Lecture Notes in Electrical Engineering

Volume 486

AF876915

**** Indexing: The books of this series are submitted to ISI Proceedings, EI-Compendex, SCOPUS, MetaPress, Springerlink ****

Lecture Notes in Electrical Engineering (LNEE) is a book series which reports the latest research and developments in Electrical Engineering, namely:

- Communication, Networks, and Information Theory
- Computer Engineering
- Signal, Image, Speech and Information Processing
- Circuits and Systems
- Bioengineering
- Engineering

The audience for the books in LNEE consists of advanced level students, researchers, and industry professionals working at the forefront of their fields. Much like Springer's other Lecture Notes series, LNEE will be distributed through Springer's print and electronic publishing channels.

For general information about this series, comments or suggestions, please use the contact address under "service for this series".

To submit a proposal or request further information, please contact the appropriate Springer Publishing Editors:

Asia:

China, *Jasmine Dou, Associate Editor* (jasmine.dou@springer.com) (Electrical Engineering)

India, *Swati Meherishi, Senior Editor* (swati.meherishi@springer.com) (Engineering)

Japan, *Takeyuki Yonezawa, Editorial Director* (takeyuki.yonezawa@springer.com) (Physical Sciences & Engineering)

South Korea, *Smith (Ahram) Chae, Associate Editor* (smith.chae@springer.com) (Physical Sciences & Engineering)

Southeast Asia, *Ramesh Premnath, Editor* (ramesh.premnath@springer.com) (Electrical Engineering)

South Asia, *Aninda Bose, Editor* (aninda.bose@springer.com) (Electrical Engineering)

Europe:

Leontina Di Cecco, Editor (Leontina.dicecco@springer.com)

(Applied Sciences and Engineering; Bio-Inspired Robotics, Medical Robotics, Bioengineering; Computational Methods & Models in Science, Medicine and Technology; Soft Computing; Philosophy of Modern Science and Technologies; Mechanical Engineering; Ocean and Naval Engineering; Water Management & Technology)

Christoph Baumann (christoph.baumann@springer.com)

(Heat and Mass Transfer, Signal Processing and Telecommunications, and Solid and Fluid Mechanics, and Engineering Materials)

North America:

Michael Luby, Editor (michael.luby@springer.com) (Mechanics; Materials)

More information about this series at http://www.springer.com/series/7818

Society of Automotive Engineers of China
(SAE-China)
Editor

Proceedings of the 19th Asia Pacific Automotive Engineering Conference & SAE-China Congress 2017: Selected Papers

Springer

Editor
Society of Automotive Engineers of China
(SAE-China)
Beijing
China

ISSN 1876-1100 ISSN 1876-1119 (electronic)
Lecture Notes in Electrical Engineering
ISBN 978-981-13-4172-4 ISBN 978-981-10-8506-2 (eBook)
https://doi.org/10.1007/978-981-10-8506-2

This Springer imprint is published by the registered company Springer Nature Singapore Pte Ltd.
The registered company address is: 152 Beach Road, #21-01/04 Gateway East, Singapore 189721, Singapore

Contents

Mechanical Performance Study of Lithium-Ion Battery Module Under Dynamic Impact Test

Fangyuan Shi, Huili Yu, Xianqing Chen, Taisong Cui, Hui Zhao and Xiangnan Shi

1 Introduction

Lithium-ion batteries have been widely used as rechargeable power for electric vehicles for the big advantages of its high energy density and lightweight. However, as the explosive growth of electric vehicle equipped with lithium-ion batteries, more and more electric vehicle fire accidents caused by lithium-ion batteries have been reported, and the safety performance of lithium-ion battery has attracted great attention. Vehicle crashes are the leading cause of catching fires in electric vehicle accidents, while mechanical intrusion caused failure of battery which is the most common cause of these fires. The severe damage of a lithium-ion battery could result in an internal short circuit condition following a thermal runaway.

In order to reduce the risk of catching fires in electric vehicle crash accidents, it is necessary to study the response of the lithium-ion battery under mechanical loading. Sahraei et al. [1–5] performed a set of loading test on different types of lithium-ion cells and developed homogenized model which captured the short-circuit point well. Avdeev et al. [6] carried out lateral loading experiments on cylindrical lithium-ion battery cells and proposed two homogenization ways for jellyroll. Greve and Fehrenbach [7] introduced stress-based Mohr–Coulomb criterion as the short-circuit criterion for jellyroll and predictions of short circuit agreed well with the indentation and bending test data. Lai et al. [8–11] performed in-plane constrained test and developed two kinds of macrohomogenized material models. Choi et al. [12] carried out different kinds of mechanical tests to study the response

F. Shi (✉) · H. Yu · X. Chen · T. Cui · H. Zhao · X. Shi
Chongqing Changan Automobile Company, Ltd., No. 589 Airport Avenue, Shuangfeng Bridge, 401120 Yubei District, Chongqing, China
e-mail: reedincqu@foxmail.com

F. Shi · H. Yu · X. Chen · T. Cui · H. Zhao · X. Shi
State Key Laboratory of Vehicle NVH and Safety Technology, Chongqing, China

Society of Automotive Engineers of China (SAE-China) (ed.), *Proceedings of the 19th Asia Pacific Automotive Engineering Conference & SAE-China Congress 2017: Selected Papers*, Lecture Notes in Electrical Engineering 486, https://doi.org/10.1007/978-981-10-8506-2_1

of lithium-ion pouch cells, investigated its mechanical behavior toward temperature, strain rate, and anisotropy.

All of those literatures mentioned above are mainly focused on the investigation of mechanical properties on lithium-ion battery cell level, and most of these test conditions are quasi-static. Moreover, failure strain of separator plays a key role in the safety performance of pouch cell. It has been studied that separator under higher strain rate resulted in smaller elongation ratio and a good elongation characteristic of separator could improve safety performance of pouch cell under mechanical loadings [13, 14]. So dynamic impact, instead of quasi-static, is chosen for acquiring the lowest failure thresholds of battery module and this failure threshold could be served as a reference for crash safety design of electric vehicle.

2 Materials and Methods

In this study, a kind of lithium-ion battery module was studied, and it is composed of three parts, which are plastic cell holder, aluminum alloy radiator, and lithium-ion battery cells. There are eight lithium-ion battery cells in this module, and they are layered arrangement with four in parallel and two in series (4P2S), which is shown in Fig. 1. The SoC of all modules was 100%, and the specifications of the battery module are listed in Table 1.

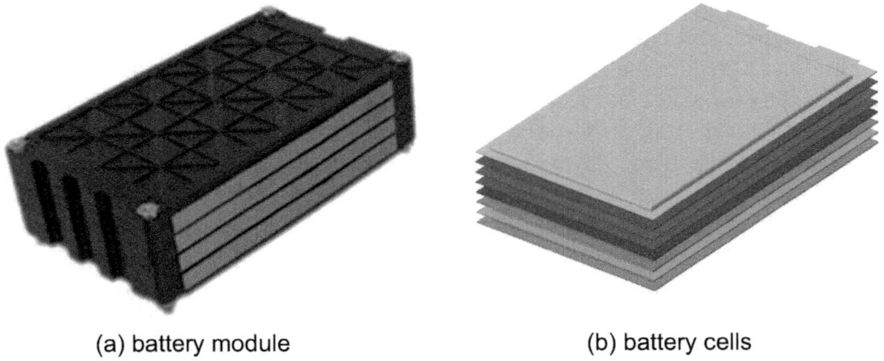

Fig. 1 Diagram of battery module

Table 1 Specifications of battery cell and battery module

Categories	Battery cell	Battery module
Nominal voltage (V)	3.7	7.4
Nominal capacity (Ah)	20	80
Length (mm)	215	269
Width (mm)	132	157
Thickness (mm)	7.4	74

These dynamic impact tests were conducted in a drop tower. The whole drop mass was 40 kg, which was released from a height of 5.1 m. A semi-cylindrical steel punch was used, the diameter and the length of the punch was 80 and 240 mm, respectively. In these tests, an ultradynamic strain gauge was used to acquire the voltage signal of the battery module. The high-speed camera and digital image correlation (DIC) method were utilized to capture the impact scene and measure battery module deformation. The experiment setup is shown in Fig. 2.

In order to obtain the punch force and voltage of the lithium-ion battery module under different dynamic impact direction, multidynamic impact tests were designed, and each test condition was repeated two times, which is shown in Table 2.

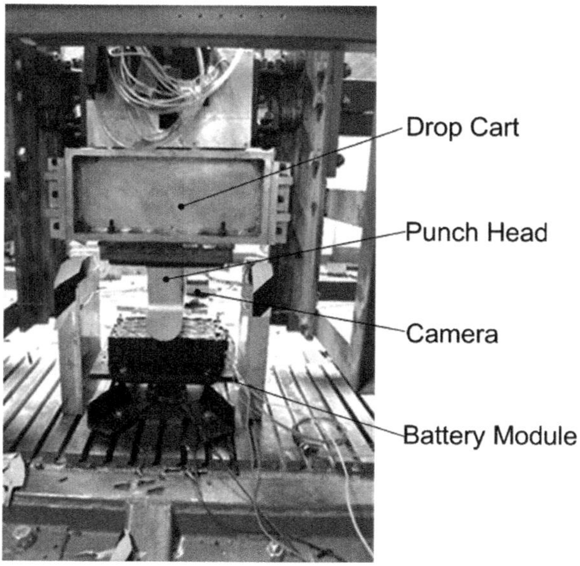

Fig. 2 Dynamic impact test setup

Table 2 Dynamic impact test matrix

Impact direction	SoC (%)	Drop mass (kg)	Drop height (m)	Repeat times
Length	100	40.0	5.1	2
Width	100	40.0	5.1	2
Thickness	100	40.0	5.1	2

3 Results

3.1 Dynamic Impact in Length Direction

When the battery modules were subjected to dynamic impact in length direction, the component parts were dispersed due to plastic parts fracture and electrolyte leakage was not observed. Figure 3 depicts the battery module impact in length orientation during impact and after impact.

The punch force, punch displacement, and battery module voltage were measured over time and plotted in Fig. 4. Fire or short circuit was not observed in both of the two tests. Voltage maintained almost constant during test. In the initial stage, the loading force increased rapidly with the increasing of battery module deformation

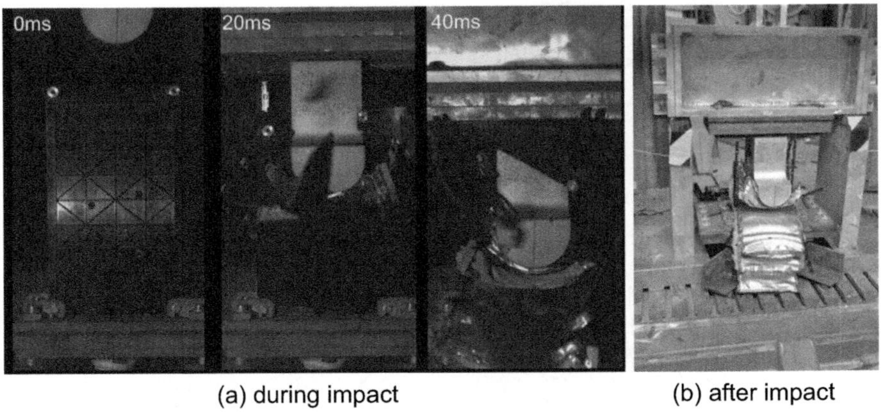

Fig. 3 Dynamic impact test in length direction

Fig. 4 Measured load/voltage–displacement curve for dynamic impact in length direction

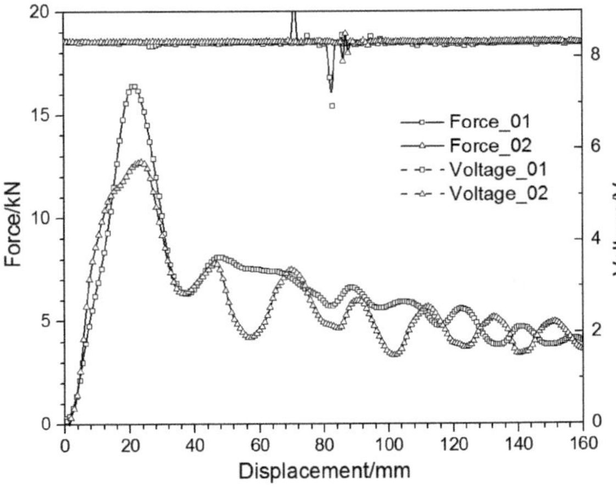

from 0 to 20 mm, and the maximum punch force and related deformations for the tests were 12.7/16.4 kN, and 22.7/20.7 mm, respectively. Then, the loading force declined rapidly due to the disassembly and fragmentation of battery module component, and the minimum impact load was only 6.3 kN. After that, the loading force showed a gradual downward trend and had some shocks. When the deformation of the battery module was over 160 mm, the impact head was in contact with rubberlike stopper.

3.2 Dynamic Impact in Width Direction

When the battery modules were subjected to dynamic impact in width direction, the component parts were also dispersed a little due to plastic parts fracture which was similar to dynamic impact in length direction, and electrolyte leakage was not observed. Figure 5 depicts the battery module impact in width orientation during impact and after impact.

All the measured parameters of the battery modules are shown in Fig. 6. Fire or short circuit was not observed like length direction impact. One can see that initially, the punch force increased rapidly with the increasing of battery module deformation from 0 to 20 mm, and the maximum impact load and associated displacement for the test were about 13 kN, and 20 mm, respectively. Then, the loading force declined rapidly, and the minimum impact load was nearly 5.6 kN. After that, the loading force showed a gradual upward trend which was different with length direction result. In the length direction impact test, battery cells were apart from battery module in the impact. While in the width direction impact test, most of components were still connected together that prevented the punch move downward which caused the punch force to increase. When the deformation of the battery module was over 85 mm, the impact head was also in contact with stopper.

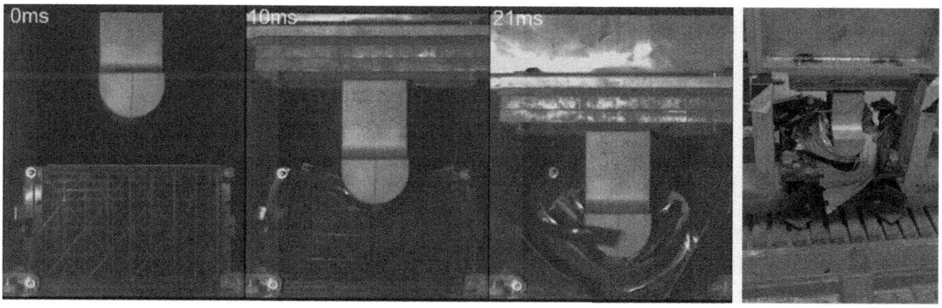

Fig. 5 Dynamic impact test in width direction

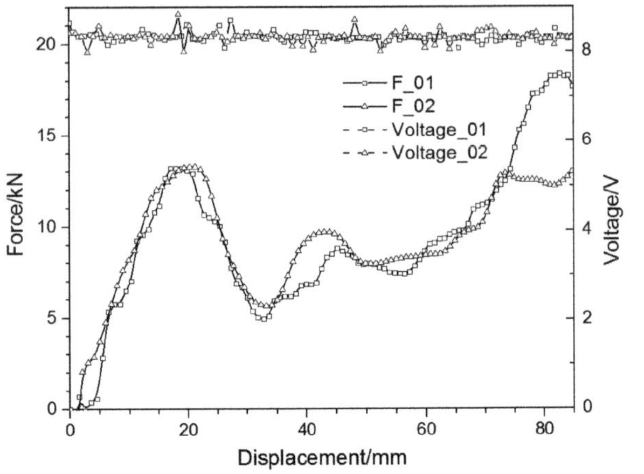

Fig. 6 Measured load/voltage–displacement curve for dynamic impact in width direction

3.3 Dynamic Impact in Thickness Direction

When the battery modules were subjected to dynamic impact in thickness direction, the component parts were not dispersed except the top plastic holder and some mist-like stuff, which was likely to be electrolyte, splashed out on the left side in the image. After the impact test, the battery caught fire as soon as a lot of smoke was sprayed out. Figure 7 depicts the battery module impact in thickness direction during impact and after impact.

The force and voltage versus displacement is shown in Fig. 8. It is easy to see from the figure that the voltage decreased as the punch force rose to its first peak. And the voltage decrease probably means there is a short circuit occurred in the battery module. In the initial deformation stage between 0 and 12 mm, the punch force experienced a sharp increase from 0 to 170 kN, and this peak force is far greater than the other two impact directions. After that, because of plastic parts and cells fracture, the force showed a gradual downward trend and had some shocks. When the deformation of the battery module was over 25 mm, the punch contacted with stopper.

4 Discussion

The test results show that impact directions have great influences on mechanical characteristics, electric voltage, and crash failure threshold of battery module.

The peak punch force in thickness direction is 170 kN, which is much higher than that measured in the other two test directions. The battery module could be regarded as an anisotropic material, and it may be possible to develop a homogenized battery module model for fulfilling engineering requirement to represent anisotropic mechanical characteristics.

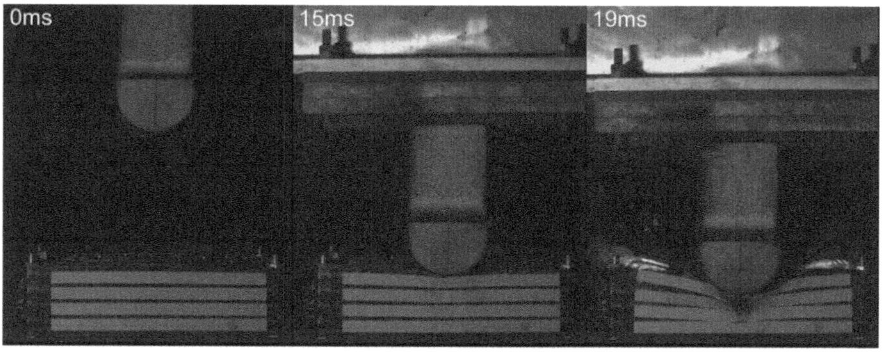

(a) during impact

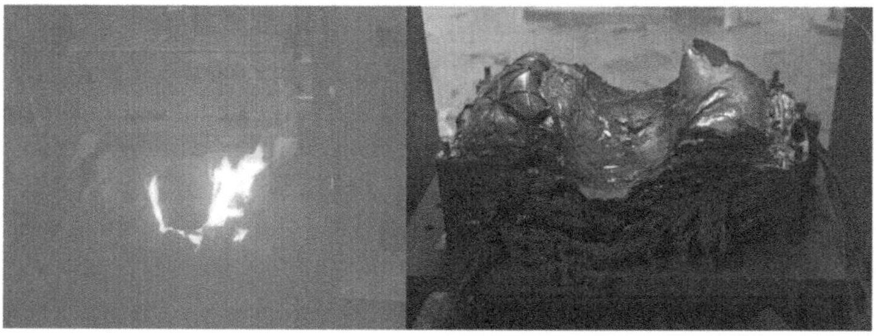

(b) after impact

Fig. 7 Dynamic impact test in thickness direction

Fig. 8 Measured load/voltage–displacement curve for dynamic impact in thickness direction

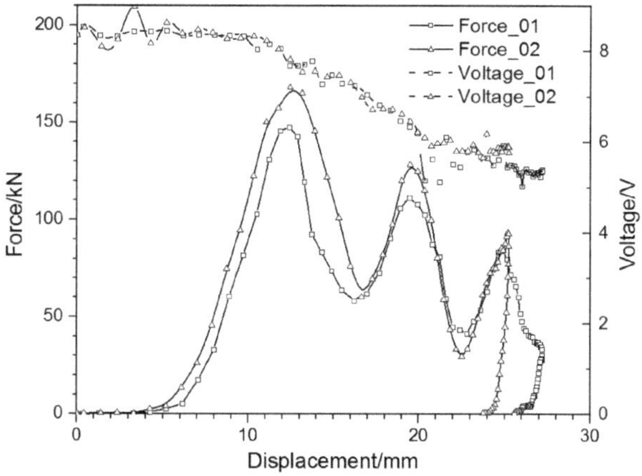

When the battery modules were subjected to dynamic impact in length and width direction, there is still no thermal runaway though the deformation is more than 50% of the initial length. However, if the battery modules were subjected to dynamic impact in thickness direction, the battery modules would be short circuit and got burning even though the deformation is only 16% of the initial length which is only 12 mm. It shows that thickness direction of such battery module is more vulnerable than the other two directions. And we'd better arrange the battery modules' thickness direction in the safest direction in the vehicle to decrease the risk of getting mechanical intrusion in an accident.

In this study, the lithium-ion battery cells in all of these modules were arranged in a layered structure. When the battery module was subjected to dynamic impact in length/width or thickness direction, it means that the battery cell was also subjected to dynamic impact in length/width or thickness direction, respectively. To further examine the potential lithium-ion battery cells mechanical fracture mechanism, several additional tests were performed. Battery cells deformed in different loading directions is shown in Fig. 9. Plots of loading force, cell voltage, and cell surface temperature are shown in Fig. 10.

It is clear that crushing test in thickness direction was associated with much less deformation when the cell experienced a voltage drop and temperature increase, indicating that the risk of such test condition is higher than lateral crushing test in width direction. The results are matching well with battery module level test results. In reality, internal short circuit occurs because of the failure of separator. The results of this study are in agreement with several previous studies examining the influence of different test conditions on failure threshold of battery cells [15–18], all of which demonstrated the challenges of managing the separator of lithium-ion battery cells.

According to the above research results, this type of battery module arrangement in the battery pack was changed in the new generation EV vehicle. In its last generation vehicle, its thickness direction was identical with vehicle's z direction (Fig. 11). There would be a high risk of getting fired when the battery experienced an intrusion from the bottom. So in order to reduce the risk of impact from the bottom of vehicle, the thickness direction of battery module was placed toward the vehicle's x direction in the new generation vehicle. In the vehicle's x direction, huge space could deform in front and rear crash to absorb crash energy, so the battery's vulnerable position (thickness direction) will be well protected.

(a) Indentation test (b) Lateral crush test

Fig. 9 Quasi-static test setup and cell deformation

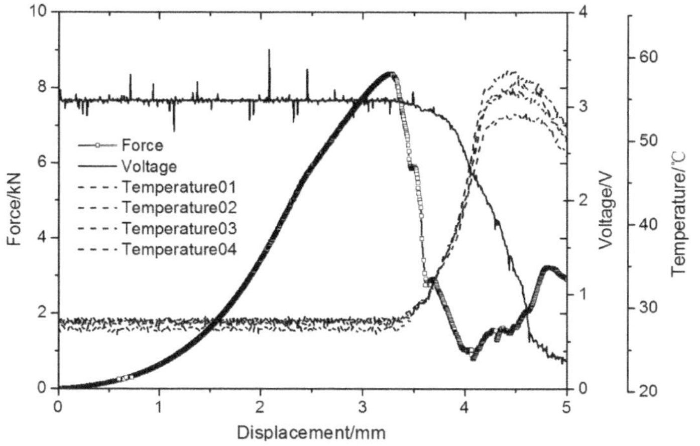

(a) Indentation test

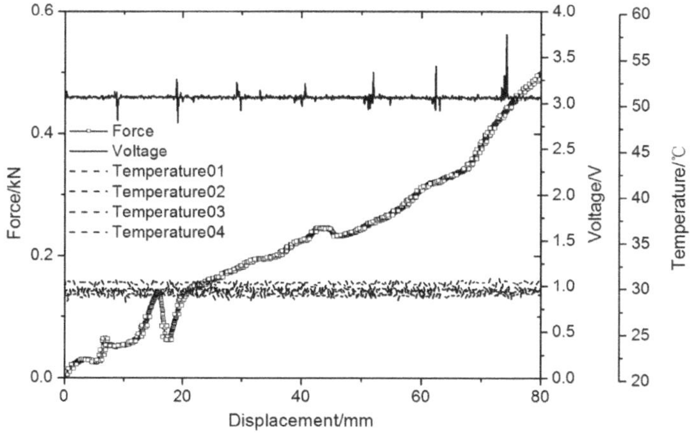

(b) Lateral crush test

Fig. 10 Results of quasi-static test

5 Conclusion

In this study, several dynamic impact tests were conducted on lithium-ion battery modules under different impact directions. Results showed that impact directions have great influences on mechanical characteristics, electric voltage, and crash failure threshold of battery module. In particular, if the battery modules were subjected to dynamic impact in thickness direction, when the deformation exceeded 12 mm which is only 16% of the initial length, these battery modules would get

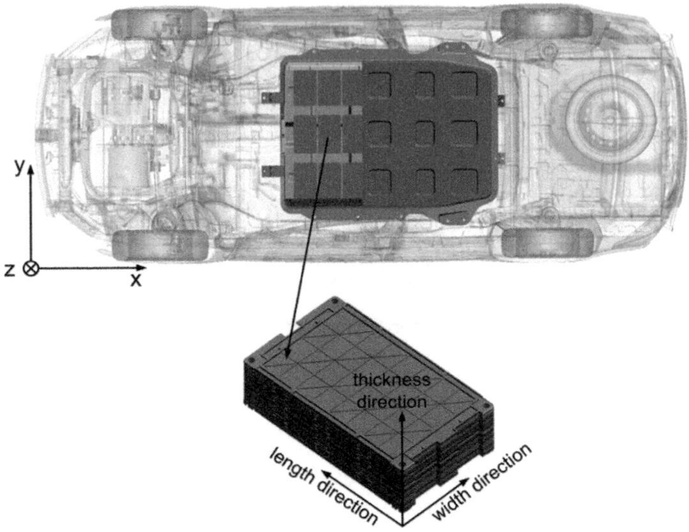

Fig. 11 Layout of battery in the EV vehicle

short circuit and burn. The conclusions from this research provide valuable information for future safety performance design of electric vehicle.

Acknowledgements The authors would like to express their sincere thanks to Mr. Guanhua Chen and Dr. Hailin Luo in TsingHua University for their help in the dynamic impact tests.

References

1. Sahraei E, Campbell J, Wierzbicki T (2012) Modeling and short circuit detection of 18650 Li-ion cells under mechanical abuse conditions. J Power Sources 220:360–372
2. Sahraei E, Meier J, Wierzbicki T (2014) Characterizing and modeling mechanical properties and onset of short circuit for three types of lithium-ion pouch cells. J Power Sources 247:503–516
3. Sahraei E, Hill R, Wierzbicki T (2012) Calibration and finite element simulation of pouch lithium-ion batteries for mechanical integrity. J Power Sources 201:307–321
4. Wierzbicki T, Sahraei E (2013) Homogenized mechanical properties for the jellyroll of cylindrical lithium-ion cells. J Power Sources 241:467–476
5. Xia Y, Wierzbicki T, Sahraei E, Zhang X (2014) Damage of cells and battery packs due to ground impact. J Power Sources 267:78–97
6. Avdeev IV, Martinsen MJ, Francis AB (2012) Materials testing of a lithium ion battery separator for use in finite element analysis. In: Proceedings of the ASME 2012 international mechanical engineering congress and exposition (IMECE 2012)
7. Greve L, Fehrenbach C (2012) Mechanical testing and macro-mechanical finite element simulation of the deformation, fracture, and short circuit initiation of cylindrical lithium ion battery cells. J Power Sources 214:377–385

8. Lai W-J, Ali MY, Pan J (2014) Mechanical behavior of representative volume elements of lithium-ion battery cells under compressive loading conditions. J Power Sources 245:609–623
9. Lai W-J, Ali MY, Pan J (2014) Mechanical behavior of representative volume elements of lithium-ion battery modules under various loading conditions. J Power Sources 248:789–808
10. Ali MY, Lai W-J, Pan J (2013) Computational models for simulations of lithium-ion battery cells under constrained compression tests. J Power Sources 242:325–340
11. Ali MY, Lai W-J, Pan J (2015) Computational models for simulation of a lithium-ion battery module specimen under punch indentation. J Power Sources 273:448–459
12. Choi HY, Lee I, Lee JS, Kim YM, Kim H (2013) A study on mechanical characteristics of lithium-polymer pouch cell battery for electric vehicle. In: Paper No. 13-0115, Proceedings of 23rd international technical conference on the enhanced safety of vehicles (ESV 2013)
13. Luo H, Jiang X, Xia Y, Zhou Q (2015) Fracture mode analysis of lithium-ion battery under mechanical loading. In: Proceedings of the ASME 2015 international mechanical engineering congress and exposition (IMECE 2015)
14. Xu J, Wang L, Guan J, Yin S (2016) Coupled effect of strain rate and solvent on dynamic mechanical behaviors of separators in lithium ion batteries. Mater Des 95:319–328
15. Halalay IC, Lukitsch MJ, Balogh MP, Wong CA (2013) Nanoindentation testing of separators for lithium-ion batteries. J Power Sources 238:469–477
16. Sheidaei A, Xiao X, Huang X, Hitt J (2011) Mechanical behavior of a battery separator in electrolyte solutions. J Power Sources 196:8728–8734
17. Lamb J, Orendorff CJ (2014) Evaluation of mechanical abuse techniques in lithium ion batteries. J Power Sources 247:189–196
18. Ren F, Cox T, Wang H (2014) Thermal runaway risk evaluation of Li-ion cells using a pinch-torsion test. J Power Sources 249:156–162

Combustion System Development of GDI-T Engine for China Stage III Fuel Consumption Regulation

Zhanfeng Wang and Haie Chen

1 Introduction

In China, with increasing of oil price and strict emissions regulations, high efficiency, low carbon and low emissions have been necessary condition for developing a new gasoline product. It means that in-cylinder direct injection turbocharged gasoline engine with high-rated power, low fuel consumption, low emission and fast response will be mainstream products in future.

In order to promote automobile factory developing energy-saving car, Chinese government has introduced corresponding fuel consumption regulations according to the actual situation of our country. Chinese government has published 《Passenger car fuel consumption evaluation method and target》 in 30 December 2011. This regulation has come into effect on 1 January 2012 in Beijing firstly. And the first year is introduction stage, then gradually strict every year, finally will become into effect in the whole nation in 2015. Table 1 is the fuel consumption requirement of passenger car. Figure 1 is the milestone of fuel consumption target.

2 Main Parameters and Design Target of CA4GB GDI-T Engine

The main parameters and performance targets are as below.

Table 2 is main parameters and performance targets. We can see that specific power of this engine is above average level, and specific torque of this engine is at first class level from Figs. 2 and 3.

Z. Wang (✉) · H. Chen
Powertrain Station, FAW Group Co.,Ltd. Research and Development Institute,
No. 8899, Dongfeng Street, 130013, Jilin Province Changchun, China
e-mail: wangzhanfeng@faw.com.cn

Society of Automotive Engineers of China (SAE-China) (ed.), *Proceedings of the 19th Asia Pacific Automotive Engineering Conference & SAE-China Congress 2017: Selected Papers*, Lecture Notes in Electrical Engineering 486, https://doi.org/10.1007/978-981-10-8506-2_2

Table 1 Fuel consumption requirement of passenger car for China Stage III

Vehicle kerb mass (CM) (kg)	Target value of fuel consumption (L/100 km)
CM $\leq$ 750	5.2
750 < CM $\leq$ 865	5.5
865 < CM $\leq$ 980	5.8
980 < CM $\leq$ 1090	6.1
1090 < CM $\leq$ 1205	6.5
1205 < CM $\leq$ 1320	6.9
1320 < CM $\leq$ 1430	7.3
1430 < CM $\leq$ 1540	7.7
1540 < CM $\leq$ 1660	8.1
1660 < CM $\leq$ 1770	8.5
1770 < CM $\leq$ 1880	8.9
1880 < CM $\leq$ 2000	9.3
2000 < CM $\leq$ 2110	9.7
2110 < CM $\leq$ 2280	10.1
2280 < CM $\leq$ 2510	10.8
2510 < CM	11.5

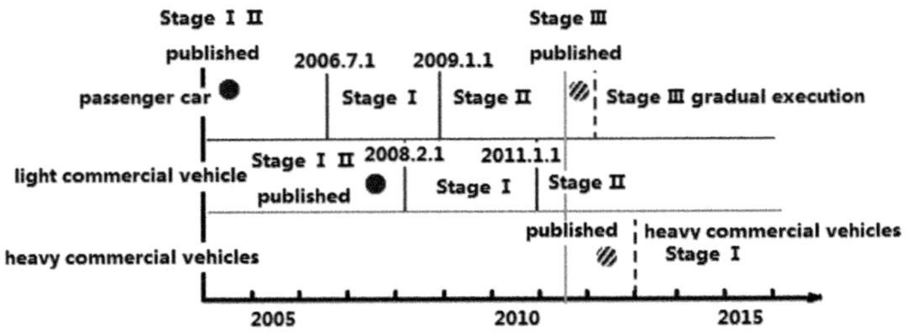

Fig. 1 Milestone of fuel consumption target

3 Combustion System Design

Figure 4 is the whole layout of CA4GB GDI-T combustion system. Fuel injector is side mounted.

We use the high tumble intake port in this engine. The intake port flow characteristic is shown in Fig. 5. The average flow coefficient of intake port is 0.44, and the tumble ratio is 2.5. The flow and tumble of each cylinder is uniformity; the difference is within 1%.

The top surface shape of piston is shown in Fig. 6.

The maximum injection pressure is 15 MPa. Hole patterns for three injectors are in Table 3.

Table 2 Main parameters and performance targets

Main parameters and performance target	
Bore (mm)	76.5
Stroke (mm)	75.6
Displacement (L)	1.39
Cylinder number	4
Firing order	1–3–4–2
Compression ratio	9.6
Max. cylinder pressure (MPa)	11
Fuel	RON 93
Power/speed (kW/r/min)	100/4500–5500
Max. torque/speed (N m/r/min)	220/1650–4350
Specific power (kW/L)	72
Specific torque (N m/L)	158.2
Emission level	EU V

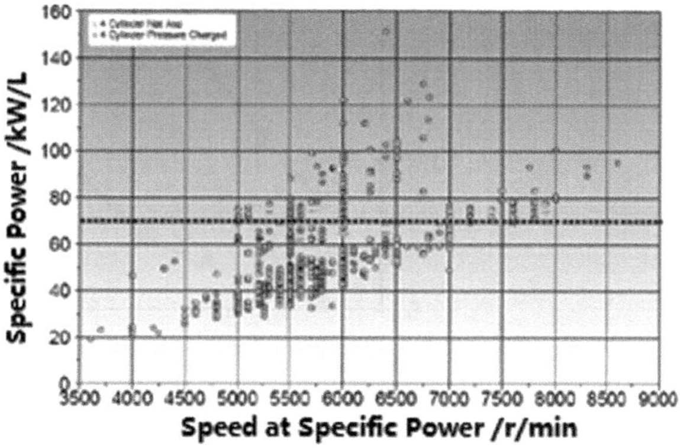

Fig. 2 Benchmark of specific power

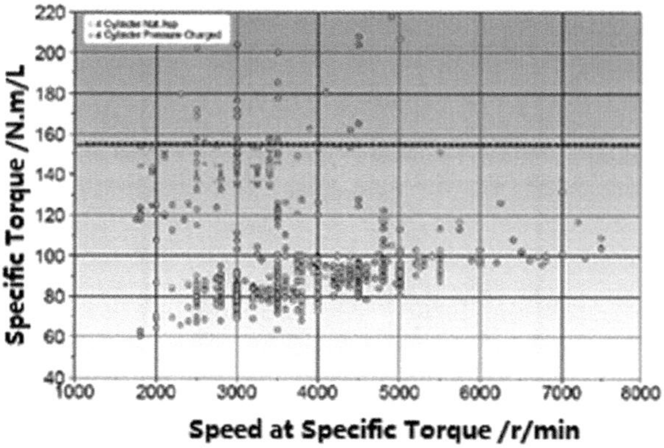

Fig. 3 Benchmark of specific torque

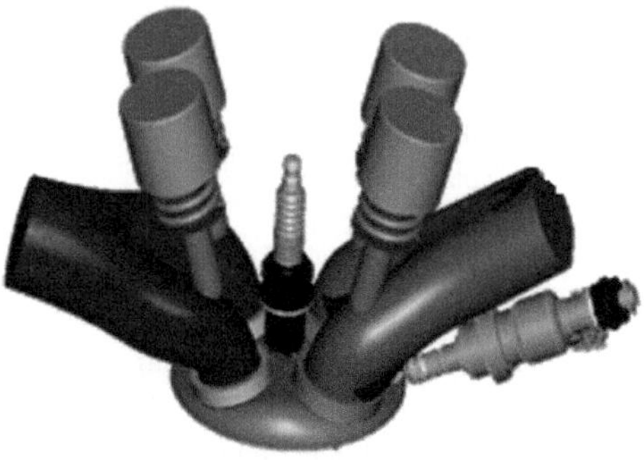

Fig. 4 Layout of combustion system

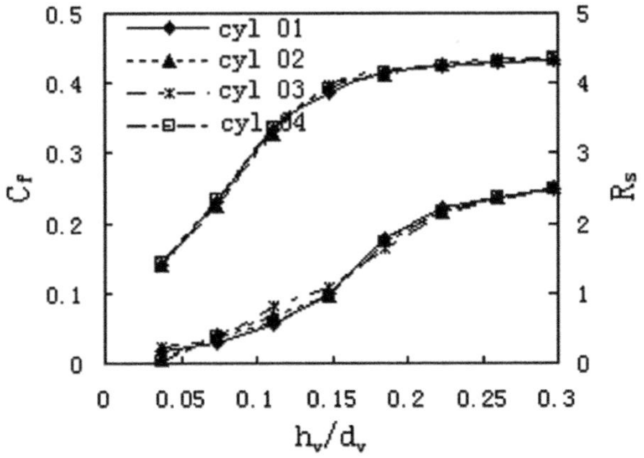

Fig. 5 Intake port flow characteristic

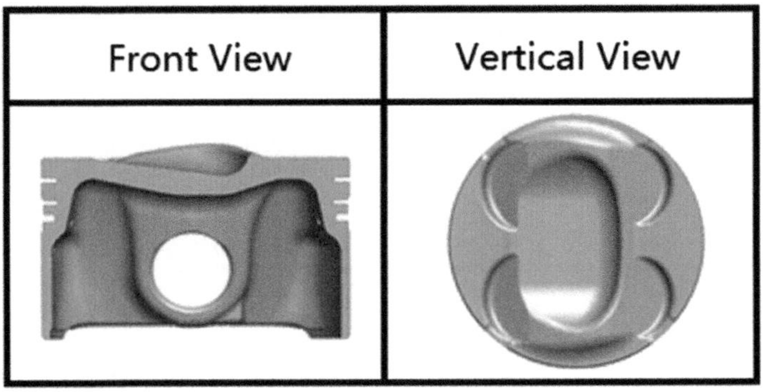

Fig. 6 Shape of piston chamber

Table 3 Hole pattern for three injectors

	Case 1	Case 2	Case 3
Spray scheme			
Hole number	6	7	6

Table 4 shows main cam parameters. Maximum lift of intake valve is 9.5 mm, and the duration of 1 mm lift is 195 °C A. Maximum lift of exhaust valve is 9.0 mm, and the duration of 1 mm lift is 181 °C A. Cam profile of intake and exhaust valve are shown in Fig. 7.

Figure 8 is engine operation line in compressor map 1, named as case 1. Figure 9 is engine operation line in compressor map2, named as case 2. These two turbochargers both can meet engine performance requirement. Case 1 has more surge margin than case 2, but case 1 should be easy to over speed than case 2 in altitude environment. So we will make comparisons about two cases in performance test.

Table 4 Main cam parameters

	Lift (mm)	VC/1 mm (°C A)	VC/1 mm (°C A)	MOP (°C A)
In base timing	9.5	382	577	478
Ex base timing	9.0	157	338	248
In max timing	9.5	332	527	428
Ex max timing	9.0	207	388	298

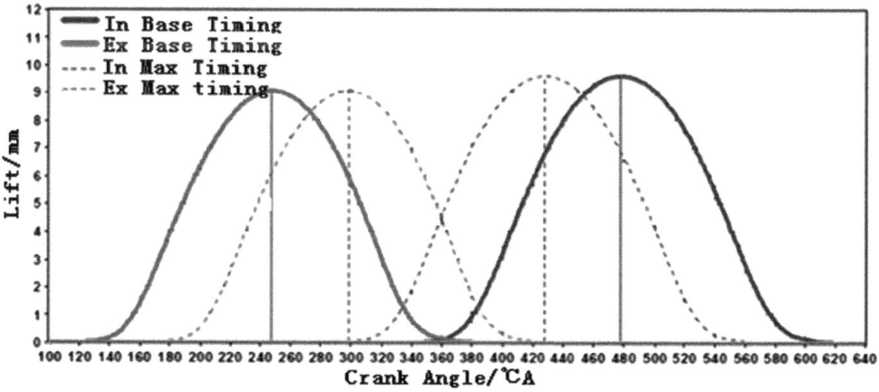

Fig. 7 Cam profile of intake and exhaust valve

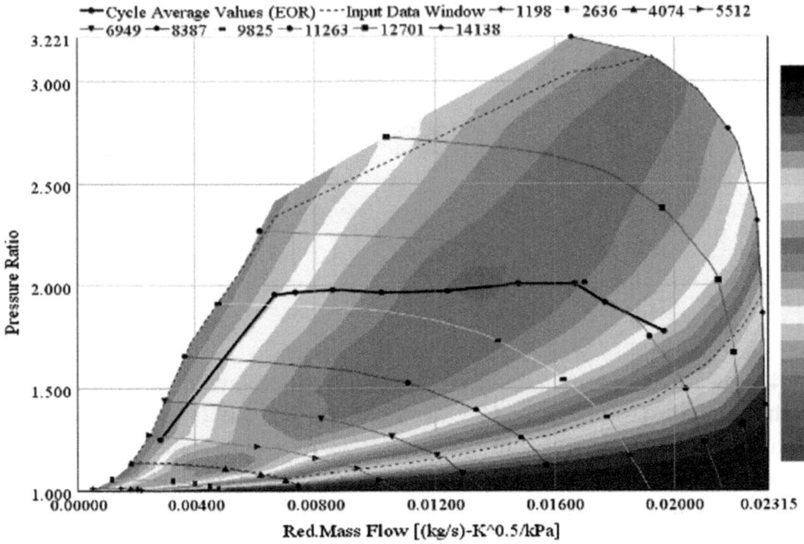

Fig. 8 Operation line in turbocharger 1

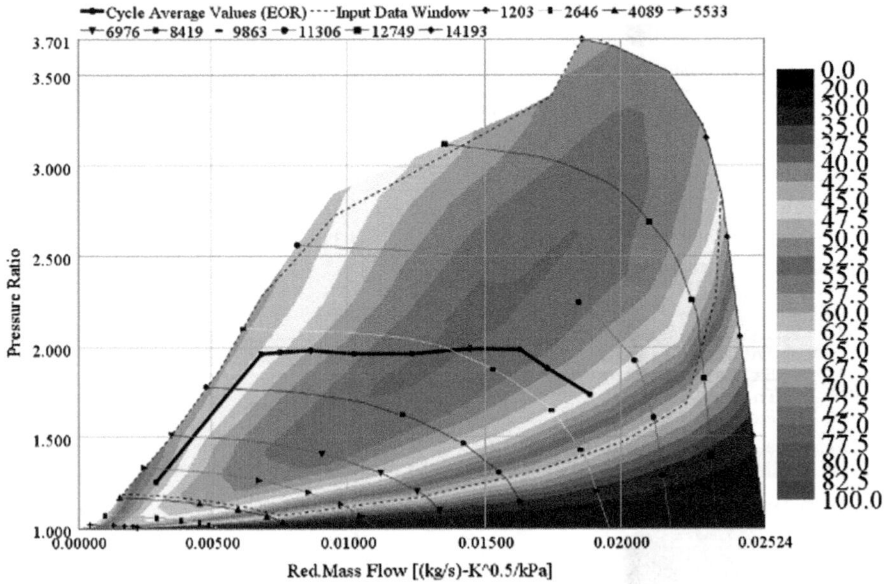

Fig. 9 Operation line in turbocharger 2

4 Optical Engine Test

In order to shorten the performance development period and choose the most proper injector, firstly we did optical single cylinder engine test.

Test was carried out at three engine operating conditions in optical engine including 1650 r/min-full load, 2000 r/min-2 bar BMEP and 1200 r/min-1 bar BMEP. We measured and recorded spray in cylinder, process of mixture formation and flame propagation and distribution after ignition on different injector scheme. We compared quality of mixture in cylinder for different timings, analyzed the effect of flame kernel shape and soot, finally determined injector scheme and injection timing.

Figure 10 is the comparison of spray at 1650 r/min-full load with three injectors. We can see that there is a wall-hitting phenomenon with injector of case 2 at −225 °C A ATDC. Wall hitting will lead to instantaneous dry friction between piston ring and cylinder wall, meanwhile increase oil dilution. There is also a wall-hitting phenomenon with injector of case 3 at −210 °C A ATDC. There is a wall-hitting phenomenon with injector of case 2 or case 3 at −200 °C A ATDC. There is no obvious a wall-hitting phenomenon with injector of case 1 in the whole process of spray.

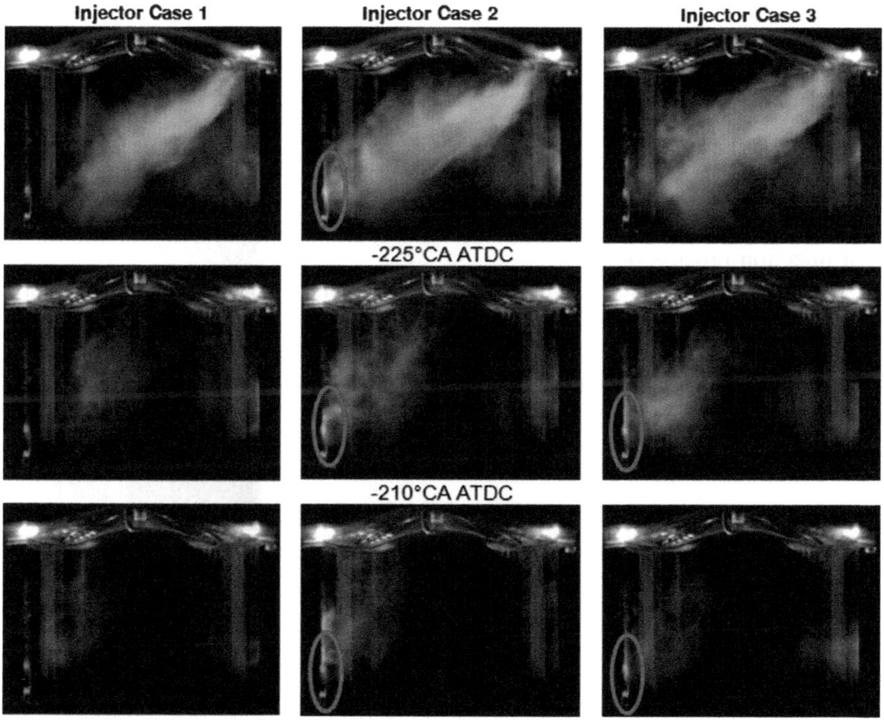

Fig. 10 Comparison of spray at 1650 r/min-full load

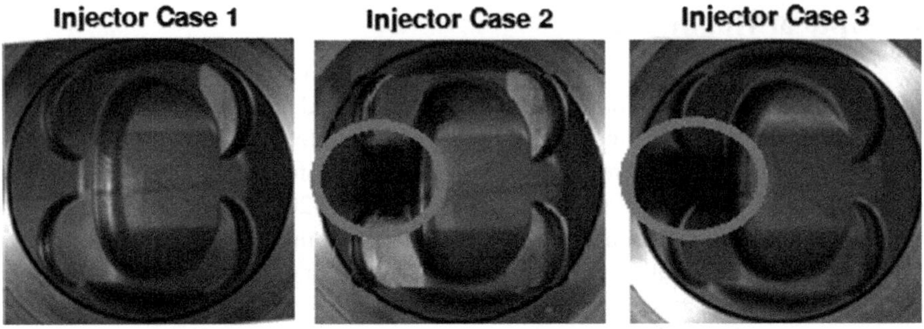

Fig. 11 Comparison of deposition on piston surface

Figure 11 is the comparison of deposition on piston surface with three injectors. We can see very clearly that there is obvious carbon deposition phenomenon on valve pit between intake and exhaust valve with injector of case 2 or case 3. It is means that fuel spreading is not good with injector of case 2 or case 3. Partial fuel stored in piston top, which lead to diffusion combustion in cylinder, then carbon deposit.

Figure 12 is the comparison of spray at 2000 r/min-2 bar BMEP with three injectors. We can see that there is no obvious wall-hitting phenomenon with three injectors in the whole process of spray. Three injectors all meet the requirements on this condition.

Figures 13 and 14 are the comparisons of spray at 1200 r/min-1 bar BMEP with three injectors. We use split injection strategy on this condition. Injection timing of first injection is −250 °C A ATDC, and duration is 1.2 ms. Injection timing of second injection is −110 °C A ATDC, and duration is 1.2 ms. In Fig. 13, we can see that there is wall-hitting phenomenon with injector of case 3 at −238 °C A ATDC. There is also wall-hitting phenomenon with injector of case 2 at −234 °C A ATDC. There is no wall-hitting phenomenon with injector of case 1 in the whole first injection process of spray.

From Fig. 14, we can see that there is wall-hitting phenomenon with injector of case 2 and injector of case 3 at −82 °C A ATDC. There is no obvious wall-hitting phenomenon with injector of case 1 in the whole second injection process of spray.

Finally, we choose injector of case 1 by optical engine test because there is no wall-hitting and carbon deposit phenomenon on various conditions. So we use injector of case 1 in the following performance development test.

5 Performance Development Test

5.1 Turbocharger Test Results

Figure 15 is full load results of the two turbochargers. We can see that both turbochargers can achieve 220 N m maximum torque target value at 1500 r/min which is lower than target speed 1650 r/min. It can reduce lagging when turbocharged

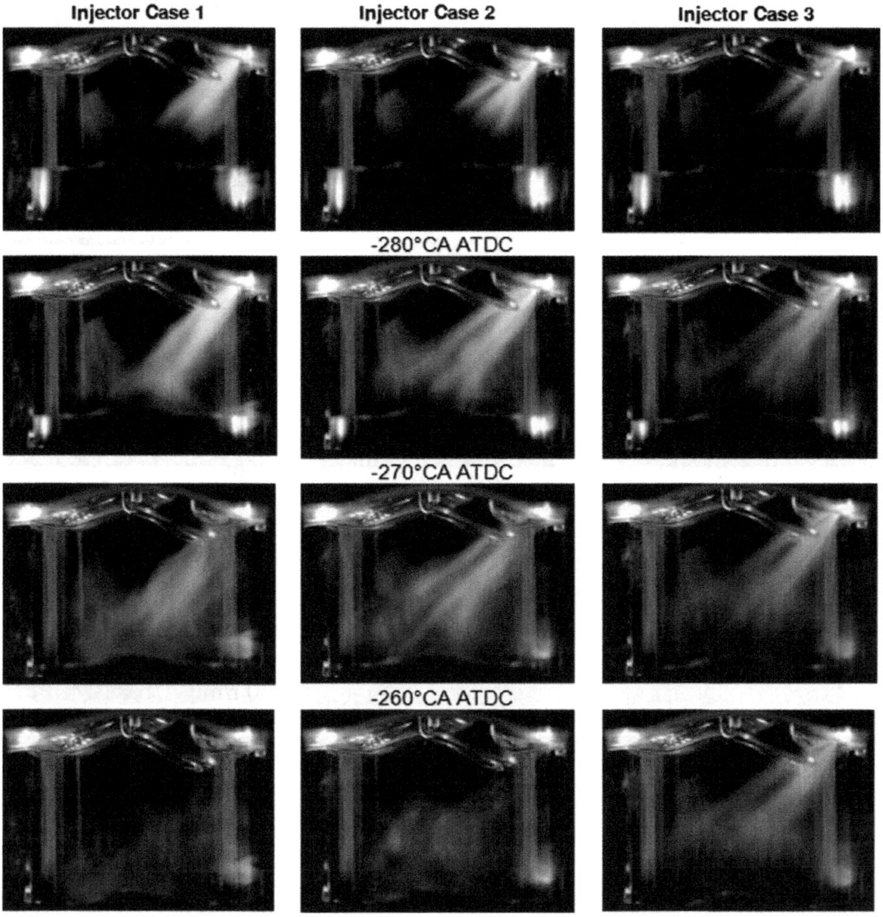

Fig. 12 Comparison of spray at 2000 r/min-2 bar BMEP

engine accelerate at low speed. Figure 16 is boost comparison of these two turbochargers. Torque of turbo case 2 from 1000 to 2500 r/min is higher than turbo case 1; torque of turbo case 1 and case 2 from 2500 to 5500 r/min are equivalent. But boost of turbo case 2 is lower than turbo case 1. Performance at low speed of turbo case 2 is better than turbo case 1.

Figure 17 is the comparison of fuel consumption at part load with two turbochargers. We choose four-part load conditions as 2000 r/min-2 bar BMEP, 2000 r/min-4 bar BMEP, 3000 r/min-3 bar BMEP, 4000 r/min-6 bar BMEP. We can see that fuel consumption of turbo case 1 is better than turbo case 2. It will help to reduce vehicle fuel consumption.

We predict operation line of altitude based on test results in the plain. Figures 18 and 19 are altitude capacity of two turbochargers. The red line is at sea level, the green line is at altitude 1000 m, the blue line is at altitude 2000 m and the purple

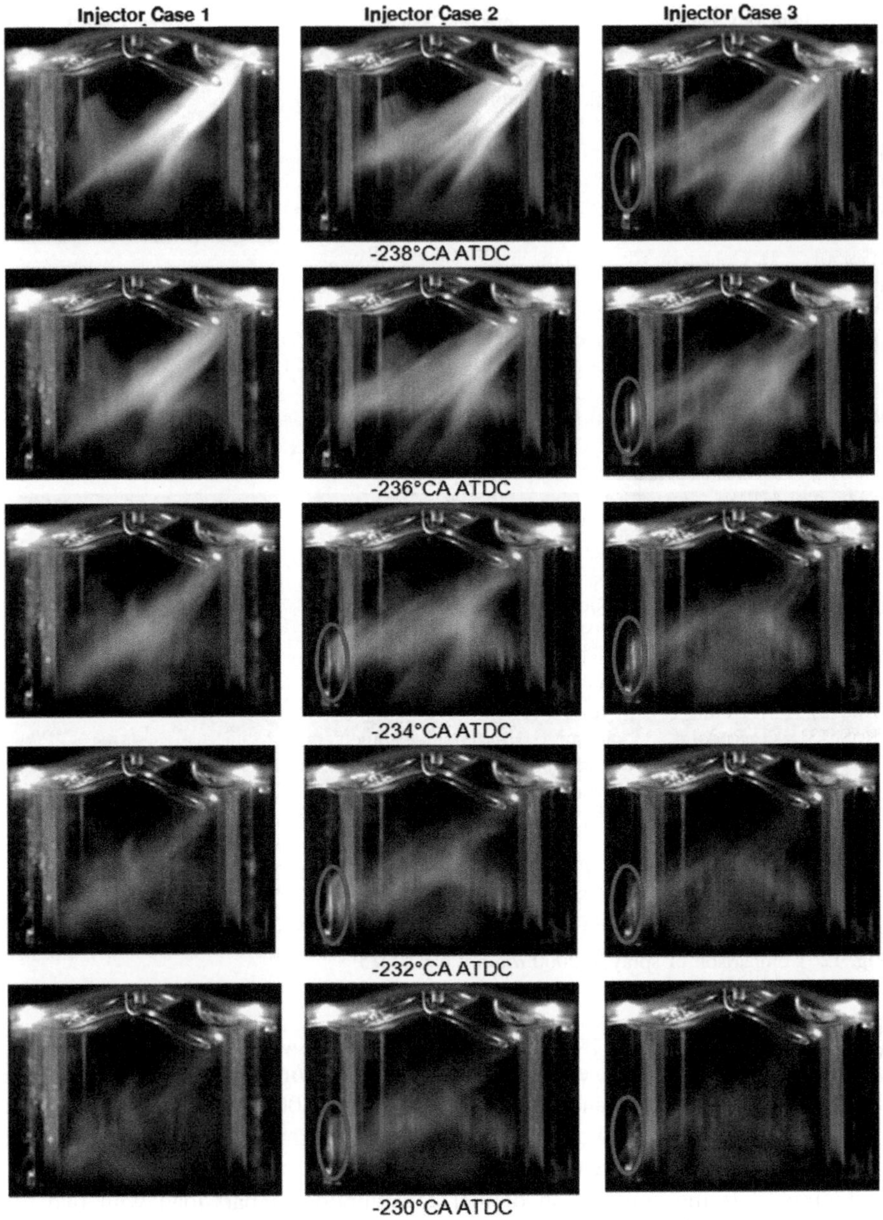

Fig. 13 First injection comparison of spray at 1200 r/min-1 bar BMEP

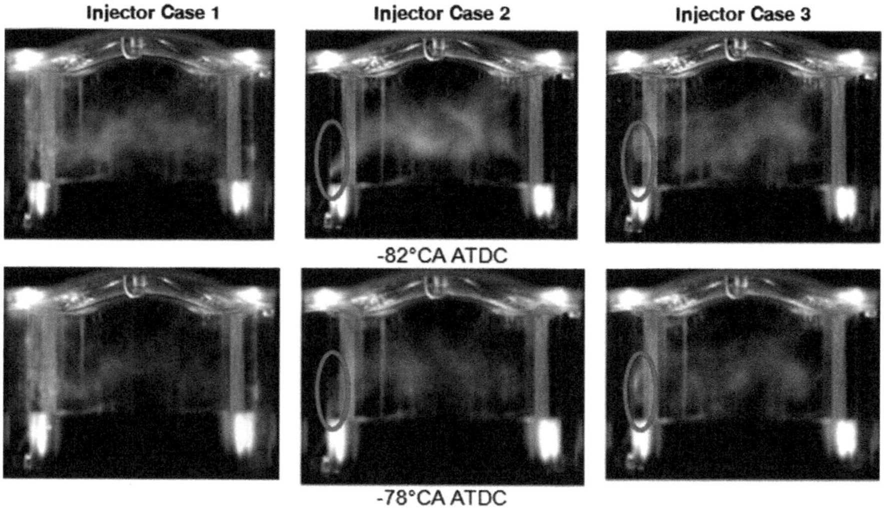

Fig. 14 Second injection comparison of spray at 1200 r/min-1 bar

Fig. 15 Full load results of turbocharger

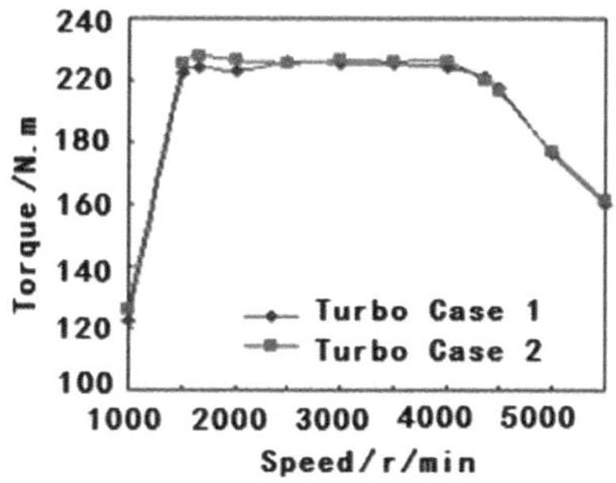

Fig. 16 Boost comparison of turbocharger

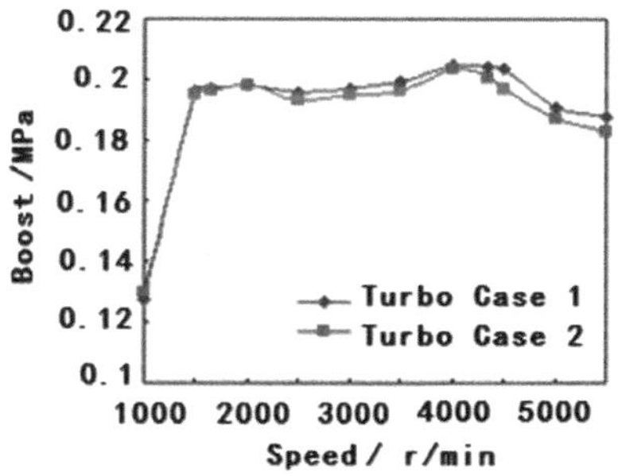

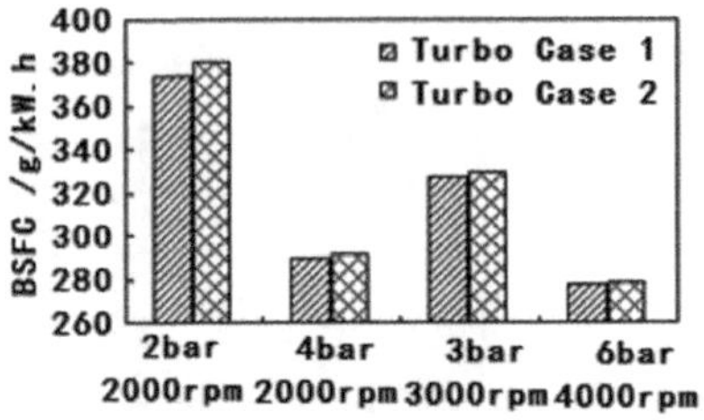

Fig. 17 BSFC at part load with two turbochargers

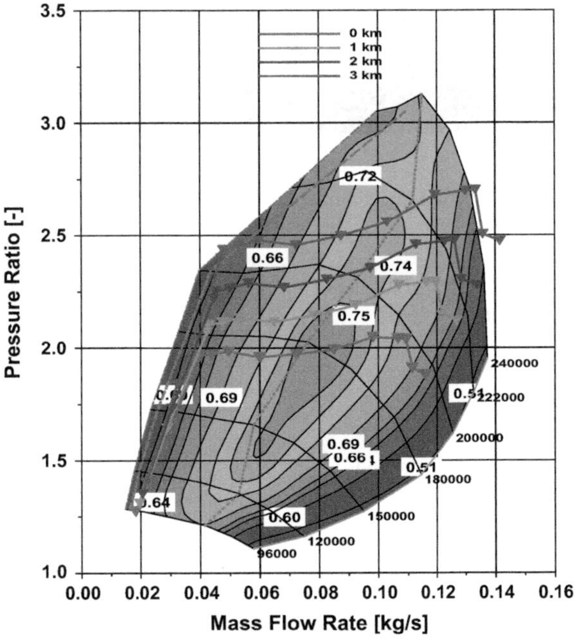

Fig. 18 Altitude capacity of turbo 1

line is at altitude 3000 m. We consider that there is 10% surge margin at sea level. Two turbochargers both meet this limit value. But turbo 2 is much easier to happening surge than turbo case 1. And turbo 1 is much easier to over speed than turbo 2 at altitude environment.

Turbocharger cannot provide air and boost pressure of engine required in time when engine changes the operating condition. There is lagging time for turbocharger. Acceleration capability of engine is worse as longer lagging time. Lagging phenomenon is more evident because of wider speed range and smaller inertia. Figures 20 and 21 are transient response of two turbochargers. Engine speeds are at 1500 and 2000 r/min. We change engine load from 10% maximum load to 90% maximum load, and then comparison response time of two turbochargers.

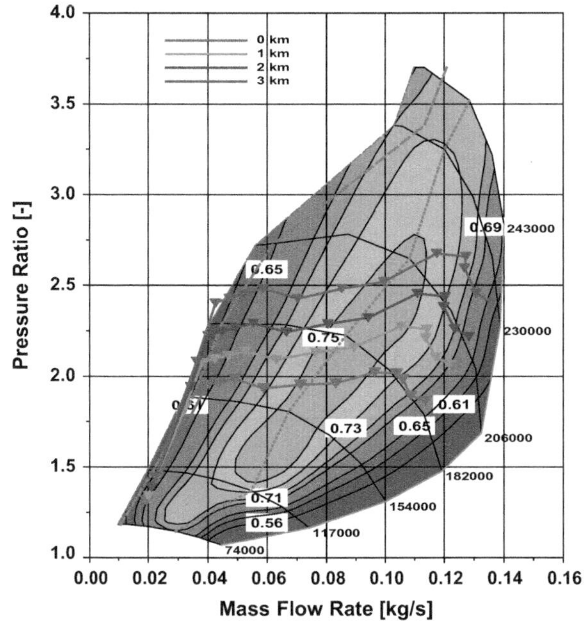

Fig. 19 Altitude capacity of turbo 2

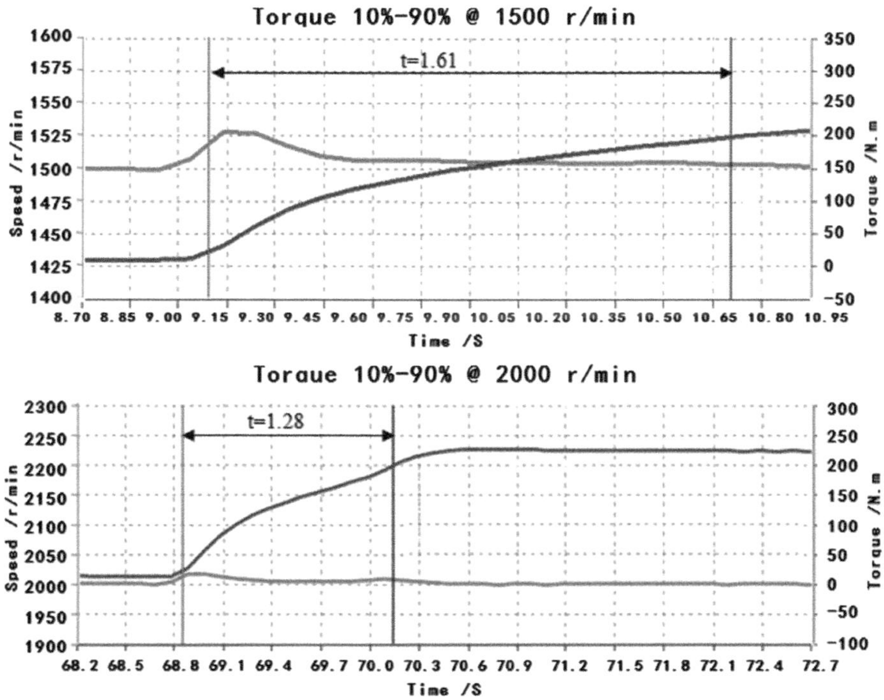

Fig. 20 Transient response of turbo 1

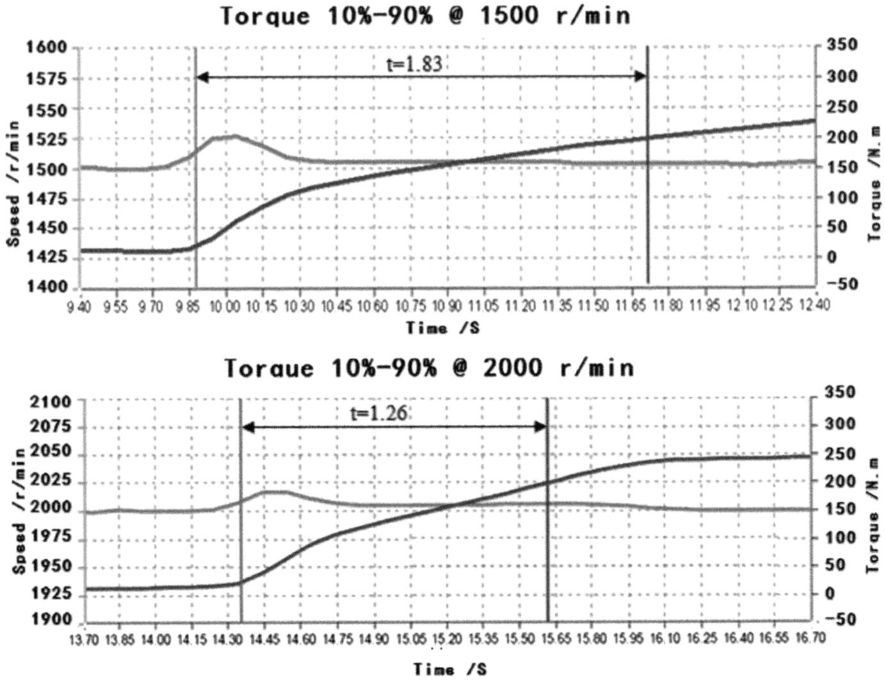

Fig. 21 Transient response of turbo 2

Table 5 is transient response time comparison results of two turbochargers. Response time at 1500 r/min of turbo case 1 is shorter than turbo case 2, but response time at 2000 r/min of turbo case 1 is longer than turbo case 2.

Figure 22 is torque slop from knee to 90% maximum torque. We can see that this engine has very good transient response capacity.

An overall consideration of full load performance, part load fuel consumption, altitude capacity and transient response capacity, we decide to choose turbo case 1.

5.2 *Catalyst Heating Test Performance*

We choose 1200 r/min-1 bar BMEP to represent catalyst heating behaviour. Lower specific HC emission and higher port specific heat flow is better for emission. Higher heat flow can heat catalytic converter faster, so that the catalyst can fast light off. We control 20 °C coolant temperature and 20 °C oil temperature. Figure 23 is results of catalyst heating. We can see that this engine has very low HC emission and very high heat capacity.

Table 5 Transient response time comparison results of two turbochargers

Speed (r/min)	1500	2000
Torque (N m)	10 → 90%	10 → 90%
Turbo case 1 (s)	1.61	1.28
Turbo case 2 (s)	1.83	1.26

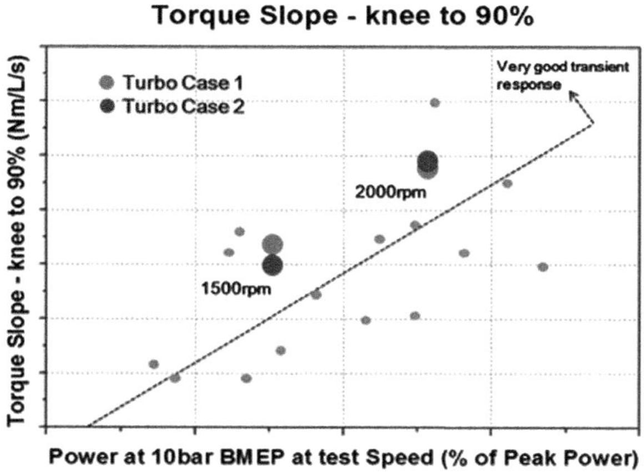

Fig. 22 Torque slope comparison of two turbochargers

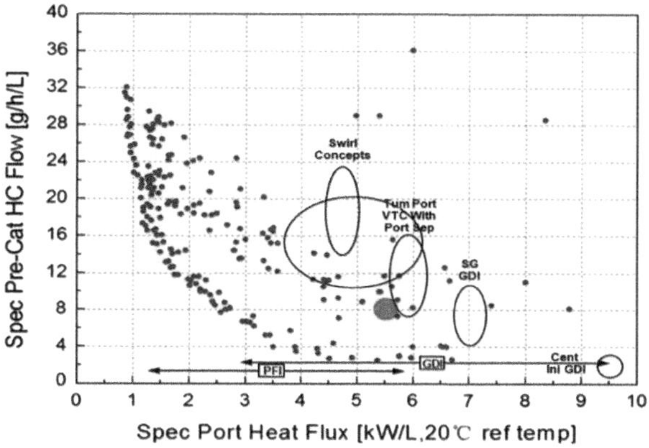

Fig. 23 Catalyst heating results

5.3 *Oil Dilution Test Results*

Oil dilution not only impacts life of lubricant oil but also thin oil film between liner and piston. Oil dilution is very important for direct injection engine. The oil dilution test including 5500 r/min-full load, 1500 r/min-full load, 2500 r/min-12 bar BMEP on engine of cold condition and a cycle test of 6 h duration. Figure 24 is the results of oil dilution test. We can see that the oil dilution ratio at every test condition is less than 6%.

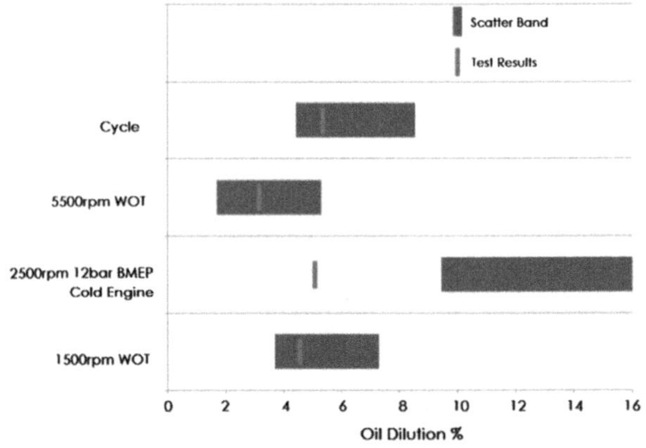

Fig. 24 Results of oil dilution test

5.4 Final Results of Performance Development

After completing whole performance test, maximum power of this engine achieves 100 kW, specific power achieves 72 kW/L. Maximum torque of this engine achieves 220 N m from 1500 to 4350 r/min, specific toque achieves 158.2 N m/L. The result is shown in Fig. 25. BSFC at 2000 r/min-2 bar BMEP achieves 373 g/kWh as Fig. 26.

6 Irregular Combustion Test Research

Per-ignition is an abnormal combustion phenomenon in cylinder before normal spark plug ignition. Super knock from pre-ignition is more destructive than conventional knock, and the mechanism is still not very clear. Pre-ignition easily appear

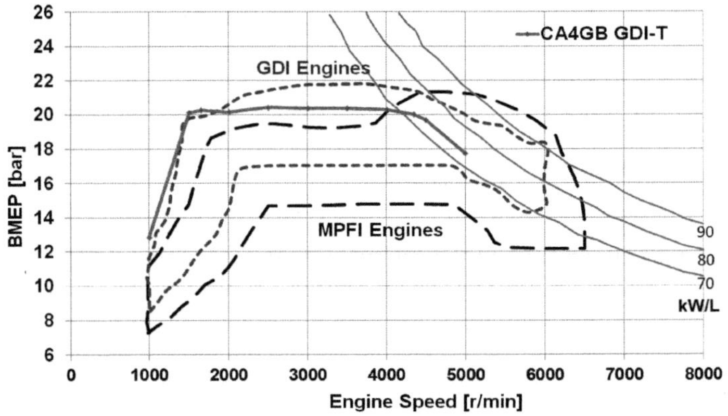

Fig. 25 Full load curve of this engine

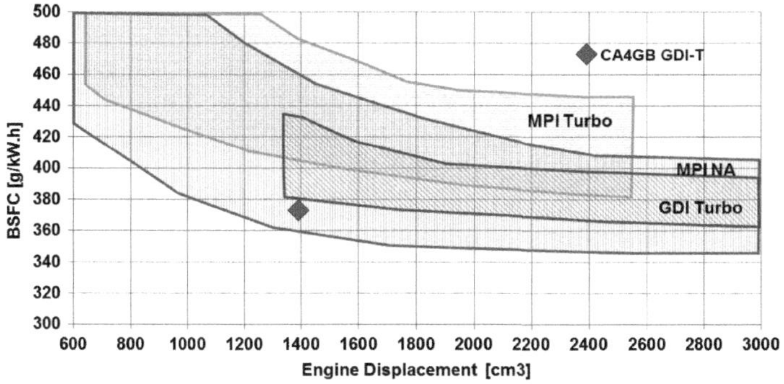

Fig. 26 BSFC at 2000 r/min-2 bar BMEP (FEV Benchmark)

at low engine speed and high load. In this research, we use optical fibre-testing system named VisioSet. This system can measure conventional knock, pre-ignition, post-ignition and flame propagation speed for direct injection engine. Fibre transducer is shown in Fig. 27. This is the key components in the test system and is integrated in spark plug. There are 70 channels in fibre transducer including four layers. It is shown in Fig. 28. There are 28 channels in the first layer and the second layer. There are seven channels in third layer and fourth layer. From first layer to third layer are mainly used for measuring knock, and the fourth layer is mainly used for flame kernel formation and flame propagation direction after ignition.

Lower diagram of Fig. 29 is the test cycle for pre-ignition test. We can see that maximum cylinder pressure is no more than 10 MPa in the first 5400 s time. So we think that there is no pre-ignition phenomenon during this period. After operating condition is changed to 1500 r/min-full load, there are 3 times five cycles that

Fig. 27 Fibre transducer integrated spark plug

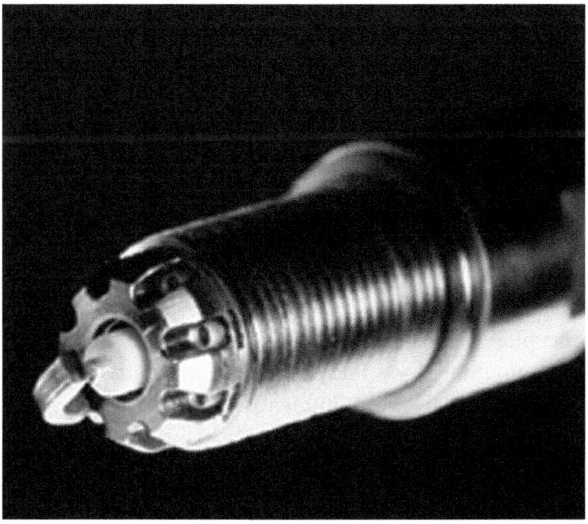

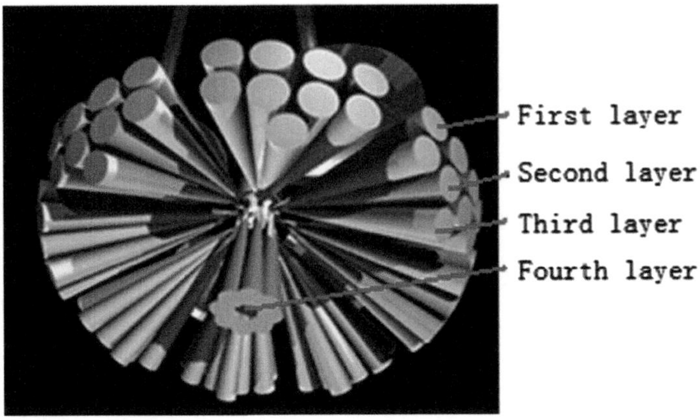

Fig. 28 Channel of transducer

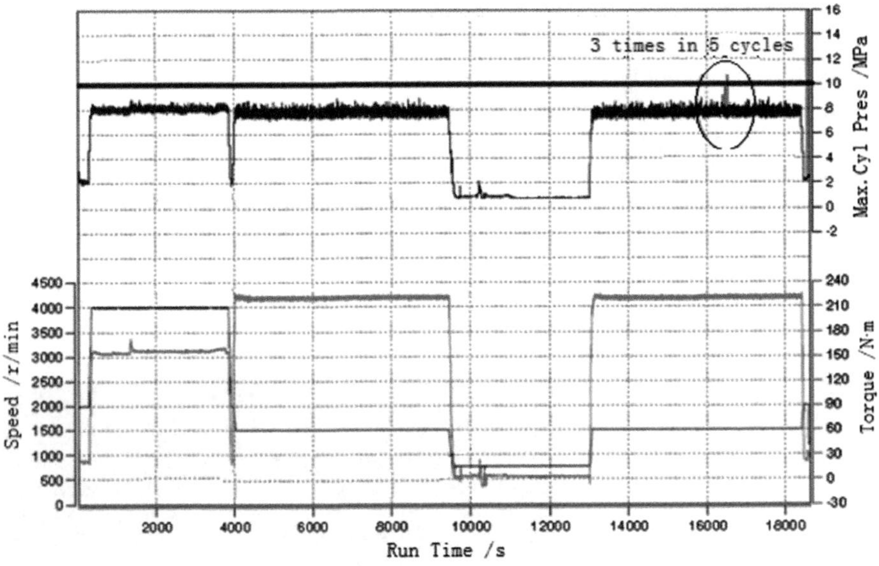

Fig. 29 Pre-ignition status with single injection strategy

maximum cylinder pressure exceed 10 MPa. Generally, we think that it is acceptable for one-time pre-ignition in 1 h. So we think that there is pre-ignition phenomenon if we use single injection strategy.

Figure 30 is light intensity distribution in cylinder. We can see that there is uneven distribution of light intensity phenomenon in cylinder obviously. The light is stronger on pre-ignition position than others position. That shows that there is

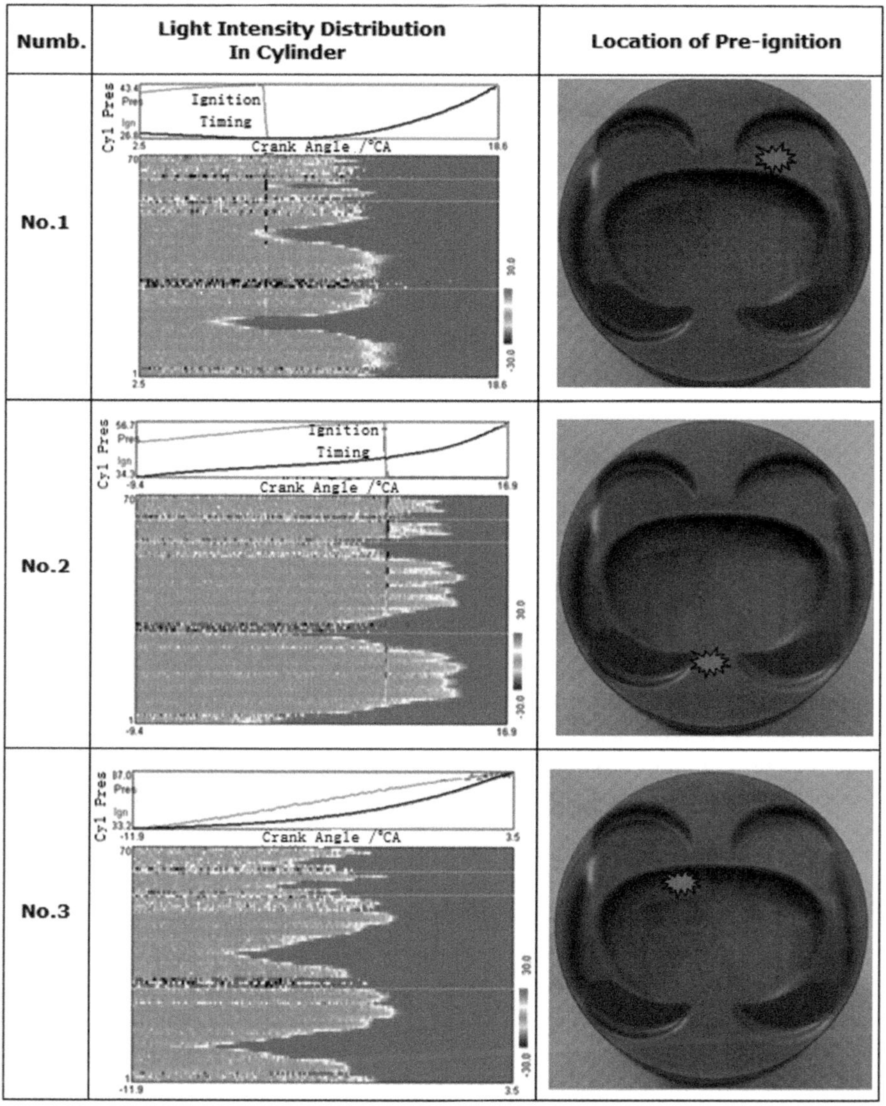

Fig. 30 Light intensity distribution in cylinder

intense burning on pre-ignition position. We also see that the position of pre-ignition is not fixed, it means that pre-ignition is sporadic and uncertainty.

Then, we change the injection strategy into split injection. The results are shown in Fig. 31. We can see that maximum cylinder pressure is no more than 10 MPa in the two 5400 s durations. It means that there is no pre-ignition phenomenon in cylinder because when we use split injection strategy, the distance of fuel penetration should be shorter. It can avoid fuel wall wetting on single injection condition.

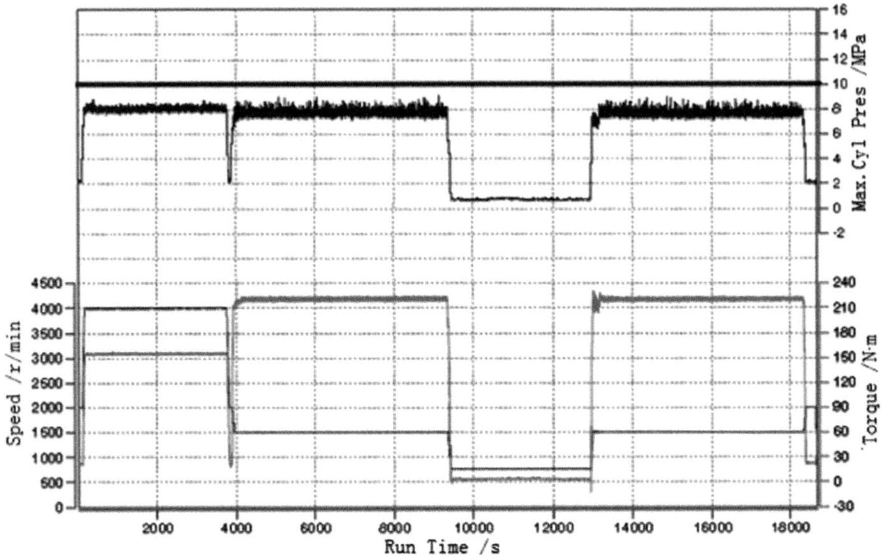

Fig. 31 Pre-ignition status with split injection strategy

Meanwhile using split injection strategy can improve air movement in spark plug, promote the mixing of fuel and air, reduce the pre-ignition tendency. But because pre-ignition is sporadic and uncertainty, we cannot control the pre-ignition by using split injection strategy completely. So split injection only can be a calibration method for restraining pre-ignition.

7 Simulation Results in Vehicle

Table 6 is simulation results of dynamic performance and fuel consumption in vehicle. The maximum vehicle speed is 210.5 km/h; acceleration time of 0–100 km/h is 10 s. NEDC integrated fuel consumption is 5.85 L/100 km which is 15% lower than China's Stage III fuel consumption regulation.

Table 6 Simulation results in vehicle

Item	Unit	Target value	Simulation results
Max. vehicle speed	km/h	208	210.5
Acceleration time [0–100 km]	s	10.2	10
Fuel consumption [NEDC cycle]	L/100 km	6.9 (China's Stage III)	5.85

8 Summary and Limitations

We have developed a new GDI-T gasoline engine by thermodynamic simulation, optical engine test and performance development test, etc. Finally, NEDC-integrated fuel consumption matched this GDI-T engine is 15% lower than China's Stage III fuel consumption regulation meanwhile meeting the dynamic performance of the vehicle.

We have finished engine calibration work and have not finished the vehicle calibration. So the vehicle performance data are the simulation data.

Crash Energy-Absorption Structure Design Based on an Innovated Leightweight Technology

Weizhao Gao, Yiwen Li, Xuewu Zhu, Zhou Yao and Ledan Liu

1 Introduction

During vehicle design process, lightweight is always viewed as an efficient way to cut off energy consume and gas emission. Besides, lightweight could either bring great contribution to the performance of vehicle in crashing safety, ride comfort, NVH property etc., while among them, the crashworthiness-based lightweight process had attracted relative more attention [1, 2]. Theoretically, the vehicle crash safety behavior can be promoted from two directions. One is to enhance the material strength to promote the stiffness of the vehicle's living cab [3, 4], while the other is to lower the sum of crash energy by weight reduction [5]. Since the material substitution may sometimes increase the research and develop cost and lower the profits, lightweight is relatively more attractive to most vehicle design engineers.

Aiming at promoting the lightweight technology into a higher level, a number of experts have invested great efforts to develop efficient lightweight methods in the passed two decades. As seen from the ever-published literature, it indicates that lightweight methods can be mainly classified into three aspects, those are to introduce lighter materials like aluminum alloy, magnesium alloy, plastic, and composite materials [6, 7], to use newly originated manufacturing process like TWB and TRB [8, 9], and to carry out structure optimization by topology or numerical optimization technology [10]. All in all, the abovementioned ways are merely limited the lightweight process into the macro-view, that means the optimal structures have no essential differences by comparison of the traditional ones. Actually, the lightweight could either be achieved through microstructural design

W. Gao (✉) · Y. Li · X. Zhu · Z. Yao · L. Liu
China FAW Co., Ltd. R&D Center, Changchun 130000, China
e-mail: gaoweizhao@rdc.faw.com.cn

Society of Automotive Engineers of China (SAE-China) (ed.), *Proceedings of the 19th Asia Pacific Automotive Engineering Conference & SAE-China Congress 2017: Selected Papers*, Lecture Notes in Electrical Engineering 486, https://doi.org/10.1007/978-981-10-8506-2_3

[11–13]. By carefully combining a number of microelements into a macro-component, the lightweight target can be controlled to a more accurate level and may sometimes generate a relative lighter task.

In this paper, an innovated lightweight technology based on the negative Poisson's ratio structure and space truss composite material is discussed. A new type of crash box and bumper assembly is designed and trial-manufactured under the crashworthiness targets. The objective is to investigate the application feasibility of this new lightweight technology and provide a new vision for the lightweight design of car body.

2 Theory and Methodology

It is widely acknowledged that the well-designed crash box and bumper assembly (Fig. 1), which serves as a significant energy-absorbing component, may greatly reduce passengers' injuries when vehicle crash accidents happen [14, 15]. Considering the crash box, since its primary function is energy absorption, the higher energy absorption capability along with lower breaking crash force behavior was always treated as the key characteristic of a better design. While for the frontal crash bumper, its key function in the crashing process is to maintain the lateral stability. Therefore, the bending stiffness of the bumper attracts relative more attention than the energy absorption. Therefore, the lightweight design of crash box and bumper should be treated differently when searching for the optimal task.

2.1 The Negative Poisson's Ratio Structure

To reconstruct a substituted crash box from the micro-view, the negative Poisson's ratio structure is introduced, and to provide a more specific view of its appearance and working principle under the axial crash load, the microstructure and

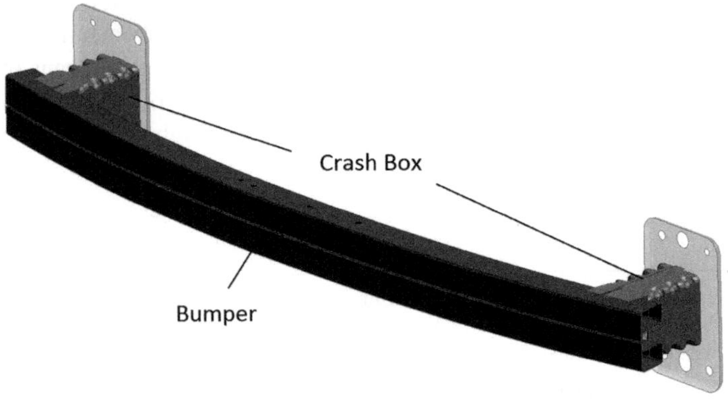

Fig. 1 Crash box and bumper assembly

deformation mode are shown in Fig. 2. As seen from it, this sort of structure is made up of a series of micro-components which is composed by two sorts of truss elements. When crash accidents happen, element 1 and element 2 can undertake stretch and compression force respectively, and the composition of the enormous micro-deformations may generate a progressive three-dimensional compression phenomenon, which can be defined as a negative Poisson's ratio deformation mode. By comparison with the traditional crash box, this kind of structure has two main advantages. Firstly, the design and optimization process are more flexible by adjusting fewer parameters, thus to sustain the crashworthiness target at a more accurate level. Secondly, the crash force may increase more steadily because of its space construction, which can further reduce the breaking force and weight on basis of the same level of crash energy.

The manufacturing process of negative Poisson's ratio structure is shown in Fig. 3. To attain the basic parts of the structure, the metal sheet has to be cut into net-shaped components. After then, the stamping process is followed to form the two sorts of elements with the input design parameters theta. The final step is to carry out a welding process, to combine the elements into a complete structure. Additionally, the manufacturing process of this negative Poisson's structure is not merely limited into that shown in Fig. 3, for that the newly originated manufacturing process like three-dimensional printing process could also be applied herein to produce higher quality structures.

2.2 *The Space Truss Composite Material*

While on the reconstruction process of the frontal crash bumper, the space truss composite material (Fig. 4), which is originated from a bionic consideration, is applied herein. Generally, this sort of structure constitutes three main parts, the outer skin, flesh, and inner bones. In this work, the glass fiber composite material panel is utilized as the skin, the foam is filled herein as the flesh, and the aluminum hollow tubes linked by steel wire are used as bones. Theoretically, when the frontal crash happens, the skin can sustain the integrity of the structure, the bones can provide an anti-force to prohibit the bending deformation, and the flesh can evaluate

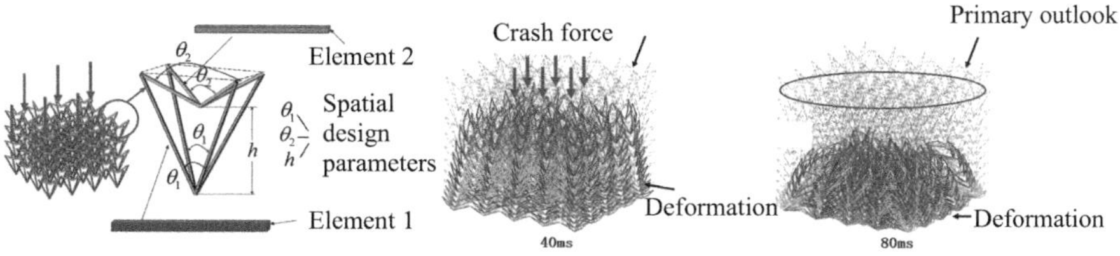

Fig. 2 Negative Poisson's ratio structure and its deformation mode

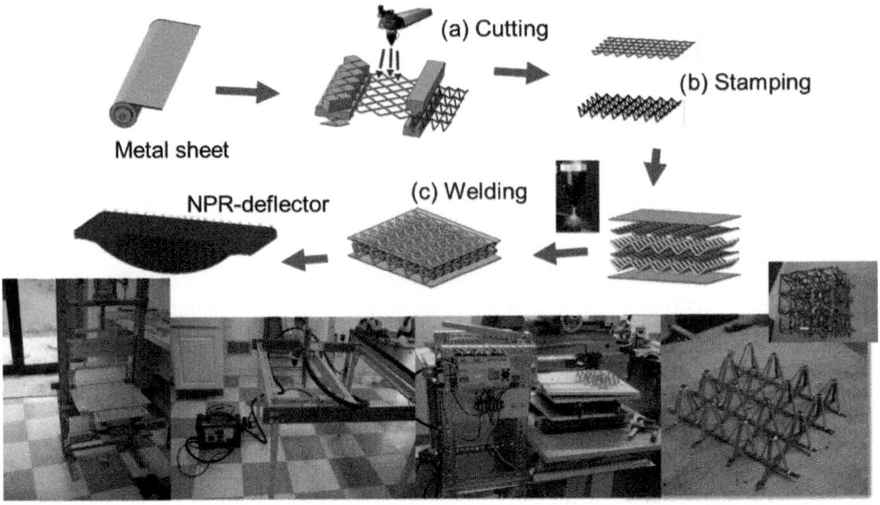

Fig. 3 Manufacturing process of negative Poisson's ratio structure

the stability of the deformation process. By combing the deformation characters of the three main components, the anti-bending capability can be greatly promoted.

The manufacturing process is shown in Fig. 5. As seen from it, the aluminum tube and steel wire has to be precisely produced firstly, which will be followed by a composition process to combine the two components into intact bones. After then, the bones have to be bonded with the ever prepared composite material panels by the structural glue. Lastly, the foam injection into the gap will help to gain the needed space truss composite materials. While among the whole process, the bonding step deserves more attention, for that the weaker adhesive connection will result in a sharp decline in the lateral stability.

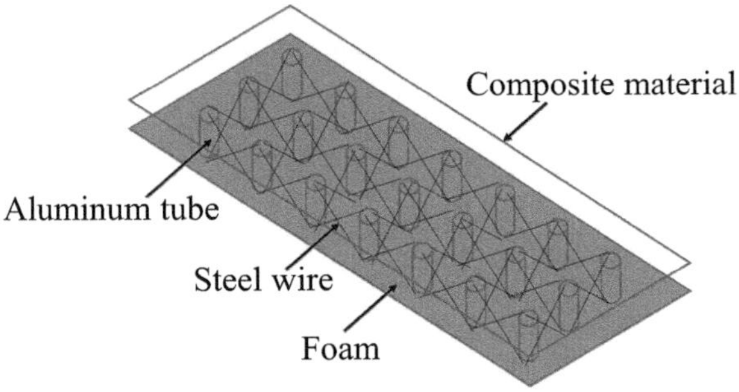

Fig. 4 Space truss composite material

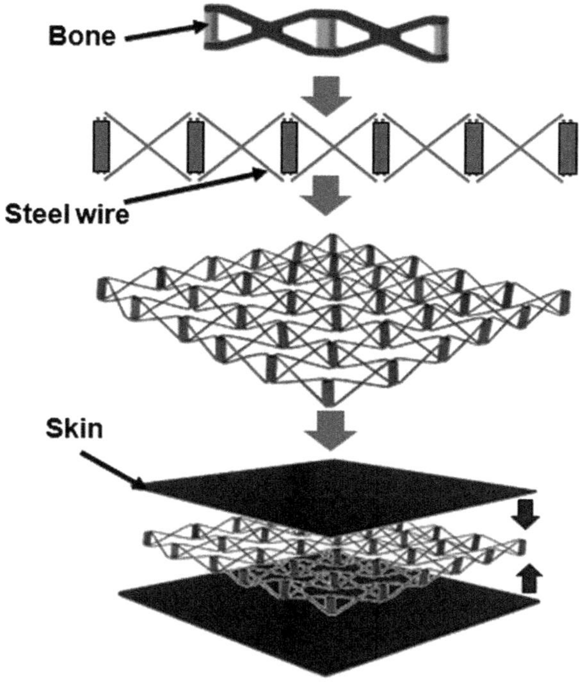

Fig. 5 Manufacturing process of space truss composite material

2.3 *The Physical Test Arrangement*

2.3.1 Test of Crash Box

To fully investigate the crashworthiness of the newly designed crash box, two kinds of physical test is applied herein. At the beginning, the quasi-static compression test, which is shown in Fig. 6, is arranged to validate the deformation mode as we predicted.

Besides, to conduct a more accurate sensitivity analysis between design parameters and crashworthiness behaviors during the high speed crash, the hammer dropping test, with a 50 kg weight and 4.3 m height, is introduced herein. The specimen of the negative Poisson's ratio structure-based crash box is shown in Fig. 7. As seen from it, two thin metal panels are served as the crush front and three layers of negative Poisson's ratio structures are filled among them. By defining the thickness with different value per layer and with or without foam injection into the microstructure, three tasks are originated and listed in Table 1, thus to dig out the correlation between design parameters and crashworthiness (Fig. 8).

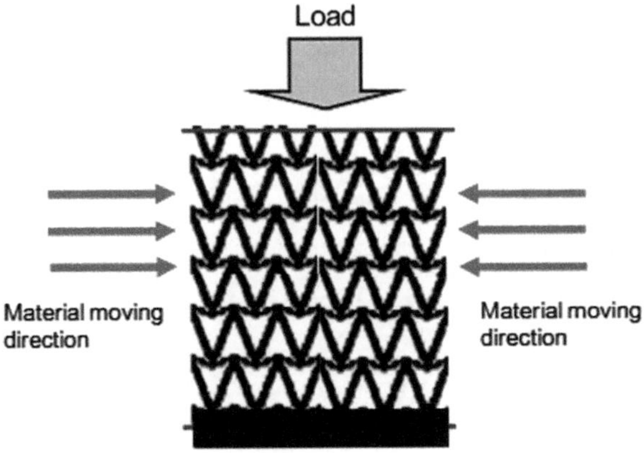

Fig. 6 Quasi-static test of crash box

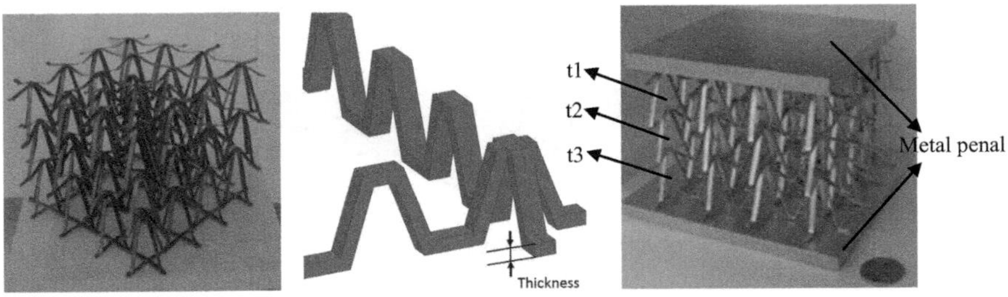

Fig. 7 Negative Poisson's ratio structure

Table 1 Tasks of crash box specimens

Task No.	t1 (mm)	t2 (mm)	t3 (mm)	Foam-filled
C1	1.0	1.2	1.8	Yes
C2	1.0	1.0	1.0	No
C3	1.0	1.0	1.0	Yes

2.3.2 Test of Crash Bumper

To prove the anti-bending capability of the space truss composite material structure, we need firstly to find evidence to demonstrate that the bone can really promote the anti-deformation ability under the axial load, which is one of the key properties for sustaining the deformation stability. Therefore, the quasi-static test (Fig. 9) is applied to validate this view.

After then, it is necessary to find out which component is highly related with the bending stiffness, thus to provide engineers a more accurate guidance in designing a space truss composite-based bumper. So that the quasi-static three-point bending test (Fig. 10) is applied to realize this purpose.

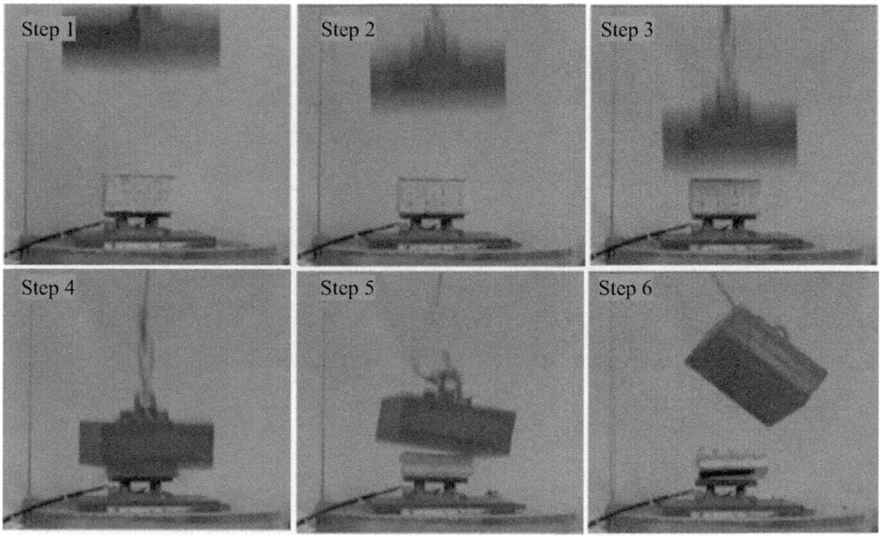

Fig. 8 Hammer dropping test of the crash box

Fig. 9 Quasi-static test of bumper

To carried out these two kinds of physical tests, five different specimens are proposed either, which are listed in Table 2. The differences among them are the foam filler and bone structure. Specifically, tasks B1 and B3 have all the components mentioned above, while tasks B2 and B5 have no bones, task B4 has no steel wire in the bones by comparison with task B1.

2.3.3 Test of the Reconstructed Assembly

Based on the abovementioned physical component test, the optimal task form of crash box and bumper can be attained. After then, the crash box and bumper

Fig. 10 Three-point bending test of bumper

Table 2 Tasks of bumper specimens

Test name	Task No.	Skin	Foam	Aluminum tube	Steel wire
Quasi-static	B1	Yes	Yes	Yes	Yes
	B2	Yes	Yes	No	No
Three-point bending	B3	Yes	Yes	Yes	Yes
	B4	Yes	Yes	Yes	No
	B5	Yes	Yes	No	No

Fig. 11 A class car of FAW

assembly will be carefully optimized according to crashworthiness behavior of the prototype, an A-class car from FAW, shown in Fig. 11. To validate its lightweight efficiency, a frontal crash sled test with a 923 kg weight and 15.6 km/h crash velocity is recommended, which is shown in Fig. 12.

Fig. 12 Sled test of crash box and bumper assembly

3 Results and Analysis

The deformation mode result of the negative Poisson's ratio crash box quasi-static test is shown in Fig. 13. It shows that the gradually compression deformation in axial direction can result in the shrink of the other two dimensions, which means that the negative Poisson's ratio deformation phenomenon is emerged. Besides, the force versus compression distance curve is shown in Fig. 14. It indicates that the breaking force of the negative Poisson's ratio structure is not the peak force, which is quite different with the traditional structures. Additionally, this curve either suggests us that the anti-load rises gradually with fluctuations along with the increase of compression displacement, which is viewed as the needed characteristic for a crash box design.

The deformation mode of crash box hammer dropping test is shown in Fig. 15, and the force versus time data, peak force, and residual length are recorded during the hammer dropping test, which are shown in Fig. 16 and Table 3, respectively. It can be seen from the results that the load of the three tasks increased with the same tendency among the first 8 ms, while the force versus time curve of task C3 became relative moderate during the left time, with only a 60% peak force by comparison with the other two tasks. Therefore, we can conclude from the comparison between tasks C1 and C3 that the equal value scattering of layer thickness perform better

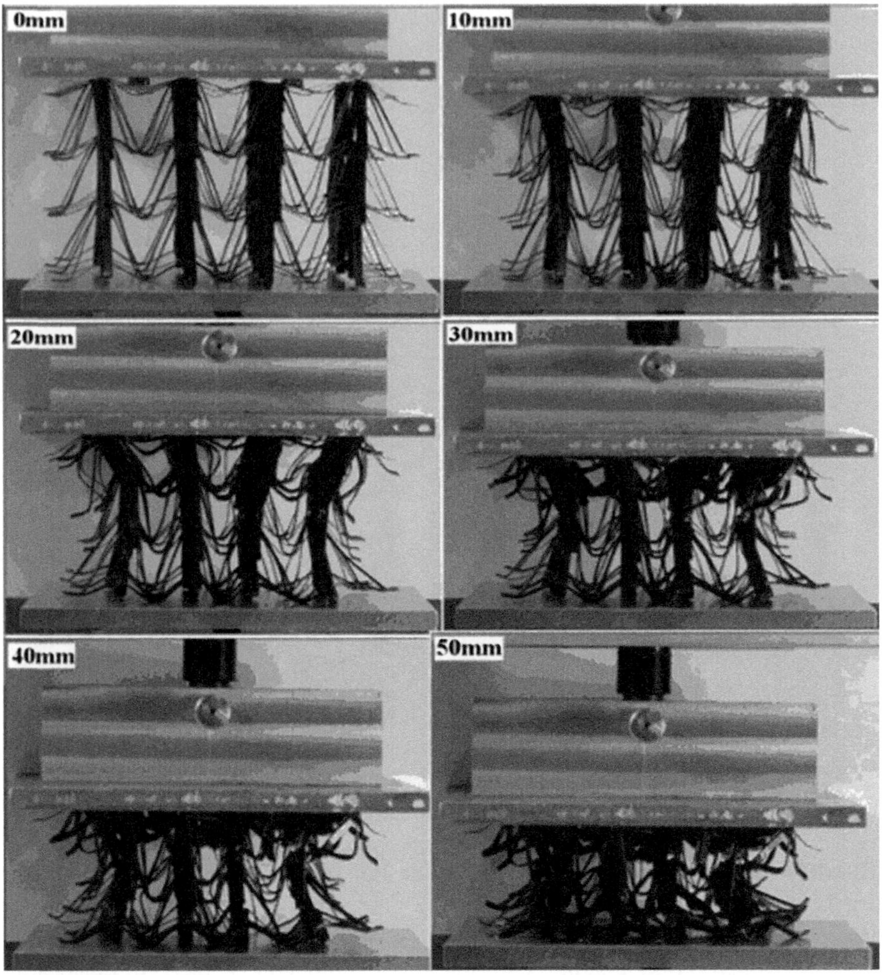

Fig. 13 Deformation mode of negative Poisson's ratio crash box's quasi-static test

Fig. 14 Results of crash box's quasi-static test

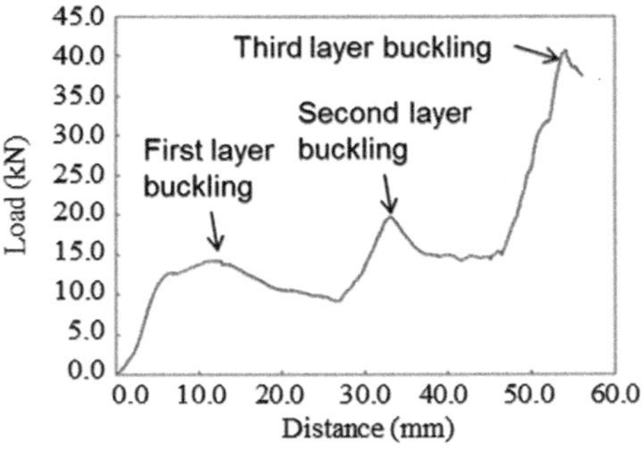

Fig. 15 Deformation of negative Poisson's ratio crash box's hammer dropping test

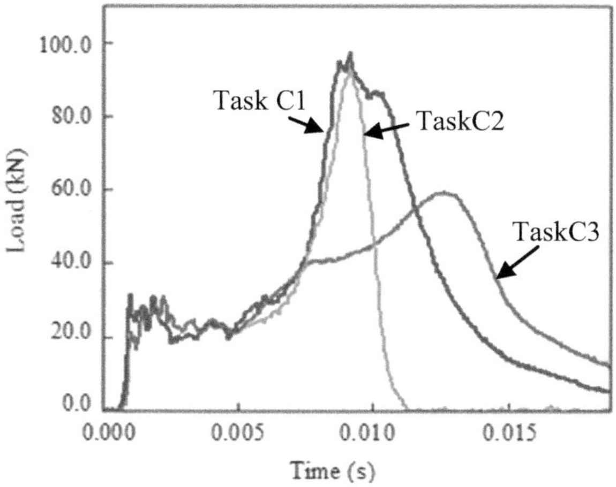

Fig. 16 Results of the crash box's hammer dropping test

Table 3 Results of the crash box's hammer dropping test

Task No.	Peak force (kN)	Residual height (mm)
C1	95.9	49
C2	90.8	13.5
C3	58.9	41.5

crashworthiness behavior than the gradually rising of thickness defining. Additionally, it is also obviously to see that the foam injection into the gap of the microstructure can greatly enhance the deformation stability and reveal an ideal crash behavior. Therefore, task C3, which has the same thickness scattering and with foam injection, is suggested to reconstruct the crash box of the prototype in this work.

The force versus compression distance curve of bumper quasi-static test is shown in Fig. 17. By comparison, we can see that the peak force of task B1 is nearly 10 times as that of task B2, which means that the bone, which is composed

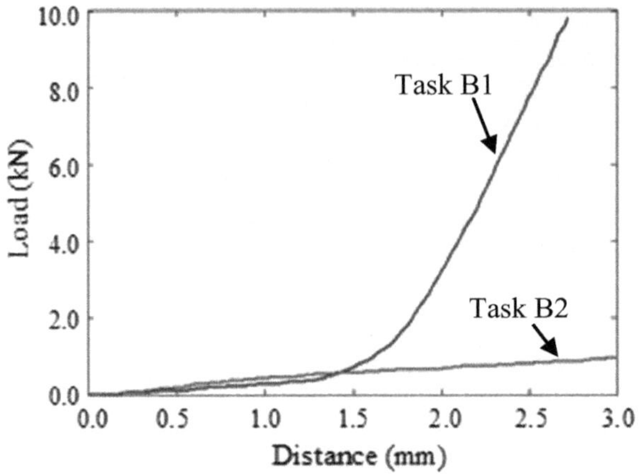

Fig. 17 Results of the bumper's quasi-static compression test

by aluminum tube and steel wire, plays an important role in elevating the axial compression stuffiness.

The force versus deformation value of the three-point bending test of crash bumper is recorded, as shown in Fig. 18. Besides, the deformation mode is shown in Fig. 19. The results show that the anti-force of task B3 is nearly three times as that of the other two tasks on the same deformation value, which indicates the bending stiffness of task B3 is relative higher either. Therefore, we can conclude that not only the aluminum tube but also the steel wire cannot be eliminated from the structure, if the bending stiffness is in the concern. Therefore, the space truss composite material of task B1 or B3 is suggested to reconstruct the crash bumper of the prototype.

Fig. 18 Results of bumper's the three-point bending test

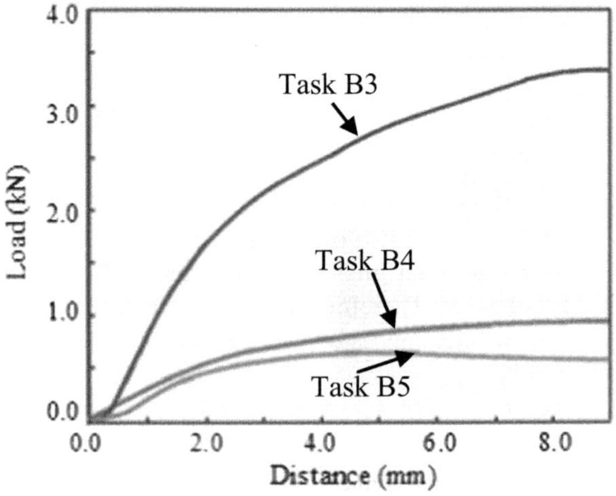

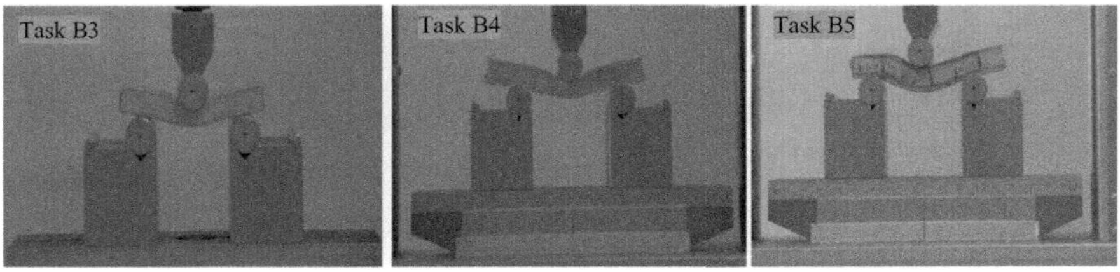

Fig. 19 Deformation of the newly constructed bumper's three-point bending test

On the basis of the above analysis, Task C3 of crash box and Task B1/B3 of bumper are recommended to reconstruct the crash box and bumper assembly herein. According to crashworthiness behavior of an A-class car from FAW, the crash box and bumper assembly based on the microstructure are redesigned and optimized through CAE analysis by using Lsdyna commercial software. Through a number of optimization processes, the optimal microstructure-based crash box and bumper assembly is attained, which is shown in Fig. 20. The deformation record of sled test by CAE and physical test is shown in Fig. 21, and the acceleration versus time curve comparison is presented in Fig. 22. As seen from the results, the deformation mode and acceleration tendency between the simulation and experimental have high consistency, which indicates that the CAE model of the crash box and assembly has high accuracy and can be utilized to drive an optimization process.

The sled tests of prototype and optimal newly designed task are both carried out to make a more detailed comparison. In the crashworthiness research, total energy absorption, average crash force, and SEA, which are listed in Eqs. (1)–(3), are widely used in many literature to value the crashworthiness of a typical structure [3]. And under the same test boundary conditions, better design always means higher value for all the three parameters mentioned above.

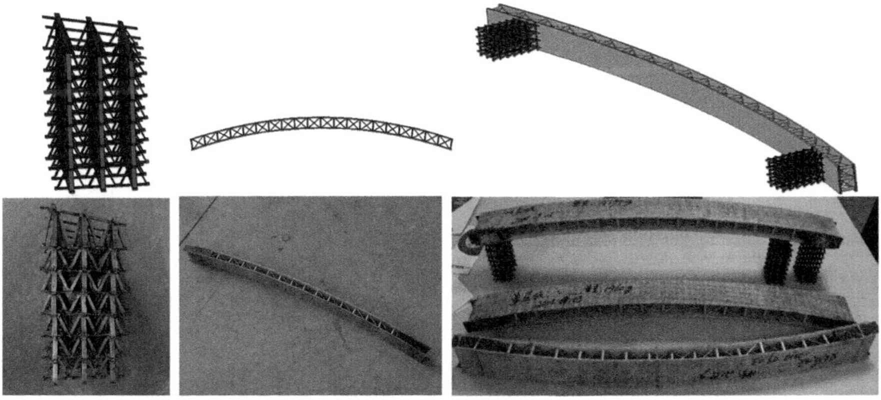

Fig. 20 Optimal reconstructed crash box and bumper assembly

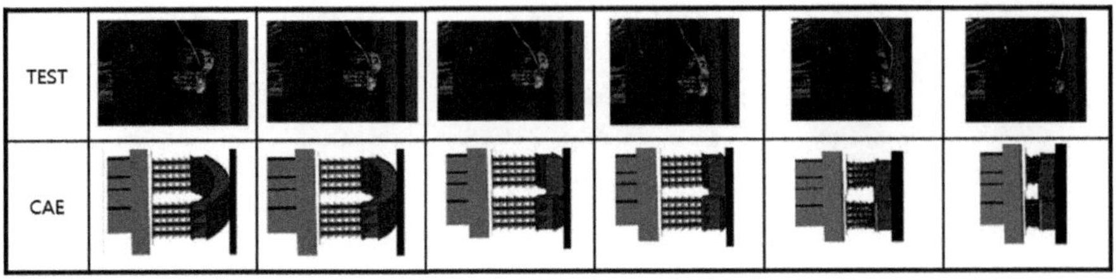

Fig. 21 Comparison of CAE and physical sled test of reconstructed crash box and bumper assembly

Fig. 22 Acceleration versus time comparison between CAE and physical sled test

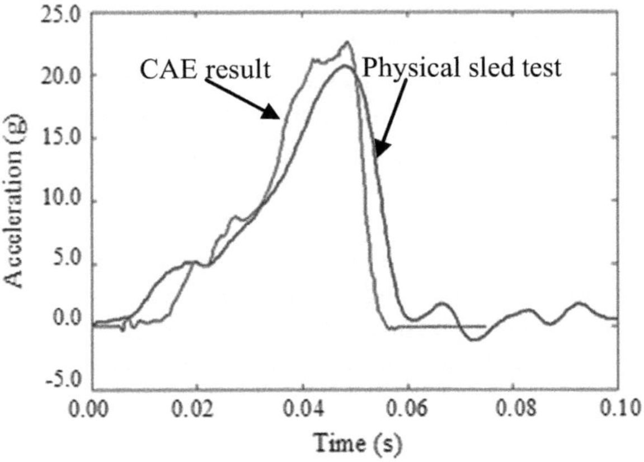

$$E = \int f \mathrm{d}x \tag{1}$$

where crash force f versus deformation x data are used herein.

$$F = \frac{E}{d} \tag{2}$$

where E and d are the energy absorption and deformation value.

$$\mathrm{SEZ} = \frac{E}{m} \tag{3}$$

where E and m are the energy absorption and mass value respectively.

The acceleration of the sled test is shown in Fig. 23. It indicates that the peak acceleration of the newly designed structure is higher than that of the prototype. The results of the sled test are listed in Table 4. As can be seen from it, the energy absorption and average crash force of the new task are 8.188 kJ and 92.0 kN, which come to a 12.39 and 2.45% increase, respectively. Besides, the SEA rises from 1.351 to 2.767 kJ/kg by comparison with the prototype, which further approves the

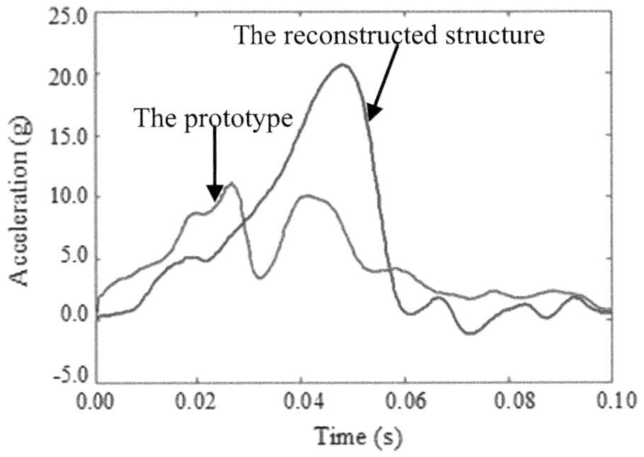

Fig. 23 Acceleration versus time comparison between the prototype and the reconstructed assembly

Table 4 Comparison of sled results

Parameters	New task	Prototype	Percentage of improvement (%)
E (kJ)	8.188	7.285	12.39
F (kN)	92.0	89.8	2.45
SEA (kJ/kg)	2.767	1.351	104.81
M (kg)	2.959	5.392	45.12

excellent capability of the new task in energy absorption capability. Furthermore, the total mass of the new task is merely 2.959 kg, with a 45% cut off by comparison with prototype. All these abovementioned evidences indicate that the potential capability of negative Poisson's ratio microstructure and space truss structure technology in the crashworthiness-based lightweight design is enormous.

4 Conclusions

The innovated lightweight method discussed in this paper brings a micro-vision of structure design. The component test results indicate that the negative Poisson's ratio structure and space truss composite material do have advantages in the crash box and bumper design. And the sled crash test results either show that the deformation mode of the new task could be under control, and the energy-absorption capability increased dramatically with a significant weight reduction compared to the prototype.

Additionally, it can be seen from this paper that the lightweight design from a micro-view will be a future trend. A number of other microstructures might be originated by researchers around the world in the next few years. By combination with the new manufacturing technologies, such as three-dimensional printing processes, the microstructure will gradually release its potential capability in lightweight area and will provide vehicle engineers with more choices to drive a lightweight design.

Acknowledgements This work is supported by the Technological Innovation Project No. kg 15034 of China FAW Co., Ltd. The authors also would like to express their thanks for the guidance and support from Ma Z. D. and Zhou Z. from University of Michigan.

References

1. Wang H, Li EY, Li GY (2009) The least square support vector regression coupled with parallel sampling scheme metamodeling technique and application in sheet forming optimization. Mater Des 30(5):1468–1479
2. Vitaly E, Jeya P (2009) Updated evaluation of size and mass effects in front-to-front crashes involving light vehicles. SAE Technical Paper, 2009-01-0375
3. Horst L, Niels P (2014) Benefit structural adhesives in full car crash applications. SAE Technical Paper, 2014-01-0811
4. Sujeet KS, Moqtik AB, Saraf MR (2014) Lightweight materials and their automotive applications. SAE Technical Paper, 2014-28-0025
5. Velayudham G, Javier R, Robert J (2014) Collaborative design optimization for light weight design. SAE Technical Paper, 2014-01-0392
6. Hirsch J, Al-Samman T (2013) Superior light metals by texture engineering: optimized aluminum and magnesium alloys for automotive applications. Acta Mater 61(3):818–843
7. Schweizerhof K, Weimar K, Munz T, Rottner T (1998) Crashworthiness analysis with enhanced composite material models in LS-DYNA merits and limits. In: LS-DYNA world conference, Detroit, Michigan, USA
8. Pan F, Zhu P, Zhang Y (2010) Metamodel-based lightweight design of B-pillar with TWB structure via support vector regression. Comput Struct 88(1–2):36–44
9. Pallett RJ (2001) The use of tailored blanks in the manufacture of construction components. J Mater Process Technol 117(1–2):249–254
10. Kim Heung-Soo (2002) New extruded multi-cell aluminum profile for maximum crash energy absorption and weight efficiency. Thin-Walled Struct 40(4):311–327
11. Zhang W, Ma ZD, Hu P (2013) Mechanical properties of a cellular vehicle body structure with negative Poisson's ratio and enhanced strength. J Reinf Plast Compos 33(4):342–349
12. Ma ZD, Wang H, Kikuchi N et al (2003) Function-oriented material design for next-generation ground vehicles. ASME 2003 international mechanical engineering congress and exposition
13. Wang H, Ma ZD, Raju B (2006) Function-oriented fastener design for composite armor under ballistic impact. 21st annual technical conference of the American society for composites, Dearborn, Michigan
14. Zarei H, Kroger M (2008) Optimum honeycomb filled crash absorber design. Mater Des 28 (1):193–204
15. Li YW, Xu T, Xu TS et al (2010) Optimal design of energy-absorbing structure of autobody under low-speed crash. Trans Beijing Inst Technol 30(10):1175–1179

Research on CFD Steady-State Simulation Analysis and Optimization for Defrosting of Passenger Vehicles

Chengsheng Zhang

Nomenclature

A	Surface area (m^2)
c_p	Specific heat [J/(kg K)]
C_{emp}	Coefficient of heat absorption for ice
h_{ext}	Heat transfer coefficient [W/(m^2 K)]
$h_s(T_{ice})$	The enthalpy at the corresponding temperature of the ice layer (J/kg)
$h_L(T_F)$	The enthalpy at the corresponding freezing temperature after melting into water (J/kg)
L	Latent heat flux (W)
Q_{ext}	Heat flux (W)
q	Heat conduction of glass into ice (W)
s	Ice thickness (m)
T_{ext}	Ambient temperature (K)
T_{ice}	Ice temperature (K)
T_F	Freezing point temperature (K)
ρ_s	Ice layer density (kg/m^3)

1 Introduction

In the cold conditions, the automobile air conditioning defrosting and demisting system plays an important role in ensuring the driver's vision and driving safety, and China's laws and regulations of GB 11555-2009 to the field of defrosting area and time also made specific requirements [1]. In view of the defrosting performance in traffic safety and laws and regulations play an important role, so it's used as an

C. Zhang (✉)
Vehicle Integration Department, AIWAYS Automobile (Shanghai) Co., Ltd., Room102, Building D, No. 1687 Changyang Road, Changyang Campus, 200082 Yangpu District, Shanghai, China
e-mail: chason.zhang@ai-ways.com; 372393810@163.com

Society of Automotive Engineers of China (SAE-China) (ed.), *Proceedings of the 19th Asia Pacific Automotive Engineering Conference & SAE-China Congress 2017: Selected Papers*, Lecture Notes in Electrical Engineering 486, https://doi.org/10.1007/978-981-10-8506-2_4

important vehicle performance indicator to deal with. Now with the help of CFD analysis, transient defrosting analysis method has been widely used in automobile factories, but with the development of the automotive industry, how to meet the objectives and requirements of air conditioning system in a shorter development cycle have become the focus of project development.

In the analysis process, if using transient method to optimize, which will spend more optimization time and computing resources, and if use the steady-state analysis optimization, there is no verification of the target wind speed evaluation standards. The industry target wind speed evaluation indicators are more, including 1, 1.5, 2, and 2.5 m/s are used by some enterprises. The reason is due to the role of air viscosity, so that the inner surface of the glass forms a velocity boundary layer; the velocity value with the distance away from the wall gradually increased. So the first layer of the boundary will cause the situation that the different thickness of the grid makes the different wind speed distribution [2].

In the defrosting performance development of a new large-scale automobile project, the lower boundary of vision area B is very close to the junction of glass and instrument panel, which is only 0.032 m. At the same time in order to avoid the head-up display (HUD) system, the length of main defrosting outlet in *Y* directions is only equal to vision area B's 35.5% in *Y* directions, and the defrosting air volume flow (350 m^3/h) and defrosting time (18 min) of the target are consistent with the original small car (but the area of vision area B is 0.6 m^2 larger than small car). The defrosting outlet and the view area are shown in Fig. 1. So it is difficult to develop the defrosting performance.

If using the transient method to optimize directly, on the optimization of time may not be able to meet project requirements. Then before this model analysis, application of STAR-CCM+ software to simulation the company existing model for exploring the relationship among the steady-state and transient analysis method and climatic test, and the steady-state model setting and evaluation method are summarized. In this way, the structural optimization of the steady-state analysis is realized, and the development cycle is shortened through the development process of transient analysis confirmation and actual vehicle climate test acceptance [3, 4]. Defrosting analysis only takes the main defrosting as an example in this paper.

2 Mathematical Model

The conventional STAR-CCM+ defrosting analysis consists of two processes: steady-state analysis establishes the initial flow field and then analyzes the transient temperature field. Both the steady and unsteady incompressible three-dimensional

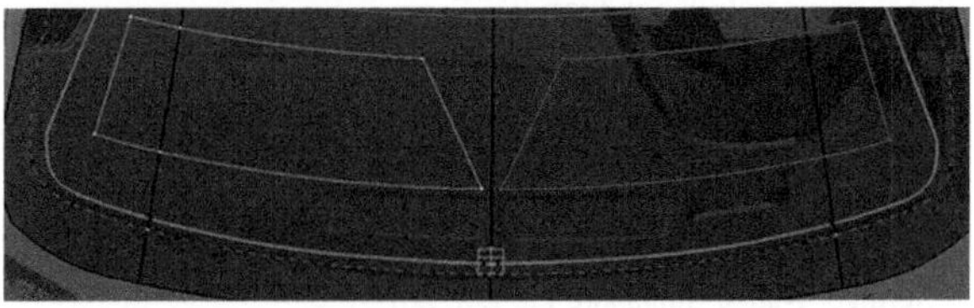

Fig. 1 Front windshield vision area and defrosting outlet of the new development large car

flow fields are used in this simulation, and the temperature field model of convective heat transfer and thermal conduction are also used; all these processes of flow and heat transfer are governed by the laws of conservation of physics. In numerical heat transfer, what we are most concerned about is the mathematical expression of these conservation laws—partial differential equations, usually called the governing equations [5]. The governing equations are as follows.

For incompressible fluids, the fluid density is constant, so the equation of mass conservation is as follows:

$$\frac{\partial(\rho u)}{\partial x} + \frac{\partial(\rho v)}{\partial y} + \frac{\partial(\rho w)}{\partial z} = 0 \tag{1}$$

Momentum conservation equation, also known as the Navier–Stokes equation, the general form is [6]:

$$\frac{\partial(\rho \emptyset)}{\partial t} + \mathrm{div}(\rho u \emptyset) = \mathrm{div}(\Gamma\,\mathrm{grad}\emptyset) + S \tag{2}$$

where ρ is the density, $\emptyset$ is the generic variable, Γ is the generalized diffusion coefficient, S is the generalized source term.

Energy conservation equation is as follows:

$$\frac{\partial(\rho T)}{\partial T} + \mathrm{div}(\rho \vec{v} T) = \mathrm{div}\left(\frac{k}{c_p}\mathrm{grad}\,T\right) + S_T \tag{3}$$

where T is temperature, k is fluid heat transfer coefficient, c_p is specific heat, S_T is source term and represents the internal heat source of a fluid and the part of fluid mechanical energy that is converted to heat by viscous action.

When calculating the transient temperature field, you need to add a thin layer model. The thin layer model considers the convective heat transfer between the interior flow of the cabin and the inner surface of the glass, the convective heat transfer from the outer surface of the ice layer to the surrounding environment, and the conduction of the glass and the ice layer itself.

Wherein the convective heat transfer between the ambient and the ice layer is as follows:

$$Q_{\mathrm{ext}} = h_{\mathrm{ext}}(T_{\mathrm{ext}} - T_{\mathrm{ice}})\mathrm{d}s \tag{4}$$

where Q_{ext} is the heat flux, h_{ext} is heat transfer coefficient, T_{ext} is ambient temperature, T_{ice} is ice temperature.

Convection heat transfer between cabin and the interior surface of glass can be deduced by Eq. 4.

Heat transfer from inner surface to outer surface of glass is as follows:

$$Q_{\mathrm{cond}} = -\lambda A \frac{(T_{\mathrm{gi}} - T_{\mathrm{ge}})}{\mathrm{d}x} \tag{5}$$

where Q_{cond} is the heat flux, λ is thermal conductivity, A is surface area, T_{gi} is internal surface temperature of glass, T_{ge} is external surface temperature of glass.

The defrosting model assumes that the melted ice becomes water flow away, do not re-coagulation into an ice. The corresponding principle is shown in Fig. 2 [7].

Which according to the quality of conservation and assumptions are as follows:

$$\dot{m}_{\mathrm{melt}} = \dot{m}_{\mathrm{run-off}} \geq 0 \tag{6}$$

The energy balance equation of the whole defrosting process is as follows:

$$\dot{m}_{\mathrm{melt}} * h_s(T_{\mathrm{ice}}) + q = \dot{m}_{\mathrm{run-off}} * h_L(T_F) + (1 - C_{\mathrm{emp}}) * q \tag{7}$$

where $h_s(T_{\mathrm{ice}})$ is the enthalpy of the solid ice at the temperature of the ice layer, and $h_L(T_F)$ is the enthalpy that is associated with the melting ice at the freezing temperature, T_F is the freezing point temperature, q is heat conduction of glass into ice, C_{emp} is coefficient of heat absorption for ice.

These enthalpies are related to temperature by the equation:

$$h_s(T) = C_pT, h_L(T) = C_pT + L \tag{8}$$

where L is a latent heat flux.

Using these relations with the mass balance and rearranging gives an explicit expression for the melting rate:

$$\dot{m}_{\mathrm{melt}} = \frac{C_{\mathrm{emp}}q}{C_p(T_F - T_{\mathrm{ice}}) + L} \tag{9}$$

If Eq. (9) predicts a negative $\dot{m}_{\mathrm{melt}}$, it is clipped to zero.

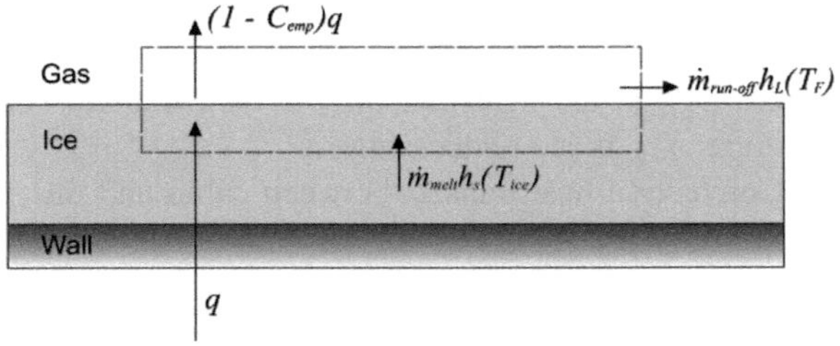

Fig. 2 Defrosting principle diagram

Therefore, the ice thickness can be solved by the following equation:

$$\frac{d_s}{d_t} = -\frac{\dot{m}_{\text{melt}}}{A * \rho_s} \tag{10}$$

where s is the ice thickness, ρ_s is the ice layer density, and A is the ice layer area.

3 Model Building and Boundary Conditions

3.1 Model Building

In order to obtain accurate analytical results, a complete model needs to be established; otherwise, the analysis results will be preferred [8]. So according to the vehicle passenger cabin data, to establish the model, surface mesh and volume mesh quality are adjusted according to enterprise standard. The grid encryption processing the region which is formed by the front windshield, dashboard, and the main and auxiliary driving side window glass is shown in Fig. 3.

where surface mesh in encryption area is 0.004 m and maximum volume mesh size is 0.01 m, the thickness of the first boundary layer is 0.001 m, and glass mesh stretching is 0.005 m. The total surface mesh number is 4583850, and total volume mesh number is 10743943. HVAC and defrosting air duct model are established by detailed data and using MRF mode processing blower.

3.2 Boundary Conditions

Physical model includes three-dimensional, steady, incompressible, K-ε turbulence model, all $y+$ wall function and separation method, which are used for the flow

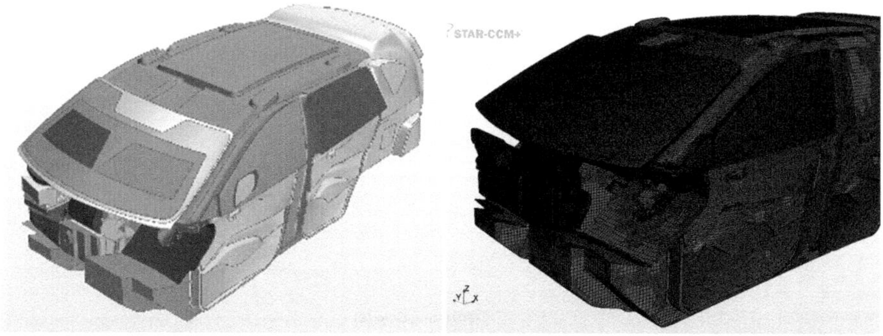

Fig. 3 Grid model diagram

field model. The second-order upwind difference scheme is used to discretize the space, and the sSimple algorithm is used for the iterative computation. The total step size of steady flow field is 3000 steps, and the convergence condition is less than 10^{-3}.

Then in the transient temperature field model, the steady model is changed to unsteady, and the energy solution equation and defrosting equation are added to the flow field model. The glass walls adopt heat convection boundary condition, and the other wall is adiabatic. The frost thickness is set to 0.0005 m, and the cabin initial and ambient temperature are set to 255.15 K. The total time of transient temperature field analysis is 1400 s, the time step is 1 s, and the step number is 10 steps per unit time step.

4 Analyses of Existing Vehicle Simulation and Test Results

Through the one method to establish the steady state and transient analysis simulation model for the existing small car and medium-sized vehicle. Two models are analyzed by 350 m^3/h air volume flow, which come from the blower performance and air duct flow-pressure loss curve. But the main defrosting duct air volume flow of the medium-sized vehicle accounted for 60% of total air volume flow; however, the small cars accounted for total volume flow of 72%.

Due to the repeated tests of different batches of engines and vehicles, there is an inevitable deviation between the test results. Therefore, we need to select the experimental data. The temperature–time curve of two model outlet input is shown in Figs. 4 and 5, which are from 1.6 L engine test data. Those data are the worst data of the temperature–time curve in the test results. By comparing the two sets of data, we can find that the trend of air temperature of two curves before 1000 s is close.

Put the data into the simulation models and to do the simulation.

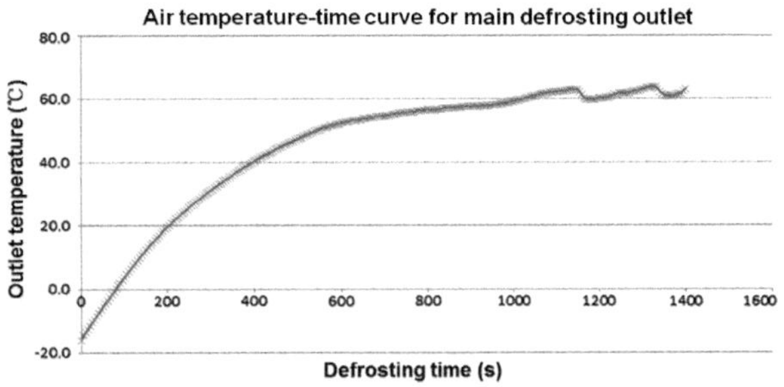

Fig. 4 Temperature-time curve of the outlet of medium-sized vehicle

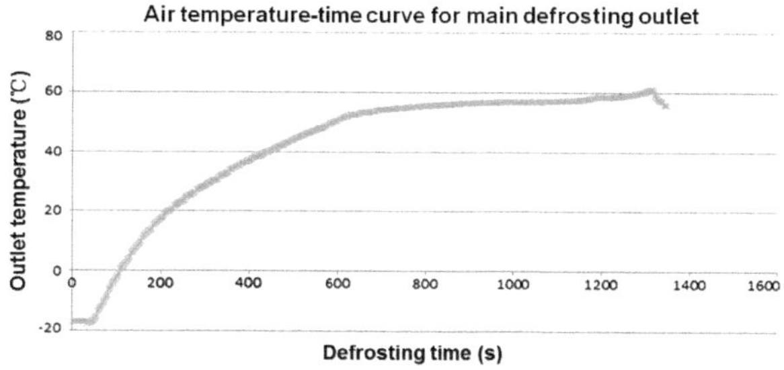

Fig. 5 Temperature-time curve of the outlet of small car

4.1 Comparison of Simulation and Test Results

The results of the analysis of the medium-sized vehicle are shown in Table 1. The test results show that the defrosting time is 20 min, and the transient results are in good agreement with the experimental results. Through the comparison of the steady-state analysis glass surface wind speed distribution results and the experimental results, initial determine the 0.5 m/s wind speed distribution can meet the 20 min front windshield frost cleaning requirements.

Taking the same model processing method, the results of defrosting analysis of the small car models are shown in Table 2 as follows. The target defrosting time is 18 min, and the transient results are basically consistent with the experimental results. Through the comparison of the glass surface wind speed distribution results, determine the initial 1 m/s wind speed distribution can meet the 18 min front windshield frost cleaning requirements. The bigger the target wind speed is, the shorter the defrosting time needed, which is mainly because the larger the target wind speed is, the stronger the convective heat transfer ability is.

4.2 Model Analysis and Experience Summary

Combining the results of steady-state, transient simulations and defrosting tests of the two existing models, it can be concluded that under the current model setting and boundary conditions, if the defrosting target value is 18 min, the wind speed coverage area of 1m/s in area B of the field of vision should reach more than 95% (actual vehicle defrosting effect than the corresponding transient analysis results better). The defrosting time and its corresponding target wind speed can be selected with reference to Table 3.

Through these models defrosting results, can be seen that the position of the air impact point in vision area is relatively lower (basically in the under edge of vision area B), resulting in the lower position of the ice breaking, which is not conducive to shorten the defrosting time.

If the target wind speed zone extends beyond the left and right side edges of vision area B, it is helpful for rapid defrosting on both sides of vision area B.

Table 1 Defrosting steady-state, transient simulation and the climate test results of the medium-sized vehicle

Defrosting time (min)	Steady-state analysis results (wind speed distribution)	Transient analysis results (frost layer thickness)	Experimental results (main defrosting target time 20 min)
10	–	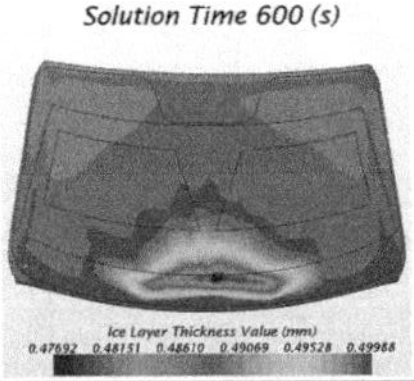	
15	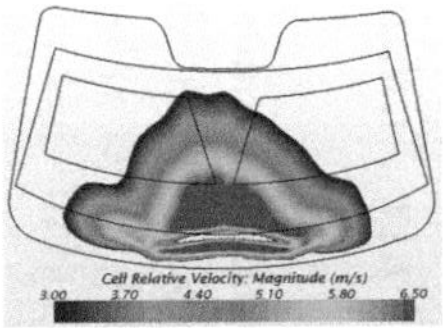	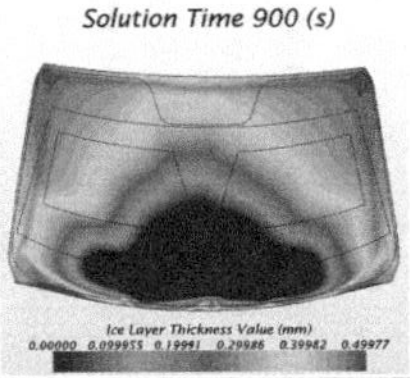	
18	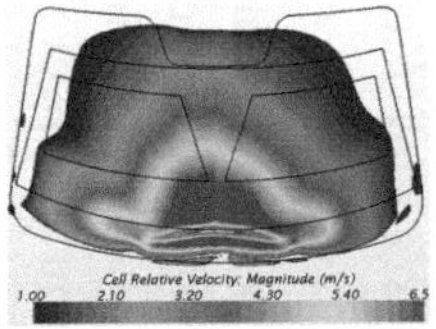	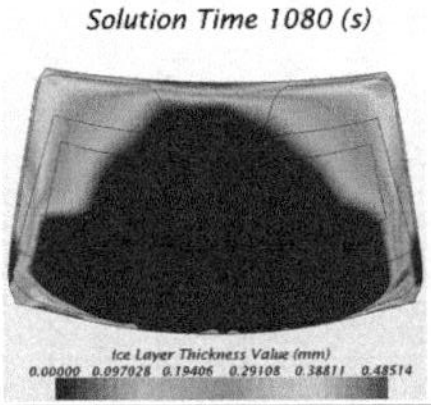	–
20	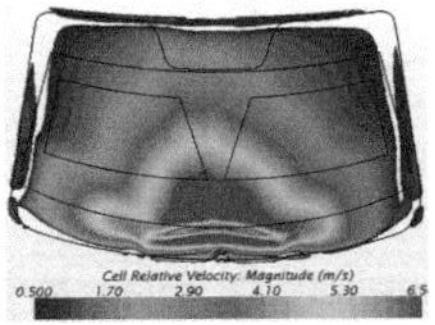		

Table 2 Defrosting steady-state, transient simulation and the climate test results of the small car

Defrosting time (min)	Steady-state analysis results (wind speed distribution)	Transient analysis results (frost layer thickness)	Experimental results (main defrosting t arget time 18 min)
10	–	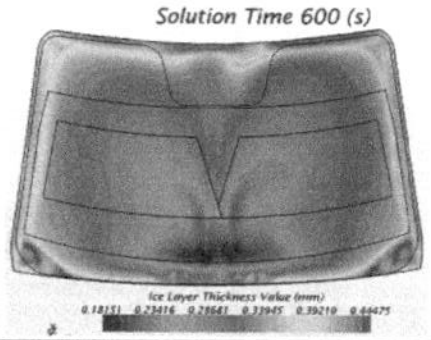	
15	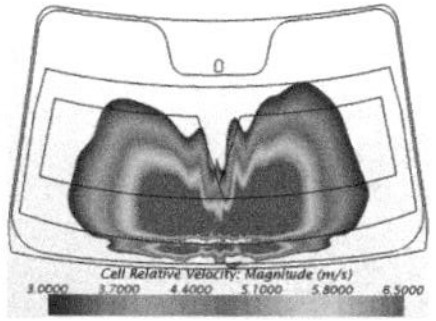	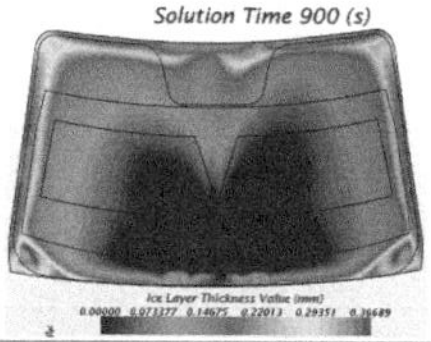	
18	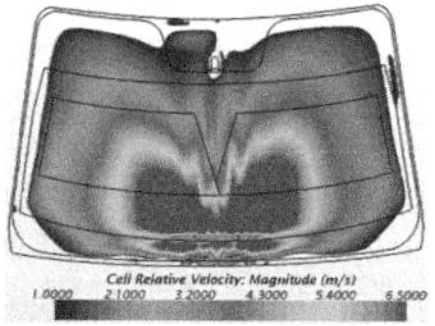	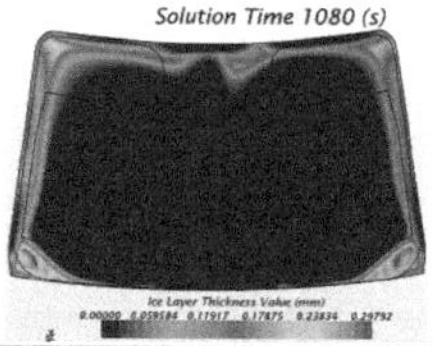	
20	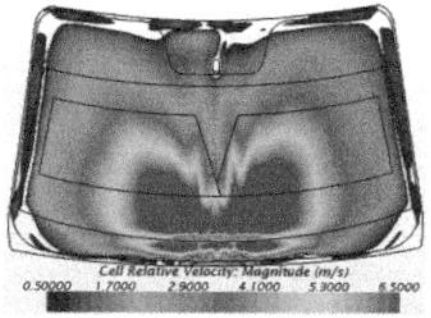	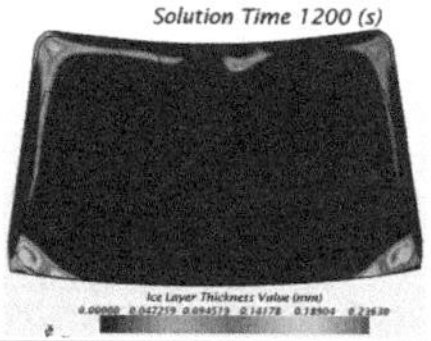	–

Table 3 Defrosting time and its corresponding target wind speed

Defrosting time (min)	15	18	20
Target wind speed (m/s)	3	1	0.5
Percentage of coverage area (%)	95	95	95

Medium-sized vehicle test found a problem in extremely cold conditions which is the existence of frost layer in the lower corner of vision area B, so combined with simulation analysis, it is found that the coverage of the wind speed by 1 m/s in the lower corner of vision area B can be solved.

5 Analyses and Optimization of a New Vehicle Defrosting

5.1 Simulation Analysis and Optimization of Defrosting Steady Flow Field

The conclusion of the defrosting steady-state analysis is obtained by the simulation, to guide the new development vehicle analysis and optimization and achieve the defrosting performance requirements. First of all, analysis of the original version air duct data (the original version of the air duct data is shown in Fig. 6), the steady-state analysis results as shown in Fig. 7.

According to the analysis of the target wind speed distribution on the original version, defrosting air flow is too concentrated, leading to the front windshield glass target wind speed distribution area is too small, and most view area is not covered, including the left and right sides.

But the main defrosting duct outlet air volume flow has accounted for 62% of the total defrosting volume flow, so consider to optimizing the air distribution in outlet, reduce the air volume flow of middle part of outlet. At the same time taking into account the installation position of sun sensor, the defrosting outlet from an air

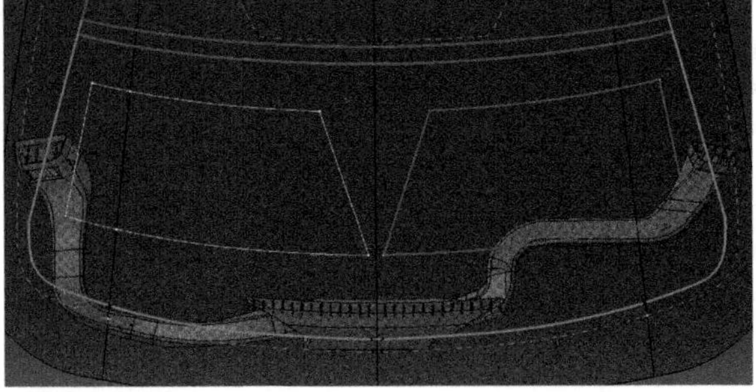

Fig. 6 Original version of the air duct data

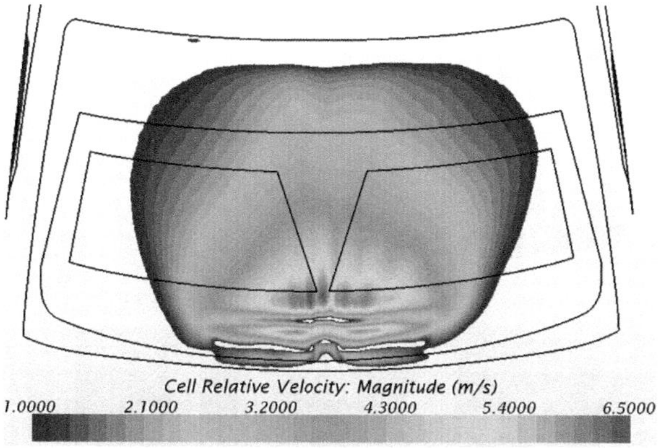

Fig. 7 Steady-state analysis results of the original version air duct

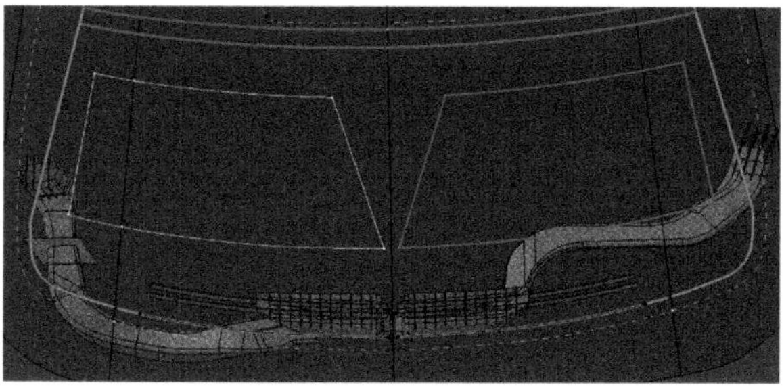

Fig. 8 Air duct data of two outlet scheme

outlet into two air outlets. And in order to facilitate the optimization of the outlet position and shape, the outlet area is increased in the directions of *X*, and adds two horizontal grilles, and the concrete scheme is shown in Fig. 8.

According to the two air outlets scheme of the main defrosting duct, set the five variables of optimization, include the outlet position, outlet shape, outlet angle, air grille area and shape. Because it involves a lots of optimization schemes, in order to reduce the space, so Table 4 only for the schemes that have obvious effect.

Based on optimization on the outlet and the shape of the air duct, scheme three has to meet the requirements; the target speed covers the entire defrosting area of 95% and above; and the main defrosting air volume flow accounts for 67% of the total air volume flow. The velocity uniformity of the air outlet is greater than 0.85, and the maximum air velocity and the total pressure loss of the air duct meet the requirements of the targets set.

At the same time, reference before the analysis conclusion, the defrosting air impact point position adjustment to the vision area A and A′ under the edge and on both sides of the vision area B region has a target velocity distribution, in order to achieve the purpose of accelerating the overall defrosting time.

Table 4 Optimization schemes and flow field results

Schemes	Schemes description	Schemes example (left half of main defrost outlet)	Wind velocity distribution on front windshield surface
Original scheme	1. A horizontal grille rotates 38, and the other is vertical 38° 2. Five vertical grilles are rotated 65°, 50°, 35°, 20°, 5°	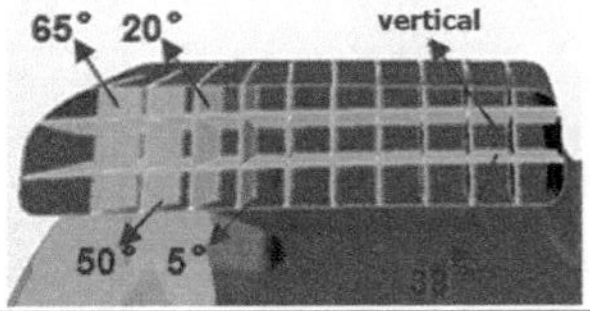	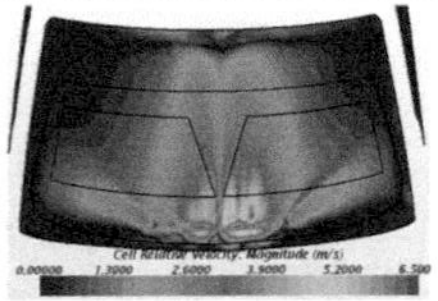
Scheme I	1. Two horizontal grilles are 38° 2. Six vertical grilles are rotated 52°, 45°, 38°, 31°, 24°, 17° 3. Block part of the grilles	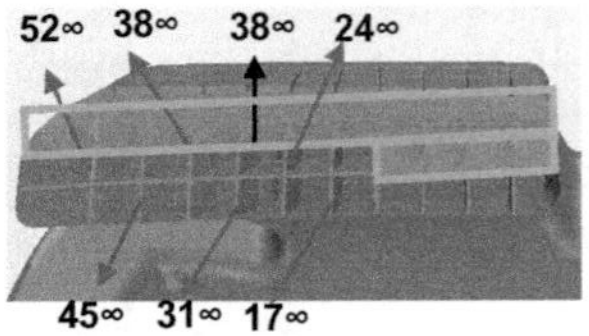	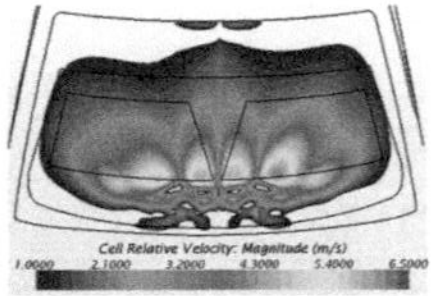
Scheme II	According to the scheme II to re-construction defrosting duct and outlet, increase the two sides of the outlet area, so that the two lower corners can be covered by the target wind speed	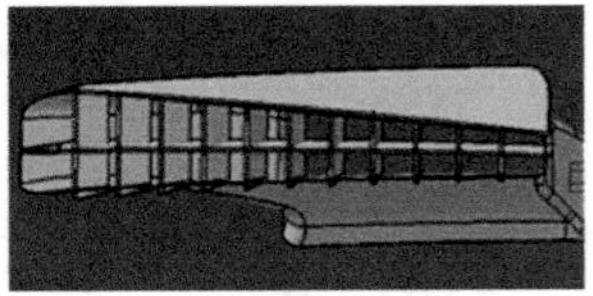	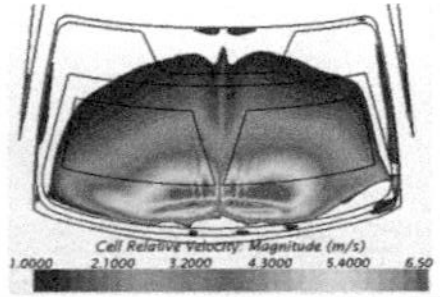
Scheme III	Adjust the main defrosting duct on the basis of the scheme II, reduce the X-way width of the duct and adjust the position of the side defrosting duct junction so that to increase the target wind speed coverage area		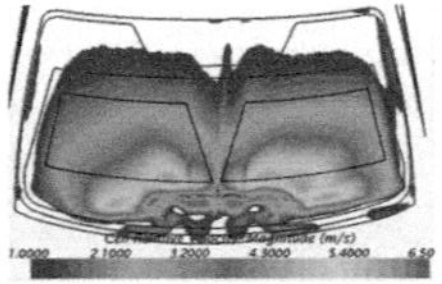

But due to the shape and arrangement reasons, project has made overall changes on HVAC. The new HVAC defrost air duct interface position is shifted 0.03 m than the original HVAC in the positive *Y* direction, and the main defrosting outlet grilles also need to be analyzed according to a given scheme, so the resulting scheme three cannot be implemented.

The defrosting air duct needs to match the new HVAC design and optimization. Therefore, the defrosting air duct and air outlet need to re-optimized; the air outlet grille shape and the junction of the main defrosting air duct are the two main optimization directions.

In the analysis and optimization process, the CAD data are modified by parameterization, the mesh is rebuilt, and the models are optimized by replace the changing parts. The air duct and outlet grilles structure and optimization results are shown in Table 5.

The air duct re-optimization based on the directions of the first optimization includes the defrosting duct shape, position and angle of the grilles. Due to the middle section in the defrosting outlet of HVAC has a large offset distance from Y=0 section, resulting in the target wind speed coverage in the visual field is obvious difference on the left and right side, defrosting process of the weak wind side will be slightly longer.

But because the new HVAC performance has increased, so the amount of air volume flow has increased, and to a certain extent, which will improve the defrosting effect. On the whole, the steady-state analysis of this new scheme shows that the 1m/s wind speed coverage area has reached 95% of vision area B, and the air volume flow accounted for 63% of the total air volume flow, the velocity uniformity of the air outlet is greater than 0.85, and the maximum air velocity and the total pressure loss of the air duct meet the requirements of the targets set. So the scheme can be used to transient analysis and verification. The new scheme is shown in Fig. 9.

5.2 *Transient Temperature Field Analysis and Verification of the Final Scheme*

Because the engine of new vehicle is a 1.6 T engine, it is developing now and has any experimental data. So the close air temperature–time curve for transient analysis is selected from the experimental database, which is a 1.5 T engine air temperature-time curve, as shown in Fig. 10.

This air temperature–time curve will also help to shorten the defrosting time than 1.6 L engine's curve (as shown in Figs. 4 and 5). The results of transient analysis are shown in Table 6; in the 18 min defrosting time required by the target, the frost layer melting area of the vision area A, A′ and B reached 95%, meeting the development target.

Table 5 Comparison of the second optimization schemes and results

Schemes	Schemes description	Schemes example (left half of main defrost outlet)	Wind velocity distribution on front windshield surface
Original scheme	1. In order to cope with the new HVAC, the duct inlet is stretched in the *Y* direction 2. The main defrosting air outlet shape and exposed grilles are changed according to the molding requirements	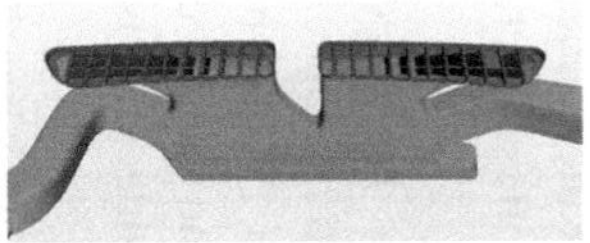	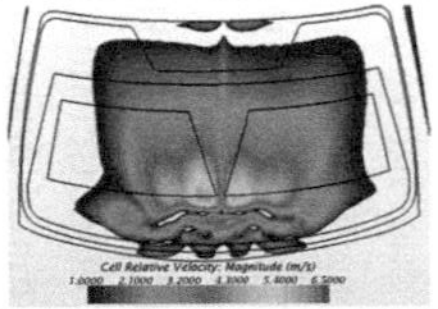
New scheme	Adjust the contour shape of the main defrosting air duct, the junction in middle of the main air duct and the junction with the side air duct	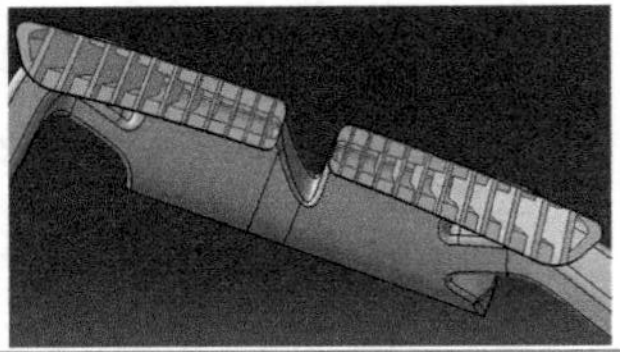	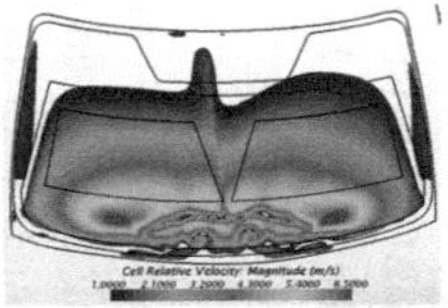

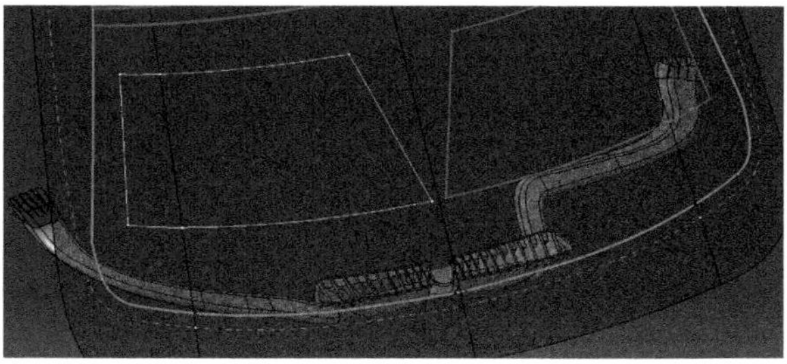

Fig. 9 Re-optimization air duct after change the HVAC

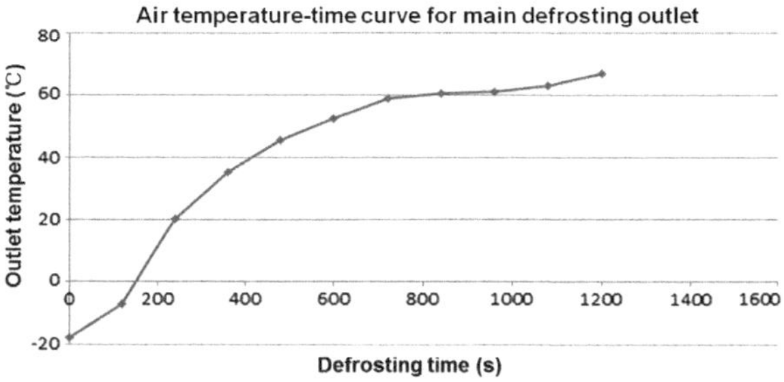

Fig. 10 Temperature-time curve for transient analysis of the new development large car

Table 6 Transient analysis results of the new development large car

Defrosting time (min)	10	15	18
Transient analysis results	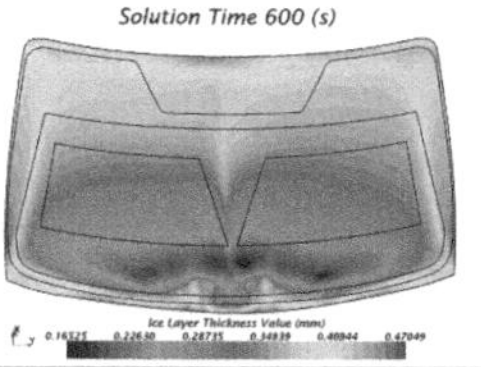	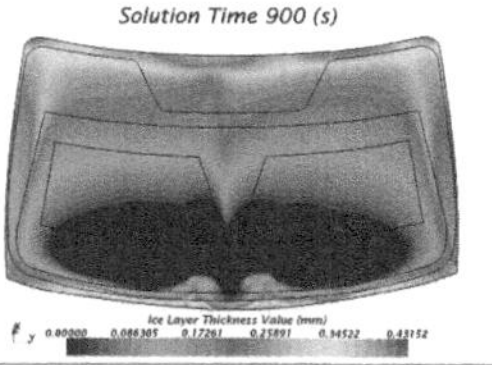	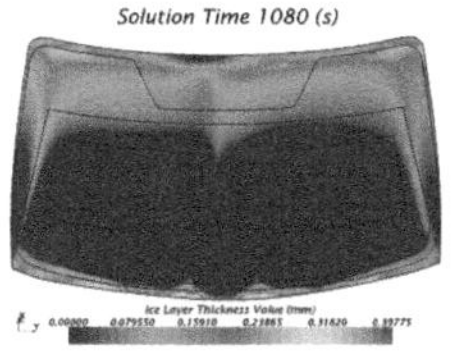

6 Conclusions

Based on the analysis of defrosting simulation and test of existing models, a method for predicting defrosting time and defrosting effect using steady state analysis was found, and applied to the project development of new vehicles, and the defrosting scheme optimization speed is improved. At the same time, this analysis also optimizes the defrosting air impact point position and wind speed distribution on the new vehicle. Specific conclusions are as follows:

(1) Determine the relationship between the coverage area of the target wind speed in the steady-state analysis and the transient defrosting time and establish the model setting and evaluation standard;
(2) Application of steady-state defrosting analysis model, to optimize defrosting duct and outlet of the new vehicle; the optimization speed of defrosting scheme has been improved by 50%, and the cost of manpower and computing resources has been shortened and forms a complete set of simulation analysis process;
(3) Through the analysis, pointed out the shortcomings of the original vehicle defrosting performance, and these problems are optimized in this model analysis, the defrosting performance of this model is improved.

Acknowledgements This work was conducted as part of the large SUV development project; the authors gratefully acknowledge the project partners for their contribution and assistance. The authors would also like to thank the environmental simulation test chamber colleagues for their help and the editorial board of *Automotive Innovation* for their guidance opinions.

References

Journals

1. Feng D, Zhengqi G, Yi Y et al (2009) Numerical simulation analysis on defrosting and defogging for automotive windshield. Automot Eng 31(2):175–179
2. Zhou A, Wang S, Liu J (2012) Computational fluid dynamics simulation of auto defrosting. Automobile Technol 12:51–53

Monograph

3. Dynamic Versteeg HK (2010) An introduction to computational fluid dynamic: the finite volume method. World Publishing Company, Beijing
4. Patankar SV (1984) Numerical calculation of heat transfer and fluid flow. Zhang Zheng. Science Press, Beijing
5. Wang F (2004) Computational fluid dynamics analysis: Principles and applications of CFD software. Tsinghua University Press, Beijing

Collected Papers

6. Hu Z, Shi D, Zhang J, Zou J The deicing analysis of vehicle based on STAR-CCM+[C]

Website

Others

7. GB 11555-2009, Motor vehicles-windshield demisting and defrosting systems performance requirements and test methods [S]
8. STAR-CCM+User Guide

A Study on a Carmaker's Brake Pedal Feel DNA

Wenfei Wang, Hua Ai, Guofan Hu, Zhengmin Liu, Hang Zhang and Xiaojun Liu

1 Introduction

As an important part of vehicle attribute, brake performance has a direct influence on customers' impression of a car [1]. However, different carlines under the same brand usually have different brake performance, which not only makes it difficult for customers to have a good brand recognition, but also causes repetitive work and thus waste of resources. Therefore, to study the brake DNA has profound effect. Brake pedal feel is a comprehensive feeling of deceleration, brake pedal movement, pedal force, etc when the driver depresses the brake pedal. It is the most direct brake feedback the customer can feel and is a necessity for a good brake. In this paper, we analyzed the difference in braking style in an automaker, and based on this analysis we've conducted a DNA study to achieve the goal of reducing cost and increasing efficiency.

2 Factors Affecting Brake Pedal Feel

2.1 Objective Specifications Affecting Brake Pedal Feel

Pedal feel is determined by brake performance; therefore, to analyze the pedal feel specifications could be regarded as the analysis of brake performance. The objective specifications that affect brake performance are as follows: foundation brake specification, vehicle specification, brake pedal specification, brake master cylinder and pipe specification [2, 3]. Take the foundation brake for example; the main

W. Wang (✉) · H. Ai · G. Hu · Z. Liu · H. Zhang · X. Liu
Chongqing Changan Auto R&D Center, Changan Automobile Co., Ltd.,
Chongqing, China
e-mail: wangwf1@changan.com.cn

Society of Automotive Engineers of China (SAE-China) (ed.), *Proceedings of the 19th Asia Pacific Automotive Engineering Conference & SAE-China Congress 2017: Selected Papers*,
Lecture Notes in Electrical Engineering 486, https://doi.org/10.1007/978-981-10-8506-2_5

specifications are diameter of the caliper oil cylinder, effective radius of the brake, friction coefficient, drag torque and braking power, clearance between brake disc and brake pad.

Looking at the brake process, the effect of foundation brake and the continuous braking effect is represented by the braking distance s, and in the equation of $s = s_1 + s_2, s_1$ is the distance when the brake is working, while s_2 is the continuous braking distance.

Let us say the braking distance within t_1 is $s_1' = v_0 t_1$, then v_0 is the vehicle speed at initial braking, vehicle speed at t_2 is $v_1 = v_0 + \frac{1}{2}kt_2^2$, braking distance is $s_1'' = v_0 t_2 - \frac{1}{6}a_{b\max}t_2^2$, and $a_{b\max}$ is the maximum brake deceleration.

The distance during which the bake is working is $s_1 = s_1' + s_1'' = v_0 t_1 + v_0 t_2 - \frac{1}{6}a_{b\max}t_2^2$, continuous braking distance is $s_2 = \frac{v_0^2}{2a_{b\max}} - \frac{v_0 t_2}{2} + \frac{a_{b\max}t_2^2}{8}$, so the ultimate braking distance is as follows:

$$s = s_1 + s_2 = \left(t_1 + \frac{t_2}{2}\right)v_0 + \frac{v_0^2}{2a_{b\max}} - \frac{a_{b\max}t_2^2}{24} \tag{1}$$

So we can see that the main factors affecting braking distance are as follows: the time when brake starts to work, the vehicle speed when depress the brake and the maximum brake deceleration

For disc brake, the pressing force of brake pad on brake disc is $F_{\max} = P_{\max}\pi\frac{d_1^2}{4}$, $P_{\max}$ is the maximum hydraulic pipe pressure, and d_1 is the diameter of brake cylinder piston.

Drag torque is as follows:

$$T = 2fFR \tag{2}$$

f stands for friction coefficient, F is the pressing force of brake pad on brake disc, R is the effective braking radius, namely the distance from the center line of brake pad to the axis of brake disc.

Maximum friction:

$$f_{\max} = fF_{\max} = fP_{\max}\pi\frac{d_1^2}{4} \tag{3}$$

Brake power:

$$\eta = \frac{fF_1 + fF_2}{F} \tag{4}$$

fF_1, fF_2 is the friction of rubbing pair on brake disc.

Based on (1)–(4) we can conclude:

(1) Bigger braking effective radius and brake cylinder diameter plus higher friction means better brake performance.
(2) Bigger brake drag torque means lower brake efficiency and worse brake performance.

2.2 Subjective Factors Affecting Pedal Feel

We could follow the same analysis logic used in foundation brake. Based on these analyses, we started from the subjective evaluation feature of pedal feel and concluded that the following factors have influence on pedal feel:

(1) Pedal travel: related to pedal lever ratio, pedal assembly clearance, pedal/including supporting longitudinal stiffness, the pressure build up characteristic of the brake main cylinder with vacuum booster/main cylinder diameter etc.
(2) Pedal effort: related to initial pre-load, pedal movement stiffness, input and output characteristic of the brake main cylinder with vacuum booster, front and rear brake starting pressure etc.
(3) Pedal damping and delay, the response speed when braking, vehicle deceleration.
(4) Modulation of vehicle deceleration, which is when the driver depresses the brake pedal to brake to reach his target speed, and then he can tip in or tip out to achieve bigger or smaller deceleration.

Based on these factors, we conclude the metrics defining brake pedal feel are showed in Chart 1.

3 Analysis of a Carmaker's Brake Pedal Feel Status

3.1 Subjective Evaluation Results for the Pedal Feel on 4 Cars

We randomly chose four cars currently under development; the subjective evaluation results are as follows:

Car 1: initial travel is long with strong and clear response, but the pedal is heavy in the later stage.

Brake pedal feel metrics	Unit
Pedal force dead travel	[mm]
Deceleration dead travel @0.1g	[mm]
Initial response	[mg]
Pedal travel @0.4g	[mm]
Peal modulation co-efficient@0.60~0.80g	[mm/g]
Pedal force modulation co-efficient@0.60~0.80g	[N/g]

Chart 1 Brake pedal feel metrics

Car 2: initial travel is a little long with strong and clear response, braking power ramps up in the later stage; the modulation is not good. Also, the pedal is heavy in the later stage.
Car 3: initial travel is OK; with overly strong initial response (easily have head-toss problem), braking power ramps up in the later stage; the modulation is not good. Also, the pedal is heavy in the later stage.
Car 4: initial travel is OK, initial response is vague (not confident); braking power is reasonable in the later stage. The braking is linear, but the pedal is heavy in the later stage.

After the evaluation and analysis, we could see this carmaker doesn't have a uniform brake behavior across its different carlines, there is no DNA. The reasons for this are as follows: product source is different, design input is different and the design target is different.

3.2 Objective Data Analysis on These Four Cars

VBOX is a intelligent test equipment that uses high-performance micro-computer as core; it is used to test the automobile performance, fuel economy, brake performance etc.

We used VBOX to recorded the pedal feel data on these four cars; by adopting data treatment methods like correction of zero drift and eleven means smoothing, we have processed the original data and compiled these data into chart 2 [4, 5].

Meanwhile, we plotted out the pedal force versus deceleration based on the data, as shown in Fig. 1; pedal travel versus brake deceleration, as shown in Fig. 2; pedal force versus pedal travel, as shown in Fig. 3.

We could see that the pedal feel on these four cars are obviously different which is in line with previous subjective evaluation.

Objective data for brake pedal feel					
Brake performance	Unit	Car 1	Car 2	Car 3	Car 4
Pedal force dead travel	[mm]	34	26	22	12
Deceleration dead travel @0.1g	[mm]	31	24	21	19
Initial response	[mg]	171	148	223	86
Pedal travel @0.4g	[mm]	49	42	32	36
Peal modulation co-efficient@0.60~0.80g	[mm/g]	35	19	22	30
Pedal force modulation co-efficient@0.60~0.8	[N/g]	150	141	168	180

Chart 2 Statistic result for brake pedal feel metrics

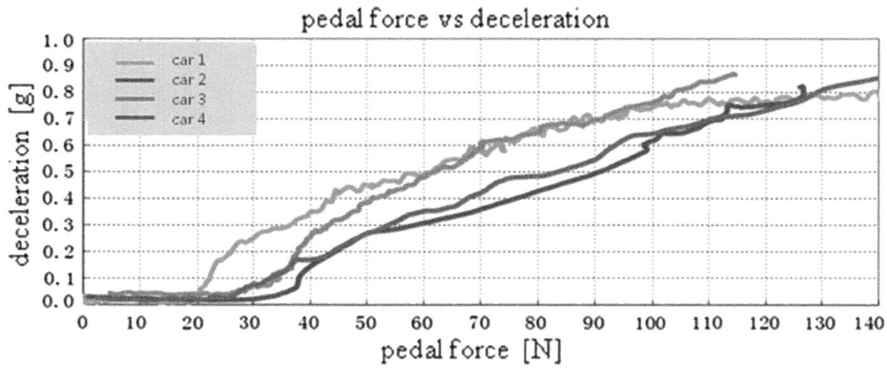

Fig. 1 Pedal force versus deceleration

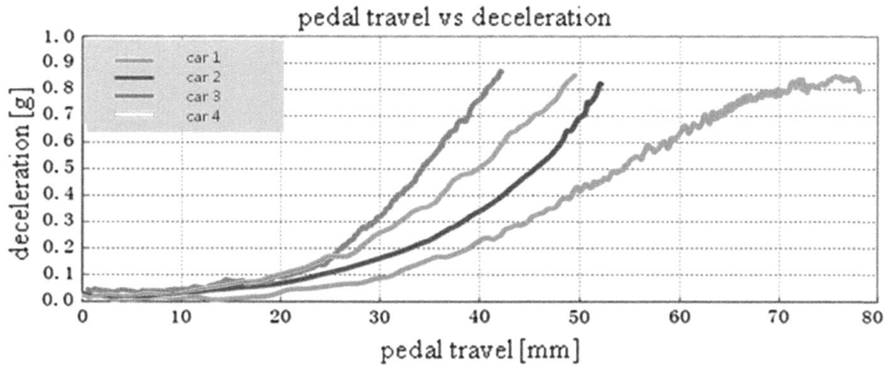

Fig. 2 Pedal travel versus deceleration

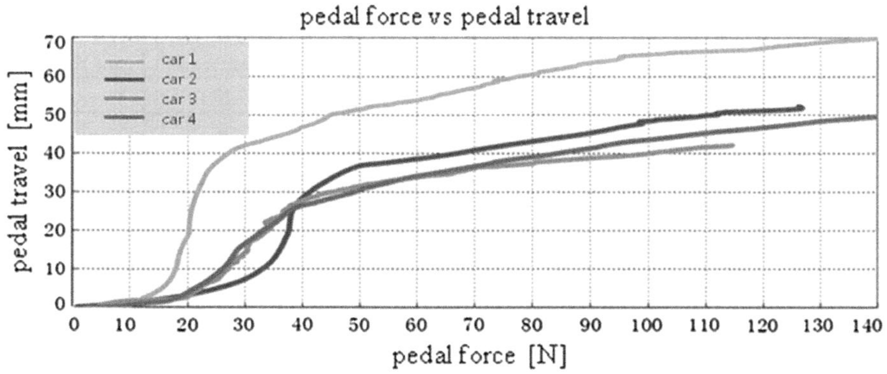

Fig. 3 Pedal force versus pedal travel

4 Brake Pedal Feel DNA Metrics Definition

4.1 DNA Metrics Definition

Before defining the brake pedal feel DNA, we need to benchmark other carmakers to make sure the DNA style is line with the market trend and meanwhile has its own distinctive features, which is not easy to be copied or surpassed. Based on the benchmarking result, we defined the DNA style for this carmaker:

(1) Comfortable: good ergonomics, reasonable pedal force, strong pedal feedback
(2) Responsive: small pedal travel input the driver could feel some deceleration response, small pedal force adjustment(increase) the driver could feel the increase in deceleration(power)
(3) Controllable: quite linear, deceleration build up is progressive and smooth, no response delay, no jump in.

Based on the style description above, we have determined the DNA metrics as showed in chart 3.

4.2 Refinement on the Four Cars' Pedal Feel DNA

Based on the brake pedal feel DNA range, we adjusted the relevant metrics for the four cars, and the pedal force versus deceleration after refinement is showed in Fig. 4, pedal travel versus deceleration is shown in Fig. 5 and pedal force versus pedal travel is showed in Fig. 6.

We could see that after implement the DNA standard, the four cars' pedal feel is more uniform, thus having a better experience and easy to be recognized.

Brake pedal feel DNA metric			
Brake pedal feel	Unit	Lower limit	Upper limit
Pedal force dead travel	[mm]	14	22
Deceleration dead travel @0.1g	[mm]	14	22
Initial response	[mg]	130	190
Pedal travel @0.4g	[mm]	29	40
Peal modulation co-efficient@0.60~0.80g	[mm/g]	25	35
Pedal force modulation co-efficient@0.60~0.80g	[N/g]	110	150

Chart 3 Brake pedal feel DNA metrics

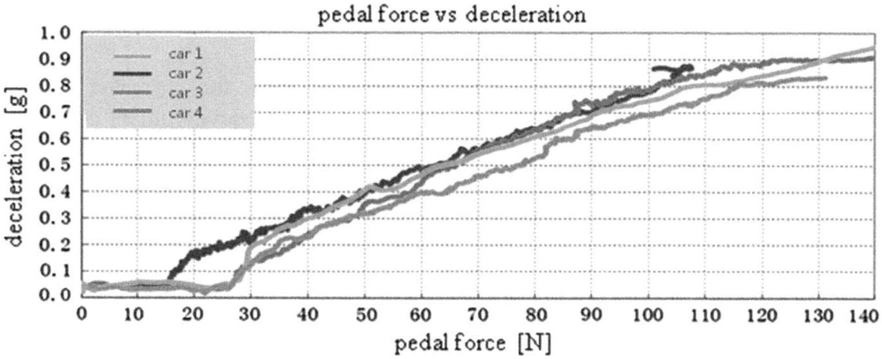

Fig. 4 Pedal force versus deceleration after refinement

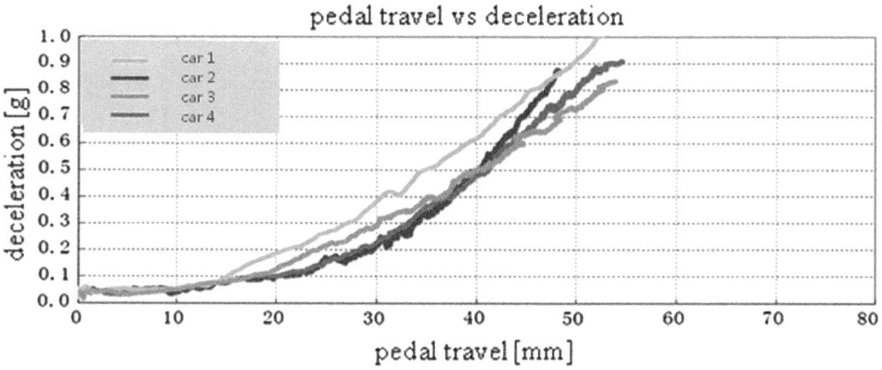

Fig. 5 Pedal travel versus deceleration after refinement

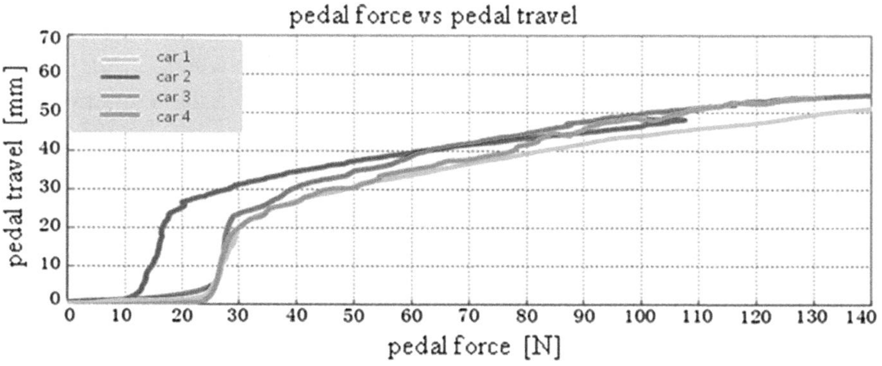

Fig. 6 Pedal force versus pedal travel after refinement

5 Conclusion

By defining and implementing a brake pedal feel DNA, the carmaker's brake performance is more uniform, which in turn reduced their development cost. And the method used in this brake DNA study can be learnt and used in the DNA work for other attributes. This paper is only a record of the initial result of this study, so it needs further verification and refinement on whether these metrics are comprehensive enough.

References

1. Yu Z (2009) Automobile theory, 5th edn. China Machine Press, Beijing
2. Xiong W (2013) ABS brake performance simulation computing and evaluation study. [Master degree thesis]. East China Jiaotong University
3. Maciuca DB, Hedriek JK Advanced nonlinear brake system control for vehicle platooning. University of California at Berkeley, CA 94720
4. Song Z, Sun Y et al (2014) Pressure sensor zero drift correction based on vague neural network. Instrum Technol Sensor 3:11–13
5. Zhou F, Fan X, Ye Z (2007) Study and application of means drift algorithms. Control Decis 22(8):0841–0847

The Dimensional Tolerance Analysis in Different Assembly Procedures Based on 3DCS

Yufeng Xia, Lai Jiang, Dong Yang and Yuting Zhou

Nomenclature

M_s	measurement, mm
M_0	nominal value of measurement, mm
λ_i	constant parameter
T_i	tolerance, mm
A_0	basic dimension for closed loop
A_i	basic dimension for composition loop
ES_i	upper deviation for composition loop
EI_i	lower deviation for composition loop
ξ	random variables
$E(\xi)$	mathematical expectation of ξ
PSBO	panel, side body outer
RCP	reinforcement, cir pillar
Est. Low	estimated low
Est. High	estimated high
Est. Range	the difference between the Est. Low and Est. High
LSL	lower specification limit
USL	upper specification limit
L-Out	the actual percentage of measurement values falling below the lower specification limit
H-Out	the actual percentage of measurement value falling above the upper specification limit
Tot-Out	the actual percentage of measurement values falling above the upper specification limit and below the lower specification limit

Y. Xia · L. Jiang (✉) · D. Yang · Y. Zhou
College of Materials Science and Engineering, Chongqing University,
400044 Chongqing, China
e-mail: 15023161316@163.com

Society of Automotive Engineers of China (SAE-China) (ed.), *Proceedings of the 19th Asia Pacific Automotive Engineering Conference & SAE-China Congress 2017: Selected Papers*, Lecture Notes in Electrical Engineering 486, https://doi.org/10.1007/978-981-10-8506-2_6

1 Introduction

Dimensional tolerance analysis has an important influence on product function, cost, assembly and quality [1]. It is obvious that all parts cannot be manufactured to an exact level of accuracy; in order to achieve higher product quality, the integration of the assembly procedure with tolerance design is vital [2]. Therefore, dimensional tolerance analysis plays a significant role in optimizing the product quality after assembly.

Currently, dimensional tolerance analysis is widely applied in the automotive, aerospace and shipbuilding industries. Boeing, Airbus and Embraer have established a perfect dimensional analyzing system. With the development of computer-aided tolerance (CAT), three-dimensional tolerance analysis has gradually replaced the traditional one-dimensional worst case method. The main analytical tools for CAT are 3DCS, VSA and Cetol6σ. The first two utilize a Monte Carlo analysis. Substantial studies on tolerance analysis and synthesis have been conducted by many researchers. Chen et al. [3] comprehensively reviewed several three-dimensional tolerance analysis methods and compared the advantages and disadvantages of several different analytical methods (i.e., the Tolerance-Map (T-Map), the matrix model, the unified Jacobian-Torsor model and the direct linearization method (DLM)). Qureshi et al. [4] used a statistical tolerance analysis approach for an over-constrained mechanism based on optimization and a Monte Carlo simulation, and they discussed how to analyze the influence of geometric deviations on an over-constrained mechanism under nonlinear conditions; Reich et al. [5] utilized 3DCS to investigate the three-dimensional tolerance of main IETR components which provided an important basis for the selection of manufacturing and assembly tolerances of IETR components. Lin [6] proposed an optimization method that integrates the genetic algorithm and Monte Carlo Simulation to search simultaneously for the optimal values of nominal values and tolerance of location of bearings. Lee et al. [7] used genetic algorithm and process capability index to optimize a lower arm, the standard deviation can be controlled more effectively according to the research. In addition, the assembly process was treated as an important factor that influenced the assembly quality. Sinanoglu and Borklu [8] developed an assembly sequence planning system based on binary vector representations and a neural network approach was utilized to optimize the assembly sequence for the assembly system. Viganò and Gomez [9] have defined an approach to extract the liaison graph from a 3D CAD model and analyzed a method to find a feasible assembly sequence for the product. Xu et al. [10] built a Dynamic Assembly Model (DAM) for planning the assembly sequence of complex products, an approach that reduced the complexity of the assembly sequence. However, these comprehensive studies did not address the means by which an assembly process influences size deviations.

In this study, the Three-Dimensional Control System (3DCS) and the Monte Carlo Method (MCM) were adopted to analyze an optimization design for an assembly process for B-pillar parts and identify the crucial tolerances to make further improvements. An urban automotive B-pillar is shown in Fig. 1. It is located between the front and rear doors extending from the roof to the bottom of the vehicle and consists of many steel stampings with complex surface structures assembled by welding [11]. The vital measurement targets are described in Fig. 2. Figure 2a depicts the planar distance between PSBO and RCP and d1 and d2 reflect the length of the gap when the two surfaces are in contact with each other. In practical applications, the inappropriate matching of the two surfaces can lead to potential failure with respect to water tightness, the overall appearance and abnormal noise. Figure 2b shows the circle interference between the holes in the PSBO and the RCP; the value of p1 and p2 should be small enough so that the locking pin of door can easily pass through. To avoid possible catastrophic failure, two types of analysis objects were required to have high accuracy when the actual assembly procedure was finished. By adopting a new assembly process and optimizing the tolerance design derived from a contribution analysis in 3DCS, the assembly quality was greatly improved.

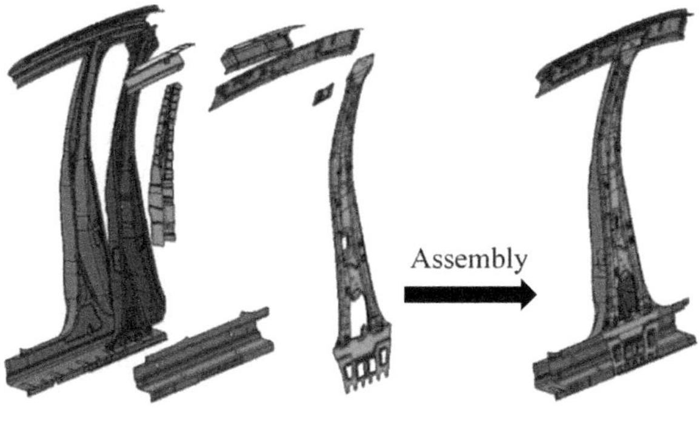

Fig. 1 The structure of a B-pillar

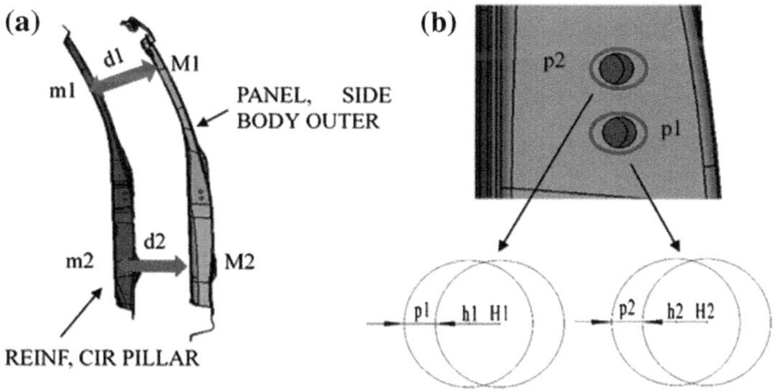

Fig. 2 Measurement targets [*Notes* **a** Measurement1: distance of two plane (d1, d2); **b** Measurement2: circle interference (p1, p2)]

2 Principle of the Monte Carlo Method

In a dimensional tolerance analysis, it is most important to calculate the final deviation, which gradually increases during the assembly process, according to the manufacturing tolerances of the parts. Currently, approaches of dimensional tolerance analysis involved Worst Case Analysis and Mont Carlo Method.

The analysis model of linearity deviation can be expressed as Eq. (1).

$$M_s = M_0 + \sum_{i=1}^{n} (\lambda_i T_i), \tag{1}$$

The analysis model of nonlinear deviation can be expressed as Eq. (2).

$$M_s = f(T), \tag{2}$$

where M_s represents a measurement, M_0 is the nominal value of a measurement, λ_i ($i = 1, 2, \ldots, n$) is a constant parameter, T_i is the tolerance, and n is the total number of tolerances.

Dimension chains play a very important role in tolerance analysis. In assembly parts design, the closed dimension group consists of a series of certain ordered sizes is called dimension chains. In the dimension chains, the indirectly ensured sizes were called closed loops, and the directly assurance sizes were called composition loops. According to the impact of the closed loop, the composition loop can be divided into the increasing loop and decreasing loop.

Worst case analysis method supposes all the dimensions reach their worst possible values, i.e. with absolute maximum size variation [12]. The basic dimension for closed loop was shown in Eq. (3).

$$A_0 = \sum_{i=1}^{n} A_i - \sum_{i=n+1}^{m} A_i, \tag{3}$$

where A_0 represents the basic dimension for closed loop, A_i stands for the basic dimension for composition loops, and m is the amount of composition loops, n is the amount of increasing loop, $m - n$ is the amount of decreasing loops.

The upper deviation for closed loop can be described in Eq. (4).

$$ES_0 = \sum_{i=0}^{n} ES_i - \sum_{i=n+1}^{m} EI_i, \tag{4}$$

The lower deviation for closed loop can be described in Eq. (5).

$$EI_0 = \sum_{i=0}^{n} EI_i - \sum_{i=n+1}^{m} ES_i, \tag{5}$$

The tolerance for closed loop can be described in Eq. (6).

$$T_0 = \sum_{i=0}^{m} T_i, \tag{6}$$

where ES_i is the upper deviation for composition loops, EI_i is the lower deviation for composition loops, T_i is the tolerance for each composition loops.

Tolerance range acquired by this method contains the most extreme dimension of each part, regardless of how changes in the dimension of those parts within its range, the final target measurement size will be distributed within the established tolerance range. However, the tolerance band assigned to every individual part was relatively small, which proposed a high demand for processing technology. Since this method is too conservative, the processing costs tended to increase in car body assembly process.

Monte Carlo Method is a typical approach to obtain the approximate solutions for mathematical, physical and engineering problems [13, 14]. Assumption the problem X needs to be solved is the mathematical expectation $E(\xi)$ of random variables ξ, the approximate approach to determine X is to establish repeated sampling tests n times for ξ, and the arithmetic mean $\overline{\xi_n}$ for those independent sequence of $\overline{\xi_n}$ (i = 1, 2, 3, …, n) can be counted according to Eqs. (7) and (8) is the law of large numbers proposed by Колмогоров. And when n tends to infinitely, Eq. (9) will be workable. That is, the arithmetic mean can substitute the unknown problem X when sufficient sampling tests were proceeded. The variance of unbiased estimates σ^2 for $\overline{\xi_n}$ can be indicated in Eq. (10).

$$\overline{\xi_n} = \frac{1}{n}\sum_{i=1}^{n} \xi_i. \tag{7}$$

$$\lim_{n\to\infty}\left(\overline{\xi_n} = x\right) = 1. \tag{8}$$

$$\overline{\xi_n} \approx E(\xi) = X. \tag{9}$$

$$\sigma^2 = \frac{1}{n-1}\sum_{i=1}^{n} \left(\xi_i - \overline{\xi_n}\right). \tag{10}$$

In dimensional tolerance analysis, the main procedures based on the MCM are as follows: firstly, the dimension distribution of each composition loop is determined; secondly, the simulation times N is depended on the required calculation accuracy; thirdly, in accordance with the distribution type and range for the dimensions, the random samples ($X1$, $X2$, $X3$, …, Xn) of composition loops are achieved after random sampling; then these samples are brought into the tolerance functions, and the dimension of closed loops are figured out as an subsample; at last, the integrated samples of closed loops are formed via N times' repeat working. With the statistical treatment for the closed loop samples, the tolerances and mean value are confirmed. The analysis flow chart of Monte Carlo Method is shown in Fig. 3.

In a word, problems that cannot be solved by general equations or experimental methods can be resolved by the Monte Carlo Method, which can analyze the dimensional chain for tolerance analysis at different confidence levels and using various probability distributions. Due to the complexity of the B-pillar components, it is preferential to perform tolerance studies with a specialized tool like Three Dimensional Control System. 3DCS uses a Monte Carlo Number Generator to assign a value for each tolerance during simulation process, and the statistical distribution-types integrated normal distribution, Weibull distribution, uniform distribution, user-defined distribution, etc.

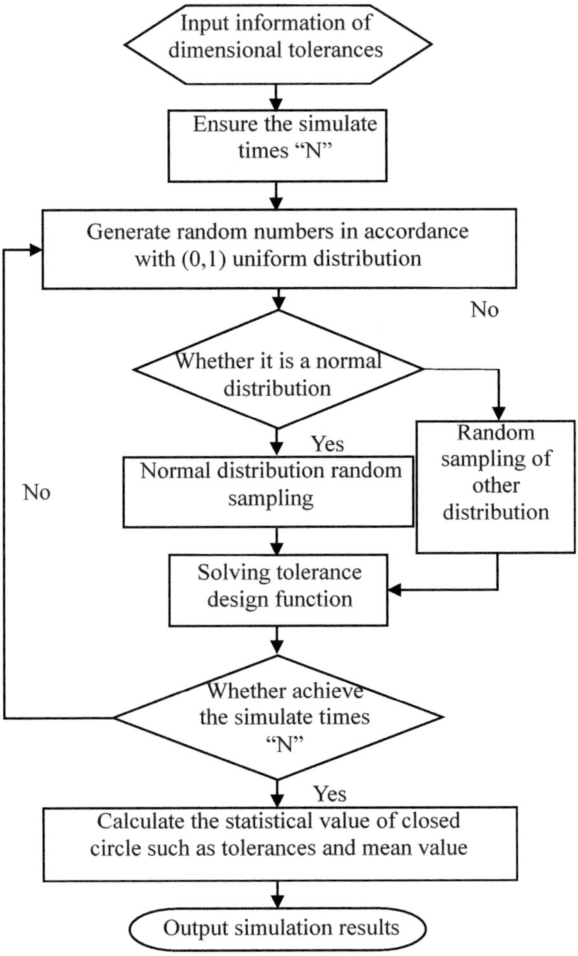

Fig. 3 Flow chart of Monte Carlo method

3 Size and Tolerance Specifications

3.1 *Size Specification of Measurement Targets*

The two measurement targets are shown in Fig. 2, and the specifications are displayed in Table 1. For the distance on the matching surface (d1, d2), the ideal situation is that the gap distance equals zero after assembly, but in practice a size deviation is present. For final products, the size deviation should be controlled within a certain range (-0.5 mm $\leq$ d1, d2 $\leq +0.5$ mm) to avoid potential failure. As shown in Fig. 4, the diameters of h1 (h2) and H1 (H2) were 13 mm, and the diameter of locking pin was 12 mm. To assure that the locking pin could pass through those holes, the circle interference (p1, p2) must be distributed within a certain range (-1.0 mm $\leq$ p1, p2 $\leq$ 0).

3.2 *Tolerance Specification*

The measurement elements of the mechanical parts require both size tolerance and form and positional tolerance, which directly affect the quality of the assembled parts. Size tolerance refers to the absolute value of the maximum size and minimum size, or the differences between the higher deviation and lower deviation and the acceptable deviations in this range. However, mechanical parts not only have a size tolerance; unavoidably, geometric features such as points, lines, surfaces, or the

Table 1 The specification of measurement targets

Measurement targets	Nominal (mm)	Low (mm)	High (mm)
d1, d2	0	−0.5	+0.5
p1, p2	–	−1.0	0

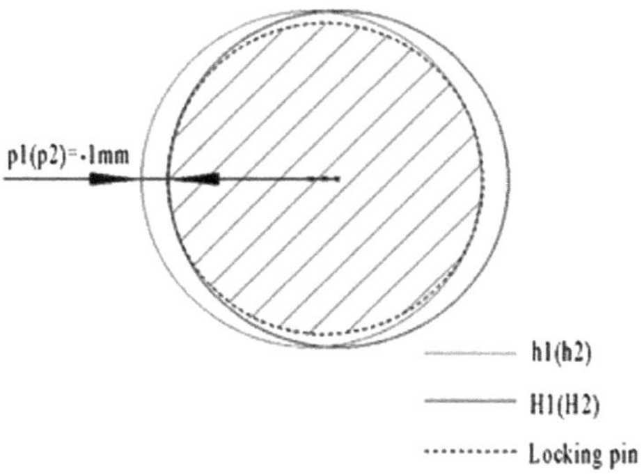

Fig. 4 The circle interference

Table 2 The tolerance specification for B-pillars (mm)

Tolerance items	Size tolerances		Form and positional tolerances	
	Higher deviations	Lower deviations	Surface profile	Position profile
Main positioning hole	+0.10	0	–	0.10
Second positioning hole	+0.10	0	–	0.10
Via holes of PSBO	+0.50	0	–	1.00
Via holes of RCP	+0.50	0	–	1.40
Reference positioning surface	–	–	0.50	–
Matching surface of PSBO	–	–	1.40	–
Matching surface of RCP	–	–	1.00	–
Positioning hole of all fixtures	+0.10	0	–	0.20
Positioning pin of all fixtures	0	-0.05	–	0.20
Positioning surfaces of all fixtures	–	–	0.40	–

position of the actual parts are different from the ideal state, and such deviations known as form and positional tolerance. The specification for those tolerances is shown in Table 2.

4 Analysis of Different Assembly Process

4.1 Successive Assembly Process

As shown in Fig. 5, the traditional method is a successive assembly method, in which all parts are welded together in succession by using a single fixture. The characteristics of this method are a relatively simple assembly process and less occupation of the welding station. However in actual practice product quality suffered significant defects when this method was used, such as a large distance between matching surfaces and severe failures in circle interference.

Figure 6 shows the simulation results of two measurement targets in 3DCS when using the traditional successive process. All simulation times in this article were 2000. In these graphs, the abscissa is the size deviation and the vertical axis is the number of products. The mean is the average value, ±3.00 STD is the Six Sigma Standard, and LSL and USL are the Lower Specification Limits and the Upper Specification Limits, respectively, which are shown in Table 1. The sections below the LSL and above the USL indicate unsatisfactory assembly; the smaller the areas, the higher the quality of the assembly of the parts. Figure 6a, b reflect the

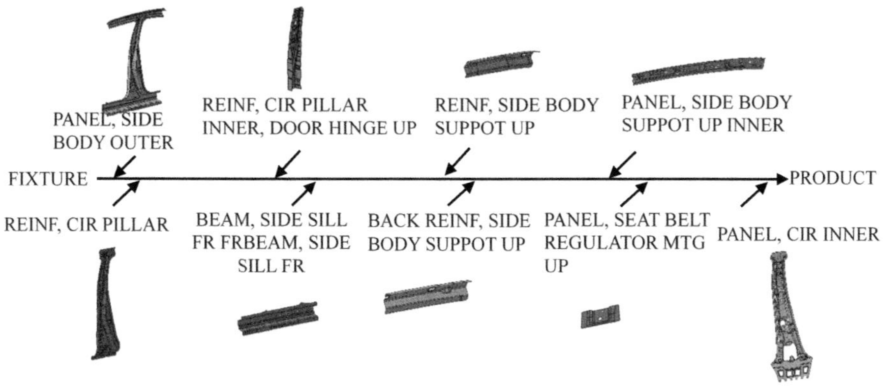

Fig. 5 Successively assembly process

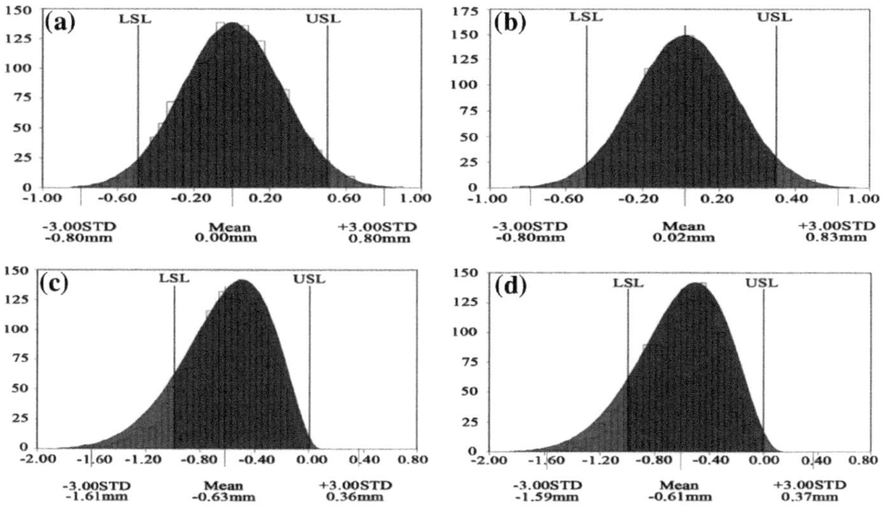

Fig. 6 Distribution of the successive assembly results (*Notes* **a** d1, **b** d2, **c** p1, **d** p2)

distribution of the gap distance and these results are summarized in Table 3. The error rates were 5.90 and 6.80%, which could result in a large number of substandard products during a mass-production process. Figure 6c, d reflect the distribution of interference of the via holes between the PSBO and the RCP. The results show that the via holes show serious interference; the error rates of via holes interference reached 13.65 and 13.40%, which far exceeds the specification requirement. These failure rates provided by the simulation are highly consistent with actual production. For this unacceptable assembly, craftsmen will spend a lot of time in secondary processing, resulting in a greater waste of time and money.

Table 3 Size deviations for the two assembly processes

Measurement	Nominal (mm)	Est. Low (mm)	Est. High (mm)	Est. Range (mm)	L-out (%)	H-out (%)	Tot-out (%)	
Successive assembly	d1	0	−0.80	+0.80	1.60	2.95	2.95	5.90
	d2	0	−0.80	+0.83	1.63	2.90	3.90	6.80
	p1	–	−1.61	+0.02	1.80	13.55	0.10	13.65
	p2	–	−1.74	+0.06	1.80	12.85	0.55	13.40
Sub-assembly	d1	0	−0.73	+0.73	1.46	2.20	2.00	4.20
	d2	0	−0.67	+0.71	1.38	1.50	2.40	3.90
	p1	–	−1.23	+0.03	1.26	1.75	0.55	2.30
	p2	–	−1.20	+0.07	1.27	1.25	0.85	2.10

4.2 Sub-assembly Process

In order to minimize the failures caused by the traditional method, a new approach was designed to improve the overall quality. This article provides a method of assembly procedure called sub-assembly, as shown in Fig. 7. Three sub-assemblies are utilized, Fixture 1, Fixture 2 and Fixture 3 were chosen to complete the sub-assembly, and Fixture 4 was used to complete the final assembly.

Figure 8a, b reflect the distribution of the gap distance by using sub-assembly procedure. Those results are shown in Table 3. The error rates about d1 and d2 were reduced to 4.20 and 3.90%. Figure 8c, d show the distribution of the via hole interference between PSBO and RCP. The results show that the via holes show slight interference—the error rate for the via hole interference was only 1.25 and 1.46%, far below the interference found with the successive assembly process.

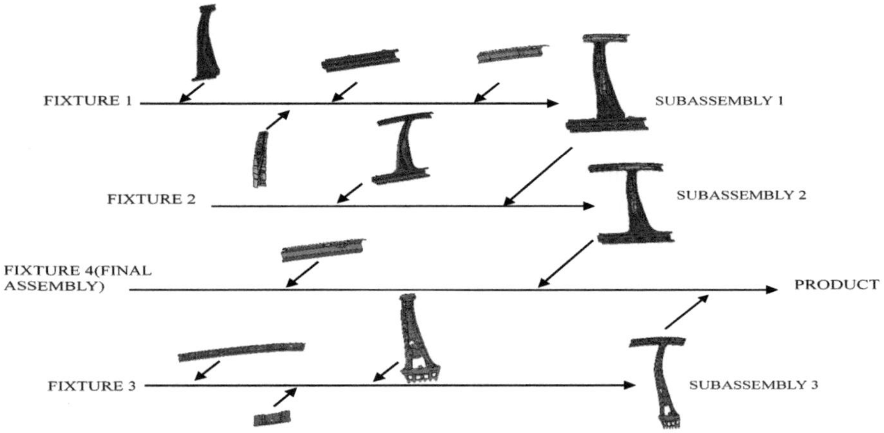

Fig. 7 Sub-assembly process

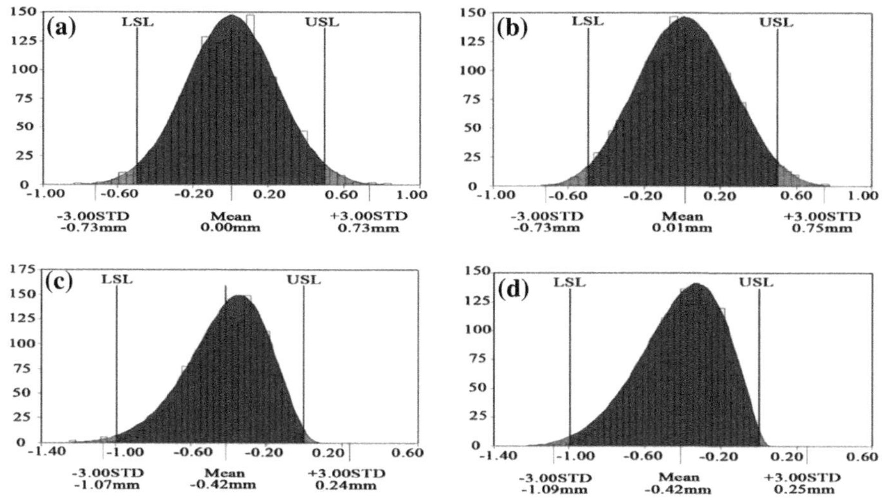

Fig. 8 Distribution of the sub-assembly process results (*Notes* **a** d1, **b** d2, **c** p1, **d** p2)

The new assembly procedure has effectively controlled the assembly variation, especially with respect to the hole interference, which has been greatly reduced.

5 Tolerance Optimization for the Sub-assembly Procedure

5.1 Tolerance Contribution Analysis

The use of the sub-assembly process can make a significant contribution in failure reduction, compared to the traditional process. To further optimize the assembly quality, a tolerance contribution analysis was crucial. It is well known that tolerance design is an essential factor that influences the accuracy of mechanical parts. However, the blind pursuit of precision could increase the difficulty of the stamping die manufacturing and the fixture design. Therefore, finding the tolerance that has the highest contribution to the measurement targets is the most efficient way for designers to optimize tolerances.

The tolerance contribution analysis results from 3DCS are shown in Fig. 9. Figure 9a shows the tolerance contribution for the plane distance between PSBO and RCP (d1 and d2)—the most influential tolerance for this measurement target was the surface profile of PSBO, which reached 75.22% (d1) and 74.95% (d2). Figure 9b displays the tolerance contribution for the via hole interference the positional tolerance between hole 1 and hole 2 had the highest impact on the accuracy of the assembly and the contribution of the positional tolerance for p1 and p2 was 56.68 and 57.09%. These critical dimensional tolerances had a significant effect on the assembly quality, so optimizing them could directly reduce the deviations. Changes were made as follows:

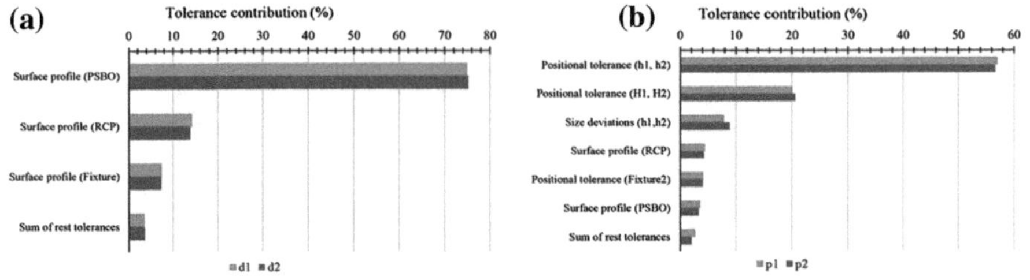

Fig. 9 Tolerance contributions of measurement targets (*Notes* **a** Tolerance contribution for d1 and d2; **b** tolerance contribution for p1 and p2)

- Surface profile tolerance for PSBO: 1.0 mm (the former value was 1.4 mm as shown in Table 2).
- Positional tolerance of hole1 and hole2 in RCP: 1.2 mm (the former value was 1.4 mm as shown in Table 2).

5.2 *Results After Optimizing*

Figure 10 shows the results for the measurement targets using the optimized tolerances. As shown in Fig. 10a, b, the errors for the surface distance were eliminated. The error rates shown in Fig. 10c, d indicate that the interferences were

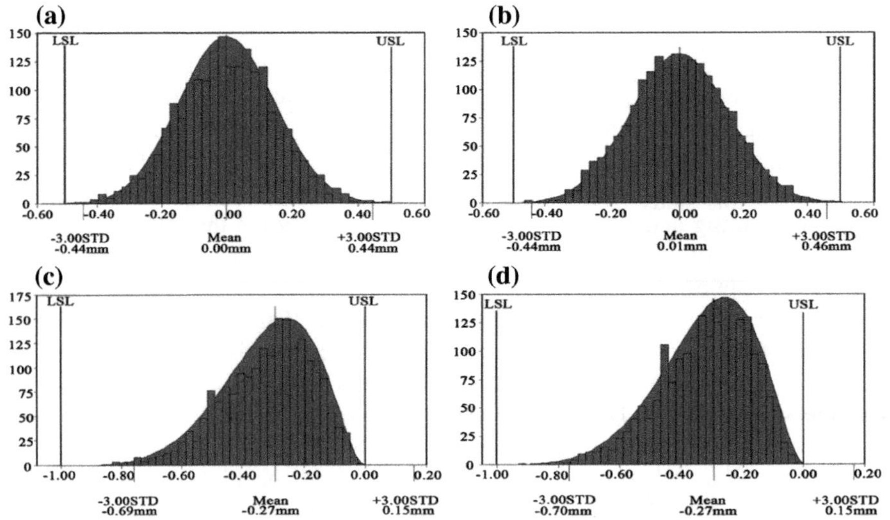

Fig. 10 Distribution of the sub-assembly process results after tolerance optimization (*Notes* **a** d1, **b** d2, **c** p1, **d** p2)

obviously reduced when the new tolerance specifications were used. All of the measurement targets were distributed in the required ranges (-0.5 mm $\leq$ d1, d2 $\leq$ 0.5 mm; -1 mm $\leq$ p1, p2 $\leq$ 0) when utilizing the sub-assembly procedure with the optimized tolerances.

6 Conclusions

Traditionally, tolerance design was dependent on the calculation of dimensional chains or on previous experience. However, it is difficult to find crucial tolerances using these methods. In addition, during the actual welding assembly process, a different assembly sequence will have an impact on the final product quality. Formerly, trial-manufacture of the product was necessary, followed by different assembly methods and a final test to determine product acceptability. This was very time-consuming and if the sample size was insufficient, the final results were questionable. In this study, 3DCS was used to investigate the specific size deviations for B-pillar parts when utilizing different assembly procedures, and the crucial tolerances were explored to optimize the final assembly quality.

In the B-pillar welding assembly process for cars, different assembly methods ultimately have a great impact on product quality. With established tolerances, the sub-assembly method for assembling the resulting product produces higher product quality than the sequential assembly method in a mass production process. The use of sub-assembly effectively reduces the number of defective products and improves product assembly accuracy.

According to the contribution analysis, designers can determine the contribution of each tolerance towards size deviations in the final product assembly, then further optimization can be accomplished. Using this method, key tolerances can be extracted from the many tolerances present. For the B-pillar parts, the most influential tolerances for measurement objectives were determined, and the quality of assembled parts was greatly improved after redesigning these important tolerances. There are numerous measurement targets for each part in the entire B-pillar product, and this study only analyzed the measurement objectives for PSBO and RCP. The remaining parts of the B-pillar are also worthy of in-depth research.

References

1. Huang GM, Shiau CS (2009) An optimal tolerance allocation model for assemblies with consideration of manufacturing cost, quality loss and reliability index. Assembly Autom 29(3):22–229
2. Ckeikh A, Mcgoldrick PF (1988) The influence of cost, function and process capability on tolerance. Int J Qual Reliab Manag 5(3):15–28
3. Chen H, Jin S, Li ZM et al (2014) A comprehensive study of three dimensional tolerance analysis methods. Comput Aided Des 53:1–13

4. Qureshi AJ, Dantan JY, Sabri V et al (2012) A statistical tolerance analysis approach for over-constrained mechanism based on optimization and Monte Carlo simulation. Comput Aided Des 44(2):132–142
5. Reich J, Chiocchio S, Cordier JJ et al (2009) Three-dimensional tolerance investigation on main ITER components. In: The 25th symposium on fusion technology. Fusion engineering and design. Elsevier Science Sa, Rostock, Germany vol 84, pp 1584–1587
6. Lin CW (2012) Simultaneous optimal design of parameters and tolerance of bearing locations for high-speed machine tools using a genetic algorithm and Monte Carlo simulation method. Int J Precis Eng Manufact 13(11):1983–1988
7. Lee KK, Roy YC, Han SH (2014) Tolerance optimization of a lower arm by using genetic algorithm and process capability index. Int J Precis Eng Manufact 15(6):1001–1007
8. Sinanoglu C, Borklu HR (2005) An assembly sequence-planning system for mechanical parts using neural network. Assembly Autom 25(1):38–52
9. Viganò R, Gomez GO (2012) Assembly planning with automated retrieval of assembly sequences from CAD model information. Assembly Autom 32(4):347–360
10. Xu ZJ, Li Y, Zhang J et al (2012) A dynamic assembly model for assembly sequence planning of complex product based on polychromatic sets theory. Assembly Autom 32(2):152–162
11. Xing ZW, Fang HS, Xu WL (2008) Simulation and formability analysis of drawing process for car B-pillar. Forging and Stamping Technol 33(2):27–30 (in Chinese)
12. Karmak S, Maiti J (2012) A review on dimensional tolerance synthesis: paradigm shift from product to process. Assembly Autom 32(4):373–388
13. Dantan JY, Qureshi AJ (2009) Worst-case and statistical tolerance analysis based on quantified constraint satisfaction problems and Monte Carlo simulation. Comput Aided Des 41(1):1–2
14. Bruyere J, Dantan J, Bigotr R et al (2007) Statistical tolerance analysis of bevel gear by tooth contact analysis and Monte Carlo simulation. Mech Mach Theory 42(10):1326–1351

The Research of Cylinder Deactivation Technology on Gasoline Engine

Mingdong Zhang, Jinqiang Ma, Hang Su, Cancan Xu, Hua Zhang, Jingwen Lu, Lianbao Li, Hong Wei, Shuangqing Li, Jun Li and Ruiping Wang

Nomenclature

BSFC	Brake specific fuel consumption
CDA	Cylinder deactivation
CAM	Camshaft
DFLHA	Dual feed hydraulic lash adjuster
DOHC	Dual overhead camshaft
DVVL	Double-variable valve lift
ECU	Engine control unit
FB	Fuel benefit
GDI	Gasoline direct injection
HLA	Hydraulic lash adjuster
NVH	Noise, vibration and harshness
OCV	Oil control valve
PWM	Pulse-width modulation
RFF	Standard roller finger follower
SRFF	Switching roller finger follower
TC	Turbocharger
TGDI	Turbocharger gasoline direct injection
TDC	Top dead centre
VVA	Variable valve actuation
VVT	Variable valve timing

R. Wang
Zhejiang Geely Royal Engine Co., Ltd., 315800 Ningbo, China

M. Zhang (✉) · J. Ma · H. Su · C. Xu · H. Zhang · J. Lu · L. Li
H. Wei · S. Li · J. Li · R. Wang
Ningbo Geely Royal Engine Components Co., Ltd., 315336 Ningbo, China
e-mail: zhangmingdong@geely.com

Society of Automotive Engineers of China (SAE-China) (ed.), *Proceedings of the 19th Asia Pacific Automotive Engineering Conference & SAE-China Congress 2017: Selected Papers*, Lecture Notes in Electrical Engineering 486, https://doi.org/10.1007/978-981-10-8506-2_7

1 Introduction

As the increasingly strict emission regulations, fuel consumption and market put forward higher requirements on engine technology for energy conservation and emissions reduction; the more energy-efficient and environmentally friendly engines are well received by consumers. To reduce the fuel consumption is one of the biggest challenges that the car manufacturers are facing currently. More and more advanced technology has been applied to reduce fuel consumption. With the rapidly development of the electrical control technology, the engine has been more advanced with the application of VVT and VVA technologies. Many OEMs and suppliers develop technologies for the low-energy and high-efficiency gasoline engines in several research projects, and many advanced technologies are applied in high-volume production. Among them, the cylinder deactivation (CDA) not only satisfies the requirement of power dynamic output, but also reduces the fuel consumption. In the past two years, GM, Chrysler, Volkswagen, Ford, Honda, Mercedes companies applied the technology on its new models. Because of the engine vibration and thermal loads' imbalance problems after cylinder deactivation, it has been used only in multi-cylinder engine. In the recent years, some domestic auto company develop the CDA engine products such as SAIC. The cars on the market are mainly used four-cylinder engine today. However, cylinder deactivation technology for four-cylinder engine is rarely applied. Now we are making some cylinder deactivation technology research on a four-cylinder gasoline engine.

2 Valve train Cylinder Deactivation

2.1 CDA Fuel-Efficient Theory

In practice, vehicle's working condition is complex and changing. The engine cannot keep operating on an economic fuel consumption at full MAP. At idle and partial load, the engine is running at low-load conditions. At this time because the throttle percentage is small and the throttling effect is increased, pumping energy loss is increased to cause the deterioration of fuel efficiency. If some cylinder is deactivated at partial load conditions, and to ensure the adequate power output, it will be necessary to increase the throttle percentage to intake more fresh air for the working combustion cylinders [1]. Due to the increased throttle opening, the pumping loss is reduced, the pressure in the intake manifold is increased, the air aerating efficiency is increased, and the combustion quality would be better [2, 3]. This can increase the working cylinders efficiency. So improve the fuel economy.

2.2 CDA Valve Train

This test uses the two-stage switchable roller finger follower to realize the range of valve lift switching from zero to maximum. Figure 1 shows the CDA mechanism schematic. Finger follower mainly consists of an inner arm and an outer arm. The inner arm roller bearing contacts with the high lift cam, and the outer arm roller bearing contacts with the cams which have 0 mm lift. There are two oil rails in the HLA. One is used for valve lash compensation, while the other one is used to control the switching of the finger follower [4].

Figure 2 shows the SRFF-3L assembly relationship with the cylinder head. Two OCV assembled in the inlet and exhaust side and respectively control the inlet and exhaust valve for each cylinder.

Figure 3 shows the SRFF-3L in the default position where the electronic signal to the OCV is absent. The OCV regulates the high pressure oil to the switching pressure port to move the latch pin. The inner rocker arm and out rocker arm are

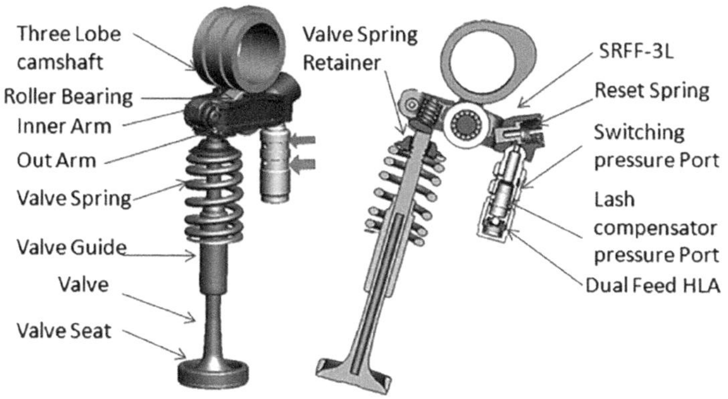

Fig. 1 Switching roller finger follower components

Fig. 2 Cylinder deactivation mechanism assembly relationship

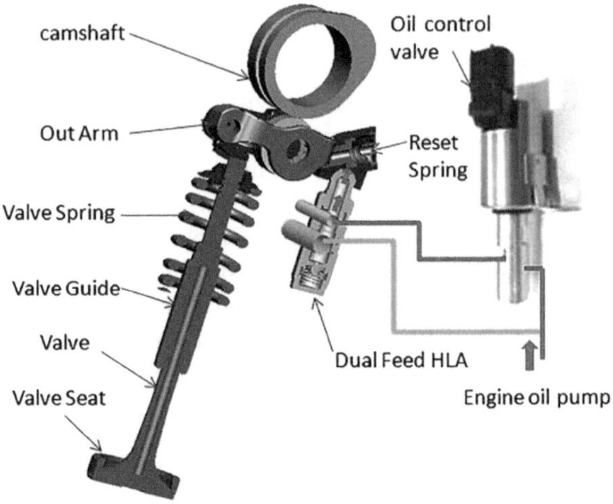

Fig. 3 Cylinder deactivation system

separated, and the inner arm is lost motion, while the no lift lobes remain in contact with the outer arm slider pads. The cylinder does not take a breath and stop the fuel injection and ignition. There is no cylinder power output. When the ECU electrical signal is absent, the reset spring moves the latch pin back. The lift lobe contacts the roller bearing, actuating the SRFF arm to open and close the valve. Now, the SRFF-3L functions similar to the standard RFF [4].

3 CAE Simulation

Before the engine test, we used the CAE tools to simulate the valve train structure and engine performance and fuel consumption.

3.1 Intensity Simulation Calculation

Figure 4 shows the half of the SRFF grid model, radial roller bearing adopted to simplify the tablet. The rocker arm used the tetrahedral mesh division and the lock pin using mixed cell division in the process of grid divides. The SRFF material elastic modulus is 2.1E5MP, and Poisson's ratio is 0.3. In the process of the simulation calculation, it applies linear radial load.

Figure 5 shows the deformation of the SRFF based on the simulation calculation. The calculation results show that the biggest displacement of rock arm along the *Y* axis is 88 μm and the static arm along the *Y* is 40 μm.

Figures 6 and 7 show the inner rock arm and out rocker arm of the first principal stress distribution; the maximum principal stress area is three-direction tensile stress

Fig. 4 Half of the SRFF grid model

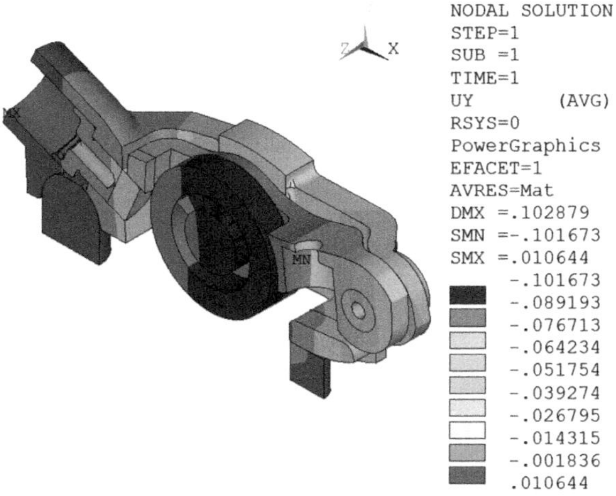

Fig. 5 Deformation of the SRFF

state. The inner arm maximum first principal stress is about 738 MPa, while outer rocker arm the first maximum principal stress is 319 MPa.

3.2 *Valve Train Dynamics Simulation Calculation*

In the process of design, AVL EXCITE Timing Drive single valve-train valve system model is established, then dynamic analysis was carried out on the CAM contour line of CDA, examining whether there will be a valve fly, bounce off spring and camshaft wear phenomenon such as to evaluate the valve dynamic characteristics.

Figure 8 shows the EXCITE TD valve train dynamics simulation model. Model mainly includes the driver, CAM, rocker arm, valve and valve spring. CDA engine

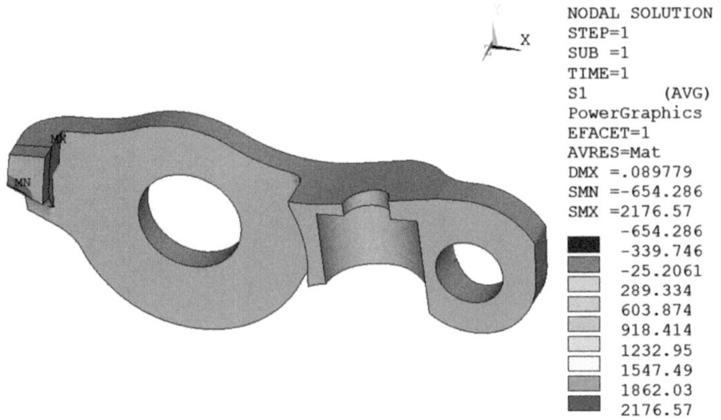

Fig. 6 First principal stress of the out rock arm

Fig. 7 First principal stress of the inner rock arm

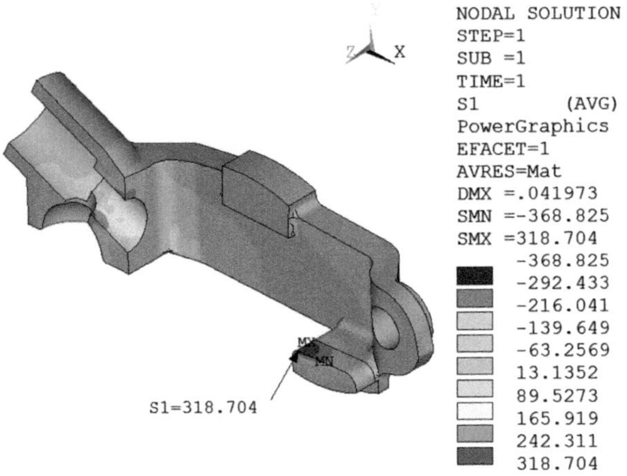

valve system's dynamic characteristics simulate computation by the simulation model, to assess whether the valve train meets the design requirements.

Figure 9 shows the simulation result that CDA engine valve closes to the valve seat speed. The result shows that the biggest speed is less than 0.6 m/s. It will not result in the valve seat and valve wear failure quickly in the life cycle of engine.

Figures 10 and 11 show the contact force of CAM and rocker arm, and the engine does not allow the CAM and rocker arm fly off phenomenon under the rated speed. It only allows slight fly off phenomenon when the engine over speed run. The results show there is no CAM and rocker arm fly off phenomenon at rated speed or over speed. The design requirements are met.

Figure 12 shows the contact stress between CAM and rocker arm. The CAM and rocker arm are steel-to-steel contact, and the maximum contact stress should be less than 1500 Mpa. The calculation results show that the maximum contact stress either

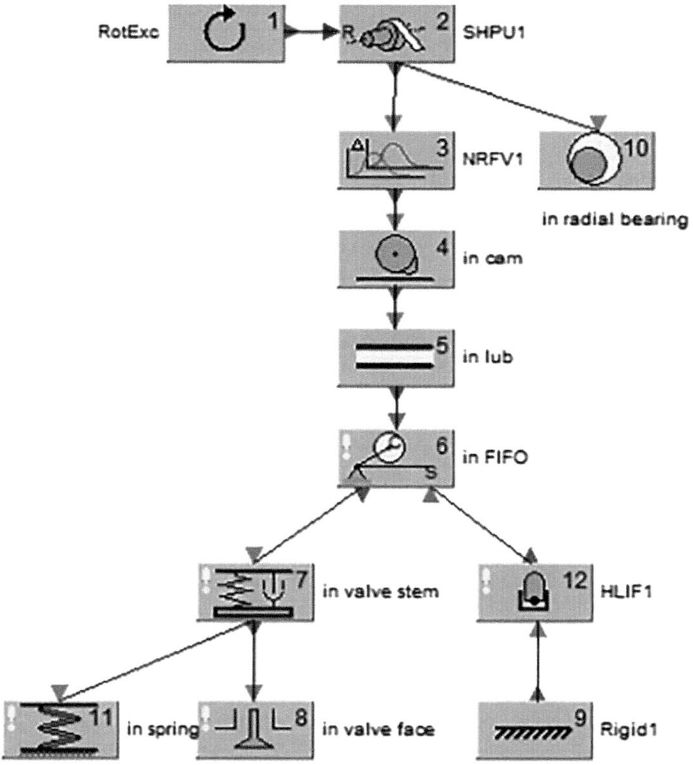

Fig. 8 Valve train dynamics simulation model

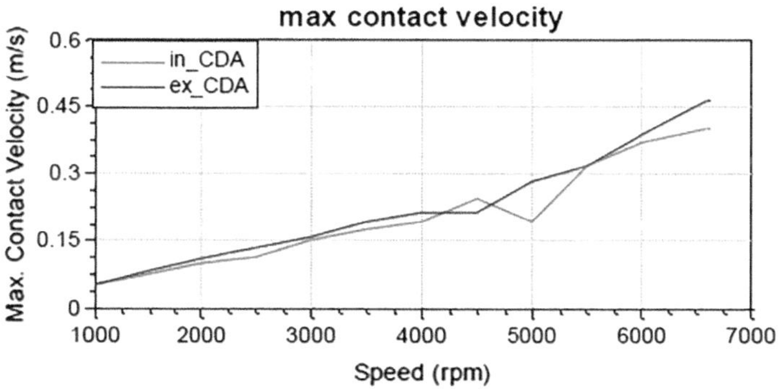

Fig. 9 Cylinder deactivation system

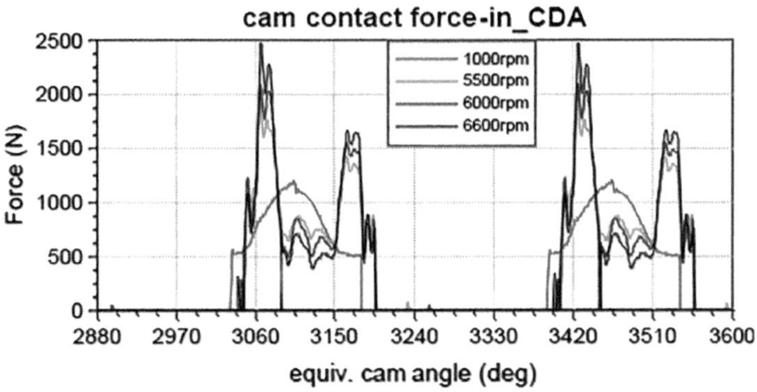

Fig. 10 Intake contact force of the CAM and rocker arm

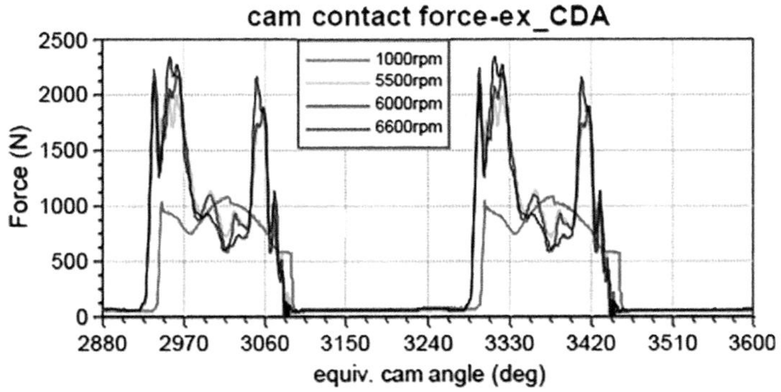

Fig. 11 Exhaust contact force of the CAM and rocker arm

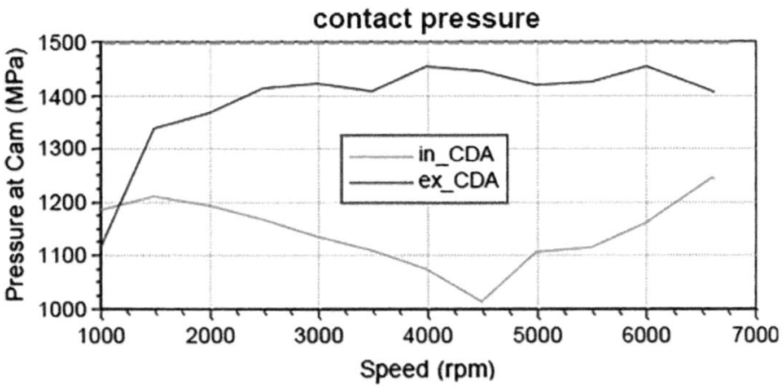

Fig. 12 Contact stress between CAM and rocker arm

on exhaust camshaft or intake camshaft is less than 1500 Mpa under different engine speed. It will not make fast damage of component due to insufficient strength in the process of engine running.

Figures 13 and 14 show the force of the intake and exhaust valve spring. In the process of engine running, the valve spring does not allow the adjacent two wire spring contact leading to abnormal changes of spring force. The results show that the engine under various working speed has no spring stress occurred at abnormal mutation.

3.3 Performance and Fuel Consumption Simulation

At the early stages of the design, we can predict the CDA engine Fuel-efficient by the GT POWER software simulate results. The rule of four-cylinder and the two-cylinder fuel consumption with the engine speed, load torque of change is simulated. The calculated results can provide theoretical reference for the engine design process.

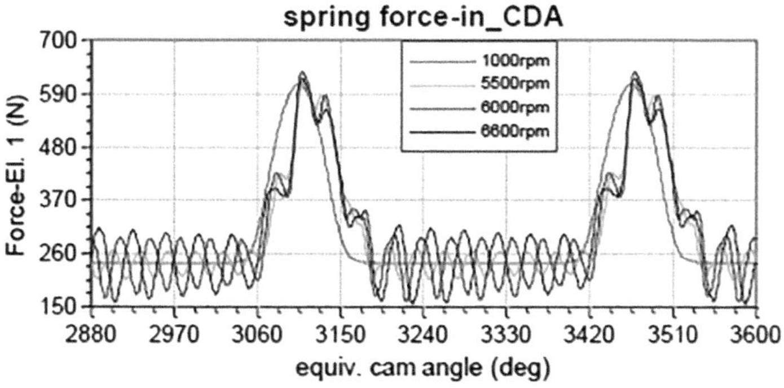

Fig. 13 Force of the intake valve spring

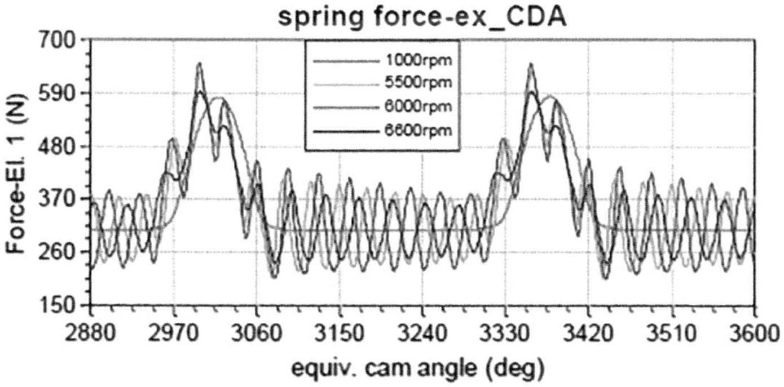

Fig. 14 Force of the exhaust valve spring

Figure 15 shows the GT-POWER model of CDA engine. The model mainly includes air intake system, exhaust system, fuel injection parts, crankcase and cylinder. The engine's main basic parameters are as shown in Table 1. To apply GT-POWER for CDA engine cylinder performance simulation, it controls two- and three-cylinder injection module no-work by changing the parameters close intake and exhaust valve of the model. It simulates the full-load torque and fuel consumption with SRFF-3L. In addition, the 2000 RPM part load torque and fuel consumption with the SRFF-3L or without is simulated.

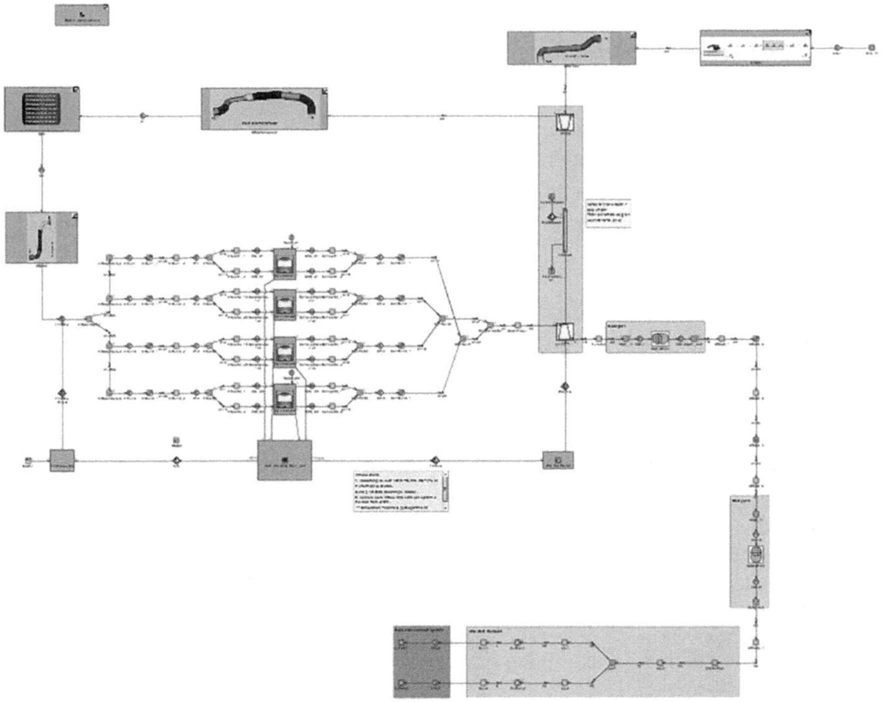

Fig. 15 GT-POWER model of CDA engine

Table 1 Basic engine parameters

Project	Parameters
Engine type and technology	Gasoline L4, four-stroke, GDI, TC, DOHC, DVVT
Engine displacement (L)	1.799
Bore and stroke (mm)	82.5 × 84.2
Compression ratio	9.7
Bore spacing	88
Engine power (kW/rpm)	135/5500
Engine torque (Nm/rpm)	300/(1750–4000)

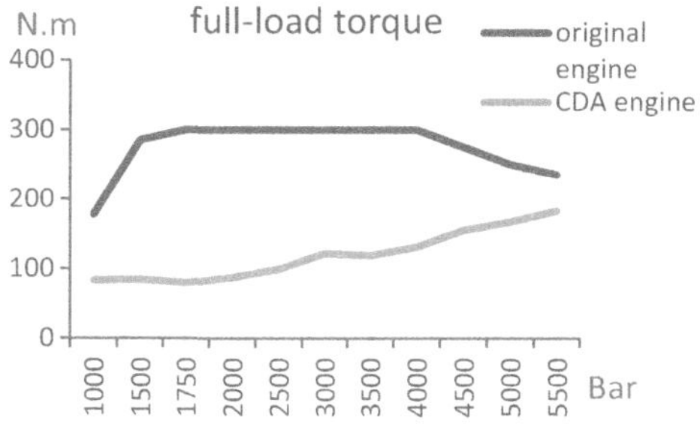

Fig. 16 Full-load torque for two-cylinder and four-cylinder states

Figure 16 shows the full-load torque calculation results of CDA engine working over two-cylinder and four-cylinder. The results show that CDA engine working at two-cylinder and four-cylinder has a bigger difference for the output torque.

Figure 17 shows the change of the engine fuel consumption rate for four- and two-cylinder engine at 2000 rpm. The results show that CDA engine has a significant fuel consumption benefits when the engine load is lower than 5 bar. With the load increasing, CDA engine fuel consumption rate is greater than four state. Particularly, 2000 rpm @ 2 bar can reduce fuel consumption rate of 12.92%, and 2000 rpm @ 3 bar can reduce fuel consumption rate of 11.45%.

Figures 18 and 19 show the vehicle fuel economy calculation point distribution of WLTC and NEDC cycles. WLTC cycle working point distribution is scattered compared with NEDC cycle. There is a more percentage that engine working in low speed and medium load in WLTC cycle. Engine works in low speed and low load in NEDC cycle. The CDA technology has a good fuel benefit on low speed and low load, so vehicle fuel consumption test using the NEDC cycle will lead to lower fuel consumption.

Figure 20 shows the NEDC and WLTC cycles' vehicle fuel consumption. In order to assess the actual benefits, vehicle fuel consumption with CDA technology simulated the vehicle fuel consumption of the NEDC and WLTC cycles. The results show that fuel consumption of the 1.8TGDI engine vehicle has a significantly reduced with CDA technology under NEDC and WLTC cycles. The CDA technology is the great way to improve vehicle fuel economy.

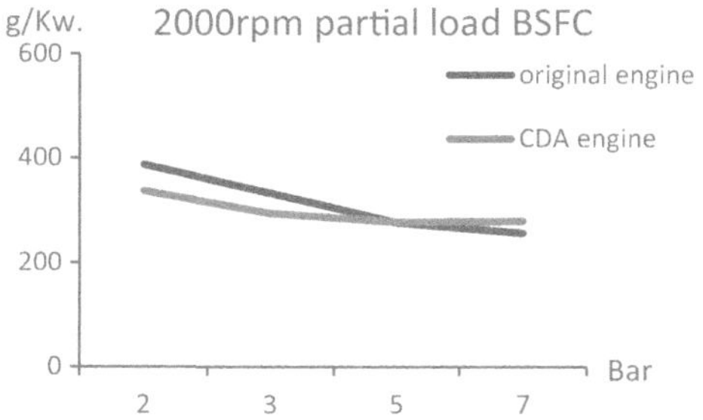

Fig. 17 2000 rpm fuel consumption for two-cylinder and four-cylinder state

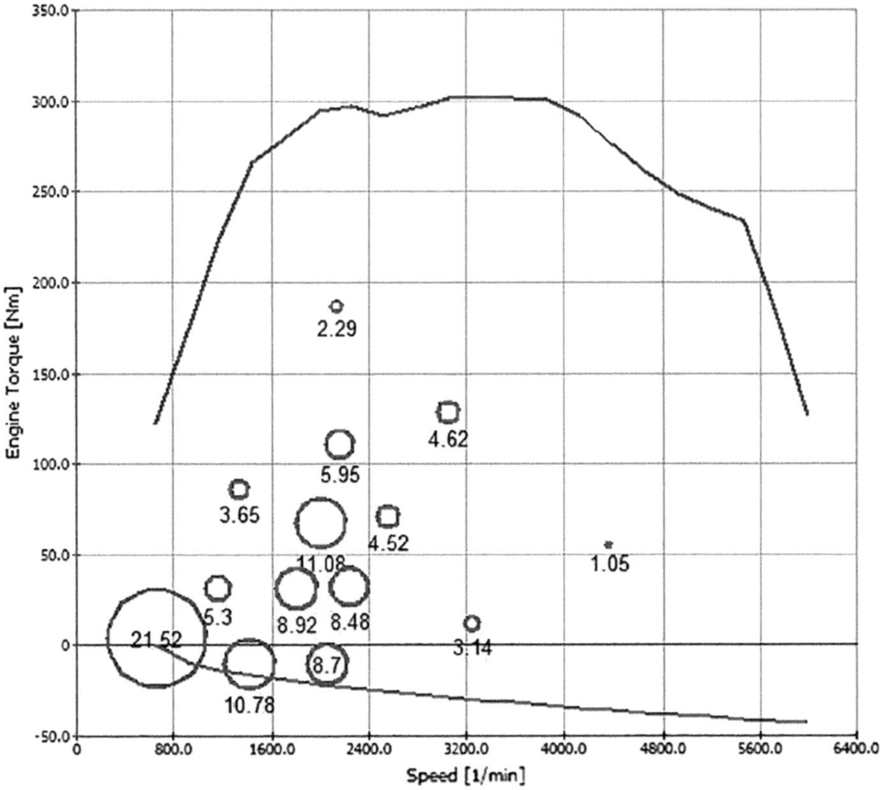

Fig. 18 WLTC cycle engine working point

4 CDA Experiment

4.1 Experimental Conditions

Cylinder deactivation experiment is based on a four-cylinder 1.8TGDI charging engine, valve timing mechanism with DOHC and DVVT technologies.

In order to get the engine BSFC benefit of cylinder deactivation, we calibrated both 2-cylinder state and 4-cylinder state. The main control boundary parameters are inlet temperature, water temperature, oil temperature and fuel temperature. During the test, the inlet temperature is (24 ± 1) °C, water temperature is (90 ± 2) °C, the oil temperature is (90 ± 3) °C, and fuel temperature is (28 ± 0.5) °C.

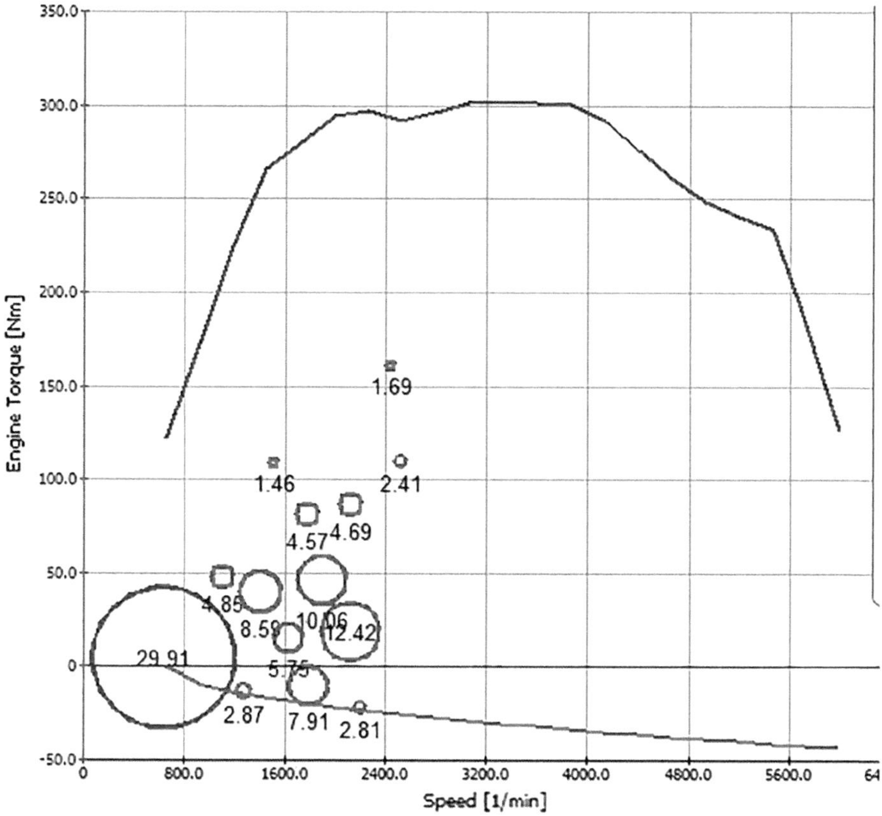

Fig. 19 NEDC cycle engine working point

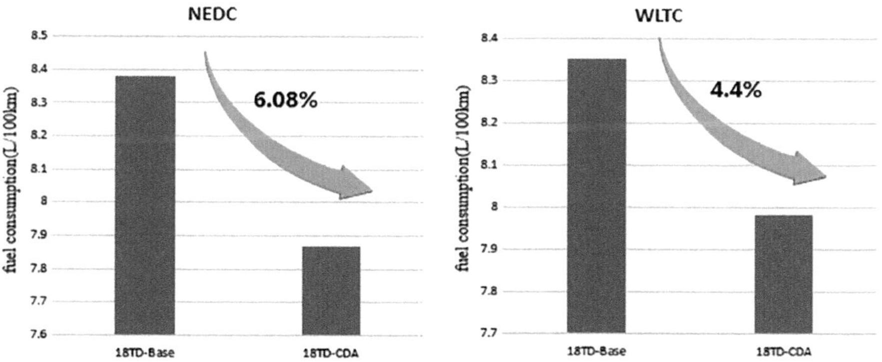

Fig. 20 NEDC and WLTC cycles vehicle fuel consumption

4.2 Experimental Results' Analysis

Figure 21 shows the results of engine fuel consumption under various loads and speed. The fuel benefit calculation is as follows:

$$\mathrm{FB} = \frac{\text{Fuel Consumption of 2 cylinder} - \text{Fuel Consumption of 4 cylinder}}{\text{Fuel Consumption of 4 cylinder}} \times 100\% \tag{1}$$

In the low load, cylinder deactivation has obvious fuel consumption advantages in this engine. While the load increases, the advantages of fuel consumption gradually reduce. The deterioration of fuel consumption is shown at the red areas on Fig. 21. Under the same load, the CDA engine fuel consumption benefit increases and with the speed increases it's gradually decreases. As the load increases, fuel consumption benefit gradually decreases under the same speed. So the best CDA working area is low load and low speed for this engine.

Figure 22 shows two-cylinder work and four-cylinder work fuel consumption at 2000 rpm. When the load is lower than 5 bar, cylinder deactivation is beneficial for fuel consumption in this engine. When the load exceeds 5 bar, cylinder deactivation technology has no fuel consumption advantages. The experimental results are similar with the simulation results as shown in Fig. 17.

Figure 23 shows the changes of intake manifold pressure between the two-cylinder and four-cylinder. When engine works in cylinder deactivation mode, the engine needs more gas into the working cylinder to ensure the same output power performance. When throttle valve opens wider, intake manifold pressure is raised and the throttling energy loss is decreased.

Figures 24 and 25 show the engine combustion data of AI50 and AI90 at 2000 rpm. When engine works in cylinder deactivation, AI50 is located in the ATDC 10 °CA. So it is better for fuel economy. As the engine load increases and

Fuel-benefit Bar

rpm

%	1	2	3	4	5	6	7
1000	-10.45	-7.52	-2.24	11.82			
1250	-12.87	-10.16	-4.83	5.08	16.52		
1500	-15.48	-9.34	-7.00	-3.61	8.38		
1750	-17.68	-11.15	-8.70	-3.55	0.70	10.60	
2000	-18.40	-12.98	-11.57	-7.57	-2.48		
2250	-18.28	-12.23	-10.62	-8.55	-4.97		
2500	-16.92	-13.15	-11.22	-7.17	-4.74		
2750	-15.11	-14.34	-10.00	-6.83	-5.27	-1.27	4.20
2900	-17.24	-14.30	-10.27	-8.11	-4.35	-1.31	3.98
3500	-13.78	-11.94	-9.64	-6.24	0.07	2.86	6.81
4000	-17.20	-11.02	-8.28	-4.83	1.22	3.88	
4500	-13.84	-9.67	-6.98	-5.13	1.64		
5000	-15.37	-7.34	-8.82	0.31	1.92		
5500	-14.95	-14.46	-11.27	-8.04	-0.28		

Fig. 21 Fuel benefit of CDA engine

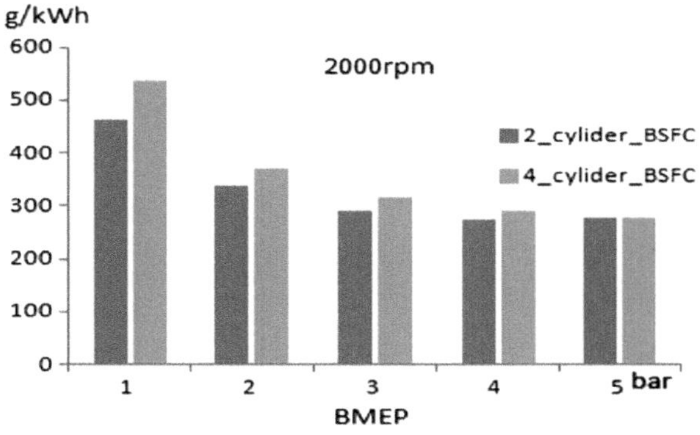

Fig. 22 Fuel-efficient MAP figure at 2000 rpm

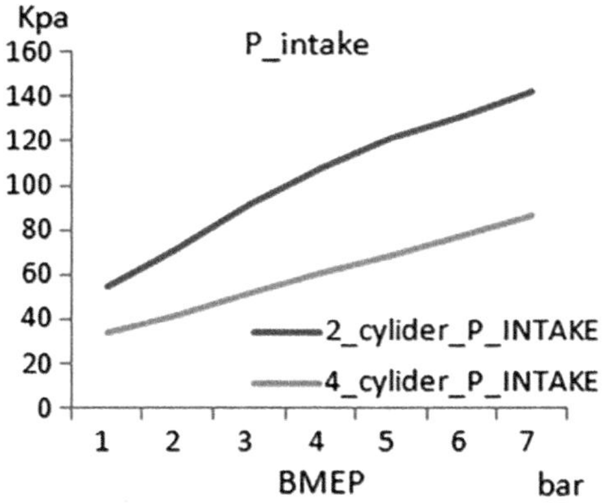

Fig. 23 Intake manifold pressure between the two-cylinder and four-cylinder

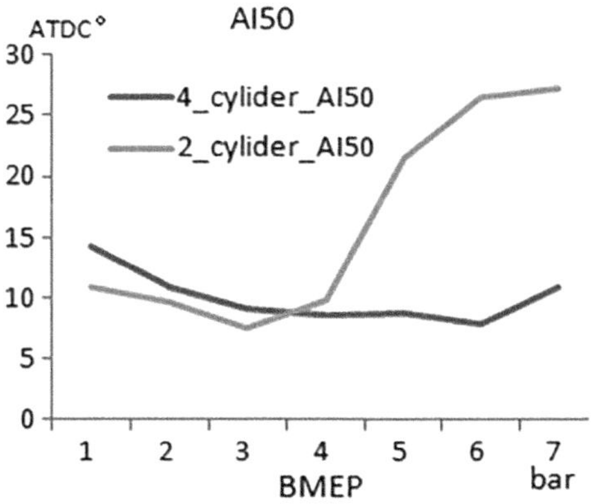

Fig. 24 AI50 at 2000 rpm

single cylinder load is heavy, AI50 value increases sharply. Engine knock tendency is increasing when using equivalent air–fuel ratio. The ignition angle is more delayed to cause deterioration of fuel economy.

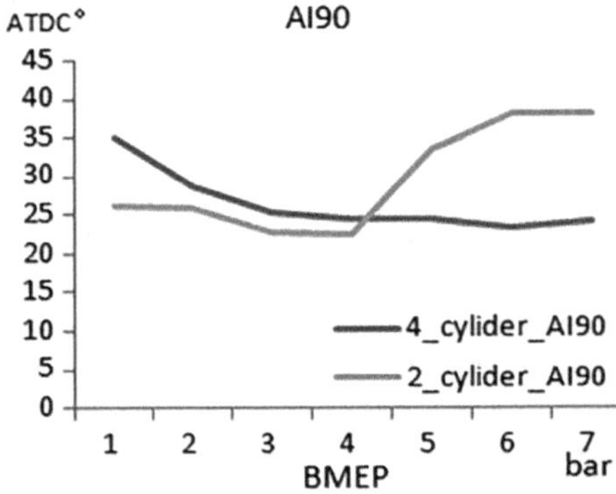

Fig. 25 AI90 at 2000 rpm

Figures 26, 27 and 28 show the emission of the engine at 2000 rpm. The CO and HC emissions are lower than four-cylinder working mode. When the engine load is increasing, throttle valve opens wider and the intake manifold pressure rises. The gas flow speed into the cylinder is faster. At the same time, the cylinder temperature is higher than four-cylinder mode. So fuel atomization is better and mixture is relatively homogeneous. When cylinder temperature is high and the quenching wall layer is thin, it may lead to the lower HC emissions. Since NO_x emissions are easy to be generated under the high temperature and oxygen enrichment in engine cylinder, the high temperature in the cylinder needs to be reduced. At the four-cylinder working condition, temperature is relatively lower. NO_x emissions of catalysts are a big challenge and may bring the increased cost of vehicle post-processing.

4.3 Challenges and Possible Solutions

Due to the CDA engine needs the same output power with four engine, the load may be increased, and cylinder firing interval angle increases for each 720 crank angle, especially after two-cylinder stop work the torque fluctuation is bigger.

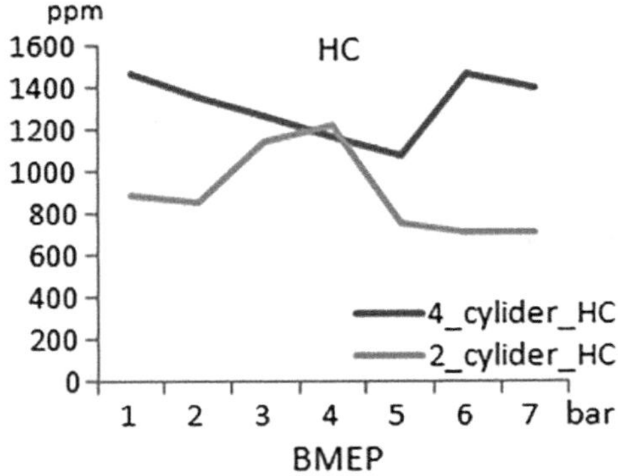

Fig. 26 HC emission of the engine at 2000 rpm/min

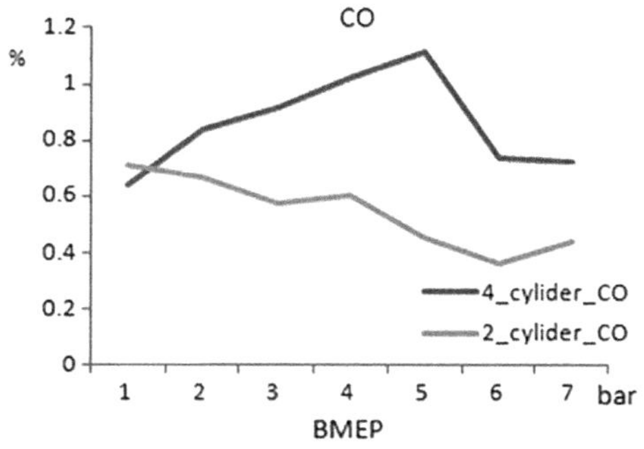

Fig. 27 CO emission of the engine at 2000 rpm

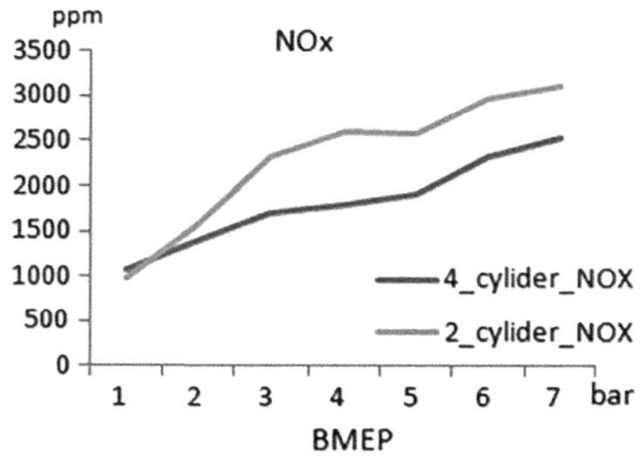

Fig. 28 NO_x emission of the engine at 2000 rpm

It makes the vibration of the engine more [2]. There are two available means to solution it. One of it reduces the stiffness of engine mounting, improves the damping of suspension, reduces the vibration of transmission, but ordinary suspension can't well meet the two requirements at the same time. It needs to adopt electro-hydraulic control suspension, which will affect the cost of the engine. The other way can increase the mass of flywheel and make use of the flywheel energy storage to reduce torque inhomogeneity, which could reduce the car responsiveness speed, which increase the flywheel mass to reduce wave vibration is limited.

When CDA engine closes the intake and exhaust valves, that cylinder temperature is low due to unworked cylinder causes engine heat load non-uniform. Each cylinder engine works strength uneven. It has an effect on the engine's reliability and life. When switching to four-cylinder working mode, there is no immediate work in high load due to the low temperature in cylinder. In the 80 s, BMW engineer developed a part that can lead high temperature exhaust gas to the cylinder in deactivation and maintain its internal temperature at a high level. It kept the engine internal heat load evenly and reduced the deformation of the cylinder block and Brush loss. It could also make rapid response of engine power output [5].

5 Control Strategy Analysis

5.1 CDA Control Discuss

For cylinder deactivation technology, one of key factors to face is the optimal cylinder deactivation operation area. It needs to ensure that the engine performance, fuel economy, emissions and a good NVH performance of power train.

Power performance: First we test the engine at both two-cylinder state and four-cylinder state to get the relationship of throttle percentage and torque output under different rotational speeds. It's determined by the experimental data that the cylinder deactivation operation area under various rotational speeds. Then calibrate the engine ECU to ensure that under the same speed after switching to cylinder deactivation mode, engine torque output is not falling.

Economy: Based on the bench test, we can get the BSFC under two-cylinder state and four-cylinder state. By comparing the BSFC under different operation mode, we can obtain the cylinder deactivation operation area.

Emissions: Ensuring the engine fuel economy and power performance unchanged, we should also guarantee the engine emissions meet the emission regulation. So during the engine cold start, warming-up, etc., the engine should work on four-cylinder mode.

5.2 Control Strategy

Figure 29. The switch sequence diagram design of cylinder deactivation control. First a reference position is used to CDA engine default sequence calculation, There is a delay time from the control signal trigger to actuator action. So CDA mechanism needs a switch delay to make up it. Switch delay is the angle between activating position and the target tooth, for CDA actuators to execute switching operation. The activating timing and the controller trigger position is revised by the first reference position and VVT position. Activating timing is calculated as below. Ultimate controller output PWM signals to control the solenoid valve movement

$$\begin{aligned}\text{Activating timing} = {} & \text{Current tooth number} + (\text{Angle between reference tooth to target} \\ & - \text{Switch delay} - \text{VVT position})\end{aligned} \tag{2}$$

Figure 30 shows the switch logic design of cylinder deactivation control. ECU detect whether meet the basic strategy conditions at the began control, for example fault codes, speed, battery voltage state, whether in the warm-up state, Slow down, Whether the transmission shift etc. If all the condition is met, engine ECU controls the engine to operate on four-cylinder mode or two-cylinder mode based on the

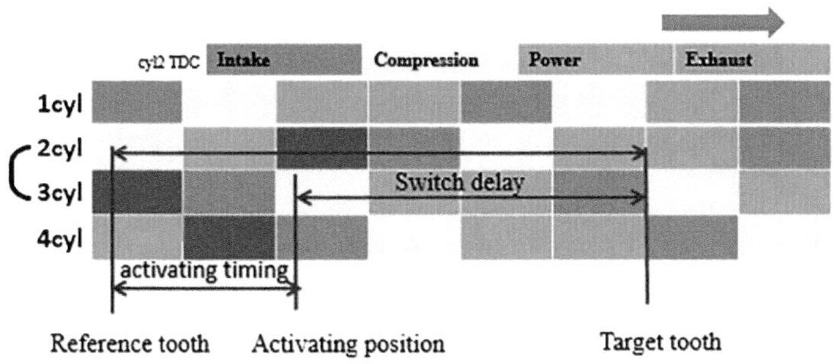

Fig. 29 Switch sequence diagram design of cylinder deactivation control

control strategy, and adjusts the control signal to the actuators such as throttle valve, fuel injector, ignition plug, VVT, etc., to ensure the power output is smooth-going. During the operation mode switching, the engine should not make the driver feel uncomfortable, and vibration and harshness (NVH) performance should be guaranteed.

Figure 31 shows the torque fluctuation at 2000 rpm@3.5 bar during operation mode switching. ECU controls the engine to operate under cylinder deactivation mode and normal mode. At each mode, engine keeps operating 25 s and the torque data is recorded. We can find that during the switching mode, there is a slight fluctuation on torque output. ECU can quickly adjust the engine control parameters to keep it working stably.

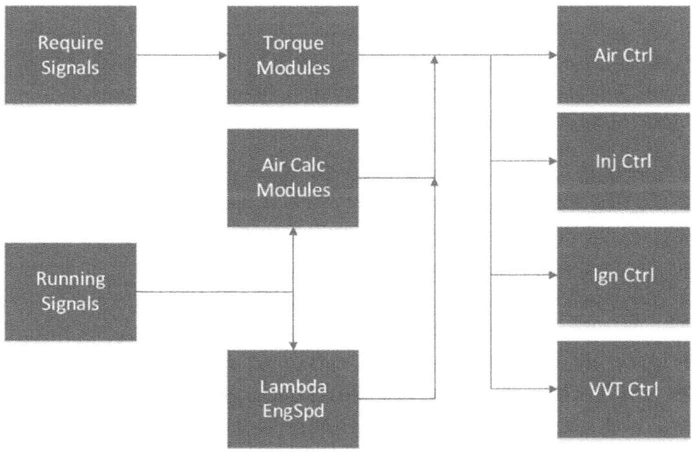

Fig. 30 Switch logic design of cylinder deactivation control

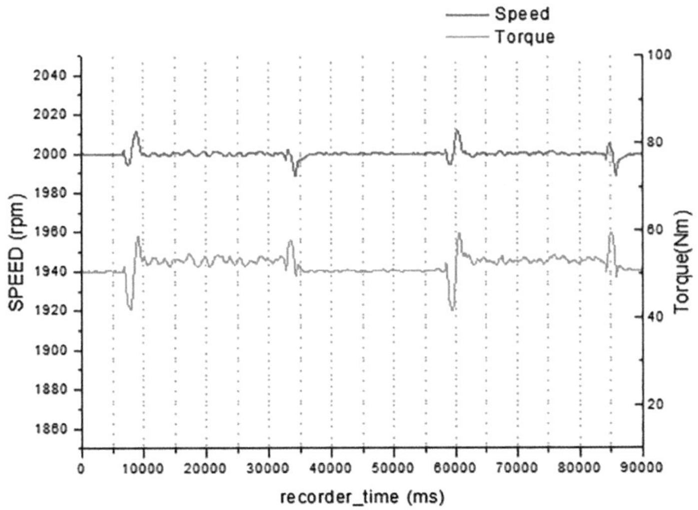

Fig. 31 Torque fluctuation at 2000 rpm@3.5 bar during operation mode switching

6 Conclusions

This paper has made a brief description of cylinder deactivation technology. After using the technology, the benefits of actual fuel consumption and changes in emissions have been achieved. The mechanism of cylinder deactivation and the test data prove that by using this mechanism, we can realize the CDA function on a 1.8L TGDI engine, and it is possible to achieve a considerably higher fuel-saving advantage at engine partial load. At the same time, this paper is also to explore the challenges and the solutions in the CDA engine design process. Finally, to introduce the demand of the cylinder deactivation technology control strategy should meet the requirements of performance, fuel economy and emission.

1. Cylinder deactivation can effectively reduce the fuel consumption under 4–5 bar, and the fuel-benefit rate reached to 18.40% at 2000 rpm @ 1 bar.
2. Cylinder deactivation reduces inlet throttling loss because of the throttle percentage increasing. The ignition advance angle is more ahead compared to the four-cylinder mode work at the same load.
3. Cylinder deactivation can effectively reduce the fuel consumption in partial engine load, but as the load increases knock tendency to be more and more obvious. Fuel economy promotion effect also gradually reduced. The HC and CO emissions are better than four-cylinder work, but NO_x emission is terrible.
4. Cylinder deactivation control strategy must consider the engine performance, fuel economy, emissions and NVH, etc.

Acknowledgements The authors would like to acknowledge the support of GEELY Corporation team colleagues that helped make this work possible. The authors acknowledge the support from the GEELY Engineering, Prototype, Metrology and Testing Departments.

References

1. Flierl R, Gollasch D, Knecht A, Pohl D, Hannibal W (2006) Perspectives of the fully variable valvetrain univalve on the basis of a 2.0 l spark-ignition engine. MTZ 67(7/8):560–567
2. Zhang D, Yuan Y, Cui Y (2007) Technology of cylinder deactivation of automotive gasoline engine. Small Intern Combust Engine Motorcycle 36:89–93
3. Zhou L (1999) Internal-combustion engine. Beijing China Machine Press
4. Radulescu A, McCarthy JR, Brownell S, Eaton Corp (2013) Development of a switching roller finger follower for cylinder deactivation in gasoline engine applications. SAE Technical Paper 2013-01-0589. https://doi.org/10.4271/2013-01-0589
5. Federenko Y, Korzhov M, Fillippov A, Atamanenko N, Method and system for power control of internal combustion engines using individual cycle cut-off. Society of Auto-motive Engineers, paper no 960462

LP EGR Influence on Performance of Turbocharged Direct Injection Gasoline Engine

Hong Wei, Lianbao Li, Lin Yang, Narendra Purania, Xuehai Qin, Huacheng Zhou, Dongya Chen, Xiaoli Tian, Yunlong Kuang and Ruiping Wang

Definitions/Abbreviations

LP	Low pressure
HP	High pressure
EGR	Exhaust gas recirculation
CEGR	Cooled exhaust gas recirculation
CR	Compressor ratio
WLTC	World light test cycle
RDE	Real driving emission
TDI	Turbocharged direct injection
DP	Delta pressure
COV	Coefficient of variation
BMEP	Brake mean effective pressure
CA50	Crank angle where 50% of FMB
BSFC	Break-specific fuel consumption
VVL	Variable valve lift
VVT	Variable valve timing
HCCI	Homogeneous charge compression ignition
ECU	Electronic control unit
TC	Turbo charger
WOT	Wide-open throttle
S415	Smoke intensity
CO	Carbon oxide
NO_x	Nitrogen oxide
HC	Hydrocarbon

R. Wang
Zhejiang Geely Royal Engine Co., Ltd., Ningbo 315800, China

H. Wei · L. Li · L. Yang · N. Purania (✉) · X. Qin · H. Zhou · D. Chen
X. Tian · Y. Kuang · R. Wang
Ningbo Geely Royal Engine Components Co., Ltd., Ningbo 315336, China
e-mail: Narendra.Purania@geely.com

Society of Automotive Engineers of China (SAE-China) (ed.), *Proceedings of the 19th Asia Pacific Automotive Engineering Conference & SAE-China Congress 2017: Selected Papers*, Lecture Notes in Electrical Engineering 486, https://doi.org/10.1007/978-981-10-8506-2_8

ZWOUT	Ignition angle
wnwsa	Intake VVT
wnwse	Exhaust VVT
Prist_w	Fuel rail pressure
Mpa	Megapascal

1 Introduction

As strict fuel consumption and emission regulation requirements of the global governments, it is an important topic to reduce the fuel consumption of the engine. Some key technologies play critical role in reducing fuel consumption and emission of TDI gasoline engine, such as Miller cycle, cooled EGR, VVL, water injection in cylinder, and HCCI. According to the analysis of technical feasibility, cost and fuel benefit potential, cooled EGR is an ideal choice, so we developed the cooled EGR technology with CR optimization to reduce the engine fuel consumption [1–3].

The main function of EGR technology is to influence the combustion by means of decreasing the combustion temperature in the cylinder to lower the NO_x emission of the engine. Faced with the strict fuel consumption regulations, the potential of EGR technology has been further exploited as it can improve the mechanical as well as the thermal efficiency of an engine. On the one hand, it was able to decrease the pumping loss to improve the engine mechanical efficiency; on the other hand, lower combustion temperature, leading to the decrease of coolant heat dissipation loss and an exhaust temperature, overall suppresses the engine knocking tendency [4–6]. In conclusion, the suppression of engine knock was able to advance the ignition angle, to increase the CR, to lower the exhaust temperature, and furthermore to reduce the fuel injection enrichment, which led to decrease in the fuel consumption. The knock suppression and lower exhaust temperature have significant contribution for fuel benefit, especially in high load, rather than in part load [7, 8].

It had been a trend to apply the EGR technology to upgrade the turbocharged gasoline engine, and the cooled EGR technology could be divided into two types: HP EGR and LP EGR. Compared to HP EGR, LP EGR has more advantage in broad operating range (LP EGR can cover the entire engine speed and load points, while HP EGR cannot operate in low speed and high load), more fuel consumption benefit and is compatible with RDE test cycle. According to the requirement of fuel consumption reduction, we chose to develop the LP EGR technology in this research. The Great Wall, BorgWarner, and Valeo have already done some researches in turbocharged gasoline engine LP EGR technology and confirmed the fuel consumption potential of LP EGR technology (Table 1) [9–12].

Table 1 Comparison of HP EGR and LP EGR

Factors		HP	LP
Effect on combustion	Reduce the knock	−	+
	Stability	−	+
	Transient response	+	−
	BSFC	−	+
	HC	+	−
	NO_x	−	+
Effect on engine	Pressure of EGR inlet	+	−
	EGR working range	−	+
	Acid gas	−	+
	Temperature	−	+
	Filter	+	−
	Difficulty of layout	+	−
	Other components influence	+	−
Cost	EGR system	+	−

Notes + pros; − cons

2 Test Equipment and Test Procedure

2.1 Test Equipment

A 1.5-L three-cylinder TDI gasoline engine was applied to the research. Compared with the base engine, the piston, camshaft, and ignition coil have been redesigned before they were added to the LP EGR system. To ensure combustion stability, we used a 90 MJ ignition coil.

Furthermore, we chose to analyze the data of CR11 because it showed the best fuel benefit of vehicle fuel consumption simulation.

As shown in Fig. 1, the LP EGR system layout was arranged in such an extent to extract EGR gas downstream of the turbine and to introduce it upstream of the

Fig. 1 LP EGR schematic diagram of TC engine

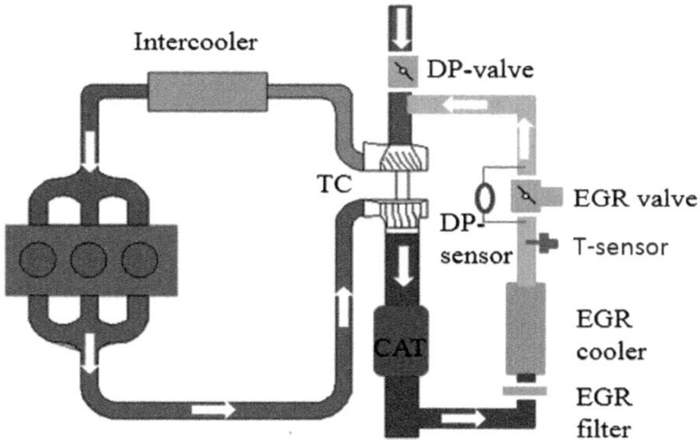

compressor. To adjust the EGR system pressure differential a delta P valve was adjusted so that the EGR flow could meet the engine requirement.

Figures 2 and 3 show the layout of the sensor and engine. In gas side, the concentration of CO_2 had measured in the position of post-intercooler and post-turbo to calculate the EGR rate. Exhaust gas analyzer had been used to analyze the emission of HC, CO, and NO_x; sensors to measure temperature and pressure were assembled at the upstream and downstream of the necessary components to

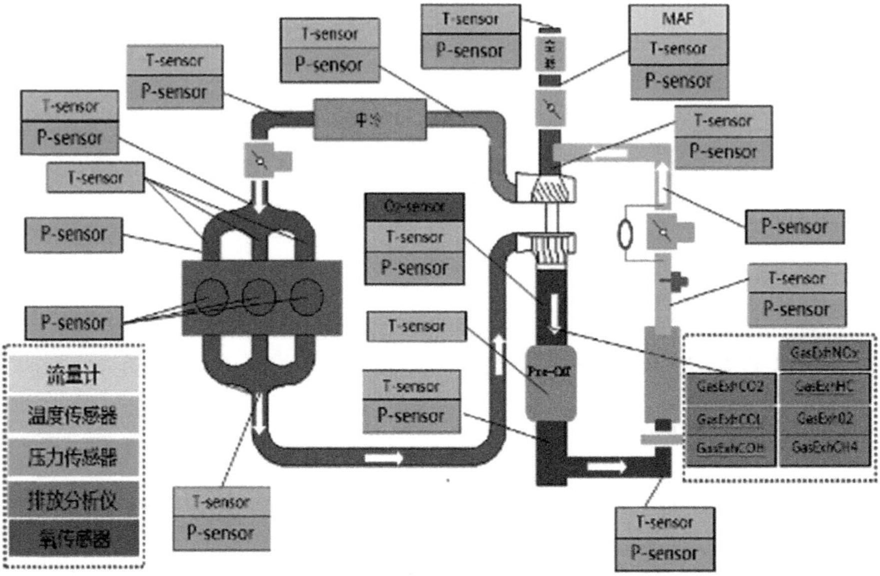

Fig. 2 Sensor schematic diagram of engine gas side

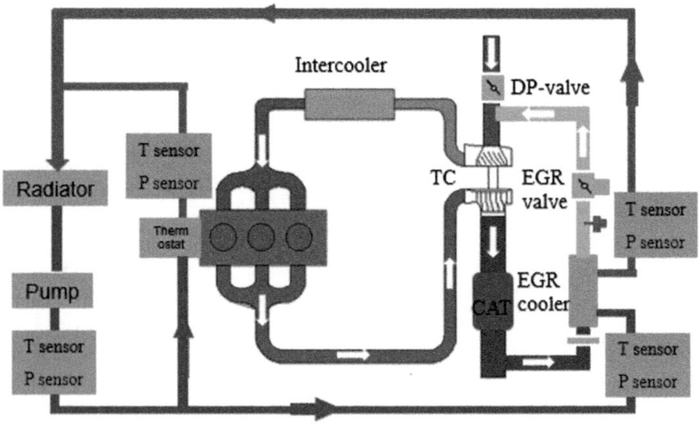

Fig. 3 Sensor schematic diagram of engine coolant side (update)

Table 2 Test equipment list

Equipment	Model
Electric dynamometer	AVL-22O-525
Fuel consumption meter	AVL735S, AVL753C
Cylinder pressure sensor	KISTLER6115CF-5DQ03
Charge amplifier	AVL-4P4G
Combustion analyzer	AVL-INDICOM
Lambda analyzer	EATS-630
Exhaust gas analyzer	HOLIBA7500DEGR
Smoke admeasuring apparatus	AVL415S
T-sensor	PT100
T-sensor	PT100
P-sensor	Accuracy: 0.3%
P-sensor	Accuracy: 0.5%

monitor the engine operation. On the coolant side, a temperature and pressure sensor was installed at the EGR cooler (coolant) inlet/outlet to monitor the EGR cooler cooling efficiency. The mixed gas temperature of intercooler outlet should be controlled based on the engine speed.

The cylinder pressure sensor and Lambda analyzer are used to detect the cylinder pressure and engine air–fuel ratio. An independent ECU was used to control the EGR system. The complete list of equipment is shown in Table 2.

2.2 *Test Procedure*

We focused on the calibration of the EGR rate, VVT, fuel rail pressure, and fuel injection phase. Variants, such as throttle, EGR valve, DP valve, TC duty cycle, A/F, ignition phase, VVT, fuel rail pressure, and fuel injection phase, were calibrated.

In the process of calibration, the first step was to sweep the EGR rate; the EGR rate is in the range from 0 to 20%, and the step length is 2%, by opening the EGR valve. The second phase was to adjust (closing direction) the DP valve to create the EGR system pressure differential and EGR flow if the EGR rate could not meet the requirement by controlling the EGR valve. We adjusted the variants during the ignition phase, and then measured the variation tendency of fuel consumption by the changing of the EGR rate. The third step was to calibrate the VVT of the intake camshaft and exhaust camshaft to get the optimal VVT phase. To avoid missing the optimal fuel consumption point, we swept the EGR rate again after confirming the VVT gets the optimal combination of EGR rate and VVT phase. Finally, we calibrated the fuel rail pressure and fuel injection phase at the optimal EGR rate and VVT phase and achieved the lowest fuel consumption point. During the full calibration, the bmep controlled at a fixed value for the each test operating

Table 3 Limit requirements of the engine test

Limit requirements of the engine test	
Fuel	Standard 95# unleaded gasoline
Intake temperature	25 ± 3 °C
Outlet temperature of intercooler	$T = n/200 + 25(\ °\text{C})$ & $T \leq 50(\ °\text{C})$ &
DP valve	ΔP $\geq$ 2.5 kPa at low load
Oil	SAE 0W-20 ACEA A5/B5, C2
Cooling system temperature	Temperature of coolant outlet 100 ± 5 °C
Coolant	ethanediol 50% + water 50%
Fuel temperature	25 ± 2 °C
Fuel pressure	4 bar
Continuous temperature of main oil channel <130 °C, limit oil temperature <142 °C, pressure of main oil channel >1 bar PT100	
Exhaust back pressure	≤47 kPa
Data acquisition conditions	The difference of engine speed and target speed should below 1% 油耗测量时间不超过20 s

point and COV limit below 3%. Air/Fuel ratio lambda controlled at 1, it is also required to enrichment at high load to keep the better engine dynamic performance.

There are some limit requirements before the engine calibration to keep the accuracy of the data. The limit requirements are shown in Table 3.

3 Experimental Results and Discussion

To achieve minimum bsfc point, engine calibration performed in following sequences 1. Sweep EGR rate 2. VVT timing optimization 3. Sweep EGR rate again to find best bsfc point 4. Fuel injection phase 5. Fuel injection pressure.

Part I Sweep the EGR rate

3.1 Influence of EGR to BSFC, Power and Torque

- EGR-specific heat capacity is higher, which leads to a lower combustion temperature in the cylinder. So NO_x is reduced and fuel consumption is improved (also, increased EGR can reduce pumping losses). For example, fuel consumption was reduced by 16.5% at 1920 rpm@18 bar shown in Fig. 4.

- At wide-open throttle (WOT) working conditions, the addition of EGR into the intake system means limit the fresh air inflow, thus reduced the engine power and torque. For example, 1500 rpm@WOT, torque is reduced by 0.9%. 4000 rpm@WOT, torque is reduced by 4.2% (Figs. 5 and 6).

3.2 *Influence of the LP EGR on Combustion Parameters*

- COV—EGR rate is less than 16%, and the combustion is relatively stable. However, the EGR rate gets higher than 16%, and COV exceeded the limit causing combustion instability at 1750 rpm@3.5 bar as shown in Fig. 7.

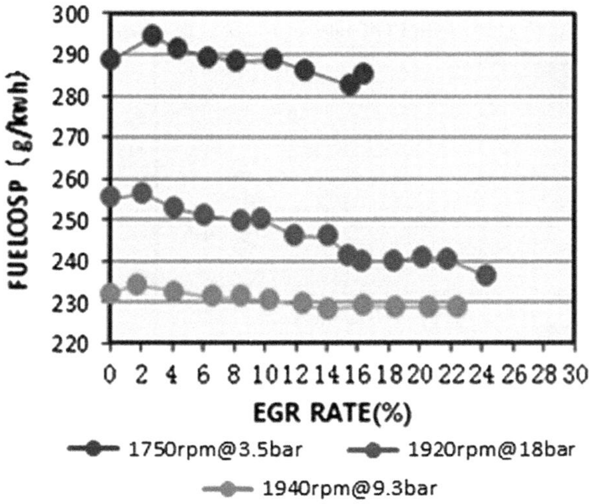

Fig. 4 BSFC with EGR rate change trend

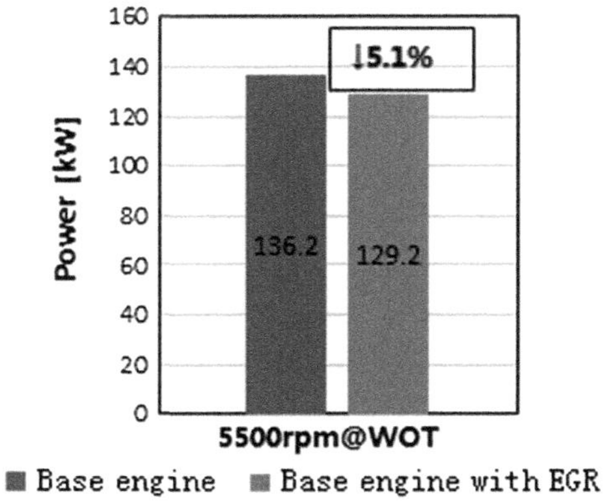

Fig. 5 External characteristic torque contrast

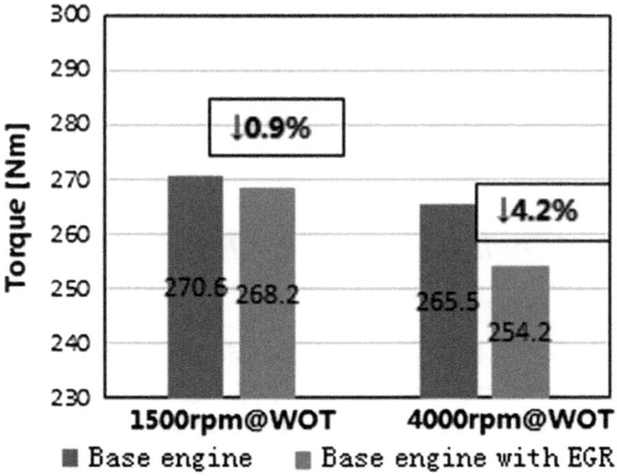

Fig. 6 External characteristic power contrast

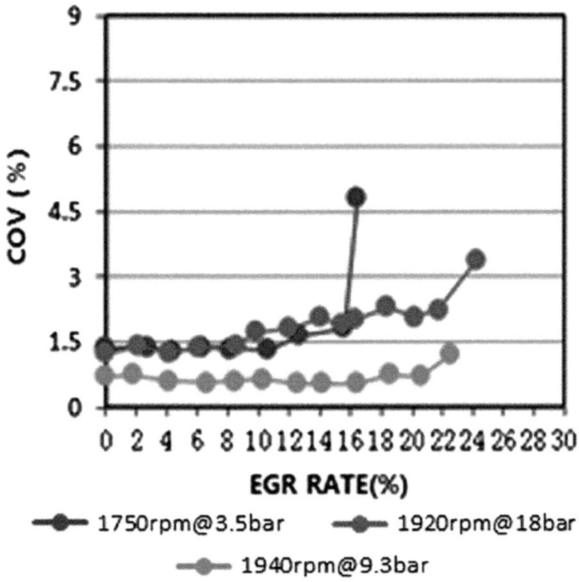

Fig. 7 COV with EGR rate change trend

- ZWOUT—Increasing EGR can reduce combustion temperature in the cylinder and the knocking tendency, which allows advancing ZWOUT as shown in Fig. 8.
- CA50—Despite an increased EGR rate and advanced ZWOUT CA50 trend is not very noticeable as shown in Fig. 9.
- Combustion Duration—The higher EGR rate the higher combustion duration as shown in Fig. 10.

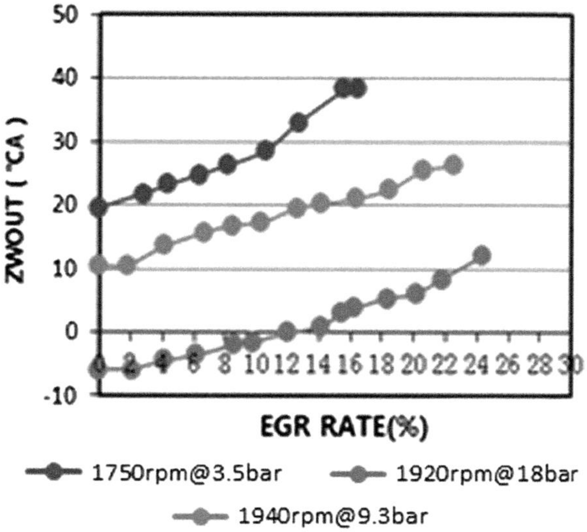

Fig. 8 ZWOUT with EGR rate change trend

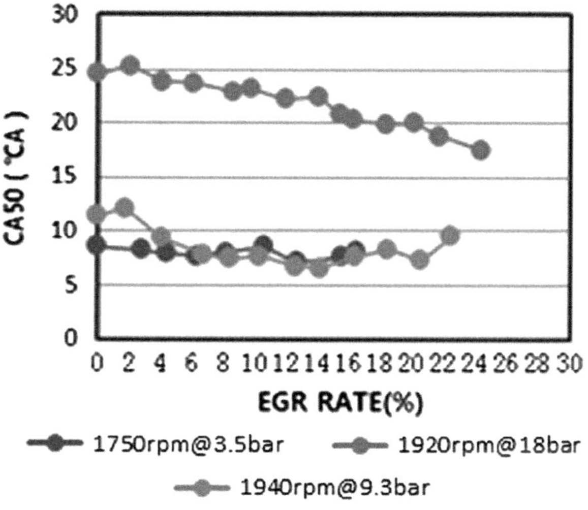

Fig. 9 CA50 with EGR rate change trend

3.3 LP EGR Influence on Emissions

- NO_x—EGR contains virtually no oxygen and reduces the oxygen contained in the cylinder. Higher specific heat capacity of EGR leads to lower combustion temperature. NO_x formation is dependent on combustion temperature. Test data show a downward trend of NO_x emissions (reduction of 50–88%) as shown in Fig. 11.

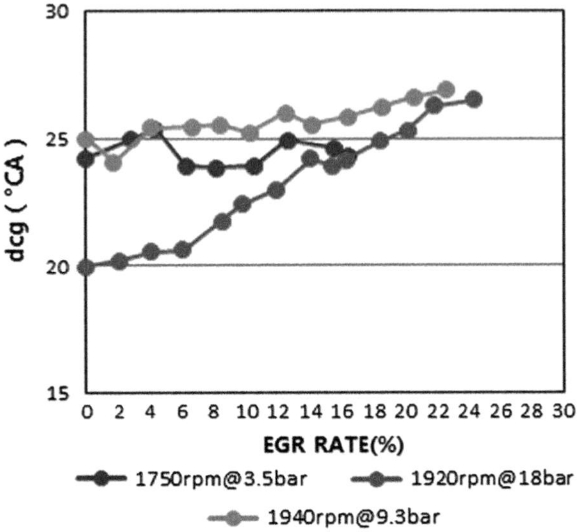

Fig. 10 Combustion duration with EGR rate change trend

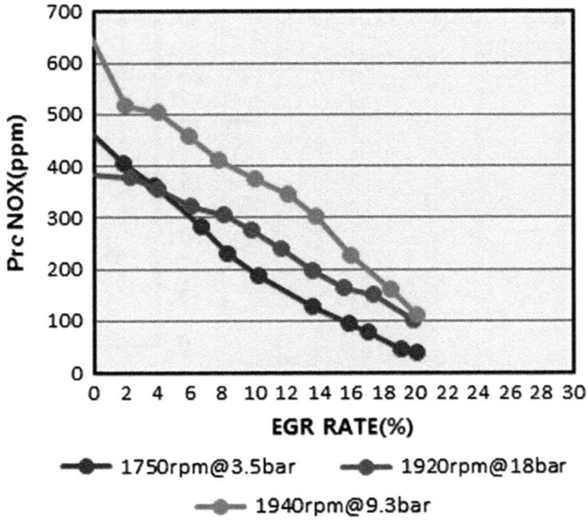

Fig. 11 NO_x with EGR rate change trend

- HC—With the increase of EGR rate, HC emission values rise (20–60% increase). Due to the increase of EGR, combustion temperature as well as combustion speed is reduced. Furthermore, combustion duration increases leading to higher HC as shown in Fig. 12.
- CO—With the increase of EGR, CO emission values fluctuate, but there is no obvious effect on them in general as shown in Fig. 13.
- S415—There is no effect of EGR on S415 emissions at all as shown in Fig. 14.

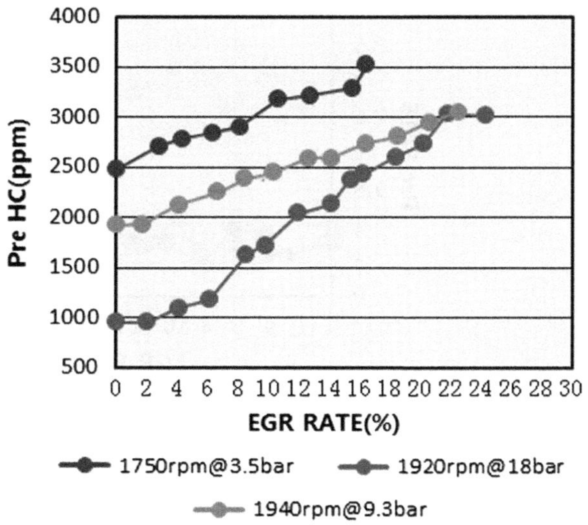

Fig. 12 HC with EGR rate change trend

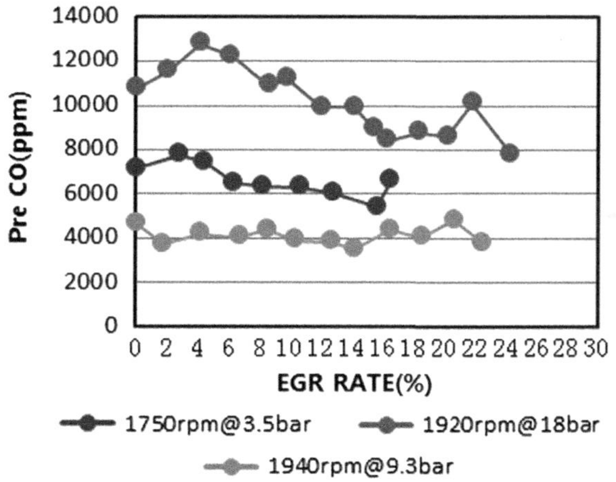

Fig. 13 CO with EGR rate change trend

3.4 Influence of the LP EGR on λ

- EGR reduces combustion temperature and lowers the exhaust temperature, which allows to reduce fuel enrichment (High speed and high load condition).
- There is no obvious influence of EGR increase regarding λ (Figs. 15 and 16).

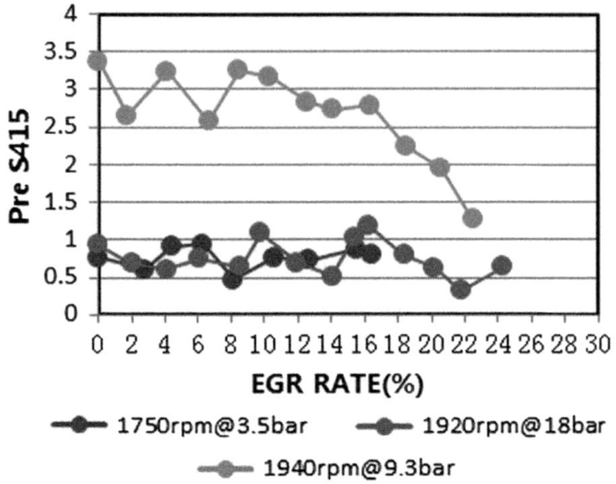

Fig. 14 S415 with EGR rate change trend

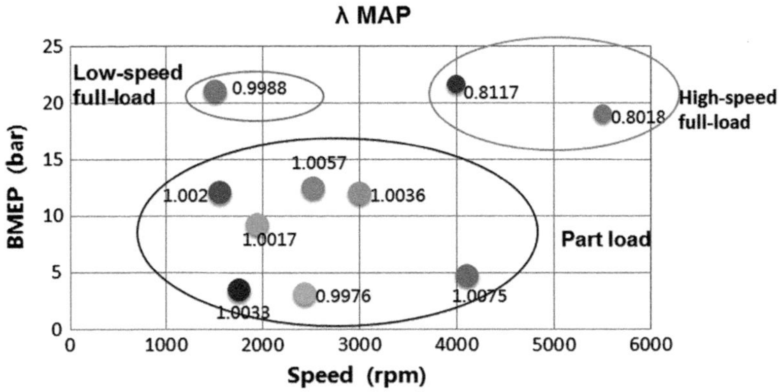

Fig. 15 λ MAP with EGR

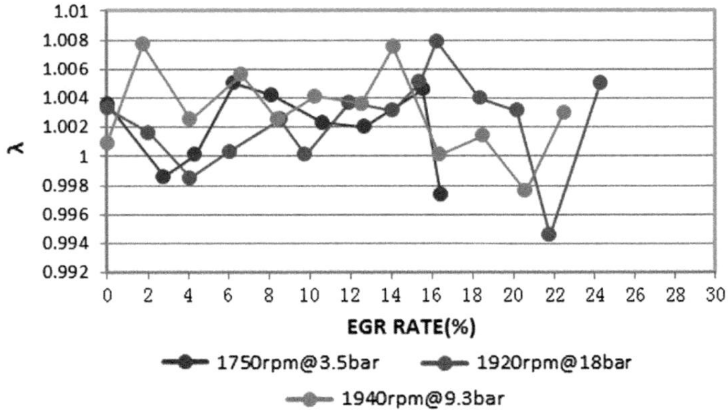

Fig. 16 λ with EGR rate change trend

3.5 Influence of the LP EGR on Pumping Gas Loss

With the increase of EGR rate, partial load showed a trend of gradual decline in the pumping loss (however, high load condition—pumping loss is on the rise) as shown in Fig. 17.

3.6 LP EGR Influence of Turbo Charger Duty Ratio

In high load condition, with increase of EGR rate, turbo charger duty ratio showed a trend of increase as shown in Fig. 18.

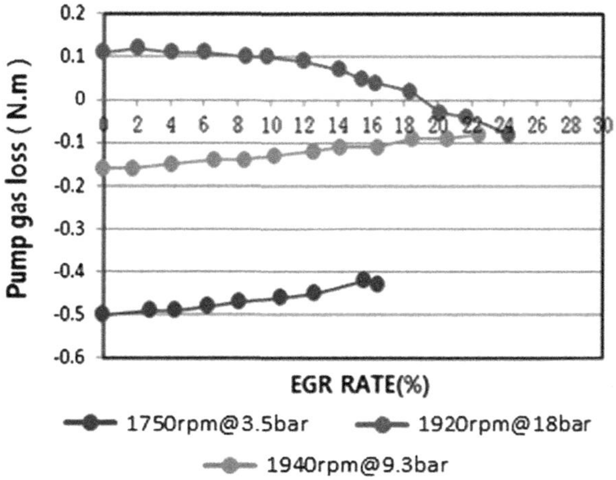

Fig. 17 Pumping gas loss with EGR rate change trend

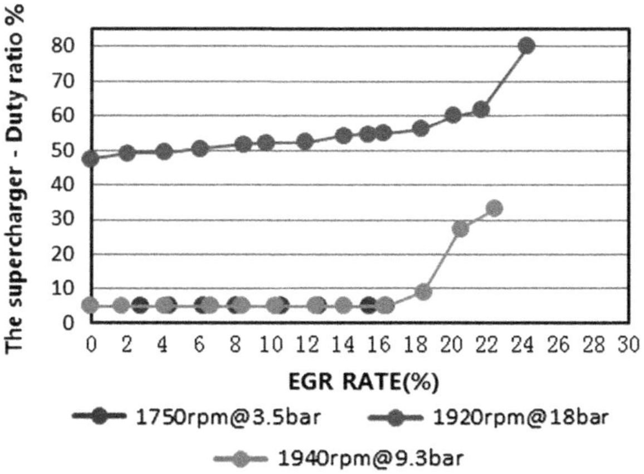

Fig. 18 Turbo charger duty ratio pumping gas loss with EGR rate change trend

3.7 LP EGR Influence on Intake and Exhaust Temperature Pressure

- T intake (before intercooler)—with the increase of EGR rate, the temperature has a rising trend.
- T exhaust (before turbo charger)—with the increase of EGR rate, the temperature has a downward trend.
- P intake—with the increase of EGR rate, the pressure has a rising trend (Figs. 19, 20, 21, and 22).
- P exhaust (before turbo charger)—with the increase of EGR rate, the pressure has a rising trend.

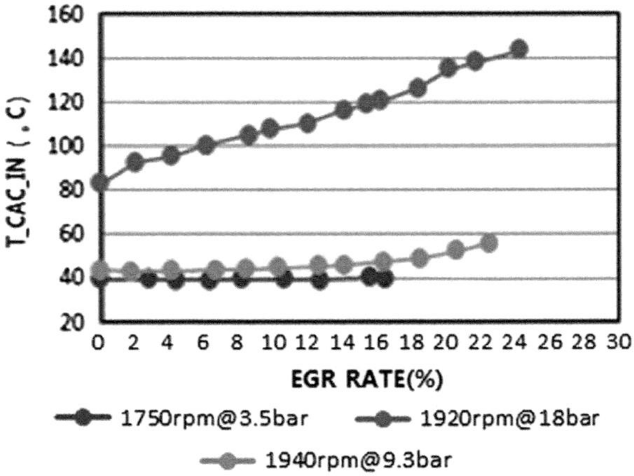

Fig. 19 T intake (before CAC) with EGR rate change trend

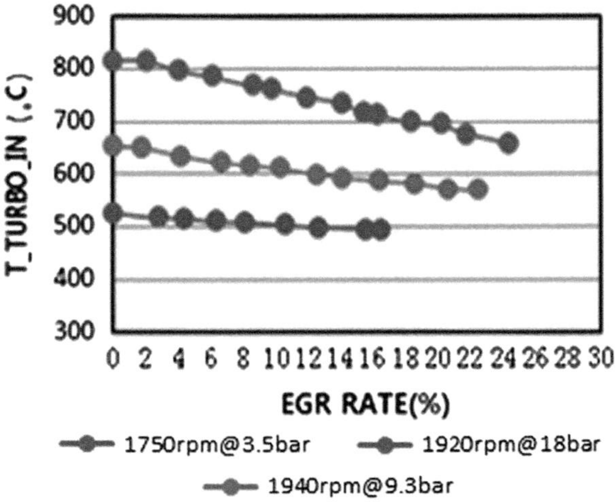

Fig. 20 Exhaust temp. (before turbo charger) with EGR rate change trend

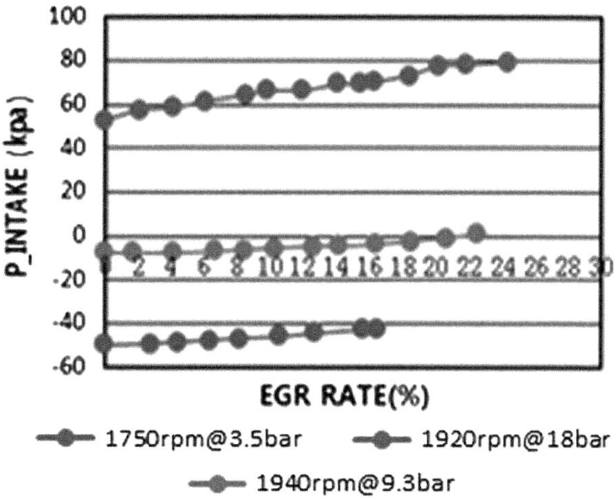

Fig. 21 P intake with EGR rate change trend

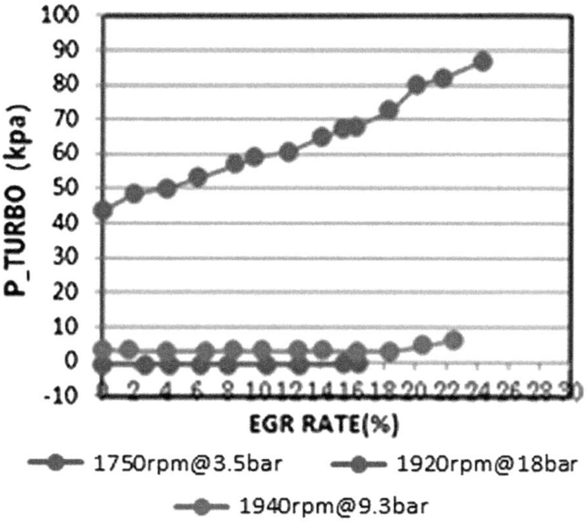

Fig. 22 P exhaust (before turbo charger) with EGR rate change trend

3.8 BSFC MAP and the EGR Rate MAP

- 1920 rpm@18 bar: Fuel consumption can be reduced by 16.5%.
- Medium load: Adding a higher EGR rate can have obvious fuel consumption benefits.
- Low load (below 4 bar), due to internal EGR, by adding an external EGR, fuel consumption benefits are not evident (Figs. 23 and 24).

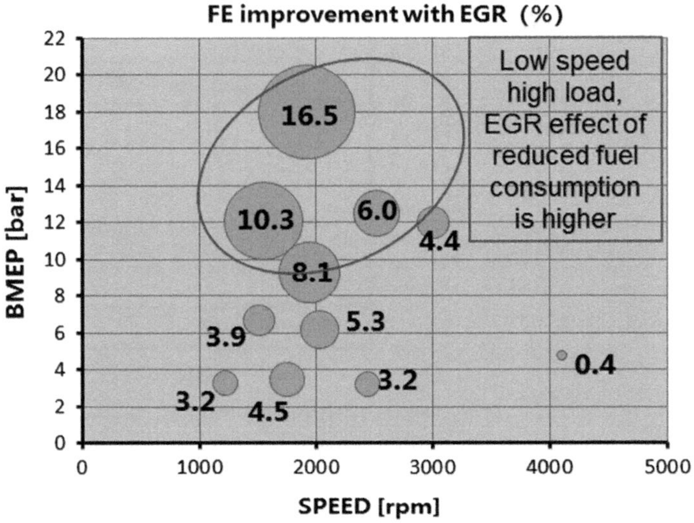

Fig. 23 FE improvement with EGR (%)

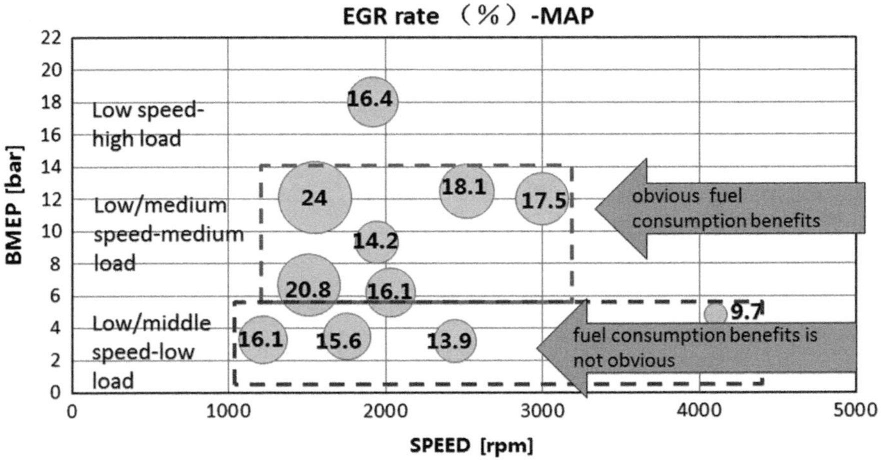

Fig. 24 EGR rate (%)-MAP

3.9 Sweep the EGR Rate Summary

The effect of EGR on combustion parameter and performance of the engine. EGR mainly applies to reduce NO_x; however, adding EGR leads to suppress knocking and helps to reduce fuel consumption by advancing ignition angle. But after the increase of EGR, engine power and torque will be reduced.

Part II Valve Timing Optimization

3.10 *Optimized Result*

Internal EGR was widely used by overlapping intake/exhaust valve. It reduces pumping loss at part load. Optimizing VVT leads further fuel consumption reduction.

At 1750 rpm@3.5 bar, optimized VVT wnwsa\wnwse (32.5, 13.5) leads to BSFC benefit of 0.56 g/kWh compared to initial point VVT wnwsa\wnwse (28.6, 6.6) as shown in Fig. 25. At 1750 rpm@3.5 bar, optimized VVT wnwsa\wnwse (32.5, 13.5) leads to BSFC benefit of 1.64 g/kWh compared to initial point VVT wnwsa\wnwse (28.6, 6.6) as shown in Fig. 26. At 1920 rpm@18 bar, this has no BSFC benefit as shown in Fig. 27.

3.11 *Valve Timing Optimization Conclusion*

VVT optimal sweeping operation point, through adjusting the valve overlap, controls the internal EGR to achieve good combustion stability; meanwhile, EGR could decrease pumping loss, after optimizing the VVT, gained good BSFC benefit.

Part III Sweep EGR Rate Again

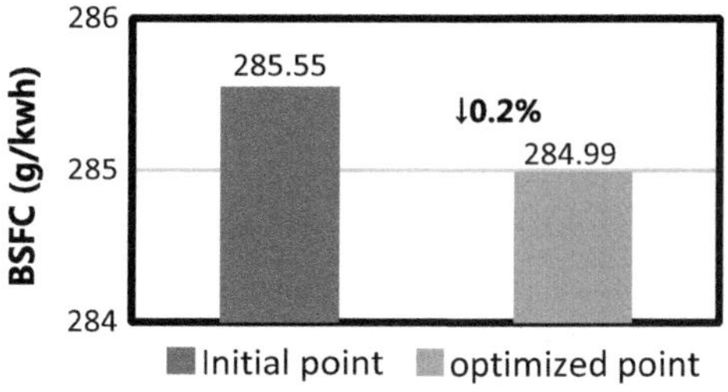

Fig. 25 Initial point BSFC versus VVT optimized point BSFC

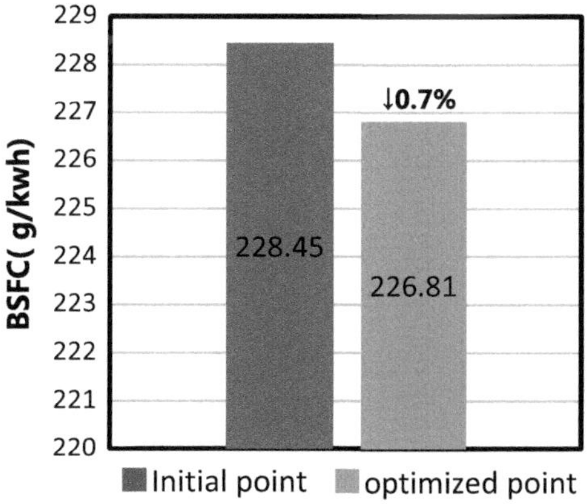

Fig. 26 Initial point BSFC versus VVT optimized point BSFC

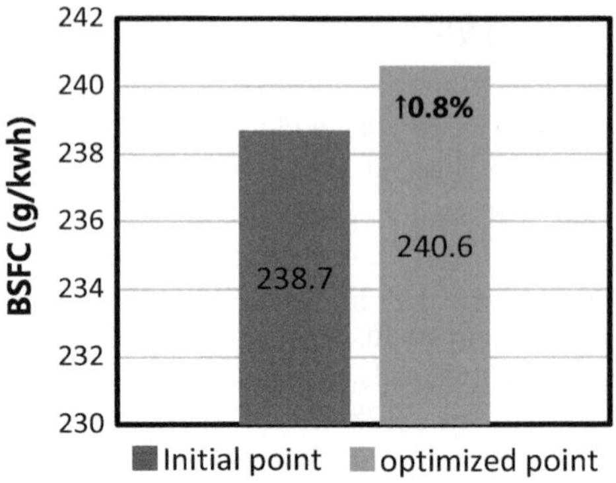

Fig. 27 Initial point BSFC versus VVT optimized point BSFC

After optimizing VVT to find minimum BSFC by valve overlap, external EGR is added to further improve BSFC. Initially set the best VVT angle and sweep the EGR rate. Combustion parameter was monitored to ensure the maximum usage.

3.12 Results of the Sweep

As shown in Fig. 28 at 1940 rpm@9.3 bar, under the condition of swept through the EGR rate, compare the VVT best, BSFC reduced by 1.3 g/kWh or 0.6%.

3.13 Summarize the EGR Rate Scan

For the entire test, in view of the three conditions, all the EGR rate of repeat point, BSFC are further optimized.

Part IV Optimization Injection Phase Angle and Fuel Rail Pressure

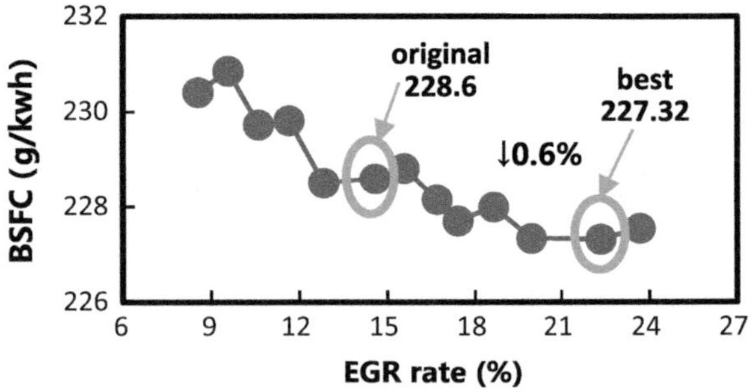

Fig. 28 VVT BSFC best versus EGR rate

Fuel atomization strongly depends on fuel pressure, spray angle, and flow calibration. Adjusting the injection pressure can achieve the high-quality atomizing effect. At different speed, load point VVT and airflow movement in the cylinders have been changed. It is necessary to adjust fuel rail pressure to have good quality atomization effect. You adjust the injection phase angle to set the fuel injection time by keeping fuel rail pressure constant. Changing the injection phase angle and fuel rail pressure could reduce the fuel consumption.

3.14 Effect of Fuel Rail Pressure and Injection Phase Angle on the BSFC

BSFC decreases with the increase of injection phase angle as shown in Fig. 29, under the working condition of 1750 rpm@3.5 bar. Minimum BSFC observed at 330° injection phase angle. BSFC decreases with the increase of injection pressure, when rail pressure is 18 Mpa and minimum fuel consumption is 282.74 g/kWh. Fuel consumption reduction of 0.8% compares to 16 Mpa fuel rail pressure.

BSFC decreases with the increase of injection phase angle as shown in Fig. 30, under the working condition of 1940 rpm@3.3 bar. Minimum BSFC was observed at 330° injection phase angle. BSFC decreases with the increase of injection pressure, but it is not necessary as shown in Fig. 30 when fuel pressure is at 20 Mpa; fuel consumption is significantly higher than others fuel pressure. When rail pressure is 18 Mpa, minimum fuel consumption is 224.18 g/kWh. Fuel consumption reduction of 1.2% compares to the initial point.

Figure 31 shows under operating condition at 1920 rpm@18 bar. With the increase of rail pressure and injection phase angle, fuel consumption has a rising trend. Best BSFC is 237.6 g/kWh corresponding to the rail pressure (10 Mpa) and injection phase angle (330), compared to the initial point of optimization of 2.48 g/kWh. Fuel consumption reduction of 1.03% compares to the original point.

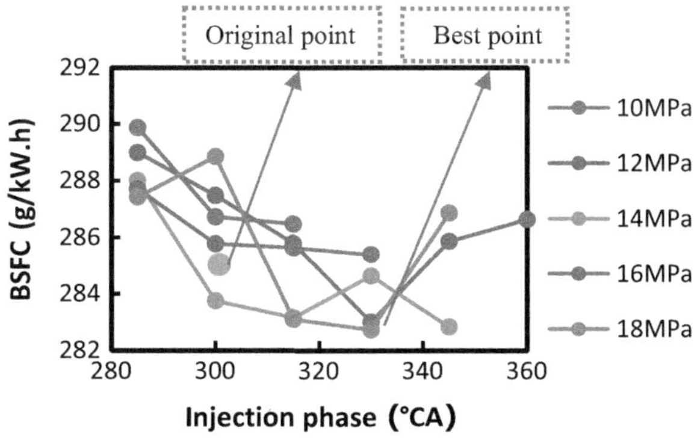

Fig. 29 BSFC best versus injection phase

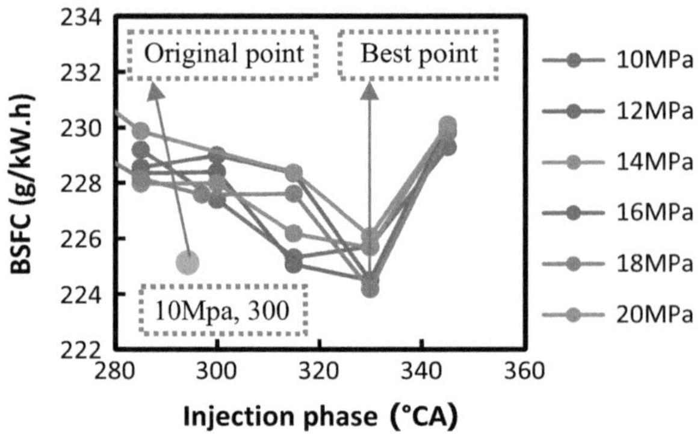

Fig. 30 BSFC best versus injection phase

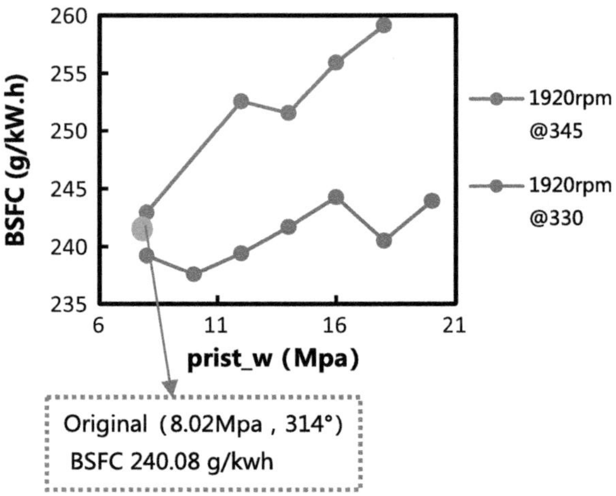

Fig. 31 BSFC best versus fuel rail pressure

3.15 *Effect of Fuel Rail Pressure and Injection Phase Angle on the Emission*

The most efficient way to reduce NO_x is exhaust gas recirculation; however, HC increases with higher EGR rate. Evaluation of the effects on emissions is done by adjusting the fuel rail pressure and injection phase angle. As shown in Fig. 32, under operating condition 1750 rpm@3.5 bar, with the increase of injection phase angle, NO_x emission showed a trend of deterioration. Compared to same injection rail pressure, NO_x emission increased with the increase of injection phase angle. As shown in Fig. 33, under operating condition 1750 rpm@3.5 bar, with the increase of injection phase angle, CO emissions are on the declining trend; the comparison in the same injection phase angle, CO emissions increased with the increase of injection pressure, is on the downward trend. As shown in Fig. 34, under operating condition 1750 rpm@3.5 bar, with the increase of injection phase angle,

Fig. 32 NO_x versus injection phase

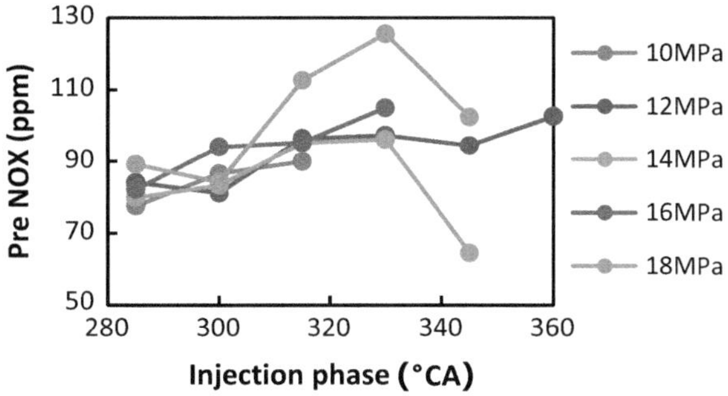

Fig. 33 CO versus injection phase

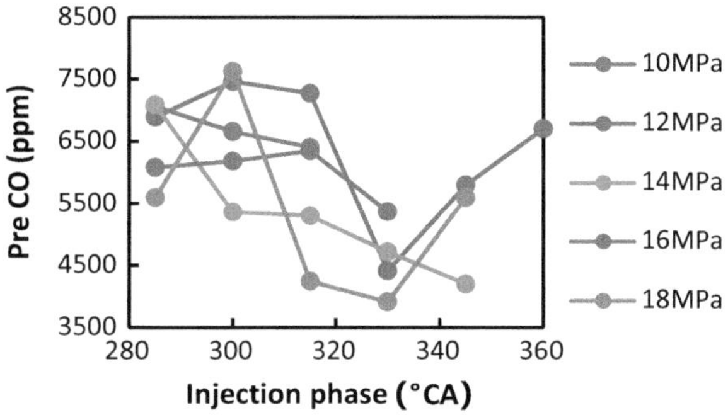

Fig. 34 HC versus injection phase

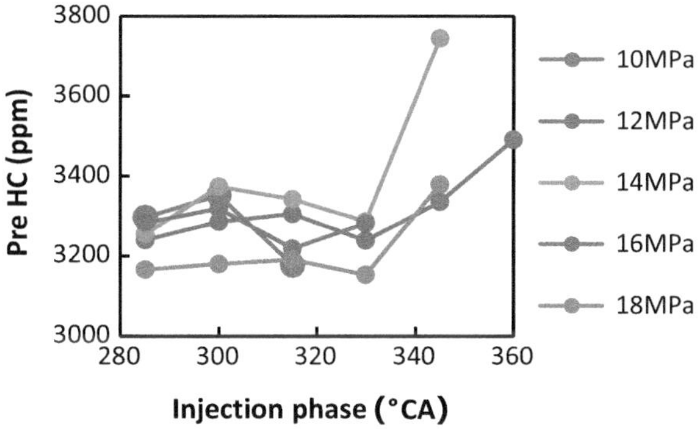

HC emissions will have deteriorating trend; compared to the same injection phase angle, HC emission decreased with the increase of injection pressure. As shown in Fig. 35, under operating condition 1750 rpm@3.5 bar, with the increase of injection phase angle, S415 emissions will be worsening; compared to same injection phase angle, S415 emission increases with the increase of injection pressure.

As shown in Fig. 36, under operating condition 1940 rpm@9.3 bar, with the increase of injection phase angle, NO_x emission showed a trend of deterioration; comparing in the same injection phase angle, NO_x emission increased with the

Fig. 35 S415 versus injection phase

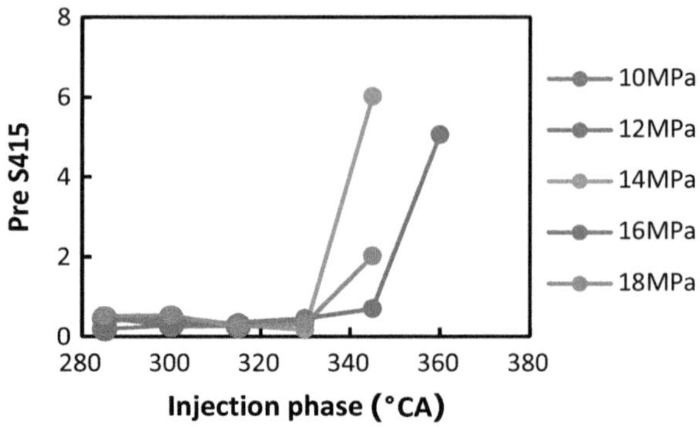

Fig. 36 NO_x versus injection phase

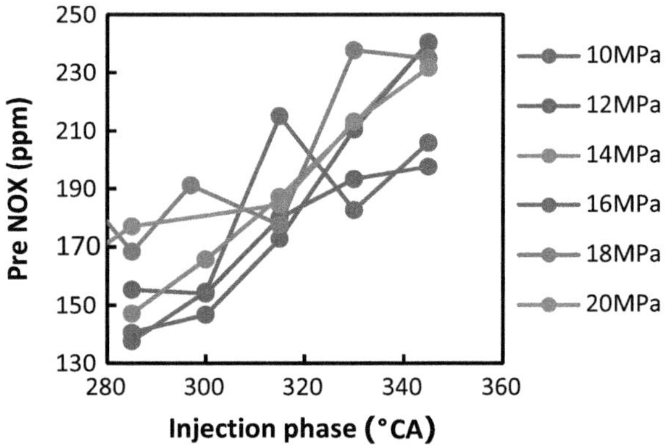

increase of injection pressure. However, when the rail pressure increases to 20 Mpa, NO_x emissions have the downward trend. As shown in Fig. 37, under operating condition 1940 rpm@9.3 bar, with the increase of injection phase angle, CO emissions are on the rise; comparing in the same injection phase angle, CO emissions increased with the increase of injection pressure. However, when the rail pressure rose to 18 Mpa, CO emissions have the downward trend. As shown in

Fig. 37 CO versus injection phase

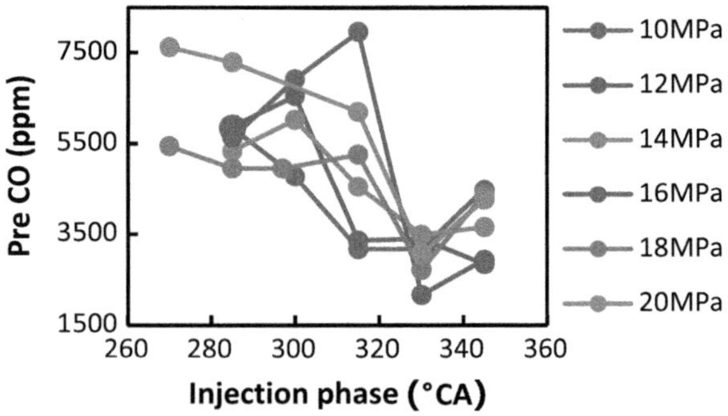

Fig. 38, under operating condition 1940 rpm@9.3 bar, with the increase of injection phase, HC emission is on the decline; comparing in the same injection phase angle, HC emissions increased with the increase of injection pressure. However, but when the rail pressure increased to 18 Mpa, HC emissions have the downward trend. As shown in Fig. 39, under operating condition 1940 rpm@9.3 bar, with the increase of injection phase angle, S415 emissions are on the rise. Comparing in the same injection phase angle, S415 emissions increased with the increase of injection pressure. However, when the rail pressure increases to 20 Mpa, S415 emissions have the downward trend.

As shown in Fig. 40, under operating condition 1920 rpm@18 bar, with the increase of injection pressure, NO_x emission is on the rise; comparing in the same injection rail pressure, NO_x emission increased with the increase of injection phase angle. In injection phase angle of 300° and 315°, NO_x values are incorrect due to emission bench error. As shown in Fig. 41, under the operating condition 1920 rpm@18 bar, with the increase of injection pressure, CO emissions are on the decline. Comparing in the same injection pressure, CO emissions decreased

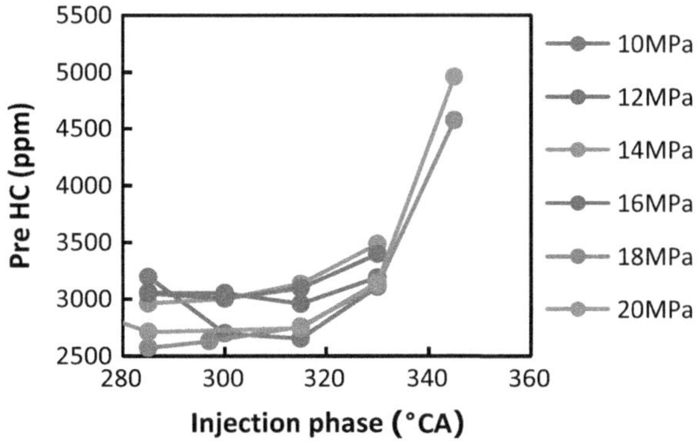

Fig. 38 HC versus injection phase

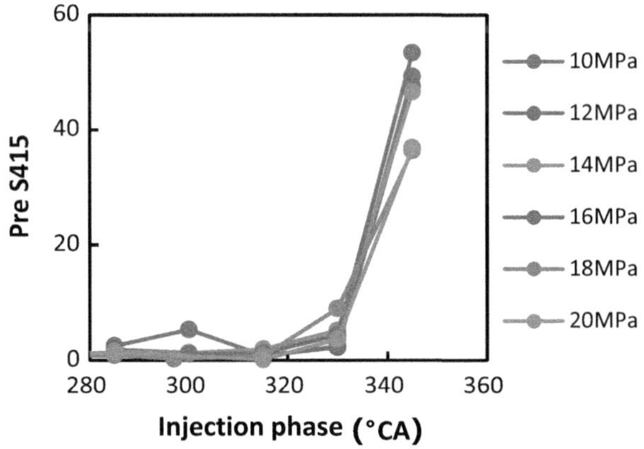

Fig. 39 S415 versus injection phase

Fig. 40 NO_x versus fuel rail pressure

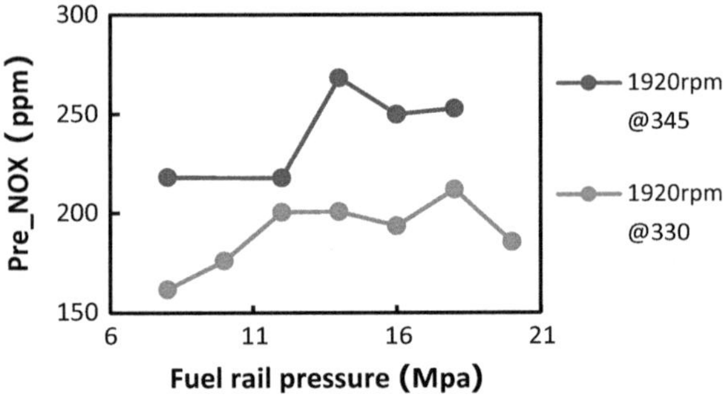

Fig. 41 CO versus fuel rail pressure

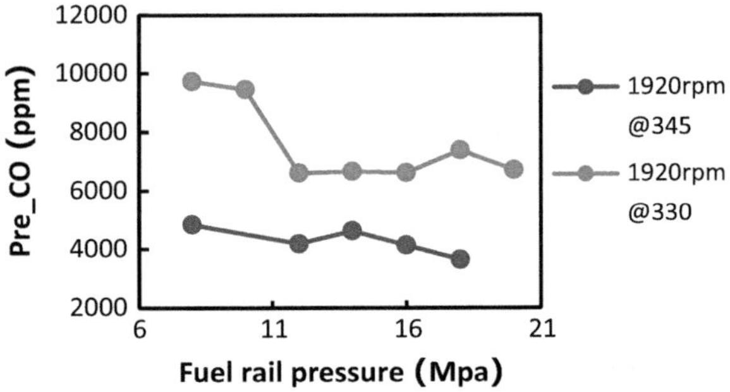

with the increase of injection phase angle. Fuel rail pressure and the increase of injection phase angle to reduce CO emissions is up to 60%. In injection phase angle of 300° and 315°, CO values are incorrect due to emission bench error. As shown in Fig. 42, under operating condition 1920 rpm@18 bar, HC has the upward trend with the increase of rail pressure and injection phase in angle; in the 300° and 315° injection phase angle, HC values are incorrect due to emission bench error. As shown in Fig. 43, under operating condition 1920 rpm@18 bar, Smoke number increases with the increase of fuel rail pressure and injection phase angle, especially in the injection phase. In injection phase angle of 300° and 315°, S415 values are incorrect due to the emission apparatus error, and the result is inconclusive.

Fig. 42 HC versus fuel rail pressure

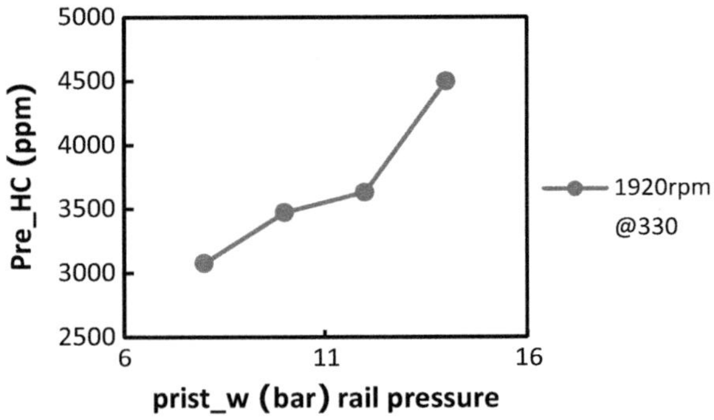

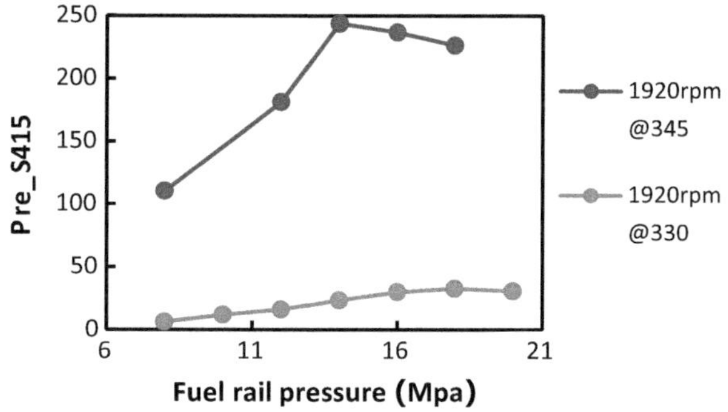

Fig. 43 S415 versus fuel rail pressure

3.16 Optimization Injection Phase Angle and Rail Pressure Conclusion

Fuel rail pressure and injection phase angle optimal setting is critical to achieving minimum BSFC. Based on test data, it is evident that the different speed and load conditions have different fuel rail pressure and injection phase angle.

During the calibration phase, it is important to evaluate the trade-off between fuel consumption and emission pollutants.

4 Summary

To maximize the benefits of LP EGR on a turbocharged direct injection engine, it is important to calibrate the engine properly, which is a critical process. To achieve the best fuel consumption, it is necessary to find the optimal EGR rate, VVT angle, injection phase angle, and fuel rail pressure. During our study, we observed that EGR influences engine combustion, pumping loss and exhaust temperature. Furthermore, the reduction of the combustion temperature and of the pumping loss led to a significant improvement of BSFC and emissions. However, engine power and torque declined. The main findings of this study are set out below:

- The addition of cooled exhaust gas to filtered air lowers the oxygen concentration of air–fuel mixture in the combustion chamber and causes a slowing combustion, which leads to the reduction in combustion temperature and reduces NO_x formation. A high load also suppresses knocking tendency, which allows to advance ignition angle and to improve the fuel consumption benefit.
- The pumping loss will decrease by adding the throttle opening at medium and low load. When the exhaust introduced into the combustion; at high load, the exhaust flow will limit the intake flow, and then the engine performance will be lower;

- LP EGR also affects the intake/exhaust temperature: After the exhaust has been introduced into the intake system, the intake temperature increases a little. On the other hand, the introduction of exhaust gas to the intake system decreases the combustion temperature, which leads to a reduction of the exhaust temperature. The intake pressure will be higher, and as the duty cycle will increase when the EGR system work, it leads to the increasing of the intake and exhaust pressure;
- Because of the optimization of the combustion parameters and pumping, the BSFC will be lower with the increasing of the EGR rate. The BSFC reduction at medium to high load is higher. We achieved 16.5% BSFC benefit at low speed and high load. Furthermore, some BSFC benefits at low load (3–4 bar) according to the dyno test results.
- Our study demonstrates the best EGR rate for minimum BSFC at different load ranks between 14 and 20%. Because of the limit of the mixture gas dilution, the engine is going to misfire if the EGR rate is too high, which will adversely affect the BSFC benefit.
- LP EGR technology has some effect on the emission, and the combustion temperature decreases significantly by adding LP EGR so that NO_x is reduced by 90%. On the downside, HC rises and CO shows just a little fluctuation but essentially remains constant.
- Finally, the air flow is limited under high load conditions, if EGR used, which leads to a performance degradation of the engine.

References

1. Zhong L, Musial M, Reese R (2013) EGR systems evaluation in turbocharged engines. SAE paper 2013-01-0936
2. Potteau S, Lutz P, Leroux S (2007) Cooled EGR for a turbo SI engine to reduce knocking and fuel consumption. SAE paper 2007-01-3978
3. Song D, Jia N, Guo X, Ma X et al (2014) Low pressure cooled EGR for improved fuel economy on a turbocharged PFI gasoline engine. SAE Technical Paper 2014-01-1240
4. Liu Z, Cleary D (2014) Fuel consumption evaluation of cooled external EGR for a downsized boosted SIDIDICP engine. SAE Technical Paper 2014-01-1235
5. Su H, Favelier S, Dutot A, Surbled K, Morello FP, Chereau C (2015) Low pressure EGR impact on condensate and product specifications. VALEO powertrain system, 5th Aachen Colloquium China automobile and engine technology
6. Potteau S, Lutz P, Leroux S, Moroz S et al (2007) Cooled EGR for a turbo SI engine to reduce knocking and fuel consumption. SAE Technical Paper 2007-01-3978. https://doi.org/10.4271/2007-01-3978
7. Bermúdez V, Luján JM, Climent H, Campos D (2015) Assessment of pollutants emission and aftertreatment efficiency in a GTDi engine including cooled LP-EGR system under different steady-state operating conditions. Appl Energy 158:459–473
8. Torres Garcia M (2009) Combustion characteristics, emissions and heat release rate analysis of a homogeneous charge compression ignition engine with exhaust gas recirculation fuelled with diesel. Energy and Fuels

9. Fontana G, Galloni E (2009) Experimental analysis of a spark-ignition engine using exhaust gas recycle at WOT operation. Appl Therm Eng 29:1131–1137
10. Potteau S, Lutz P, Vogel JB, Rochette C, Surbled K, Tomas E, Vitiello N, Bernard JS (2011) Turbo SI engine fuel consumption improvements by using cooled EGR. SIA congress of December 2011
11. Hu E (2009) Experimental investigation on performance and emissions of a spark-ignition engine fuelled with natural gas-hydrogen blends combined with EGR. Int J Hydrogen Energy 200901
12. Tang C, Guan L, Feng Z, Zhan C, Yang K, Huang Z (2017) Effect of di-n-butyl ether blending with soybean-biodiesel on the near-nozzle spray characteristics. Fuel 191:300–311

Reconstruction of the Management System of Automobile Manufacturing Enterprise in the Future

Yu Xiang

1 Background

In the last ten years, automobile manufacturers have been trying to understand and adapt to the great changes brought by information technology and Internet development. The three laws of network—Moore's law, gilder's law and Metcalfe's law maximize the digitization, which reduces the cost of manufacturing and expands the size of the market. The steady development of information technology has not only had a profound impact on the management system of automobile manufacturers, but also brought serious challenges.

2 Challenges and Trends

The automobile manufacturing enterprise can only survive by satisfying the customer's needs, and develop by keeping good relationship with the partners. The development of technology is changing the relationship between enterprises and customers, enterprises and partners. There are two remarkable trends associated with the needs. The first trend is the trend of socialization. The second trend is customer experience has received more recognition. Those two trends are gradually affecting the management system of automobile manufacturers.

Y. Xiang (✉)
Changan Automobile Management Innovation and IT Center,
No. 260, Jianxin East Road, Jiangbei District, Chongqing 400023, China
e-mail: xiangyu@changan.com.cn

Society of Automotive Engineers of China (SAE-China) (ed.), *Proceedings of the 19th Asia Pacific Automotive Engineering Conference & SAE-China Congress 2017: Selected Papers*, Lecture Notes in Electrical Engineering 486, https://doi.org/10.1007/978-981-10-8506-2_9

2.1 Challenges

2.1.1 Enterprises Have Difficulty in Identifying Information Quickly

New information, new knowledge is being generated all the time, and it is difficult for automobile manufacturing enterprises to quickly identify information that is valuable.

2.1.2 The Effectiveness Requirement of the Enterprise Management Decision Is Higher

In the information age, all kinds of information are gushing, thus the enterprise managers are required to make effective decisions in the short time, or the opportunities will be lost, allowing competitors to gain competitive advantages.

2.1.3 It Is Getting Harder and Harder for Companies to Master New Technology

There is a growing number of new industrial technologies, both in products and in machines. Mastering the new technology is getting more and more difficult. At the same time, the massive information of production and management has also increased the management requirements of information system and the risk of information security.

2.1.4 New Requirements for the Enterprise Organization

Enterprises are no longer solo "warriors". Collaboration, union, integration, and the cross-industry reorganization are more and more chosen by enterprises. Traditional enterprise management mode cannot adapt to the new organization form.

2.1.5 The Management of Human Resources Is Getting Harder

The Internet is a relatively fair place. In the vertical? Network time, there are not so many rules, employees can enjoy their imagination and creativity. Mobile connectivity flattens the communication between people. Traditional vertical communication is outdated. The scope and depth of the information that the employees receive have been greatly extended Enterprises have gradually increased reliance on human resources due to the knowledge competition between enterprises, accordingly increased the staff management difficulty.

In addition, car manufacturers are in an awkward position in the battle for talents with Internet companies: the old backbone remains anemic and the new elite can absorb nothing. Car manufacturers often value systems, regulations, while Internet companies encourage innovations, flat communication, and personal values in the corporate vision. More and more people prefer Internet companies.

2.2 *Three Remarkable Trends*

Despite many challenges, companies can make breakthroughs in the management system by analyzing industry characteristics and finding trends. Automobile manufacturing enterprises survive by meeting the needs of their customers, and develop by maintaining good relationships with their partners. The development of technology has changed the relationship between business and customers, businesses and partners. There are three trends that need to be focused on: one is social trends, the other is the trend of customer experience, and three is the data driven deep personalization.

These three trends are gradually influencing the management system of automobile manufacturers.

2.2.1 Social Trends

With the rapid development of the Internet, big data, cloud computing, the emergence of various instant messaging software and tools, makes the information more and more dynamic and interconnected. Communication mode has changed from vertical to flat communication. Organization structure of traditional enterprises timely adjusts of the original "dinosaur", changes the original bureaucratic approval process, and also creates an ecosystem including suppliers, distributors, consumers and other related parties. It requires companies to break through the traditional management interface to think. Companies like Haier, millet, Letv, have made certain exploration. Automotive manufacturers should actively follow up.

2.2.2 Attention to the Trends of Customer Experience

Customer experience becomes more diverse, fragmented. The diversification of customer experience refers to the customer experience is composed of more and more types of experience, from the original sensory experience, emotional experience to interactive user experience, the overall user experience and other. Fragmentation of customer experience refers to what kind of experience cannot be estimated for the customers. Sometimes a spark of inspiration can ignite the enthusiasm of customers, bring them the ultimate experience Qihoo 360 chairman Zhou Hongyi made comments on micro innovation He believes that as long as the user can move the hearts of the most sweet point, a good solution to the problem, and sometimes accomplish a great task with little effort by clever maneuvers, this single point of breakthrough called micro innovation. Therefore, the value proposition of the enterprise needs to be changed from "what kind of product I produce" to "what kind of experience can I bring to the customer". Good customer experience requires auto manufacturing enterprise to redefine the channel, users' way of life and to the products and services, which will also become the key to the success of future automobile products.

2.2.3 Data Driven Deep Personalization Trends

It is understood and realized that personalized service is based on consumer's consistent demand. However, the so-called "personalized push" in reality often becomes "junk information bombing". Ninety-eight percent of social media users had received advertisements on social media, but only 18% said they thought they were "good enough." For the offline shopping experience, only 10% of consumers said they received personalized service or advice from the store. In the Internet age, car manufacturers should focus on large data analysis, enhance understanding of consumers, and provide targeted recommendations and full channel services to create greater values.

These three trends will drive car manufacturers to changes in strategy, organization, human resources, and processes to reshape the management system. Because enterprise to support operation, one of the core of innovation, change, or the management of the enterprise, even the so-called Internet changed the minds of thinking things, difficult to keep up with the evolution of management also will become the enterprise transformation and survival.

3 Reconstruction of Enterprise Management System

3.1 Strategy

The traditional management theory to guide enterprises in the formulation of strategies, generally follow the steps. The first step is to determine the mission and vision, followed by analysis. The third step is to determine the scope of business. The next step is to determine the dominant source and design value system, lists the growth ladder. The last step is to set financial goals and determination and implementation plan, which stands for the perspective of enterprises' development strategy. Moreover, corporate strategy is based on a static business environment: whether it is positioning strategy or enterprise competition strategy. Professor Michael porter put forward "five competitive model", which is described by the competitive environment of enterprise, to determine the corresponding competitive strategy, which holds that any enterprise can choose competition strategy based on its industry competitors. But Internet has made these static corporate strategies difficult to follow.

Especially Internet has been penetrated into various industries. Traditional enterprise transformation involves the crossover, across the industry chain, in a rapidly changing business environment to establish the static enterprise strategy, and the result is the same as making his mark.

In the Internet era, the automobile manufacturing enterprises cannot be limited into a system. From the beginning of the strategic formulation, we need to interconnect a variety of resources, to create a win-win new platform and achieve win-win cooperation with partners. There are three levels of strategy that car manufacturers can consider.

3.1.1 Traditional Manufacturing Plus Internet Connection Strategy

Automobile manufacturers can use the Internet as a platform and a tool to continue provide products and services, but it requires Internet innovation. Because Internet is not superimposed on the traditional automobile manufacturing industry. It is also not a simply tool. The traditional industry and Internet industry, according to the gene to their mutual integration not only involves the culture, management, more reveal the path of the methods, tools and implementation. There are two more types to choose from. Firstly, the traditional industrial chain intelligence and Internet optimization. The Internet is a tool for improving operational efficiency. Automobile manufacturers can use the Internet to optimize channels and manufacturing, integrate the Internet into traditional automobile manufacturing. Secondly, the Internet has led the transformation of traditional manufacturing industries, which is leading the transformation of automobile manufacturing enterprises. Although SuNing has a huge loss in e-commerce reform. It is a problem in the reform plan and implementation rather than a problem in strategic direction.

Besides the connectivity strategy, car manufacturers may also use strategic approaches such as guerrilla, flanking, and cost leadership, although they are no longer a static strategy.

3.1.2 Platform Based Cross-Boundary Strategy

Automobile manufacturing enterprises in the use of connection strategy if accumulated a large number of user groups, form a cohesive community, you can through the “+” Internet connection, providing enterprises and car users value exchange or business trading platform, similar to a huge commercial organisms, began to cross-border marketing.

3.1.3 Enterprise Strategy of Business Ecosystem

In 1993, an American scholar James Moore in the Harvard Business Review put forward the concept of "business ecosystem" for the first time, breaking the traditional on the premise of industry division of competition strategy and business model thinking, and put forward the ecological concept "co-evolution" business. Companies, societies and nature are equally ecological. An industry, an enterprise, a region, a sector, are mutual balance and maintain the ecological chain. It is a reasonable structure, complete functions and stable ecological system. An important feature of ecology is interdependence, interdependence, and even interdependence. Once the ecology is established, once the ecological chain is formed, the whole ecology will be very united and will instinctively block or delay the decline of the whole ecology. The highest level strategy of automobile manufacturing enterprises is also through the "Internet plus" serving as the means of expansion, with traditional enterprises as the center, forming an industrialized business ecosystem.

In terms of the allocation of resources related to the implementation of the strategy, the traditional way of allocating resources to a project will also be changed. Based on the changes brought by socialization, in the future, the companies will create value and experience to design the configuration of resources, and to optimize the efficiency of resource allocation based on large data analysis.

3.2 Organization

The organization of the traditional automobile manufacturing enterprise is in series, from product planning, research and development, manufacturing, marketing, service to the final user. There are a lot of driving wheels between the product planning and the user. But those driving wheels don't know where the customers are. The middle layer of the enterprise, includes the social suppliers and vendors. The middle layer is far away from the enterprise and the user. The organizational structure of the hierarchy reduces the response efficiency to customer needs and customer problems.

In "the out of control," Kevin Kelly mentioned the Internet represents several characteristics: distributed management is not mandatory at the center of the control, secondary units had the characteristics of autonomy, highly connected to each other between secondary units, between point to point through the network to form the influence of the nonlinear cause-and-effect relationship. These characteristics are in line with all the trends of the Internet age: to innovate and evolve faster than competitors.

In the future, the enterprise is no longer the center of power. Internet has pushed companies to adjust their organizational structure. The organization is becoming more and more platform, network, ecology, will be able to quickly docking customer needs, customer interaction and interactive ecosystem. The information inside the organization will become more transparent, symmetrical, and more efficient. To adapt to society, more and more attention is paid to the organizational structure of the customer experience, which will show the following types.

3.2.1 Reticular Tissue

In Internet era, auto company will become a node in the network interconnection. The hierarchical organization will evolve into a network organization. Traditional vertical communication, vertical reporting methods have been slowly flattened. The original single point of contact customers will become more than a sector of the customer. The traditional pyramid is the organization of the inverted pyramid. The top layer is the interface of the customer team, the middle is a variety of resources platform, and the bottom is the functional service layer. Leadership is changed from the original commander into a resource provider.

3.2.2 Reticular Tissue

After the organization has been changed, with the change of the market and customer value creation, the organization resources will adapt the new organization structure and cooperation mechanism according to the new node.

Reticular tissue has no specific strategic objectives. It is a strategic development direction and makes its organization into a strategic state. In the self-organizing state, the authority and decentralization will naturally occur. According to the demands of the mission, everyone can be the center, and everyone can become the CEO. It's important to note that everyone can be a center, does not mean just that everyone is necessarily a center, but has the possibility to become a center. In other words, "de-centralization" is not a complete center, it simply changes the center of centralization and becomes a control center. In self-organization, it must be highly authorized, so that everyone is automatically responsible and automatically pursuing synergy. Self-organization emphasizes the value creation of employees.

3.2.3 Online Organization

In the future, auto companies' online communication and collaboration will become the norm, not only within the enterprise, as well as between customers. Car networking and electronic business platform will accelerate the process. Online communication improves the communication efficiency and facilitate the sharing of information on the one hand, to facilitate the management of the team all over the world; on the other hand, online is the precipitation of internal interactive data, is conducive to fine management, reduces management costs.

The three different forms of organization common features is flattening and networking, vertical, decentralization and democratization, socialization, internationalization, not only ensure the highly efficient and flexible operation, and greatly stimulated the innovation ability of organization members.

In the optimization of the organizational structure of the enterprise, we should start with the following aspects: first, do a good job in the analysis of key activities to determine the organizational structure. Second, use decentralized organization. Third, change the organization management, make the organization structure to flat and specialized. In order to adjust the structure of the organization, the stability of traditional automobile manufacturing business and the flexibility of Internet transformation should be ensured. One solution is to set up separate business units for the transformation. Traditional businesses and internet-based businesses are not interfering with each other. Traditional businesses continue to carry out business under the original structure to ensure efficiency and output. In addition to the new independent business units, the "Internet plus" business carries out the exploration and innovation of uncertainty with the resources of traditional business. A mature, handover of a transformation to the traditional business, curing to the traditional business, and at the same time the culture, processes and systems gradually form a complete set of matching, the two complement each other and fusion, jointly improve the efficiency and flexibility of the organization.

3.3 Human Resources

Human resource management principles of traditional automobile manufacturing enterprises include "selection, education, and stay", employer providing jobs, employees receiving work tasks from supervisor arrangement. The main role of an employee is to carry out which lacks of innovation and creativity of employees in the long-term structured work. Quality of work and efficiency cannot be guaranteed. The enterprise will increase the supervision and inspection of the post to check the staff performance, thereby causing the organization overstaffed. There will also be an atmosphere of mistrust between enterprises and employees.

Social change are based on the fact that the employees in the company will no longer just simply performing the task. The market competition and the pressure through the flat organization will be transferred to everyone inside the enterprise. Everyone will focus on improving the customer satisfaction and engage in the related work, Employee from the functional departments no longer feel far away from the market, from the customers. It requires the establishment of a mechanism to achieve the transformation. A good example is Haier's competing single posts, "elected officers" mechanism, because they ensure the dynamic optimization, embodies the equality of opportunity.

In terms of human resource training, enterprises need to attach the positions of the forefront market with the nearest customers, because innovation and creativity will be produced mostly here. Enterprises should pay attention to the discovery and selection of interdisciplinary talents with innovative genes and Internet thinking. Under the Internet environment, enterprise training should reflect two kinds of thinking ability and systematically carry out human resource training: one is the ability of community operation, and the enterprise cannot get fans to get high; Second, multimedia production ability. Enterprises need to make the training content into visual, audible new media products.

Incentives should change, too. Mobile Internet era of employees is no longer the carrot and the stick under the screws completely. They have enough information to master the core of the enterprise, have enough ability to aggregate a number of consumers, have enough speed to respond to market changes, and have enough knowledge to carry on the independent innovation. The incentive and restraint mechanism represented by salary bonus has been degenerated and replaced with KPI cancellation, challenge, professionalism and game survival mechanism.

3.4 Process

The traditional automobile manufacturing enterprise series process—from product planning, research and development, manufacturing, marketing, service may be changed, mainly in: (1) social change based on social network formed a strong network of human resources market, (2) research and development of automobile manufacturing enterprises can have the innovation ability from outsourcing that is now facing customers, suppliers, research institutions, universities and other individuals or organizations, so called Crowd-sourcing mode. This series process will change the traditional flow, and Crowd-sourcing party data sharing will be more and more. In this model, the planning time will be shorter. In addition, because of the car sharing, the users are no longer directly interfacing with the automobile manufacturing enterprises. Car sharing companies will consider all kinds of users in the scene to provide car sharing services, receiving user feedback and experience, as the window of automobile enterprise. The manufacturing, marketing and service process of automobile manufacturing enterprises will change. Internet thinking will be shown in the traditional automobile manufacturing enterprises.

3.4.1 Product Planning and Development Process

The impacts of technological development on product and business models, especially for the Internet, Internet of things, and robotics change the business environment, while customers are always fundamental. Automobile manufacturers should think about how enterprises can seize customers and market opportunities under the new technology and economic environment.

Successful demand management has seven key activities. First, the requirements should be confirmed according to the expectations of stakeholders, the appointment objectives and constraints of the mission, the operational objectives and the success criteria of mission tasks. The second is to obtain requirements from the system design process and organize them into hierarchical tree structures. Third, establish a plan for managing requirements and keep current project plans and requirements consistent. The fourth is to establish a two-way traceability between requirements, which can track the corresponding design, manufacturing and test work through demand tracking. The fifth is to establish a control baseline for the requirements, to track the requirements state and to change the situation throughout the project. The sixth is the change request that evaluates the baseline of requirement control in the life cycle of the project and implements the change request approved by the change committee. To estimate the impact of the change and negotiate new conventions on this basis to incorporate requirements change into the current project in a manageable way. Seven is to define validation methods for each requirement, maintain consistency between requirements, application scenarios and architecture/design, and take action to eliminate inconsistencies.

It's certainly a great idea to get users involved in product development. Customers need to be engaged, with the rapid development of network. by opening fast communication channels between automobile manufacturing enterprise with the customer, the users will be gradually involved in the factory in the future new product development. The user and the automobile manufacturing enterprises will both be the winners.

In the process of product design, automobile manufacture enterprises give full consideration to the flexible design in advance, after the listing, by user feedback information, timely to users are not satisfied with some fixed trough point (such as fuel consumption, the exterior, inside and outside decoration, etc.), optimizing, update, iterative production, rapid production of customer satisfaction products.

3.4.2 Manufacturing Process

Information technology will be more embedded in the manufacturing process. In the process of vehicle production, automobile manufacturing enterprises realize the lifetime monitoring and management of vehicles through the "pre-loading" application of vehicles. Whether there is the use of installed on-board wireless smart computer in the car in front of the vehicle, whether real-time product parameters can be acquired after the vehicle sales under the running state are important. Through the wireless data transmission module in the on-board computer, the information will be collected real-time and information will be transferred back to the database. By using platform cloud computing system, unified data analysis for each car, can be timely found out to prevent the potential threat and reasonably to guide product upgrades, and establish reasonable product delivery routes with available GPS module to reduce business operating costs, reduce the model car loans and the risks of automobile manufacturing enterprises and influence.

By built-in electronic chips in the vehicle, dedicated wireless hand-held terminals can collect vehicle information at any time, any place through the wireless network. They can identify the car manufacturing enterprise, realize the convenient and quick query vehicle information, including receiving vehicle orders, confirming the deposit, arranging production, parts distribution, vehicle assembly, car, debt collection, hand in a car, after-sales service, maintenance, spare parts supply, etc. Through the application of informatization in the whole production process, the production of vehicles has truly implemented with intelligent management.

The users can assemble their most favorite configuration, and customize the color/structure of the body, interior and exterior colors, and various electronic functions and so on. Car products will be exclusive to every customer. Through the deep personal customization of products, it also indirectly realizes the "production of sales and production", maximizes industry value and minimizes the process cost.

3.4.3 The Marketing Process

The process of selling cars in traditional bricks-and-mortar stores will change. The traditional physical store marketing channel construction cost is high, the period is long, and the coverage scope of individual stores is limited.

E-commerce platform will become the main vehicle marketing tool. In this process, the automobile manufacturers need to use the Internet mentality, based on the rules and regulations of the automobile industry, and they can embody the characteristics of the industry on the basis of commercial logic. When building an e-commerce platform, automobile manufacturers need to determine the differences between e-commerce platforms and other automobile manufacturers. There are eight factors to consider. One is customer segmentation, who is the target user of e-commerce platform. The second is the value proposition. The difference between the products and services of e-commerce platform versus the e-commerce platform of other automobile enterprises. Third, the user obtains the channel. E-commerce platform is to introduce users to online from the offline, or direct network to seek users. Fourth, customer relations. How to build customer relationships with online users and how different from the customer relationships maintained by 4S stores. Fifth, profit and capital flow. The e-commerce platform is self-built, or based on a third-party platform or a social investment and partnership. Sixth, the core resources, capital and talent of e-commerce platform. Seven is the core activity that the e-commerce platform creates value. Eight is what the cost structure of the e-commerce platform looks like.

Once the e-commerce platform is established, the automobile manufacturers can take the advantages of the Internet through e-commerce platform to display the comprehensive sales promotion policies of automobile manufacturers. At the same time, auto manufacturers also allow distributors to improve their product promotion policies. Using various Internet enterprises and self-owned platforms, automobile manufacturers can make new car sales lead and facilitate sales through diversified forms of activities. The customer can personalize product configuration through e-commerce platform, and clinch a deal, which not regularly assists in the inventory of enterprises and dealers, reduces the operating costs of enterprises and dealers.

3.4.4 Service Process

Through mobile interconnection, the connection between online and offline users and services is established, and the interaction between enterprises and users is broadened and the interaction mode is easier. Enterprises, through intelligent equipment, complete vehicle route set. Vehicles will secured, and timely share other needs of customers. If there is need of repair, the car system actively push notifications. After Customer confirm the repair, vehicles will automatically arrange with the customer designated service center or the nearest service center for repair. When vehicle maintenance is finished, the vehicle will automatically drive to the customer designated locations. The customers can remotely monitor all the time, all the customers can complete shopping, order a meal, work and perform other daily needs through the electric business platform carried by the vehicle itself.

Automobile manufacturing enterprises set up maintenance service network. In the network, all units are on standby to repair his near vehicles, in order to provide more timely-rescued repair service to our customers. In order to organize the scattered social workers, the system is organized, the network is formed, and the whole service is provided to the "auto race" in a timely manner.

4 Conclusions

The progress of science and technology drives the evolution of the management system of automobile manufacturing enterprises. In the future, the thinking and methods of the Internet will be reflected in the design and operation of the enterprise management system. The transformation of the Internet and the "Internet plus" are based on the management principle and the transformation of the system, which is the combination of the traditional industry and the genes of the Internet industry, and the co-evolution of technology and management.

Development Integrated Roof Rack of Nylon and Continuous Fiber Composite

Hyun-Kyung Kim, Jeong-Ho Lee, Chun-Ho Park and Jae-Wha Kim

1 Introduction

The latest automotive safety regulations have tightened automotive safety standards and performance is a major concern vis-à-vis vehicle maintenance and automotive insurance cost. In particular, vehicular weight reduction is a primary focus as it enables ready reduction in fuel consumption, which is of major concern given depletion of fossil fuels and environmental concerns. Therefore, there is a need for an innovative technology in automotive engineering, capable of satisfying new safety regulations, performance with weight reduction. During the last 20 years, technical development has focused upon material optimization and dimension into preloading engineering through computer aided analysis and the optimization management strategies, while existing structure and production method does not changed. The roof rack is mounted on the uppermost car body of the vehicle and provides two or more structures on the left and right sides to mount the loop crossbar necessary for the cargo loading of the consumer. It provides the maximum load energy of the cargo and the kinetic energy Absorbing to prevent deformation of the car body and to minimize the deformation of the car body and to support the stability of the cargo by exhibiting the maximum strength and energy absorbing ability even when operating the vehicle excessively. The roof rack consists of front/rear stance, front/rear stance cover, roof pad, left/right side of the center side rail, and the front/rear stance and side rail are bolted to the top of the roof panel.

H.-K. Kim (✉) · J.-H. Lee
Advanced Body Engineering and Trim Engineering Team, HMC R&D, Seoul, Korea
e-mail: caar@hyundai.com

C.-H. Park
Automotive Part Design, LG-Hausys, Seoul, Korea

J.-W. Kim
Auto TS&D, LG-Chemicals, Seoul, Korea

Society of Automotive Engineers of China (SAE-China) (ed.), *Proceedings of the 19th Asia Pacific Automotive Engineering Conference & SAE-China Congress 2017: Selected Papers*, Lecture Notes in Electrical Engineering 486, https://doi.org/10.1007/978-981-10-8506-2_10

It carries the function of representing the high-grade design character through the support of the load of the load cargo through the carrier mounted on the loop crossbar and the polished surface treatment of the stance cover and the side rail of the roof rack. In the conventional automobile design, the central side rail is attached to the top of the loop while maintaining a certain distance from the top of the loop. The roof rack mounted on the roof panel uses aluminum as the side rail to withstand the load of the load cargo. Each automobile company has its own design goal by adding more evaluation and quality items. While using aluminum, the material is very expensive, but the weight is low and the exterior appearance of the vehicle is made in the shape of a drawer, so it has many quality problems such as manufacturing and surface treatment. Recently, researches on the light weight of the bumper back beam and roof rack have been carried out by improving the heat treatment method of 7000 series aluminum.

In this study, to realize fuel efficiency enhancement, surface treatment diversification and material cost reduction by light weighting, a new concept roof loop with a new concept loop top and close design to make various appearance and high rigidity material not simple change of aluminum material Respectively. Based on the load test of the vehicle, the material and the molding method that can satisfy the permissible physical properties and the environmental conditions required at the upper part of the loop are specified and the nylon based material Glass continuous fibers were selected. In order to realize appearance gloss, we have developed the coating surface treatment of nylon material and confirmed the minimum physical property standard to overcome the deterioration of material properties and physical property durability due to heat and environmental temperature drop. The design and components for the material development and system configuration for roof racks were fabricated and the performance of the single part was confirmed through material environment test, surface treatment test, loading performance test, and user condition loading performance test. We confirmed the technical performance of the car by loading the cargo loading condition and the vehicle driving condition by the actual car test. The load test was carried out under the condition of maliciousness against the general use condition of the customer and further verification was made. Through this study, we achieved light weight, material cost saving and various color effect compared to existing method while satisfying basic performance of cargo loading performance and appearance gloss.

2 Body

2.1 Concept Design

The roof racks provide the ability to engage with an article carrier or anchoring device where the baggage is seated via load transfer of the load weight to the roof panel and surrounding components, as well as the position fixation of the roof top load cargo. In order to improve the fuel efficiency of the manufacturer, there is a lot of application of the roof rack in the vehicle. The weight of the roof rack is 2.4–4.1 kg and the weight ratio of the side rail is 55%. It is required to develop the technology to improve the fuel efficiency by developing the alternative method of aluminum. Figures 1 and 2 show the design changes of the roof rack and the crossbar mounting structure.

The basic performance of the roof rack is the main function of the load carrying performance of the vehicle driving basis, the fastening or slipping prevention performance of the goods loop carrier, and the appearance color performance, and the additional function for preventing the driving noise of the roof rack mounting hole when the roof carrier is not mounted Have.

The basic performance of the roof rack is the main function of the load carrying performance of the vehicle driving basis, the fastening or slipping prevention performance of the goods loop carrier, and the appearance color performance, and the additional function for preventing the driving noise of the roof rack mounting hole when the roof carrier is not mounted Have. In this study, it is aimed to reduce

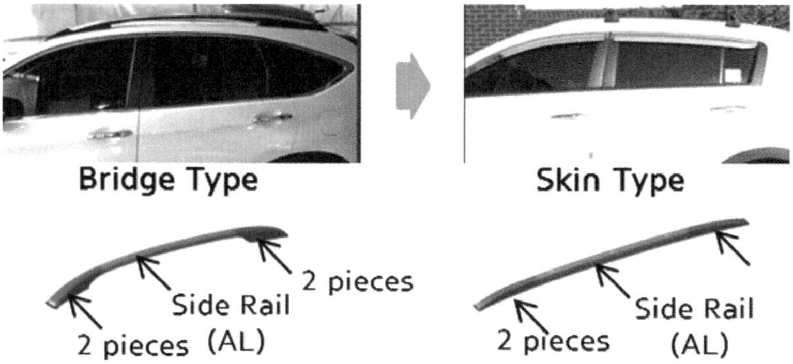

Fig. 1 Design type of roof rack

Fig. 2 Installed roof rack with crossbars

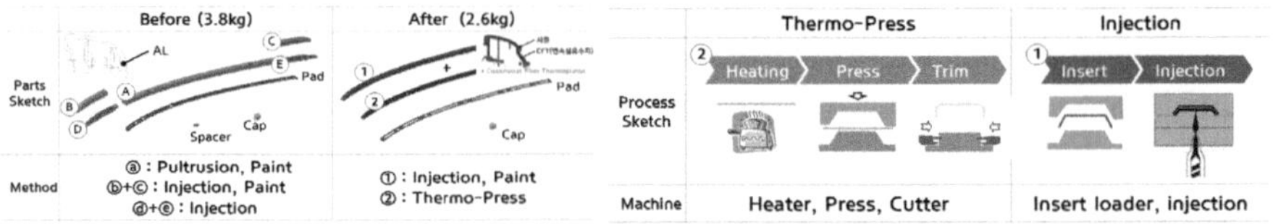

Fig. 3 Concept design of AL and integrated skin type plastic roof rack

the weight of the aluminum rail, stance and stance cover of the roof rack by replacing the high rigid plastic and continuous fiber with the insert injection method to secure the load performance. Figure 3 shows the pre- and post-design and molding methods proposed in this study. We will verify the performance and function through the concept development for the lightweight part and the basic method plan to realize it.

The choice of composite material of roof rack is an important technical factor of roof rack loading performance. The minimum material strength standard for plastics sensitive to environmental conditions on the roof panel is a high temperature condition, and the minimum material strength standard for humidity is the WET condition. Continuous fiber selection was made of glass fiber with low material cost and excellent matrix impregnation, and injection plastic material was selected as nylon which has high rigidity among engineering plastics and excellent environmental resistance compared to material cost. Matrix of glass fiber was selected as nylon with low physical property under high temperature WET condition. Through this, we confirmed that the basic Matrix is valid for both the hot press/injection nylon. The material composition ratio of injection nylon is set to 40% (GF+MD) which is used for existing stance. The lowering of the physical properties at higher temperature than the lower temperature leads to the necessity of high temperature verification at the actual vehicle condition. The physical properties of the material are shown in Figs. 4 and 5.

Material			DRY			WET 80%		
			-40℃	23℃	80℃	-40℃	23℃	80℃
Resin	PA6+(GF+MD)40	Strength (MPa)	168.5	127.7	56.0	160.8	100.6	56.3
		Elongation (%)	2.0	2.9	9.7	1.6	3.0	11.1
Fiber	CFT (PP+GF50)	Strength (MPa)	-	480	200	-	280	160
	CFT (PA6+GF50)	Strength (MPa)	-	620	420	-	350	300

Fig. 4 Material properties of resin and fiber

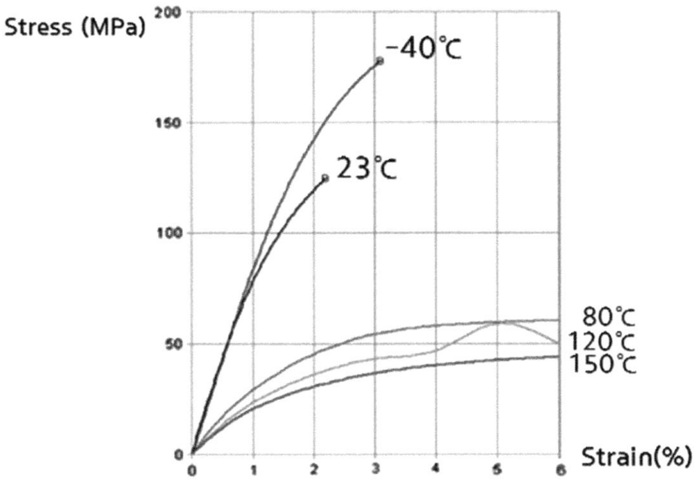

Fig. 5 Environmental property of resin

3 Performance Simulation

Recently, we have tried to compare the performance of the manufacturer verification criteria with the customer use condition due to the change of the roof rack design without providing the crossbar. The roof racks with close-coupled roofs conform to the same test method as specified by the manufacturer and the national standard, but the discrepancy between the manufacturer's evaluation condition and the customer's usage condition as a standard member for the crossbar fixation is shown in Fig. 6. The goods crossbar has a clamping pin w/o non-dissimilar structure on the part fixed to the roof rack. In the clamping fin type structure, it can be seen that the roof top load is more secure through the hole of the crossbar pin and the roof rack. The clamping jig was fabricated to confirm the difference in loading performance between the bolt fixing method and the left/right clamping method, which is the condition of customer use, and the front lifting test was performed on the roof rack of vehicle. It is confirmed that the evaluation method of the manufacturer is 3 times harder than the deformation amount of the roof rack compared with the customer use condition and it is difficult to confirm the slip condition of the crossbar. The clamping equipment and results are shown in Fig. 7.

The loading performance of the roof rack was evaluated by the performance analysis of the new roof rack concept for the roof panel constraint and room temperature conditions in vehicle in four evaluation modes (Fig. 8). It was confirmed that the roof panel of the roof rack is advantageous with 5 points compared to the 4 points and the glass fiber reinforcement of the material is advantageous. In addition, the roof panel constraint takes precedence over the fiberglass reinforcement, and when the roof panel restraint conditions change, the analysis results are satisfactory for the loading performance without roof-rack new concepts of fiberglass reinforcement.

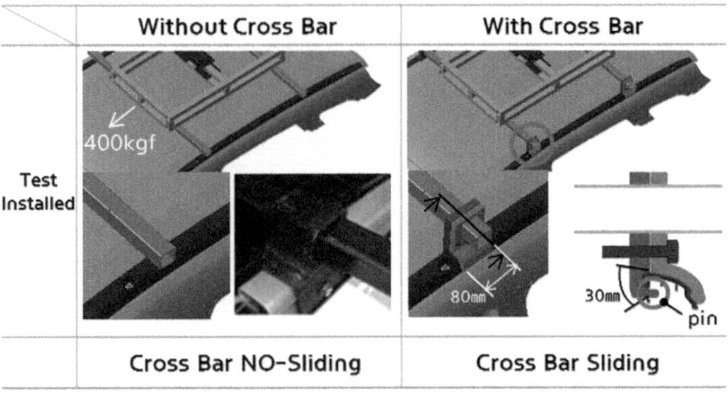

Fig. 6 Installation crossbars on roof rack

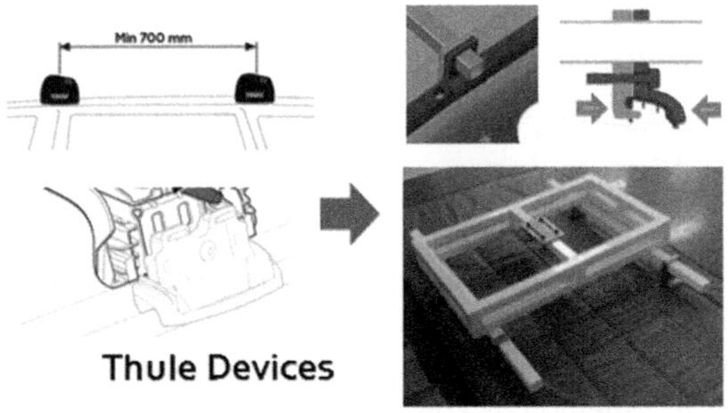

	-	Manufacture Deflection(mm)	Customer Deflection(mm)
Frt. UP (Front Lift)	5mm ↓	3.6	1.2

Fig. 7 Performance differences with crossbars

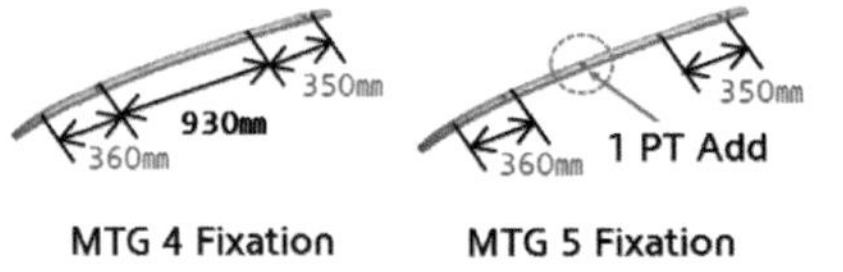

Loading CAE (23℃, DRY)	MTG 4		MTG 5	
	Def.(mm)	Load(MPa)	Def. (mm)	Load(MPa)
Frt. UP	1.9	254	2.1	229
Rr. UP	4.5	126	2.1	170
Frt. Pull	1.2	32	0.2	63
Frt. 20	1.2	36	0.3	67

Fig. 8 Simulation result of mounting fixation and carrier loading

Since the crossbar of the product should satisfy the condition of the clamping pin of the roof rack, the type and the design structure of the composite material are set as control factors by using the L9 (2^3) orthogonal table for the continuous fiber

+ PIN	23℃ WET			80℃ WET		
	Def. (mm)	CFT (MPa)	PA6 (MPa)	Def. (mm)	CFT (MPa)	PA6 (MPa)
Frt. UP	0.64	90	47	0.98	120	45
Rr. UP	0.16	41	26	0.21	51	23
Frt. Pull	0.51	168	33	0.60	214	15
Frt. 20	0.54	160	31	0.65	225	16

- PIN	23℃ WET			80℃ WET		
	Def. (mm)	CFT (MPa)	PA6 (MPa)	Def. (mm)	CFT (MPa)	PA6 (MPa)
Frt. UP	0.64	91	47	0.97	120	**45**
Rr. UP	0.16	41	26	0.21	52	23
Frt. Pull	Slide 8	26	14	Slide 6	31	12
Frt. 20	Slide 7	37	21	Slide 5	30	14

Fig. 9 Simulation result of carrier loading with/without pin of crossbar

reinforced concept of the present invention. The results of the loading performance analysis for the maximum deformation of the roof rack are shown in Fig. 9.

The analysis results show that the load of the continuous fiber is higher than the load of the injection material at high temperature in the case of the clamping pin, but it does not exceed the maximum value of the injection material strength and satisfies the load carrying load. In the case of the front/rear lift test without clamping pin, the result is the same as the case with clamping pin. As a result, it was confirmed that there was no difference between the front and rear lift tests. In case of front/side load test of clamping pin, slip phenomenon occurred 5–6 mm between the product crossbar and roof rack, and the load of the injection material and the continuous fiber both become smaller due to slipping phenomenon. It can also be seen that the forward/lateral load test of the clamping pin oil is concentrated on the continuous fibers. In this way, the development of the roof rack of this development concept is satisfied with the result of the analysis of the mounting performance and the loading performance of the product crossbar. The lift test of the manufacturer of the analysis result is the same as the customer's use condition, and the slip test of the manufacturer implies the possibility of inconsistency with the customer's actual condition of using the clamp cross pin without condition, so that the slip test of the manufacturer is supplemented.

4 Development and Verification

Molding of glass fiber composites is done through fabric heating, press molding through mold press and end cutting. The process conditions and development improvement items are shown in Fig. 10. The average thickness of the basic fabric is 0.3 mm, and a method of pre-lapping or stacking on the top of the bottom mold is generally used in accordance with the maximum thickness. In this development, a full charge technique was used instead of the flow charge technique for laminating continuous fibers on the mold, it was verified through a microscopic photograph that it is a good way to prevent peeling of the joint interface part of the fabric through the evaluation of Roof rack are shown in Fig. 11.

The loading performance of the roof rack is divided into lift resistance and slip resistance test according to the evaluation method prescribed by the regulations of

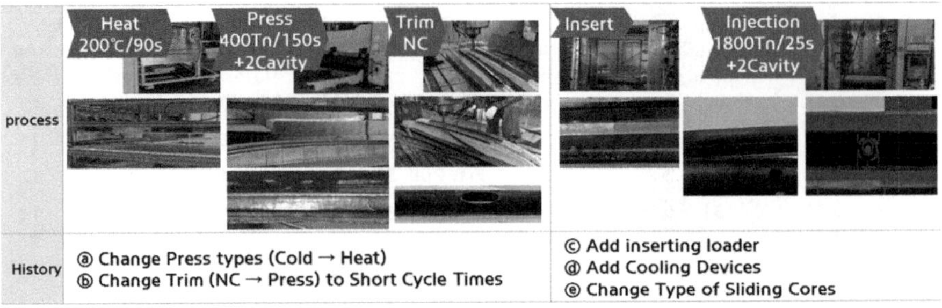

Fig. 10 Producing process in composite fiber and injections of integrated roof rack

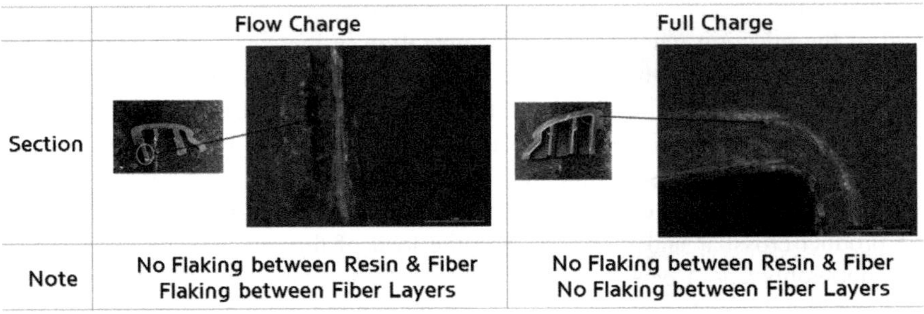

Fig. 11 Producing process in composite fiber and injections of integrated roof rack

KS4084-1, DIN75-305. The slip resistance is a sudden stop and a sudden start mode of the vehicle. The lift resistance refers to the load transferred to the roof crossbar according to the dynamic load value due to the lift effect of the stacked cargo during the running of the vehicle. Basic Test Results are shown in Fig. 12.

Continuous Fiber 2 M Manufacture of Cross-Bar Clamping Equipment's equipped with 4-point Mounting Criteria and Cross-Bar Equipment's provided by the Manufacturer in the Regulation for the manufactured Part-1. The results of the evaluation of the actual vehicle and evil condition are shown in Fig. 12. The performance of the developed product satisfies the performance of the part-1 under the conditions of the single-item standard evaluation condition and the real-vehicle

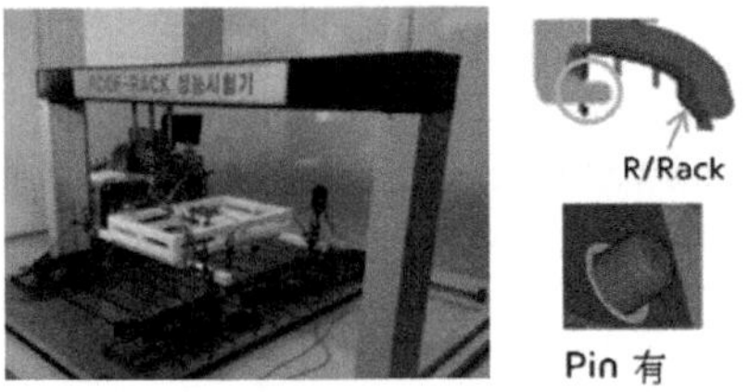

+Pin Cross Bar MTG	Cross Bar 4 MTG		Cross Bar 2 MTG	
	CAE Maximum Def.(㎜)	TEST Maximum Def.(㎜)	TEST Maximum Def.(㎜)	TEST Permanent Def.(㎜)
Frt. UP	0.64	1.54	0.70	0.05
Rr. UP	0.16	0.09	0.50	0.0
Frt. Pull	0.51	0.17	1.9	0.0
Frt. 20	0.54	0.71	0.0	0.0

Fig. 12 Test result of loading performance of roof rack

Fig. 13 Test result with evil condition

Fig. 14 Test result of crossbar with evil condition

standard evaluation (2 points). It is confirmed that it is a developed part which is superior in weight saving effect compared with aluminum. The Thule product crossbar, which is generally used by the user, is subjected to front lift evaluation for the condition 1 of malfunction in the actual vehicle equipped with part respectively. Deformation and breakage of parts occurred at 400 kgf of loading weight and 300 kgf of product crossbar, and deformation and damage of the developed roof rack did not occur. Those results are shown in Figs. 13 and 14. Through this, we verified the feasibility of stacking performance and customer safety of the developed technology and products, and the feasibility of future products.

5 Conclusion

In this study, we developed and evaluated nylon and continuous fiber composite roof racks for weight reduction of manufacturers to provide fuel economy improvement information for consumers to purchase their cars. We evaluated the degree of severity according to the loading performance evaluation equipment of the manufacturer, We conducted the evaluation of single and actual car with respect to importance of glass continuous fiber laminating method, continuous fiber function and article crossbar by developing parts for technology selection and performance development. The loading performance and the safety of use were verified and verified. The conclusions obtained through this development are as follows.

(1) The integrated roof rack using nylon and continuous fiber composite material has satisfied the loading performance of the roof rack against aluminum roof rack for vehicles.
(2) The manufacturer's bolt clamping for the roof rack performance test, which is equipped with the product crossbar, is 3 times more severe than the customer's crossbar clamping.
(3) The minimum requirements of the integrated roof rack are that nylon material property is more than 120 Mpa and Fixation Constraint of Loop panel is 450 mm or less.
(4) The actual crossbar safety load is 300 kgf or less in maximum upward dynamic force.

Parameter Sensitivity Analysis and Uncertainty Optimization of Suspension K&C Characteristic Response

Chengyuan Ren, Kai Li, Shuangshuang Li and Song Gao

1 Introduction

Because of the advantage of simple structure, small footprint, and low unsprung mass, McPherson suspension is widely used in modern cars. When the vehicle is running, the wheel alignment parameters change with the beating of the wheel, only with reasonable geometric parameters to ensure its good K&C characteristics. K&C characteristics of the suspension directly affect the vehicle's handling stability, ride comfort, and tire life. In the design of suspension, generally think that the smaller the wheel alignment parameters change with wheel jump is optimal [1, 2]. This theory is used as an optimization method in the paper [3, 4]. But in the real vehicle design, the theory is not completely applicable. Sometimes in order to meet some of the performance of the car, we need to increase the amount of certain changes. So in the vehicle design, usually make a specific characteristic of K&C as the standard and make the design of the suspension K&C characteristics as close as possible to the standard K&C characteristics. In the real vehicle design, due to the presence of geometric error, installation error, the actual hard point has certain fluctuation and uncertainty which cause the suspension K&C characteristic fluctuation bigger compared with the standard K&C characteristic. The small changes may bring about great changes for car performance. The traditional optimization design has neglected the uncertain factors in the design [5, 6]. Aiming at this problem, this paper establishes the mathematical model of Mcpherson suspension. Based on this model, Sobol method [7] is used to analysis the suspension K&C characteristics sensitivity and find out the hard point of impact. Considering the uncertainty factors in the suspension design, the optimization design method based on the interval analysis is constructed which makes the K&C characteristic of the target suspension have strong robustness.

C. Ren (✉) · K. Li · S. Li · S. Gao
Beijing Electric Vehicle Co., Ltd, Beijing, China
e-mail: rcy5177@163.com

Society of Automotive Engineers of China (SAE-China) (ed.), *Proceedings of the 19th Asia Pacific Automotive Engineering Conference & SAE-China Congress 2017: Selected Papers*, Lecture Notes in Electrical Engineering 486, https://doi.org/10.1007/978-981-10-8506-2_11

2 Mathematical Model of Suspension

Based on the rigid body kinematics theory the mathematical model of Macpherson suspension is established. Geometric model is shown in Fig. 1; A_i is the connection point between the suspension and the body. Spherical Joint A_5 is A_3, A_4 two spherical joints equivalent connection point; B_i is the initial points of the steering knuckle, and C is the center of the wheel. The origin of the coordinate system is located at the center of the vehicle where *x*-axis points to the direction of the vehicle; *z*-axis pointing up; *y*-axis is perpendicular to the plane *xoz* and points to the left side of the vehicle.

2.1 Steering Knuckle Coordinates

Steering knuckle relative to the location of the car body can be made of three consecutive finite rotation around the *x*-, *y*-, *z*-axis rotation angle of α, β, γ. Under these rotations, the steering knuckle coordinates will be same with the wheel center $\boldsymbol{C}(C_x, C_y, C_z)$. Coordinate transformation matrix of the steering knuckle is as follows:

$$R = \begin{bmatrix} c\alpha c\beta & -s\alpha c\beta & s\beta & m \\ s\alpha c\gamma + c\alpha s\beta s\gamma & c\alpha c\gamma - s\alpha s\beta s\gamma & -c\beta c\gamma & n \\ s\alpha s\gamma - c\alpha s\beta c\gamma & c\alpha s\gamma + s\alpha s\beta c\gamma & c\beta c\gamma & l \\ 0 & 0 & 0 & 1 \end{bmatrix} \tag{1}$$

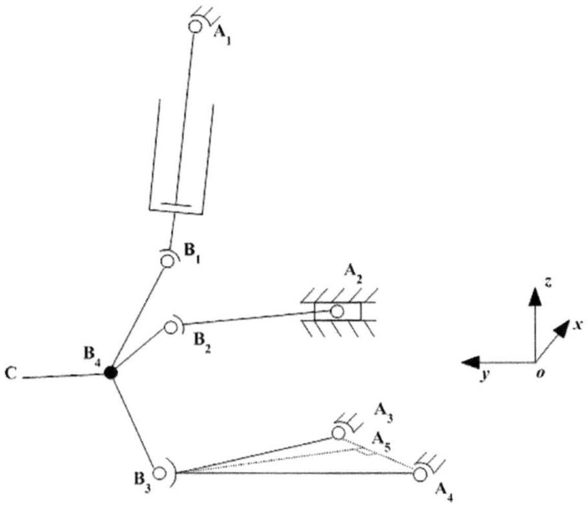

Fig. 1 Macpherson suspension geometry model

$$m = C_{tx} - [R(1,1)C_x + R(1,2)C_y + R(1,3)C_z];$$
$$n = C_{tx} - [R(2,1)C_x + R(2,2)C_y + R(2,3)C_z];$$
$$l = C_{tx} - [R(3,1)C_x + R(3,2)C_y + R(3,3)C_z];$$

Here, $c\alpha = \cos\alpha, s\alpha = \sin\alpha$. The rest can be done in the same manner. $C_t = [C_{tx}, C_{ty}, C_{tz}]$ is the coordinate of wheel center C after its rotation. When the wheel is jumping, the z-coordinate of the wheel center C has a dz input. C_{tz} can be calculated from Eq. (2).

$$C_{tz} = C_z + dz \tag{2}$$

All points of the steering knuckle should satisfy Eq. (3).

$$\begin{cases} B_{it} = RB_i \\ C_t = RC \end{cases} \tag{3}$$

There are five unknown parameters in the formula (3). When the C is used as the input, the four unknown parameters need four constraint equations. Due to the connecting rod is a rigid body, the length of the connecting rod of the suspension is unchanged, and the angle between the steering knuckle is unchanged; four constraint equations can be written as follows:

$$\begin{cases} (x_{A_2} - x_{B_{2t}})^2 + (y_{A_2} - y_{B_{2t}})^2 + (z_{A_2} - z_{B_{2t}})^2 = l^2_{A_2B_2} \\ (x_{A_5} - x_{B_{3t}})^2 + (y_{A_5} - y_{B_{3t}})^2 + (z_{A_5} - z_{B_{3t}})^2 = l^2_{A_5B_3} \\ l^2_{B_1B_4} + l^2_{B_1B_3} - 2l_{B_1B_4}l_{B_1B_3}\cos\angle B_1B_4B_3 = l^2_{B_1B_3} \\ l^2_{B_1B_4} + l^2_{B_1B_2} - 2l_{B_1B_4}l_{B_1B_2}\cos\angle B_1B_4B_2 = l^2_{B_1B_2} \end{cases} \tag{4}$$

When the C_{tz} is in different positions, the coordinate of each point can be obtained by the formula (3) and (4). According to the definition of the alignment parameters of Mcpherson suspension, the caster angle θ, inclination angle β, toe angle τ, and camber angle γ can be written as follows:

$$\theta = \arctan\frac{B_{3tx} - A_{1x}}{B_{3tz} - A_{1z}} \tag{5}$$

$$\beta = \arctan\frac{B_{3ty} - A_{1y}}{B_{3tz} - A_{1z}} \tag{6}$$

$$\tau = \arctan\frac{C_{tx} - B_{4x}}{C_{ty} - B_{4y}} \tag{7}$$

$$\gamma = \arctan\frac{C_{tz} - B_{4z}}{C_{ty} - B_{4y}} \tag{8}$$

2.2 Verification of Mathematical Model

Following assumptions are made in the establishment of the mathematical model: All parts are rigid, and all parts are connected by rigid hinges. According to the assumption, the simulation model of the Mcpherson suspension in ADAMS/Car is obtained as shown in Fig. 2.

According to the mathematical model of the suspension, the MATLAB program is written to calculate the change of the position parameters of the suspension with the wheel jump, and the simulation results are compared with the ADAMS/CAR. Results are shown in Figs. 3 and 4. The simulation results of the toe angle, camber angle, inclination angle, caster angle are in agreement with results of the mathematical model. It shows that the mathematical model of the suspension is reliable and can provide guarantee for the further study.

3 Sensitivity Analysis of Suspension K&C Characteristic

The purpose of sensitivity analysis of suspension K&C characteristic is to find out the key hard points that affect the alignment parameters of the wheel during the wheel jump. Because there are many factors that affect the K&C of the suspension, their sensitivities cannot be resolved at the initial stage of the study. Therefore, it is necessary to select the factors which are more sensitive. Here, we choose the maximum variation of alignment parameters in the process of wheel jump as objective function. The equation can be written as follows:

Fig. 2 Simulation model of Mcpherson suspension

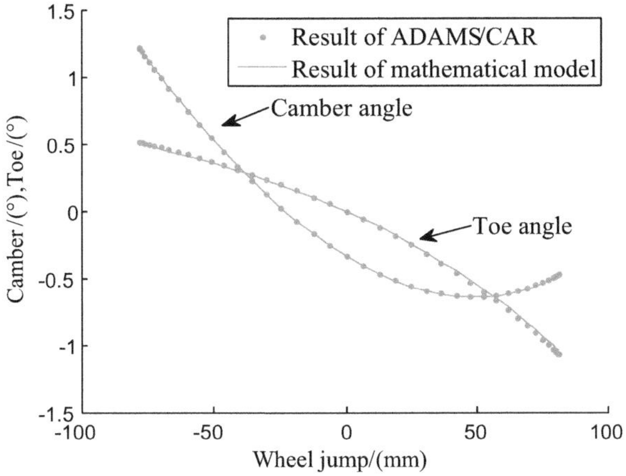

Fig. 3 Camber angle and toe angle

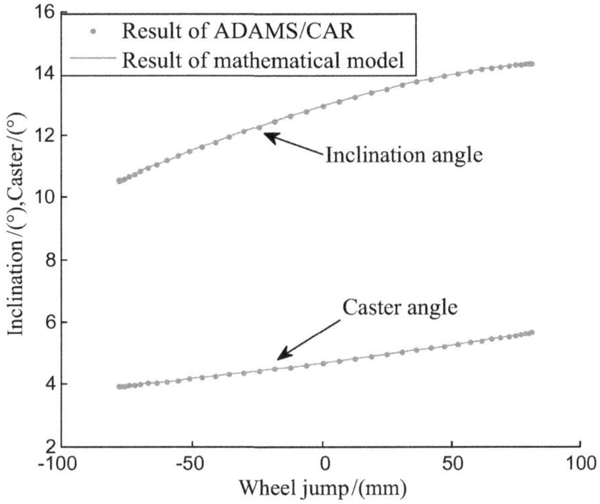

Fig. 4 Inclination angle and caster angle

$$\begin{aligned} \theta_t &= \max(\theta) - min(\theta) \\ \beta_t &= \max(\beta) - min(\beta) \\ \tau_t &= \max(\tau) - min(\tau) \\ \gamma_t &= \max(\gamma) - min(\gamma) \end{aligned} \quad (9)$$

3.1 Design Variable

The selected suspension hard point parameters have a decisive influence on the suspension alignment parameters. In this paper, B_1, B_2, B_3, A_2, and A_5 are selected as the design variables. The perturbation range of each variable is −10–10 mm. Coordinates of each hard point are shown in Table 1.

3.2 Sensitivity Analysis

Latin hypercube sampling of the 15 parameters is carried out according to Table 1 in ISIGHT. Take two capacities of 3000 samples, expressed in matrix $P_{m\times n}$ and $Q_{m\times n}$, respectively. Each row in a matrix is a set of specific input combination of the objective function. m = 3000, n = 15.

We Assume H_i is the matrix which is replaced by the column i $(i \leq 15)$ of the matrix P into the column i of the matrix Q; H_{-i} is the matrix which is replaced by the column i $(i \leq 15)$ of the matrix Q into the column i of the matrix P. The output response value of design parameters can be obtained when each group design parameters in the matrix were put into the objective function calculation. Then the variance estimation and parameter sensitivity of the system response can be obtained by the Monte Carlo method [8] as follows:

$$\hat{f}_0^2 = \frac{1}{n}\sum_{r=1}^{n} f_P(x_{r1}, \cdots, x_{rm}) f_Q(x'_{r1}, \cdots, x'_{rm}) \tag{10}$$

$$\hat{V}(y) = \frac{1}{n}\sum_{r=1}^{n} f_P^2(x_{r1}, \cdots, x_{rm}) - \hat{f}_0^2 \tag{11}$$

$$\hat{U}_i = \frac{1}{n}\sum_{r=1}^{n} f_P(x_{r1}, \cdots, x_{rm}) f_{H_i}(x'_{r1}, \cdots, x_{ri}, \cdots, x'_{rm}) \tag{12}$$

$$\hat{U}_{-i} = \frac{1}{n}\sum_{r=1}^{n} f_P(x_{r1}, \cdots, x_{rm}) f_{H\_i}(x_{r1}, \cdots, x'_{ri}, \cdots, x_{rm}) \tag{13}$$

S_{x_i} is the main effect index of input parameters x_i, S_{x_i}: $\hat{S}_{x_i} = \frac{\hat{U}_i - \hat{f}_0^2}{\hat{V}(y)}$ is the effect of a single parameter for output, which reflects the size of the first-order sensitivity. $S_{x_i}^T$ is the full effect index of input parameters x_i, $S_{x_i}^T$: $\hat{S}_{x_i}^T = 1 - \frac{\hat{U}_{-i} - \hat{f}_0^2}{\hat{V}(y)}$ contains the main

Table 1 Initial hard point coordinates

	B_1	B_2	B_3	A_2	A_5
x	881.24	1006.85	866.81	1061.91	872.22
y	−618.53	−704.81	−729.82	−359.36	381.14
z	429.22	345.18	244.35	355.27	267.97

effect of the parameters and the interaction of the parameters with the other parameters. At the same time, the magnitude of global sensitivity was also reacted.

3.3 Results Analysis

Based on the mentioned objective function and the Monte Carlo calculation method for global sensitivity of Sobol, the sensitivity of the fifteen parameters can be obtained by entering the sample points, as shown in Figs. 5, 6, 7 and 8.

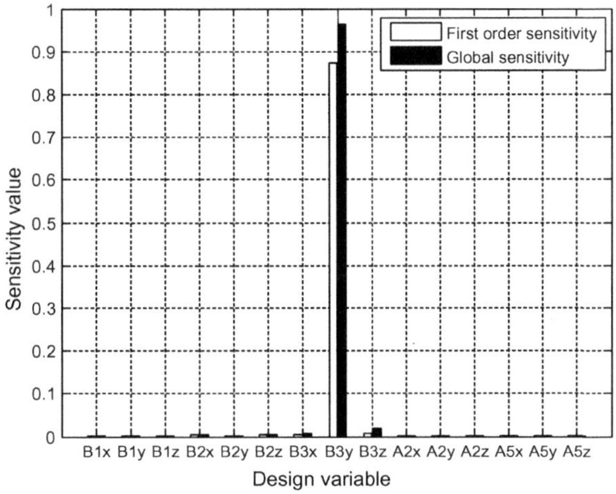

Fig. 5 Inclination angle sensitivity

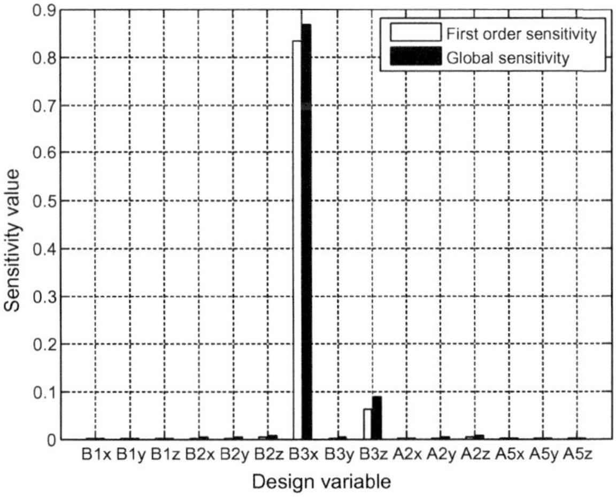

Fig. 6 Caster angle sensitivity

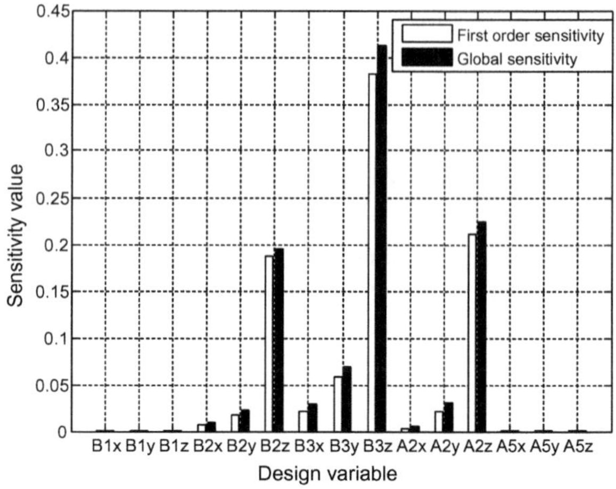

Fig. 7 Camber angle sensitivity

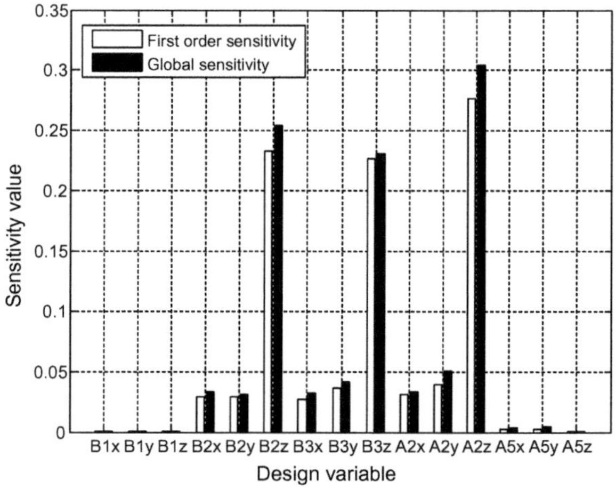

Fig. 8 Toe angle sensitivity

From Fig. 5, parameter B_{3y} has the maximum inclination angle sensitivity during the wheel jump; the rest of the parameters sensitivity can be negligible. From Fig. 6, the sensitivity value of B_{3x} is the largest, followed by B_{3z}, the rest can be ignored. From Fig. 7, the global sensitivity value of the camber angle is sorted as follows: $B_{3z}> A_{2z}> B_{2z}> B_{3y}> A_{2y}> B_{3x}> B_{2y}$; from Fig. 8, the global sensitivity value of the toe angle is sorted as follows: $A_{2z}> B_{2z}> B_{3z}> A_{2y}> B_{3y}> A_{2x}> B_{3x}$.

Based on the sensitivity analysis, parameter B_{3x}, B_{3y}, B_{3z}, A_{2y}, A_{2z}, B_{2x}, B_{2z} are more sensitive to the objective function. Therefore, these parameters are selected as optimization design variables and record as x_1, x_2, x_3, x_4, x_5, x_6, and x_7, respectively.

4 The Uncertainty Optimization Method Based on Interval Analysis

4.1 Mathematical Transformation Model of Nonlinear Interval

Based on interval mathematics [9], the interval is defined as a pair of ordered real numbers:

$$A^I = [A^L, A^R] = \{x | A^L \leq x \leq A^R, x \in R\} \tag{14}$$

where I represents the interval, L and R are lower and upper bound of interval.

The mathematical method for interval comparison is generally divided into two categories: One of them called "interval probability" is used to quantitatively describe the degree of the interval is better than the other one. Another kind is called "interval order relation", which is used to qualitatively judge the merits and defects of the two intervals. In this paper, we choose "interval order relation" as interval comparison method. For the minimization optimization problem, the order relation of interval B is better than interval A which can be expressed as follows:

$$\begin{cases} A^I \leq_{CW} B^I & A^C \geq B^C \text{且} A^W \geq B^W \\ A^I <_{CW} B^I & A^I \leq_{CW} B^I \text{且} A^I \neq B^I \end{cases} \tag{15}$$

where A^C and A^W respectively represent the midpoint and the radius of the interval.

$$A^C = \frac{A^R + A^L}{2}, \quad A^W = \frac{A^R - A^L}{2} \tag{16}$$

The order relation expressed its preference for interval midpoint and radius. By formula (15) we can know that when the interval B^I is better than the interval A^I, it must meet that the midpoint and radius of the interval B^I is less than A^I. So we hope to find an optimal design variables which will let the objective function of the interval has a minimum of midpoint and minimum interval radius. Based on this idea, the uncertain optimization problem in the formula (14) can be converted to the following deterministic optimization problem as follow:

$$\min\left(f^C(\boldsymbol{X}), f^W(\boldsymbol{X})\right) \tag{17}$$

4.2 The Optimization Process of Suspension Hard Point Based on Interval Analysis

The deterministic optimization problem which is obtained from the nonlinear interval model is a typical two layer nesting optimization problem. The inner layer optimization is used to calculate the interval of uncertain objective function and constraints, and the outer layer is optimized for the design variables. Because of the existence of nested, the deterministic optimization problem is often discontinuous and non differentiable. In this paper, multi-objective particle swarm optimization algorithm based on crowding distance [10] (MOPSO-CD) is used to solve this problem in the optimization of inner and outer layers. MOPSO-CD is a stochastic optimization and search algorithm, which can only know the design variables information, so it is a good choice to solve the inner and outer layer optimization. When choosing the optimal solution, the least distance method [11] is used, which can avoid the subjectivity of the optimal solution. It can be expressed as follows:

$$\min d = \sqrt{\sum_{i=1}^{n} \left[\frac{f_i(x) - \min f_i(x)}{\max f_i(x) - \min f_i(x)} \right]^2} \tag{18}$$

where n is the number of components in the target vector; f_i is the i target component.

4.3 Constraint Interval Setting for Design Variables and Objective Function

Design variables and the uncertainty values range as shown in Table 2. Constraint interval is from 85% standard K&C characteristics to 115% standard K&C characteristics.

4.4 Optimization Process and Result Analysis

Taking the change between K&C characteristic and standard K&C as objective function,

Table 2 Range of design variables and uncertain quantity

Design variable	Design interval	Uncertain quantity
x_1	[858.142, 875.478]	[−2, 2]
x_2	[-737.118, -722.522]	[−2, 2]
x_3	[241.906, 246.793]	[−1, 1]
x_4	[-357.291, -362.611]	[−1,1]
x_5	[351.717, 358.823]	[−1, 1]
x_6	[996.782, 1016.92]	[−2, 2]
x_7	[341.728, 348.632]	[−1, 1]

$$\min D_i = \sqrt{\frac{1}{n}\sum_{j=1}^{n}\left(y_{ij} - Y_{ij}\right)^2} \tag{19}$$

where D_i is the i objective function; D_1–D_4 refers to the inclination angle, caster angle, camber angle and toe angle; y_{ij} is the response values for the j sampling point; Y_{ij} is the standard value of K&C for the j sampling point; n is the number of samples.

Based on the interval order relation, the uncertain optimization model with interval is converted to a deterministic optimization model, and the optimized hard point information is obtained, as shown in Table 3. The optimization results are shown in Table 4.

From Table 4, we can see that the midpoint of the objective function is reduced after optimization. This means that the change between the optimized K&C characteristic and standard is reduced and the optimized K&C characteristic is more close to the standard; on the other hand, the objective function radius is relatively reduced, which indicates that the K&C characteristic fluctuation range is reduced, and the optimized K&C characteristic is more robust.

In order to verify the robustness of the optimized hard points, 30 sets of data are randomly selected in the uncertain domain of the initial hard point and optimized hard point, respectively, and the simulation test is carried out on the 30 sets of data. Formula (19) is used to calculate the change of the suspension K&C in each group, and the test results are shown in Figs. 9, 10, 11, and 12.

Table 3 Comparison of variables before and after optimization design

Variable	Initial	Results	Final
x_1	866.81	866.8182	866.82
x_2	−729.82	−729.7184	−729.72
x_3	244.35	244.7974	244.80
x_4	−359.36	−361.0009	−361.00
x_5	355.27	353.6275	353.63
x_6	1006.85	1007.5775	1007.58
x_7	345.18	343.9905	343.99

Table 4 Results of optimization

Object	Objective function interval	Midpoint of the objective function			Objective function radius		
		Before	After	Rate	Before	After	Rate
D_1	[0.0063,0.2183]	0.2230	0.1123	49.6%	0.1873	0.1060	43.4%
D_2	[0.0589,0.2865]	0.2726	0.1727	36.6%	0.1779	0.1138	36.0%
D_3	[0.0120,0.0394]	0.0327	0.0236	27.8%	0.0278	0.0158	43.1%
D_4	[0.0203,0.2511]	0.2264	0.1357	40.1%	0.2096	0.1154	44.9%

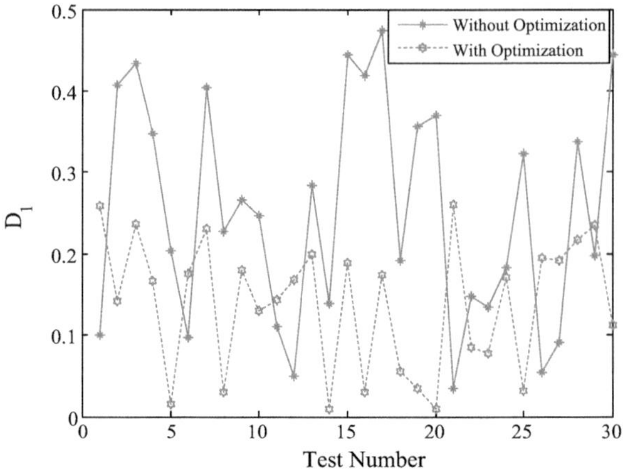

Fig. 9 Variation curve of inclination angle

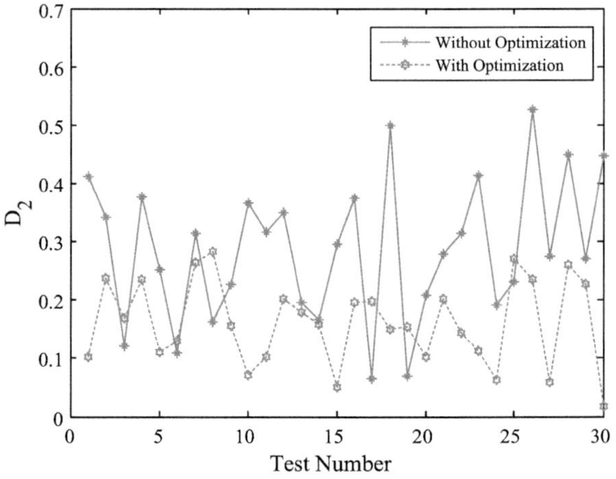

Fig. 10 Variation curve of caster angle

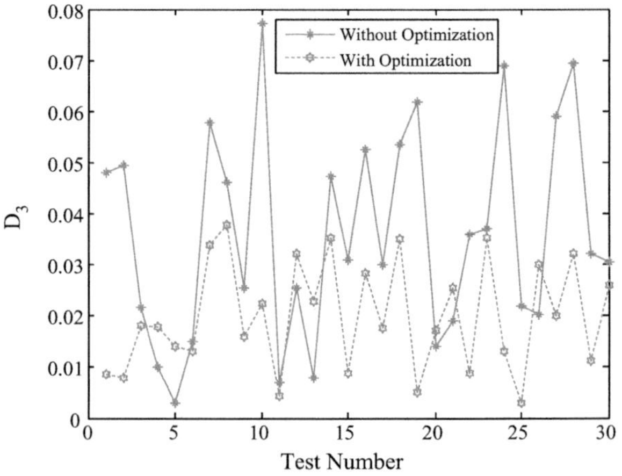

Fig. 11 Variation curve of camber angle

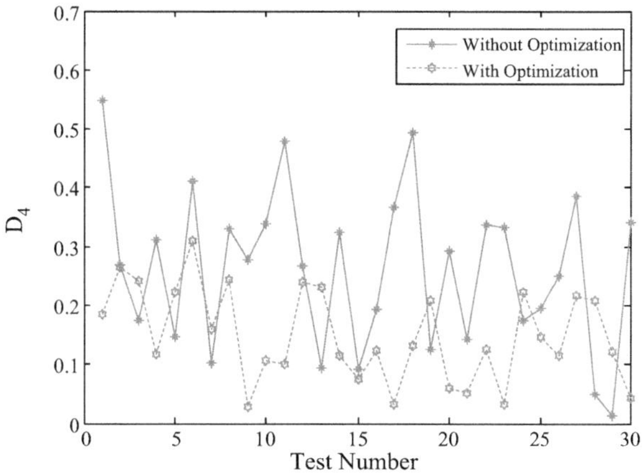

Fig. 12 Variation curve of toe angle

From Figs. 9, 10, 11, and 12, we can find that the mean value of variation and fluctuation range is significantly reduced compared with the optimized K&C and the initial K&C. It can be seen that the optimized K&C is more close to the standard K&C and has stronger robustness.

5 Conclusions

1. In this paper, the mathematical model of Mcpherson suspension is established and compared with the Mcpherson suspension in the ADAMS/CAR module, and the calculation results are consistent with the simulation results, which prove the correctness of the mathematical model.
2. And based on this model, the global sensitivity is calculated by the Sobol method. Through applying wheel jumping input of the model, the sensitivity of the suspension hard point is obtained. By comparing the sensitivity values of each parameter, we found that: y-coordinates of lower-control-arm outer hard point have great influence on inclination angle; x-coordinates of lower-control-arm outer hard point have great influence on caster angle; z-coordinates of lower-control-arm outer hard point, tierod inner, and outer hard point determine the size of camber angle and toe angle.
3. Robustness is an important index to evaluate the quality of products. In this paper, uncertain factors are considered in the design of the suspension. Based on the nonlinear interval order relation and the multi-objective particle swarm optimization algorithm, an optimal design method based on interval analysis is proposed. The results show that the method is robust and practical.

References

1. Jornsen R (1992) Automotive chassis element. Translated by Zhang Hongxin, Yu Zhuoping. Popular Science Press, Beijing, 136–176
2. Donald B, Geoffrey H, John PW (1987) Car suspension and handling. British Library Cataloguing in Publication Data, London
3. Lu J, Zhou K, Guo L, Hou Y (2012) Design and parametric optimization of mcpherson front suspension of electric vehicle. J Mech Eng 48(8):98–103
4. Xiang T, Shen L (2014) An optimization of the front suspension linkages of a racing car with ADAMS. Automot Eng 36(4):486–490
5. Feng T, Zhong Z, Yan X, Chen Y (2012) Multi-objective optimization for multi-link suspension based on NSGA-II algorithm. Automot Eng 32(12):1063–1066
6. Feng J, Chen X, Zheng S, Hou S (2013) Optimization for multi-link suspension hard-points based on PDJI-MOPSO algorithm. China Mech Eng 24(13):1841–1845
7. Sobol IM (1993) Sensitivity estimates for nonlinear mathematical models. Math Model Comput Exp 1(4):407–414
8. Zhang X (2010) Research on the global sensitivity analysis method of air defense missiles weapon system-of-systems combat effectiveness. National University of Defense Technology in Changsha
9. Moore RE (1979) Methods and applications of interval analysis. Prentice-Hall Inc, London
10. Raquel CR, Naval Jr. PC (2005) An effective use of crowding distance in multi-objective particle swarm optimization. In: Proceedings of Congress evolutionary computation. ACM Press, Washington DC, USA, pp 257–264
11. Junning Chen, Chaiy Rungsiyakull, Wei Li, Yuhang Chen, Michael Swain, Qing Li (2013) Multiscale design of surface morphological gradient for osseointegration. J Mech Behav Biomed Mater 20:387–397

Simulation Study on Engine Cylinder Block Machining Line Based on Plant Simulation

Wenbin Tao, Dong Wang, Qijun Hu, Shaochun An, Yunping Zhang and Shuwen Xie

1 Introduction

The system of a production line is a highly dynamic, highly changing and complex system, so it is very difficult to analyze and optimize the entire system. At present, in the process of the production line construction, people always use the traditional mathematical methods and statistical methods to abstract, simulate, and analyze the concrete problems of the production system, but that is difficult to describe, analyze, and judge [1]. With the leaping development of virtual modeling and simulation technology, it is very possible to solve these problems. As a system modeling and experimental analysis method, computer simulation technology can combine the resources of the production line, the route of the process, and the data of the process in the actual production line. The expression of simulation technology is easy to understand, and it can reflect the actual dynamic process of production and features with simulation technology [2].

2 Theoretical Foundation

2.1 Simulation Definition

VDI directive 3633 defines: "Simulation as the emulation of a system, including its dynamic processes, in a model one can experiment with. It aims at achieving results

W. Tao (✉) · D. Wang · Q. Hu · S. An · Y. Zhang · S. Xie
Chongqing Changan Automobile Co., Ltd, Chongqing 401120, China
e-mail: 490802611@qq.com

D. Wang
e-mail: wangdong1@changan.com.cn

Society of Automotive Engineers of China (SAE-China) (ed.), *Proceedings of the 19th Asia Pacific Automotive Engineering Conference & SAE-China Congress 2017: Selected Papers*, Lecture Notes in Electrical Engineering 486, https://doi.org/10.1007/978-981-10-8506-2_12

that can be transferred to a real-world installation. In addition, simulation defines the preparation, execution, and evaluation of carefully directed experiments within a simulation model."

2.2 Simulation Application

In general, simulation helps us to determine the design of process planning, investigate the utilization of the resources and of the machines, determine the number of parts and of the tool, optimize the control strategies you devised, evaluate different alternatives by running a number of simulation experiments.

2.3 Simulation Process

The system of a production line itself has the complexity, diversity, and other characteristics. In order to ensure the orderly development of the simulation analysis and the effectiveness of simulation results, developing our simulation is necessarily by a cyclical and evolutionary process; after several cycles, we will arrive at the final model. When we implement our simulation project, we will proceed like in Fig. 1.

2.4 Simulation Analysis Tool

When the production system is running, the system is affected by random factors, which show the characteristics of discrete, random, concurrency, and recursion. Therefore, it is a typical system of discrete event [3]. In this paper, Siemens' Plant

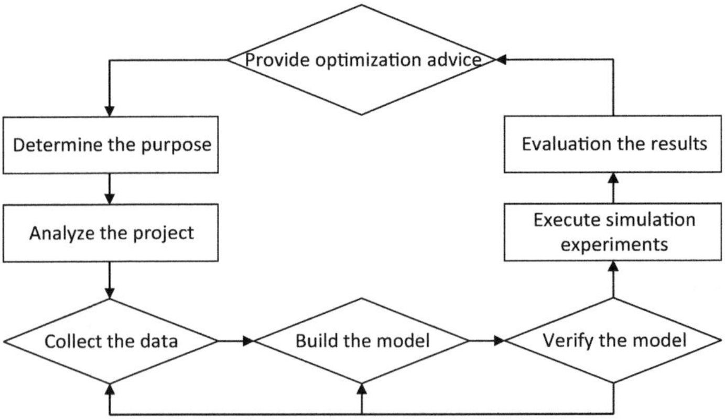

Fig. 1 Process of simulation

Simulation is used as the simulation platform to simulate the production system and the operation of the system. Plant Simulation provides an object-oriented simulation environment for the establishment of discrete time flow process, through simulation, experiment, and optimization to improve the production efficiency of dynamic system [4]. Through the real-time modeling of the production line, you can dynamically analyze the capacity and quota of the production line and provide an objective and optional graphical analysis and data for the construction and improvement of the production line.

At present, the most commonly used simulation algorithms are event-oriented event scheduling; activity-oriented activity scanning; and process-oriented process interaction. Plant Simulation uses the process interaction, the simulation clock to advance, which will move all events which meet the time conditions from the future events list (FEL) to current events list (CEL), and then scan each record of the CEL by activity scanning. When the condition is true, the activity is executed, the state of the system is modified, and the next event is determined. When all the records of the CEL are completed, the scan of the CEL is finished, the simulation clock is advanced, and the earliest occurrence event in FEL is moved to CEL until the simulation ends [5]. The advantage is that we can skip the time between events.

3 Simulation and Experimental Verification

3.1 Production Line Analysis

3.1.1 Production Line Analysis Content

In this paper, an automobile enterprise engine cylinder block machining line is simulated and analyzed, for the cylinder machine production line designing to provide support and decision-making reference. We will verify the capacity of the production line, evaluate the utilization rate of the gantry manipulator, and optimize the number of the zero-point clamping and of the dummy head.

3.1.2 Production Line Main Strategy

In this paper, an automobile enterprise engine cylinder block machining line is simulated and analyzed. The machining line consists of cutting equipment, auxiliary equipment, and testing equipment; the whole line is ring layout, and the process route is shown in Fig. 2. The layout of cutting equipment is clustered, the layout of auxiliary equipment is arranged in series, and the layout of testing equipment utilizes online.

Among them, the cutting equipment adopts the gantry manipulator to load, unload, locate, and clamp machined workpieces; the auxiliary equipment adopts the robot automatic to load and unload machined workpieces; the automatic conveyor

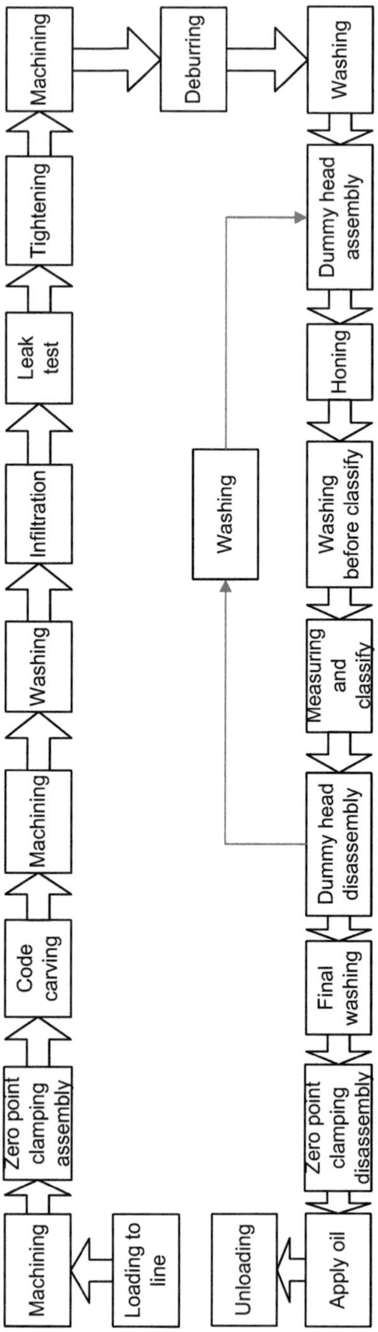

Fig. 2 Engine cylinder block machining line process

should be equipped between the process equipment; and the tactics of gantry manipulator follows the logic of actual production. The online inspection station is set behind the cutting process. The water pressure inspection station is set behind the leakage testing station. The SPC station is set behind the finishing process. The parts are located and clamped with the zero-point clamping. The dummy head is applied in the rear section of the machining line, the backflow of which utilizes the automatic conveyor.

3.1.3 Production Line Main Parameters

The number and layout of production line equipment and automatic conveyor are arranged according to the layout drawing. The cycle time of the single equipment is 68 s/piece; the cycle time of the multiple equipment = the designed cycle time of the whole line × the number of equipment.

The equipment involved in the utilization rate and the mean time to repair (MTTR) includes manual equipment, automatic domestic equipment, automatic import equipment, and the gantry manipulator. And the utilization of the equipment and the mean time to repair (MTTR) of the equipment are shown in Table 1.

The parameters of the equipment used for logistics: The speed of the automatic conveyor is 0.23 m/s, the speed of the gantry manipulator is 3.14 m/s, and the acceleration is 1.26 m/s^2.

The parameters of the equipment used for detection: Three-coordinate detection and online sampling, and the frequency and the time of detection are shown in Table 2.

The waste rate of the production line is 2% of machining waste and 3% of casting material waste.

3.2 Assumption and Simplification of Model

The modeling method cannot reflect the whole of the real system which is a complex system. The simulation model is not the repetition or rehearsal of the real system. Therefore, the model must be assumed and simplified. In this paper, the following assumptions and simplifications are made to the simulation model [6]:

Table 1 Utilization rate and MTTR

	Manual equipment	Automatic domestic equipment	Automatic import equipment	The gantry manipulator
The utilization rate (%)	100	99.40	99.60	99.60
MTTR (min)	0	45	45	30

Table 2 Parameters of detection

No.	Process ID	Three-coordinate detection			Online sampling detection	
		Detection frequency	Detection time (min)	Number of equipment	Detection frequency	Detection time (s)
1	OP20	1/station/day	17	5	1/40	500
2	OP40	1/station/day	30	5	1/40	300
3	OP50	1/station/day	50	10	1/40	2200
4	OP60	1/station/day	30	8	1/40	1280
5	OP130	1/station/day	25	5	1/10	100
6	OP170	1/station/day	25	3	1/10	90

(1) In the case of setting MTA of the equipment, the equipment must be a failure and MTTR of the equipment is a certain time.
(2) The waste is concentrated in two stations to be removed from the machining line, the machining waste is removed at the SPC station of the finishing process, and the casting material waste is removed at the water pressure inspection station. The parts of the remaining processes are considered only to be inspected and are not considered to be removed from the machining line.
(3) When multiple steps are included in a process, the multi-step setting is achieved by increasing the number of single processing units connected in series in the process and keeping the processing time of each step consistent with the cycle time of the process.
(4) Manual loading and unloading process: Manual loading and unloading time are incorporated into the cycle time of the equipment and is not considered separately.
(5) The manual equipment of the production line does not take into account the failure.

3.3 *Simulation Model Establishment*

Plant Simulation uses the method of object-oriented modeling. By dragging the graphics and adding additional programs, you can quickly build a system model, which makes the modeling process more quickly [7]. Based on the library of typical equipment provided by Plant Simulation, we customize the equipment that matches the actual situation. By programming with SimTalk, we achieve the control of the simulation process. Then according to the collected information of the production line, the assumptions, and simplification of model, we build the simulation model of the engine cylinder machine line (shown in Fig. 3) and map the strategy and parameter of production line, the cycle time of equipment, and other information completely into the simulation model.

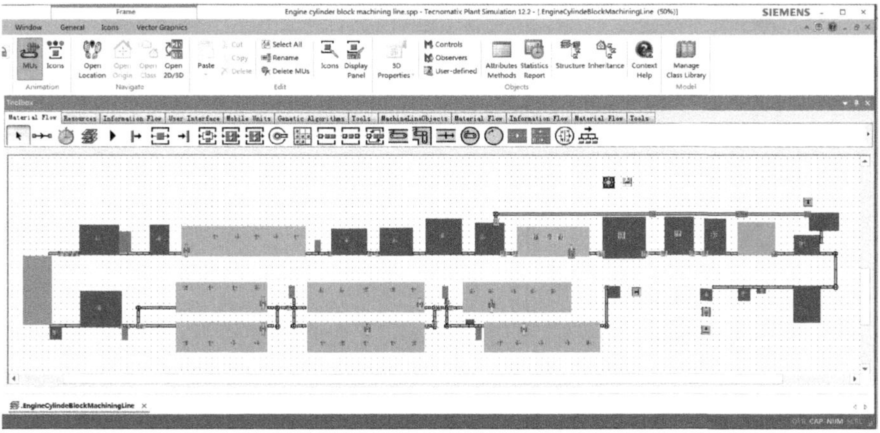

Fig. 3 Engine cylinder block machining line simulation model

Among them, the controls of the production line, including the basic data, shift management, experimental manager, the collection, and the analysis of the data are shown in Figs. 4, 5, 6 and Table 3. And the experimental manager is used when we optimize the number of the zero-point clamping and of the dummy head.

The robot, the gantry manipulator, and other automatic equipment involve logic programming, according to their design strategy, respectively, on the control of the loading and unloading, the control of looking for target, the control of mobile and other process programming (shown in Figs. 7 and 8).

3.4 Simulation Model Validation

After finishing the first draft of the model, we have to verify it. We will test each and every object we created and all parameters we set. Executing the model, we will arrive at the data. The verification between the results and the design scheme and the comparison between the results and the production line which has been

Fig. 4 Controls of the production line

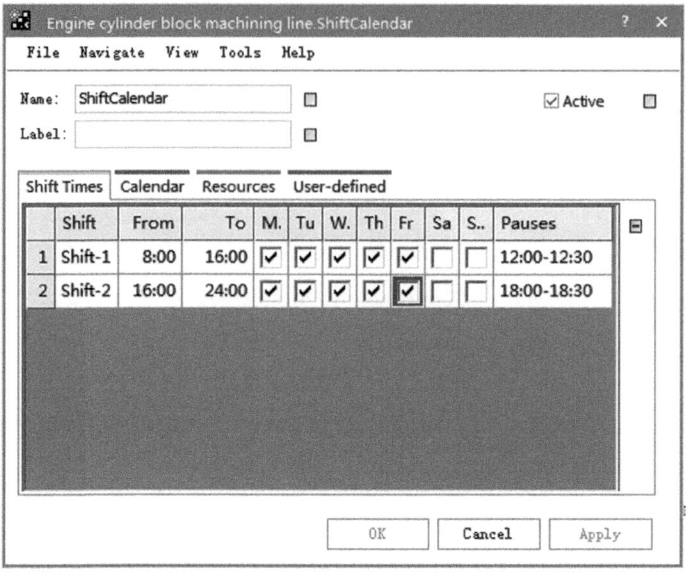

Fig. 5 Shift management

Fig. 6 Part of the program

```
is
    tlb:object;
    countTime:time;
do

    tlb:=T_ExportDataTlb;
    if tlb.ydim>=1 then
        countTime:=0;
        for local i := 1 to tlb.ydim loop
            countTime:=CountTime+tlb["DTDTime",i];
        next;
        DTD:=countTime/tlb.ydim;
    end;
end;
```

completed are carried out. The existing model is optimized and perfected. The aim is to make the model closer to the real production line, making the simulation results more accurate.

4 Key Findings and Achievements

This paper takes the task of the machining line as a database for one month and sets the simulation time for 30 days in the model. Then run the model and observe the work of the machining line. By the statistical module provided by the simulation

Table 3 Part of the parameters

Process ID	Name	Type of process	Parallel	Cycle time	Availability(%)	MTTR
OP10	Loading	Manual station	No	15	100	00:00.0
OP20_1	Cutting	Automatic station	Yes	05:19.0	99.4	45:00.0
OP20_2	Cutting	Automatic station	Yes	05:19.0	99.4	45:00.0
OP20_3	Cutting	Automatic station	Yes	05:19.0	99.4	45:00.0
OP20_4	Cutting	Automatic station	Yes	05:19.0	99.4	45:00.0
OP20_5	Cutting	Automatic station	Yes	05:19.0	99.4	45:00.0
OP30	Assembly	Manual station	No	50	100	00:00.0
OP40_1	Cutting	Automatic station	Yes	05:19.0	99.6	45:00.0
OP40_2	Cutting	Automatic station	Yes	05:19.0	99.6	45:00.0
OP40_3	Cutting	Automatic station	Yes	05:19.0	99.6	45:00.0
OP40_4	Cutting	Automatic station	Yes	05:19.0	99.6	45:00.0
OP40_5	Cutting	Automatic station	Yes	05:19.0	99.6	45:00.0

Fig. 7 Controls of the gantry manipulator

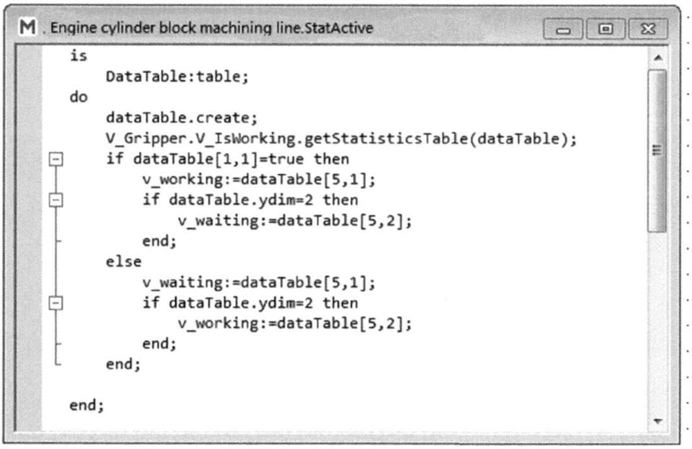
.Engine cylinder block machining line.StatActive

```
is
    DataTable:table;
do
    dataTable.create;
    V_Gripper.V_IsWorking.getStatisticsTable(dataTable);
    if dataTable[1,1]=true then
        v_working:=dataTable[5,1];
        if dataTable.ydim=2 then
            v_waiting:=dataTable[5,2];
        end;
    else
        v_waiting:=dataTable[5,1];
        if dataTable.ydim=2 then
            v_working:=dataTable[5,2];
        end;
    end;

end;
```

Fig. 8 Program of the StatActive

software and the relevant statistical method obtained by programming with SimTalk, the capacity of the whole machining line, the utilization rate of the gantry manipulator, the number of the zero-point clamping, and the number of the dummy head are simulated and output in the form of graph and table.

4.1 Production Line Capacity

In this paper, the capacity of the machining line includes production line JPH, OEE, the completion rate of the cycle time, comprehensive yield, the whole line rate, and other indicators. Overall equipment effectiveness (OEE)refers to the overall efficiency of the equipment used in the production line. Jobs per hour (JPH) refers to the unit hour output of the production line. Among them, the design indicators of the machining line are shown in Table 4.

By the database module of the model, the results of the simulation are output as dialog box, and the output values of the indicators are shown in Fig. 9.

Combined with the simulation results, the capacity of the machining line meets the design requirements. Before applying the simulation software, there is no effective way to analyze the above-mentioned capacity of the machining line, only be judged by experience. Application simulation fills the vacancy well.

Table 4 Engine cylinder block machining line design indicators

JPH	OEE	The whole line rate	Comprehensive yield	The completion rate of cycle time
40	≥ 80	$\geq 93.5\%$	$\geq 95\%$	100%

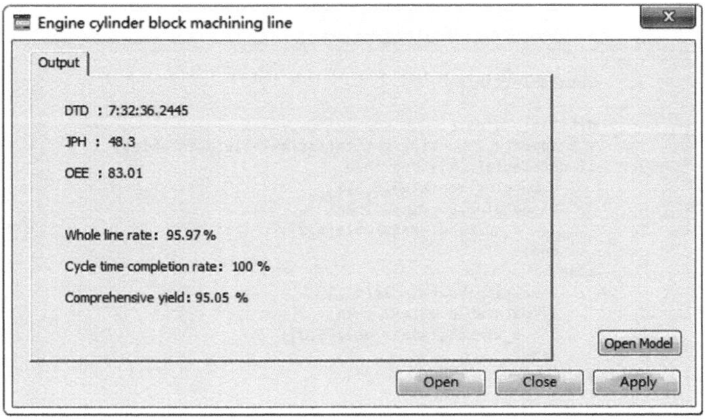

Fig. 9 Output values of the indicators

4.2 Gantry Manipulator Utilization

The simulation model of the gantry manipulator with the most cycle time in the production line is selected. The simulation model is shown in Fig. 10.

Often, the traditional way of assessing the capability of gantry manipulator is in the static state. The time at which the gantry manipulator completes a cycle is counted and calculated as T, and the cycle time is t; the utilization rate of the gantry manipulator can be calculated from the above data and expressed as V.

$$V = \frac{T}{t} \times 100\% \tag{1}$$

Among them, the main working process of the gantry manipulator is as follows (Fig. 11). By calculating the time T required to complete a cycle which is 55.03 s (shown in Table 5), the utilization rate of the gantry manipulator is calculated by the above equation $V \approx 81\%$.

The chart statistic module of the simulation model is used to output the time histogram of the working time and the waiting time of the gantry manipulator, shown in Fig. 12. According to the figure, the rate of the gantry manipulator waiting time is 16% and the working time is 84%. The results of the simulation and calculation are basically the same. Because many situations will happen in actual production, such as the gantry manipulator waiting for the equipment, the equipment waiting for the gantry manipulator and so on, the method of the simulation is closer to the actual production than the traditional static calculation.

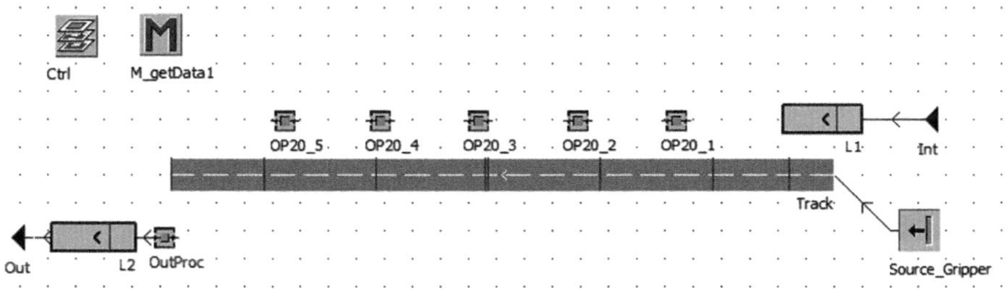

Fig. 10 Simulation model of gantry manipulator

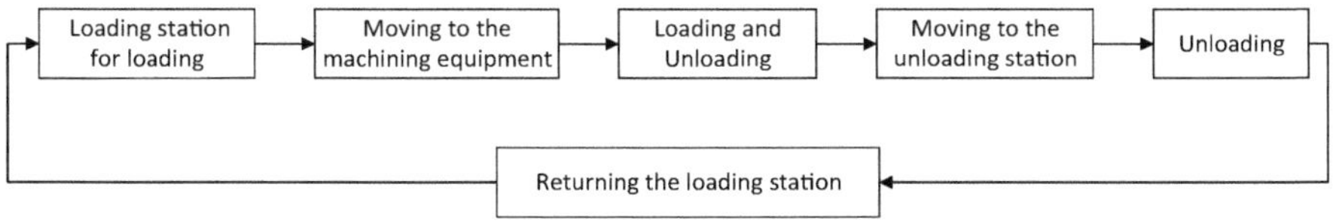

Fig. 11 Working process of the gantry manipulator

Table 5 Cycle time calculation table

Process	X-direction		X-direction return		Z-direction (loading or unloading)
	Distance	Time	Distance	Time	Time
Loading					6.95 s
↓	20500 mm	12.96 s	21400 mm	13.5 s	
First Machine					0 s
↓					
Second Machine					0 s
↓					
Third Machine					0 s
↓					
Fourth Machine					0 s
↓					
Fifth Machine					12.26 s
↓	900 mm	2.68 s			
Unloading					6.68 s
Total Time	15.64 s		13.5 s		25.89 s

Fig. 12 Histogram of utilization rate of gantry manipulator

4.3 *Number of the Zero-Point Clamping and Dummy Head*

The number of the zero-point clamping is the number of online parts multiplied by the corresponding wide release factor, so the number of online parts can be calculated to design the number of the zero-point clamping. At present, there is no accurate statistical calculation for the number of online parts. In the simulation model, the calculation method is to count the number of loading parts on the loading station, the number of unloading parts of the finished product on unloading station and the number of unloading parts unqualified on the middle station of the machining line, and the corresponding calculation logic is written into the control methods of the simulation model. Running the model, the results are shown in Fig. 13. According to the figure, the maximum number of online parts is 435, and we can get the number of zero-point clamping multiplied the maximum number by the value by the allowable factor.

The machining line JPH and the number of online parts of the dummy head section are taken as the index, and the number of the dummy head is taken as the influencing factor. And the number of the dummy head is simulated by designing the experiment manager (shown in Fig. 14) module of the software. By designing the multi-factor test in the experimental module, the number of the dummy was studied. The influence of the number of different dummy head on the number of JPH and online parts is obtained. The simulation results are shown in Fig. 15.

Considering JPH and the number of online parts of the dummy head section, it is recommended that the minimum number of dummy head needs to be 55 for the machining line. Before applying the simulation software, there is no effective calculation method to calculate the above-mentioned digital of the production line, only through calculating the number of the workpiece filling the entire production line. In practice, the production line is not completely filled by the workpiece. Therefore, through the traditional method, we will get a larger number, resulting in a lot of waste and idle. So, application of simulation can solve the problem.

In summary, application of simulation software on the whole line for statistics, we can analyze the data of the production line, predict whether the production line meets the design requirements, find problems, and optimize them. Applying the simulation software programming function, we can analyze the cycle time and the ability of the gantry manipulator, and dynamically evaluate whether the cycle time meets the design value. Applying the simulation software experiment manager, we can match the number, reasonably determine the number of the tray on the production line and reduce unnecessary waste.

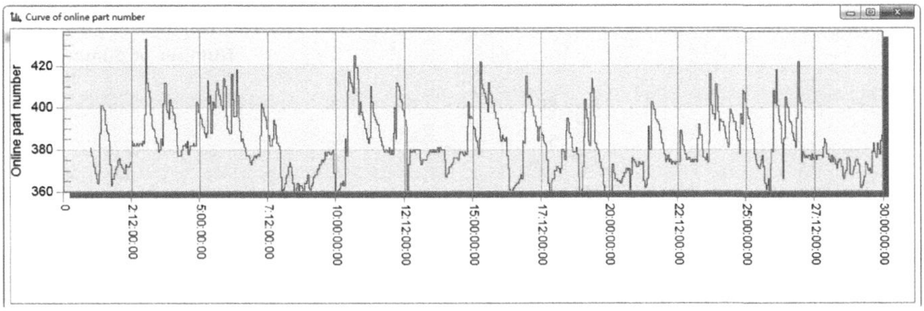

Fig. 13 Online part number curve

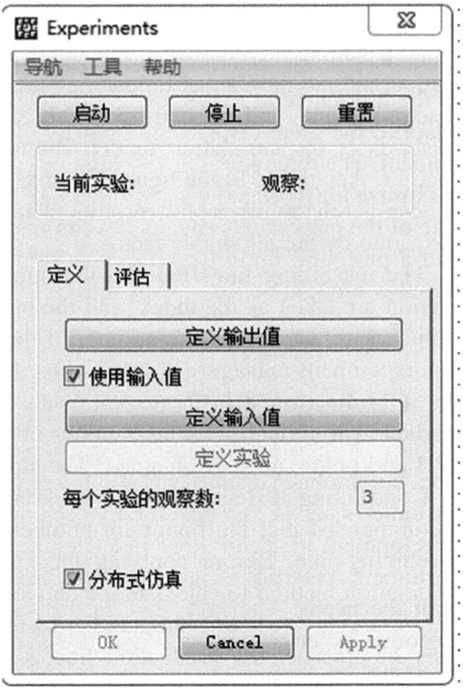

Fig. 14 Manager experiment

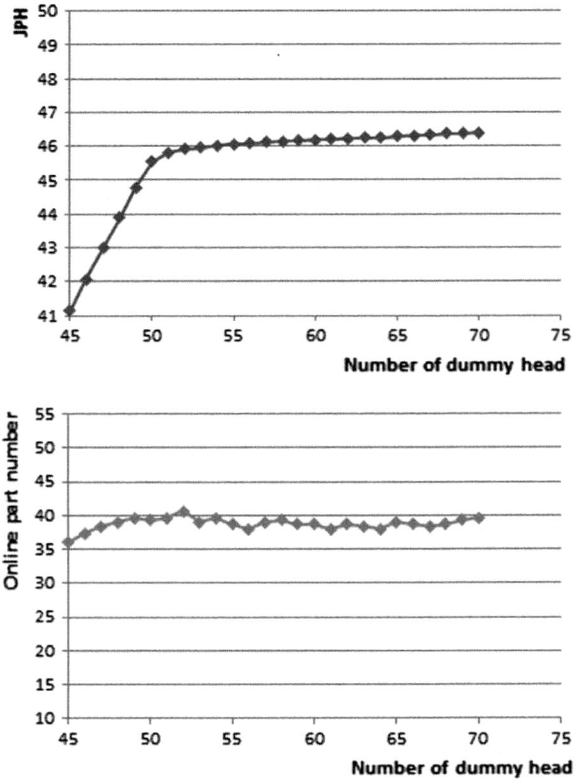

Fig. 15 Number of dummy head trend

5 Conclusions

In this paper, the simulation analysis of the production line is carried out with Plant Simulation as the platform. And the material flow unit, the information flow unit, and the statistical analysis unit which are provided by Plant Simulation are combined with the parameters of the process and the strategy of the engine cylinder block machining line. To propose modeling method and carry out two-dimensional modeling, we achieve the assessment of the capacity of the production line, verification of the results of the simulation, and the measures proposed. Thus, effective evaluation and decision can be presented for the construction of the production line.

References

Journals

1. Minghui Li, Yuqiang Shi, Junjia Wang, Bingwan Cai (2015) Simulation of packaging manufacturing system based on Em-plant. Manuf Autom 37(12):24–27
2. Guanglun Xiong, Xin Wang (1999) Application and development of simulation technology in manufacturing industry. J Syst Simul 3:145–151
3. Jian Mao (2011) Simulation study on production logistics system of automobile parts based on witness. Ind Eng 14(3):125–127

Monograph

4. Zhang X (2003) Modern warehousing logistics technology and equipment. Chemical Industry Press, Beijing, 301–430
5. Jerry B (2008) Discrete time system simulation. In: Xiao T, Fan W (Eds), Biography. Mechanical Industry Press, Beijing, 87–93
6. Zhang X (2000) Modern production logistics and simulation. Qinghua University Press, Beijing, 211–292
7. Xiaoping Zhang, Wei Shi, Yukun Liu (2008) Logistics system simulation. Qinghua University Press, Beijing, pp 205–261

Platform-Based Evolution and Optimization of Vehicle Body in White Using Implicit Modeling Technology

Zhongcai Qiu, Bo Liu, Ke Wang, Jinsheng Zhang, Bo Lu, Huili Yu, Bo Dian, Qin Yang and Xiaoqing Li

1 Introduction

With increasing competition in global automotive market, how to reduce research and design costs, shorten development cycle, develop more reliable products has become key to success for overall OEMS. Under this background, most of OEMS from Europe, Japan, USA, etc., have proposed platform-based development strategy; i.e., diverse vehicles with different appearance, functions, and market positioning are developed from the same platform, including under body, chassis, powertrain. For example, the MQB platform from Volkswagen is shared by sedan (JETTA, etc.) and SUV (TIGUAN, etc.). Since most of components of underbody are shared, the costs of development and manufacturing will be reduced. Moreover, most of CAE and experimental validations of the shared components are eliminated; thus, the development cycle will be shorten remarkably [1–3].

In general, the most advanced and key technologies of an OEM are integrated in its most recent platforms, one of which is lightweight. A vehicle developed from a lightweight platform will have advantages on fuel efficiency, carbon emissions, handling, etc. Thus, considering costs, many lightweight technologies, e.g., structure optimization, advanced bonding process, lightweight material such as hot forming steel, aluminum are applied in MQB from Volkswagen, DELTA from GM, etc.

Massive iterative design optimization, CAE, and experimental validations are very necessary for developing a reliable platform. The key of a platform is modularization of chassis and powertrain, while BIW is a supporter for them. So when

Z. Qiu (✉) · K. Wang · J. Zhang · B. Lu · H. Yu · B. Dian · Q. Yang · X. Li
Chongqing Changan Automobile Co., Ltd, Globe R&D Center,
Chongqing 401120, China
e-mail: qiuzc@changan.com.cn

B. Liu · H. Yu · B. Dian
Chongqing Changan Automobile Co., Ltd, Commercial Vehicle R&D Center,
Chongqing 400023, China

Society of Automotive Engineers of China (SAE-China) (ed.), *Proceedings of the 19th Asia Pacific Automotive Engineering Conference & SAE-China Congress 2017: Selected Papers*, Lecture Notes in Electrical Engineering 486, https://doi.org/10.1007/978-981-10-8506-2_13

schemes of chassis or powertrain are changed frequently, the BIW structure should be changed correspondingly. In addition, the performance of the changed BIW, e.g., safety, NVH, needs fast assessment. If CAE analysis is advanced to early development phase without CAD data, structural design validations and optimization could be enabled. Thus, more schemes of the platform could be validated and the BIW structure will be more reliable and has a higher lightweight level [4, 5].

SFE CONCEPT is an implicit modeling software integrated CAD and CAE. SFE CONCEPT model is controlled by the position of influence points, curvature of link lines and shape of base sections. These geometry elements are assembled by mapping relations, as shown below.

All geometry parameters are interrelated logically. Hence, once a parameter is changed, all parameters associated with it will be changed correspondingly and the finite element mesh will be generated automatically. During platform-based vehicle development process, the BIW could be constructed efficiently with SFE CONCEPT. Furthermore, major performance, e.g., safety, NVH, could be assessed quickly. For example, SFE CONCEPT parametric modeling without CAD data enabled development of the new MAZDA 2 (launched in 2014) to be advanced by 8 months. In this paper, a SUV BIW is established using SFE CONCEPT. After that, according to platform-based evolution rules specified by Changan Automobile Co., Ltd, the SUV BIW is evolved into MPV BIW and Sedan BIW respectively. Then, the evolved BIWs are optimized to have a higher lightweight level comparing with the reference vehicles which have the same geometry dimension.

2 Implicit Modeling of BIW of Platform Basic Vehicle

For a long time, CAE analysis must be carried out after complete detailed CAD data are finished. Moreover, it is time-consuming to clear up some unnecessary CAD details for CAE model preparing. Fortunately, even without detailed CAD data, all CAD schemes could be constructed and validated fast with SFE CONCEPT.

2.1 Theory of SFE CONCEPT

SFE CONCEPT model is based on implicit parametric modeling technology. When the parametric of influence points, curvature of link lines, and shape of base sections is changed, all associated geometry will be changed correspondingly and the finite element mesh could be generated automatically, which enables any complicated CAD schemes changed easily. Since using implicit parametric modeling and multiflange autoidentification technology, high quality of finite element mesh could be generated with SFE concept [6]. Hence, more design space is available and more schemes could be validated in early design phase so that the body structure will have better performance, high lightweight level, and low costs. Moreover, less or no design changes will occur during later phase due to lack of performance. Besides, shape parametric for design of experiment (DOE) and multidiscipline optimization can be provided by SFE CONCEPT model.

2.2 Modeling of BIW of Platform Vehicle

A compact SUV with 2580 mm wheelbase is selected as the platform basic vehicle. New vehicles will derive from this SUV. In general, wheelbase, wheel-track, front and rear overhang of the platform should be adjustable for different evolved vehicles from it, as Fig. 1 shows. For underbody of the BIW, to satisfy these requirements, the length of front crash boxes, heel kick, and the rear end of rear floor rail needs to be changed. Thus, the number of shared underbody components can be maximized, as Fig. 2 shows. To fulfill these changes flexibly and generate CAD/CAE data correctly, the BIW SFE COPNCEPT model should be divided into several modules as Fig. 3 shows. The cut-off locations marked with red lines share the same influence points and base sections. The divided modules are defined as some logical groups, some of which are fixed, while the others are flexible. So the

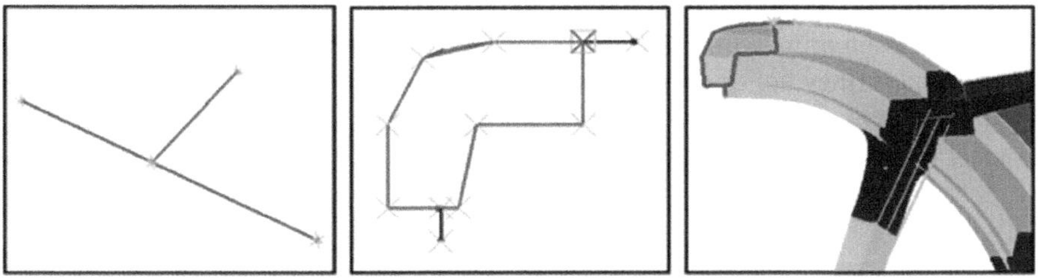

Fig. 1 Structure parameterize description

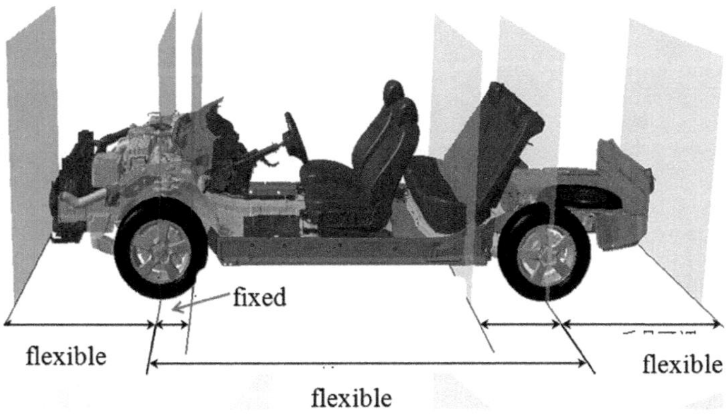

Fig. 2 Flexible requirements for the platform

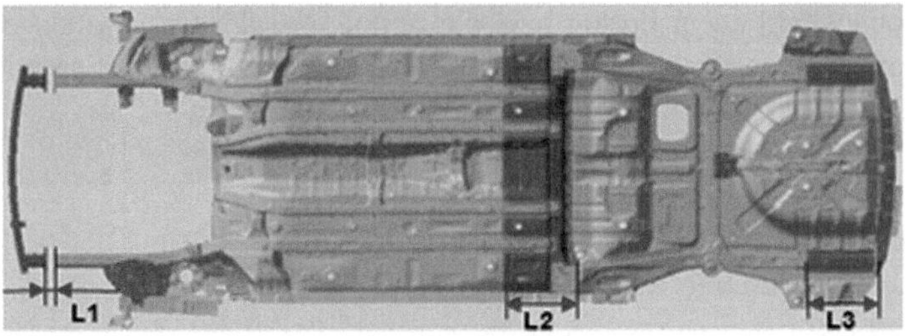

Fig. 3 Adjustable underbody components for platform

BIW dimension can be changed through translating coordinates of these flexible logical groups quickly. Generally, less than 5 min is needed for the whole process, which is far more efficient than traditional CAD/CAE method (at least 1 week). High quality mesh will be generated within extra 5 min automatically.

3 Platform-Based Evolution of BIW from SUV to MPV and Optimization

Generally, comparing with SUV, MPV has larger roominess such as RAV4 contrast with Alphard (they are developed on the same platform, i.e., TNGA from Toyota). In addition, the rear two rows of seats can slide forward and backward. Therefore, to meet these requirements, wheelbase of the SUV is lengthening by 180 mm and

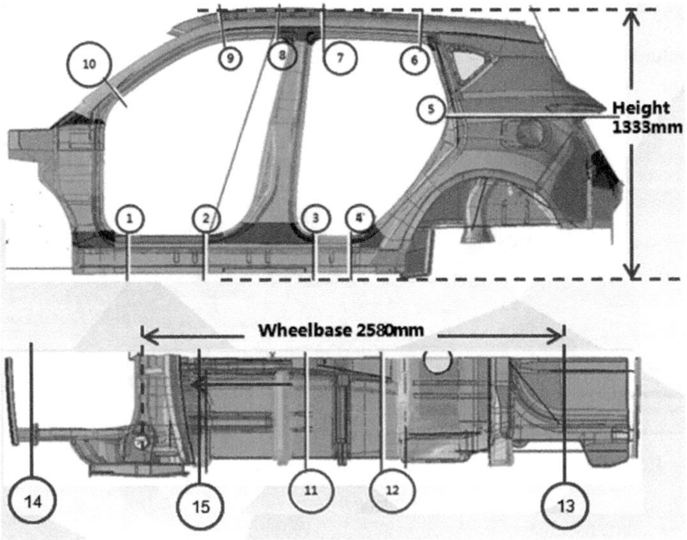

Fig. 4 Divided modules of SFE CONCEPT BIW model

the height is increased by 50 mm, as Fig. 4 shows. In addition, the backdoor frame is leaned back by 50 mm. And then, the evolved MVP BIW has the same dimension as the reference MPV, M1.

To enable the rear two rows of seats slide forward and backward, the heel kick and the rear floor panel where the spare tire is placed should be flatten. Besides, the upper cross member on the rear floor also should be removed, as Fig. 5 shows, which will cause rigidity of BIW decrease.

Finally, an initial MPV BIW is achieved. The whole evolution process only needs two weeks. Afterward, stiffness and modal analysis is conducted for the evolved MPV BIW and the result is shown in Table 1, where T and B represent for torsion and bending. The units of mass, frequency, bending stiffness, and torsion stiffness are Kg, Hz, N/mm and KN.m/rad, respectively.

According to Changan's specification of NVH targets, torsion stiff of BIW with similar size as M1 should be above 1100 KN.m/rad. While bending stiff, first torsion, and bending freq should be above 10000 N/mm, 35, 45 Hz, respectively. From Table 1, it can be seen that bending stiffness is less than the target value and the first bending freq decreases too much comparing with the SUV. Thus, the evolved MPV has to be optimized. As the definition of strain energy U shows in

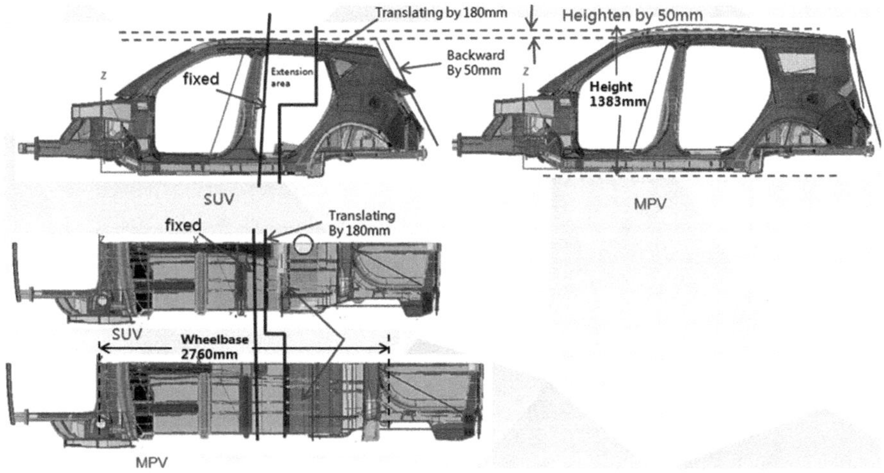

Fig. 5 Dimension changes from SUV to MPV

Table 1 Stiffness and frequency of the evolved MPV BIW

	SUV	Evolved MPV	M1
Mass	321.8	333.5	385.2
First T freq	40.83	38.73	32.51
First B freq	53.40	48.23	44.56
B stiffness	11393	8476	11052
T stiffness	1304	1102	946

Eq. 1, if the strain energy of an area is high, it means that the stress σ or train ε in the area is high and the stiffness of the area is weak. So the area needs to be strengthened.

$$U = \int_V \delta\sigma^T \varepsilon \mathrm{d}V \tag{1}$$

On the basis of strain energy distribution as Fig. 6 shows, the rocker, upper end of B pillar, low end of C pillar, side window frame, and backend of rear floor should be strengthened emphatically.

According to Fig. 6, 15 optimization schemes, including adjusting location of beams, optimizing section shape, adding reinforcement and ribs, thickening beams, are carried out, as Fig. 7 shows. And then, optimization result is shown in Table 2 (Fig. 8).

As Table 2 shows, after optimization, both stiffness and frequency reach beyond target values. But the mass increases by 14.9 kg. Therefore, 19 lightweight schemes, including removing unnecessary parts, reducing thickness, and optimizing section shape, are carried out, as Fig. 4 shows. Through lightweight schemes, there is a decrease in weight by 13.8 kg. Also, the stiffness and frequency are as shown in Table 3.

To explore lightweight potential of the evolved MPV, multidiscipline parametric optimization (MDO) is conducted using SFE combined with ISIGHT, NASTRAN, etc. Thirteen variables extracted from SFE CONCEPT model of evolved MPV BIW. 100 analysis sample points are generated by Latin Hypercube method.

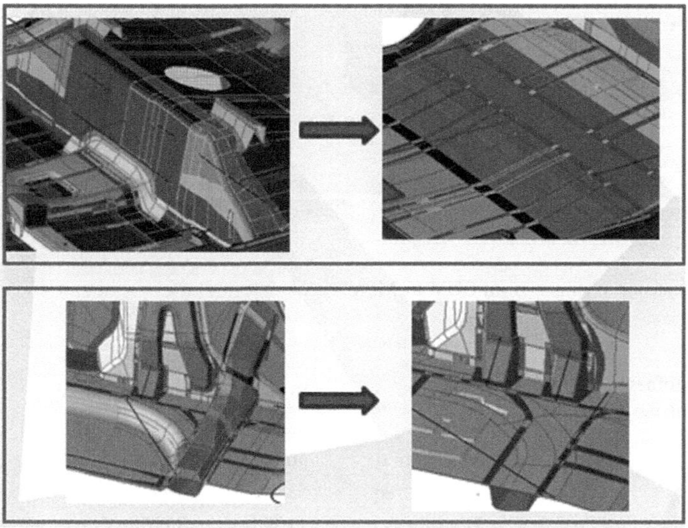

Fig. 6 Local structure changes from SUV to MPV

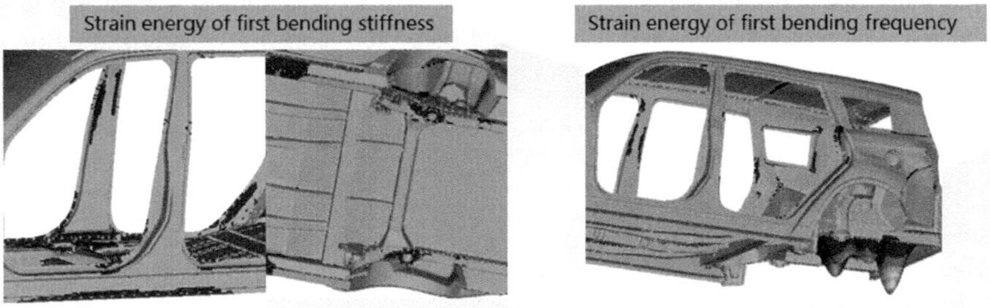

Fig. 7 Strain energy distribution of evolved MPV

Table 2 Optimization results of evolved MPV

	Evolved MPV	Optimization	Ratio	Target	M1
Mass	333.5	348.4	4.5%↑	/	385.2
First T freq	38.73	38.21	1.3%↓	35	32.51
First B freq	48.23	49.90	3.5%↑	45	44.56
B stiffness	8476	10228	20.7%↑	10000	11052
T stiffness	1102	1125	2.1%↑	1100	946

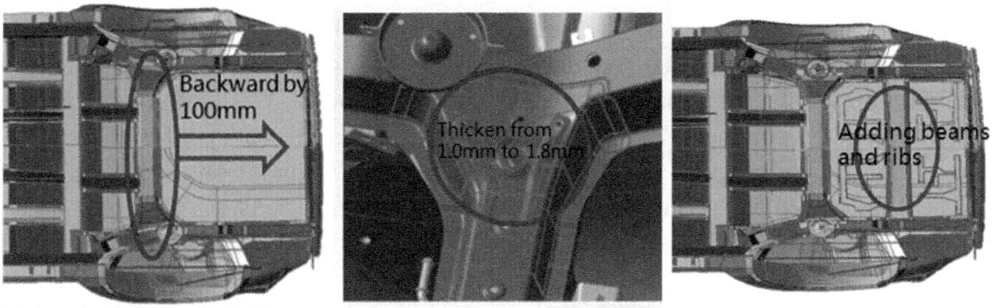

Fig. 8 Partial optimization schemes of the evolved MPV

Table 3 Results of evolved MPV after lightweight

	Evolved MPV	Optimization	Lightweight	Target	M1
Mass	333.5	348.4	334.6	/	385.2
First T freq	38.73	38.21	38.22	35	32.51
First B freq	48.23	49.90	49.95	45	44.56
B stiffness	8476	10228	10063	10000	11052
T stiffness	1102	1125	1069	1100	946

And then, based on these 100 points, response surfaces, including weight, stiffness, and frequency, are established using radial basic function (RBF) approximation as Eq. 2 shows.

$$F(x) = \sum_{p=1}^{p} d_p \|X - X^P\| \tag{2}$$

where X is variable X_1, X_2 … X_m, P is the number of sample points, $P > m$, d_p is the radial from X_m to X_p.

Set stiffness, frequency as constraints and weight as targets, MDO is run with mixed integer sequential quadratic programming (MISQP). After MDO, the weight reduces by 1.8 kg furthermore. Also, the stiffness and frequency are as shown in Table 4. As seen from Table 4, the BIW of the evolved MVP after MDO is lighter than M1 by 42.4 kg, while stiffness and frequency are almost the same. It is emphasized that all optimization, including lightweight and MDO, also only takes two weeks.

4 Platform-Based Evolution of BIW from SUV to Sedan and Optimization

Usually, comparing with SUV, sedan has lower body height larger luggage roominess which means that the rear overhang of sedan is longer. Thus, to meet these requirements, wheelbase and rear overhang of the SUV is lengthening by 105 and 100 mm, respectively. Meanwhile, the height is decreased by 80 mm, as Fig. 4 shows. And then, a sedan BIW, which has the same dimension as the reference sedan, M2, is evolved from the USV, as Fig. 9 shows.

The different modules between sedan and SUV, including rear upper part of body-side, skirts, package tray, etc., are derived from M2. These different modules and common modules with SUV are assembled together. Hence, a new sedan is generated, as Fig. 10 shows.

The whole evolution process only needs two weeks. Afterward, stiffness and modal analysis is conducted for the evolved sedan BIW and the result is shown ins Table 5.

Table 4 Results of evolved MPV after lightweight

	Evolved MPV	Lightweight	MDO	Target	M1
Mass	333.5	334.6	332.8	/	385.2
First T freq	38.73	38.22	38.60	35	32.51
First B freq	48.23	49.95	49.66	45	44.56
B stiffness	8476	10063	10130	10000	11052
T stiffness	1102	1069	1089	1100	946

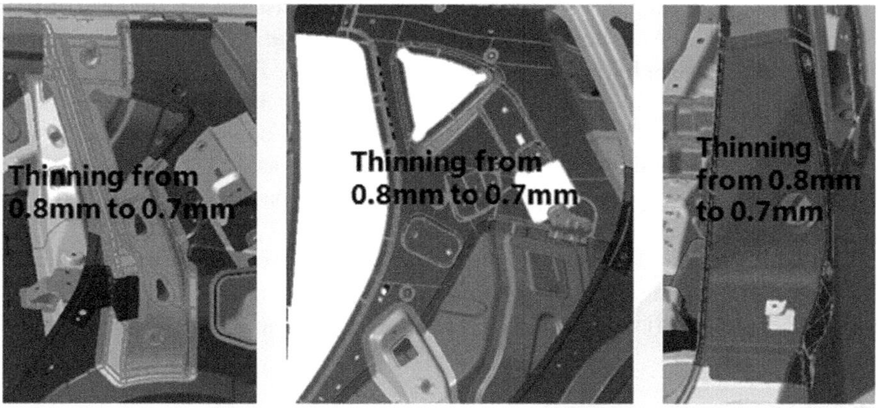

Fig. 9 Partial lightweight schemes of the evolved MPV

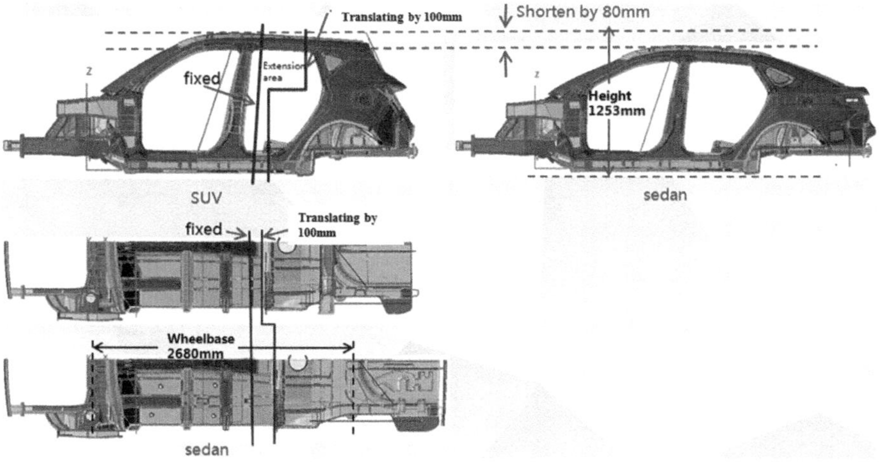

Fig. 10 Dimension changes from SUV to sedan

Table 5 Stiffness and frequency of the evolved sedan BIW

	SUV	Evolved sedan	M2
Mass	321.8	327.5	328.1
First T freq	40.83	48.28	47.6
First B freq	53.40	54.67	62
B stiffness	11393	10070	12121
T stiffness	1304	1228	1248

According to Changan's specification of NVH targets, torsion stiffness of BIW with similar size as M2 should be above 1100 KN.m/rad. While bending stiffness, first torsion, and bending frequency should be above 10000 N/mm, 45, 50 Hz, respectively. From Table 5, it can be seen that the first bending frequency is less than the target value. Thus, the evolved sedan has to be optimized. On the basis of strain energy distribution as Fig. 11 shows, the cowl assembly, upper end of B pillar, upper end of C pillar, mapping location between package tray, and body-side should be strengthened emphatically (Figure 12).

According to Fig. 11, seven optimization schemes, including adjusting location of beams, optimizing section shape, adding reinforcement, thickening beams, adding ribs, are carried out, as Fig. 11 shows. And then, optimization result is shown in Table 6.

Fig. 11 Evolution process of a new sedan

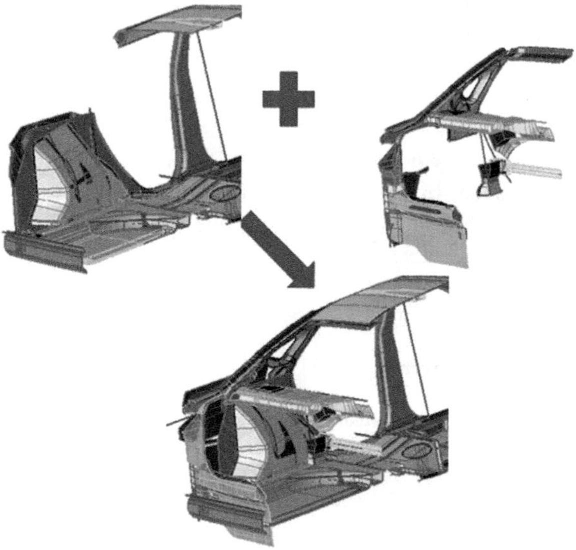

Fig. 12 Strain energy distribution of evolved sedan

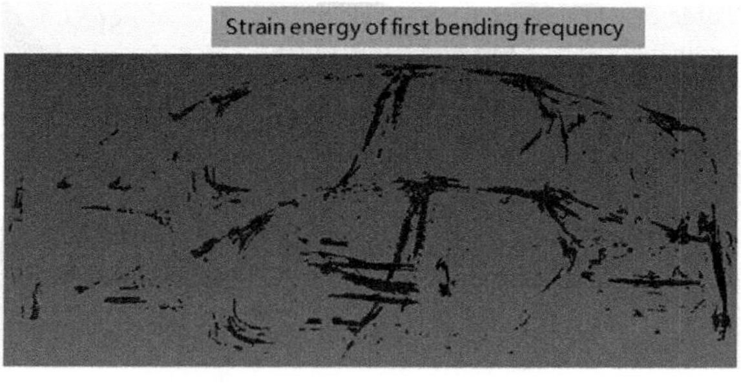

Table 6 Optimization results of evolved sedan

	SUV	Evolved sedan	Optimization	M2
Mass	321.8	327.5	332.3	328.1
First T freq	40.83	48.28	48.55	47.6
First B freq	53.40	54.67	60.85	62
B stiffness	11393	10070	11270	12121
T stiffness	1304	1228	1284	1248

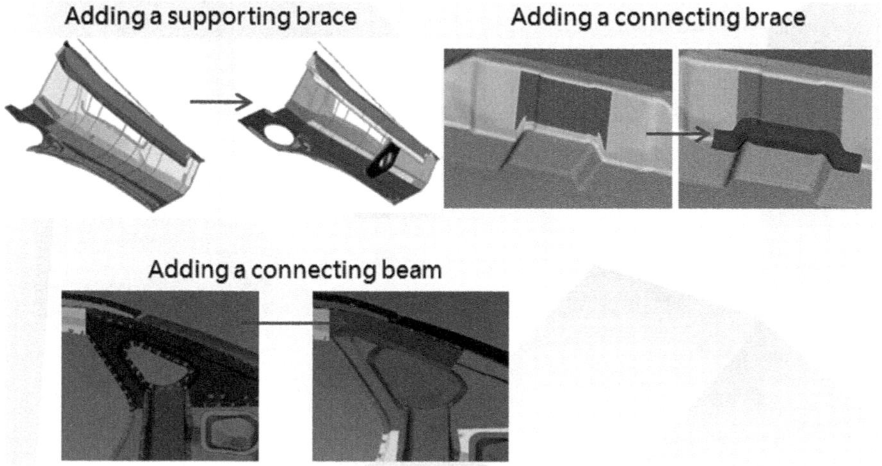

Fig. 13 Partial optimization schemes of the evolved sedan

As Table 6 shows, after optimization, the first bending frequency reaches beyond 60 Hz. But the weight increases by 4.8 kg and the evolved sedan is heavier than M2 by 4.2 kg. Therefore, 25 lightweight schemes including removing unnecessary parts, reducing thickness, and optimizing section shape are carried out, as Fig. 13 shows. Through lightweight schemes, there is a decrease in weight by 13.6 kg and the evolved sedan is lighter than M2 by 7.1 kg. Also, the stiffness and frequency are as shown in Table 7. It is emphasized that all optimization including lightweight also only takes two weeks (Fig. 14).

Table 7 Results of evolved sedan after lightweight

	SUV	Evolved sedan	Optimization	Lightweight	M2
Mass	321.8	327.5	332.3	318.7	328.1
First T freq	40.83	48.28	48.55	49.19	47.6
First B freq	53.40	54.67	60.85	60.18	62
B stiffness	11393	10070	11270	10796	12121
T stiffness	1304	1228	1284	1264	1248

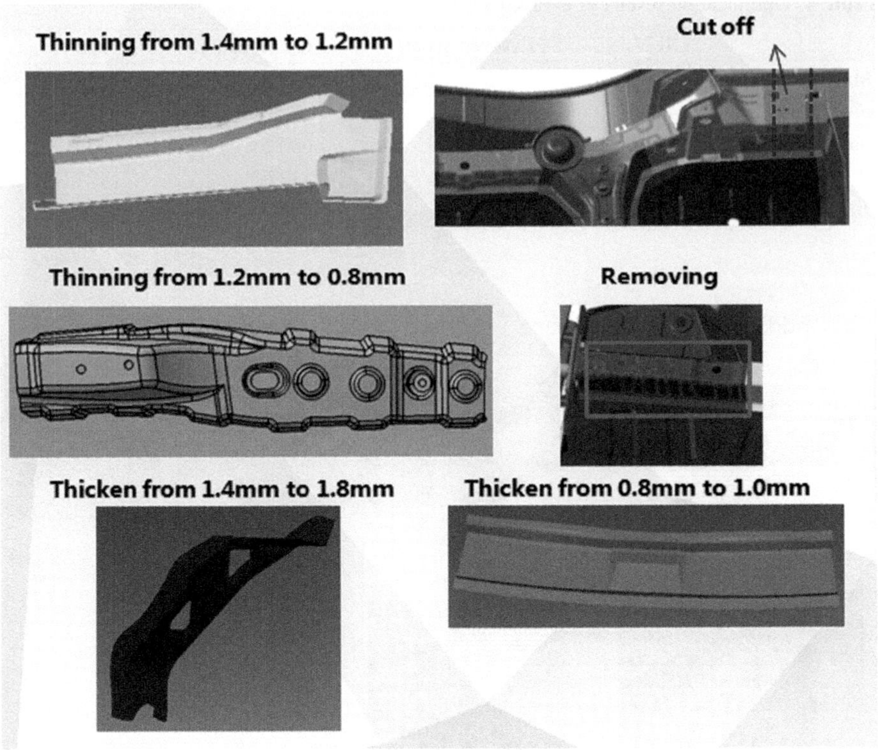

Fig. 14 Partial lightweight schemes of the evolved sedan

5 Conclusions

1. According to platform-based rules, A SUV BIW is evolved into A MPV and A sedan with implicit modeling software SFE CONCEPT. It is proved that SFE CONCEPT has higher efficiency than traditional CAD/CAE software.
2. After fast optimization, the evolved BIW has a higher lightweight level comparing with the reference vehicles which have the same geometry dimension.
3. During platform development phase, with the same or less time, more optimization design could be validated through SFE CONCEPT model. Thus, the developed vehicle will be more reliable and has a higher lightweight level.

Acknowledgements The authors sincerely thank the financial supports from The National Key Research and Development Program of China (No:2016YFB0101703, No:2016YFB0101601, No:2016YFB0101604, No:2016YFB0101606,) and Chongqing Research Program of Distinguished Youth Foundation (NO:cstc2014jcyjjq50002).

References

1. Xiaofeng Ju (2012) Study of BIW platform develop strategy. Automobile Technol 2:7–10
2. Shen J, Wang D (2013) Study on integrated development and application of vehicle body platform architecture. Automobile Technol 1:34–37
3. Yuan H, Chen D, et al (2015) Previous optimization of car body based on platform and modularization. J Mech Electr Eng 32(9):1181–1185, 1217
4. Donald R, Jones GM (2001) A taxonomy of global optimization based on response surfaces. J Global Optim 21(4):345–383
5. Rousselet B, Hang EJ (1983) Design sensitivity analysis in structural mechanics, III, effects of shape variation. J Struct Mech 10(3):273–310
6. Guohong SHI, Jinyan WU, Zhengchao Song (2014) Optimization design in developing a vehicle body in forward engineering. J Automotive Saf Energy 5(3):238–243

Preparation Technology of Silicon–Carbon Composite Anode Material Based on Expanded Graphite for Lithium-Ion Battery for Vehicles

Huiming Chen, Tao Jiang, Changru Rong, Dan Wang, Xinyan Mi and Kejin Zhang

1 Introduction

Electric vehicles (EVs) are considered an inexorable trend of ground transportation due to their energy conservation and environmental protection. The power density of batteries becomes growing important. Li-ion batteries are the most successful energy storage devices invented in the past two decades. However, the endurance mileage deficiency of EVs is still the main obstacle which restricts the development of electric vehicle industry, so it is significant to improve the operation energy efficiency of electric vehicle in the case of limited vehicle energy. To meet these requirements, it is necessary to find higher-capacity electrode materials. At present, the most widely used anode material is graphite, whose theoretical capacity is only 372 $mAhg^{-1}$ [1, 2]. In recent years, some new high-capacity anode materials have aroused widespread concern, such as Si [3], Sn [4], Sb [5] and their oxides [6, 7] are promising alternative candidates due to their markedly higher capacities. Especially, silicon has the highest capacity of 4200 $mAhg^{-1}$. Nevertheless, the huge volume expansion in lithiation/de-lithiation caused a sharp drop in the cycling performance of silicon. Therefore, the research of silicon anode is focused on improving the cycle performance. The first promising way is the use of nano-scaled silicon [8]. Another way is to disperse the active silicon particles into one or two elastic matrices or some active materials with low volume change to form a composite structure [9, 10].

In this study, a new term, silicon-rich expanded graphite, was proposed. The silicon-rich expanded graphite was prepared by in situ reduction methods using the silicon source produced by TEOS hydrolysis. The silicon–carbon composite anode material was prepared by compositing silicon-rich expanded graphite with

H. Chen (✉) · T. Jiang · C. Rong · D. Wang · X. Mi · K. Zhang
China FAW Co., Ltd., Changchun, China
e-mail: chenhuiming@faw.com.cn

Society of Automotive Engineers of China (SAE-China) (ed.), *Proceedings of the 19th Asia Pacific Automotive Engineering Conference & SAE-China Congress 2017: Selected Papers*, Lecture Notes in Electrical Engineering 486, https://doi.org/10.1007/978-981-10-8506-2_14

commercial graphite anode. Our results showed that the cyclability of the composite anode material can be significantly improved, and after 100 cycles at 70 mA/g, the capacity retention rate can still be more than 90%.

2 Experimental

The fabricated process of silicon–carbon material contained two steps. The first step was the synthesis of silicon-rich expanded graphite. Then, compositing silicon-rich expanded graphite with commercial graphite anode was the last step. The silicon-rich expanded graphite was fabricated as follows. Prepare 0.75 mol/L ammonia solution. Then mix expanded graphite into the solution. The mixture was stirred slowly at 45 °C for 1.5 h. During this period, the TEOS solution (0.3 wt%) was added into the mixture at a rate of 1 ml/min. The weight ratio of expanded graphite and Si contained in TEOS was 3:1. Dry the moisture in the mixture quickly after TEOS was hydrolyzed completely. Then, we had got the precursor of silicon-rich expanded graphite. Some Mg powder was mixed into the precursor for reductant. The resultant composite precursor was heat-treated at 750 °C for 4 h in an argon atmosphere to obtain the silicon-rich expanded graphite. The products were washed with dilute acid and deionized water. The silicon–carbon composite material was prepared via ball-milling of silicon-rich expanded graphite and commercialized graphite material (weight ratio 1:2).

The phase components of silicon-rich expanded graphite were obtained by X-ray diffraction (XRD) with Cu Kα radiation. The morphology and element distribution of the material were examined using scanning electron microscopy with energy-dispersive X-ray analysis (SEM-EDX). Electrochemical workstation was used to test the electrochemical performance of the materials in the button cell (vs. Li).

3 Results and Discussion

The SEM-EDX photographs of the silicon-rich expanded graphite are shown in Fig. 1. The silicon is homogeneously distributed between the expanded graphite layers. Expandable graphite can provide an excellent buffer space for volume expansion of silicon.

The XRD pattern of the silicon-rich expanded graphite is shown in Fig. 2. There are only the crystalline diffraction peaks of the silicon and graphite. Notably, no other impurity phases are detected. This result shows that the silicon-rich expanded graphite is pure in crystallography. The SiO_2 from hydrolysis of TEOS is completely reduced.

The charge–discharge curves after the first lithium intercalation and cycling test have been shown in Fig. 3. Fig. 3a shows that the charge and discharge capacity are 736.2 and 761.6 mAhg^{-1}. The coulombic efficiency has reached 103.5%. At the

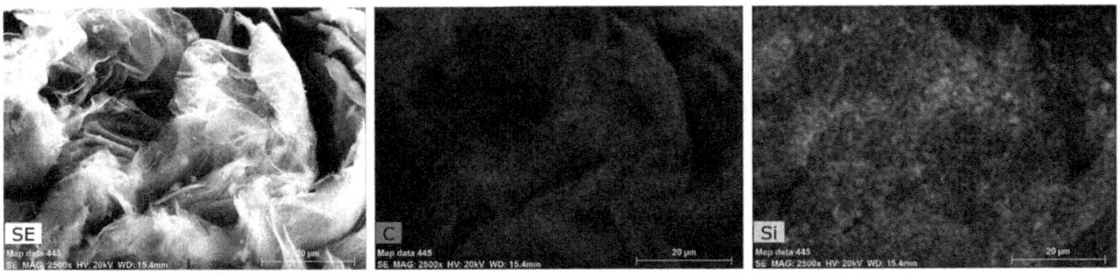

Fig. 1 SEM-EDX photograph of the silicon-rich expanded graphite

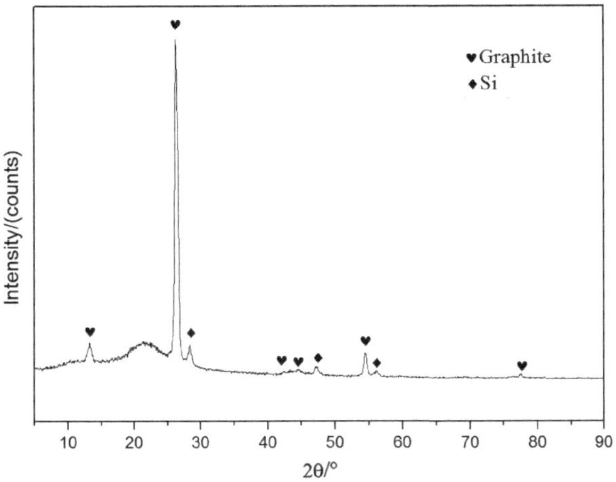

Fig. 2 XRD pattern of the silicon-rich expanded graphite

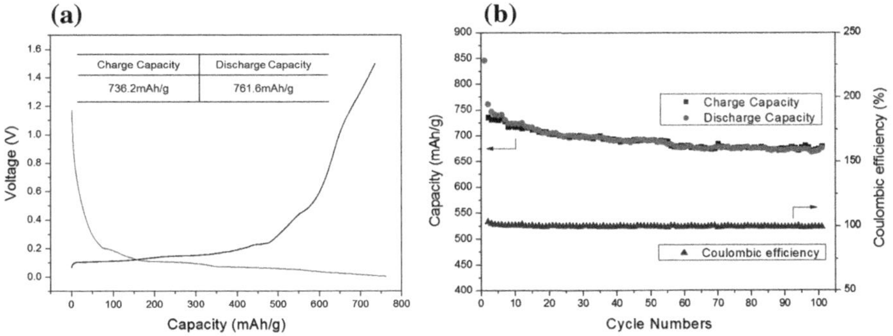

Fig. 3 **a** Voltage versus capacity curves and **b** cycling performance (vs. Li/Li^+) at 70 mA/g at room temperature

initial cycles, the capacity decreased faster. As a consequence of the gradual stability of silicon structure, the capacity has gradually stabilized for the subsequent cycles. In the 100 times cycles, the coulombic efficiency remains between 99 and 103%. The discharge capacity remains 690.2 $mAhg^{-1}$, and the retention efficiency of capacity can keep above 90% at the 100th cycle.

The CV curves of the initial four cycles are shown in Fig. 4. There is a reduction peak near 0.7 V in the curve of first cycle. The irreversible reaction between the electrode material and the electrolyte occurs in this voltage [11]. The peak disappeared in the subsequent cycles. The two reduction peaks at 0.25 and 0.17 V correspond to the phase transformation of Li_xSi [12]. The peaks corresponding to lithiation of graphite can be found below 0.1 V [13]. In the other side, the occurrence of lithium ion from graphite occurs at 0.2 and 0.25 V [14]. The oxidation peaks exist at the corresponding voltage. At the same time, the oxidation peaks between 0.5 and 0.6 V show that the de-lithiation from silicon react at this voltage. The first cycle has a lower current and higher oxidation voltage because of the structural instability of SEI. The peak value of current decreases gradually with the increase of cycle times, which indicates that the silicon structure is stable gradually.

The impedance spectra of the composite anode in de-lithiated state after different cycles are shown in Fig. 5. The impedance profile shows a semicircle in the high-frequency region, another semicircle in the middle frequency region and a sloping straight line in low-frequency region. The SEI film decomposed by electrolytes shows a semicircle at high frequency in the Nyquist plots, and the straight line at low frequency can be attributed to the Li-ion diffusion [15]. The second semicircle at the medium-frequency region shows the insertion of lithium ion into silicon or carbon [16]. The fitting impedance data of the EIS is shown in Table 1 according to the equivalent circuit (top left corner in Fig. 5). The solution resistance (R_S) associated with the concentration of components in the electrolyte solution. R_S is almost constant in 100 cycles. The resistance of passivating film (R_{film}) and charge–transfer resistance (R_{ct}) of the composite anode decrease significantly along

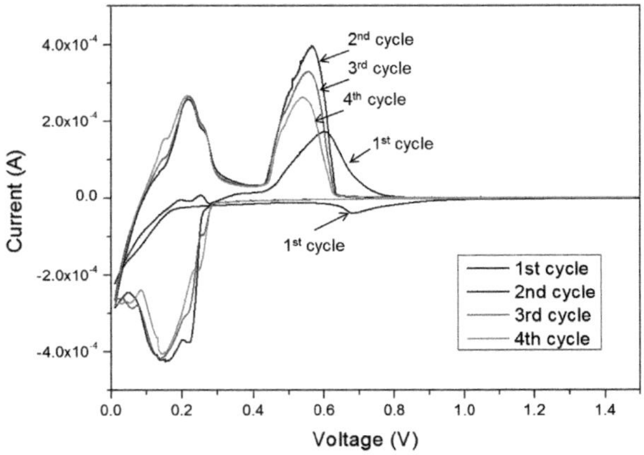

Fig. 4 CV profile of the silicon–carbon composite material

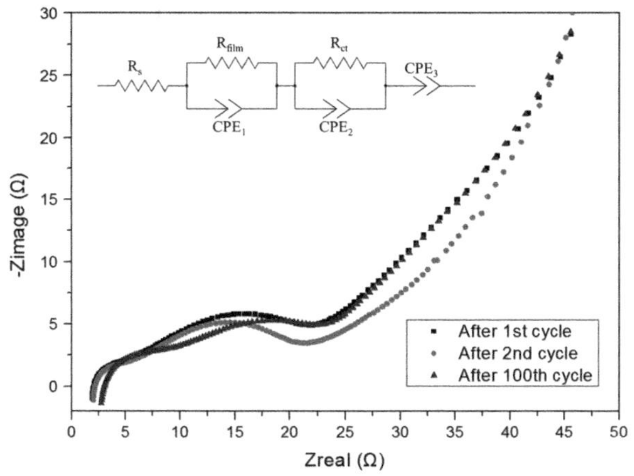

Fig. 5 Nyquist plots of the composite anode in de-lithiated state after different cycles (R_S, ohmic resistance; R_{film}, resistance of passivating film; R_{CT}, charge–transfer resistance; CPE, constant phase element)

Table 1 Fitting impedance data of the EIS according to the equivalent circuit

	After first cycle	After second cycle	After 100th cycle
$R_s(\Omega)$	1.629	2.214	2.887
$R_{film}(\Omega)$	19.93	2.946	4.535
$R_{ct}(\Omega)$	1.12×10^4	15.34	14.54

with cycling. As a result, the transport channel of Li-ion is not completely built in the first cycle. In the subsequent cycles, the structure of the Li-ion channel is gradually stable, and almost unchanged.

4 Conclusions

A concept of silicon-rich expanded graphite was obtained in this paper. The silicon–carbon composite material was prepared via ball-milling of silicon-rich expanded graphite and commercialized graphite material. The composite material has both the high capacity of silicon and the stable cycleability of graphite. In addition, the preparation process is simple and suitable for industrial production. In this research, we got the silicon–carbon composite material with a reversible capacity with the capacity more than 700 mAhg^{-1} for Li-ion batteries for electric vehicles.

References

1. Wu Y, Jiang C, Wan C, Tsucida E (2000) Effects of catalytic oxidation on the electrochemical performance of common natural graphite as an anode material for lithium ion batteries. Electrochem Commun 2(4):272–275
2. Johnson BA, White RE (1998) Characterization of commercially available lithium-ion batteries. J Power Sources 70(1):48–54
3. Kasavajjula U, Wang C, Appleby AJ (2007) Nano- and bulk-silicon-based insertion anodes for lithium-ion secondary cells. J Power Sources 163(2):1003–1039
4. Dou P, Cao Z, Wang C, Zheng J, Xu X (2017) Ultrafine Sn nanoparticles embedded in shell of N-doped hollow carbon spheres as high rate anode for lithium-ion batteries. J Power Sources 404(15):342–349
5. Zhang F, Liu S, Wang J, Du Y, Sun L (2017) Experimental investigation and thermodynamic assessment of the Li-Sb system. Calphad 57:28–36
6. Yuan X, Xin H, Qin X, Li X, Liu Y, Guo H (2015) Self-assembly of SiO/Reduced Graphene Oxide composite as high-performance anode materials for Li-ion batteries. Electrochim Acta 155(10):251–256
7. Tian Q, Tian Y, Zhang W, Huang J, Zhang Z, Yang L (2017) Impressive lithium storage of SnO_2@TiO_2 nanospheres with a yolk-like core derived from self-assembled SnO_2 nanoparticles. J Alloy Compd 702(25):99–105
8. Chan Candace K, Peng Hailin, Liu Gao, McIlwrath Kevin, Zhang Xiao Feng, Huggins Robert A, Cui Yi (2008) High-Performance Li battery Anodes using silicon nanowires. Nature Nanotech 3(1):31–35
9. Yoshio M, Wang H, Fukuda K, Umeno T, Dimov N, Ogumi Z (2002) Carbon-Coated Si as a lithium-ion battery anode material. J Electrochem Soc 149(12):A1598–A1603
10. Hasegawa T, Mukai SR, Shirato Y, Tamon H (2004) Preparation of carbon gel microspheres containing silicon powder for lithium ion battery anodes. Carbon 42(12–13):2573–2579
11. Wu Junxiong, Qin Xianying, Liang Gemeng, et al (2016) A binder-free web-like silicon-carbon nanofiber-graphene hybrid member for use as the anode of a lithium-ion battery. New Carbon Mater 31(3):321–327
12. Huang X, Yang J, Mao S, Chang J et al (2014) Controllable synthesis of hollow Si anode for long-cycle-life lithium-ion batteries. Adv Mater 26(25):4326–4332
13. Wang C, Zhao H, Wang J, Wang J, Lv P (2012) Electrochemical performance of modified artificial graphite as anode materia for lithium ion batteries. Ionics 19(2):221–226
14. Dong A, Shen L, Qiao Y (2014) Preparation and performance of Si-carbon composite. Chinese Battery Ind 13(3):119–122
15. Li H, Huang X, Chen L (1999) Electrochemical impedance spectroscopy study of SnO and nano-SnO anodes in lithium rechargeable batteries. J Power Sources 81–82:340–345
16. Zuo P, Yin G, Ma Y (2007) Electrochemical stability of silicon/carbon composite anode for lithium ion batteries. Electrochim Acta 52:4878–4883

Robust Design of Active Vehicle Interior Noise Control System Based on Adaptive Notch Filter

Yinglong Zhao, Lijun Zhang, Dejian Meng and Pinjie Zhang

1 Introduction

Active noise control (ANC) technology, based on sound wave interference theory, is better than passive noise control in reduction of low-frequency noise, which mainly comes from engine harmonic operation under daily driving conditions [1]. Adaptive notch filter is quite often used in ANC system and has significant effect of reducing single-frequency noise with small computation load. In ANC system using adaptive notch filter, the secondary path is often regarded as a pure delay [2] and only its amplitude–frequency and phase–frequency characteristics [3] are considered when the performance of adaptive notch filter is estimated. This is not quite suitable for complicated acoustic system in vehicle cabins. The variety of driving conditions causes the potential modelling mismatch of secondary path, which might result in robust problems of ANC system.

There are usually two ways to deal with the adverse effects of modelling mismatch in the secondary path. The first way is to model the secondary path online [2, 4, 5] so that the model can make appropriate real-time changes, but the complexity of the algorithm increases largely and extra stochastic noise might also appear. The other way is to develop model library [6], corresponding to various possible circumstances of secondary path; therefore, extra sensors are demanded and acquiring signals from window motor or ECU are essential. In view of the defect of these two methods above, robust design of ANC system was proposed in this paper aiming at optimizing key parameters in adaptive notch filtering algorithm, without increasing extra online computation or extra sensors.

Y. Zhao (✉) · L. Zhang · D. Meng (✉) · P. Zhang
School of Automotive Studies, Tongji University, Shanghai 201804, China
e-mail: Lukaslong@163.com

D. Meng
e-mail: mdj0218@163.com

Society of Automotive Engineers of China (SAE-China) (ed.), *Proceedings of the 19th Asia Pacific Automotive Engineering Conference & SAE-China Congress 2017: Selected Papers*, Lecture Notes in Electrical Engineering 486, https://doi.org/10.1007/978-981-10-8506-2_15

2 Performance Prediction of Single-Channel Adaptive Notch Filter

This study adopted an adaptive notch filter, which consists of a notch filter and adaptive algorithm. The reference signal is filtered through the notch filter, and secondary acoustic signal is then generated. Using the adaptive algorithm, the coefficient of notch filter is adjusted adaptively so that the notch filter can keep updating with the change of noise. The notch filter is low order and simple structured so that the sinusoidal signal of a specific frequency can easily be synthesized, which is suitable for eliminating the harmonic noise from the engine. Besides, adopting the LMS algorithm is of many advantages, including the simplified algorithm, less computation, and better convergence characteristic.

The ANC system in our study was equipped with one microphone and one loudspeaker, and the secondary path could be expressed as a (K+1)-order finite impulse response (FIR) filter. The system was required to simultaneously eliminate the noise in O different frequencies. The block diagram of the system is shown in Fig. 1, and the system equations are shown in Eq. 1.

In the diagram, $\boldsymbol{x}$ is the reference signal, y is the output of the ANC system, $\boldsymbol{H}_s$ is transfer function of the secondary path, d is the primary signal, and e is error signal obtained from the microphone.

$$\begin{aligned} &\boldsymbol{x} = [\sin\omega_1 nt, \cos\omega_1 nt, \ldots, \sin\omega_O nt, \cos\omega_O nt]^T \\ &y(n) = \boldsymbol{x}^T(n)\boldsymbol{w}_{2O}(n), \quad s = \sum_{i=0}^{K} h_{si}\boldsymbol{y}(n-i) \\ &d(n) = \boldsymbol{x}^T\boldsymbol{A}_{2O}, \qquad \boldsymbol{R}_{2O}(n) = \sum_{i=0}^{K} h_{si}\boldsymbol{x}(n-i) \\ &e(n) = d(n) + s(n), \quad \boldsymbol{w}_{2O}(n+1) = \boldsymbol{w}(n) - 2\mu e(n)\boldsymbol{R}_{2O}(n) \end{aligned} \tag{1}$$

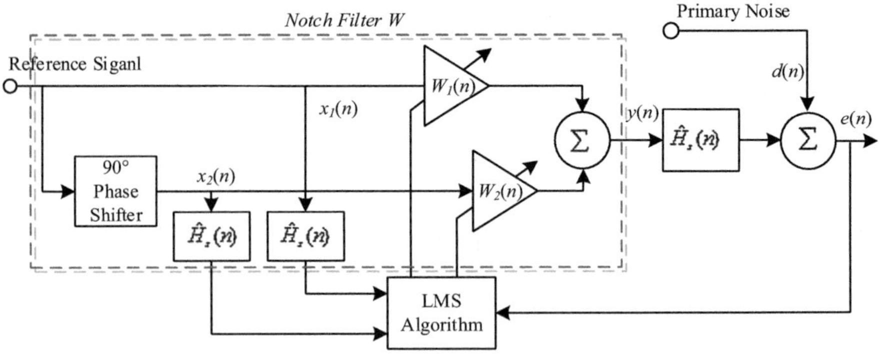

Fig. 1 Block diagram of the single-channel adaptive notch filter

where $w = [w_1, w_2, \ldots, w_{2O-1}, w_{2O}]^T$ is the weight coefficient adjusted in the adaptive algorithm and h_{si} represents the ith coefficient of the FIR sequence of the transfer function $\boldsymbol{H}_s$.

The rotation matrix is created as follows:

$$
\boldsymbol{c} = \begin{bmatrix} \begin{matrix} \cos\omega_1 t & \sin\omega_1 t \\ -\sin\omega_1 t & \cos\omega_1 t \end{matrix} & 0 & 0 \\ 0 & \ddots & 0 \\ 0 & 0 & \begin{matrix} \cos\omega_O t & \sin\omega_O t \\ -\sin\omega_O t & \cos\omega_O t \end{matrix} \end{bmatrix} \tag{2}
$$

Then, the object function J of the adaptive notch filter can be expressed as follows:

$$
\begin{aligned}
J = E[e^2(n)] &= E[\boldsymbol{A}^T\boldsymbol{xx}^T\boldsymbol{A} + 2\sum_{i=0}^{L} h_{si}\boldsymbol{w}^T\boldsymbol{c}^{-i}\boldsymbol{xx}^T\boldsymbol{A} + \sum_{i=0}^{L}\sum_{j=0}^{L} h_{si}h_{sj}\boldsymbol{w}^T\boldsymbol{c}^{-i}\boldsymbol{xx}^T\boldsymbol{c}^{j}\boldsymbol{w}] \\
&= \frac{1}{2}\left\|\boldsymbol{A} + \boldsymbol{A}_{\mathbf{mp}}^T\boldsymbol{w}\right\|_2
\end{aligned} \tag{3}
$$

where $\boldsymbol{A}_{\mathrm{mp}} = \sum_{i=0}^{K} h_{si}\boldsymbol{c}^{-i}$ contains all the Fourier transform results of the secondary path. From Eq. 3, it can be shown that when:

$$
\boldsymbol{w} = \boldsymbol{w}_{\mathrm{best}} = -\left(\boldsymbol{A}_{\mathbf{mp}}^T\right)^+\boldsymbol{A} \tag{4}
$$

J reaches its minimal value. Specially, when $\boldsymbol{A}_{\mathbf{mp}}^T\left(\boldsymbol{A}_{\mathbf{mp}}^T\right)^+\boldsymbol{A} = \boldsymbol{A}$, $J = 0$. And $\boldsymbol{A}_{\mathbf{si}} = h_{si}\boldsymbol{c}^{-i}$ is defined in order to simplify the equation. Furthermore, it is specified that $\Delta\boldsymbol{w} = \boldsymbol{w} - \boldsymbol{w}_{\mathrm{best}}$. Therefore, the iteration formula of weight vector in Eq. (1) can be written as follows:

$$
\begin{aligned}
\Delta\boldsymbol{w}(n+1) = \Delta\boldsymbol{w}(n) &- 2\mu\boldsymbol{A}_{mp}\boldsymbol{xx}^T\sum_{i=1}^{L}\boldsymbol{A}_{si}^T\Delta\boldsymbol{w}(n-i) \\
&- 2\mu\boldsymbol{A}_{mp}\boldsymbol{xx}^T\left(\sum_{i=1}^{L}\boldsymbol{A}_{si}^T\boldsymbol{w}_{best} + \boldsymbol{A}\right)
\end{aligned} \tag{5}
$$

If $J = 0$, the last term in Eq. 5 becomes 0. Even if $J = 0$, it can be observed from simulations that the last term can also be neglected. Additionally, we define $\boldsymbol{W} = [\Delta\boldsymbol{w}^T(n), \ \ldots, \ \Delta\boldsymbol{w}^T(n-K)]^T$, and the preceding Eq. 5 can be expressed as follows:

$$W(n+1) = \tilde{C}^n \tilde{A} \tilde{C}^{-n} W(n) \tag{6}$$

where $\widetilde{A} = \widetilde{A}_1 - 2\mu \widetilde{A}_2$, and

$$\tilde{A}_1 = \begin{bmatrix} E_{2O\times 2O} & 0 \\ E_{2OK\times 2OK} & 0 \end{bmatrix}, \tilde{A}_2 = \begin{bmatrix} A_{\mathrm{mp}} x x^T A_{s0}^T & \cdots & A_{\mathrm{mp}} x x^T A_{sK}^T \\ & 0 & \end{bmatrix}, \tilde{C} = \begin{bmatrix} c & & \\ & \ddots & \\ & & c \end{bmatrix} \tag{7}$$

Equation 6 is the kinetic equation of the single-channel adaptive notch filter. Convergence factor α can be used to evaluate the convergence performance as follows:

$$\alpha = -f_s \cdot \ln \lambda \tag{8}$$

where f_s is the sampling frequency of ANC system and λ is the spectral radius of $\widetilde{A}\widetilde{C}^T$.

When $\alpha > 0$, the system is convergent, and as α becomes bigger, the system converges faster, while $\alpha < 0$ indicates that the system is unable to converge.

3 Secondary Path's Modelling Mismatch

Since there are a variety of driving conditions, the interior acoustic environment is quite likely to change a lot, while the model of secondary path does not make corresponding adjustment in time. This causes error between model of secondary path and real circumstance, which is called modelling mismatch of secondary path. The factors causing this modelling mismatch include different states of windows' opening or closing, which is investigated in this paper.

3.1 Model of Secondary Path

As for the ANC system equipped in a car, secondary path is mainly composed of peripheral circuits, loudspeaker (secondary acoustic source), the acoustic in passenger compartment, and microphone (also serves as error sensor). The model of secondary path and signal transmission is shown in Fig. 2.

In the algorithm of ANC system, secondary path is often modelled using unit impulse response function. In order to obtain unit impulse response function, we set the ANC to output band-limited white noise signal (y), which is the input signal of secondary path, and then we can get the microphone's output signal (e), namely

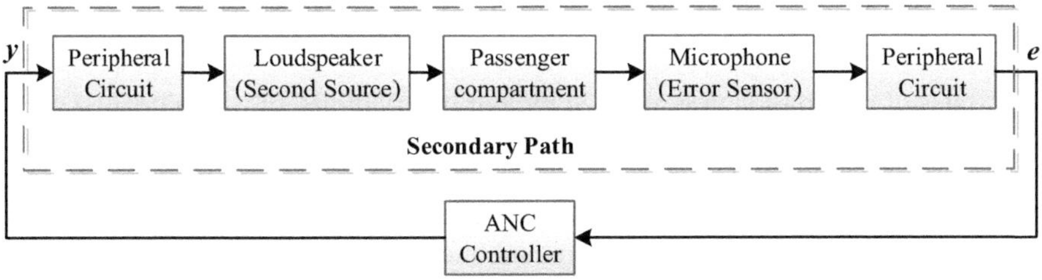

Fig. 2 Model of secondary path and signal transmission

output of secondary path. With input and output signals, using correlation identification method (8), the unit impulse response function can be then computed. The placement of microphone and loudspeaker is shown in Fig. 3. The vehicle loudspeaker was adopted as secondary acoustic source, and the microphone was placed above the head of driver.

Under different states of windows' opening or closing and different distributions of passengers, the secondary paths are also different. Here, we take one circumstance to illustrate the model of secondary path: All windows are closed. The unit impulse response function is shown in Fig. 4.

Fig. 3 Locations of microphone and loudspeaker

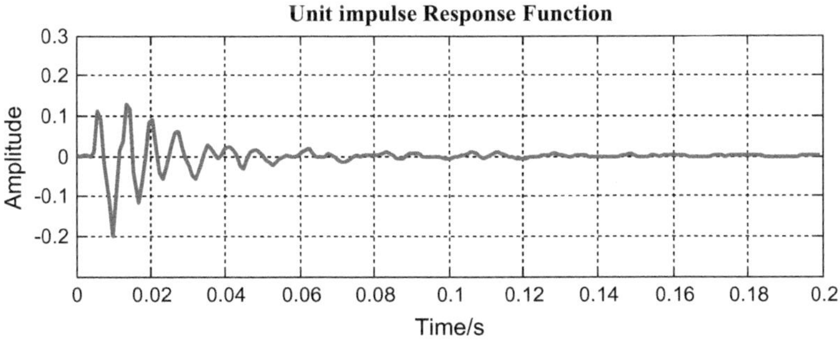

Fig. 4 Unit impulse response function of secondary path

3.2 Influence of Modelling Mismatch in the Secondary Path

The states of windows are represented by four binary digits, of which the first one to the fourth one indicates the left front window's state, the right front window's state, the left rear window's state, the right rear window's state, respectively. '0' stands for closing state while '1' for fully opening state. For example, '1000' indicates that the left front window is fully open and the other three windows are closed. Considering the spatial symmetry of windows, 0010 and 0001 are regarded as the same circumstance, so 0001 is ignored. Similarly, 0101, 0110, 1101, 0111 are also ignored, and the circumstances which are concerned in this research are 0000, 1000, 0010, 1100, 1010, 1001, 0011, 1110, 1011, and 1111.

'0000' (all windows were closed) was chosen to be reference state, while other circumstances were compared with the reference state to illustrate the change of function transfer in secondary path. The comparison with reference state includes changing rate of amplitude and the amount of phase change, which are defined in Eq. 9. The subscript '0' stands for reference state while '1' for new state. And Fig. 5 shows those changes of secondary path. Since the variation of windows' state leads to considerable change in secondary path, it is important to investigate the influence of ANC system when there is modelling mismatch in secondary path.

$$\Delta \text{Amp}(f) = \left|1 - \frac{\text{Amp}_1(f)}{\text{Amp}_0(f)}\right|, \quad \Delta\varphi(f) = |\varphi_1(f) - \varphi_0(f)| \tag{9}$$

In this paper, the nominal value of state '0000' was regarded as identification value, which remained unchanged in the secondary path, while the true value in secondary path was replaced with the nominal value under different windows' opening or closing states. By computing convergence factor α under each window's opening or closing state, the convergence performance of ANC system can be

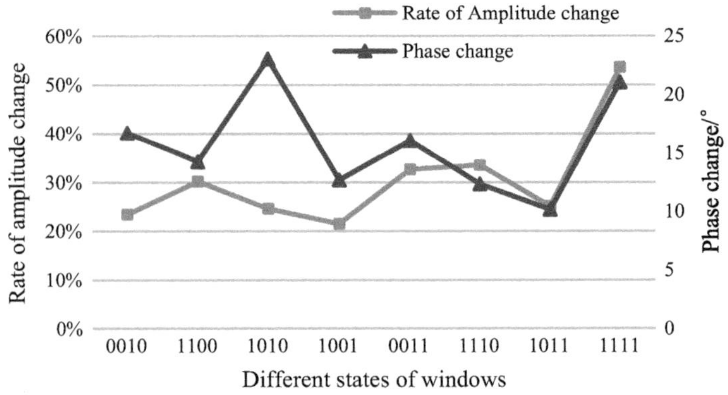

Fig. 5 Changing rate of amplitude and the amount of phase change of secondary path under different windows' opening or closing states

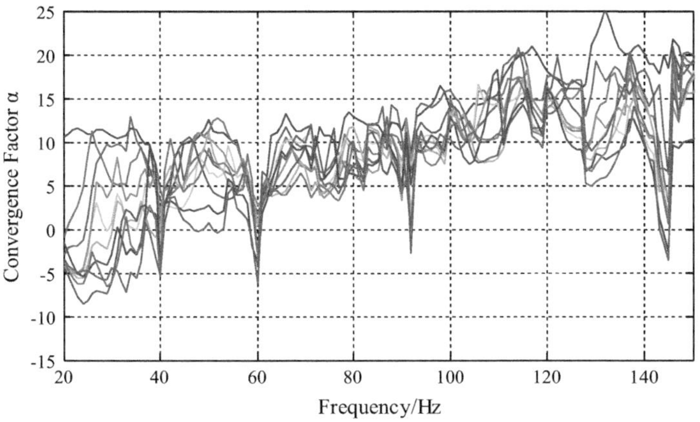

Fig. 6 Convergence performance in different circumstances when there was modelling mismatch in secondary path. Each color curve stood for one circumstance of windows' opening or closing state

evaluated. Figure 6 shows the curve of convergence factor α in different circumstances. The working frequency of ANC system is usually in low-frequency range, so only the frequency between 20 and 150 Hz is investigated. When convergence factor was less than 0, especially in the 20–40 Hz range and near 60, 90, and 145 Hz, the system did not converge, which shows that the ANC system is not robust to the change of windows' opening and closing.

4 Robust Design of ANC System

4.1 Designed Variables

In the algorithm of adaptive notch filter, convergence coefficient and the model of secondary path could be adjusted, which were considered controllable. In practice as long as the phase difference between the secondary path model and real circumstance is less than 90°, the ANC system will not diverge. Hence, the idea of robust design in this research was that the phase of the secondary path model used in practice was in the middle of the phases of the secondary path in each circumstance, in the meantime choosing appropriate convergence coefficient μ also needed to be done.

The iterative formula of the weight vector for a single-channel ANC system is as shown in Eq. 10 (the system was here assumed to eliminate noise in only one frequency, $O = 1$). Since the reference signal is sinusoidal signal, the role of secondary path can be described by the amplitude A and phase φ as shown in Eq. 10.

$$\begin{bmatrix} w_1(n+1) \\ w_2(n+1) \end{bmatrix} = \begin{bmatrix} w_1(n) \\ w_2(n) \end{bmatrix} - 2\mu \begin{bmatrix} h_s * \sin \omega n\Delta t \\ h_s * \cos \omega n\Delta t \end{bmatrix} e(n) \begin{bmatrix} w_1(n+1) \\ w_2(n+1) \end{bmatrix} \\ = \begin{bmatrix} w_1(n) \\ w_2(n) \end{bmatrix} - 2\mu \begin{bmatrix} h_s * \sin \omega n\Delta t \\ h_s * \cos \omega n\Delta t \end{bmatrix} e(n) \tag{10}$$

Besides, the combined effect of the convergence coefficient μ and the secondary path h_s on the reference signal is reflected by the changes of the amplitude and phase of the reference signal. Hence, the iterative formula can be further rewritten as follows:

$$\begin{bmatrix} w_1(n+1) \\ w_2(n+1) \end{bmatrix} = \begin{bmatrix} w_1(n) \\ w_2(n) \end{bmatrix} - 2 \begin{bmatrix} c_1 & c_2 \\ -c_2 & c_1 \end{bmatrix} \begin{bmatrix} \sin \omega n\Delta t \\ \cos \omega n\Delta t \end{bmatrix} e(n) \tag{11}$$

where the product of the convergence coefficient μ and the amplitude A of the transfer function can be represented with $\sqrt{c_1^2 + c_1^2}$ and c_1/c_2 is then tangent of the phase angle. Hence, c_1, c_2 can be regarded as design variables, so that convergence coefficient and characteristic of the secondary path can be considered simultaneously.

4.2 Criterion Function

The purpose of robust design is to make the performance of ANC system remain high level and has small fluctuation, when there are changes in the ANC system (states of windows' opening and closing). Here, convergence factor α was not expected to be a specific value, but was expected to be as large as possible. Referring to the method of determining the criterion function in the tolerance model (7), the criterion function in this paper is defined as follows:

$$f(c_1, c_2) = -\mu_\alpha + \beta\sigma_\alpha \tag{12}$$

where μ_α = mean of convergence factor α under different windows state, σ_α = standard deviation of convergence factor α, and β = weight coefficient of standard deviation.

The value of criterion function is expected to be as small as possible. Robust design in this research can be formulated as follows:

$$\min \quad f(\boldsymbol{c}) = -\mu_\alpha + \beta\sigma_\alpha - k\min(\alpha) \tag{13}$$

where $\boldsymbol{c} = (c_1, c_2)$ and $k = \begin{cases} p, \min(\alpha) < 0 \\ 0, \min(\alpha) \geq 0 \end{cases}, \ (p > 0)$

In Eq. 13, $k\min(\alpha)$ acts as penalty term. The addition of penalty term allows divergence to be taken into account when looking for the optimal design variables so that the ANC system after robust design can converge in any circumstance.

4.3 Evaluation of Robust Design

According to Eq. 13, robust design here can be regarded as a nonlinear optimization problem with multivariable and no constraints, which can be solved using downhill simplex algorithm (9). By taking $\beta = 1$ and $p = 50$, the optimized c_1, c_2 in different frequencies can be obtained as shown in Fig. 7.

Figure 8 (left) shows the curve of convergence factor in different circumstances after robust design. The ANC system after robust design has better performance, including small fluctuation and no divergence, which indicates that the system is robust even when there are changes of windows' states. To better illustrate the performance of the system, the mean of convergence factors in different circumstances was computed, including before and after robust design. Besides, the standard deviation was also computed to evaluate the stability of the system. These statistical values are shown in Fig. 8 (right). It was found that the mean of convergence factors after robust design was larger and the standard deviation was also smaller.

5 Active Noise Reduction Experiment

In order to evaluate the real performance of the ANC system, an active noise reduction experiment based on rapid control prototyping (RCP) was built. Under steady-state condition, experiments were carried out in different circumstances of

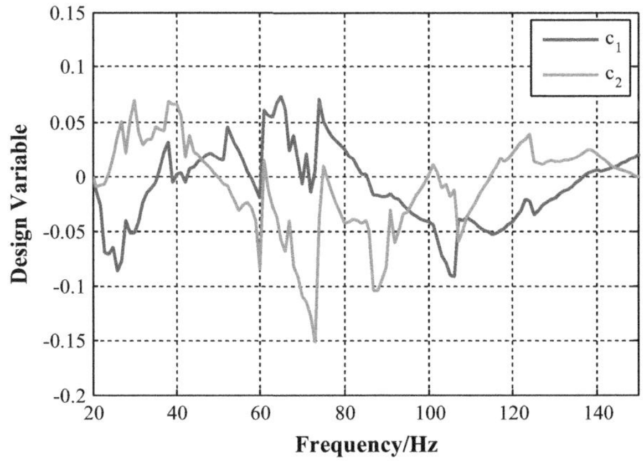

Fig. 7 Optimized design variables in robust design

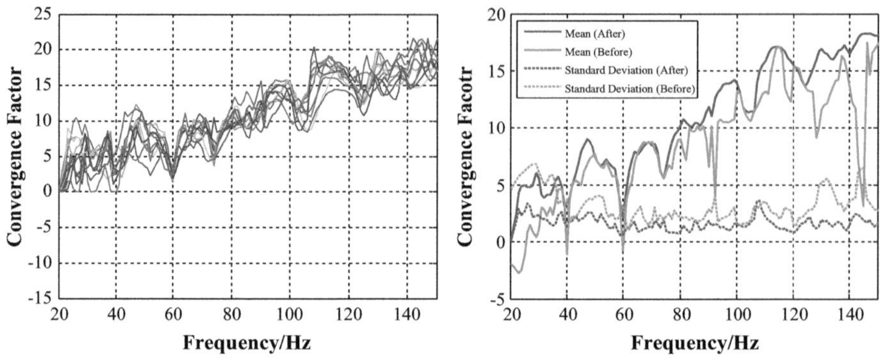

Fig. 8 Left: Convergence performance in different circumstances after robust design. Each color curve stood for one circumstance of windows' opening or closing state. Right: Statistical values of convergence factors before and after robust design

windows' opening and closing states, so that the effectiveness of the robust design could be tested by comparing the performance of the ANC system before and after robust design.

5.1 Construction of Experiment Platform

Based on RCP and using the dSPACE simulation system, the experiment platform was built as shown in Fig. 9 (left). The engine speed signal was regarded as reference signal and was connected to the CAN interface of the dSPACE system using CAN bus. The engine speed obtained from CAN interface was able to update in real time and was also more accurate, which can almost avoid the impact of frequency mismatch. The error signal was connected to the A/D channel of dSPACE system through an external conditioning circuit, and the secondary acoustic signal was connected with the power amplifier through the D/A channel so as to drive the vehicular loudspeaker, which served as secondary acoustic source. In the experiment, MicroAutoBox, signal acquisition instrument, power supply device, and other equipment were all put in the luggage compartment, as shown in Fig. 9 (right).

5.2 Settings of Verification Experiments

The relationship between the frequency of harmonic noises caused by the unbalanced inertia force of the engine and the engine speed can be formulated: $f_i = iNn/60\tau$, where i = harmonic order, f_i = frequency of harmonic noises, N = number of cylinders, n = engine speed, and τ = stroke coefficient (four-stroke $\tau = 2$).

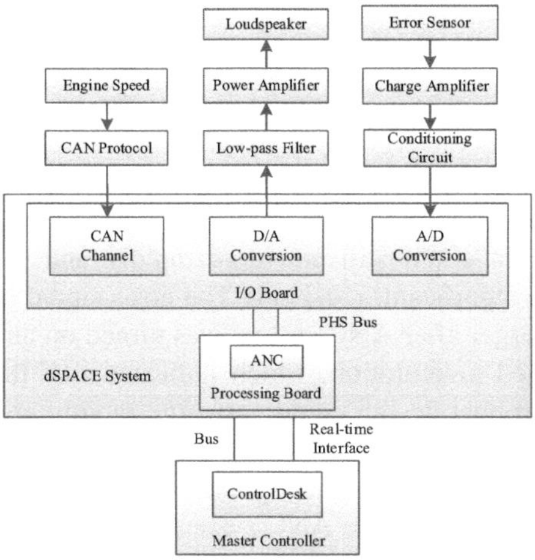

Fig. 9 Left: Experiment platform in this research based on RCP. Right: Equipment used in the experiment, which was placed in the luggage compartment

A four-cylinder four-stroke engine was used in this research. In the experiment, several representative circumstances of windows' opening and closing states were selected to carry out active noise reduction experiment to test the effectiveness of robust design. These circumstances are 0000, 1000, 1100, 1110, and 1111, corresponding from all windows' closing to all windows' opening. During the experiment, the car was set in neutral and the engine kept at a constant speed. According to Fig. 5, the system diverges in the frequency range of 20–40 Hz and near 60 and 145 Hz, and using $f_i = iNn/60\tau$, corresponding engine speeds could be computed. Hence, three group experiments were carried out, and the engine speeds are 700 rpm (idle conditions), 1000 rpm, and 2700 rpm. Each working condition lasted 15 s, and in the first 5 s the ANC system was turned off, while in the later 10 s it was turned on. And the amplitude of the primary noise was set to 1.

The error signal from error sensor could reflect the convergence characteristics of the system, and the size of the error signal could also reflect the residual sound pressure. Hence, the error signal in time domain was used as the evaluation index. The secondary acoustic signal from the controller was also recorded to check the system convergent or divergent. Since the range of output signal from MicroAutoBox controller was 0–1, a DC offset of 0.5 for the output signal was set in the control algorithm, which could be filtered by the power amplifier. In order to prevent the secondary acoustic signal from becoming too large, a saturation module was added to the secondary acoustic signal to limit the output in the range of 0.05–0.95. Hence, when the system diverged, the secondary acoustic signal reaches saturation, which could be used to check the convergence of system.

5.3 Results of Verification Experiments

Figure 10 shows experiment results under idle conditions. It was found that (1) 0000, 1000, and 1110: The system could converge rapidly both before and after robust design, and effectively reduce the error signal; (2) 1000: The system before robust design converged slowly and the reduction of error signal was not significant, while the system after robust design could still converge rapidly and the reduction of error signal was much more significant; (3) 1111: The error signal of the system before robust design became larger after ANC system was turned on and the secondary acoustic signal also reached a saturation, which indicated that the system diverged, while the system after robust design could converge rapidly and reduce error signal effectively.

Similarly when the engine speed was 1000 rpm, corresponding results were obtained. For lack of space, only circumstances that the ANC system before robust design diverged were shown in Fig. 11. When the states of windows were 1100,

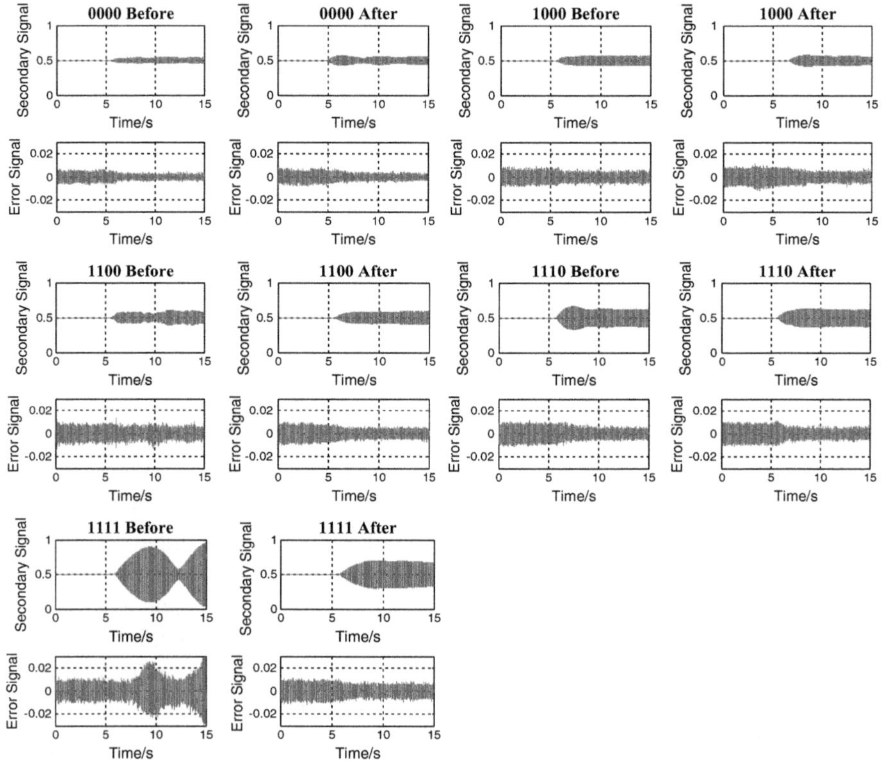

Fig. 10 Secondary acoustic signal and error signal under idle conditions (engine speed 700 rpm). 'Before' stands for the system before robust design while 'After' for the system after robust design

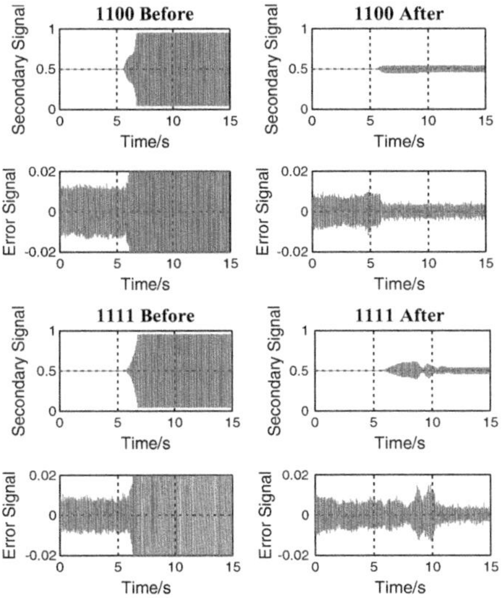

Fig. 11 Secondary acoustic signal and error signal when engine speed was set to 1000 rpm

1110, and 1111, the system before robust design diverged, while the system after robust design could converge rapidly and reduce the error signal effectively.

And when the engine speed was set to 2700 rpm, in the circumstances of 1100 and 1110 windows' state, the system before robust design diverged, while the system after robust design could converge rapidly, as shown in Fig. 12.

The results of the experiment show that with the robust design suggested in this research, the ANC system could rapidly converge and reduce the primary noise in each typical window's opening and closing states. The system after robust design had better robustness comparing with the ordinary ANC system.

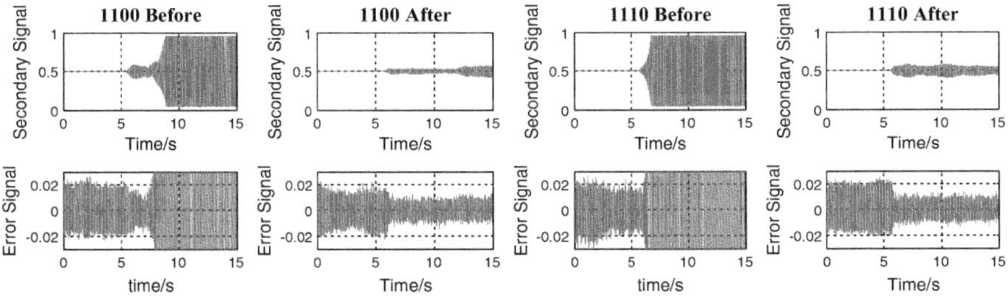

Fig. 12 Secondary acoustic signal and error signal when engine speed was set to 2700 rpm

6 Conclusion

As an effective means of suppressing low-frequency noise from engine, the technology of active noise reduction has a wide application in the field of passenger cars with increasing command for comfort, while the variability of driving conditions and the complexity of acoustic environment in the passenger compartment raise a high requirement on the robustness of the ANC system. The results in this research could be summarized as follows:

- The coefficient factor α established in this paper could indicate the convergence of the system, and the size of coefficient factor could describe the convergence speed.
- The system diverged in several frequency ranges when there was modelling mismatch in secondary path, which means the system was not robust to the changes in the states of windows' opening and closing.
- The convergence coefficient in the adaptive algorithm and characteristic of the secondary path could be considered at the same time in the robust design using a vector $\boldsymbol{c}$. And the addition of penalty term allowed divergence to be taken into account when looking for the optimal design variables. And through simulation and experiment, the robust design in this research was proven to be effective.

References

1. Jin X, Zhang L (2004) Prediction and control of noise in the car. Tongji University Press, Shanghai
2. Chen K (2014) Active noise control. National Defense Industry Press, Beijing
3. Qian Y (2008) Current status and trends of active control of vehicle interior noise. Sci Technol Inf 03:4–6
4. Hua C (2009) Research on hardware-in-loop simulation of active noise control in automobile cabin. Tongji University, Shanghai
5. Shahsavari B, et al (2015) Adaptive repetitive control design with online secondary path modeling and application to bit-patterned media recording. IEEE Trans Magn 51.4:1–8
6. Deng H (2014) Active control of vehicle interior noise based on adaptive notch filter. Tongji University, Shanghai
7. Chen L (2000) Robust design. China Machine Press, Beijing
8. Fang C (1988) Progress identification. Tsinghua University Press, Beijing
9. Lagarias JC, Reeds JA, Wright MH et al (1998) Convergence properties of the Nelder-Mead simplex method in low dimensions. SIAM J Optim 9(1):112–147

A Study on the Relationship Among Several Friction-Induced Instability Mechanisms Based on Multi-point Contact Nonlinear Dynamical Friction-Induced Vibration Model

Lijun Zhang, Zhuang Zhang and Dejian Meng

1 Introduction

Frictional squeal is widely existed in various mechanical systems such as brake system, clutch, train wheel/rail and so on [1, 2]. To date, the mechanism of friction, vibration and noise has not been fully understood, nor an effective control method has been established. Therefore, it has always been the focus of attention in academia and industry.

Up to now, the mechanism of frictional vibration and noise mainly includes stick-slip, sprag-slip, friction coefficient velocity negative slope theory, modal coupling theory and so on, which can be seen in many review literature in detail [1–18]. Among them, mode coupling instability is wildly used [19]. Negative friction-velocity slope instability theory used to be a widely accepted theory. However, it cannot explain the dependence of friction noise on friction coefficient and squeal also occurs when friction-velocity slope is positive. So this theory is challenged [18, 20, 21, 22]. Stick slip is often regarded as the main reason for the occurrence of low frequency squeal [23, 24]. Review history discoveries, it can be seen that researchers seem to be "excessively" focused on their own theory, and try to explain all the phenomena of frictional vibration and noise with only one theory. At the same time, they seriously neglected the strict definition of the conditions for the proposed theory, and they lacked mutual understanding with other theories. This may be the reason why there is no consensus on the mechanism of frictional vibration and noise. In paper [25], a lumped-parameter model with soft-rigid friction pairs was established and through the formula derivation and numerical calculations, the three theories were discussed.

In order to make the model closer to the real system, a three-degree-of-freedom contact friction dynamics model is established to explore the mechanism of fric-

L. Zhang · Z. Zhang (✉) · D. Meng
Tongji University, Shanghai, China
e-mail: zhangzhuang951@126.com

Society of Automotive Engineers of China (SAE-China) (ed.), *Proceedings of the 19th Asia Pacific Automotive Engineering Conference & SAE-China Congress 2017: Selected Papers*, Lecture Notes in Electrical Engineering 486, https://doi.org/10.1007/978-981-10-8506-2_16

tional squeal and the relationship between mode coupling instability theory, negative friction-velocity slop instability theory, and stick-slip theory. At the same time, Bifurcations of friction coefficient for mode coupling of the system under asymmetrical pressure distribution is also investigated. Through this study, the applicable conditions of different theories and their dynamic response characteristics are clarified and the theoretical system of frictional vibration and noise is improved.

2 Mode Coupling Instabilities

2.1 Physical Model of Friction Vibration

According to the topology and contact mode between pad and disk, a two-point contact, two-mass, and three-degree-of-freedom nonlinear dynamic model are established for a kind of friction vibration system which composed of a flexible body–rigid body friction pair with lumped parameters with reference to [25]. The model is shown in Fig. 1. In Fig. 1, the soft body with lumped parameter consists of two particles with mass m_1, m_2 and three freedoms of x, y and θ, a massless rigid rod of length L which connect to two particles, a spring with stiffness k_{11} and a damper with coefficient of c_{11} both of which have a degree of α_{11} with the x-axis, a spring with stiffness of k_{12} and a damper with coefficient of c_{12} both of which have a degree of α_{12} with the x-axis, a spring with stiffness of k_{21} and a damper with coefficient of c_{21} both of which have a degree of α_{21} with the x-axis, a spring with stiffness of k_{22} and a damper with coefficient of c_{22} both of which have a degree of α_{22} with the x-axis. The rigid body is a massless rigid band translated at speed V_0. There are two points of contact between the flexible body and the rigid body, and the normal contact stiffness is k_{13} and k_{23}. In order to keep them in a frictional contact state, the normal forces P_1 and P_2 are applied to the particles.

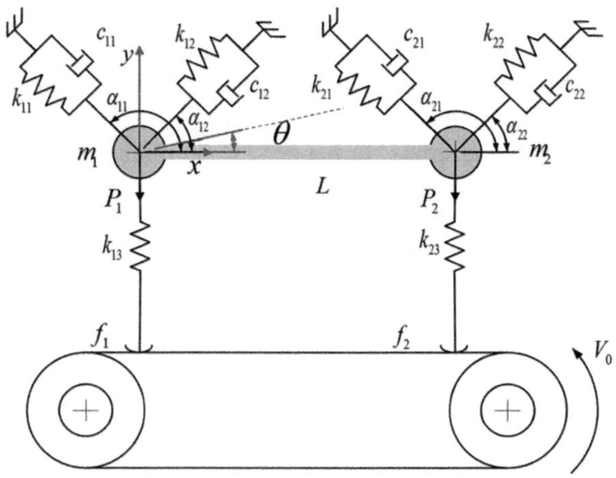

Fig. 1 3DOF nonlinear friction vibration model

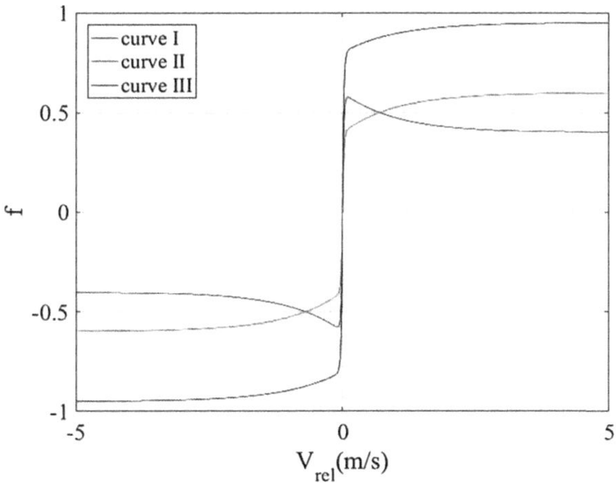

Fig. 2 Friction coefficient-velocity curve

2.2 Friction Coefficient Model

The friction coefficient between flexible body and rigid body is f_1,f_2, the mathematical expression is:

$$f_i = \text{sgn}(V_{\text{reli}})\mu_i \quad i = 1, 2 \tag{1}$$

$$\mu = \left(1 - e^{-d|V_{\text{reli}}|}\right)\left\{\mu_k - (\mu_k - \mu_s)e^{-h|V_{\text{reli}}|}\right\} \quad i = 1, 2 \tag{2}$$

$$V_{\text{reli}} = V_o + \dot{x} \quad i = 1, 2 \tag{3}$$

where μ_k, μ_s are the control parameters of dynamic and static friction coefficient, d, h are the control parameters of the value of the positive slope near the origin and the slope of the dynamic friction area. V_{rel} is the relative velocity between particle and belt.

The friction characteristic parameter combination I ($\mu_s = 0.6, \mu_k = 0.4,\ d = 50, h = 1$) and the parameter combination II ($\mu_s = 0.4, \mu_k = 0.6, d = 50, h = 1$) are defined to obtain the friction coefficient characteristic with negative friction-velocity slope and positive friction-velocity slope, which are used in the subsequent numerical calculation and analysis. The characteristic curves are shown in Fig. 2.

2.3 Dynamic Model of Friction System

Assuming that the massless rigid rods rotate at a small angle, so $\sin\theta = \theta$, $\cos\theta = 1$.

Then, the dynamics equation of the system can be expressed as Eq. (4):

$$\mathbf{M}'\dot{\mathbf{X}}' + \mathbf{K}'\mathbf{X}' = \mathbf{F}' \tag{4}$$

where $\mathbf{M}' = \begin{bmatrix} m_1 + m_2 & 0 & -m_2L\theta \\ 0 & m_1 + m_2 & m_2L \\ -m_2L\theta & m_2L & m_2L^2 \end{bmatrix}$, $\mathbf{X}' = \begin{bmatrix} \frac{d^2x}{dt^2} \\ \frac{d^2y}{dt^2} \\ \frac{d^2\theta}{dt^2} \end{bmatrix}$, $\mathbf{X}' = \begin{bmatrix} x \\ y \\ \theta \end{bmatrix}$,

$\mathbf{F}' = \begin{bmatrix} f_1k_{13}y + f_2k_{23}(y + L\theta) \\ -P_1 - P_2 \\ -f_2k_{23}L\theta(y + L\theta) - P_2L \end{bmatrix}$, The expression for $\mathbf{K}'$ is given in Appendix A.

Define: $T = \sqrt{m_1/k_{11}}, t = T\tau$, stiffness ratio $\kappa_1 = k_{12}/k_{11}, \kappa_2 = k_{13}/k_{11}$, $\kappa_3 = k_{22}/k_{11}, \kappa_4 = k_{21}/k_{11}, \kappa_5 = k_{23}/k_{11}$, mass ratio $\mu = m_2/m_1$, the dimensionless transformation of Eq. (4) is Eq. (5):

$$\mathbf{M}''\ddot{\mathbf{X}}'' + \mathbf{K}''\mathbf{X}'' = \mathbf{F}'' \tag{5}$$

where $\mathbf{M}'' = \begin{bmatrix} 1 + \mu & 0 & -\mu L\theta \\ 0 & 1 + \mu & \mu L \\ -\mu L\theta & \mu L & \mu L^2 \end{bmatrix}$, $\ddot{\mathbf{X}}'' = \begin{bmatrix} \frac{d^2x}{d\tau^2} \\ \frac{d^2y}{d\tau^2} \\ \frac{d^2\theta}{d\tau^2} \end{bmatrix}$, $\mathbf{X}'' = \begin{bmatrix} x \\ y \\ \theta \end{bmatrix}$, $\mathbf{F}'' =$

$\begin{bmatrix} 0 \\ -\frac{P_1 + P_2}{k_{11}} \\ -\frac{P_2}{k_{11}}L \end{bmatrix} + \begin{bmatrix} 0 \\ 0 \\ -f_2\kappa_5L\theta(y + L\theta) \end{bmatrix}$, $f_i = \mathrm{sgn}(V_{\mathrm{reli}})\mu_i$, $V_{\mathrm{reli}} = \sqrt{m_1/k_{11}}V_0 + \frac{dx}{d\tau}$,

$\mu_i = \left(1 - e^{-d\sqrt{\frac{k_{11}}{m_1}}|V_{\mathrm{reli}}|}\right)\left\{\mu_k - (\mu_k - \mu_s)e^{-h\sqrt{\frac{k_{11}}{m_1}}|V_{\mathrm{reli}}|}\right\} i = 1, 2,$ the expression for $\mathbf{K}''$ is given in Appendix B.

The equilibrium point can be adjusted to the coordinate origin by the translation transformation, so make $x = u^* + u, y = v^* + v, z = w^* + w$ in the Eq. (5):

$$\begin{cases} (1 + \mu)\ddot{u} - \mu L(w + w^*)\ddot{w} + \kappa_{11}u + (\kappa_{12} - f_1\kappa_2 - f_2\kappa_5)v + \kappa_{13}w = \left(f_1\kappa_2 - f_1^*\kappa_2 + f_2\kappa_5 - f_2^*\kappa_5\right)v^* \\ (1 + \mu)\ddot{v} + \mu L\ddot{w} + \kappa_{21}u + \kappa_{22}v + \kappa_{23}w = 0 \\ -\mu L(w + w^*)\ddot{u} + \mu L\ddot{v} + \mu L^2\ddot{w} + \kappa_{31}u + \kappa_{32}v + \kappa_{33}w = f_2\kappa_5L(w + w^*)(v + v^* + Lw + Lw^*) \end{cases} \tag{6}$$

The expressions for $u^*, v^*, w^*, \ddot{u}, \ddot{v}, \ddot{w}, \kappa_{11}, \kappa_{12}, \kappa_{21}, \kappa_{22}, \kappa_{23}, \kappa_{31}, \kappa_{32}, \kappa_{33}, f_1, f_2, f_1^*, f_2^*$ are given in Appendix C.

3 Influence of Friction Characteristics on Mode Coupling Instability

3.1 System Linearization and Stability Analysis

Equation (7) can be obtained by performing a Taylor series expansion of the friction coefficient formula and then ignoring the high order term in the expansion:

$$f_i(V_{reli}) = f(\sqrt{m_1/k_{11}}V_0 + \dot{u}_i) = f(\sqrt{m_1/k_{11}}V_0) + \frac{\partial f}{\partial V_{\text{reli}}}|_{\dot{u}=0}\dot{u} + o(\dot{u}^2) = f^* + \beta\dot{u} \tag{7}$$

where $\beta = \partial f / \partial V_r|_{\dot{u}=0}$ represents the slope of the friction coefficient at the dimensionless speed.

From Eq. (7) is easy to prove $f_1 = f_2 = f^* + \beta\dot{u}$.

Equation (7) is substituted into Eq. (6), and only the vibration near the equilibrium position is taken into account. Therefore, Eq. (8) can be obtained by neglecting the higher order terms $\dot{u}\dot{v}, \dot{u}v, \dot{u}w, w\ddot{w}, w\ddot{u}, wv, w^2, \dot{u}w, \dot{u}v$.

$$\begin{cases} (1+\mu)\ddot{u} - \mu L w^* \ddot{w} + \kappa_{11}u + [\kappa_{12} - f^*(\kappa_2+\kappa_5)]v + \kappa_{13}w - \beta(\kappa_2+\kappa_5)v^*\dot{u} = 0 \\ (1+\mu)\ddot{v} + \mu L\ddot{w} + \kappa_{21}u + \kappa_{22}v + \kappa_{23}w = 0 \\ -\mu L w^*\ddot{u} + \mu L\ddot{v} + \mu L^2\ddot{w} + \kappa_{31}u + (\kappa_{32} - f^*\kappa_5 L w^*)v + (\kappa_{33} - f^*\kappa_5 L v^* - 2f^*\kappa_5 L^2 w^*)w - \beta\kappa_5 L^2 w^*(v^* + L w^*)\dot{u} = 0 \end{cases} \tag{8}$$

The linearized system (9) can be obtained by transforming Eq. (8) into a matrix form:

$$\mathbf{M}\begin{pmatrix}\ddot{u}\\ \ddot{v}\\ \ddot{w}\end{pmatrix} + \mathbf{C}\begin{pmatrix}\dot{u}\\ \dot{v}\\ \dot{w}\end{pmatrix} + \mathbf{K}\begin{pmatrix}u\\ v\\ w\end{pmatrix} = \mathbf{0} \tag{9}$$

where $\mathbf{M} = \begin{bmatrix} 1+\mu & 0 & -\mu L w^* \\ 0 & 1+\mu & \mu L \\ -\mu L w^* & \mu L & \mu L^2 \end{bmatrix}$, $\mathbf{C} = \begin{bmatrix} -\beta(\kappa_2+\kappa_5)v^* & 0 & 0 \\ 0 & 0 & 0 \\ -\beta\kappa_5 L w^*(v^* + L w^*) & 0 & 0 \end{bmatrix}$,

$\mathbf{K} = \begin{bmatrix} \kappa_{11} & \kappa_{12} - f^*(\kappa_2+\kappa_5) & \kappa_{13} \\ \kappa_{21} & \kappa_{22} & \kappa_{23} \\ \kappa_{31} & \kappa_{32} - f^*\kappa_5 L w^* & \kappa_{33} - f^*\kappa_5 L v^* - 2f^*\kappa_5 L^2 w^* \end{bmatrix}$.

It can be seen from Eq. (9) that due to the existence of friction–velocity slope, the system produces a damping matrix **C**; that is, the equivalent damping in the u direction is $-\beta(\kappa_2+\kappa_5)v^*$, the equivalent damping in the v direction is 0, and the equivalent damping in the w direction is $-\beta\kappa_5 L w^*(v^* + L w^*)$.

For the linearized system (9), the eigenvalues can be calculated. The linearized system (9) is stable when the real part of all the eigenvalues is negative, and the linearized system (9) is unstable when at least one real part of eigenvalue is positive.

As the three-degree-of-freedom system is more complex, in order to facilitate the study, we can start from a simple case, assuming that the corresponding parameters in the two units are the same, that is $m_1 = m_2, k_{11} = k_{21}, k_{12} = k_{22}, k_{13} = k_{23}$, $\alpha_{11} = \alpha_{21}, \alpha_{12} = \alpha_{22}, P_1 = P_2$.

In the following, the influence of friction characteristics on mode coupling instability is analyzed under the friction-velocity slope of $\beta = 0, \beta < 0$ and $\beta > 0$.

3.2 $\beta = 0$

The eigenvalues can be calculated and the mode coupling instability is determined by the sign changes of the real part, so that bifurcation point of friction coefficient can be determined according to the derivation of Appendix D.

When the friction coefficient f^* is small, the real part of the eigenvalues of the system is 0 and the system has three stable modes. With the increase of the friction coefficient f^*, the real part of the eigenvalues of the system are positive and four eigenvalues are negative, the system has one unstable mode and two stable modes. Thus, the bifurcation point of the friction coefficient f^* is: when $\kappa_{12} > 0$, the bifurcation point is $f_a^* = (1 + \kappa_1^2 + \kappa_2^2 + 2\kappa_1 \cos(2\alpha_1 - 2\alpha_2) - 2\kappa_2 \cos 2\alpha_1 - 2\kappa_1\kappa_2 \cos 2\alpha_2)/2\kappa_2\kappa_{12}$; when $\kappa_{12} < 0$, the system has no bifurcation point.

Define the parameter combination I ($m_1 = 0.5\,kg, m_2 = 0.5\,kg, k_{11} = 50\,N/m$, $k_{12} = 50(2+\sqrt{3})/(2-\sqrt{3})\,N/m., k_{13} = 100/(2-\sqrt{3})\,N/m$, $k_{21} = 50\,N/m$, $k_{22} = 50(2+\sqrt{3})/(2-\sqrt{3}), k_{23} = 100/(2-\sqrt{3}), \alpha_{11} = 150°, \alpha_{12} = 30°, \alpha_{21} = 150°$, $\alpha_{22} = 30°, P_1 = 7N, P_2 = 7N, L = 0.1m$)

Using the friction coefficient combination I, keeping $\beta = 0$ and changing f^*, the change of the real part and the imaginary part of the linearized system (9) with different friction-velocity slope can be calculated. As shown in Fig. 3. The red '+' indicates that the real part of the eigenvalue is greater than 0 and the black '.' indicates that the real part of the eigenvalue is less than or equal to 0.

It can be seen from Fig. 3:

(1) When $\beta = 0$, only when the friction coefficient $f^* > f_a^* \approx 0.75$, the real part of the eigenvalues of the system is positive, which means that the system has mode coupling instability.

3.3 $\beta < 0$

The real part and the imaginary part of the eigenvalues are calculated for the linearized system (9) by the same method with a negative β. As shown in Fig. 4a, b.

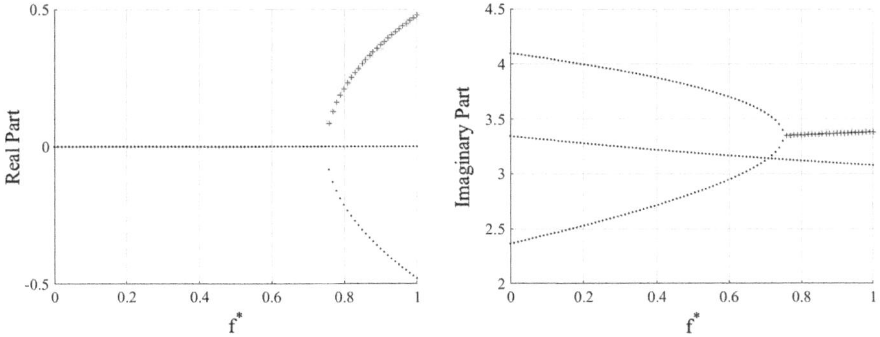

(a) The locus of real part with respect to friction coefficient.

(b) The locus of imaginary part with respect to friction coefficient.

Fig. 3 Locus of eigenvalue with respect to friction coefficient. **a** The locus of real part with respect to friction coefficient. **b** The locus of imaginary part with respect to friction coefficient

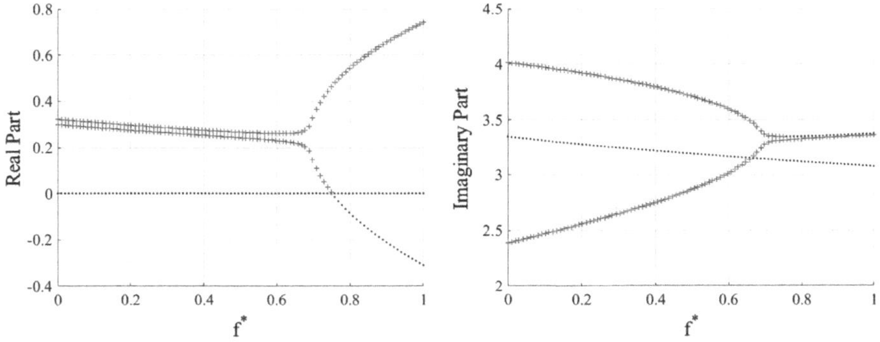

(a) The locus of real part with respect to friction coefficient.

(b) The locus of imaginary part with respect to friction coefficient.

Fig. 4 Locus of eigenvalue with respect to friction coefficient. **a** The locus of real part with respect to friction coefficient. **b** The locus of imaginary part with respect to friction coefficient

It can be seen from Fig. 4:

(1) When $\beta < 0$, regardless of the value of the coefficient of friction, the system always has eigenvalues with positive real part and the system is always unstable;

(2) When $\beta < 0$ and $f^* < f_b^* \approx 0.75$, the system only has the instability caused by the negative slope of the friction-velocity; when $f^* > f_b^* \approx 0.75$, the mode coupling occurs, and the negative friction–velocity slope instability and mode coupling instability exist simultaneously;

(3) It can be seen from Figs. 3b, 4b that the change of friction-velocity with or without negative slope is not significant for the imaginary part of the eigenvalue, which means that the negative slope of friction-velocity does not significantly change the unstable frequency.

Therefore, the negative friction-velocity slop will increase the instability of the system, but will not significantly change the value of the unstable frequency.

3.4 $\beta > 0$

In the same way as before, using the system parameter combination I and the friction coefficient parameter combination II, we can get the change of the real part and imaginary part of the eigenvalues with a positive β. As shown in Fig. 5a, b.

It can be seen from Fig. 5:

(1) When $\beta > 0$, only when $f^* > f_c^* \approx 0.75$, the system has eigenvalues with positive real part, and the system has mode coupling instability;
(2) When $\beta > 0$, when $f^* < f_c^* \approx 0.75$, the real part of the system eigenvalues is always negative, which indicates that the system is always stable;
(3) By comparing Figs. 3b and 5b, it can be found that the change of friction-velocity with or without negative slope is not significant for the imaginary part of the eigenvalue, which means that the negative slope of friction-velocity does not significantly change the unstable frequency.

In summary, the positive friction-velocity slope will enhance the instability of the system, but will not significantly change the value of unstable frequency.

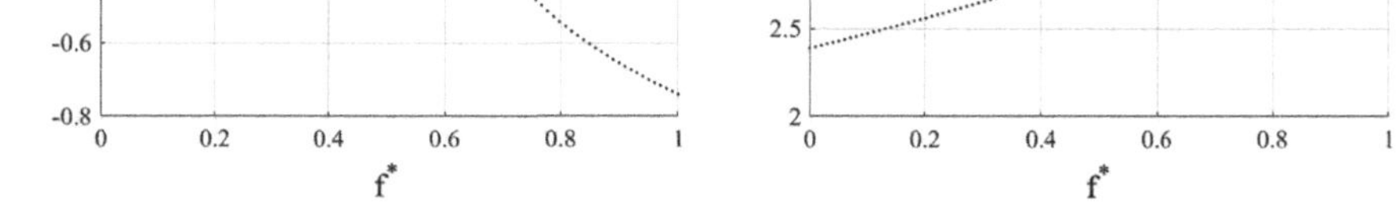

(a) The locus of real part with respect to friction coefficient.

(b) The locus of imaginary part with respect to friction coefficient.

Fig. 5 Locus of eigenvalue with respect to friction coefficient. **a** The locus of real part with respect to friction coefficient. **b** The locus of imaginary part with respect to friction coefficient

3.5 Summary of Friction Coefficient and Friction-Velocity Slope

Summarizing the above analysis, the system instability under different friction-velocity slopes is shown in Table 1. It can be seen from the table that system stability is closely related to the value of friction coefficient and the sign of friction-velocity slope.

Referring to Table 1, using the system parameter combination I, we randomly select the friction coefficient f^* and the friction-velocity slope β to calculate the eigenvalues and determine the stability. As shown in Fig. 6.

According to the value of bifurcation point and the friction-velocity slope, the figure is divided into four areas: (1) positive slope no mode coupling stable area (I); (2) positive slope mode coupling unstable area (II); (3) negative slope no mode coupling unstable area (III); (4) negative slope mode coupling unstable area (IV). The boundaries between the four areas are $\beta = 0$ and $f^* = f_a^* = f_b^* = f_c^*$ (the value of bifurcation point may not be equal). As shown in Fig. 7:

Table 1 Bifurcation point and mode stability with respect to different kinds of slope

Sign of slope	Mode stability characteristics	
	Coefficient of friction < Bifurcation point	Coefficient of friction > Bifurcation point
$\beta = 0$	3 stable modes	2 stable modes and 1 unstable mode
$\beta < 0$	1 stable mode and 2 unstable modes	2 stable modes and 1 unstable mode
$\beta > 0$	3 stable modes	2 stable modes and 1 unstable mode

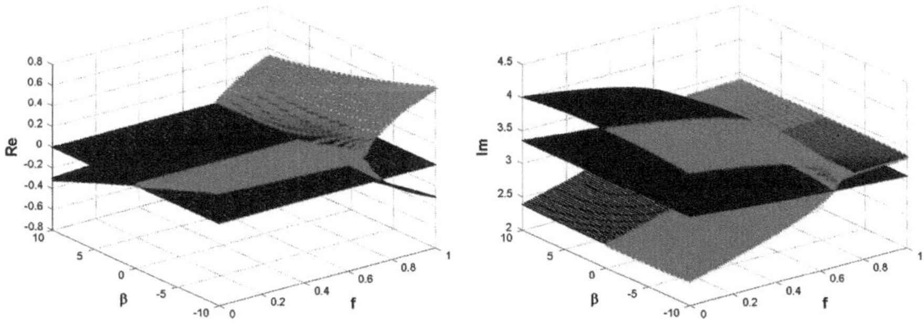

(a) The locus of real part with respect to friction coefficient.

(b) The locus of imaginary part with respect to friction coefficient.

Fig. 6 Locus of eigenvalue with respect to friction coefficient. **a** The locus of real part with respect to friction coefficient. **b** The locus of imaginary part with respect to friction coefficient

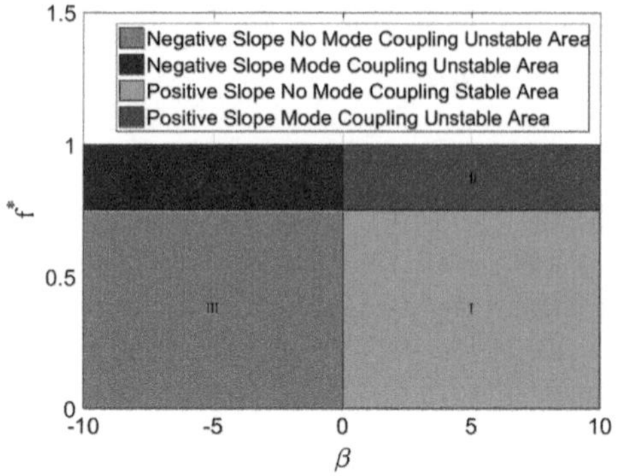

Fig. 7 Four areas when friction coefficient and slope change

It can be seen from Fig. 7 that the instability due to the negative friction-velocity slop and the instability caused by mode coupling are two completely independent mechanisms of frictional vibration and noise generation, and they are independent of each other and can co-exist.

4 Effect of Friction Characteristics on Stick-Slip Motion

On the following, the influence of friction characteristics on the stick-slip motion of the system and its dependency on the working conditions are systematically investigated in four areas.

Considering that there may be limit cycles, the Bendixson–Dulac criterion for the existence of limit cycles is introduced firstly [26].

Define $\mathbf{X}(\tau) = (p(\tau) \quad q(\tau) \quad r(\tau) \quad s(\tau) \quad t(\tau) \quad u(\tau))^T$, and Eq. (10) is obtained by convert Eq. (6) into state space equation:

$$\frac{d\mathbf{X}(\tau)}{d\tau} = \begin{pmatrix} p_1(a,b,c,d,e,f) \\ p_2(a,b,c,d,e,f) \\ p_3(a,b,c,d,e,f) \\ p_4(a,b,c,d,e,f) \\ p_5(a,b,c,d,e,f) \\ p_6(a,b,c,d,e,f) \end{pmatrix} = \mathbf{P}(\mathbf{X}) \tag{10}$$

where $\dot{a} = d = p_1(a,b,c,d,e,f), \dot{b} = e = p_2(a,b,c,d,e,f), \dot{c} = f = p_3(a,b,c,d,e,f)$, $\dot{d} = p_4(a,b,c,d,e,f)$, $\dot{e} = p_5(a,b,c,d,e,f)$, $\dot{f} = p_6(a,b,c,d,e,f)$.

Substituting the system parameter combination I into Eq. (10), the divergence can be obtained:

$$\nabla \cdot \mathbf{P} = \frac{\partial p_1}{\partial a} + \frac{\partial p_2}{\partial b} + \frac{\partial p_3}{\partial c} + \frac{\partial p_4}{\partial d} + \frac{\partial p_5}{\partial e} + \frac{\partial p_6}{\partial f} = 7.464\beta v \quad (11)$$

Assuming that the belt and the particles are never disengaged, the dimensionless normal force is $n_v = -\kappa_2(v + v^*) \geq 0$.

According to the Bendixson–Dulac criterion [26], the necessary condition for the generation of limit cycles in the plane is that $\nabla \cdot \mathbf{P}$ changes the sign. Therefore, a limit cycle only occurs when β changes the sign.

4.1 Analysis on the Characteristics of Stick-Slip Motion in Aera III

It can be seen from the red line in Fig. 2 that when $|V_{\mathrm{rel}}|$ is very small, $\beta > 0$ and the difference between the particle velocity and the belt velocity is small, and the system is in the stick stage. When $|V_{\mathrm{rel}}|$ is relatively large, $\beta < 0$ and the difference between the particle velocity in x direction and the belt velocity is small, the system is in the slip stage. This indicates that when the limit cycle occurs, the friction coefficient is bound to change between the positive slope area and negative slope area, which means that the appearance of the limit cycle will be accompanied by stick-slip phenomenon. The calculation and verification analysis is carried out at different belt speeds.

(1) $$V_0 = 1 \quad m/s$$

At this time, because of the negative friction-velocity slope, the system is always unstable and the particle velocity in x direction reaches the belt speed. So the intermittent stick-slip phenomenon is produced, and finally it enters the stable limit cycle (Fig. 8). It can be seen from Fig. 9 that there is no separation between the particles and the belt.

(2) $$V_0 = 3.5 \quad m/s$$

It can be seen from Fig. 10 that when the belt speed increases, the positive real part of the eigenvalues decreases and the system tends to be more stable. And the higher the belt speed, the harder it is to achieve the belt speed. As can be seen from Fig. 11, since only the slip motion is generated, the friction coefficient is also stable and no abrupt change occurs.

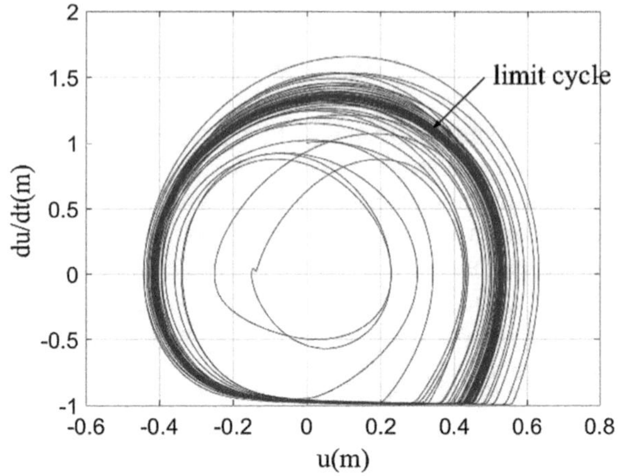

Fig. 8 Phase diagram of the mass in x direction

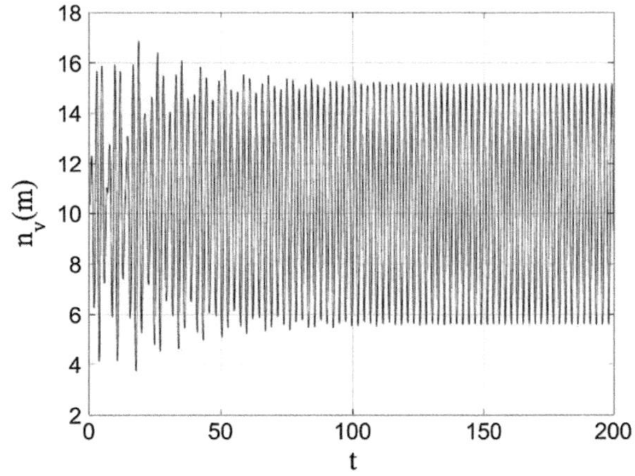

Fig. 9 Time-domain diagram of the non-dimensional normal force

4.2 Analysis on the Characteristics of Stick-Slip Motion in Aera I

It can be seen from Eq. (11) that when the friction-velocity slope is positive, the sign of $\nabla \cdot \mathbf{P}$ does not change, so the limit cycle cannot be generated. Using the friction coefficient parameter combination II and the system parameter combination I, the system is in the positive slope mode coupling stable area of Fig. 7. The simulation results of the nonlinear system (6) at low belt speed are shown in Figs. 12 and 13. Obviously, because the eigenvalues of the system have a negative real part, the system tends to slide steadily and it is very difficult to produce stick-slip motion and the friction coefficient is stable.

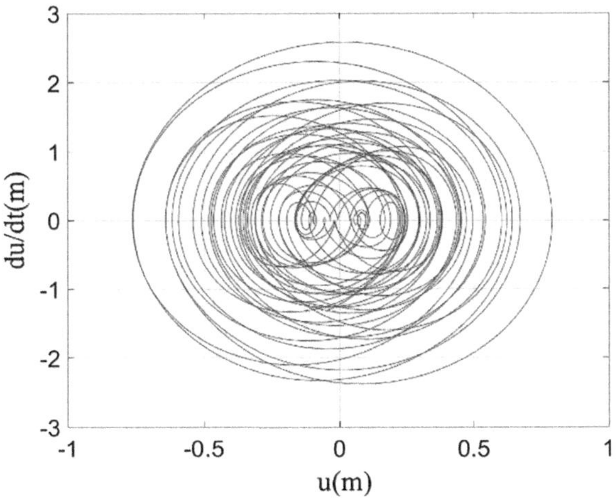

Fig. 10 Phase diagram of the mass in x direction

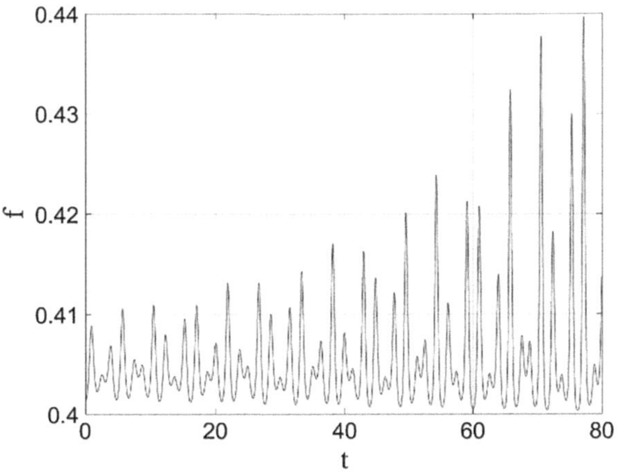

Fig. 11 Time-domain diagram of friction coefficient

4.3 Analysis on the Characteristics of Stick-Slip Motion in Aera II and IV

At the moment, the system is unstable and the instability of the system is aggravated. The velocity of the particle is more likely to reach the belt speed. Now we use $\mu_s = 0.8, \mu_k = 0.95, d = 50, h = 1$, system parameter combination I and $V_0 = 2\ m/s$ to calculate. It can be seen from Fig. 14 that the system tends to generate stick-slip after $t = 6$, however, it can be seen from Fig. 15 that after $t = 5$ there is a detachment of the particle and the belt. In other words, the occurrence of stick-slip motion is usually accompanied by the intermittent separation of particle and band, which is consistent with the literature [24]. Since n_v is less than zero, which violates the assumptions of the model, further analysis of this problem will not be carried out.

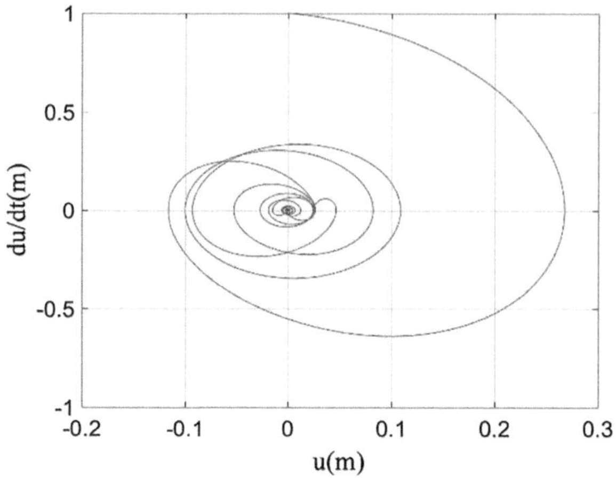

Fig. 12 Phase diagram of the mass in x direction

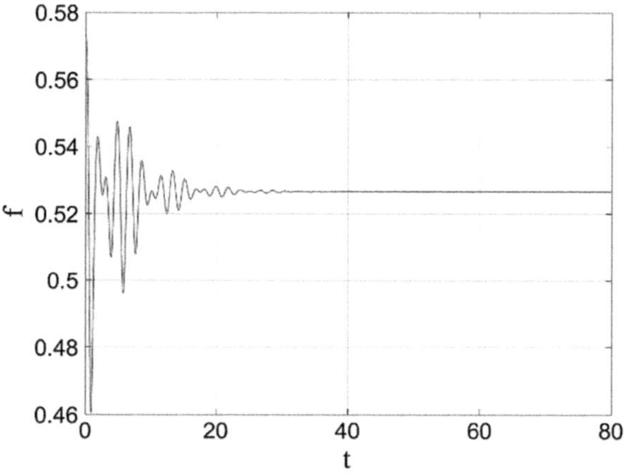

Fig. 13 Time-domain diagram of friction coefficient

5 Summary

In this paper, according to the topology and contact mode between pad and disk, a nonlinear friction and vibration dynamic model with dual mass and three-degrees-of-freedom is established for a kind of friction and vibration system composed of flexible body–rigid body friction pairs with lumped parameters. By means of mathematical formula derivation and numerical calculation, the following conclusions are obtained:

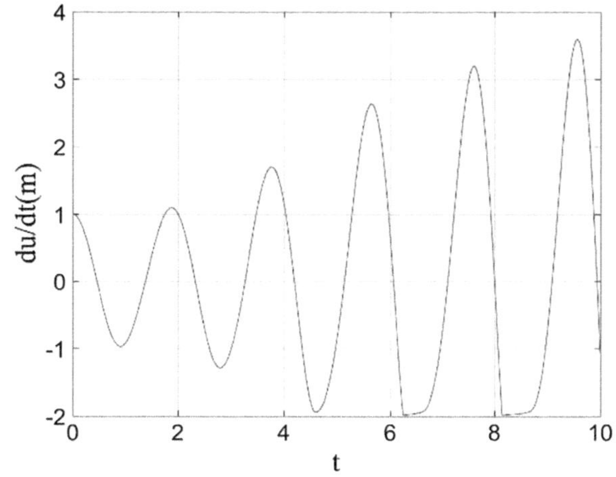

Fig. 14 Time-domain diagram of the velocity of the mass in x direction

Fig. 15 Time-domain diagram of the non-dimensional normal force

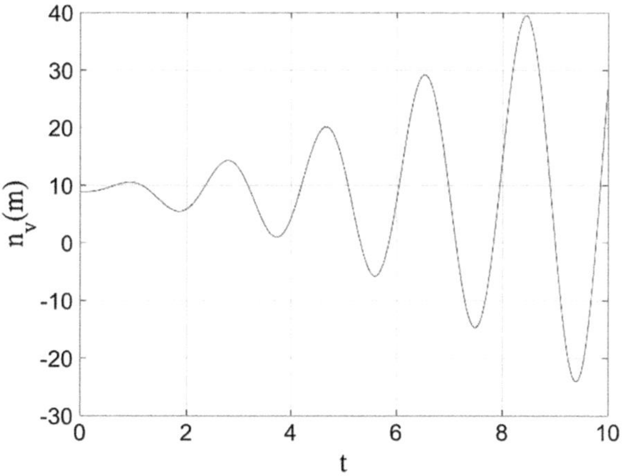

(1) The damping matrix produced by the negative friction-velocity slope exacerbates the instability of the system; the mode coupling can also be produced when the friction coefficient-velocity slope is positive; the sign of the friction coefficient-velocity slope does not change the area of mode coupling and the effect on the imaginary part of the eigenvalue is not significant;

(2) According to the friction coefficient and its slope, the system is divided into four areas: (1) positive slope no mode coupling stable area (I); (2) positive slope mode coupling unstable area (II); (3) negative slope no mode coupling unstable area (III); (4) negative slope mode coupling unstable area (IV);

(3) When the system is in negative slope no mode coupling unstable area, whether the system is able to produce stick-slip phenomenon is dependent on the belt speed to a certain extent. When the speed of belt is small, the system can produce intermittent stick-slip phenomenon, and finally enter a stable limit cycle movement, but when the speed of belt is large, it is more difficult to produce stick-slip phenomenon. When the system is in positive slope no mode coupling stable area, the stick-slip phenomenon is difficult to occur even with very small belt speed because the system stabilizes quickly. When the system is in positive slope mode coupling unstable area and negative slope mode coupling unstable area, the stick-slip is accompanied by the separation of the particles and the belt.

The above conclusions explain the generation of brake noise from the mechanism level, help researchers understand the generation of brake noise, and also guide the design of the brake.

Appendix A

$$\mathbf{K}' = \begin{bmatrix} k_{11}\cos^2\alpha_{11} + k_{12}\cos^2\alpha_{12} + k_{21}\cos^2\alpha_{21} + k_{22}\cos^2\alpha_{22} & k_{11}\sin\alpha_{11}\cos\alpha_{11} + k_{12}\sin\alpha_{12}\cos\alpha_{12} + k_{21}\sin\alpha_{21}\cos\alpha_{21} + k_{22}\sin\alpha_{22}\cos\alpha_{22} & k_{21}L\sin\alpha_{21}\cos\alpha_{21} + k_{22}L\sin\alpha_{22}\cos\alpha_{22} \\ k_{11}\sin\alpha_{11}\cos\alpha_{11} + k_{12}\sin\alpha_{12}\cos\alpha_{12} + k_{21}\sin\alpha_{21}\cos\alpha_{21} + k_{22}\sin\alpha_{22}\cos\alpha_{22} & k_{11}\sin^2\alpha_{11} + k_{12}\sin^2\alpha_{12} + k_{13} + k_{21}\sin^2\alpha_{21} + k_{22}\sin^2\alpha_{22} + k_{23} & k_{21}L\sin^2\alpha_{21} + k_{22}L\sin^2\alpha_{22} + k_{23}L \\ k_{21}L\sin\alpha_{21}\cos\alpha_{21} + k_{22}L\sin\alpha_{22}\cos\alpha_{22} & k_{21}L\sin^2\alpha_{21} + k_{22}L\sin^2\alpha_{22} + k_{23}L & k_{21}L^2\sin^2\alpha_{21} + k_{22}L^2\sin^2\alpha_{22} + k_{23}L^2 \end{bmatrix}$$

Appendix B

$$\mathbf{K}'' = \begin{bmatrix} \cos^2\alpha_{11} + \kappa_1\cos^2\alpha_{12} + \kappa_3\cos^2\alpha_{21} + \kappa_4\cos^2\alpha_{22} & \sin\alpha_{11}\cos\alpha_{11} + \kappa_1\sin\alpha_{12}\cos\alpha_{12} + \kappa_3\sin\alpha_{21}\cos\alpha_{21} + \kappa_4\sin\alpha_{22}\cos\alpha_{22} - f_1\kappa_2 - f_2\kappa_5 & \kappa_3L\sin\alpha_{21}\cos\alpha_{21} + \kappa_4L\sin\alpha_{22}\cos\alpha_{22} \\ \sin\alpha_{11}\cos\alpha_{11} + \kappa_1\sin\alpha_{12}\cos\alpha_{12} + \kappa_3\sin\alpha_{21}\cos\alpha_{21} + \kappa_4\sin\alpha_{22}\cos\alpha_{22} & \sin^2\alpha_{11} + \kappa_1\sin^2\alpha_{12} + \kappa_2 + \kappa_3\sin^2\alpha_{21} + \kappa_4\sin^2\alpha_{22} + \kappa_5 & \kappa_3L\sin^2\alpha_{21} + \kappa_4L\sin^2\alpha_{22} + \kappa_5L \\ \kappa_3L\sin\alpha_{21}\cos\alpha_{21} + \kappa_4L\sin\alpha_{22}\cos\alpha_{22} & \kappa_3L\sin^2\alpha_{21} + \kappa_4L\sin^2\alpha_{22} + \kappa_5L & \kappa_3L^2\sin^2\alpha_{21} + \kappa_4L^2\sin^2\alpha_{22} + \kappa_5L^2 \end{bmatrix}$$

Appendix C

$$\ddot{u} = \frac{d^2u}{d\tau^2}, \ddot{v} = \frac{d^2v}{d\tau^2}, \ddot{w} = \frac{d^2w}{d\tau^2}, \dot{u} = \frac{du}{d\tau},$$

$$\kappa_{11} = 1 + \kappa_1\cos^2\alpha_1 + \kappa_3 + \kappa_4\cos^2\alpha_2, \kappa_{12} = \kappa_{21} = \kappa_1\sin\alpha_1\cos\alpha_1 - \kappa_4\sin\alpha_2\cos\alpha_2,$$

$$\kappa_{13} = \kappa_{31} = -\kappa_4 L\sin\alpha_2\cos\alpha_2$$

$$\kappa_{22} = \kappa_1\sin^2\alpha_1 + \kappa_2 + \kappa_5, \kappa_{23} = \kappa_{32} = \kappa_5 L, \kappa_{33} = \kappa_5 L^2$$

$$f_i = \mathrm{sgn}(V_{\mathrm{reli}})\mu_i, \mu_i = \left(1 - e^{-d\sqrt{\frac{k_{11}}{m_1}}|V_{\mathrm{reli}}|}\right)\left\{\mu_k - (\mu_k - \mu_s)e^{-h\sqrt{\frac{k_{11}}{m_1}}|V_{\mathrm{reli}}|}\right\},$$

$$V_{\mathrm{reli}} = \sqrt{m_i/k_{11}}V_0 + \dot{u}$$

u^*, v^*, w^* solved by the equation:

$$\begin{cases} \kappa_{11}u^* + \kappa_{12}v^* + \kappa_{13}w^* = f_1^*\kappa_2 v^* + f_2^*\kappa_5 v^* \\ \kappa_{21}u^* + \kappa_{22}v^* + \kappa_{23}w^* = -\frac{P_1+P_2}{k_1} \\ \kappa_{31}u^* + \kappa_{32}v^* + \kappa_{33}w^* = f_2^*\kappa_5 w^* L(v^* + Lw^*) \end{cases}$$

Appendix D

When $\beta = 0$, the conjugate eigenvalues of the system are:

$$\Omega_{1,2}''^2 = \frac{1}{2}\left[(1+\kappa_1+\kappa_2) \pm \sqrt{1+\kappa_1^2+\kappa_2^2+2\kappa_1\cos(2\alpha_1-2\alpha_2)-2\kappa_2\cos(2\alpha_1)-2\kappa_1\kappa_2\cos(2\alpha_2)-2f^*\kappa_2\kappa_{12}}\right]$$

Define
$\Delta = 1+\kappa_1^2+\kappa_2^2+2\kappa_1\cos(2\alpha_1-2\alpha_2)-2\kappa_2\cos 2\alpha_1 - 2\kappa_1\kappa_2\cos 2\alpha_2 - 2f^*\kappa_2\kappa_{12}$,
therefore:

(1) When $\Delta > 0, \Omega_1$ and Ω_2 are conjugate pure imaginary numbers, real parts are zero。
(2) When $\Delta = 0, \Omega_1$ and Ω_2 are conjugate pure imaginary numbers, real parts are zero。
(3) When $\Delta > 0, \Omega_1$, and Ω_2 are complex conjugate pairs with one real part greater than zero and one less than zero.

Obviously
$1+\kappa_1^2+\kappa_2^2+2\kappa_1\cos(2\alpha_1-2\alpha_2)-2\kappa_2\cos 2\alpha_1-2\kappa_1\kappa_2\cos 2\alpha_2 > 0$, so we can determine:

When $\kappa_{12} > 0$ and $f^* > \frac{1+\kappa_1^2+\kappa_2^2+2\kappa_1\cos(2\alpha_1-2\alpha_2)-2\kappa_2\cos 2\alpha_1-2\kappa_1\kappa_2\cos 2\alpha_2}{2\kappa_2\sin 2\alpha_1+2\kappa_1\kappa_2\sin 2\alpha_2}$, the real part of the eigenvalues of the system is positive.

References

1. Watson FR (1939) The journal of the acoustical society of america. J Acoust Soc Am 11 (1):15–20
2. Ouyang H, Nack W, Yuan Y et al (2005) Numerical analysis of automotive disc brake squeal: a review. Int J Veh Noise Vib 1(3–4):207–231
3. Hoffmann NP, Gaul L. Friction induced vibrations of brakes: research fields and activities (2008). SAE Technical Paper.
4. Kinkaid NM, O'reilly OM, Papadopoulos P (2003) Automotive disc brake squeal. J Sound Vib 267(1):105–166
5. Nishiwaki M. Review of study on brake squeal[J]. JSAE Review, 1990, 11(4).

6. Crolla DA, Lang AM (1991) Paper VII (i) brake noise and vibration-the state of the art. Tribology Series 18:165–174
7. Ibrahim RA (1994) Friction-induced vibration, chatter, squeal, and chaos, Part I: mechanics of contact and friction. Appl Mech Rev 47(7):209–226
8. Ibrahim RA (1994) Friction-induced vibration, chatter, squeal, and chaos—part II: dynamics and modeling. Appl Mech Rev 47(7):227–253
9. Yang S, Gibson RF (1997) Brake vibration and noise: reviews, comments, and proposals. Int J Mater Prod Technol 12(4–6):496–513
10. Papinniemi A, Lai JCS, Zhao J et al (2002) Brake squeal: a literature review. Appl Acoust 63(4):391–400
11. Chen F, Quaglia RL, Tan CA. On automotive disc brake squeal Part I: mechanisms and causes (2003). SAE Technical Paper.
12. Dessouki O, Drake G, Lowe B, et al. Disc brake squeal: diagnosis and prevention (2003). SAE Technical Paper.
13. Ouyang H, Nack W, Yuan Y, et al. On automotive disc brake squeal part II: simulation and analysis (2003). SAE Technical Paper.
14. Chen F, Abdelhamid MK, Blaschke P, et al. On automotive disc brake squeal part III test and evaluation (2003). SAE Technical Paper.
15. Chen F, Tong H, Chen S E, et al. On automotive disc brake squeal part IV reduction and prevention (2003). SAE Technical Paper.
16. Bhushan B (1980) Stick-slip induced noise generation in water-lubricated compliant rubber bearings. J Lubr Technol 102(2):201–212
17. Eriksson M, Bergman F, Jacobson S (1999) Surface characterisation of brake pads after running under silent and squealing conditions. Wear 232(2):163–167
18. Bergman F, Eriksson M, Jacobson S (1999) Influence of disc topography on generation of brake squeal. Wear 225:621–628
19. Hoffmann N, Fischer M, Allgaier R et al (2002) A minimal model for studying properties of the mode-coupling type instability in friction induced oscillations. Mech Res Commun 29(4): 197–205
20. Ibrahim RA (1994) Friction-induced vibration, chatter, squeal, and chaos—part II: dynamics and modeling. Appl Mech Rev 47(7):227–253
21. Carr J (2012) Applications of centre manifold theory. Springer Science & Business Media, New York, NY
22. Chen GX, Zhou ZR (2003) Correlation of a negative friction–velocity slope with squeal generation under reciprocating sliding conditions. Wear 255(1):376–384
23. Kang J, Krousgrill CM, Sadeghi F (2009) Oscillation pattern of stick–slip vibrations. Int J Non-Linear Mech 44(7):820–828
24. Meziane A, D'Errico S, Baillet L et al (2007) Instabilities generated by friction in a pad–disc system during the braking process. Tribol Int 40(7):1127–1136
25. Zhang L, Wu J, Meng D (2015) Relationship among mode coupling, friction-velocity slope and stick-slip motion. J Tongji Univ (Nat Sci) 43(12):1850–1859
26. Chen WC (2008) Nonlinear dynamics and chaos in a fractional-order financial system. Chaos, Solitons Fractals 36(5):1305–1314

Six-Sigma Robust Optimization on the NVH Performance of the CRFM Based on Isight

Lin Liu, Songbo Zhang, Xueliang Li and Bo Zhang

1 Introduction

In engineering design, there are many uncertainties and random fluctuations that lead to the reliability distribution of product performance. When one deterministic optimal solution is close to the constraint boundary, product performance may not satisfy constraints if the random fluctuations of uncertain factors exist [1]. Thus, reliability and robustness design methods have been proposed to solve such problem by guiding optimal solutions away from the boundary of feasible region [2]. Robust design requests that while fluctuating in six-sigma range, product performance meets design requirements. Through the combination of reliability design and robust design based on tolerance model, the reliability and robustness of design are improved by keeping mean response away from constraints and taking sigma level as the evaluation index of quality.

Six-sigma robust optimization was firstly proposed by Motorola, and then it became a popular method in the electronic industry and benefited many transnational corporations. At the beginning of twenty-first century, because of the intense competition in the international automobile market, six-sigma robust optimization was widely adopted by automobile factories and component suppliers such as GM, Ford Motor, Volkswagen, Delphi. At the same time, six-sigma robust optimization became hot topic among electronic industry in China. It is obvious that six-sigma robust optimization will soon be fully applied as the automobile industry in China develops.

In this paper, a six-sigma robust optimization method based on CAE technology using IDDOV robust design process is developed. At the beginning of design process, sensitivity factors that influence the NVH performance of the CRFM are sought and robust design space is defined. Uncertain factors are transformed by

L. Liu (✉) · S. Zhang · X. Li · B. Zhang
CAE Department, Changan Automotive Engineering Institute, Chongqing, China
e-mail: liulin66662008@163.com

Society of Automotive Engineers of China (SAE-China) (ed.), *Proceedings of the 19th Asia Pacific Automotive Engineering Conference & SAE-China Congress 2017: Selected Papers*, Lecture Notes in Electrical Engineering 486, https://doi.org/10.1007/978-981-10-8506-2_17

random models that obey the reliability distribution. The reliability and robustness of product performance are improved by decreasing the effects caused by the random fluctuation of uncertain factors. It can both decrease development costs and improve users' experience in terms of the NVH performance by using this method.

2 Six-Sigma Robust Design

A common deterministic optimization problem can be formulated as follows:

$$\left.\begin{array}{ll}\min & F(x_1, x_2, \ldots, x_n,) \\ s.t. & g_j(x_1, x_2, \ldots, x_n,) \leq 0 \\ & x_{L,i} \leq x \leq x_{U,i}\end{array}\right\} \tag{1}$$

where $x_{U,i}$ is upper bound of design variables and $x_{L,i}$ is lower bound of design variables.

A common robust design problem can be formulated as follows [3]:

$$\left.\begin{array}{l}\min \quad F(\mu_y(x_i), \sigma_y(x_i)) = \mu_y(x_i) + \omega_1 \sigma_y(x_i) \\ \quad + \omega_2 \sum_{i=1}^{m} \sigma_{g_j}(x_i) \\ s.t. \quad \mu_{g_j}(x_i) - n\sigma_{g_j} \geq 0 \\ \quad x_{L,i} + n\sigma_{x_i} \leq \mu_{x_i} \leq x_{U,i} - n\sigma_{x_i}\end{array}\right\} \tag{2}$$

where m is the amount of the constraints. ω_1 and ω_2 are weight factors. μ_{x_i} and σ_{x_i} refer to mean value and standard deviation of random parameter. $x_i.\mu_y$ and σ_y refer to mean value and standard deviation of response. σ_{g_j} represents the standard deviation of constraint. n is sigma level which is depended on the requirement of designers and when $n = 6$ means six-sigma robust optimization.

Compared with deterministic optimization, the standard deviation of constraints and responses is added to the objective function in six-sigma robust optimization in order to seek the most robust performance of objective function and reduce the sensitivities of the uncertain factors.

3 Six-Sigma Robust Design for the NVH Performance of the CRFM

The flowchart of six-sigma robust design is shown in Fig. 1.

(1) The NVH simulation model of the CRFM is set up.
(2) Parametric identification: Performance objective and design parameters are verified.

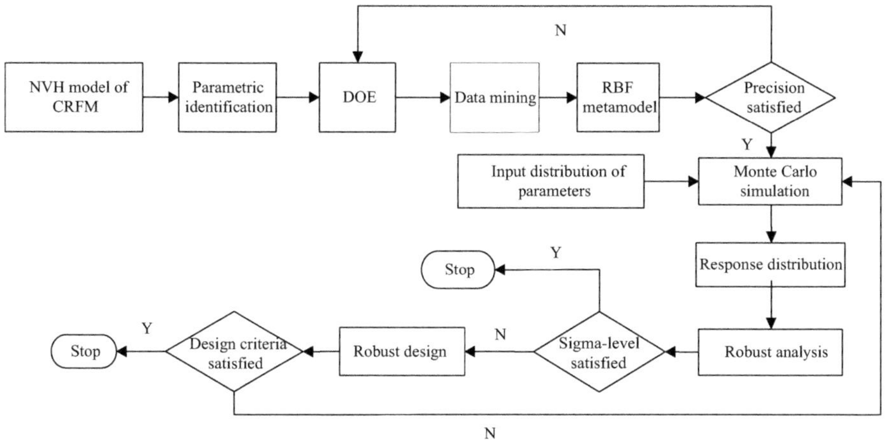

Fig. 1 Flowchart of robust design process for NVH performance of the CRFM

(3) Design of experiment: Samples are generated in the entire design space by choosing a specific DOE method which is suitable to the design problem. The responses of the objective function at these samples are evaluated by calling the CRFM simulation model, and then data mining is accomplished.
(4) Metamodel construction: In order to save the time cost to run an expensive NVH simulation model, RBF metamodels are constructed using all the samples.
(5) Monte Carlo simulation: According to the distribution of design parameters, the distribution of responses is evaluated based on RBF metamodels.
(6) Robust analysis: The sigma level of responses is checked to determine whether the robust optimization should be continued or stopped.

Based on Isight, Fig. 2 shows the flowchart of six-sigma robust design for the NVH performance of the CRFM. Through the first three subprocesses, 32 samples are generated and the responses at these samples are evaluated. The latter two subprocesses complete the data mining, Monte Carlo simulation and six-sigma robust optimization.

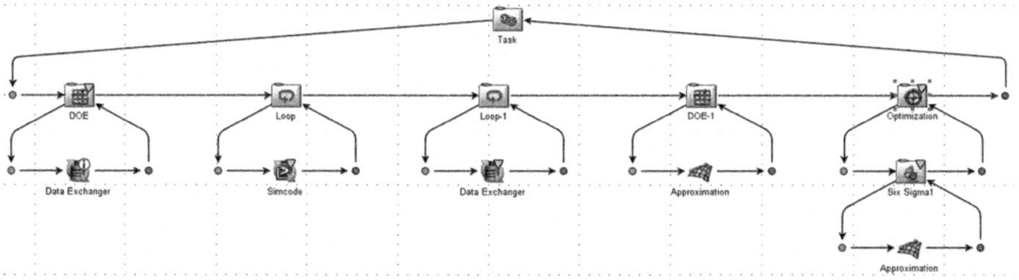

Fig. 2 Flowchart of six-sigma robust design process based on Isight

3.1 Establish the Analysis Model of NVH Performance of the CRFM

Establish the detailed analysis model of NVH performance of the CRFM, as shown in Fig. 3, with the fan imbalance as the excitation which is loaded at the axis of the fan. The responses are accelerations of the point at 12 o'clock on the steering wheel in all three directions (XYZ) of the vehicle coordinate system. The curves of acceleration responses of the point at 12 o'clock on steering wheel are shown in Fig. 4.

Fig. 3 Simulation model of the vehicle

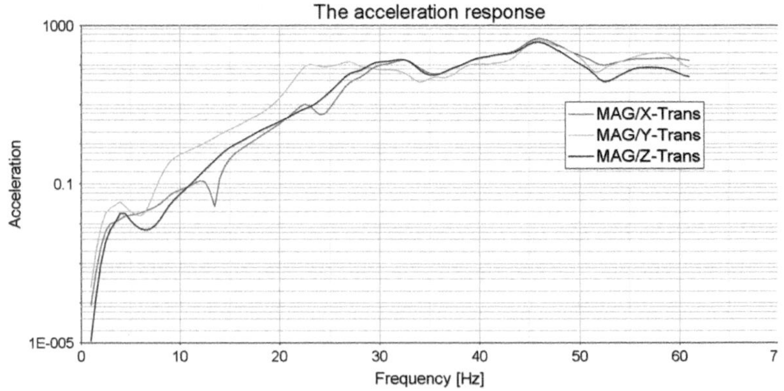

Fig. 4 Curves of acceleration responses at 12 o'clock on the steering wheel

3.2 Parametric Identification

According to the feedback of the customers and the real car tests, the vibration problem of the steering wheel caused by CRFM will be analyzed with the P graph. Three inputs and two outputs of the case are identified in order to get the relevant parameters for the steering wheel vibration caused by CRFM as shown in Fig. 5.

3.3 Design of Experiments

Through the data exchanger component in the Isight, seven relevant design parameters including fan imbalance, pad stiffness of CRFM in different directions causing steering wheel vibration are specified. In order to ensure the impartiality of experiment level, the DOE matrix is designed by orthogonal experimental method [4], and the designed matrix selects the orthogonal array named $L_{32}(4^7)$, which means three-factor, four-level, and 32 times experiments, as shown in Table 1.

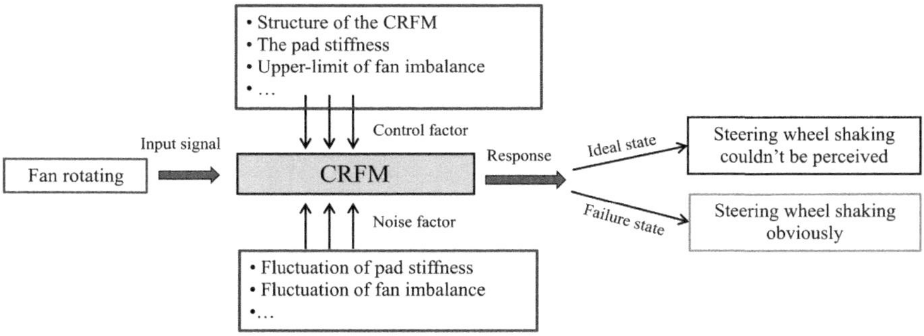

Fig. 5 P graph analysis for NVH performance of the CRFM

Table 1 Design parameters and level

Number	Parameters	Level			
1	bush_up_x	20	80	140	200
2	bush_ up _y	20	80	140	200
3	bush_ up _z	20	80	140	200
4	bush_down_x	20	80	140	200
5	bush_down_y	20	80	140	200
6	bush_down_z	20	80	140	200
7	imbalance	5.0E-06	1.5E-05	3.0E-05	5.0E-05

Sample points are obtained as follows: (1) Design the matrix through the DOE component in Isight, and obtain 32 simulation models corresponding to the 32 sample points. (2) With the application of the LOOP component in Isight, the 32 simulation models could be solved. (3) Through the data exchanger component in Isight, the acceleration responses of the point at 12 o'clock on the steering wheel in different frequencies corresponding high-low fan speed is parameterized and the response results are extracted in batch.

3.4 Metamodel Construction

RBF metamodel [5] is constructed based on the DOE sample points. In order to guarantee the robust analysis and optimization accuracy, RBF metamodel is effective if the accuracy is greater than 90%. Some of the modal accuracy evaluations are shown in Figs. 6 and 7, and RBF metamodel accuracy statistics are shown in Table 2.

Fig. 6 Evaluation of precision for high_x

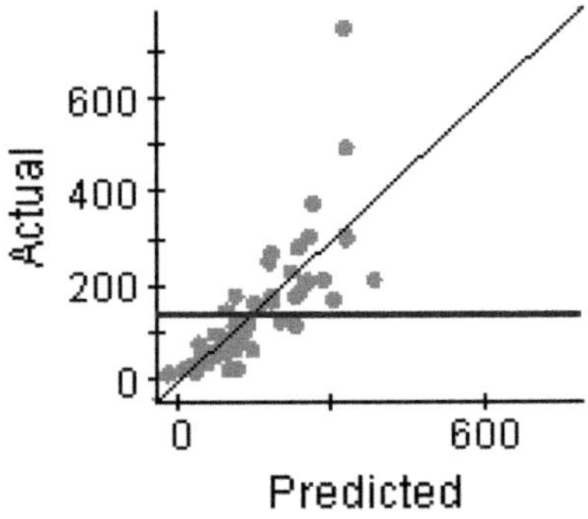

Fig. 7 Evaluation of precision for high_y

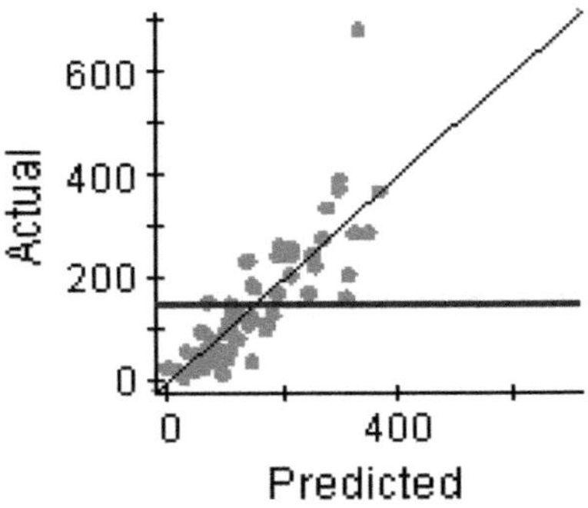

Table 2 Precision for RBF metamodels of acceleration responses

Number	Parameters	Precision (%)
1	high_x	93.8
2	high_y	93.5
3	high_z	93.6
4	low_x	91.9
5	low_y	93.9
6	low_z	92.3

3.5 *Monte Carlo Simulation*

Monte Carlo simulation is a numerical computation [6] based on the theory of reliability and statistics, which use random sampling as the main ways of simulation. The relative functions are used to create random numbers that obey the corresponding reliability distribution and then start the simulation.

According to the multi-batch test results of fan imbalance and pad stiffness, study the parameters and identify the distribution by using statistics method, as shown in Fig. 8. Based on the RBF metamodel sampling method, the Monte Carlo model is built. Then, obtain the random features like mean, variance, range, distribution of acceleration response of the point at 12 o'clock on the steering wheel, as shown in Fig. 9.

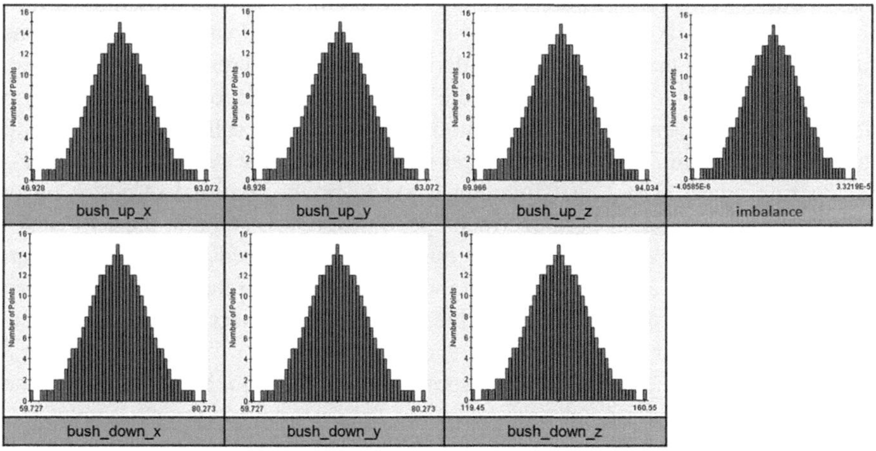

Fig. 8 Design parameters distribution of the CREM

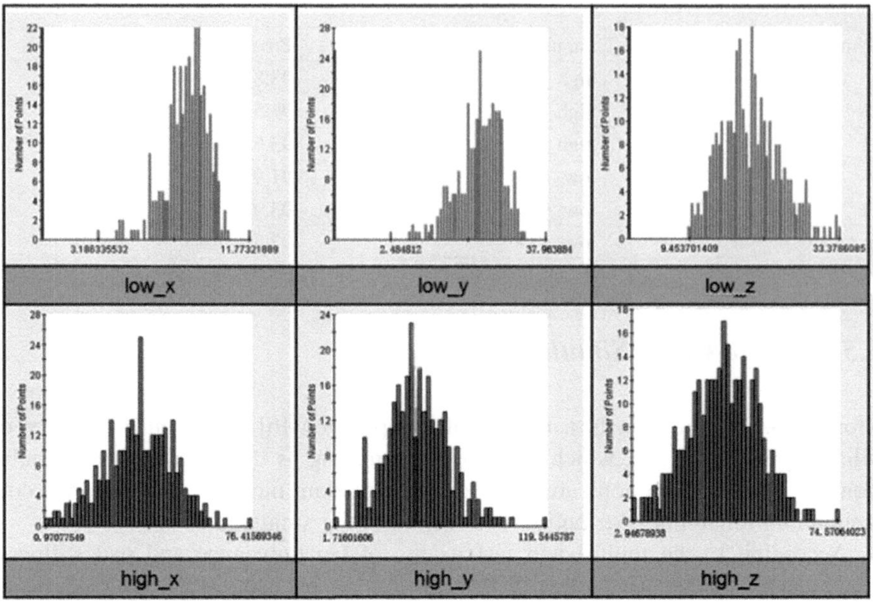

Fig. 9 Acceleration response distribution at 12 o'clock on the steering wheel

3.6 *Six-Sigma Robust Analysis and Optimization*

Deterministic optimization problem is transformed into the reliability optimization problem through six-sigma robust design. Set the target value of the response variables, and then the distribution regularity of acceleration response and six-sigma level at the point at 12 o'clock on the steering wheel is obtained through the simulation. Establishment of optimization model: Using self-adaptive simulated annealing algorithm. The response target is set to 100 at the point at 12 o'clock on the steering wheel. The target function is set to be the minimal response on the steering wheel. No matter high or low range, the weight is equal in the x-, y- and z-directions. Multiple optimization schemes [7] are obtained by Isight software. Taking into account all the factors like the design and manufacture feasibility, one optimization plan is chosen to satisfy the need for the project. The design parameters before and after optimization are shown in Table 3.

By optimizing, the acceleration response sigma level in Y-direction is raised from 1.35 to 8.0 at high range and the rate of the performance target completion is increased from 82 to 100%, as given in Table 4. The response distribution interval before and after optimizing is shown in Figs. 10 and 11.

Table 3 Comparison of design parameters

Number	Design parameters	Initial value	Optimal design
1	bush_up_x	55	32
2	bush_ up _y	55	32
3	bush_ up _z	82	40
4	bush_down_x	70	86
5	bush_down_y	70	86
6	bush_ down_z	140	112

Table 4 Comparison of response sigma level and reliability

Number	Responses	Initial design		Optimal design	
		Sigma level	Reliability (%)	Sigma level	Reliability (%)
1	high_x	8.0	100	8.0	100
2	high_y	1.35	82	8.0	100
3	high_z	8.0	100	8.0	100
4	low_x	8.0	100	8.0	100
5	low_y	8.0	100	8.0	100
6	low_z	8.0	100	8.0	100

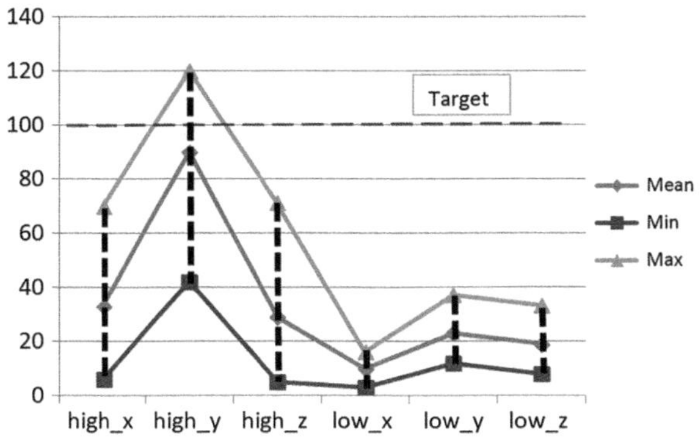

Fig. 10 Response distribution in the initial design

4 Test

In order to verify the performance improvement by using optimal design instead of initial design, four sets of vibration-isolating pad samples are made, and four vehicles are chosen randomly to test the NVH performance of the CRFM. Then, the four sets of pad samples are changed to do the same test once more. The test method is presented as follows: (1) The engine is turned off, and the fan is

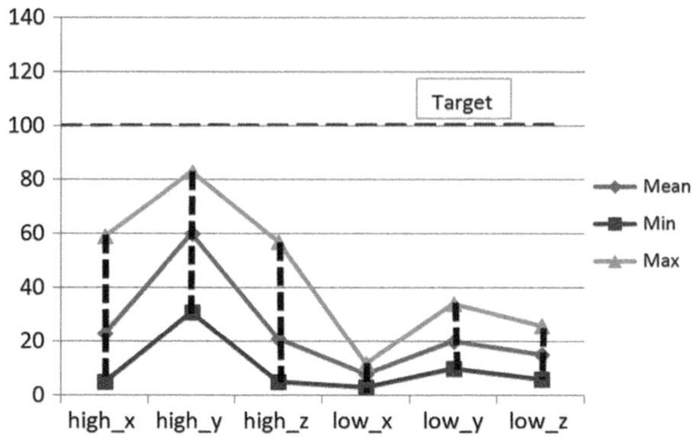

Fig. 11 Response distribution in the optimal design

Fig. 12 Test point of acceleration response

connected to the extra power. (2) The acceleration response is measured at 12 o'clock on the steering wheel as shown in Fig. 12.

According to the test results, (1) the regularities of response distribution are the same between test and simulation. (2) As can be seen from Fig. 13, by using robust optimal design of vibration-isolating pads, the acceleration responses at 12 o'clock on the steering wheel decrease from 29 to 35%.

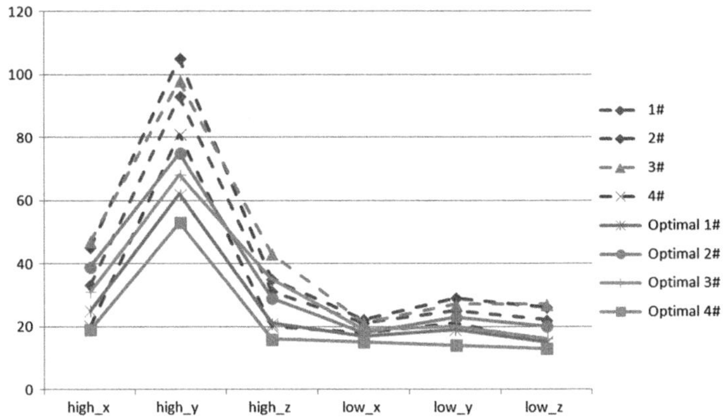

Fig. 13 Comparison of the test results

5 Conclusion

In this paper, the NVH performance of the CRFM is analyzed and optimized for robustness based on Isight that includes many methods such as DOE, RBF metamodel, Monte Carlo simulation, six-sigma robust analysis, and optimization. The paper concludes as follows.

(1) By using the robust design method based on CAE, acceleration response of the point at 12 o'clock on steering wheel under the fluctuation of fan imbalance and pad stiffness has been simulated by the random model under the certain reliability distribution, the most robust design parameters and sensitive factors have been found, the low cost and high robustness designs have been implemented, the cost of quality improvement has been reduced, the vehicle performance has been improved from the source of design, and the NVH experience of customers has become better.
(2) Taking into consideration the influence of the fluctuation of fan imbalance and pad stiffness at 12 o'clock on steering wheel, the reliability that response reaches the target has been increased from 82 to 100% through the optimization of this method.
(3) The regularities of response distribution are the same as simulation according to the test results, which shows the accuracy of the simulation. The acceleration responses at 12 o'clock on the steering wheel decrease clearly with the robust optimal design of vibration-isolating pads, proving the validity of the scheme.

References

1. Maglaras G, Ponslet E, Haftka RT et al (1996) Analytical and experimental comparison of probabilistic and deterministic optimization. AIAA J 34(7):1512–1518
2. Nikolaidis E, Stroud WJ (1996) Reliability-based optimization: a proposed analytical-experimental study. AIAA J 34(10):2154–2161
3. Koch PN, Yang RJ, Gu L (2004) Design for six-sigma through robust optimization. Struct Multidiscip line Optim 26:235–248
4. Ye KQ (1998) Orthogonal Column Latin hypercubes and their application in computer experiments. Am Stat. J Am Stat Assoc Theor Methods 27(11):1430–1439
5. Liping W, Don B. A Comparis on of meta-modelling methods using practical in dustry requirements. AIAA 2006–1811
6. Siebert B, Sommer K (2004) New development of the GUM and Monte Carlo techniques. Tech Mess 71(2):67–80
7. Shimoyama K, Oyama A, Fujiii K (2008) Development of multi-objective six-sigma approach for robust design optimization. J Aerospace Comp, Inf Comm 5(8):215–233

Automobile Corrosion Mechanism and Cases Analysis

Dapeng Zhao, Minglei Liu, Guofeng Li, Chunyan Tan and Bo Feng

1 Introduction

According to the statistic data, automobile holding quantity in China had reached 193 million in 2016 and will exceed 200 million in 2017. Annual losses brought by metal corrosion is about 150–250 dollars for one single vehicle, and thus it can be estimated that total losses in China incurred by corrosion will be 30–50 billion dollars in 2017. For vehicle enterprises, appearance, performance, and security of vehicles will be directly influenced by metal corrosion, which can severely damage the enterprise brand value as well as customers' economic profit and security. Therefore, vehicle corrosion has got more and more attention in recent years. This paper made a detailed analysis on causes of vehicle metal corrosion based on corrosion theory and proposed corresponding anticorrosion measures in combination with actual corrosion cases.

2 Corrosion Mechanisms

Electrochemical corrosion refers to the phenomenon that a short circuit galvanic cell that consists of anode and cathode is formed by impure metal in corrosive medium, and anode with lower negative potential will be dissolved [1]. Main reactions are as follows:

D. Zhao (✉) · G. Li · C. Tan · B. Feng
Brilliance-Auto Researching & Development Center, Shenyang 110000, China
e-mail: zdp19880616@163.com

M. Liu
China Petroleum Pipeline Engineering Co. Ltd, Langfang 065000, China

Society of Automotive Engineers of China (SAE-China) (ed.), *Proceedings of the 19th Asia Pacific Automotive Engineering Conference & SAE-China Congress 2017: Selected Papers*, Lecture Notes in Electrical Engineering 486, https://doi.org/10.1007/978-981-10-8506-2_18

$$\text{Anode: } M - xe^- = M^{x+} \tag{1}$$

$$\text{Cathode: } O_2 + 2H_2O + 4e^- = 4OH^- \tag{2}$$

$$2H^+ + 2e^- = H_2 \tag{3}$$

where the formula (2) describes reaction of oxygen absorption corrosion while formula (3) for hydrogen evolution corrosion. Oxygen absorption corrosion will control the corrosion in case of enough oxygen on the reaction surface; otherwise hydrogen evolution corrosion will prevail.

As shown in Fig. 1a, metal A and C with different activity have corresponding natural corrosion potential $\varphi_{e,a}$ and $\varphi_{e,c}$ when put into corrosive medium. Owing to unequal electrode potential, metal A and C will be inter-polarized and get a coupling potential φ_{corr} finally. Therefore, potential of both electrodes, anode and cathode will get a deviation from its own natural potential to generate potential difference, where:

$$\text{Anode: } \Delta E_a = \varphi_{\text{corr}} - \varphi_{e,a} \tag{4}$$

$$\text{Cathode: } \Delta E_c = \varphi_{\text{corr}} - \varphi_{e,c} \tag{5}$$

The current will be generated due to the electric potential deviation, and the current on the anode surface is corrosion current, of which value could be read from Tafel Curve. Tafel Curve describes the relationship between electrode potential deviation and surface current density, as shown in Fig. 1b, in which vertical coordinate of middle horizontal part is natural corrosion potential, top part as anodic polarization curve, and lower part as cathodic polarization curve. With the Tafel Curve, current density generated by potential gradient can be acquired.

Faraday's law describes the relationship between electrode reaction current density and electrode weight losses [2], as shown in the formula below. And the value of electrode reaction current density can reflect corrosion progress.

$$i = z \cdot F \cdot \frac{m}{Mst} \tag{6}$$

where

Z is element valence;
F is Faraday's constant, 96485 C mol^{-1};
M is molar mass, g mol^{-1};
s is electrode surface area, cm^2;
t is electrode reaction time, year.

Fig. 1 E-chemical corrosion mechanism diagram

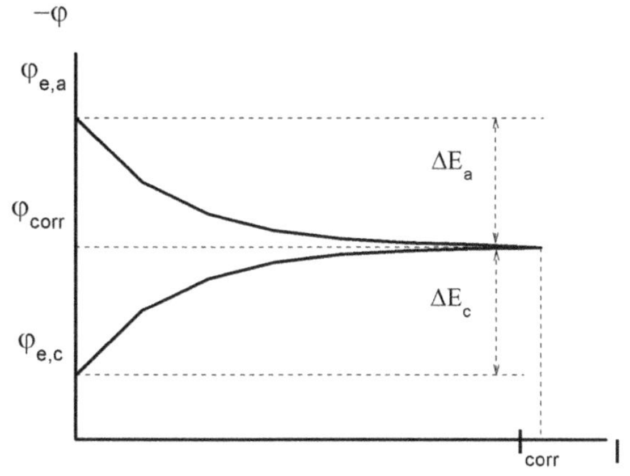

a. Inter-polarized curve

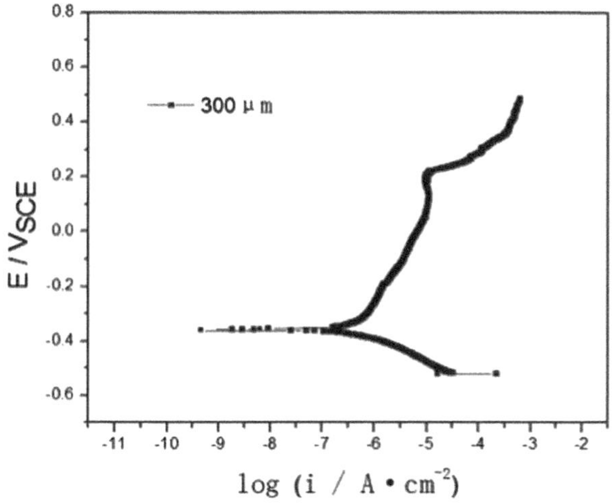

b. Tafel-Curve

3 E-Chemical Corrosion

Corrosion is caused by metal oxidation in water or moisture circumstance of which most are e-chemical process. Currently, normal vehicle corrosion types include micro-difference corrosion, pitting, bubbling corrosion, crevice corrosion, galvanic corrosion, SCC, and corrosion fatigue.

3.1 Micro-difference Corrosion

Due to micro-difference of material component, atom array, phase and tissue, various micro-zones have corresponding electrochemical activity and will form galvanic cell when in contact with corrosive medium. Micro-differences mainly include: phase and tissue difference generated during phase thermo-transferring, such as the difference of martensite, ferrite, pearlite, retained austenite; grain and grain boundary difference generated during crystallization by composition segregation; stainless steel chrome-depleted region generated by compound Cr23C6 yielding under thermal affection [3], shown in Fig. 2. As most metal used in structure or machinery is alloy, micro-differences widely exist on metal surface. Most of them can cause general corrosion of which some could develop into pitting.

3.2 Pitting

As shown in Fig. 3, pitting is commonly found at flat position. The protecting layer, such as oxidation membrane or metal coating, is not uniform, and then the weak point would be corroded, and the corrosion product will cover these points and hinder oxygen diffusing to electrode surface, so the oxygen is reduced at surrounding area and the covered metal which acts as anode will be continually dissolved into a pit. Therefore, the sealed region is formed, and positive ion Fe^{2+} is accumulated in it. For balancing the positive charge, smaller negative ion Cl^- could diffuse into the pit and make it acidic. The localized acidic circumstance will

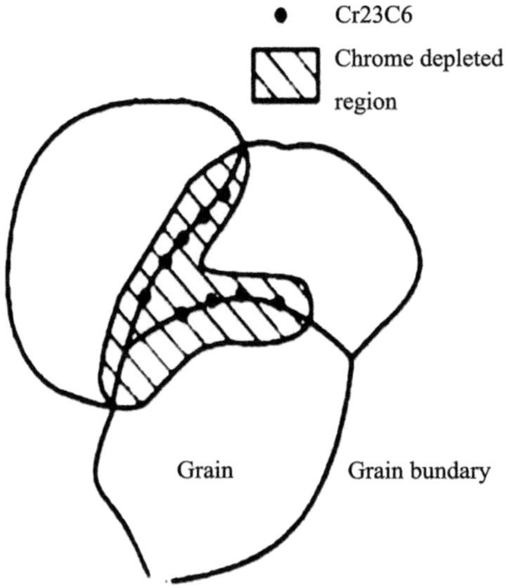

Fig. 2 Cr23C6 yielded in chrome-depleted region

accelerate metal dissolving, and its dissolving rate is much higher than the average corrosion [4]. Owing to its high dissolving rate, difficulty in detection and prediction, pitting is one type of the most hazardous corrosions.

3.3 Crevice Corrosion

When immersed into corrosive medium, crevice corrosion has the similar mechanism with pitting, and the difference is that the product will be accumulated at the crevice opening at initial, and the sealed region will be formed in the crevice. But for the most of automobile crevice corrosion, they are not in accordance with above electrochemical mechanism. Vehicles almost are serving under atmospheric environment, if vehicle gets waded, the water will travel into crevices from the opening under the capillary action. The inside water is hard to expel outside and makes crevice long-term moist. Besides the strict crevice, closed angle corner and fastener have similar mechanism to make them vulnerable against corrosion.

3.4 Galvanic Corrosion

Galvanic corrosion is the most typical electrochemical corrosion. Two kinds of different electrochemical activity metal will be inter-polarized when contact with each other and immersed into corrosive medium, and the active metal will be anode and be corroded [5], shown in Fig. 4. For vehicle, the galvanic corrosion of bulk metal is not common and cases are mainly occurred between anodic coating and base metal, and the anodic coating will be rapidly dissolved, such as galvanized fasteners and tube connectors, fasteners with Dacromet coating.

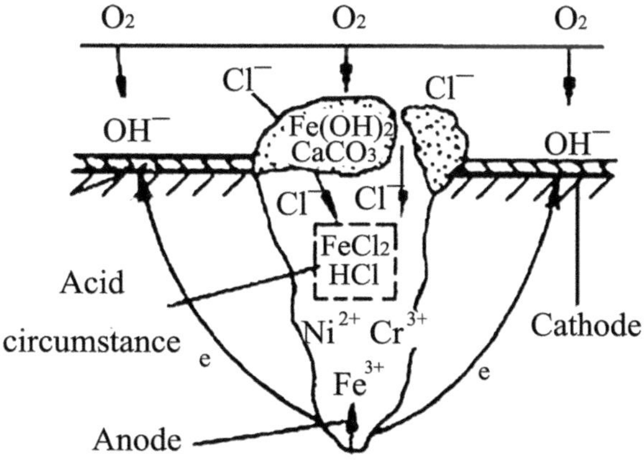

Fig. 3 Pitting mechanism diagram

Fig. 4 Galvanic corrosion diagram

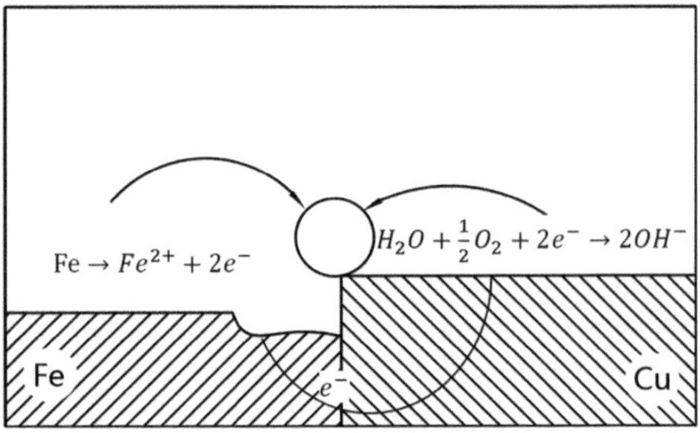

3.5 Stress Corrosion Cracking (SCC) and Corrosion Fatigue

As shown in Fig. 5, micro-cracks exist on the loaded metal structure surface, which break the material continuity and redistribute the stress, and the stress concentration will be generated on crack tip.

Higher stress will improve the metal activity, accelerate crack dissolving and extension rate, until brittle fracture occurs because of the residual metal strength cannot resist the load [6]. SCC is one failure form contributed by both stress and corrosive medium, and its dissolving rate is much higher than the simple superposition of two factors affecting separately. As there is no obvious deformation observed on the metal structure, the fracture is poor in predictability and will be catastrophic accident in case of occurrence. Similar with SCC, but corrosion fatigue is brought by alternating stress and dynamic load of which failure stress is much lower. The crack extension includes active crack tip dissolution and corrosion fatigue typical crack extension.

Fig. 5 Stress corrosion cracking mechanism diagram

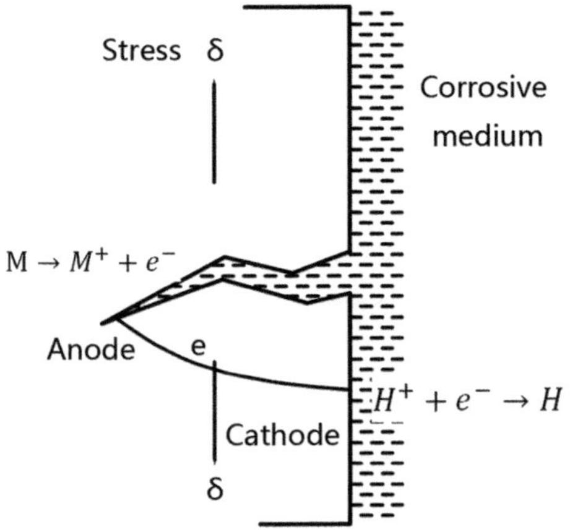

4 Corrosive Factors and Cases Analysis

4.1 Auto-Corrosion Complaints Statistics

866 pieces of automobile corrosion complaints in recent 2 years are collected in this paper, and the result is shown in Fig. 6.

Fig. 6 Automobile corrosion complaints statistics

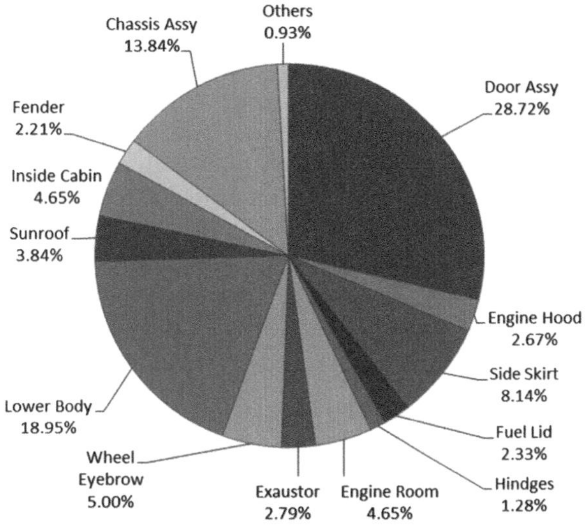

(a). Auto-corrosion distribution

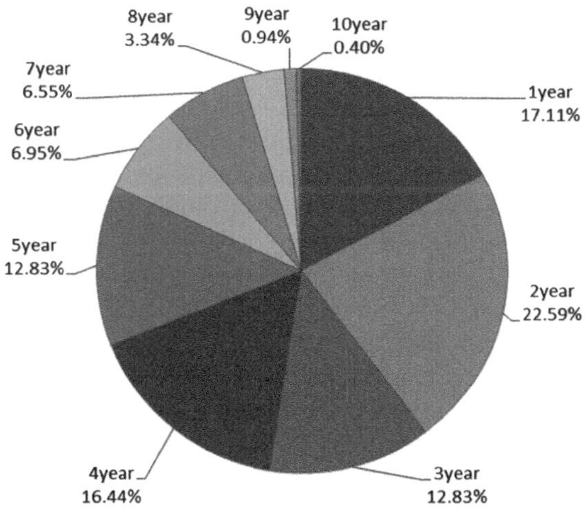

(b). Corrosion occurring time after purchasing

From the pie chart (a) in Fig. 6, corrosion distribution could be read, and most of them are grouped in door assy, side skirt, lower body, and chassis. Common problems, including visible rust, coating peeling, cannot cause serious consequences, but it can make a ugly out looking of vehicles which the owner do not want to accept. 7.16% of the complaints are material failure, including corrosion perforation and fracture. These problems would bring about terrible result or catastrophe if the failure parts are located at key structure position.

Pie chart (b) shows corrosion occurring time after purchasing. As the chart shows, corrosion mostly occurred in first 6 years. Automobile industry widely identified that there is no visible rust in first 6 years and no corrosion perforation for economic vehicle in first 10 years, and there is no visible rust in first 10 years and no corrosion perforation in first 12 years. Therefore, it can conclude that these complaints are quality problems.

So anticorrosion problems are existed, and it can come out that automobile corrosion could be affected by various factors, and those factors could be classified into two categories, which include internal factor dominated by materials and processes and external factor dominated by corrosive environment. Combining the statistics and two projects, automobile road accelerated corrosion tests, the following test will detail influencing factors.

4.2 External Factors

4.2.1 Hydrographic Environment Effect

Hydrographic environment includes relative humidity, rainfall, car washing, road surface gathered water. Water acts as an essential factor in corrosion. The relative humidity represents the water membrane formation ability on a metal surface, and according to relevant research, corrosion rate will increase dramatically when humidity is beyond 60% [7]. Car washing, rainfall, and road surface gathered water can moisten the metal structure in the form of spraying or plashing, which makes protective coating damage and corrosion is more liable to occur.

In automobile corrosion complaint statistics, as shown in Fig. 6a, 28.72% cases were found at door drainage outlet, and 18.95% cases were found at lower body, and 8.14% cases were found at lower side skirt, and 13.84% cases were found at chassis part. It is known that these areas above have more chance to contact with water. When water intrudes into door panel cavity, it will gather and flow at bottom of cavity, and finally drain out from the lower hole. As the water has wetting ability, drainage outlet would keep long-term moisture, and organic coating will fail. Lower body, chassis, wheel bow, and fender are closer to the ground, so water is more likely spraying or plashing and attaching to them during car running.

According to statistics, in these areas, metal plate corroded really heavy, and in some cases large area holes were formed or massive material lost on metal plate, as shown in Fig. 7. Though corrosion is caused by comprehensive factors, and rich

Fig. 7 Automobile corrosion complaint statistics

(a). Left rear side lower body corrosion perforation

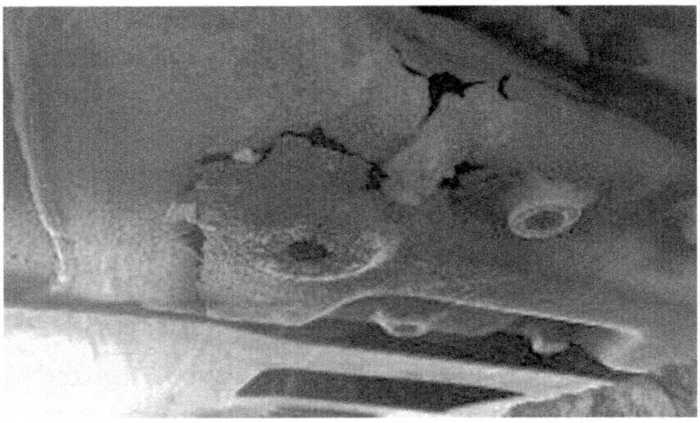

(b). Rear beam corrosion failure

moisture environment is not the only factor of corrosion failure, water plays a really important role in the reaction.

4.2.2 Temperature and UV Light

Chemical reaction rate has a strong correlation with temperature, and the relationship had described by Arrhenius, as shown in formula 7, so it can come out that corrosion rate will improve as temperature increases. Furthermore, metal material will cool faster for its high thermal conductivity, and water vapor in air is liable to condense into liquid on metal surface to form corrosion conditions.

$$k = Ae^{-\frac{Ea}{RT}} \tag{7}$$

where

k	is reaction rate;
A, E, R, a	are related with chemicals;
T	is temperature.

The combined effect of high temperature and UV light contributes to organic coating aging. High temperature or UV light can enhance polymeric segmental relaxation, and then chemical bonds are more risky to break. The accumulation of bonds breaking will cause atom losing, and then porosity will rise and corrosion resistance will decrease. In addition, owing to different thermal expansion rate between coating and base metal, long-term high- and low-temperature cycling will further accelerate the peeling off. Though it does not bring about metal losing, it does damage to metal coating and makes metal contact corrosive with media directly.

4.2.3 Chloride Effect

Chloride ion has high migration rate and penetrability because of its tiny diameter, and chloride ion could destroy passivation membrane on metal surface and expose fresh metal to corrosive medium. In pit and crevice zones, acidic corrosive medium will be generated by accumulated chloride ion. Chloride ion has various sources mainly comes from dust, mud, coast atmosphere, road deicing salt, etc.

Related research has proven that chloride density determines the relative influence of corrosion rate, shown as in Table 1. The corrosion rate will improve with chloride density rising [8].

The chloride is more likely to cause localized corrosion, so weight loss cannot reflect corrosion degree comprehensively. Electrochemical impedance spectroscopy (EIS) can describe corrosion reaction procedure from the mechanism. The value of electron transfer resistance R_{ct}, which represents the Fe oxidation reaction rate, could be read in fitting circuit.

Table 2 shows the Q235 steel R_{ct} value soaking in snow water and snow water with chloride for series period of time. From the date in Table 2, it can conclude that chloride decreases the resistance value sharply [9].

Table 1 Relationship between chloride density and corrosion rate

$w(Cl^-)$ (%)	0	1	2	3
Bare steel mm/year	0.09741	0.10415	0.13656	0.014199
Galvanized steel mm/year	0.05256	0.06660	0.07097	0.07310

Table 2 Influence of chloride on electrotransfer

Soaking period (h)	24	72	120	169
Snow water $R_{TC}/\Omega \cdot cm^{-2}$	6564	7161	9862	7687
Snow water with chloride $R_{TC}/\Omega \cdot cm^{-2}$	1837	1394	1488	1178

4.2.4 Silt and Gravel Spattering

When organic protective coating is damaged by spattered gravel or silt kicking on surface of running vehicles, which has the largest influence on cold-rolling non-galvanized metal plate, the bared metal will contact corrosive environment directly and corrosion will happen. Such kind of corrosion can be more easily found at chassis and side apron, because electrophoresis coating is the only anticorrosion procedure of most chassis parts, and electrophoresis coating is really fragile under spattered gravel or silt attack, as shown in Fig. 8. For such kind of corrosion, the most common methods could be adopted at present are PVC armor spraying and chassis shield to prevent gravel attacking electrophoresis coating.

4.3 Internal Factors and Cases Analysis

4.3.1 Welding Surface Quality

Welding quality can severely affect the surface forming quality [10], which will also directly influence the follow-up coating quality. Currently, main welding technologies adopted in automobile enterprises are spot welding and arc welding. For spot welding, columnar copper electrode can conduct welding current as well as provide restraint stress at welding point. In case of wide crevice between two metal sheets, the electrode will provide much higher restraint stress, and then large deformation will appear at the welding point owing to metal strength decline caused by heating. Columnar copper is softer than metal plate and its geometry size will decrease by abrasion during welding process, so proper welding spot cannot be got.

Fig. 8 Chassis corrosion without armor or shield

Fig. 9 Welding quality influence on coating

a. Spot welding

b. Gas shield welding

Otherwise, high electrode columnar pressure, circuit value or welding time period can cause spot burning through; metal sheet surface contamination and electrode columnar also could affect the surface quality and anticorrosion ability, shown as in Fig. 9a.

Gas shielded welding and filling metal are widely adopted in arc welding technology. The parameters in gas shielded welding, such as circuit value, surface contamination, arc length, purity and component of shielding gas, humidity would affect the surface quality. If the parameters' value cannot satisfy wielding process specification, splash will easily produce around welding pass. Therefore, the surface formed will not be smooth enough to prepare a good coating, shown in Fig. 9b.

4.3.2 Organic Coating

Leng·A had pointed out that water will travel through the organic coating to the coating/base metal in really short time [11], and the ion density is five times than the surface. Besides, coating peeling also needs electric conductive oxidation layer and active water on interface. Coating quality mainly refers to macro- and micro-defects of organic coating, coating thickness, etc. Coating defects play a key role in coating failure. It works as the water and ions travel channel, when coating contacts with corrosive medium, oxygen, water, and other ions would diffuse to coating/base metal interface and form continuous or discontinuous liquid phase. Diffusion kinetics is mainly provided by concentration gradient, osmotic pressure, temperature gradient, etc.

At the interface, metal atoms will oxidized into metal ions, and the negative ions will diffuse to the interface to balance the metal ions positive charge. The negative ions accumulation will produce acidic corrosive environment and boost the corrosion. Metal ions and negative ions form corrosion product and it will peel off protective coating [12], as shown in Fig. 10a. Wapner·K found the water and ions diffusion rate at the coating/base metal interface is much higher than in polymer coating [13], and it is irrelevant with polymer components. So the corrosive medium will penetrate to the coating/metal interface from coating defects, and then diffuse forward along the interface to cause bubbling, as shown in Fig. 10b. Zinc galvanizing would suppress this problem properly. Zinc coating forms a thin layer of passive film, which can decrease corrosion rate and delay bubbling, shown in Fig. 10c.

Therefore, organic coating quality is really important in automobile anticorrosion, and it is better combining with zinc galvanizing technology.

4.3.3 Anodic Coating

Anodic coating is the metal coating with higher electrochemical activity than metal base, and usually has passivation property. As protective, anodic coating is made continuously on metal surface, it will separate metal and corrosive medium, and it has certain anticorrosion property owing to its passivation layer. If anodic coating is damaged, coating and metal will form a natural galvanic cell, where coating is the anodic electrode due to its higher negative potential [14]. The cathodic protection mechanism is shown in following diagram. Zinc will release electrons in procedure of oxidation, and the electrons will transfer to the bare base metal, so Fe cannot turn to ion by losing electrons, shown as Fig. 11.

As the anodic electrode has much larger area than cathodic electrode, corrosion rate is really slow, and anodic electrode will provide efficient electrons for metal base to prevent it from oxidation. Anodic coating has excellent corrosion resistance in atmospheric or low moisture environment, but in heavy moisture, especially with chloride ion environment, corrosion resistance will decline sharply, and it is much obvious to be found in zinc galvanic coating, as shown in Fig. 12a. Dacromet coating has excellent stain resistance performance when it keeps completed. But if it is damaged and contact with other negative metal, galvanic corrosion will cut down its service life [15], as shown in Fig. 12b, and it is much more serious when is serviced with stainless steel. For this kind of corrosion prevention, fasteners or fittings galvanized separately should not be adapted in heavy moisture environment [16], or some of them, such as screws, could be replaced by nonmetal ones. Dacromet coating has relatively higher corrosion resistance, but should be isolated from other bare metal.

Fig. 10 Coating failure

(a). Organic coating peeling

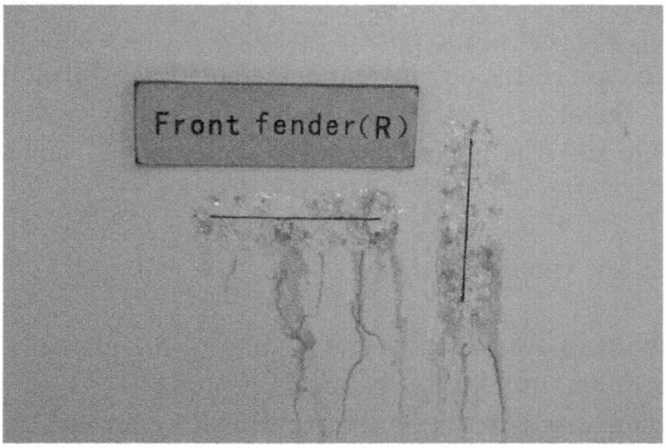

(b). Organic coating bubbling on cold rolling sheet

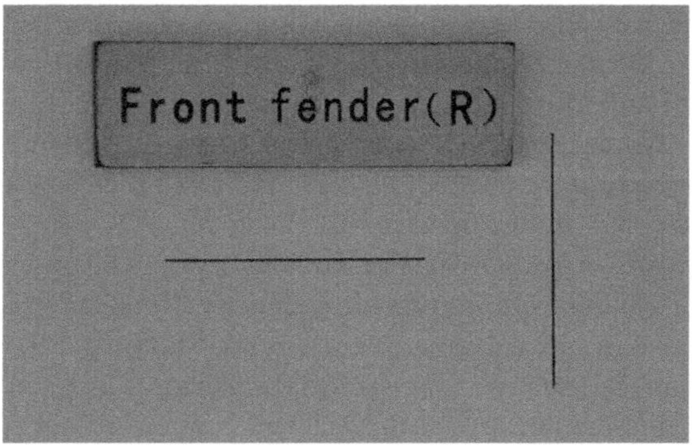

(c). Salt spray test on zinc galvanized sheet

Fig. 11 The mechanism of cathodic protection

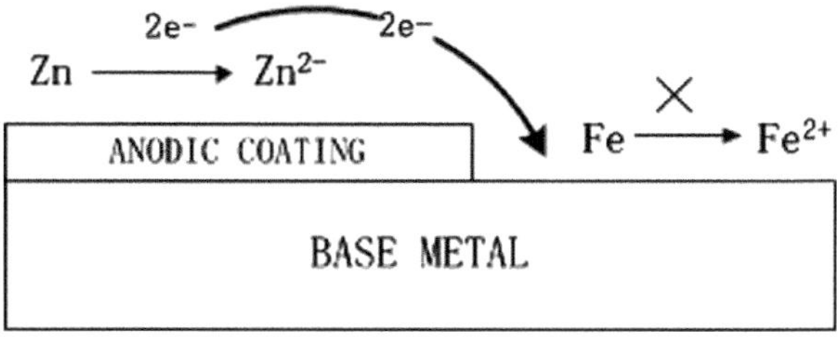

Fig. 12 Anodic coating corrosion

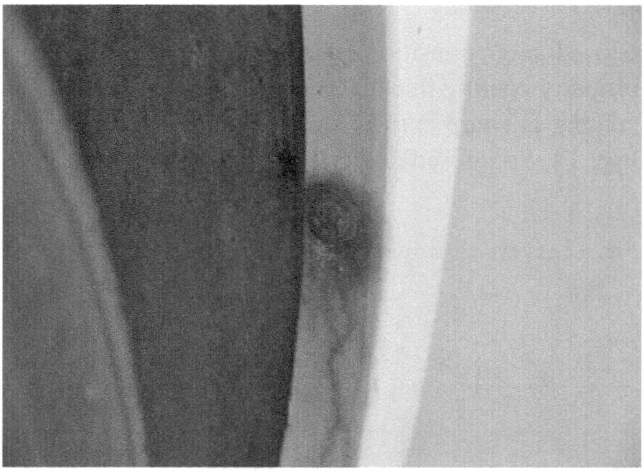

(a). Zinc galvanizing screw corrosion

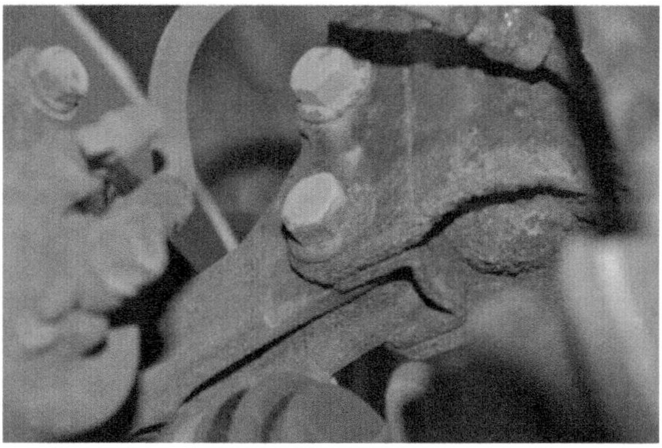

(b). Dacromet coating corrosion

4.3.4 Crevice and Fastener

The main reason for corrosion occurred at crevice or fastener position is that those places are easy to accumulate water and keep its inner side moisture, shown as in Fig. 13. Crevice avoiding is the most effective measure to prevent this corrosion, such as getting accurate geometric size of metal sheet, improving spot wielding sealant spraying quality, sealing metal plate crevice by structure glue [17], folding and sealing the edge in multilayer stamping parts, spraying grease and taking rubber jacket at screwed fittings like rack end and tie rod end.

4.3.5 Closed-Angle Transition Effect

Closed-angle transition effect means that the transition line angle is too acute to form property coating thickness in galvanizing, electroplating, painting, etc. Besides that, coating at sharp corn is easily broken, and metal is more active [12], as shown in Fig. 14. So, closed angle transition should be grained into obtuse.

Fig. 13 Crevice corrosion

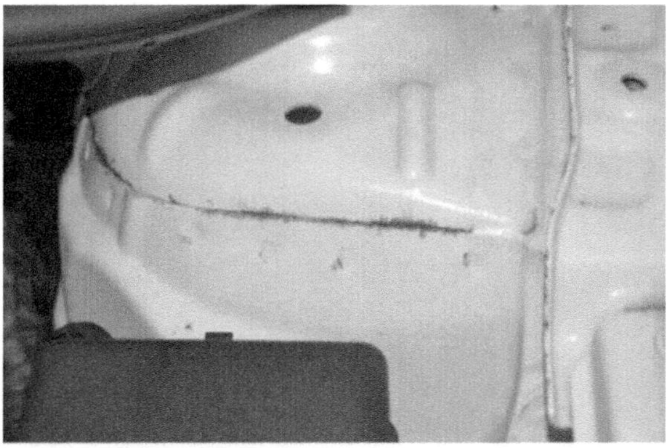

(a). Crevice corrosion in engine room

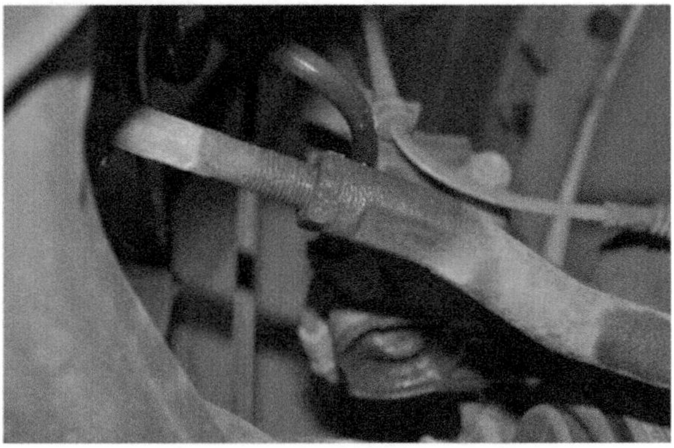

(b). Fastener corrosion at rack end

Fig. 14 Sharp corner transition corrosion

4.3.6 Impeded Drainage

Drainage is the principle concern in structure designing. Drainage hole should be set at lowest point, but it cannot guarantee good drainage. Runoff water will flow outside from the hole after it comes into cavity, but water will still exist in forms of membrane and crevice water as the effects of wetting action and metal plate crevice. Long-term moisture will damage protective coating and corrode base metal. Furthermore, porous corrosion products will contribute to water retention and intensify corrosion [13], shown in Fig. 15a. Hydrophobic coating, structure glue filling in metal crevice structure, and cavity ventilation improvement can help to inhibit the corrosion process.

Muffler pitting is another typical case. With lower temperature at end part of muffler, moisture in exhaust gas will be condensed into liquid, and nitrogen oxides, sulfur oxides and chloride will dissolve into it. As there is no drainage path, the condensed liquid could only be vapored out by residual heat and liquid salinity rises up during evaporation process to become much more corrosive, to produce muffler pitting damage [3, 4], shown in Fig. 15b. Stainless steel is the main anticorrosive measure adopted for muffler at present. SUH409 stainless steel is commonly used in mid-and-low class vehicles owing to its lower price. But generally, stainless steel will form effective membrane when the functional element content reaches 12%, while in SUH409 the value is 11%, so its anticorrosion performance is relatively worse when compared with SUS410, SUS430, SUS430 M, SUS436L, etc. With increasingly fierce market competition, as well as customers' demand for high-quality products, the profit margin is gradually compressed and products with high quality and lower price will be more favored. So low-cost material combined with drainage structure designing to inhibit corrosion will be the mainstream in future muffler anticorrosion development.

Fig. 15 Impeded drainage corrosion

(a). Door panel pitting

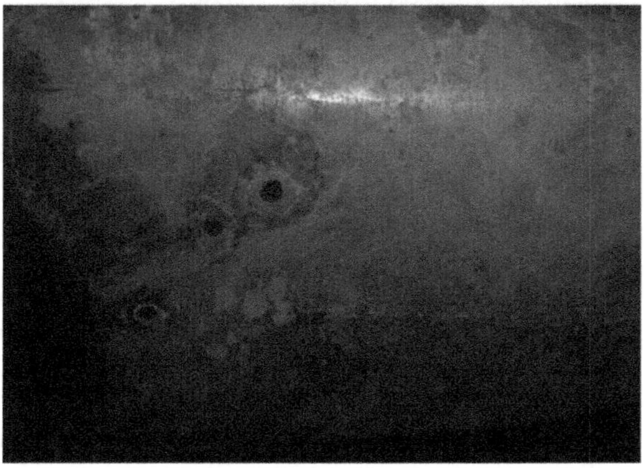

(b). Muffler pitting

5 Conclusions

Automobile metal corrosion is a complicated electrochemical procedure, which corrosion mechanisms are various due to different material, structure, and corrosive medium. Therefore, protective measures should be chosen as per corrosion mechanism rather than simply adopt single or similar measures based on salty spray test strength. Furthermore, corrosion is one process influenced by multiple factors, which shall be considered comprehensively during anticorrosion designing in order to obtain low-cost and high-quality product. Finally, some effects could be generated when multiple measures are adopted simultaneously, the positive effects should be utilized, while the negative ones should be avoided.

References

1. Cao C (2008) Principles of electrochemistry of corrosion. Chemical Industry Press, Beijing, pp 12–31
2. Perez (2003) Electrochemistry and corrosion science. Chemical Industry Press, Beijing, 68–73
3. Li H, Jiang Z (2013) Corrosion behavior of ferritic stainless steel with 15wt% chromium for the automobile exhaust system. Metall Mater 20(9):850–859
4. Tzanakis I, Bolzoni L, Eskin DG, Hadfield M (2017) Evaluation of cavitation erosion behavior of commercial steel grades used in the design of fluid machinery. Metall Mater Trans A 5:2193–2206
5. Xiaohong Z, Quanzhong G (2015) Galvanic corrosion behavior of couples of hot rolled steel SS400 and cold rolled steel ST12 with two coating. J Chin Soc Corrosion Prot 1:86–90
6. Zhao J, Chevil K, Eadie R (2017) Crack growth modeling and life prediction of pipeline steels exposed to near-neutral pH environments: dissolution crack growth and occurrence of crack dormancy in stage I. Metall Mater Trans A 4:1629–1640
7. Burgstaller W, Schimo G, Hassel AW (2017) J Solid State Electrochem, 1–12
8. Shi Y, Liang P, Liu F (2014) Automobile zinc galvanized steel corrosion behavior in deicing salt solution. Mater Prot 47(3):53–55
9. Ning S, Shi Y, Liang P, Lin L (2015) The corrosion behavior of automobile steel in the snow water with snow melting salt. J Liaoning Shihua Univ 35(1):58–61
10. Serebryakov An V, Shulin EL, Serebryakov Al V, Bogatov AA (2004) Metallurgist (9):487–490
11. Leng A, Streekel H, Stratmann M (1999) The delamination of polymeric coating from steel. Part 2: First stage of determination, effect of type and concerntration of cations on delamination. Corros Sci 41:579
12. Choi Y-S, Kim J-G, Kim Y-S, Huh J-Y (2008) Corrosion characteristics of coated automotive parts subjected to field and proving ground tests. Int J Automotive Technology, 2008, No.5 p. 625–631
13. Wapner K, Stratmann M, Grundermier G (2006) Electrochim Acta 51(16):3303–3315
14. Zhao D (2014) Study on hydrogen permeation and hydrogen Embrittlement of X80 Pipeline Steel and its HAZ Caused by Cathodic Protection. Master 3–12
15. Zhu X, Chen X (2013) Dacromet fastener corrosion analysis and improvement measure. Locomotive & Rolling Stock Technology 41–42
16. Wang Y, Cai Y (2013) Discussion of common corrosions from vehicel corrosion test and anti-corrosive optimization. In: The 7th Sino-American Academic Symposium on environmental corrosion and degradation tests of material, 2013, May 20–21, Guanzhou, pp 31–37
17. Jie W-z, Peng W, Zhao Z (2015) Representative corrosion problems analysis of automobile accelerated corrosion test. Experience Exchange 7(29):43–45

Capacity Allocation of Driving System and Braking System for Distributed Drive Electric Vehicle

Dejian Meng, Zhuoyu Li and Lijun Zhang

1 Introduction

Safety, energy conservation, environmental protection, and comfort are timeless themes in the development of automobile industry, and it is well-recognized that electrification and intellectualization are frontier domains of automobile technology [1]. A distributed drive electric vehicle driven by wheel-hub motors or wheel-driving motors has advantages of compact structure and high transmission efficiency, while the driving force as well as the braking force of each wheel can be controlled independently, rapidly, and precisely by controlling driving motors, which possesses the technical feature of both electrification and intellectualization and has become a focus and hot topic of automobile industry [2].

The mainstream inter-axis capacity allocation method for the driving and braking systems of distributed drive electric vehicle adopts uniform distribution among four wheels, neglecting the effect of load transfer during acceleration and deceleration, and the technological superiorities of distributed drive vehicles are not fully utilized to improve vehicle's dynamic property and braking performance. Meanwhile, existing research papers mainly concentrate on inter-axis dynamic allocation strategies of driving and braking force [3, 4], neglecting the effect of inter-axis capacity allocation [5, 6]. Therefore, it is of great necessity and importance to carry out research on the optimal method of capacity allocation for distributed drive electric vehicle so as to attain better dynamic and braking performance.

Optimized Design and Coordinated Control of Electrified Chassis System equipped with In-Wheel or Close-to-Wheel Motors (U1564207).

D. Meng · Z. Li (✉) · L. Zhang
Collaborative Innovation Center for Intelligent New Energy Vehicle,
Tongji University, Shanghai 201804, China
e-mail: lzhy0512@163.com

Society of Automotive Engineers of China (SAE-China) (ed.), *Proceedings of the 19th Asia Pacific Automotive Engineering Conference & SAE-China Congress 2017: Selected Papers*, Lecture Notes in Electrical Engineering 486, https://doi.org/10.1007/978-981-10-8506-2_19

This thesis aims at improving vehicle's dynamic property and braking performance. A novel inter-axis capacity allocation method of driving system and braking system is proposed, analyzed, and verified.

2 Total Capacity Matching for Driving and Braking Systems

The research is carried out based on a compact car parametric model of which the design and performance specifications are shown in Table 1.

In the following two sections, the total capacity of driving and braking systems will be matched in accordance with the vehicle dynamics and braking performance targets, respectively.

2.1 Total Capacity Matching for Driving System

The dynamic performance targets are shown in Table 2.

The maximum speed of vehicle corresponds to the motor continuous operating region [7], therefore the rated power P_e and the maximum rotate speed $n_{\max}$ can be matched as follows:

$$P_e = \left(mgf + \frac{C_D A u_{\max}^2}{21.15}\right)\frac{u_{\max}}{3600\eta_T} = 35.63\ \text{kW} \tag{1}$$

$$n_{\max} = \frac{u_{\max} i}{0.377r} = 1416\ \text{rpm} \tag{2}$$

In the equations above, m is the total mass; f is the rolling resistance coefficient; C_D is the air resistance coefficient; A is the frontal area; $u_{\max}$ is the maximum speed; η_T is the transmission efficiency; i is the transmission ratio, which is assigned to be i = 1 0 here; r is the wheel radius.

Table 1 Specifications of the vehicle

Specifications	Value
Total mass m/kg	1535
Wheel radius r/m	0.281
Rolling resistance coefficient f	0.013
Air resistance coefficient C_D	0.31
Frontal area A/m^2	2.0
Wheelbase L/m	2.540
CM to front axle distance L_f/m	1.172
Height of center of mass h_g/m	0.5
Wheel rotational inertia I_w/kg m^2	3.6

Table 2 Dynamic performance targets

Target	Value
Maximum speed u_{max}/km h^{-1}	150
Maximum gradability α/%	30
0–100 km/h acceleration time t/s	9.5

The maximum gradability of vehicle corresponds to the motor short-time operation region, thus the maximum torque T_{max} is written as

$$T_{max} = \frac{\left(mgf\cos\alpha + mg\sin\alpha + \frac{C_D A u_s^2}{21.15}\right)r}{i_g \eta_T} = 1268\ \text{Nm} \quad (3)$$

The 0–100 km/h acceleration time also corresponds to the short-time operation region:

$$\delta m\left[\int_0^{\frac{u_e}{3.6}} \frac{du}{\frac{T_{max}}{r} - \left(mgf + \frac{C_D A(3.6u)^2}{21.15}\right)} + \int_{\frac{u_e}{3.6}}^{\frac{100}{3.6}} \frac{du}{\frac{T_{max} u_e}{r(3.6u)} - \left(mgf + \frac{C_D A(3.6u)^2}{21.15}\right)}\right] = t \quad (4)$$

It is calculated that the maximum torque T_{max} = 1419 Nm. Comparing the size and taking the bigger one as the final result, i.e., T_{max} = 1419 Nm. Meanwhile, assign the value of speed ratio as 1.875 [8].

By applying the steps above, the total capacity parameters of the driving motor are obtained. Round the numbers off and the parameters are shown in Table 3.

2.2 *Total Capacity Matching for Braking System*

According to the national standard GB21670-2008 and combining the vehicle specifications, the braking performance targets can be easily obtained (Table 4).

The total braking capacity can be matched in accordance with the braking distance and average deceleration targets, and the hydraulic braking capacity demanded should be the difference of total capacity and regenerative braking capacity [9]. The braking distance is calculated by the following formula [10]:

Table 3 Total capacity parameters of driving system

Parameter	Value
Maximum rotate speed/rpm	1420
Rated rotate speed/rpm	760
Rated power/kW	36
Maximum power/kW	114
Maximum torque/N·m	1420

Table 4 Braking performance targets

Initial speed/km h^{-1}	Braking distance/m	Average deceleration/m s^{-2}
150	165.7	/
120	108.4	5.76
100	77.0	/
80	50.8	/

$$S = \frac{1}{3.6}\left(\tau' + \frac{\tau''}{2}\right)u_0 + \int_0^{u_c} \frac{1}{25.92a_{\max}(u)}\mathrm{d}u^2 + \int_0^{u_c} \frac{1}{25.92a_{\max}(u)}\mathrm{d}u^2 \quad (5)$$

where τ' + τ''/2 represents the action duration of brake, which is assigned to be 0.4 s. u_0 represents the initial speed and u_c is the conventional speed.

The average deceleration is calculated by

$$d_m = \frac{\left(u_b^2 - u_e^2\right)}{25.92(s_e - s_b)} \quad (6)$$

in which $u_b = 0.8u_0$, $u_e = 0.1u_0$. s_b is the distance that vehicle covers from u_0 to u_b; se is that from u_0 to u_e.

Substitute different initial speed values from Table 3 to calculate the hydraulic braking capacity demanded and the final result matched is T_h = 1535 Nm. The varying pattern of the maximum braking force with vehicle speed is shown in Fig. 1.

On the basis of total capacities we have matched, the following sections will concentrate on the influence and optimization of inter-axis capacity allocation method.

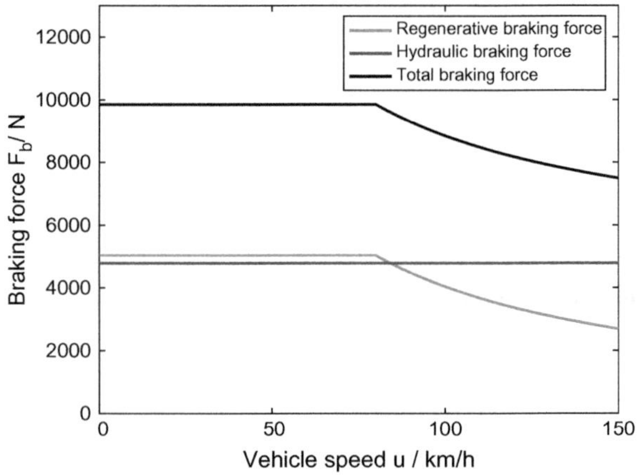

Fig. 1 Maximum braking force curve

3 Capacity Allocation Method of Driving System

Define the motor capacity allocation ratio, denoted as β_D, as the ratio of the maximum torque of front motor to the total maximum torque:

$$\beta_D = \frac{T_{f,\max}}{T_{f,\max} + T_{r,\max}} \tag{7}$$

Its value range is $0 \le \beta_D \le 1$.

Since the variation of β_D changes the value of the total maximum torque and the variation of μ changes the road maximum adhesive force, it is necessary to compare the numerical size of motor driving force and road adhesive force under the simultaneous variation of β_D and μ so as to obtain the corresponding actual driving force and then derive the analysis formula of maximum acceleration, maximum gradability, 0–100 km/h acceleration time, and so on.

In the next section, the derivation process of analysis formula of dynamic specifications will briefly demonstrate, taking the maximum acceleration, for example. In addition, in this essay we use the complete *I*/*f*/*r* formula for calculation, which takes grade resistance, air resistance, etc., into consideration [11]:

I curve:

$$F_{dr} = \frac{1}{2}\left[\begin{array}{l} -\frac{G\cos\alpha}{h_g}\sqrt{(L_r + p)^2 - \frac{4h_g L}{G\cos\alpha}F_{\mathrm{df}}} \\ + \left(\frac{G(L_r + p)\cos\alpha}{h_g} - 2F_{\mathrm{df}}\right) \end{array}\right] \tag{8}$$

f line group:

$$F_{dr} = -\frac{(\mu h_g + L)}{\mu h_g}F_{\mathrm{df}} + \frac{G(L_r + p)\cos\alpha}{h_g} \tag{9}$$

r line group:

$$F_{dr} = \frac{\mu h_g}{L - \mu h_g}F_{\mathrm{df}} + \frac{\mu G(L_f - p)\cos\alpha}{L - \mu h_g} \tag{10}$$

where $p = f(h_g - r)$. α is the road slope angle; h_g is the height of center of mass; L_f is the center of mass-to-front axle distance; L_r is the center of mass-to-rear axle distance; F_{df} is the driving force of front wheels and F_{dr} is that of rear wheels; μ is the road adhesion coefficient.

3.1 *Influence to Maximum Acceleration*

3.1.1 Analysis Formula Derivation

Under the circumstance of achieving the maximum acceleration, the vehicle speed and slope angle are both assigned to be zero, and the value of other parameters are assigned according to the Tables 1 and 3.

Since the total motor capacity is already decided, the total driving force of two axles is a constant value. Plot the motor driving force line and *f*/*r* line groups of different μ value, as shown in Fig. 2.

It can be seen in Fig. 2 that with the variation of μ value, the relative location of *f*/*r* line group and motor driving force line is also changing, and the second group of *f*/*r* line, of which $\mu = 0.337$, is considered to be the critical condition, whose μ value is defined as the critical value. Therefore, we can make a partitioned discussion following these steps:

(1) Calculate all the critical values: μ_0, μ_1, μ_2 ... and demarcate intervals accordingly;
(2) In each μ interval, demarcate sub-intervals according to the critical value of β_D. For example, in the interval $\mu_0 < \mu < \mu_1$, it can be demarcated into sub-intervals I, II, III by β_D as shown in Fig. 3;
(3) In each sub-interval, compare the numerical size of motor driving force and road adhesive force and derive the analysis formula of corresponding actual driving force, and then plot the curves as shown in Fig. 4;
(4) Derive the analysis formula of the maximum acceleration in each sub-interval by using that of actual driving force.

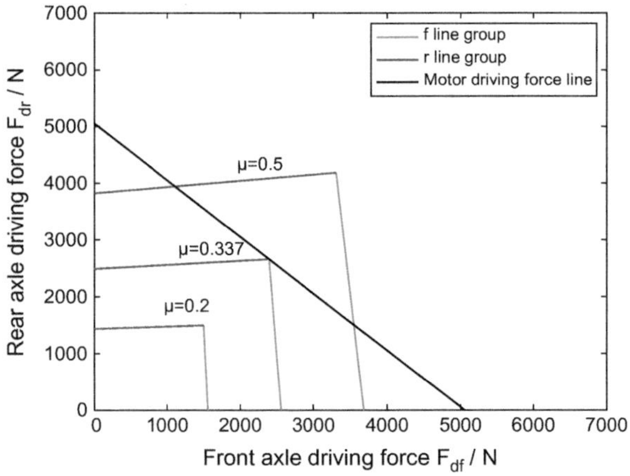

Fig. 2 *f*/*r* line groups of different μ value

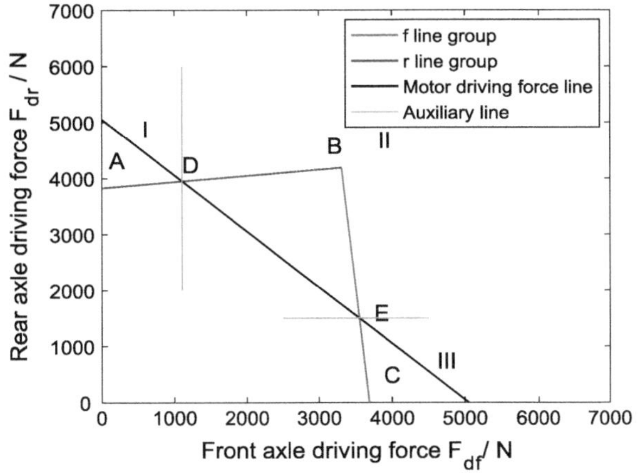

Fig. 3 Demarcating sub-intervals by β_D

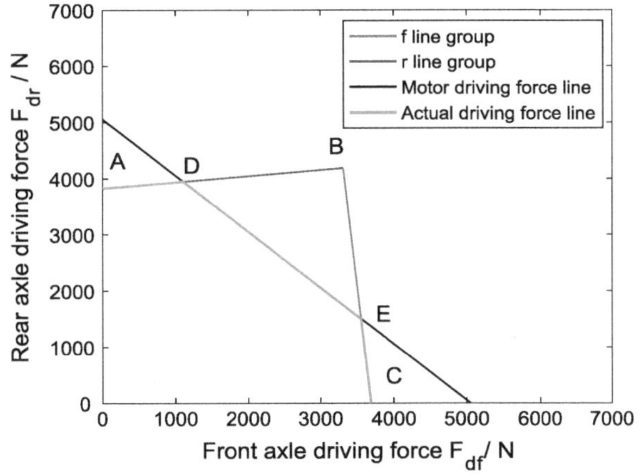

Fig. 4 Actual driving force

By applying the steps above, we can accomplish the formula derivation of all conditions under simultaneous variation of β_D and μ, and the complete formula group is obtained.

3.1.2 Result Analysis and Optimized Method Selection

According to the maximum acceleration formula group, the 3D figure of maximum acceleration can be plotted, as shown in Fig. 5.

This figure shows the maximum acceleration that the distributed drive electric vehicle can achieve under all possible capacity allocation conditions. The 3D surface has two intercept curves with $\mu - \alpha_{\max}$ plane, which correspond to $\beta_D = 0$ and $\beta_D = 1$, showing RWD and FWD conditions, respectively. When $\beta_D = 0$, i.e., in the RWD condition, the intercept curve is shown in Fig. 6.

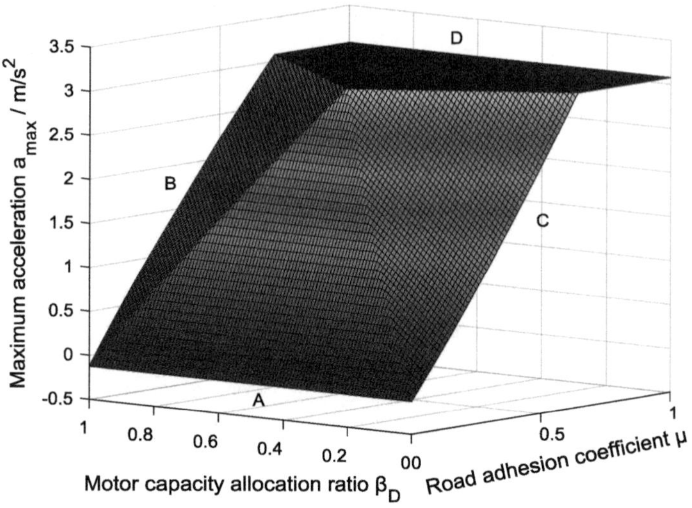

Fig. 5 Maximum acceleration

Fig. 6 Maximum acceleration in RWD condition

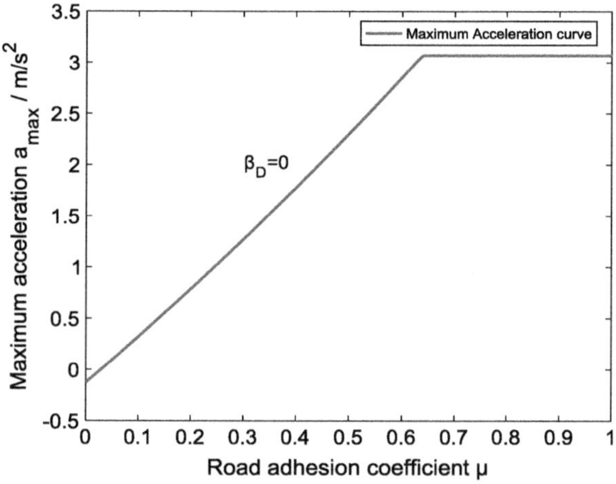

It is shown clearly in the figure that under this condition, the maximum acceleration value keeps raising with the increase of μ, achieving its maximum value 3.07 m/s^2 (about 0.31 g) until $\mu = 0.63$, and the utilization of road adhesion force is 49% only, which has indicated the weakness and limitation of RWD.

On the other hand, the curvature of this curve proves that the maximum acceleration does not increase strict linearly, but with its slope growing slowly. This is because the axle load moves backward with the increase of acceleration and driving strength, thus the vehicle load allocated to the rear axle is increasing. Since the whole motor capacity is allocated on the rear axle, its accelerating ability also increases in the same pace, which results in the growth of slope. Similarly, C and D surfaces are not planes.

By observing the variation trend of the maximum acceleration in Fig. 5, it is obvious that when the road adhesion coefficient μ is small ($\mu < 0.33$), the whole surface is nearly symmetrical on the both ends of β_D axle (C and D surfaces), and the maximum acceleration value is small, meaning that there are two wheels slipping or the motor driving force is greater than road adhesion force. In the middle section of β_D axle (A surface), the maximum acceleration value is larger, but the motor driving force does not work completely due to the limitation of road adhesion condition. When $\beta_D = 0.48$, the acceleration value achieves its maximum value. When μ is larger ($0.33 \leq \mu \leq 0.68$), the acceleration value achieves the maximum value in the middle section of β_D axle (D surface). However, when μ is a large enough, the acceleration value always maximizes.

Take the utilization of adhesion coefficient into consideration, we can obtain Fig. 7.

As shown in Fig. 7, the utilization is large near $\beta_D = 0.48$. While the D surface indicates that once the motor driving force works completely, larger the road adhesion coefficient is, smaller the utilization will be.

In conclusion, for the distributed drive electric vehicle with specifications above, the capacity allocation ratio for two axles should be 0.48:0.52, which can optimize the maximum acceleration performance.

3.2 Influence to Maximum Gradeability

Under the circumstance of achieving the maximum gradability, the vehicle speed is assigned to be u_s = 10 km/h constant, thus the accelerating resistance doesn't exist

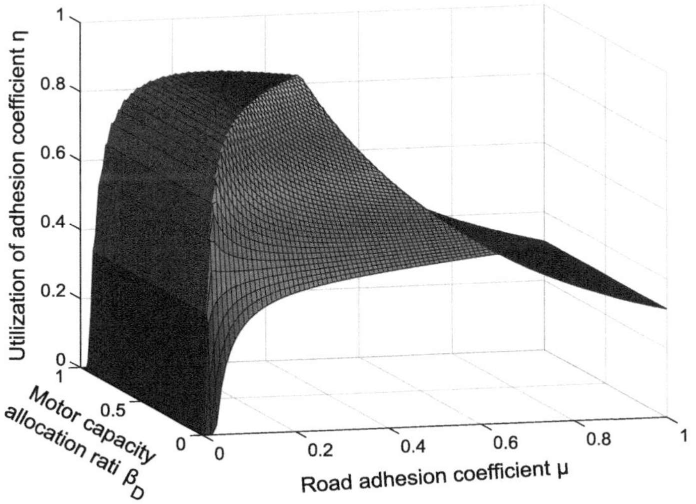

Fig. 7 Utilization of adhesion coefficient

and the air resistance is constant, too. Denote the maximum driving force of the vehicle as F_d, thus

$$F_d = Gf\ \cos\alpha + G\ \sin\alpha + F_w \tag{11}$$

Then the maximum gradability can be obtained. Under the simultaneous variation of β_D and μ, the derivation of maximum gradability formula is almost exactly the same as that of the maximum acceleration and so are the variation trend of their 3D figures, so I won't repeat it.

Notice that we have to presuppose a value of slope angle α because it may be included in the analysis formula of some interval boundaries and that results in a bit inevitable error in the interval boundary value. But it only has minimal effect on the maximum gradability value and its variation trend.

3.3 *Influence to 0–100 Km/H Acceleration Time*

The research process of 0–100 km/h acceleration time is similar to that of the maximum acceleration. However, since the normal speed is 80 $\leqslant$ 100 km/h, the maximum torque is not constant during the whole acceleration process, but there is a turning point at u = 80 km/h after which the maximum torque $T_{\max}$ begins to decrease as a function of speed u. When u = 100 km/h, $T_{\max}$ = 1136 Nm. In other words, there is more than one motor driving force line, therefore the partitioned discussion is of more complexity.

In the following section, we focus on the distinction between the derivation process of 0–100 km/h acceleration time and that of the aforementioned two parameters.

3.3.1 Analysis Formula Derivation

Under the circumstance that the minimum 0–100 km/h acceleration time is achieved, the slope angle is assigned to be zero. Plot two motor driving force lines and *f*/*r* line groups of different μ value, we can get a plurality of μ critical values and demarcate intervals by them. In each μ interval, demarcate sub-intervals according to β_D is shown in Fig. 8.

In addition to this, the vehicle speed u is also variable in each β_D interval, leading to changes of the actual maximum driving force. Hence the sub-sub-interval ought to be demarcated in each β_D interval. In a word, we have to do the partitioned discussion according to '$\mu \rightarrow \beta_D \rightarrow u$' sequence and, finally, derive the analysis formula of 0–100 km/h acceleration time in every sub-interval.

3.3.2 Result Analysis and Optimized Method Selection

According to the 0–100 km/h acceleration time formula group, the 3D figure of acceleration time can be plotted, as shown in Fig. 9.

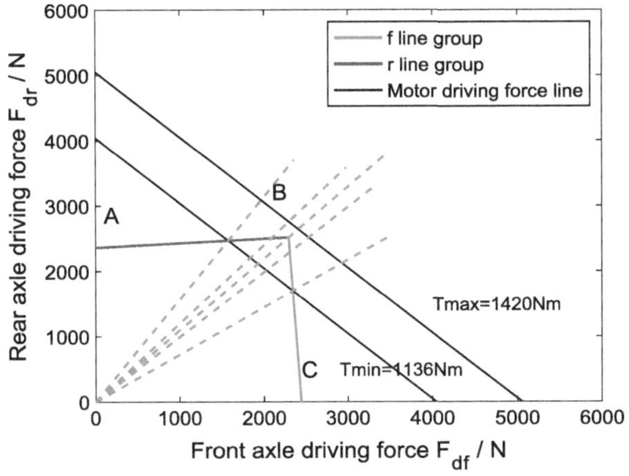

Fig. 8 Demarcate sub-intervals by β_D

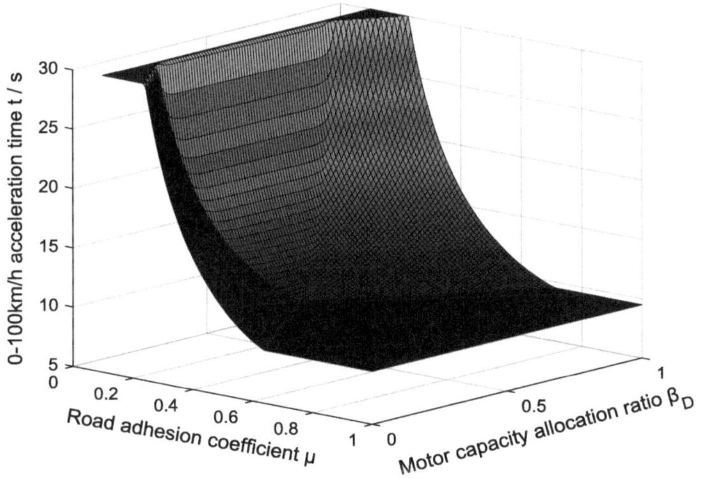

Fig. 9 0–100 km/h acceleration time

The variation trend of Fig. 9 is basically corresponding to the figures of maximum acceleration and maximum gradability, thus the optimized capacity allocation ratio is still 0.48:0.52. When μ is large enough, the maximum value of 0–100 km/h acceleration time converges to 9.5 s, which meets the design requirement.

3.4 Optimized Allocation Method

Synthesizes each kind of situation, we can conclude that for the compact car parametric model used, when the motor capacity allocation ratio β_D equals 0.48, that is, the capacity allocation for two axles is 0.48:0.52, three dynamic performance parameters mentioned are optimized. When different parametric models are adopted, we can obtain newly optimized allocation ratios by following this method. With this, a novel capacity allocation method of driving system for distributed drive electric vehicle is proposed, which is more reasonable than four-wheel uniform distribution.

4 Capacity Allocation Method of Braking System

Following the definition of motor capacity allocation ratio β_D defines the hydraulic brake capacity allocation ratio β_B. Now there are three independent variables in the research of braking process: (1) the motor capacity allocation ratio β_D; (2) the hydraulic brake capacity allocation ratio β_B; (3) road adhesion coefficient μ. Since the braking effect of motor regenerative braking and hydraulic braking are regarded to be of no difference because the coordinate control of these two braking systems is not considered, we define the total brake capacity allocation ratio β_T to simplify things. β_T is the ratio of the braking capacity of front axle to the total braking capacity of both axles:

$$\beta_T = \frac{\beta_D T_{\max} + \beta_B T_h}{T_{\max} + T_h} \tag{12}$$

Its value range is $0 \leqslant \beta_T \leqslant 1$.

In the next section, we will discuss the influence of simultaneous variation of β_T and μ to braking distance and average deceleration [12].

4.1 Influence to Braking Distance

The derivation of braking distance formula is similar to that of 0–100 km/h acceleration time. Do the partitioned discussion according to '$\mu \rightarrow \beta_D \rightarrow u$' sequence and finally derive the analysis formula in each sub-interval.

There is one point to notice that after the vehicle speed exceeds 80 km/h, the regenerative braking capacity will decrease, but the hydraulic braking capacity remains constant, hence the term $T = 80T_{\max}/u + T_h$ should be used. The 3D figure of braking distance with initial speed 100 km/h is shown in Fig. 10.

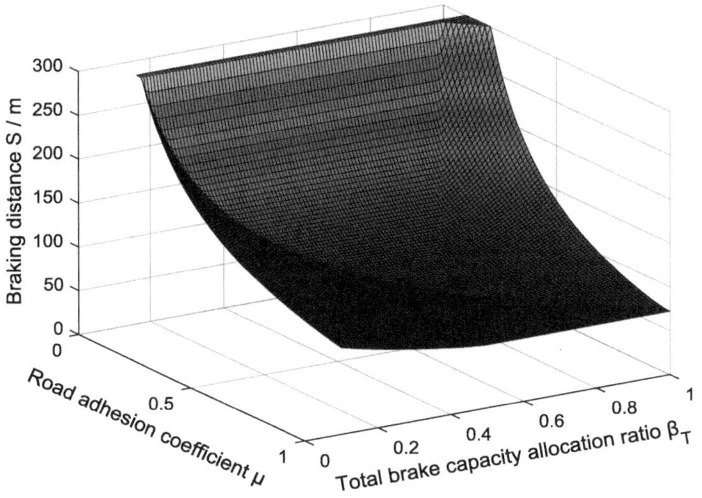

Fig. 10 Braking distance of 100 km/h

Its variation trend is also similar to that of 0–100 km/h acceleration time, and the optimized value of β_T is about 0.66. Then assign the value of β_D to be its optimized value, or $\beta_D = 0.48$ and convert the horizontal axis of Fig. 10 from β_T to β_B, hence the correlativity of braking distance to β_B is obtained, as shown in Fig. 11. It is observed that the vehicle braking performance is optimized when $\beta_B = 0.85$.

4.2 Influence to Average Deceleration

The derivation process of average deceleration formula is similar to that of the braking distance. The 3D figure is presented in Fig. 12. Similarly, the optimized value of β_T is still 0.66.

4.3 Optimized Allocation Method

In conclusion, for the compact car parametric model used, when the total braking capacity allocation ratio β_T equals 0.66; that is, the capacity allocation of hydraulic braking system for two axles is 0.85:0.15, and the two braking performance parameters are optimized. Similarly, this capacity allocation method can be applied to various kinds of distributed drive electric vehicles.

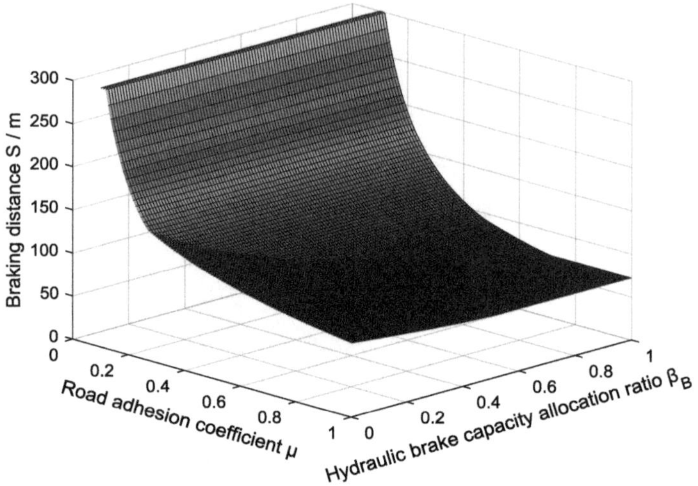

Fig. 11 Braking distance of 100 km/h to β_B

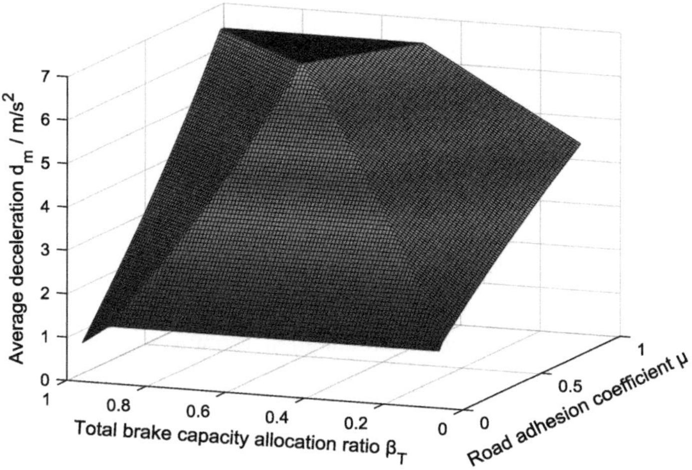

Fig. 12 Average deceleration of 100 km/h

5 Summary

In this thesis, quantitative research is carried out on capacity allocation of driving system and braking system for distributed drive electric vehicle. Taking regenerative braking into account and on the basis of vehicle parametric model, the influence of inter-axle capacity allocation of driving and braking systems to vehicle dynamic performance and braking performance is studied by analytical method. For dynamic performance, the targets are maximum acceleration, maximum gradability, and 0–100 km/h acceleration time. For braking performance, the targets are braking distance and average deceleration. After that, a novel capacity allocation method is proposed, which has certain guiding significance to engineering application.

References

1. Ning G (2006) Development of wheel driving system of the electric vehicle. Shanghai Auto 11:2–6
2. Yu Z, Feng Y, Lu X (2013) Review on vehicle dynamics control of distributed drive electric vehicle. J Mech Eng 49(8):105–114
3. Li L (2010) Control system design of four-wheel Independent drive electric vehicle. Wuhan University of Technology, China
4. Zhou Y, Li S, Zhou X, et al (2010) The control strategy of electronic differential for EV with four in-wheel motors. Chinese Control and Decision Conference. 4190–4195
5. Lang W (2013) Power system modeling and driving control of distributed drive electric vehicles. Jilin University, Changchun
6. Li Y, Zhang J, Guo K et al (2015) Driving and braking force distribution between front and rear axles for 4WD electric vehicle. J Jilin Univ (Engineering and Technology Edition) 45 (3):703–710
7. Sun X (2015) Study on elctro-mechanical braking control strategy for a distributed driving electric vehicle. Beijing Institute of Technology, Beijing
8. Ehsani M, Gao Y, Emadi A (2010) Modern electric, hybrid electric, and fuel cell vehicles: fundamentals, theory, and design, Second edn. China Machine Press, Beijing
9. Gu J (2012) Vehicle control of four-wheel driven micro electric vehicle. Tsinghua University, Beijing
10. Yu Z (2006) Automobile theory, 4th edn. China Mechine Press, Beijing
11. Yu F, Lin Y (2005) Dynamics of vehicle system. China Machine Press, Beijing
12. Zhang Y, Yu Z, Zhang L, et al (2007) Related-problem analysis of electric vehicle hybrid braking system. In: Proceedings of the annual meeting of SAE-China

The Effect of Driving Cycle and Shifting Pattern on Vehicle Emissions

Ni Zhang, Linyun Wei, Xing Wang and Yongsheng Long

1 Introduction

To fulfill the Chinese government's 13th Five-Year Plan, China will implement emission regulation VI and relative gasoline criterion countrywide in January 2019. In Beijing and Shanghai, the regulation will be implemented in advance in 2017. Therefore, it is a big challenge for all the test institutions and car manufacturers to update the test equipment and technology.

Compared with China V [1] emission regulation, one of the most significant differences in China VI is to test the vehicle emission with WLTC instead of NEDC. The intention of this change is that WLTC more truly reflects the level of vehicle emissions in real drive situation, and more transient engine running conditions will be covered in the WLTC [2].

2 Comparison Between NEDC and WLTC Cycles

The following Fig. 1 shows the NEDC and the WLTC, and the main difference in two cycles is listed in Table 1.

Figure 2 shows the engine work area of the two test cycles, and compared to NEDC cycle, the engine work area tested in WLTC is much wider, especially in medium- and high-load engine working areas which means the WLTC tests the engine's emission in a more complete scope.

N. Zhang (✉) · L. Wei · X. Wang · Y. Long
Dongfeng Motor Corporation Technical Center, 430058 Wuhan, China
e-mail: zhni@dfmc.com.cn

Society of Automotive Engineers of China (SAE-China) (ed.), *Proceedings of the 19th Asia Pacific Automotive Engineering Conference & SAE-China Congress 2017: Selected Papers*, Lecture Notes in Electrical Engineering 486, https://doi.org/10.1007/978-981-10-8506-2_20

Table 1 Differences between NEDC and WLTC

Cycle	Duration (s)	Total mileage (km)	Total idle time (s)	Maximum acceleration (m/s2)	Steady condition total duration (%)
NEDC	1180	11.04	24.8	1.04	24.8
WLTC	1800	23.23	13.2	1.67	13.2

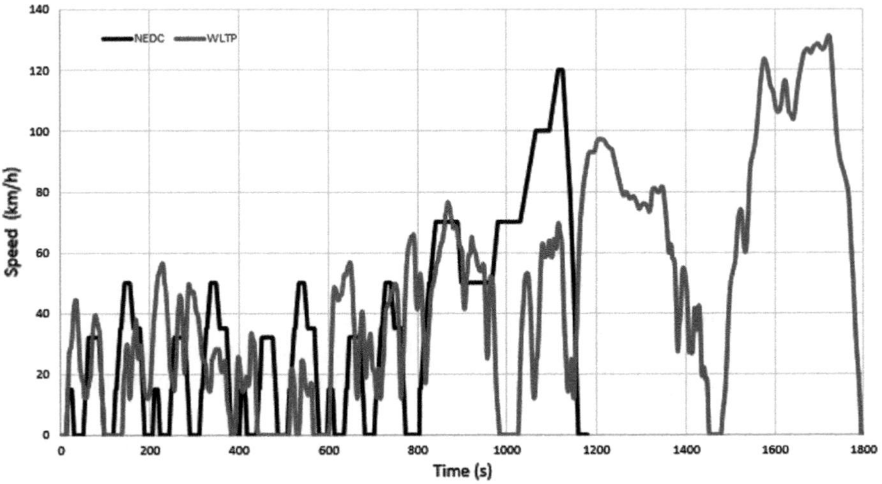

Fig. 1 NEDC and WLTC cycles

Besides, the shifting timing and gear are changed with vehicle configuration in WLTC cycle. The shifting timing is decided by rated power, engine speed, idle speed, gear number, speed ratio and road load. The details are in China VI emission regulations [3]. However, the shifting timing is fixed in NEDC cycle, so there is a big influence on vehicle fuel consumption and emissions.

3 Test Equipments and Vehicle

3.1 Test Reparation

The study is carried out in emission laboratory in DFMTC. The test vehicle is normal production vehicle from Dongfeng Corporation. The test gasoline vehicle has a manual (5-speed) transmission, and the test mileages are larger than 3000 km. The specifications of test vehicle and engine are listed in Tables 2 and 3.

The tests are implemented on ambient temperature 25 °C. Before testing, the tire pressure, oil temperature, and battery storage will be checked. The chassis dynamometer will be set up before the formal test. According to driving cycle, the

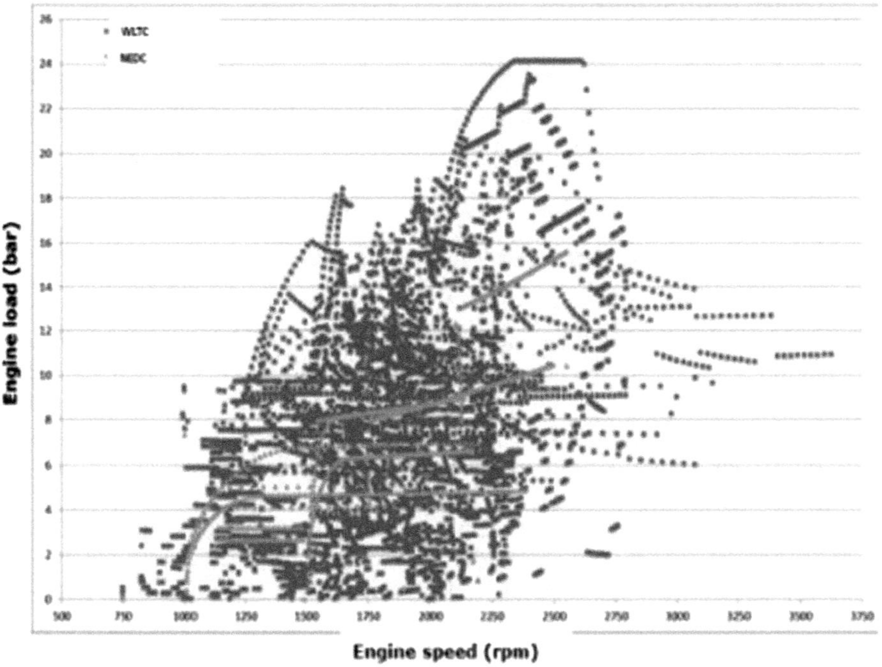

Fig. 2 Engine working area in NEDC and WLTC

Table 2 Specification of test vehicle

Weigh	1960 kg
Tire pressure	210 kPa
Road load	0.03346 + 1.6745v + 102.54vv
Transmission type	MT
Catalyst type	TWC

Table 3 Specification of test engine

Engine type	4 cylinder in line, 4 strokes, water-cool, turbo
Bore/Stroke	73/83.4 mm
Volume	1.396 l
Compress ratio	8.9
Ignition sequence	1-3-4-2

pretest will be carried out. And then, the test vehicle will be soaked in ambient temperature 25 °C for at least 6 h. The WLTC cycle used for this study is the four-phase test (low-phase, medium-phase, high-phase, and extra high-speed phase), which is found in the typical European vehicles test. All the tests are conducted three times.

3.2 Test Equipment

The gaseous pollutants including CO, THC, NOx and NMHC are measured in bags per phase by CVS system. And the bag results and continuous results will be analyzed [4]. The specification of test equipment is listed in Table 4.

4 Test Results of Different Driving Cycle

The road-load determination was carried out in real road and then set on the chassis dynamometer for the cold-start tests. One remarkable difference between NEDC and WLTC is the setting of test inertia. Both driving cycles use the same vehicle mass, and the vehicle inertia is used according to the China VI Type-Approval procedure. This option is to directly comparing the effect on emission of the different driving cycles.

Emissions are measured per phase in two set of bags on the cold-start NEDC and in four set of bags on the cold-start WLTC. But, our test cell is only equipped with three set of bags, so the 4th bag analysis on WLTC is obtained by analyzing emissions from low-speed phase bag as soon as sampling in this bag was over. Figure 3 shows gathering procedure of sample gases.

The emission results are illustrated as the average of three test repeats in this paper. Figure 4 shows all the gaseous pollutants measured on NEDC and WLTC cycles. The total hydrocarbon emissions are below the ChinaVI (a) regulatory limits in both driving cycles. Because of transient acceleration in WLTC cycle, the catalyst temperature is increased faster in WLTC cycle than that in NEDC cycle. Thus, the hydrocarbon emission in WLTC cycle is less than that in NEDC cycle. The study reported by Cecile Favrel shows the similar THC emission result with us [5]. As shown in Fig. 2, there is no significant different in NEDC and WLTC cycles, for NOx and NMHC.

There is significant difference between two driving cycles in CO emission results. CO emission result on WLTC cycle is above the regulatory limit of China VI (a). This can be explained that air–fuel ratio enters enrichment area in extra high-speed phase of WLTC. As we can see in Fig. 5, the lambda is larger than 1, when the 4th-phase driving cycle begins. In order to protecting the three-way catalyst, the engine works in the enrich area.

Several technical methods are carried out to decreasing the emission of CO. First, the air–fuel ratio is adjusted, and consequently enriching conditions are broadened. Figure 6a shows the variation of temperature before the turbo at

Table 4 Specifications of test equipment

Emission analyzer	HORIBA 7400CVS
Chassis dynamometer	MAHA AIP-ECDM48 M
Climatic cabine	WEISS WK554/10-35/R0

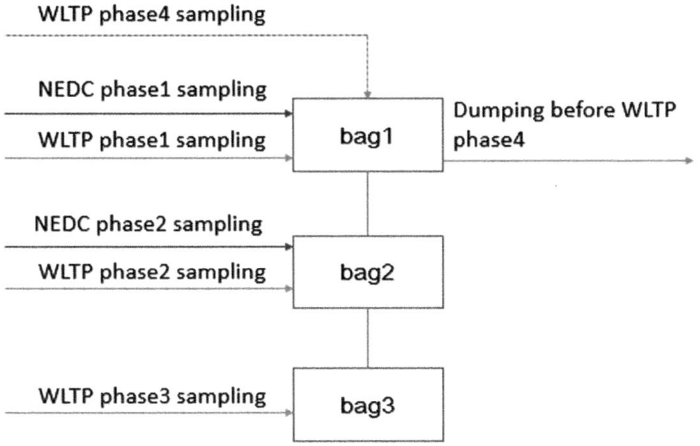

Fig. 3 Sampling procedure

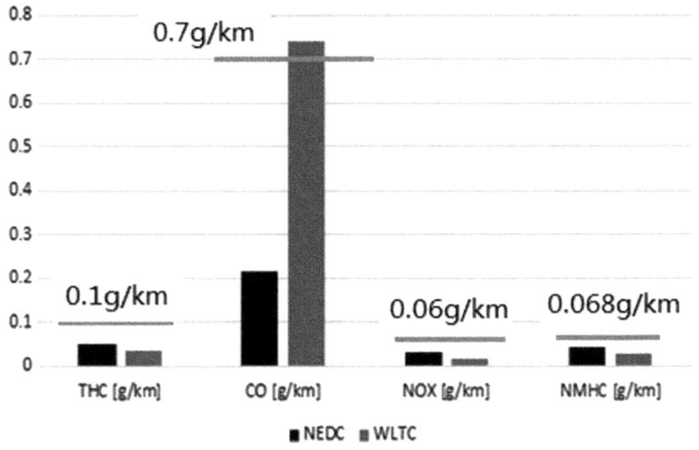

Fig. 4 Comparison of gaseous pollutants emissions in NEDC and WLTC

different air–fuel ratio conditions. Figure 6b illustrates the impact of temperature on CO emission.

The peak value area of CO emission is zoomed as illustrated in Fig. 7. It is clear that adjusting enriching has a prominent impact on CO emission. If the temperature is controlled at 980 °C, the maximum value of CO emission is reduced to approximate 250 ppm from 890 ppm. And through calculating, final CO emission is 0.43 g/km, which is under the limit of regulation.

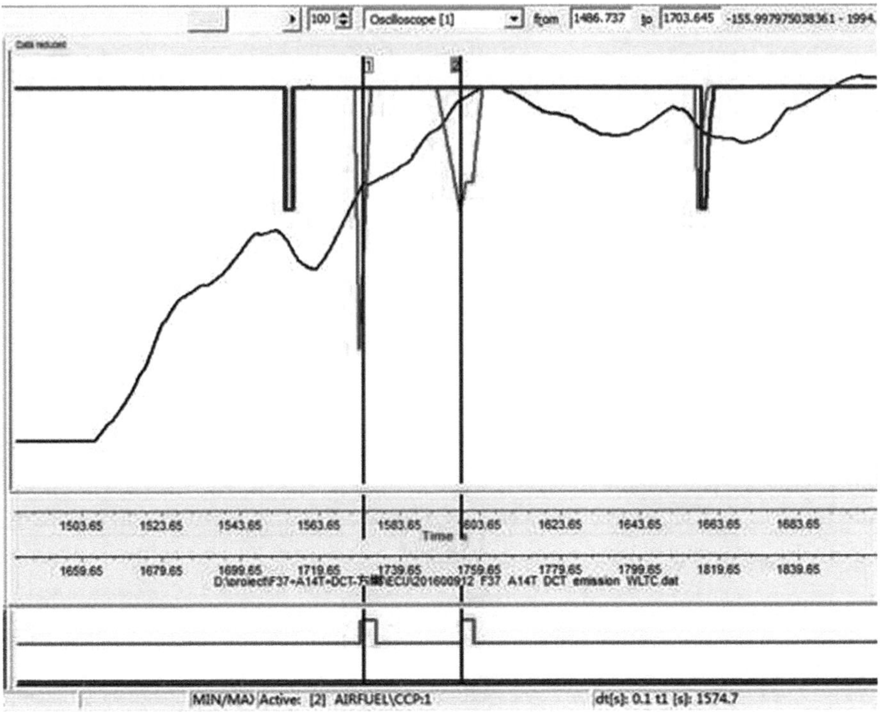

Fig. 5 Lambda in WLTC

There is a shifting criterion when running the WLTC cycle according to vehicle mass, engine speed and so on. On the basis of analysis, the engine works on high-load area on shifting pattern 1, so that ECU enters enrich control. So the second method is to modify shifting pattern. The shifting pattern is modified to making engine working in high-speed and low-load area in line with China VI regulation. Through this method, it is possible to decrease CO emission. Figure 8 shows that the CO emission is decreased to 0.39 g/km by using shifting pattern 3. But vehicles are shifted at a very high engine speed, so there is an effect on driving comfort, which needs a further estimation.

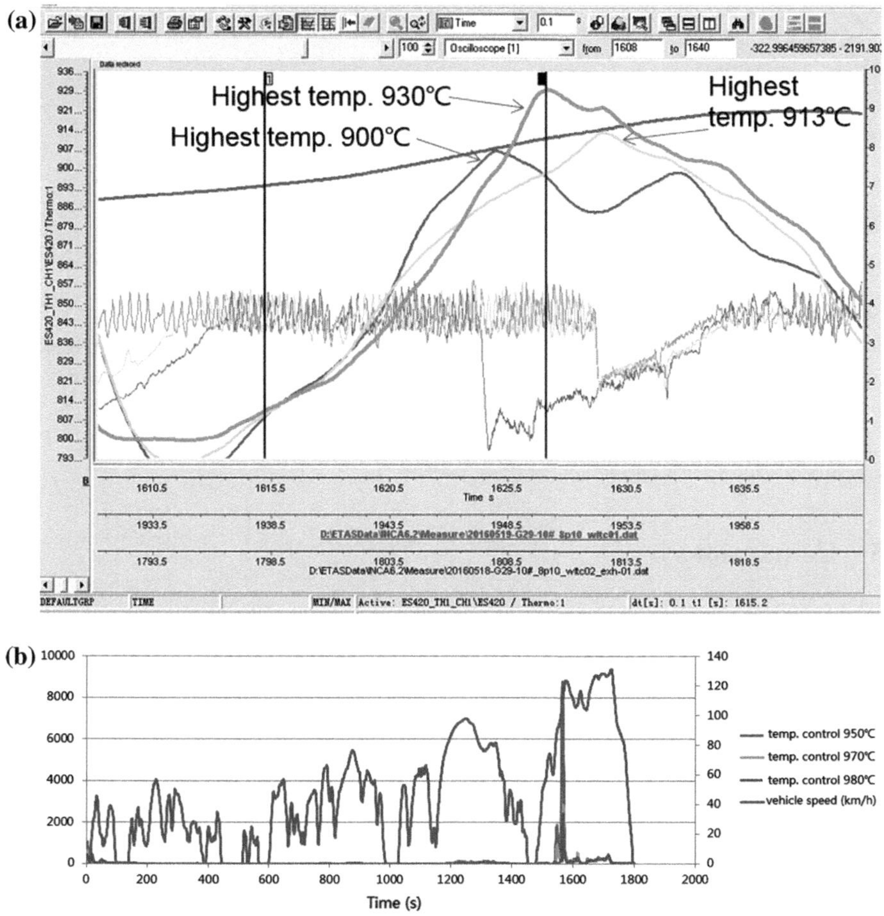

Fig. 6 **a** Highest temperature at different air–fuel ratio and Fig. 6. **b** The impact of temperature on CO emission

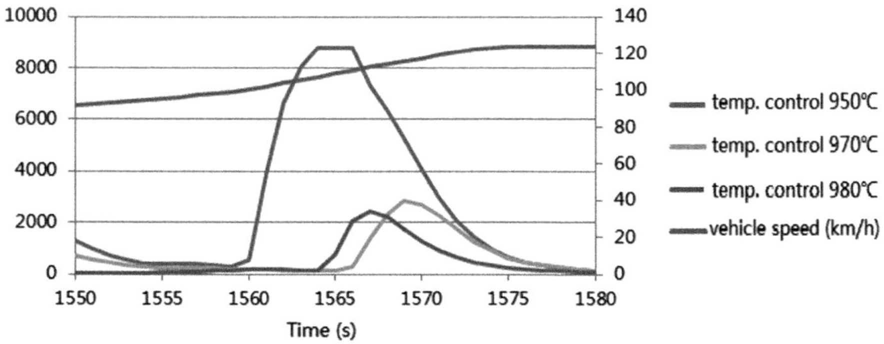

Fig. 7 Peak values of CO emission at different temperature

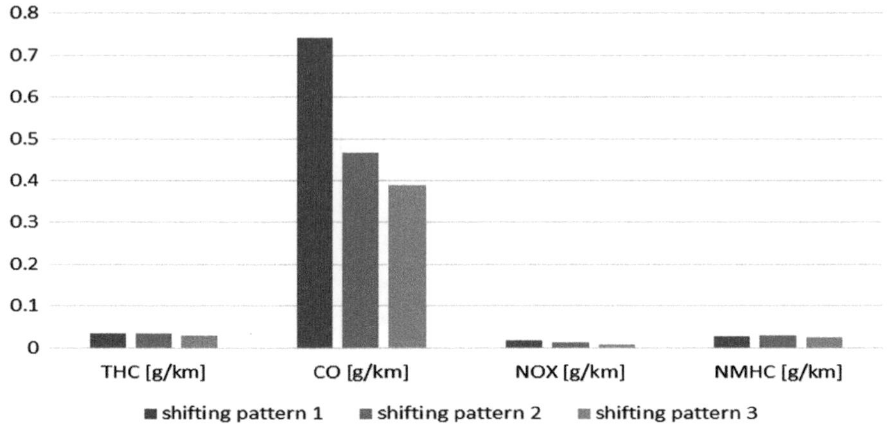

Fig. 8 Emission results on different shifting patterns

5 Conclusions

As the China VI regulations are published, vehicle testing institutions and car manufacturers are carrying out pre-studies. The effect of driving cycles on emission has been studied in DFMTC laboratory. The results are summed as follows:

In order to excluding the effect of inertia, the test inertia is set according to China V regulations in this paper. But further tests are needed to carry out to find out how TEST MASS affect.

2. The driving cycles have small effects on emissions of THC, NOx, and NMHC.

3. Comparing with NEDC cycle, the engine runs in high-load area in WLTC cycle, which makes air–fuel ratio entering enrich state and CO emission out of regulatory limit. The CO emission could be reduced to regulatory limit through air–fuel calibrating and choosing suitable shifting pattern.

References

1. GB18352-2013 轻型汽车污染物排放限值及测量方法(中国第五阶段),附录C 常温冷启动后排气污染物排放试验(I型试验),P33-P80
2. 轻型汽车污染物排放限值及测量方法(中国第六阶段),5.4 标准制定依据,P34
3. 轻型汽车污染物排放限值及测量方法(中国第六阶段),附录C 常温冷启动后排气污染物排放试验(I型试验),P37-P122
4. Piotr Bielaczyc, Andrzej Szczotka, Joseph Woodburn. Regulated and Unregulated Exhaust Emissions from CNG Fueled Vehicles in Light of Euro 6 Regulations and the New WLTC/GTR 15 Test Procedure. SAE International, 2015-01-1061
5. Cecile Favre, Dirk Bosteels, John May. Exhaust emission from European market-available passenger cars evaluated on various drive cycles. SAE International, 2012-24-0154

A New Fuel Pump Driver Using ePower

Shinian Li, Zuoyun Huang and Kilian Martin

Nomenclature

L_{mk} Average inductance
ΔL_k Maximum inductance variation due to the flux
ε Coefficient relative to the phase current flux
I_k Excitation current

1 Introduction

Nowadays, the emission limitation is more and more strict and many methods are used for the purpose, such as reduction weight, HEV, start-off system. At the same time, the performance of autos is also required higher and higher. In this background, BLDC motors are increasingly popular in automotive driving applications because of their high torque to inertia ratio, small mass and size, low maintenance, superior power density, and high efficiency [1].

In most speed driving systems, some type of shaft position sensors such as an optical shaft position encoder or resolver is fitted to provide a signal that is used to maintain an appropriate space angle between the stator and rotor fields in the motor [2][2]. In the fuel pump application, the BLDC motor is assembled in the fuel tank. The fuel flows through the motor, and it can also cool the motor when it is running. It will take a lot of cost to set up the installation line for the position sensor as well as the sensor itself is costly. At the same time, the sensor may reduce the reliability of the whole system.

S. Li (✉) · K. Martin
Infineon Integrated Circuit (Beijing) Co. Ltd, 18F, Tower B, Zhao Lin Plaza, No.19 Rong Hua Middle Road, Yi Zhuang Economic-Technological Development Area, 100176 Beijing, China
e-mail: shinian.li@infineon.com

Z. Huang
Keboda Technology Corp, Shanghai, China

Society of Automotive Engineers of China (SAE-China) (ed.), *Proceedings of the 19th Asia Pacific Automotive Engineering Conference & SAE-China Congress 2017: Selected Papers*, Lecture Notes in Electrical Engineering 486, https://doi.org/10.1007/978-981-10-8506-2_21

The ePower is smart and powerful microcontroller and driver that is fully match the fuel pump using case. It integrates a BEMF comparator that can provide the rotator information and save the cost on the external comparison circuit. A three-phase bridge driver (BRDV) simplifies the MOSFET pre-driving circuit and makes the driver system more safe and reliable, as well as saves the pre-driving cost.

2 Solution Introduction

The following figure shows the solution overviews. The power supply (6–16 V) can connect the power management unit (PMU) through a filter circuit. The PMU is responsible for generating all required voltage supplies for the embedded MCU (VDDC, VDDP) and the external supply (VDDEXT). It provides two communication interfaces, LIN interface and PWM interface. In order to debug the system easily, a J-Link interface is added. Three terminals of the motor connect to the ePower device through a simple filter and the 3-phase voltages compares each other in BEMF comparators then provide the rotator position information. The rotator position information is expressed as the hall signals (called virtual halls). The virtual halls are captured by the compare and capture unit 6 (CCU6), and it can generate suitable block communication logics. Due to the hall signals captured by CCU6, the general-purpose timer 1/2 (GPT12) submodule can count and timing the event and then provide the speed information. Due to the block commutation logics from CCU6 and speed information from GPT12, the core computes the controlling signals—three-phase comparison values. These comparison values are sent to the CCU6, and it can generate the three-phase PWMs. The PWMs can drive motor after going through the bridge driver (BRDV). In this solution, DC current is sampled by a low side single shunt, integrated current sense amplifier, and ADC1. The current is not used for the current control but for over-current protection (Fig. 1).

The motor control software is shown in Fig. 2. After the ePower device setup, the peripherals are initialed, such as time T2, timer T3, GPT12, CCU6. Then the off-state diagnoses (include short to ground, short to battery voltage, motor three-phase ends opened) are implemented. In the infinite loop, the communication interface includes receiving the speed command and feeds back the system status. There are three interrupts related to the motor control function. In CCU6 T12 period interrupt service, the motor control strategy is handled. CHE interrupt is for the block commutation setting, and the GPT1 T2 overflow interrupt is for the speed control.

2.1 *Current Measurement*

The DC link current was sampled by a low side single shunt through external low-pass filter, integrated current sense amplifier (CSA), and ADC. The amplifier gain could be programmable during 10, 20, 40, and 60 which are decided by the

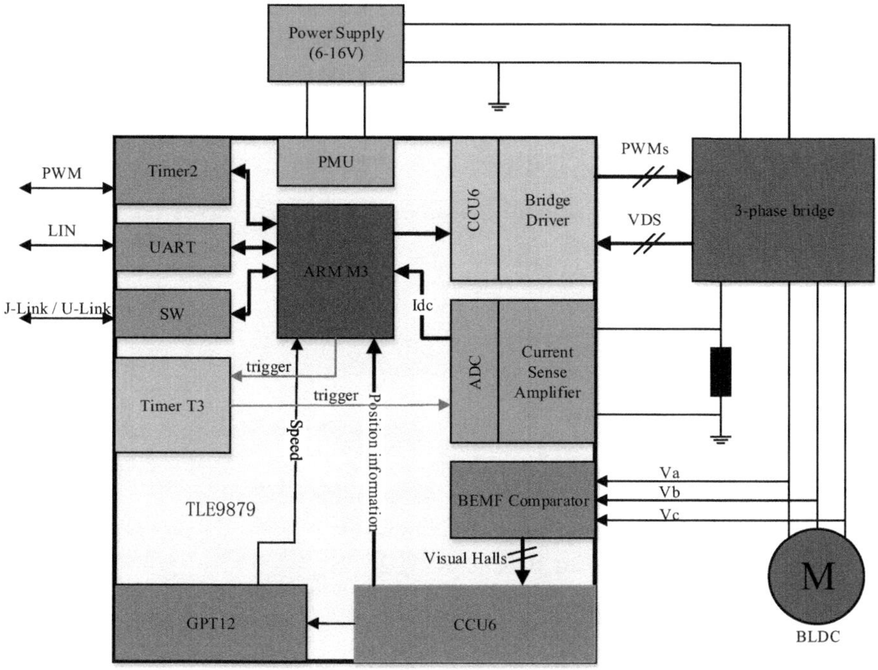

Fig. 1 BLDC driver based on ePower TLE9879

maximum current in the real application [3]. There are two ADCs in ePower TLE987x: ADC2 is used for device monitor function and ADC1 for users' application. Here, ADC1 works in the EMI mode triggered by the Timer 3 that is synchronized to motor driving PWMs. In initial stage (before start motor), the current was sampled for 16 times in software start mode and the average value is the zero offset value which should be removed when read DC current value. In motor running stage, the current was sampled and compared to the configured maximum current value then avoid the system working over-load too long time (Fig. 3).

2.2 *Initial Position Detection*

For fuel pump application, one strict requirement is the fast startup from standstill. At the same time, the startup should be very reliable like the DC motor. In traditional sensor-less startup method, BLDCM rotator needs to be fixed to a position and then starts up by I/F or V/F method. In this method, it needs a long time to fix the position because of the rotor inertia. The other method is starting the motor directly without position alignment. In this method, the failure to startup often happens. Therefore, initial position detection is needed and should be finished in a short time.

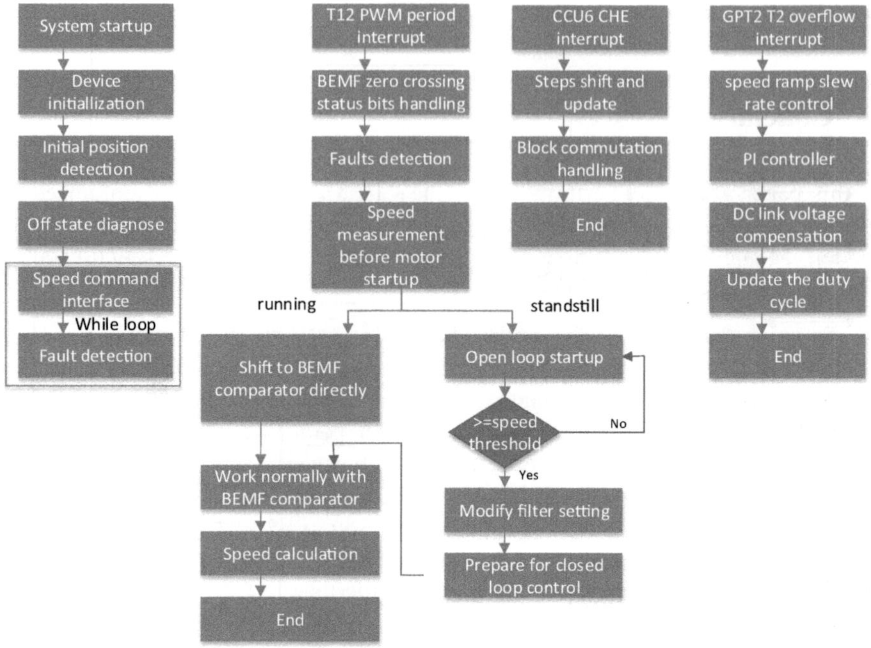

Fig. 2 Software flowchart

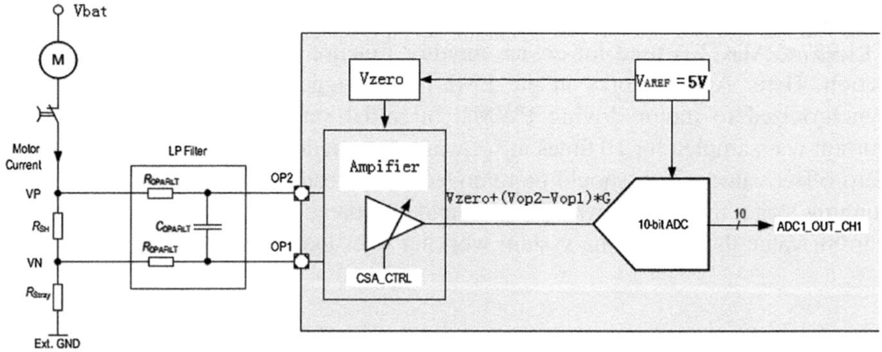

Fig. 3 CSA block diagram

The BLDC motor inductance variation could be approximated by the following formula.

$$L_k = L_{\text{mk}} + \Delta L_k \cos(2\theta) + \varepsilon I_k \sin(\theta) \tag{1}$$

where L_{mk} is the average inductance, ΔL_k is the maximum inductance variation due to the flux, ε is the coefficient relative to the phase current flux, and I_k is the excitation current. Then, we can get the inductance distributed on the motor electrical angle like the following figure (blue line). Normally, ΔL_k and ε are more significant in injection BLDC motor in comparison with the surface type. From the figure, we see the inductance varies with the angle and the current varies with the inductance shown by the red line. Therefore, we can estimate the rotor angle by detecting the induction current [4].

From Fig. 4, once we inject a rotating voltage to the three-phase bridge and the voltage is small enough so that the motor will maintain standstill state, the DC current will be different due to different vector voltage and there is a biggest one. The simplest rotating vector is six block vectors (AH-BL, AH-CL BH-CL, BH-AL, CH-AL, CH-BL). When rotating the motor shaft, the vector according to the biggest DC current is different (Table 1).

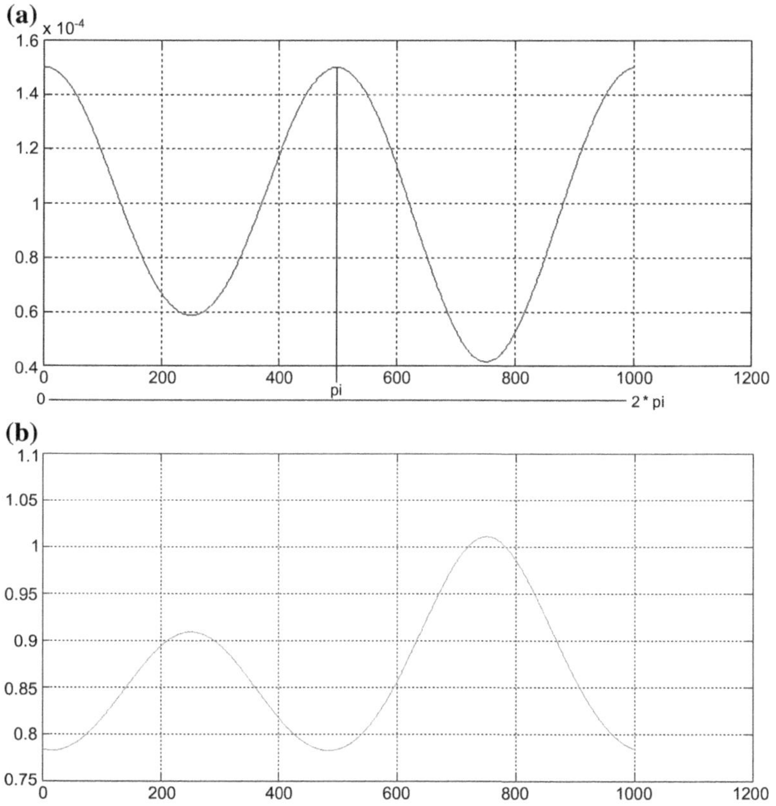

Fig. 4 **a** BLDC inductance with rotator angle. **b** BLDC current with rotator angle

Table 1 DC currents due to different sectors

Voltage vector	Sector	Measured DC current					
AH-BL	1	347	345	326	297	*346*	301
AH-CL	2	304	339	346	357	315	*355*
BH-CL	3	*359*	305	308	358	345	344
BH-AL	4	346	*349*	332	299	310	349
CH-AL	5	299	340	*347*	358	314	295
CH-BL	6	332	296	303	*362*	317	346

2.3 Virtual Hall Signals Generation

A BEMF comparator is integrated into ePower TLE987x to detect the rotator position. In block commutation method, there is always a phase opened and then its BEMF voltage could be detected. The BEMF voltage is compared to the natural point voltage, and the output logic bit is the virtual hall signal that is same as hall element output. In TLE987x, the natural point is simulated by the middle point of the other two on-state phases. Take the C-phase opened for instance, A-phase and B-phase are on state, then one end connected to battery voltage and another connected to ground, its middle point could be reckoned as natural point. In order to remove the commutation current pulse and noise, a low-pass filter is integrated as well. The filter time could be set according to the speed and hardware case (Fig. 5).

The filtered virtual hall signals are sent to CCU6. Any edges of virtual signals will trigger T13 in CCU6 to start to count. Users can set the T13 period value which is determined by the motor speed. The product of the period multiplied by the speed should be 30° which needs to be right shifted to do block commutation. The period event will trigger hall signals in comparison to see if correct hall event (CHE) happened. Then, the CHE will trigger the updating multi-channel PWM pattern (MCMP) (Fig. 6).

2.4 Speed Detection and Control

As mentioned before, the position could not be detected in low-speed range and the motor starts by V/F method. In this method, speed control is open loop. A speed is

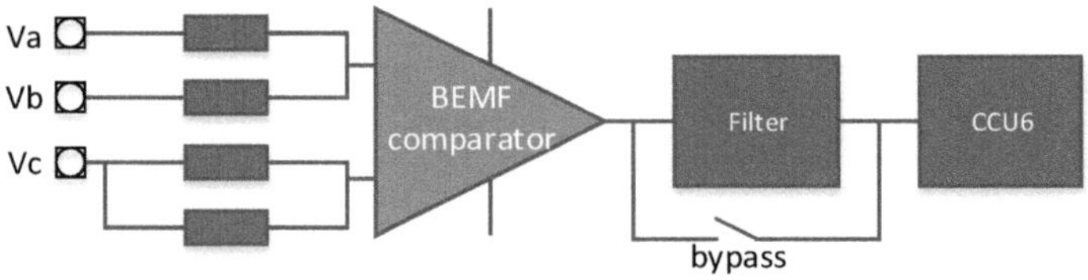

Fig. 5 BEMF comparator

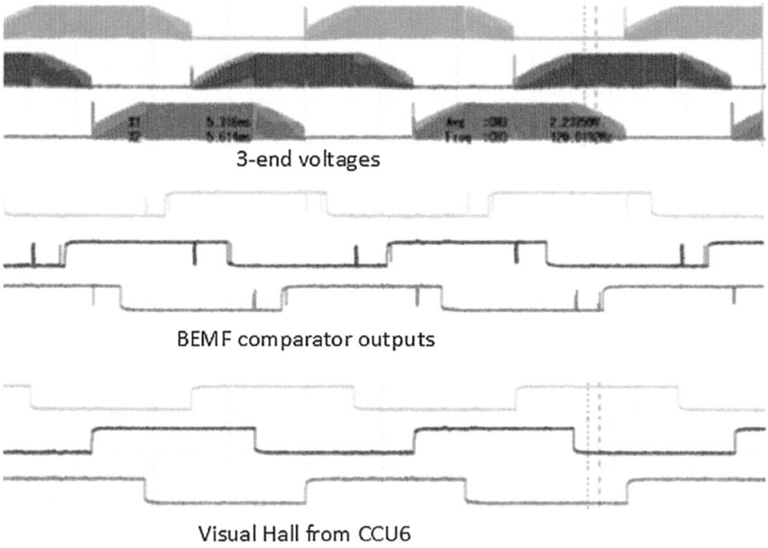

Fig. 6 Position detection signals

given in a ramp slew rate and the slew rate could be adjusted for stable startup. At the same time, there is no position feedback, the angle is the integration of the given speed, and the six blocks could be derived from the given angle. In order to maintain the given speed, the reference voltage is proportional to the given speed and then the motor can run up.

The open-loop control is a very short time, and it will turn into the closed loop when the speed is over a threshold (about 5% of the normal maximum speed). In closed loop, the virtual hall signals are obtained through the BEMF comparator described in the last sector. In CHE interrupt service, the GPT12 T5 timer value was read and the speed is calculated by the following formula.

$$\text{Speed} = \frac{f_{T5}}{N} \cdot \frac{1}{6} \cdot 60 = 10\frac{f_{T5}}{N} \tag{2}$$

where N is the T5 counter value and f_{T5} is the T5 frequency (Fig. 7).

The reference speed is given by a ramp controller, and then a PI control is implemented. The output is the compared value for PWM generation.

2.5 PWM Generation

In this solution, current detection is for the initial position measurement and over-load protection but not for the current loop control. The speed PI controller output is send to the CCU6 for PWM generation. CCU6 has two timers, T12 and

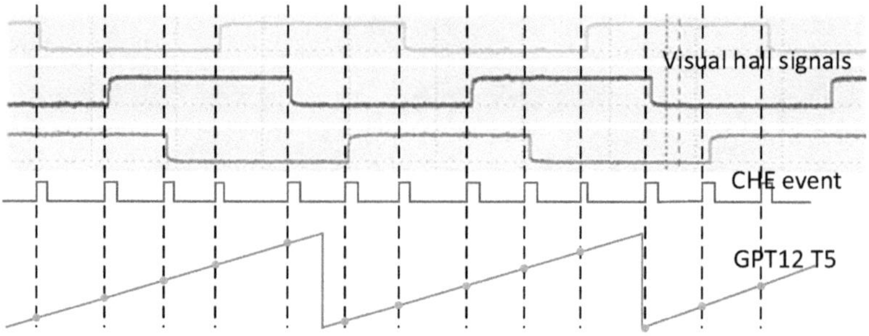

Fig. 7 Speed detection

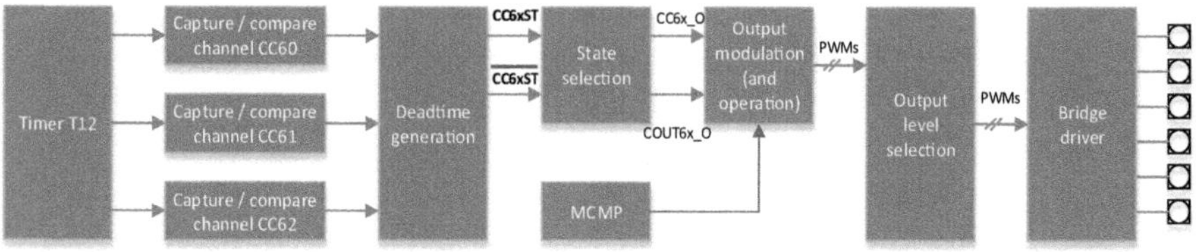

Fig. 8 PWMs generation by CCU6

T13. T13, as mentioned before, is used for the position detection. T12 has three compare–capture channels for three-phase PWMs. These PWMs are logic level and generated by Multi-channel PWM pattern (MCMP) mode which shown in Figs. 8 and 9.

2.6 3-Phase Bridge Driver

In ePower TLE987x, the three-phase bridge driver is integrated to save the space and cost when we design the driving system. One biggest feature of the pre-driver is that the charge and discharge current of the MOSFET gate can be adjusted by the slew rate control. There are four current steps that can be programmed, and there are eight stages (3 bits) that could be used in each step. Additionally, there are 32 levels (5 bits) available for each stage. By this feature, users do not need to design charge/discharge path and do not need to change the gate resistance for better driving performance or smaller EMI property.

The second feature is that the three-phase bridge was monitored a lot to ensure the safe and reliability of driving system. The off-state diagnoses include the open-load diagnose and short to ground or battery voltage diagnose. It is implemented in the initial process or after the over-current fault to ensure the driver is safe before working. In working state, the VDS, VCP, and VDH are always monitored. The over-voltage and under-voltage thresholds could be configured due

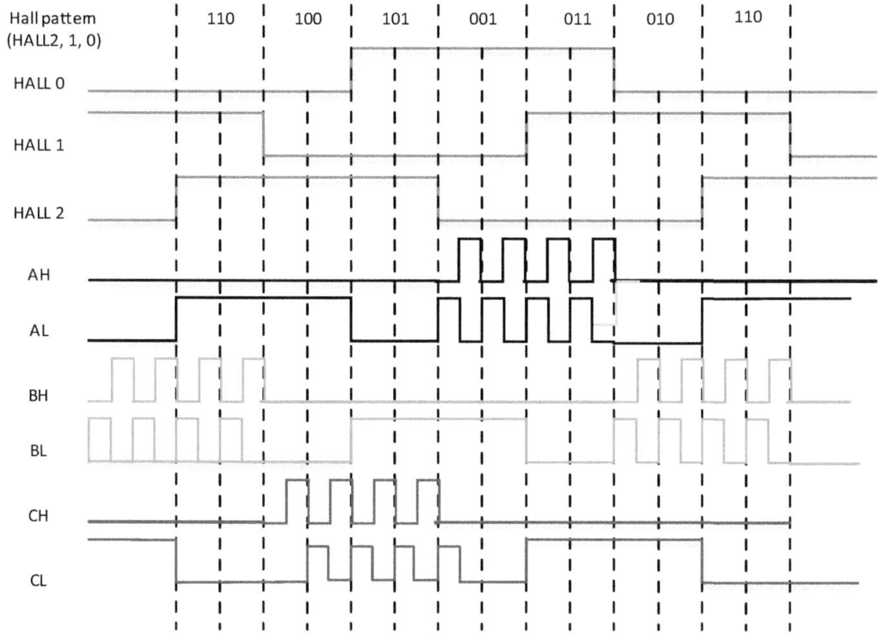

Fig. 9 Hall signals and its corresponding driver signals

to the application case. The VDS is monitored to see if there is over-current happened caused by short circuit. By this, the MOSFETs in bridges and driver could be protected. The threshold of VDS is determined by formula $I * R$, where I is the allowed maximum current of MOSFET and R is the DC-on resistance (Fig. 10).

3 Experiments and Results

To certify the solution, a simplified fuel tank was set up which is shown in the following left figure. The controller is shown in the right figure, which is based on TLE9879. The pump motor speed is about 6000 RPM@12 V.

3.1 Startup

The following figure shows the motor runs from standstill up to the reference speed. In standstill, the V/F method is used to run the motor. When it reaches a speed threshold, the BEMF signals could be detected clearly. After that, the control system shifts to the closed speed loop, and the position information is from the BEMF comparators. In the demo application, the threshold speed is set to

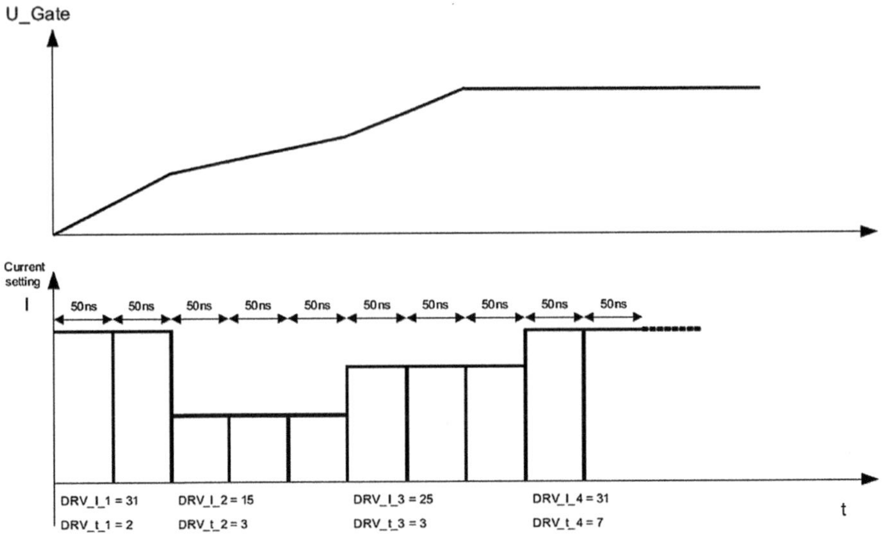

Fig. 10 Bridge driver slew rate control

300 RPM. From the figure, the rising time is about 150 ms and there is nearly no overshoot. The startup test is repeated 3600 times in 2 h, and there is no failure to startup. Therefore, it is believed that the startup is so reliable that can be used in fuel pump system (Fig. 12).

3.2 *Speed Regulation*

The following testing is for the performance of the speed regulation. The test is like that the speed dropped from the maximum speed 6000 RPM to minimum speed 2000 RPM in 50 ms and increased from 2000 to 6000 RPM in 50 ms. The first target is that the system should maintain stable status. Second, the response time should be as short as possible. In the following figures, the green line is the DC link voltage, the yellow line is the DC link current, the sky-blue pulse is the A-phase virtual hall signal, and the purple-blue pulse is the speed command—PWM communication. From the two figures, the system is very stable and the response time is less than 100 ms when there is speed jump (Fig. 13).

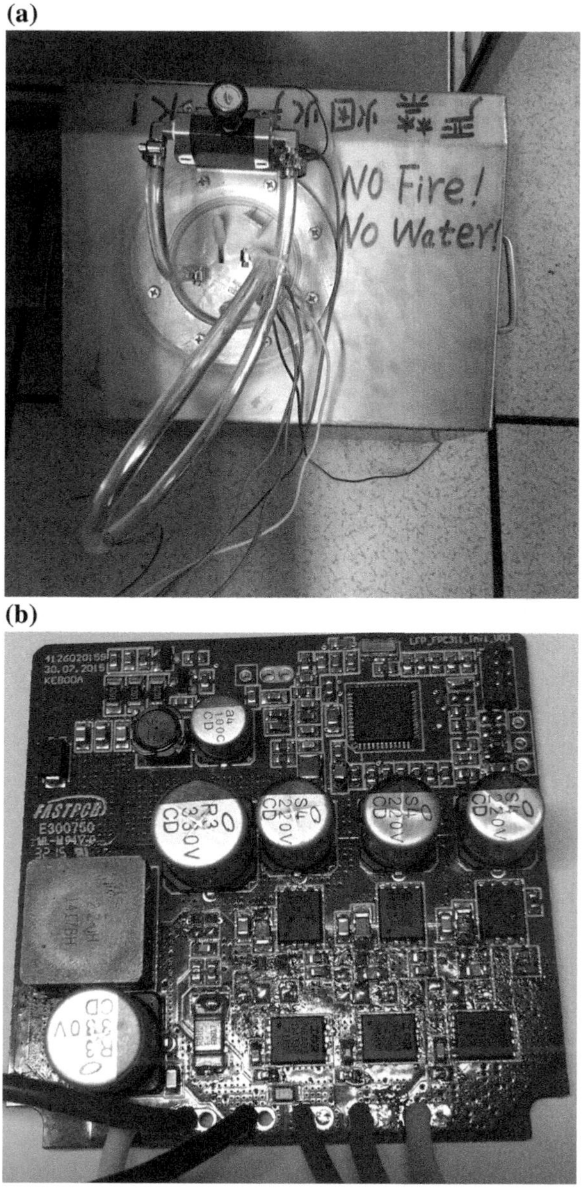

Fig. 11 **a** Fuel pump. **b** The driver based on ePower

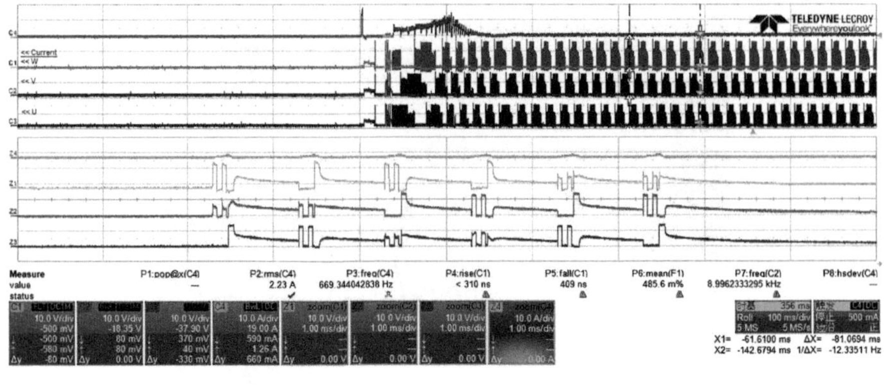

Fig. 12 Startup

3.3 Voltage Response

In order to make the system more stable, the voltage compensation is added to the system. The following figures show the results when the voltage jumped from 8 to 16 V and from 16 to 8 V in 20 ms. Because the DC link voltage determines the maximum speed of motor, when the voltage jumps from 8 to 16 V, the speed also increases to higher speed and the process is stable. When the voltage dropped from 16 to 8 V, the maximum speed decreases to lower speed and the whole process is also very stable (Fig. 14).

3.4 Pressure Response

Pressure response is a very important requirement in fuel pump application. In Fig. 11, there is a valve to change the oil pressure. In this experiment, we adjust the value to change the oil pressure continuously and the pressure range is during 2 and 6 bar. The following figure shows that the system is very stable, and the response is about 50 ms (Fig. 15).

(a)

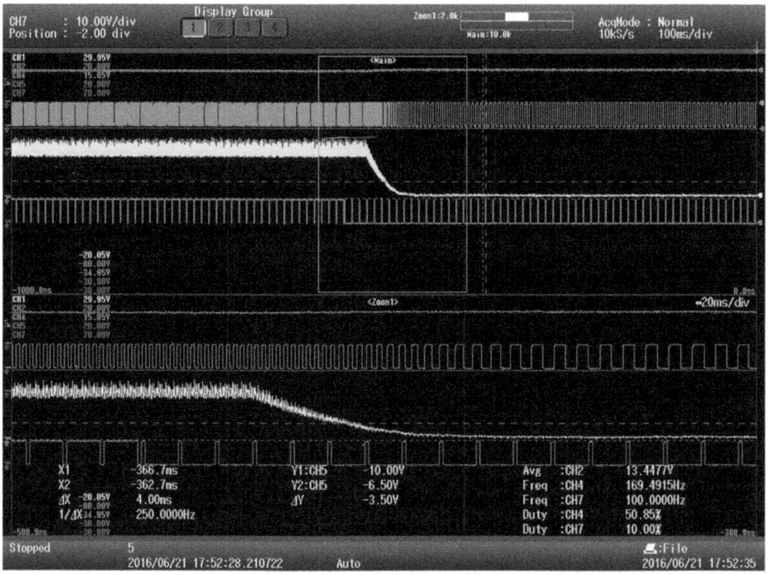

(b)

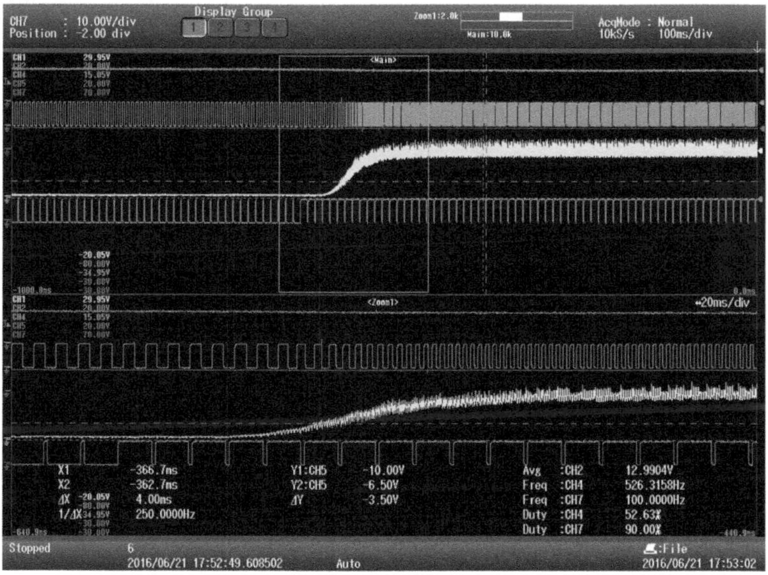

Fig. 13 **a** Deceleration. **b** Acceleration

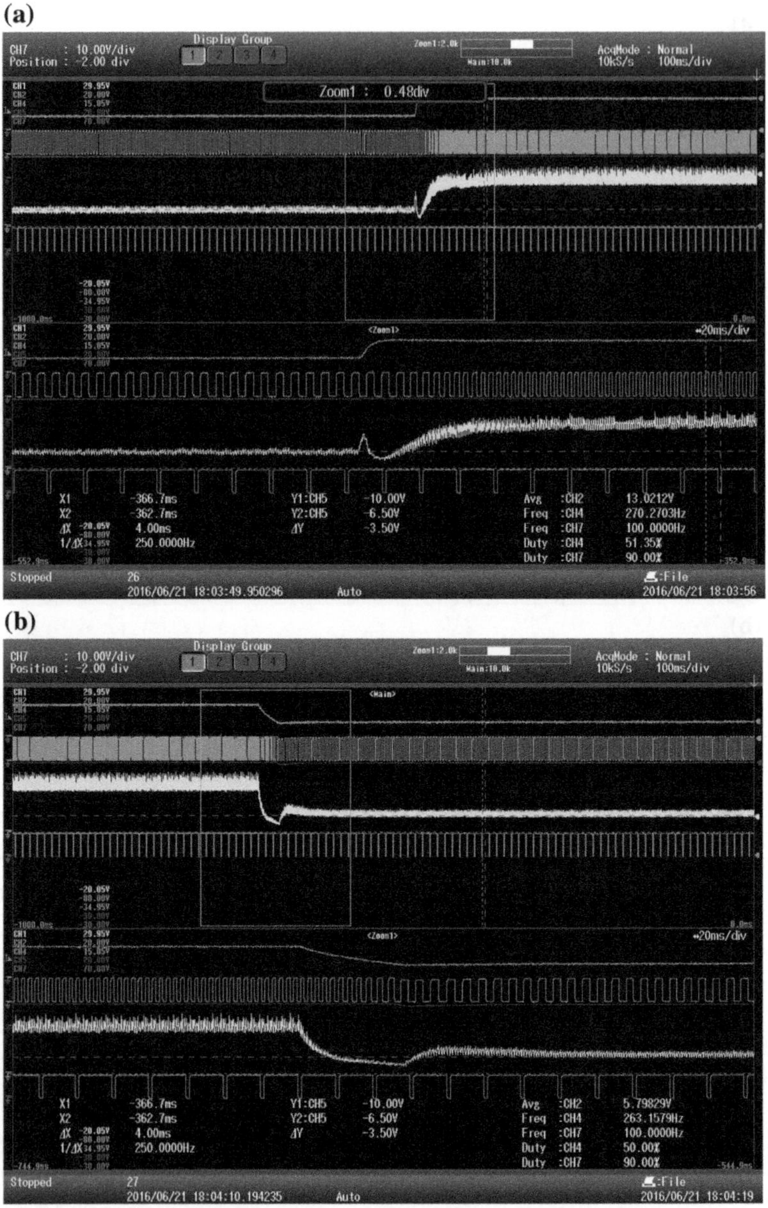

Fig. 14 **a** Voltage varieties from 8 to 16 V. **b** Voltage varieties from 16 to 8 V

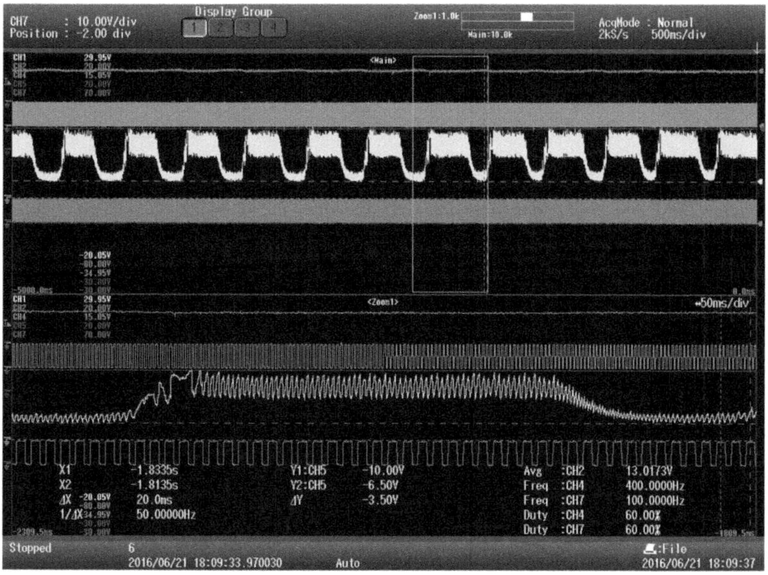

Fig. 15 Pressure variety

4 Conclusions

Infineon ePower TLE987x integrated voltage regulator, three-phase bridge driver, BEMF comparator, and other useful peripherals. The serial experiments show that the fuel pump driver based on ePower can perfectly overcome the key points in fuel pump system such as reliable and fast startup, quick response for speed regulation, much stability in pressure variety, and voltage fast changing. Therefore, ePower TLE987x is very powerful and smart that can fully handle the fuel pump application.

Acknowledgements I want to take this chance to thank the team leaders Ma Pan and Richard Zhu. They greeted the project and gave strong support in the proceedings. At the same time, I am very grateful to our Germany colleagues Martin Kilian and Mysliwitz Daniel who offered me the confidence and discussed with me when there are problems.

References

1. Sebastian T, Slemon GR, Rahman MA (1986) Modelling of permanent magnet synchronous motors. IEEE Trans MAG-22(9):1069–1108
2. Wu R, Slemon GR (1990) A permanent magnet motor drive without a shaft sensor. 90/CH 2935-5/90/0000-0553$01.00, 1990, IEEE
3. Infineon-TLE9879QXA40-UM-v01_00-EN, www.infineon.com
4. Cardoletti L, Jufer M (1992) Brushless DC motor standstill indirect position sensor. Intelligent Motion, April 1992 Proceedings, 99–109

Road Noise Prediction Based on FRF-Based Substructuring Method

Jingchao Xu, Chao Ren, Ran Xie and Jianfeng Huang

1 Introduction

Reducing vehicle inner noise level has become a top issue for our vehicle engineers. In the meantime, with the population of new energy vehicles, the road-induced noise would take up the main component of vehicle inner noise. The FBS hybrid model method is introduced in this paper to analyze and solve this problem. Both simulation model and test result are contained in the model in order to maintain the perfect balance between the size of model and result precision. This is an effective way to predict and solve road noise problem.

2 Methodology

2.1 Vibro-Acoustic Transfer Path Analysis

The vibration and acoustics of machinery system were summed by many excitation sources. In order to better analyze and optimize vibration and noise in the vehicle, different excitation and transfer paths should be considered, respectively. Transfer path analysis [1] was invented to tackle these issues. A "source-path-receiver" model was established to quantitative the contribution of every path, and then measures can be taken to reduce the forces transmitted through the main paths.

J. Xu (✉) · C. Ren · R. Xie · J. Huang
GAC Automotive Research & Development Center, Guangzhou, China
e-mail: xujingchao@gaei.cn

Society of Automotive Engineers of China (SAE-China) (ed.), *Proceedings of the 19th Asia Pacific Automotive Engineering Conference & SAE-China Congress 2017: Selected Papers*, Lecture Notes in Electrical Engineering 486, https://doi.org/10.1007/978-981-10-8506-2_22

The math equation [2] is written as follows:

$$r(\omega) = \sum_{i=1}^{n} \frac{R(\omega)}{S_i(\omega)} \cdot S_i(\omega)$$

$r(\omega)$ denotes receiver spectrum;

$\frac{R(\omega)}{S_i(\omega)}$ denotes FRF between the receiver and the source applied at the transfer path i;

$S_i(\omega)$ denotes the excitation force i.

Generally speaking, the transfer path analysis involves two factors: firstly, the excitation forces acting upon every transfer path under operating condition, and secondly, the transfer functions of every path between the receivers and sources. The transfer functions can be obtained by test or analytic methods, and the excitation forces can also be obtained through direct measurements and indirect measurements. This process is illustrated in Fig. 1.

2.2 *Hybrid Simulation of FRF-Based System Synthesis*

Transfer function is of main importance during operating transfer path analysis, which can be generally obtained by FE model numerical analysis or test method. By defining the rigid or elastic connection property of the coupling points, FRF-based system synthesis is based upon the dynamic property of every sub-component to predict the assembled structural dynamic character. Either experimental or analytic FRFs should be fine.

Assuming the subsystem A and B (see Fig. 2) are represented by transfer function matrix $\{A\}$ and $\{B\}$, respectively. They are connected by an elastic connector to form the assemble C. The stiffness matrix of the elastic connector can be represented by $\{K_s\}$. Then, the transfer function matrix [3] of assemble C can be written as follows:

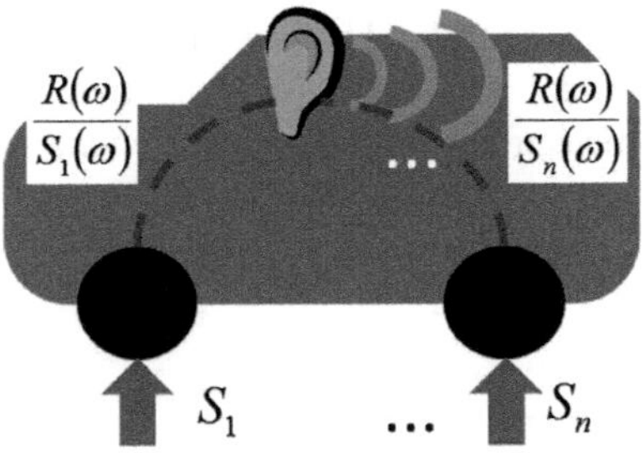

Fig. 1 Transfer path analysis process of road noise

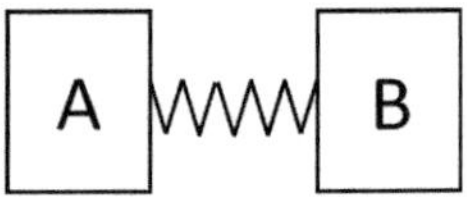

Fig. 2 Substructure A and B are connected by elastic connector

$$\begin{bmatrix} [H_C]_{RR} & [H_C]_{RS} & [H_C]_{RT} \\ [H_C]_{SR} & [H_C]_{SS} & [H_C]_{ST} \\ [H_C]_{TR} & [H_C]_{TS} & [H_C]_{TT} \end{bmatrix} = \begin{bmatrix} [H_A]_{RR} & [H_A]_{RS} & 0 \\ [H_A]_{RS} & [H_A]_{SS} & 0 \\ 0 & 0 & [H_B]_{TT} \end{bmatrix} - \begin{Bmatrix} [H_A]_{RS} \\ [H_A]_{SS} \\ -[H_B]_{TS} \end{Bmatrix} \left[[H_A]_{SS} + [H_B]_{SS} + [K_S]^{-1} \right]^{-1} \begin{Bmatrix} [H_A]_{RS} \\ [H_A]_{SS} \\ -[H_B]_{TS} \end{Bmatrix}^T$$

Subscript S denotes the coupling connector DOF A and B.

Subscript R and T, respectively, denote the inner DOF of A and B.

With this method, we can use transfer functions obtained by analytic and experimental methods to create the whole analytic model, where some FE components are hard to create or the prototypes are unavailable.

2.3 Inverse Matrix Method

The loads acting on the spindle heads can be measured through direct method. However, the equipment should be placed in the transfer path, which should bear high strength. Another method of inverse matrix [4] was introduced to identify the loads acting on the spindle head, while the knuckle accelerations can be viewed as responses. The theory equations were listed below:

$$\begin{Bmatrix} f_1 \\ \cdots \\ f_n \end{Bmatrix} = \begin{bmatrix} \frac{\ddot{X}_{11}}{F_1} & \frac{\ddot{X}_{12}}{F_2} & \cdots & \frac{\ddot{X}_{1n}}{F_n} \\ \cdots & & & \\ \frac{\ddot{X}_{m1}}{F_1} & \cdots & & \end{bmatrix} \begin{Bmatrix} \ddot{x}_1 \\ \cdots \\ \ddot{x}_m \end{Bmatrix}$$

$\begin{Bmatrix} \ddot{x}_1 \\ \cdots \\ \ddot{x}_m \end{Bmatrix}$ denotes the acceleration responses;

$\frac{\ddot{X}_{ij}}{F_j}$ transfer function between spindle head and knuckle.

The loads $\{f\}$ can be calculated through this method.

3 Full Vehicle Modeling

3.1 Trimmed Body Model

This paper mainly concerns about the effect of chassis component. We can use experimental transfer function to replace the FE trimmed body model, in order to avoid the error which introduced from the establishment of FE model. The test was conducted in the semi-anechoic room. All of the chassis components were removed, and the trimmed body was placed on four air springs to create the free–free condition. The transfer function was tested between exciting points of chassis mounts and receiver points of the human ear location, the seat rails, and the steering wheel. The measurement is shown in Fig. 3.

The exciting hammer was used as excitation. The transfer function showed below is between left trailing arm front mount and steering wheel 12'o clock point, and the coherence of frequency domain is of good quality (see Fig. 4).

Fig. 3 Trimmed body placed on four air springs

Fig. 4 Acoustic transfer function and coherence

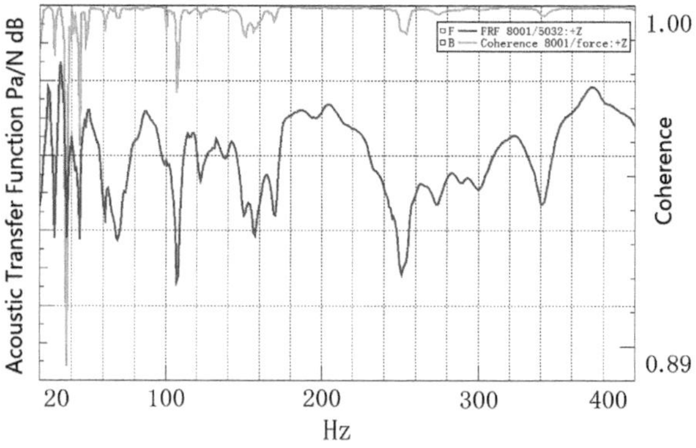

3.2 Shock Absorber Model

For automobile companies, the damper seems like a black box system. It is almost impossible to predict its actual dynamic property through detailed physical geometric modeling. This paper regards the damper as a mass-spring-damp system, on the premise of assuming the small displacement system as a linear system. Under the help of experiment, we can build a parameterized model based on the test data [5].

A test bench was built to conduct this test. Firstly, the damper was fixed rigidly upon the bench vertically. Secondly, a thin elastic cord passing through the top of damper inflicts prestressing force upon the damper to make sure the damper maintains its length at ride height, when the vehicle is at curb weight. A large vibration generator was hung upside down above the damper to exert operation forces. The image was shown (see Fig. 5).

The impedance data was obtained through the sensors and data acquisition, which will be used later to modify the simulation model. The x- and y-direction bending modes were also measured to adjust the CAE models (see Fig. 6).

3.3 Sub-frame Model

Modal analysis was used to identify nature frequencies and modal shapes of certain component. It can also be used to correlate the simulation models with actual

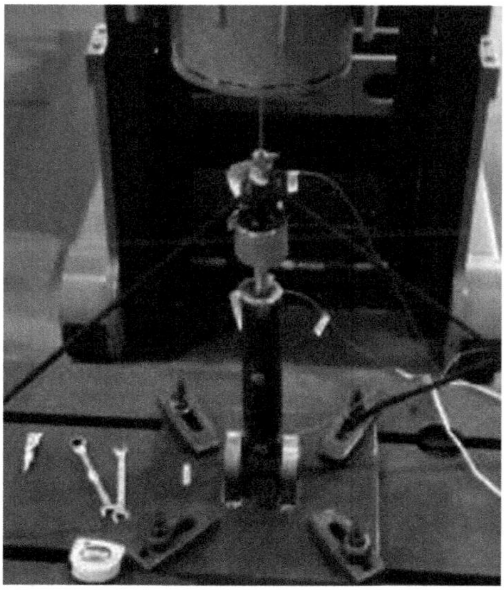

Fig. 5 Test bench of damper

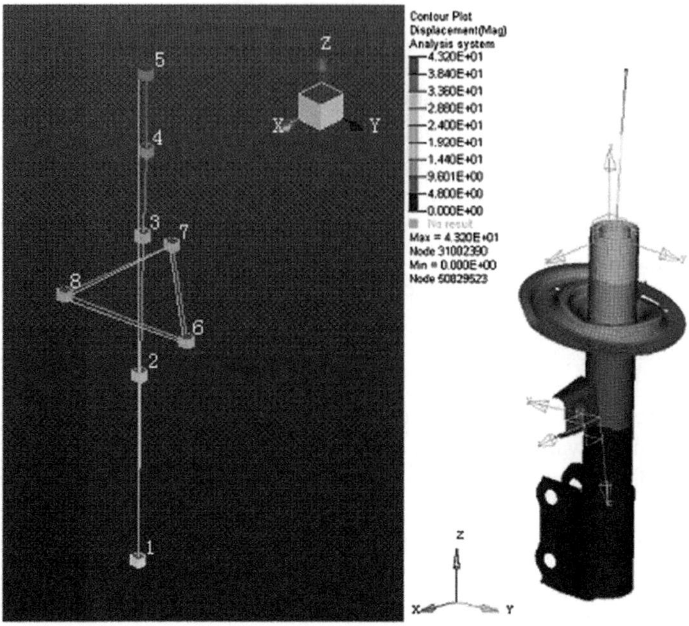

Fig. 6 Bending mode of damper

components. For example, the rear sub-frame modal test was conducted. After the verification of the mass and size, the correlation analysis was used to adjust this model. It can be seen through the MAC matrix graph [6] that the simulation model corresponded well with the actual entity (see Fig. 7).

3.4 Other Models

Special benches were built to measure the constraint mode of the steering column and links of suspension. Constraint mode is much easier to reflect the actual condition. It is also easier to correlate and adjust the simulation model. The test benches were shown (see Figs. 8 and 9).

The power train system containing engine, transmission exerts effective influence on inner noise within low- and medium-frequency domain. The transfer functions can also be obtained through experimental method under free–free condition (see Fig. 10).

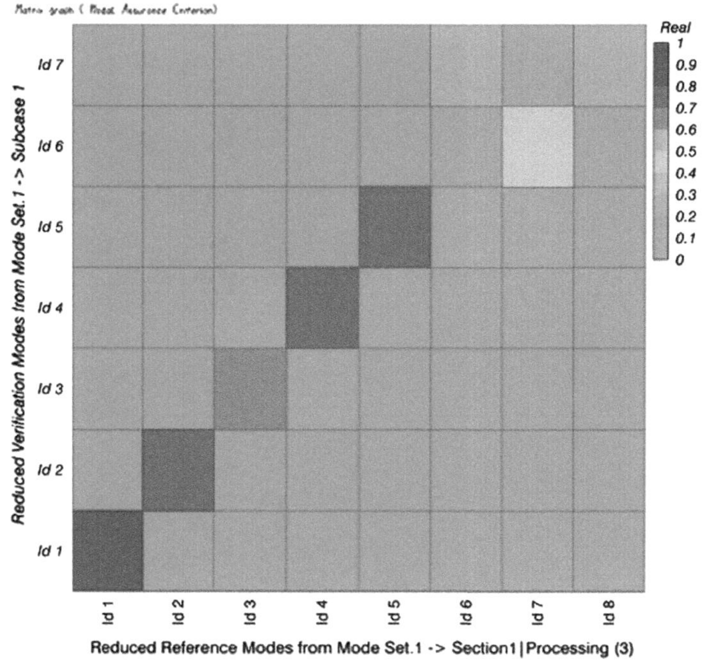

Fig. 7 MAC matrix graph of rear sub-frame

Fig. 8 Steering wheel test bench

3.5 Rubber Bushing

In order to alleviate the excitation transmitted through the suspension, rubber bushing mounts are of extensive use in chassis system. As is described above, the stiffness matrix of connection is of key importance to FRF-based system synthesis.

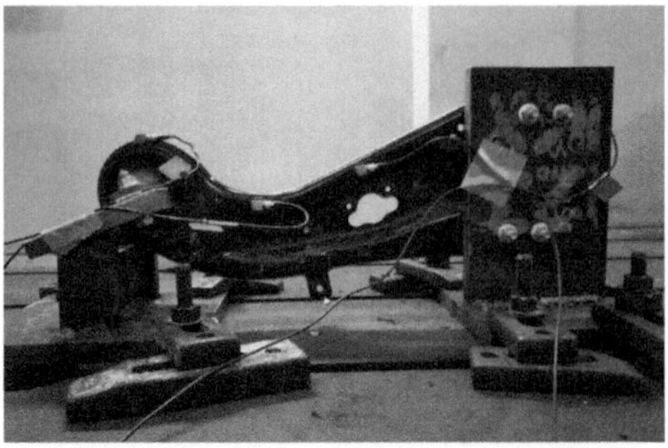

Fig. 9 Trailing arm test bench

Fig. 10 Power train hanged with elastic ropes

So, we need to get the exact dynamic property of these connecting mounts. They are preloaded during the dynamic stiffness test. Three directions were tested, and the frequency of excitation force was exerted up to 400 Hz. The equivalent dynamic stiffness was calculated via the curve of transfer function [7]. The test bench is illustrated in Fig. 11.

3.6 Assembling of Full Vehicle FBS Model

Because of the nonlinearity of tires, it is of great difficulty to establish accurate models. All the transfer functions and IPIs of every FE component were extracted on free–free condition. They are synthesized along with the trimmed body transfer functions to establish the full vehicle analytic model without tires (see Fig. 12).

The assemble transfer function between inner receivers and hub centers was calculated via FBS full vehicle model. These transfer functions were validated by means of full vehicle transfer function experiment. The test was conducted in the

Fig. 11 Rubber bush test bench

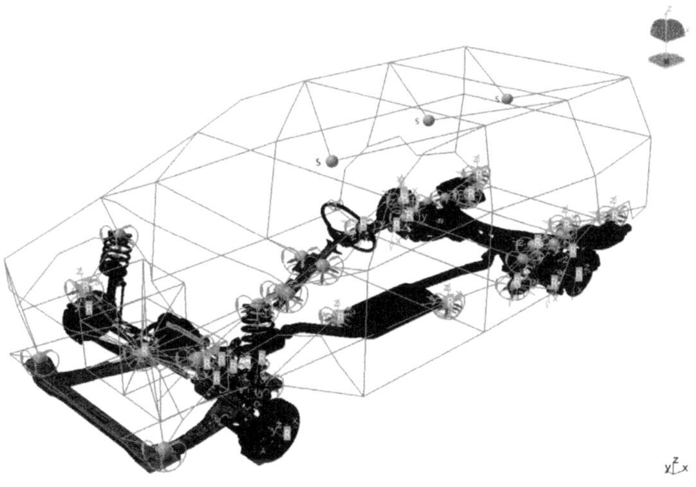

Fig. 12 FBS full vehicle model

semi-anechoic room, with one tire removed from the vehicle, and the hub was lifted by an elastic rope. The excitation was exerted on the brake disk using a hammer. Both transfer functions were compared below, and the results matched well from 20 to 250 Hz within frequency domain, which proved the FBS model was of enough accuracy for engineering application (see Figs. 13 and 14).

Fig. 13 Full vehicle with one tire removed

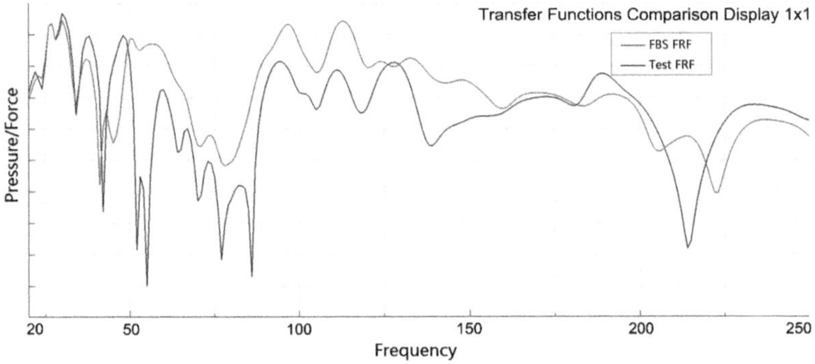

Fig. 14 Comparison of FRFs

4 Calculation and Analysis

4.1 Load Identification

In NVH issue, the full vehicle can be viewed as a multi-exciting linear system. Four acceleration transducers were placed on every knuckle to measure the responses under certain operational condition. The transfer function between spindle head and knuckle was calculated by using the FBS full vehicle analytic model. Then, the excitation loads were obtained by using inverse matrix method. The number of acceleration transducers placed on the knuckle should be at least two. Three translational forces on every knuckle were calculated by using inverse matrix method (three rotational forces on every spindle head were neglected).

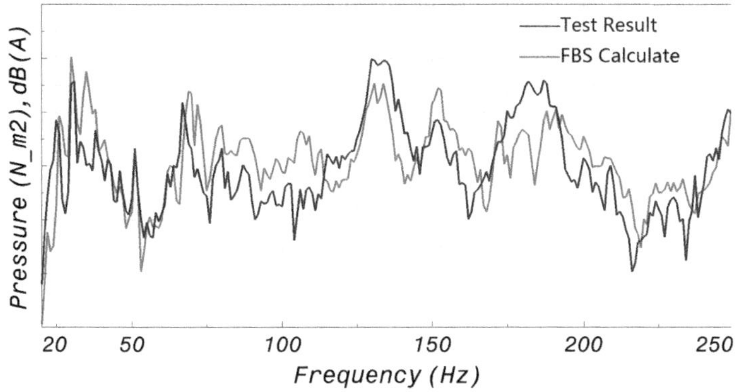

Fig. 15 Comparison of driver's right ear road noise level (A)

4.2 Verification of Road Noise Result

The loads obtained above were imported in the full vehicle hybrid FBS model. Then, the road-induced noise level was calculated and compared with the actual measured data. The overall sound pressure level between the calculated results with actual noise was within approximately 2 dB(A) for the concerned frequency range and the tendency showed perfect match, which meant the model was high in precision (see Fig. 15).

5 Conclusion

In this paper, the hybrid model method of establishing the full vehicle was introduced to estimate the road-induced noise at initial development stage. The principle based on the synthesis of both analytic and experimental transfer function improved the efficiency and precision outstandingly. The inverse matrix method to identify the load was also introduced in this paper and the road noise was validated by importing the identified load in the full vehicle FBS model, which is of key importance to discern potential risk and modify faulty parts at development stage.

References

1. Padilha PEF (2006) Comparison of estimation techniques for vibro-acoustic transfer path analysis. Shock Vib 13:459–467
2. Kim Y (2015) Application of experimental transfer path analysis and hybrid FRF-based substructuring model to structure-borne noise. SAE Technical Paper, https://doi.org/10.4271/2015-01-2267

3. K. Wyckaert. Flexible Substructuring Coupling Based on Mixed Finite Element and Experimental Models: a Step Ahead of Transfer Path Analysis[J]. Proceedings, 1996, 12–15
4. Park J (2001) Operational spindle load estimation methodology for road NVH applications. Noise & Vibration Conference & Exposition
5. Rao MD (2002) Measurement of equivalent stiffness and damping of shock absorbers. Exp Tech 26(2):39–42
6. Sakai T (2001) Development procedure for interior noise performance by virtual vehicle refinement, combining experimental and numerical component models. Sae Noise & Vibration Conference & Exposition 9:283–284
7. QJGAC 1220.009-2015. Technical conditions of rubber bush of passenger vehicle Chassis. 2015

The Analysis of a Sound-Generating Mechanism for Tire Pattern Groove Air Pumping by CFD

Haichao Zhou, Zhen Jiang, Guolin Wang, Hao Qian and Lei Qiao

1 Introduction

Traffic noise is a key source of noise pollution, and the noise generated by the rolling of tires is dominant when the vehicle speed exceeds 70 km/h. Tire noise includes tire vibration noise and aerodynamic noise, but the pattern groove air-pumping noise is an aerodynamic mechanism that generates noise at medium- and high-frequency ranges, compared with tire surface vibration, which radiates in medium and low-frequency ranges [1, 2]. The air-pumping noise is usually considered to be caused by the tread pattern volume fluctuations that are related to tire deformations. This noise is generally associated with all airflow fluctuations and acoustic resonances near the contact area; therefore, the tire tread pattern shape and parameters determine the type and volume of the air-pumping noise [3, 4]. Using tire label (EC) No 122/2009 and the norms for green tire technical specifications provided by the China Rubber Industry Association, we consider tire noise performance in terms of overall tire performance levels. Research about the mechanism that produces the noise of air pumping and release is significant.

Kim [5] used the CFD approach to compute the air-pumping noise of a single transverse groove with both ends closed and air pressure distributed. Wang et al. [6] combined Lighthill acoustic analogy method and the vortex sound theory analysis to analyze the air-pumping noise mechanism of a single transverse pattern with both ends open and determined that the noise level has a positive relationship with air vortices. Based on the volume change of the rolling grooves of the tire, it was found that the air-pumping noise was generated according to the model of acoustic

H. Zhou (✉) · Z. Jiang · G. Wang · H. Qian · L. Qiao
School of Automotive and Traffic Engineering, Jiangsu University,
Zhenjiang 212013, China
e-mail: hczhou@ujs.edu.cn

Society of Automotive Engineers of China (SAE-China) (ed.), *Proceedings of the 19th Asia Pacific Automotive Engineering Conference & SAE-China Congress 2017: Selected Papers*,
Lecture Notes in Electrical Engineering 486, https://doi.org/10.1007/978-981-10-8506-2_23

monopole and the superposition of a formula that described the acoustic source by Wang [7], and her analysis results show that the pumping noise increased with the increase in the groove depth or circumferential dimension, where the depth is more significant. Using the tire lab drum noise test, Ge [8] obtained the effects of transverse groove geometric parameters on air-pumping noise, and pointing out that the length of the transverse groove had an obvious effect on noise. In order to determine the sound spectral characteristics of pattern noise and precisely predict tire rolling noise, a semi-empirical predict model for tire air-pumping noise was built by Wei [9, 10]. The noise prediction results obtained by this method were compared with those obtained by a carved tire indoor noise test, and the prediction results using the established model were in good agreement with the test results. Shu [11] build a single lateral pattern groove air-pumping noise model, which included roads and the surrounding air domain, and used a two-way fluid-structure interaction method to simulate the flow field and the noise. Upon analyzing the effect of the geometrical parameters of the lateral pattern of the groove cross section on the air-pumping noise pattern, it was found that the air-pumping noise increased as the groove width increased, and it also increased as the groove depth increased; however, it decreased as the draft angle of the groove wall increased. In order to control air flow and reduce noise, Zhou [12] arranged a bionic V-riblet non-smooth structure in the bottom of the tire groove and analyzed the tread aerodynamic noise. In that study, by comparing the Reynolds stress and turbulent kinetic energy, the noise reduction mechanism of the bionic V-riblet non-smooth structure on tread aerodynamic noise was revealed.

From the review of the various studies that have been conducted on air-pumping noise, we may see that previous research has focused on the single lateral pattern groove and the effect of geometrical parameters; however, the mechanism that generates noise in response to air flow fluctuations, which is an important element to understand, has not been studied.

This paper is organized as follows: In the first section, the tire FEM model of complex tread patterns is established by Abaqus software, and then the contact shape and groove deflection are used to validate the model. The second section describes the analysis method for the air-pumping noise using Lighthill acoustic analogy method and FW-H equation. A comparison of the prediction results and test results shows that the analysis method is appropriate. In the third section, the effects of tire load, rolling speed and inflation pressure on air-pumping noise are analyzed. From exhaustive research on the air flow field and variation characteristics, the mechanism of the air-pumping noise pattern is determined.

2 Tire FEM Model Building and Verification

The main function of tire tread grooves is to provide grip performance and drainage water. The deformation of tread under various loads affects the volume of the groove and results in air-pumping noise. The contact shape and pattern grooves deformation under normal load must be first obtained to analyze air-pumping noise. Therefore, the FEM model of the tire (205/50R16) was established. The Yeoh model was adopted in the finite element analysis (FEA) using for rubberlike elastomers. The cord-rubber composite, belts, and tire carcasses were modeled with rebar layers. The rubber materials were modeled with CGAX3H and CGAX4H elements. The contact characteristics used involved Coulomb friction. The build process of the tire model is presented in Fig. 1. First, the tire is inflated state with no load, no inflation pressure, and no tread pattern; this step is arranged on a 2D tire around the rim; next, the 2D tire is revolved in a signal pitch using *symmetric model generation, and then the signal pitch tread pattern is tied to the 2D tire. After completion of the tying process, the 2D signal pitch tire is subjected to a inflate pressure of 200 kPa; third, symmetric model generation is used to generate the three-dimensional (3D) tire model shown in Fig. 1. Meanwhile, the results from the axisymmetric analysis are transferred to the 3D model. The 3D model of the tire is characterized by 205,379 nodes and 166,439 elements. Vertical deformation of the tire is achieved under a load of 4000 N, with the inflation pressure of 200 kPa.

The tire vertical deformation analysis includes four steps, namely:

1. Build the tire 2D section and a pitch tread pattern model, respectively.
2. Make the tread pattern tie the tire 2D section and form a pitch tire FEM model; at the same time, establish surfaces contact between tire and rim, and apply inflation pressure on the inner surface (200 kPa in this study).

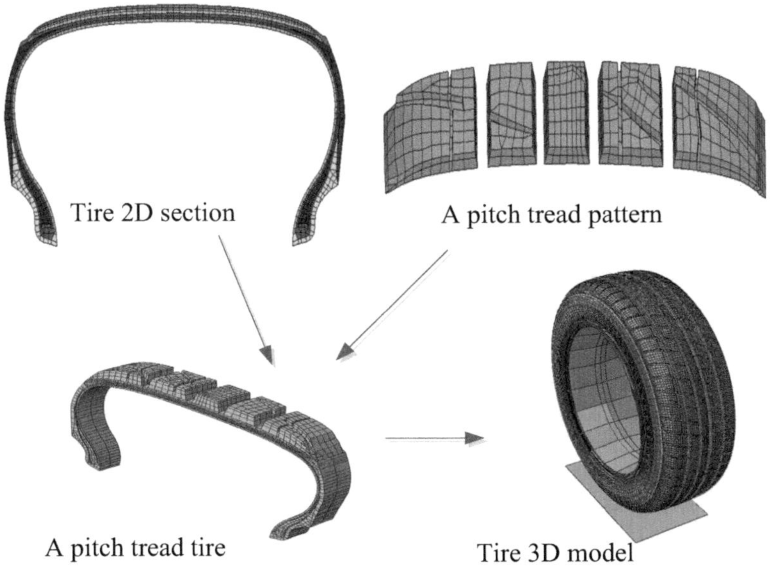

Fig. 1 Tire FEM model

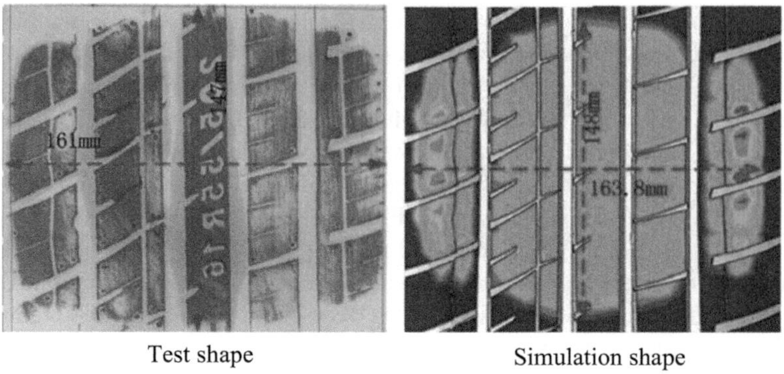

Fig. 2 Comparison of tire contact shape between test and simulation

3. Revolve the single pitch model into a 3D tire FEM model; establish surfaces contact between tire and road. The penalty method was invoked for this purpose.
4. Simulate the static deflection and move the road surface to a specific position until a given force, say the rated load, is exerted on the tire.

The static tire contact shape was simulated under the above assumptions, which are shown in Fig. 2. The tire contact shape is tested using the static loading machine (the MTM-2 tire testing machine), with a pressure-sensing pad. The detailed test method was provided by Zhou et al. [13]. As shown in Fig. 2. the contact shape of the simulation was consistent with the test shape. Also, the contact long axis and contact short axis of the test results are, respectively, 161 and 147 mm; the simulation prediction results are almost the same as the test results, in which the contact long axis and short axis were 163.8 and 147 mm, respectively.

3 Groove Air-Pumping Noise Pattern Analysis and Method Verification

3.1 Aeroacoustics Theory

The acoustic wave equation of Lighthill reflects the relationship between acoustic wave motion and flow field parameters. The FW-H equation is derived directly from the N-S equation by means of Lighthill acoustic analogy, and no assumption is made in the derivation. The integral surface can be located in the nonlinear region of sound propagation [14]. The sound problem of a moving object in a fluid can be expressed by the FW-H equation:

$$\frac{1}{c_0^2}\frac{\partial^2 p'}{\partial t} - \nabla^2 p' = \frac{\partial^2}{\partial x_i \partial x_j}(T_{ij}H(f)) - \frac{\partial}{\partial x_i}((p_{ij}n_j + \rho u_i(u_n - v_n))\delta(f)) + \frac{\partial}{\partial t}((\rho_0 v_n + \rho(u_n - v_n)\delta(f)) \tag{1}$$

where the f is the groove surface, $H(f)$ is the step function, and $\delta(f)$ is the Dirac function:

$$H(f) = \begin{cases} 1 & f(\overline{x,t}) > 0 \\ 0 & f(\bar{x},t) < 0 \end{cases}, \quad \delta(f) = \frac{\bar{\partial} H(f)}{\partial f} \tag{2}$$

where n_j is the unit normal vector that points to the outer surface of surface ($f > 0$); u_n is the normal velocity on the f = 0 surface; v_n is the normal component of surface velocity; i, j, k are representative Descartes coordinate system x- , y- , z-axis positive directions, respectively; P_{ij} is the compressible fluid stress tensor:

$$P_{ij} = p\delta_{ij} - \mu\left(\frac{\partial u_i}{\partial x_j} + \frac{\partial u_j}{\partial x_i} - \frac{2}{3}\frac{\partial u_k}{\partial x_k}\delta_{ij}\right) \tag{3}$$

Lighthill stress tensor:

$$T_{ij} = \rho u_i u_j + P_{ij} - c_0^2(\rho - \rho_0)\delta_{ij} \tag{4}$$

It can be seen that the three parts on the right side of the equation are the sound radiation sources. The first represents the turbulence induced by the fluid motion, which is equivalent to the quadruple source. The second part is the dipole source of the force acting on the surface of an object. The third includes the unsteady mass flow into the fluid, which is equivalent to the monopole source.

3.2 A Method for Calculating the Volume Change of a Single Pattern Groove

The fundamental mechanism of air-pumping noise is caused by the deformation of the groove. Hence, it is necessary in order to obtain the deformation characteristics of the groove patterns to further study air-pumping noise. In view of the tire in this paper, an experimental study on the sinkage of the grooves was carried out. During the test, the sinkage of the grooves under different loads was measured by a displacement sensor. The sensor model is a SKR resistance—mechanical sensor, the accuracy of which is ±0.5%. The sensor was installed on a steel plate with a hole in the middle and stuck into the groove of the middle of the contact area; load was then applied to the tire drum. Figure 3 shows the instruments used for the test. The experiment was conducted to test the sinkage of the shoulder groove and the crown groove. In the test,

Fig. 3 Sample tire and tire static-loaded test machine

the sensor and the groove bottom contact position was not easy to ascertain, so the test data processing was based on groove sinkage under an initial load of 500 N, so as to prevent errors caused by contact problems.

During the test, the data of the sinkage under different loads were obtained, as shown in Fig. 4. It can be seen from the Fig. 4 that groove deflection increased corresponding to increase load; however, the deflection of the crown groove was much larger than that of the shoulder groove. This should be beneficial for passenger car tires, because the shoulder contact pressure is greater than that of the crown, and the tire therefore will have improved handling performance. All the grooves deflection tends to be stable at 3500 N. The fact that the simulated results are so close to the test demonstrates the validity of the tire FEM model. Due to errors in the testing parameters of the rubber and the constitutive model of the model (the hyperelastic model and the real rubber tire with viscoelasticity), there are some differences between the simulation and the experiment.

Fig. 4 Pattern grooves deflection between test and simulation

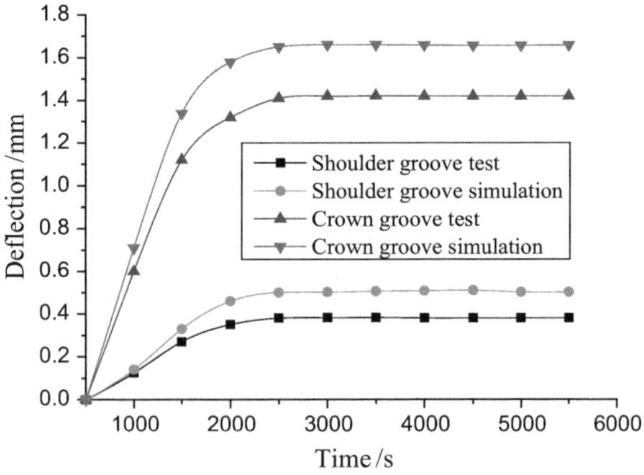

When the tire is rolling, the grooves volume change has an important influence on the air-pumping noise. The grooves volume deformation occurs mainly at the bottom of the groove sinkage, and at the same time, the surface of both sides of the groove is raised outwards, as shown in Fig. 5. Since the rubber can be considered as incompressible, it may be presumed that the total volume of the block is unchanged before and after deformation [15]. Therefore, this paper presents a geometric analysis method to calculate the volume deformation of a pitch pattern.

When the pattern block is subjected to load, the bottom of the groove of the pattern has a certain amount of sinkage, and the side wall of the groove is raised outwards. In Fig. 5, the dashed line represents the shape of the block and the groove before deformation, and the solid line represents the shape of the pattern after deformation caused by the load. Because the rubber is considered incompressible, we have the following:

$$S_1 = 2S_2 \tag{5}$$

where S_1 is the area of the block before deformation; this area is reduced by the area that overlaps with the deformed block. S_2 is half of the tire after the deformed area is reduced to the area of the original pattern.

Then, the volume deformation of a pitch groove can be expressed as follows:

$$\mathrm{V} = (2S_2 + 2S_3 + bh)\Delta l \tag{6}$$

where S_3 is the triangular area composed of the end point of the front end of the deformation block, the end point of the groove after the deformation of the pattern block, and the intersection of the bottom of the deformed tire with the un-deformed part of the front side wall; b is the groove width; and Δl is the length of a pitch.

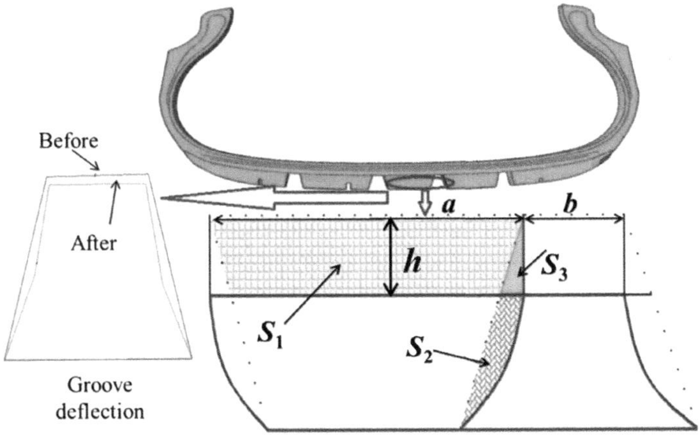

Fig. 5 Geometry diagram of pattern deformation

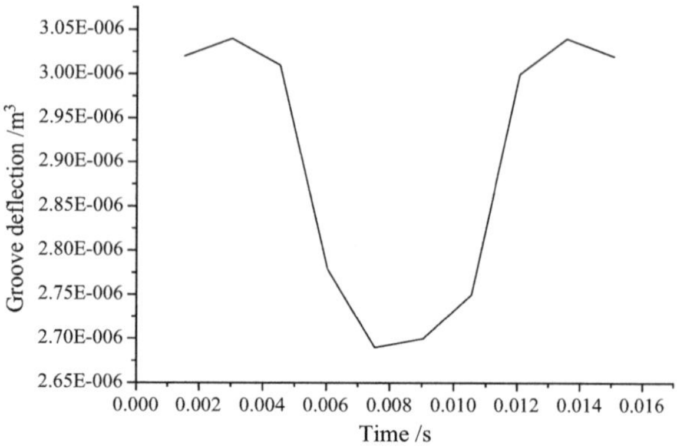

Fig. 6 Volume change of a single pitch pattern groove

Also because

$$2S_2 + 2S_3 = S_1 + 2S_3 = ah \tag{7}$$

Formula (6) can be changed to

$$\mathrm{V} = h(\mathrm{a} + \mathrm{b})\Delta l \tag{8}$$

where a is the block bottom width, and h is the pattern sinkage. Therefore, Formula (8) can be used to calculate the volume change of a pitch groove.

The volume change of a single pitch groove can be obtained by subtracting the deformation amount of the original groove. The analysis of a single pitch pattern groove can be used to obtain the volume change characteristics of the pattern groove from the ground to up. As shown in Fig. 6, the distance from the beginning of the tread pattern to the ground constitutes the entire tread process, and the groove deflection is first reduced and then gradually increased.

3.3 *The Calculation Method of Air-Pumping Noise*

It is assumed that the tire is not rolling, and that the relative motion of tire and air is simulated by the relative velocity of the given air, as shown in Fig. 7. At the same time, with the help of the user-defined function (UDF), the deformation of the grooves was applied as the boundary conditions for the model, recreating the characteristics of the volume change of the pattern grooves. The flow field includes the inside and outside of the groove pattern.

The flow field analysis is carried out by ANSYS-FLUENT, which includes two parts: steady and unsteady calculations. The realizable $k - \varepsilon$ model is solved by steady calculation, and it serves as the initial flow field for the unsteady calculation. The large eddy simulation LES method is utilized for the unsteady calculation, and the simulation time step is 1e-5s. Pressure and velocity coupling using SIMPLE algorithm is adopted, and the spatial and temporal parameters are improved by using the two-order discretization method. In the calculation of the sound field, the groove surface is defined as the noise source, and FW-H equation is used to predict the sound pressure level.

In order to verify the effectiveness of the analysis method of air-pumping noise, the rolling noise of the tire was tested using the laboratory drum, and the test device is shown in Fig. 8. The comparison of 1/3 octave band sound level of the test and simulation is shown in Fig. 9.

The air-pumping noise energy is mainly concentrated above 800 Hz the middle- and high-frequency band, so we need only analyze the frequency above 800 Hz. It can be seen in Fig. 9 that when the frequency band is greater than 800 Hz, the frequency spectrum characteristics of the test and simulation have a very high degree similarity. Also, there is a significant noise peak around 1200 Hz. However, the sound pressure level in the test is higher than that of the predicted. This is due to the model that only considers air-pumping noise in the simulation, the resonant

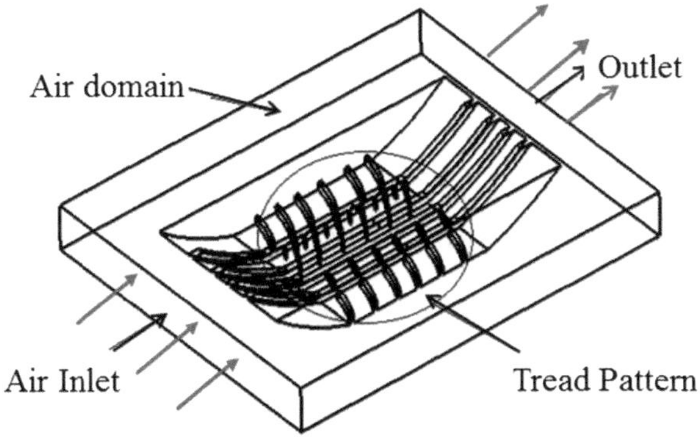

Fig. 7 CFD model of air-pumping noise analysis

Fig. 8 Tire noise drum testing device

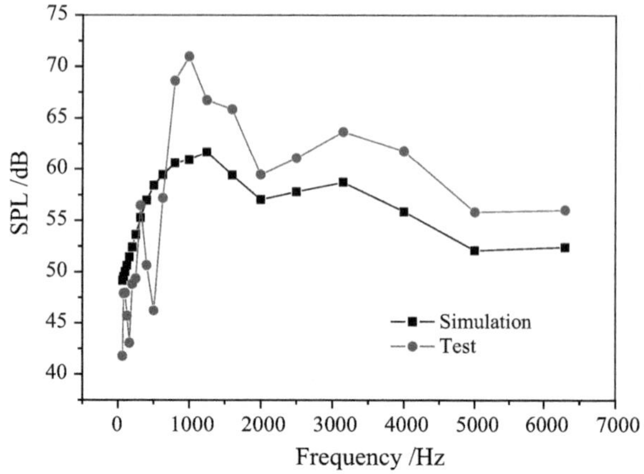

Fig. 9 Comparison of 1/3 octave band sound-level pressure between simulation and test

frequency of the cavity formed by the transverse groove and the surface, of approximately 1200 Hz, which causes the noise to increase, but in the test results, not only the cavity resonant and horn effect, but also the tire vibration noise component are included in all the frequency. This shows that the sound pressure level predicted by simulation is less than the test results.

4 Influence of Tire Maneuvering Factors on Air-Pumping Noise

4.1 Influence of Load on Air-Pumping Noise

In order to analyze the influence of load on air-pumping noise, the pattern volume deformation rate under three loads of 75, 100, and 125% was analyzed with a speed of 80 km/h and a tire pressure of 240 kPa. Based on the computational fluid dynamics (CFD) method and the acoustic equation, the farfield 1/3 octave band sound level of the test and simulation is shown in air-pumping noise under the three loads which were calculated and analyzed, as shown in Fig. 10.

Figure 11 shows that the noise spectrum characteristics of the three loads are basically consistent, and the noise peak frequency is approximately 1200 Hz. As the load increases, the air-pumping noise increases, so we may understand that the tire load and air-pumping noise are positive correlation.

The main mechanism of aerodynamic noise is caused by fluctuating pressure, which is sourced from the impact of airflow on the groove surface, and this action leads to the pressure pulsation on the groove surface. Therefore, by investigating the dynamic pressure conditions of the airflow in the groove, we can explain the performance of air-pumping noise. Figure 11 shows the dynamic pressure cloud in

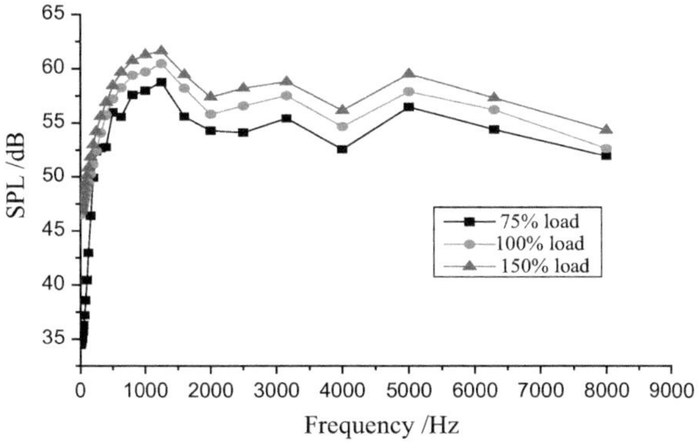

Fig. 10 Spectrum of 1/3 octave band sound level pressure under different loads

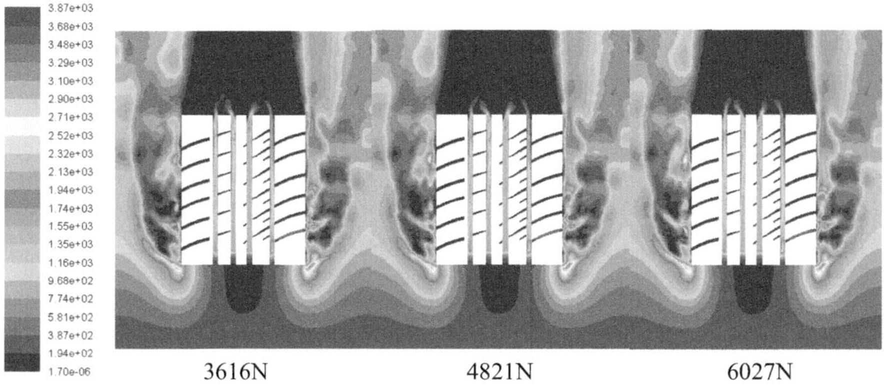

Fig. 11 Dynamic pressure contour under different loads

the pattern grooves under the three loads, and we find that the noise increases as the dynamic pressure increases caused by the flow field in the groove. The airflow mainly flows in the longitudinal groove, while a small amount of airflow is discharged along the transverse groove, and the interaction between the inside airflow and the outside airflow will produce turbulence around the outlet of the transverse grooves. Figures 12a–c are the scatter plot of the dynamic pressure acted on the pattern surface in the transverse direction under three loads, and it can be seen from Fig. 12 that as the tire load increases, the dynamic pressure gradually increases. Also, it is shown that the increase of load will lead into the groove deformation seriously. The tire roll will extrude more airflow, and the influence of airflow on the groove surface is more severe. Consequently, reducing the tire load can greatly inhibit the groove deformation, reduce the airflow dynamic pressure acts on the groove surface, and decrease air-pumping noise.

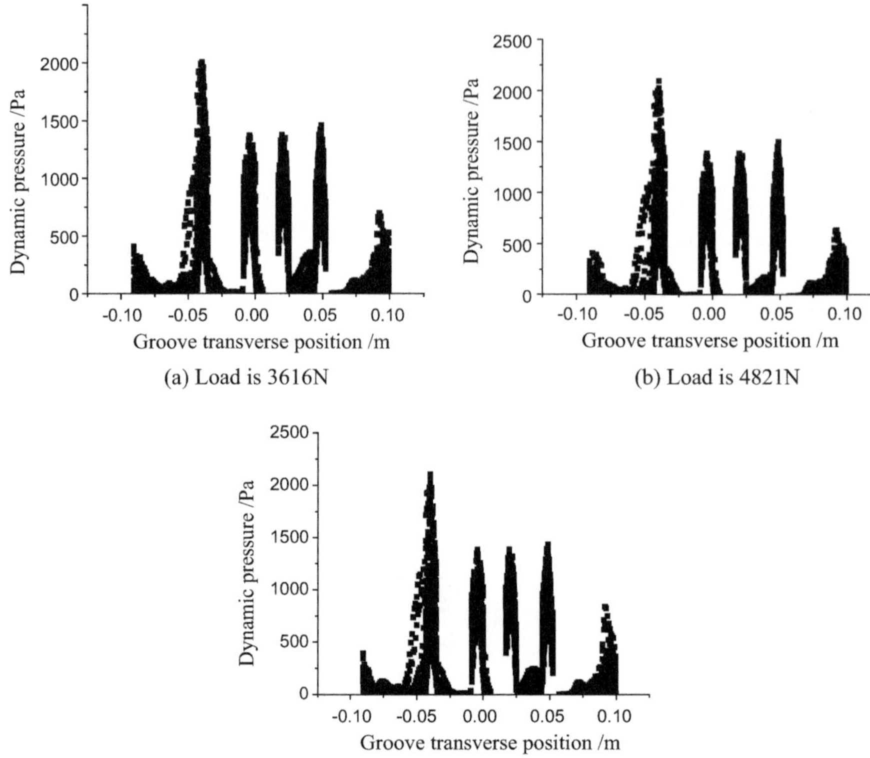

(a) Load is 3616N

(b) Load is 4821N

(c) Load is 6027N

Fig. 12 Dynamic pressure scatter plot of groove surface under different loads

4.2 Influence of Tire Rolling Speed on Air-Pumping Noise

In order to analyze the influence of speed on the pattern pumping noise, the influence of 60, 80 and 120 km/h on air-pumping noise was analyzed under the 4000 N load and a tire pressure 240 kPa. Based on the computational fluid dynamics (CFD) method and the acoustic equation, the conditions of the air-pumping noise were calculated and analyzed, and the farfield 1/3 octave band sound level of air-pumping noise under three speeds, as shown in Fig. 13.

Figure 13 shows that the noise spectrum characteristics of the three speeds are basically consistent, and the noise peak frequency is approximately 1200 Hz. As the speed increases, the air-pumping noise increases, so we confirmed that the tire speed and air-pumping noise are positive correlation.

Figure 14 shows the dynamic pressure cloud in the grooves flow field under the three speeds. The airflow mainly flows along the longitudinal groove, but a small amount of airflow is discharged along the transverse groove, and it has a significant

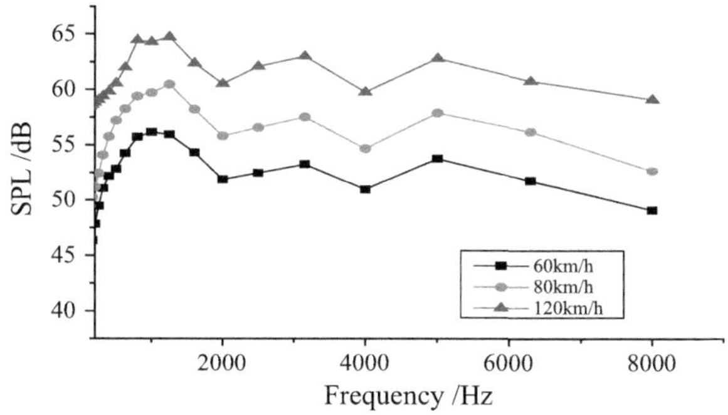

Fig. 13 Spectrum of 1/3 octave band sound-level pressure under different speeds

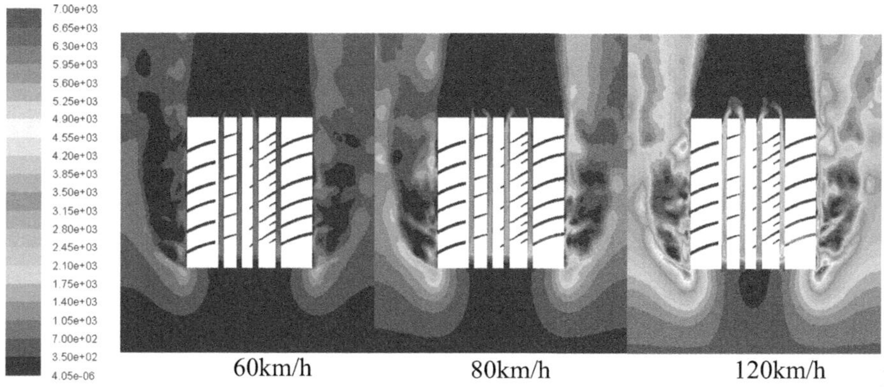

Fig. 14 Dynamic pressure contour under different speeds

effect on the external flow field at the exit of the transverse groove. Figure 15a–c is the scatter plot of the dynamic pressure acts on the pattern surface in the transverse direction under three speeds, and it can be seen from Fig. 15 that as the tire speed increases, the dynamic pressure gradually increases. It is shown that the increase in tire speed causes the airflow increase, the amount of airflow exhausted in the pattern grooves is great lager in a unit time, and the impact effect of the air on grooves surface is more severe. Therefore, reducing the speed can greatly reduce the fluid dynamic pressure acted on the grooves surface and reduce the impact of airflow on the grooves surface, thus also decrease the air-pumping noise.

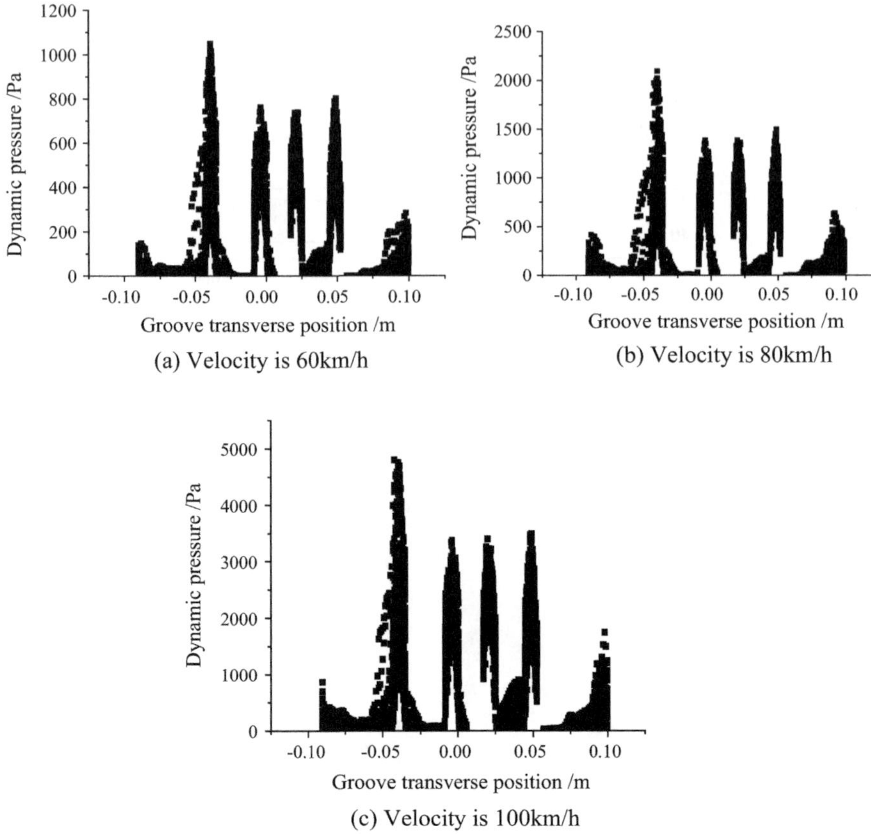

Fig. 15 Dynamic pressure scatter plot of groove surface under different speeds

4.3 Influence of Tire Inflation Pressure on Air-Pumping Noise

The tire pattern pumping noise under the three inflation pressures of 180, 240 and 300 kPa was analyzed under the speed 80 km/h and the load 4000 N. Based on the computational fluid dynamics (CFD) method and the acoustic equation, the change of air-pumping noise was calculated and analyzed, and the farfield 1/3 octave band sound pressure of air-pumping noise under three inflation pressures, as shown in Fig. 16.

Figure 16 shows that the noise spectrum characteristics of the three tire pressures are basically consistent, and the noise peak frequency is approximately 1200 Hz. As the inflation pressure increases, the air-pumping noise decreases, so we confirmed that the inflation pressure and air-pumping noise are negative correlated.

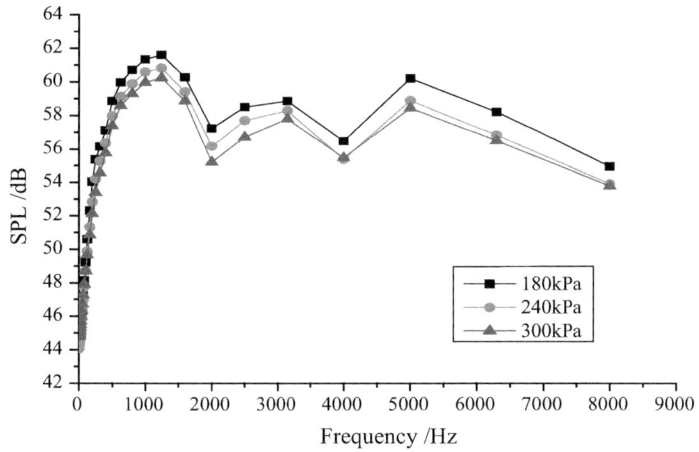

Fig. 16 Spectrum of 1/3 octave band sound level under different inflation pressures

Figure 17 shows the dynamic pressure cloud in the groove under the three pressures. The airflow mainly flows in the longitudinal groove, while a small amount of airflow is discharged along the transverse groove, and the interaction between the inside airflow and the outside airflow will produce turbulence around the outlet of the transverse grooves. Figures 18a–c is the scatter plot of the dynamic pressure acts on the pattern surface in the transverse direction under three pressures, and it can be shown from Fig. 18 that as the tire pressure increases, the dynamic pressure gradually decreases, and the pattern volume deformation rate also decreases. It is shown that increase tire pressure can suppress the grooves deformation, and reduce the amount of airflow exhausted during the tire rolling process, and decrease the impact of airflow on the groove surface. Therefore, the air-pumping noise is small. However, increasing the tire pressure will cause the tire vibration noise to deteriorate, and increase tire stiffness; this will have a bad effect on vehicle ride comfort.

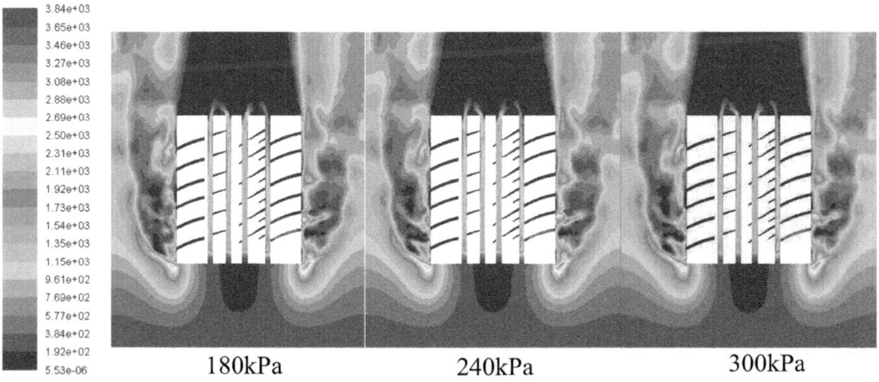

Fig. 17 Dynamic pressure contour under different pressures

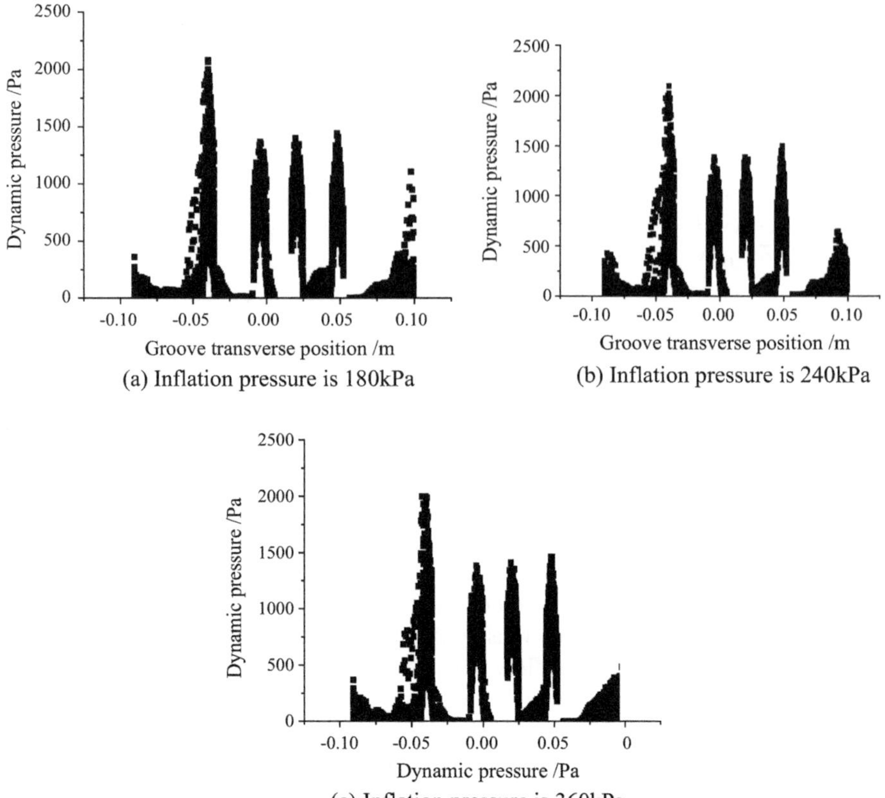

(a) Inflation pressure is 180kPa
(b) Inflation pressure is 240kPa
(c) Inflation pressure is 360kPa

Fig. 18 Dynamic pressure scatter plot of groove surface under different pressures

5 Conclusions

(1) The FEM model of the complex tread pattern tire is established by Abaqus software, and contact shape and pattern groove deflection were used to validate the model. The results of a tire rolling analysis show the characteristics of pattern groove volume change is from small to large.
(2) The air flow inside the pattern groove characters is analyzed using CFD, and the air-pumping noise is evaluated used Lighthill acoustic analogy method and FW-H equation. From the noise spectrum comparison between the prediction result and test result, it is showed that the analysis method and model for air-pumping noise are effective and accurate.
(3) The effects of tire load, rolling speed, and inflation pressure on air-pumping noise are analyzed. The detailed research of the air flow filed and variation characteristics discloses the mechanism of the pattern groove air-pumping

noise. It is found that dynamic pressure, which results from unsteady airflow, acted on the groove surface is the main generate source of the pattern air-pumping noise.

Acknowledgments The National Natural Science Foundation of China (Grant No. 51605198, 51675240); Jiangsu Province Funds for Young Scientists (Grant No. KB2016042722); Six talent peaks project in Jiangsu Province (Project no. JXQC-011).

References

1. Plotkin KJ, Montroll ML, Fuller WR (1980) Generation of tire noise by air pumping and carcass vibration. In: Proceedings of the international conference noise control engineering, noise control for the 80's, Inter-Noise 80, vol 1, Miami, Florida, 8–10 Dec 1980
2. Gautam P, Chandy AJ (2016) A three-dimensional numerical investigation of air pumping noise generation in tires. J Vib Acoust 138(6):061005
3. Wei Y, Feng Q, Wang H et al (2016) A hybrid numerical-experimental analysis for tire air-pumping noise with application to pattern optimization. Noise Control Eng J 64(1):56–63
4. Gautam P, Chandy AJ (2016) A three-dimensional numerical investigation of air pumping noise generation in tires. J Vib Acoust 138(6):61–66
5. Kim S, Jeong W, Park Y et al (2006) Prediction method for tire air-pumping noise using a hybrid technique. J Acoust Soc Am 119(6):3799–3812
6. Wang G, Mao Z, Zhou H et al (2012) Method research of calculation of air-pumping noise of single transverse groove of tire. J Mech Eng 48(4):116–119
7. Wang L, Li Z, Xia Y (2014) Numeircal simulation of tire pumping noise. J Univ Sci Technol China 44(6):483–487
8. Liu H, Ge J (2015) Noise features of tires with variable pitch tread. Tech Acoust 34(6): 540–541
9. Wei Y, Feng QZ (2016) PCR tire air-pumping noise prediction and optimization method. China J Highw Transp 29(2):130–135
10. Wei Y, Feng X, Zheng X et al (2015) A hybrid experimental-numerical analysis for radial tire air pumping noise generation mechanism. J Vib Shock 34(11):166–172
11. Shu Y, Zhou B, Zhang C et al (2017) The research of air-pumping noise of a lateral pattern groove tire[J]. Mach Des Manuf 1:242–244
12. Zhou H, Yang J, Liang C et al (2015) Effect of bionic V-riblet non-smooth structure on aerodynamic noise of tire tread pattern. J Huazhong Univ Sci Technol (Natural Science Edition) 43(11):27–32
13. Zhou H, Wan G, Ding Y, Yang J, Liang C, Fu J (2015) Effect of friction model and tire maneuvering on tire-pavement contact stress. Adv Mater Sci Eng. https://doi.org/10.1155/2015/632647
14. Zhang N, Xie H, Wang X, Wu BS (2016) Computation of vortical flow and flow induced noise by large eddy simulation with FW-H acoustic analogy and Powell vortex sound theory. J Hydrodyn 28(2):255–266
15. Stevenson K, Stallwood B, Hart AG (2008) tire rubber recycling and bioremediation: a review. Bioremediat J 12:1–11

Study of Intake Charge Control Method on TGDI Engine with CVVL System

Benzhong Han, Liguang Li, Ning Liu and Lihua Zhang

Nomenclature

TGDI	Turbocharged gasoline direct injection
VVT	Variable valve timing
CVVL	Continuously variable valve lift
Psr	Intake manifold pressure
TOA	Throttle opening angle
IL	Intake valve lift
BSFC	Brake specific fuel consumption
BMEP	Brake mean effective pressure
Rip	The pressure ratio after and before throttle valve
ZWOUT	Ignition advance angle

1 Introduction

Automotive OEMs are developing various technologies to fulfill the requirements of more and more stringent fuel consumption and emission regulations, especially due to the reduction of CO_2 emissions. China government published phase IV fuel consumption regulation in 2015, which requires companies to achieve target 5L/100 km of average fleet fuel consumption in the year 2020 for OEMs. It is estimated that the average fuel consumption targets could be 4L/100 km and lower after 2025. Therefore, OEMs are taken various technical measures to reduce ICE fuel consumption level, e.g., turbocharger, GDI, downsizing, VVT, smart components, electrification. Continuously variable valve lift (CVVL) system is well

B. Han (✉) · L. Li
College of Automotive Engineering, Tongji University, Shanghai 201804, China
e-mail: Benzhong.han@UAES.com

N. Liu · L. Zhang
United Automotive Electronic Systems Co., Ltd., Shanghai 201206, China

Society of Automotive Engineers of China (SAE-China) (ed.), *Proceedings of the 19th Asia Pacific Automotive Engineering Conference & SAE-China Congress 2017: Selected Papers*, Lecture Notes in Electrical Engineering 486, https://doi.org/10.1007/978-981-10-8506-2_24

known by automotive engineers as one of the ideal solutions, which can improve engine performance, reduces fuel consumption and emissions at the same time instead of compromising, so that CVVL system has been introduced by more and more automotive OEMs, such as global OEMs BMW, Nissan, Toyota, Hyundai [1–4], meanwhile Chinese OEMs are under development, such as Greatwall.

Turbocharged GDI engine with CVVL system requires not only complex mechanical design, but also complex electronic control system. Figure 1 shows the relationship among intake valve lift (IL), throttle opening angle (TOA), and intake manifold pressure (Psr) at a typical operating point, engine speed 2500 rpm, Brake Mean Effective Pressure (BMEP) 0.4 MPa with fixed VVT position. More variation means more complexity for system control. With increasing intake valve lift from 2.7 mm to maximum 8.64 mm, TOA decreases from 100 to 14% and Psr falls from 1120 to 730 hPa. The operating point 2500 rpm, 0.4 MPa can be ensured by many combinations of IL and TOA. The question is whether there is an optimal combination.

Three charge control methods are defined as shown in Fig. 2 to realize one constant operating point, e.g., 2500 rpm, 0.4 MPa with fixed VVT position: intake valve lift, throttle valve opening, or combination of both.

Intake valve lift control is the charge control method just through intake valve lift at the condition that throttle valve is fully opened and intake manifold pressure reaches the ambient pressure. Throttle valve opening control is the charge control method only via throttle valve at condition of maximal intake valve lift. Combination control is the charge control method with combination of intake valve lift and throttle opening. The aim of this study is to verify which charge control method can obtain the best torque output, lowest brake specific fuel consumption (BSFC) and emissions, meanwhile to discover the application boundary condition of the charge control method.

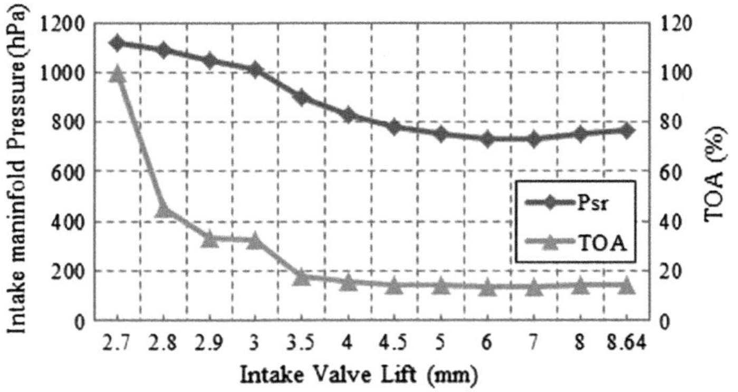

Fig. 1 Relationship of throttle opening angle, intake manifold pressure versus intake valve lift at 2500 rpm, 0.4 MPa

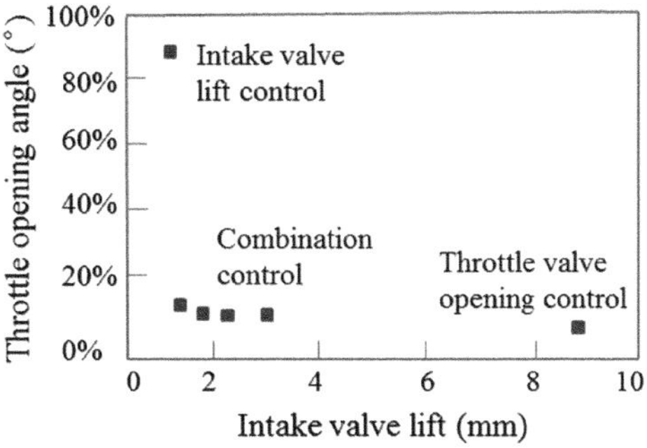

Fig. 2 Definition of three charge control methods

2 Test Setup

A 1.5L TGDI engine with CVVL system, which is equipped with all necessary measurement sensors and instruments for fuel consumption and emission measurement, is employed in this test.

The bench test operating points cover the whole engine working conditions. Engine speed sweeps from 1000 to 5600 rpm. BMEP sweeps from minimal 120 kPa to maximal 1200 kPa. For each operating point, best fuel economy with different intake valve lift is optimized by optimal VVT position, best ignition angle, and excess air ratio 1. For example, Fig. 3 indicates that at engine speed 2500 rpm, BMEP 0.4 MPa, intake valve lift 3 mm, the lowest fuel consumption is attained when intake VVT is advanced to −30 to −32° CA and exhaust VVT is retarded to 25–35° CA.

3 Analysis of Charge Control Methods

3.1 Measurement Result Analysis

Figure 4 shows the relationship of BSFC, TOA, and IL characteristics versus Psr on different operating points from above measurements. With the decrease of IL, TOA

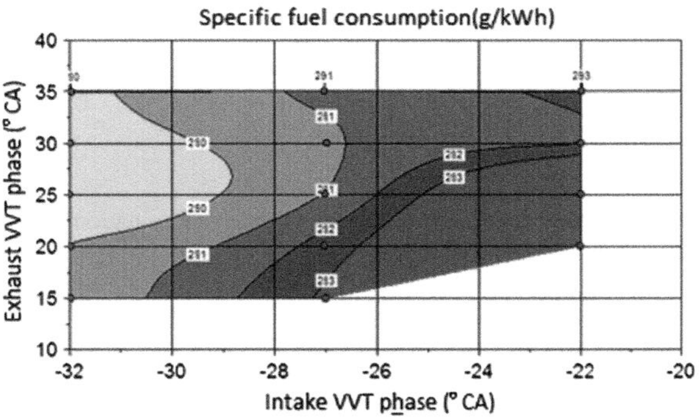

Fig. 3 Influence of intake and exhaust VVT on fuel consumption

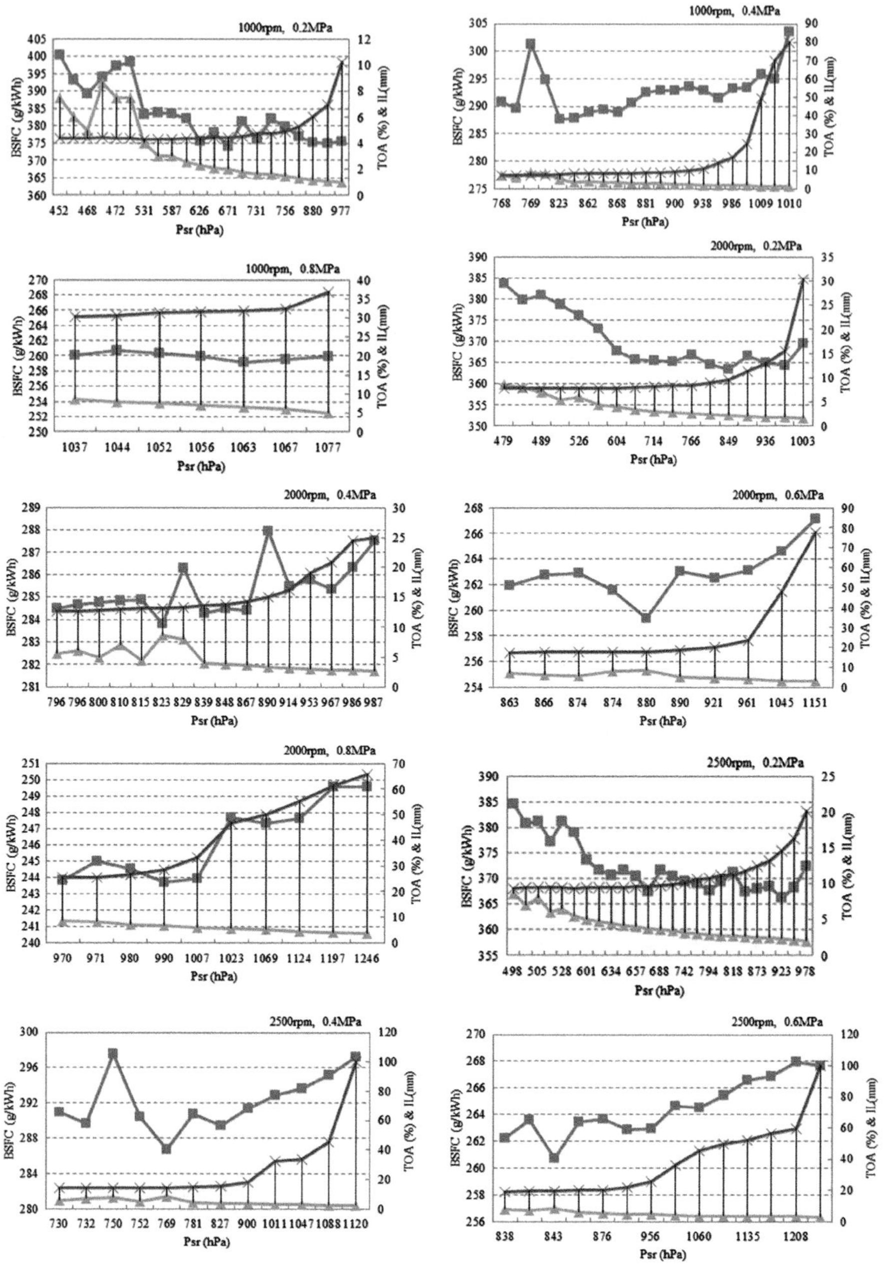

Fig. 4 Effect of valve lift and throttle combination control to manifold pressure and fuel consumption

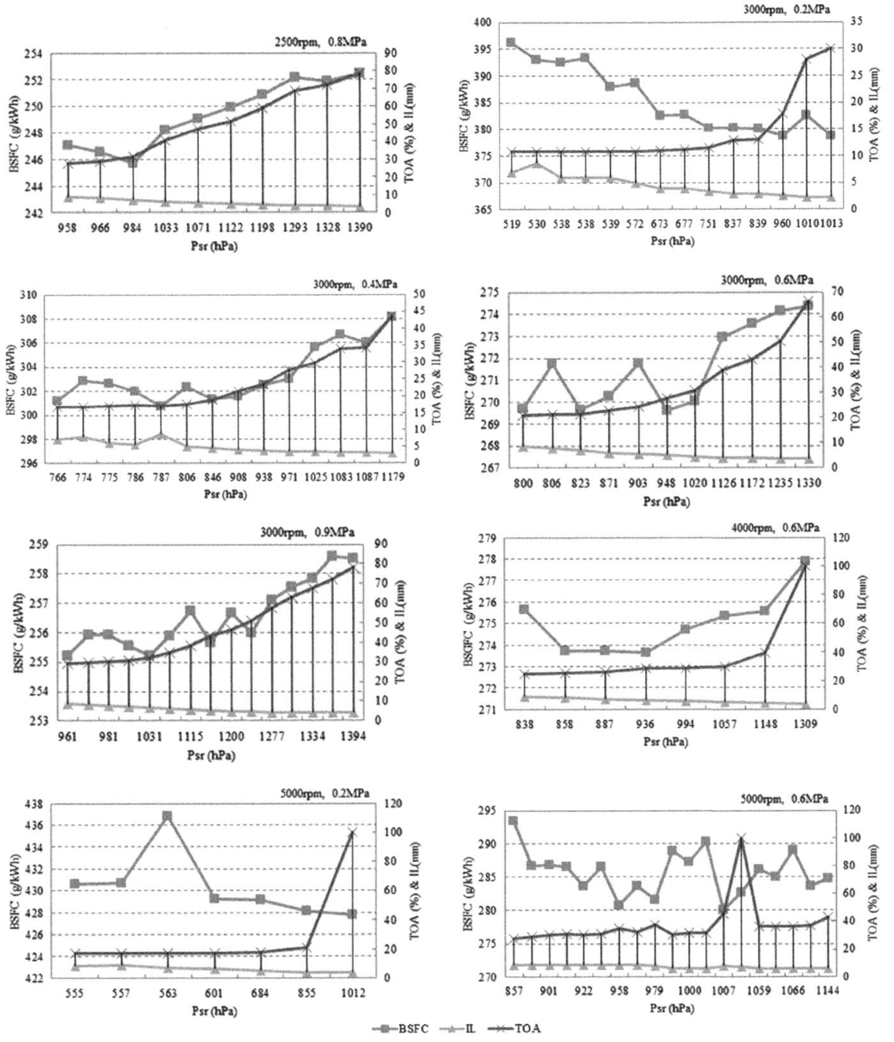

Fig. 4 (continued)

has to be increased to ensure constant engine speed and load. TOA keeps relatively smooth at high valve lift and rises steeply at low valve lift. Following phenomena are observed from the figures:

- Lowest BSFC is achieved with combinations of intake valve lift and throttle opening instead of pure throttle control or intake valve lift control for all operating points.
- At each operating point, there is a range of intake manifold pressure where the lowest BSFC can be found. For example on 1000-rpm variation of BMEP

0.2 MPa, the lowest BSFC appears at the range of intake pressure 626–977 hPa; 0.4 MPa, at 823–881 hPa; 0.8 MPa, at 970–1077 hPa. Similar conclusion can be drawn for 2000, 3000 rpm, and other engine speeds.

- The difference of maximum and minimum BSFC is about 7% on low speeds and low to middle loads; 3% on high speeds and low load; BSFC has less change on high loads than low loads, only 1% difference and lowest BSFC occurs normally at highest intake valve lift on high loads. BSFC reduction is much lower than the results reported in other articles [5–7].
- Lift dithering can be found between full valve lift to 6 mm, even to 4 mm at some operating points while engine performance remains the same, for example, 2000 rpm, 0.4 MPa; 2500 rpm, 0.2 MPa; 0.4 MPa; 3000 rpm, 0.2 MPa, 0.4 MPa.

Explanation to the phenomenon why there is a range of intake manifold pressure where the lowest BSFC can be found: Pumping loss occurs mainly at throttle valve when TOA is very small; or at intake valve when IL is very low, since throttle valve and intake valve both are throttling device. Pumping loss is higher, and corresponding BSFC is higher when intake valve at very low lift or throttle valve at very small opening; otherwise, BSFC is lower. This is also the reason why the lowest BSFC occurs with combination control.

Under high loads, BSFC has less change than under low loads, because pumping loss is lower at high loads due to the fact that throttle valve is opened widely and valve lift is quite high. Lowest BSFC occurs normally at highest intake valve lift, because the lower the valve lift is, the higher the pumping loss is. At high speeds and low loads, BSFC gain is very less and the pumping loss is quite high, that may be caused by mechanical loss of CVVL mechanism of the test engine.

For the phenomenon of lift dithering on special operating points shown in Fig. 4, it is just the airflow characteristics, that the air mass flow is insensitive to valve lift when the pressure ratio after and before the intake valve exceeds 0.95. Therefore, CVVL system should avoid frequently adjusting in this area to prolong actuators lifetime.

3.2 *The Optimal Charge Control Method and Application Boundary Condition*

In order to understand the relationship of TOA, IL, and Psr deeply, a dimensionless parameter intake pressure ratio (Rip) is introduced, which stands the pressure ratio after and before throttle valve, to make the test result more universal. As illustrated in Fig. 5, further observation of the pressure before and after throttle valve and Rip indicates that the air pressure before throttle valve remains at ambient pressure and Rip rises linearly with the air pressure in manifold when the air pressure in manifold is lower than 960 hPa. Rip keeps constant 1 and the air pressure before throttle

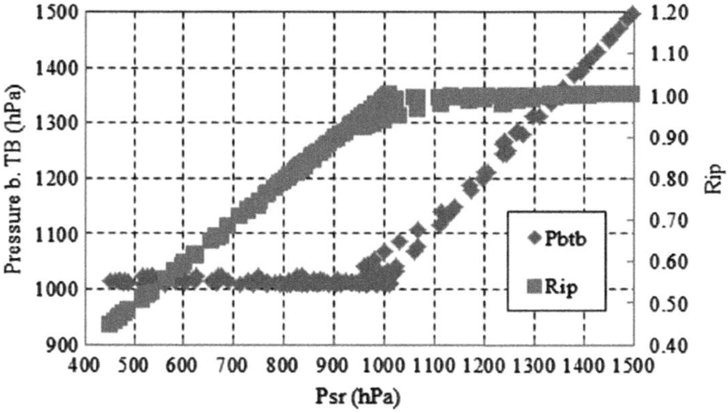

Fig. 5 Relationship of manifold pressure, pressure before throttle valve and Rip

valve goes up linearly with the air pressure in manifold when the air pressure in manifold is higher than 1013 hPa. Transient pressure from 950 hPa to ambient pressure 1013 hPa is nonlinear.

Following conclusions can be drawn from Fig. 5:

(1) It is one-to-one correspondence between Rip and manifold pressure when Rip is less than 0.95 or Psr is less than 960 hPa, and this correlation is not affected by engine speed and load. Adjusting throttle valve opening means Psr change and Rip change. Certainly to keep constant speed and load, IL must be modified accordingly.
(2) Turbocharger is activated after Rip exceeds 0.95. Proper TOA must be remained to ensure Rip over 0.95 when Psr lies between 960 and 1013 hPa.
(3) Psr and the pressure before throttle valve become equal when Rip reaches 1; i.e., ambient pressure exceeds 1013 hPa.

From system control view, the optimal combination of throttle valve opening and intake valve lift should meet following criteria: 1. lowest BSFC; 2. best emissions; 3. avoidance of insensitive area at pressure ratio after and before intake valve 0.95 to 1.

The best Psr on each operating point shown in Fig. 4 is sorted out according to criteria of lowest BSFC and avoidance of insensitive area. The impact on emissions will be analyzed later in Sect. 4. The data map of best Psr is smoothed and derived the result in Fig. 6.

Figure 6 shows that at all operating points, there are best Psr, which can achieve lowest BSFC and ensure system stability. Best Psr increases with engine load and has less change with engine speed.

The measurement result of IL and TOA at best Psr is sorted out further and plotted on Figs. 7 and 8.

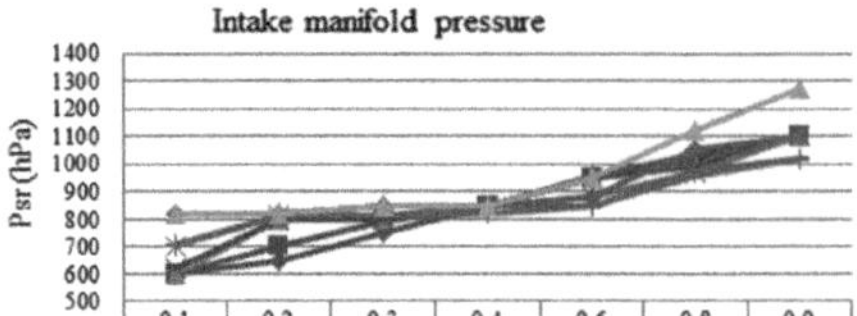

BMEP (MPa)	0.1	0.2	0.3	0.4	0.6	0.8	0.9
1000	600	650	750	850	960	1030	1100
1250	600	700	790	850	950	1020	1100
1500	600	800	800	850	950	990	1100
1750	620	800	800	850	950	990	1100
2000	700	810	810	850	880	970	1100
2500	710	810	810	820	850	980	1020
3000	820	820	850	850	850	960	1020
4000	820	820	850	850	880	1050	1100
5000	820	820	850	850	950	1120	1270

BMEP (MPa)

(a) Best intake manifold pressure

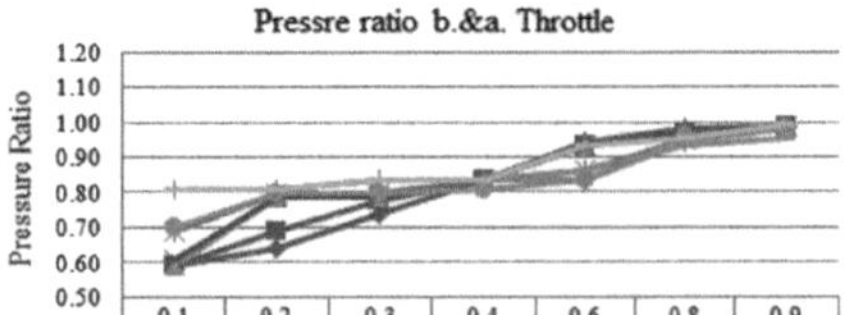

BMEP (MPa)	0.1	0.2	0.3	0.4	0.6	0.8	0.9
1000	0.59	0.64	0.74	0.83	0.94	0.98	0.99
1250	0.59	0.69	0.78	0.83	0.93	0.97	0.99
1500	0.59	0.79	0.79	0.83	0.93	0.97	0.99
1750	0.61	0.79	0.79	0.83	0.93	0.97	0.99
2000	0.69	0.80	0.80	0.83	0.86	0.95	0.99
2500	0.70	0.80	0.80	0.81	0.83	0.96	0.98
3000	0.81	0.81	0.83	0.83	0.83	0.94	0.98
4000	0.81	0.81	0.83	0.83	0.86	0.93	0.96
5000	0.81	0.81	0.83	0.83	0.93	0.96	0.99

BMEP (MPa)

(b) Best intake pressure ratio

Fig. 6 Best intake pressure and ratio

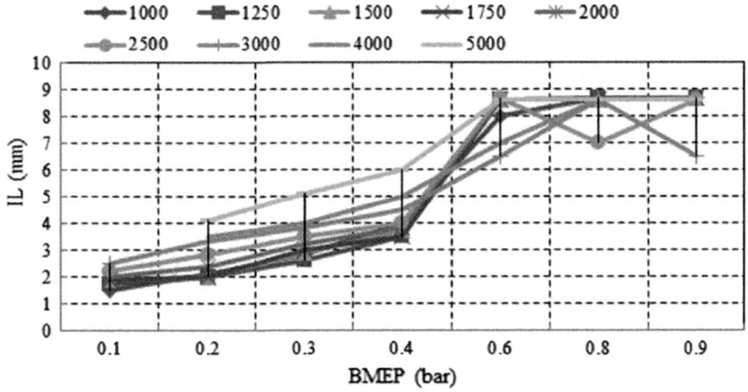

Fig. 7 Intake valve lift at best Psr

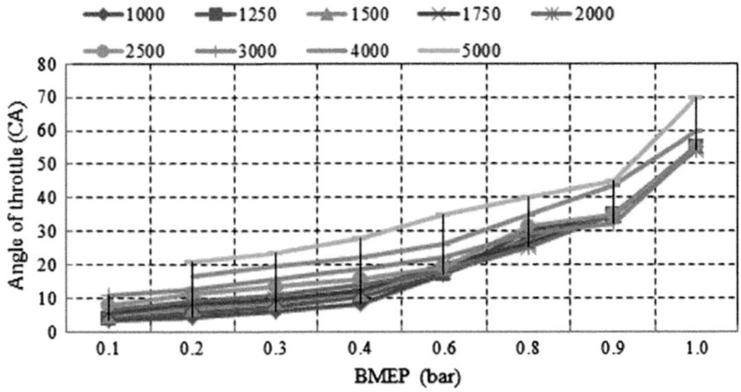

Fig. 8 Throttle opening at best Psr

Following conclusions can be obtained from the observation of Figs. 6, 7 and 8:

(1) At low loads, shown in Fig. 7, BMEP rises almost linearly with IL at all engine speed when BMEP is less than 0.6 MPa; i.e., Rip is less than 0.95 that derives from Fig. 6. Therefore, charge control can be realized through intake valve lift at this area while best Psr is achieved through TOA, shown in Fig. 8.
(2) At middle to high loads, shown in Fig. 8, BMEP goes up further via TOA till Rip exceeds 0.95 when intake valve reaches maximum lift 8.64 mm and BMEP lies between 0.6 and 0.8 MPa.
(3) At high loads, BMEP increases as conventional engine mainly through turbocharger with assistance of throttle valve when Rip is bigger than 0.95, which is shown in Fig. 6. Throttle valve opens fully after Psr exceeds ambient pressure 1013 hPa.

4 The Impact of Best Psr on Emissions

Figure 9 takes typical operating points at 2000 and 3000 rpm as example to illustrate the test result that HC and NO_x emissions change with Psr, which is put into yellow frame on each operating point. Following observation conclusions are reached:

(1) HC emissions have little change with Psr at all speeds and loads.
(2) NO_x emissions increase significantly with Psr rising at lower to middle loads, i.e., 0.4–0.6 MPa, while NO_x emissions have little change with Psr at high loads area, i.e., over 0.8 MPa, and have almost no change with Psr at very low loads, i.e., 0.2 MPa.
(3) NO_x emissions at best Psr in yellow frame are the relatively optimal result for all operating points.

This phenomenon can be explained from engine intake flow and combustion process [8].

HC emissions have little change with Psr at all speeds and loads because Psr has almost no effect on mixture formation for GDI engine.

NO_x emissions have significant change with Psr at lower to middle loads, i.e., 0.4–0.6 MPa; direct reason is that the ignition preparation timing is reduced greatly and best ignition advance angle (ZWOUT) is retarded with deceasing valve lift and rising Psr. Root cause is that in-cylinder airflow velocity increases a lot with valve lift deceasing and Psr rising, which improves air–fuel mixing in cylinder, increases combustion velocity and temperature, leading to increased NO_x emissions. Figure 10 demonstrates the influence of intake valve lift on combustion velocity at engine speed 2000 rpm and BMEP 0.4 MPa. It is shown that in commonly used valve lift region (2.5–8.5 mm), the combustion duration decreases when intake valve lift decreases, benefiting from the increased in-cylinder flow velocity due to low lift.

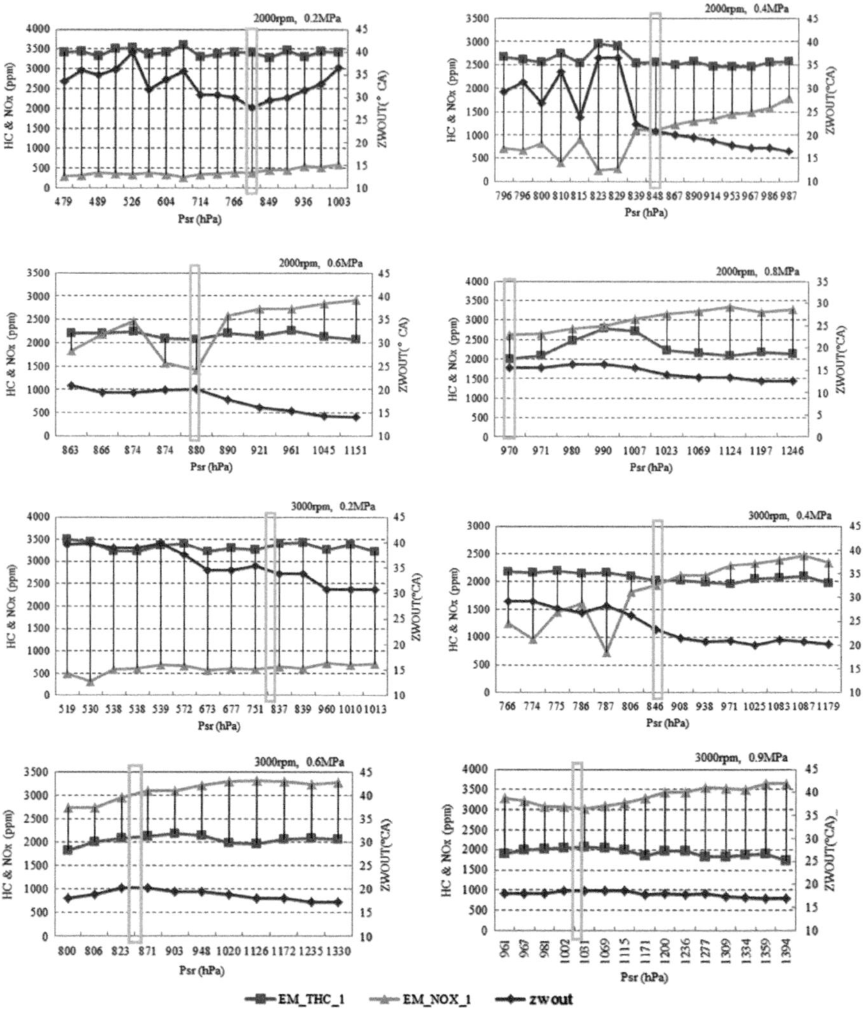

Fig. 9 Effect of Psr on emissions

NO_x has little change with Psr on high loads, because ignition advance angle and in-cylinder flow speed has very less change with Psr in this condition.

NO_x emissions have almost no change with Psr at very low loads 0.2 MPa, because on one hand, small lift causes the flow velocity at intake valve in critical state; on another hand, too small lift leads to early intake valve closing which has a strong inhibiting effect on in-cylinder flow.

In summary, lowest BSFC and relatively optimal emissions both are achieved at best Psr simultaneously because of the best compromise of pumping loss and appropriate intake airflow and in-cylinder flow velocity at the best Psr.

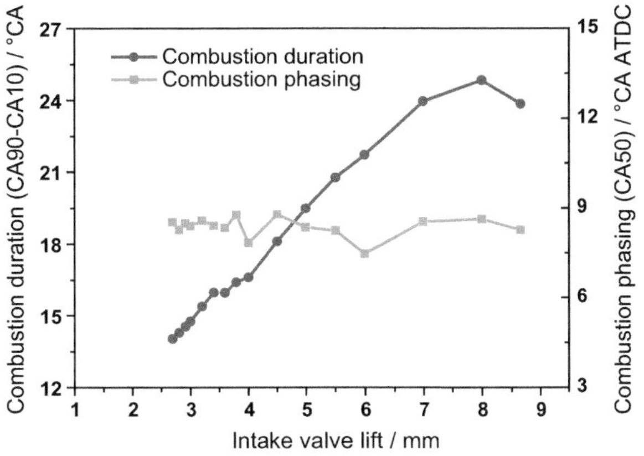

Fig. 10 Influence of intake valve lift on combustion

5 Conclusion

(1) There are best Psr on all operating points for TGDI engine with CVVL system, at which lowest BSFC and relatively optimal emissions both are achieved at the same time.
(2) The combination of intake valve lift and throttle opening is the best charge control method at low to middle loads when Rip is smaller than 0.95. Intake valve lift is the dominant control method and throttle opening is assistance method to achieve best Psr.
(3) Throttle opening is the main charge control method when intake valve lift reaches 8.64 mm, the maximum lift position, till Rip exceeds 0.95.
(4) The combination of turbocharger and throttle opening is the best charge control method on high loads after Rip exceeds 0.95. Turbocharger is the dominant control method and throttle opening is assistance method to ensure Rip greater than 0.95.
(5) Turbocharger pressure is the sole charge control method after throttle is fully opened and Rip reaches 1.

This work gives a design foundation for control software functions development of CVVL engine and helps improving CVVL engine control quality and calibration efficiency.

References

1. Ha K, Han D, Kim W (2010) Development of continuously variable valve lift engine. SAE Technical Paper 2010-01-1187
2. Unger H, Schneider J, Schwarz C, Koch K-F VALVETRONIC—Experience from 7 years of series production and a look into the future. Paper presented at 29th international Vienna motor symposium, Austria, 24–25 Apr

3. Kiga S, Mae Y, Akasaka Y, Tomogane K (2007) Development of innovative Variable Valve Event and Lift (VVEL) system. SAE Technical Paper 2007-01-3548
4. Harada J, Yamada T, Watanabe K (2008) The new L4 gasoline engines with VALVEMATIC system. Paper presented at 29th international Vienna motor symposium, Austria, 24–25 Apr 2008
5. Grimaldi F, Gervais D, Marchal A, Floch A (2007) Single-cylinder experiments for downsizing-oriented SI concepts: GDI and VVL thermodynamic comparison. SAE Technical Paper 2007-24-0013
6. Wang T, Zhang Y, Wang L, Wu Q, Liu D, Zhao H, Liu S (2009) Effect of throttle-free load control strategies on a si engine performance. Trans CSICE 1000- 0909, 03-0224-07, 27(3)
7. Tagami H, Yasui Y, Sato M, Ito H (2007) Management system for continuously variable valve lift gasoline engine. SAE Technical Paper 2007-01-1200
8. Takemura S, Aoyama S, Sugiyama T, Nohara T et al (2001) A study of a continuous Variable Valve Event and Lift (VEL) system. SAE Technical Paper 2001-01-0243

Analysis and Detection Methodology of Knock Phenomenon in Gasoline Engines Based on Cylinder Pressure Sensors

Wentao Zhang, Tong Wu, Lidong Dong and Wei Hao

1 Introduction

Knock is a major phenomenon of abnormal combustion for gasoline engines. The acoustic, optical, mechanical, and thermodynamic characteristics are studied by researchers. The occurrence of knock affects the improvement of power output and fuel economy of gasoline engines. Severe knocking combustion even causes the engine damage. Many investigations about knock estimating algorithm, knocking signal processing, knock control strategies, and modeling and simulation of knocking combustion were conducted by researchers. Nevertheless, very little investigations about detection methodologies of knocking combustion were conducted. Recently, the knocking combustion evaluation is based on time–frequency analysis with short-time Fourier transform, discrete wavelet transform, fuzzy clustering algorithm, and neural network algorithm, which are too sophisticated for engineering applications.

In this paper, the investigations, which include the selection of optimized knock window, the determination of knocking signal frequency, and the methodologies of evaluating knocking intensity, were conducted based on the measurement result by cylinder pressure sensors.

W. Zhang (✉) · T. Wu · L. Dong · W. Hao
China FAW Corporation Limited R&D Center, Changchun 130000, China
e-mail: zhangwentao@rdc.faw.com.cn

Society of Automotive Engineers of China (SAE-China) (ed.), *Proceedings of the 19th Asia Pacific Automotive Engineering Conference & SAE-China Congress 2017: Selected Papers*, Lecture Notes in Electrical Engineering 486, https://doi.org/10.1007/978-981-10-8506-2_25

2 The Test Setup

2.1 Experimental Apparatus

As shown in Fig. 1, the measuring system consists of an AVL 740 instantaneous fuel consumption meter, Horiba electric dynamometer, a Horiba MEXA-7500DEGR emission analyzer, an ETAS-ES630 AFR analyzer, a TGDI (turbocharged gasoline direct injection) engine, and an engine management system (EMS).

Main specifications of the 1.0-liter turbocharged GDI engine in the investigation are summarized in Table 1.

In order to conduct a sufficient investigation about knocking intensity and background noise at different engine operating conditions, the margin of operating conditions is wide and is shown in Table 2.

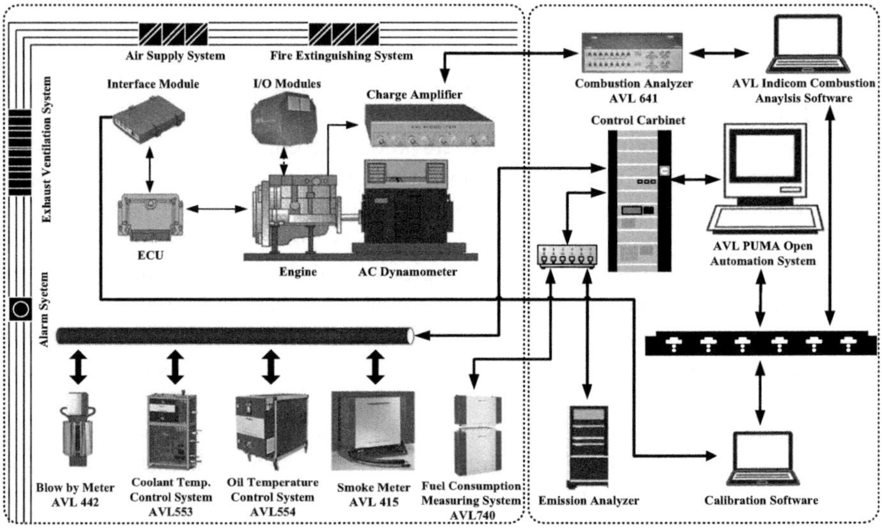

Fig. 1 Measuring system

Table 1 Main specifications of 1.0L TGDI engine

Items	Parameters
Displacement/L	0.999
Maximum net power/kW	88.0 (5500 r/min)
Maximum torque/N m	170 (1500–4500 r/min)
Compression ratio	10.0
Fuel	93#
Ignition order	1-2-3
Technical features	TGDI, DVVT, variable oil pump
Electric control system	FAW-EMS

Table 2 Engine operating conditions

Engine speed (r/min)	1000, 1500, 2000, 2500, 3000, 3500, 4000, 4500, 5000
Engine load	1 bar (BMEP), 20, 40, 60, 80% (relative air charge) …, Full load
Time length of measurement	1000 combustion cycles

Cylinder pressure signal was measured and processed by Kistler (6115BFD35Q05) cylinder pressure sensor and AVL Indicom combustion analysis system.

2.2 *Data Acquisition and Processing Method*

The original signal, acquired by knock sensor, includes many noises such as the noise signals from the background, the engine block, valve drive mechanism, and the impact between the piston and cylinder wall, which makes the original signal not valid and need be processed.

The data processing procedure, as shown in Fig. 2, includes the signal high-pass filtering, rectifying, integrating, and removing the noise signals. After the processing procedure, the knock intensity is estimated and the frequency of knock occurrence is calculated.

3 Selection of Knock Window

The selection of knock window is significant to the knock detection. The knock detection based on the cylinder pressure signal during the whole combustion cycle is influenced by the noises from ignition signal, valve closing, etc. The detection with high resolution during a proper time window prevents the influence of noises and achieves a more accurate test result with relatively lower requirement for hardware performance. In previous investigation [1–5], the selected knock detection window starts from 3°CA before the first peak of high-pass filtered cylinder pressure signal and ends when the difference between neighboring peak and valley of high-pass filtered cylinder pressure signals is below 30 kPa. As is illustrated in Fig. 3, the detection window starts from 4°CA when the cylinder pressure decreases sharply and ends in 20°CA where the pressure curve is smooth. In the case shown in Figs. 4 and 5, the detection window starts from 4°CA when the cylinder pressure increases sharply and ends in 20°CA where the curve of heat release rate and total heat release are both smooth. But due to the selection methodology in the reference paper [1–5], 40°CA should be selected as the end of the knock window, which extends the detection window and affects the knocking intensity evaluation.

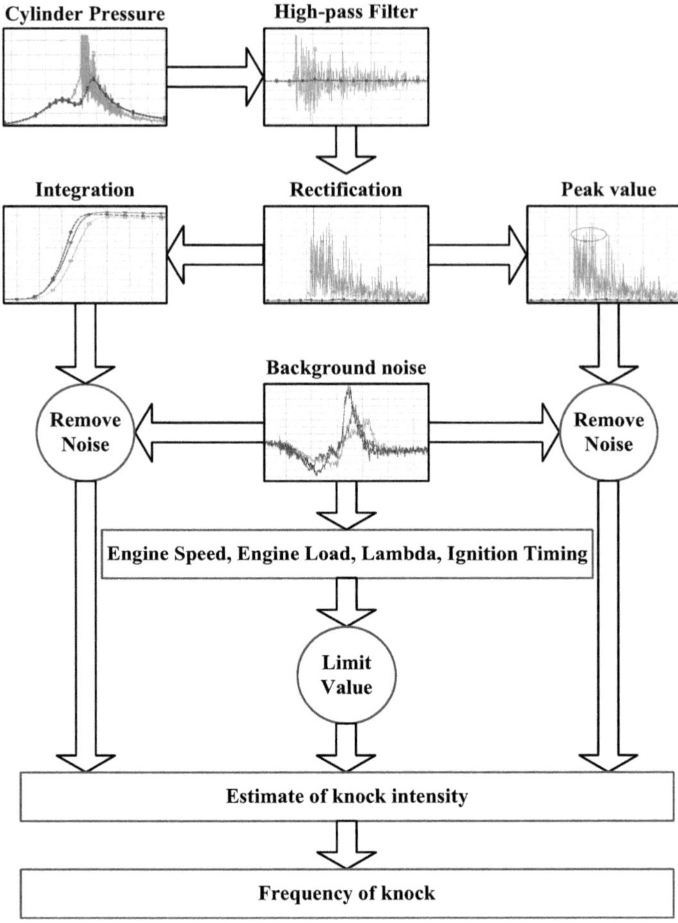

Fig. 2 Data processing procedure of cylinder pressure signal

As shown in Figs. 6 and 7, the knock combustion starts from 2–25°CA and ends in 34–63°CA; therefore, the whole knock window is about 2–63°CA. The recommended detection window of conventional knocking combustion is 10–70°CA for naturally aspirated and turbocharged gasoline engines. Super knock could occur in highly turbocharged gasoline engines, and the recommended window is from 0 to 70°CA. Furthermore, for the engines with obvious pre-ignition, the detection window extends from −10 to 70°CA and should be divided into two sections (one section for knock detection and the other for pre-ignition). For the engine in this paper, 0–30°CA is selected as the detection window in the case knocking intensity is not investigated.

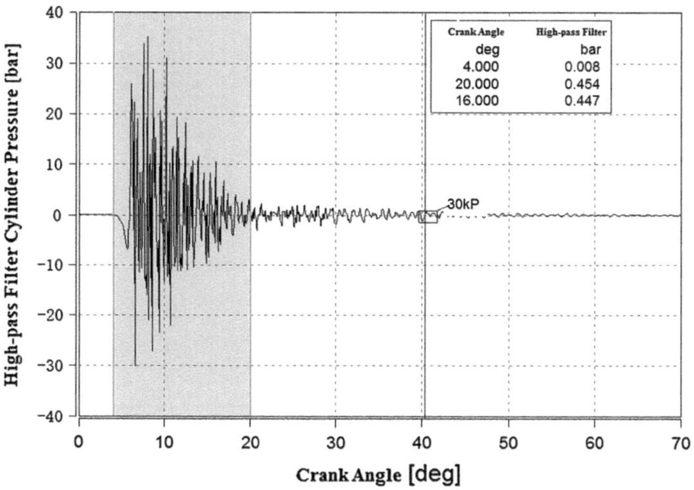

Fig. 3 Selection of knock window based on cylinder pressure signal

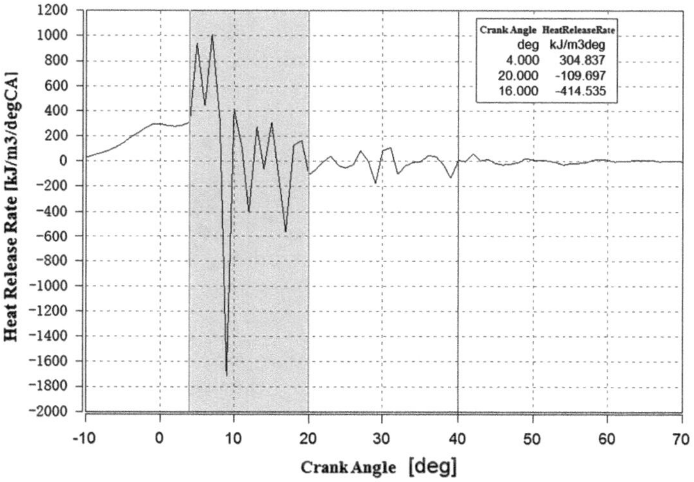

Fig. 4 Selection of knock window based on heat release rate

4 Determination of Knock Frequency

The velocity of the pressure wave is 100–300 m/s when the slight knocking combustion occurs, and the velocity is 800–1000 m/s when the intense knocking combustion occurs. The knock could make a strong impact force inside the cylinder, and the knock is divided into vibrations in circumferential direction, radial direction, and axial direction. The vibrational frequency is 4–25 kHz. The knocking frequency and

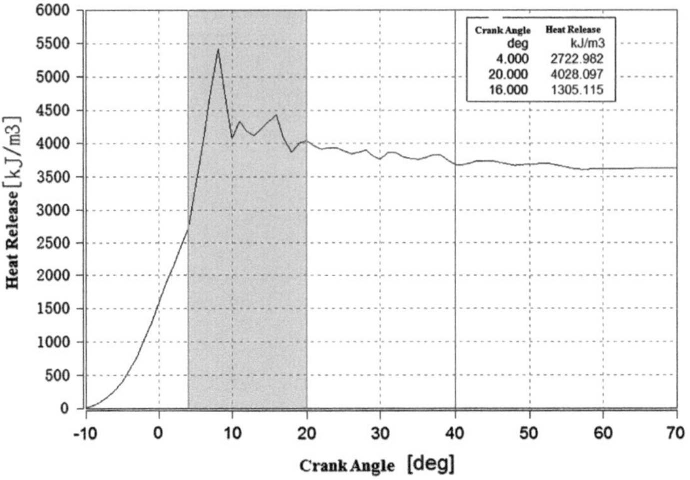

Fig. 5 Selection of knock window based on total heat release

Fig. 6 Start and end crank angle of knocking combustion

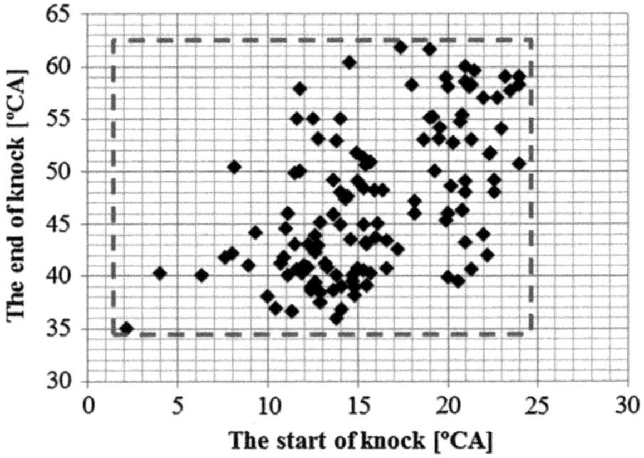

Fig. 7 Duration of knocking combustion

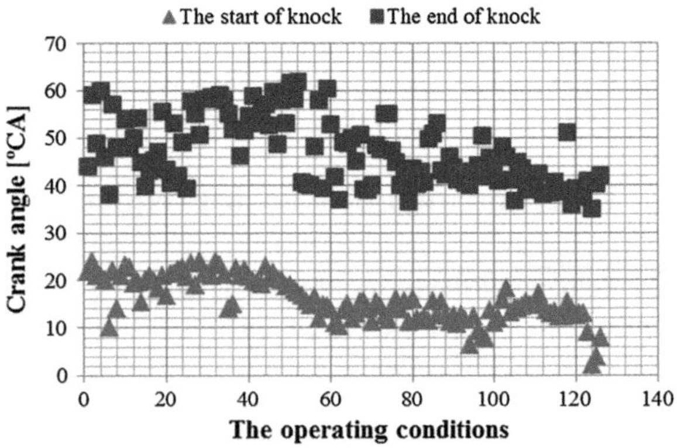

the order of the knocking vibration are confirmed based on the power spectral density result of cylinder pressure signal by FFT methodology. As shown in Fig. 8, the frequency of the first-order vibration is 8.2 kHz, the frequency of the second-order vibration is 15.13 kHz, and the frequency of the third-order vibration is 19.8 kHz.

The Bessel function to estimate the frequency is as follows:

$$f_{m,n} = \frac{c \times a_{m,n}}{\pi \times B} \tag{1}$$

where $f_{m,n}$ is the frequency of the vibration (Hz); C is the sound velocity (m/s); $a_{m,n}$ is the factor of vibration mode; B is the diameter of cylinder (m); m is the circumferential vibration number; and n is the radial vibration number. The spark plug cylinder pressure sensor is applied and arranged in the top central section of the combustion chamber. So the radial vibration amplitude is mainly detected in the test.

As shown in Table 3, the predicted value of frequency is 10% higher than actual value. There are several reasons for this difference:

(1) The combustion chamber is simulated as simply cylinder-shaped, but the actual shape is more sophisticated.
(2) The sound velocity is constant as 1000 m/s in the calculation, but the actual sound velocity is varied by different local temperature. The local sound velocity and temperature decrease when the piston is moving down.
(3) The frequency decreases and the wavelength increases when the piston moves downward and the volume of the chamber increases.
(4) The axial direction vibration is ignored in the calculation, and the influence of axial vibration becomes more significant when the piston moves downward.

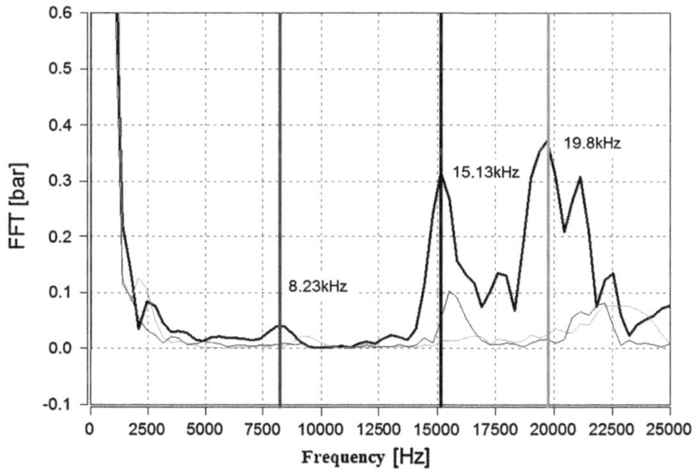

Fig. 8 Power spectral density result after FFT

Table 3 Vibration mode

Vibration mode	1,0	0,0	1,1
Vibration type			
$a_{m,n}$ [3]	1.8412	3.8317	5.3314
Calculating frequency (kHz)	8.14402	16.94843	23.58192
Actual frequency (kHz)	8.2	15.13	19.8

5 Evaluation of Knock Intensity

At present, there is no acknowledged methodology for the evaluation of knock intensity. Among the existing methodologies, MAPO and IMPO are commonly applied, which are defined as follows:

$$\mathrm{MAPO} = \max\left(\left|P_{\mathrm{hpf}}\right|\right)_{\theta_1}^{\theta_2} \tag{2}$$

$$\mathrm{IMPO} = \frac{1}{N}\sum_{1}^{N}\int_{\theta_1}^{\theta_2} \left|P_{\mathrm{hpf}}\right| \mathrm{d}\theta \tag{3}$$

where N is the cycle number; θ_1 is the start angle, and θ_2 is the ending angle of the knock window; P_{hpf} is the high-pass filtered cylinder pressure value; MAPO is the max absolute value of the high-pass filtered cylinder pressure within the knock window of 0–70°CA; and IMPO, as the power of filtered cylinder pressure signal, is the integrated absolute value of the high-pass filtered cylinder pressure within the knock window of 0–70°CA.

Actually, evaluation based on the value of MAPO and IMPO cannot effectively identify slight knock or intense knock. Therefore, the BGNO is defined as the background noise in the selected knock window as follows:

$$\mathrm{BGNO} = \max\left(\left|P_{\mathrm{hpf}}\right|\right)_{-30}^{0} \tag{4}$$

The SNR is defined as the ratio of knock intensity to background noise. The calculation of SNR is as follows:

$$\mathrm{SNR} = \frac{\mathrm{MAPO}}{\mathrm{BGNO}} \tag{5}$$

The knock intensity is influenced by the ignition timing and lambda and has a strong relationship with the engine speed and engine load. The variation of MAPO, BGNO, and SNR values with different engine speed is illustrated in Figs. 9, 10, and 11. The values of MAPO and BGNO increase when the engine speed increases. But the value of SNR decreases slightly with the increasing engine speed because

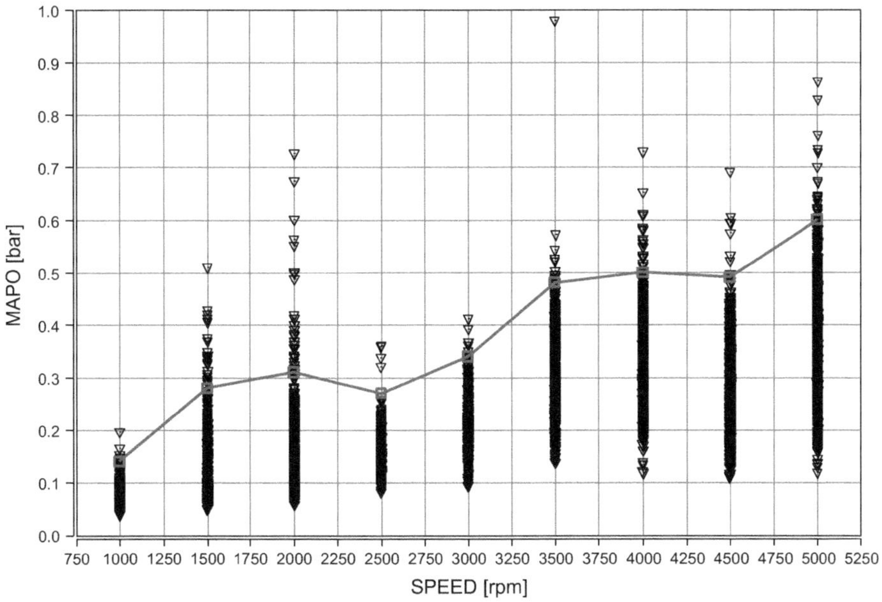

Fig. 9 Variation of MAPO value with different engine speed

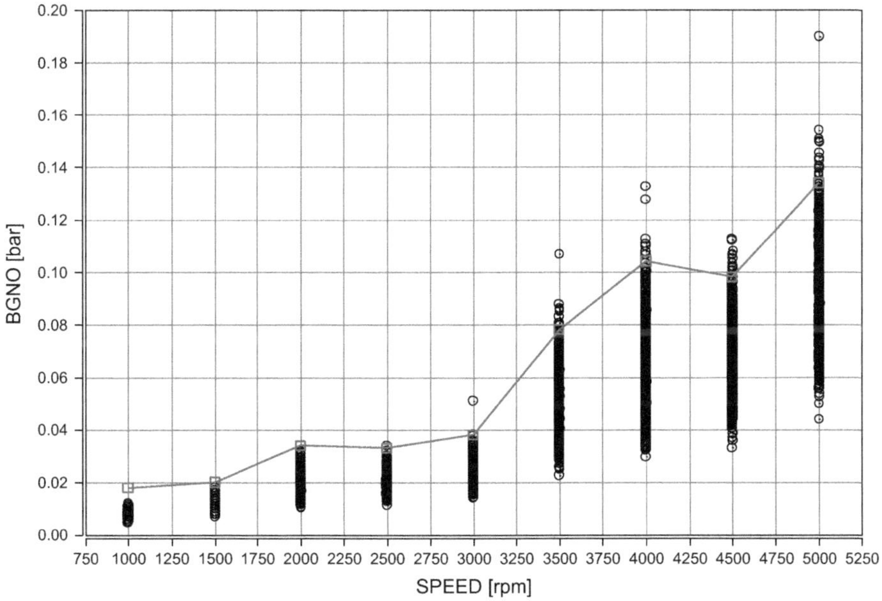

Fig. 10 Variation of BGNO value with different engine speed

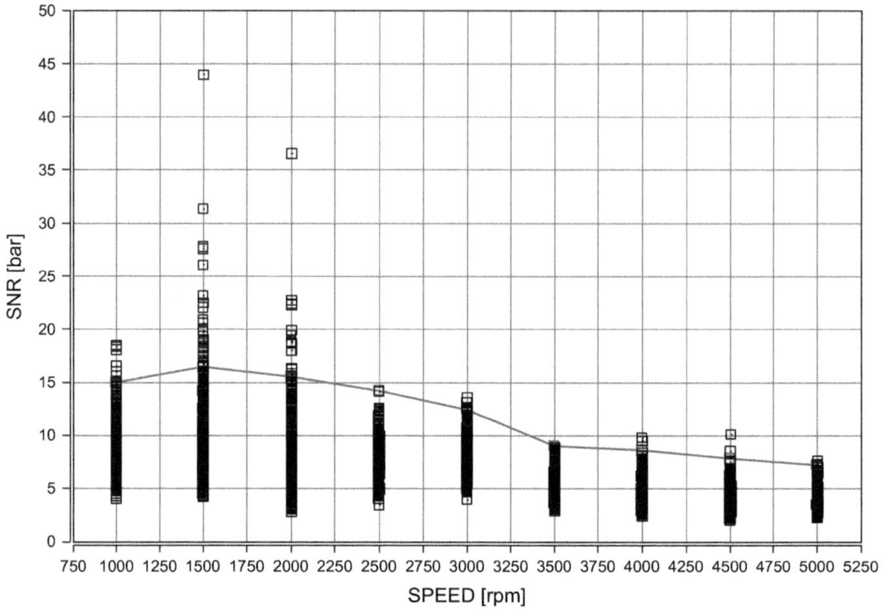

Fig. 11 Variation of SNR value with different engine speed

the background noise from the water pump, electric generator, power steering pump, and air condition compressor increases with the increasing engine speed.

The variation of MAPO, BGNO, and SNR values with different engine load is illustrated in Figs. 12, 13, and 14. The values of MAPO and SNR increase when the engine load increases. And the value of BGNO increases slightly with the increasing engine load.

The variation of MAPO and BGNO values with different ignition timing is illustrated in Figs. 15 and 16. The value of MAPO increases with the advancing ignition timing, and there is a linear relationship between MAPO and ignition timing. But the relationship would become exponential if the ignition timing continues advancing. Besides, the value of BGNO increases rapidly with the advancing ignition timing because the knock window extends with the advancing ignition timing.

The variation of MAPO and BGNO values with different value of lambda is illustrated in Figs. 17 and 18. Due to the influence of different combustion rate and combustion phase, the values of MAPO and BGNO increase when the air-fuel mixture becomes richer.

The evaluation of knock intensity based on MAPO cannot prevent the influence of background noise. And the evaluation based on BGNO eliminates part of the influence of background noise, which cannot represent the actual knock intensity. In order to represent the actual knock intensity and eliminate the influence of background noise, the KI is applied in the evaluation and defined as follows:

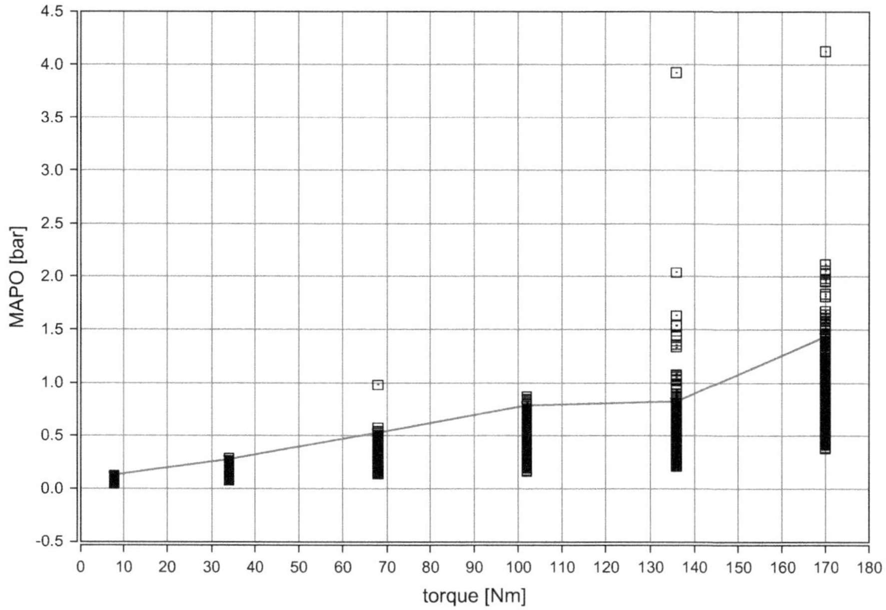

Fig. 12 Variation of MAPO value with different engine load

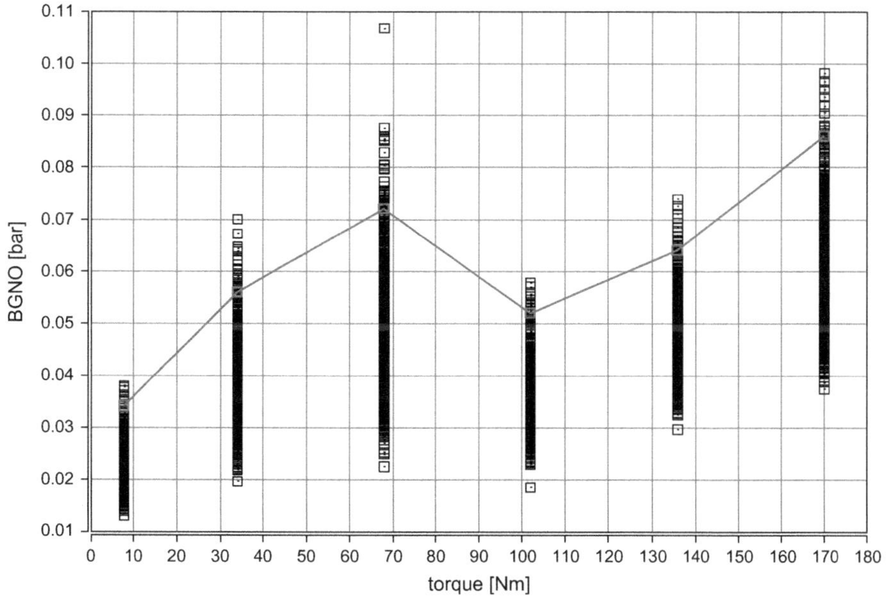

Fig. 13 Variation of BGNO value with different engine load

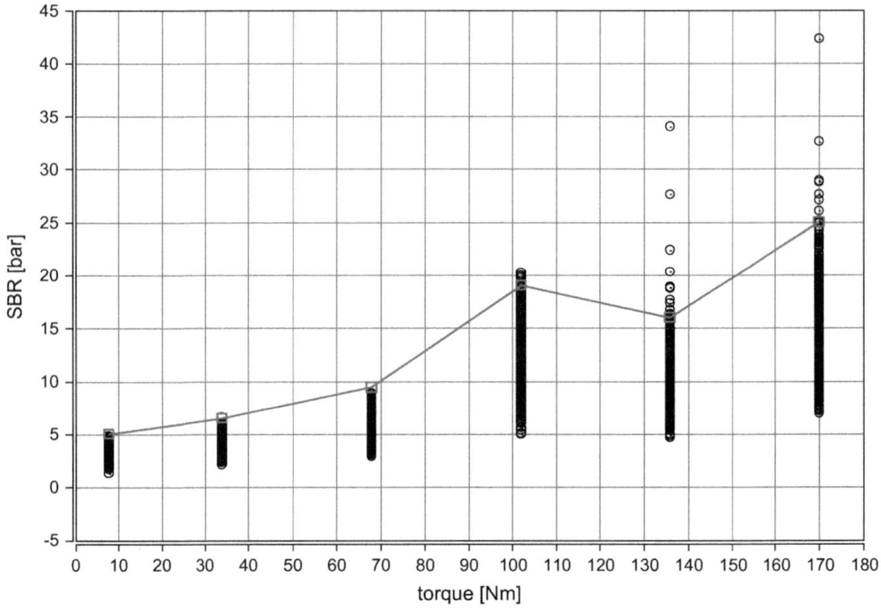

Fig. 14 Variation of SNR value with different engine load

Fig. 15 Variation of MAPO value with different ignition timing

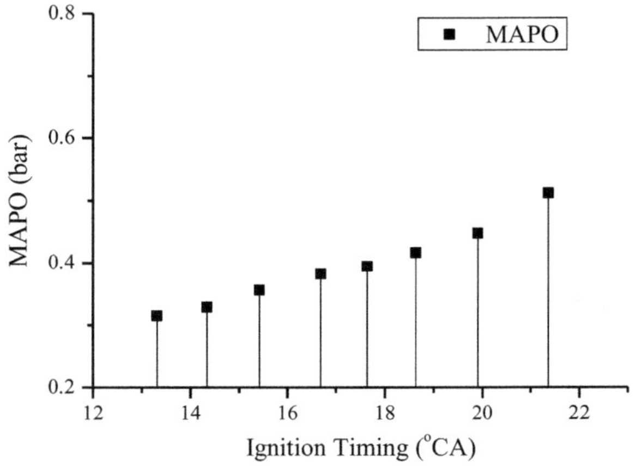

$$\mathrm{KI} = \frac{\sum_{1}^{N} (\text{resolution}) \cdot \sum_{\theta_1}^{\theta_2} |P_{\text{hpf}}|}{\mathrm{N} \cdot \max(|P_{\text{hpf}}|)_{\theta_1}^{\theta_2} \cdot (\theta_2 - \theta_1)} \tag{6}$$

where the resolution is the sampling resolution.

As illustrated in Fig. 19, the KI is the ratio of the integrated power of high-pass filtered cylinder pressure to the integrated power of high-pass filtered peak cylinder

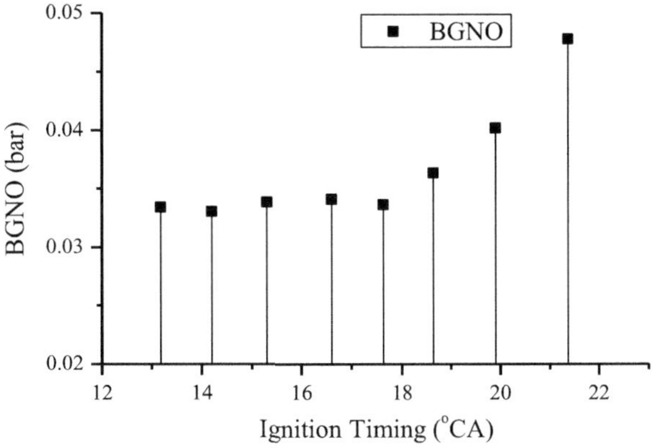

Fig. 16 Variation of BGNO value with different ignition timing

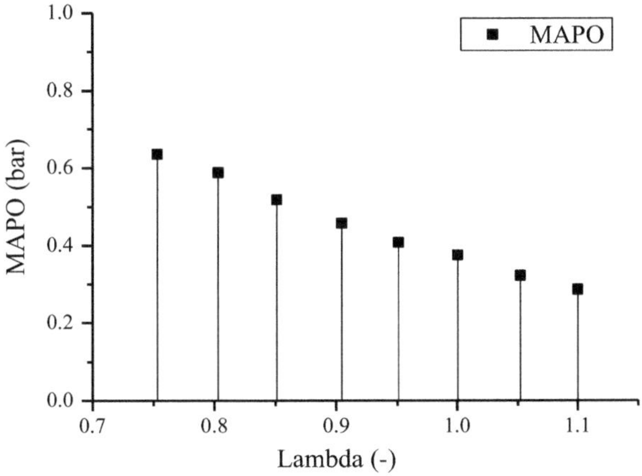

Fig. 17 Variation of MAPO with different lambda value

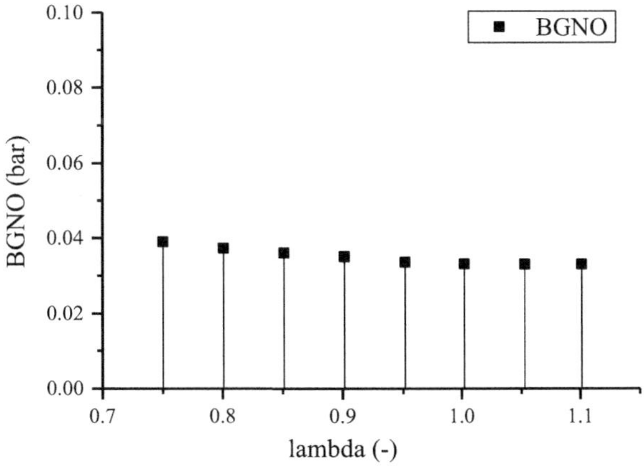

Fig. 18 Variation of BGNO with different lambda value

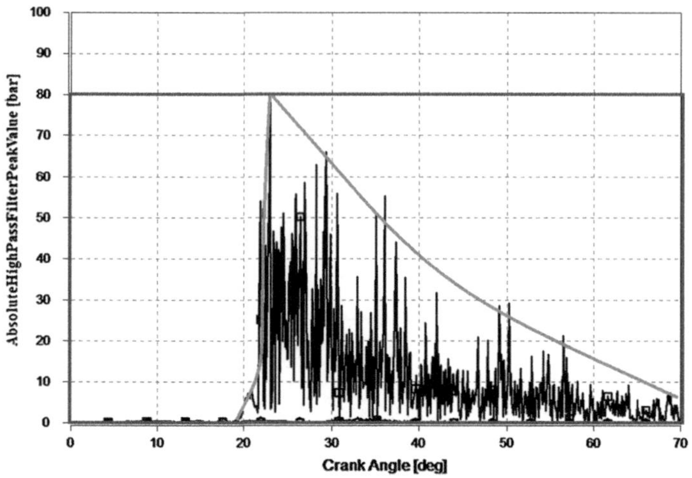

Fig. 19 Definition of KI

pressure within the knock window. The value of KI decreases when the knocking intensity increases or the knock window extends. The evaluation of knocking intensity based on KI value reflects the influence of different background noise and different sampling resolution. The recommended value of the start angle of knock window is 20°CA after knock occurs.

The methodology of the knock intensity evaluation should change in different engine operating conditions. The influence of different ignition timing and lambda should be considered. The value of knocking intensity is related to combustion rate and combustion phase, but is not related to whether the knock occurs. The value of knocking intensity is related to ignition timing and lambda even though there is no knock combustion inside the cylinder.

6 Conclusions

In this paper, the analysis about the cylinder pressure signal was conducted. Furthermore, the optimized knock window was selected. And the frequency of knocking signal was determined. Besides, evaluation methodology of knock intensity was investigated and confirmed. The conclusions are as follows:

1. The selection of knock window is significant to the knock detection. The optimized pre-ignition and knock window should be selected and confirmed separately. The knock window of 20°CA after the knock occurrence is recommended based on the investigation results.

2. The value of knock frequency is significant in the knock sensor calibration. The frequency of knocking signal is estimated by Bessel function, FFT, and power spectrum density analysis
3. The methodology of the knock intensity evaluation should change in different engine operating conditions. Having eliminating the influence of background noise, the knocking intensity (KI) is defined based on different sampling resolutions, operating conditions of various ignition timings, and Lambda values.

References

1. Heywood JB (1988) Internal combustion engine fundamentals. McGraw-Hill International editions, New York
2. Lee J-H, Hwang S-H, Lim J-S, Jeon D-C A new knock-detection method using cylinder pressure, block vibration and sound pressure signals from a SI engine. SAE Technical Paper 981436
3. Brunt MFJ, Pond CR, Biundo J Gasoline engine knock analysis using cylinder pressure data. SAE Technical Paper 980896
4. Carstens-Behrens S, Urlaub M, Böhme JF, Förster Jet al (2002) FEM approximation of internal combustion chambers for knock investigations. SAE Technical Paper 2002-01-0237
5. Scholl D, Davis C, Russ S, Barash T The volume acoustic modes of spark-ignited internal combustion chambers. SAE Technical Paper series 980893

Fuel Consumption Contribution Research on Map-Controlled VDOP and PCJ by Simulation

Lifen Wu, Huibin Qing, Qingqiang Zeng, Kun Yang and Xiaodong Chen

Nomenclature

VDOP	Variable displacement oil pump
PCJ	Piston cooling jet
HLA	Hydraulic lash adjusters
VCT	Variable cam actuation
OCV	Oil control valve
TC	Turbocharger

1 Introduction

A significant effort in current automotive research is oriented to identify areas where a potential exists for fuel consumption and emissions reduction. Map-controlled variable displacement oil pump together with map-controlled PCJ can make a contribution in this direction owing to their intrinsic capability of matching oil consumers demand at various operating conditions besides the critical situation at hot engine idling [1–3].

Oil pumps were normally designed for the least favorable case, which means high oil temperature with the largest clearances. Further, design criteria are to supply sufficient oil supply at low speeds (hot idle) to hydraulic lash adjusters (HLA) and to variable cam actuation (VCT) system which is at the farthest point on the oil circuit, and at high speeds it has to ensure adequate supply for the highly loaded connecting rod bearings. For a conventional oil pump, the oil displacement is proportional to the speed and the oil flow increases with engine speed increasing, so it will exceed the lubricant system's demand at mid speeds and high speeds.

L. Wu (✉) · H. Qing · Q. Zeng · K. Yang · X. Chen
Power Train Research Institute, Chang an Automobile Co., Ltd,
Chongqing 401120, China
e-mail: 15011603@163.com

Society of Automotive Engineers of China (SAE-China) (ed.), *Proceedings of the 19th Asia Pacific Automotive Engineering Conference & SAE-China Congress 2017: Selected Papers*, Lecture Notes in Electrical Engineering 486, https://doi.org/10.1007/978-981-10-8506-2_26

Normally, an internal bypass valve is installed in the oil pump to relieve the excess flow; however, in the conventional fixed oil pump there will be a huge parasitic losses because of this. Figure 1 shows a typical engine actual oil pressure versus the requirements [4–6]. It shows that the actual oil pressure is much higher than the required oil pressure at mid speeds.

However, map-controlled VDOP can run at any target pressure point. The oil pressure is actively controlled for various running conditions via a solenoid according to a map in the engine control module. The feedback oil gallery ensures the pump delivery is automatically matched to the engine requirements at any condition. There will be little excess flow supply, so a mass of fuel consumption will be saved.

The traditional piston cooling jet (PCJ) is employed in the past years to avoid the piston temperature to rise too high. PCJs will be turned on when the oil pressure in main gallery meets the PCJ's opening pressure, so it often happened that oil jet to the underside of each piston when the engine is cold start or at low load, namely the piston temperature is low. So the spraying oil will lead the temperature of combustion chamber to be much cooler, the combustion deteriorates, and the fuel consumption and the exhaust increased.

In order to decrease the losses caused by traditional PCJ, map-controlled PCJ is used. Map-controlled PCJ spraying oil onto the underside of each piston is controlled by a single oil control valve (OCV), which is the layout at an independent oil gallery. This valve has on/off control only, it will be closed by control module when engine is cool or the engine load is low. And when the OCV is closed, there will be no oil flow into jets, so the temperature of the piston will rise rapidly, combustion efficiency of engine will improve, and emissions will decrease. At the same time, when the PCJ is closed, the resistance of piston scraping oil down to crankcase will decrease also.

While it is usual that map-controlled VDOP and map-controlled PCJ technology will be applied on the same engine. They should match up well with each other in order to save energy as much as possible. At the cold start condition, the temperature of piston is low, it is not necessary to cool it, so map-controlled PCJs are all closed by OCV. Then temperature of piston will rise quickly, the combustion in the

Fig. 1 Typical engine actual oil pressure versus the requirements

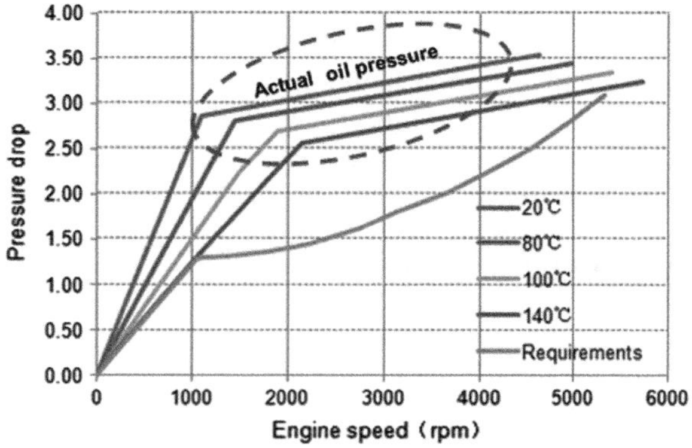

chamber will be wealthier and less emission will be produced. But at this condition, oil filling speed of the lubrication system is fatal to ensure the engine safety and decrease the friction loss, so it works with the maximum displacement of map-controlled VDOP. At low speed together with low load and middle speed with low load conditions, temperature of piston is not so high, in order to save energy, map-controlled PCJ is closed and the map-controlled VDOP will work with smaller displacement. It will save energy for a engine at these conditions. At middle speed with high load and high speed with high load conditions, temperature of piston is high, to avoid combustion knocking and piston be damaged, it needs to spray oil to piston to cool it, at the same time the map-controlled VDOP worked with the maximum displacement.

In this paper, a lubrication circuit is researched, which is equipped with HLA, turbocharger (TC), two VCT units, and especially with a map-controlled VDOP and map-controlled PCJs, etc. A lubrication circuit CAE model has been built up to assess the fuel consumption of map-controlled VDOP together with map-controlled PCJ. And three steps were done. First to confirm the maximum displacement of VDOP, and then to determine the oil pressure control strategy of map-controlled VDOP. And final fuel consumption contribution of two technologies is simulated.

In this paper, the oil pressure control strategy of map-controlled PCJ is known, and it needs to determine the oil pressure control strategy for map-controlled VDOP only. The fuel consumption contribution involved in this paper is due solely to hydraulic power of map-controlled VDOP used to deliver oil to every part of lubrication system, not including dry friction power caused by mechanical motion [7–10].

2 The Model

A lubrication circuit model equipped with HLA, TC, two VCT units, and with map-controlled VDOP and map-controlled PCJs was built up by GT-ISE showed in Fig. 2.

When the model is built, the control signal of VDOP and map-controlled PCJ is significant. There are two main oil galleries laid on the cylinder body. One is used to supply oil to main bearings and the turbocharger. The other is used to supply oil to PCJs, and the OCV is layout at the front of the gallery to control the oil gallery through or not.

2.1 Boundary Conditions

Figures 3 and 8 show the boundary conditions for the simulation model, including the main components' single test date at different temperatures, clearance for main bearings, and the control strategy of PCJ.

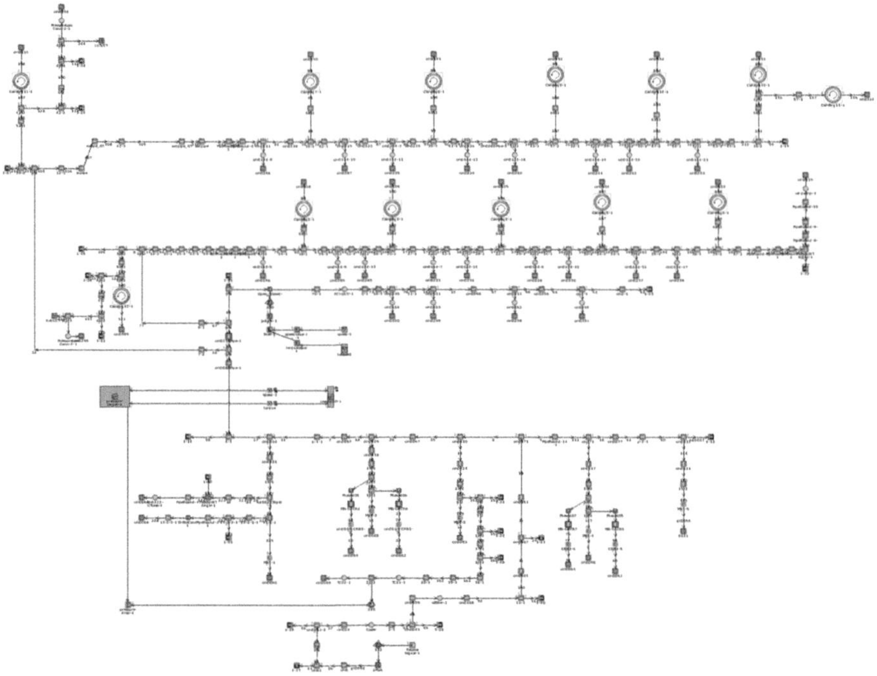

Fig. 2 Lubrication circuit model

Fig. 3 Flow characteristic for oil cooler

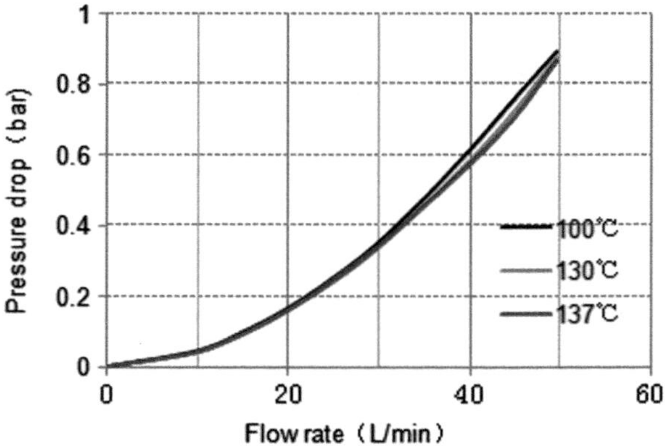

2.1.1 The Flow Characteristic for Oil Cooler

As shown in Fig. 3, it is the flow characteristic for oil cooler, and is derived from experienced supplier rig. The pressure drop from oil cooler inlet to outlet increase with oil flow rate increase, but there is little pressure drop changes with temperature rise.

2.1.2 The Flow Characteristic for Oil Filter

As shown in Fig. 4, it is the flow characteristic for oil filter and is derived from supplier rig. The pressure drop of oil filter from inlet to outlet increase with oil flow rate increase, and the pressure drop will decrease with temperature rise.

2.1.3 The Flow Characteristic for PCJ

As shown in Fig. 5, the opening pressure of PCJ is about 2 bar, before reaching the opening pressure, there is a little flow rate leakage. The pressure drop of PCJ from inlet to outlet will increase with oil flow rate increase, and the pressure drop will decrease with temperature rise.

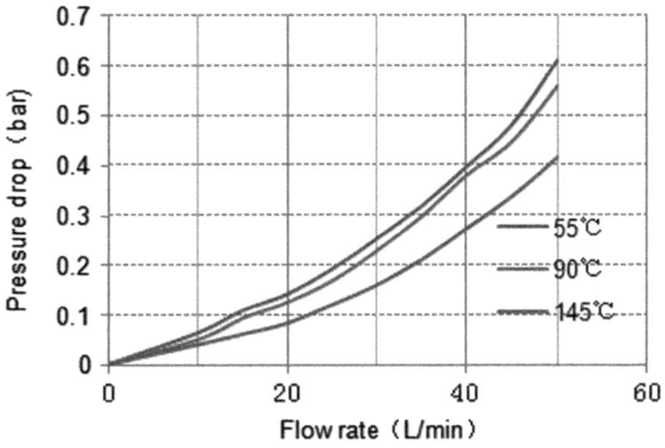

Fig. 4 Flow characteristic for oil filter

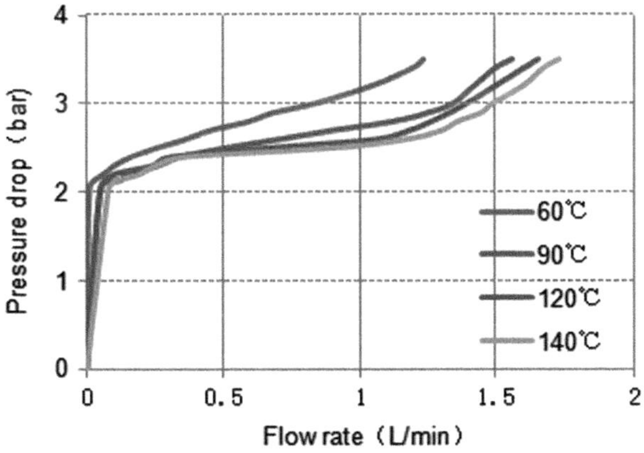

Fig. 5 Flow characteristic for PCJ

2.1.4 The Flow Characteristic for HLA

As shown in Fig. 6, it is leakage curves derived from supplier rig. The pressure drop of HLA from inlet to outlet increase with oil flow rate leakage increase, and the pressure drop will decrease sharply with temperature rise.

2.1.5 The Clearance for Main Bearings

As shown in Fig. 7, it is the clearance of main bearings, which are derived from simulation results.

2.1.6 The Control Strategy of PCJ

In order to avoid combustion knocking and piston be damaged, meanwhile save energy as much as possible, we proposed a PCJ control strategy showed in Fig. 8.

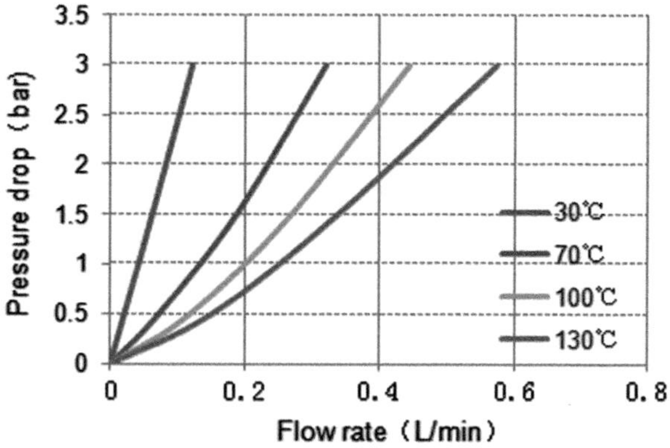

Fig. 6 Flow characteristic for HLA

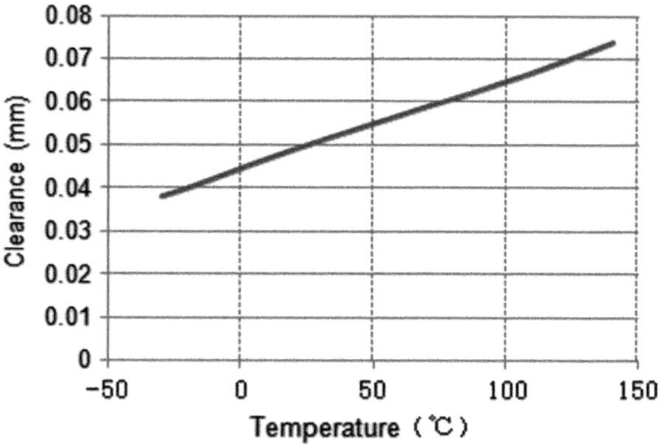

Fig. 7 Clearance of MB

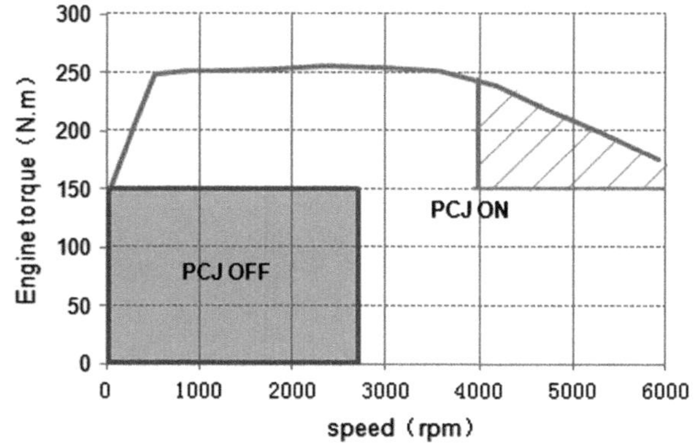

Fig. 8 Control strategy of PCJ

As showed, the OCV activation is decided by engine speed and actual load. During low speeds from 750 to 2750 rpm, meanwhile the engine torque is low than 150 N m, the ECU control deactivates the OCV solenoid, so PCJs are closed in that operating zone, there is no flow rate to PCJ. When the engine speed exceeds 4000 rpm, at the same time, the engine torque is higher than 150 N m, the oil flow rate of PCJ is required to be the maximum. At other conditions, the solenoid gets actuated, and the PCJs are opened in that operating zone. The flow rate jetting to piston depends on the oil pressure.

The simulation of oil pressure control strategy of VDOP will go on to match up the control strategy of PCJ and all hydraulic pressure requirements include VCT mechanisms, turbochargers, hydraulic lash adjusters, main bearings, etc., under all engine operating conditions (Fig. 9).

2.2 *Hydraulic Pressure Requirement*

Figures 10, 11, and 12 show the hydraulic pressure requirements. Several factors have large impact in increasing the demanded of the oil pump include expansion of the VCT operating region to lower engine speeds for improving fuel economy and higher oil temperature. The engine also adopts piston oil jets for suppressing engine knocking. The hydraulic pressure needed by these devices made it is necessary to increase the discharge rate of the oil pump substantially. The VDOP was therefore adopted to satisfy all of the requirements above at any engine condition.

2.2.1 Operating VCT Units

To support VCT to shift as quickly as required when it needs to improve engine full load performance, part load fuel consumption, and emissions at transient operations. Sufficient P-Port oil pressure is needed. In this paper, at 750 rpm engine

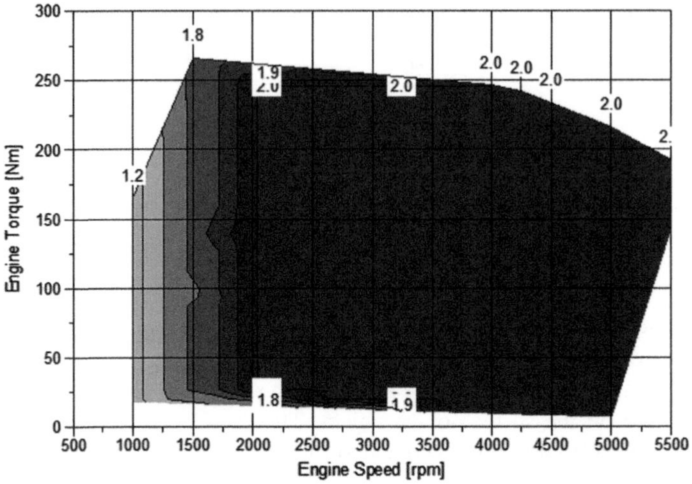

Fig. 9 Oil pressure needed by VCT

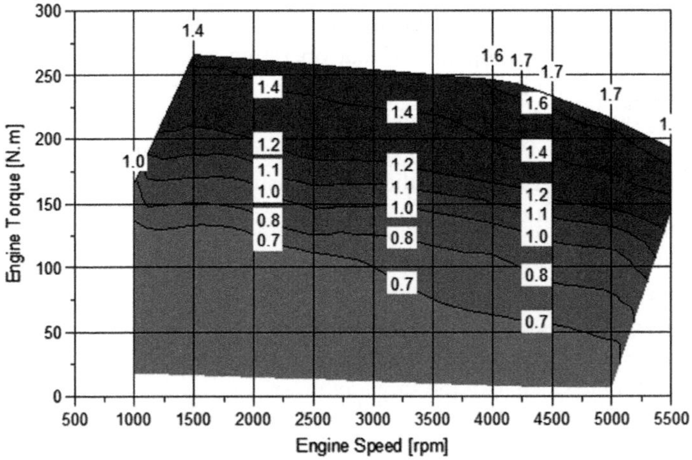

Fig. 10 Oil pressure needed by TC

speed with 100 C oil temperature, 1.5 bar VCT P-Port oil pressure is needed. The oil pressure needed at other speeds is shown in Fig. 9.

2.2.2 TC Requirement for Cooling and Lubricate

To satisfy the cooling and lubricate requirements of the turbocharger, the oil pressure needed in the paper is shown in Fig. 10. From the picture, we can see that the oil pressure changes little with speed changes, while it will change much faster with load changes. It is suitable for any oil temperature.

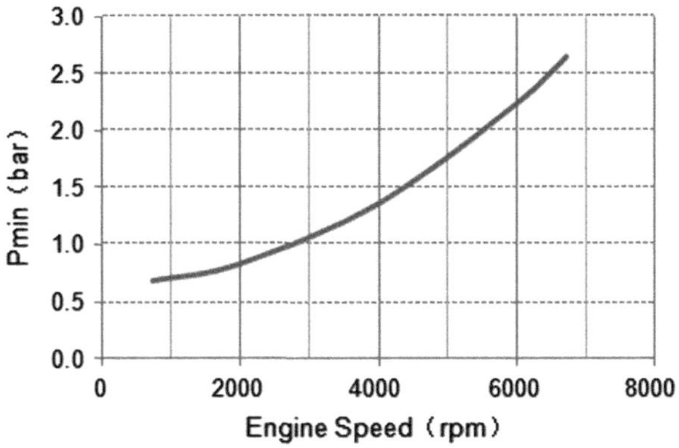

Fig. 11 Minimum oil pressure for journal bearings

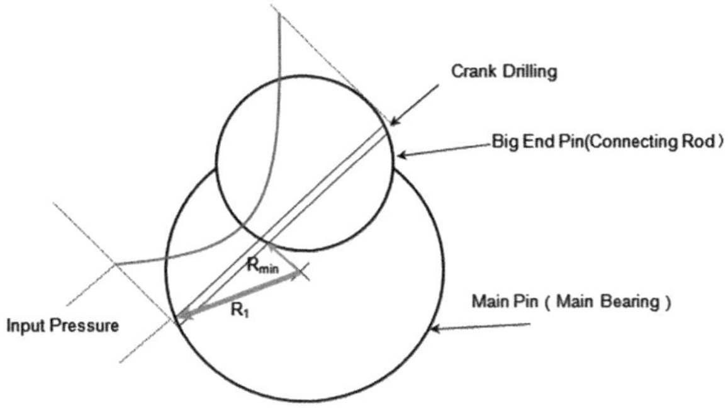

Fig. 12 Oil pressure drop in the crank drilling

2.2.3 Operating HLA Devices

At 750 rpm engine speed, 140 C oil temperature, at least 0.5 bar oil pressure is needed for operating HLA devices. There is no special oil pressure requirement at other conditions. It is offered by HLA supplier.

2.2.4 The Minimum Oil Pressure for Journal Bearings

In order to generate the lifting capacity in journal bearings, such as crankshaft, connecting rod, camshaft, the minimum oil pressure for journal bearings is needed. The critical lifting capacity needed is connecting rod bearings. So in order to satisfy oil pressure requirements of all bearings, the centrifugal pressure drop in the crank drilling should be overcome.

The oil pressure drop in the crank drilling between a crankshaft main pin and a big end pin is illustrated in the Fig. 12.

The oil pressure along the crank drilling, the purple curve shown above, begins at the main pin end with an input pressure. With a rotating crankshaft, the oil pressure drops as it moves toward the central rotating axis of the crankshaft as a result of the centrifugal force acting in the opposite direction. Once the oil has passed the central axis, the pressure increases as the centrifugal force adds to the existing oil pressure.

The centrifugal pressure drop is calculated with the following formula:

$$\Delta P_C = \frac{0.5\rho n^2 (R_1^2 - R_{\min}^2)}{100000} \quad (1)$$

where ρ is the oil density (kg/m^3), n is the engine speed (rad/s), R_1 the main bearing radius (m), and $R_{\min}$ the minimum perpendicular distance of the drilling to the crankshaft rotating axis (m).

3 Simulation Results

3.1 The Maximum Displacement of Map-Controlled VDOP

As we know, the maximum displacement of VDOP is confirmed to supply sufficient oil to bearings and hydraulic components, including VCTs, HLA, TC, and other hydraulic components. In this paper, in order to satisfy the VCT's shifting speeds, TC and HLA's oil pressure requirements. The demand of VCT P-Port is the most severe requirement for pump sizing. So the design point for pump size is going on at this condition.

After simulation, at 750 rpm engine speed with 100 C oil temperature, the engine will see 1.73 bar pump exit oil pressure with 11.6 l/min oil flow. When the volumetric efficiency of the pump at 750 rpm engine is 80% with 1.73 bar pump exit pressure and pump performance loss is 5% at the end of life, then the maximum displacement for VDOP will be 20.3 cc/rev engine speed.

3.2 The Oil Pressure Control Strategy of VDOP

As the oil pump is VDOP, this means at any engine operating condition other than the sizing point, the pump will have spare capacity, so adjustment or variable control is needed. However, to fully utilize the benefit of this VDOP, the engine lubrication system mapping is needed together with on-demand control strategy.

According to the PCJ's control strategy and the hydraulic pressure required. We simulate the control strategy of the map-controlled VDOP at certain conditions showed in Fig. 13. Simulations under fully warm engine conditions were carried out over a range of engine speeds and loads, the oil temperature in main gallery was around 85–110 °C. For each condition, the feed pressure to the VCT or turbocharger was maintained constant by adjusting the pump delivery.

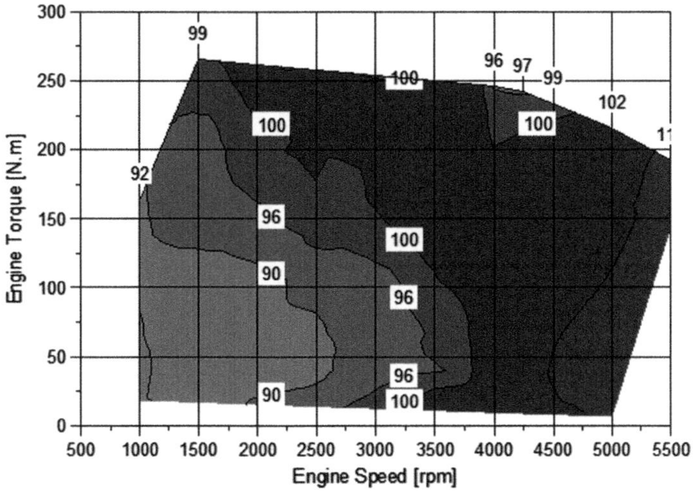

Fig. 13 Simulation conditions

At the temperature above, the pump outlet flow rate will cover the range from 7 to 24 L/min, showed in Fig. 14.

As shown in Fig. 15, the general trends show oil pressure increasing with engine speed and load. The pressure at which PCJs opened and closed marked an important point in the sweep of gallery pressure. This pressure was determined by monitoring the response of oil flow into the gallery as the feed pressure was changed. Because the PCJs take a high proportion of the total flow, switching between jets on and jets off produces a marked change in the sensitivity of flow rate to feed pressure. During low speeds from 750 to 2000 rpm, the VDOP operates with a delivery pressure from 1.22 to 2.1 bar. When the engine speed exceeds 4000 rpm, the OCV of PCJ was triggered. The oil pump capacity increase will allow higher oil pressure, high up to 3.39 bar. When speed is between 2750 and 4000 rpm, and the engine load is up 150 N, the ECU control activates the PCJ solenoid, so the oil pressure in feedback oil gallery is about 3.3 bar.

When PCJ is needed or when we need to shift the VCT, oil pump capacity could be increased in advance before PCJ switching on signal or VCT shifting signal was triggered. The oil pump capacity increase will allow higher oil pressure so faster VCT response rate to be achieved. It also allows PCJs to be turned on when they are needed.

3.3 *Fuel Consumption Contribution of VDOP*

Usually, power loss caused by oil pump comes from two parts, one is dry friction power caused by mechanical motion, and the other is hydraulic power used to

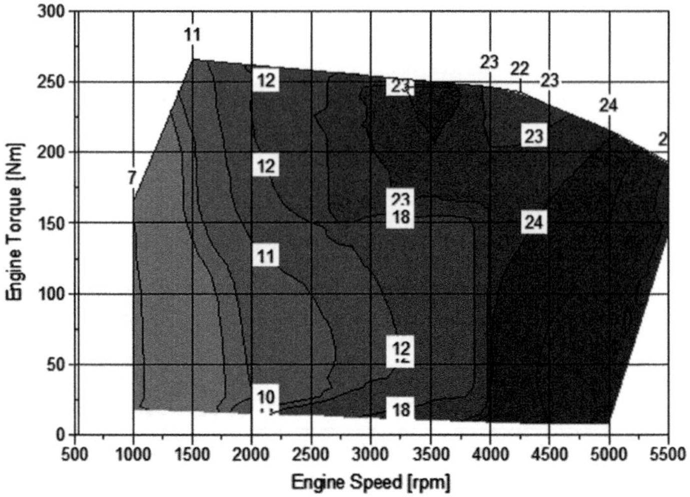

Fig. 14 Pump outlet flow rate for VDOP

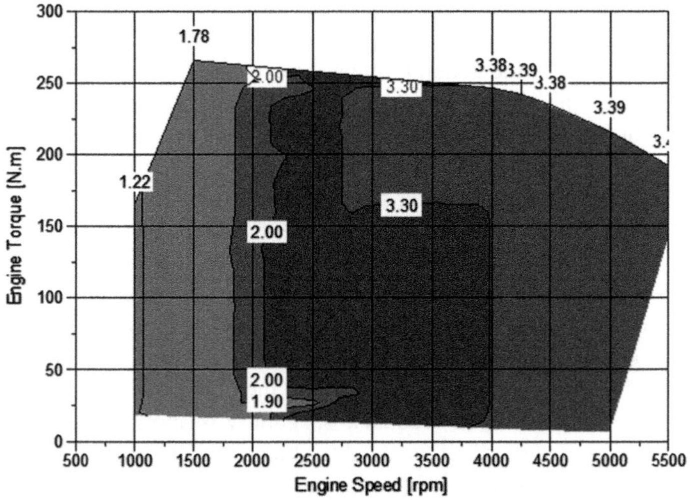

Fig. 15 Pump outlet oil pressure for VDOP

deliver oil to every part of lubrication system. The fuel consumption contribution in this paper is due solely to hydraulic power, not including dry friction power. Hydraulic power often gets from the formula showed below.

$$P_t = a \times \Delta p \times q \tag{2}$$

P_t is hydraulic power, Watt. Δp is the oil pressure drop between pump inlet and pump outlet, bar. q is the oil pump flow rate, L/min. a is a factor used to revise the units difference while q and P_t come from simulation data.

Figure 16 shows the hydraulic power map (the unit is watt) of the VDOP. When the engine speed is below 2000 rpm with any engine load condition, the hydraulic power is about 45 W. When the engine condition is in the regime that the speed is between 2000 and 4000 rpm, and engine load is below 150 N m, the hydraulic power is around 45 W too. At the other conditions, the hydraulic power is 150 W approximately.

For reflect the fuel consumption contribution of map-controlled VDOP together with map-controlled PCJ, a traditional fixed oil pump whose pump size is 20.3 cc/rev, the relief valve is opened at 4 bar and with ordinary PCJs are introduced in the same lubrication system model. The simulation results show that in all conditions, the hydraulic power exceeds 100 W; it is high up to 200 W at high engine speed with high load showed in Fig. 17 (the unit is watt).

Figure 18 is fuel consumption contribution fraction of VDOP compare to the fixed oil pump, it is showed that the fuel consumption saving ratio is very attractive. At lower speed, the power consumption reduces up to 73–85%, yet it saves 20% when high speeds with high loads. It is because when engine is work at high speeds with high loads, the oil pump delivery of fixed and variable oil pumps all meets the requirement well.

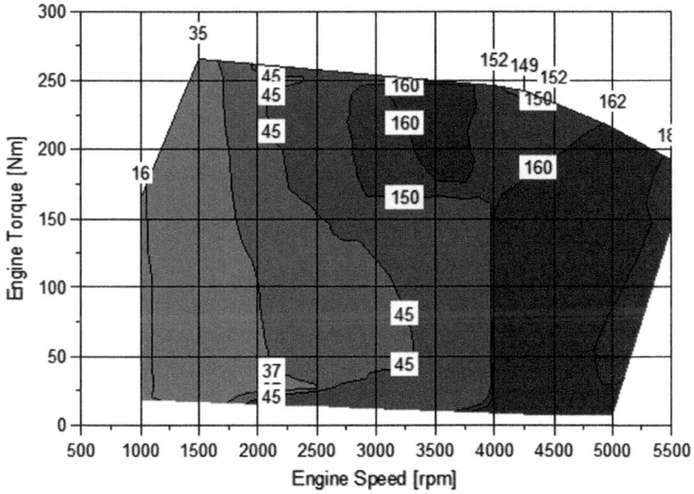

Fig. 16 Hydraulic power of VDOP

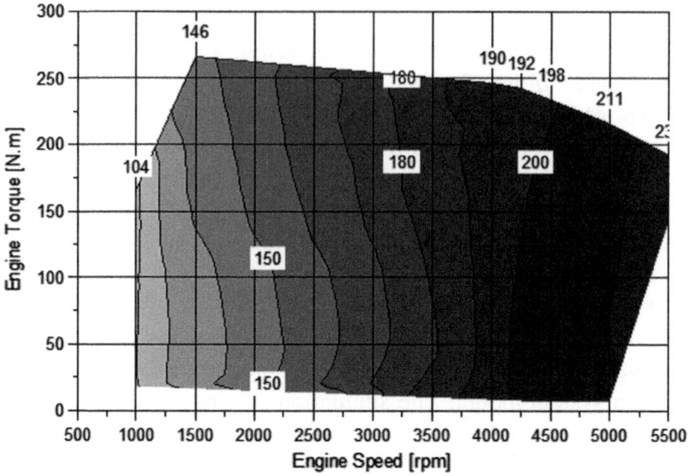

Fig. 17 Hydraulic power of fixed oil pump

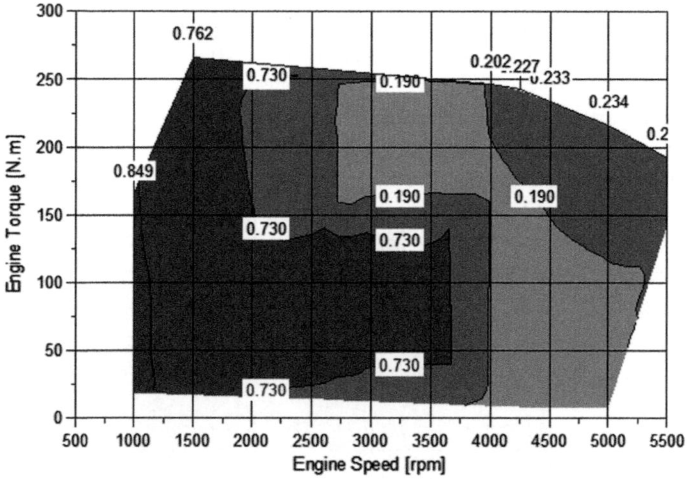

Fig. 18 Fuel consumption contribution of VDOP compare to the fixed oil pump

4 Conclusions

The paper describes the procedure to assess fuel consumption contribution of map-controlled VDOP together with map-controlled PCJ.

Firstly, the maximum pump size was confirmed to satisfy the most severe requirement, the method is the same as a fixed oil pump.

Then, according to the PCJ's control strategy and the hydraulic pressure required. The control strategy of the map-controlled VDOP is set up on certain condition which the engine often works at. It is showed that when PCJ was triggered, the oil pump delivery increases dramatically, at the same time the oil pressure in feedback oil gallery will increase in order to feed the PCJ as much as it needs.

Finally, in comparison with a fixed oil pump whose pump size is same as the map-controlled VDOP, it shows that the power consumption saving ratio is very attractive at lower speed and also has decrease space when high speed and high load.

Acknowledgements I would like to appreciate my team members who co-work for developing the CI engine with map-controlled VDOP and map-controlled PCJ.

References

1. Kandavalli PB, Karthi R, Kumar SS (2017) Benefits of variable discharge oil pump on performance of 3 cylinder SI engine. SAE Technical Paper 2017-26-0051. https://doi.org/10.4271/2017-26-0051
2. Shayler PJ, Cheng L, Li Q (2016) A modified oil lubrication system with flow control to reduce crankshaft bearing friction in a litre 4 cylinder diesel engine. SAE Technical Paper 2016-01-1045. https://doi.org/10.4271/2016-01-1045
3. Wang D, Ding H, Jiang Y (2012) Numerical modeling of vane oil pump with variable displacement. SAE Technical Paper 2012-01-0637, Published 04/16/2012
4. Arata T, Novi N, Ariga K, Yamashita A, Armenio G (2012) Development of a two-stage variable displacement vane oil pump. 2012-01-0408, Published, 04/16/2012
5. Mancò S, Nervegna N, Rundo M, Armenio G (2004) Modelling and simulation of variable displacement vane pumps for IC engine lubrication. SAE Technical Paper, 2004-01-1601
6. Klingebiel F, Kahlstorf U (2000) Simulating engine lubrication systems with 1-D fluid flow models. SAE Technical Paper, 2000-01-0284
7. Chun SM, Park YW, Jang S (2000) A study on engine lubrication system by optimized network analysis—Part II: parametric study. SAE Technical Paper, 2000-01-2923
8. Neukirchner H, Kramer M, Ohnesorge T (2002) The controlled vane-type oil pump for oil supply on demand for passenger car engines. SAE Technical Paper, 2002-01-1319
9. Mancò S, Nervegna N, Rundo M, Armenio G (2004) Displacement vs flow control in IC engines lubricating pumps. SAE Technical Paper, 2004-01-1602
10. Loganathan S, Govindarajan S, Suresh Kumar J, Vijayakumar K, Srinivasan K (2011) Design and development of vane type variable flow oil pump for automotive application. SAE Technical Paper, 2011-28-0102

The Development and Application of the Front Axle Assembly Line

Yi Zhang, Jun Qian, Hongying Qin, Peng Teng, Zhu Li and Yongchao Wu

1 Introduction

The model is the first senior car developed by Dongfeng independent brand, which has the front axle with an independent suspension. In order to make the assembly of the front axle meet 33 V/H production capacity of Dongfeng passenger car new energy plant, need to re-design and construct a new front axle assembly line to ensure its industrialization [1].

The production line includes front shock absorber subassembly unit, the production line of lower half bridge and the main dock of upper and lower half bridge, which arranged in the side of front power assembly line. The front shock absorber subassembly unit is composed of electric servo-cylinder, press and various sensors. The production line of lower half bridge is mainly composed of press-fit equipment, front brake disc end beating detection equipment, positioning fixture, rotating rack, roller table and so on. The overall development ideas of the paper are as follows.

By using servo-automatic transitional device to realize the automatic tightening of the equipment. Through the use of automatic detection of jumping and flip technology, automatic feeding mechanism and roller conveyor technology to improve the production capacity and the level of automation, we achieve the production cycle [2].

Optimize quality control by using a networked intelligent system to automatically identify vehicle parameters and adapt the equipment parameters to production [3], while ensuring traceability and controllability of assembly quality.

Control the investment in the budget with the minimum manufacturing costs while ensuring the production and quality.

Y. Zhang · J. Qian · H. Qin (✉) · P. Teng · Z. Li · Y. Wu
Dongfeng Motor Corporation Technology Centre, Economic and Technological Development Zone, 663, Zhu Shanhu Avenue, Wuhan, Hubei, China
e-mail: qinhy@dfmc.com.cn

Society of Automotive Engineers of China (SAE-China) (ed.), *Proceedings of the 19th Asia Pacific Automotive Engineering Conference & SAE-China Congress 2017: Selected Papers*, Lecture Notes in Electrical Engineering 486, https://doi.org/10.1007/978-981-10-8506-2_27

In the case of the existing workshop area is scarce, arrange the workshop compactly. The assembly line is designed according to the optimization of the line edge arrangement and the shortest path of the logistics route. With annular arrangement, the area of the line is optimized greatly while the connection is reasonable and compact. The actual occupying area of the front axle assembly line is 36 m^2. The layout is shown in Fig. 1.

2 Upper Half Bridge Front Shock Absorber Spring Press Machine

The spring press machine, which is used for the former shock absorber spring pressure and nuts automatically tightened [4], mainly consists of frame, clamping and arming device, pressing device, automatic safety door, tightening device and other components, as shown in Fig. 2.

The structural features of upper half bridge front shock absorber spring press machine are shown in Fig. 3. Achieve the horizontal and vertical movements of the moving device by using two single-acting (horizontal and vertical) cylinders respectively [5–7]. The horizontal cylinder pushes the connecting plate in horizontal direction, the connecting plate through the guide rod pushes the mounting seat of the tightening machine horizontal movement, and the tightening machine mounting seat is integrated with the tightening device. In order to ensure the horizontal movement of the tightening equipment's stability and accuracy, double-track support is used under the connecting plate. The vertical cylinder is fixed on the connecting plate while connected to the mounting seat of the tightening machine to control the movement of the tightening machine in the vertical direction.

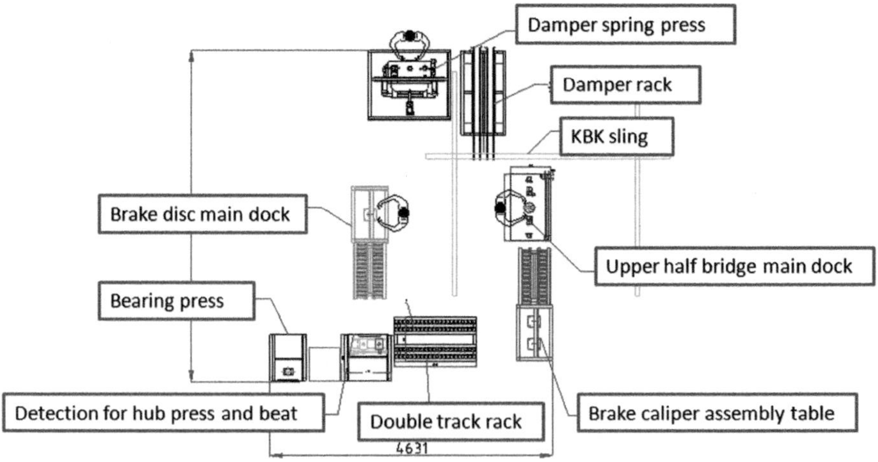

Fig. 1 Overall layout of the front axle assembly line

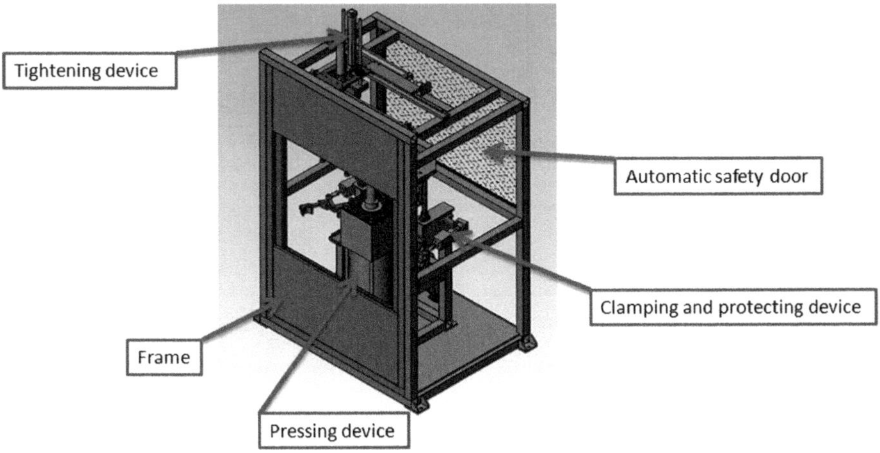

Fig. 2 Structure of the front shock absorber spring press machine

The guide rod is fixed on the connecting plate and is connected to the mounting seat of the tightening machine while vertically movable in the guide sleeve to realize the guiding effect in the vertical direction of the tightening machine. The operation process of the upper half bridge front shock absorber spring press machine is shown in Fig. 4.

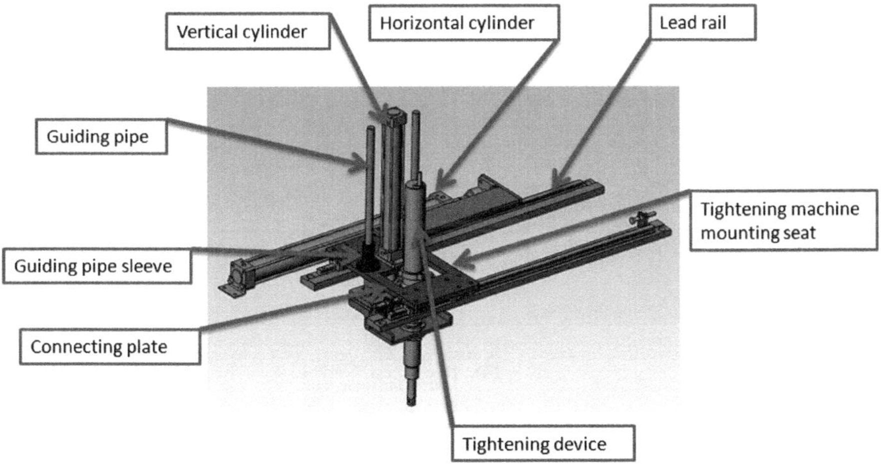

Fig. 3 Structure of the automatic moving device of the servo-tightening machine

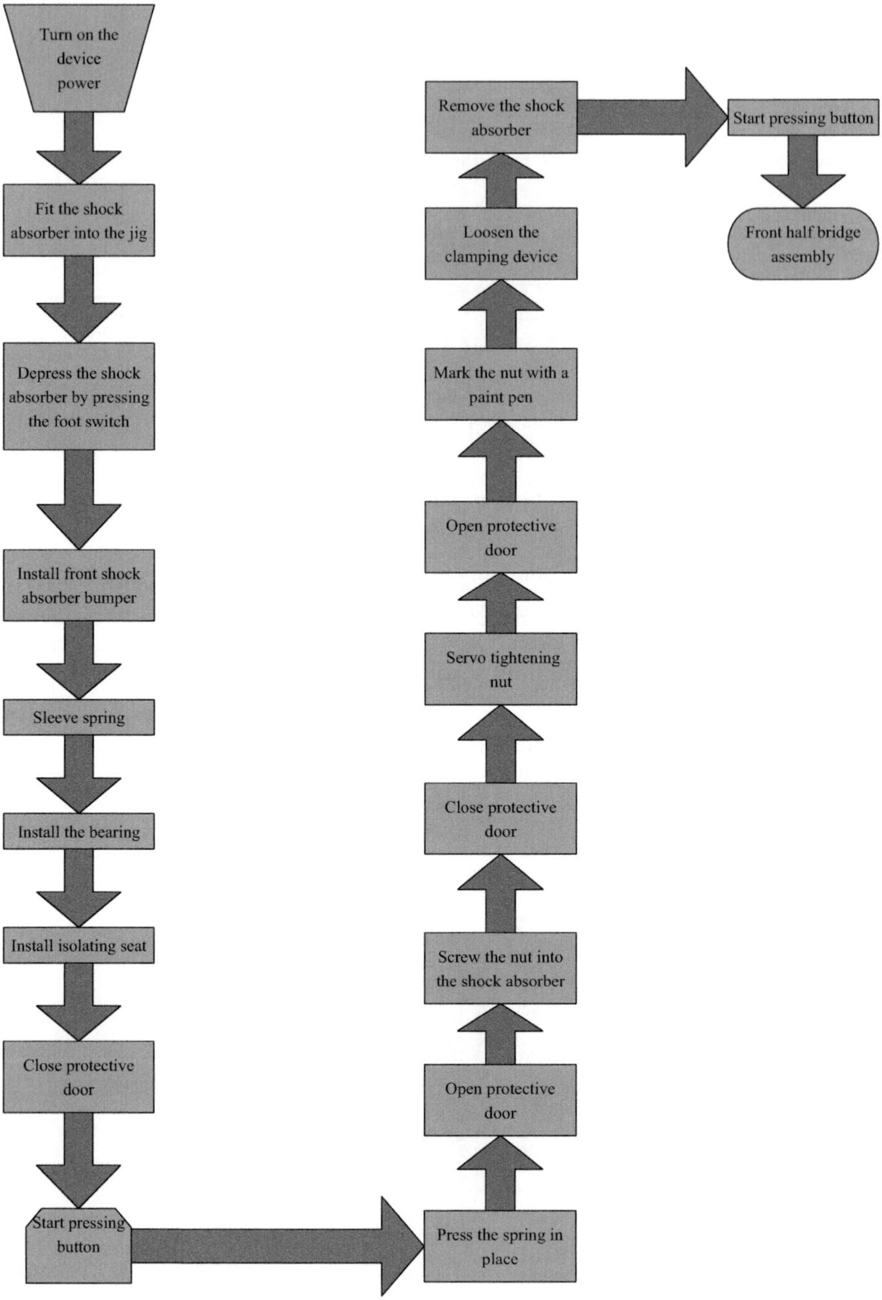

Fig. 4 Operation process of the upper half bridge front shock absorber spring press machine

3 The Assembly Line of Lower Half Bridge

The assembly line of lower half bridge consists of bearing ring press machine, subassembly platform of the front hub-front brake disc, the stand-alone device used for pressing the front wheel to the steering knuckle, disc brake vibration measurement equipment and brake calliper subassembly platform. The assembly line is designed according to the optimization of the line edge arrangement and the shortest path of the logistics route. It greatly optimizes the line edge area and improves the productivity of labour. The overall layout of the assembly line of lower half bridge is shown in Fig. 5.

3.1 *Bearing Ring Press Machine*

The structure of the bearing ring press machine is shown in Fig. 6. As ring press is followed by the bearing press-fitting process, it is very tight with the other equipment while controlling the beat accurately [8, 9]. To this end, calculate the beat accurately according to the other equipment to achieve the synchronization of the beat. The sorting of the retaining ring is precisely controlled by the PLC to ensure its accuracy. The retaining ring is easy to flip when placed and causes the

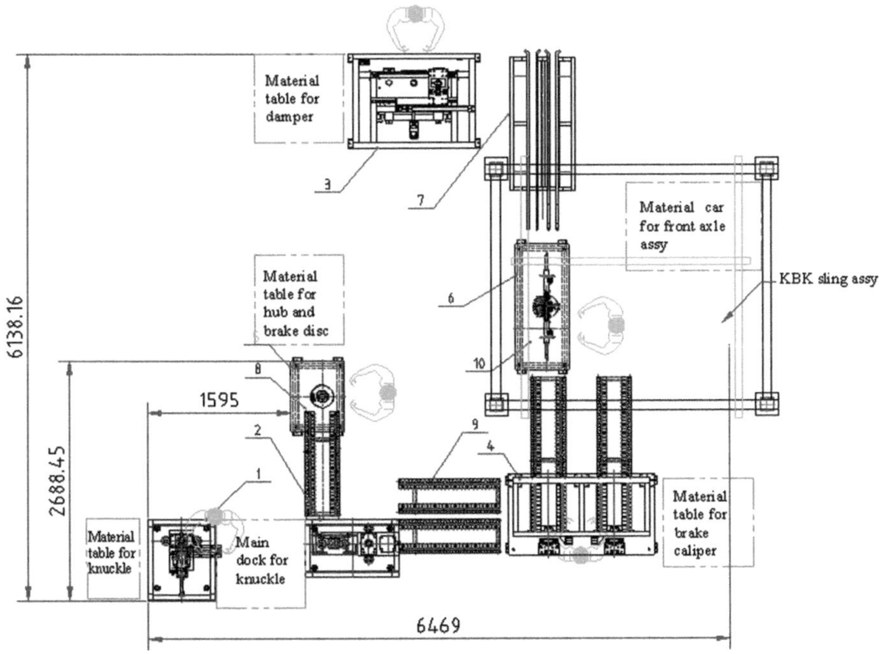

Fig. 5 Overall layout of the assembly line of lower half bridge

operating time increase itself lighter. Then design the profiling tooling to position the retaining ring precisely and pressing accurately. The features of the bearing ring press machine are as follows. The operation process of the bearing ring press machine is shown in Fig. 7.

3.2 Device for Pressing the Front Wheel and Testing the Brake Disc Vibration

The features of the device for pressing the front wheel and testing the brake disc vibration are as follows [10, 11]. It is controlled by PLC to realize the integration of the hub press fit and the beating detection automatically, while shifting the workpiece to the flipping mechanism after detection by the straight rail. The flipping mechanism holds the hub steering knuckle assembly and turns it 180°. Then pulls the workpiece out of the machine automatically through the slide cylinder and enters the feeding roller, as shown in Fig. 8.

This equipment is facing various problems, such as the same equipment integrated hub bearing pressing and the automatic detection of the brake disc vibration, the brake disc flip 180° after press, the quality of the assembled parts lager, handling difficulty and so on [12]. The design integrated the equipment for the hub bearing pressing and the brake disc vibration detection, multi-frequency detect the

Fig. 6 Picture of the bearing ring pressing machine

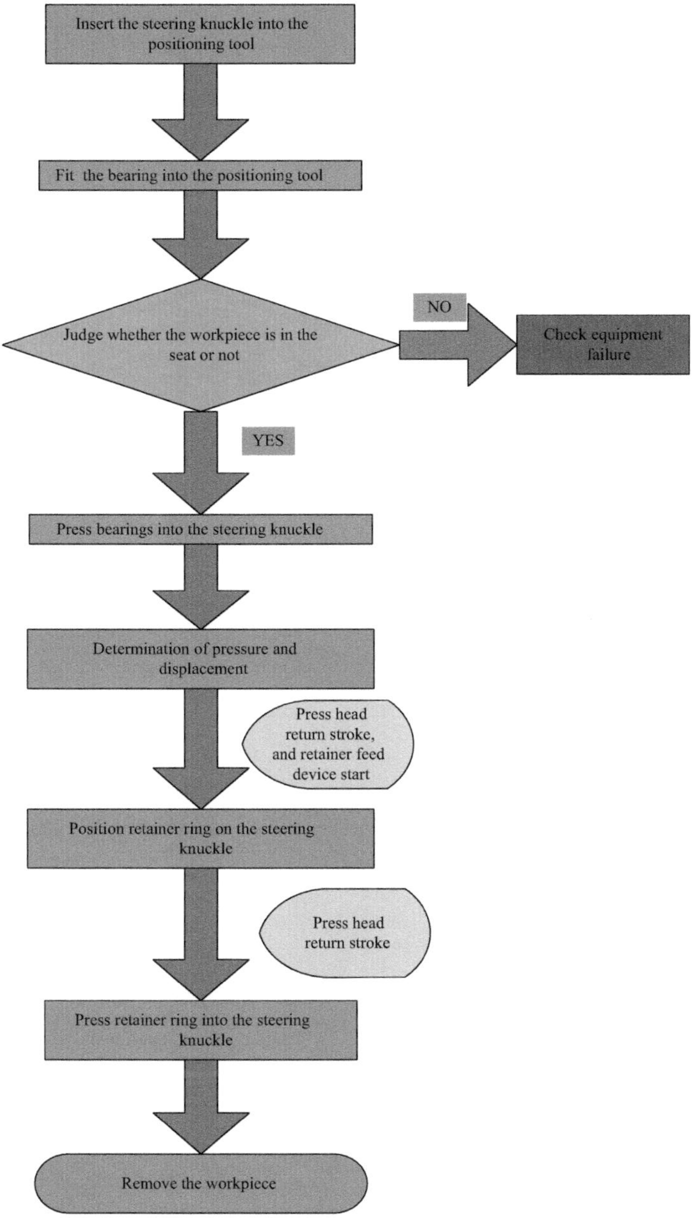

Fig. 7 Operation process of the bearing ring press machine

brake disc vibration automatically after press fitting [13, 14]. The flipping mechanism can clamp brake disc automatically, flip the workpiece 180° driven by the motor and place the workpiece in the feeding roller by the vertical cylinder

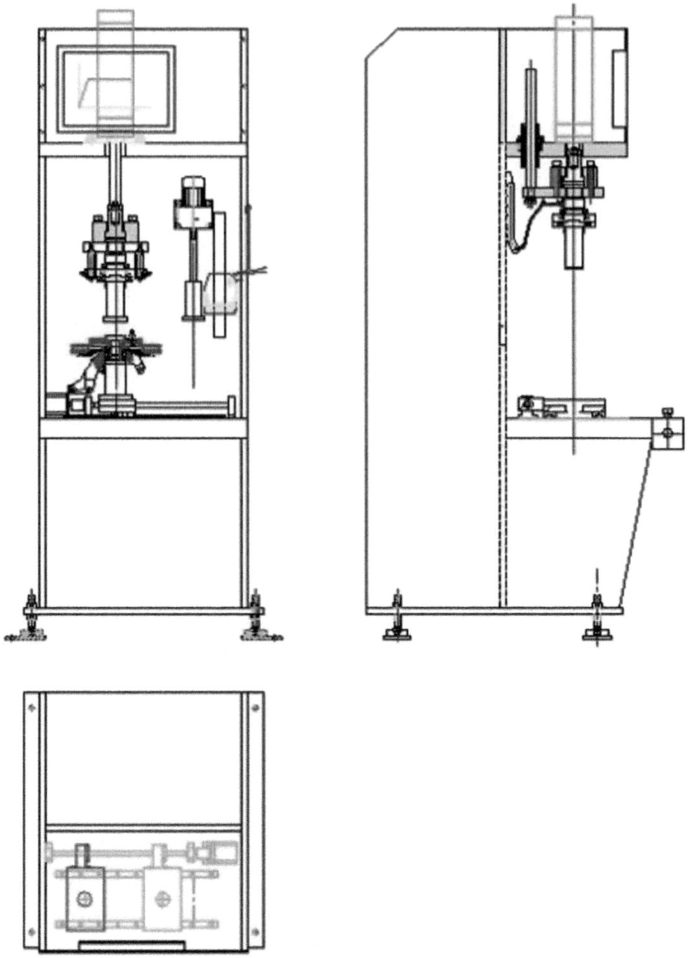

Fig. 8 Device for pressing the front wheel and testing the brake disc vibration

automatically. The operation process of the device for pressing the front wheel and testing the brake disc vibration is shown in Fig. 9.

3.3 Brake Calliper Tightening Equipment

The structural characteristics of servo-brake calliper tightening device are as follows [15]. Two bolts have a certain angle with the vertical line used to tighten the brake calliper. In order to ensure the accuracy of the two bolts fastening, biaxial tightening structure was used [16]. As shown in Fig. 10, a slide rail and a movable

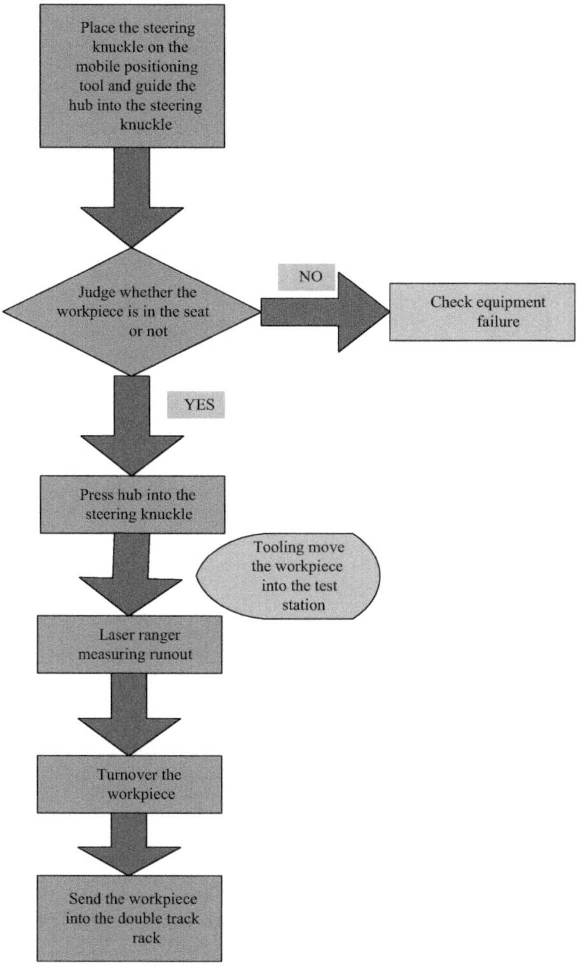

Fig. 9 Operation process of the device for pressing the front wheel and testing the brake disc vibration

cylinder above the biaxial tightening shaft can guarantee the spindle flexible movement effectively and fasten conveniently. The equipment ensures the accuracy of the bolt fastening while also saving man-hours to facilitate operation.

The body is automatically fed through a two-track rack. Control the brake disc positioning device to rise and fall through the lifting cylinder, in order to achieve the brake disc and roller separation and bonding. Then solve the problem of larger quality and manpower transfer difficultly for the brake disc. For the high-accuracy of the calliper positioning, the calliper positioning device was been designed. The operation process of the servo-brake calliper tightening device is shown in Fig. 11.

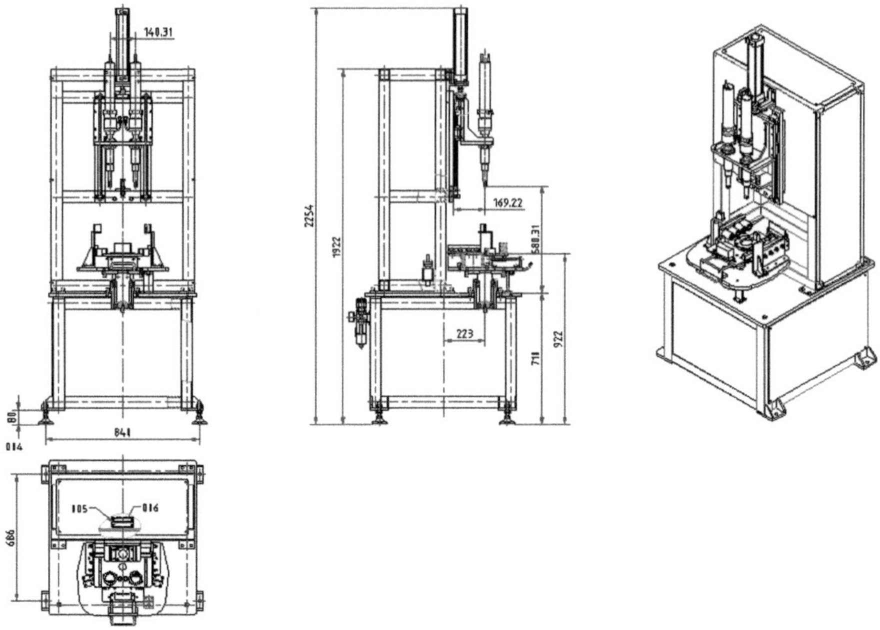

Fig. 10 Brake calliper tightening equipment

3.4 *Automatic Feeding Structure for the Lower Half of the Bridge Assembly*

As shown in Fig. 12, this mechanism is mainly composed of a positioning device, lifting cylinder assembly, conveying roller, etc. It jacks up the hub brake disc assembly by the lift cylinder components and installs the brake calliper. The cylinder drops after the bolts are fastened by biaxial servo-tightening machine [17]. Then the lower half of the bridge assembly slid down to the double roller rack and transported to the next station automatically.

4 The Main Dock of Upper and Lower Half Bridge

The main dock of upper and lower half bridge is mainly composed of frame, positioning fixture, tightening device and other components, and the schematic diagram is shown in Fig. 13.

The automatic feeding device structural features of upper half bridge are as follows. Left and right front suspension spring (upper half bridge) can be automatically transported to the main dock, and distinguish the left and right to prevent the possibility of wrong. The feeding device can be automatically adjusted

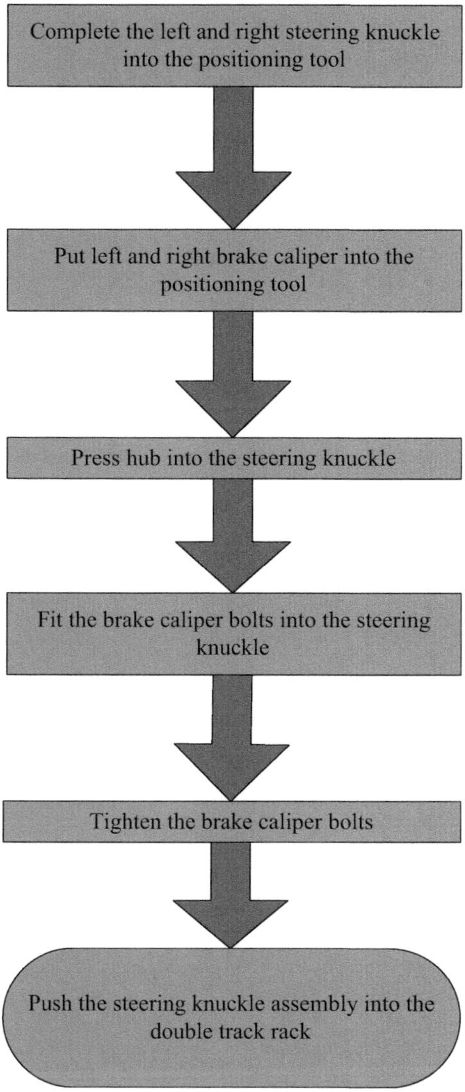

Fig. 11 Operation process of the brake calliper tightening equipment

according to the height of the main dock, to ensure that the workpiece can be accurately delivered to the main dock [18, 19]. All track rollers are made of materials that are harmless for the workpiece to protect the workpiece effectively.

In the assembly, fastening torque of the device is up to 165 Nm. As the rebound force is too large, the device will cause safety problems. In order to prevent rebound, design of special fixture to fix it after tightening. As shown in Fig. 14, tooling has a certain position to ensure the accuracy of tightening.

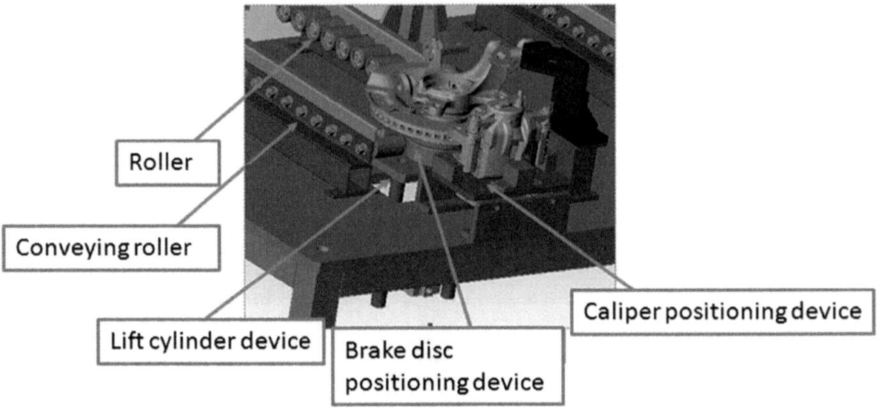

Fig. 12 Automatic feeding structure for the lower half of the bridge assembly

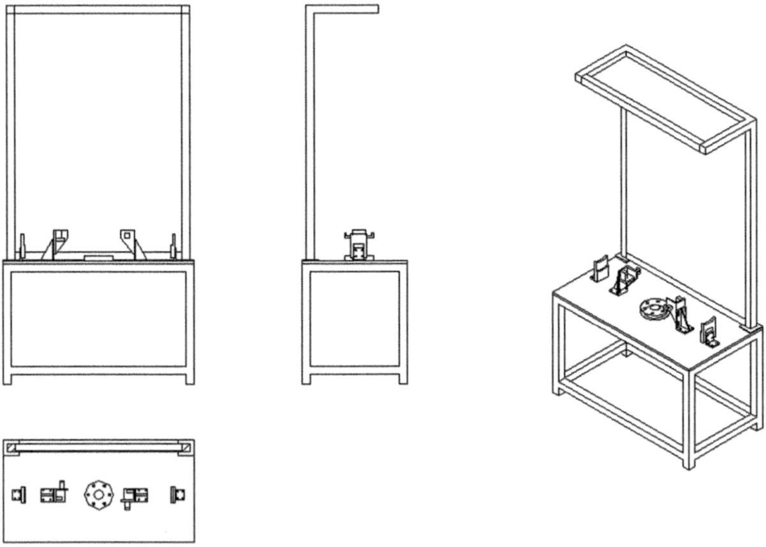

Fig. 13 Main dock of upper and lower half bridge

5 Intelligent Assembly System

Automobile manufacturing industry ANDON system is ideally suited to continuous production line; currently in the automotive industry it is widely used. ANDON system is an effective method to improve manufacturing quality and productivity [20]. The intelligent assembly system obtains the traffic flow sequence through the ANDON system, establishes traffic queue matrix of the production line and

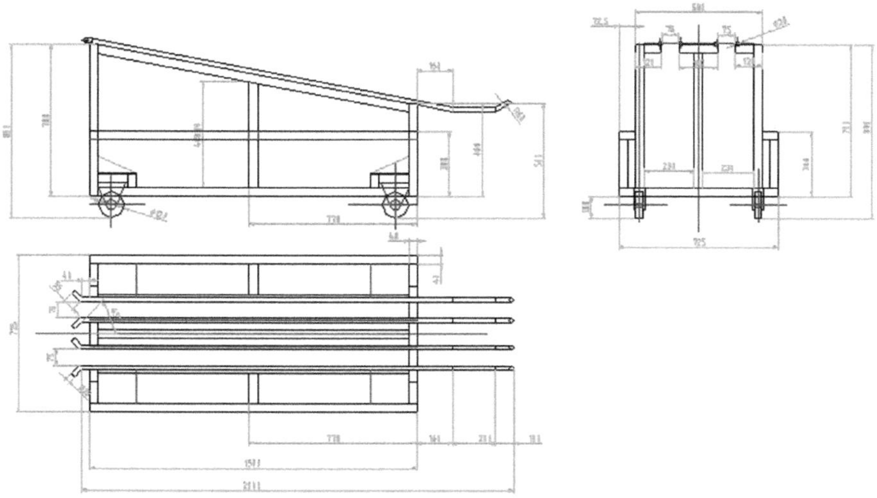

Fig. 14 Automatic feeding device of upper half bridge

transmits the vehicle information that is to be assembled to the demand equipment [21, 22]. The equipment automatically selects the assembly parameters and assembles the equipment. The equipment selects the assembly parameters automatically and assembles [23–26].

The station shall be provided with a monitoring guide screen to guide the production. As shown in Fig. 15, and guide the screen should be able to display the process documents, equipment parameters, assembly results and other information of the vehicle which is current assembly.

6 Applications and Achievements

The front half bridge assembly line has been put into use. Through the production verification, the design indexes have been reached, and the process time has reached 33 JPH. The production line is equipment multi-functional integration, process layout and intelligent control technology, while reducing the process area, the number of operators, inventory and equipment investment, reducing production costs. The assembly line has high integration, high automation and stable operation. The difference between the assembly line and the similar technology at home and abroad is shown in Table 1.

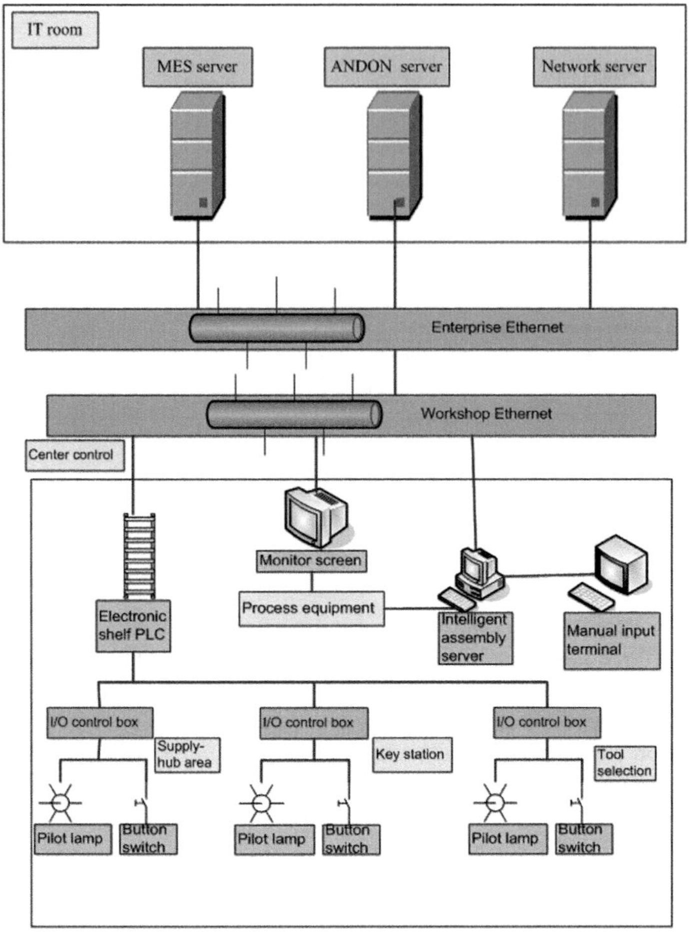

Fig. 15 Schematic diagram of the intelligent assembly system

Table 1 Difference between the assembly line and the similar technology at home and abroad

Comparison program	Mainstream independent brands	Domestic joint venture	Dongfeng independent brand
Servo-automatic screwing and moving technology	Single-axis hand-held servo-tightening machine fastening (or pneumatic tools)	Two-axis servo-tightening machine simultaneously fastening	The single-axis servo-tightening machine is used for fastening, and the automatic tightening and shifting technique of the servo-motor is applied for the first time

(continued)

Table 1 (continued)

Comparison program	Mainstream independent brands	Domestic joint venture	Dongfeng independent brand
Automatic detection run-out of brake disc and brake disc turnover technology	Pressure equipment, testing, flip equipment independent operation, the use of manual handling mode to achieve parts transport	Press-fit testing equipment integration, the use of manual handling mode to achieve the parts flip transport	Pressing, testing, flip equipment integration, the application of brake disc beating the amount of automatic detection technology, the development of automatic flip technology
A half bridge assembly installation and automatic feeding mechanism	The use of manual handling mode to achieve the workpiece transport	Spreader auxiliary mode to achieve the workpiece transport	The first application of the assembly and automatic feeding technology for the lower half bridge
Intelligent system control technology	Vehicle information match: job instructions; assembly instructions: paper process files	Vehicle information match: job instructions; assembly instructions: monitor the guide screen	Vehicle information matching: MES system obtains vehicle information automatically; assembly instructions: monitor the guide screen

7 Conclusions

The assembly line can be produced to 33 V/H ability to ensure the front axle assembly of industrialization. Subsequent front axles of new models can be designed based on synchronous engineering for the structure of the assembly line and servo-tightening equipment, a variety of models sharing this line will greatly reduce the equipment investment. At the same time, we obtained two national patent technologies as follows; automatic detection and turnover device, the patent number was 201320775615.2; assembly of the lower half bridge and the automatic feeding mechanism, the patent number was 201320775107.4.

Acknowledgements On the completion of my thesis, I would like to express my deepest gratitude to all whose kindness and advice have made this work possible.

References

1. Andersen GK, Anderson BK, Yakes CK et al (2003) Integrated vehicle suspension, axle and frame assembly. U.S. Patent: 6516914, 2003-2-11
2. Bruhn H, Felske A (1985) A new optical axle measuring instrument for wheel alignment in assembly-line production. Schweizer Archiv Für Tierheilkunde 109(10):537–538
3. Resano Lázaro A, Luis Pérez CJ (2008) Analysis of an automobile assembly line as a network of closed loops working in both, stationary and transitory regimes. Int J Prod Res 46 (17):4803–4825
4. Li X, Wu W, Ban S et al (2017) Third-order natural frequency combined-type shock absorber, has metal spiral spring fixed on outer side of rubber cylinder, and free rolling plate downwardly moved to determine natural frequency of damping system. 中国专利: CN106678267-A, 2017-05-17
5. Zhang H, Yuan XJ (2016) About the shock absorber damping torque analysis of the problem. Automot Pract Technol 7:240–242
6. Schorn M, Isermann R (2004) Improving the performance of an active steering system by variation of the antiroll bar stiffness at the front and rear axle. IFAC Proceedings Volumes 37 (22):154–156
7. Ryabov IM, Chernyshov KV, Pozdeev AV (2016) Energy analysis of vehicle suspension oscillation cycle. Procedia Eng 150:357–362
8. Wang SB (2015) Based on TRIZ theory of elastic bearing retainer safety disassembling tool innovation design. Dev Innov Mach Electr Prod 4:16–18
9. Rong LS (2013) The design principle and application of single wheel bearing inner ring pressing machine. Popular Sci Technol 6:157–158
10. Xu SH, Ding GQ (2008) Design of servo control system for vehicle wheel bearings push-mounting machine. Electron Meas Technol 8:20–23
11. Lu H, Yu DJ, Chen N et al (2014) Analysis of automotive disc brake systems vibration stability by introducing uncertain parameters. J Vib Eng 6:900–906
12. Feng XL, Li BN (2013) Development of performance test bed for vehicle disc brake. Automobile Parts 9:91–93
13. Kang J, Krousgrill CM, Sadeghi F (2009) Comprehensive stability analysis of disc brake vibrations including gyroscopic, negative friction slope and mode-coupling mechanisms. J Sound Vib 324:387–407
14. Liang HJ (2015) Development of an equipment for detecting brake disc ending run-outs. South China University of Technology
15. Liu P, Huang J (2012) Screw and data acquisition system for fasteners of brake caliper unit. Railway Locomotive Car 1:58–62
16. Ge Q (2011) Study on car float caliper disc brake caliper assembly technique. Equip Manuf Technol 11:104–105
17. Zhang YB, Shu ZJ, Shi LH et al (2005) The decoiling auto-line material sending system based on servo system. Manuf Technol Mach Tool 8:43–47
18. Wu Z (2016) Simulation analysis of automatic loading robot. Mech Eng 7:70–71
19. Xiu Q (2015) Design of mechanical structure of automatic feeding machine and its servo drive control. Mach Build Autom 3:155–157
20. Li ZH, Man ML (2015) Design and application of ANDON system in automobile production line. J Chengde Petrol Coll 3:39–43
21. Xu HG (2013) The research of the auto assembly plant ANDON system that adopts PLC integrated fail-safe system. Microcomput Appl 10:35–37
22. Wang Q (2012) The application of ANDON system in mixed production of automobiles with different body structures. Autom Appl 10:37–38
23. Guo ZX, Wong WK, Leung SYS et al (2009) Intelligent production control decision support system for flexible assembly lines. Expert Syst Appl 36:4268–4277

24. Zha XF, Du H, Lim YE (2001) Knowledge intensive Petri net framework for concurrent intelligent design of automatic assembly systems. Robot Comput Integr Manuf 17:379–398
25. Liu MZ, Ma J, Lin L et al (2017) Intelligent assembly system for mechanical products and key technology based on Internet of things. J Intell Manuf 28(2):271–299
26. Deng HQ (2002) NKR front axle assembly line rearrangement and modification. Xi'an University of technology

Prediction of Vehicle Population of Wuhan Based on Model of Econometrics

Mingrui Wang, Li Fu, Qiguang Xie, Yabo Du and Fan Dong

1 Introduction

For automotive industry practitioners, accurate prediction of vehicle population is an important job to plan production capacity precisely and adjust future development strategy. Inaccurate prediction brings unrealistic production plan, leading to resource waste and predicament of production and management. Besides that, vehicle population is also a crucial index for urban construction department and traffic administration to allocate public parking resources reasonably, and to effectively plan traffic network. Nowadays, urban administrators realize that it is not sustainable to solve urban transportation problems through satisfying transportation demand ceaselessly. They need to present reasonable transportation planning and management solutions, and the first step is to predict the vehicle population. Therefore, it is of great significance to predict the vehicle population exactly, whether for automotive industry or for overall economic development of the whole society.

In the 1950s, Marc Nerlove built up a prediction model based on preceding data of the USA, which was actually time series equations between vehicle demand, amount of end-life vehicle, and vehicle population, and then, he gave a simple and short-term prediction. Due to prevailing of balance concept, he also introduced a balance state of vehicle market to the model, which became a personal error. On all accounts, it initiates the researches on vehicle population prediction. Later researchers usually use regression model to do prediction, and some researchers such as Dargay and Gately use logistic regression model especially. Their model gave good results in accordance with product life cycle theory. However, when researchers start to predict vehicle population, vehicle is already broadly used in Western countries, especially the USA. Thus, the models they use are not of universal versatility in undeveloped countries such as China and India which are at a

M. Wang (✉) · L. Fu · Q. Xie · Y. Du · F. Dong
Dongfeng Motor Corporation Technical Center, Wuhan 430058, China
e-mail: wangmr@dfmc.com.cn

Society of Automotive Engineers of China (SAE-China) (ed.), *Proceedings of the 19th Asia Pacific Automotive Engineering Conference & SAE-China Congress 2017: Selected Papers*, Lecture Notes in Electrical Engineering 486, https://doi.org/10.1007/978-981-10-8506-2_28

phase that vehicle population is in rapid increase. The national conditions and development phase of them are completely different.

Contraposing vehicle population of China, researchers commonly choose different methods when doing short-term and long-term prediction. In short-term prediction, time series method is a fundamental method, and Wang Xiaohong and Chen Daoping both did productive work on that. Thanks to improvement of computer science, neural network is applied to this area. Chen Yong built a neural network model comprised with feature of time series, and his model outputs a good prediction result. In long-term prediction, researchers mainly conduct their analysis based on assumptions of China's GDP growth. Wang Yini found Gompertz model could explain the development trend of vehicle population in various countries to some extent. Assuming GDP growth rate of China varies from 3 to 10% and popularizing rate would reach and remain at 0.52, she gave a mid-term and long-term prediction through Gompertz model. Wei Baoren and Yagita Hiroshi built up a regression model of per capita GDP assuming vehicle growth rate of China is at high speed, intermediate speed, and low speed, and eventually, they calculate that the vehicle population of China is 188 million in 2020.

Predicting vehicle population is no longer a new subject, time series method; neural network method and some other methods are used to do this job. They all have their own limitations. For example, the advantages of time series method are less data required and thus the prediction method is simple, but the shortage is only the vehicle population is taken into account statistically as a time-varying variable, ignoring other economic factors that may affect it, such as people's purchasing power, price of gasoline, road construction. There also exists mutual affect among them. Time series method could not reflect or remove this mutual affect, and the prediction result could only describe the trend qualitatively. Neural network method is an actually very appropriate method in prediction area due to its nature of black box which does not need any empirical formulas and could capture the internal law from abundant data. Its advantage is very evident that it gives a prediction of high precision; moreover, the model could be a multi-input–multi-output (MIMO) model, making it possible to predict on the circumstance of multi-factors and multi-targets. However, the more accurate it could be, the more expenses it needs to undertake; for instance, it takes a long time to train the model, and there is no definite guideline for users to select control variables. In most of the time, selection of control variables depends on users' experience and even scientific intuition, which affects its generalization ability. For those complicated networks, the error function could be a multi-dimensional curved surface meaning that there might exist some local ravines in it and obviously local minimum values could be found. Hence, the error gradients at the local minimum value points would increase no matter which direction it will trace. As a consequence, the training could not step over this value. This is another shortcoming of neural network [1].

Compared with above methods, the advantage of the econometrics model is that economic factors are introduced as the explanatory variables, which improves the prediction accuracy but not at expense of calculation convenience. Therefore, this paper uses econometrics model to predict the vehicle population.

2 Econometrics Model

Econometrics, which formed in the 1930s of last century, is an interdiscipline of economics, statistics, and mathematics. One using it analyzes statistical data based on mathematical and statistical methods and establishes models to explore and verify the economic laws. After more than a half of century's development, the application of econometrics in various fields has become more and more extensively.

2.1 *Steps of Establishing Econometrics Model*

Steps using econometrics to establish the prediction model of vehicle population are indicated as follows:

(1) Design of theoretical model. One selects appropriate variables based on research objective. The research objective in this paper is the vehicle population of Wuhan. The variables are selected in Chap. 3.1.
(2) Collection of sample data. One collects relevant statistics after determining explanatory variables.
(3) Establishment of mathematical model. Multiple linear regression model is deployed, and the expression is [2]:

$$y_i = \beta_0 + \beta_1 \cdot x_{1i} + \beta_2 \cdot x_{2i} + \cdots + \beta_k \cdot x_{ki} + u_i, \quad i = 1, 2, \cdots, n \quad (1)$$

β is regression parameters of model, and u is error and disturbance term.
(4) Estimation of regression parameters. MATLAB is used to calculate the parameters in this paper.
(5) Test of model. The steps are introduced in Chap. 2.2.
(6) Validation and prediction after applying the model.

2.2 *Modeling Check*

After establishing the model, it is necessary to check whether it can be put into use. In this paper, one mainly uses the R^2 test, F test, t test, and multi-colinearity test for hypothesis conditions.

(1) R^2 test is actually a test on fitting goodness. In multiple linear regression model, the coefficient of R^2, between 0 and 1, is a measure of the fitting degree of the model. The closer it approaches to 1, the better the fitting degree is.

(2) F test is a significance test on the overall linearity of the regression equation. It examines whether the explanatory variables have significant effects on the explained variable.
(3) t test is a test on regression parameters. It aims at judging whether each parameter has a significant effect on explained variable.
(4) Multi-colinearity test is to judge whether linear relationship exists among variables, and one should eliminate variables that have strong colinearity so that the independent effect on explained variable by explanatory variables is ensured.

3 Establishment of Model

3.1 Variable Selection

The vehicle population is selected as explained variable y in this paper. According to economic theory and actual situation of economic development in Wuhan, one selects the following indicators as explanatory variables of the model:

(1) Gross domestic product (GDP) $\boldsymbol{x}_1$. GDP is an important indicator to measure the overall economic situation of a country or region.
(2) Above-scale industrial output value $\boldsymbol{x}_2$. Above-scale industrial output value refers to output value of all state-owned industrial enterprises and non-state-owned industrial enterprises with prime operating revenue over 20 million yuan [4]. It is the embodiment of the industrial strength of a country or region and is also the foundation of the national economy. Wuhan is the National Central City of China and the country's major industrial base, especially the automotive industry. Therefore, there is a close relationship between above-scale industrial output value and vehicle population.
(3) Disposable income of urban residents $\boldsymbol{x}_3$. Disposable income of urban residents refers to the income of residents that can be freely controlled. This statistics reflects the level of people's ability to buy a car. In recent years, the vehicle population in Wuhan has been in growth, which is directly related to the rapid increase in disposable income of Wuhan residents.
(4) Total retail sales of consumer goods $\boldsymbol{x}_4$. The total retail sales of consumer goods are the most direct reflection of people's consumption demand. With the improvement of living quality, people's demand for vehicles is also growing. Therefore, it is reasonable to select this statistic as an explanatory variable.
(5) Total investment in fixed assets $\boldsymbol{x}_5$. Through the investment of fixed assets, the industrial structure of a country or region can be optimized continuously, and the economic strength can be improved. Wuhan municipal government is always encouraging and developing the automobile industry, so there is a close relationship between this statistic and vehicle population.

(6) Vehicle population of last year $\boldsymbol{x}_6$. Vehicle population of last year also has an impact on consumer psychology. Before 2012, the average increase in vehicle population of Wuhan is about 80,000. However, after the announcement that 2.3 million vehicles are the "purchase red line" by municipal government of Wuhan [3], vehicle population in 2014, 2015, and 2016 increased rapidly by 320, 400, and 410 thousand because of consumers' concern that whether the municipal government would release vehicle purchase restriction order. This rapid growth of vehicle population is the direct embodiment of consumer psychology of panic. Therefore, this statistic should be considered in the model as a variable to represent the psychology of consumers Table 1.

3.2 Model Test

According to *Wuhan Statistical Yearbook-2015* [4], official Web site of bureau of statistic of Wuhan municipal government [5] and *2015 Annual Report of Wuhan Vehicle Emission Control* [6], one can conduct calculation via MATLAB and get the results in Table 2.

(1) R^2 test. Fitting goodness R^2 and adjusted fitting goodness $\bar{R}^2$ are both close to 1, indicating fitting degree is high.
(2) F test. On the condition of significance level $\alpha = 0.05$, F test value is larger than critical value, suggesting linear relationship of model holds significantly.
(3) t test. On the condition of significance level $\alpha = 0.05$, absolute value of t test values of explanatory variables are not entirely larger than critical value, demonstrating existence of colinearity of model, which needs to be adjusted.
(4) Multi-colinearity test. In order to investigate the degree of multi-collinearity, the correlation coefficient matrix of explanatory variables is obtained, and explanatory variables with strong multi-colinearity are excluded. One establishes single-explanatory-variable models for explained variable and replaces variables with low fitting goodness by variables with high one. Through stepwise regression method, the adjusted model is finally determined. According to the above strategy, one can finally determine the model as below:

$$y = \beta_0 + \beta_3 \cdot x_3 + \beta_5 \cdot x_5 + \beta_6 \cdot x_6 \quad (2)$$

Regression analysis of final model is described in Table 3.

Referring to Table 3, the adjusted model passes R^2 test and F test. t test value shows existence of a certain degree of multi-colinearity of the mode, but one can see that only absolute value of t_5 is smaller than the critical value and is actually close to it. Thus, the multi-colinearity has been reduced substantially after adjustment. The D.W. value of the model is close to 2, suggesting the self-correlation degree is low. Hence, this model is still acceptable.

Table 1 Statistical data of 2003–2015 [4, 5]

Year	Vehicle population y	Gross domestic product x_1 (100 million yuan)	Above-scale industrial output value x_2 (100 million yuan)	Disposable income of urban residents x_3 (Yuan)	Total retail sales of consumer goods x_4 (100 million yuan)	Total investment in fixed assets x_5 (100 million yuan)	Vehicle population of last year x_6
2003	560,649	1622.18	1994.49	8524.56	885.69	645.06	506,010
2004	623,302	1882.24	2402.34	9564.05	996.22	822.20	560,649
2005	652,713	2261.17	2674.41	10849.72	1128.64	1055.18	623,302
2006	703,481	2679.33	3162.06	12359.98	1293.33	1325.29	652,713
2007	759,874	3209.47	4010.30	14357.64	1518.30	1732.79	703,481
2008	818,626	4115.51	6251.79	16712.44	1850.05	2252.05	759,874
2009	907,644	4620.86	6317.94	18385.02	2464.09	3001.10	818,626
2010	1,071,697	5565.93	7004.96	20806.32	2570.40	3753.17	907,644
2011	1,225,208	6762.20	8461.21	23738.09	3031.79	4255.16	1,071,697
2012	1,350,903	8003.82	10195.88	27061.00	3467.37	5031.25	1,225,208
2013	1,544,096	9051.27	11654.64	29821.22	3916.60	6001.96	1,350,903
2014	1,816,498	10069.48	11764.59	33270.00	4369.32	7002.85	1,544,096
2015	2,223,577 [6]	10905.60	12564.58	36436.00	5102.24	7725.26	1,816,498

Table 2 Statistical description of data

Correlation coefficient		t test value	Critical value of t test
β_0	100,350	−1.2209	2.4469
β_1	−7.9290	−0.1410	
β_2	−188.1146	−2.6231	
β_3	−23.8414	−1.3244	
β_4	−204.0553	−2.1092	
β_5	316.5907	4.3599	
β_6	1.3708	15.3998	
Fitting goodness R^2	0.9994	F Test Value	Critical value of F test
Adjusted fitting goodness $\bar{R}^2$	0.9987	1569	4.2839

Table 3 Statistical description of adjusted model

Correlation coefficient	t test value		Critical value of t test
β_0	96,213	0.8573	2.2622
β_3	−38.0097	−2.6606	
β_5	118.1343	2.0268	
β_6	1.4108	9.8386	
Fitting goodness R^2	0.9974	F test value	Critical value of F test
Adjusted fitting goodness $\bar{R}^2$	0.9947	1132.9	3.3738
D.W. value	1.4750		

Adjusted equation:

$$y = 96213 - 38.0097x_3 + 118.1343x_5 + 1.4108x_6 \quad (3)$$

4 Validation and Prediction

Referring to related news and Report on the Work of Government of Wuhan [8], one can validate and predict the vehicle population of Wuhan in 2016 and 2017. Results are filled out in Table 4.

Table 4 Validation and prediction results of vehicle population of Wuhan

Year	Disposable income of urban residents x_3 (Yuan)	Total investment in fixed assets x_5 (100 million Yuan)	Vehicle population of last year x_6	Vehicle population y (10 thousand)
2016	40,590 [7]	5610.6 [8]	2,223,577 [6]	273.86
2017	43,837 [8]	6171.66 [8]	2,637,577 [9]	331.42

On the basis of related news, the vehicle population of Wuhan increases by 414,000 [9] by the end of 2016; thus, one can calculate that the actual total population is 2.6376 million in 2016 [9]. By using the modeling equation, one can calculate that the validation result is 2.7386 million. It is close to the actual population whose error is 3.8%, suggesting that prediction of the model is accurate. Eventually, one can predict the vehicle population of Wuhan in 2017 is over 3.3 million.

5 Conclusion

Due to their availability, 13 years' statistical data are merely collected, and some data can only be obtained through news reports because statistical department has not published accurate data. All of these factors affect the accuracy of the model. Notwithstanding the inaccuracy brought by those factors mentioned above, the validation result indicates the model holds and it can predict vehicle population of Wuhan effectively.

There are some papers which predict vehicle population based on econometrics, but very few selects the population of last year as a variable into analysis as a representation of the consumer psychology; nevertheless, this paper conducts some discussions. Taking into account the economic factors and consumer psychology, the econometric model is established and modified withstanding sufficient test and outputs validated result. Therefore, this method is precise enough for short-term prediction of vehicle population.

Acknowledgements We would like to express our deepest gratitude to Prof. Wang Xuejun and Prof. Fan Ruguo, two respectable, responsible, and amiable professors from Economics and Management School of Wuhan University. Without their enlightening instructions, impressive kindness, and patience, we could not complete this paper. Their keen and vigorous academic observation motivates us not only in this paper but also in our future work.

References

1. Chen Y. The methodology and implementation of car ownerships forecast technique. South China University of Technology. Master Thesis, U491.14 (in Chinese)
2. Li Z, Pan W (2004) Econometric. Higher Education Press, Beijing (in Chinese)
3. Wang L, Zhen G. 40 thousand vehicles bought by Wuhan locals at expense of 60.2 Billion, close to purchase red line [EB/OL]. http://hb.qq.com/a/20160107/014023.htm (in Chinese)
4. Bureau of Statistic of Wuhan Government (2015) Wuhan statistical yearbook-2015. China Statistic Press (in Chinese)
5. Bureau of Statistic of Wuhan Municipal Government (2016) Major Indicators of national economy of Wuhan in 2015. [EB/OL]. http://www.whtj.gov.cn/details.aspx?id=2814, 2016/02/22 (in Chinese)

6. Wuhan Environmental Protection Bureau. 2015 Annual report of Wuhan vehicle emission control [EB/OL]. http://www.whepb.gov.cn/jdcGzdt/20012.jhtml (in Chinese)
7. Guan X, Zhou T. GDP of Wuhan grows by 7.6% in first half year, increase of disposable income of urban residents higher than rest of the country. [EB/OL]. http://hb.people.com.cn/n2/2016/0722/c194063-28711047.html (in Chinese)
8. Wuhan Municipal People's Government. Report on the work of Government of Wuhan [EB/OL]. http://old.whtv.com.cn/zt/2017/lh/bg/201702/t20170215_718923.shtml (in Chinese)
9. Zeng Q. Vehicle population of Wuhan close to 2.7 Million, 410 thousand of increase in 2016 [EB/OL]. http://wuhan.auto.qq.com/a/20170106/005547.htm (in Chinese)

Establishment and Application of Dynamic Characteristics Prediction Model for Variable Valve Timing System

Cheng Liu, Mingxi Liang, Junjun Wang, Cheng Meng, Fei Lin and Pucheng Pei

Non-real-time changing of valve overlap has restricted the improvement of engine mechanical efficiency for a long time. In 1980s, company Alfa Romeo firstly applied variable valve timing technology to their engines. After that, the application of this technology rapidly grew in OEMs worldwide for its significant reduction in fuel consumption and ease of implementation.

Recently, the amending of emission legislation makes the requirements of fundamental functions and dynamic characteristics of VVT system to a higher level and promotes the constantly upgrading of VVT products.

For instance, company Hilite integrated trigger wheel with rotor at the front of traditional vane-type phaser, which, to a certain extent, improved the system integration. Company INA developed central valve products which integrated oil control valve with central bolt, this modification successfully decreasing the length of oil gallery, and obviously improved system responsiveness. Company BorgWarner developed camshaft torque actuated phaser which applied 2 check valves inside to collect assistant torque from camshaft. By using aforementioned assistant torque, CTA phaser is able to operate without oil supply [1]. And in order to make system base position and power source more flexible, Denso and Hitachi developed phaser products which can be locked at multiple position and driven by electric motor [2].

In the field of VVT dynamic characteristics simulation, researchers from Jilin University proposed that cam phaser could be simplified as hydraulic cylinder

---

C. Liu (✉) · M. Liang · J. Wang · C. Meng · F. Lin
Dongfeng Motor Corporation Technical Center, Wuhan 430058, China
e-mail: liuc@dfmc.com.cn

M. Liang
e-mail: liangmingx@dfmc.com.cn

C. Liu · P. Pei
State Key Laboratory of Automotive Safety and Energy,
Tsinghua University, Beijing 100084, China

Society of Automotive Engineers of China (SAE-China) (ed.), *Proceedings of the 19th Asia Pacific Automotive Engineering Conference & SAE-China Congress 2017: Selected Papers*, Lecture Notes in Electrical Engineering 486, https://doi.org/10.1007/978-981-10-8506-2_29

model during simulation in 2011. However, the simplified model neglects the impact of moment of inertia and cannot load cam torque to rotor accurately. Therefore, it is very difficult to achieve accurate result by using such method [3]. In 2013, a vane-type model was used to analyze the causes that affect the system responsiveness by researchers of Chongqing University and Changan Automobile Co., Ltd. However, the simulation model they used is not able to realize closed-loop control. In addition, they did not take the system leakage and unlock process into consideration, which made the model difficult to reflect the actual working process of variable valve timing system [4].

This paper contents the establishment of a more practicable VVT model by software AMESim, the simulation of system dynamic performance and the influence of the design parameters and engine operating conditions are discussed in combination with rig test results. This prediction method can help to predefine VVT parameters at early stage of project. Makes it possible to check system dynamic characteristics before dynamic test and provides new solutions for solving development issues.

1 Establishment of VVT System Model

1.1 System Working Process Analysis

The phase shifting function of VVT system is implemented via different torques which acted on cam phaser rotor, at the time VVT remains stable. The relationship of different torques is [5–7] (Fig. 1, Table 1):

$$I\ddot{\theta} + b\dot{\theta} = M_{\mathrm{Op}_1} - M_{\mathrm{Op}_2} + M_{\mathrm{Sp1}} - M_{\mathrm{Sp2}} - M_{\mathrm{Cam}} \quad (1)$$

where

$$I = \frac{1}{2} * m_1 * (r_1^2 + r_2^2) + \frac{1}{4} * \frac{1}{2} * m_2 * (r_1^2 + R^2) \quad (2)$$

$$M_{\mathrm{Op1}} = i * P_a * l * (R - r_1) * \frac{R + r_1}{2} \quad (3)$$

$$M_{\mathrm{Op2}} = i * P_r * l * (R - r_1) * \frac{R + r_1}{2} \quad (4)$$

$$M_{\mathrm{Sp1}} = M_{\mathrm{pre1}} + k_1 * \theta \quad (5)$$

$$M_{\mathrm{Sp2}} = M_{\mathrm{pre2}} + k_2 * \theta \quad (6)$$

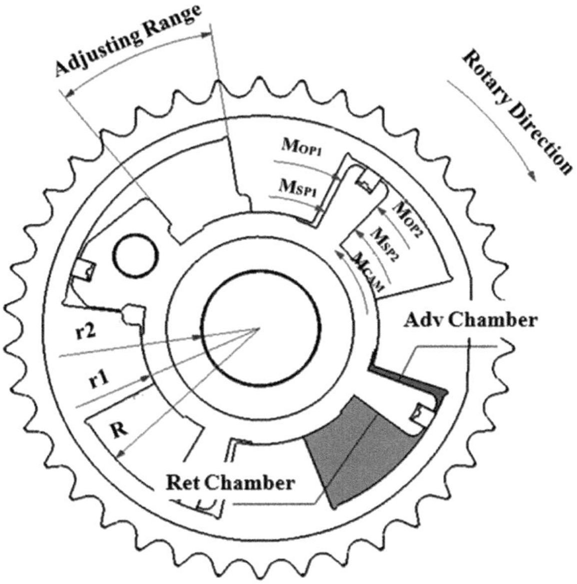

Fig. 1 Torques on cam phaser

Table 1 Description of parameters

Parameters	Description
I	Inertia of rotor and camshaft
b	Viscous damping coefficient
M_{Op1}	Hydraulic torque at advance chamber
M_{Op2}	Hydraulic torque at retard chamber
M_{Sp1}	Assistant spring torque
M_{Sp2}	Resistant spring torque
M_{Cam}	Camshaft torque
θ	Angular movement of rotor
φ	Angular position of camshaft
α	Angle between fuel pump cam lobe and 1st cam lobe

$$M_{\text{Cam}} = 2M_c(\varphi+\theta) + 2M_C(\varphi+\theta+120) + 2M_c(\varphi+\theta+240) + M_{\text{Hpc}}(\varphi+\theta+\alpha) \tag{7}$$

In addition to evaluate the torques relationship, another critical characteristic of VVT system is flow balance, which leads to equations:

$$Q_a = \frac{\pi}{360} \times \left(R^2 - r_1^2\right) \times L \times \Delta\theta + Q_{Li} + Q_{Lo} \tag{8}$$

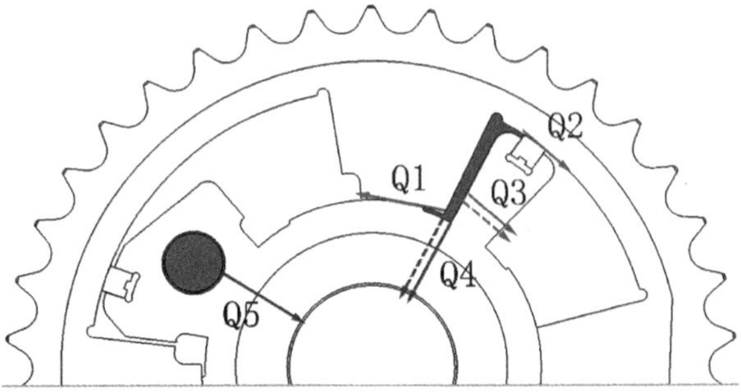

Fig. 2 Internal and external leakages of cam phaser

$$Q_r = \frac{\pi}{360} \times \left(R^2 - r_1^2\right) \times L \times \Delta\theta + Q_{Li} + Q_{Lo} \tag{9}$$

In these equations, Q_{Li} is the internal leakage caused by difference of oil pressure on both sides of vanes, leaked oil flow from working chambers to non-working chambers through axial and radial clearances internally. Q_{Lo} represents system external leakage rate which shows the allowance of the oil drain out of cam phaser through axial clearance and locking pin groove during engine operation (Fig. 2).

As a result of leak point analysis, here [8]:

$$Q_{Li} = iQ_1 + iQ_2 + 2iQ_3 = \left(\frac{\mathrm{L}h_1^3}{2B_1} + \frac{\mathrm{L}h_2^3}{2B_2} + \frac{(R - r_1)h_3^3}{B_3}\right) \times \frac{i\Delta p}{6\mu} \tag{10}$$

$$Q_{Lo} = 2iQ_4 + Q_5 = \left(\frac{i\beta h_3}{360}\ln\left(\frac{R}{r_1}\right) + \frac{dh_5^3}{4B_5}\right) \times \frac{\Delta p\pi}{3\mu} \tag{11}$$

1.2 System Simulation Model

As a typical electrohydraulic servo system, VVT and its related systems can be modeled and analyzed by mature commercial software AMESim [9].

Base on its working principles, a VVT system can be abstracted into five interrelated subsystems: phaser, oil control valve, oil supply unit, valve train unit, and control unit. Oil paths of OCV are connected with chambers in phaser, p port of OCV is connected with oil supply unit, and solenoid receives the PWM signal from control unit which is used to real-time change the oil paths and system flow rate. Rotor, stator, and different oil chambers are included in model of phaser. Advance and retard chambers are connected by each other's internal leakage model, and all chambers have external leakages to the environment. In addition, there are two angular position sensors attached at rotor and stator. The angular deviations of actual and theoretical position are sent to control unit instantaneously to achieve function of closed-loop control.

2 Dynamic Simulation and Test

2.1 Simulation Inputs

During the operation of engine, a bidirectional torque acts on rotor because of the periodic open and close of valve. Commonly, when a valve opens with camshaft's rotation, the valve spring resists the force and impedes the movement of the shaft. Similarly, when a valve closes under valve spring force, it also pushes on the cam lobe in opposite direction to the valve opening and promotes the rotation of the shaft.

When hydraulic torque is greater than the resistance one, rotor will move toward target position and the volume of pressure chamber increases. Otherwise, rotor will stop moving toward target position. If there is a check valve at p port of OCV, phase will keep the position at the end of last movement due to the incompressibility of oil. And, if there is no check valve at p port of OCV, the resistance torque will push the rotor move backward and bring about influences on system response speed and phase precision.

Therefore, it can be seen that the phase shifting of VVT system is a discontinuous process. The input of camshaft torque and oil pressure will directly affect the accuracy of calculation. Camshaft torque data can be obtained by running valve train sub-model in Fig. 3 and will be used in following calculations.

The red, green, and blue curves respectively represent camshaft torques at 1000, 3000, and 5000 rpm. Figure 4 indicates that camshaft torque is dominated by spring force at low engine speed. And it shows good accordance to valve

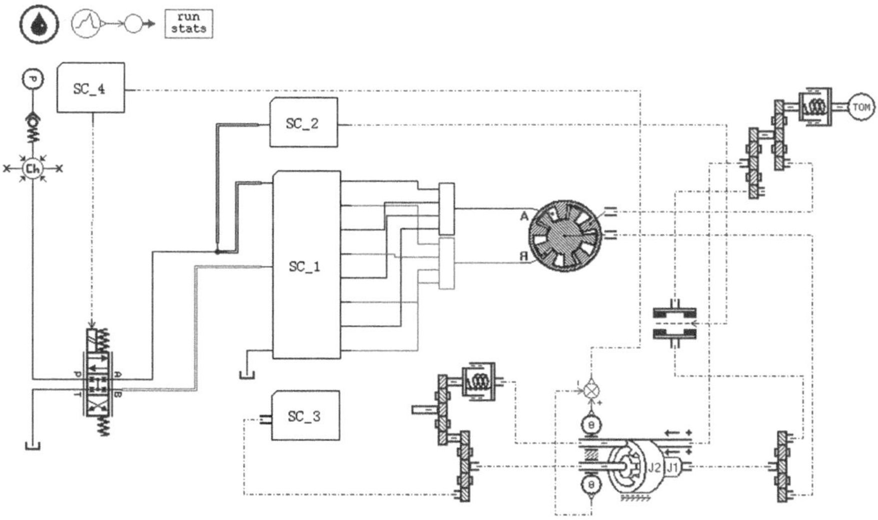

Fig. 3 AMESim model of VVT system

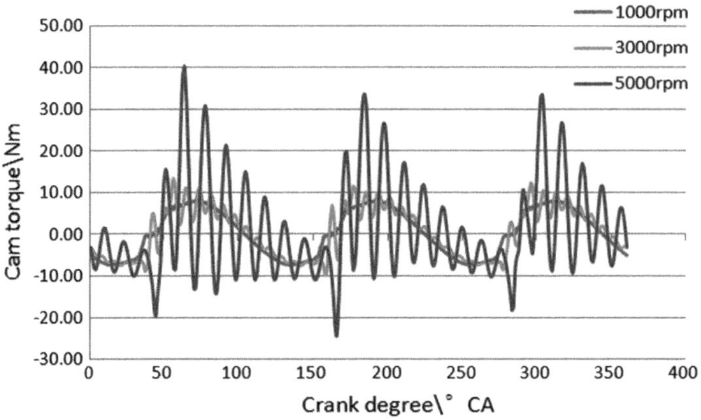

Fig. 4 Camshaft torques input

movement. With the increase of the inertial force of valve train components, peak value of the camshaft torque increases and an optimal dynamic fluctuation occurs at high-speed conditions. However, the torque still well consists with valve movement.

Figure 5 shows the oil pressure map obtained from 1D lubrication simulation under 10–140 °C oil temperature and 750–5500 rpm engine speed. These data will attach to oil supply sub-model as simulation inputs. It can be seen that the oil pressures are at relatively low level under low speed and high-temperature conditions due to the limit of oil pump output capability and low oil viscosity.

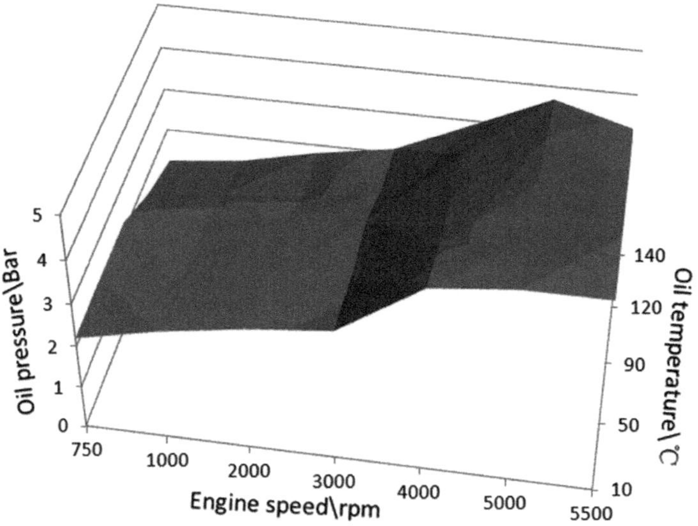

Fig. 5 Oil pressures input

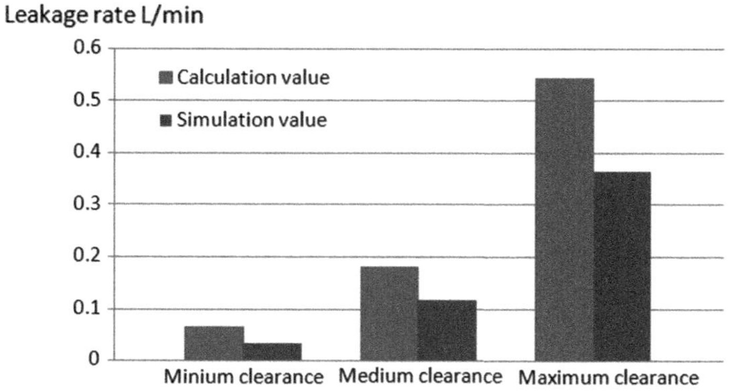

Fig. 6 Total leakages under different leakage clearance value

In order to further ensure the theoretical accuracy of prediction model, the leakage module of cam phaser should be checked before calculation. The results show that the maximum deviation between simulation results and calculation results is 0.18 L/min. Calculation value of total leakage is larger than simulation results under 90 °C, 2.3 bar condition regardless of the values of leakage clearance (Fig. 6).

Main reason of the deviation here is that the internal leakage at sealing blades has been ignored in simulation model. Therefore, the leakage module has been modified base on experience to better reflect the real working situation of VVT system.

2.2 *Simulation Results of Response Speed*

Response speed is one of the most critical parameters for VVT system. Practically, target phase of engine will change with different road conditions as well as driver's operation. If VVT system cannot follow the change of PWM signal in time, the engine will be impossible to get ideal overlap. And finally, it affects the precision of air flow, which against the original design intention for VVT to provide engine suitable timing phase under different conditions.

For the sake of avoiding impact of overshoot at the beginning of adjusting process, the response speed is usually defined as the ratio of the phase angle difference and the time of adjustment, at the range of 10–90% angular position, that is:

$$v = \frac{2(\theta_{90\%} - \theta_{10\%})}{t_{90\%} - t_{10\%}} \tag{12}$$

Using the aforementioned model, the calculation results of system response speed at low, medium, and high oil temperatures (10, 90, 140 °C) are shown in the following figure.

As it is shown in Fig. 7, system response speed increases with engine speed and trends to vary gently after 5000 rpm. It is because that the hydraulic torque becomes larger as the speed increases, which enhance the angular speed of vanes and raise the percentage of movement toward target direction. At high-speed conditions, output pressures of oil pump gradually reach the rated pressure and no longer rise. At this time, hydraulic torque absolutely exceeds the peak of maximum negative cam torque, which makes the system lose the profits of percentage of movement toward target direction. Meanwhile, high oil pressure brings larger internal and external leakages to cam phaser. That is why the system response speed increases gently even drops a bit at high engine speed.

In these three sets of curves, response speeds are limited by fluidity of oil at low oil temperature and affected by output pressure at high oil temperature. Conditions around 90 °C are ideal for fast adjusting.

2.2.1 Influences of Coil Spring

The coil springs installed at the front of cam phasers are classified into assistant spring and resistant spring according to the different directions of its output torque. A basic function of resistant spring is to force the rotor back to original position when engine stops and oil supply cuts off, which guarantees a constant relative angular relationship between camshaft and crankshaft every time engine starts (Fig. 8).

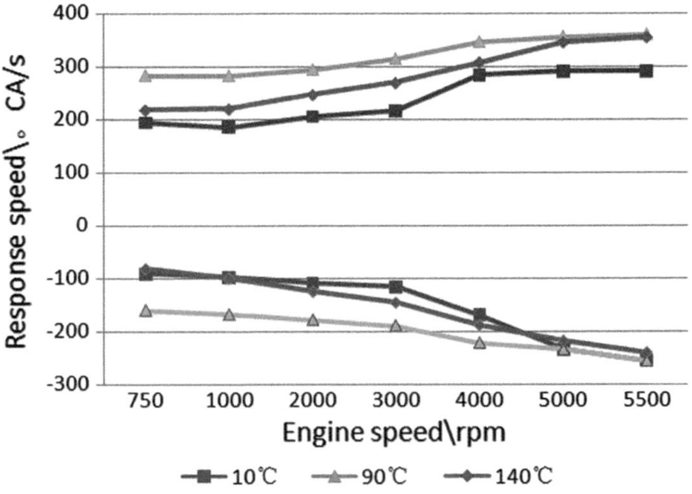

Fig. 7 Response speeds under different oil temperatures

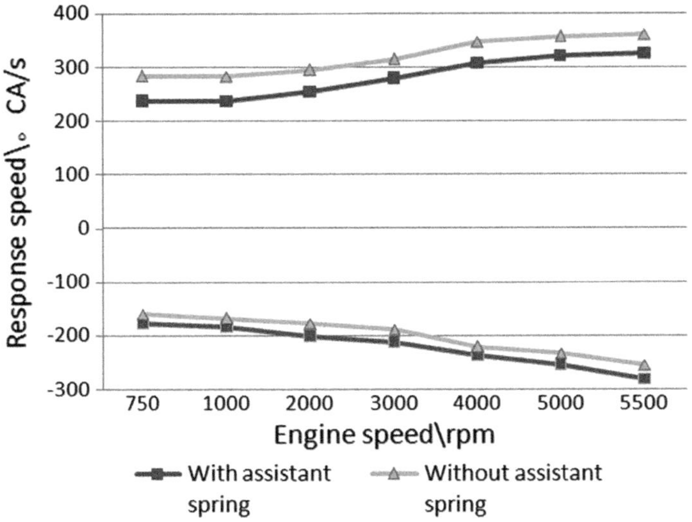

Fig. 8 Response speeds with and without assistant spring

Table 2 Settings of coil springs

Settings	Without spring	With spring
Stiffness	0 Nm/°	0.018 Nm/°
Preload	0 Nm	2.2 Nm
Damping coefficient	0.0001 Nm/rpm	

The difference between two groups of simulations is the settings of preload and stiffness of coil springs. Refer to Table 2.

Comparison shows that system response speed is improved by 14.8% toward advance direction and drops by 9.3% toward retard direction. Curve of adjusting speeds move upwards after the application of assistant spring.

2.2.2 Influences of Pressure Ratio

Pressure ratio is a parameter representing the pressure output capability of vane-type phaser under per unit length. The pressure ratio is determined once the number of oil chambers, the inner and outer diameters of rotor, and the thickness of rotor are determined. In order to balance the pressure output capability and arrangement spaces at axial and radial direction, the pressure ratio of 4-vanes VVT is usually chosen from 3.0 to 3.5.

The geometric parameters of rotor are adjusted during calculation to evaluate the influence brought by different pressure ratios (Table 3).

Table 3 Settings of rotor parameters

	Low PR	Medium PR	High PR
Number of chambers	4		
Inner diameter (mm)	23		
Outer diameter (mm)	38		
Thickness (mm)	15.4	16.5	17.6
Pressure ratio	2.82	3.02	3.22

The results are that:

As it is shown in Fig. 9, system response speeds are improved by 16.1 percent toward advance direction and 13.0% toward retard direction. If there is no package issue regarding system layout, a design with higher pressure ratio can be applied to improve system responsiveness.

2.2.3 Influences of Oil Leakage

Oil leakage affects the pressure building-up process of hydraulic systems. Due to wear and tear, the leakage rates of some aluminum alloy phasers will significantly increase comparing with its initial state. In addition, when there is a big distinction between material thermal expansion coefficient of rotor and stator, a potential risk of leakage will be expected especially at extreme temperature.

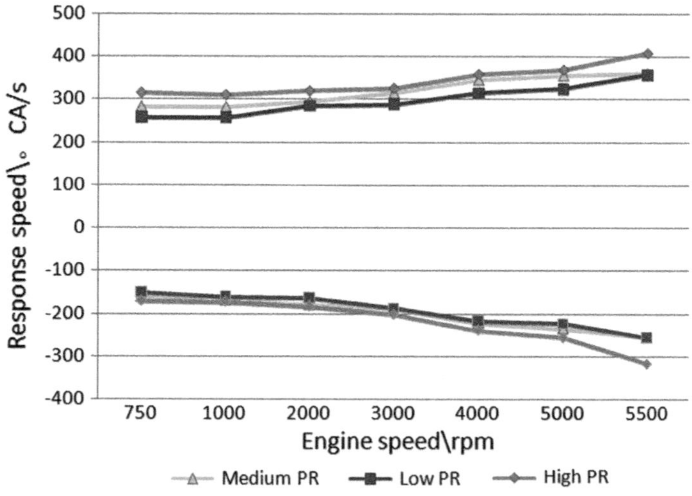

Fig. 9 Response speeds under different pressure ratio

Once the leakage clearances exceed particular thresholds, the function of sealing will be invalid. Pressure inside oil chambers will be not established easily. Significant fluctuations of timing phase will occur and cam phaser will be difficult to achieve the target position in time.

As what we can see in Eqs. 10 and 11, leakage clearance has a decisive effect on leakage results. Extremely large clearance will lead to excessive leakage, which affect the basic functions of VVT system. Extremely small clearance will make troubles to assembly and manufacturing process, which affect the cost of product finally.

The values of clearances are adjusted during calculation to evaluate the influence brought by different total leakages (Table 4).

The results are that:

As it is shown in Fig. 10, the influence of different oil leakages on system responsiveness can be neglected within a particular range of clearance. Common wear during engine service life will not cause too much impact on its behavior.

Table 4 Settings of leakage clearances

	Low leakage	Medium leakage	High leakage
Radial clearance (mm)	0.016	0.051	0.086
End clearance (mm)	0.030	0.040	0.050
Pin diameter (mm)	8		
Pin &hole clearance (mm)	0.05		
Leakage (%)	22	100	339

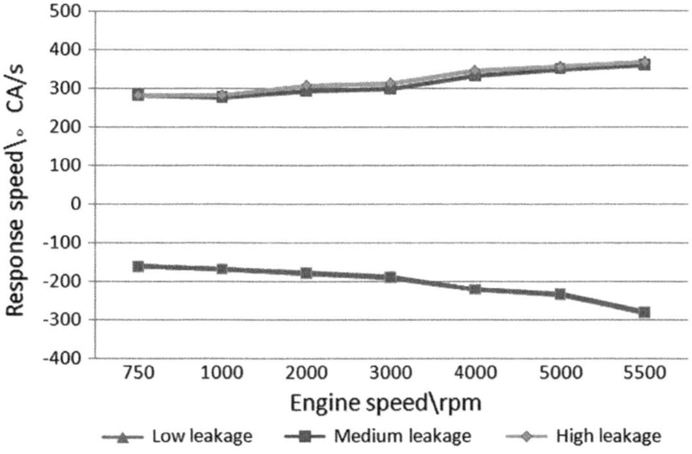

Fig. 10 Response speeds under different system leakage

2.3 *Verification of Simulation Results*

To verify the consistency of simulation results and actual performance, a mock-up test rig is used to measure the response speed of the central valve VVT system (Table 5).

As it is shown in Fig. 11, the prototype includes cylinder head, valve train, and timing system. Encoders are installed at camshafts and crankshaft to reflect angular position real-timely. During the test, the crankshaft is driven by a motor at constant speed. Oil is heated outside and enters cylinder head through main gallery. Once the test condition is stable, this system will follow the preset program to adjust the phaser moving toward advance direction and retard direction. During mock-up test, angular positions, oil temperatures, oil pressures, and other data are automatically recorded for subsequent analysis.

5w-30 lubricating oil is used in this test. To verify the worst case conditions, two cam phasers with excessive leakage are used in this prototype. Here we compare test results at 10 and 140 °C with simulation results which have shown in 2.2 this paper (Figs. 12 and 13).

The simulated and measured results are in good agreement with each other and show that this model is effective to predict the values and trends of VVT response speed.

Table 5 Major devices of test rig

Device name	Parameters
Angular encoder	Accuracy < 0.15 °CA
Motor	Range of Rotary Speed: 0–8000 rpm
External oil pump	Range of Pressure: 1.5–6 bar

Fig. 11 VVT system dynamic test bench

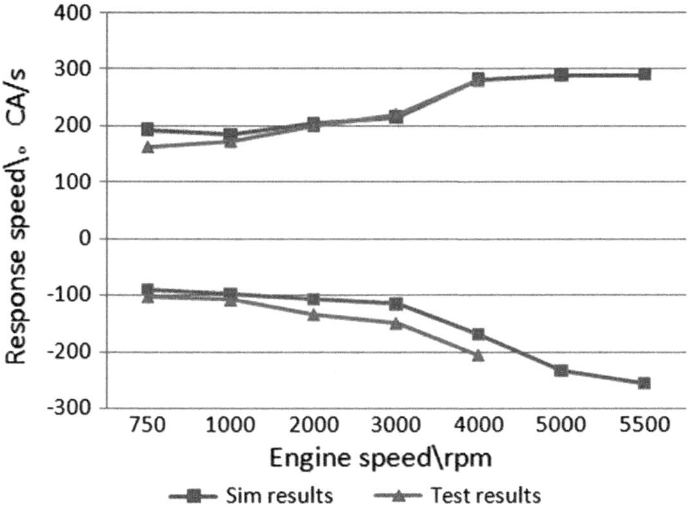

Fig. 12 Data comparison at 10 °C

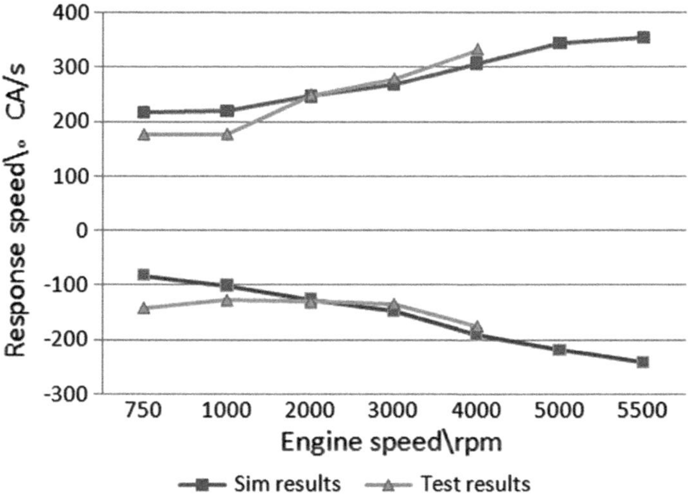

Fig. 13 Data comparison at 140 °C

2.4 Simulation Results of Phase Precision

Phase precision, the stability of phase, is defined as the difference between actual and theoretical timing phase when system is completely stable. According to experience, accuracy of phase is mainly affected by the performance of hydraulic oil (oil temperature, air content) and OCV characteristics (PWM frequency, range of holding point current).

2.4.1 Influences of Oil Temperature

Viscosity is the basic properties of oil. Different temperatures make great differences in oil viscosity. Based on the simulation results at 3000 rpm, phase precision is much higher under low temperature (±0.29 °CA) than under medium temperature(±0.61 °CA) and high temperature (±0.72 °CA). The accuracy is getting worse along with the oil temperature rise and oil pressure decrease. When the accuracy is over than ±1.5 °CA, some adjustments for system parameters are necessary to avoid negative impacts on engine performance and durability due to the excessive fluctuations of system timing phase (Fig. 14).

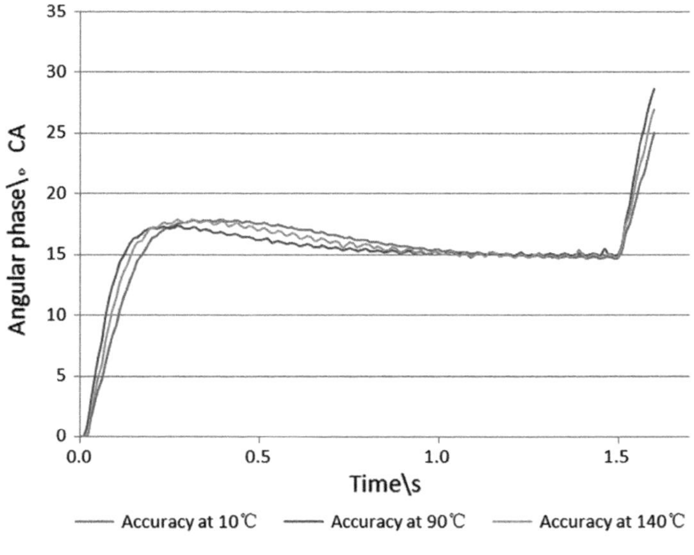

Fig. 14 Phase precision under different oil temperature

2.4.2 Influences of Air Content

Increasing air content significantly lowers the bulk modulus of oil, which makes the incompressibility of the oil cannot be neglected at several extreme conditions. In addition, the air will result in pressure fluctuations and affect the reliability of hydraulic system. As it is shown in Fig. 15, the phase precisions are steadily deteriorated from ±0.29 °CA to ±1.07 °CA when the air content increases from 5% to 15%. Due to its great influences on stability of timing phase, air content is necessary to be limited during practical use.

2.5 Verification of Simulation Results

Test device described in 2.3 is able to measure timing phase accuracy as well. During the test, the engine keeps stable at operating conditions and target phase is set manually. Value of phase deviations will be obtained in 4–5 s period after the cam phaser reaches target position (Fig. 16, Table 6).

The simulated and measured results are in good agreement with each other and show this model is effective to predict values and trends of VVT phase precisions.

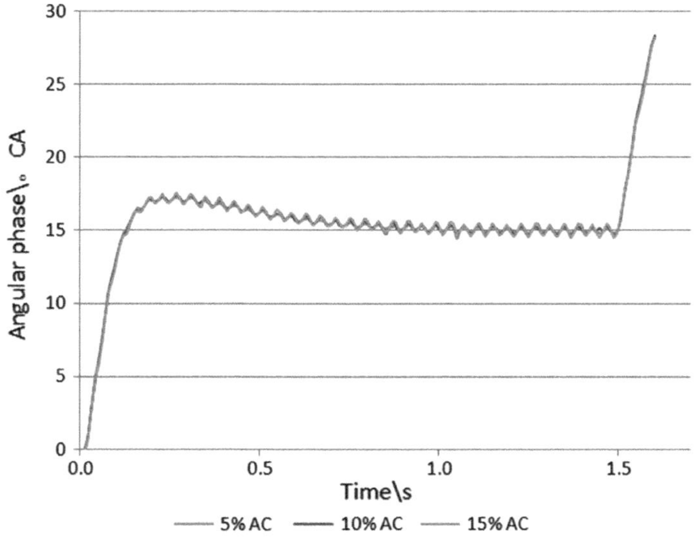

Fig. 15 Phase precision under different air contents

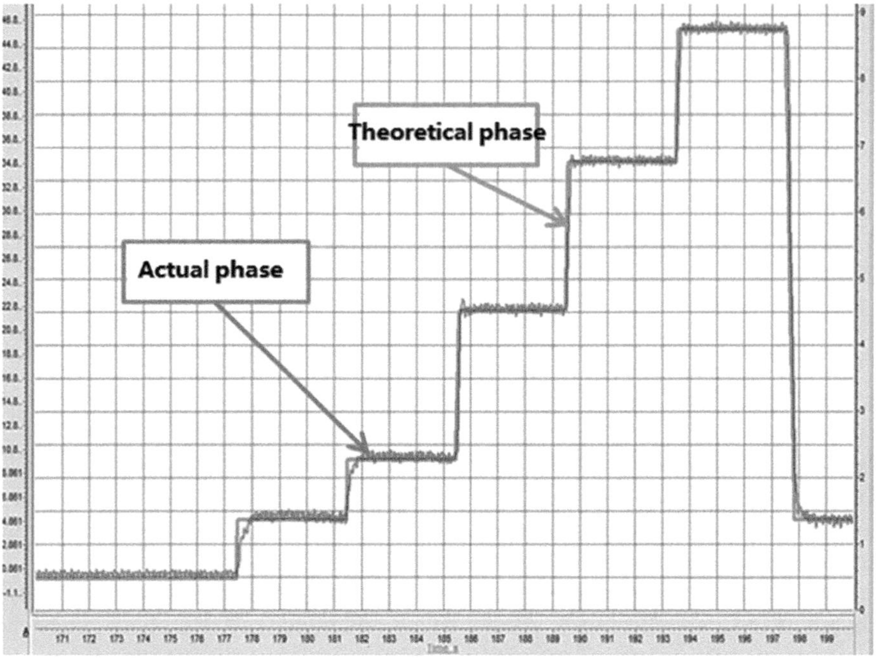

Fig. 16 Measurement results of timing phase accuracy

Table 6 Data comparison

	10 °C	90 °C	140 °C
Speed	3000 rpm		
Simulation results	±0.29	±0.61	±0.72
Test results	±0.40	±0.63	±0.64

3 Conclusions

In this paper, a model has been established to predict system dynamic characteristics of a 3-cylinder TGDI engine. Combined with the results of mock-up rig test, the impacts of different engine operating conditions and design parameters on VVT dynamic characteristics are investigated systematically. The conclusions are drawn as follows:

(1) Spring characteristics have significant impacts on VVT response speed. Curves of adjusting speeds move upwards after the application of assistant spring. The responsiveness improvement in advance direction is greater than the decline in retard direction.

(2) System responsiveness will be improved along with the increase of pressure ratio.

(3) The influence of different oil leakage on system responsiveness can be neglected within a particular range of leakage clearance.
(4) Phase precisions are obviously deteriorated with the increase of oil temperature and air content, but all simulated and tested results are still within reasonable range of ±1.5 °CA.

It takes little imagination to foresee, with the trend of widespread applications of VVT, and the higher level requirements of its dynamic characteristics, prediction method described in this paper will become a necessary tool to bring more conveniences on engine developing and system issue analyzing.

References

1. Smith F (2005) A camshaft torque actuated vane style VCT phaser. SAE Paper, 2005-01-0764
2. Hatorri M (2008) Development of variable valve timing system control by electric motor. SAE Paper, 2008-01-1358
3. Dong R (2011) Co-simulation study on continuous variable cam phasing system. Jilin University, Chang Chun
4. Zhang L (2013) Testing and numerical analysis on the dynamic response characteristics of vane-type induction CVCP. Automotive Engineering, vol 35, no 10
5. Astarita M (2004) Virtual testing and design of experiments for functional validation of cam phaser control. SAE Paper, 2004-01-0785
6. Zheng L (2004) Characterization of engine variable cam phaser fluid dynamics and phaser's ability to reject system disturbances. SAE Paper, 2004-01-1389
7. Zhang Y (2016) Calculating of the camshaft load torque and research on virtual check valve control strategy. I.C.E & Powerplant, vol 33, no 2, Apr 2016
8. Hou G (2006) Engineering fluid mechanics. China Machine Press
9. Liang Q (2016) Advanced tutorial on hydraulic system simulation by software AMESim. China Machine Press

Steering Mode Switch Control of Four-Wheel-Independent-Steering Electric Vehicle

Xinbo Chen, Fengmei Luo, Peng Hang and Jie Luo

1 Introduction

Compared with traditional front-wheel steering (FWS) vehicle, four-wheel-independent-steering (4WIS) electric vehicle (EV) has the advantages of good mobility and maneuverability. 4WIS EV can make all wheels move around the same instantaneous steering center in the procedure of turning, to realize zero sideslip angle of vehicle to improve handling stability [1]. Since each wheel of the vehicle can be individually controlled, 4WIS EV can realize FWS, rear-wheel steering (RWS), four-wheel steering (4WS), oblique driving, pivot steering, and so on. Switch from FWS or RWS to 4WS can enhance stability of vehicle at high speed and improve maneuverability by reducing the turning radius to adapt narrow space at low speed.

Many scholars have done a lot of researches on steer-by-wire (SbW) 4WIS EV in recent years. Min et al. [2] have manufactured FABOT concept car which can go leftward or rightward without turning and can turn in standing state, and examined the effect of the converted presence of Ackerman geometry in steering performance. Lam et al. [3] have solved instantaneous turning center in real time using Ackermann geometry and calculated each wheel angle based on the least square method. Lai et al. [4] have obtained the dynamical relationship among wheels using Ackerman geometry and steady state of sideslip angle and yaw rate. Ma et al. [5] have proposed an algorithm of steering angle assignment based on sliding mode

X. Chen
Key Laboratory of Ministry of Education in the Field of Road and Transport Engineering, Tongji University, Shanghai, China

X. Chen · F. Luo (✉) · P. Hang · J. Luo
School of Automotive Studies, Tongji University, Shanghai, China
e-mail: 1763267858@qq.com

X. Chen
Clean Energy Automotive Engineering Center, Tongji University, Shanghai, China

Society of Automotive Engineers of China (SAE-China) (ed.), *Proceedings of the 19th Asia Pacific Automotive Engineering Conference & SAE-China Congress 2017: Selected Papers*, Lecture Notes in Electrical Engineering 486, https://doi.org/10.1007/978-981-10-8506-2_30

control. Liang [6] has analyzed the characteristics of six movement modes and several switches. She has got varying curves during the switch process and determined the control strategy based on MATLAB, adopted by simulation experiment and real vehicle test. But the curves are not smooth in the process.

In this paper, the dynamical models of the 4WIS EV and SbW system are established and two typical switches between different steering mode conditions are selected to achieve smooth switch without parking, which is verified by real vehicle test.

2 Structure of 4WIS EV

In order to explore structure design of 4WIS EV and verify the control strategy of switch between different motion modes, a novel SbW 4WIS prototype EV is manufactured. As shown in Fig. 1, the prototype is composed of four same and independent drive/steer-by-wire systems [7, 8]. As shown in Fig. 2, it is an integrated design of steering system, drive and brake system, and suspension system, which consists of tire, hub motor, brake disk, suspension supporting platform, worm and gear reducer, steering motor, shock absorber, and other parts.

1 tire; 2 hub motor; 3 brake disk; 4 suspension supporting platform; 5 worm and gear reducer; 6 wheel steering angle sensor; 7 steering motor; 8 upper sliding column; 9 undercarriage upper swing arm; 10 shock absorber; 11 spring; 12 undercarriage lower swing arm; 13 brake caliper; 14 lower sliding column.

Fig. 1 4WIS prototype EV

Fig. 2 SbW system

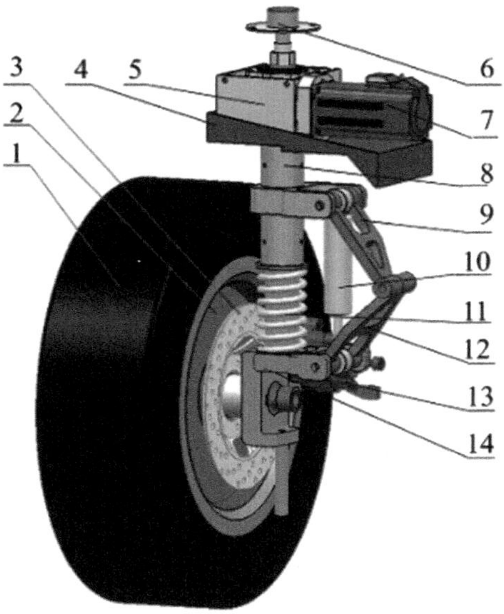

3 Modeling

3.1 Single Trace Model

4WIS EV can realize different steering modes, i.e., FWS, RWS, 4WS, pivot steering, oblique driving. Switch from one mode to another must start after the former has been in steady state. During the switch process, the EV should automatically adjust the wheel angle and remain stable, and the switch time should not be too long. Stability is that vehicle can resist external disturbances to maintain stable [9, 10] and drive following the given direction, which is controlled by driver through the steering system and wheels. Vehicle stability control is divided into two categories [11]: One is path holding, described by sideslip angle, namely, the whole vehicle is seen as a particle, and path holding is mainly decided by integrated force on tires; the other is stability, described by yaw velocity, and it is decided by yaw moment on the force on the tire. These two sides are interrelated.

Firstly, the dynamical model of 4WIS EV should be established to solve the switch problem. The model is simplified as a single trace model with 2 degrees of freedom (DOF). As shown in Fig. 3, only lateral and yaw motions are taken into consideration to minimize the complexity of modeling with the assumption of constant longitudinal velocity [12].

The vehicle dynamical equations can be derived according to Newton's second law.

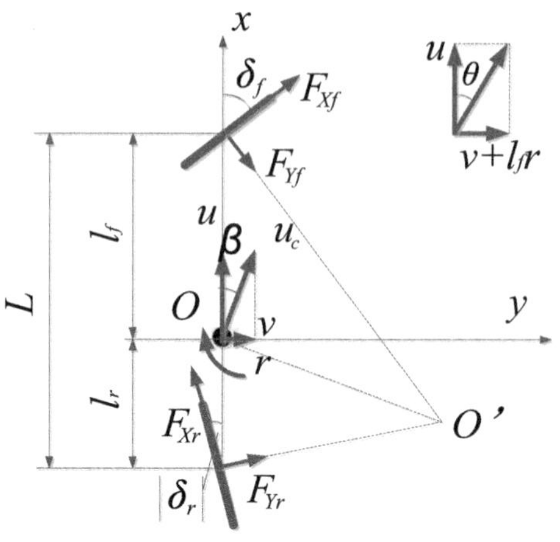

Fig. 3 A linear single trace model with 2 DOF

$$\begin{cases} \sum F_y = m(\dot{v} + ur) \\ \sum M_z = I_z \dot{r} \end{cases} \tag{1}$$

The lateral forces of the front and rear axial can be expressed as:

$$\begin{cases} F_{Yf} = -C_{\alpha f}\alpha_f = -C_{\alpha f}\left(\arctan\frac{v + l_f r}{u} - \delta_f\right) \\ F_{Yr} = -C_{\alpha r}\alpha_r = -C_{\alpha r}\left(\arctan\frac{v - l_r r}{u} - \delta_r\right) \end{cases} \tag{2}$$

In Eq. (2), α_f is front wheel sideslip angle (rad); α_r rear wheel sideslip angle (rad); δ_f front wheel angle (rad); δ_r rear wheel angle (rad); $C_{\alpha f}$ front tire cornering stiffness (N/rad, two tires, positive); $C_{\alpha r}$ rear tire cornering stiffness (N/rad, two tires, positive); u longitudinal velocity of vehicle (m/s); v lateral velocity of vehicle (m/s); l_f distance from center of gravity (CG) to front axle (m); l_r distance from CG to rear axle (m); r yaw rate of vehicle (rad/s); F_{Yf} lateral force of front axial (N); F_{Yr} lateral force of rear axial (N); and u_c resultant velocity of vehicle.

Substituting Eq. (2) into Eq. (1), it follows that:

$$\begin{cases} \dot{v} = -\frac{C_{\alpha f}}{m}\left(\arctan\frac{v + l_f r}{u} - \delta_f\right)\cos\delta_f - \frac{C_{\alpha r}}{m}\left(\arctan\frac{v - l_r r}{u} - \delta_r\right)\cos\delta_r - ur \\ \dot{r} = -\frac{l_f C_{\alpha f}}{I_z}\left(\arctan\frac{v + l_f r}{u} - \delta_f\right)\cos\delta_f + \frac{l_r C_{\alpha r}}{I_z}\left(\arctan\frac{v - l_r r}{u} - \delta_r\right)\cos\delta_r \end{cases} \tag{3}$$

Further, the sideslip angle of vehicle can be written as:

$$\beta = \arctan\frac{v}{u} \tag{4}$$

If the steering radius is defined as the distance between CG and instantaneous center of the vehicle, the steering radius is as follows.

$$Radius = \frac{l_f \cos\left(\arctan\frac{v + l_f r}{u}\right)}{\sin\left(\arctan\frac{v + l_f r}{u} - \beta\right)} = \frac{l_r \cos\left(\arctan\frac{v - l_r r}{u}\right)}{\sin\left(\arctan\frac{v - l_r r}{u} - \delta_r\right)} = \frac{\sqrt{u^2 + v^2}}{r} \tag{5}$$

Here is a clear definition of the switch process: Start time is when the former mode comes to a steady state, and the initial state is a set of all the motion parameters at the moment; terminal time is when input angles of each wheel maintain unchanged, and terminal state is a set of all the motion parameters at the moment.

From Eqs. (3) to (5), it can be seen that there are only two independent variables among yaw rate, sideslip angle, and steering radius. As mentioned before, stability control of vehicle can be divided into two categories: the sideslip angle and yaw rate control. Consequently, the input angles of front wheels and rear wheels of vehicle can be solved through Eqs. (3)–(5),

$$[r, \beta, Radius] = f\left(\delta_f, \delta_r\right), \text{s.t.}\begin{cases} r(t) \\ \beta(t) \end{cases} \tag{6}$$

When sideslip angle is very large, the driver cannot make vehicle produce yaw moment through steering wheel, which is why vehicle is hard to control with large sideslip angle [13]. Therefore during the switch of different modes, the absolute value of sideslip angle should be controlled to come to zero gradually, and if not, it needs to be within a certain range. At the same time, path holding requires yaw rate should maintain unchanged. And the paper adopts that sideslip angle comes to zero when the steering state is stable.

3.2 *Model of SbW System*

For controller design, the simplified model of SbW system is shown in Fig. 4.

1 motor rotor; 2 worm gear reducer; 3 kingpin and knuckle; 4 tire and rim; 5 brake system.

Dynamical equation of the motor rotor and worm is described by Eq. (7).

$$J_m\ddot{\delta}_m + B_m\dot{\delta}_m + T_{w2m} = T_m \tag{7}$$

where J_m is the moment of inertia of motor rotor (kg m^2); δ_m the rotational angle of motor rotor (rad); B_m the vicious friction of motor (N m s); T_{w2m} the torque from the wheel assembly to the motor (N m); and T_m the torque generated by the steering motor (N m);

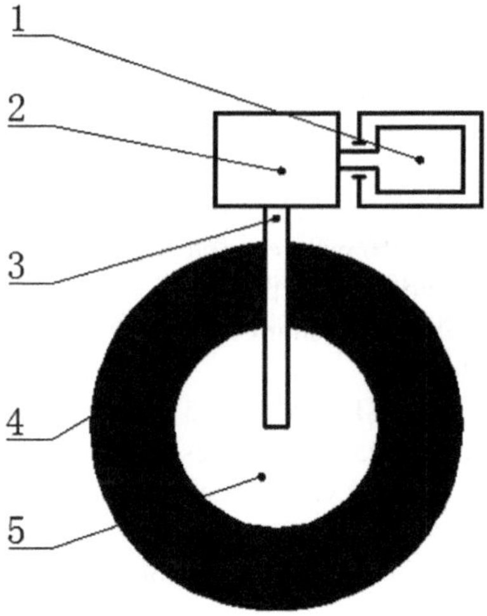

Fig. 4 Simplified structure of SbW system

$$T_m = T_{\mathrm{m}}^{*} + \Delta T_{\mathrm{pert}} \tag{8}$$

where T_m is the control torque (N m) and ΔT_{pert} is the motor perturbation (N m).

Dynamical equation of the turning wheel is described as:

$$J_{\mathrm{w}}\ddot{\delta}_w + B_w\dot{\delta}_w + T_F + T_{\mathrm{e}} = T_{m2w} \tag{9}$$

where J_w is the moment of inertia of the wheel assembly (kg m^2); δ_w the steering angle of the wheel assembly (rad); B_w the vicious friction of the wheel assembly (N m s); T_F the Coulomb friction of the system (N); T_e the self-aligning torque (N); and T_{m2w} the torque from the motor to the wheel assembly (N m).

As k is the gear ratio of worm gear reducer, the following relationship holds:

$$\frac{\ddot{\delta}_{\mathrm{m}}}{\ddot{\delta}_w} = \frac{\dot{\delta}_{\mathrm{m}}}{\dot{\delta}_w} = \frac{\delta_{\mathrm{m}}}{\delta_w} = \frac{T_{m2w}}{T_{w2m}} = k \tag{10}$$

According to Eqs. (7)–(10), the dynamical equation of whole SbW system can be defined as:

$$J_{\mathrm{w}}\ddot{\delta}_w + B_w\dot{\delta}_w + T_F + T_{\mathrm{e}} = k(T_m - J_m\ddot{\delta}_m - B_m\dot{\delta}_m) \tag{11}$$

Rewritten as:

$$J_{\mathrm{eq}}\ddot{\delta}_w + B_{\mathrm{eq}}\dot{\delta}_w + T_l = T_{\mathrm{eq}} \tag{12}$$

$$J_{\mathrm{eq}} = J_{\mathrm{w}} + k^2 J_{\mathrm{m}}, B_{\mathrm{eq}} = B_{\mathrm{w}} + k^2 B_{\mathrm{m}}, T_l = T_e + T_F + k\Delta T_{\mathrm{pert}}, T_{\mathrm{eq}} = kT_m^{*} \tag{13}$$

where J_{eq} is the moment of inertia of the equivalent system (kg m^2); B_{eq} the vicious friction of the equivalent system (N m s); T_{eq} the torque exerted on the equivalent system (N m); and T_1 the lumped disturbance (N).

4 Control Algorithm

4.1 Control Algorithm of Steering Mode Switch

The control algorithm of the whole vehicle is shown in Fig. 5, where δ^* is the reference wheel angle and δ is real wheel angle.

It has been assumed that the tire-cornering properties are in a linear range and the lateral acceleration of the vehicle is limited to 0.4 g. Accordingly, the condition under low speed and small wheel angle is selected. Assuming that the vehicle longitudinal velocity is 20 km/h, the stable state is described as the front wheel angle $\delta_f = 5°$ and the rear wheel angle $\delta_r = 0°$. Yaw rate r and lateral velocity v can be calculated according to Eq. (3). The steady-state value of yaw rate r_0 = 0.3106 rad/s and lateral velocity v_0 = 0.09635 m/s.

Two typical switches from FWS to RWS or to 4WS are selected to research.

4.1.1 Switch from FWS to RWS

Researching on the switch process from FWS to RWS. During the switch, unchanged yaw rate is the control strategy without change of sideslip angle. The constraint equation is defined as:

$$[\beta, Radius] = f(\delta_f, \delta_r) \tag{14}$$

where $r(t) = r_0$.

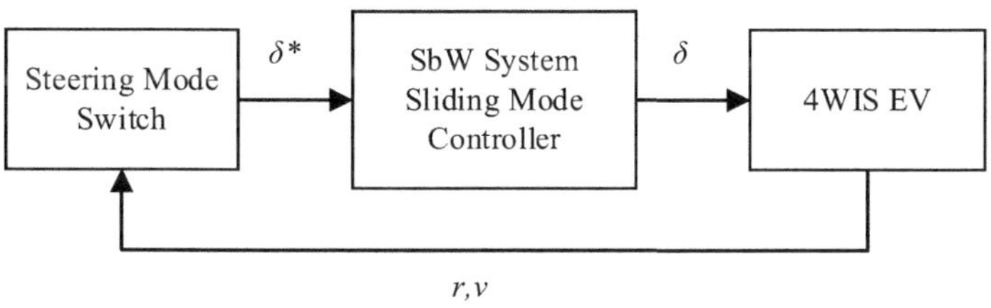

Fig. 5 Control system of 4WIS EV for angle tracking

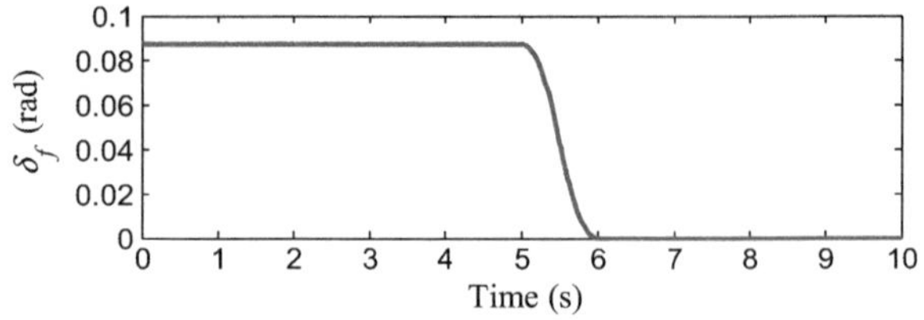

Fig. 6 Change of front wheel angle δ_f during switch

Also known as:

$$\begin{cases} \dot{v} = \frac{l_f C_{\alpha f}}{l_r C_{\alpha r}} \left(\arctan \frac{v + l_f r_0}{u} - \delta_f \right) \cos \delta_f \\ \beta = \arctan(v/u) \\ Radius = \sqrt{u^2 + v^2}/r_0 \end{cases} \tag{15}$$

To solve Eq. (15), one of the variables must be constrained, and the change of front wheel angle is chosen as the constraint condition. In Fig. 6, when $t = 0$ s, $\delta_f = 5°$ (0.0873 rad), which is the step angle input, and then it goes into FWS mode. When $t = 5$ s, it switches to RWS mode, which continues for 1 s and then 4 s. The fitted curves during the switch time from $t = 5$ s to $t = 6$ s are given by:

$$\delta_f = \begin{cases} 5 - 10(t-5)^2, & 5 < t \leq 5.5 \\ 2.5 - 10(t - 5.5) + 10(t - 5.5)^2, & 5.5 < t \leq 6 \end{cases} \tag{16}$$

4.1.2 Switch from FWS to 4WS

Researching on the switch process from FWS to 4WS. During the switch, unchanged yaw rate is the control strategy without change of sideslip angle. The constraint equation is written as:

$$[\beta, Radius] = f\left(\delta_f, \delta_r\right) \tag{17}$$

where $r(t) = r_0, \beta(t_1) = 0$.

Equation (15) can be derived from the constraint Eq. (17) as well.

One of the variables needs to be limited to solve Eq. (15). After switch, sideslip angle is zero, which means lateral velocity is zero. From the formation of Eq. (15), lateral velocity is easier to control than sideslip angle. In Fig. 7, during the time $t = 0 \sim 5$ s, the lateral velocity is stable. When $t = 5$ s, it switches to 4WS mode, which continued for 1 s. When $t = 6$ s, lateral velocity is zero. The fitted curves during the switch time from $t = 5$ s to $t = 6$ s are

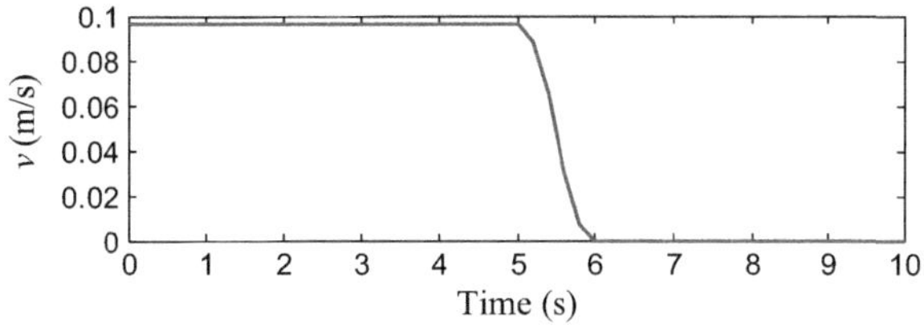

Fig. 7 Change of lateral velocity v during switch

$$v = \begin{cases} 0.09635 - 0.09635 \times 2(t-5)^2, & 5 < t \leq 5.5 \\ 0.09635/2 - 0.09635 \times 2(t-5.5) + 0.09635 \times 2(t-5.5)^2, & 5.5 < t \leq 6 \end{cases} \tag{18}$$

4.2 Sliding Mode Control Designed for SbW System

To solve the problem of controller design for perturbed model, the sliding mode controller is designed.

Considering the second-order nonlinear system [14]

$$\ddot{\theta} = f\left(\theta, \dot{\theta}\right) + g\left(\theta, \dot{\theta}\right)u + d(t) \tag{19}$$

where f and g are known as nonlinear functions; u and $y = \theta$ are the control input and the measurement output, respectively; and $d(t)$ is disturbance, and $|d(t)| \leq D$.

Taking the reference angle instruction as θ_d, the error is defined as:

$$e = \theta_d - \theta \tag{20}$$

Define sliding mode function as $s = \dot{e} + ce$, where $c > 0$. And then

$$\dot{s} = \ddot{e} + c\dot{e} = \ddot{\theta}_d - \ddot{\theta} + c\dot{e} = \ddot{\theta}_d - f - gu - d + c\dot{e} \tag{21}$$

The sliding mode controller is designed as:

$$u = \frac{1}{g}\left(-f + \ddot{\theta}_d + c\dot{e} + \eta \operatorname{sgn}(s)\right) \tag{22}$$

Taking *Lyapunov* function as $L = \frac{1}{2}s^2$

$$\dot{L} = s\dot{s} = s\left(\ddot{\theta}_d - f - gu - d + c\dot{e}\right) = s\left(\ddot{\theta}_d - f - \left|-f + \ddot{\theta}_d + c\dot{e} + \eta \mathrm{sgn}(s)\right| - d + c\dot{e}\right)$$
$$= s(-d - \eta \mathrm{sgn}(s)) = -sd - \eta|s| \tag{23}$$

Taking $\eta \geq D$, it follows that:

$$\dot{L} = -sd - \eta|s| \leq 0 \tag{24}$$

$\mathrm{sgn}(s)$ is sign function, which is given by:

$$\mathrm{sgn}(s) = \begin{cases} 1 & s > 0 \\ 0 & s = 0 \\ -1 & s < 0 \end{cases} \tag{25}$$

While modeling, large uncertainties and disturbances need large gain η, which results in great chatter. In order to prevent chatter, the controller uses saturation function sat(s) instead of the sign function sgn(s), i.e., (where Δ is boundary layer.)

$$\mathrm{sat}(s) = \begin{cases} 1 & s > \Delta \\ ks & |s| \leq \Delta \quad , k = 1/\Delta \\ -1 & s < -\Delta \end{cases} \tag{26}$$

The nature of sliding mode control using saturation function is that outside the boundary layer, the system goes quickly to sliding mode state; inside the boundary layer, feedback control is used to reduce the chatter produced in the fast switch of sliding mode.

For SbW system, according to Eqs. (12) and (13), take $x_1 = \theta$ as wheel angle, u as the ideal output torque of the motor, and the angle instruction as the output angle $\theta_d(t)$ of steering mode switch. It follows that:

$$\begin{cases} \dot{x}_1 = x_2 \\ \dot{x}_2 = \ddot{\theta} = f + gu + d \end{cases} \tag{27}$$

$$f = -\frac{B_{\mathrm{eq}}}{J_{\mathrm{eq}}}, g = \frac{k}{J_{\mathrm{eq}}}, d = -\frac{T_l}{J_{\mathrm{eq}}} \tag{28}$$

5 Simulation Analysis

During the switch process from FWS to RWS and from FWS to 4WS, the response curves of front wheel angle δ_f, rear wheel angle δ_f, sideslip angle of vehicle β, steering radius *Radius,* and lateral velocity v are illustrated in Figs. 8, 9, 10, and 11, where "→" denotes switch direction, and FW and RW denote the abbreviation of front wheel and rear wheel, respectively.

From Figs. 8, 9, 10, and 11, it can be seen that during the switch process from FWS to RWS, the control strategy is shown in Fig. 6, including the changing curves of yaw rate and front wheel angle input. During the switch, the changing trends of rear wheel angle and front wheel angle are same, sideslip angle and steering radius change a little, and steering radius changes slightly irregularly. Due to the unchanged longitudinal velocity, lateral velocity exists change in a certain

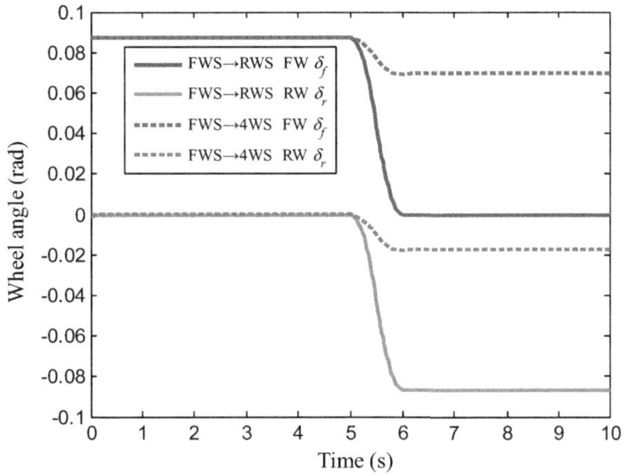

Fig. 8 Response curves of wheel angle δ_f/δ_r

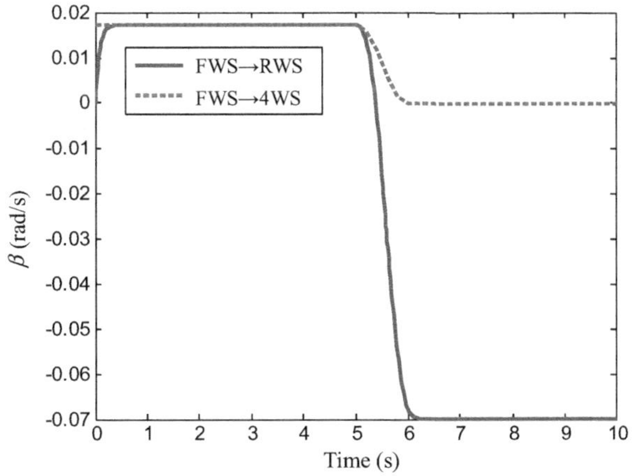

Fig. 9 Response curves of sideslip angle β

Fig. 10 Response curves of steering radius *Radius*

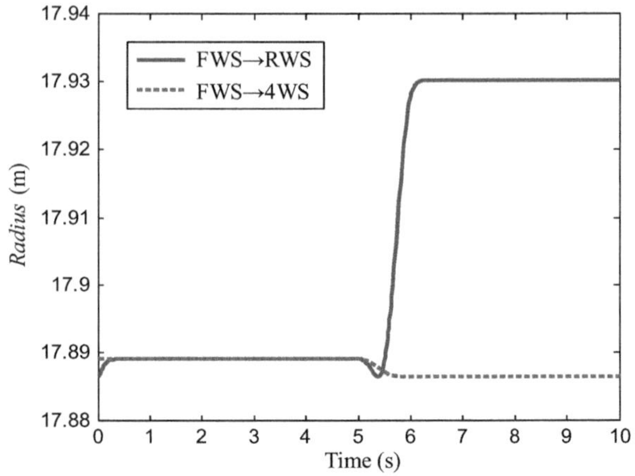

Fig. 11 Response curves of lateral velocity *v*

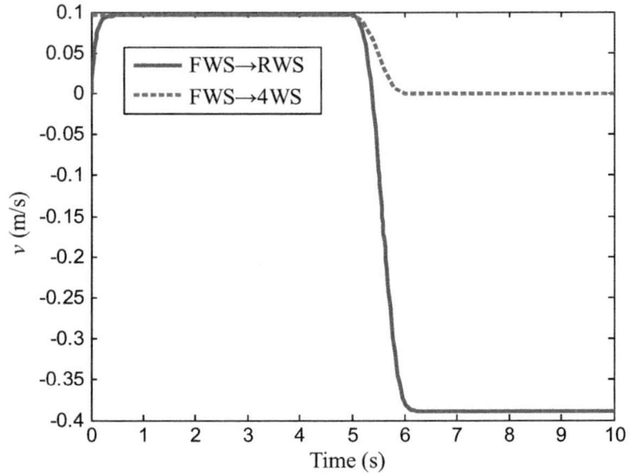

range, and consequently the velocity of CG will vary, which will result in the change of steering radius. In real vehicle tests, the wheel angles should change smoothly, and accordingly the changing curve of front wheel angle is practical as shown in Fig. 6.

Similarly, in the switch process from FWS to 4WS, the control strategy is shown in Fig. 7, including unchanged yaw rate, sideslip angle which is zero finally, and the changing curves of lateral velocity. It is proved that the changing curve of lateral velocity is also practical as shown in Fig. 7.

During the switch process from FWS to RWS, the sliding mode controller is used to track the front and rear wheel angles, using sign function sgn(s) and saturation function sat(s), respectively, $(\Delta = 0.01)$, and the simulation results are shown in Figs. 12, 13, 14, and 15.

Fig. 12 Front wheel angle tracking

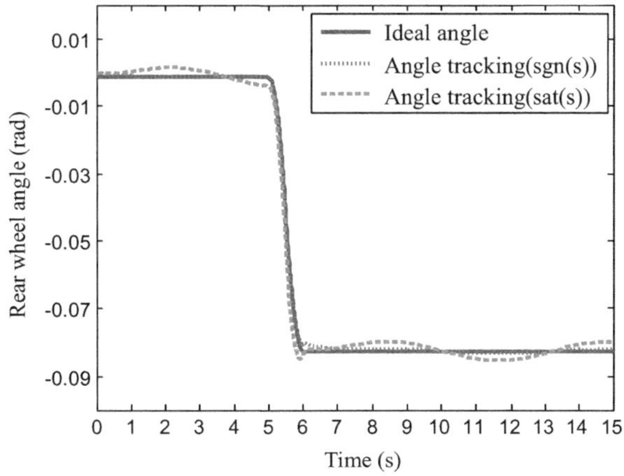

Fig. 13 Rear wheel angle tracking

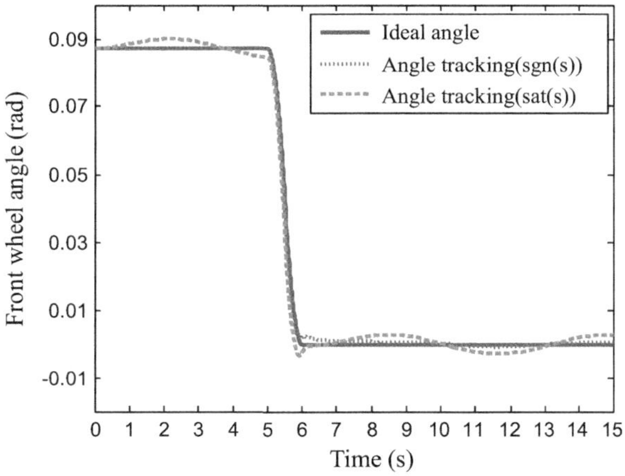

It can be seen from Figs. 12, 13, 14, and Fig. 15 that the chatter can be effectively prevented at the expense of the tracking accuracy if using saturation function instead of sign function. The simulation results of angle tracking during the switch process from FWS to 4WS are similarly available and no more details here.

6 Real Vehicle Tests of Steering Mode Switch

The steering system of 4WIS EV of SbW can achieve steering mode switch in driving. To verify the control strategy as mentioned above, real vehicle tests have been conducted. Several quintessential examples should be cited that switch from FWS to RWS or to 4WS.

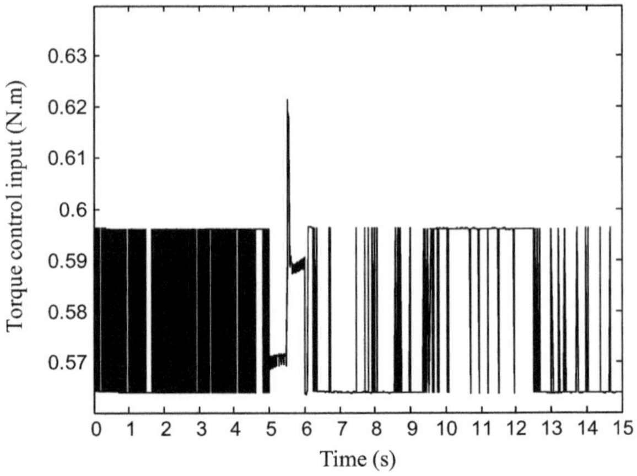

Fig. 14 Control torque input (sign(s))

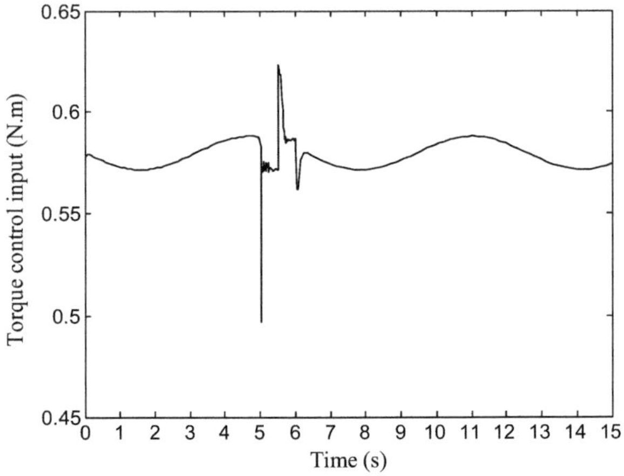

Fig. 15 Control torque input (sat(s))

6.1 Switch from FWS to RWS

Considering the clearance of worm and gear reducer, the allowable error between the object angle and the actual wheel angle is set to 1.5°, which is also the main factor affecting the response time of steering system. In order to see the switch process clearly, the frequency of steering motor is set to 500 Hz, under which the response time of the motor is also increased. The test data of switch process from FWS to RWS are shown in Fig. 16 where FR and FR* mean real angle and object angle of front right wheel, respectively, FL and FL* mean front left wheel, RR and RR* mean rear right wheel, and RL and RL* mean rear left wheel.

Figure 16 demonstrates the object wheel angles and real-time angles during the switch process from FWS to RWS. It can be seen that wheels can track object angles well. When wheels get into stable state, the error is within 1.5°, which meets

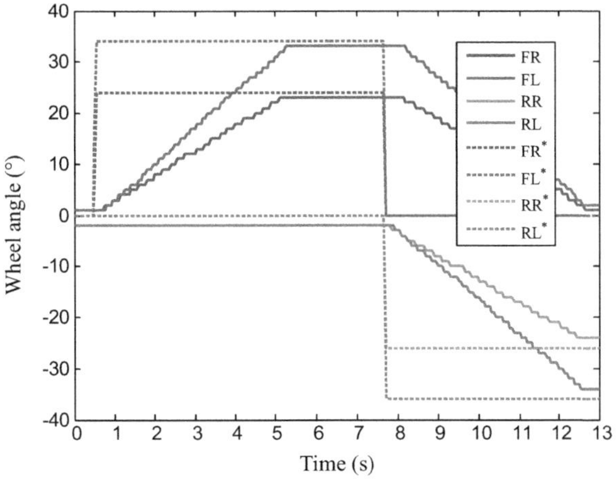

Fig. 16 Object wheel angles and real-time angles during the switch process from FWS to RWS

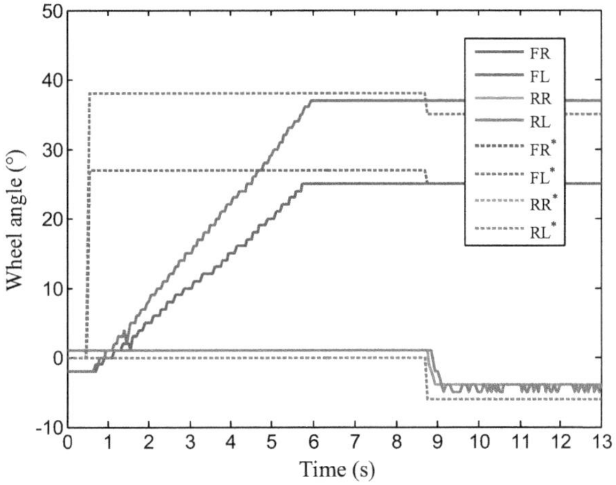

Fig. 17 Object wheel angles and real-time angles during the switch process from FWS to 4WS

the requirement. When the real angles of rear wheels come to their expected values, the switch process is finished. At this time, the angles of front wheels are about zero. One can find that there are small fluctuations in the process, which are mainly caused by the clearance of worm and gear reducer. What's more, it is proved that this system has faster response with higher frequency of steering motor.

6.2 Switch from FWS to 4WS

The test data of switch process from FWS to 4WS are shown in Fig. 17, which demonstrates the object wheel angles and real-time angles during the switch process. It can be seen that wheels can track object angles well. When wheels get into stable state, the error is within 1.5°, which meets the requirement. When the real angles of four wheels come to their expected values, the switch is over. And front wheels and rear wheels are turning in opposite directions, which can improve the steering flexibility of the vehicle.

7 Conclusions

In this paper, a novel 4WIS EV with SbW system is proposed, and the dynamical model of the whole vehicle system and SbW system is established. The control algorithm of steering mode switch is designed considering relative constraints, and controller for SbW system is designed using sliding mode control algorithm. Two typical conditions of steering mode switch are chosen: switch from FWS to RWS or to 4WS. The changing curves of vehicle motion parameters during the switch process are calculated based on MATLAB, and the control algorithm adopted by simulations and real vehicle tests is finally determined. The tests show that the sliding mode controller for SbW has good tracking accuracy. Besides, the test results of switches among FWS, RWS, and 4WS indicate that the prototype can realize various steering mode switches. Further, the control system and algorithm are effective through data acquisition and analysis.

Acknowledgements National Nature Science Foundation of China (51375344), (U1564207) and (5150051692).

References

1. Sun LL (2008) Dynamics and control of four-wheel-independent-steering mobile robot. Harbin Institute of Technology, Harbin
2. Min WC, Park JS, Lee BS et al (2008) The performance of independent wheels steering vehicle (4WS) applied Ackerman geometry. In: International conference on control, automation and systems. IEEE, pp 197–202
3. Lam TL, Qian H, Xu Y (2010) Omnidirectional steering interface and control for a four-wheel independent steering vehicle. IEEE/ASME Trans Mechatron 15(3):329–338
4. Lai X, Chen XB, Wu XJ et al (2014) A study on control system for four-wheel independent driving and steering electric vehicle. Appl Mech Mater 701–702:807–811
5. Ma L, Gong SJ, Li B et al (2015) Steering control scheme for electric vehicles with four-wheel independent steering. Automot Eng 37(9):1029–1034
6. Liang D (2015) Research on the controlling strategy of motion mode switch of all wheel independent steering electric vehicle. Tongji University, Shanghai
7. Chen XB, Fang SD, Hang P et al (2016) Four wheel steering and driving by wire integrated suspension. China: 201610832731.1
8. Lai X, Chen XB, Wu XJ et al (2015) Study on motion control systems and algorithms for four-wheel independent driving and steering electric vehicles. Chin J Automot Eng 5(5): 334–340

9. Yu ZS (2011) The theory of automobile. The China Machine Press, Beijing
10. Thomas D (2006) Gillespie. Fundamentals of vehicle dynamics. Tsinghua University Press, Beijing
11. Chen J (1999) The simulation study on vehicle dynamic control. Automot Eng 2(2):199–205
12. Wu XJ (2015) The research and application of four wheel independently steering-by-wire system. Tongji Univerity, Shanghai
13. Fu H (2008) Research on sideslip angle estimation and control algorithm for vehicle electronic stability system. Jilin University, Changchun
14. Liu JK (2015) Sliding mode control design and MATLAB simulation, 3rd edn. Tsinghua University Press, Beijing, pp 51–56

The Innovation and Application for Analysis Measurement in Insurance of Completely Car Quality

Pan Liu, Zhang Qiang, Mi Fuzheng and Zhang Shuai

1 Overview for Analysis Measurement

The application of the dimension engineering is one of the important indicators of the technical capability of auto plant, the production department using technical countermeasure to tightening ensure the dimension accuracy, in case of the fault happen. For the issue car which been defined out of tolerance of the dimension accuracy by auditor, quality department should take action at first time to analyse the root cause and offer the optimize solution.

The dimension deviation of the completely car normally is the combination from three aspects: deviation of the car body, deviation of the supplier part, and deviation of the assembly process. Analysis measurement using the technology solution to carry out the inspection for completely car and supplier part creates the data link for root cause analysis, successfully locates the main source for deviation, and further supports the production department to optimize the dimension accuracy or technic process, insures the compliance and stability of the completely car quality status. We will roughly discuss the function, equipment, and strategy according to the process during the analysis measurement.

1.1 Classified According to the Function of Analysis Measurement

The function of the analysis measurement includes three main domains: complete car measurement, matching measurement, and purchase part measurement. How to

P. Liu (✉) · Z. Qiang · M. Fuzheng · Z. Shuai
Quality Management Department, Beijing Benz Automobile
Company, Beijing 100176, China
e-mail: liupan@bbac.com.cn

Society of Automotive Engineers of China (SAE-China) (ed.), *Proceedings of the 19th Asia Pacific Automotive Engineering Conference & SAE-China Congress 2017: Selected Papers*, Lecture Notes in Electrical Engineering 486, https://doi.org/10.1007/978-981-10-8506-2_31

further more improve the reliability of the measurement result and the analysis efficiency based on the traditional strategy? The paper will discuss the technical innovation of the above three main domains and the optimization of the measurement strategy, based on the optimization, emphasis the system integration, it make lots of essential research at several directions and acquire some positive fruit.

Completely car measurement: The analysis measurement mainly focusses on the completely car which has gap or transition issue during the quality audit after the assembly shop finishes all the working steps. As we discussed above, the deviation of gap or transition may come from body in white, supplier part, or assembly process, as a final status of the combined deviation, the completely car analysis measurement should strip the deviation from the three aspects one by one. There are two key factors which are needed to be considered during the completely car analysis measurement; first one is which alignment should be used for the specific fault analysis, second one is how to eliminate the distort of the car body if the completely car alignment has been used finally. The two factors we will discuss detail in the following. As Fig. 1 is the picture for complete car analysis measurement.

Fig. 1 Completely car analysis measurement

Matching analysis measurement: The purpose for matching analysis measurement is to use the master car body, Cubing, as reference to analysis the dimension accuracy of the supplier part after been assembly to completely car, due to the Cubing have high accuracy, the tolerance should be controlled within ±0.15 mm, the deviation of gap and transition after the part been assembled to the cubing should been contributed by the part itself or the assembly process.

The benefit for the matching analysis measurement is the deviation of car body will not affect the quality status any more, due to cubing's high accuracy, but the biggest challenge is how to insure the accuracy of the cubing during the manufacture process, assembly process, and maintenance process after high-frequency matching test. As Fig. 2 is the picture for matching analysis measurement.

Supplier part measurement: As one of the main contribution from three deviation aspects, the accuracy of purchase part has affection to the finial quality status of completely car, due to the huge number of supplier parts for one model car, normally the part only been measured and audited at supplier's plant, car plant cannot control the quality directly, therefore, the feedback and countermeasure for quality issue been affected. In order to improve quality control level of the supplier part, two strategies have been figure out; one is using flexible fixture to speed up the analysis measurement process and increase the number which can be measured at car plant; second one is to unify the measurement specification for supplier at the beginning of the project, furthermore to carry out audit and SPC analysis for supplier's daily measurement result at car plant.

Fig. 2 Matching analysis measurement

1.2 Classified According to the Equipment of Analysis Measurement

Equipment for contactile measurement: The traditional contactile measurement equipment has high accuracy and high reliability; it is still the main metrology equipment that is being widely used. Due to the improvement of the technic requirement of passenger car and the trend for streamline design of car profile, there have more and more requirement for gap and transition issue, the efficiency of the contactile measurement is lower to analysis this kind of dimension issue.

Equipment for optic measurement: A kind of useful complementation for contactile equipment, the optic equipment already been widely used for analysis measurement working, instead of measure single point, the color map report and portable equipment greatly improve the analysis efficiency. For the quality issue as: waist line matching measurement; simulation assembly; gap and transition issue. The optic measurement has great (active) advantage.

How to choose contactile or optic equipment for specific quality issue according to the character of the dimension fault, planning a reasonable measurement strategy, fully using the advantage of the equipment itself should been carefully considered, it will greatly improve the analysis efficiency.

1.3 Classified According to the Coordination of Measurement

Coordination of completely car: The alignment which also been used during the manufacture process, all the welding fixtures which been use by body shop are calibrated the position using the completely car alignment, it have significant meaning for the accuracy insurance of the car body. Based on the principle of unify the reference for design, manufacture and measurement audit. The deviation which been measured with the completely car alignment is the only evidence for body shop engineer to adjust the welding fixture, the deviation on the report can be directly use to adjust the fixture. If the purpose for measurement is to guide body shop modify the accuracy of welding fixture, all the deviation report should be measured with completely car coordination (Fig. 3).

Coordination of local alignment: If the quality issue only related with the local area matching accuracy, then need to consider using the local alignment for evaluation. For example, the transition and gap issue between fuel tank cover with sidewall, except the deviation of the sidewall and cover part itself, the root cause mainly come from the relative position of the cover to the tank opening of the sidewall, if using the completely car alignment to measure the shape and position, the deviation combine the position and shape of tank opening, cannot directly get the conclusion for the quality issue for this kind of issue, it should be using the local alignment of the tank opening to evaluate the position of the cover, the result can

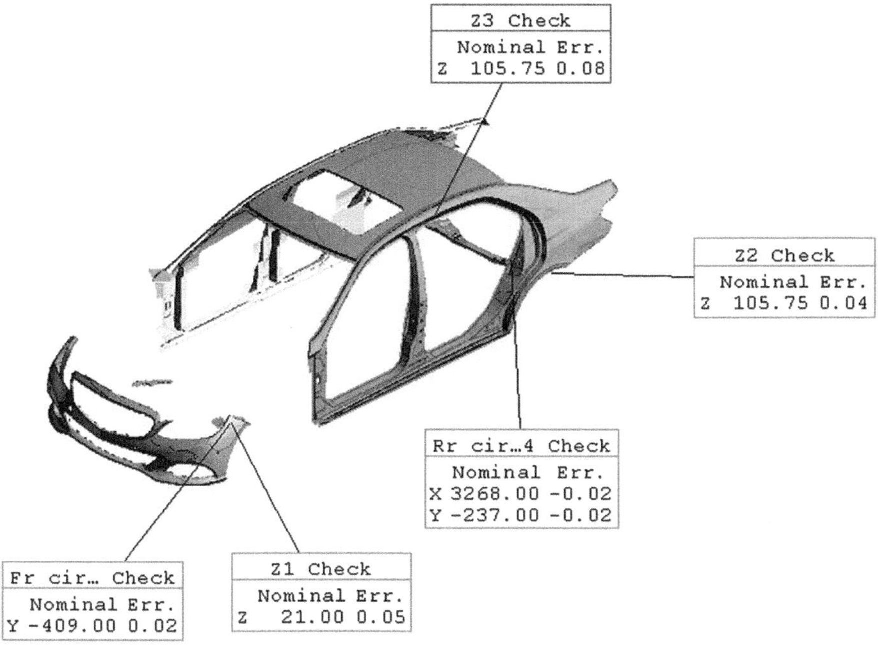

Fig. 3 Create completely car alignment

directly been used for production to optimize the assembly process. As Fig. 4 is the picture for tank opening local alignment, the Z4 and Z6 are constructed points by edge surface.

Coordination of part alignment: The accuracy of the manufacture process of the part itself has significant influence to the quality status after the part been assembled to the car body. The part alignment which uses the manufacture RPS point should be considered if only the part's deviation need to be evaluated. The result of this alignment can be directly used for supplier to optimize the accuracy of the production (Fig. 5).

2 The Innovative Application for Each Function Area

2.1 *The Development of Automatic Adjustable Equipment for Completely Car Measurement*

The static torsion stiffness of the car body needs to be considered during the completely car measurement: compare with the body in white, the status of car body of the completely car will be changed due to the weight of the suspension system, power train system, and so on. Although this kind of torsion of car body

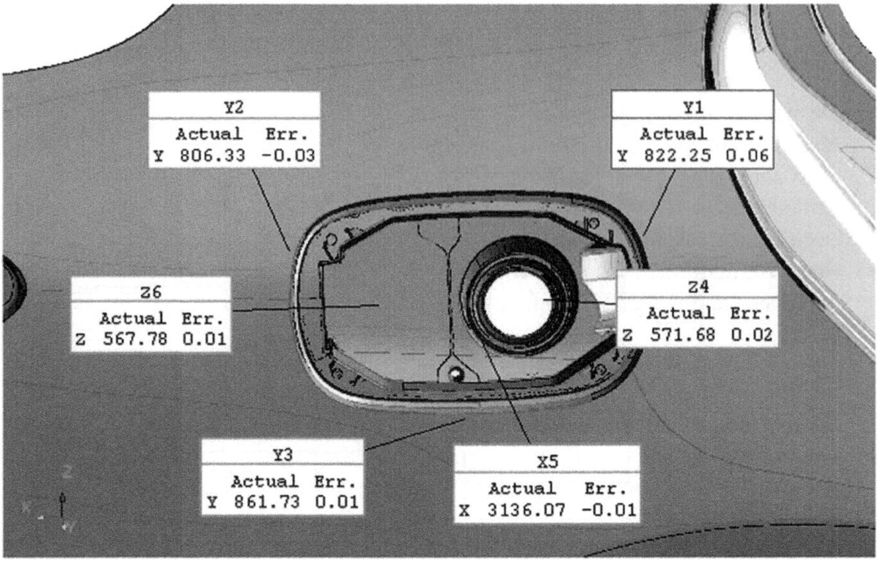

Fig. 4 Local alignment of tank opening

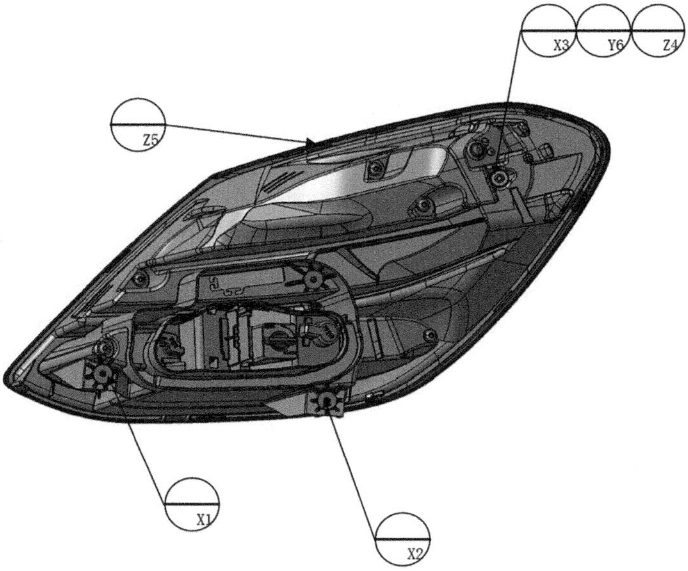

Fig. 5 Part alignment

had already been considered in the conception, except some special parts, for example the door frame, normally the design data do not have two statuses: one version for body in white, another version for completely car after loaded. It means for completely car analysis measurement, if we use the same RPS point to create alignment, the distort of the car body caused by the weightiness will be combined in the report and cause misunderstanding. On the other hand, the result of analysis measurement report should be directly used for related department to optimize the product process and eliminate the deviation of the complete car. In order to avoid this situation, it is necessary to develop one set of equipment which can compensate the distortion of the completely car (Figs. 6 and 7).

The development of the automatic adjustable equipment successfully solves the issue; the whole system is a closed-loop control equipment combined with support unit, elevate unit, self-motion height sensor, and control unit' the software which is specifically written using C++ will automatically control the system to adjust the car body according to pre-setting parameter. The experiments verify that the set can ensure the accuracy of the alignment within 0.1 mm between body in white and completely car body.

2.2 *The Innovation Application for Cubing Accuracy Insurance*

The function of cubing is to verify the dimension accuracy of the supplier part and the hanging on part of the body shop during the matching test. The purpose is during the test do not need to consider the influence of the deviation of the car body,

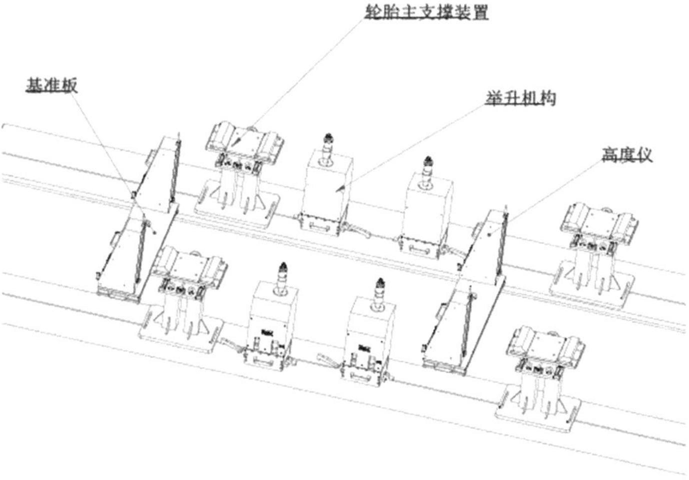

Fig. 6 Automatic adjustable equipment

Fig. 7 Example for measurement

only the (dimension) deviation of the supplier parts and the assembly process need to be considered in the measurement result. The accuracy of the cubing will have significant affection to the matching result. The completely cubing was build up by over hundred subunits; the accuracy of the cubing in each subunit after been assembled should be controlled within ±0.15 mm.

In order to insure the high accuracy of the cubing, it is necessary to develop the innovation technology from the conception, material, manufacturing, and assembly. The aluminum material of Cubing easy to be deformed during high frequency testing of assembly/disassembly matching, the connector part is the main part and if it been deformed, the deviation will misleading the customer thinking the matching supplier part have deviation. Figure 8 shows the design drawing of Ccubing. Due the aluminum easy to be manufactured, and less inner stress, it been widely used for Cubing design, but the disadvantage is rigidity lower, for example, the rigidity of aluminum model ZL-AC4CH-SVW only 60-68HB, if it been used for connector of the Cubing body, the connector will be deformed after several times try out.

There have significant meaning to improve the life circle time and the measurement accuracy of the Cubing via technical innovation to optimize the rigidity of the key part of Cubing, especially the connector. Using the stainless steel 3Cr13 to manufacture the connector of Cubing, the rigidity can be reach HRC28-32; it can be designed as a embody part to assemble with cubing body after calibration, the accuracy can fulfil the requirement.

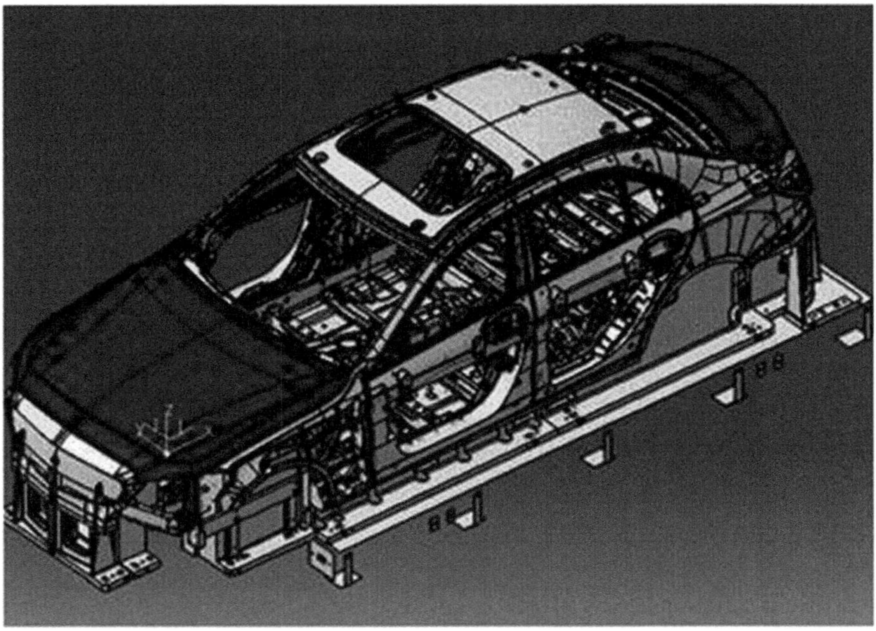

Fig. 8 Cubing conception

The surface of Fig. 9 is one of the key connectors which are used for rear bumper assembly to sidewall rear; it decide the accuracy in *X* direction after the rear bumper had been assembled. The area been damaged after frequently try out test before SOP production; the accuracy cannot reach the target. The CMM measurement result shows the deviation of the bumper after been assembled over 1.2 mm in *X* direction, and the connection apparently has clearness. The innovation technology module designs the connector using material 3Cr13 stainless steel, as the Fig. 10, the new connector been assembled after calibrated, after testing, it been verified fulfil the accuracy requirement, the new design extend the life circle of Cubing and save the running cost. Figure 10 show the final status after been improved.

2.3 *The Innovation Technology of Purchase Part Measurement*

The quality status of the supplier part is one of the main factors for completely car quality insurance. There have been more than several hundreds of supplier part for one model. In order to control the dimension accuracy and improve the analysis efficiency for quality issue of supplier part, it need innovate the measurement method, and the measurement strategy during the analysis process.

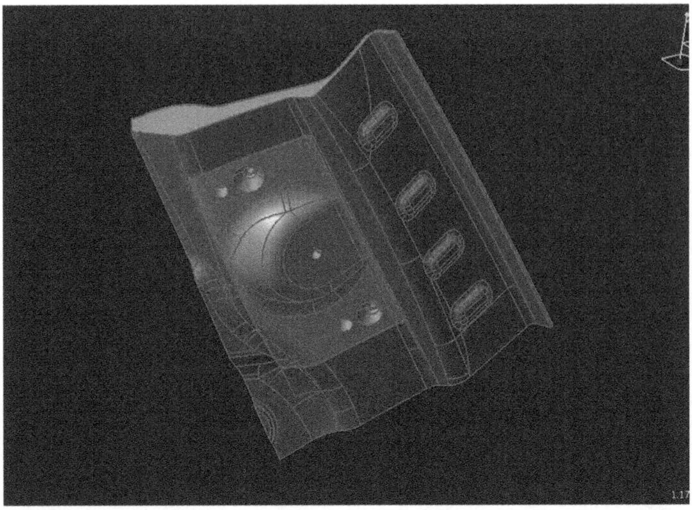

Fig. 9 Module design

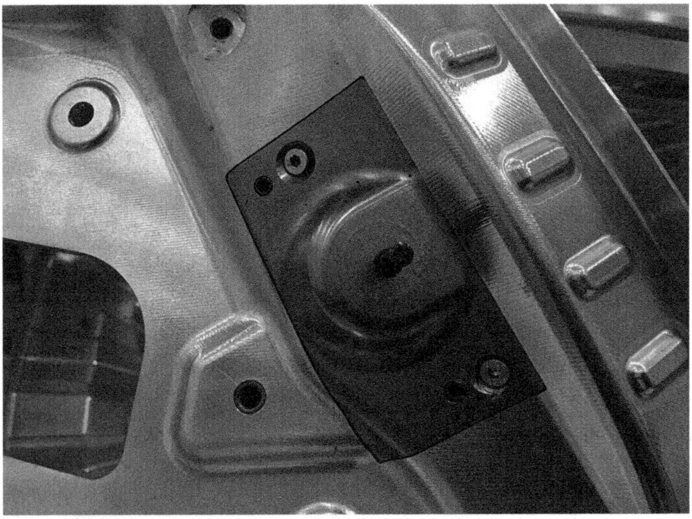

Fig. 10 Stainless part been assembled

Except the normal traditional way to measure the supplier part using designed fixture, the innovation strategy which uses the high accuracy flexible fixture to construct the measurement fixture been used more and more widely. When the supplier part has dimension issue for production quality, it can be quickly respond via flexible fixture according to the part design drawing and the measurement specification, after adjusting the holding position and calibrate the accuracy; it can solve the bottleneck of measurement efficiency and control large volume supplier part. Figure 11 is one of the examples to measure supplier's BIW part using flexible fixture.

The "Audit Measurement of Supplier Part" is the innovation strategy for quality insurance of supplier part. The measurement team of car plant regulates the supplier's conception of dimension control, including specification of the fixture design and manufacture, unify the measurement standard, plan the measurement strategy, edit the measurement element, and so on. This significant meaning of this strategy is to extending the supplier's daily measurement to be one of the segments for auto plant quality audit planning. After the part beginning the series production, the CMM operator will carry out the daily result checking and SPC analysis according to the measurement result which is measured by supplier and send to auto plant to insure acquire the first-hand information about the part quality status after been produced (Fig. 12).

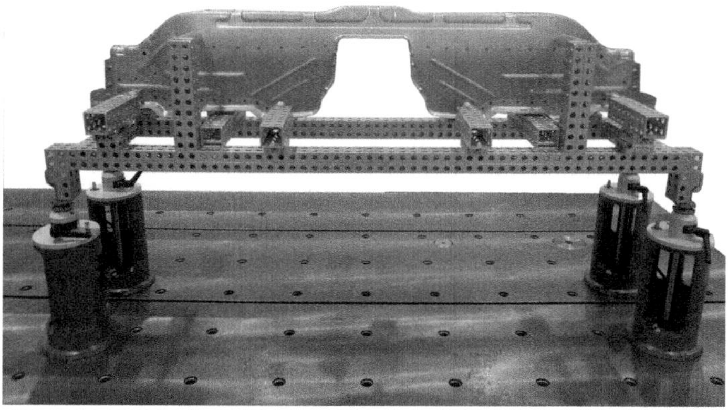

Fig. 11 Flexible fixture

Fig. 12 Measurement of supplier part

The efficiency and reliability of the analysis measurement has been greatly improved via the technic innovation of each function area. Innovation offers the powerful support to quality control for production of the auto plant.

References

1. Li M (1996) The application of coordination measurement machine and quality control. Combined machine tool and automatic machining 6
2. Wang ZM (2009) Research on automatic measurement technology of coordinate measuring machine. Mach Tool Hydraul 37

Study on Lightweight Optimization of Low Control Arm for Vehicle Suspension Based on Kinematic Envelope Analysis

Yongzhou Lin and Shaoliang Guo

1 Introduction

With the wide application of digital prototype and finite element simulation techniques, high-performance suspension systems with lightweight have been developed in the past decades. Weight reduction of suspension components can effectively decrease fuel consumption and carbon emissions and improve vehicle's driving performance. Among all the vehicle structural components, the low control arm (LCA) is particularly important to minimize roll angle of the vehicle and improve control stability when driving on rough road [1, 2]. Therefore, lightweight structure design and optimized method are essential in product development process.

Great deals of research on structural design and optimization application in engineering development have been carried out in past years [3, 4]. For example, Sekulski [5] used genetic algorithm optimization technique to minimize the weight of ship structure variables. Marco Cavazzuti [6] established topology optimization model to develop aluminum control arm according to the predetermined requirements. What is more, Wu [7] described a methodology to optimized vehicle components using parallel chaos optimization algorithm. Hoe [8] presented shape optimization of lower control arm considering multi-disciplinary constraint conditions.

Despite the work mentioned above, there are still limited studies concerning the acquisition of design domain for structure optimization; what is more, comparative performance simulation and bench tests are seldom conducted for the structure verification. Thus in this study, kinematic envelope technique was implied to determine the original design domain of LCA using digital prototype methods, and

Y. Lin (✉) · S. Guo
Guangzhou Automobile Group Co., Ltd, Automotive Engineering Institute,
Guangzhou 511434, China
e-mail: linyongzhou@gaei.cn

Society of Automotive Engineers of China (SAE-China) (ed.), *Proceedings of the 19th Asia Pacific Automotive Engineering Conference & SAE-China Congress 2017: Selected Papers*,
Lecture Notes in Electrical Engineering 486, https://doi.org/10.1007/978-981-10-8506-2_32

then, topology and size optimization was conducted to obtain the optimal design. Finally, fatigue analysis and bench test were performed to verify the optimal structural scheme.

2 Kinematic Envelope Analysis

The difficulty in the design domain acquisition is that the distances between low control arm and other suspension components keep changing during suspension traveling and steering. Thus, kinematic envelope analysis in CATIA digital prototype module will be introduced and applied for this purpose, to determine the design domain technologically in this section.

2.1 Suspension Modeling

The three-dimension (3D) model of multi-link suspension was first imported into CATIA software, as shown in Fig. 1, to simulate the movement of suspension components during wheels traveling and steering; here, the developed LCA is presented as point-line model.

The mainly topological joints and degree of freedom were firstly defined according to the connecting relationship and load transfer paths of multi-link suspension. Each type of joints and degree of freedom were shown in Fig. 2. Driven pair of vertical translation between knuckle and test rig was implied to the model representing the movement of suspension travel.

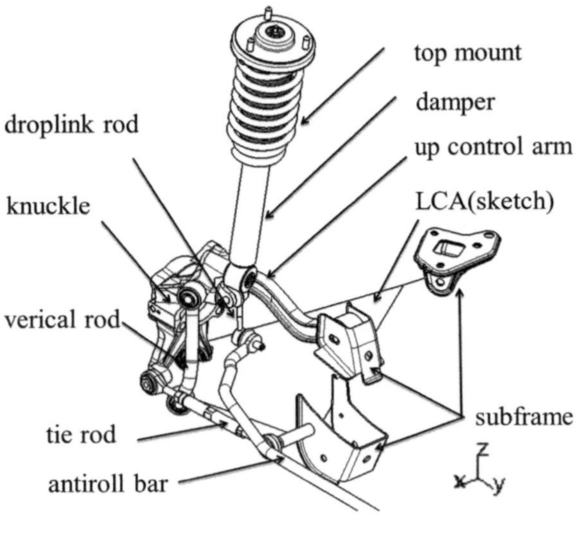

Fig. 1 Suspension modeling of multi-link suspension

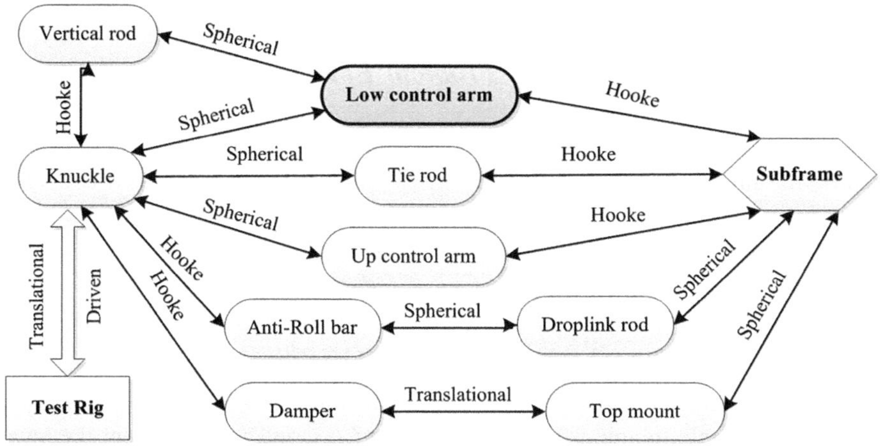

Fig. 2 Components and mechanics joints

2.2 *Acquisition of Design Domain*

Based on the digital prototype model of the developed suspension system, the simulation was driven and the dynamic envelopes of each related components referred to the LCA were generated. Then, all these envelopes were assembled into a new model to form the limit geometrical boundary for LCA, as shown in Fig. 3. According to the standard for multi-link suspension system, the dynamic distances between LCA and other suspension components should not be less than 10 mm, considering tolerance within manufacture and assembly processes, and thus, the 3D design domain of LCA was determined technologically.

Fig. 3 Dynamic envelopes of each related component

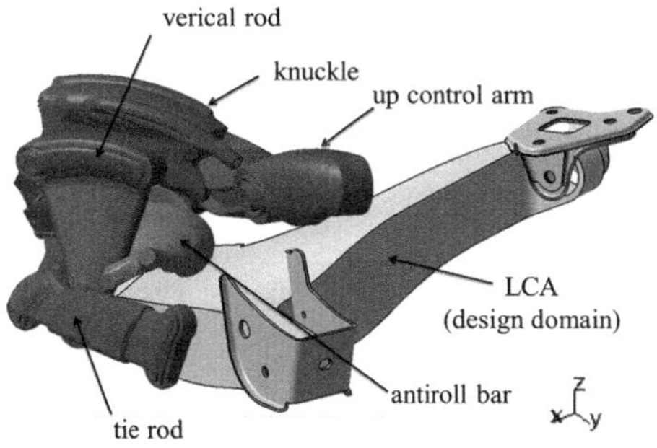

3 Topology Optimization

3.1 Topology Optimization Problem Formulation

The mathematical model of topology optimization using variable density method can be formulated as follows:

$$\begin{aligned} &\text{Minimize}\, M = M(den) \\ &\text{Subject to:} \quad C_{\text{Low}} \leq f(den) \leq C_{\text{Up}} \end{aligned} \tag{1}$$

$$0 \leq den_i \leq 1.0, \quad i = 1, \ldots, n. \tag{2}$$

Here, $M = M(den)$ represents the objective function; *den* denotes the density of the design variable, ranging from 0 to 1, while C_{Low} and C_{Up} represent the lower and upper limit values of the response function described as *f*(*den*).

3.2 Finite Element Modeling

The design domain of the LCA was imported into OptiStruct module of HyperWorks software for finite element modeling. In general, the model was characterized by 18107 3D solid elements for design region, and 325 elements for connection areas defining as non-design region. QSTE380 was considered with the following properties: Young's modulus 205 GPa, Poisson ration 0.28, and yield strength 380 MPa.

According to the boundary condition, the LCA was linked to the chassis at the inside points A and B, which were fully constrained, and loads coming from the wheel hub were applied at the outside points C and D through the knuckle and vertical rod, as shown in Fig. 4. Load cases of the model were defined by

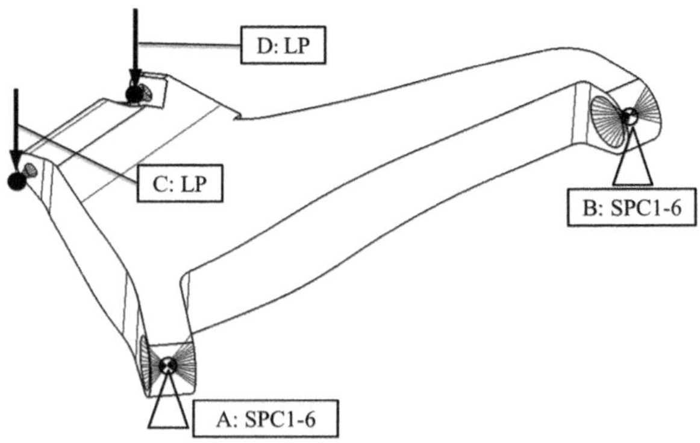

Fig. 4 Boundary and loading condition of LCA

considering the most typical working conditions, including accelerating, steering, braking, and vertical loading conditions.

3.3 Topology Optimization Simulation and Results

The topology optimization problem of LCA has been expressed as a mass minimization problem, respecting to multiple static working conditions. Weight coefficients of 25% have been allocated for each loading case, first natural frequency of 300 Hz has been imposed as constraint functions, and convergence criterion was based on the relative volume when change of 70% has been achieved.

According to the definition of convergence criterion, the optimization process completed after 13 iterations, as shown in Fig. 5. The final density distribution layout of the design domain was shown in Fig. 6, which turned out to be L-shaped, linking the knuckle and vertical rod to the chassis.

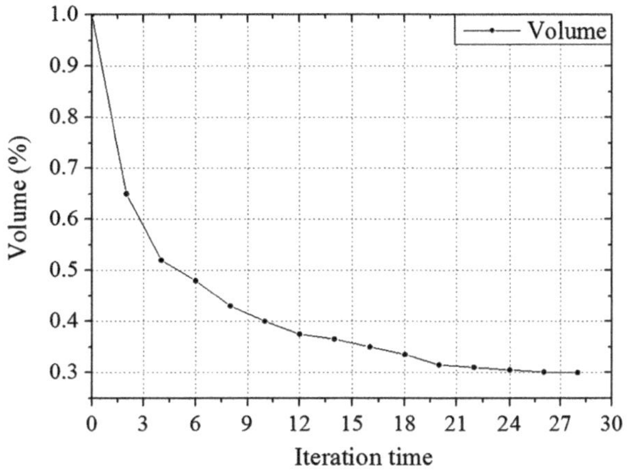

Fig. 5 Variation of volume during optimization process

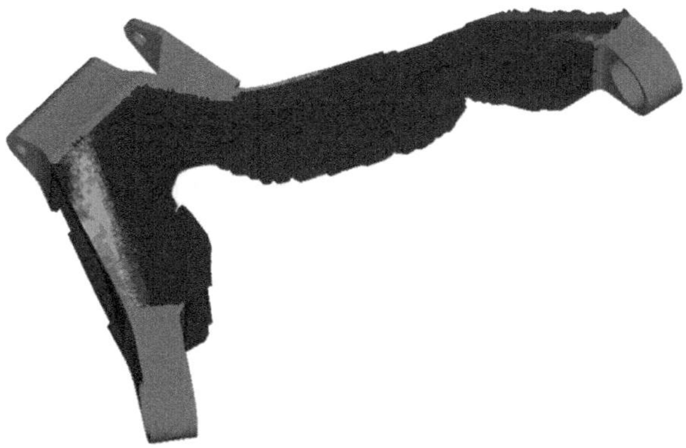

Fig. 6 Optimal layout of the LCA

4 Sizing Optimization

4.1 Sizing Optimization Formulation

The mathematical model of sizing optimization based on first-order optimization algorithm can be formulated as follows:

$$\begin{aligned} &\text{Minimize}\, M = M(x_i) \\ &\text{Subject to:} \quad D_{\text{Low}} \le f(x) \le D_{\text{Up}} \end{aligned} \tag{3}$$

$$x_{\text{Low}} \le x_i \le x_{\text{Up}}, \quad i = 1, \ldots, n \tag{4}$$

Here, $M = M(x_i)$ represents the objective function; x_i denotes the design variable, each ranging from x_{Low} to x_{Up}, while C_i^{Low} and C_i^{Up} represent the lower and upper limit values of the response function described as $f(x)$.

4.2 Sizing Optimization Simulation and Results

The optimal layout of design domain obtained in topology optimization was export to CATIA using OSSMooth tool for further detailed design. Considering the manufacturing requirement and cost efficiency of stamping process, the structure of LCA was divided into six parts made of steel plate (as shown in Fig. 5). Finite element model of the new scheme was established in OptiStruct, which consisted of 15,918 shell elements and 204 rigid elements, and shared the same material definition and boundary conditions as discussed above. The finite element model of new scheme was shown in Fig. 7.

Sizing optimization was then conducted in OptiStruct module to achieve the best combination of thickness for each steel plate, which basically fell into three parts. Firstly, the objective function was set to minimize the weight of the LCA. Secondly, the thicknesses of all plates of the LCA were defined as design variables, each ranging from 1.5 to 5.0 mm. while the mean von Mises stress 380 MPa was defined as limit of response parameter. Finally, the input parameters of objective function, optimization algorithm, and loop control method were defined.

The optimal result has been obtained after six iterations, as shown in Fig. 8. Thus, the optimal design of the lower control arm has been achieved, as shown in Table 1, which met the predetermined requirements of strength.

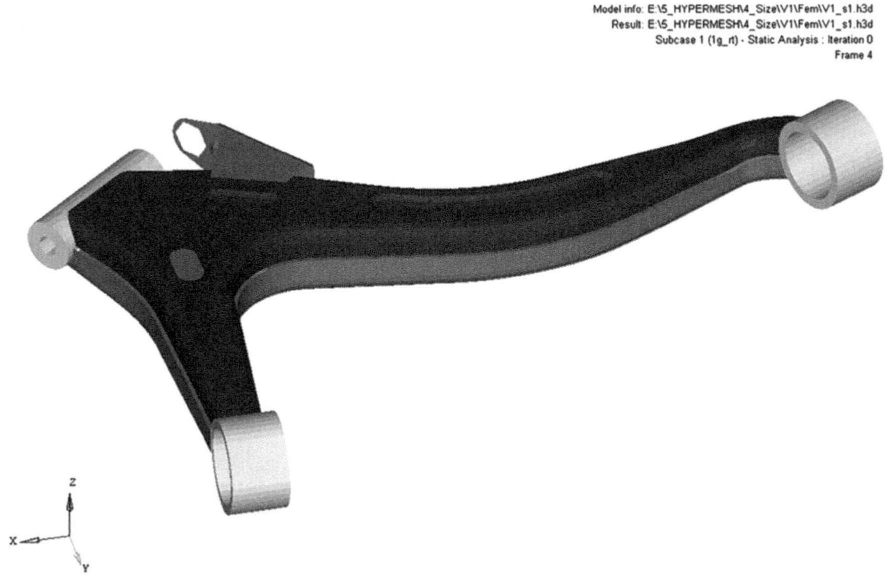

Fig. 7 New finite element model

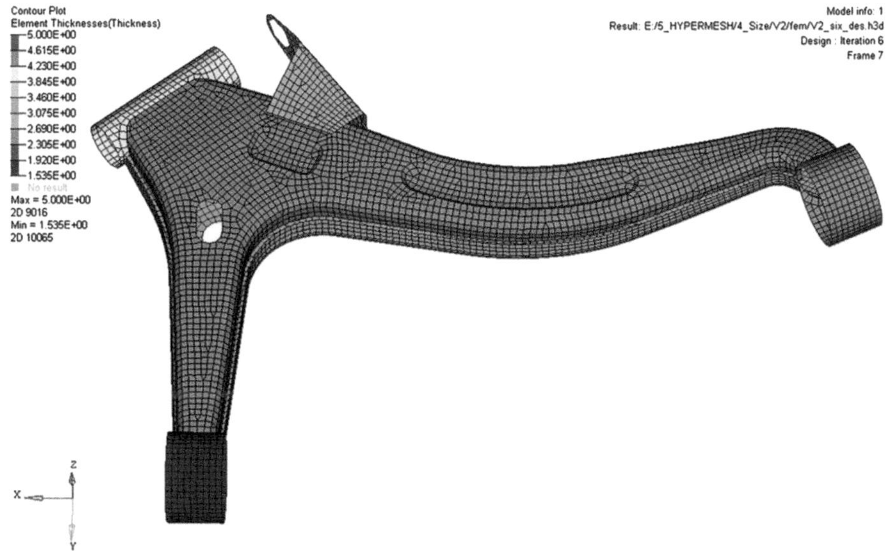

Fig. 8 Optimal result of the LCA

Table 1 Sizing optimization process of variables

Design variable	Bush-1	Bush-2	Bush-3	Plate-1	Plate-2	Plate-3
Initial value/mm	5	5	10	3	3	3
Range/mm	3–15	3–15	3–15	1–10	1–10	1–10
Optimal solution/mm	3.0	3.0	4.5	3.0	3.5	3.0

5 Structure Verification

Fatigue analysis of the optimal design has been carried out in ncode software using strain-life approach and rain-flow counting method. Based on the six-axis wheel loads recorded on the proving ground, loading spectra of single complete durability cycle, as shown in Fig. 9a, were calculated and imported into ncode for simulation [9, 10].

The results showed that the LCA has survived for more than six life cycles without exceeding the failure limit of material, as shown in Fig. 10.

To validate the results of strength and fatigue analysis, sample of LCA was manufactured for half suspension durability bench test, as shown in Fig. 11. Three-axis limit loads were imposed on the LCA for 1E + 6 cycles. The results of bench test showed that the LCA survived its target life period without any damage or significant deformation.

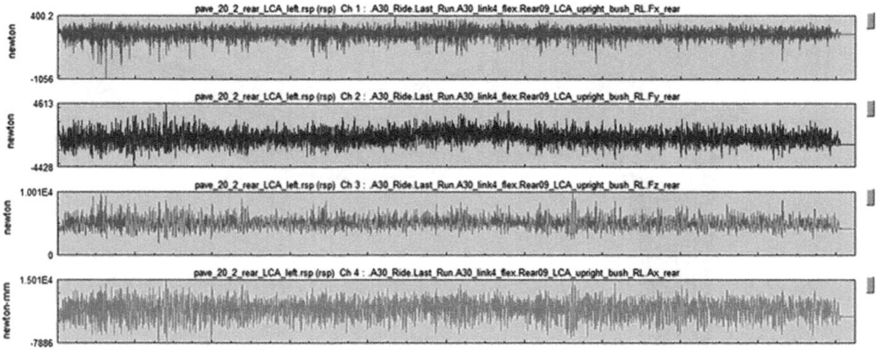

Fig. 9 Loading spectrum of single durability cycle

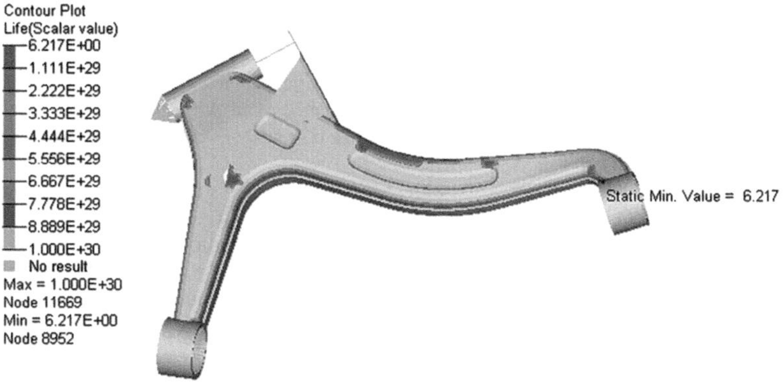

Fig. 10 Fatigue analysis result of the LCA

Fig. 11 Durability bench test of half suspension for structure verification

6 Conclusion

Lightweight design of low control arm for multi-link suspension based on kinematic envelope technique and finite element optimization methods has been conducted in this paper. Original design domain for optimization has been technologically determined based on kinematic envelope technique, and then, topology and sizing optimization has been conducted to obtain the optimal structural scheme.

Fatigue analysis and bench tests have been performed to verify the feasibility and reliability of the optimal design. The forward structural design and optimization methods proposed in this paper provided a reference for lightweight design of other automotive suspension parts.

References

1. Gobbi M, Haque I, Papalambros P et al (2006) A critical review of optimization methods for road vehicles design. American Institute of Aeronautics and Astronautics
2. Hui-ye L, Qi F et al (2011) Lightweight design for rear suspension's upper arm of Roewe750. Mach Des Res 27(2):111–118
3. Hai-bo Z, Rui-jun Z et al (2014) A finite element analysis of automobile suspension control arm base on ANSYS. Modular Mach Tool Automat Manuf Tech 3:150–155
4. Gao D, Wang D et al (2013) Topology optimization of conditioner suspension for mower conditioner considering multiple loads. Math Comput Model 58:489–496
5. Sekulski Z (2009) Least-weight topology and size optimization of high speed vehicle passenger catamaran structure by genetic algorithm. Mar Struct 22(4):691–711
6. Cavazzuti Marco, Baldini Andrea, Bertocchi Enrico et al (2011) High performance automotive chassis design: a topology optimization based approach. Struct Multidisciplinary Optim 44(1):45–56
7. Wu XL, Cao BG, Li XY et al (2011) Component sizing optimization of plug-in hybrid electric vehicles. Appl Energy 88(3):799–804
8. Heo S, Kang D, Lee J et al (2013) Shape optimization of lower control arm considering multi-disciplinary constraint condition by using progress meta-model method. Int J Automot Technol 14(3):499–505
9. Palma ES, Petracconi CL, Ferreira SE (2009) Fatigue behavior analysis of a rear tow hook pin of a passenger vehicle. Eng Fail Anal 16(7):2408–2416
10. Lijun Q, Daojun W, Nianjiong Y (2012) Multiaxial fatigue analysis of vehicle lower control arm based on simulated road excitation. Automot Eng 34(3):249–254

Effect of Engine Oil on Fuel Consumption and Durability of Heavy-Duty Vehicle

Zhuo Wang, Yanan Zhu, Chunsheng Tao, Qingguo Wang, Mo Zhao and Zhenfang Chen

1 Introduction

1.1 Fuel Economy Standards

Improving fuel economy has been a key focus across the automotive industry since many years ago. For heavy-duty commercial vehicles, they accounted for 36% of oil consumption and GHG emissions from the transportation sector in 2010, and the benefits from minor gains in fuel economy can lead to significant savings for fleets as well as owners and operators. Setting heavy-duty vehicle (HDV) efficiency standards is significantly more challenging, because HDV fleets are extremely diverse in terms of vehicle size and configuration as well as usage patterns.

Governments have driven the improvement of diesel engine fuel economy in transportation through tightening legislation all over the world. These legislations began to be implemented in 2014–2015; the percentage reduction in limits of fuel consumption is 10–20%. It is expected that stricter limits of fuel consumption will be implemented by 2020 (Table 1).

1.2 Oil Specifications

The development trend of diesel engine oil specification also reflects the effect of diesel engine oil on fuel consumption. The need for diesel engine fuel economy

Z. Wang · Y. Zhu (✉) · C. Tao · Q. Wang
China FAW Corporation Limited R&D Center, Changchun, China
e-mail: zhuyanan@rdc.faw.com.cn

M. Zhao · Z. Chen
Lubrizol Management (Shanghai) Co., Ltd, Shanghai, China

Society of Automotive Engineers of China (SAE-China) (ed.), *Proceedings of the 19th Asia Pacific Automotive Engineering Conference & SAE-China Congress 2017: Selected Papers*, Lecture Notes in Electrical Engineering 486, https://doi.org/10.1007/978-981-10-8506-2_33

Table 1 Global comparison of fuel economy/GHG standards for heavy-duty vehicle [1]

Country/ Region	Target year	Vehicle type	Regulation type	Energy saving	Next phase
USA	2014	GVW > 3.86 t	GHG/Fuel efficiency	6–23%	2020 (Phase 2)
Canada	2014	GVW > 3.86 t	GHG/Fuel efficiency	6–23%[a]	2020 (Phase 2)
China	2014	GVW > 3.5 t	Fuel consumption	10.5–14.5%[b]	2019 (Phase 3)
Japan	2015	GVW > 3.5 t	Fuel economy	12.1–12.2%[c]	2020 (Phase 2)
EU	–	–	GHG	–	2016–2017
India	–	–	Fuel efficiency	–	2018–2019 (Phase 1)

[a]Implemented beginning with MY2014, CO_2 emission reductions ranging from 6 to 23% in the MY2017 timeframe (as compared to a MY2010 baseline)
[b]Compared to the limits with Phase 1
[c]Improvement over 2002 performance

improvement resulted in the establishment of API's proposed category 11 (PC-11) program which aimed at establishing a new standard to ready it for the market in 2016. European Automobile Manufacturers Association (ACEA) is also considering the introduction of similar requirements in the future specification update. Compared with other means of reducing fuel consumption, the use of energy-saving engine oil presents a high performance/price ratio (Table 2).

1.3 Measurement Methods

There are three measurement methods for fuel consumption: simulating calculation method, chassis dynamometer method, and proving ground test. Typical examples of light-duty vehicle (LDV) drive cycles on chassis dynamometer include the Environmental Protection Agency (EPA) Federal Test Procedure 75 (FTP-75) [3], the New European Drive Cycle (NEDC), the Japanese JC08 Cycle, and the Worldwide harmonized Light vehicles Test Cycles (WLTC); examples of HDV

Table 2 Specification of diesel engine oil and the deed of fuel economy improvement

Specification	ACEA future specification	PC-11 [2]
Characteristic	Series E is considering to introduce a specification of HTHS <3.5 to meet the need of fuel economy improvement	CK-4: high HTHS ($\geq$ 3.5 mPa s) FA-4: low HTHS (2.9–3.2 mPa s): fuel economy improvement

drive cycles include the Urban Dynamometer Driving Schedule (UDDS), the European Transient Cycle (ETC), the Japanese JC05 Cycle, and the World Transient Vehicle Cycle (WTVC) on chassis dynamometer. Proving ground test doesn't have a fixed drive cycle, with SAE J 1321 as an example.

1.3.1 Fuel Economy Measurement Cycle (Chassis Dynamometer Method)

Measurement cycle is a scientific and practical method to evaluate the influence of engine oil on fuel consumption. The cycle of fuel economy measurement is mostly carried out by means of emission measurement cycle. Every government establishes the measurement cycle according to every real-world driving condition, and there is a big difference between different cycles as shown in Table 3.

China heavy-duty vehicles measurement cycle (C-WTVC) which adjusts accelerations and decelerations is based on the World Transient Vehicle Cycle (WTVC). Drive cycle is the reference and foundation of fuel economy measurement. It is effective to evaluate the influence of engine oil on the fuel consumption by establishing a practical and reasonable measurement cycle which is in the fixed dynamic condition.

1.3.2 Proving Ground Test

American SAE J1321 is a test method for fuel consumption by the actual operation of vehicles. SAE J1321 testing involves at least a control vehicle and a test vehicle which are identical. After selecting the representative route, "weigh tanks" for accurate gravimetric measurement of fuel or fuel flow meter is used to calculate fuel consumption (Table 4).

As mentioned previously, it is found that fuel economy measurements carried out on the chassis dynamometer with fixed measurement cycle show good reproducibility; however, test results have deviation due to the difference between measurement cycle and real-world driving conditions. Proving ground test is more similar to the real-world driving conditions; however, if scientific and reasonable test methods are not specified clearly, it is difficult to distinguish between test reproducibility and experimental error. Because of the difference between China and other countries of geographical positions and urban traffic situations, it is of great significance to establish drive cycle in accordance with the real-world driving

Table 3 Selected operating parameters by test cycle [2, 3]

Test cycle	WTVC	Japan JE05	US UDDS	US Manhattan cycle
Running time (s)	1800	1800	1060	1090
Average speed (km/h)	41.00	26.94	30.40	11.00
Maximum speed (km/h)	87.6	88.0	93.3	40.9

Table 4 SAE J1321 [4]

Item	Requirement
Vehicle	Test vehicle + Control vehicle
Test route selection	Long-haul operation >64.4 km
	Other operation 30% of the tank capacity or 22.7 L fuel consumption, greater
Vehicle test speed	Representative of the average operation
Cargo weight	Representative of the fleet operations
Driver selections	Sufficiently skilled and excellence of test procedure conduct
Fuel measuring	Portable weigh tank method or flow meter method
Test accuracy	1%

conditions in China to differentiate fuel efficiency benefits from oil to oil. The test cycle is carried out by the equipment with good test precision, the vehicle is driven on a proving ground test track, and the test conditions are relatively controllable.

2 Data Acquisition System and Measuring System

Before establishing drive cycle, the vehicle was instrumented with data acquisition systems for recording multiple key parameters; the collected data was processed, analyzed, and combined to create drive cycle. The cycle was assessed to be representative of the real driving condition of EMS vehicle in Beijing. Before measuring fuel consumption on proving ground test, the vehicle was instrumented with fuel consumption measuring system for monitoring fuel consumption in real-time and driver aid device, etc.

2.1 Data Acquisition System

The data acquisition system was installed at the corresponding position of the vehicle, and sufficient key parameters were collected such as GPS data, CAN data, oil temperature data, and atmospheric temperature data. The data acquisition system is mainly divided into three parts: the Sierra part, the temperature acquisition part, and the SOMAT part. Circuit connection of data acquisition system is shown in Fig. 1.

2.2 Fuel Consumption Measuring System

The fuel consumption rate is the mass of the fuel consumed by the engine for per 1 kW effective power in 1 h. Brake specific fuel consumption (BSFC) is commonly

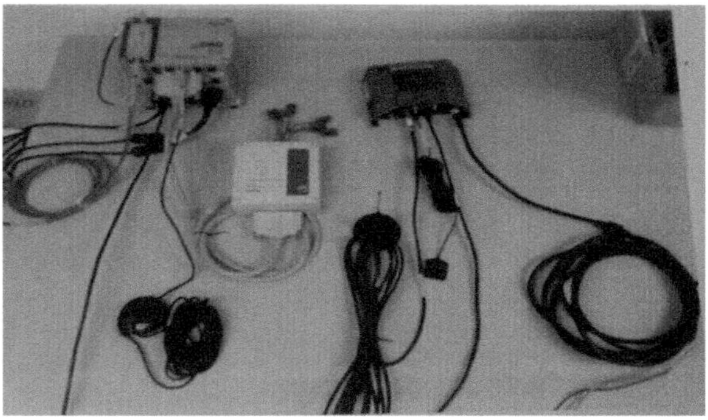

Fig. 1 Circuit connection of data acquisition system

used as a measure of how efficiently a given amount of fuel is being converted by the engine into a specific output torque [5]; the unit is g/(kW h). By accurately measuring the fuel flow rate, engine output torque, and engine speed, BSFC can be recorded.

$$\mathrm{BSFC} = \frac{\gamma}{\mathrm{Pe}} = \frac{\gamma}{\omega \cdot \tau} \tag{1}$$

Pe is the power produced in watts where (kW)
γ is fuel consumption in grams per second (g/s)
τ is engine output torque in newton meters (Nm)
ω is engine speed in radians per second (rads/s)

2.2.1 Fuel Consumption Measurement Device

The fuel supply system was designed and improved to accurately measure fuel consumption. Coriolis mass flow meter was an independent fuel consumption measurement device; the advantage was mass flow rather than volume flow which needs calculation as temperature changes. The flow accuracy was 0.1%, and the repeatability was better than 0.05%; the calibrated Coriolis mass flow meter was to measure the fuel flow rate from 0.24 to 300 kg/h. Figures 2 and 3 show the interior structure and the installation position of Coriolis mass flow meter.

2.2.2 Engine Output Torque Measurement Device

The torque transducer in engine output torque measurement device was suitable for measuring the torque of the long-term continuous high-speed rotating components. On account of engine output torque being generated by rotation, the non-contact torque transducer was long-term maintenance free. The torque transducer consisted of two separate parts: the rotor and the stator. The rotor comprised the measuring

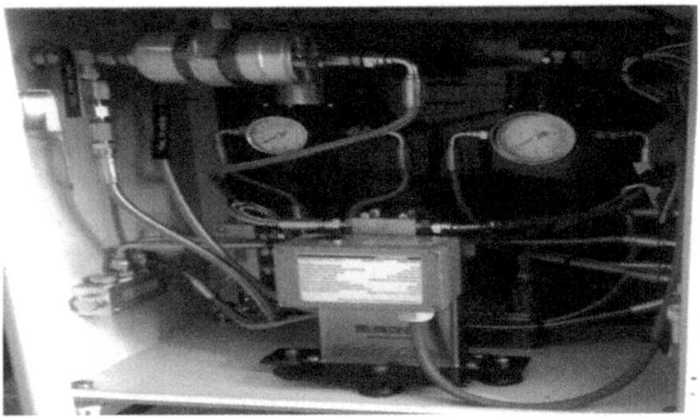

Fig. 2 Interior structure of Coriolis flow meter

Fig. 3 Installation position of Coriolis flow meter

body and the signal transmission elements. The rotor electronics generated the bridge excitation voltage and transmitted the measuring signal to the evaluation unit (stator) and was located at the outer circumference of flywheel, and the stator electronics was located at the flywheel cover. The installation positions of the rotor and the stator are seen in Figs. 4 and 5.

2.2.3 Engine Speed Measurement Device

Engine speed is monitored by and accessible from the vehicle J1939 Control Area Network Binary Unit System (CANBus). This system is the broadcast serial bus standard for connecting electronic control units (ECUs) to send and receive messages to control the vehicle's electronic systems [6].

2.3 *Driver Aid Device*

The control box of driver aid device was installed in cab, and the tablet of driver aid device which was connected to the control box via Ethernet was mounted on the

Fig. 4 Installation positions of the stator

Fig. 5 Installation positions of the rotor

windscreen, as shown in Figs. 6 and 7. The driver drove the vehicle in accordance with drive cycle shown on the screen and voice prompt.

After installation of the data acquisition systems, raw data from over 10 km of driver operation was collected through a computer connected to the SOMAT part and was analyzed to detect the correlation of various parameters. According to the correlation of various parameters and the driver's driving experience, by adjusting the parameters of data acquisition systems, the data acquisition systems could operate normally and the vehicle was in a best running state to give reliable results.

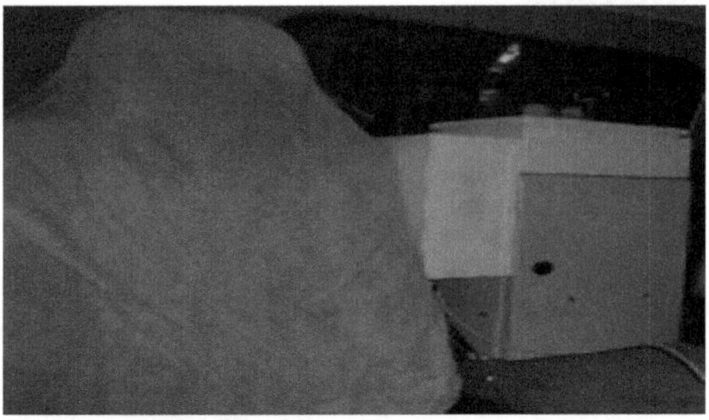

Fig. 6 Installation positions of control box of driver aid device

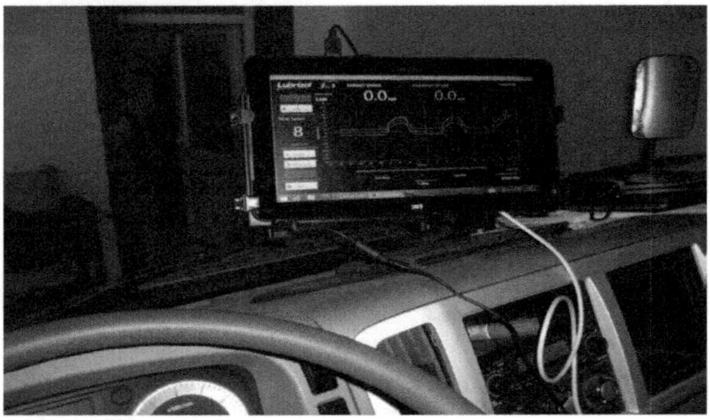

Fig. 7 Tablet of driver aid device

3 Drive Cycle Development

With the development of urbanization and logistics industry, the inter-city logistics industry has developed rapidly, and statistics show that the sales of HDV increase year by year. Beijing is the most typical city in China which has a relatively well-developed road network and large population density, and road traffic in Beijing is characterized by more traffic jams, longer vehicle retention time, and more express roads, etc. Therefore, selected vehicle was a FAW J6F configuration, and selected operation conditions were EMS logistics fleets in Beijing. To comprehensively generate road condition in Beijing, the test vehicle was given to EMS logistics fleets that used it on normal logistics delivery routes in Beijing urban and suburb.

First of all, multiple key parameters, which was about 3047 km in urban and 15504 km in suburb, were generated from the test vehicle during its normal operation, and the specific route is shown in Fig. 8. The multiple key parameters were processed, analyzed, and calculated by mathematical statistics. Based on these statistics, a drive cycle of EMS vehicle in Beijing shown in Fig. 9 was established; the characteristics of the drive cycle of EMS vehicle in Beijing are seen in Table 5. The drive cycle consisted of two distinct parts: part 1 to simulate urban operation and part 2 to simulate suburb driving.

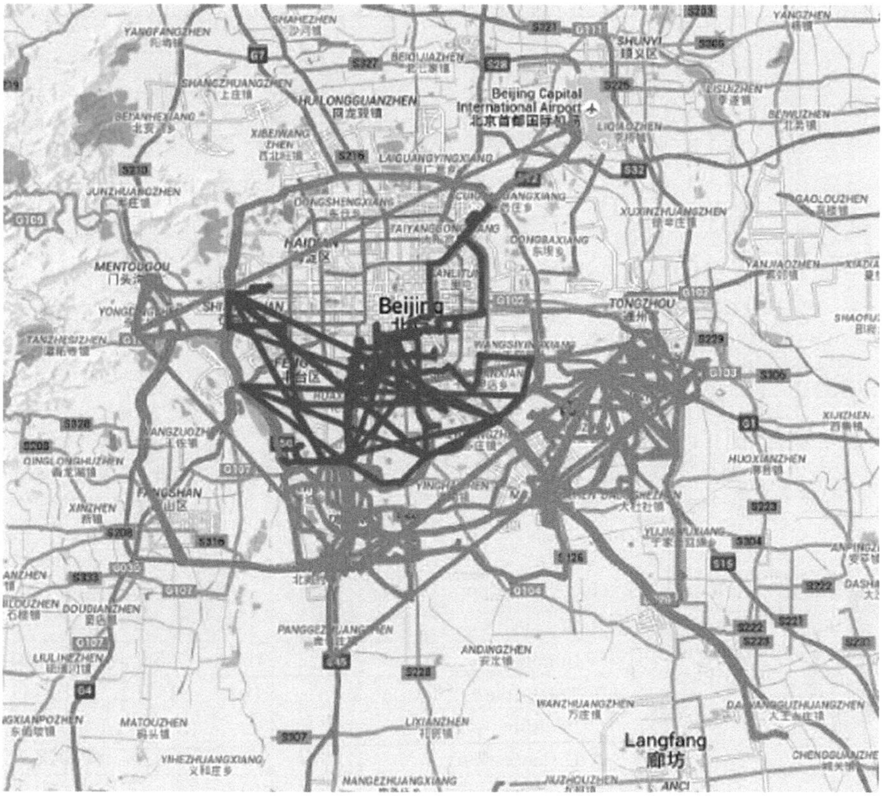

Fig. 8 Specific route of EMS vehicle

4 Fuel Consumption Measurement

In order to investigate fuel economy, tests on chassis dynamometer and proving ground were implemented on a FAW J6F to compare the fuel consumption of the candidate oil 5 W-30 diesel engine oil (using Lubrizol fuel economy formulation) with that of the reference oil CI-4 15 W-40 diesel engine oil.

4.1 Chassis Dynamometer Test

Chassis dynamometer test of fuel consumption was performed according to GB/T 27840-2011 “Fuel Consumption Test Methods for Heavy-Duty Commercial Vehicles,” using C-WTVC drive cycle.

An engine oil fuel economy evaluation test of the candidate oil relative to the reference oil was undertaken on chassis dynamometer at China Automotive

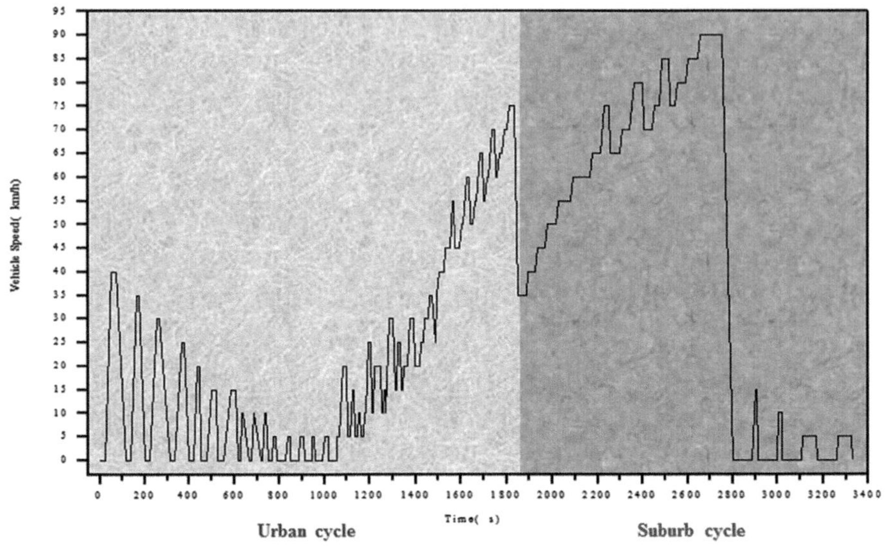

Fig. 9 Drive cycle of EMS vehicle in Beijing

Table 5 Characteristics of the drive cycle of EMS vehicle in Beijing

Condition	Urban	Suburb	Total
Total time (s)	1803	1562	3365
Idle time (s)	–	–	23.3
Distance (km)	9.8	18.8	28.7
Average speed (km/h)	19.6	43.4	30.7
Maximum acceleration (m/s^2)	–	–	0.35
Maximum deceleration (m/s^2)	–	–	1.67
Proportion (%)	64	36	100

Technology and Research Center (CATARC), which should be able to accurately simulate the vehicle road resistance, accelerating and decelerating condition, equivalent inertia, etc. During the test, data acquisition system also collected the data (Fig. 10).

The result showed a clear differentiation between the averages of fuel consumption in L/100 km with a total of 3.8% fuel saving on the candidate oil compared to the reference oil.

4.2 *Proving Ground Test*

The proving ground test was carried out at proving ground under the Ministry of Transport from July 2015 to November 2015. The selected track field was a high-speed circuit, as shown in Figs. 11 and 12 (Table 6).

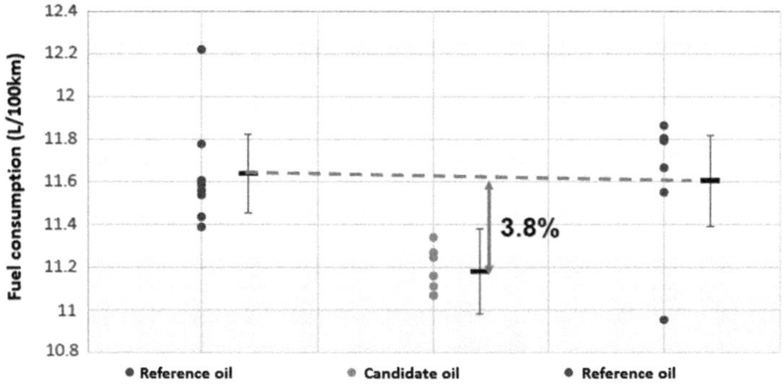

Fig. 10 Results of chassis dynamometer method test

Fig. 11 Proving ground under the ministry of transport

Fig. 12 Vehicle running on high-speed circuit

The un-laden weight of FAW J6F is 2.8 tons, and laden weight is 4.5 ton. Barrels of oil and water were added as a temporary solution to increase the weight. The laden FAW vehicle was tested with the drive cycle of EMS vehicle in Beijing on the proving ground under the Ministry of Transport. The candidate oil showed a total of 1.7% less BSFC compared to the reference oil in total drive cycle. Comparison of the corrected BSFC data can be found in Fig. 13.

Table 6 Test cycle number of proving ground test

Sample	Effective test cycle number	Invalid test cycle number	Total test cycle number
Reference oil	20	2	22
Candidate oil	20	9	29
Reference oil	24	25	59

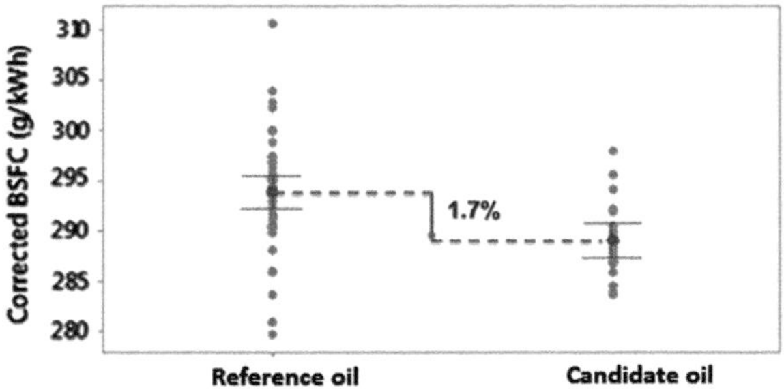

Fig. 13 Corrected BSFC data (total cycle)

It can be seen in Figs. 14 and 15 that using the candidate oil relative to the reference oil, 2.5% BSFC could be saved in urban cycle of the EMS vehicle drive cycle in Beijing; the candidate oil measured 1.5% improvement over the reference oil in suburb cycle of the drive cycle.

The drive cycle of EMS vehicle in Beijing was required to differentiate the fuel economy benefits from oil to oil while remaining representative of the real-world data recorded. The candidate oil showed a decrease of 1.5–2.5% BSFC compared to the reference oil.

5 Field Test

The field test began from August 2015; the test vehicle information is in Table 7. Before the test, the vehicle had driven more than 110,000 km. The engine was flushed twice with the candidate oil to minimize the contamination from the used

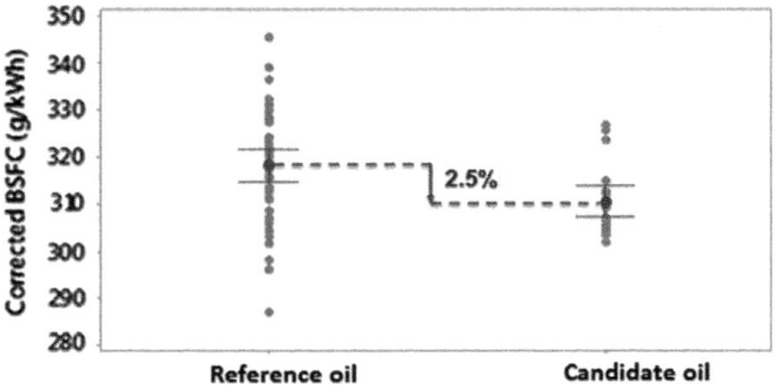

Fig. 14 Corrected BSFC data (urban cycle)

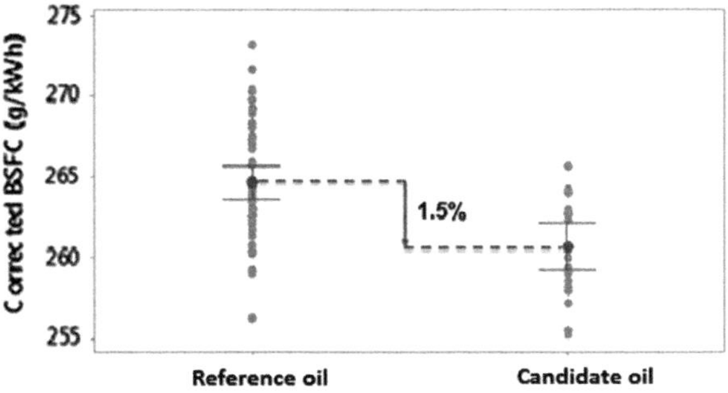

Fig. 15 Corrected BSFC data (suburb cycle)

Table 7 Test vehicle information

Information	
Vehicle number	吉H39367
Engine style	CA4DD1-15E4R
Vehicle style	FAW J6F, platform truck
Initial mileage	112,603 km
Carrying capacity	Usually 7 t–8 t, occasionally more than 10 t
Running route	Rural road in Yanji, cement paving road

oil. During the test, 200 ml oil sample was taken for analysis. When it reached 150,000 km, the engine was taken back to FAW R&C to disassemble for measurement (Fig. 16).

5.1 Oil Sample Analysis

5.1.1 Viscosity

The viscosity of used oil is influenced by oxidation, light component evaporation, fuel dilution, water contamination, and mechanical shear. From Fig. 17, it was found that the viscosity changed slightly with the mileage increase, and it did not exceed the limited value of GB/T 7607-2010. The oil could still provide good lubricating protection.

5.1.2 TAN & TBN

The oil property can be influenced by low and high temperatures, and the oxidation of oil usually leads to TAN increasing and TBN decreasing. As is shown in Fig. 18,

Fig. 16 Field test vehicle

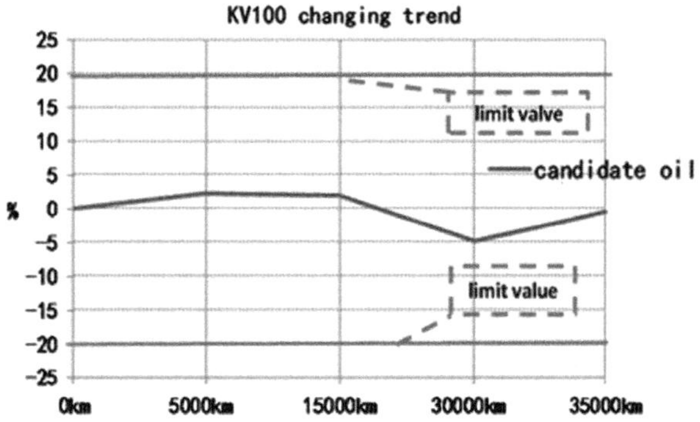

Fig. 17 Viscosity changing trend at 100

TAN increased slightly with the mileage increasing while TBN decreased; neither of them exceed the limited value of GB/T 7607-2010. The oil could still provide good acid neutralization ability.

5.1.3 Oxidation, Nitration, and Sulfonation

The oil aging extent can be reflected by oxidation, nitration, and sulfonation. As is shown in Fig. 19, oxidation, nitration, and sulfonation increased slightly with the mileage increasing.

5.1.4 Metallic Element

Iron in the engine oil mainly comes from wear of cylinder and piston ring, copper comes from liner and bearing, and aluminum comes from piston and bearing. Silicon mainly comes from contaminant of dust from the external environment. As is shown in Figs. 20, 21, 22, and 23, it was found that iron, copper, aluminum, and silicon increased with the mileage increasing, which did not exceed the limited value of GB/T 7607-2010. The oil could still provide good anti-wear ability.

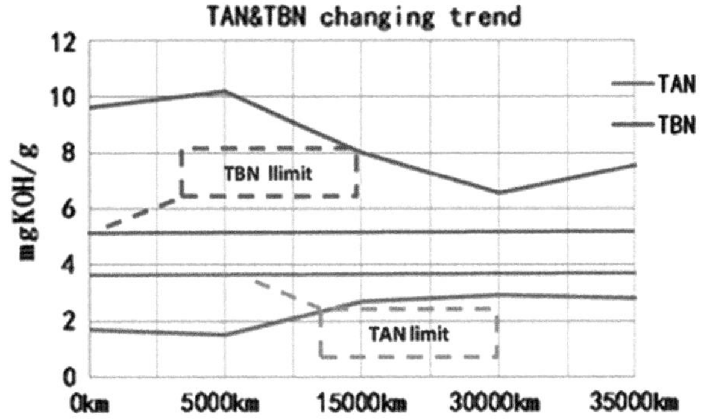

Fig. 18 TAN & TBN changing trend

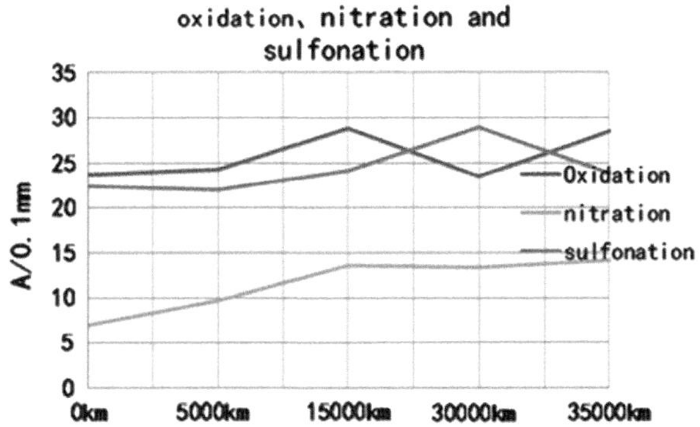

Fig. 19 Oxidation, nitration, and sulfonation changing trend

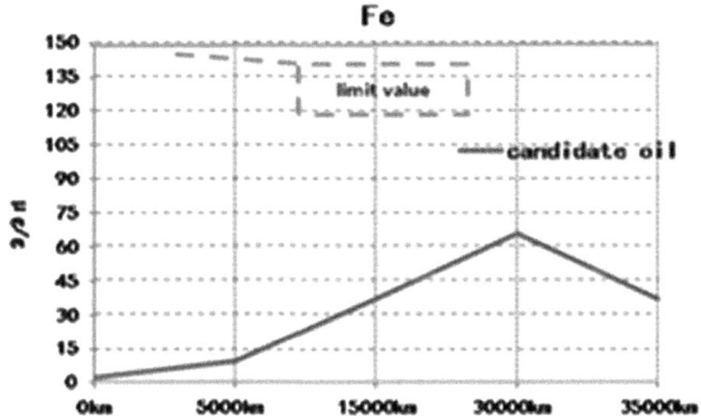

Fig. 20 Iron content

5.2 *Engine Disassemble*

Figures 24, 25, 26, 27, 28, and 29 are pictures of the disassembled engine. Carbon deposition on the head of piston was normal, reticulate pattern of cylinder was good, and there was no severe polishing. The wearing area of the middle part of the

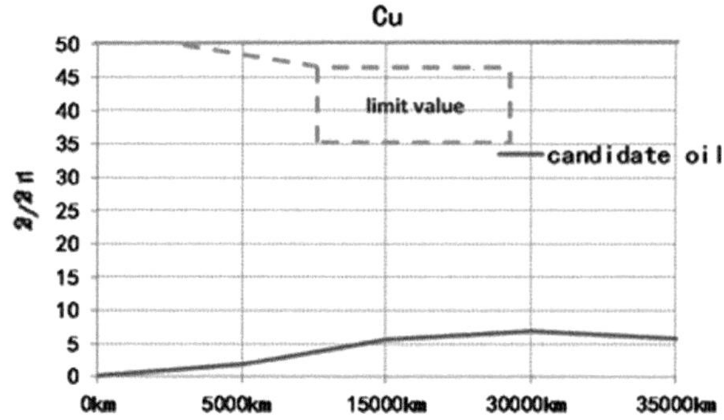

Fig. 21 Copper content

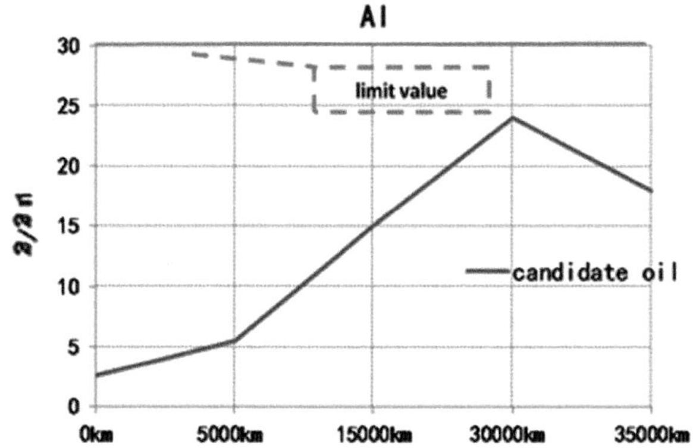

Fig. 22 Aluminum content

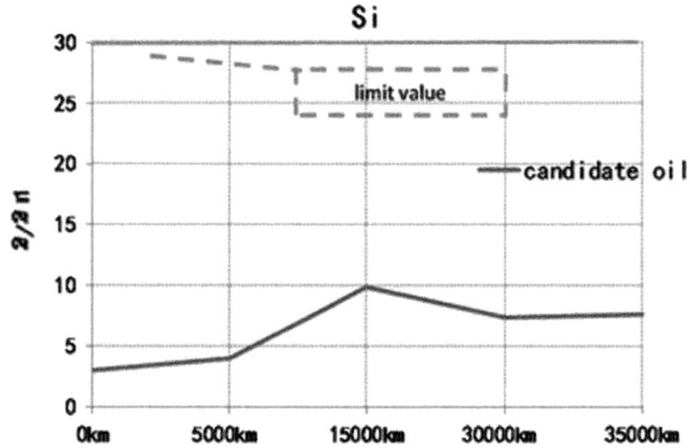

Fig. 23 Silicon content

first piston ring was 80–90%; the second piston ring was 75–80%. The lower bearing of the main bearing had obvious wear due to the worst lubricant condition; two upper bearings of connecting rod had obvious wear; the rest of wearing components were good. In view of the 150,000 km running mileage, the result was acceptable.

Fig. 24 Piston

Fig. 25 Reticulate pattern of cylinder

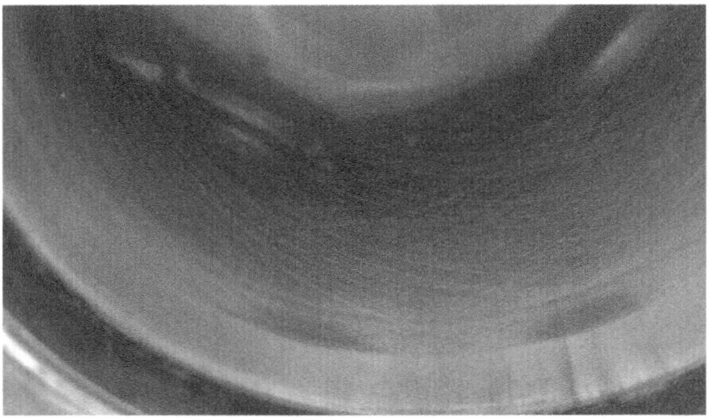

Fig. 26 First piston rings

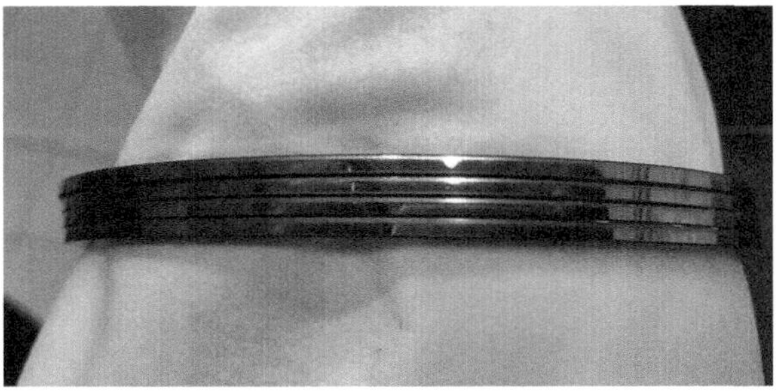

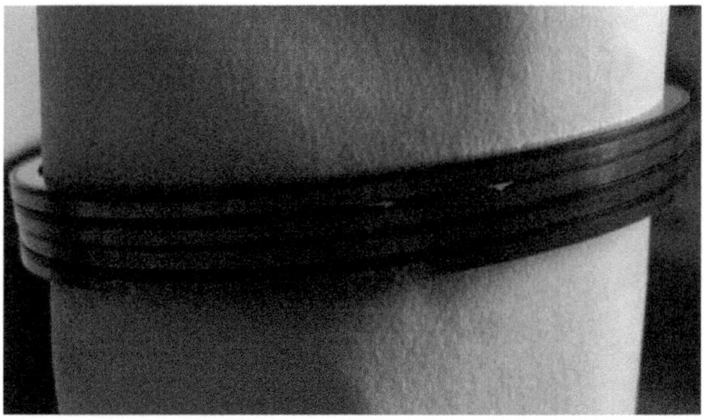

Fig. 27 Second piston rings

Fig. 28 Main bearings

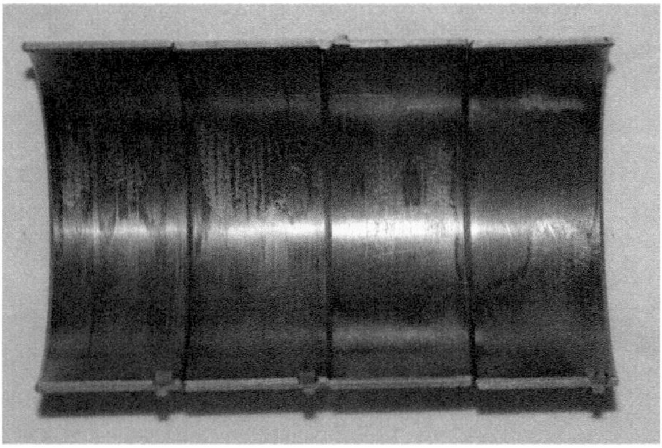

Fig. 29 Connecting rod bearings

6 Conclusions

Based on the analysis of the test results, the following conclusions can be drawn:

Compared with the reference oil CI-4 15 W-40 diesel engine oil, the candidate oil 5 W-30 diesel engine oil (using Lubrizol fuel economy formulation) had the certain fuel efficiency benefits. The result of chassis dynamometer test showed a clear differentiation between the averages of fuel consumption in L/100 km with a total of 3.8% less fuel consumed on the candidate oil compared to the reference oil.

Using the drive cycle of EMS vehicle in Beijing, the proving ground test showed that the candidate oil showed a decrease of 1.5–2.5% BSFC compared to the reference oil.

During the field test, the low-viscosity candidate oil provided good lubricant protection. The oil monitoring result showed that the oil properties were normal. The inspect result of the disassembled engine was acceptable. Due to the long-running mileage of the vehicle before the test and the changing of oil, it was difficult to determine the real reason for engine wear. More research work and tests need to be carried out.

Acknowledgements The authors would like to acknowledge support from Lubrizol and FAW R&D Center colleagues: Bin, Zhang from Lubrizol for project coordination, Fei, Gu and Jifeng, Han from FAW for vehicle and engine technical support, Li, Yu and Weinan, Jiang from FAW for analysis of the field trial data.

References

1. Transportpolicy (2017). http://transportpolicy.net/index.php?title=Global_Comparison:_Heavy-duty_Fuel_Economy_and_GHG. Accessed 12 April 2017
2. API (2017) http://www.api.org/products-and-services/engine-oil/eolcs-categories-and-documents/latest-oil-categories. Accessed 12 April 2017
3. DieselNet (2017) https://www.dieselnet.com/standards/cycles/#us-ld. Accessed 12 April 2017
4. SAE International Surface Vehicle Recommended Practice "Fuel consumption test procedure —type II," SAE Standard J1321, Rev Feb 2012
5. Moore W, Sutton M, Donnelly K (2013) Development of long haul heavy duty vehicle real world fuel economy measurement technique. SAE Technical Paper 2013-01-0330. https://doi.org/10.4271/2013-01-0330
6. Moore W, Finch T, Sutton M (2013) Development of heavy duty diesel real world drive cycles for fuel economy measurements. SAE Technical Paper 2013-01-2568. https://doi.org/10.4271/2013-01-2568

A Simple Theory of Fuel Escape from Combustion Chamber

Xuan Huang, Jianjiao Zhuo, Yongming Zheng, Hongyu Wang and Zhaoshuo Wu

Nomenclature

C_1	Tyre damping coefficients
H_i	Perturbation input
DOF	Degree of freedom

1 Introduction

Turbo gasoline direct-injection (TGDI) engine has been put on the Chinese market in recent years to improve fuel economy and provide higher power for vehicle. However, compared to the normally aspirated engine, direct-injection and turbo technology resulted in shorter time for fuel atomization and higher blow-by flow, respectively. It gives rise to higher quantity of fuel escape from combustion chamber and further resulted in higher fuel consumption, oil consumption, fuel dilution, liner wore problems, and so on.

Study of fuel escape theory has been presented from the point of fuel dilution mechanism in this paper. Primarily, there are two questions for fuel dilution mechanism study:

(A) How and how much fuel escaped into crankcase.
(B) How and how much fuel came out from crankcase.

This paper mainly focuses on the first question and proposed a simple model to study approaches and mass of fuel escape into crankcase during one engine cycle; furthermore, test has been proposed to study quantity of fuel mass which comes out from combustion room. This model and corresponding experiments were applied to

X. Huang (✉) · J. Zhuo · Y. Zheng · H. Wang · Z. Wu
Powertrain Department of Technical Center, Jiangxi Changhe Automobile CO., LTD, Jingdezhen 333000, China
e-mail: huangx_01@changhe-suzuki.com

Society of Automotive Engineers of China (SAE-China) (ed.), *Proceedings of the 19th Asia Pacific Automotive Engineering Conference & SAE-China Congress 2017: Selected Papers*, Lecture Notes in Electrical Engineering 486, https://doi.org/10.1007/978-981-10-8506-2_34

one TGDI engine; good agreement was obtained on blow-by flow and fuel mass escape from combustion chamber prediction and experiments. However, though fuel came out from crankcase and fuel dilution theory was beyond this paper, it should be noticed that the test result of fuel mass came out from crankcase of this TGDI engine indicated even different mass of fuel escaped into crankcase at different engine speed and torque, while fuel mass out from it was quite stable (stable oil temperature resulted in stable evaporation of fuel in oil). Suggestions about reduction of fuel mass escape into crankcase and also fuel dilution have been provided based on the calculation and test results.

2 Fuel Escape Theory

Before the fuel escape study, primarily sources of fuel in combustion chamber should be defined. For TGDI engine, there are two ways for fuel into combustion chamber; one is fuel injection, and the other is fuel back from CCV system and intake system. It is reasonable to assume fuel injection is the only source for fuel into combustion chamber because the very small quantity of fuel mass back from intake system compare with mass of fuel injection. On the other hand, for the fuel escape into crankcase, generally, there are two approaches:

(A) Fuel gas escape with blow-by flow.
(B) Fuel wetting on liner and escape with liner oil film by ring scrape.

In the first case, fuel gas is part of blow by flow, which can be considered as idea gas and homogeneous mixed in combustion chamber gas, which means same percentage fuel and other gas of total combustion chamber gas escaped into crankcase. Otherwise, fuel particle distribution in combustion chamber involved in very complex gas flow changes inside it which is beyond the purpose of engineering application. In the second case, fuel wetting has great influence on dynamic viscosity of liner lubricants and further affect lubricant film thickness on liner which cannot be simply ignored. Under this circumstance, influences of lubricants dynamic viscosity which result from different extent of fuel wetting have been considered by quote Vogel equation and fitting function. However, fuel wetting on liner has very complex changes which are caused by fuel evaporation, thermal variation, and gas flow motion in combustion chamber. So, in order to simplify calculations, it will be assumed that the fuel wetting on liner oil film is homogeneous, and further thermal variation and evaporation will be ignored.

2.1 Fuel Escape with Blow-by

Significant amount of work has been done through engine tests, experiments, and theoretical modeling to simulate the blow-by mass during engine circle [1, 2]. The modeling work of blow-by is advanced by considering more and more detailed physics and realistic descriptions for different purposes such as bore distortion, ring flutter, ring and groove interaction, and so on. For fuel escape study, it is acceptable that construct model by coupling ring dynamics, blow-by through crevice and lubrication of liner models. For blow-by calculation, liner and piston rings have been divided into regions according to the different cavity regions as shown in Fig. 1; the gas pressures in these regions are assumed to be uniform.

Additionally, fuel mass in blow-by was considered in this paper. After fuel injected into combustion chamber, some part of fuel mass is mixed with air in combustion chamber and others wetting on liner. Fuel mixed in air would escape with blow-by flow, and it can be considered as four conditions according to the engine stroke and fuel injection timing setting as Fig. 2.

(A) Intake stoke before SOI: no fuel in combustion chamber; blow-by is assumed as 100% air.
(B) Intake stoke after SOI and compression stoke before spark: fuel injected into combustion chamber and blow-by has different percentage of fuel inside it according to the fuel mass corresponding to the crankshaft angle. It should be mentioned that in this case fuel was assumed to be uniformly distributed in combustion chamber.
(C) Power stoke before EVO: gas and unburned fuel in combustion chamber, blow-by has very small percentage fuel inside it which can be ignored.

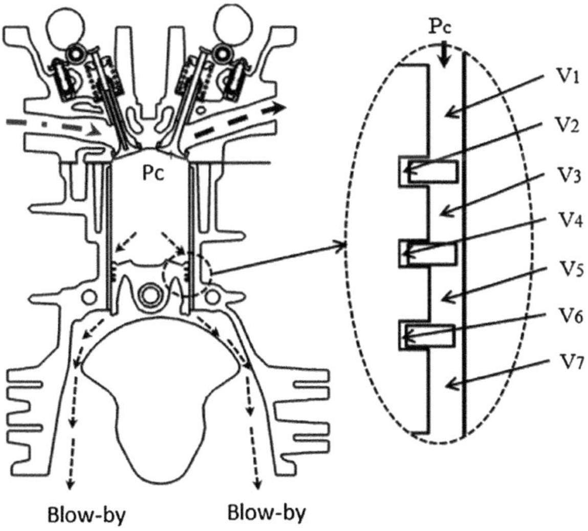

Fig. 1 Cavity regions

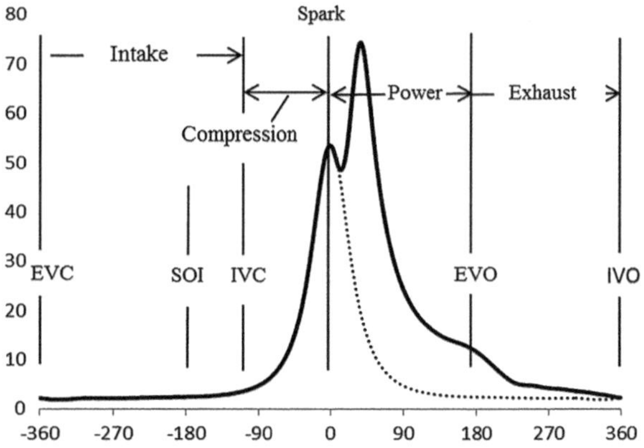

Fig. 2 Stoke and fuel injection timing setting [3]

(D) Exhaust stoke: exhaust gas and very samll amount of unburner fuel in combustion chamber, very small amount of unburned fuel in blow by which can be ignored and assume blow by is 100% exhaust gas.

Since correlation between fuel mass in blow-by and crankshaft timing is defined, it can calculate fuel mass escape with blow-by by combining fuel gas flow model with blow-by model.

Three sub-models constructed for blow-by calculation are as below.

(A) Gas flow through crevice.
(B) Lubrication of liner oil film.
(C) Ring dynamic.

2.1.1 Gas Flow Through Crevice

Many researches have been done on simulation of gas flow through crevice; the simplest model only includes gas flow through ring gaps; further, gas flow through clearance between ring and its groove has been proposed, and ring flutter and liner wear model have been applied to blow-by simulation model. However, for fuel escape calculation purpose, it is acceptable to use blow-by model with gas flow through ring gaps and clearances between ring and its groove.

Firstly, flow of gas through clearance between a ring and its groove basic theory is as below.

The total gas flow between each ring and its groove is given by [2].

$$\dot{m}_{\mathrm{cle}} = \int_{0}^{2\pi} \dot{m}_{\mathrm{cle},n}\, \mathrm{d}\theta \tag{1}$$

where

$$\dot{m}_{\text{cle},n} = \rho\left(-\frac{h^3}{12\mu}\frac{\partial P}{\partial x} + U\frac{h}{2}\right) \tag{2}$$

where

ρ: density of gas, kg/m^3
h: nominal film thickness, m
μ: viscosity, Pa s
P: gas pressure, Pa
U: ring speed, m/s

Secondly, flow of gas through the ring gap theory is as below.

The gas flow through piston ring gap is assumed to be the isentropic orifice flow as [1].

$$\dot{m}_{\text{gap}} = \frac{c_D A_{\text{gap}} P_U}{\sqrt{RT_U}} f_m \tag{3}$$

where

$$f_m = \gamma^{\frac{1}{2}}\left(\frac{2}{\gamma+1}\right)^{\frac{\gamma+1}{2(\gamma+1)}} \text{ for } \frac{P_D}{P_U} \le \left(\frac{2}{\gamma+1}\right)^{\frac{\gamma}{\gamma-1}} \tag{3.1}$$

Or

$$f_m = \left(\frac{P_D}{P_U}\right)^{\frac{1}{\gamma}}\left\{\frac{2\gamma}{\gamma-1}\left[1-\left(\frac{P_D}{P_U}\right)^{\frac{\gamma-1}{\gamma}}\right]\right\}^{\frac{1}{2}} \tag{3.2}$$

for $\frac{P_D}{P_U} > \left(\frac{2}{\gamma+1}\right)^{\frac{\gamma}{\gamma-1}}$

And

$$C_D = 0.85 - 0.25\left(\frac{P_D}{P_U}\right)^2 \tag{3.3}$$

where

C_D: the discharge flow coefficient.
A_{gap}: ring gap area, mm^2.
R: gas constant of ideal air.
T_U: temperature of gas on the higher pressure side, K.
P_U: pressure of gas on the higher pressure side, Pa.
P_D: pressure of gas on the lower pressure side, Pa.
γ: coefficient of heat

Equation of C_D is obtained by fitting the experimentally measured data [4].

Till now, all gas flow models have been built, and equations based on mass conservation and Newton's second law are as follow.

$$\frac{\mathrm{m}_{V2}}{P_{V2}}\frac{\mathrm{d}P_2}{\mathrm{d}t} = \dot{m}_{12} - \dot{m}_{23} \tag{4}$$

$$\frac{\mathrm{m}_{V3}}{P_{V3}}\frac{\mathrm{d}P_3}{\mathrm{d}t} = \dot{m}_{13} + \dot{m}_{23} - \dot{m}_{34} - \dot{m}_{33} \tag{5}$$

$$\frac{\mathrm{m}_{V4}}{P_{V4}}\frac{\mathrm{d}P_4}{\mathrm{d}t} = \dot{m}_{34} - \dot{m}_{45} \tag{6}$$

$$\frac{\mathrm{m}_{V5}}{P_{V5}}\frac{\mathrm{d}P_5}{\mathrm{d}t} = \dot{m}_{35} + \dot{m}_{45} - \dot{m}_{56} - \dot{m}_{57} \tag{7}$$

$$\frac{\mathrm{m}_{V6}}{P_{V6}}\frac{\mathrm{d}P_6}{\mathrm{d}t} = \dot{m}_{56} - \dot{m}_{67} \tag{8}$$

where,

m_{Vi}: initial gas mass in region V_i, g
V_{i}: volume of region i divided as shown in Fig. 1., mm^3.
P_{i}: pressure of region i divided as shown in Fig. 1., Pa.
$\dot{m}_{ij}$: mass flow from region i to region j, g/s.

2.1.2 Fuel Mass Escape with Blow-by

It is important to understand air and fuel mass inside combustion chamber at different crankshaft angle; however, as mentioned before, fuel escape only happens under two conditions as follow.

(A) Intake stoke after SOI.
(B) Compression stoke before spark.

In both cases, air mass can be calculated by combustion chamber pressure and gas temperature, and fuel mass can be calculated by the injection fuel mass flow. Because the fuel inject pressure is significantly higher than combustion chamber pressure under these two conditions, and fuel gas is considered as ideal gas in this case, it is feasible to calculate air mass according to the gas state equation directly as given by the formula below.

$$M_{\mathrm{air}} = \frac{P_C V_C}{RT_{\mathrm{gas}}} \tag{9}$$

M_{air}: total air mass in combustion chamber.
V_C: combustion chamber volume at given crankshaft angle

where

$$V_C = V_{\text{cylhead}} + V_{\text{piston}} + \left(S + h_{\text{gasket}}\right) * \pi \left(\frac{B}{2}\right)^2 \quad (9.1)$$

where

V_{cylhead}: combustion room volume on cylinder head, m^3.
V_{piston}: combustion room volume on piston, m^3.
S: piston travel distance, m.
h_{gasket}: thickness of cylinder head gasket after tighten, m.
B: bore diameter, m

And,

$$\mathrm{S} = r_{\text{crank}}(1 - \cos\theta) + \frac{r_{\text{crank}}}{l_{\text{rod}}}(1 - \cos 2\theta) - \frac{r_{\text{crank}}}{l_{\text{rod}}}\frac{\mathrm{e}}{r_{\text{crank}}}\sin\theta \quad (9.2)$$

where

P_C: combustion chamber pressure at given crankshaft angle Pa.
T_{gas}: gas temperature at given crankshaft angle K

$$M_{\text{fuel}} = (\dot{m}_{\text{finlet}} - \dot{m}_{\text{foutlet}} - \dot{m}_{\text{fl}})\Delta t \quad (10)$$

M_{fuel}: fuel mass in combustion chamber at given crankshaft angle, g.
$\dot{m}_{\text{finlet}}$: fuel mass flow injection into combustion chamber at given crankshaft angle, g.
$\dot{m}_{\text{foutlet}}$: fuel mass came out from combustion chamber at given crankshaft angle, g.
$\dot{m}_{\text{fl}}$: fuel mass flow on liner at given crankshaft angle, g

Fuel mass escape with blow-by flow can be calculated by combine the equations as follow.

$$\dot{m}_{\text{foutlet}} = \left(\frac{\dot{m}_{\text{gap}} + \dot{m}_{\text{cle}}}{M_{\text{air}}}\right) \times (M_{\text{fuel}} - \dot{m}_{\text{fl}}) \quad (11)$$

2.2 *Fuel Escape with Liner Lubricant (Fuel Wetting)*

Piston ring lubrication is influenced by liner and ring surfaces roughness, ring surface characteristics, bore distortion, liner wear, piston tilt, ring flutter, lubricant effects, and so on. For fuel escape study, fuel wetting on liner mainly resulted in the lubricant characteristics change and further affects the piston ring lubrication condition, especially the inlet condition of up-stokes. Because during up-stokes, lubricant inlet condition depends on the lubricant film left on liner after down-stokes which effective viscosity and shear significantly influenced by the continuing fuel injection and further wetting on liner. Under this circumstance, formulas for lubricant viscosity change at given crankshaft angle calculation which have been proposed as follows based on the assumption of fuel uniform distribution on liner.

$$\log \mu_e = A \log \mu_{\text{oil}} + B \log \mu_{\text{fuel}} \tag{12}$$

where

A: percentage of oil mass on liner lubricant film.
B: percentage of fuel mass on liner lubricant film
μ_{oil}: dynamic viscosity of engine oil, Pa s.
μ_{fuel}: dynamic viscosity of fuel, Pa s

Furthermore, the Vogel equation is used for the lubricant temperature/viscosity relation

$$\mu_0(T) = \rho \alpha_0 \exp\left(\frac{\theta_1}{\theta_2 + T}\right) \tag{13}$$

where,

ρ: density of lubricant, kg/m^3.
$\alpha_0, \theta_1, \theta_2$: Vogel correlation parameters.
T: temperature of lubricant, K

Combining Eqs. (12) and (13),

$$\log \mu_2 = A \log\left(\rho \alpha_{\text{oil}0} \exp\left(\frac{\theta_{\text{oil}1}}{\theta_{\text{oil}2} + T_{\text{oil}}}\right)\right) + B \log\left(\rho \alpha_{\text{fuel}0} \exp\left(\frac{\theta_{\text{fuel}1}}{\theta_{\text{fuel}2} + T_{\text{fuel}}}\right)\right) \tag{14}$$

In this research, fuel mass on liner as an input data has been given by experimental results directly as shown in Fig. 3, which shows the liner fuel film mass percentage at given crankshaft angle.

On the hand of ring lubrication and oil film thickness on liner, as researches done before [1, 2], three lubricating conditions have been applied for rings and different ring regions, respectively.

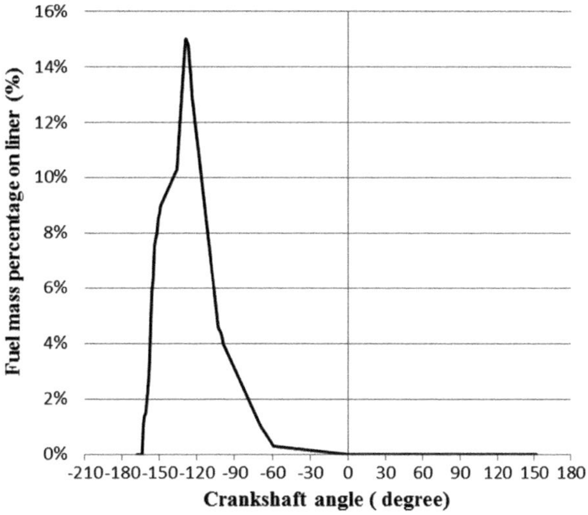

Fig. 3 Fuel mass percentage on liner

2.2.1 Pure Hydrodynamic Lubrication Model

For the rough surface, the Reynold's equation with flow factor proposed by Patir and Cheng [5] is used as below:

$$\frac{\partial}{\partial \mathrm{x}}\left(\phi_x \frac{h^3}{\mu}\frac{\partial \overline{P}}{\partial x}\right) + \frac{\partial}{\partial \mathrm{y}}\left(\phi_y \frac{h^3}{\mu}\frac{\partial \overline{P}}{\partial y}\right) = \frac{U_1 + U_2}{2}\frac{\partial \bar{h}_T}{\partial x} + \frac{U_1 - U_2}{2}\sigma\frac{\partial \phi_S}{\partial x} + \frac{\partial \bar{h}_T}{\partial t} \tag{15}$$

where

$\bar{h}_T$: averaged local film thickness, m.
ϕ_S: shear factor.
ρ: lubricant density, kg/m^3.
ϕ_x: flow factor.
ϕ_y: flow factor.
μ_e: oil effective viscosity, Pa s.
U_1: ring speed, m/s.
U_2: liner speed, m/s

The oil film thickness between the piston ring and liner is much smaller than piston ring radius. It can be calculated as an infinite flat bearing. The average Reynold's formula is simplified as follow.

$$\frac{\partial}{\partial \mathrm{x}}\left(\phi_x \frac{h^3}{\mu}\frac{\partial \overline{P}}{\partial x}\right) + \frac{\partial}{\partial \mathrm{y}}\left(\phi_y \frac{h^3}{\mu}\frac{\partial \overline{P}}{\partial y}\right) = \frac{U_\mathrm{r}}{2}\frac{\partial \bar{h}_T}{\partial x} + \frac{U_r}{2}\sigma\frac{\partial \phi_S}{\partial x} + \frac{\partial \bar{h}_T}{\partial t} \tag{16}$$

The mean hydrodynamic shear stress is given by:

$$\bar{\tau} = -\frac{\mu_e}{\bar{h}}(\phi_f + \phi_{fs}) + \phi_{fp}\frac{\bar{h}}{2}\frac{\partial \overline{P}}{\partial y} \tag{17}$$

The mean hydrodynamic force on Y-axis is given by:

$$\overline{F}_{y,\mathrm{oil}} = \bar{\tau} dA_{\mathrm{oil}} \tag{18}$$

2.2.2 Pure Boundary Lubrication Model (Asperity Contact)

When oil supply is insufficient, pure boundary lubrication model is used according to Greenwood and Tripp's [6] theory:

$$P_{y,\mathrm{asp}} = \frac{16\sqrt{2}}{15}\pi(\sigma\beta\eta)^2 E' \sqrt{\frac{\sigma}{\beta}} F_{2.5}\left(\frac{\bar{h}(h)}{\sigma}\right) \tag{19}$$

η: surface density of asperity peaks, m.
β: radius at asperity summit, m.
E': composite elastic modulus of contact, Pa.
$F_{2.5}$: Gaussian distribution factor [7]

$$F_{y,\mathrm{asp}} = P_{y,\mathrm{asp}}\, dA_{\mathrm{asp}} \tag{20}$$

2.2.3 Mixed Lubrication Model

For mixed lubrication condition, force on y-axis direction contains both asperity contact force and hydrodynamic force.

$$F_{y,\mathrm{mix}} = F_{y,\mathrm{asp}} + F_{y,\mathrm{oil}} \tag{21}$$

2.3 *Piston Ring Force Model*

Previously, all force generated by gas pressure and lubricants has been established as shown in Fig. 4, and the force balancing equation is introduced in this section.

$$F_y = F_{y,\mathrm{asp}} + F_{y,\mathrm{oil}} + F_{y,e} + F_{y,\mathrm{gas}} + F_{y,a} = 0 \tag{22}$$

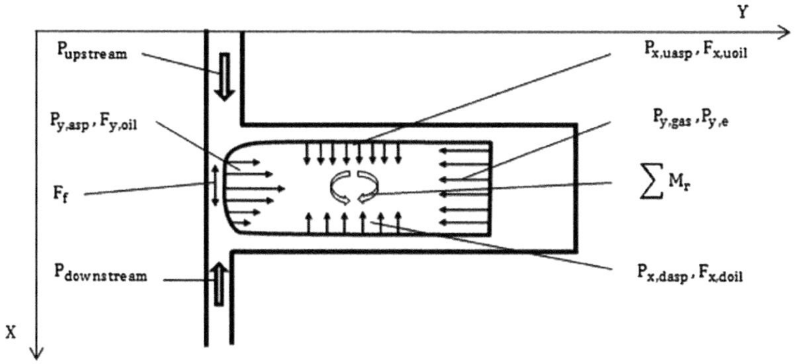

Fig. 4 Ring force definition

$$F_x = F_{\mathrm{x},u\mathrm{asp}} + F_{x,u\mathrm{oil}} + F_{x,d\mathrm{asp}} + F_{x,d\mathrm{oil}} + F_{x,\mathrm{upstream}} + F_{x,\mathrm{downstream}} + F_f + F_{x,a} = 0 \tag{23}$$

$$\sum M_r = 0 \tag{24}$$

3 Simulation Algorithms

(1) Define initial gas pressure in different regions.
(2) Calculate gas flow at given crankshaft angle.
(3) Define initial lubricant thickness for every given crankshaft angle.
(4) Calculate ϕ_x, ϕ_s and other relative arguments.
(5) Solve average Reynolds equation.
(6) Calculate asperity contact force and iteration lubricant thickness according to the force balance.
(7) Set circulation condition based on 720° crankshaft angle.

$$|(h_0 - h_{720})/h_0| < \xi$$

where

h_0: minimum lubricant thickness at 0° crankshaft angle.
h_{720}: minimum lubricant thickness at 720° crankshaft angle.
ξ: convergence tolerance for lubricant thickness iteration.

4 Experiments

In order to measure flow of blow by, and mass of fuel escaped into crankcase, and furtherly fuel mass escaped out from crankcase, experiments have applied to one three cylinder TGDI engine, related engine parameters as Table 1.

In order to get accurate blow-by data, it is necessary to block all air make up ports for engine before blow-by measurement, and for TGDI engine with two pressure control valves for boost and unboost states, corresponding two blow-by meters are used for test at some special engine running conditions. In this case, AVL blow-by meter has been applied to measure blow-by flow, and results of blow-by map is shown in Fig. 5.

In this paper, experiment have done on the TGDI engine and engine setting for fuel mass in oil and fuel mass out crankcase survey as follow.

It is a new engine for test, built history, measurement data and performance check shown good consistence with other engines. After about 15 h normal

Table 1 Engine basic parameters

Parameters	Values
Bore size	73 mm
Stroke	81.77 mm
Liner surface roughness	3.2 μm
Ring surface roughness	1.6 μm
*T*TDC	460.16 K
*T*BDC	422.16 K
Lubricants	SAE-0 W-20
Fuel	92#

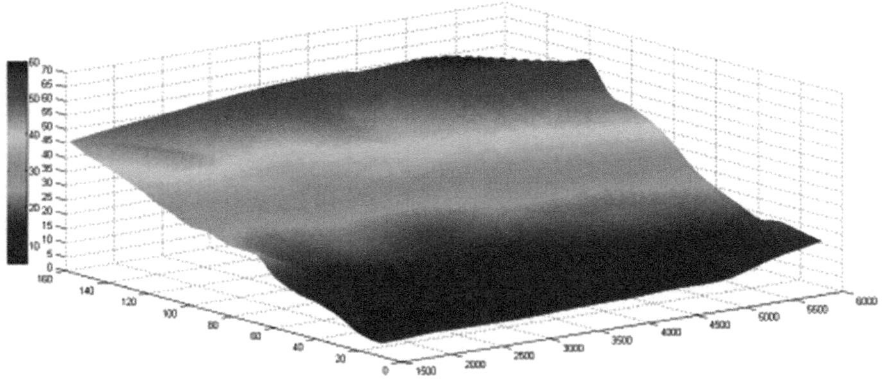

Fig. 5 Blow-by map

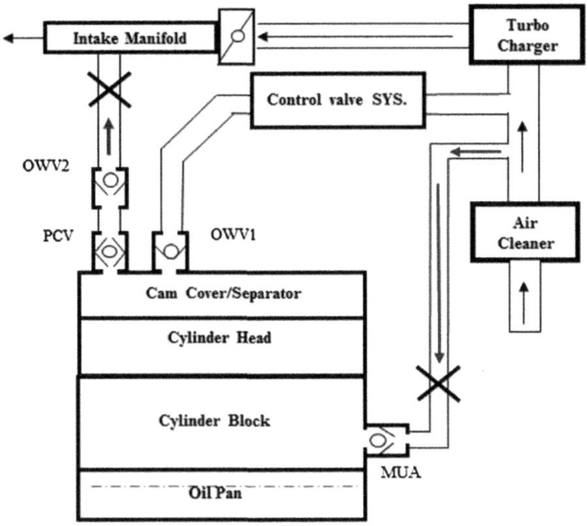

Fig. 6 Engine setting on test bed

running-in test procedure, engine is set at the state for fuel mass measuring as shown in Fig. 6.

where OWV1 means the crankcase pressure control on boost side, and × means port has disconnected and both sides have physically blocked, MUA means make up valve; because only full load condition test is required in this case, full load valve port is the only one gas outlet port used here. However, it is a feasible change to other ports according to different purpose and test conditions.

Details of control valve system are shown in Fig. 7.

As shown in Fig. 7, switch 1 and switch 2 are electrical control valves to separate the fuel and oil mass escaped during warm up and real test. Filter in Fig. 7 is designed for oil and fuels escaped from crankcase capture which mass has been measured before test. However, small amount of fuel will evaporate into blow-by meter in reality, while it is acceptable for engineering application. Because mass of fuel out from crankcase is quite small, in order to get accurate test result, 30 min or 1 h steady engine running condition is performed. Oil temperature and coolant temperature are controlled to 90 and 105 °C, respectively.

As shown in Fig. 8, fuel mass out from crankcase is quite stable (around 70 g) through all 1 h tests with different engine speed and fuel dilution percentage.

On the other hand, fuel mass in oil can be measured by fuel dilution percentage measurement and total oil mass measurement. For fuel dilution percentage measurement, oil sample should be taken at required oil temperature such as 90 °C.

According to Fig. 9, the highest fuel dilution occurred at engine speed 4000 rpm and full load condition. And as mentioned before, stable fuel mass out from crankcase under different running conditions, and 4000 rpm and full load shown highest fuel dilution percentage according to the Fig. 9, so 4000 rpm and full load

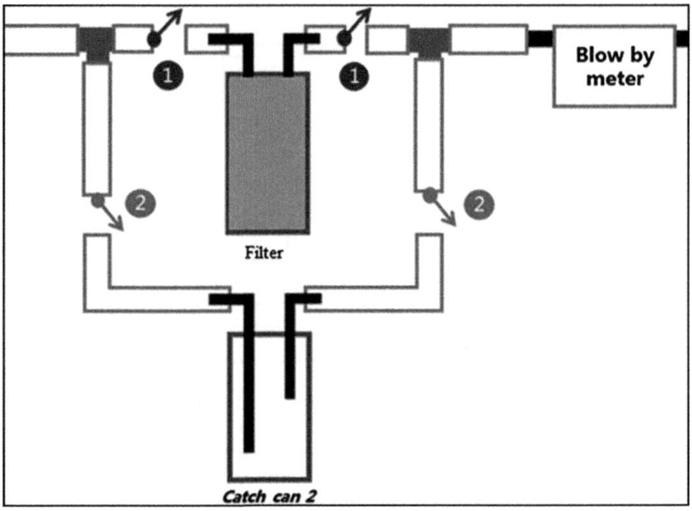

Fig. 7 Control valve system setting on test bed

Fig. 8 Test result of fuel mass out crankcase

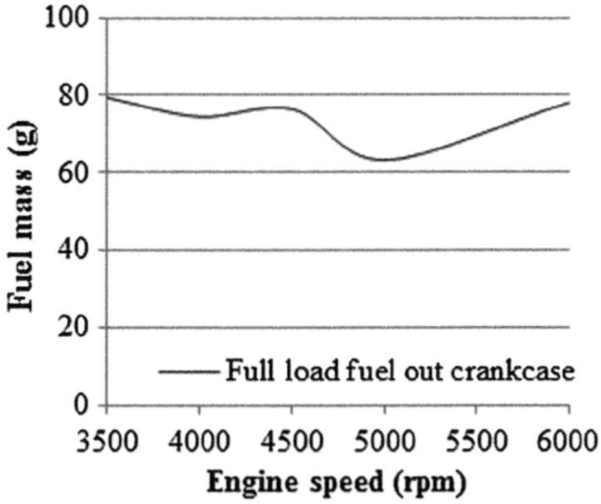

can be defined as the highest fuel escape condition for this engine. Under this circumstance, cylinder pressure at 4000 rpm full load is measured and corresponding calculation of escaped fuel mass is performed based on the reality cylinder pressure (Fig. 10).

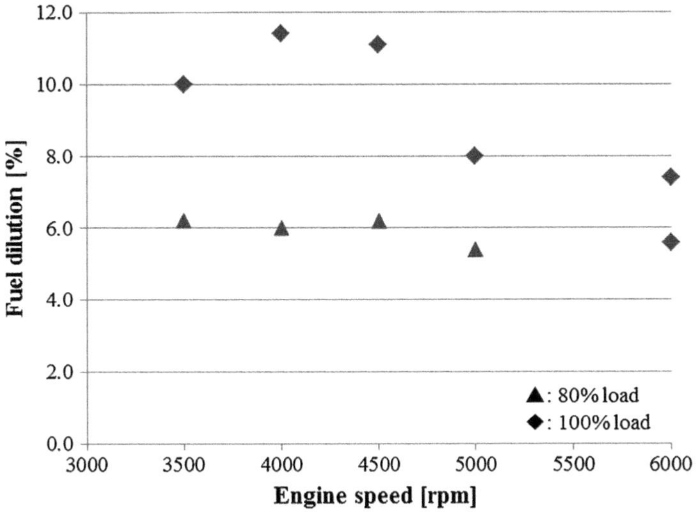

Fig. 9 Test results of fuel dilution percentage

Fig. 10 Cylinder pressure data at 4000 rpm full load

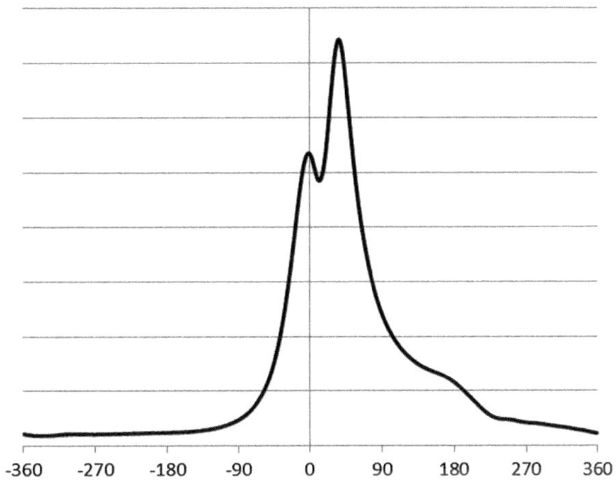

5 Results and Discussions

Flow of blow, fuel mass escaped with blow by and liner film thickness have calculated on the TGDI engine, here below introduced different engine speed but all full load condition results corresponding to the fuel escape experimental data. Furthermore, particularly introduced liner film thickness calculation result at 4000 rpm full load engine running condition, which is the worst case according to the engine experiment results.

5.1 Blow-by Flow and Fuel Mass Escaped with Blow-by

As shown in Fig. 11, blow-by test and calculated results are close enough for engineering application. For this engine, the maximum blow-by is 52 L/min at 6000 rpm and full load engine running condition.

Calculated result of fuel mass escaped into crankcase with blow-by at 4000 rpm full load is shown in Fig. 12.

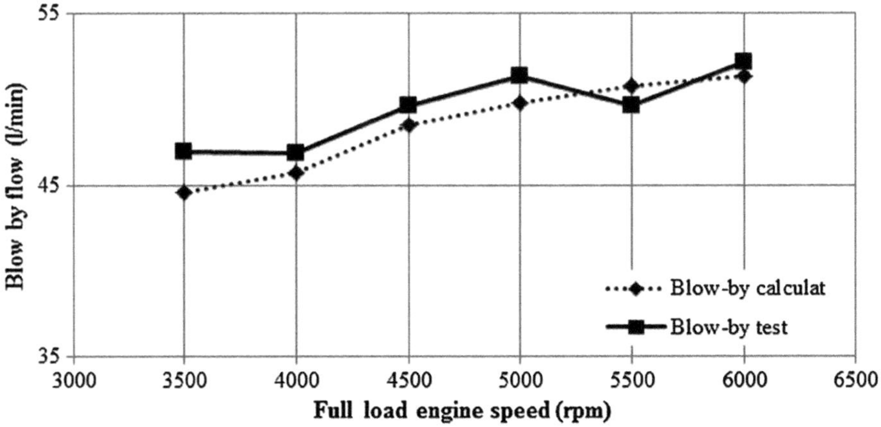

Fig. 11 Blow-by flow comparison

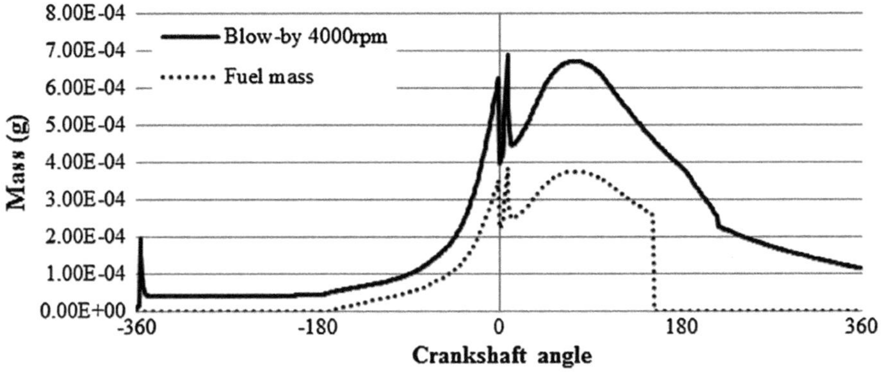

Fig. 12 Fuel mass escape with blow-by

5.2 Liner Lubricant Film and Fuel Mass Escaped with Liner Lubricants

Calculated results of minimum liner lubricant film thickness are shown in Fig. 13, which indicated significant influence on liner oil film thickness during fuel injection timing resulted by fuel wetting on liner.

Fuel mass escaped into crankcase with liner lubricants at 4000 rpm full load is shown in Fig. 14.

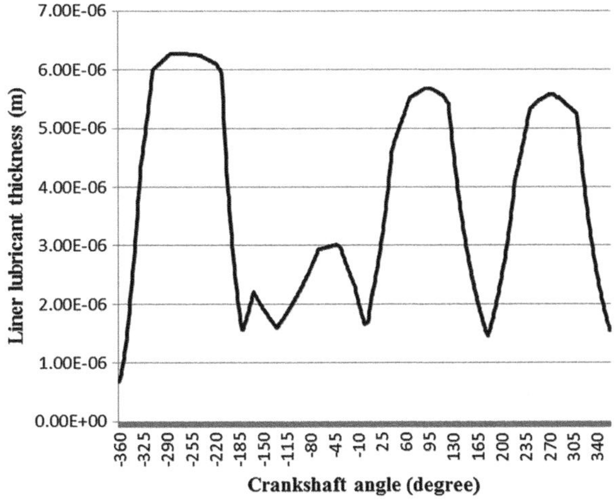

Fig. 13 Liner lubricant film thickness

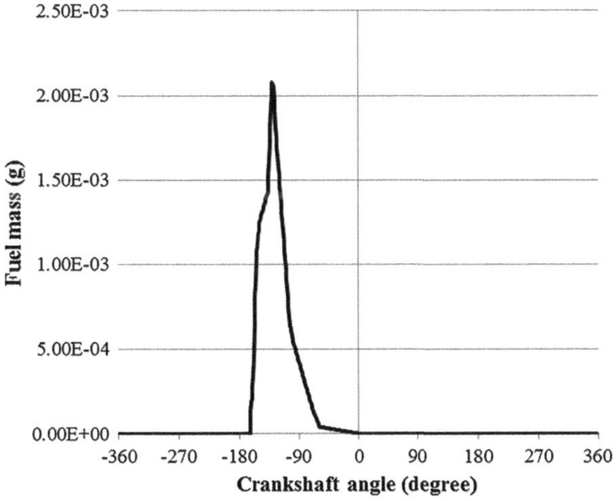

Fig. 14 Fuel mass escape with liner lubricant

5.3 Total Fuel Mass Escaped into Crankcase

Total fuel mass escaped into crankcase as Fig. 15 about 5% difference can be found through the comparison result between test and calculation at 3500 rpm and full load condition, while generally it is acceptable for prediction.

As shown in Figs. 15 and 16, fuel mass escaped into crankcase with liner lubricant and blow-by is about 60 and 40%, respectively, according to the calculated results, which means fuel wetting on liner is the main way for fuel escape into crankcase in this case, and decrease of fuel wetting percentage and blow-by flow will be the effective ways to reduce fuel mass escape into crankcase. While on the other hand, it is essentially to find the mechanism of fuel in blow-by mix into oil and further caused fuel dilution issue.

5.4 Total Fuel Mass Escaped from Crankcase

Experimental data of fuel mass remain in oil and out crankcase have been shown as Figs. 17 and 8. And comparison with fuel mass in oil and out crankcase is shown in Fig. 18, which indicated that most parts of fuel remained in the crankcase and further resulted in the high percentage of fuel dilution.

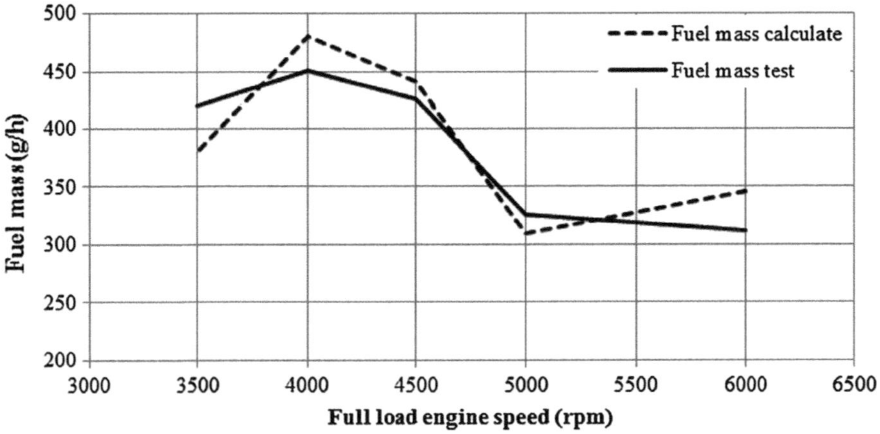

Fig. 15 Fuel mass escape comparison

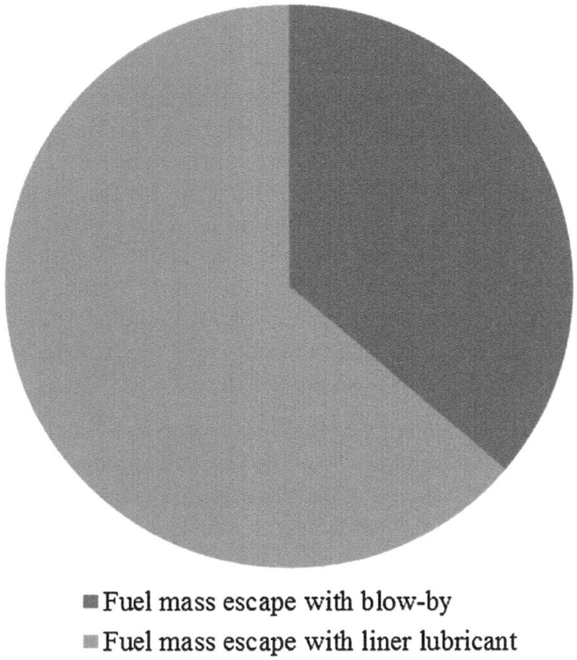

Fig. 16 Comparison between fuel mass escape with blow-by and liner lubricant at 4000 rpm full load

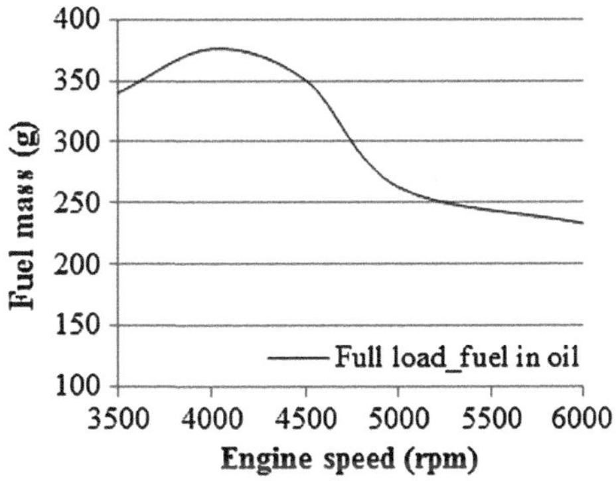

Fig. 17 Test result of fuel mass in oil

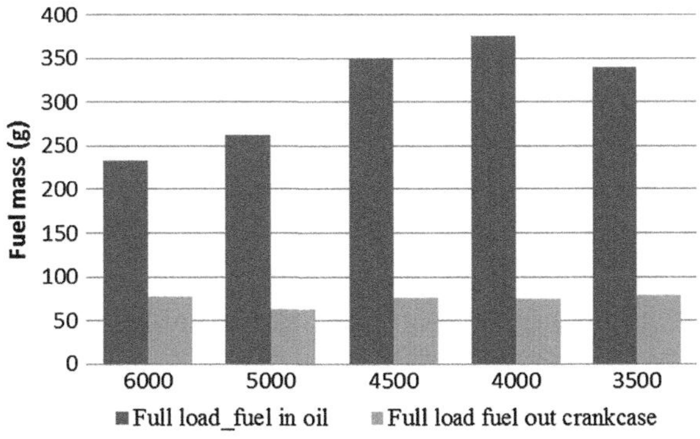

Fig. 18 Comparison with fuel mass in oil and out crankcase

6 Conclusions

The calculated results and combination of these two models indicated that fuel wetting on liner and blow-by both are the primary approach for fuel that came into crankcase. However, during engine running in the condition with low fuel wetting percentage, blow-by is the main way for fuel escape.

For fuel dilution phenomenon, till now there is no further study that explained the principle of how fuel and oil are mixed together in the crankcase and oil pan or in the oil separation system of engine, while this paper and calculated results indicated that the formation of fuel dilution should be the main concern on the mechanism of fuel gas and oil vapor mixed together.

Other conclusions can be inferred as follow:

(1) Higher quantity of fuel mass came into crankcase on engine running cycle with higher speed and load. For engine introduced in this paper, severe fuel dilution phenomenon resulted by fuel into crankcase have found on engine.
(2) Fuel dilution percent is a constant number under stable engine running condition because of oil evaporation surface area and oil temperature; fuel mass injection and fuel timing are under constant condition. This means fuel inlet and outlet condition are all constant.
(3) Suggestion for the fuel escapes reduction.
   (A) Coolant temperature: higher coolant temperature can improve liner temperature and further improve fuel evaporation ability on liner and reduce fuel mass on liner.
   (B) Fuel injection: right SOI and timing and multiple injection can reduce the fuel mass that came into crankcase with blow-by, and the possibility of fuel mix with liner oil film and better burning of fuel also have significant influence on this phenomenon.
   (C) Blow-by: smaller blow-by reduces the fuel mass that came into crankcase, while it may result in higher fraction power.

References

1. Tian T, Sonin AA, Wong V, Heywood JB (1997) Modeling the performance of the piston ring-pack in internal combustion engines
2. Wannatong K (2007) Simulation algorithm for piston ring dynamics. Simul Modell Pract Theory:127–146
3. Heywood JB (1988) Internal combustion engine fundamentals. McGraw-Hill, New York
4. Shapiro AH (1953) The dynamics and thermodynamics of compressible fluid. The Roland Press Company, New York
5. Patir N, Cheng HS (1979) Application of average flow model to lubrication between rough sliding surfaces. ASME J Lubr Technol 101:220–230

6. Greenwood JA, Tripp JH The contact of two nominally flat rough surfaces. In: Proceedings of the institution of mechanical engineers, 185(1970-71) 625
7. Yun JE, Chung YC, Chun SM, Lee KY (1995) An application of simplified average Reynolds equation for mixed lubrication analysis of piston ring assembly in an internal combustion engine. SAE Paper 952562

Strategy Research on Key Technologies of Misfire Detection on Gasoline Engine

Long Qin, Chunjiao Zhang, Yongyi Huang, Feng Xu, Xue Lei, Liuchun Yang, Kai Wang, Junpeng Wang, Fengmin Tian and Fanwu Zhang

1 Introduction

The laws and regulations on car emission get more and more restrictive. The State Department of Environmental Protection and General Administration of Quality Supervision jointly released Limits and Measurement Methods for Emissions from light-duty vehicles (VI) at the end of 2016, which described that the OBD systems, through the use of an onboard computer(s), shall monitor emission systems in-use for the actual life of the vehicle and shall be capable of detecting malfunctions of the monitored emission systems, illuminating a malfunction indicator light (MIL) to notify the vehicle operator of detected malfunctions, and storing fault codes identifying the detected malfunctions [1]. Misfire detection is a mandatory of the critical OBD issues. Misfire is defined as, because of no ignition, too lean or rich air fuel mixture, low cylinder pressure at compression stroke, mixture cannot be combusted, when engine is running. From the federal emissions test, a certain percentage of misfire will cause catalyst damage and emissions obviously increase [2]. In OBD systems, the percentage of misfire, which cause catalyst overheating, and irreversible damage, and emissions exceeding OBD limits shall be monitored. So, it is essential to research misfire detection of OBD systems. In this paper, a kind of misfire detection strategy, by instantaneous power and torque in a high-performance 4 cylinder engine ECU, is developed [3, 4].

L. Qin (✉) · C. Zhang · Y. Huang · F. Xu · X. Lei · L. Yang · K. Wang
J. Wang · F. Tian · F. Zhang
Dongfeng Motor Corporation Technical Center, Wuhan, China
e-mail: qinlong@dfmc.com.cn

Society of Automotive Engineers of China (SAE-China) (ed.), *Proceedings of the 19th Asia Pacific Automotive Engineering Conference & SAE-China Congress 2017: Selected Papers*, Lecture Notes in Electrical Engineering 486, https://doi.org/10.1007/978-981-10-8506-2_35

2 Principles of Gasoline Misfire Detection

When engine is under normal combustion and a certain percentage of misfire, engine speed shape, is shown in Fig. 1. It is easily obtained that engine speed behaves abnormal with unstable shape when engine is under misfire (the red region).

The basic misfire monitoring algorithm is based on evaluating the combustion time during which crankshaft covers 90° angle, and the choice to locate the starting point can be selected experimentally, as reported in the following section. In order to put forward a sophisticated and robust method of monitoring each combustion cycles and determine efficient misfire events, in this paper, monitoring misfire by instantaneous power and torque is applied.

$$M = M_L + M_F \tag{1}$$

$$M_F = J\dot{w} \tag{2}$$

$$w = \frac{d\theta}{dt} \tag{3}$$

$$P_F = CM_F w = CJ\,\dot{w}\,w \tag{4}$$

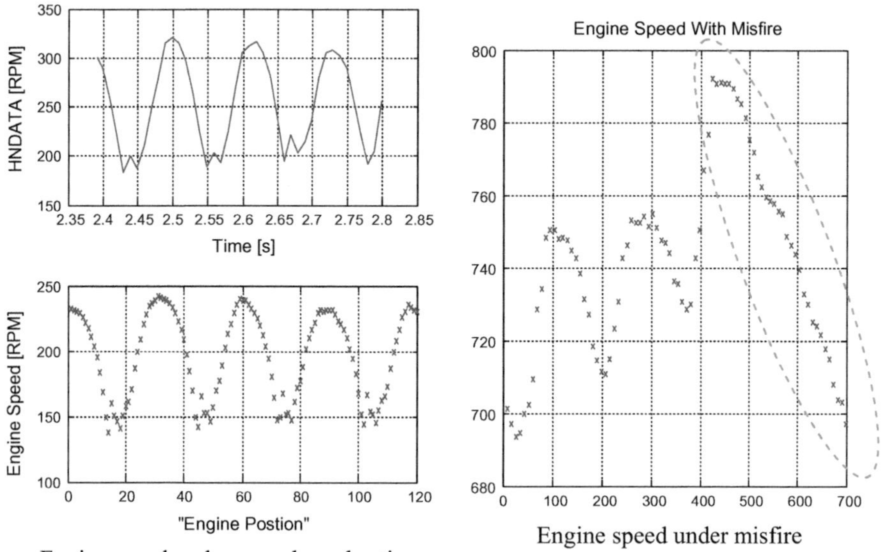

Engine speed undernormal combustion

Engine speed under misfire

Fig. 1 Engine speed behaviors under normal combustion and misfire where M is engine indicated torque; M_L is electric load torque; M_F is flywheel torque; $\dot{w}$ is angular acceleration; J is engine rotation inertia; P_F is flywheel power, C is constant.

If misfire happens, whatever the reason is, the engine instantaneous torque and power will fluctuate [5, 6]. Integrate and evolve engine instantaneous torque and power near combustion top dead center (TDC) with crankshaft covering 90° angle $\theta_n(90)$. The starting point of the optimal angular will be selected depending on engine speed and spark timing. n Indicates engine is in the n time ignition. For simplified presentation, $\theta_n(90)$ showed in the equations is θ_n.

$$\int_0^{\theta_n} M_F \times d\theta = \int_0^{\theta_n} J\dot{w} \times d\theta = \int_0^{\theta_n} J\dot{w} \times d\theta/dt \times dt = J\int_0^{\theta_n} \left(\frac{d\theta}{dt}\right)^2 \approx C_1 \frac{1}{T^2(\theta_n)} \tag{5}$$

$$\int_0^{\theta_n} P_F \times d\theta = \int_0^{\theta_n} CJ\dot{w}w \times d\theta/dt \times dt = J\int_0^{\theta_n} \left(\frac{d\theta}{dt}\right)^3 \approx C_2 \frac{1}{T^3(\theta_n)} \tag{6}$$

where $T(\theta_n)$ is the generic consecutive combustion time interval in the n time ignition.

Then, implement the difference between these integration values both in n and $n+1$ time ignitions, and the following will be received:

$$D_M(n+1,n) = \int_0^{\theta_{n+1}} M_F \times d\theta - \int_0^{\theta_n} M_F \times d\theta \approx C_1\left(\frac{1}{T^2(\theta_{n+1})} - \frac{1}{T^2(\theta_n)}\right) \tag{7}$$

$$D_F(n+1,n) = \int_0^{\theta_{n+1}} P_F \times d\theta - \int_0^{\theta_n} P_F \times d\theta \approx C_2\left(\frac{1}{T^3(\theta_{n+1})} - \frac{1}{T^3(\theta_n)}\right) \tag{8}$$

Further, three consecutive times $D_M(i+1,i)$ and $D_F(i+1,i)$ will be obtained, where $i = n-1, n, n+1$.

Four engine running roughness values, $R_{M1}(n)$, $R_{M2}(n)$, $R_{F1}(n)$, $R_{F1}(n)$, to detect misfire, are proposed, defined as:

$$R_{M1}(n) = D_M(n+1,n) - \text{median}\{D_M(n,n-1), D_M(n+1,n), D_M(n+2,n+1)\} \tag{9}$$

$$R_{M2}(n) = D_M(n+1,n) - \text{mean}\{D_M(n,n-1), D_M(n+1,n), D_M(n+2,n+1)\} \tag{10}$$

$$R_{F1}(n) = D_F(n+1,n) - \text{median}\{D_F(n,n-1), D_F(n+1,n), D_F(n+2,n+1)\} \tag{11}$$

$$R_{F2}(n) = D_F(n+1,n) - \text{mean}\{D_F(n,n-1), D_F(n+1,n), D_F(n+2,n+1)\} \tag{12}$$

where '*median*' is the algorithm of the median value of these three members; '*mean*' is the algorithm of the median value of these three members.

Furthermore, the mean values of engine speed, engine load, and spark timing of three consecutive combustions intervals are calculated. Then these four roughness threshold values will be calibrated after coolant temperature and spark efficiency compensation in the bench [7].

Finally, in the bench, which one(s) of these four roughness and its or their corresponding roughness threshold value(s) will be calibrated and selected, for the purpose to optimize the misfire detection accuracy, robustness, and complexity, depending on different engine speed and engine load. Thus, misfire which happened or not will be determined [8].

3 Realization of Misfire Detection Control Strategy

Those factors that disturb the misfire detection accuracy, robustness, and complexity should be evaded. Enable conditions of misfire detection are as follows [9]:

(a) Engine operating conditions range fulfilled, what laws and regulations claim;
(b) No fuel cut;
(c) Not in neutral gear;
(d) Clutch engaged;
(e) No rough road;
(f) Higher coolant temperature;
(g) Small fluctuation of throttle opening.

Misfire detecting stops if enable conditions do not meet the requirements until enable conditions would be fulfilled in the next time.

The misfire detection control strategy will be proposed in detail, as follows:

(1) Different spark timing will affect engine combustion performance, which will cause different torque and power behavior. According to spark timing and engine speed, the starting point of the optimal angular $\theta_n(90)$ will be determined.
(2) Calculate four engine running roughness values $R_{M1}(n)$, $R_{M2}(n)$, $R_{F1}(n)$, $R_{F1}(n)$, then identify the ignition cylinder number for evaluating.
(3) The mean values of engine speed, engine load, and spark timing of three consecutive combustions intervals are calculated in three consecutive ignition

times $i = n-1, n, n+1$. Then these four roughness threshold values $R_{M1}(n)^0_{\lim}$, $R_{M2}(n)^0_{\lim}$, $R_{F1}(n)^0_{\lim}$, $R_{F2}(n)^0_{\lim}$ will be received after coolant temperature and spark efficiency compensation in the bench. Roughness threshold values will self-adapt during the total engine running cycle. The purpose to self-adapt the roughness threshold values is to compensate them which maybe shift, because of the differences of engine manufacturing, abrasion, fatigue, and aging of air intake and combustion system, by continuously updating the roughness thresholds. Ultimately, the effective and accurate judgment of roughness thresholds will be realized. The following is the roughness thresholds self-adapting flowchart in Fig. 2.

Take one of four roughness values $R_{M1}(n)^0_{\lim}$ self-adapting method as an example, because the left three roughness values self-adapting methods are the same. The way to self-adapt is: When engine is in stable driving conditions and misfire is not detected, store the reasonable roughness values $R_{M1}(n)^j$ in current

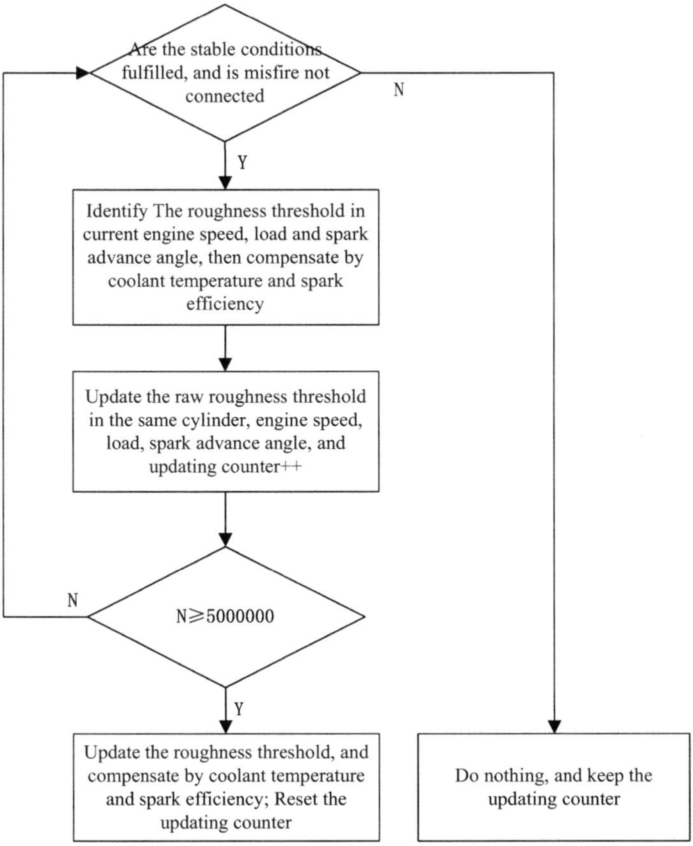

Fig. 2 The roughness threshold self-adapting flowchart

spark timing, engine speed, engine load, and discard the illogical values which have big differences with $R_{M1}(n)^0_{\lim}$, where j indicates that the current cylinder is in j combustion time. And store the raw value of roughness threshold after the 1st time self-adapting with the following algorithm:

$$R'_{M1}(n)^1_{\lim} = (1-\alpha) \times R_{M1}(n)^0_{\lim} + \alpha \times R_{M1}(n)^j \tag{13}$$

where $\alpha \in [0, 1]$.

By such analogy, the raw value of roughness threshold $R'_{M1}(n)^m_{\lim}$ will be determined after m times self-adapting about the same cylinder.

When $m = 5 \times 10^6$, the new value after compensating $R'_{M1}(n)^{5\times10^6}_{\lim}$ with coolant temperature and spark efficiency, as the new roughness threshold $R_{M1}(n)^1_{\lim}$. That is to say, update a new roughness threshold $R_{M1}(n)^s_{\lim}$ in every 5×10^6 self-adapting times, where s indicates that system is in s roughness threshold updating time. At the same time, the other three roughness threshold values $R_{M2}(n)^s_{\lim}$, $R_{F1}(n)^s_{\lim}$, $R_{F2}(n)^s_{\lim}$ can be determined by the same method.

(4) Which one(s) of these four roughness values and its or their corresponding roughness threshold value(s) will be calibrated and selected based on engine speed and engine load of the same cylinder in three consecutive ignition times $i = n-1, n, n+1$. If the selected roughness value(s) is/are lower than the corresponding threshold values, misfire has definitely happened. So the misfire flag and misfire cylinder number will be recorded and saved.

The simplified misfire detection strategy flowchart is demonstrated in Fig. 3. Finally, count the ideal combustion times and misfire times for every cylinder, thus engine misfire percentage will be obtained.

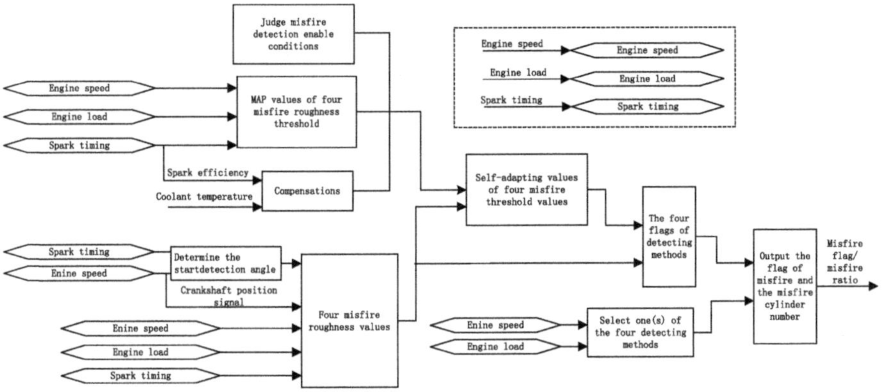

Fig. 3 The simplified flowchart misfire detection strategy flowchart

4 Strategy Verification

To verify the sufficient accuracy of misfire detection strategy, car test is ultimately implemented in the car. The parameters of the engine in the car are as follows:

- Engine type: 4 cylinders in line, MPI
- Displacement: 1994 cc
- Compression ratio: 11.59:1
- Maximum torque: 190 N m at 4000 rpm
- Maximum power: 95 kW at 4500 rpm
- Maximum engine speed: 6000 rpm.

The method to generate misfire in this paper is by a misfire generator, which is very common, flexible, and convenient in the car factory.

As we know, engine speed fluctuation has great difference in variability when engine is running high speed, compared normal combustion with misfire. And when engine speed is in idle, it is most difficult to distinguish the difference in variability. So it is critical to verify the accuracy, reliability, and robustness of misfire detection algorithm when engine is in idle [10].

The following Picture 4 presents the engine running behaviors when engine is in normal combustion. The engine speed steadily fluctuates between 710 and 730 rpm (Fig. 4).

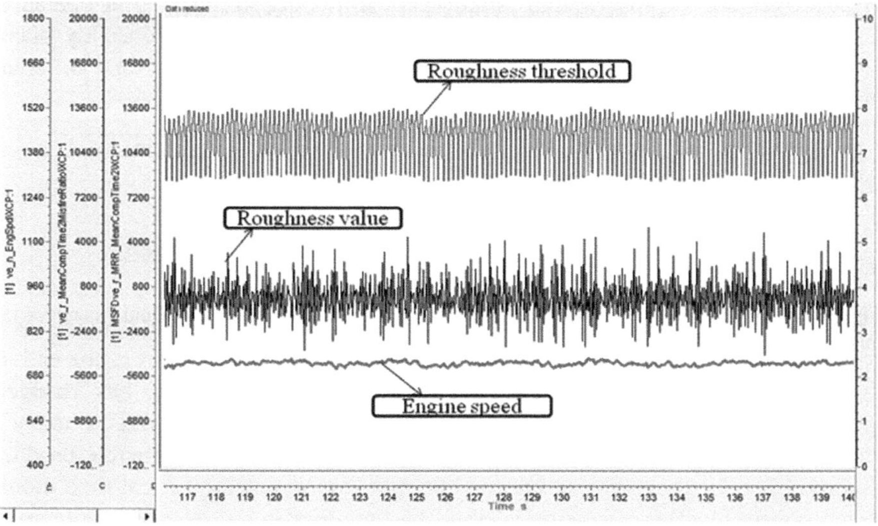

Fig. 4 The engine running behaviors when engine is under normal combustion

When engine is in idle, the starting point of the optimal angular of the evaluating angle $\theta_n(90)$ is equal to 20° after the TDC of the cylinder under combustion. And the final algorithm to select to detect misfire, calibrated by engine speed and loads, is based on Eq. (12). It is reported in Fig. 5 that when only one cylinder misfire happens, and its misfire percentage is equal to 1%, the engine speed will drop approximately 35 rpm.

When each cylinder has the same 1% of misfire, in Fig. 6, it is indicated that engine speed has larger fluctuation. The engine speed will drop approximately between 35 and 60 rpm.

From the result of engine running behaviors when only one cylinder is under misfire, as well as when each cylinder has the same 1% of misfire. The misfire detection algorithm is enough efficient and accurate to distinguish misfire event(s) and the corresponding misfire cylinder(s).

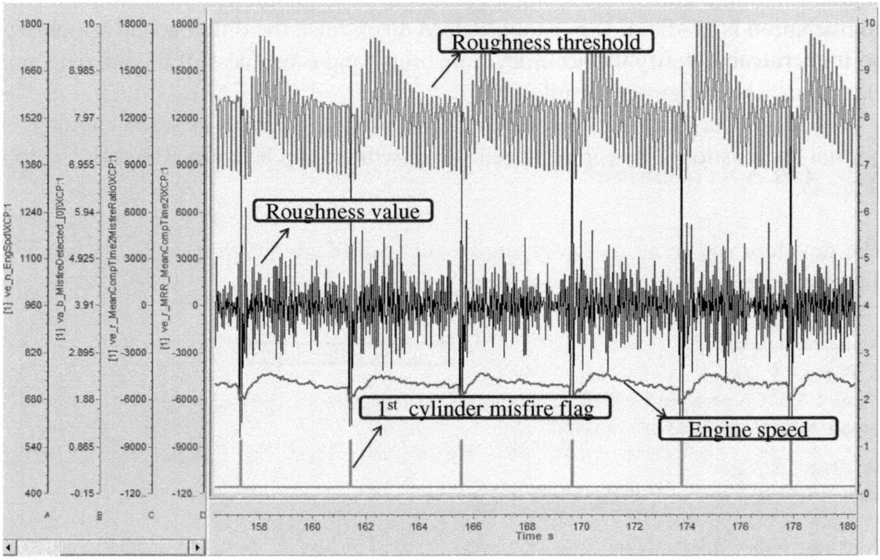

Fig. 5 The engine running behaviors about all four cylinders when only one cylinder is under 1% misfire

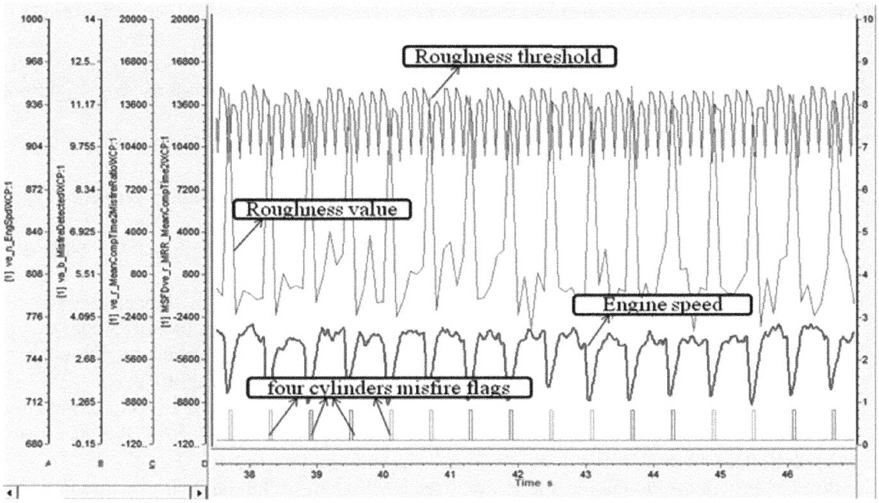

Fig. 6 The engine running behaviors when each cylinder has the same 1% of misfire

5 Conclusion

In the paper, an analysis of the variability of the engine speed fluctuation when engine is under normal combustion and misfire is presented. A misfire detection algorithm, based on instantaneous power and instantaneous torque, is put forward, and verified in a high-performance 4-cylinder engine.

These misfire detection algorithms evaluate the change of power and torque near combustion TDC with crankshaft covering 90° angle. The starting point of the optimal angular is calibrated by engine speed and spark timing.

The basic four engine roughness thresholds are determined by engine speed, load, and spark timing, which are modified by spark efficiency and coolant temperature. At the meantime, roughness thresholds continually self-adapt to recognize the difference in different cylinders when engine is running.

Ultimately, the optimal algorithm of the four ones is selected, according to the calibration of engine speed and load.

This final algorithm is checked and verified in the car, and the result shows that all events of misfire and cylinders are absolutely reported, which meet the requirements of engineering design.

References

1. GB18352.6-2016 (2016) Limits and measurement methods for emissions from light-duty vehicles (III, IV). China Environ Sci Press, Beijing (in Chinese)
2. Ribbens WB, Rizzoni G Onboard diagnosis of engine misfires. SAE Technical paper 901768 1990-09-01
3. Klenk M, Moser W, Muller W, Wimmer W Misfire detection by evaluating crankshaft speed —a means to comply with OBD II. SAE paper 930399 1993-03-01
4. Mahieu V, Duponcheele P, Leduc B et al Misfire detection on S.I. engines, by instantaneous torque analysis. SAE Technical paper 2000-01-0367
5. Moro D, Ponti F, Leduc et al An approach for misfire diagnosis in critical zones of the operating range of a high performance engine. SAE Technical paper 2003-01-1058
6. Merkisz J, Tomaszewski F et al Application of the time-frequency selection of the vibration signal for misfire sensing in diesel engines. SAE Technical paper 2009-01-0242
7. Alves MZ et al Practical evaluation and computational simulation of the effect of engine front end accessory drive (FEAD) dumpers on OBD misfire monitor signal. SAE Technical paper 2012-36-0329
8. Ponti F Instantaneous engine speed time-frequency analysis for on-board misfire detection and cylinder isolation in a V12 high performance engine. 2002 Fall Technical Conference ASME-ICE Division, ICEF2002-538
9. Cavina N, Poggio L, Sartoni G (2011) Misfire and partial burn detection based on ion current measurement. SAE Int J Engines 4(2):2451–2460
10. Williams J An overview of misfiring cylinder engine diagnostic techniques based on crankshaft angular velocity measurements. SAE Technical Paper 960039 1996-02-01

A Study on Finite Element Analysis and Optimization of Rear Bush of Vehicle Lower Control Arm

Dengmu Yuan, Bin Hu, Yong You and Yonghui Qi

1 Introduction

Rubber bushes are widely used in the connection structure of the chassis components. Due to the unique mechanical properties of rubber, this super-elastic material has an important effect on the performance of the vehicle, including low-frequency handling stability, ride comfort, and high-frequency NVH performance. The rubber bush should have an enough life during the period of use. In Fig. 1 for the rubber bush of the lower control arm in road test, there was fatigue damage. The main part of the destruction of the bush is in opening part of the transition fillet. This phenomenon is mainly due to local stress concentration and the average stress caused by a larger value.

The finite element modeling of the cracked rubber bush is carried out. The Mooney–Rivlin constitutive model of rubber is selected, and the stiffness simulation of the finite element model is built by using the ABAQUS. The result is evaluated by the stiffness of test to ensure that the model is accurate. Through the static analysis of the common conditions of the vehicle, it is found that the regional results with the largest stress and the easiest concentration are consistent with the damage in the road test. By changing the arrangement direction of the opening angle of the bush, the approximate model of the simulation result is established by ISIGHT. The optimal opening angle of bush is obtained, and the stress of distribution is reasonable.

D. Yuan (✉) · Y. You · Y. Qi
Commercial Vehicle Business Department, Chongqing Changan Automobile Co., Ltd., 260 East jianxin Road, jiangbei District, Chongqing, China
e-mail: 21454423@qq.com

B. Hu
CAE Department, Chengdu Konghui Automobile Science & Technology Co., Ltd., Chengdu, China

Society of Automotive Engineers of China (SAE-China) (ed.), *Proceedings of the 19th Asia Pacific Automotive Engineering Conference & SAE-China Congress 2017: Selected Papers*, Lecture Notes in Electrical Engineering 486, https://doi.org/10.1007/978-981-10-8506-2_36

Fig. 1 Cracked bush of the lower control arm

2 Finite Element Modeling and Analysis of Rubber Bushing

2.1 Establishment of Finite Element Model

According to the geometry, a first-order tetrahedral element is established for the rubber bushing, with a cell size of 1 mm and a number of 204401. The inner ring of the bush is rigidly coupled, and the outer ring of the bush restrains the degree of freedom in six directions. The surface of the rubber is self-contacting. The initial direction of the bush opening passes through the center of the ball of the lower control arm, the model shown in Fig. 2.

Unlike normal metal materials, the deformation of rubber materials is very complex, accompanied by large displacement and strain. The rubber is a nonlinear material. Different to normal metal material, which can be described with a few linear coefficients, the super-elastic materials like rubber need the 17 continuum mechanics theory. Therefore, the strain energy density function is used in this paper to describe the mechanical properties of rubber. The finite element solver software is needed to input the corresponding strain energy density function parameters. The parameters are determined by many experiments, including the uniaxial tensile compression test, biaxial tensile compression test, flat test, volume test and so on. In the simulation, the commonly used strain energy function constitutive model contains Mooney–Rivlin, Ogden, Yeoh, neo-Hookean model. Each model is a special form of the strain energy function [1]. Once the form of the strain energy function w is determined, the Cauchy stress tensor can be determined:

Fig. 2 .

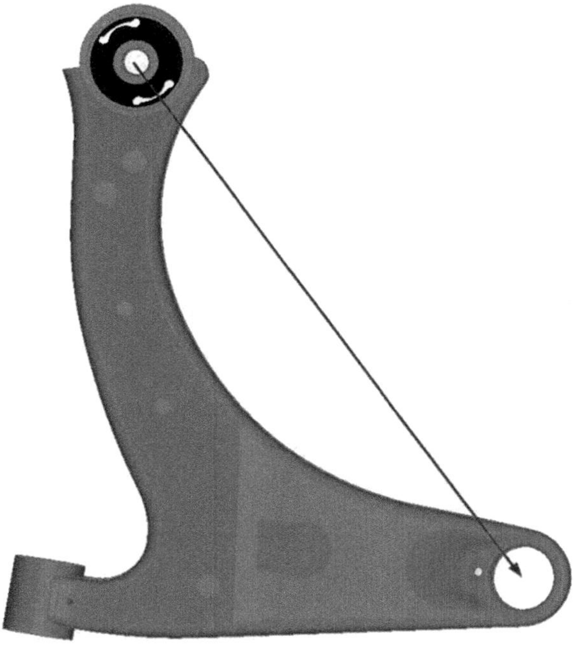

$$\sigma = -pI + 2\frac{\delta w}{\delta I_2}B - 2\frac{\delta w}{\delta I_2}B^{-1} \tag{1}$$

σ Stress tensor
I Unit tensor
B Cauchy–Green deformation tensor
p Due to incompressible assumptions introduced by the hydrostatic pressure
w Strain energy

The invariant of deformation tensor is:

$$I_1 = B \tag{2}$$

$$I_2 = \frac{1}{2}\left[I_1^2 - tr(B^2)\right] \tag{3}$$

$$I_3 = \det B \tag{4}$$

The relationship between the invariant of B and the principal elongation is

$$\begin{aligned} I_1 &= \lambda_1^2 + \lambda_2^2 + \lambda_3^2 \\ I_2 &= \lambda_1^2\lambda_2^2 + \lambda_2^2\lambda_3^2 + \lambda_1^2\lambda_3^2 \\ I_3 &= \lambda_1^2\lambda_2^2\lambda_3^2 \end{aligned} \tag{5}$$

λ is principal elongation.

The constitutive model of the material is selected as the Mooney–Rivlin model. The model is often applied to the elastic deformation of the strain within 150% of the simulation, which is an accepted constitutive model and is more suitable for bush simulation [2]. There are two preconditions for the application of this model: 1) Isotropic incompressible material. 2) The shear deformation follows Hooke's law, even when the shear is superimposed in different directions. The strain energy function W is:

$$W = C_{10}(I_1 - 3) + C_{01}(I_2 - 3) \tag{6}$$

$$\begin{aligned} I_1 &= \lambda_1^2\lambda_2^2\lambda_3^2 \\ I_2 &= \lambda_1^2\lambda_2^2 + \lambda_1^2\lambda_3^2 + \lambda_2^2\lambda_3^2 \end{aligned} \tag{7}$$

λ_1, λ_2, and λ_3 are the main elongation of the material in three directions.

C_{01} and C_{10} material parameters.

In Mooney–Rivlin constitutive model, parameters C_{01} and C_{10} are worth the accuracy of the simulation. Generally, these can be obtained through the relevant test of rubber material. However, due to the high cost, this paper adopts the fitting method of empirical relation. Compared to Stiffness the hardness of bushings is easy to obtain and therefore widely used in automobile industry. The hardness of Bushings is usually determined by applying a pressure on the rubber surface. In the work of this paper a cylindrical intender is used to create the pressure. There is a certain correspondence between the ratio of C_{01} and C_{10} of the bushing and the hardness, and simplified curve of empirical relationship is shown in Fig. 3.

The hardness of the bush is 70, from the curve 3 to take $C_{01}/C_{10} = 0.0143$.

Due to the incompressibility of the vulcanized rubber material, Poisson's ratio is approximately 0.5. The formula between rubber material elastic modulus E and hardness is:

$$G = \frac{E}{2(1+u)} = \frac{1}{3}E \tag{8}$$

$$G = 2(C_{01} + C10) \tag{9}$$

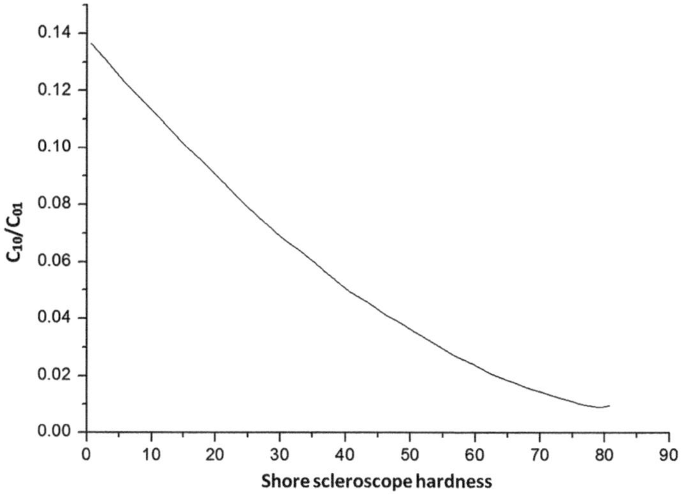

Fig. 3 Bush shore hardness and C_{01}/C_{10} relationship

The test data fitting formula of rubber between material elastic modulus E and hardness is:

$$\lg E = 0.0198Hs - 0.5432 \quad (10)$$

Simultaneously $C_{01} + C_{10} = 1.1614$ is obtained. Therefore, the input parameters of the material are:

$C_{01} = 0.0164$ and $C_{10} = 1.145$.

2.2 *Stiffness Simulation and the Standard*

The purpose of stiffness on the standard is to ensure that the finite element modeling is consistent with the actual state, material model selection, and material parameter input which is reasonable. To adjust the stiffness parameters in the simulation, the bushing stiffnesses in three mutually perpendicular directions are measured according test standard. Based on the results of the experimental measurements, the parameters in the simulation are modified to ensure the consistency of the simulation and experiment. The curves of experiment and simulation are shown in Figs. 4, 5, and 6.

From the results of the subject, the three directions of the bushes on the results consistent with the ideal can be used for subsequent stress analysis.

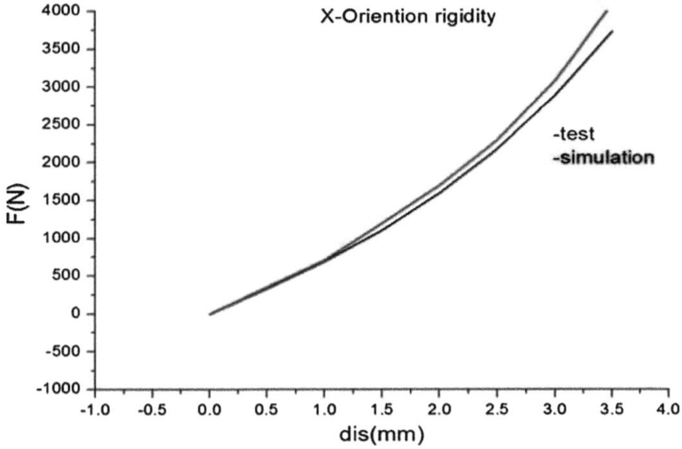

Fig. 4 Bush opening direction stiffness on the standard results between simulation and test

2.3 *Stress Analysis*

For the finite element analysis of the common conditions of the vehicle for the model of the bush, this paper chooses ten typical operating conditions of the vehicle in the running process. By establishing the multi-body dynamics model of the vehicle, the bush is subjected to force and torque when the corresponding working conditions are extracted. According to the working condition and force of the lower control arm, the main concern is the Max Bump, the Max Rebound, the Longitudinal shock, the Reverse Shock, the Pothole Braking. From the results of the analysis, the area where the maximum stress concentration is present is in the transition fillet of the bush opening area. The results of the analysis are as follows (Figs. 7 and 8; Table 1).

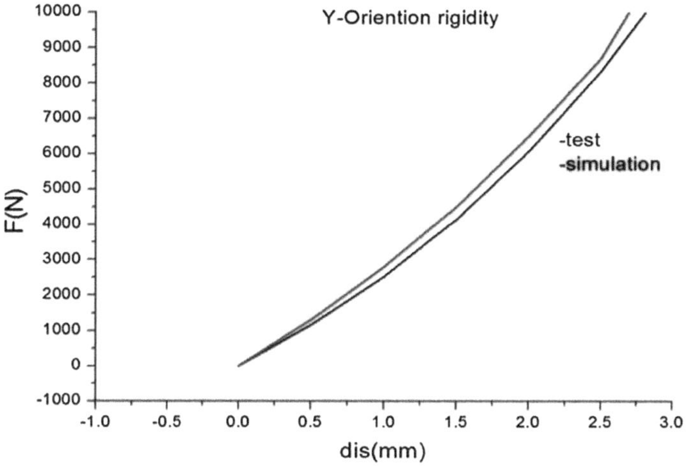

Fig. 5 Bush solid direction stiffness on the standard results between simulation and test

Comparing the results of the analysis and the results of the actual bushing failure, the fillet stress in the transitional regions of these openings is larger and it is necessary to optimize the maximum stress.

3 Optimization

The opening angle of the bush is optimized without changing the structure. In other words, the bush is rotated at the central axis of the inner sleeve, and the peak of the stress is found to have an ideal opening angle. The software ISIGHT is applicated to

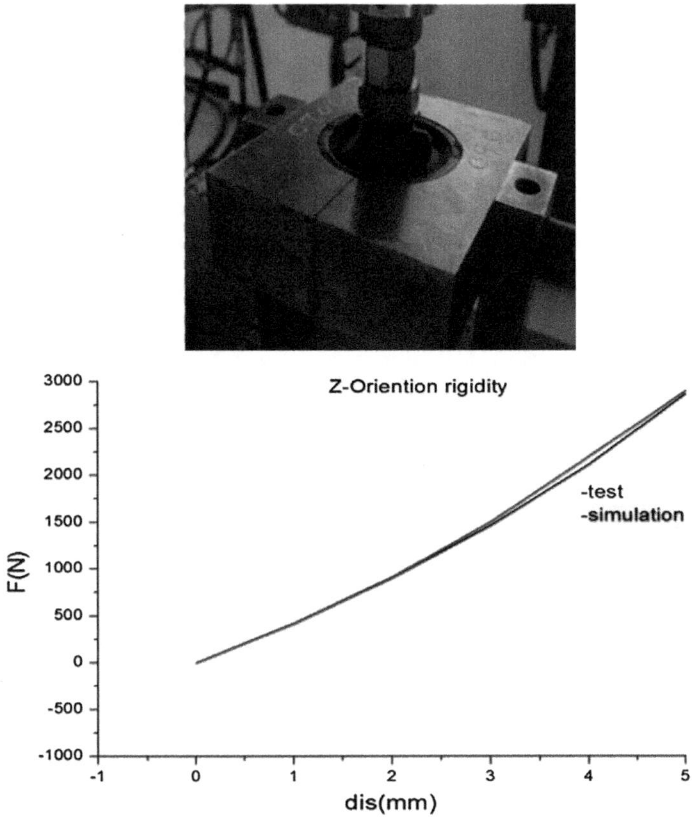

Fig. 6 Bush axial stiffness on the standard results between simulation and test

perform the optimization. The opening direction as the single optimization variable rotates around the bushing axis. The optimal orientation angle archives the minimal stress.

Since the load input of the bush's force and direction vary at different angles, and for the bush, the nonlinear material is costly to solve. Therefore, this paper uses ISIGHT to establish an approximate model to optimize the results to find the best angle of the bush opening [3]. Approximate model can reduce the number of high-intensity computer simulations, there are many algorithms, including the response surface model, radial basis function neural network model, kriging model, orthogonal polynomial model. The approximation model approach is a method of approximating a set of input variables (independent variables) and output variables (corresponding variables) by means of mathematical models. The ISIGHT neural network structure consists three layers: input layer, middle layer and output layer. An input vector is transformed to a non-linear form and

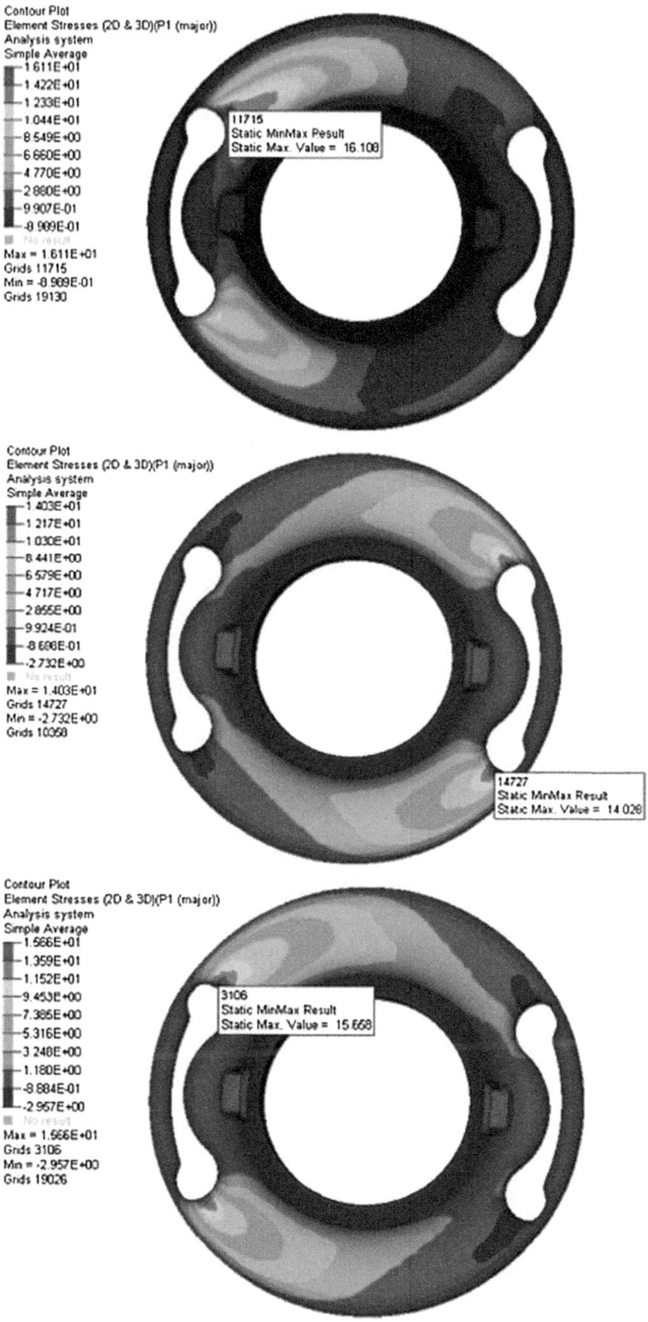

Fig. 7 Analysis of the results of cloud (1–5 conditions)

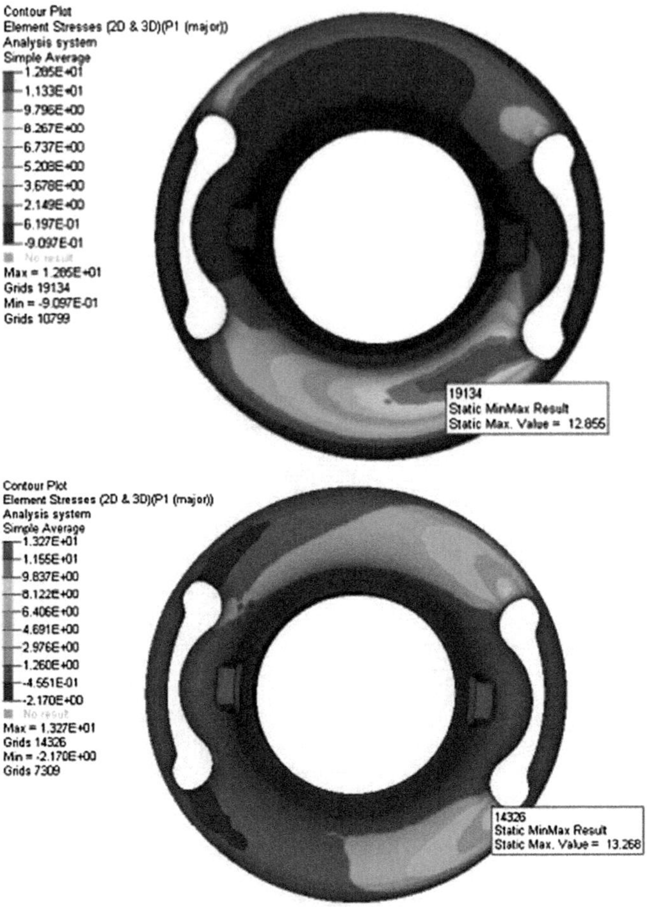

Fig. 7 (continued)

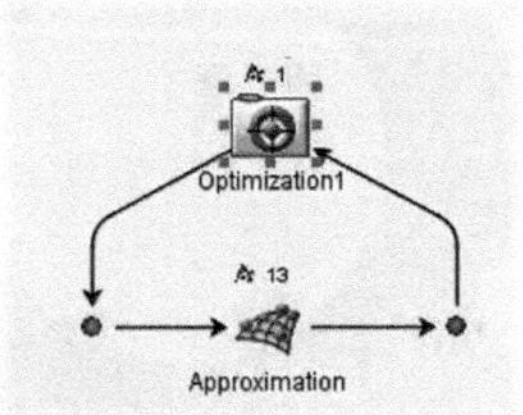

✓			Angle	Forward_Braking	Left_Longitudinal_Shock	Left_Reverse_Shock	Max_Bump	Max_Rebound
✓	1	1	120.0	11.477	8.5774	8.8038	10.453	13.257
✓	1	2	120.12	11.474	8.5746	8.8016	10.447	13.251
✓	1	3	120.16	11.473	8.5737	8.8008	10.445	13.249
✓	1	4	120.28	11.471	8.5709	8.7986	10.439	13.243
✓	1	5	120.96	11.455	8.5551	8.786	10.407	13.211
✓	1	6	121.08	11.452	8.5523	8.7837	10.401	13.205
✓	1	7	123.33	11.401	8.5	8.742	10.294	13.097
✓	1	8	123.46	11.398	8.4971	8.7398	10.288	13.091
✓	1	9	123.33	11.401	8.5	8.742	10.294	13.097

Fig. 8 Approximation model optimization flowchart

Table 1 Bush stress results statistics

Number	Load conditions	Max stress (MPa)
1	Max bump	16.108
2	Max rebound	14.028
3	Longitudinal shock	15.658
4	Reverse shock	12.855
5	Pothole braking	13.268

mapped into a new space. The mapping of the middle layer to the output layer is linear, and the output layer implements a linear weighted combination in the new linear space, where the weight is the network adjustable parameter. The Radial basis network measures the positions and the Euclidean distances of the independent variables.

Approximate model needs to be based on the experimental results' data through fitting, regression, interpolation, and other methods to create an approximate simulation model simulation program. In this paper, ten angles are selected to analyze the strength of the bush, and the results are taken as the modeling parameters of the approximate model. This approach has chosen Radial Basis Function Neural Network, which can approximate the nonlinear function. It does not need the mathematical hypothesis to have the characteristics of black box. The learning speed is fast and has good ability of normalization. The bushing or the direction of the opening rotates counter clockwise around its axis. The rotation angles and results of the stress are as follows (Table 2).

The results of the analysis are as follows. When the bush is rotated 123° counterclockwise, it is best to see that the peak stress of the lower control arm is the best. On the 123° to re-solve, the results are as follows (Table 3).

Table 2 Bush stress results statistics at different angles

Angle (degree)/max stress	Max bump (MPa)	Max rebound (MPa)	Longitudinal shock (MPa)	Reverse shock (MPa)	Pothole braking (MPa)
0	16.108	14.028	15.65	12.855	13.268
30	14.635	17.477	14.56	11.427	11.135
45	17.713	18.551	13.98	10.838	10.518
60	13.5	18.13	12.82	9.711	9.415
75	12.636	17.746	10.77	9.228	9.141
90	9.129	17.64	10.04	6.281	6.2
105	7.587	14.89	9.409	5.812	5.551
120	7.845	13.3	9.424	7.938	7.671
135	10.705	10.164	11.82	9.744	9.519
150	13.21	8.363	15.21	10.998	12.846

Table 3 Optimized post-stress results

Number	Load cases	Max stress (MPa)	Decrease the percentage (%)
1	Max bump	11.5	40.6
2	Max rebound	8.68	61.6
3	Longitudinal shock	9.1	72
4	Reverse shock	10.24	25.5
5	Pothole braking	12.99	2.1

From the results' point of view, the five operating conditions significantly, in addition to the Pothole Braking conditions, and other conditions of the peak stress have been greatly reduced to improve the high-peak local stress concentration. Of course, the final results should also refer to the dynamics and other aspects of the comprehensive consideration.

4 Conclusions

The finite element model of bush is established by simulation. After the simulation analysis, the angle of the bush opening direction is obviously optimized without changing the structure, and the stress peak of the bush in the process is obviously reduced, which provides guidance for the design and arrangement of the bush.

References

1. Zhao Z Automotive suspension and rubber elastic components theory and design. National Defense Industry Press
2. Wang W, Liu T, Zhao S (2004) Determination of common parameters of rubber Mooney-Rivlin model. Characteristic Rubber Prod 8
3. Lai Y ISIGHT parameter optimization theory and examples of detailed. Beijing University of Aeronautics and Astronautics Press

Dimensional Variation Simulation Analysis of Front Wheel Camber for Macpherson Suspension

Kun Ju, Jing Ye and Xiaohai Wang

Nomenclature

σ_j	Standard Deviation
$C_j\%$	Contribution Percentage
PV	Physical Verification

1 Introduction

Macpherson suspension system is easy to guarantee the vehicle performance requirements with the advantages of simple and compact structure that takes up less space, and the variation of wheel camber and kingpin inclination are small compared with other suspension system. So it has been widely applied on front-drive cars, but in the process of modern car manufacturing, there are many factors that cause the camber variation to make it deviate from the value of the design.

Front wheel camber, one of the four-wheel alignments of Macpherson suspension on the car's performance, is particularly important, which could make the car appear the following problems if its tolerance is not well controled:

(1) The tire is abnormal wear;
(2) Steering wheel is overweight or floating trembling when driving;
(3) The car running deviation to the left or right when going straight;
(4) Steering wheel is too heavy or cannot automatically back.

So, it is quite necessary to analyze Macpherson suspension's design dimensional tolerance and manufacturing process and deeply research for the factors affecting dimensional tolerance of front wheel camber.

K. Ju (✉) · J. Ye · X. Wang
Process Engineering Division, GAC Engineering, 668 East of Jinshan Road, Panyu District, Guangzhou 511434, China
e-mail: jukun@gaei.cn; jkzjfdaq@126.com

Society of Automotive Engineers of China (SAE-China) (ed.), *Proceedings of the 19th Asia Pacific Automotive Engineering Conference & SAE-China Congress 2017: Selected Papers*, Lecture Notes in Electrical Engineering 486, https://doi.org/10.1007/978-981-10-8506-2_37

2 The Structure and Camber of Macpherson Suspension

Macpherson suspension consists of shock absorbers, steering knuckles, steering drag links, control arms, steering machine, sub-frame, and other parts, as shown in Fig. 1.

The camber is the angle between the wheel centerline and the vertical line viewing from the rear car; outward is positive, and inward is negative, as shown in Fig. 2.

3 The Problem of the Camber

The camber design value is $-20' \pm 30'$ in a car, but some production data is outside the scope of design, and it has the following characteristics by statistic:

(1) Right camber ranges between $-20'$ and $-50'$ (deviating from the median to negative camber trend) when left camber is outside the design scope, as shown in Fig. 3.
(2) Left camber ranges between $0'$ and $10'$ (deviating from the median to positive camber trend) when right camber is outside the design scope, as shown in Fig. 4.

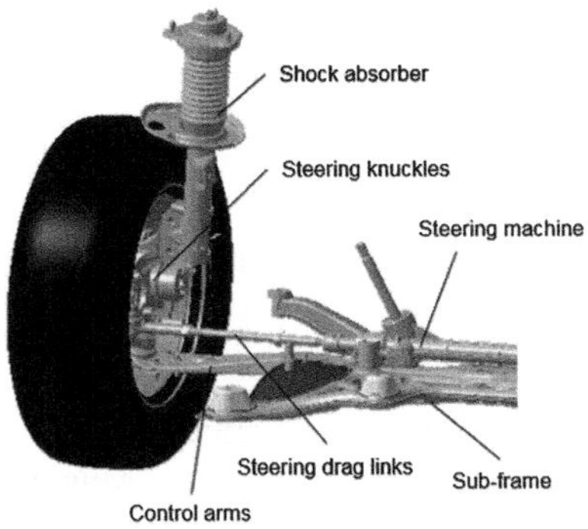

Fig. 1 Structure of Macpherson suspension

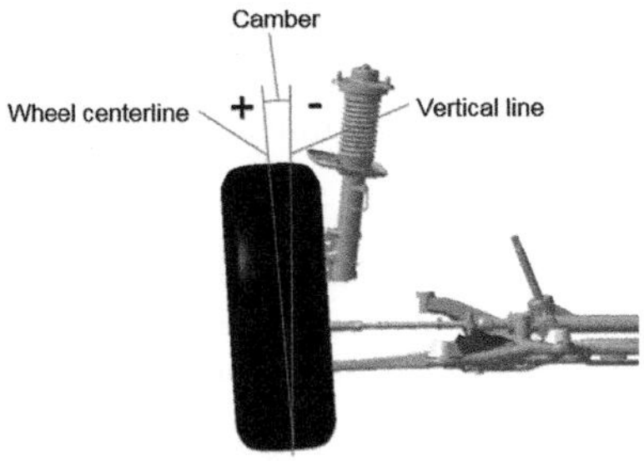

Fig. 2 Camber of McPherson suspension

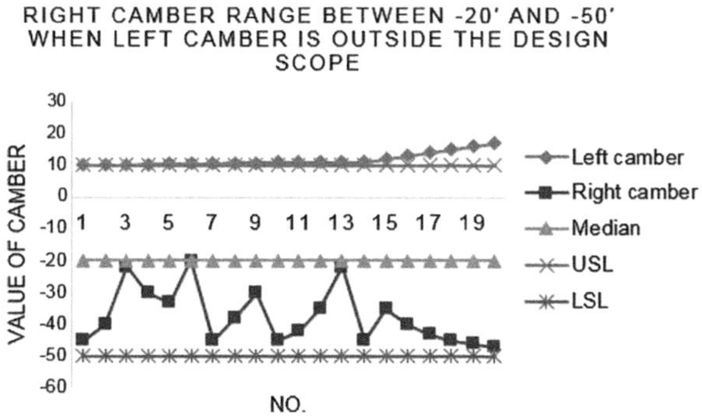

Fig. 3 Left camber statistical data

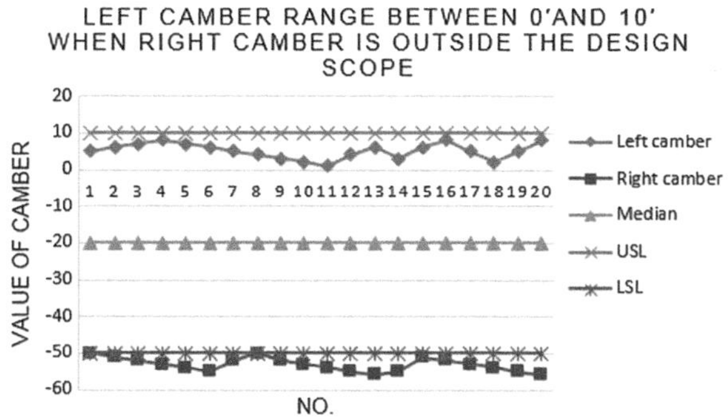

Fig. 4 Right camber statistical data

(3) The value of right camber gets bigger in the negative direction when the value of left camber deviating from the median to positive limit by the statistical data, the values of both affect each other.

4 Problem Analysis

Comprehensive analyses show that main reasons for the problem are in the following three aspects:

(1) Design: whether the design tolerances of parts are reasonable or not;
(2) Manufacturing: if the design tolerances are no problem, whether the accuracy of parts and components are good or not;
(3) Assembly process: if the two above is OK, check assembly process.

Therefore, the way to solve the problem also contains three aspects: tolerance simulation, physical verification, and measures establishment.

4.1 Tolerance Simulation

The purpose of the tolerance simulation is to judge the rationality of the design tolerances and assembly process and to find the factors affecting dimensional deviation and determine the contribution of each factor. Monte Carlo simulation and sensitive calculation are included.

4.1.1 Monte Carlo Simulation Method

Dimensional engineering is the study of variation accumulation for assembly products; dimensional variation analysis is a key component of dimensional engineering. Given any assembly product, a variation model can be generally represented as follows [1].
where *M* is a set of measurements, *T* is a set of variation inputs, and *f*(*T*) is a set of functions.

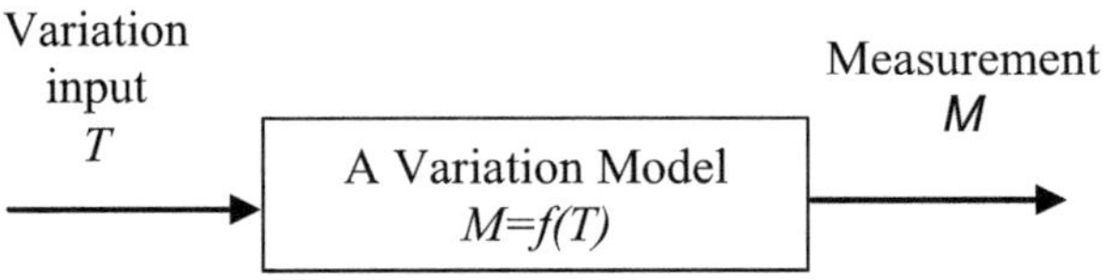

Because the variation inputs are random variables, a Monte Carlo simulation is a natural choice for dimensional variation analysis. Required inputs for the analysis include:

(**Input 1**) Tolerance definitions as variation model inputs *T*— these tolerances define the ranges and probability distributions of component tolerances.
(**Input 2**) Assembly operation definitions as variation model functions *f*(*T*)—the assembly operations dictate the component locating methods, fixture, locations, and tooling methods.
(**Input 3**) Measurement definitions as variation model outputs *M*—the measurements describe the output of a variation model.

A general flowchart describing the model for Monte Carlo tolerance analysis is shown in Fig. 5. A stepwise explanation of the algorithm is given [2]:

1. Monte Carlo simulation is a method for performing statistical variation analysis. The output distribution is a function of the distributions of the input variables and the assembly function. Thousands of samples of the input variables are combined to get a reliable measure result of the output distribution. So, a sample should be decided before the simulation, but the sample size should be sufficiently large in order to ensure a stable and consistent result. To assure accurate, as well as narrow confidence intervals, 5000 of the sample be used in this paper.
2. No part can be manufactured perfectly. A certain amount of variation is inherent in any manufacturing process. Tolerances are set to specify the limits which

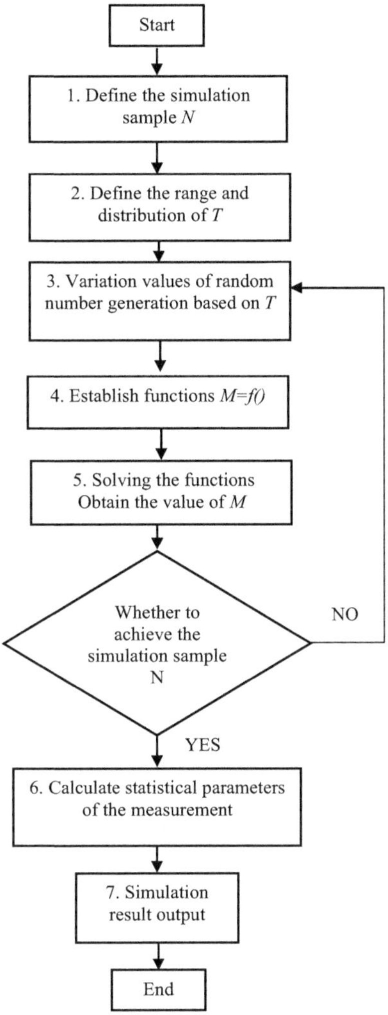

Fig. 5 Flowchart of Monte Carlo tolerance simulation

control the dimensions of important features on a part. Individual parts are manufactured and then put together to form an assembly. Each of these individually manufactured parts can have several different tolerances. It is the effect of these tolerances in the assembly that makes assembly tolerance simulation studies necessary. Proper specification of tolerances in the simulation model is essential. Tolerances that are specified correctly show the exact effect they have on an assembly when simulations are carried out. The input variables of *T* represent the process capability of manufacture; it contains range, distribution, sigma level, etc.

Range: the tolerance value;
Distribution: type of distribution that describes the tolerance region; different distributions represent different processes; the following are the different types of distribution available in 3DCS, as shown in Fig. 6.
Sigma level: This value relates the range to the number of standard deviations the range includes. Half of the range is divided by this number to determine the standard deviation of the tolerance. The sigma level is $\pm 3\sigma$ without other specified in this paper. This specification is required for the normal, Weibull, step, left skew, and right skew distributions.

3. Variation values for the tolerances to be analyzed are generated by the random number generator based on the input variables of *T*. Every variation value is with the tolerance range and compliance probability distribution and $\pm 3\sigma$ level.
4. To establish the functions $M = f(T)$, moves are used to not only define how a part locates to another part, but also decide a specific order to describe an assembly or manufacturing sequence. It means to assemble the variation parts with each other by constraints and a certain sequence.
5. After assembling every variation part of the product with each other, the analysis target *M* will be calculated; repeat the five steps above until M_1, M_2, M_3, ... M_N obtained.
6. According to the central-limit theorem, the sum of the sample values of some independent random variables is always approximately equal to the normal distribution no matter what the distribution of the random variables is. Therefore, the corresponding statistical parameters of *M*, such as average, standard deviation and interval estimation, are calculated by the normal distribution theory.
7. After performing a Monte Carlo simulation, a variation analysis provides the following information:

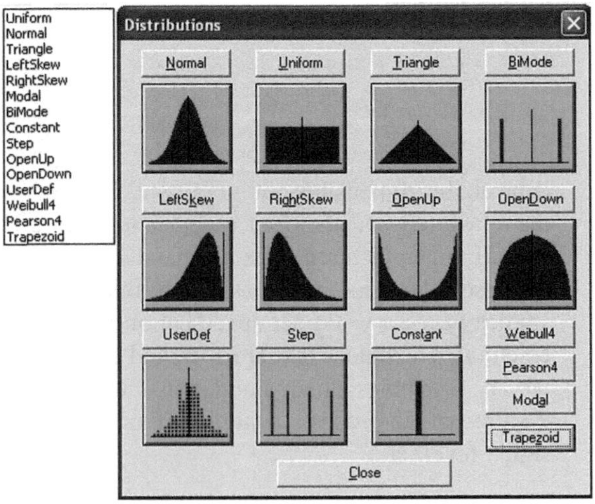

Fig. 6 Distribution available in 3DCS

Descriptive statistics—calculations made directly from the sample data such as mean, minimum, maximum, range, standard deviation, confidence limits, skewness, kurtosis, etc.
Inferential statistics—estimations based on a curve fitting algorithm such as estimated low, high, range, percentage out-of-spec, etc.

4.1.2 Sensitive Analysis Theory

For a measurement, there can be any number of contributors (tolerances); sensitivity analysis uses a methodology to determine the effect of individual tolerances on a given measure.

The relationship between a measurement M and contributors T_j is as follows [3]:

$$M = f\left(T_j, j = 1, 2, \ldots, N_T\right) \quad (1)$$

where f is a generic function and N_T is the number of contributors. By using Taylor series expansion, Eq. (1) can be approximated by

$$M = M_0 + \sum_j \frac{\partial f}{\partial T_j} \Delta T_j + \frac{1}{2} \sum_{jk} \frac{\partial^2 f}{\partial T_j \partial T_k} \Delta T_j \Delta T_k + \cdots \quad (2)$$

where M_O is the nominal measurement value, $\Delta T_j = \left(T_j - T_{j0}\right)$ is the change of the contributor T_j, and T_{j0} is the initial value. In 3DCS, the initial value, T_{j0}, is set as zero.

Contribution analysis is used to find out how big each contributor T_j affects the measurement M, and the analysis assumes that: (a) all contributors are independent of each other, and (b) the random distribution for each contributor T_j is normal. Measurement M is based on the first-order derivative, that means,

$$M = M_0 + \sum_j \frac{\partial f}{\partial T_j} \Delta T_j \quad (3)$$

The standard deviation relationship based on (3) can be presented as follows:

$$\sigma_m^2 = \sum_j \left(\frac{\partial f}{\partial T_j}\right)^2 \sigma_{T_j}^2 \quad (4)$$

If all first-order derivatives and all stand deviations, $\left(\frac{\partial f}{\partial T_j}\right)^2 \sigma_{T_j}^2$ are known, the contribution percentage ($C_j\%$) is calculated by

$$C_j\% = \frac{\left(\frac{\partial f}{\partial T_j}\right)^2 \sigma_{T_j}^2}{\sum_j \left(\frac{\partial f}{\partial T_j}\right)^2 \sigma_{T_j}^2} \times 100\% \qquad (5)$$

The calculation of sensitive holds all variables (tolerances) at their nominal value and then moves one tolerance at a time to determine the effect that it has on the total variation at a specified output location. Once this is done for each tolerance, a list of the top contributors to variation is given in descending order. It can be assumed at this point that changing a given tolerance will directly affect the output condition per the contribution percentage listed for that tolerance in the list.

4.1.3 Variation Simulation of Camber Based on 3DCS

3DCS is a software tool for performing tolerance simulation analysis, that can be executed Monte Carlo simulation and contribution calculation based on the two algorithms above. After the simulation has been run, the simulation window displays a report in the form of a histogram and statistical data. This output contains simulation output and sensitivity output.

Create a Tolerance Simulation Model

Dimensional tolerance simulation is the computer's emulation of an assembly to predict the amount of variation resulting from a process sequence. In addition, dimensional simulations help us determine the key tolerances that contribute to a particular variation. The basic components of tolerance simulation are as follows.

Part Geometry: The basic requirement of a model is the creation of part geometry. All part components that are analyzed have to be defined in the model. This is done by selecting the appropriate features from the CAD geometry and/or defining points representing part components. The CAD geometry in this case contains shock absorbers, steering knuckles, steering drag links, control arms, steering machine, sub-frame, BIW, and tire as shown in Fig. 1.

Assembly Sequence: This information defines how the multiple parts sequentially locate in an assembly. Within 3DCS, a move defines how one part locates to another. Moves must be added to the model to represent position of each part in the assembly. The assembly sequence of Macpherson suspension is shown in Fig. 7.

Tolerances: The 3DCS analyst model is built and simulated to study the effects of tolerances in an assembly. Tolerances are of great significance in a model. Proper definition and application of tolerances to appropriate areas of the model are required. The key tolerances inputted in the simulation model are shown in Table 1.

Measurements: To understand the effects of tolerances in a model, measurements are defined to predict the parameters of interest. These measurements quantify the

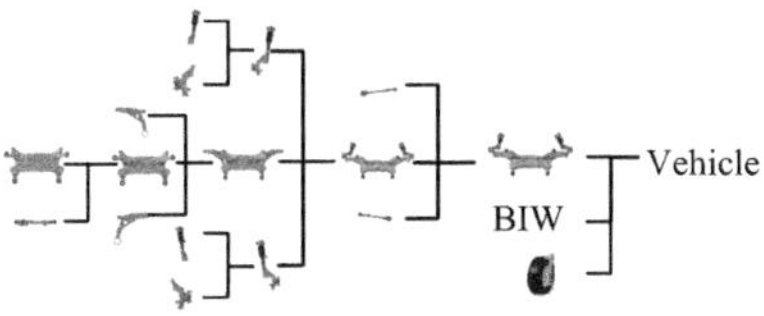

Fig. 7 Assembly sequence of Macpherson suspension

Table 1 Tolerance input in 3DCS model

No.	Tolerance (mm)
T_1	0.28
T_2	0.20
T_3	0.21
T_4	0.45
T_5	2.0
T_6	2.0
T_7	0.45
T_8	0.45
T_9	2.0
T_{10}	0.45
T_{11}	0.6

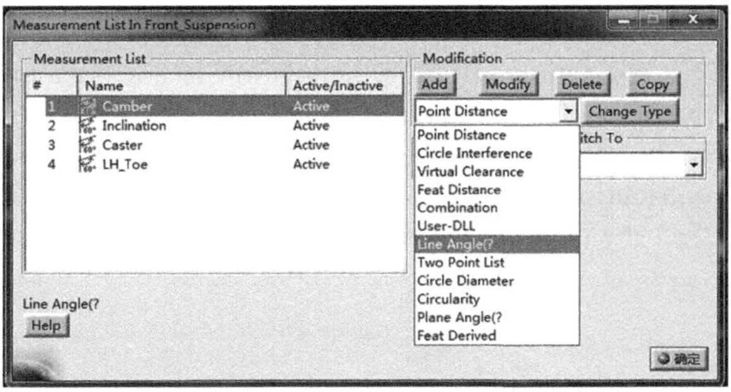

Fig. 8 Use line angle measurement to define the analysis target in 3DCS

variation induced in the desired areas of the model. The value of the camber is $-20' \pm 30'$; use the line angle measurement in 3DCS to create the analysis target to define this value in the model, as shown in Fig. 8.

Simulation Output

The value of the camber is $-20' \pm 30'$ according to the design document, use 3DCS to simulate 5000 virtual assemblies without considering gravity, temperature and deformation, and assume all parts tolerance distribution are normal, process capabilities for all parts satisfied with 6σ. Figure 9 shows that the simulation result of 6σ is 0.55°, 0.55° converted to deviation of minute is $\pm 16.5'$, less than the

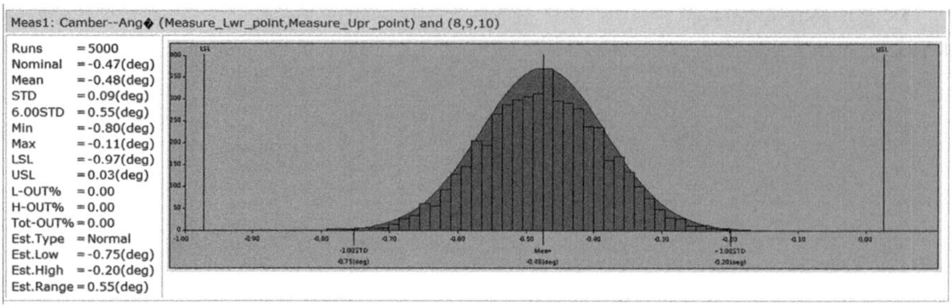

Fig. 9 Simulation output by 3DCS

design object value ±30′, this means that all parts tolerance and process capability is satisfied with design requirement. The reason for the problem may be manufacturing process that some parts beyond their tolerance range or some assembly processes are not right, so it is necessary to monitor the deviation accuracy of parts and assembly processes, but how many deviations should be monitored and how big are these factors affecting the accuracy of the camber angle? The sensitivity (contribution) output will tell you the truth.

4.1.4 Sensitivity (Contribution) Output

The destination of 3DCS simulation is not only to do 3D-stacking up, to check whether the parts tolerance and processes capability meet the design requirement or not, but also to find the factors affecting dimensional deviation and determine the contribution of each factor for the camber. Figure 10 shows four main joints that affect the camber and row from big to small by their contribution.

Camber (Ang◆ (Measure_Lwr_point,Measure_Upr_point) and (8,9,10))

Index	Tolerance	Point	Part	Range	Percent	Graph
1	Ctrl_Arm_sphere_position_composite	Shock_absorber_XYZ2(FRTZF)	Knuckle3610101ASV0000_1	M:0.280(mm)	25.76%	
2	Ctrl_Arm_sphere_position_composite	Shock_absorber_XYZ1(FRTZF)	Knuckle3610101ASV0000_1	M:0.280(mm)	18.69%	
3	Shock_absorber_upr_panel__to_BIW	Shock_absorber_upr_panel__to_BIW	Front_Suspension	M:2.000(mm)	10.42%	
4	Shock_absorber_XYZ_position_composite	Shock_absorber_XYZ2(FRTZF)	Shock_absorber3610121ASV0000_	M:0.210(mm)	7.86%	
5	Shock_absorber_to_Knuckle&Bracke_Asm_Pattern_rigid	Shock_absorber_to_Knuckle&Bracke_Asm_Pattern_rigid	Front_Suspension	M:0.200(mm)	7.20%	
6	Shock_absorber_XYZ_position_composite	Shock_absorber_XYZ1(FRTZF)	Shock_absorber3610121ASV0000_	M:0.210(mm)	5.70%	
7	Steering_Gear_XY	Steering_Gear_XY	Sub_Frame3710055ASV0000_1	M:2.000(mm)	5.47%	
8	Left_rod_sphere_position_Y	Left_rod_sphere_XYZ1	Steering_Gear_1	M:2.000(mm)	4.96%	
9	Shock_sphere_position	Shock_sphere	Shock_absorber_upr_panel	M:1.000(mm)	3.70%	
10	Ctrl_Arm_sphere_position	Ctrl_Arm_sphere	Ctrl_Arm	M:2.000(mm)	3.12%	
11	Ctrl_Arm_bolt_position_composite	Ctrl_Arm_front_bolt_XY(PLTZF)	Sub_Frame3710055ASV0000_1	M:1.732(mm)	2.25%	
12	Shock_absorber_XYZ_position_composite	Shock_absorber_XYZ1(PLTZF)	Shock_absorber3610121ASV0000_	M:0.530(mm)	1.10%	
13	Ctrl_Arm_bolt_position_composite	Ctrl_Arm_rear_bolt_XY(FRTZF)	Sub_Frame3710055ASV0000_1	M:1.000(mm)	0.96%	
14	HP_Sub_Frame&Steering_Gear_to_BIW	Sub_Frame_XY	Front_Suspension	C:0.450(mm)	0.79%	
15	HP_Steering_Gear_to_Sub_Frame	Steering_Gear_XY	Front_Suspension	C:0.750(mm)	0.70%	
16	Steering_gear_profile	Steering_Gear_Z1(Group)	Sub_Frame3710055ASV0000_1	M:2.000(mm)	0.40%	
17	HP_Ctrl_Arm_rear_bolt_to_sub_frame	Ctrl_Arm_rear_bolt_XY	Front_Suspension	C:0.700(mm)	0.39%	
18	Ctrl_Arm_sphere_position	Ctrl_Arm_sphere	Knuckle3610101ASV0000_1	M:0.600(mm)	0.27%	
19	HP_Ctrl_Arm_to_SubFrame	Ctrl_Arm_rear_point	Front_Suspension	C:0.400(mm)	0.13%	
20	HP_Sub_Frame&Steering_Gear_to_BIW	Sub_Frame_X	Front_Suspension	C:0.450(mm)	0.07%	

Sum of Rest 67 Contributors = 0.08%

Fig. 10 Sensitivity output by 3DCS

No. 1 The joint between Knuckle and Shock absorber of which contribution contains three factors (T_1–T_3) is 58.21%;

T_1. Hole position for absorber matching on knuckle.
T_2. Hole pin floating between knuckle and absorber.
T_3. Hole position for knuckle matching on absorber.

No. 2 The joint between absorber and BIW of which contribution contains two factors (T_4–T_5) is 15.12%:

T_4. Hole pin floating between absorber and BIW.
T_5. Hole position for absorber on BIW.

No. 3 The joint between sub-frame and control arm of which contribution contains three factors (T_6–T_8) is 11.72%:

T_6. Hole position for matching control arm on sub-frame.
T_7. Hole pin floating between the bolt and sub-frame.
T_8. Hole pin floating between the bolt and control arm.

No. 4 The joint between control arm and knuckle of which contribution contains three factors (T_9–T_{11}) is 10.43%:

T_9. Ball joint position of control arm assembly.
T_{10}. Hole pin floating between steering knuckle and ball joint.
T_{11}. Hole position for ball pin on steering knuckle.
All above is the first step to analyze this problem, and the second step is to validate simulation result of the four main factors physically.

4.2 Physical Verification (PV)

In order to find each contribution of the four main joints, single variable method is used in the physical verification, four steps included as follows:

1. Select 26 cars randomly measure their values of the camber and calculate statistical parameters:
   $\sigma_m = \sqrt{\frac{1}{25}\sum_{i=1}^{26}(x_i - u)^2} = 0.097°$, $6\sigma_m = 0.58°$, as shown in Table 2;
2. Choose one car to be a base, then change factor of T_1 for 26 times by replacing part from the 26 cars of T_1, and keep other factors the same; measure 26 values of the camber, and calculate σ_1, using the same method to calculate σ_2–σ_{11}, as shown in Table 3;
3. Calculate the contribution of every factor, as shown in Table 3:

$$C_i\% = \frac{\sigma_i^2}{\sigma_m^2} \tag{6}$$

Table 2 PV statistical data for camber of left wheel

No.	Nominal value for camber of left wheel (′)	Measured value for camber of left wheel (′)	Variation (′)
1	−20	−3.5	16.5
2	−20	−10.4	9.6
3	−20	−7.2	12.8
4	−20	−12.8	7.2
5	−20	−1.7	18.3
6	−20	−24.8	−4.8
7	−20	1.2	21.2
8	−20	−13.7	6.3
9	−20	−7.9	12.1
10	−20	−9.8	10.2
11	−20	−5.6	14.4
12	−20	−4.3	15.7
13	−20	−9.8	10.2
14	−20	−7.3	12.7
15	20	−17	3
16	−20	−12	8
17	−20	−17.8	2.2
18	−20	−13.3	6.7
19	−20	−18.4	1.6
20	−20	−17.7	2.3
21	−20	−15.6	4.4
22	−20	−8.2	11.8
23	−20	−7.6	12.4
24	−20	−12.9	7.1
25	−20	−11.1	8.9
26	−20	−12.5	7.5
σ_m (′)	/	5.80	5.80
σ_m (°)	/	0.097	0.097
$6\sigma_m$ (′)	/	34.81	34.81
$6\sigma_m$ (°)	/	0.58	0.58

4. Calculate the contributions of the four joints:

$$\text{No. 1}: C_{T_1-T_3}\% = \frac{\sigma_1^2 + \sigma_2^2 + \sigma_3^2}{\sigma_m^2} \tag{7}$$

$$\text{No. 2}: C_{T_4-T_5}\% = \frac{\sigma_4^2 + \sigma_5^2}{\sigma_m^2} \tag{8}$$

Table 3 PV statistical data of contribution analysis

No.	Contribution value (°)	$C_i\%$	$C_{T1}-C_{T3}\%$	$C_{T4}-C_{T5}\%$	$C_{T6}-C_{T8}\%$	$C_{T9}-C_{T11}\%$
σ_1	0.31	28.53	52.30	/	/	/
σ_2	0.23	15.17		/	/	/
σ_3	0.17	8.60		/	/	/
σ_4	0.17	8.23	/	16.20	/	/
σ_5	0.16	7.97	/		/	/
σ_6	0.12	4.10	/	/	12.50	/
σ_7	0.13	5.18	/	/		/
σ_8	0.10	3.22	/	/		/
σ_9	0.10	2.90	/	/	/	9.40
σ_{10}	0.11	3.30	/	/	/	
σ_{11}	0.10	3.20	/	/	/	
σ_{m}	0.58	/	/	/	/	/

$$\text{No. 3}: C_{T_6-T_8}\% = \frac{\sigma_6^2 + \sigma_7^2 + \sigma_8^2}{\sigma_m^2} \tag{9}$$

$$\text{No. 4}: C_{T_9-T_{11}}\% = \frac{\sigma_9^2 + \sigma_{10}^2 + \sigma_{11}^2}{\sigma_m^2} \tag{10}$$

Contribution of No. 1–No. 4 can be calculated by Eqs. (6)–(10), as shown in Table 3.

4.3 *Comparison of Result Between 3DCS Simulation and Physical Verification*

3DCS simulation calculates the statistical parameter 6σ, contribution of main joints $C_i\%$ by Monte Carlo theoretically, and physical verification also calculates these parameters. How different between the two methods? The following is the comparison of the result.

4.3.1 Discrepancy of Standard Deviation

The first parameter shows in Table 4, that 6σ (3DCS) is 0.55°, $6\sigma_m$ (PV) is 0.58°, the distinction between the two is within 5%, which shows that 3DCS simulation is close to PV.

Table 4 Discrepancy of standard deviation

6σ (3DCS simulation)	$6\sigma_m$ (PV)	Discrepancy
0.55°	0.58°	5%

4.3.2 Discrepancy of Sensitivity Result

The second parameter, contribution results in Fig. 11, shows that the distinction between the two is close to 10%; if we think 10% is an acceptable level, it proves that 3DCS simulation is a reliable method to forecast manufacturing problem, to find and solve the problem in design stage.

4.4 Production Problem Investigation and Measures Be Taken

After tolerance simulation, the contributions of factors are confirmed, and it gives a clear direction to solve problems.

4.4.1 Investigation of Components Accuracy and Assembly Process

(1) Investigation of components accuracy: According to the results of the simulation analysis, the key parts with larger contribution are preferred to investigate.
(2) Confirmation of assembly process: Engineers investigated the assembly process in the plant and found that actual assembly process is same as the design; assembly process problem can be ruled out.

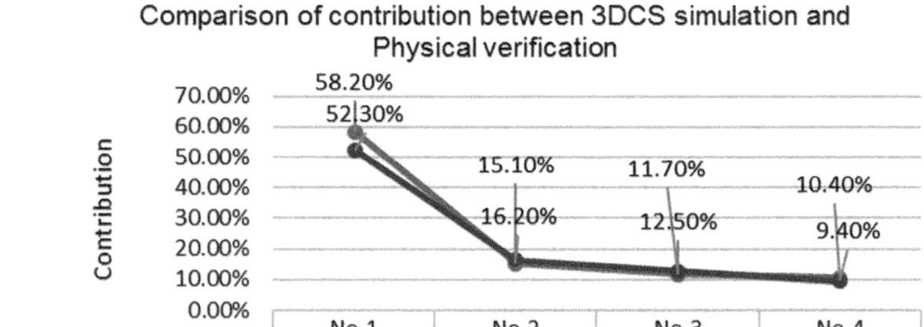

	No.1	No.2	No.3	No.4
Contribution(3DCS)	58.20%	15.10%	11.70%	10.40%
Contribution(PV)	52.30%	16.20%	12.50%	9.40%

Fig. 11 Comparison of contribution between 3DCS simulation and PV

The problem is own to the poor manufacturing accuracy of some parts. As shown in the following Table 5.

4.4.2 Measures Be Taken

(1) Repairing: Pick out these cars with camber problem, and replace parts with higher contribution by high-precision parts, reducing their effect to the camber.
(2) Because the contribution of shock absorber and steering knuckle joint is the largest, it is very important to ensure the joint between shock absorber and steering knuckle, but the accuracy of the two parts is not good, and it is very difficult to improving the precision in a short time; a temporary fixture is developed to assemble the two parts with each other for improving the accuracy of assembly. Figure 12 shows that the defect rate of the camber is reduced largely with the fixture used, without improving the precision of the parts.
(3) The permanent solutions are to improve the process of the key parts, to strengthen the dimensional precision monitoring of parts with higher contribution, guaranteeing their manufacturing accuracy same as the design.

Table 5 Accuracy of key components

No.	Part name	Accuracy (OK/NOK)
1	Steering knuckle	NOK
2	Shock absorber	NOK
3	Sub-frame	NOK
4	BIW	OK
5	Control arms	OK
6	Steering drag links	OK
7	Steering machine	OK

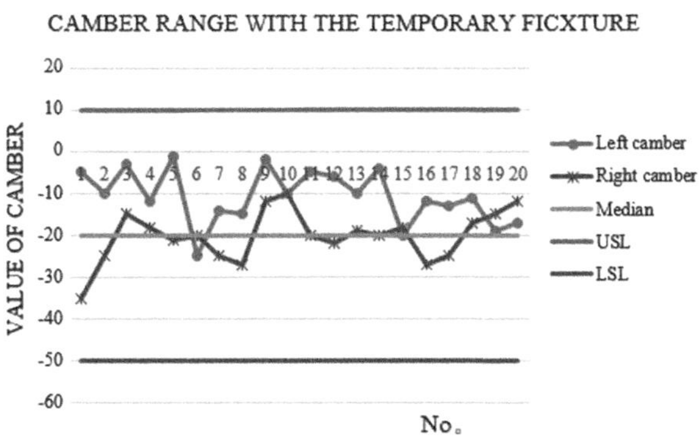

Fig. 12 Camber range with the temporary fixture

5 Conclusions

3DCS simulation shows that there are many factors affecting the size of camber, among which the contribution of shock absorber and steering knuckle joint is the largest, and the most sensitive factors are from this joint, secondly BIW, thirdly sub-frame. So, it is very important to know how to distribute tolerances of components in the design stage and control them in manufacturing and assembling to ensure the size of camber, and it is one of big problems that automobile manufacturers needed to conquer. But as long as the main factors are certain by dimensional tolerance simulation, engineers can distribute them reasonable tolerances, and measures can be taken to guarantee the size of camber that reduces the probability of a problem in manufacturing stage.

However, there are many assumptions in this simulation model:

(a) Linear or nonlinear model

The model is assumed to be a linear model; all parts are treated as rigid bodies, without regarding the deformation, but some parts is over constraints, and it means that they could be deformed by applying of force, and the model would change to a nonlinear one in this case. Thus, to improve the accuracy of simulation, a more in-depth study of nonlinear model is needed.

(b) Tolerance distribution and process capability

All tolerance distributions are assumed normal, but some tolerance distribution may be other types of distribution due to its different manufacturing process, and sigma level also does not obey to 6 sigma. The research of process capability is also very important and can be perfected by establishing a database.

(c) The rationality of the simulation sample

In the particular case of the confidence interval for percent out of spec, 3DCS uses the following statistical results: For large sample sizes—in the thousands—the sampling distribution of percent out of spec is normally distributed with mean = (samples out of spec/total samples) (A) and standard deviation = $v\ (pq/n)$ (B) where p equals the right hand side of (1), $q = 1 - p$, and n = the sample size. The 90, 95, and 99% confidence intervals for percent out of spec will therefore be given by mean $\pm$ 1.645 $v\ (pq/n)$, mean $\pm$ 1.960 $v\ (pq/n)$, and mean $\pm$ 2.326 $v\ (pq/n)$, respectively. Because the convergence to normality may be very slow, it is recommended that sample sizes in the 20,000 range be run whenever possible. Five thousand samples are used in this paper; while for the consistent and reliable application of the Monte Carlo simulation to the statistical tolerance analysis, the number of samples is the key of precision. By a large number of samples, the precision can be improved, but the computational cost will be increased. The improvement of this approach should be an area for some intense research on stochastic methods coupled with worse case methods. How to balance calculation precision and calculation resource cost is also worth studying [4].

Acknowledgements The authors would like to thank Dimensional Control System Inc. (DCS), Engineering Technology Associates (China) Inc. (ETA-China), and Pan-I Information Technology (Shanghai) Co., Ltd (PAN-I) for their support.

References

1. Zhou YQ, Thagu V Monte Carlo simulation and variation analysis. Dimensional Engineering News, DEnews_06_2009
2. Dantan JY, Qureshi AJ (2009) Worst-case and statistical tolerance analysis based on quantified constraint satisfaction problems and Monte Carlo simulation. Comput Aided Des 41(1):1–12
3. Zhou YQ, Thagu V Theory for HLM analysis. Dimensional Engineering News, DEnews_09_30_07
4. Qureshi AJ, Dantan JY, Gayton N et al (2012) A statistical tolerance analysis approach for over-constrained mechanism based on optimization and Monte Carlo simulation. Comput Aided Des 44(2):132–142

The Real-Time Measurement Function of the Random Measuring Instrument Improves the Manufacturing Quality of the Workpiece Effectively

Zhengde Zhu

The random meter refers to those instruments and meters (collectively referred to as "detection devices") attached to the process equipment in the production line. They can perform simultaneous or simultaneous detection of the processing operations during the machining process. Because this type of detection device is always attached to a machine or by a number of machine tools composed of short-term, in fact, It has become an automatic process control system components, so we call it the random meter. Since the eightieth of last century, "online testing, process control" concept has been popular at China and abroad manufacturing industry, as an important part of the implementation of SPC, in the production line "online gauge" has also been rapid development in company, from Simple special gauge to the complex CNC multi-parameter automatic measuring machine, a wide range with multi-function[1–3].

1 Development of Random Measuring Instrument: Grinder Active Measuring Instrument → Combined Machine Tool Automatic Correction Measurement → Processing Center Random Measurement System

Until the end of the last century, the random meter mainly includes the following two types:

(1) widely used in the grinding machine as the main machine of the active measuring instrument is a pioneer of random meter. As early as the early fiftieth of last century, by the famous instrument manufacturers, Italy's MARPOSS

Z. Zhu (✉)
Shanghai Volkswagen Powertrain Co., Ltd., Shanghai, China
e-mail: 13564011215@163.com

Society of Automotive Engineers of China (SAE-China) (ed.), *Proceedings of the 19th Asia Pacific Automotive Engineering Conference & SAE-China Congress 2017: Selected Papers*, Lecture Notes in Electrical Engineering 486, https://doi.org/10.1007/978-981-10-8506-2_38

company launched for the raceway grinder internal grinding machine active measuring instrument, after more than half a century of continuous efforts in the major machine tool plant. Automotive industry is the representative of the user's cooperation, the current performance of all kinds of excellent active measuring instrument has become a grinder or special grinder's standard configuration. Although theoretically, it is possible to use active measurement techniques for lathes, especially for highly specialized equipment, but for some reason, there are few application examples in domestic and foreign [1].

(2) Another type of random meter is mainly used for special machines and machine tools automatically adjust the measuring instrument, especially in the boring process. The early years there have been a measuring instrument known as the "automatic warning". Since it is not possible to act as active measuring instruments and automatic tamper measuring instruments to form closed loop system of instruments, tool and workpiece, is gradually replaced by the former. [2, 4].

From the beginning of this century, modern automobile manufacturing industry has changed from the monotonous mass production, to a variety of small and medium-size production mode, resulting in more and more powertrain plant and components plant in the process planning, and resulting in choosing flexible high-performance machine tools and processing center to compose automatic production lines [5]. In fact, for those complex parts, such as engine cylinder, cylinder head, gearbox housing, and other box parts, not only with complex shape, not only is the shape complex and processing requirements are high, but also the waste will cause great losses. Therefore, how to improve the quality of the manufacturing center has a great significance, feedback compensation in the device, and other functions of the random measurement system is a very effective method [6, 7].

So what is the difference between the random meter and the online gauge? In the broad sense, the online gauge is located in the field, so the random meter should be attributed to the field of online inspection. Because its frequency of detection is high, usually 100%. And with its function improving, it is good for simplifying the equipped online inspection tools, and is of great significance for the adjustment of process planning and reduction of equipment investment while improving the manufacturing quality of workpieces. Since the 1980s, due to the promotion of SPC (Statistical Process Control) in automobile plants, reliable quality can be acquired through stable production process, and the information to deal with is only the data of low frequency random check from the online inspection equipment [8], which reflects the modern management concept "Quality comes from manufacturing". It is clear, however, that the accuracy of the sampling data depends on the machining accuracy of the relevant machine tool, and it is clear that the random meter is closely related to the on-line inspection (consistent target). To this end, it is recommended that in the case of involving random meters, try not to use the word "online" to avoid conceptual confusion.

2 Processing Center the Basic Composition and Main Function of the Random Measurement System

2.1 The Composition of the Random Measurement System

Compared to other manufacturers of measuring instruments, production and processing center random measurement system is much less professional factory, which shows that the production needs of their business is far from the application has been used for many years of online gauge and active measuring instrument, etc. In fact, even the introduction of its application of such article is also rare.

The detection device for the random measurement of the machining center is mainly composed of contact probe, signal receiver, and output cable (or interface device). According to the nature of the transmission signal, it is divided into infrared and radio. In contrast, the latter signal transmission capability is stronger. In addition to the long transmission distance, the radio signal is not affected by the obstacles. But in fact, in the enterprise, former type is used more often. Figure 1 shows the composition and working procedure of a typical system. The test result of the trigger probe is send to the receiver installed in the machining center by infrared signal. The receiver passes the output cable (or via the interface unit). Signal is transmitted to the machine control system. The contact-type probe, as an infrared signal transmitter, can send signals within 360°. The interface device in Fig. 1 can deal with the signal and send it to the CNC system of the processing center. But in most cases, the test program is programed by machine tool manufacturers according to the actual needs of the preparation, according to the input signal to achieve the corresponding function.

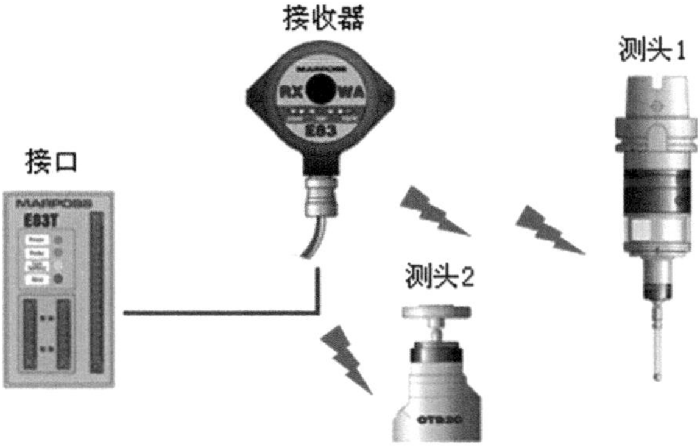

Fig. 1 The basic composition and working process of the machining center stochastic instrument

2.2 *The Main Function of the Random Measurement System*

The measuring object can be a workpiece, a fixture, or a tool, and it is designed and implemented according to the needs of different users. When the object is a workpiece and a fixture, the probe 1 shown in Fig. 1 will be used. As the contact probe is like a tool, it is usually stored in the tool magazine of the machining center. According to different requirements, it is called out before or after a machining process, and then the automatic detection is performed according to the program to achieve a certain function. And when the detection object is a tool, use the probe 2 in Fig. 1, when the "seat" of the probe is fixed in the machining center of the machine table. In summary, real-time measurements performed by random meters can be achieved primarily for the following purposes:

- Tool status detection

The detection of the tool status is also referred to as "tooling", see figure. In this case, the tool is set by the measuring device (probe) provided on the work surface of the machine tool, and the tool in the magazine is set to the tool according to the presetting process, and then the judgment is made after comparing with the predetermined value. At the same time, through the detection of the tool can also achieve the tool wear, damage or installation of the correct type or not. Figure 2 is a few examples of "knife measurement": a tool is being long (high) detection, b is the direction of the tool radius measurement, c detection of the tool at this time has been damaged, through the knife measurement can be found and alarm. The use of a random meter to perform in-machine tool setting not only saves the manpower and material resources when setting the tool outside the machine, but also improves the working efficiency. In addition, the environment of tool setting and cutting is the same, which can reduce the effect of temperature to maximum extent.

In the "tool state detection" of this application, the detection signal is used in the cable transmission mode input interface device, or directly connected with the machine numerical control system. There are two types of contact and non-contact (optical), and Fig. 2 is an example of a more commonly used contact.

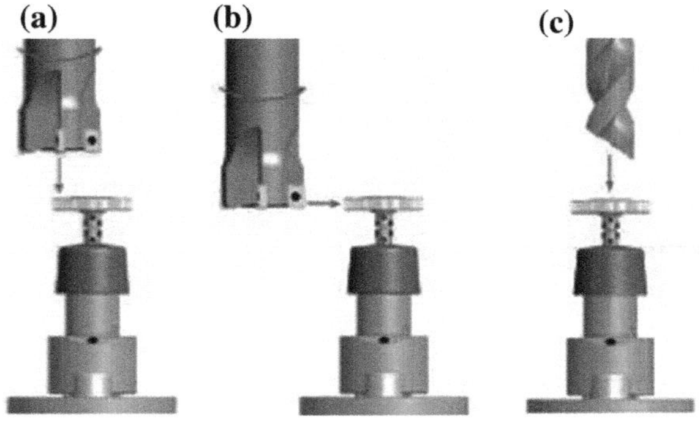

Fig. 2 Random instrument of the knife detection

- Machine tooling parameters set

By means of on-line measurement by a random meter, indirect or direct access to the most suitable processing parameters of the machining center in the execution of the next operation step can be achieved, so that the manufacturing quality of the workpiece can be greatly improved. This type of targeted and intelligent work method is used in many occasions where there are cooperation relations or special requirements, such as the machining of the upper surface of the cylinder block with a cylinder liner, the machining of the combustion surface of the cylinder head, and the like.

- To ensure the correct processing status: the workpiece, fixture "correction" and compensation

The so-called correct refers to the workpiece in order to ensure the correct installation, positioning, and take the appropriate measures. As for the existence of inconsistencies, there are not only the reason of the fixture, but also the impact of the workpiece itself. Undoubtedly, the alignment in processing is the basis for ensuring the quality of workpiece processing. As for the deviation measured during the alignment of the fixture, and due to temperature changes and tool wear and other gradient factors, the stability of the processing of the state may affect the quality of the finished product changes, if necessary, need to take some Compensation measures. Random measurement systems also play an important role.

- Automatic detection of the workpiece

After a process is completed, or after all processes have been completed, the workpiece is automatically measured, i.e., the inspection of the manufactured product is performed directly on the machine tool, which is another very important function of on-line measurement in the machine. At this point, a coordinate measuring machine moved to the machine, obviously, which can greatly reduce the auxiliary time of offline measurement, reduce the cost of quality. In fact, today's on-machine measurement capabilities are indeed very powerful, in addition to a variety of geometric elements can be a rapid detection, the use of specially developed software can also be completed offline programming, through the computer simulation, can also avoid Machine interference may occur in the interference, collision etc.

3 Application Examples and Their Analysis

As mentioned earlier, machining centers have been used in the domestic machinery manufacturing industry, including the automotive industry, many years ago, but the equipment random detection system has only appeared in the past decade. As a result of significant improvements in manufacturing quality, efficiency and reduce errors, in the enterprise, especially automotive engines, gearbox and other applications are increasing. Here are some examples from production.

3.1 Temperature Compensation and Tool Wear Compensation

About 10 years ago, an engine plant was in the process of accepting a flexible cylinder head automatic production line, and the machine capacity index was found to meet the requirements when several key linear dimension parameters of the machined parts of the two machining centers were evaluated. However, when the performance evaluation of the process, that is, the continuation of two or more shifts of sampling data for statistical analysis, there is a large deviation, process capability index, the value is low, that is, the quality of the process does not meet the requirements, After a more comprehensive analysis of the possible causes, it is confirmed that it is caused by changes in the ambient temperature. Obviously, it is difficult to eliminate the resulting error without taking compensation measures. Finally, by adding a random detection function to the machine tool, the problem was solved completely. The method is: equip a trigger type probe in the magazine, and according to the pre-set frequency (such as once //10 pieces), like a tool-like take out, hit a fixed position on the fixture to install the workpiece. Since it is judged correctly that the change in this position is linearly related to the controlled key size, it is possible to adjust the amount according to the change of the measured value, thereby effectively compensating.

Likewise, another method may be used for temperature compensation or tool wear compensation. Not long ago, in order to ensure the dimensional accuracy of the cylinder head after machining, an automotive engine plant in the south took the random probe's probe on the surface of the workpiece after milling and measured it at an interval of once every 10 pieces. If there is a large deviation, the processing parameters are automatically adjusted according to the set compensation mode. In general, the impact of temperature changes or tool wear is regularity, and corresponding compensation can be determined accordingly.

3.2 Machine Tooling Parameters Set

The aluminum cylinder in Fig. 3 adopts a mosaic cylinder liner, which is the purchased part, whose mounting plane (see the green arrow in the figure) is below the plane of the cylinder (see the red arrow in the figure) One of the processes in the machining center is to process the cylinder liner. In order to ensure that the axial distance from the plane on the cylinder to the mounting plane can be controlled within the specified tolerance range, the on-line detection system is provided in the machine. The h value to be controlled in this process is determined by the height of the bottom plane of the cylinder to the height of the upper plane H_i and the height of the cylinder liner; i.e., $h = H_i - L$.

Since the bottom of the cylinder is fixed to the bearing surface of the machine tool holder, the latter is the machined reference and L is the set value. Therefore, in

Fig. 3 Random detection is used to determine the processing parameters

order to ensure a consistent h value, the height value of each workpiece (H_i) must be obtained by online detection. Determine the corresponding cutting amount m: $m = H_i - (h + L)$.

The specific method is as follows: In Fig. 3, the touch probe is used to measure 4 points in sequence on the cylinder, and according to the data obtained by the average, and then compare the previous set value, to find the corresponding amount of cutting, As the basis for processing cylinder liner mounting surface.

3.3 Fixture Correction

The machining center shown in Fig. 4 has a huge turntable called "switch", with two "trays" in its diametrical direction.

Two worktables are fixed in the radial direction, which are in fact two rotation worktables to hold component parts, used to install cylinder head cover. The position of the two tables always corresponds to the "up and down" and "processing" stations in front and rear of the machine; that is, when a group of workpieces are processed in the processing station, the operator is engaged in the upper and lower feed stations. After the completion of the process of unloading and re-feeding. "Switch" and "tray" of the rotation accuracy is very high, but the former in the exchange of two table position, you must first lift the whole turntable lift, and then turn 180°, and then fall by the four cone, Taper holes, depending on the cone to match the positioning device. Due to poor working conditions, it will inevitably have the cooling liquid into the aluminum, debris, and other adhesion on the positioning surface, which will cause the turntable trace offset, and passed to the table ("tray") and its fixture. But the positioning device is determined by the principle and structure, but also to ensure that the micro offset can only be translated and can not be skewed. As can be seen from Fig. 4, the cylinder head cover is machined to be upright, and if it does not monitor the factors that cause the error, it

(a)
(b)

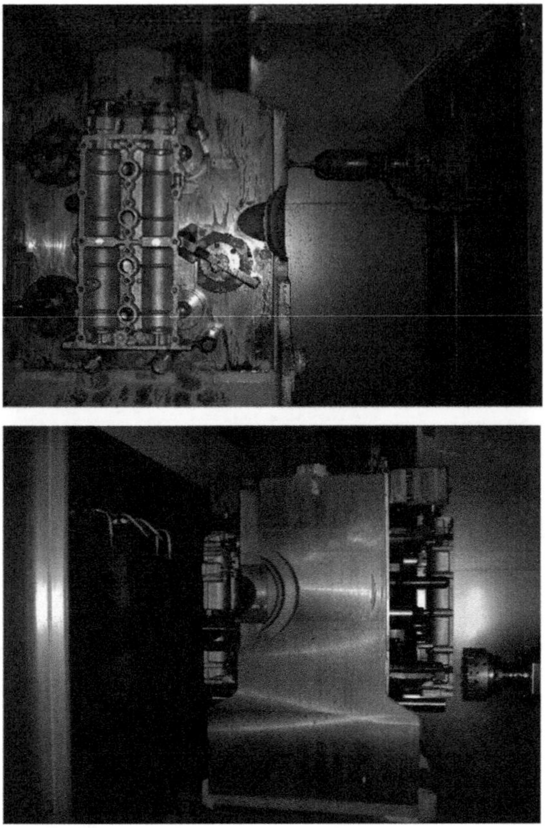

Fig. 4 Random detection for fixture correction

will be detrimental to the quality of the workpiece. To this end, the random test is arranged. For this purpose, a random inspection link was arranged, and it was said that compensation was made through “correction”. The upper part of the fixture is provided with a reference block. When the table is placed in the machining station of the machine tool, before the cutting of the workpiece, the power head is called out and hit the small plane of the reference block (see Fig. 4). The state of the jig is determined by comparison with a preset value, and when the deviation exceeding the allowable range occurs, the operator or the operator is informed of the processing. According to the actual situation of the workpiece to be processed as shown in Fig. 4, the tolerance range is set to ±0.2 mm; that is, when the result of the on-line measurement is less than ±0.2 mm, it is considered that the machining error caused by the fixture offset can be solved by compensation. And in the subsequent processing, through the t-cutting parameters in the introduction of the corresponding compensation value, in order to eliminate the impact of fixture offset, so as to ensure the quality of the workpiece manufacturing.

3.4 Workpiece Correction

The machined part is a large aluminum casting on a new car engine the sprocket hood, which is extremely important for the processing of four of these machines in many of the machine tools. In Fig. 5, these holes are shown from left to right, with the fourth and the rightmost one being located (corresponding to the "machined") position. In order to ensure the processing quality of the hole, it is necessary to keep the center of rotation of the tool in line with the center of the workpiece blank hole. However, it can be seen from the figure that the four holes are radiated, scattered, and the pore size and center are high. In this case, if the processing of each workpiece is still a constant process, then even in the fixture of the nuances, or some differences in the casting itself, will affect the hole manufacturing quality. To this end, the workpiece must be "correct", that is, the use of machine tool random detection system before processing each one by one for each blank hole to measure. The method is to precisely determine the coordinate position of the hole center by up and down, left and right of the circle, totally 4 points. According to it, execute the process of each hole targetedly. Obviously, after the "workpiece alignment" manufacturing quality has been fully guaranteed. In addition, the "correct" at the same time, can also get the rough roof of the blank, if the further use of variable programming, and thus can achieve the automatic allocation of rough blank, so that both can ensure that the hole during the cutting force is not too large, in order to avoid damage to the machine and tool, and improve the durability of the tool as well to achieve the highest efficiency.

4 The Application of Random Measurement Improves the Quality of the Process

Use the random meter set in the machining center to perform on-machine tool setting, complete the automatic setting of relevant machining parameters by on-line measurement before machining, or perform "correction" on fixtures and workpieces

Fig. 5 Random detection for workpiece alignment

to make corresponding corrections accordingly and compensate due to the temperature and abrasion through the online inspection after processing. All these, not only to ensure the quality of processing parts, but also can effectively improve the quality of the production process to run. Figure 6 is the result of the process capability analysis in Example 4 (3.4). The evaluation item selected is the hole in Fig. 5 where the probe is being corrected: $\varphi 23_{0}^{+0.05}$ is also the highest one of the four holes being found. In order to verify the quality of the physical processing and the quality of the production process, according to a month during the normal production standardized of sampling data, a statistical analysis. In Fig. 6, a is a single-value process, also known as "scatter plot", reflecting the changes in the period of processing items, b is a histogram. According to this, it is possible to calculate the index value—process capability index C_P and C_{pk} for evaluating the running quality of the production process during this period. The result is $C_P = 3.24$, $C_{pk} = 2.95$, it is clear that the quality of the process center has reached quite high level.

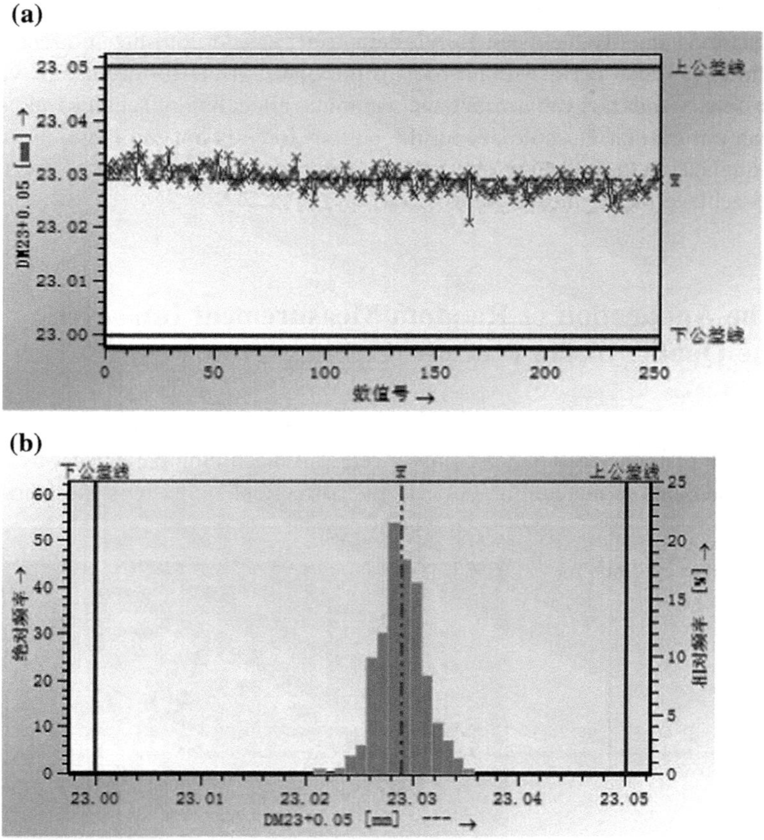

Fig. 6 Example 4 of the statistical analysis results

Figure 7 is a process capability analysis of Example 3 (2.3) in the previous section. From Fig. 4b, it can be seen that the "correction" of the fixture is measured by the on-line measurement. With the corresponding compensation, the workpiece size parameter that directly improves the accuracy is the same height as the axis of the power head, because the value of this value depends entirely on the milling amount of the tool on the vertical surface of the workpiece. Similar to the above-mentioned statistical analysis of Example 4, in order to verify the actual manufacturing quality of the workpiece and the quality of the process of the production process, but also for nearly a month to standardize the sampling of the data obtained by the assessment of the object is a fixed surface 4b in the working surface of the distance: 20.4 ± 0.2. From the obtained single-valued process map (Fig. 7a) and histogram (Fig. 7b), and thus the calculated value of the process capability

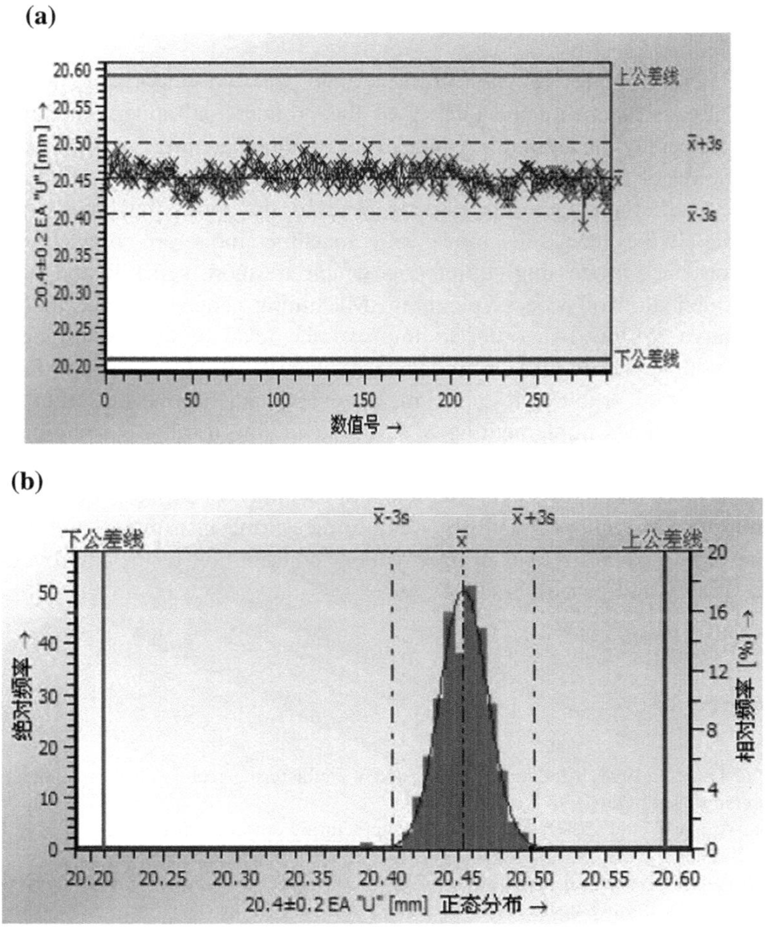

Fig. 7 Example 3 of the statistical analysis results

index, which reflects the level of the operation quality during this period: $C_P = 3.33$, $C_{pk} = 3.01$, Also shows that the quality of the process is quite high.

In fact, from the perspective of the role of the random meter, the user is more inclined to choose "correct", "compensation" and in some processes for each processed parts, automatically generate certain items of cutting costs. In other words, it is expected that after the random meter is applied to a machining center with excellent versatility, it can effectively improve the manufacturing quality of the processing equipment located at the production site, truly embodying the contemporary manufacturing industry, especially the most representative car Factories, following the rules of modern quality concept: "product quality comes from manufacture, the effective action has been taken.

5 Conclusion

Random measuring device and on-line gauge are two important and effective technical means to ensure the quality of the product for modern manufacturing industry, especially the automotive industry. From the beginning of this century, the transformation of the production model brought about by the economic development and market changes has directly lead to the trend that such traditional equipments like special machines and machine tools are replaced by the high-performance processing equipments which are more versatile and flexible in the implementation of process planning. Machining centers are one of the most representative [6]. And in order to improve the level of the machining center, equipped with a random measuring device is the only way. Over the past few years, there are a lot of articles on active measurement and on-line detection devices, including companies implementing SPC and achieving results, but few have been concerned with random measurement systems for machining centers. Through the analysis of its function, especially with examples analysis, explains the necessity of machining center equipped with the random measuring instrument, and combined with the online gauge, It will bring excellent contribution to the improvement of product quality and process quality.

References

1. Gu Y, De ZZ (1998) Modern car advanced manufacturing technology. Shanghai Jiaotong University Press, Oct 1998
2. Wang Z, Zhu Z et al (2000) Automobile manufacturing detection technology. Beijing Institute of Technology Press, Nov 2000
3. China Automotive Engineering Society 2006 (2007) World automotive technology development research. Beijing Institute of Technology Press, Aug 2007
4. Zhu Z (2008) Compensatory boring tool in the modern automobile engine factory fine boring process application. Automotive Engineer Tianjin (No. 10)

5. Zhu Z (2012) Online detection of intelligent, simple, flexible has become the basis of enterprise process planning. Automotive & Accessories Shanghai (No. 19)
6. Zhu Z (2011) To adapt to more varieties of mixed production line of online detection technology and equipment. Metal Processing Beijing (No. 12)
7. China Automotive Engineering Society (2016) Automotive advanced manufacturing technology development tracking research. Beijing Institute of Technology Press, Oct 2016
8. Luo X et al (2010) Online testing and SPC in the process of quality control in the application of machinery manufacturing. Shanghai (No. 10)

Research on the Application of Real-Time Monitoring System for Manufacturing Quality of Industrial Production Based on Industrial 4.0

Zhengde Zhu

1 Preface

As everyone knows, the Industrial 4.0 is the core of industrial intelligence-based manufacturing plant. In order to ensure the product quality, we must follow the modern quality concept, namely "quality comes from design and manufacturing, not from inspection." To reach the target of ensuring product quality, while reducing cost and improving efficiency, it is of great significance for such a large automobile manufacturing country like China to strengthen the stability of the production process, so it is very necessary to continue the implementation of statistical process control (SPC) which is a mature technology. Obviously, in order to be adapted to the rapid, small batch, customized production of intelligent manufacturing demand, we need to make full use of information technology, to optimize and expand the function of traditional SPC system.

Shanghai Volkswagen Powertrain Co., Ltd (VWPT), which was founded in 2005, has expanded from a plant to three plants (name as project 1, project 2, project 3), and has also experienced a series of production, promotion from EA111 to EA211 in a short term of 10 years (2006–2016). Varieties have increased from 3 to 5; in 2016, the volume reached 1.6 million units. In fact, VWPT is one of the most advanced worldwide engine production bases, both in production technology and manufacturing process. At present, the modern automotive industry as a representative of the manufacturing industry is carrying out Industrial 4.0 and "Chinese manufacturing 2025," and creating a transparent plant to achieve the enterprise intelligent manufacturing is a correct path, the so-called transparency, in essence, is integrated technology based on network, to make the various functional

Z. Zhu (✉)
Shanghai Volkswagen Powertrain Co., Ltd, Shang Hai, China
e-mail: 13564011215@163.com

Society of Automotive Engineers of China (SAE-China) (ed.), *Proceedings of the 19th Asia Pacific Automotive Engineering Conference & SAE-China Congress 2017: Selected Papers*, Lecture Notes in Electrical Engineering 486, https://doi.org/10.1007/978-981-10-8506-2_39

departments, operate with high effective interaction, which reflects the basic characteristics of the intelligent manufacturing.

As early as 4, 5 years ago, VWPT has worked together with Q_DAS company, which is a high level of professional statistical analysis software (DE) enterprises, and applied a real-time monitoring system in a plant network covering the production site, which has taken an important step to create a transparent plant, and to realize enterprise intelligent manufacturing. The "transparent" project is to establish a high level of quality information system according to the enterprise demand, with production site quality data visualization, data centralized management, backstage automatic analysis and assessment (CAMERA), and a series of advanced features. After several years of efforts, the implementation and completion of the transparent plant project, has not only greatly improved the production quality and operation level of the enterprise, also has provided a good example for the assembly plant, and many component plants in the domestic automobile industry, and different industries and enterprises which have similar needs.

2 The Change of Production Mode Request New Requirements for the Real-Time Monitoring System

The automobile engine, especially the small displacement automobile engine, as a typical product of modern manufacturing industry, since last 10 years at the end of the last century, its manufacturing process, mode of production, has changed greatly. As mentioned earlier, in order to meet the changing needs of today's automotive consumer market, and to be adapted to multi-variety, flexible production model, it is a necessity to improve the production process of real-time monitoring system functions. Taking the crankshaft production line as an example, over the years, the shaft grinding process is mainly consist of special grinding wheel cutting-type grinding machine, including the basic combination for processing line 5 gear shaft neck and neck connecting shaft for (respectively stage 1, 4 and stage 2, 3) equipment for each one. In the first line of the crankshaft in VWPT plant (see Fig. 1a), the three parts of the grinding machine are in the red circle.

The workpieces are transported in air through manipulator and get into the machine one by one. But in Project II, the tracking function of CNC grinder has occupied a dominant position in crankshaft grinding. Through the control program adjustment, it is convenient to be used for grinding various types of crankshafts, which has good flexibility. Due to the adoption of a single installation, all the main journal and connecting rod neck can be processed, and the efficiency and the precision can be improved. At the same time, since the technology scheme and layout use the fishbone which makes it easy to expand the production capacity (see many arrows in Fig. 1b), compared with the traditional linear layout shown in

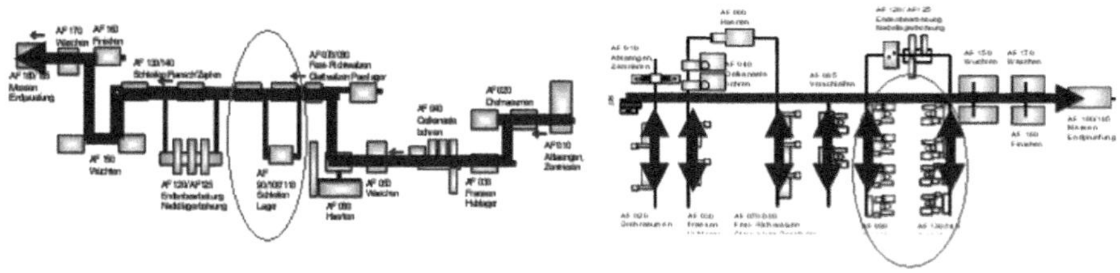

Fig. 1 Examples of the traditional and flexible curved axes

Fig. 1a, fishbone scheme improves the flexible level of the production line. It can be seen from Fig. 1b, in the past few years, the crankshaft has been built in the second line; the grinding process has reserved space to put another 8 sets of the same machine tool, and the graph shows that the actual number of grinding machine now is 6.

At this point, the process monitoring system is facing the problem: Through the grinding process in Fig. 1b, the sampling workpieces sent to the production side for checking are likely to have 14 different states, and if the check does not make a distinction and go on measuring, and deals with the workpieces' data according to the traditional approach, it has no means. Because at this time, if you want to judge and analyze whether the process is in control and stable measurement must be refined to one specification of the above over 10 conditions, only after recognizing the specific condition the targeted data processing is meaningful. Therefore, we must uniform the data format settings of those testing equipment that are related and provide quality information. This requires that need to be included in the quality data for evaluation in addition to the measurement value. Of course, depending on that the specific circumstances of the content will be different, but the modes of the data format are the same, then we will get the graph shown in Fig. 2 as a reflection of the operating process of the analysis of the situation. Different from the traditional SPC, the graph contains a number of curves, each corresponding to a specific situation, such as a machine tool or even more refined to a power head, or a corresponding part of the processing site. Having the refine dI monitoring system, we can recognize some abnormal phenomena that cannot be recorded in detail and targeted and take some actions according to the analysis and judgement. For example, tool wear occurs in the processing, only when a specific power head of a machine tool is targeted, analysis and measures can be worked out targetly (see Fig. 3), and if you do not use the data format set, we will not be able to timely adjust. As we see, the function of SPC needs to expand correspondingly with the change of technology and production methods.

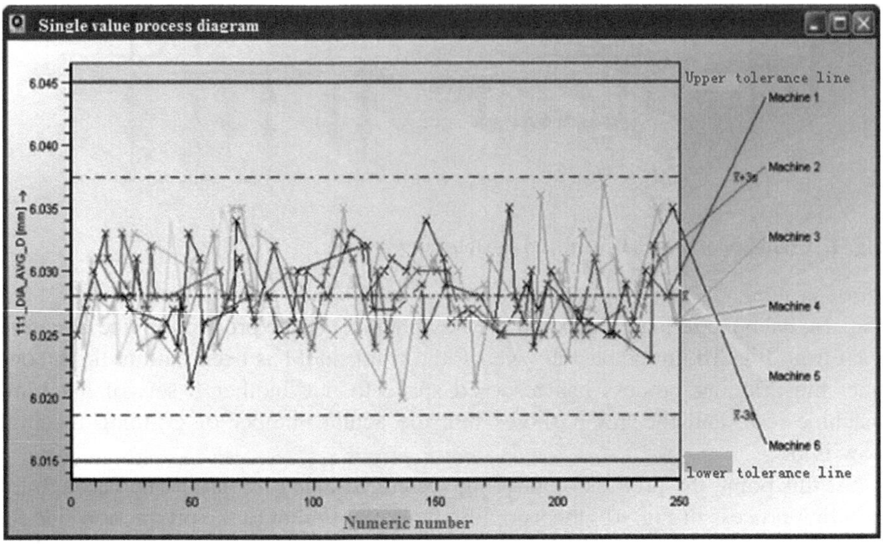

Fig. 2 It shows the detailed information related to the real-time monitoring of the process after the data format

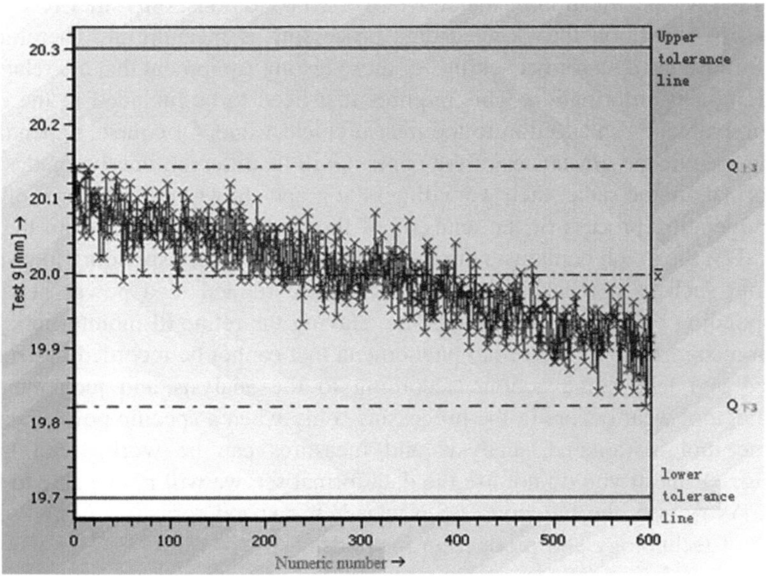

Fig. 3 Analysis and processing of the tool wear

3 The Necessity and Main Configuration of the High-Level Network Real-Time Monitoring System

Shanghai Volkswagen Powertrain Co., Ltd is a representative mainstream enterprise in the past 10 years with the rapid development of the domestic automobile industry, and the continuous expansion of production capacity, its number of measuring points, and online inspection devices on the floor is increasing rapidly. But since these measurement points (and the online inspection tools) are added along with the construction of the production workshop, it brings the following problems:

- The source of measurement data is fragmented: Measurement points are spread across the entire enterprise production area of three plants (in the three large combined plants), and it takes a lot of manpower and time to get all the data on the site;
- The measurement data are isolated from each other: The data from each measurement point is stored in a different computer format in its own computer;
- The measurement data cannot be shared: Since there is no established network communication between the existing measurement points, so if you need to access data, the only way is to collect the data on each measurement point. It is very troublesome, and also it is of low efficiency and easy to make mistakes;
- The measurement data information cannot be timely feed backed to the management departments (or supervisor): Due to the scattered measured points, once the measured value exceeds the control line, or the instability of the data occurs. It takes a long time for on-site quality staff to analyze the data and take countermeasures and feedback.
- To solve the actual situation, lack of operation personnel quality consciousness: Because the field operators are not able to see the fluctuation of the measured data visually and real-time process capability index, it is difficult to change the trend of the production process to make prediction and judgment. It is more difficult to adjust in time.

In order to improve the above problems, and based on the Volkswagen global transparent factory building suggestion, Shanghai Volkswagen Powertrain Co., Ltd plans to set up a real-time network monitoring system for manufacturing quality, automatic analysis, and evaluation system with field data visualization, centralized management, background data automatic evaluation system (CAMERA system). The function of the system should be able to meet the following requirements:

1. to provide advanced statistical process control theory and methods, to reduce the deviation in the production process, so as to improve production efficiency;
2. by better quality data visualization and human–computer interaction, to improve engineers and relevant personnel's working efficiency, to improve the sensitivity of field personnel for quality data, and to make timely feedback on production possible;

3. to achieve the centralized storage, management, and analysis of quality data from the production site in the whole plant;
4. to provide a standard statistical analysis and evaluation strategy for the management system, processing and assembly equipment acceptance, and process capability evaluation of the measurement system;
5. to be able to provide a comprehensive solution for automatic data entry, visualization, statistical analysis, data management, data display, data distribution model of automatic selection, evaluation, and quality information archiving.

So, what are the characteristics of the mass production data from the production site? The data from the production site have two kinds: One kind is the production site inspection; for example, Marposs measuring device placed next to the production line in three different plants in the workshop, they are divided into manual Marposs measure and important station automatic Marposs measuring machine (most of them placed near GROB machining center). The other is all kinds of precision measurement room testing equipment, such as Hexagon's CMM three coordinate measuring machine, Mahr's MFK form and position tolerance meter, Hommel's roughness meter, etc. (Fig. 4 left three/data sources) the goal of the project is to measure all of these results of the measuring instrument and to save production data not only as reference paper, but also save data to the database, the engineer in responsibility can do real-time analysis, while the manager can get to know the real-time quality of products, such as OEE data through the Web page (Fig. 4 in the database connection), meanwhile the measurement results on the Web page will be displayed on the big screen in the production field, which makes

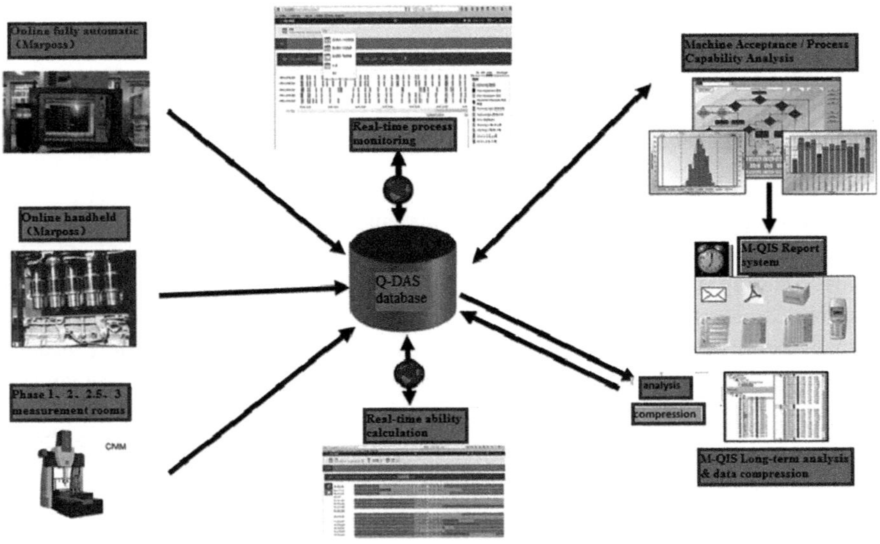

Fig. 4 Schematic diagram of data transmission in real-time monitoring system for networked manufacturing quality

production staff have a more intuitive understanding of the overall production quality and also, can make a more quick response (such as shutdown, tool change). If there is any doubt about the results of the field measurement, they can use the qs-STAT software interface to retrieve the data from the database to analyze (Fig. 4 on the right) for long-term production, not only the current quality data need analysis, but also these “big data” are required to predict the future trend of production quality, to avoid problems. So in the design of the whole system, we need to set up all kinds of automatic reporting tasks by means of M-QIS (lower right), and through these regular reports, the quality trend production process can be controlled by enterprises management departments, to guide the future direction of our continuous improvement.

Apparently, by the data acquisition and processing in this way, it is not enough to only rely on measuring check value provided by line. It cannot fully reflect the running status of the flexible production process, and surely, it is difficult to implement effective monitoring. In addition, as mentioned above, now enterprises pay more attention to product quality tracking and tracing. After taking a series of measures, marking two-dimensional code on the key parts has been more and more common (see Fig. 5), which contains a batch, type, blank manufacturer plant, mold tool number information, etc. In many enterprises of Volkswagen system, since many code marking stations are placed in the final inspection of the workpiece, the two-dimensional code will also contain some information obtained through inspection; for example, two-dimensional code marking on the crankshaft in the contains the journal grouping results obtained, and the measurement data of inspection can provide all measurement data of the workpiece under a certain state. It is not only necessary for workpiece evaluation, but also very valuable in the tracking of product quality by enterprises. In recent years, chip technology of RFID RF (based on the application of automatic identification device for contactless electronic data) is also becoming more and more common, and RF signal automatically recognizes the target and gets access to relevant data; identification does not need manual intervention. It is applicable to all kinds of environment. The system consists of an electronic chip, data read–write device, and control management software: Electronic data chip which is attached to a data carrier identification object on the part has small volume and lightweight, waterproof, anti-oil, high-temperature and high-pressure resistance characteristics. The electronic data chip is usually mounted on the machined workpiece by means of threaded fasteners, and Fig. 6 is a typical example. When it enters the working magnetic field, it can accept the RF signal sent by the reading and writing device. Similarly, the RF signal emitted by the read–write device is provided with a carrier, and the information can be written into the electronic data chip. Under normal circumstances, the basic information from the production line parts is as follows:

(1) The basic information—including the part number (sometimes the drawing number), blank factory code, production plant code, blank batch, on-line time, etc.;

Fig. 5 Two-dimensional code on the surface of the workpiece

(2) The manufacturing information—the workpiece is reflected by machining state and each process information, which is actually the most important content that record and storage, including information such as machine tools, equipment number, the spindle (fixture) number: processing information, i.e., parts in the procedure (state qualified or not is rework or not, processing date and time, if it belongs to the sampling parts, test results are OK or not); if it is online measurement procedures, such as connected to the cylinder block and cylinder head production line's sealing test (leakage test assay) process, the measured results will be written to the electronic data chip at this time.

Fig. 6 Workpiece in the feeding station with electronic

4 Based on qs-STAT Enterprise Edition Statistical Analysis Software to set up the Whole Plant Network Process Monitoring System

Companies follow the Volkswagen “Transparent Factory” concept, in the implementation of the project, based on qs-STAT Enterprise Edition statistical analysis software from q-DAS company, the establishment of manufacturing quality monitoring system networking site in the whole company, and then making own personalized customization, optimization, generates quality information real-time monitoring manufacturing data network. The system covers all the production lines in the three plants (the 3 connecting rods, the 4 curved axes, the 3 cylinder block, the 3 cylinder head, the 1 axis of the cams, and 2 housing lines). As mentioned earlier, the quality information from the production site contains:

(1) production line process;
(2) high precision instrument in the measurement room;
(3) at the end of line 100% final inspection.

According to the requirements of VWPT, the system will meet the needs of three levels (line operators, responsible engineers, and relevant managers), as shown in Fig. 7.

The line operator (located in the left side Fig. 7) can observe the fluctuation data through the monitor at any time, control the production process of the station, and all production measurement room can quickly feedback measure data to the production site through the network after checking the parts.

After getting the feedback data, the line operator can make adjustments in time, and if there are special fluctuation, it can feedback to quality management personnel immediately, which reduces the TP and TQ sectors workload, and also improves the data transmission efficiency and timeliness. The data will also be sent to the central quality database after recognized by the line operator, for engineers to query and analyze.

Refer to the middle part of Fig. 7. Since the whole system support analysis of different stage of production process, engineers in responsibility on site in the early stage, according to production process and process, make the appropriate directory file and edit the corresponding detection device, so that the quality assurance and technical engineers in responsibility can easily reach the aim of filtrating and tracing

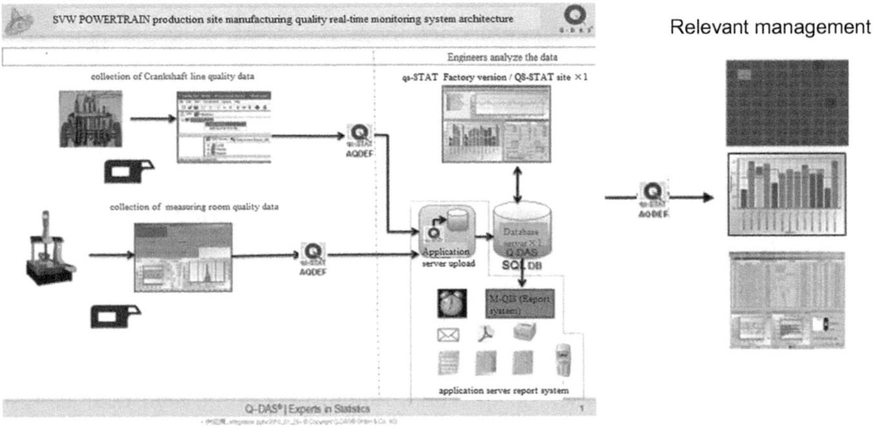

Fig. 7 Framework of real-time monitoring system for manufacturing quality with three layers of data visualization

data. At the same time, the system also allows engineers quickly analyze the corresponding short-term and long-term process capability by analyzing the quality data acquired from measure points; of course, the site supervisor engineers can also monitor production quality trend through the monitoring module, so that any problem can be treated in time.

Right side of Fig. 7 corresponds enterprise management; in accordance with the site supervisor engineers present, automatic report system can periodically obtain measure data from the corresponding automatic central quality database and generate standard SPC report. And the report can be viewed through e-mail or the form of public SkyDrive. Chy-STAT WEB web system works out a query page, as long as the relevant management leaders of the enterprise have a certain understanding of the system. They can monitor the quality of the whole plant at any time through any network-connected PC terminal.

5 Based on Whole Process of the Network Monitoring System, in the Whole Plant Improve the Quality of Real-Time Monitoring

After three years of hard work, the measurement point in the entire workshop production line and the four production measuring room has been equipped with network, using qs-STAT statistical analysis software as monitoring system, to achieve the effective control of the whole production process manufacturing quality. This set of new monitoring system uses Firefox Web browser as a software platform, to avoid the tedious software installation and configuration process, so that everyone can easily use the Q-DAS software to observe and monitor the whole plant's quality data. And it also has the following advantages: (1) to report regularly, to avoid complex preparation on product scrap, process and equipment capacity, OEE and stability; (2) online KPI; (3) easy and intuitive user interface; (4) with Q-DAS statistical core, to ensure the perfect statistics.

Further study also shows that for a long-term and continuous production, it not only requires analyzing the quality of current data, but also handling the "big data" for the future production process of quality trend forecast, in order to avoid the quality problems of products manufacturing. So in the developed system, through the M-QIS (see Figs. 4 and 7 in the lower right below) all kinds of automatic reporting tasks are set up, through which the production process trend can always be grasped and controlled, and the direction of the enterprise's continuous quality improvement in the future is guided. Today, the most used in Powertrain Assembly is the "monitoring module" and "capability module" of the whole system. The former can monitor real-time measurement result in a specific time, a specific production line, as well as a specific production equipment and so that it can statistic and display the actual production indicators. In the monitoring module, the data and varieties of charts of the measured parts can be retrieved at any time (see Fig. 8a–c) and can be read by a click on the report, and you can also send the

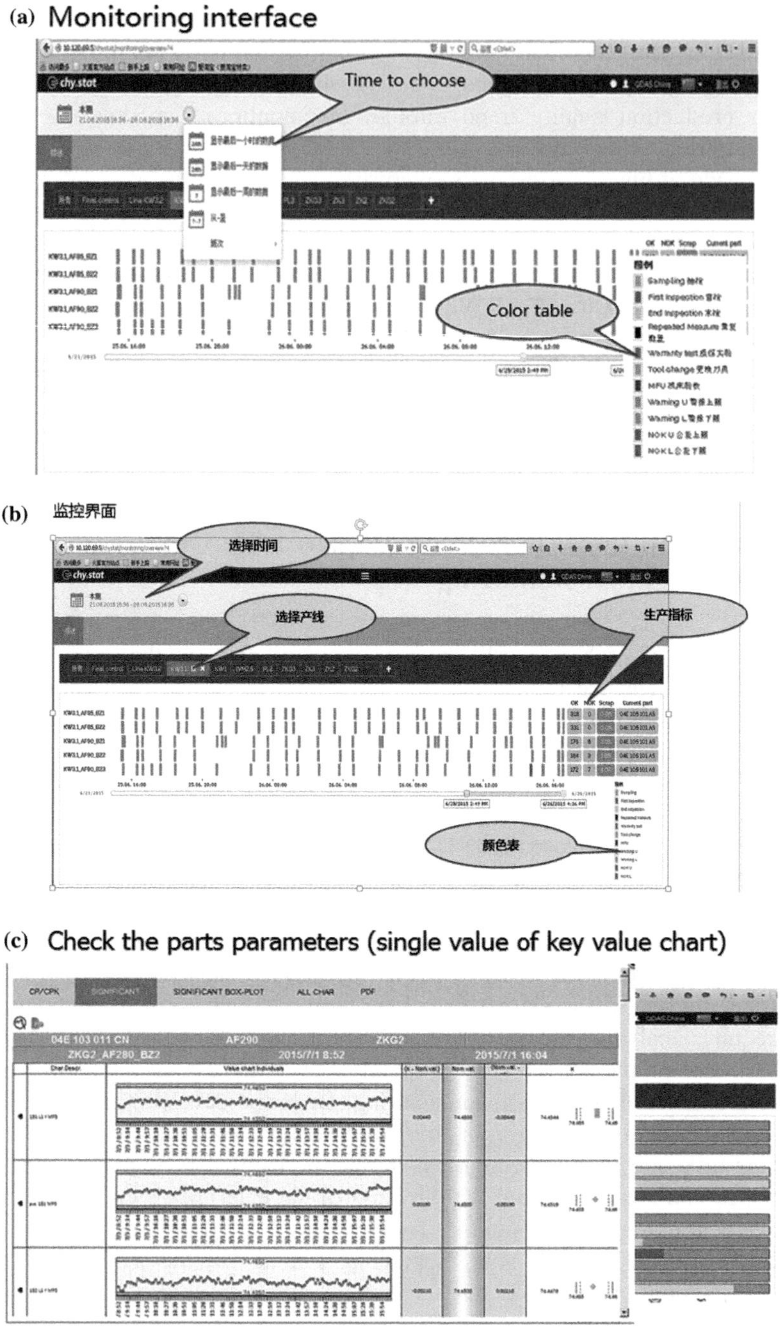

Fig. 8 **a** from the monitoring module in the system work monitoring interface 1, **b** from the monitoring module in the system work monitoring interface 2, **c** from the monitoring module in the system work monitoring interface 3

above interface to other related people. Either the basic line operator or the managers can easily distinguish between these colors to detect whether the work is in accordance with the requirements, and whether the results can fulfill the requirements. If the engineers on site need to do further analysis, you can click on a specific color column in Fig. 8 page, that is, to get more detailed information. Of course, you can also save the data in the form of DFQ alone for further analysis.

"Module" can take a shift as a unit, and calculate and evaluate the ability of each shift, each production line each process, so that each line foreman and the director has an intuitive understanding of the quality of the actual production capacity. Through the analysis of the capacity of these processes, we can adjust the production line or shift, etc., targetedly. Of course, the above two modules will eventually combine with other modules of the Q-DAS system to work together, such as the O-QIS module that ensures the complete input of additional information of measured data, the qs-STAT module used to do more statistical analysis on the data. After all, in the Web system you can only see a relatively simple icon; more icons need to be displayed on a more professional tool (Fig. 9).

6 Real-Time Monitoring and Automatic Reporting System for the Whole Plant

By Use of the previously described optimized, recognizable data processing and analysis system, the department managers, and even responsible leaders, can easily observe and monitor the production process according to their own needs,. and can handle it when necessary. The main steps are the relevant professionals with qs-STAT statistical analysis software, use quality database, to obtain the overall monitor on the process after data processing from the production line and assembly line through comparison and filtering, etc., and then judge the operation situation of the production process, and analysis, to find out the problems in the process. It depends on the specific circumstance to decide which method to use to complete the steps. In normal condition, it is up to production site engineers and full-time quality assurance personnel, to be responsible for the implementation of the daily monitoring, using a targeted state of real-time monitoring. In addition to the monitor of the field operator through the control box in the test table at any time, the monitor can be realized from the system terminal set on site selectively.

Figure 10 on the left shows the needs to call out the required instructions, and after that, the monitoring page and content can be selected as needed, the right is an example. As for the use of M-QIS automatic reporting system, daily, weekly, and monthly reports and other customized reports can be sent to relevant departments and relevant functions through the electronic documents and can be designed or expressed in various forms according to the actual needs. The daily report and the weekly report can be used as shown in Fig. 11a as a process analysis report, and weekly and monthly reports summarize through the tolerance utilization and

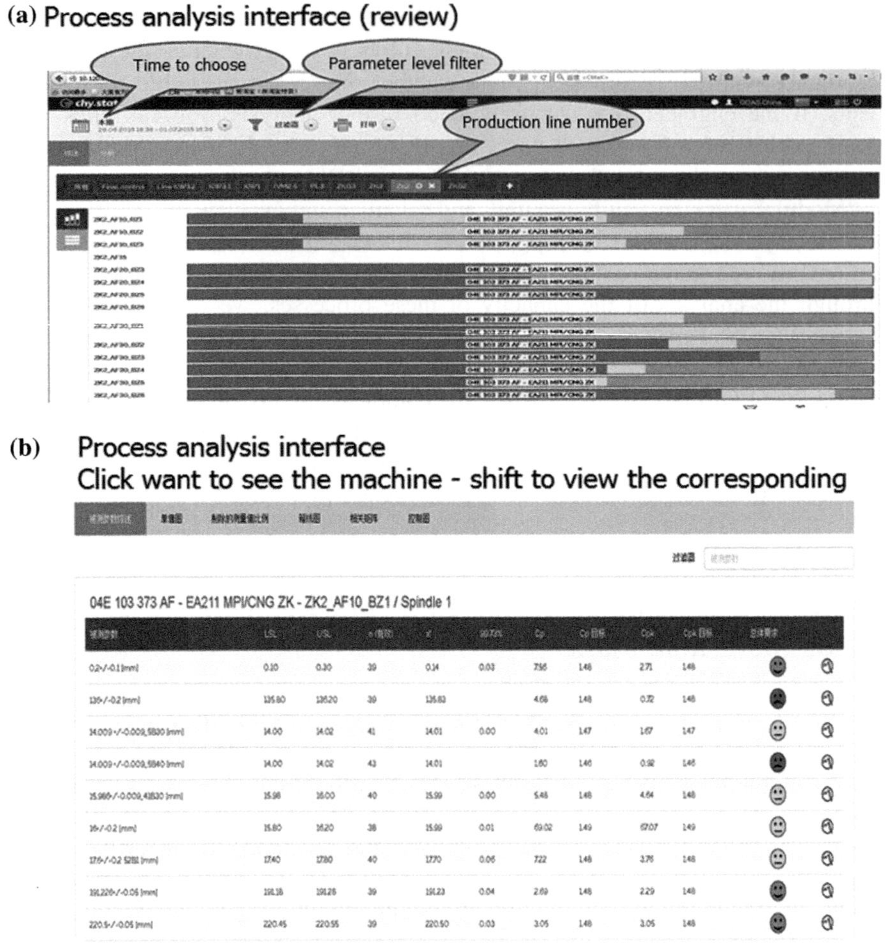

Fig. 9 **a** from the system ability module work monitoring interface 1 **b** from the system ability module work monitoring interface 2

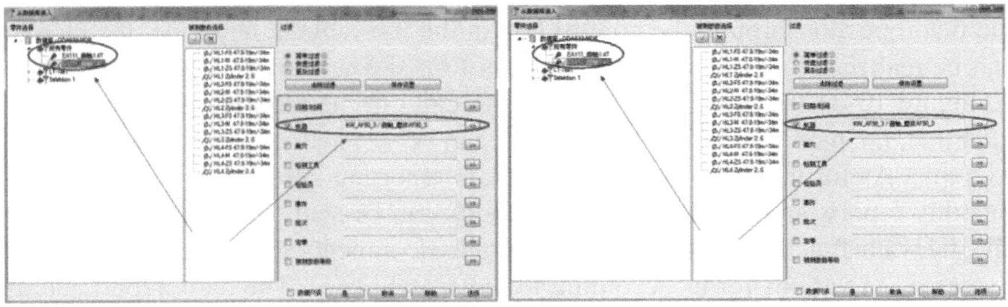

Fig. 10 Production site monitoring system terminal has a selective representation of the page pattern

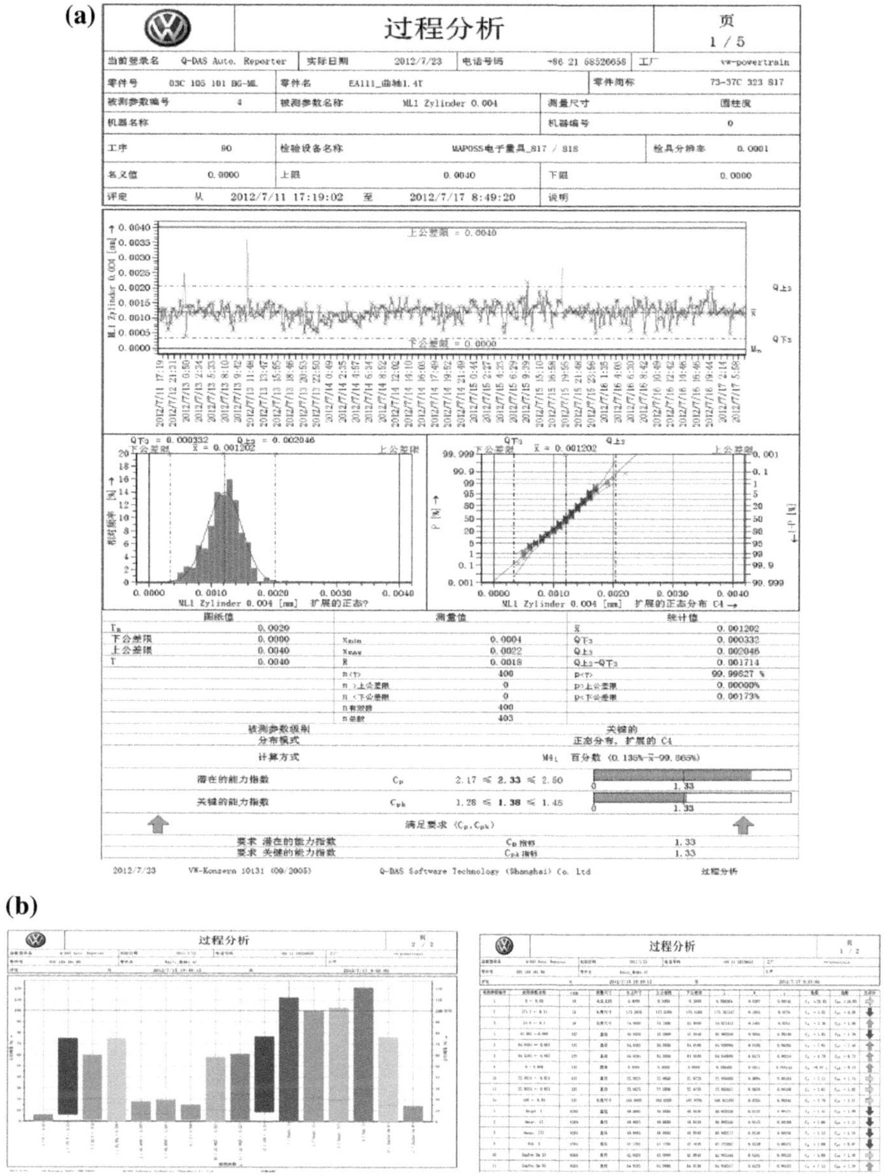

Fig. 11 **a** Example of analysis report selected in daily, weekly 1. **b** Example of analysis report selected in daily, weekly, and monthly 2

process analysis results (see Fig. 11b left and right), and Cp/Cpk trend charts and other forms, so that it can intuitively reflect and state the production process of a period of time, as the base to optimize production and improve process.

7 Conclusion

Shanghai Volkswagen Powertrain implements the "transparent plant" project, as an example of the successful application of the enterprise version qs-STAT statistical analysis software. As a result of the combination of automobile, especially cars, the growing demand for personalized products for the consumer market, and thus in the implementation of the project, it both talk around the greatest degree of adaptation and to meet the multi-types, flexible production targets, And further use the contemporary information and network technology, to optimize and expand the traditional SPC system. And then through the improvement of "qs-STAT Enterprise Edition statistical analysis software" function, a whole plant network and real-time manufacturing quality monitoring system is completed, which not only includes function of automatic data entry, data visualization three-tier structure, statistical analysis, data management, but also provides a data distribution model's automatic selection, a variety of capacity assessment, and the quality of information archiving the requirements of enterprises. It can be called as a comprehensive solution.

The last thing to mention is that the German Volkswagen (headquarters) hoped to copy the Czech Republic's Volkswagen Group's engine plant's lately completed, "transparent plant" project content which shows good results, intactly to Shanghai Volkswagen Powertrain. They may also be out of "good intentions." They hope that we can avoid some detours, since our domestic understanding of this ability is one-sided. After communication with Czech factory (by sending engineers out, and inviting them in), we recognized that there are big differences between the two plants, and "copy" is impossible and meaningless; we must do it based on our own feature. The final fact is after closely and unremitting efforts made by the Chinese engineers of both sides, a satisfying solution is finally provided to all sides after three years.

References

1. Luo Xiaolin and other online testing and SPC in the process of quality control in the application of machinery manufacturing Shanghai 2010 No.10
2. Xiao Fan Engine Production Line Parts Management Model Shanghai Automotive Shanghai 2012. No.9
3. Zhu Zhengde to improve the quality of the manufacturing process of the product is the conduct of the enterprise's vision of Chengdu 2012. No.4

4. Guo Jin Shu and other Chinese intellectual development under the tide of enterprise information development stage of mechanical design and manufacturing engineering Nanjing 2015. No.9
5. ZHU Zheng-de Research on the Development of Modern Automobile Manufacturing Industry Based on "Industrial 4.0" Changchun 2015. No.5
6. Zhou Benhua and other field data acquisition technology and intelligent manufacturing system in the application of manufacturing technology and machine tools Beijing 2016. No.6

Development of a Water-Cooled Condenser

Isao Tamada, Yasuhiro Mizuno, Hiroyasu Shimanuki, Takaki Okochi and Zong Xiumei

1 Introduction

The development of plug-in hybrid vehicles (PHVs) and electric vehicles (EVs) has recently accelerated in response to the fuel economy regulations enforced by various countries, and there is a rising need for high-efficiency heat pump (HP) systems to secure a low heat source during heating and to increase the cruising distance of EVs. The following two approaches to securing a heating source are used for HP systems because in electric vehicles exhaust heat from the engine is unavailable during heating: (1) using a compressor to produce hot, high-pressure refrigerant gas and warm the air inside the cabin by means of a heat exchanger and (2) producing hot water from hot, high-pressure refrigerant gas with the aid of a heat exchanger and supplying it to the cabin to warm the inside air. Approach 1 offers good efficiency because it can directly warm the air, but drastic changes must be made to the air conditioning unit, which contains an evaporator and a heater, inside the cabin. Approach 2, on the other hand, can use the existing air conditioning unit without modification because the hot water is produced by a heat exchanger; this approach could be used for a larger number of vehicles as a relatively inexpensive HP system. The heat exchanger used for approach 2 is mounted downstream of the compressor and is designed to condense and liquefy the hot refrigerant gas by cooling it with water (hereafter, "water-cooled condenser"). The refrigerant entails a phase change from gas to liquid, and as a result, a large heat exchange rate can be expected. A water-cooled condenser must provide the following functions to work with the HP system and to meet vehicle needs:

- High performance: High-efficiency heat exchange and low pressure loss to ensure quick heating performance,

I. Tamada · Y. Mizuno · H. Shimanuki · T. Okochi · Z. Xiumei (✉)
Denso Corporation, 1-1, Showa-Cho, Kariya, Aichi Prefecture 448-8661, Japan
e-mail: xiumei_zong@dich.denso.com.cn

Society of Automotive Engineers of China (SAE-China) (ed.), *Proceedings of the 19th Asia Pacific Automotive Engineering Conference & SAE-China Congress 2017: Selected Papers*, Lecture Notes in Electrical Engineering 486, https://doi.org/10.1007/978-981-10-8506-2_40

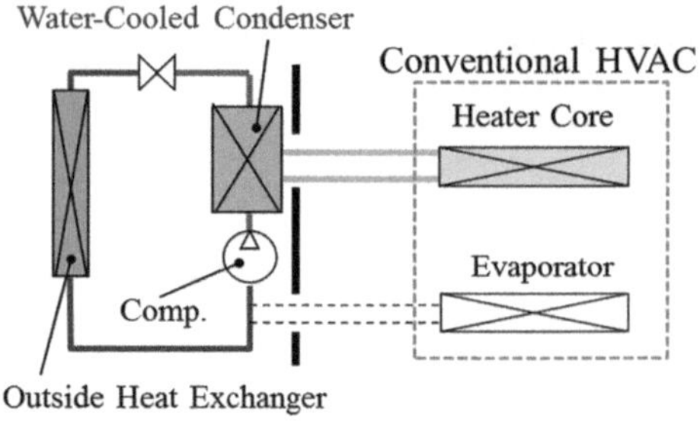

Fig. 1 Heat pump system layout using WCDS

- Smaller mounting space: Compactness for mounting in a limited space,
- High reliability: High strength because it uses a high-pressure vessel,
- This paper describes the development of a water-cooled condenser that can meet the above needs (Fig. 1).

2 Product Concept

The water-cooled condenser is a heat exchanger designed to produce hot water by exchanging heat between the refrigerant gas and the cooling water. The structure of the developed product is shown in Fig. 2. It includes a tube consisting of a case plate and an inner plate, and both the refrigerant and watersides are provided with an inner fin. This tube is stacked in multiple layers to form a core, and each inner plate is provided with a partition to allow the refrigerant and the cooling water to flow in opposite directions and thereby ensure high performance. In addition, conventional refrigerant-side passages, as represented by existing air-cooled condensers, use a flat extruded multiway tube or straight fin. Our water-cooled condenser, on the other hand, adopts an offset fin on both the water and refrigerant sides to increase performance (details are described in Sect. 3.2). The water-cooled condenser achieves high performance, low pressure loss, high pressure durability, and fin workability at the same time by optimizing not only the general specifications of the fin, such as height and thickness, but also its offset amount and other detailed specifications (details are described in Sect. 3.3).

3 High Performance and Miniaturization of the Core

To increase the performance and reduce the size of the water-cooled condenser, we optimized the specifications of the refrigerant-side fin focusing on the large contribution of the heat transfer characteristic of the refrigerant side. Generally, the performance of a heat exchanger between two fluids is determined by the thermal resistance (inverse of heat transfer coefficient × heat transfer area) ratio between the

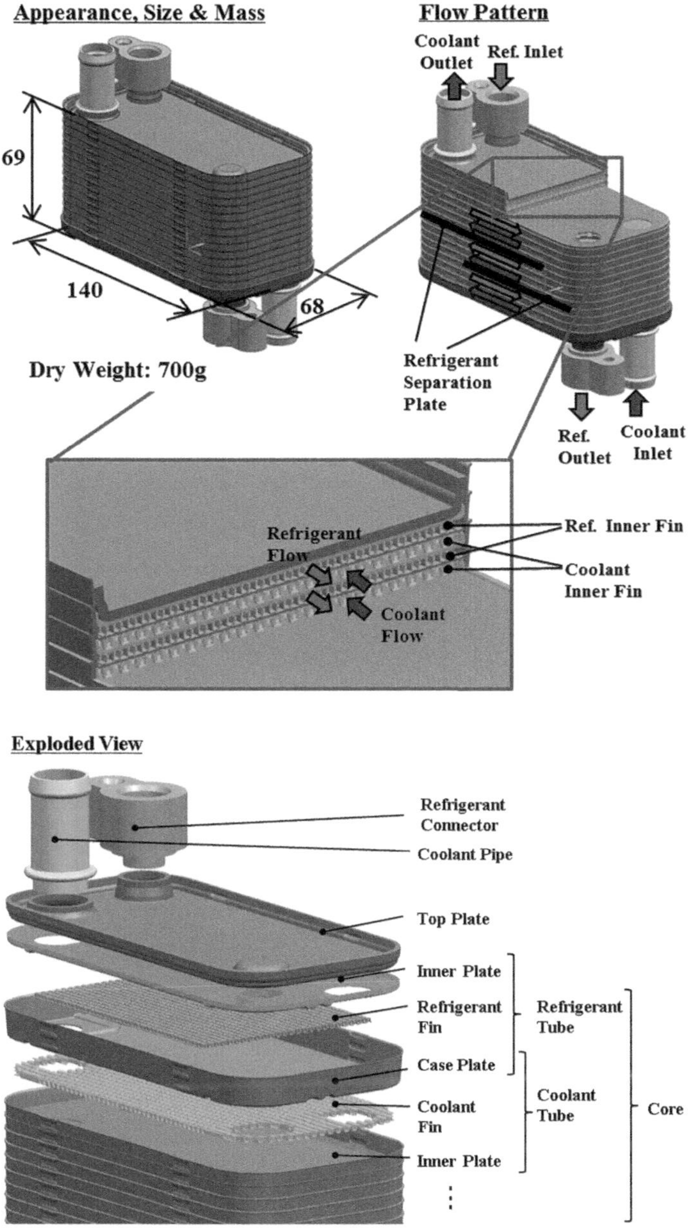

Fig. 2 Structure of the developed product

fluids. Here, let us consider a straight fin for the refrigerant side, a louver fin for the air side, and an offset fin for the waterside as standard fins for the respective fluids used in typical air-cooled condensers or oil coolers. Based on this assumption, the thermal resistance ratio of a conventional air-cooled condenser used for air conditioning systems is shown in Fig. 3. Since the air-side thermal resistance is larger than the refrigerant-side thermal resistance due to the physical properties of air, improvement of the heat transfer performance of the air-side fin greatly improves the overall performance of the air-cooled condenser. On the other hand, the rate of the refrigerant-side thermal resistance in the thermal resistance ratio of the water-cooled condenser (Fig. 4) is smaller than that of the air-cooled condenser due to the heat transfer physical properties of water. For this reason, to improve the performance of the water-cooled condenser it is essential to reduce the refrigerant-side thermal resistance by improving the condensing heat transfer coefficient.

To find the optimum specifications of the water-cooled condenser, we followed the approach:

3.1 Refining the shape of the refrigerant-side fin.
3.2 Establishing an equation for predicting the condensing heat transfer coefficient.
3.3 Optimizing the specifications of the core of the water-cooled condenser.

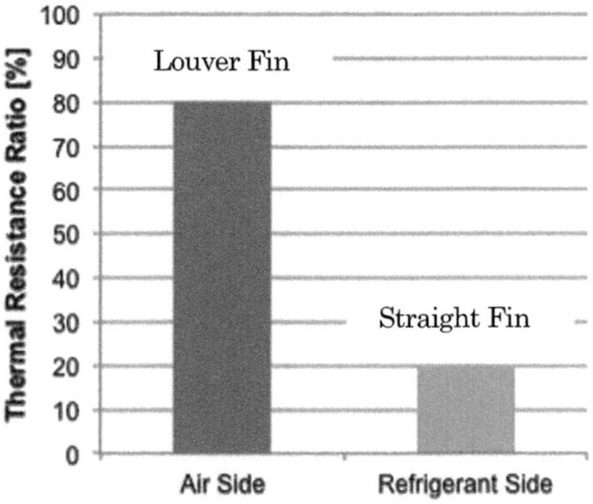

Fig. 3 Thermal resistance breakdown of conventional air-cooled condenser

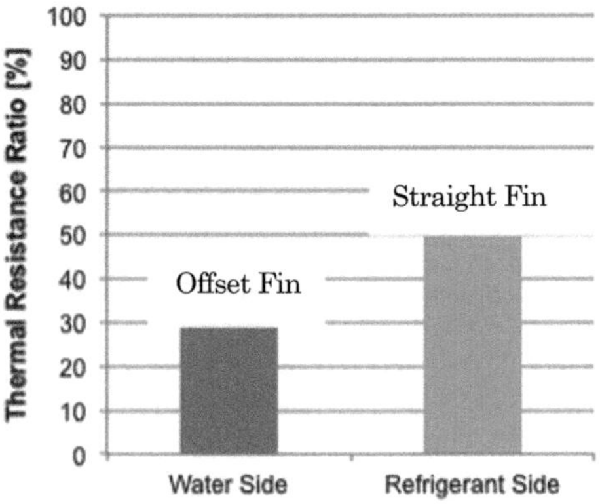

Fig. 4 Thermal resistance breakdown of water-cooled condenser

3.1 Refinement of the Shape of the Refrigerant-Side Fin

The refrigerant inside the condensation heat exchanger is usually further condensed from the inlet to the outlet, and its state changes from gas to liquid (Fig. 5). The state at the inlet where only gas exists is defined as quality 1 (x), and the state at the outlet where only liquid exists is defined as quality 0. The heat transfer coefficient also changes in response to the quality.

The heat transfer coefficient during condensation is basically expressed by the filmwise condensation model shown in Fig. 6. The model shows that it is important to keep the liquid film thin to improve the condensing heat transfer coefficient. The liquid film thickness δ in this two-phase flow is extremely close to the boundary layer thickness in a single-phase flow, and therefore, it is presumed that heat transfer can also be accelerated in the two-phase flow by a leading-edge effect as in the case of a single-phase flow. The basic structures of the fins are shown in Table 1. Of these, the wavy fin and the offset fin can more greatly amplify the leading-edge effect than the straight fin. Considering the pitch of an edge providing a leading-edge effect based on past mass production data, the offset fin can be formed with a shorter pitch than the wavy fin, and so we selected the offset fin for this development project.

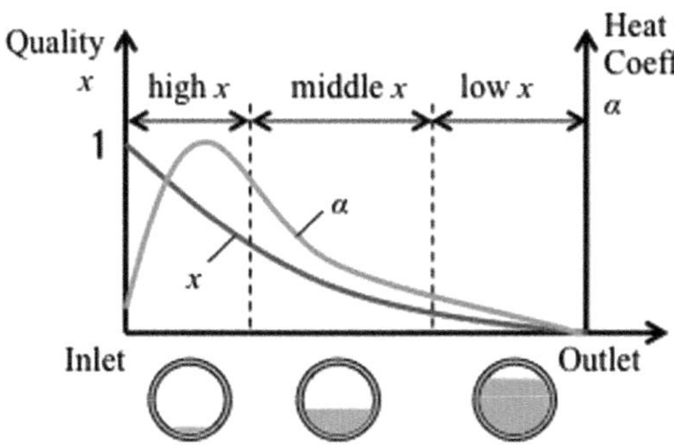

Fig. 5 Transition of quality and heat transfer coefficient in condensation heat exchanger

Fig. 6 Filmwise condensation model

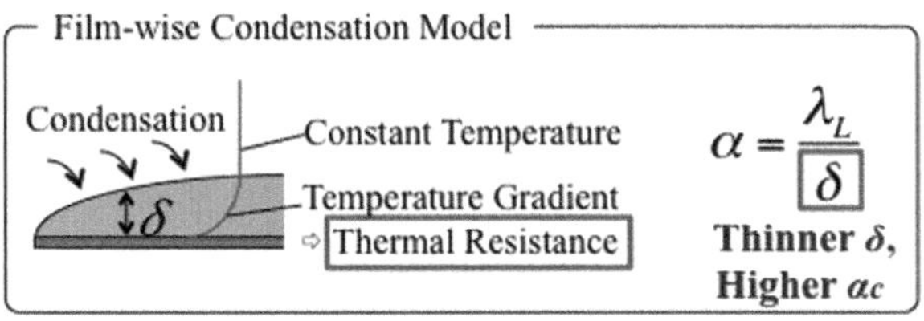

Table 1 Performance comparison of basic fin structure for condensation

	Straight	Dimple	Wavy	Offset
Structure			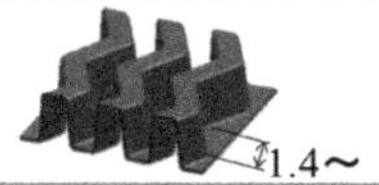	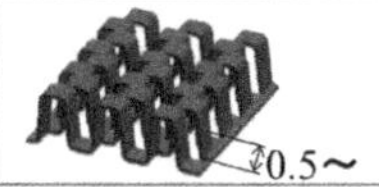
Edge pitch	Tube length	Tube length	1.4 mm~	0.5 mm~
Leading-edge effect	×	×	Δ	O

3.2 *Establishment of an Equation for Predicting the Condensing Heat Transfer Coefficient*

In this development project, we formulated a calculation model, shown in Fig. 7, capable of determining the amount of transferred heat of the whole heat exchanger to eventually optimize the specifications of the core, including the specifications of the fin. The calculation of the amount of transferred heat using this model requires an equation for estimating the refrigerant-side condensing heat transfer coefficient.

As an empirical equation for estimating the condensing heat transfer coefficient, the Shah equation [1] is used (Eq. 1). This equation is based on the Dittus-Boelter

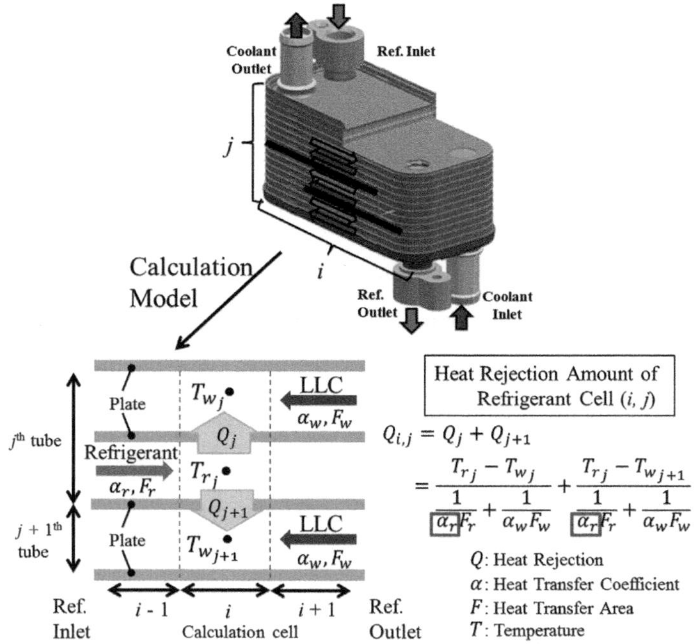

Fig. 7 Overview of calculation model for WCDS

equation [2] (Eq. 2), which is the equation for the Nusselt (*Nu*) number during turbulence, and which makes it possible to estimate the condensing heat transfer coefficient by multiplying the correction factor depending on the quality. However, this equation is applicable only to flows inside circular pipes and straight fins.

$$Nu_{\mathrm{Shah}} = 0.023Re^{0.8}Pr^{0.4}\Phi(x) \tag{1}$$

$$Nu_{\mathrm{Dittus-Boelter}} = 0.023Re^{0.8}Pr^{0.4} \tag{2}$$

On the other hand, an equation is established for estimating the heat transfer coefficients of offset fins only in single-phase flows, only of gases or liquids [3] (Eq. 3). Heat exchangers designed based on this equation include oil coolers and EGR coolers.

$$Nu = j \cdot Re \cdot Pr^{0.4}$$

When $Re < 1000$:

$$j = 0.48\left(\frac{F_{\mathrm{seg}}}{de}\right)^{-0.16}\left(\frac{F_p}{F_h}\right)^{-0.18}Re^{-0.54}$$

When $Re > 2000$:

$$j = 0.24\left(\frac{F_{\mathrm{seg}}}{de}\right)^{-0.32}\left(\frac{F_t}{de}\right)^{0.09}Re^{-0.37} \tag{3}$$

Thus, to design the heat transfer of a water-cooled condenser using an offset fin, an equation for estimating the condensing heat transfer coefficient according to the quality of the refrigerant suitable for the offset fin is additionally required. In this development project, we tried to formulate a model equation based on the Shah equation by multiplying the coefficient j depending on the quality by the improvement of performance achieved with the application of the offset fin (Eq. 4).

We decided to include the coefficient j as the variable for the quality x because the margin of performance improvement was expected to change. Since the region of the high quality consists almost only of gases, the expected specific margin of performance improvement is almost equal to that in single-phase flows. On the other hand, there were no findings or data on the margin of performance improvement in a region of low quality where liquids and gases coexist at almost the same level, and we quantified it by performing a test.

$$Nu_{\mathrm{ref\_offset}} = j\left(x, F_{\mathrm{seg}}, F_p, \ldots\right)Nu_{\mathrm{Shah}} \tag{4}$$

We placed a test section, which simulated a single layer of the tube of the water-cooled condenser, in three positions and constructed a test system capable of measuring the heat transfer coefficient from high to low quality (Figs. 8 and 9).

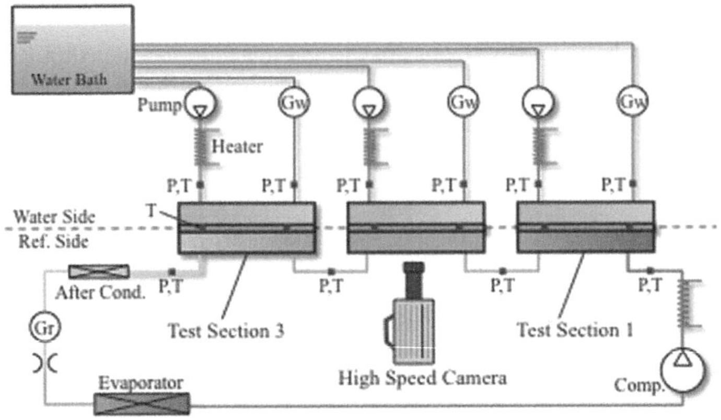

Fig. 8 Test cycle layout

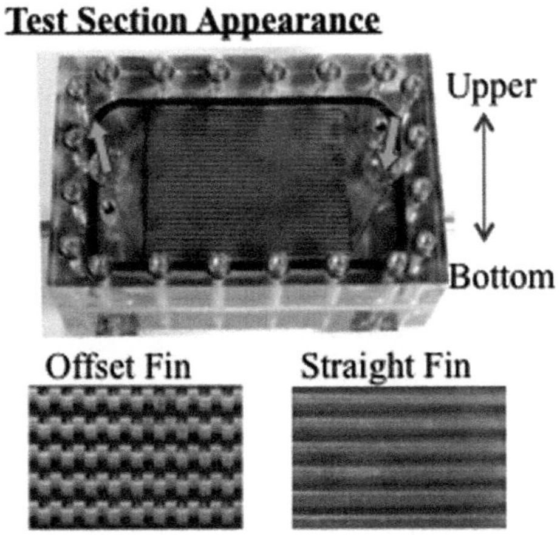

Fig. 9 Appearance of test section

In addition, as the test conditions, the heat exchange conditions per tube layer in the practical region are set. The condensing heat transfer coefficient α was calculated based on the performance Q of the test section and the heat transfer area F of the refrigerant-side fin, and the degree of contribution of the quality x to the coefficient j was quantified by comparing α of the straight fin with that of the offset fin. To calculate α with high accuracy in this step, the pitch and other specifications of the fins were unified so that the heat transfer areas F of both the fins would become equal to each other. In addition, to consider the difference between the heat transfer coefficients resulting from the difference between the fins based on the liquid film thickness and the flow pattern, we visualized the flow field using a high-speed camera (Table 2).

Table 2 Test conditions

	Test condition
Ref. inlet pressure [Mpa]	1.0
Ref. flow rate [Kg/h]	20
Coolant flow rate [L/min]	1.0

$$\alpha_{\mathrm{ref}} = \frac{Q_{\mathrm{water}}}{F_{\mathrm{ref}}\left(\overline{T_{\mathrm{ref}}} - \overline{T_{\mathrm{wall}}}\right)} \tag{5}$$

The test results are shown in Fig. 10. The solid line in the graph indicates the margin of improvement of the condensing heat transfer coefficient attributable to the application of the offset fin. In the high-quality region, the margin of improvement of the condensing heat transfer coefficient was the margin of performance improvement +20% as expected before (the liquid *Re* under this condition was approximately 50, and the margin of improvement was calculated assuming that *Nu* for the offset fin was $Re < 100$ of Eq. 3 and that for the straight fin was 3.66 of the laminar flow *Nu*), whereas the margin of performance improvement was +60% in the low-quality region, which was higher than the effect of performance improvement in the single-phase region mentioned earlier. Let us consider the reason for this effect from the results of visualization. In the high-quality region, the main difference in the flow field was the liquid film thickness. Judging from the fact that the offset fin kept the liquid film thin, its performance was presumably improved. In the low-quality region, on the other hand, turbulence was clearly observed due to the collision of liquid against the offset fin in addition to the difference in liquid film thickness. The margin of performance improvement in the low-quality region is attributable to this turbulence. These facts reveal that the application of the offset fin to a condensing flow was more effective in improving

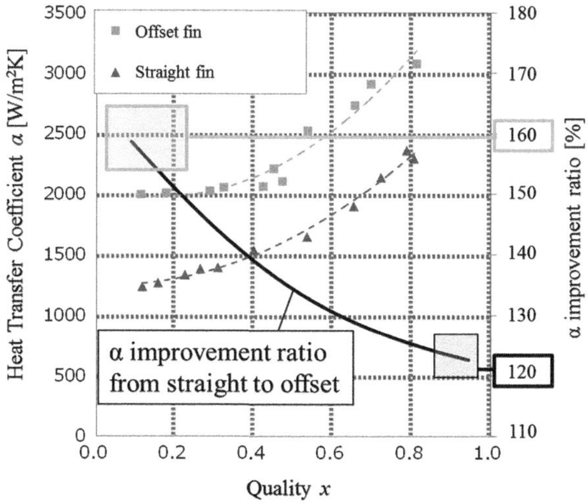

Fig. 10 Test results of condensation heat transfer coefficient for offset fin and straight fin

performance than its application to a single-phase flow, and so we used it for the developed product. In addition, by combining the coefficient j with the margin of improvement of the heat transfer coefficient obtained from this test and integrating them into the calculation model for the whole heat exchanger shown in Fig. 7, we made it possible to design the specifications of the core of the heat exchanger provided with an offset fin on the refrigerator side.

3.3 Optimization of the Specifications of the Core of the Water-Cooled Condenser

The important point when optimizing the specifications of the core is to achieve strength and workability capable of withstanding the high pressure of the refrigerant while pursuing high performance and low pressure loss. Thus, we used the formulated calculation model, FEM, and the following approach to optimize the specifications of the core (Fig. 11).

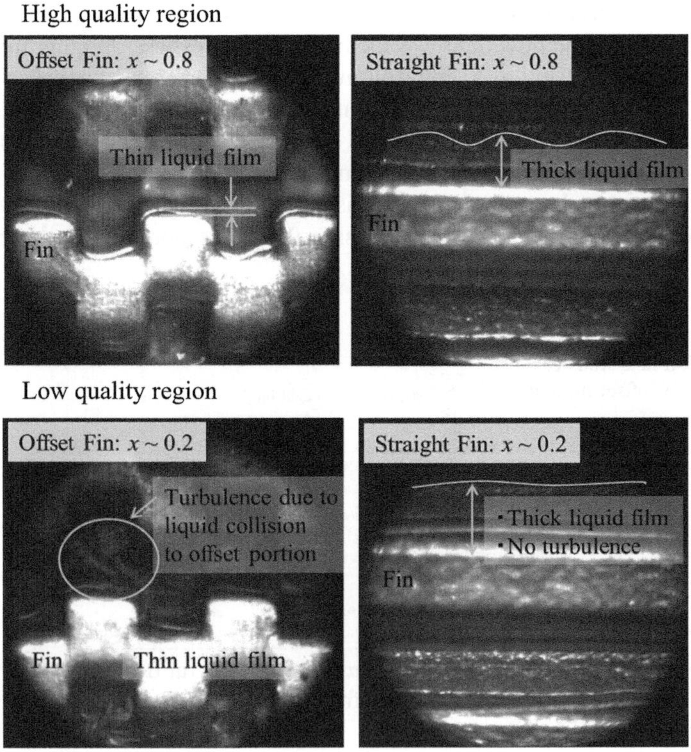

Fig. 11 Visual test results of each flow pattern

3.3.1 Determining the specifications of the refrigerant-side offset fin.
3.3.2 Determining the specifications of the water-side offset fin.
3.3.3 Optimizing the height of the refrigerant-side and water-side tubes.

3.3.1 Determination of the Specifications of the Refrigerant-Side Offset Fin

As described in Sect. 3.1, the specifications of the refrigerant-side fin (Fig. 12) are important for performance and also extremely important to secure pressure durability. The concept of securing pressure durability is to secure the cross section of the column and keep the stress imposed on the fin at or below a given level by making the fin pitch finer and increasing the plate thickness (Fig. 13). A finer pitch is also desired to secure performance. However, making the fin pitch finer and increasing the plate thickness reduce workability. The forming limit of conventional offset fins, the specifications of currently mass-produced fins, and the fin pitch and fin thickness that can meet the target performance and pressure durability of the water-cooled condenser are shown in Fig. 14. It was necessary to resolve the issues confronting conventional technologies because the target performance and pressure durability regions of the water-cooled condenser were outside of the range

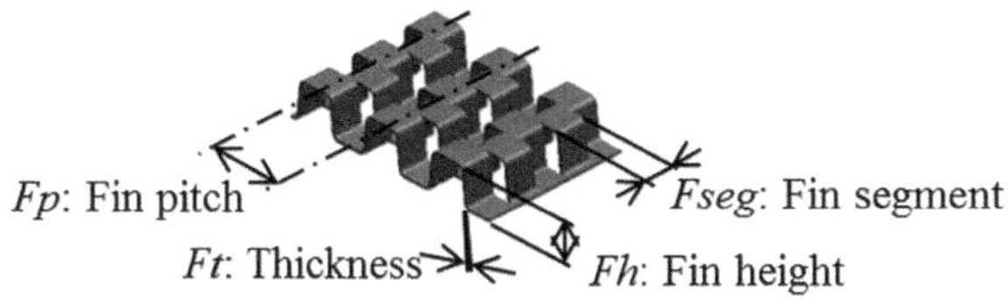

Fig. 12 Offset fin parameters

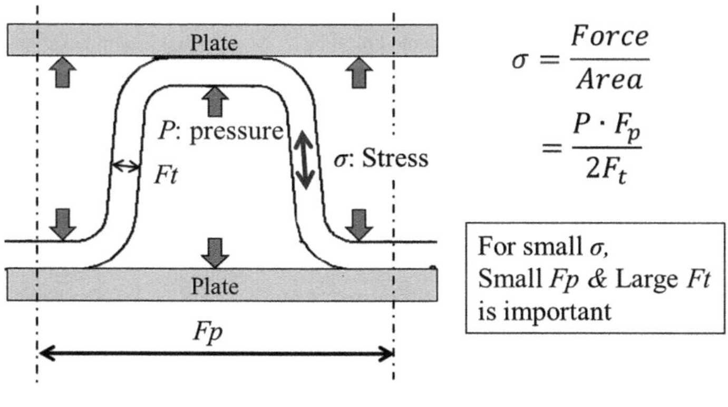

Fig. 13 Design concept of pressure durability

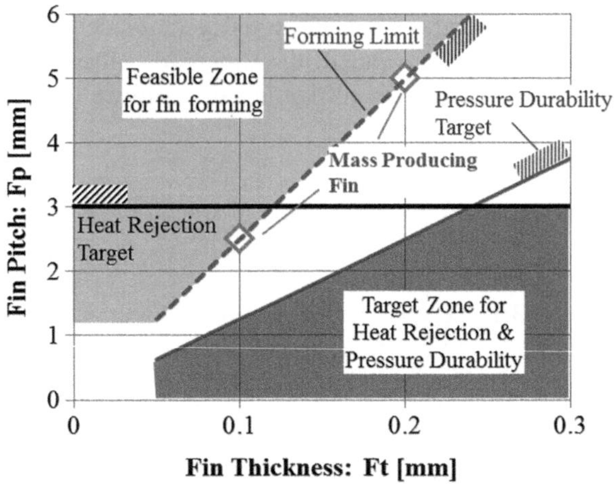

Fig. 14 Fin specifications for heat rejection, pressure durability, and forming limit

achievable with conventional offset fins (shaded area in the upper left part of the figure).

The first issue was that conventional offset fins are designed with an offset ratio (amount of offset/half the fin pitch) of 0.5 to optimize their performance (Fig. 15a). Considering the workability of an offset fin based on this assumption, the problem of fin breakage occurs in fine processing of a thick fin because excessive stress is applied to the joint of the offset fin (Fig. 15b). Thus, we considered achieving both pressure durability and workability by reducing the offset ratio to help improve workability (Fig. 15c).

However, reduction of the offset amount causes a decrease in the fin performance, and in this project, we quantified the degree of contribution of the offset amount to performance (Fig. 16) through analyses and tests and tried to achieve performance, pressure durability, and workability simultaneously.

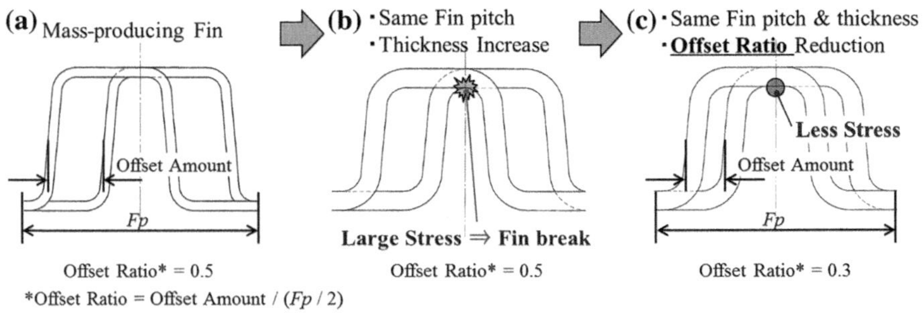

Fig. 15 Relation between offset fin forming and offset amount

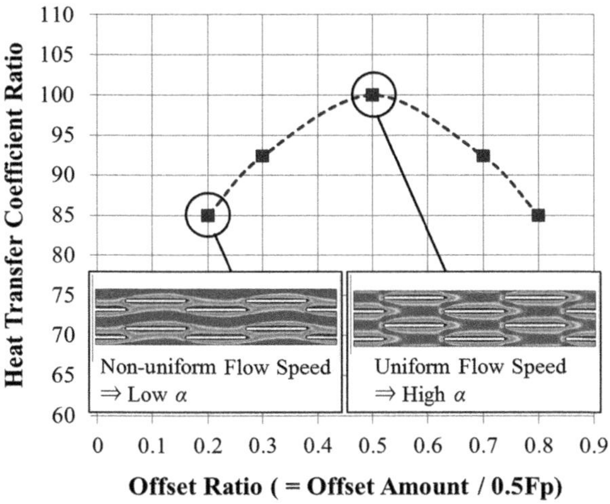

Fig. 16 Relation between offset ratio and heat transfer coefficient

The feasibility of achieving the target performance and pressure durability and workability with the offset amount taken into account is shown in Fig. 17. By reducing the offset ratio from 0.5 to 0.3, we successfully found a region in which performance, pressure durability, and workability could be realized (shaded area in the figure), and attained the desired performance, pressure durability, and workability by setting the pitch and plate thickness of the fin to 2.5 and 0.2 mm, respectively, and the offset ratio to 0.3.

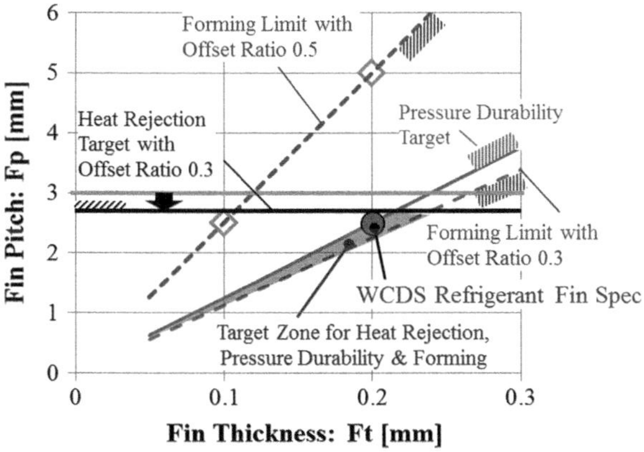

Fig. 17 Fin specifications with consideration of offset amount

3.3.2 Determination of the Specifications of the Water-Side Offset Fin

We also adopted an offset fin on the waterside to accelerate heat transfer in the working region of the water-cooled condenser (*Re* to 500). Since the water-side passage can become clogged with foreign substances, we determined the fin pitch taking into account the limit of clogging.

3.3.3 Optimization of the Height of the Refrigerant-Side and Water-Side Tubes

Using the abovementioned presumed fin and with the height per layer of tube (height of the refrigerant-side tube Fhr + height of the water-side tube Fhw) fixed, we optimized the height of the refrigerant-side tube Fhr and that of the water-side tube Fhw so as to minimize the pressure loss on both sides and maximize the performance. Figure 18 shows a graph with the height of the refrigerant-side tube per layer of tube plotted on the horizontal axis and performance per pressure loss on the vertical axis. As the value increases along the horizontal axis, the height of the refrigerant-side tube per layer of tube increases, and as the value increases along the vertical axis, performance per pressure loss increases. Thus, we performed optimization to maximize the vertical axis. As the height of the refrigerant-side tube increases, the pressure loss on the refrigerant side decreases, but the reduction of the heat transfer area or the cross section of the passage on the waterside causes a decrease in the performance of the core or an increase in the pressure loss on the water side. We determined the height of the tubes of the product so as to maximize the performance of the core while keeping the pressure loss on both sides low [4].

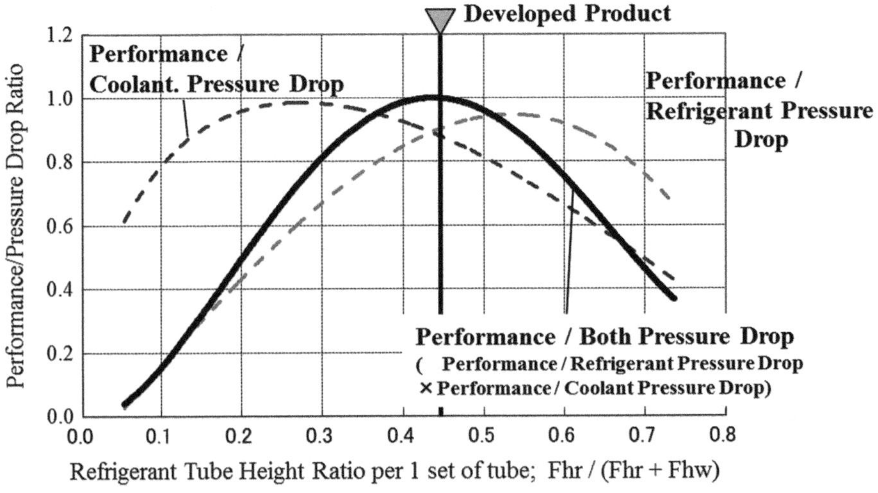

Fig. 18 Tube height optimization

4 Conclusions

We have developed a compact, high-performance, high-reliability water-cooled condenser. The product can improve heating efficiency and reduce space in heat pump systems. Mass production of the product was started in 2017.

We will use the findings and the calculation model of this study to develop other water-cooled condensers with even greater heat exchange performance and smaller footprint.

References

1. Shah MM (1979) Int J Heat Mass Transf 22(4):547–556
2. The Japan Society of Mechanical Engineers (2009) JSNE data book: heat transfer, 5th edn. The Japan Society of Mechanical Engineering, Tokyo, p 332
3. Wieting AR (1975) Trans ASME Ser C 97:3
4. JP Patent: JP2013191695, Isao Tamada

Optimization for Qualified Structure Intension of Sliding Door Based on Regulation and a Study About Key and Important Factors Affecting Simulation Accuracy

Qiuping He and Wufeng Zhou

1 Introduction

Nowadays, many vehicle manufactures pay more attention to passive safety performance, especially the sliding door retention components whose property is a vital index of the sliding door system. The sliding door is an important part of the body, which needs to satisfy ingress and egress of people and goods, and makes the inner body isolated from the outer. In addition, it must ensure the door will not be opened during crashing or driving, and the door will be opened normally after crashed.

To meet the requirements, body of sliding door requires reasonable structures, enough strength and reliable lock system. Thus, it is important that how to place B pillar, C pillar, lock stiffener for performance of sliding door retention components. Xiao [1] solved the problem that separation distance is too large between sliding door inner plate and door frame through structure optimization of the inner door and B-pillar, and the performance of sliding door retention components met the requirements of regulation.

With the development of CAE, the vehicle manufactures have realized its importance and gradually increase the investment for CAE team construction. It can decrease the times of experiments, shorten the research time and quickly reach the design requirements of performance.

The accuracy is a key point in the process of simulation for the sliding door retention components performance. Therefore, it has an important significance for doing the research of stimulation accuracy before the analysis of passing national regulations.

Q. He (✉) · W. Zhou
Changan Oushang Automobile Institute, Chongqing Changan Automobile Co., Ltd., Chongqing 400023, China
e-mail: qiupinghe@126.com

Society of Automotive Engineers of China (SAE-China) (ed.), *Proceedings of the 19th Asia Pacific Automotive Engineering Conference & SAE-China Congress 2017: Selected Papers*, Lecture Notes in Electrical Engineering 486, https://doi.org/10.1007/978-981-10-8506-2_41

2 Regulation Requirements

In September 2013, based on the world's first unified technical regulations [2], China promulgated the GB 15086-2013, motor vehicle's door locks and door retention components performance requirements and test methods, enforced on January 1, 2015 [3].

(1) Sliding door and other support are in the position of full lock off.
(2) According to the stipulation of manufacturer, the force devices are moved to reach the force of 9000 N or the total displacement of 300 mm between the speed of 20–90 mm/min.
(3) If one device force reaches the target of 9000 N before another one, then the former is maintained at the total force of 9000 N until the later device force achieves the target of 9000 N.
(4) Keeping the value at least 10 s if both of the two forces have reached the target.
(5) Measuring the separation distance between sliding door and frame if the devices have satisfied the requirements above all.
(6) The sliding door does not reach the requirements if the separation distance is beyond 100 mm.
(7) Another evaluation criterion is that the total displacement should be in the scope of 300 mm; otherwise, the sliding door does not reach the requirements.
(8) The devices are shown in Fig. 1.

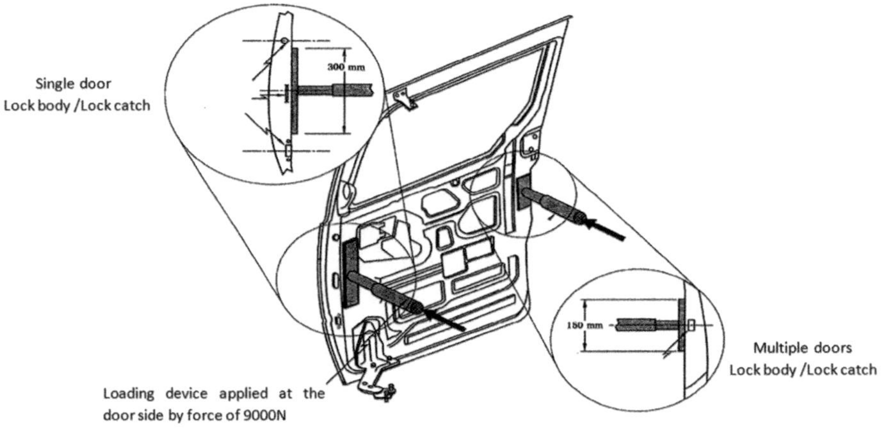

Fig. 1 National regulation

3 The Simulation Analysis of Initial Model

3.1 Simulation Model

In this paper, LS-DYNA [4] software is used to build the simulation model. The whole vehicle CAE model which contains front door and middle door is established based on the CATIA model, and the average size of grid is 10 × 10 mm. The sheet metals are meshed by 2D grids, while the locks, the pothooks, and the guide pulleys are 3D grids. There are also some connections which need to be noticed, such as the material type of welding spot is MAT100 and the bolt is RBE2 (Fig. 2).

3.2 Load Method

According to the experimental regulation requirements (as Fig. 3), the loads are put on the inner planets of sliding door approaching the B-pillar and the C-pillar,

Fig. 2 Simulation model

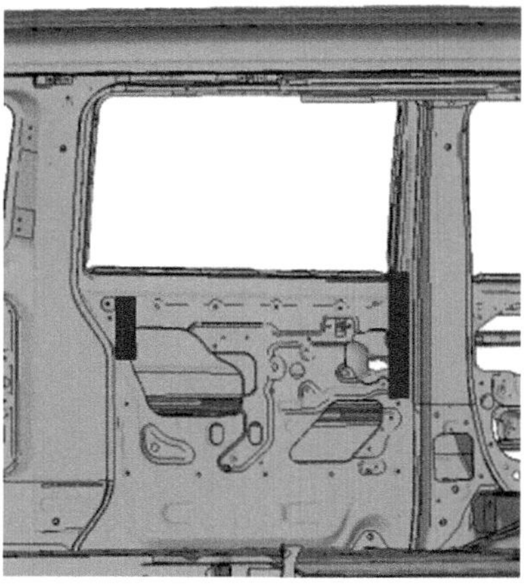

Fig. 3 Load method of the simulation model

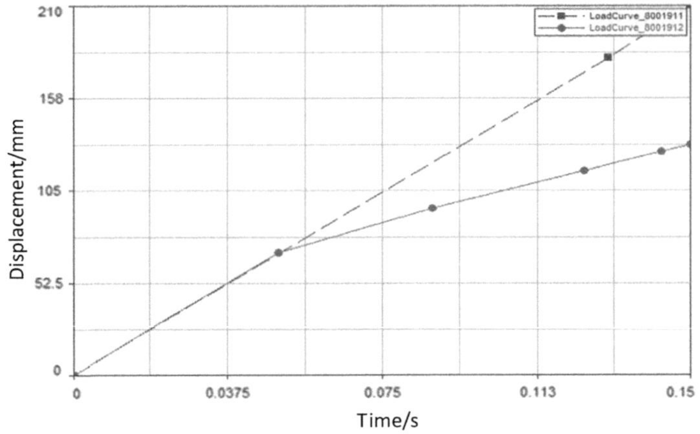

Fig. 4 Load curves of the simulation model

respectively. The enforced displacement method is adopted in the process of loading. Both two sides of load curves are consistent at first, while the side of strength larger arrives at 9000 N, adjusting the loading rates to make it fluctuate upon 9000 N (as broken line). The time–displacement curves are shown in Fig. 4.

3.3 The Simulation Results of Initial Model

The maximum separation distance of initial model is 82/184 mm in the upper and lower ends, respectively. Thus,distance of the upper and lower ends is over 100 mm, which does not satisfy the regulation requirements (Fig. 5).

Fig. 5 Simulation results of initial model

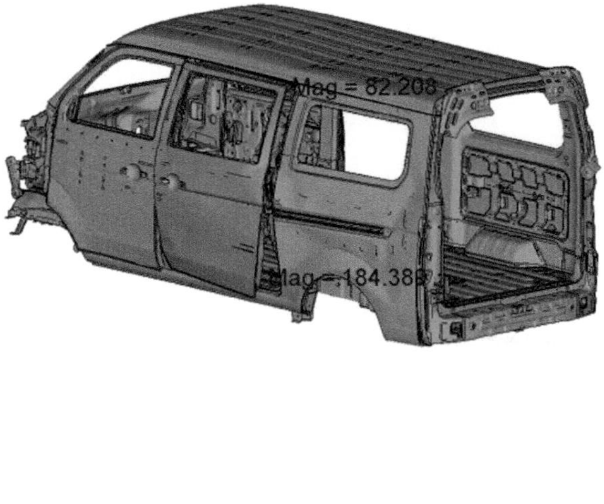

3.4 *The Reasons for Analysis*

As shown in Fig. 6a, b, it is necessary to increase the length of inner B-pillar and the strength of C-pillar stiffener because both of them have a large deformation. In addition, the sliding door inner panel strength also needs to be improved.

4 The Study for Important Factors on Simulation Accuracy

A series of important factors are summarized by comparing experiments results with simulation acquired through many calculations.

4.1 *Material: The Choice of Materials Curve*

Minamoto [5] found that the material strain rate sensitivity had remarkable effect on the impact of two identical spheres even in the low-speed impact. Sun HT [6] showed that the higher flow stress and the lower fractured strain of hot forming part of AA5182 aluminum alloy occur with the increasing of strain rate.

It is a quasi-static condition that the load speed of sliding door retention components is 50 mm/min. Considering the material strain rate, it will make the stiffness of the model higher and the separate distance smaller than the actual condition. Therefore, we should use the material curves that acquired in the condition of quasi-static (Fig. 7).

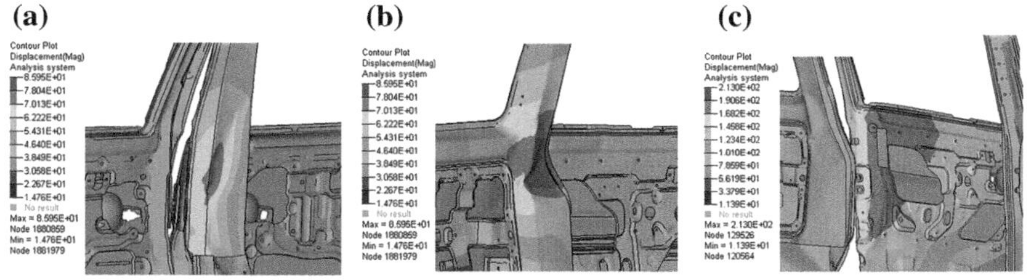

Fig. 6 Deformation pattern of parts. **a** The deformation pattern of B-pillar; **b** the deformation pattern of C-pillar; **c** the deformation pattern of sliding door inner panel

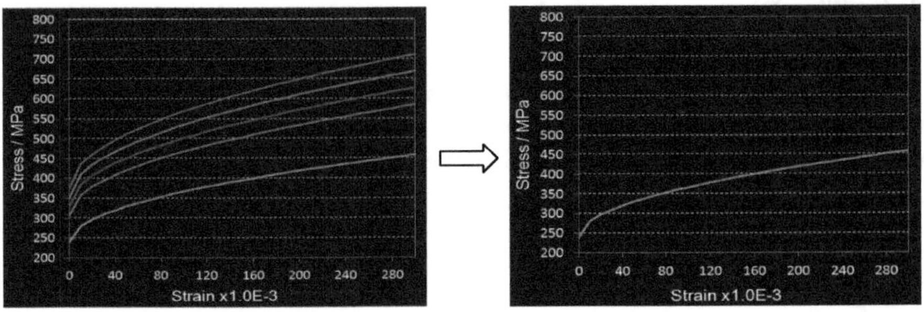

Fig. 7 Choice of materials curve

4.2 The Choice of the Vehicle Body Length

The simulation results of different length of vehicle body are shown in Fig. 8. We can find that vehicle body length has a significant influence on deformation. The disadvantage of short body is unable to express the compatibility of deformation when loading, which eventually makes the separate distance small relatively.

In chart 1, the separation distance difference of short model is −48.1% compared with the result of the experiment, while the whole vehicle model is only −7.4%, within the scope of 10%. In conclusion, the stimulation result of the whole vehicle model is more closer to the experiment, and the simulation accuracy is more higher.

Chart 1: The comparison between the simulation and experiment

Simulation model	Short model	Long model	Whole model	Experiments
Result of separation distance	28/50.8	64.5/80.2	50/87	54/93
Difference comparing with the result of experiment	−48.1/ −45.4%	+19.4/ −13.8%	−7.4/ −6.5%	/

Fig. 8 Choice of vehicle length. **a** The short model, **b** the long model, **c** the whole model

4.3 Glues

The influence of glues in experiment of whole vehicle is not consistent with the simulation model because the former is used to buffer components, while the latter has the function of connection. Therefore, the glues near to the loading plate in the simulation model are removed off.

4.4 The Assembly Relationship of Lock, Lockpin, Pothook, Nut Plate with Sliding Door

The assembly relationship of latch mechanism with sliding door should not be ignored. Besides, the type of latch mechanism is elastic-plastic material instead of rigid material, and the solid mesh is more than three layers.

4.5 The Distribution of Welding Spot

The welding spot distribution of B-pillar, C-pillar, and lock/lockpin stiffener has a huge influence on simulation results. For example, the lacking of welding spot and incomplete welding have a bad influence on the result of experiment, which is another reason about the result difference between simulation and experiment (Fig. 9).

(a) (b)

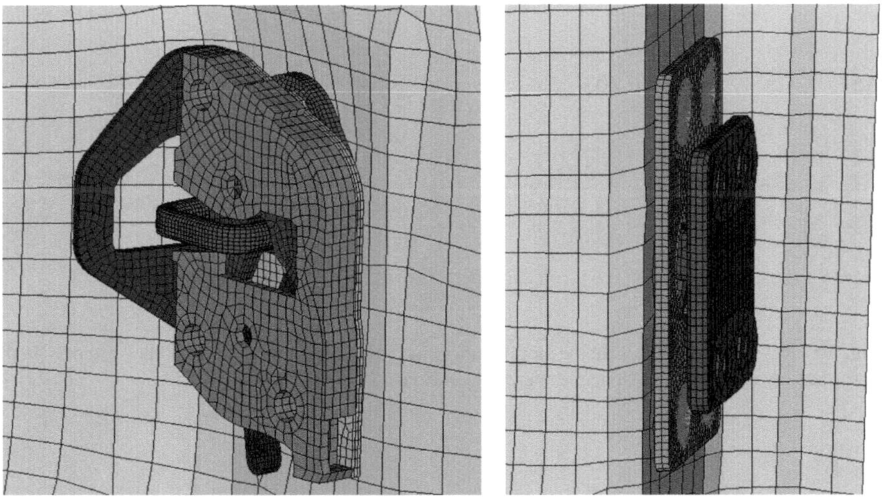

Fig. 9 Assembly relationship of sliding door and vehicle body. **a** lock/lockpin, **b** pothook/nut plate

5 Optimization

5.1 The Solutions of Optimization

In Fig. 10a, b, there are some optimization solutions for key structures as following. Firstly, the length of B-pillar stiffener and C-pillar stiffener is increased, the material is B280VK, and the thickness is added to 2 and 1.5 mm, respectively. Secondly, in Fig. 10c, the structure of lock stiffener is changed to increase the strength of lock zone, and the property of lock stiffener is also B280VK and 2.0 mm.

5.2 The Result of Optimization

Figure 11 shows that the separation distance of the final optimization solution is 50/87 mm, respectively, which satisfies the requirements of national regulation, as the showing.

5.3 Experiments

The optimization result is used to the vehicle experiment completed in Tianjin research institution. It is indicated that the final result is 54/93 mm in the upper and lower ends, respectively. The loading method and the deformation pattern are shown in Fig. 12.

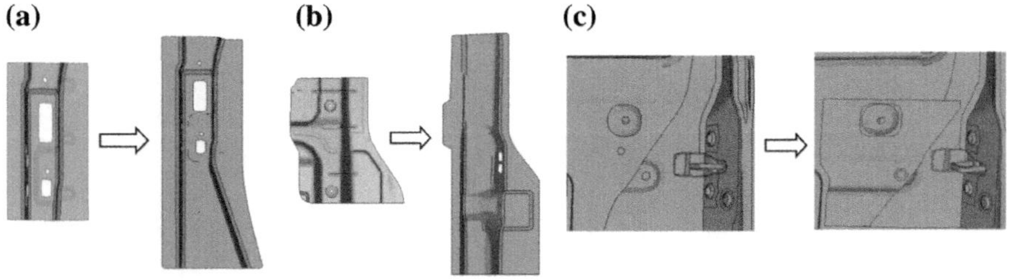

Fig. 10 Key structures of scheme optimization. **a** The optimization of B-pillar stiffener; **b** the optimization of C-pillar stiffener; **c** the optimization of lock stiffener

Fig. 11 Deformation pattern of whole vehicle after optimization

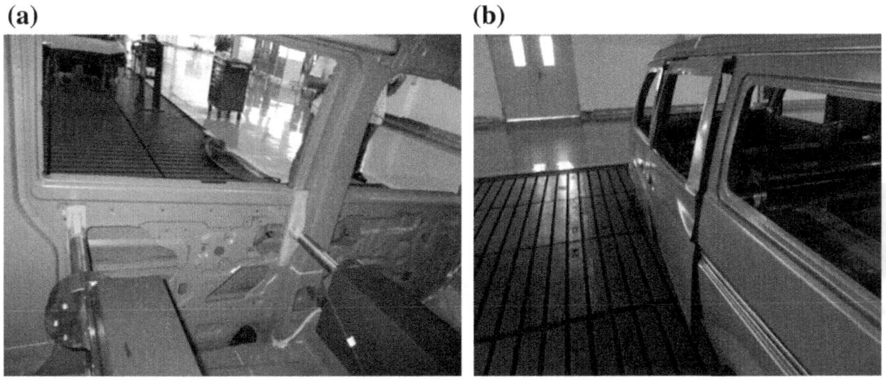

Fig. 12 Load method of experiment and the deformation pattern of whole vehicle. **a** The loading method of experiment, **b** the deformation pattern of whole vehicle

6 The Comparison of the Simulation Result and Experiment Result

6.1 The Comparison of Loading Curves

In Fig. 13, the time–force curve of B-pillar and C-pillar in simulation results is in accordance with the experiment.

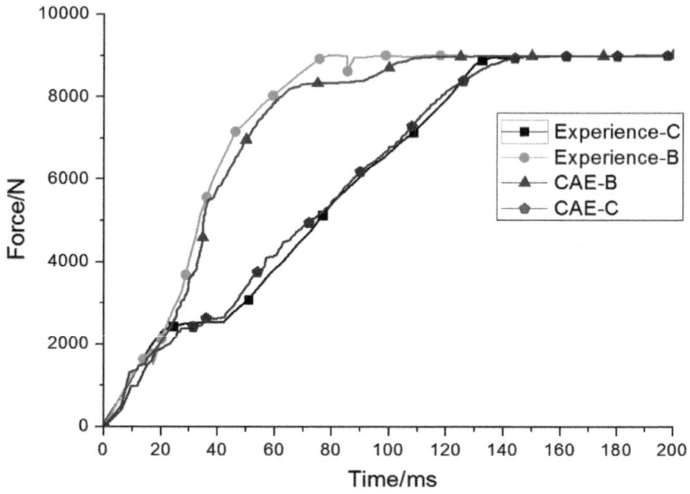

Fig. 13 Comparison of the simulation curves and experiment curves

6.2 The Comparison of Deformation Pattern in Key Portion

In Fig. 14, the deformation pattern and deformation regions of C-pillar inner panel and sliding door are in accordance with experiment.

6.3 The Comparison Between Simulation Results and Experiment Result

The results of the simulation and experimental are 50/87 and 54/93 mm, respectively, and its difference is −7.4/−6.5%.

7 Conclusions

(1) Based on the initial model, B-pillar, C-pillar, and sliding door stiffeners are optimized. The result shows that the maximum distance is lesser than 100 mm, meeting the requirements of the national regulation.
(2) There are many factors affecting the simulation accuracy of sliding door retention components, including the choice of material curves, the length of the vehicle body, gules, the distribution of welding spots, the assembly relationship between lock and sliding door, and so on.

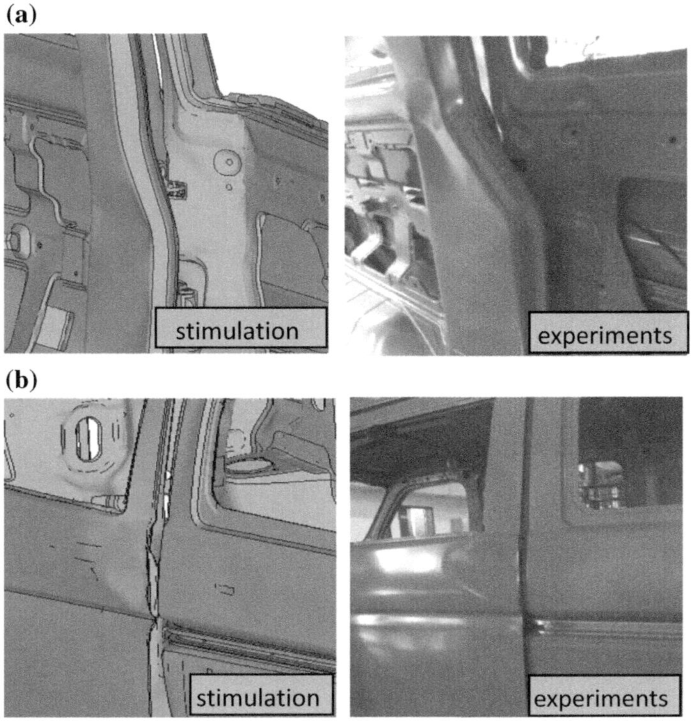

Fig. 14 Comparison of deformation pattern in key regions. **a** inner panel of C-pillar; **b** outer panel of sliding door

(3) By adjusting the key points affecting simulation accuracy, the result of the simulation model is 50/87 mm and the experiment result is 54/93 mm, and the difference of them is within the scope of 10%.
(4) This paper is of great engineering application value.

References

1. Chunyan Xiao (2010) A study on regulations of qualified structure of minibus of sliding door. Enterp Technol Develop Filed Vision 22:292
2. Zhu Y (2005) The arrival of the world's first unified technical regulations. Automobile Parts Technol
3. GB 15086-2013 (2013) Motor vehicle's door locks and retention components performance requirements and test methods. Standardization Administration of China. China Standard Press, Beijing
4. John OH (2006) LS-DYNA theory manual. Livermore Software Technology Corporation, California

5. Minamoto H, Kawamura S (2009) Effects of material strain rate sensitivity in low speed impact between two identical spheres. Int J Impact Eng 36(5):680–686
6. Sun HT, Wang J, Shen GZ et al (2013) Application of warm forming aluminum alloy parts for automotive body based on impact. Int J Automotive Technol 14(4):605–610

A Study on Thermal Runaway of Commercial Lithium-Ion Cells: Influence of SOC, Cell Chemistry and Ageing Status on Safety Performance

Peng Wu, Jan Romberg, Xiaolu Cheng, Weijian Hao, Huinan Si, Huiyu Li and Xingping Qiu

1 Introduction

Since the lithium-ion battery (LIB) has been applied in automobile applications such as plug-in hybrid vehicles and electric vehicles in recent decades, there are growing concerns about the safety issue of LIB0 [1, 2].

To predict and prevent the thermal runaway timely, a lot of work has been carried out on thermal modelling of battery and battery pack [3–5]. To make sure the model can simulate the thermal runaway behaviour correctly, several data and parameters especially some critical temperatures from the real thermal runaway situation are needed. Here, we employed the Accelerating Rate Calorimeter (ARC) to heat the cell to trigger thermal runaway and measured the critical temperatures including self-heating start point of exothermic reactions (T_s) and onset temperature of thermal runaway (T_o). Considering cells that service in vehicles are under various circumstances, we tested cells in different state of charge (SOC), types (different anode and cathode materials; different energy densities) and ageing status (SOH) to establish the relationship between the T_o and SOC, types, and SOH. In this work, the capacity fading is achieved by accelerated ageing in high temperature (50 °C).

The work belongs to the cooperative research project between BMW and Tsinghua University.

P. Wu (✉) · J. Romberg
BMW China Services Ltd, 19th Floor, Tower A, Gateway Plaza,
No. 18 Xia Guang Li, Chaoyang 100027, Beijing, People's Republic of China
e-mail: peng.wu@bmw.com

X. Cheng · W. Hao · H. Si · H. Li · X. Qiu
Advanced Power Source Lab, Department of Chemistry, Tsinghua University,
Beijing, China

Society of Automotive Engineers of China (SAE-China) (ed.), *Proceedings of the 19th Asia Pacific Automotive Engineering Conference & SAE-China Congress 2017: Selected Papers*, Lecture Notes in Electrical Engineering 486, https://doi.org/10.1007/978-981-10-8506-2_42

2 Experimental

2.1 Cell Samples

18,650-type cells with graphite anode and different cathode materials were employed in our work. Different energy density and capacity were also considered. Six types of cells were selected for our test. Their cell chemistry and types are shown in Table 1.

2.2 Accelerating Rate Calorimeter (ARC) for Heating Test

Before ARC test, the cells were charged to certain SOC (defined by cut-off voltage) with CC/CV method. Four SOCs were selected: 60, 80, 100 and 120% (over-charged). In this work, PHI-TEC I adiabatic reaction calorimeter that was manufactured by UK HEL Inc. was carried out with heat–wait–search (HWS) mode. The temperature of cell was measured by a K-type thermocouple that is attached on the surface of cell. The cell was heated to set temperature, followed by wait and search mode to see whether the self-heating reactions occurred. If the exothermic reactions were detected (defined as the d*T*/dt reached 0.03 K/min), the sample heater is shut down, and the guard heater (on top and bottom) keeps working to maintain the adiabatic circumstance until 240 °C. If no exothermic reactions occurred, another heating period starts. The test temperature range was 60–220 °C. In our work the definition for T_s is the first time ARC detected the exothermic signal, and the definition for T_o is the very temperature at which the temperature rate (d*T*/dt) reaches 10 K/min.

2.3 High-Temperature Ageing

Since the cell ageing processes are time consuming, we used high-temperature accelerated method to obtain the aged cell samples. All cells prepared to age were

Table 1 Cell types and chemistry

Type	Cathode materials	Anode materials	Energy density (mAh/cm^2)	Capacity (Ah)
Type 1	LMO/NCM523 (3:7)	Synthetic graphite	1.55	1.5
Type 2	NCM523	Nature graphite	2.14	2.0
Type 3	NCM523	Nature graphite	3.15	2.2
Type 4	NCM523	Nature graphite	4.05	2.6
Type 5	LCO/NCM523 (2:1)	Synthetic graphite	3.15	2.5
Type 6	LCO/NCM523 (2:1)	Nature graphite	3.71	2.5

Remark LMO = $LiMn_2O_4$, NCM523 = $Li(Ni_{0.5}Co_{0.2}Mn_{0.3})O_2$, LCO = $LiCoO_2$

put in the high- and low-temperature test box with 50 °C and cycled with 1C rate CC/CV method. After every 100–500 charge/discharge cycles, a capacity check-up test was carried out at 25 °C. The check-up test contains standard capacity check and a direct current internal resistance (DC-IR) check. The capacity check is defined by the manufacturer, that is three standard CC/CV cycles with 0.5C. For the DC-IR check, a pulse discharge for 30 s with I = 3C is performed at 60% SOC.

Considering the fact that cells which have a capacity of $\leq 80\%$ could be still used in a second life for stationary applications, we chose a large ageing area (100, 90, 80, 70 and 60%) for our tests. The larger area of ageing makes it more possible to investigate the ageing safety relation from a brighter scientific basis.

3 Results and Discussion

3.1 Dependence of Onset Temperatures with SOC for Fresh Cells

Type 1–6 cells without capacity fading had been respectively charged to 60, 80, 100 and 120% SOC. To make sure of the credibility of results, every SOC needs two pieces of cells for each type to test.

Table 2 and Fig. 1 show the valid data of onset temperature (T_o) for all fresh cells.

As shown in Fig. 1, the T_o of thermal runaway generally decreased with increasing SOC. The fully charged and overcharged cells tended to trigger thermal runaway at lower temperatures except for types 1 and 6.

Type 1 cells, with LMO/NCM523 (3:7) cathode and synthetic graphite anode, detected thermal runaway during the external heating test at 80 and 100% SOC. Generally, fully charged and overcharged cells show pronounced thermal runaway because the lithiated graphite and delithiated cathode material have high activity to reaction with electrolyte and release more heat when heated [6–8]. However, in the test of overcharged cells, the exothermic reactions were not strong enough to trigger a distinct thermal runaway. That probably result from the phase transfer of

Table 2 Onset temperatures (°C) of fresh cells (#1/#2)

Type	SOC (%)			
	60	80	100	120
Type 1	–/–	228.9/233.8	227.2/228.5	–/–
Type 2	248.5/251.5	247.0/246.8	216.6/212.9	203.9/198.6
Type 3	–/–	243.2/234.8	203.3/205.3	198.7/203.5
Type 4	–/–	217.7/220.2	208.9/208.0	204.9/201.3
Type 5	199.3/195.7	196.5/192.6	173.4/174.4	178.4/182.4
Type 6	202.2/203.7	200.6/201.3	180.9/179.0	201.0/203.5

Remark –/– indicates no thermal runaway detected during tests

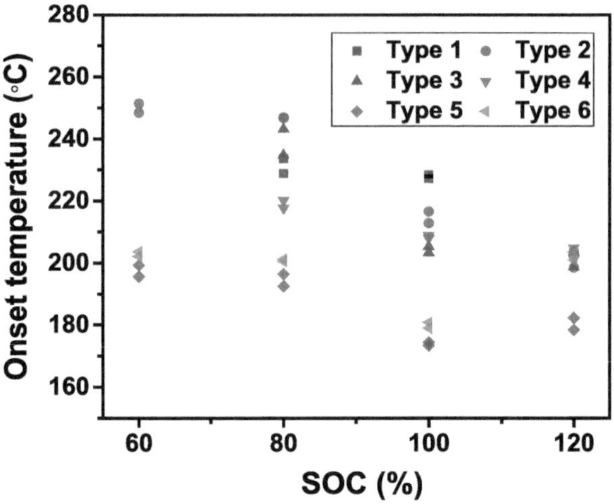

Fig. 1 Onset temperatures (°C) of fresh cells

$LiMn_2O_4$ when over delithiated. During overcharging the cell, the structure of spinel $LiMn_2O_4$ transfer to defective spinel and some Mn^{2+} dissolve into electrolyte [9], which probably makes the material exhibit better thermal stability.

Type 2–4 cells have the same chemistry but different capacity and energy density. Their onset temperatures of thermal runaway ranged from 198 to 249 °C. Different from type 3 and type 4, type 2 cells in 60% SOC showed thermal runaway upon heated up to ~250 °C. 100 and 120% SOC of type 2–4 showed lower onset temperatures than the 80% SOC (and 60% SOC for type 2) cells. Temperature versus time profiles and d*T*/dt versus temperature profiles of type 3 are given in Fig. 2.

The initial exothermic reaction of charged type 3 cells was detected at ~130 °C, and some of them were accompanied by severe venting, which resulted from the generation of large amount of gas.

Figure 3 displays the photographs of cell before and after the external heating tests.

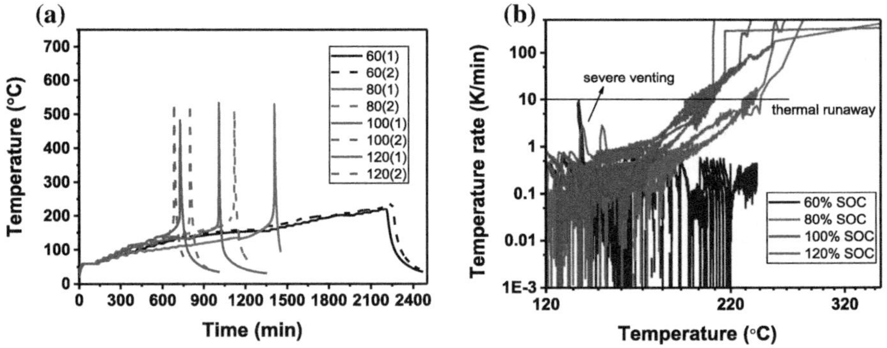

Fig. 2 Temperature versus time profiles **a** and d*T*/dt versus temperature profiles **b** of type 3

Fig. 3 Photographs of cell before **a** and after **b–d** the external heating tests

Figure 3a shows the photograph of cell before heating test. Figure 3b shows cells that did not trigger a thermal runaway during test. However, some electrolyte leakage could be seen from the surface around the cathode cover, indicating the open of the vent valve on cathode. Figure 3c, f displays cells after thermal runaway. Figure 3c shows the majority situation: the plastic cover was burned off, and a little black powder was spilled on cathode. Some fully charged and overcharged cells show extremely thermal runaway as shown in Fig. 3d. The content inside the shell spurted out, suggesting there was a very large pressure when thermal runaway occurred.

3.2 Dependence of Onset Temperatures with Cell Types for Fresh Cells

Considering the safety behaviour of fully charged cells receives more attention in automobile applications. We make a comparison of the safety behaviour of fully charged cells in different types. Figure 4 shows the temperature versus time profiles and dT/dt versus temperature profiles.

The temperature versus time curve of type 1 cell showed mild thermal runaway compared to other types. Its maximum temperature was no more than 300 °C, and the T_{max} of other types exceeded the maximum detection temperature of equipment, 500 °C. The temperature rate profiles differed in types as shown in Fig. 3b.

Types 5 and 6 started self-heating at about 140 °C, which were earlier than other types. Thermal runaway of types 5 and 6 was detected around 175 °C. The low onset temperatures of them result from the LCO content, which usually triggers decomposition when heated up to 130 °C and releases large amount of oxygen [10]. In addition, type 5 had slightly lower T_o than type 6 as shown in Table 2, even though type 5 has higher energy density. The main reason is that the synthetic graphite has slightly poor thermal stability than nature graphite.

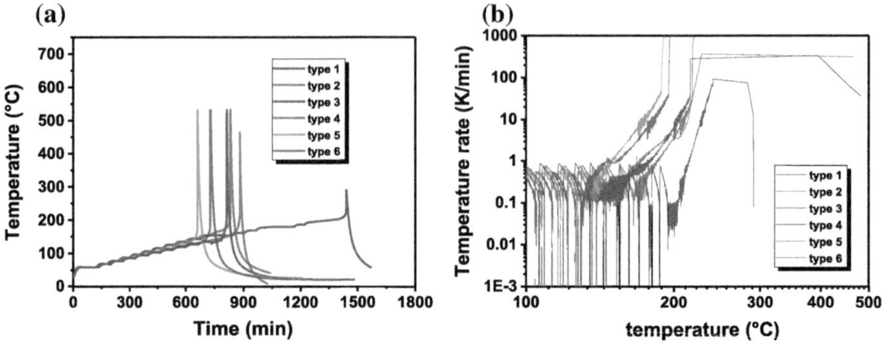

Fig. 4 Temperature versus time profiles **a** and d*T*/dt versus temperature profiles **b** of fully charged cells in different types

The energy density of types 2, 3 and 4 increases by types, but their onset temperatures were much closed and their temperature rate curves are similar as well. Their onset temperatures of thermal runaway ranged from 205 to 217 °C.

Obviously, the cathode and anode materials have a big effect on the onset temperature of thermal runaway for fresh cells, especially the cathode material. Cells that contain LCO tend to trigger thermal runaway at low temperature, and cells that contain LMO show mild thermal runaway than cells that only contain NCM523 cathode. The energy density makes little difference to the safety behaviour of cells in external heating test.

3.3 Ageing Performance

We chose the type 3 cells for ageing and obtained the capacity retention (SOH) and DC-IR increase as functions of the cycle number, as demonstrated in Fig. 5.

As shown in Fig. 5a, the capacity retention of type 3 cell decreased with cycling. At first ~300 cycles, the SOH decreased to 90%, and then at ~600 decreased to 80%. The results showed that the capacity retention decreases linearly with the increase of cycle numbers. On the contrary, the DC-IR of cells increased with the ageing depth. The DC-IR of fresh type 3 cell is ~65 mOhm, and it increased ~2.5 times after 1300 cycles.

3.4 Dependence of Onset Temperatures with Ageing Status

Type 3 cells at 90, 80, 70 and 60% SOH were tested, and their onset temperature of thermal runaway was given in Table 3. The scatter plot of thermal runaway was displayed in Fig. 6.

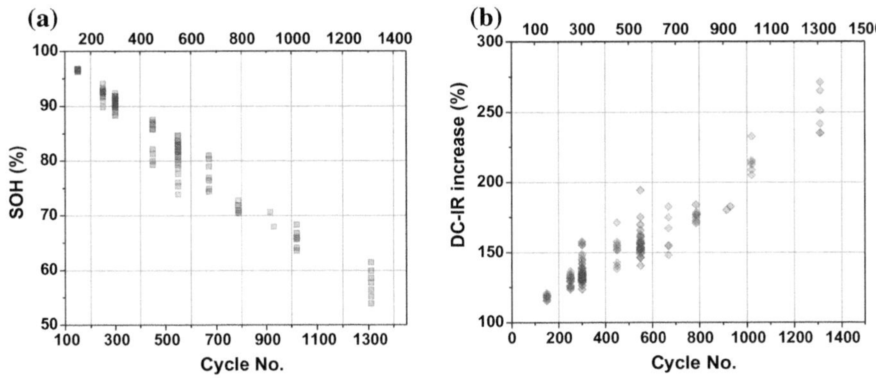

Fig. 5 SOH **a** and DC-IR increase **b** versus cycle number

Table 3 Onset temperatures (°C) of aged cells (#1/#2) for type 3

SOH(%)	SOC (%)			
	60	80	100	120
100	–/–	243.2/234.8	203.3/205.3	198.7/203.5
90	–/–	–/–	226.8/231.8	213.7/205.3
80	–/–	–/–	–/–	210.7/216.5
70	–/–	–/–	–/–	–/–
60	–/–	–/–	–/–	211.5/212.3

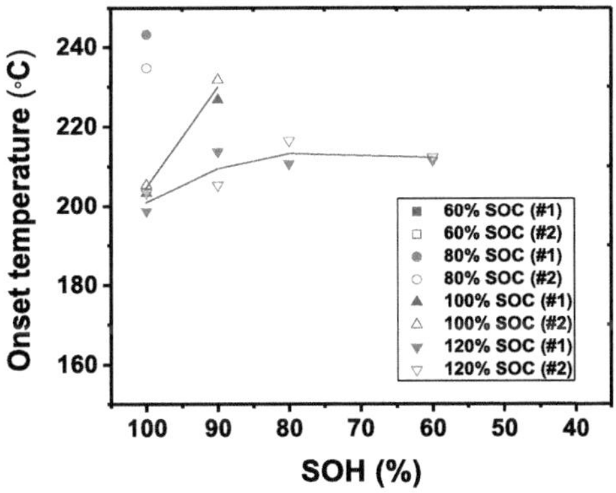

Fig. 6 Onset temperatures (°C) of aged cells for type 3

As shown in Table 3 and Fig. 6, cells harder to trigger thermal runaway with the SOH decreased in most situations. In fixed SOC, the onset temperatures of thermal runaway increased with SOH decrease except for the 120% SOC of 60% SOH. In the initial of ageing, electrolyte reacted with active material and generated solid electrolyte interface (SEI) layer. More and more active Li^+ and electrolyte was consumed with the increasing degree of the ageing, which made the cell release fewer heat and harder to trigger thermal runaway (higher T_o). But if cells were aged much deeper, for example 60% SOH in our test, some other negative factors became dominant. The negative factors contained lithium dendrite formation and lithium plating on the anode especially when overcharged, which made the cell trigger exothermic reactions at lower temperature and release more heat in tests. That's the reason why 60% SOH cells in 120% SOC triggered thermal runaway as well.

4 Conclusion

This paper presents the cooperative research between BMW and Tsinghua University. The influence of cell chemistry, state of charge and ageing status on safety performance of commercial cell was studied. The results show that the onset temperatures of thermal runaway depend on cell chemistry. Cells that contain LCO have lower onset temperatures than NCM and LMO cells. Energy density contributes little to the safety performance. The fully charged and overcharged cells usually have lower onset temperatures than cells at 80 and 60% SOC. The SOH has effect on the onset temperatures. T_o increases with SOH decrease in the initial of ageing because of the consumption of active material and electrolyte, but Li dendrite and Li plating could be brought to anode and make the cell have potential to trigger thermal runaway at lower temperatures. Our future work will focus on understanding the ageing mechanism of lithium-ion batteries.

References

1. Feng X, Lu L, Ouyang M, Li J, He X (2016) A 3D thermal runaway propagation model for a large format lithium ion battery module. Energy, Part 1 115:194–208
2. Spotnitz R, Franklin J (2003) Abuse behavior of high-power, lithium-ion cells. J Power Sour 113:81–100
3. Kim G-H, Pesaran A, Spotnitz R (2007) A three-dimensional thermal abuse model for lithium-ion cells. J Power Sour 170:476–489
4. Shah K, Chalise D, Jain A (2016) Experimental and theoretical analysis of a method to predict thermal runaway in Li-ion cells. J Power Sour 330:167–174
5. Huang P, Ping P, Li K, Chen H, Wang Q, Wen J, Sun J (2016) Experimental and modeling analysis of thermal runaway propagation over the large format energy storage battery module with $Li_4Ti_5O_{12}$ anode. Appl Energy 183:659–673
6. Haik O, Ganin S, Gershinsky G, Zinigrad E, Markovsky B, Aurbach D, Halalay I (2011) On the thermal behavior of lithium intercalated graphites. J Electrochem Soc 158:A913–A923
7. Jouanneau S, MacNeil DD, Lu Z, Beattie SD, Murphy G, Dahn JR (2003) Morphology and Safety of Li $[Ni_x\ Co_1 - {}_{2x}Mn^x]\ O_2$ $(0 \leq x \leq 1/2)$. J Electrochem Soc 150:A1299–A1304

8. Lang M, Darma MSD, Kleiner K, Riekehr L, Mereacre L, Perez MA, Liebau V, Ehrenberg H (2016) Post mortem analysis of fatigue mechanisms in $LiNi_{0.8}Co_{0.15}Al_{0.05}O_2$ − $LiNi_{0.5}Co_{0.2}Mn_{0.3}O_2$ − $LiMn_2O_4$/graphite lithium ion batteries. J Power Sour 326:397–409
9. Chung KY, Lee HS, Yoon WS, McBreen J, Yang XQ (2006) Studies of $LiMn_2O_4$ capacity fading at elevated temperature using in situ synchrotron X-ray diffraction. J Electrochem Soc 153:A774–A780
10. MacNeil DD, Christensen L, Landucci J, Paulsen JM, Dahn JR (2000) An autocatalytic mechanism for the reaction of Li_x CoO_2 in electrolyte at elevated temperature. J Electrochem Soc 147:970–979

BCRLS-EKF-Based Parameter Identification and State-of-Charge Estimation Approach of Lithium-Ion Polymer Battery in Electric Vehicles

Zhifu Wang, Zhaojian Liu and Zhi Li

1 Introduction

As one of the core components of electric vehicles, battery management system (BMS) plays an important role. SOC estimation is an important parameter for battery equalization, thermal management, and vehicle control, so exact estimation of SOC of lithium-ion battery (LiB) is crucial for the BMS in electric vehicles [1, 2].

Estimating battery parameters is the basis of SOC estimation. At present, the main methods of parameter identification can be divided into two categories: the non-time series-iterative calculation-based fitting or optimal solving (off-line) method and time series format-based iterative calculation (online) method. The former method cannot achieve the desired estimation accuracy when the battery aged or the operating condition of the battery changed, because the parameters of the battery are very sensitive to the state of the battery. The latter method can calculate the parameters of the model by measuring the voltage and current in real time, so that it has the potential to achieve an accurate prediction performance against different battery aging levels and operating conditions [3, 4].

The main methods of battery SOC estimation can be generally classified into following categories: (1) the ampere-hour counting method, (2) internal resistance method, (3) open circuit voltage method, (4) neural network and fuzzy inference method, (5) Kalman filtering method. Kalman filtering algorithm has the advantage of (a) good real-time performance, able to estimate SOC online. (b) The estimation accuracy is high. In addition, the error of the most BMS is not more than 3% due to

Z. Wang (✉)
Collaborative Innovation Center of Electric Vehicles, Bejing, China
e-mail: wangzhifu@bit.edu.cn

Z. Liu · Z. Li
National Engineering Laboratory of Electric Vehicles,
Beijing Institute of Technology, Beijing 100000, China
e-mail: liu_zhaojian@126.com

Society of Automotive Engineers of China (SAE-China) (ed.), *Proceedings of the 19th Asia Pacific Automotive Engineering Conference & SAE-China Congress 2017: Selected Papers*, Lecture Notes in Electrical Engineering 486, https://doi.org/10.1007/978-981-10-8506-2_43

its good correction effect. (c) The vehicle does not accumulate errors during driving. (d) It has good error correction capability and is not affected by the initial error [5, 6].

In this paper, based on the above research, a 3.7 V/2.4Ah LiB is chosen as the research object. Based on the Thevenin model, combining the BCRLS method with EKF algorithm is used to estimate the parameters and SOC of the battery. It can not only solve the noise problem of the unknown system, but also modify the state value in real time. In this paper, based on the analysis of a large number of SOC experiments, it is proved that the joint estimation algorithm has a great advantage in SOC estimation.

2 Establishment of Equivalent Circuit Model for LiB

2.1 *The Establishment of the Battery Model*

The battery model was established to estimate the state of the battery. Considering the battery model precision, parameter identification complexity, operability, and other factors, the Thevenin model is chosen as the battery model in this paper, which can reflect the battery dynamic and static characteristics and battery polarization phenomenon well. The schematic diagram of the model is shown in Fig. 1.

Where I is the load current (assumed positive for discharge, negative for charge), U_t is the terminal voltage, and U_p describes the diffusion voltage arising from the diffusion resistance. The electrical behavior of battery model can be expressed by Eq. (1).

$$\begin{cases} U_t = U_{oc} - IR_0 - U_p \\ \dot{U}_p = I/C_p - U_p/(C_pR_p) \end{cases} \tag{1}$$

2.2 *Identification of Model Parameters*

In this paper, the process of parameter identification was introduced on the basis of the off-line identification of the 10 °C LiB. We selected 10 points of SOC from 10

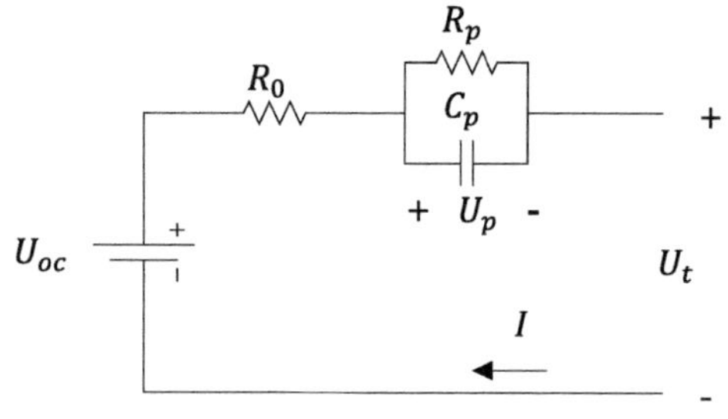

Fig. 1 Schematic diagram of the Thevenin battery model

Fig. 2 SOC-OCV graph

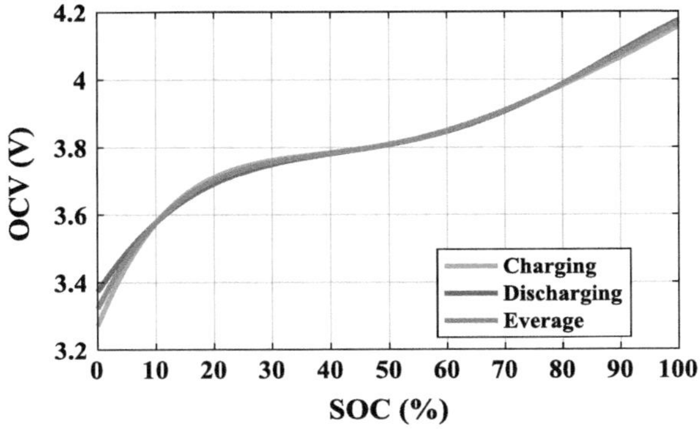

to 100% as the reference point. We used the MATLAB toolbox cftool to fit the SOC-OCV curve of the battery charge and discharge, and fitted the curve of the average. The SOC-OCV graph is shown in Fig. 2. The equation of SOC-OCV is obtained after averaging:

$$U_{oc} = 2.159s^5 - 9.192s^4 + 14.58s^3 - 10.06s^2 + 3.36s + 3.33 \quad (2)$$

where U_{oc} is the open circuit voltage, and s is state of charge.

Combining with the model state equation and the standard equation of the least square (LS) method, we can realize the programming of LS off-line parameter identification algorithm in MATLAB and estimate the discharge parameters of different SOC points at 10 °C.

2.3 *Verification of Battery Model*

The model parameters were validated by the Thevenin model as shown in Fig. 3. This model takes into account the parameter identification results under different factors, including multiplying power, temperature. The identification results were taken in the form of two-dimensional table. The parameters of the model are extracted by lookup table every time. At the same time, the SOC of the ideal condition can be calculated by the ampere-hour counting method as the reference value. Finally, the parameters were introduced into the Thevenin model, and the output voltage of the model was calculated. The error of parameter identification is obtained by comparing the output voltage of the model with the actual terminal voltage [7].

Figure 4a is the comparative profiles between the model output voltage and the actual voltage when the LiB is in SOC = 90%, at 10 °C under HPPC operating conditions. The voltage estimation error is shown in Fig. 4b. It can be seen from the graph that the coincidence degree is high, and the maximum error is not more than 10 mV in HPPC operating conditions. Figure 5 shows the test results and error

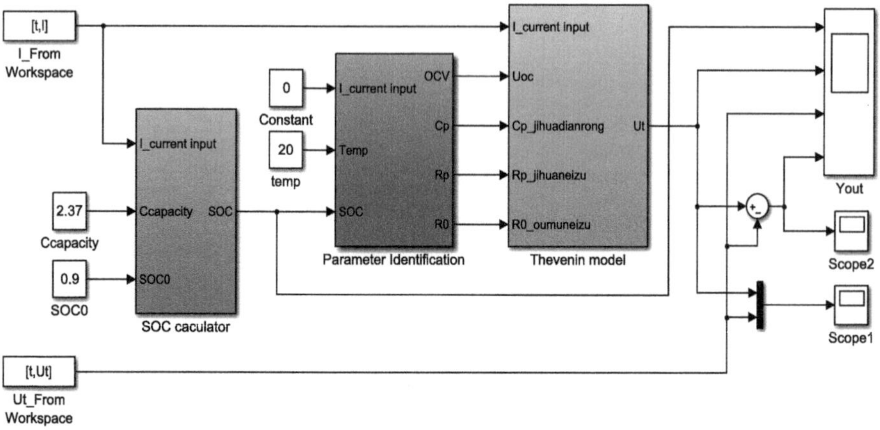

Fig. 3 Thevenin model for verifying model parameters

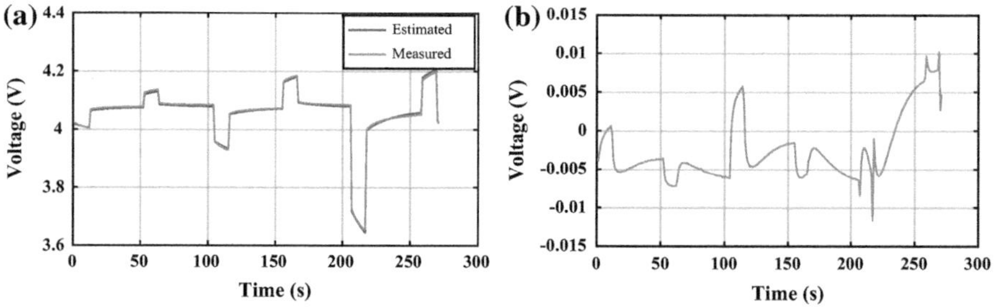

Fig. 4 Comparative profiles between the model output voltage and the actual voltage at 10 °C for HPPC operating conditions. **a** The comparative profiles, **b** the voltage estimation error

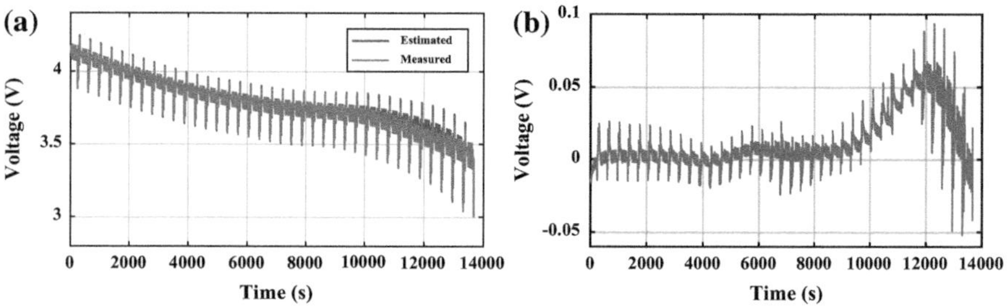

Fig. 5 Comparative profiles between the model output voltage and the actual voltage at 10 °C for DST operating conditions. **a** The comparative profiles, **b** the voltage estimation error

curves at 10 °C for DST operating conditions. As the battery SOC is less than 20%, the battery becomes extremely unstable, so the error suddenly increases. However, when the SOC is greater than 20%, the error of the battery model parameter identification is relatively small.

3 Online Identification of Model Parameters Based on BCRLS

The off-line identification method has some use in laboratory conditions. However, the SOC, temperature, and rate of the battery are often changed due to the change of the actual environment. But the off-line identification model is difficult to track the dynamic performance of the battery. So, the online identification method of model parameters is proposed. The so-called online identification is based on real-time measurement of the battery current, voltage, and temperature data to update the parameters of the model in real time [8].

The RLS online parameter identification method is asymptotically unbiased in the laboratory and the ideal white noise condition, and able to ideally identify the parameters of the system. However, in practical applications, the input and output noise of the system is colored noise rather than white noise. In this case, the RLS identification results are biased, which makes decreases in model accuracy. In this paper, the RLS method is improved according to the colored noise problem in the parameter identification of the model, and the BCRLS method is used to identify the parameters of the battery model [9, 10].

Considering the model of Eq. (1), the $e(t)$ is the colored noise uncorrelated with the output. And the recursive procedure of the algorithm is as follows:

The output of predicting system and the calculation of estimated error:

$$e(k) = z(k) - \boldsymbol{h}^T(k)\hat{\boldsymbol{\theta}}_{LS}(k-1) \tag{3}$$

Calculation of gain matrix:

$$\boldsymbol{K}(k) = \boldsymbol{P}(k-1)\boldsymbol{h}(k)\left[1 + \boldsymbol{h}^T(k)\boldsymbol{P}(k-1)\boldsymbol{h}(k)\right]^{-1} \tag{4}$$

Estimation of parameters:

$$\hat{\boldsymbol{\theta}}_{\mathrm{LS}}(k) = \hat{\boldsymbol{\theta}}_{\mathrm{LS}}(k-1) + \boldsymbol{K}(k)e(k) \tag{5}$$

Calculation of error criterion function:

$$J(k) = J(k-1) + e(k)^2\left[1 + \boldsymbol{h}^T(k)\boldsymbol{P}(k-1)\boldsymbol{h}(k)\right]^{-1} \tag{6}$$

Estimation of noise variance:

$$\hat{\sigma}^2(k) = \frac{J(k)}{k\left(1 + \hat{\theta}_C(k-1)\mathbf{D}\hat{\theta}_{LS}(k)\right)} \tag{7}$$

The value of correlation matrix D:

$$\boldsymbol{D} = \begin{bmatrix} I_{n_a} & 0 \\ 0 & 0_{n_b} \end{bmatrix} \tag{8}$$

Update covariance matrix:

$$\boldsymbol{P}(k) = \left[\boldsymbol{I} - \boldsymbol{K}(k)\boldsymbol{h}^T(k)\right]\boldsymbol{P}(k-1) \tag{9}$$

Estimated value of deviation compensation:

$$\hat{\boldsymbol{\theta}}_C(k) = \hat{\boldsymbol{\theta}}_{LS}(k) + k\hat{\sigma}^2(k)\boldsymbol{P}(k)\hat{\boldsymbol{\theta}}_C(k-1) \tag{10}$$

4 SOC Estimation Method Based on EKF Algorithm

Power battery is a nonlinear dynamic system. KF is only suitable for the estimation of linear system, so it is necessary to linearize the nonlinear system, which is approximated as a linear time-varying system, and then KF is used to estimate the state of the system. This is called the extended Kalman filter (EKF) algorithm.

When EKF is used to estimate the SOC of the battery, the state variables are selected as the SOC and the voltage U_p of the resistance-capacitance network, and the current $i(t)$ is the input quantity, and the terminal voltage U_t is the output. Then, the equation of state:

$$\begin{cases} \frac{\partial u_p t}{\partial t} = -\frac{1}{C_p R_p} U_p t + \frac{1}{C_p} i(t) \\ \frac{\partial st}{\partial t} = -k_c i(t) \end{cases} \tag{11}$$

where $k_c = k_T k_I / Q_N$ and Q_N is the rated capacity of the battery. k_T and k_I is the temperature correction factor and the rate correction factor.

The output equation of the model is:

$$\begin{cases} U_{OC} = f(s,t) \\ U_t = f(s,t) - U_p - R_i i(t) \end{cases} \tag{12}$$

The state equation is estimated by Taylor expansion around the estimated value, and the items of second-order above are omitted. The state equation and observation equation are obtained by linearization and discretization of the output equation:

$$\begin{bmatrix} S(k) \\ U_c(k) \end{bmatrix} = \begin{bmatrix} 1 & 0 \\ 0 & 1-\frac{T_s}{R_i C} \end{bmatrix} \begin{bmatrix} S(k-1) \\ U_p(k-1) \end{bmatrix} + \begin{bmatrix} -k_c T_s \\ \frac{1}{C}\left(T_s - \frac{T_s^2}{2R_p C_p}\right) \end{bmatrix} I(k-1) + \begin{bmatrix} w_1(k-1) \\ w_2(k-1) \end{bmatrix} \tag{13}$$

$$[U(k)] = \begin{bmatrix} \frac{\partial f(s,t)}{\partial s}|_k & -1 \end{bmatrix} \begin{bmatrix} S(k-1) \\ U_p(k-1) \end{bmatrix} + R_i I(k-1) + \begin{bmatrix} v_1(k-1) \\ v_2(k-1) \end{bmatrix} \tag{14}$$

where T_s is the sampling time, and w and v are process noise and observation noise. Coefficient matrix:

$$X_k = \begin{bmatrix} S(k) \\ U_c(k) \end{bmatrix}, \phi(t) = \begin{bmatrix} 1 & 0 \\ 0 & 1-\frac{t}{R_i C} \end{bmatrix}, G(t) = \begin{bmatrix} -k_c T_s \\ \frac{1}{C}\left(T_s - \frac{T_s^2}{2R_p C_p}\right) \end{bmatrix},$$

$$H(k+1) = \begin{bmatrix} \frac{\partial f(s,t)}{\partial s}|_k & -1 \end{bmatrix} \tag{15}$$

According to the EKF algorithm, the coefficient matrix is introduced to obtain the recursive process of SOC estimation:

$$\text{State prior estimate update}: \bar{X}_k = \phi_{k-1}\hat{X}_{k-1} + G_{k-1} i_{k-1} \tag{16}$$

$$\text{Kalman gain matrix}: K_k = P_{k,k-1} H_k^T \left(H_k P_{k,k-1} H_k^T + R_k\right)^{-1} \tag{17}$$

$$\text{State error covariance prior estimate}: P_{k,k-1} = \phi_k P_{k-1} \phi_k^T + Q_k \tag{18}$$

$$\text{State covariance update}: P_k = (I - K_k H_k) P_{k,k-1} \tag{19}$$

$$\text{State estimate update}: \hat{X}_k = \bar{X}_k + K_k (Y_k - \bar{Y}_k) \tag{20}$$

where the variance of process noise and measurement noise:

$$Q_k = \int_0^{\mathrm{T}} \phi(\tau) \mathrm{Q} \phi^T(\tau) \mathrm{d}\tau = \int_0^{\mathrm{T_s}} \left(\begin{bmatrix} 1 & 0 \\ 0 & 1-\frac{\tau}{C_p R_p} \end{bmatrix} \times \phi_s \begin{bmatrix} k_{w1} & 0 \\ 0 & k_{w2} \end{bmatrix} \times \begin{bmatrix} 1 & 0 \\ 0 & 1-\frac{\tau}{C_p R_p} \end{bmatrix}^{\mathrm{T}} \right) \mathrm{d}\tau$$
$$= \phi_s \begin{bmatrix} k_{w1}\mathrm{T_s} & 0 \\ 0 & k_{w2}\mathrm{T_s} - \frac{T_s^2}{C_p R_p} + \frac{T_s^3}{3(C_p R_p)^2} \end{bmatrix} \tag{21}$$

$$R_k = \sigma_k^2 \tag{22}$$

5 Joint Estimators of Model Parameters and States Based on BCRLS-EKF

When BCRLS is used for parameter identification, U_{rc} is used as the output to participate in the LS calculation after the model is discretized. While $U_{rc} = U_t - U_{oc}$, U_t is the measured value. The relationship curve between U_{oc} and SOC was measured by experiment. It can be seen that U_{rc} is not only affected by the measured value U_t, but also by the impact of SOC. In practical application, the U_{oc} changes with the SOC value of the battery, which leads to the change of the output U_{rc}. Because the SOC value cannot be obtained in the BCRLS calculation, we use EKF to estimate the state and provide the data for the online identification of the parameters [11].

In the online identification, each cycle identification process can provide relatively accurate parameters for the state estimation process, meanwhile the estimation results U_p and SOC of EKF can be used as reference data for parameter identification, so we use the identification method of BCRLS-EKF, in order to adapt to different working conditions and improve the accuracy and efficiency of the parameter identification and the state estimation. The identification process is shown in Fig. 6.

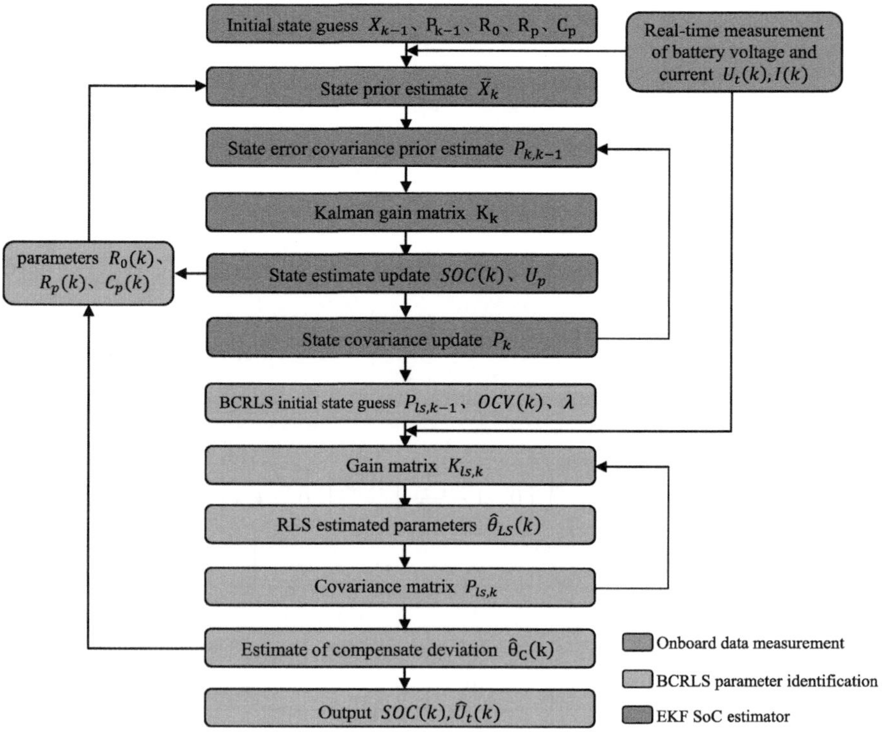

Fig. 6 Flowchart of the joint estimator with BCRLS and EKF algorithm

The off-line identification data of the corresponding SOC points are used as the initial parameters of the joint identification. The forgetting factor, the process noise covariance Q, the measurement noise covariance R, and the initial covariance P_0 were selected according to different working conditions.

In the identification process, BCRLS provides new parameters for EKF estimation, and EKF provides a relatively accurate OCV value for BCRLS. In addition, EKF estimates that the state variable U_p can be used as a reference to the BCRLS output value, and the output U_{rc} of the BCRLS can also be used to correct the measured value of the next cycle of EKF. That is to say, the output values of the two algorithms can be compared with the measured values to improve the identification accuracy of the next cycle [12, 13].

6 Experiments and Result Analysis

In order to verify the feasibility of the BCRLS-EKF algorithm, RLS-EKF and BCRLS-EKF algorithm model are built in MATLAB, and verify its accuracy. Figure 7a is the comparative profiles between the estimated voltage U_t calculated by the RLS algorithm and the measured voltage, and their error is shown in Fig. 7b. Figure 8a is the comparative profiles between the estimated voltage U_t calculated by the BCRLS algorithm and the measured voltage, and their error is shown in Fig. 8b.

Fig. 7 U_t estimation results calculated by the RLS algorithm: **a** U_t; **b** the U_t estimation error

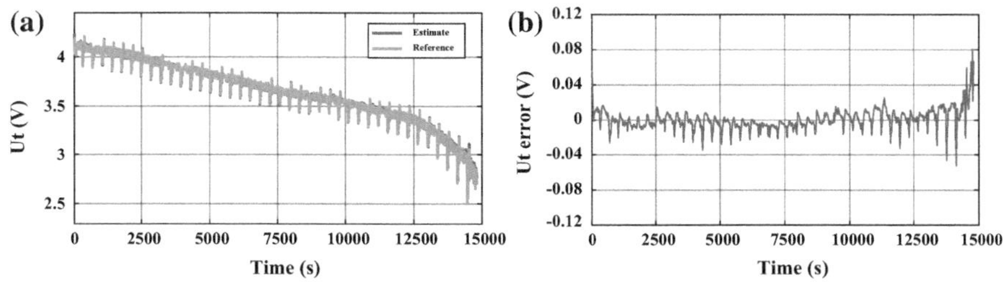

Fig. 8 U_t estimation results calculated by the BCRLS algorithm: **a** U_t; **b** the U_t estimation error

It can be seen from the figure, compared to the RLS algorithm, the error is smaller in BCRLS for DST conditions. The maximum error of U_t estimated by BCRLS is only 80 mV, and the precision is improved greatly, which provides the basis for the accurate estimation of SOC.

Validation of the SOC estimator is then performed. The input variables were obtained after processing the experimental data of battery, including time, current, voltage, and the number of real-time ampere-hour. The output variable is the estimated battery terminal voltage and SOC. Off-line experiments discharged the battery from full charge, so $SOC_0 = 100\%$ [14]. The initial values of the system are as follows:

$$X_0 = \begin{bmatrix} 1 \\ 0 \end{bmatrix}, P_0 = \begin{bmatrix} 10^5 & 0 \\ 0 & 10^5 \end{bmatrix}, Q = \begin{bmatrix} 0.05 & 0 \\ 0 & 1 \end{bmatrix}, R = 100 \quad (23)$$

In this paper, the accuracy of the joint estimation algorithm is verified in the DST cycle condition, and the accuracy of the algorithm is evaluated by absolute error. Figure 9a is the comparative profiles between the estimated SOC calculated by the RLS-EKF algorithm and the measured value, and their error is shown in Fig. 9b. Figure 10a is the comparative profiles between the estimated SOC calculated by the BCRLS-EKF algorithm and the measured value, and their error is shown in Fig. 10b.

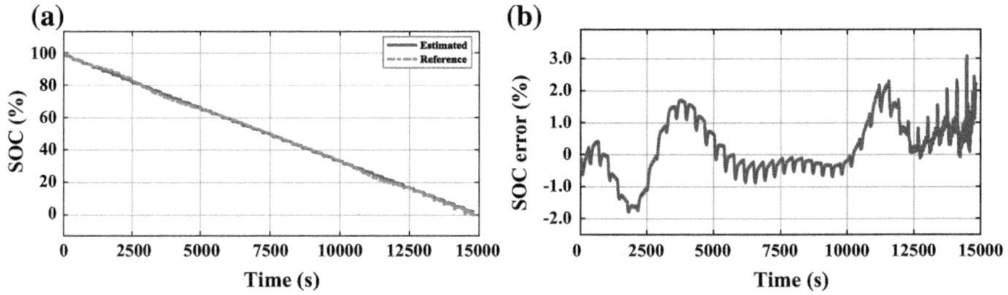

Fig. 9 SOC estimation results calculated by the RLS-EKF algorithm: **a** SOC; **b** SOC error

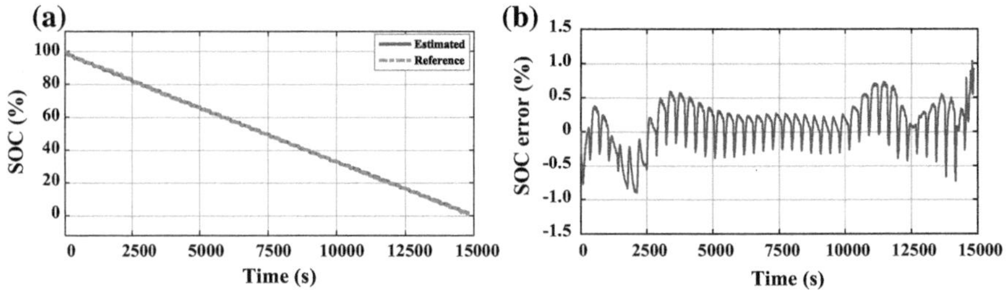

Fig. 10 SOC estimation results calculated by the BCRLS-EKF algorithm: **a** SOC; **b** SOC error

It can be seen in the DST condition, the RLS-EKF algorithm estimates the absolute error of SOC is 3%, while the BCRLS-EKF algorithm estimates the absolute error of SOC is not more than 1%. Compared with RLS-EKF algorithm, the accuracy of SOC estimation is improved obviously by using BCRLS-EKF algorithm. It can be seen that the estimation error increases obviously when the SOC is lower than 20%, which is because the battery becomes extremely unstable at that moment. In general, the SOC estimated by BCRLS-EKF algorithm based on the Thevenin model can meet the requirements of the BMS of electric vehicle.

7 Conclusion

This paper built the Thevenin battery model in MATLAB/Simulink, and to determine the ohmic resistance of the battery polarization resistance, polarization capacitance through off-line parameter identification method. The model was verified by continuous charging and discharging, HPPC and DST operating condition. The results show that the model has a high precision, and the terminal voltage error is only 10 mV, which fully meets the actual demand.

In order to solve the problem that the traditional RLS algorithm has a large error in the identification of the battery parameters under colored noise, we propose to use BCRLS algorithm to identify the parameters of the battery, which can effectively eliminate the colored error and improve the recognition accuracy. The BCRLS algorithm is combined with the EKF algorithm to estimate the SOC accurately based on the accurate estimation of the model parameters. In order to verify the reliability of the algorithm, the corresponding algorithm model was built in MATLAB/Simulink. The comparison between the DST and off-line results shows that the absolute error of the estimated SOC value is less than 1%, which can fully meet the needs of the BMS.

References

1. Finna P, Fitzpatrick C, Connolly D (2012) Demand side management of electric car charging: benefits for consumer and grid. Energy 42:358e63
2. Xiong R, Sun F, Chen Z, He H (2014) A data-driven multi-scale extended Kalman filtering based parameter and state estimation approach of lithium-ion polymer battery in electric vehicles. Appl Energy 113:463e76
3. Hariharan KS, Kumar VS (2013) A nonlinear equivalent circuit model for lithium ion cells. J Power Sources 222:210e7
4. Xiong R, Sun F, Gong X, He H (2013) Adaptive state of charge estimator for lithium ion cells series battery pack in electric vehicles. J Power Sources 242:699e713
5. Zhang C, Sun F, Zhao H et al (2000) Development of management system for electric vehicle battery series. J Beijing Inst Technol 9(1):35–40

6. Wang JP, Guo JG, Ding L (2009) An adaptive Kalman filtering based state of charge combined estimator for electric vehicle battery pack. Energy Convers Manag 50(12): 3182–3386
7. Li C (2007) Study on parameter identification and SOC estimation of Ni/MH battery for EV. Tianjin University, China
8. Xiong R, Sun F, He H et al (2013) A data-driven adaptive state of charge and power capability joint estimator of lithium-ion polymer battery used in electric vehicles. Energy 63 (1):295–308
9. Li Y, Mao Z, Wang Y (2010) BCRLS Identification method for Hammerstein-Wiener model. Int Conf Measur Technol Mechatron Autom, 745–748. IEEE Computer Society
10. Yang WL, Yu S, Cao WG (2015) Bias compensation recursive least squares identification algorithm for MISO stochastic systems. Info Comput Autom—Int Conf, 659–662
11. Xiong R, He H, Sun F, Zhao K (2012) Online estimation of peak power capability of lion batteries in electric vehicles by a hardware-in-loop approach. Energies 5(5):1455e69
12. Xiong R, Sun F, Gong X et al (2014) A data-driven based adaptive state of charge estimator of lithium-ion polymer battery used in electric vehicles. Energy 113(1):1421–1433
13. Xiong R, He H, Sun F, Zhao K (2013) Evaluation on state of charge estimation of batteries with adaptive extended Kalman filter by experiment approach. IEEE Transac Veh Technol 62 (1):108e17
14. Wu G, Zhu C, Chan CC (2010) Comparison of the first order and the second order equivalent circuit model applied in state of charge estimation for battery used in electric vehicles. J Asian Electric Veh 8(1):1357e62

Design of Triaxial Platform Used on Automobile Parts' Testing

Chengcheng Xiao and Tai Liu

Nomenclature

DOF	Degree of freedom
n	Number of components
g	Number of motion pairs
f_i	DOF of motion pair i
l_x, l_y, l_z	Length of the rods
α, β, γ	Deviation angles
$\boldsymbol{D}$	Input displacement of the platform
$\boldsymbol{X}$	Output displacement of the platform
α_X, β_Y, γ_Z	Projections of α, β, γ to X-, Y-, Z-coordinates, respectively
$[x_A, y_A, z_A]$	Coordinate value of point A
M_l^i	Mass of rod i
M_p	Mass of platform and fixture
M	Mass of moving section of specimen
$\boldsymbol{a}_i$	Acceleration of rod i
$\boldsymbol{a}$	Acceleration of platform
a_X, a_Y, a_Z	Component of platform acceleration on X,Y, Z axis, respectively
$\boldsymbol{C}$	The damping coefficient of the specimen
$\boldsymbol{K}$	Stiffness of the specimen
$\boldsymbol{F}$	Driven force
$\boldsymbol{F'}$	Force load on the specimen
F_{out}	Load cell indication
F_{real}	Real force loaded on specimen
DS	Driven signal matrix
DES	Desired signal matrix of road data
FRF	Frequency response function matrix
$\mathbf{FRF}^{-1}$	Inverse of **FRF**

C. Xiao (✉) · T. Liu
China Automotive Engineering Research Institute, Chong Qing 401122, China
e-mail: xiaochengcheng@caeri.com.cn

Society of Automotive Engineers of China (SAE-China) (ed.), *Proceedings of the 19th Asia Pacific Automotive Engineering Conference & SAE-China Congress 2017: Selected Papers*, Lecture Notes in Electrical Engineering 486, https://doi.org/10.1007/978-981-10-8506-2_44

1 Introduction

Automobile parts are widely bench tested before mounting to a prototype car or even after mounting into real car. Generally, performance and durability of automotive parts can be obtained by bench testing. Because of this, a large numbers of servo-hydraulic linear actuators which have excellent performance and long service life are applied to the bench testing. It is easy to realize one direction force loading to specimen by using linear actuator, while loading in multiple directions to parts of chassis is more comprehensive of the real condition.

During a long time in the past, bench testing procedures are only required to apply standard load forces which have a constant amplitude and frequency (usually sinusoidal waveform, single-axis loading) to specimen. Test bench can easily realize constant amplitude and single-axis loading. Nowadays, a lot studies on road load spectra loading have been introduced to get more accurate fatigue testing result [1–3]. To achieve the road load simulation purpose, not only loads data should be more accurate to the real service road load, but also the testing benches need to be sophisticated design. In this paper, a platform which can apply three direction forces or motions to specimen is introduced.

2 Mechanism Model of Triaxial Platform

As Fig. 1 shown, the mechanism model is composed of three similar subassemblies which placed along *X-/Y-/Z*-coordinate, respectively. Each subassembly driven by a linear motion performs one axis movement. Linear motion can be easily realized by either hydraulic or electrical actuator.

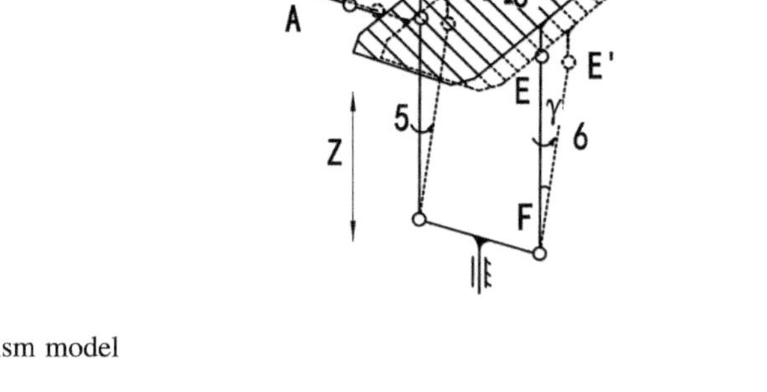

Fig. 1 Mechanism model

The mechanism model consists of 3 slide pairs, 6 connection rods, 12 connection joints, and 1 platform. Each slide pair is placed along the *X-/Y-/Z*-axis of the coordinates. There are two equaling rods which have ball joints of both ends between every slide pair and platform.

When platform is on the initial position, rods of the three directions are perpendicular to each other. All these rods, platform, and slide pairs compose to a parallel mechanism indeed.

3 Kinematical Analysis of Triaxial Platform

3.1 DOF Calculation

According to Kutzbach–Grübler criterion [4, 5],

$$\mathrm{DOF} = 6(n - g - 1) - \sum_{i=1}^{g} f_i = 9 \tag{1}$$

n Number of components.
g Number of motion pairs.
f_i DOF of motion pair i.

Each rod has own rotation DOF, subtract 6, and then, the platform has three DOF indeed which are three translational motions along *X-/Y-/Z*-axis in space.

Consider the length of the rods in *X*, *Y*, *Z* directions is l_x, l_y, l_z, respectively.

Platform is moved from initial point O to point O′. Point A, B, C, E are moved to point A′, B′, C′, E′, respectively. The deviation angles of rod AB, CD, EF are α, β, γ, respectively.

In the motion mentioned above, input displacement of *X*-axis is D_X. Input displacement of *Y*-axis is D_Y. Input displacement of *Z*-axis is D_Z.

$$\mathrm{A'B'} = \begin{bmatrix} x_A - x_B \\ y_A - y_B \\ z_A - z_B \end{bmatrix} + \mathrm{AA'} - \begin{bmatrix} D_X \\ 0 \\ 0 \end{bmatrix} \tag{2}$$

$$\mathrm{C'D'} = \begin{bmatrix} x_C - x_D \\ y_C - y_D \\ z_C - z_D \end{bmatrix} + \mathrm{CC'} - \begin{bmatrix} 0 \\ D_Y \\ 0 \end{bmatrix} \tag{3}$$

$$\mathrm{E'F'} = \begin{bmatrix} x_E - x_F \\ y_E - y_F \\ z_E - z_F \end{bmatrix} + \mathrm{EE'} - \begin{bmatrix} 0 \\ 0 \\ D_Z \end{bmatrix} \tag{4}$$

$$\mathbf{OO'} = \mathbf{AA'} = \mathbf{CC'} = \mathbf{EE'} \tag{5}$$

The input displacement of the platform position can be described as vector $\boldsymbol{D} = \begin{bmatrix} D_X \\ D_Y \\ D_Z \end{bmatrix}$, and the output displacement of the platform position can be described as vector $\boldsymbol{X} = \begin{bmatrix} x \\ y \\ z \end{bmatrix}$.

In fact, $|\mathbf{A}'\mathbf{B}'|, |\mathbf{C}'\mathbf{D}'|, |\mathbf{E}'\mathbf{F}'|$ are the length of rods.

$$|\mathbf{A}'\mathbf{B}'| = l_X \tag{6}$$

$$|\mathbf{C}'\mathbf{D}'| = l_Y \tag{7}$$

$$|\mathbf{E}'\mathbf{F}'| = l_Z \tag{8}$$

According to these equations, the relationship between input and output can be precisely described. The system is nonlinear. The three inputs and three outputs are coupled to each other. It is not easy to get the direct and inverse solution of position through analytic way. Numerical analysis would be right choice to get direct and inverse solution [4, 6].

This paper discusses an approximate analytic solution when the displacement of the platform and the length of the rods meet some qualifications.

3.2 Approximation and Error Analysis

Considering the input of the platform position

$$\boldsymbol{D} = \begin{bmatrix} D_X \\ 0 \\ 0 \end{bmatrix}, \text{and the output positon } \boldsymbol{X} = \begin{bmatrix} x_O \\ y_O \\ z_O \end{bmatrix}.$$

$$x_O = D_X - 2l_X \sin^2 \frac{\alpha_X}{2} \tag{9}$$

$$y_O = l_Y(1 - \cos \beta_Y) = 2l_Y \sin^2 \frac{\beta_Y}{2} \tag{10}$$

$$z_O = l_Z(1 - \cos \gamma_Z) = 2l_Z \sin^2 \frac{\gamma_Z}{2} \tag{11}$$

where

$\alpha_X, \beta_Y, \gamma_Z$ are projections of α, β, γ to *X*-, *Y*-, *Z*-coordinates, respectively.

In conditions of $l_X, l_Y, l_Z \gg D_X$, $\alpha_X, \beta_Y, \gamma_Z$ are all small angle. The second-order quantities can be ignored [5].

$$x_O \cong D_X \tag{12}$$

$$y_O \cong 0 \tag{13}$$

$$z_O \cong 0 \tag{14}$$

Because of the symmetries of the X-, Y-, Z-axis, displacement of the platform in X, Y, Z directions x_O, y_O, z_O is approximately equal D_X, D_Y, D_Z, respectively.

The error of the approximation in X direction can be described as follow.

Assumption: $D_X < 0.05 l_X$

$$D_X - x_O = 2 l_X \sin^2 \frac{\alpha_X}{2} < 0.0012 l_X < 0.012 D_X \tag{15}$$

Maximum error of the full stroke is 1.2%F.S., which meet the most requirement in testing engineering.

Y- and Z-axis can be derived similarly approximation of X-axis.

The displacement of the specimen in any direction can be measured by the driven displacement, respectively.

4 Kinetics Analysis

To simplify analysis, ignore the friction forces of the slide pairs and the friction moments of the join pairs. Ignore the gravity.

4.1 Driven Force and Force on Specimen

As mentioned previously, the mechanism model has three input forces, which are the driven forces of X, Y, and Z direction. The driven forces load on each slide pairs, respectively.

Set driven force: $\boldsymbol{F}$, the force load on the specimen via the platform: $\boldsymbol{F}'$

$$\boldsymbol{F} = \begin{bmatrix} F_X \\ F_Y \\ F_Z \end{bmatrix} \tag{16}$$

$$\boldsymbol{F}' = \begin{bmatrix} F'_X \\ F'_Y \\ F'_Z \end{bmatrix} \tag{17}$$

$$\boldsymbol{F} - \boldsymbol{F}' = \sum\nolimits_{i=1}^{6} \left(M_l^i \boldsymbol{a}_i\right) + M_{\mathrm{p}} \boldsymbol{a} \tag{18}$$

where:

M_l^i Mass of rod i
M_{p} Mass of platform and fixture
$\boldsymbol{a}_i$ Acceleration of rod i
$\boldsymbol{a}$ Acceleration of platform

Advanced mathematical tool is needed to get analytic solution of the mechanism model [4–7]. According to the previous approximation of the kinematic, if: $D_X < 0.05 l_X$, when F_X is applied, the platform motion in Y and Z directions can be ignored.

There

$$\begin{bmatrix} F_X \\ F_Y \\ F_Z \end{bmatrix} - \begin{bmatrix} F'_X \\ F'_Y \\ F'_Z \end{bmatrix} \cong \begin{bmatrix} M_l^1 + M_l^2 + M_P & 0 & 0 \\ 0 & M_l^3 + M_l^4 + M_P & 0 \\ 0 & 0 & M_l^5 + M_l^6 + M_P \end{bmatrix} \begin{bmatrix} a_X \\ a_Y \\ a_Z \end{bmatrix} \tag{19}$$

where

a_X, a_Y, a_Z Component of platform acceleration on X-, Y-, and Z-axis, respectively

It can be assumed that motion of X-, Y-, and Z-axis is independent of each other. In mostly case, the force loaded on the specimen can be described as Eq. (20).

$$F' = M\ddot{X} + C\dot{X} + KX \tag{20}$$

$\boldsymbol{C}$ The damping coefficient of the specimen
$\boldsymbol{K}$ Stiffness of the specimen
M Mass of specimen moving section

The relation between input force $\boldsymbol{F}$ and the specimen acceleration $\ddot{X}$ or displacement $\boldsymbol{X}$ can be described according to Eqs. (18) and (20).

In automobile parts' testing, the relevant vibration amplitude, force, and velocity are all the important parameters. Equations (18) and (20) are the fundamental mathematical model to perform the triaxial platform. Based on this, control theory model can be established.

4.2 Force Measurement

Many experiments utilize a standard actuator to carry out testing. Standard test method consists of an actuator, a load cell, and fixtures. The load cell is always placed on the end of cylinder rod to measure the loading force. As shown in Fig. 2, this method has enough accuracy to meet requirements in the quasi-static testing. While in high dynamic testing, this force measurement method has obvious defects. The mass between the load cell and specimen would produce inertia force which cannot be ignored. Load cell would give a higher output than the real force loaded on specimen.

Take a certain frequency f_0 into account.

Set mass between the load cell and the specimen as M_s, which includes the mass of fixture, ball joint, and moving section of the part.

The error of the output force:

$$F_{out} - F_{real} = -(2\pi f_0)^2 A M_s \sin(2\pi f_0 t + \varphi) \tag{21}$$

where

F_{out} Output of load cell
F_{real} Real force loaded on specimen

So the inertia force depends on frequency, amplify, and mass between load cell and specimen. Error calculation should be done before test.

In mostly automobile parts' testing cases, specimen can be set between actuator and load cell. One end of the load cell is fixed, and the other end is connected with specimen. In this condition, inertia force effect is eliminated.

When consider applying triaxial platform into bench testing, to accurately measure the multiple direction forces, the best solution is to use a triaxis load cell to measure three direction forces. For example, in the case of testing a connection rod with ball joints on both ends, connect one end of ball joint to the triaxial load cell, while connect the other end to the triaxial platform. The three direction loads can be measured without any error of inertia force.

The example of the intermediate bearing of drive shaft (IBDS) gives more detail of force measurement.

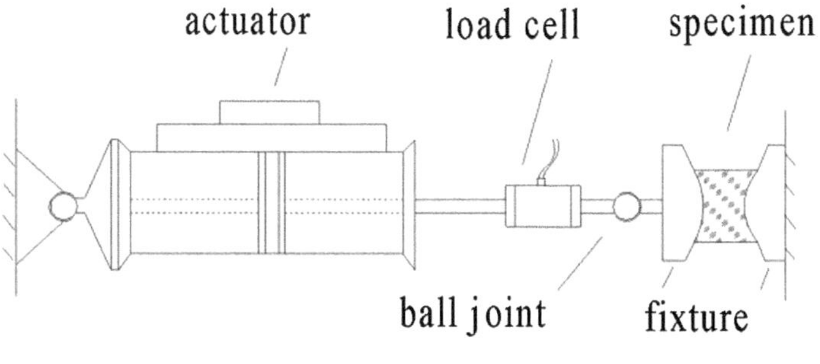

Fig. 2 One axis standard test

5 Example in Intermediate Bearing of Drive Shaft Testing

An example of intermediate bearing of drive shaft (IBDS) testing is introduced in this paper.

5.1 Instruction of IBDS

A type of IBDS is shown in Fig. 3. The inner ring of IBDS is a rolling bearing. The outer ring is elastomeric covered with metal plate. In real vehicle state, IBDS can bear drive shaft weight and compensate displacements of the drive shaft in *X*, *Y*, *Z* directions.

The triaxial platform can be applied to IBDS test requirements. Fix the mounting point to triaxial load cell and connect the inner ring to triaxial platform. The triaxial platform can perfectly reproduce the movement of the inner ring. Meanwhile, multi-axis load cell can measure the loads in different directions.

5.2 Testing Bench Design

Test bench is designed as shown in Fig. 4. Multi-axis load cell is set right below triaxial platform. Multi-axial load cell can measure the three direction forces and three moments.

As Eq. (17), F'_X, F'_Y, F'_Z are the forces loaded on specimen which should be measured.

M_X, M_Y, M_Z : bending moment of X, Y, Z direction.

The triaxial platform of the bench is driven by three linear hydraulic servo-actuators, and the actuator is equipped with a displacement sensor (LVDT), which can measure the relative displacement of the three directions of the inner ring of the test specimen. The force of the three directions is measured by the multi-axial load cell.

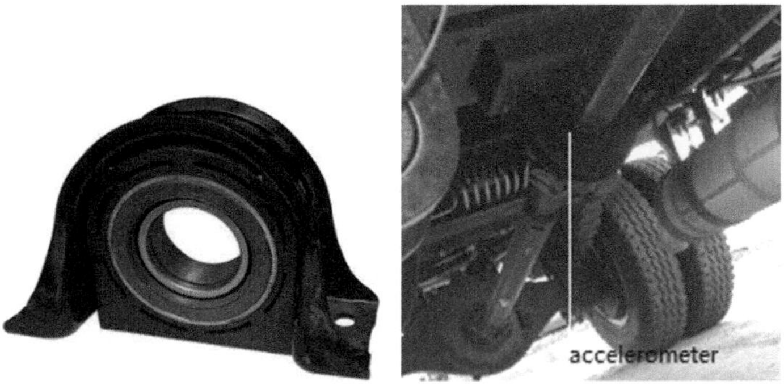

Fig. 3 Intermediate bearing of drive shaft (IBDS)

Fig. 4 Triaxial platform test bench

Each actuator can conduct load closed-loop control and displacement closed-loop control.

Actuator *X*, actuator *Y*, and actuator *Z* are, respectively, controlled by an independent controller; each displacement and force signal are fed back to the corresponding controller. To perform the load feedback control, *X-/Y-/Z*-axis controller utilizes *X/Y/Z* forces feedback signal of the multi-axis load cell, respectively. To perform the displacement feedback control, *X-/Y-/Z*-axis controller utilizes the LVDT signal of the *X-/Y-/Z*-axis actuator, respectively.

The schematic diagram of the control system is shown in Fig. 5.

The three controllers use the same external clock to generate the clock signal, which can ensure the relative phase of the signal. The closed-loop control operation is carried out in controllers, and controllers exchange data to PC computer through Ethernet. Not only the constant amplitude fatigue test but also service road load simulation can be done in this system.

Conventional PID control algorithm can be used to conduct the constant amplitude loading.

As shown in Fig. 6, sinusoidal waveform loads are simultaneously applied to specimen in *X*, *Y*, and *Z* direction, respectively. 120° of phase difference was set between *X* and *Y* direction load, while 90° of phase difference was set between *Y* and *Z* direction load.

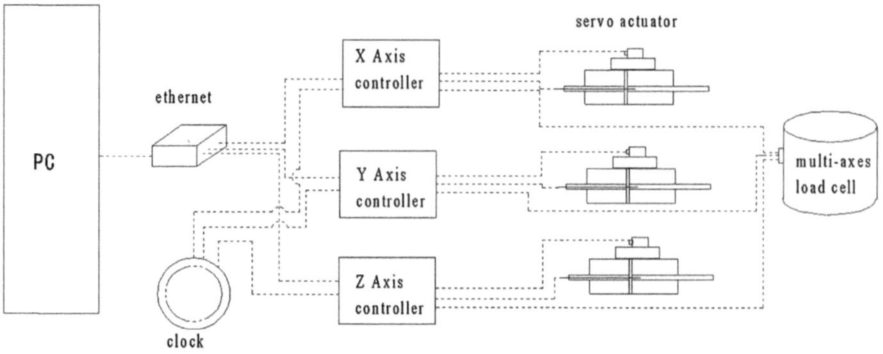

Fig. 5 Diagram of the control system

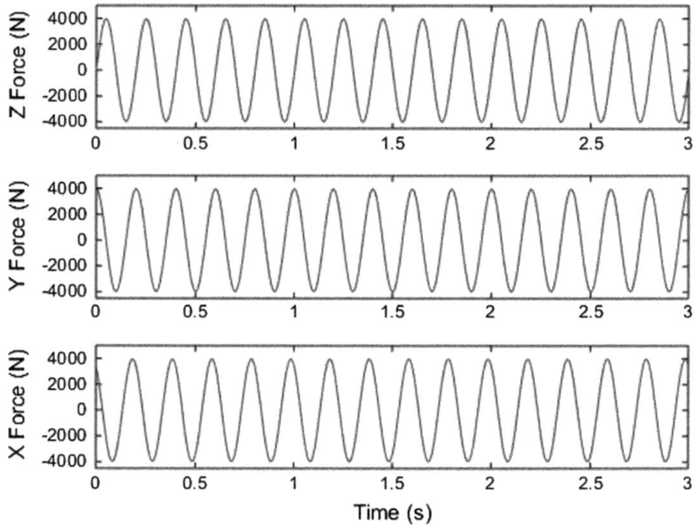

Fig. 6 Constant amplitude load

5.3 Road Loads Simulation

This test bench could carry out road load simulation. Because of the high-frequency response characteristic of the hydraulic servo-actuator, a large band of spectrum signal can be reproduced through the test bench. Meanwhile, the load cell arrangement eliminates the inertia force effect, and displacement transducer can directly measure the piston displacement of the actuator. Both the load cell and the displacement transducer (LVDT) can give an accurate indication. This would absolutely be a perfect solution for the IBDS (other similar elastomeric parts) testing.

As mentioned previously, nonlinearity exists in the control loop. Besides, *X*-/*Y*-/*Z*-axis movements are coupled to each other. It is difficult to conduct the service load simulation by the conventional PID control algorithm. To achieve the accurate simulation, the iteration control algorithm is introduced.

There are four steps to perform the iteration control algorithm.

Firstly, measure the system frequency response function (FRF). Exciting the system with a random noise signal input can obtain the relationship between the input and output across frequency band. The FRF can be built with data output and data input.

Secondly, estimate the initial driven signal. According to Eq. (22)

$$\mathbf{DS}_1 = \mathbf{FRF}^{-1} \times \mathbf{DES} \tag{22}$$

$\mathbf{DS}_1$ Initial driven signal matrix.
DES Desired signal of road data matrix.

Thirdly, conduct iteration. The second driven signal $\mathbf{DS}_2$ is obtained as Eq. (23).

$$\mathbf{DS}_2 = \mathbf{FRF}^{-1} \times (\mathbf{DES} - \mathbf{DS}_1) + \mathbf{DS}_1 \tag{23}$$

In order to get enough accuracy, repeat the iteration process until the error of the desired signal of the road data is acceptable. Iterations are used to incrementally correct the driven signal.

Fourthly, use the last driven signal of the iteration process to conduct the fatigue testing.

To simulate the road load of the IBDS part, it is needed to acquire the road data in testing field firstly. Displacement transducer, accelerometer, and strain gage are often used to collect road data. There are lots of differences among these sensors. Displacement transducer has decent accuracy in low frequency, while often not easy to mount into the place where the relative displacement is the predominant parameter [8–10]. Accelerometer should be mounted on where the acceleration is the predominant parameter. Accelerometer has sound high-frequency accuracy, while less sensitivity in very low frequency (usually below 0.6 Hz). Strain gage is used where the force is the predominant parameter.

To test the durability of the IBDS, reproducing the real motion is the priority. Use triaxial accelerometer mounted on the outer ring of the shaft rolling bearing (shown in Fig. 3) to collect the relative motion of the IBDS. The motion is combined with displacement compensation of the drive shaft and vibration of the rolling bearing. These data are high-frequency data predominant. By utilizing the acceleration data, the relative motion of the IBDS can be reproduced on testing bench.

When accelerometer is mounted on IBDS of the tested vehicle, make sure that coordinates of accelerometer coincide coordinates of tested vehicle as much as possible. Drive tested vehicle through block road to acquire the service road load.

The IBDS housing is fixed directly to the frame of tested vehicle. The frame has low-frequency vibration, which is less sensitivity to the acceleration sensor. Besides, during signal process, low-frequency acceleration signal is filtered to get an accurate reproducing. According to this mounting position, acceleration data can generate road load simulation in order to do fatigue testing.

Mount the specimen and the accelerometer to the test bench with the same condition compared to the tested vehicle state. Preload the IBDS in *Z*-axis with the calculated drive shaft weight. Use the processed road data to carry out iteration. The deviation of the output signal compared to the desired signal can be described as RMS error. After five iterations, the error of the output signal compared to the desired signal is acceptable. All the *X*-/*Y*-/*Z*-axis RMS are below 8% which can be accepted in testing engineering [8].

A segment of data of *X*, *Y*, *Z* motion and the reproducing data are shown in Fig. 7.

The output data show the reproducing in time domain. The output curves match the desired curves well. As the RMS result, the test bench can reproduce road load data with good accuracy.

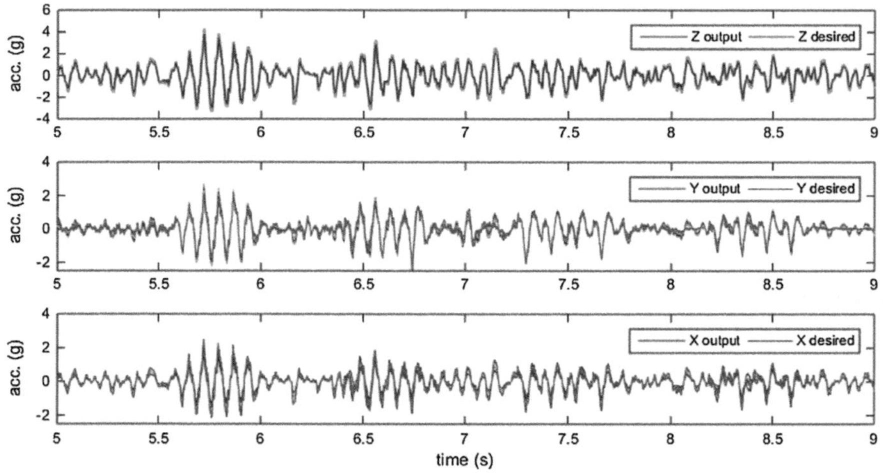

Fig. 7 Triaxial acceleration data and reproduced data

6 Conclusion

In this paper, a triaxial platform which can be applied to kinds of automobile parts' testing is introduced. Mechanism model of triaxial platform is analyzed from kinematics and kinetics. To achieve the solution quickly, the math model is simplified and the error is acceptable. The platform can be widely used in durability testing of connecting rods, elastic parts, and other important chassis components. Finally, an example of IBDS durability test solution is introduced, which has achieved good results.

Both the constant amplitude fatigue test and road load simulation can be conducted on this platform. Both conventional PID control algorithm and iteration control algorithm can be used on the triaxial platform test bench. Triaxial platform testing solution is proved to be successful.

References

1. Zhu J, Sun H, Yin H et al (2013) Application of hydraulic servo actuator to fatigue test of automobile parts. Eng Test
2. Sonsino CM (2007) Fatigue testing under variable amplitude loading. Int J Fatigue 29 (6):1080–1089
3. Karbassian A, Bonathan D, Katakami T (2010) Method and system for testing road load durability of truck rear bed: US, US 7677094 B1
4. Huang Z (2006) Adv Spatial Mech
5. Liu J, Li et al (2011) Mobility analysis of altmann overconstrained linkages by modified Grübler-Kutzbach Criterion. Chin J Mech Eng 24(4):638
6. Yang C, Huang Q, Han J (2012) Decoupling control for spatial six-degree-of-freedom electro-hydraulic parallel robot. Robot Comput-Integr Manuf 28(1):14–23
7. Ren R, Ye CC, Fan GB (2012) Analyses of error and precision of 6-DOF platform. Adv Mater Res 591–593:2081–2086
8. Sakai Y, Watanabe I, Nakamaru T (2008) Road-load input contribution analysis for suspension durability using a multi-axial road simulator
9. Wang W, Li H, Hu S et al (2016) Research on suspension durability test based on road load spectrum in proving ground. Automobile Parts
10. Tian Y, Jianhai WU, Fang S et al (2015) Data processing of fatigue test of automobile control arm bush based on road load data of blockcycle test. J Shanghai Dianji University

Application and Analysis of Laser Brazing in DPCA

Wei Wang, Heng Wang, Hongfei Ruan, Peiyong Wang and Tao Ye

1 Introduction

Resistance spot welding is the most common method in automobile manufacturing industry, but it is difficult to solve the problems such as large welding spatter, short service life of the electrode head, and a large amount of joint edge. The CO_2 protection welding, MIG welding, copper brazing, and other welding methods by the current and voltage, the stability is also difficult to be guaranteed. In recent years, as a kind of high-quality, high-precision, and high-efficiency welding method, laser brazing is becoming an important method of metal material processing and manufacturing. In 2007, DPCA first applied the laser brazing technology in the model C5 and gradually accumulated valuable experience through the application and practice of a large number of new models.

At present, DPCA has been applied to the laser welding technology in almost all the models, such as Peugeot 508, Peugeot 301, C4L, and New Elysee, and so on (Excel 1).

W. Wang (✉)
Dong Feng-Peugeot Citroen Automobile Co., Ltd.,
Hubei University of Economic, Wuhan 430000, China
e-mail: wangwei8@dpca.com.cn

H. Wang (✉)
Dong Feng-Peugeot Citroen Automobile Co., Ltd.,
Hubei University of Automotive Technology, Wuhan 430000, China
e-mail: wangheng@dpca.com.cn

H. Ruan
Dong Feng-Peugeot Citroen Automobile Co., Ltd.,
Hua Zhong Agriculture University, Wuhan 430000, China

P. Wang · T. Ye
Dong Feng-Peugeot Citroen Automobile Co., Ltd., Wuhan 430000, China

Society of Automotive Engineers of China (SAE-China) (ed.), *Proceedings of the 19th Asia Pacific Automotive Engineering Conference & SAE-China Congress 2017: Selected Papers*, Lecture Notes in Electrical Engineering 486, https://doi.org/10.1007/978-981-10-8506-2_45

Excel 1 Application of laser welding technology in DPCA

Type area	C5 (2007)	Peugeot 508 (2009)	C4L (2007)	New Elysee (2013)	Peugeot 301 (2013)
Body + roof	/	Laser brazing	Laser brazing	Laser brazing	Laser brazing
Door inner panel + door window frame	/	/	Laser welding	/	/
Trunk out + inner panel	Laser brazing	/	/	/	/

2 Fundamentals, Classification, and Equipments of Laser Brazing

2.1 Fundamentals of Laser Brazing

Laser brazing is one of the three most important parts of modern welding technology. It has the essential difference with melting welding and pressure welding. Laser brazing or laser wire filling brazing welding, its principle is used as heat source of laser beam, the focused beam is irradiated on the surface of the filler wire by an optical system. In a very short time, the welding wire is melted under the continuous heating of the beam energy to form a high temperature liquid metal, the liquid metal infiltration to be connected at the welding parts. Under appropriate external conditions, a good metallurgical bond is formed between the solidified liquid metal and the workpiece; at the same time, the material itself is not severely damaged by laser ablation. Laser brazing can be divided into soldering and brazing according to the liquidus temperature of solder. The liquidus temperature of solder is higher than 450 °C and is lower than that of the parent metal melting point of base metal, and it is called brazing. When the liquidus temperature of solder is lower than 450 °C, it is called soldering.

Laser soldering technology is commonly used in welding of electronic components on printed circuit boards, assembly of sheet originals, and so on. It already belongs to the traditional field of laser applications. The laser beam can be realized on a small area of the parts in high-speed heating with small heat affected zone, radiation time and the transmission power is easy to control and easy to realize multi-point light symmetric welding, with great flexibility.

Laser brazing is a new joining technology developed in recent years. It is especially applied to the connection of high strength steel and coated metal sheet. In recent years, laser brazing has been used in the automotive industry for the welding of galvanized steel sheets. The problem of excessive porosity and poor formation of weld seams due to the production of zinc vapor has been solved. The welding seam obtained by laser brazing not only meets the requirements of surface quality, but also greatly improves the efficiency of production and reduces the requirement of cleaning work after welding.

2.2 Classification of Laser Brazing

In accordance with the process method, laser brazing can be divided into the following categories:

(1) Single beam laser brazing

In the process of laser brazing, the solder can be supplied by presetting or by automatic filling. The laser power density is very high in the focal point, the brazing process requires only lower power densities for melt filling materials, the base metal remains solid, and wider heating range is beneficial to uniform heating of filling material. Therefore, the defocusing spot heating is usually used in the laser brazing process, and the spot diameter is adjusted according to the required brazing seam width. The semiconductor laser beam energy is evenly distributed, the spot shape can be manufactured into rectangular or strip, it is more advantageous to the stability of brazing process and the wetting spreading of solder, and the way light beams are transmitted through an optical fiber is very suitable for laser brazing.

Roll butt joint and lap joint are the two most commonly used joint forms of sheet metal laser brazing. In the process of laser automatic filling and brazing of coil butt joint, the welding wire is usually sent from the front of the laser beam, and the welding process is stable; In the process of laser automatic filling and brazing of lap joints, the welding wire is fed from the side with the lower plate, and the feeding process is more stable. As shown in Fig. 1.

(2) Double beam laser brazing

The spread of brazing at the joints is largely determined by the time and space distribution of the temperature. One of the greatest advantages of brazing process is the precise controllability of laser energy input. In order to utilize the laser energy distribution to control the distribution of temperature in time and space, a double beam and multi-beam brazing processes have been developed. More than two laser beams are used to radiate different regions, the hole or part of laser beam overlaps and radiates a region to obtain different temperature field distribution of the joints, so as to be applicable to different brazing materials and brazing processes.

Figure 2 is the laser radiation of lap joint and roll lap joint, butt joint and roll butt joint in double beam laser brazing. In the case of overlap, the gap between the upper plate and the lower plate cannot be directly heated by the beam, and the solder is difficult to be melted and spread. By using dual-beam laser brazing, a light beam heating wire, another one heating plate,the surface temperature of the upper plate is increased by the heat conduction of the upper plate,which is beneficial to spread the solder into the gap and improve the strength of the joint. Under the condition of the lap joint and roll lap joint, two beams of light overlap radiation heating solder, and heating docking of two boards at the same time, it is advantageous to evenly distribute the temperature in the position where the solder needs to be spread and to obtain satisfactory results.

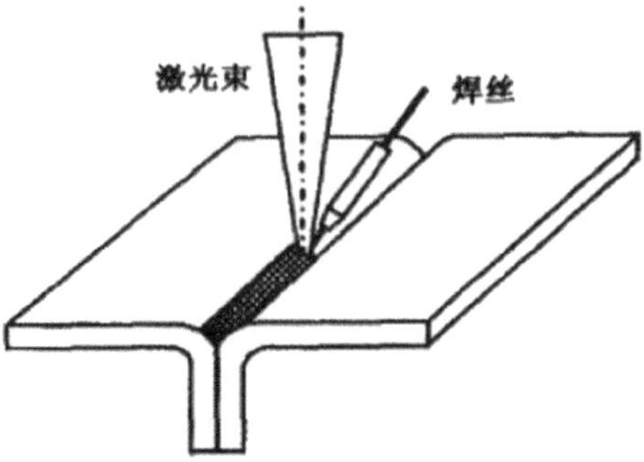

(a) Roll butt joint

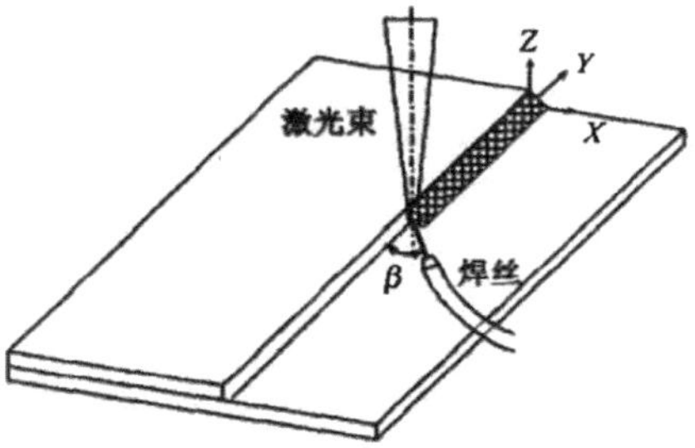

(b) lap joint

Fig. 1 Wire feeding method of laser brazing sheet metal

(3) Laser brazing with Auxiliary current

High-speed brazing requires higher laser power and wire feed speed, which can easily lead to instability in brazing process. In addition, due to the high cost of laser, it is not conducive to the economy of production by simply increasing the laser power to increase the speed of brazing. By simply applying an additional heat source, the brazing speed can be increased under the existing laser power. A DC or AC power supply is lapped between the welding wire and the base metal, and the contact position between the base metal and the welding wire is heated by contact discharge. With this method, the brazing speed and wire feeding speed can be increased by about 20–50%. Figure 3 shows the schematic diagram of the auxiliary current laser brazing system.

(4) Laser beam scanning brazing

In the roll butt joint, the joint gap at the joint is large, and the bridge capacity is usually required in the brazing process. CO_2 and YAG lasers are commonly used in industrial production; their laser power density distribution is uneven. The power density of the low order mode laser is very high, and it decreases rapidly along the radial direction of the spot. In the large gap joint, it is necessary to ensure that the

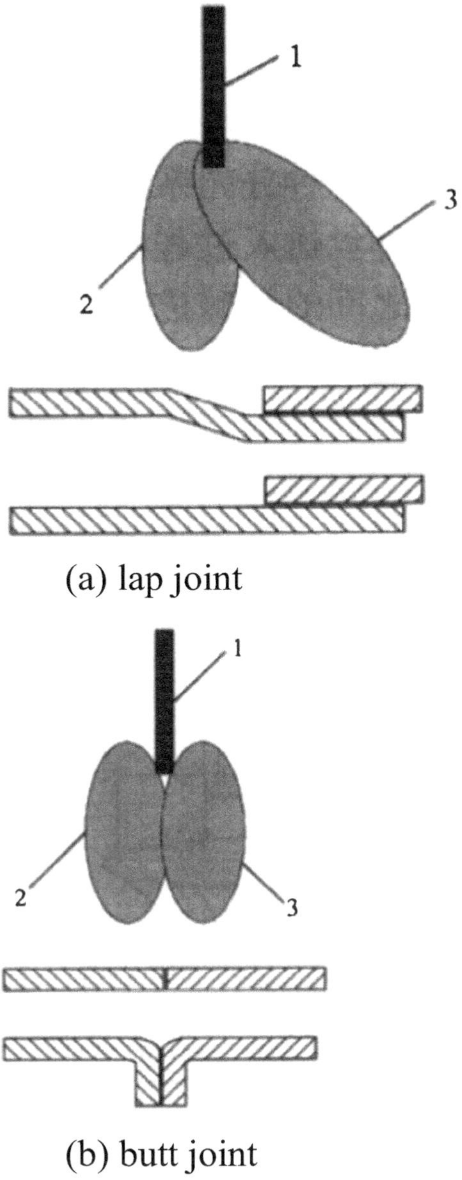

Fig. 2 Distribution mode of double beam laser brazing. 1. Filler metal, 2. heating point 1, 3. heating point 2

higher temperature is located at the interface between the solder and the base metal far away from the center of the spot. In order to ensure the good spread of the filler metal on the base metal, it is necessary to make a uniform heating of the joint position. In this case, the laser beam scanning brazing technique can be used. as shown in Fig. 4.

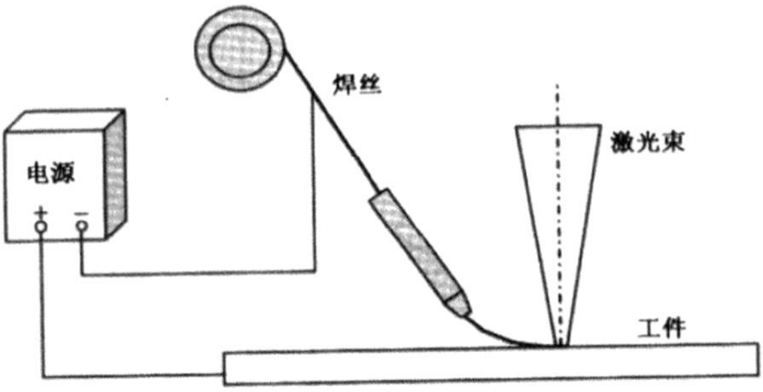

Fig. 3 Laser brazing with auxiliary current

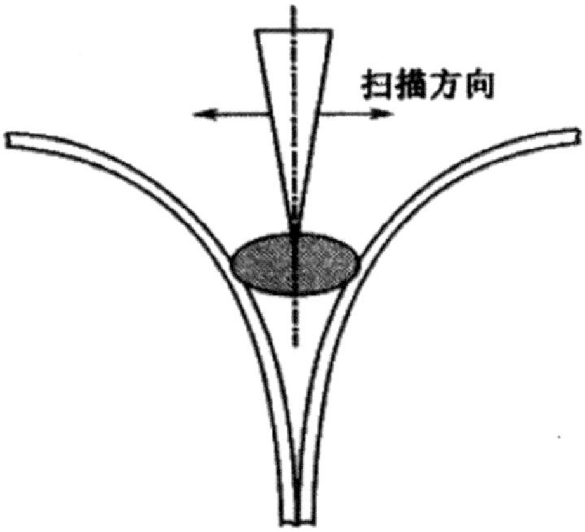

Fig. 4 Laser beam scanning brazing

(5) Hot wire laser brazing

In order to increase the production efficiency, the wire can be heated to a certain temperature and then sent to the laser radiation area. Thus, the required laser power is reduced, the brazing speed is increased, and the temperature of the welding wire heated before welding is even, which is favorable to the spreading after melting. Figure 5 is a schematic diagram of hot wire brazing. Before welding the wire into the laser heating area, the wire is heated to a point near the melting point by a heater and immediately fed into the brazing zone.

2.3 *Equipment of Laser Brazing*

Laser welding equipment is mainly composed of laser device, optical system, welding jig, welding robot, and welding control system.

Fig. 5 Hot wire laser brazing

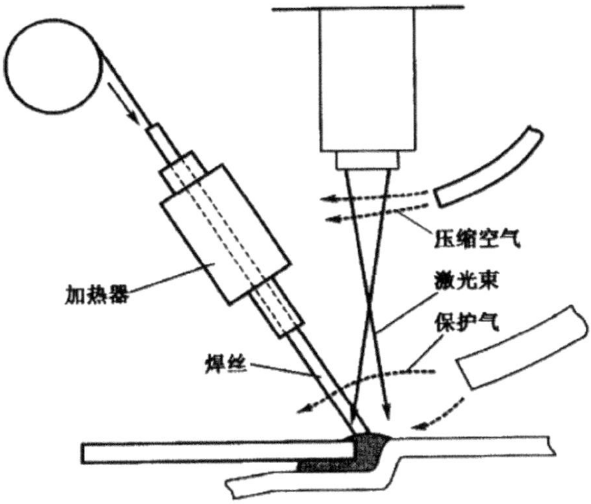

(1) Laser device: The laser device can be divided into gas laser, solid state laser, liquid laser, and semiconductor laser according to different luminous medium, and the most important performance of laser is the output power and quality of beam. At present, DPCA is mainly used in gas (CO_2) laser and solid state (Nd: YAG) laser with an average output power of 4000 W.
(2) Light guide system and focus system: The optical focusing system is composed of a circular polarizer, a beam expanding mirror, a reflecting mirror or an optical fiber and a focusing lens, which can realize the function of changing the polarization state, the direction of the light beam, the transmitted beam, and the focusing function. So the status of these optical components has an important influence on the quality of laser welding. At present, the laser head used in DPCA mainly includes brazing head and welding head.
(3) Welding robot: Due to the different methods of laser welding and laser, MIG hybrid welding, as well as the different forms of welded joints, the assembly precision of welding joints is also different. Laser welding of lap joint and laser brazing of fillet weld can be carried out by ordinary welding robot. For the laser brazing and laser welding of butt weld joints, it is necessary to use the arm welding robot, which is different from the conventional one, and the seam tracking system is usually designed. The robot used in DPCA mainly includes ABB, MOTOMAN, and KUKA.
(4) Welding jig: The welding jig can ensure the precise positioning of the connected plate or the assembly of the laser welding, ensure the weld gap, and prevent the welding deformation, so as to improving the quality of the laser welding joint. DPCA fully applied the design concept of the floating tool, which can effectively avoid the influence of the bad positioning of the fixture on the quality of the weld.

(5) Control system of laser welding: The laser welding control system mainly includes the video monitoring system of the welding process, the welding seam automatic tracking system of the robot and the correction system, the wire feeding control system, and so on. The composition of the control system is different with different laser welding methods. For example, laser welding of lap joint does not require a wire feeding system, an automatic seam tracking system, and a running track correction system. Laser cutting technology often needs to be used in conjunction with laser on line inspection system.

3 Technological Characteristic of Laser Brazing

Laser brazing is a local brazing technology, and wire feeding is usually used to provide brazing filler metal. At present, the application of laser brazing is to connect steel plates, especially those with metal coating, such as galvanized steel sheet. Because the boiling point of the coating metal is generally lower than the melting point of the base metal, the method of welding and melting is used. During the welding process, the evaporation of the coating metal can easily affect the stability of the welding process, and there will be more porosity.

The technological characteristics of laser brazing are discussed from the following aspects: process parameters, filling material, brazing flux, and shielding gas.

3.1 Laser Brazing Parameters

The basic parameters affecting the laser brazing process include laser type and power, spot diameter, brazing speed, wire feed speed, wire feed, laser incidence, wire shape and size.

(1) Laser type and laser power: CO_2 and Dd: YAG lasers, which are most widely used in industrial production, can be used as laser soldering heat sources.
(2) Diameter of spot: Defocusing heating is usually used in the process of laser brazing; the focus position determines the diameter of the spot when the beam quality is certain. The size of the spot used depends on the width of the weld. In thin plate welding, the width of the heating zone usually ranges from 2 to 3 mm, which meets the requirements.
(3) Brazing speed and wire feeding speed: The brazing speed is related to the laser power, and the laser power and the brazing speed are comprehensively considered by using the unit length energy input. The welding results show that the welding speed can be increased by one meter per minute, and the welding speed can be controlled from 1.5 to 1.6 m/min. The feeding speed is related to the filling requirement of the weld. When the filling requirement of the weld can be

satisfied, the wire feeding speed is the same as the brazing speed, which is helpful to the stability of the brazing process.

(4) Wire feed and laser incidence: The wire feeding angle can be between 0° and 90°, and the actual welding results show that a better seam forming can be obtained within 30°.
(5) Shape and dimension of welding wire. As laser brazing usually uses filler wire to provide filler metal, considering the convenience of wire feeding, the wire is generally round and linear, and its diameter can be chosen in the range of 0.4–2 mm. When the preset solder is brazed, its shape and size can be chosen according to the actual joint shape.

3.2 *Filling Material*

In the laser brazing of the connecting plate, hard material generally melts at temperatures between 650 and 1000 °C. Taking galvanized steel sheet laser brazing as an example, the materials usually include copper silicon brazing filler metal, brass filler metal, copper aluminum solder, and copper-tin solder.

3.3 *Protective Gas and Flux*

The brazing temperature field can be controlled and adjusted by adjusting the laser power, brazing speed, shape, and size of the spot, and heating mode. If the proper temperature field can be obtained, there is no need to use brazing flux to aid the wetting spread of solder. At present, laser welding brazing can achieve flux-less brazing in industrial production and can also be used for reducing protective gas to replace the flux, such as nitrogen, argon, helium, or carbon dioxide as protective gas, the air isolation effect, so as to avoid the further oxidation of the heating region.

4 Industrialized Debugging Method for Laser Brazing of Top Roof

Although the laser welding of top roof is formed in the laser welding process, the welding quality is not only decided by welding parameters, but also in the process of body forming welding. The gap value of the lap joint of the top roof and the body side is mainly guaranteed by the forming tool. If the gap value cannot meet the requirements, it can directly lead to the quality problem of the laser welding seam. Therefore, during the process of industrial debugging, we attach great importance to the debugging of forming tooling. The following contents will explain the working principle of the roof forming tooling and the method of industrial debugging and explain how to ensure the gap between the top cover and the side lap area and also how to guarantee the quality of laser welding.

4.1 Working Principle of Forming Tool for Roof Laser Brazing

The forming tool of roof is divided into two parts: Prefabricated platform for roof and forming tool, the prefabricated platform provides pre-positioning for part and ensure the initial position of the part is unique. The orientation is implemented in three directions: *XYZ*.

Work principle of prefabricated platform.

The *Z* plane is positioned by the roller according to the volume of the top cover.

- The side of roof positioned *Y*
- The posterior parts of roof positioned *X*
- The gravity of roof must pass between the two rollers rather than the other rollers. The origin of the vector is the center of gravity of the part, and the direction of support must be compatible with all types of roof.

Work principle of forming tool:

The forming tool can be divided into two parts: the floating part and the positioning part. The fixed part is connected with the robot, and the floating part keeps the positioning relation with the roof; the whole work process is as follows:

First step: When the floating tool is taken, the floating mechanisms ① and ④ are locked by the locking system of the positioning pin to ensure that the whole process of the placing of the grab piece is not floating, and the track is unique.

Second steps: The roof ② is parallel ③ to the body, the positioning pin is released, and the floating tooling is free at *XY*.

Third steps: The top is fitted with the body, the *Y* floats to fit the body, and the *X* is positioned on the body beam or side by means of the tooling.

4.2 Debugging Method for Laser Brazing of Top Roof

The method of industrial debugging is mainly divided into two parts: the debugging of track and the debugging of jigs. Both of them have an impact on the gap between the roof and the body side. The running track mainly involves the debugging of *Y* to centering. When there is a deviation between the top roof and the body side, the position of the top roof on the body can be changed by adjusting the trajectory *Y* or rotating around the center. When the position of the top roof is already OK in *Y* orientation, but the regional gap between the rooftop and the body side is still relatively large, we can debug the jig to solve the problem of partial gap, in general. In general, adjust the pressure to overpressure in Z orientation to decrease the gap between the top roof and the body side. The quality of parts also needs to be taken into account, when all the debugging of fixture and track cannot solve the problem; it is necessary to pay attention to the conformity of the parts, because the local contour of parts cannot be solved by jig debugging; it cannot eliminate the defects of large gap at the top roof and the body side. The Y position of joint between the body frame and the roof is equally important. When the Y orientation of the body is shifted toward the overall, the floating tooling will follow the offset, and the top positioned by the floating tooling is also offset. Although the gap value between the top and the side can be guaranteed, the position of Y cannot be guaranteed. Because the welding tolerance of laser welding gun track is certain, when the Y deviates too far from the position, the quality of the welding seam will also be a problem. Therefore, the vehicle size control is also particularly important. We do not expand in detail here. (there are too many aspects of body size adjustment).

The following is the requirements for the debugging process and results:

a. The gap between the side and the roof should be less than 0.3 and uniform.
b. Starting and stopping point: $X \pm 1$ mm, $Y \pm 1.5$ mm, $Z \pm 2$ mm.
c. Offset of the entire weld in the Y direction $\pm$ 1.5 mm, Z direction $\pm$ 1.5 mm.
d. The surface of roof and body side should not have obvious edges and bulges.

The welding torch track should be close to the weld edge as much as possible, and the pressure size needs to be adjusted and optimized.

5 The Common Defects and Solutions of Laser Brazing

Through a large number of practical work and research, we summed up the common defects and solutions of laser brazing (Excel 2).

The weld seam should meet the flowing requirement when it is repaired: The function dimension and design requirement should not be influenced; after repairing the weld seam, it can ensure the dimensional stability of the car body assembly and also can ensure the better surface precision and better corrosion resistance of the joint. In addition, the gumming area is not allowed to weld in the repair work, so as to avoid unnecessary welding deformation. In the area where the flange is shorter (less than or equal to 8 mm) or cannot be repaired by drilling, we can use a fillet weld at the lap joint or use MIG or plasma brazing at the butt joint. The weld length which be repaired must be equal to the length of the weld defect. It is necessary to analyze the weld ability of the base metal before repaired by arc brazing (MIG, WIG, plasma). For example, the yield strength and fracture strength of base metal can be analyzed by drawing experiment. In some special circumstances which cannot be repaired, such as in the connection due to heat causes the assembly deformation of the body, according to the standard which the parent material heating repair approach is not allowed, must cold glue or other cold processing method considering.

Excel 2 Common defects and solutions of laser brazing

Description	Defect definition	Figure of defect	Reason	Scheme
Weld offset	The weld is completely offset to a base material	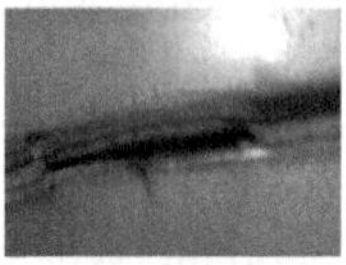	The trajectory of the robot is inconsistent with the position of the clamping part	Adjust the trajectory of the robot and check the assembly status of parts
Weld seam sag	Local material or missing	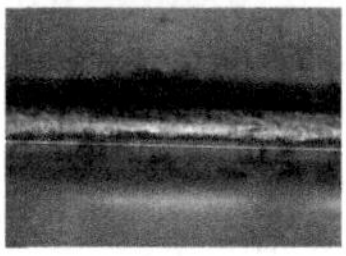	Bad spot center	The trajectory of the robot should be adjusted to match the center position of the red spot and the center of the weld
Accumulation of weld seam	Weld filler material	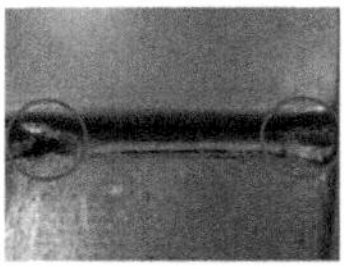	The wire feeding speed is too fast, and the diffusion of the molten solder is insufficient	The trajectory of the robot should be adjusted to match the center position of the red spot and the center of the weld
Shrinkage cavity	The welding seam presents irregular pores, which directly affects the appearance quality of the weld	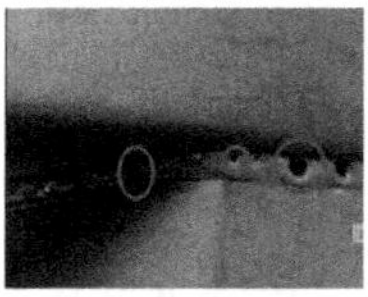	The surface of the steel plate is greasy, and galvanized steel plate is bad	Control the surface oil pollution to ensure the cleanliness of parts
Weld interrupt	The weld seam is not continuous, and the local area is not fused	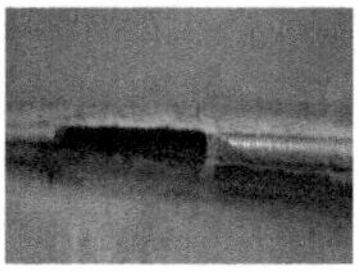	The normal welding specification is disordered, the wire feeding is not smooth, and the optical output power is unstable	Check that whether the wire feed mechanism and the light output device are normal

Excel 3 Technical requirements for weld position tolerance of laser brazing

Technical requirements for weld position tolerance (IT) of laser brazing				
Location description	*X* orientation (mm)	*Y* orientation (mm)	*Z* orientation (mm)	Ability of process
Arcing point	±1	±1.2	±1.2	CAP > 1.33
Weld seam	/	±1.5	±1.5	CAP > 1.33
End	±1.2	±1.2	±1.2	CAP > 1.33

6 Constrains of Product for Laser Brazing of Roof

6.1 Constrains of the Position for Laser Brazing Seam of Roof

In order to ensure the function requirements of the roof and body welding, the surface of roof and body side should not have obvious edges and bulges. At the same time, the gap between the roof and the body side must be <0.3 mm and uniform; the shape of weld fusion zone is uniform "V" shape. In addition, in order to meet the requirements of the flush between the front windshield and the roof, the flush between the roof and the side surface of the body side must be <1.5 mm; the flush between the left and right surfaces is in the same *X* section <1.5 mm; according to the above requirements, we have developed the following technical specifications: (Excel 3) (Fig. 6).

6.2 Constrains of Tolerance for Outer Panel of Body Side and Roof Par

The fit position of the outer panel of body side and the roof of the *Y* orientation, the width of the fitting width 8 mm, the tolerance of the requirements of the position of the opposite position is 0/+1 mm, Offset to Y_0 direction. The profile requirements of the opposite side are 0.6 mm, as smooth as possible; the transition is uniform; there can be no bump. The surface of the trace level difference is less than 0.2 mm, to avoid the excessive bulge in the welding area of the mold surface; the requirements for the fit position of the roof and body side outer panels is ±0.5 mm, the surface profile degree requirements of 0.6 mm, there should be no bump. The surface of the trace level difference is less than 0.2 mm; the requirements for the fit position of the roof and body side outer panels are 0/+1.5 mm, offset to the direction of Z+ (Fig. 7).

Fig. 6 Structural sketch

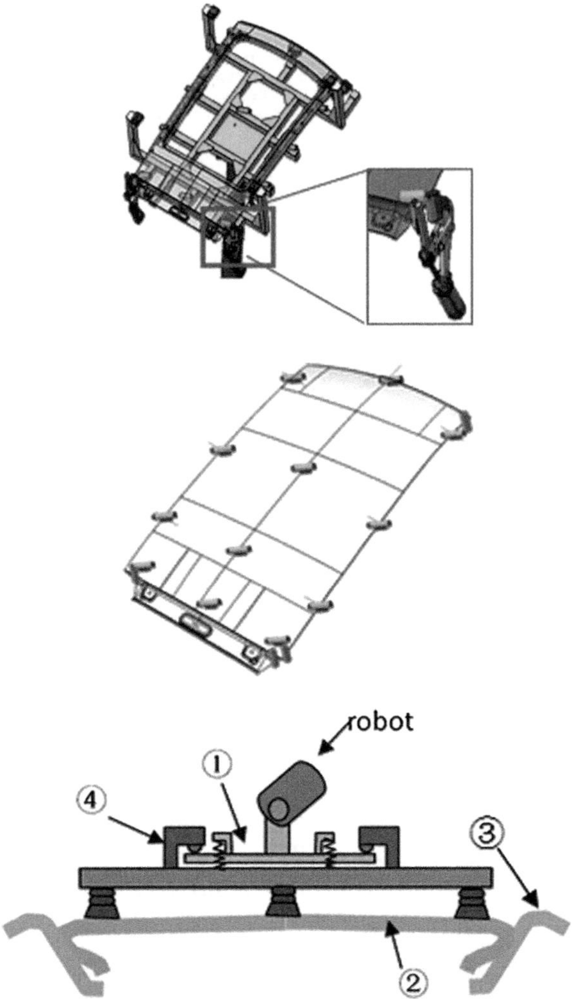

7 Optimization Scheme of the Process of Roof Laser Brazing

7.1 Analysis of Boundary Conditions for Stamping Process

According to the process scheme of body side die from seven existing models, we have analyzed the tolerance guarantee ability of traditional process for outer panel of body side. By comparison, it is found that (A + C) and B as two regions, there is only A (C) or B region in each body side of the model to meet the tolerance, but it cannot guarantee that these two areas to meet the tolerance at the same time. To meet the technical requirements of laser brazing, the process of the existing technical conditions must be optimized (Fig. 8).

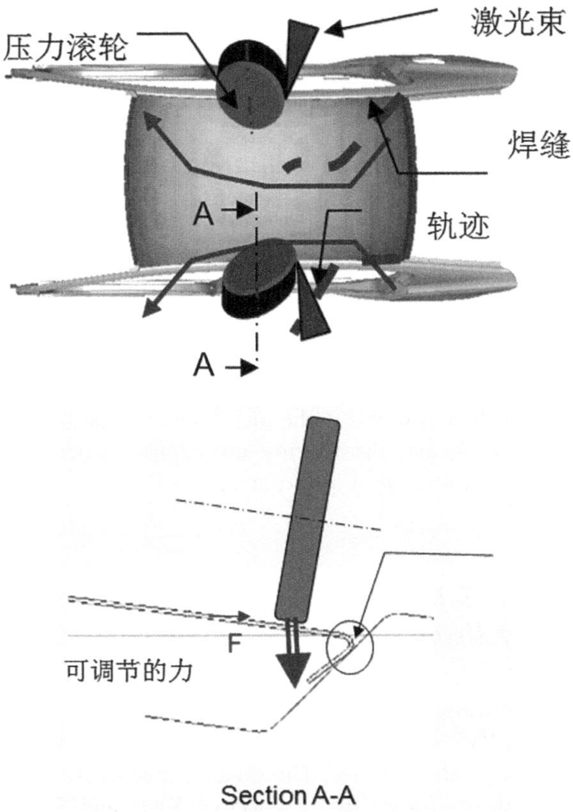

Fig. 7 Schematic diagram of laser welding

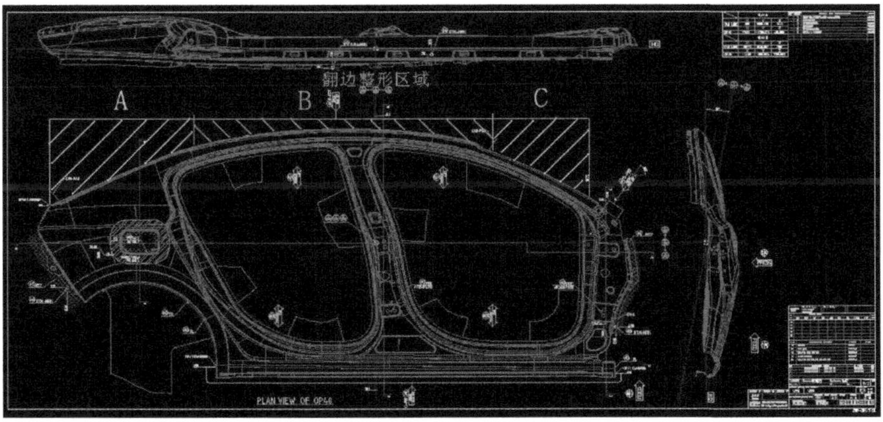

Fig. 8 Analysis of process scheme of body side die

7.2 Optimization Scheme of Welding Process (1)—Ensure the Effective Positioning of the Roof Jig by Using Zero Point Positioning System

Through the analysis and comparison of the five existing models, we find that the fixture of the zero positioning system can meet the technical requirements of laser brazing. The structure of the system is shown in the following figure (Fig. 9), the positioning lock head is a male side, and the lower part of the whole locking mechanism is a female side. When the system works, the lock head is inserted into the cavity of the positioning system, the mechanism allows the 12 mm level offset and ±5° of tilt angle to enter the lock head, when the lock head enters into the air cavity, the pressure is generated in the cavity, the top of the lock head is turned on, and the pressure is pushed upwards to the lock body, so that the inner cavity of the steel ball, the lock head, and the locking mechanism generates three points of contact; the lock head is accurately positioned.

7.3 Optimization Scheme of Welding Process (2)—Ensure Parts Fit by Using Floating Roof Jig

In order to ensure the formation of V cavity on the *Y*/*Z* planes of the body side, the roof tooling forms a "V" shaped cavity through the *X* positioning hole with the *Y*/*Z* planes of the body side and roof. The floating mechanism drives the roof and the body to automatically locate the position of *X*, *Y*, and *Z* automatically. The floating mechanism should ensure the roof and the body of naturally fit, but it cannot bear too much internal stress, so as to ensure the correct position of the roof.

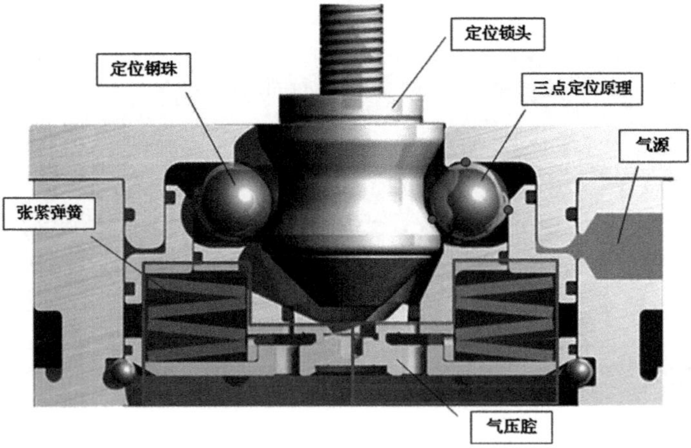

Fig. 9 Zero point positioning system

7.4 *Optimization Scheme of Welding Process (3)—Ensure the Floating Function of Tooling by Using Concentric Guide Wheel and Eccentric Guide Wheel*

The floating process in the *Z* orientation of the jig, in order to ensure that it does not rotate around the *X* and *Y* axis (rotation will lead to tilt the parts that cannot be fitted smoothly), 16 sets of guide wheel mechanism are added to the four sides of the cylinder's moving direction to limit the rotation of the tool. The guide wheel is divided into two types: concentric guide wheel and eccentric guide wheel, and all wheels can roll on the adjustable guide surface. The concentric guide wheel is fixed

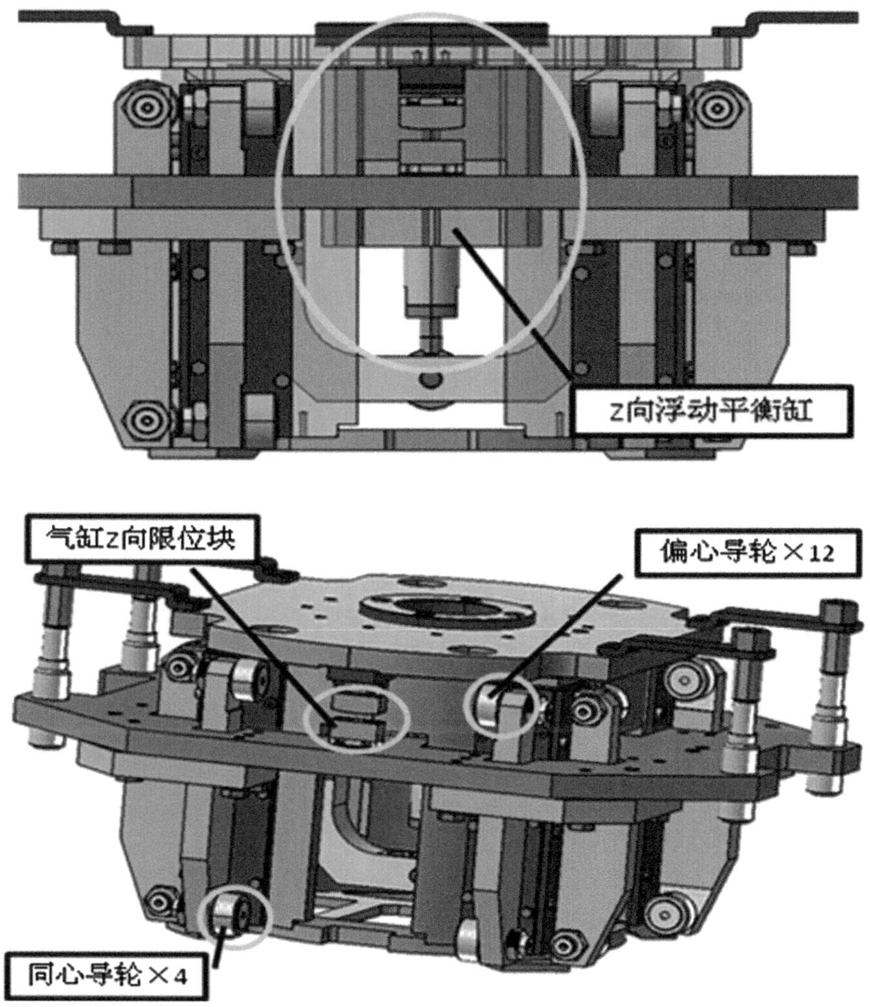

Fig. 10 Theory of the floating function of the tooling

on a theory guide surface which is used to limit the rotation of the jig. The eccentric idler wheel is used to compensate the manufacturing error and ensure the alignment of guide wheel and guide surface (Fig. 10).

7.5 Optimization Scheme of Welding Process (4)—Switch Between Floating and Locking by Using Universal Wheel Mechanism

The *XY* floating function of the tooling is completed by the universal wheel mechanism. There are four groups of universal balls at the connection part of the floating part and the positioning part. Before grasping the parts, both sides of the locking pin will lock the *XY* floating, so that the robot can pick up parts in a fixed position. After grabbing the parts, the tooling starts to drive the parts and the body to fit together. When the tool *Z* is floating, the friction force between the universal ball and the positioning part of the utility model has been very small, and the locking pin exits. The entire positioning part of the tool is in a free state in the *XY* direction, and the contact between the top cover and the vehicle body can trigger the movement of the tooling with the body position, and finally, the utility model can achieve the adaptive floating function of the tooling.

8 Conclusions

As a well-known automobile manufacturing enterprise, DPCA has been guiding the development direction of domestic automobile manufacturing technology. The development and application of laser welding technology are an important symbol for the upgrading of the manufacturing level. Over the years, the experience of the product and process constraints accumulated provides a scientific basis for the development of new models. Through the popularization and application of the B72, M3M4, T93, T91 project, which effectively improve the quality of laser brazing and the rework rate decreased from 20 to 3%, cumulative savings equipment renovation costs and manual repair costs more than 4 million RMB. At the same time, with the application and promotion of advanced technology, we have won the favor of more customers and praise and enhanced the company's brand image and market competitiveness.

References

Monograph

1. Li LJ (1993) Modern laser processing and equipment. Beijing Institute of Technology Publisher, pp 194–204
2. Laser technology in automotive industry 2005
3. Chen YB Technology of modern laser wielding. Science Publisher

Collected papers

4. Austauschkonzepte HL4006D TRUMF 2009.09.08

Simulate Design and Experimental Analysis of Commercial Vehicle Cab Suspension

Song Zhang, Kaiqiang Li, Guolin Wang and Xinyu Zhu

1 Introduction

With the higher requirement of riding comfort, the full-floating commercial vehicle cab suspension is being widely used. This research of air spring cab suspension contains a series of advantages which including low and stabilized system natural frequency, good vibration isolation effect, light weight, low noise, and that are the reasons for being widely applied in middle- and high-class commercial vehicles [1]. However, this kind of cab suspension can only bear radial force, and guide mechanism is thus required to be attached to strengthen its lateral stability, which may unavoidably make the system much more complicated. As the cab suspension system plays a key role in the vehicle system, researchers from home and abroad are also devoting themselves to exploring its performance of every aspect [2–7]. At the beginning of the product design, designers generally adopt simple empirical formulas or simplified simulation models [8–10] which are of great advantages for mature products with fixed design process. But it may have problems in complicated new products. In later developing stages of cab suspension, one possible way to solve the design defect is repeatedly revising design according to the result of prototype vehicles' repeated road test, or doing rectification and modification according to feedback from the market, which greatly increases procedure and cost of design, and reduces market competitiveness [11–14]. It can be seen that some problems are still there in the research process of cab suspension system.

The method that adopting simulation design and bench scale test in this research process of air spring cab suspension can ensure the accuracy of design, and greatly shorten the design cycle, and cut the cost. To explore the vibration performance,

S. Zhang · K. Li (✉) · G. Wang · X. Zhu
School of Automotive and Traffic Engineering, Jiangsu University,
Zhenjiang 212013, China
e-mail: mclee2@sohu.com; 514125398@qq.com

Society of Automotive Engineers of China (SAE-China) (ed.), *Proceedings of the 19th Asia Pacific Automotive Engineering Conference & SAE-China Congress 2017: Selected Papers*,
Lecture Notes in Electrical Engineering 486, https://doi.org/10.1007/978-981-10-8506-2_46

firstly ADAMS/View was adopted to build the dynamic model and the cab suspension's vibration isolation performance was analyzed. The vibration test of cab suspension in four-channel road simulator was conducted to verify the effectiveness of the built dynamic model. Secondly, a cab suspension finite element model was built by using RADIOSS according to the boundary condition of dynamic model, and experimental verification of full-load working condition was also conducted. Accordingly, the strength of cab suspension in various limiting condition was analyzed. Finally, the system fatigue test on fatigue testing machine was conducted to check the fatigue resistance of both structure and elastic parts of the cab system based on which corresponding design problems could be modified. The method mentioned above is of high value in engineering application and promotion.

2 Cab Suspension Dynamic Performance

Cab suspension dynamic performance is strongly related to the spatial structure arrangement. So it needs to explore and check the rationality of particular spatial structure. Air spring cab suspension system in this paper is consisted of front suspension and rear suspension. The front air springs are arranged vertically, and mainly bear vertical force; the rear air springs are arranged obliquely which can bear both vertical and lateral forces. The cab suspension system consists of considerable number of bushes which greatly increase the system freedom. In this way, the cab suspension can bear the pitching moment, yawing moment and rolling moment from the cab within the limit travel. Also, the large number of Elastic elements may improve the system vibration isolation if they are adjusted.

2.1 Cab Suspension Bench Test

To explore the vibration isolation effect of elastic elements (air springs and bushes) on air spring cab suspension system, the cab suspension was launched between the developed equivalent cab and four-channel road simulation testing bench (as shown in Fig. 1). The equivalent cab is developed according to the mechanical dynamics conversion principle, so the equivalent one has the same mass, center-of-mass coordinate, and moments of inertia as the original one. The four-channel road simulation testing bench is mainly consisted of servo-linear actuators, hydraulic control system, dynamic and static test software, load holders, relevant attachments, and so on. Hydraulic control system controls linear movement of actuators. The movement signals are measured and received by corresponding sensors, and then fed back to the computer after processed to complete a closed-loop control. The four servo-linear actuators can move independently with different input road spectrum (displacement spectrum) which can simulate the real working condition of the running vehicle.

Fig. 1 Bench test

Fig. 2 Acceleration sensors' position

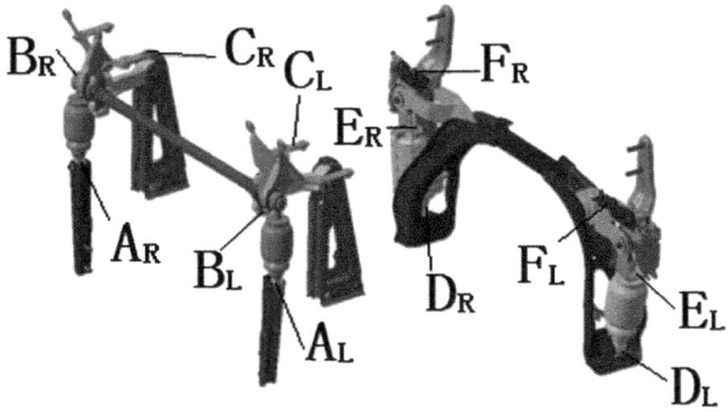

Every air spring of the cab suspension system works with one altitude valve which connected to center plenum chamber so it can keep every air spring's inner pressure as its setting value by inflation and bleeding. And it makes sure that the running vehicle can turn bad working condition into quite good work condition. In another word, the stiffness value of the air spring can always stays in the adjacent area of the inflection point of the inverse *S* curve (air spring's stiffness curve).

In order to investigate dynamic characteristics of the cab suspension system, the air spring's inner pressure was set to a normal working pressure (5 bar) and the acceleration sensors were placed on the input terminal and output terminal of relevant elastic elements (each sensor must be vertical arranged) to measure the vertical acceleration (as shown in Fig. 2). The filtered acceleration signals can be translated into displacement signals by twice integrals so that they are of the same formats as the input signals.

2.2 Cab Suspension Dynamic Model

Cab suspension bench test is restricted. In some harsh conditions, therefore simulation method is quite suitable. In ADAMS/View, the rigid-flexible coupling method was adopted to build cab suspension dynamic model (as shown in Fig. 3).

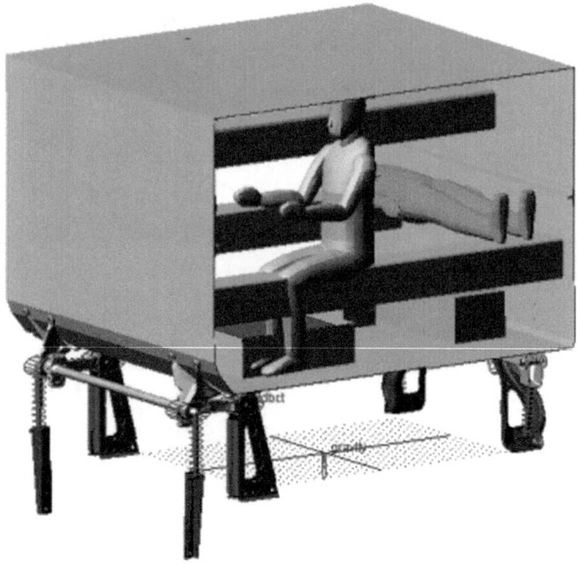

Fig. 3 Dynamics model of cab suspension

The horizontal stabilizer bar of the front suspension and portal frame of the rear suspension were modeled as elastic bodies, and the other structures as rigid body. The air springs were simulated with Spring-Damper model, and the bushes with Bushing model. In the cab, one driver and two passengers along with one toolbox were fixed to the specified location which characterized the full-load condition. The sprung mass is up to 1600 kg, much larger than the unsprung mass. The whole degree of freedom of the dynamic model is 102 which is large enough to reflect the dynamic characteristics.

2.2.1 The Stiffness Characteristic of Air Spring

Diaphragm-type air spring is wildly used in the cab suspension system. The stiffness curve seems like an inverse *S* curve which means the stiffness value is relatively small near the equilibrium point while much larger when stretched or compressed a large stroke. The stiffness can be fitted by the following formula [14]:

$$K = \frac{mD^2(p_r + p_a)}{V}x^2 + \frac{2mDS_0(p_r + p_a)}{V}x + \frac{mS_0^2(p_r + p_a)}{V} + p_r D \tag{1}$$

where K is the stiffness value, p_r is the inner air spring pressure, p_a is the atmospheric pressure, m is poly-tropic index, S_0 is the effective area of air sac on the equilibrium position, D is the changing rate of air sac' s effective area, V is the volume of air spring, and x is the displacement of air spring.

As it is difficult to fit the stiffness curve directly from the tension and compression test, formula (1) was integrated, and the relation between the air spring's load and internal pressure was obtained, which is a cubic equation with one unknown. The air spring's load is proportional to internal pressure as shown in formula (2), where F_0 is a constant.

$$K = \frac{mD^2(p_r + p_a)}{3V}x^3 + \frac{mDS_0(p_r + p_a)}{V}x^2 + \frac{mS_0^2(p_r + p_a)}{V}x + p_rDx + F_0 \tag{2}$$

The diaphragm-type air spring's load-displacement curve was tested under the normal temperature environment by adopting electro-hydraulic servo-vibration system made by the MTS company from the USA, and it was fitted by formula (3) under the normal working pressure (5 bar). The unit of load is N, while displacement is mm. As discussed above, the air spring cab suspension system can always work on the initial set position (see Fig. 4) which means the displacement x is relevantly small. Owning to the quite small coefficient of the two and three party of x, formula (3) could be substituted with (4). Then, the linear stiffness of air spring could be identified as 12.5 N/mm.

$$F = 0.0077x^3 + 0.1311x^2 + 12.50x + 3486 \tag{3}$$

$$F = 12.5x + 3486 \tag{4}$$

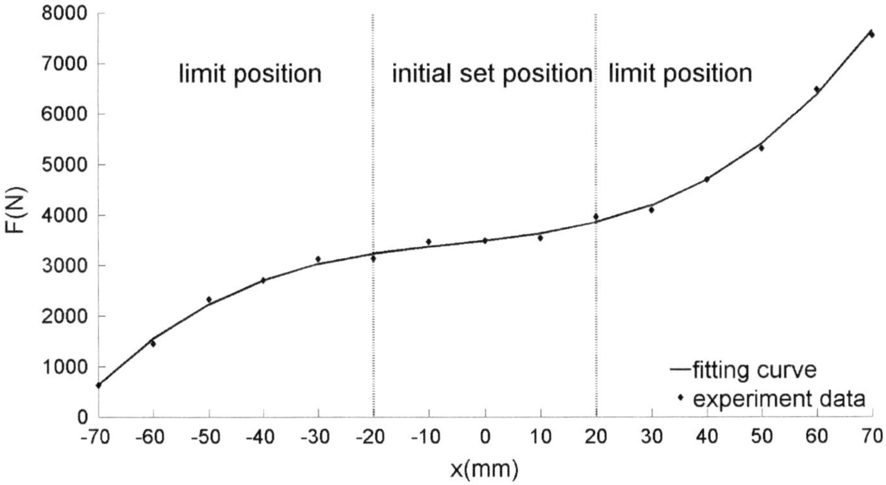

Fig. 4 Load-displacement curve

2.2.2 The Damping Characteristic of Air Spring

The damping characteristic of diaphragm-type air spring is equivalent to the viscous damping, so the velocity-load curve of the front and rear air spring suspension was tested by adopting WDTS-IV electric dynamometer under normal temperature condition. Its damping curve can be fitted with formula (5) according to the linear viscous damping model.

$$F = c \cdot v \tag{5}$$

where F is the damping force, c is the viscous damping, and v is the testing speed.

The air spring damping can be divided into compression damping and restoring damping (as shown in Fig. 5), and most of the energy is consumed on restoring stroke, so the restoring damping is much larger than the compression one. As the centroid of the cab is more closer to the front suspension whose air spring has much lager vertical stroke, the air spring's damping from the front suspension is much larger than the rear one. With comprehensive consideration of the vibration effect, the restoring damping was chosen as the simulation input value. Thus, the front air spring's damping value is 8051 N s/m, and the rear one is 4959 N s/m, respectively.

2.2.3 The Stiffness Characteristic of Bush

The main role of bushes in the cab suspension is increasing the elastic degree of freedom and supporting the cab, and due to the small volume, the damping value of bush is so small that can even be ignored. Figure 6 shows the location of bear bushes, some of the bushes have holes in the direction without extra load, so the stiffness may be much smaller.

The stiffness of cab suspension's various bushes was tested by using DJW dynamic and static stiffness of rubber fatigue testing machine under normal temperature. The result of bush stiffness test is shown in Table 1. The bush from the front suspension has much larger stiffness than the rear one with consideration of emergency braking or other exceptional case.

2.3 Comparison Between the Results of Simulation and Experiment

The same road spectrum of rub garment board (as shown in Fig. 7) was put into the corresponding channel of the test bench and the dynamic simulation model at the same time, and the displacement of the corresponding position (Fig. 2) was exported. Table 2 shows the contrast results of experiment and simulation.

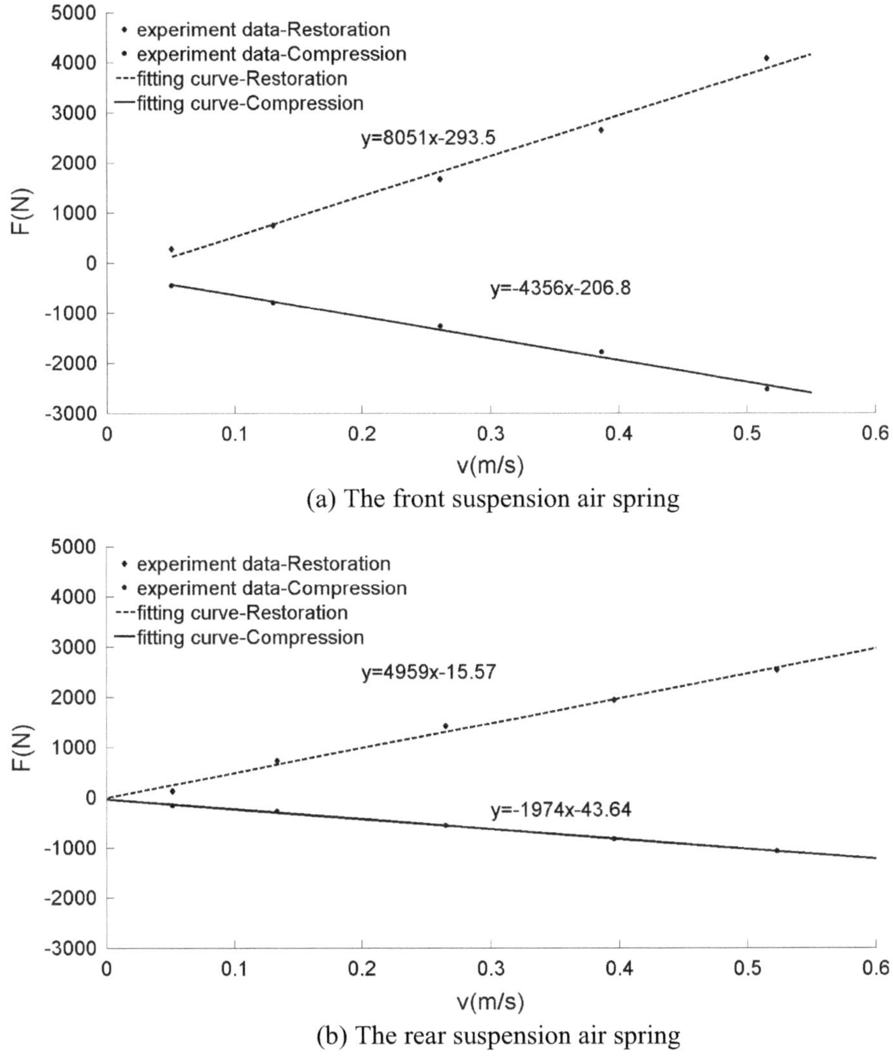

Fig. 5 Velocity-load curve

From the table, it can be seen that the max error of the simulation result is no more than 13%, and the min error is less than 4%, which meets the engineering requirement. So, the dynamic model has certain reliability.

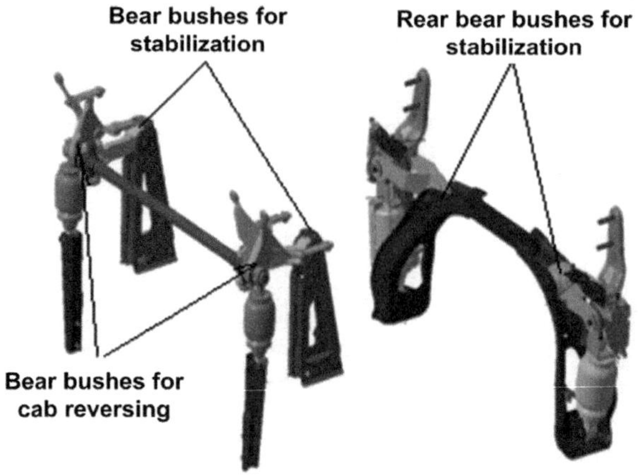

Fig. 6 Location of bear bushes

Table 1 Stiffness of bushes

Bush	Stiffness (kN/m)
Bear bushes for cab reversing (radial)	23.2
Rear bear bushes for stabilization (radial empty)	0.25
Rear bear bushes for stabilization (radial entity)	1.67
Bear bushes for cab reversing (radial entity)	44.67
Bear bushes for cab reversing (radial empty)	1.86
Bear bushes for cab reversing (axial)	0.59
Bear bushes for cab reversing (reverse)	71.64

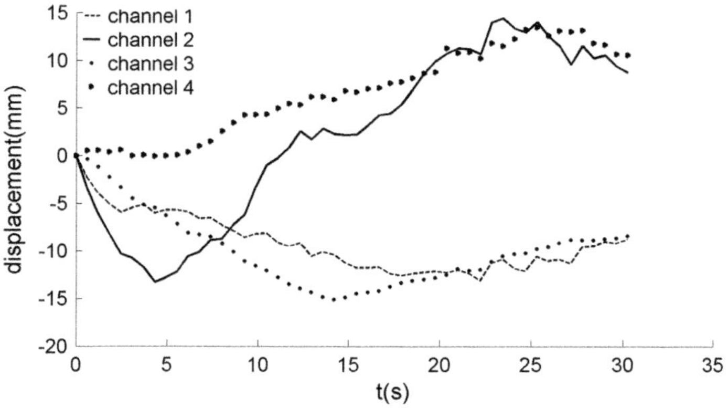

Fig. 7 Road spectrum of rub garment boards

Table 2 Experimental and simulation data

Point number	Experimental vibration amplitude (mm)	Simulation vibration amplitude (mm)	Error (%)
A_L	29.0	27.9	−3.8
A_R	15.0	13.1	−12.6
B_L	29.3	27.8	−5.1
B_R	14.6	12.8	−12.3
C_L	25.0	22.6	−9.6
C_R	11.3	10.2	−9.7
D_L	16.3	15.1	−7.4
D_R	14.5	13.4	−7.5
E_L	16.8	17.9	6.5
E_R	11.9	12.4	4.2
F_L	17.8	16.6	−6.7
F_R	13.0	12.3	−5.3

2.4 *Optimization of Cab Suspension Vibration Isolation Performance*

In the cab suspension dynamic model, the cabin seat vertical acceleration power spectrum density function was set as the optimization goal, and the air spring's linear term stiffness value and the damping values were set as optimization variables, and the pitch angle and rolling angle of the cab were set as the constraints, and adopted the generalized reduced gradient method to optimize the system parameters. Results show that the stiffness value contributes much less to goal compared to the damping value, and the rear damping has the highest sensitivity. With comprehensive consideration, the linear term stiffness value of the optimized air spring is set as 18.8 N/mm, the damping values of the front and rear suspension are 8214 and 3879 N s/m, respectively.

3 Cab Suspension Static Performance

Once the spatial structure and elastic element's dynamic performance parameters are set, the dynamic behavior of the cab suspension system is then determined. However, that is only the beginning of the R&D process, as the static performance (such as static strengthen and static stiffness) also plays a very important role in the design of cab suspension.

3.1 Static Strength Test with Full-Loaded Cab

The equivalent cab was installed on the cab suspension, and several bags of sand (a total of 300 kg) were put to the specified location to simulated the working condition when the cab is full-loaded. The equivalent cab was slowly lifted along the vertical direction, until there were no external forces affecting on the cab suspension. According to the engineering experience, the strain gauges were attached to the smooth surface where there may have more possibility to generate stress concentration (as shown in Fig. 8). The lateral stabilizer bar (Point O) was attached with a unidirectional strain gauge, which was 45° to the axis. The turnover supports of the cab (Point A_L and A_R) suffer forces from all directions, so the three-direction strain rosette was attached to the bottom plan. The other positions were attached with unidirectional strain gauges, and they were all laid out along the vertical direction. All the strain gauges remained stationary for 24 h before the strain gauge was demarcated and reset. Then the equivalent cab was landed as slowly as possible to make sure that the cab suspension suffered steady force, at the same time the strain values were recorded until the cab landed completely. The test was repeated three times, and the average values were taken as the final test results.

3.2 Static Strength Simulation with Full-Loaded Cab

HYPERMESH was used to mesh the cab suspension. Sheet metal parts were meshed as shell elements, and complex casts parts were replaced with tetrahedron solid element. The bushes were simulated with CBUSH element and the air spring with CELASI element to characterize the linear stiffness and damping. Bolted connection and welding were substituted with RBE2 element. All the parts in the cab suspension were set with proper element sizes. The cab suspension finite element model is shown as Fig. 9.

The lower ends of cab suspension were fixed and the upper ends which connected to the cab suffered vertical forces and corresponding moments derived from the dynamic model. The finite element model was submitted to the solver after all were set up.

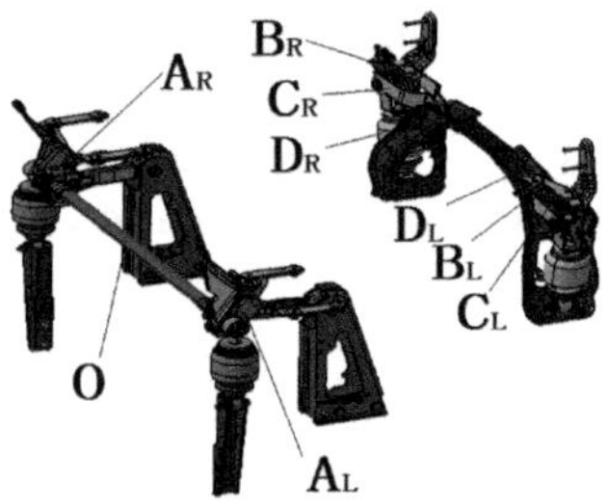

Fig. 8 Location of the strain gauges

Fig. 9 Finite element model of cab suspension

3.3 *Comparison Between the Results of Simulation and Test*

The stress values in the static strength test of the relevant parts could be calculated by the strain values and corresponding modulus of elasticity. Table 3 shows the results of simulation and experiment.

From the table, it can be seen that the error of simulation value and testing value is below 10%, and it has conformity with the error direction which shows that finite element model is effective and meets the need of engineering analysis.

Table 3 Stress value

Point number	Simulation value (MPa)	Experimental value (MPa)	Error (%)
O	17.4	16.2	7.4
G_L	186.3	173.6	7.3
G_R	159.4	151.6	5.1
H_L	12.6	11.6	8.6
H_R	11.3	10.8	4.6
I_L	12.4	12.1	2.5
I_R	10.8	10.2	5.9
J_L	53.8	49.8	8.0
J_R	45.4	43.6	4.1

3.4 Modal Analysis of Cab Suspension

The inherent characteristics of the cab suspension system can be obtained by modal analysis. On the basis of finite element model, the cab suspension system's free mode was calculated (as shown in Table 4). As the first-order bending frequency of car body is about 5–6 Hz, and the resonance frequency of unsprung weight is about 8–15 Hz, each order's frequency of the cab suspension is far away from the above scope. Therefore, the vibration of the cab suspension system could be reduced to a proper level which satisfies the driver.

3.5 Analysis of the Cab Suspension's Limiting Working Condition

It may always be confronted with all kinds of limiting conditions such as severe traffic fluctuation turbulence, emergency braking, and rapid turn in the course of vehicle running. And that is usually quite difficult to repeat with a physical prototype. However, the finite element simulation analysis is a perfect alternative at this time. In the finite element model, the 3.5 g vertical impact working condition (3.5 times of vertical gravity applied at the cab center-of-mass) was adopted to simulate passing the rough roads with large pit holes, 1 g rolling working condition (1 time of gravity applied at cab center-of-mass toward the lateral side) to simulate sharp turn, and 1 g pitching working condition (1 time of gravity applied at the cab center-off-mass toward the driving direction) to simulate the emergency brake (the system stiffness was not under consideration because cab suspension system has a large amount of elastic parts such as air spring, which has strong nonlinearity). On 3.5 g vertical impact working condition, the local stress of the cab's turnover support was larger than expected (as shown in Fig. 10a), and this can be solved by local-thickening, adding some stiffeners and so on. On 1 g braking working condition, excessive braking force results in local stress concentration of the connecting plate of the rear suspension (as shown in Fig. 10b), which can be dealt with adding stiffeners, changing material, adjusting bolts' pretension, and so on. On 1 g rolling working condition, cab suspension system's stabilizing bearing has local stress concentration (as shown in Fig. 10c) because of large rolling moment, which can be handled with enlarging the fillets, thickening the sheet metal, and so on.

Table 4 Mode of cab suspension

Order	Frequency (Hz)	Vibration mode
1	1.06	Pitch
2	1.37	Torsion
3	1.86	Vertical translation
4	2.21	Roll

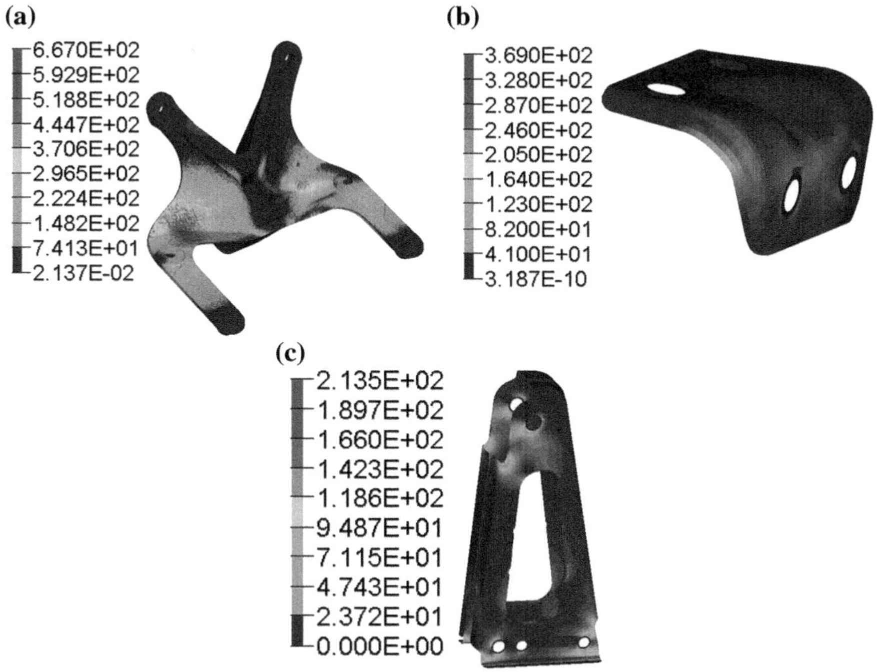

Fig. 10 Stress concentration parts under different working conditions

Though some parts of the cab suspension exist yield under limiting working condition, the stress concentration can be solved by local minor adjustment, and these small changes have little influence on dynamic characteristic of the cab suspension system.

4 Fatigue Performance of Cab Suspension

According to the statistics, the fatigue failure accounts for more than 80% among the failure of most mechanical parts. That means, the cab suspension system satisfying the requirement of static limiting strength design does not guarantee its reliability due to its large amounts of mechanical parts, and it usually needs to be tested about the fatigue resistance performance.

The cab suspension in this research contains a large number of elastic elements, which makes it difficult to conduct the fatigue simulation because of the strong nonlinearity in this system. Therefore, a cab suspension fatigue testing machine was developed to study its fatigue characteristics.

4.1 Cab Fatigue Testing Machine

To explore the vertical, longitudinal, and lateral fatigue resistance properties of cab suspension, the fatigue testing machine for cab suspension system was developed, as shown in Fig. 11.

It mainly consists of hydraulic vibration head, gantry, equivalent cab, base, etc. The equivalent cab has the same mass, center-of-mass coordinates, and the moment of inertia as the original one. It is welded by square tubes and steel plates, with high stiffness, which makes sure that the deformation is small enough. The equivalent cab has three card slots along the vertical, longitudinal, and lateral direction respectively at its centroid position. The hydraulic vibration head can connect to one of the three card slots to test the cab suspension fatigue properties of the relevant direction. With a closed-loop force control system, the hydraulic vibration head can apply stable vibration forces. The hydraulic vibration head is connected to the gantry with joint bearings, which can release the rotational degrees of freedom, avoiding over-constraints of the vibration head. The gantry linked to the vibration head is composed of I-beams and square tubes with high stiffness, and the maximum deformation of working vibration head is no more than 0.1 mm when the exciting force is more than 20,000 N.

4.2 The Cab Suspension Fatigue Test

The vibration head was connected to the vertical card slot (as shown in Fig. 12a), the longitudinal card slot (as shown in Fig. 12b), and the lateral card slot (as shown in Fig. 12c) of the cab respectively to investigate the fatigue properties in the relevant direction.

Fig. 11 Cab suspension fatigue testing equipment

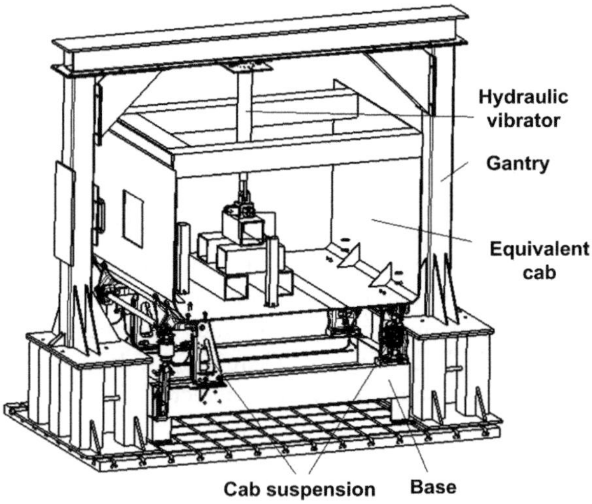

Fig. 12 Fatigue test

In the vertical test, the input force was a sine wave with an amplitude twice as the weight of cab. The balance value of force wave was 0, and the vibration frequency was 1 Hz. The vibration lasted for 50,000 cycles (about 14 h without pause). In the longitudinal and lateral tests, all the input conditions were the same as in the vertical test, except that the amplitude was 0.5 times as the cab weight. As the stress level is relatively low, the mechanical parts in the cab suspension system may encounter high-cycle fatigue problems. The fatigue assessment criteria of the cab suspension are as follows: (1) There are no cracks on the mechanical parts of the system even under the microscope; (2) there are no excessive wear or cracks on the elastic parts; (3) there are no functional damages on the cab turnover structures and locking devices, etc. If there are any cracks on the mechanical parts, the structure could be modified or reinforced on the positions where are likely to appear stress concentration; if there is excessive wear or crack on the elastic parts, they could be dealt with changing the formulation of compound without obviously changing the stiffness characteristics to enhance the wear performance; if the cab turn over structures or locking devices are functionally damaged, they can be solved by either improving the wear resistance or adjusting the stress state of the system.

5 Analysis and Discussion

Aiming at developing the cab suspension, a method that combines simulation design and workbench test has been proposed, which can find the defects of design and correct them in the early stage. The dynamic model of air spring cab suspension has been built and validated by corresponding workbench test. Since there are plenty of elastic components in the cab suspension system, the matching optimization of which can be conducted by a dynamic model, to determine the optimal vibration characteristics of the system. If the corresponding channels of the cab suspension were inputted with complex spectrum obtained from the actual measurement, the status under different working conditions can be simulated, which may be helpful to the dynamic optimization. Although the dynamic model with simplified parameters has high credibility on some working conditions, the nonlinear characteristic still needs to be considered if focusing on the nonlinearity of the system, and that is the focus of future work. The finite element model of the cab suspension with the optimized dynamic boundary conditions has been built and validated by full-load static strength test, and with the help of FE model, the modes of the cab suspension were analyzed, and possibility of resonance between sprung and unsprung components was eliminated. Meanwhile, the FE model was also used to predict the stress distribution of cab suspension under limiting conditions, with which help the structures could be strengthened in time. While the cab suspension has met the requirement of static strength (without considering the static stiffness), it is still necessary to verify the fatigue resistance properties of the system. To this end, the triaxial fatigue testing machine was developed, and the vertical, longitudinal, and lateral fatigue properties of the cab suspension were tested. Then corresponding solutions could be found to solve the problems appeared in the test. And this procedure may be repeated if necessary to make sure that the system works well. Similarly, the method can be applied to other design of systems or vehicles; it is of certain engineering application value.

6 Conclusions

(1) In this paper, dynamic model and finite element model were built and validated by corresponding workbench test. The results of simulation and test are consistent with each other. The simulation results can meet the requirement of engineering analysis.
(2) Based on the dynamics model, the dynamic properties of cab suspension system could be optimized. With the FE model, the stress state of cab suspension under limiting conditions could be predicted. And the three-dimensional fatigue characteristics of cab suspension were tested using the fatigue testing machine. The results show that this method can greatly shorten the design cycle and reduce the design cost.

(3) The proposed design method can also be applied to other vehicle systems or other full vehicles.

Acknowledgements Thanks to Jiangyin Shengshijie Machinery Manufacture Co., Ltd for applying the test field.

Thanks to JAC MOTORS for technical support of cab suspension fatigue test.

References

1. Wang K (2008) The study of dynamic simulation optimization and test of commercial vehicle cab suspension system. Jiling University, Changchun
2. Soliman A, Gazaly N, Kadry F (2014) Parameters affecting truck ride comfort. SAE Tech Pap. https://doi.org/10.4271/2014-01-0147
3. Lee G, Kim J, Kim G (1996) Improving ride quality on the cab suspension of a heavy duty truck. SAE Tech Pap. https://doi.org/10.4271/962151
4. Pang H, Li HY, Fang ZD et al (2010) Experimental study on effects of cab suspension system on heavy vehicle ride comfort. Automobile Technol 11:52–56
5. Huh K, Choi J, Yoo H (2003) Development of a multi-body dynamics simulation tool for tracked vehicles. JSME Int J Ser C Mech Syst Mach Elem Manufact 46(2):550–556
6. Sim K, Lee H, Yoon JW et al (2016) Effectiveness evaluation of hydro-pneumatic and semi-active cab suspension for the improvement of ride comfort of agricultural tractors. J Terramech 69:23–32
7. He L, Zhu SH, Zhu HL (2012) Modeling and simulation of full-float tractor cab suspension system based on ADAMS. Appl Mech Mater 1523(141):364–369
8. Wang KY, Shi WK, Yang CH et al (2010) Commercial vehicle cab suspension system vibration modes and transmission characteristics by means of ADAMS. J Jilin Univ Eng Technol Ed 40(2):330–334
9. Chen L, Ding WP, Yang ML et al (2013) The vibration transmission rate matching research of cab mounting system. Mach Des Manufac 11:76–81
10. Yan X, Jiang D, Wang ZX et al (2014) Vibration isolation performance research of commercial cab suspension system based on road load data. Lect Notes Electr Eng 328:425–432
11. He L, Zhou SH, Zhu HL (2011) Modeling and simulation of full-float cab suspension system based on ADAMS. Appl Mech Mater 141:364–369
12. Rajan V, Deshmukh C (2013) Cab suspension optimization using Matlab. SAE Tech Pap. https://doi.org/10.4271/2013-26-0147
13. Dai C, Liang JH (2012) Experimental research on a certain light truck cab suspension system effect on ride comfort. Appl Mech Mater 1500(128):1460–1463
14. Ye FH, Xu K, Zhang YP et al (2010) Development of full-floating suspension system for a commercial vehicle cab. Automobile Technol 4:33–39

Optimization Analysis on Full Vehicle Road Noise Based on Substructure Frequency Response Function

Yan Zhang, Jun Song and Yanghui Xu

1 Introduction

With the improvement of people's living standards, there are more stringent requirements for the vehicle noise, vibration, and harshness (NVH) performance. The NVH performance is mainly evaluated based on road noise. Driving road is considered as the main excitation source. The excitation energy transfers through chassis and vibration isolations input to the body eventually, generating noise, which affect the passengers comfort. Therefore, reducing the road noise is the effective way to make the improvement of NVH performance [1].

Many scholars studied the road noise, but they did research just on the body attachments, subframe, suspension, tire, and other structures of the dynamic characteristics match. Because the number of vehicle parts is particularly large, the workload of study is also very large. And the standard optimization with manual tuning parameters is tentative. It does not from the perspective of numerical analysis to solve the actual engineering problem. Kim et al. [2] suggested that the dynamic stiffness of the body attachment points on the vehicle road noise have an important impact and demonstrated by test data. Park et al. [3] investigated the links between NVH of a car such as booming, road noise, and dynamic characteristics of the front subframe system. Douville et al. [4] presented a methodology for the analysis of the noise transfer paths for an automotive suspension assembly. Kido I et al. [5] thought that tire is a key to good performance of road noise, and the coupled vibration analysis between suspension and tire is executed to understand and to improve their vibration.

In this paper, the 40–200 Hz road noise of a car is optimized based on the substructure frequency response function (FRF) method, so that all the road noise

Y. Zhang (✉) · J. Song · Y. Xu
Automotive Engineering Institute, Guangzhou Automobile
Group Co., LTD, Guangzhou, China
e-mail: zhangyan1@gaei.cn

Society of Automotive Engineers of China (SAE-China) (ed.), *Proceedings of the 19th Asia Pacific Automotive Engineering Conference & SAE-China Congress 2017: Selected Papers*, Lecture Notes in Electrical Engineering 486, https://doi.org/10.1007/978-981-10-8506-2_47

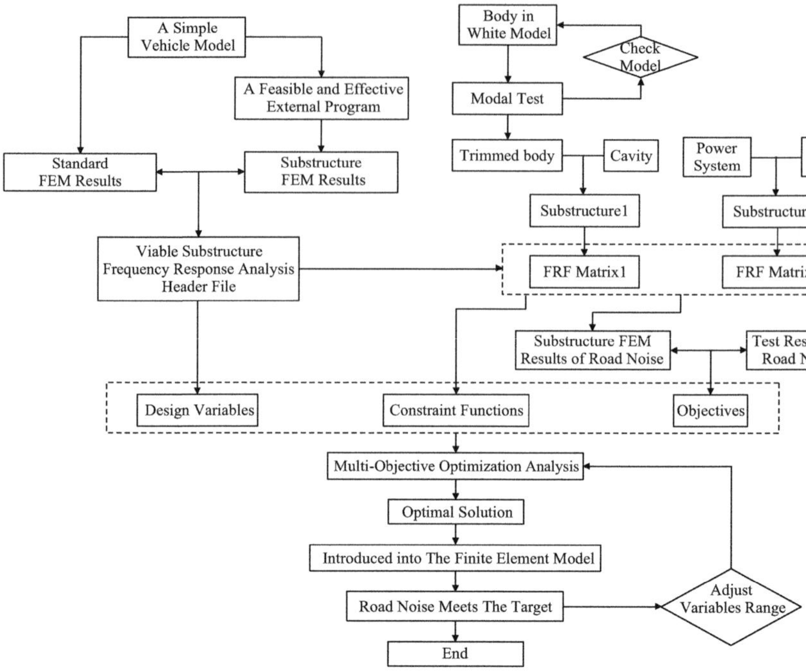

Fig. 1 Research route

peaks in the frequency band meet the target. And the research route, as shown in Fig. 1. Firstly, an efficient method for frequency response is formulated with a simple vehicle model. And then, the full vehicle finite element model of an actual car is established and analyzed by the substructure FRF method for its road noise. It is worth mentioning that the substructure FRF method increases the computational efficiency, because the method establishes the numerical relationship between input parameters and output results, and it simplifies the solving process of bulky finite element matrix. Finally, the multi-objective optimization model is established, and the optimal Latin hypercube method is used to optimize the road noise.

2 Theory

To reduce the simulation optimization time, many scholars carried out a lot of research. Gholizadeh et al. [6] proposed combination genetic algorithm with neural networks to find the optimal solution subject to multiple natural frequencies constraints, and the natural frequencies are evaluated by properly neural networks. But construction of neural network needs to take some time to simple.

Comparison with the neural network, substructure FRF is more efficient, because it establishes the numerical relationship between input variable and output variable in each substructure. The numerical relationship is only used for the calculation of dynamic analysis, which represents the frequency response characteristics of the structure. The frequency response characteristic is only related to its own structure. For a complex system, FRF is used to characterize the response characteristics of structures with simple function relations instead of complex finite element matrix reduction operations. The FRF only needs one calculation to obtain the FRF of each substructure, which is equivalent to a set of samples in the construction of the neural network. It shortens the computational time of the dynamics more significantly and becomes more convenient.

2.1 Substructure FRF

A whole structure S consists of substructures M and N, as shown in Fig. 2. The substructure M is a system with $(a + c)$ degrees of freedom (DOFs), and the substructure N is a system with $(b + c)$ DOFs. There are c DOFs [7, 8] between M and N interface.

Steady-state frequency response equation is:

$$\left(-\omega^2[M] + i\omega[B] + [K]\right)\{X\} = [Z]\{X\} = \{F\} \tag{1}$$

where ω is excitation frequency; $[M]$ is the mass matrix; $[B]$ is the damping matrix; $[K]$ is the stiffness matrix; $\{X\}$ is the displacement response; $[Z]$ is the dynamic stiffness matrix; $\{F\}$ is the external excitation. Solve Eq. (1):

$$\{X\} = [Z]^{-1}\{F\} = [H]\{F\} \tag{2}$$

where $[H]$ is the mobility matrix, as well as FRF matrix.

The equation of motion for each substructure in the frequency domain for the complex displacement is:

$$\{X\}_n = [H]_{nn}\{F\}_n \tag{3}$$

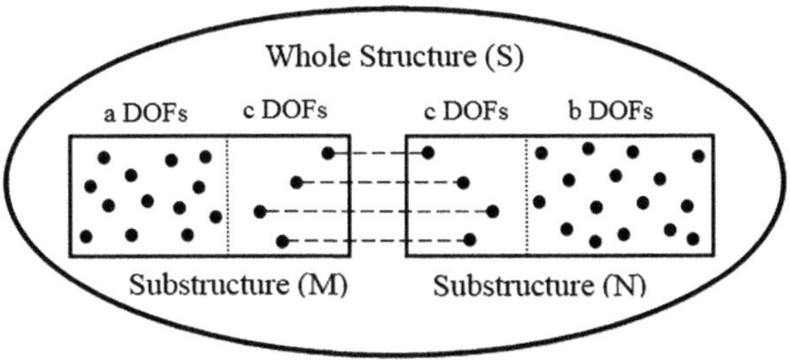

Fig. 2 Description of substructure

$[H]_{nn}$, the FRF of structure system, is a function of excitation frequency ω. The element H_{ij} in matrix $[H]_{nn}$ means the displacement response of the ith DOF when the unit excitation is loaded at the jth DOF.

Subsystem M is partitioned as:

$$\begin{Bmatrix} \{X^M\}_a \\ \{X^M\}_c \end{Bmatrix}_n = \begin{bmatrix} [H^M]_{aa} & [H^M]_{ac} \\ [H^M]_{ca} & [H^M]_{cc} \end{bmatrix}_{nn} \begin{Bmatrix} \{F^M\}_a \\ \{F^M\}_c \end{Bmatrix}_n \tag{4}$$

Subsystem N is partitioned as:

$$\begin{Bmatrix} \{X^N\}_b \\ \{X^N\}_c \end{Bmatrix}_n = \begin{bmatrix} [H^N]_{bb} & [H^N]_{bc} \\ [H^N]_{cb} & [H^N]_{cc} \end{bmatrix}_{nn} \begin{Bmatrix} \{F^N\}_b \\ \{F^N\}_c \end{Bmatrix}_n \tag{5}$$

Subsystems M and N are rigidly connected at c DOFs. It requires:

$$\{X^M\}_c = \{X^N\}_c = \{X^S\}_c \tag{6}$$

$$\{F^M\}_c = \{F^M\}_c = \{F^M\}_c \tag{7}$$

Equations (4) and (5) are expanded, and the displacement response of the c DOFs is obtained:

$$\{X^M\}_c = [H^M]_{ca}\{F^M\}_a + [H^M]_{cc}\{F^M\}_c \tag{8}$$

$$\{X^N\}_c = [H^N]_{cb}\{F^N\}_b + [H^N]_{cc}\{F^N\}_c \tag{9}$$

Assuming $\{\hat{F}^M\}_c$ and $\{\hat{F}^N\}_c$ are the transfer forces at the c DOFs, plug them into Eqs. (8) and (9):

$$\{X^M\}_c = [H^M]_{ca}\{F^M\}_a + [H^M]_{cc}\{\hat{F}^M\}_c \tag{10}$$

$$\{X^N\}_c = [H^N]_{cb}\{F^N\}_b + [H^N]_{cc}\{\hat{F}^N\}_c \tag{11}$$

Plug Eqs. (10) and (11) into Eq. (6), and the following equation is obtained:

$$[H^M]_{cc}\{\hat{F}^M\}_c = [H^N]_{cb}\{F^N\}_b + [H^N]_{cc}\{\hat{F}^N\}_c - [H^M]_{ca}\{F^M\}_a \tag{12}$$

The balance force at the c DOFs in the coupled structure S is associated with $\{\hat{F}^M\}_c$ and $\{\hat{F}^N\}_c$:

$$\{F^S\}_c = \{\hat{F}^M\}_c + \{\hat{F}^N\}_c \tag{13}$$

It is described as:

$$\{\hat{F}^M\}_c = \{F^S\}_c - \{\hat{F}^N\}_c \tag{14}$$

$$\{\hat{F}^N\}_c = \{F^S\}_c - \{\hat{F}^M\}_c \tag{15}$$

Plug Eqs. (14) and (15) into Eq. (12), then the vibration transfer force is obtained:

$$\{\hat{F}^M\}_c = \left[[H^M]_{cc} + [H^N]_{cc}\right]^{-1}\left\{[H^N]_{cb}\{F^N\}_b + [H^N]_{cc}\{F^S\}_c - [H^M]_{ca}\{F^M\}_a\right\} \tag{16}$$

$$\{\hat{F}^N\}_c = \left[[H^M]_{cc} + [H^N]_{cc}\right]^{-1}\left\{[H^M]_{ca}\{F^M\}_a + [H^M]_{cc}\{F^S\}_c - [H^N]_{cb}\{F^M\}_b\right\} \tag{17}$$

Moreover, the FRFs of the system S are represented as:

$$\begin{Bmatrix} \{X^S\}_a \\ \{X^S\}_c \\ \{X^S\}_b \end{Bmatrix}_n = \begin{bmatrix} [H^S]_{aa} & [H^S]_{ac} & [H^S]_{ab} \\ [H^S]_{ca} & [H^S]_{cc} & [H^S]_{cb} \\ [H^S]_{ba} & [H^S]_{bc} & [H^S]_{bb} \end{bmatrix}_{nn} \begin{Bmatrix} \{F^S\}_a \\ \{F^S\}_c \\ \{F^S\}_b \end{Bmatrix}_n \tag{18}$$

Taking an example of $[H^S]_{aa}$ describes the derivation process of the coupled FRF equation.

Expand the first row of matrix Eq. (18):

$$\{X^S\}_a = [H^S]_{aa}\{F^S\}_a + [H^S]_{ac}\{F^S\}_c + [H^S]_{ab}\{F^S\}_b \tag{19}$$

Recall from Eq. (4) for the uncoupled subsystem M:

$$\{X^M\}_a = [H^M]_{aa}\{F^M\}_a + [H^M]_{ac}\{F^M\}_c \tag{20}$$

For the coupled system S, Eq. (20) becomes:

$$\{X^S\}_a = [H^M]_{aa}\{F^S\}_a + [H^M]_{ac}\{\hat{F}^M\}_c \tag{21}$$

Through building the Eq. (19) and (21):

$$[H^S]_{aa}\{F^S\}_a = [H^M]_{aa}\{F^S\}_a + [H^M]_{ac}\{\hat{F}^M\}_c - [H^S]_{ac}\{F^S\}_c - [H^S]_{ab}\{F^S\}_b \tag{22}$$

Substitute Eq. (16) into Eq. (22). $[H^S]_{aa}$ is just related to the DOFs, so $\{F^S\}_b$, $\{F^S\}_c$, and $\{F^N\}_b$ are all 0. The item $[H^S]_{aa}$ in the FRF of whole structure S is simplified as:

$$[H^S]_{aa} = [H^M]_{aa} - [H^M]_{ac}\left[[H^M]_{cc} + [H^N]_{cc}\right]^{-1}[H^M]_{ca} \tag{23}$$

Referring to the solution process of $[H^S]_{aa}$, other items in the FRF of whole structure S are obtained:

$$[H^S]_{ac} = [H^M]_{ac}\left[[H^M]_{cc} + [H^N]_{ac}\right]^{-1}[H^N]_{cc} \tag{24}$$

$$[H^S]_{ab} = [H^M]_{ac}\left[[H^M]_{cc} + [H^N]_{cc}\right]^{-1}[H^N]_{cb} \tag{25}$$

$$[H^S]_{ca} = [H^M]_{cc}\left[[H^M]_{cc} + [H^N]_{cc}\right]^{-1}[H^N]_{ca} \tag{26}$$

$$[H^S]_{cc} = [H^M]_{cc}\left[[H^M]_{cc} + [H^N]_{cc}\right]^{-1}[H^N]_{cc} \tag{27}$$

$$[H^S]_{cb} = [H^M]_{cc}\left[[H^M]_{cc} + [H^N]_{cc}\right]^{-1}[H^N]_{cb} \tag{28}$$

$$[H^S]_{ba} = [H^M]_{bc}\left[[H^M]_{cc} + [H^N]_{cc}\right]^{-1}[H^N]_{ca} \tag{29}$$

$$[H^S]_{bc} = [H^M]_{bc}\left[[H^M]_{cc} + [H^N]_{cc}\right]^{-1}[H^N]_{cc} \tag{30}$$

$$[H^S]_{bb} = [H^M]_{bb} - [H^M]_{bc}\left[[H^M]_{cc} + [H^N]_{cc}\right]^{-1}[H^M]_{cb} \tag{31}$$

By now, the entire FRF matrix of the whole structure S is calculated according to the FRF matrixes of substructure M and N. The displacement response of each DOF is obtained according to Eq. (18), and its derivative is velocity response.

2.2 Optimal Latin Hypercube Design

In the optimization, the vast majority cases of the optimization objectives are multiple, and it is necessary to coordinate and compromise between the various objectives, so that each objective meets the target value as much as possible. For example, the road noise in the frequency domain is also a multi-objective optimization: for example, frequency F_1, frequency F_2 and frequency F_n corresponding noise peaks need to meet the design target at the same time. When using the common single-objective optimization, the optimization of the noise at frequency F_1 may cause deterioration of the frequency F_2 or other frequency peaks, which is accompanied by complex uncertainties that cause some trouble in engineering. It is necessary to use multi-objective to constrain and optimize all peaks in interested frequency range to meet the target.

The optimal Latin hypercube design is used for optimization analysis, which provides a reasonable and effective way to obtain variable information data. It can be used to identify key parameters, build empirical formulas and get the best design. Latin hypercube design principle: each dimension coordinate interval $\left[x_k^{\min}, x_k^{\max}\right]$, $k \in [1, n]$ is evenly distributed in m intervals and n-dimensional space. Each interval is denoted by $\left[x_k^{i-1}, x_k^{i}\right]$, $i \in [1, m]$. Constructing the n-dimensional space, the number of samples of m Latin hypercube design calls $m \times n$ Latin hypercube design. Its advantages are manifested in spatial filling and non-linear response: a 2 factors, 9 levels of research, full factorial design requires 81 points, while the Latin hypercube design requires only 9 points. In addition, the optimal Latin hypercube design [9] improves the uniformity of the random Latin hypercube design, and the factors and responses are more equalization in numerical space. An optimal Latin hypercube sampling with 2 factors, 9 levels, and 9 test design points, as shown in Fig. 3. For the same number of design variables, the number of samples of the optimal Latin hypercube method is smaller, the spatial distribution is more uniform, and the iterative solution is shorter.

3 Simple Vehicle Model

In the early stage of research on substructure FRF method, considering the convenience of model building and the quickness of solving, a simple vehicle model is studied. The simple model similar to the actual car structure is established, and the standard finite element method is used to analyze the road noise. At the same time, the substructure analysis external program is written tentatively, which contains

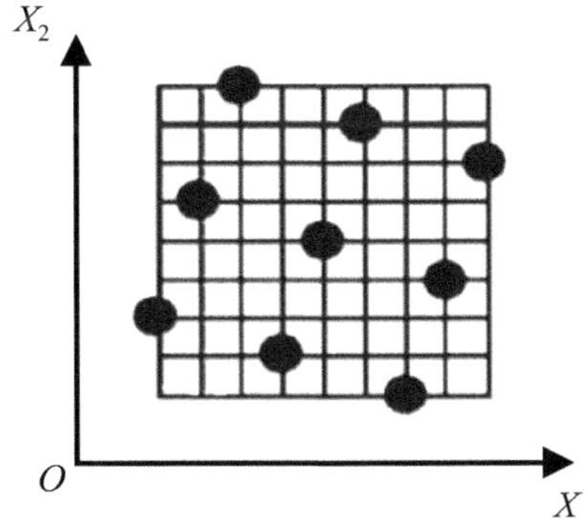

Fig. 3 An optimal Latin hypercube sampling

connecting parameters between substructures. And then, the road noise is analyzed by substructure method, repeated comparison with the standard finite element method results, and debugging the related parameters, until an effective external program used for substructure FRF analysis is obtained.

3.1 Standard Method of Frequency Response Analysis

A 3D vehicle model is built and simplified into four parts: body, acoustic cavity, power train, and chassis. The body structure is built with shell elements CQUAD4. The acoustic cavity model is built with solid element CHEXA. The chassis model is built with CBEAM elements. Taking the characteristics of mounting structure into consideration, the power train model is built with concentrated mass element CONM2, dummy element PLOTEL, and elastic element CBUSH. The body and chassis are connected through four simplified CBUSH elements. A simplified vehicle finite element model is shown in Fig. 4. Point *R* is the response point of the road noise.

The unit sinusoidal force is excited at the center of each tire in the *Z* direction, the frequency sweeps from 40 to 200 Hz, and the increment is 1 Hz. The noise characteristics are studied by standard method and substructure method, respectively. The road excitation is transferred to body though four tires and isolators, and the radiation noise is generated by vibrations of body structure.

3.2 Substructure Method of Frequency Response Analysis

As shown in Fig. 5, the vehicle model is simplified to two substructures. The substructure 1 is composed of body structure and acoustic cavity, while substructure 2 is composed of power train and chassis. Substructures 1 and 2 are connected

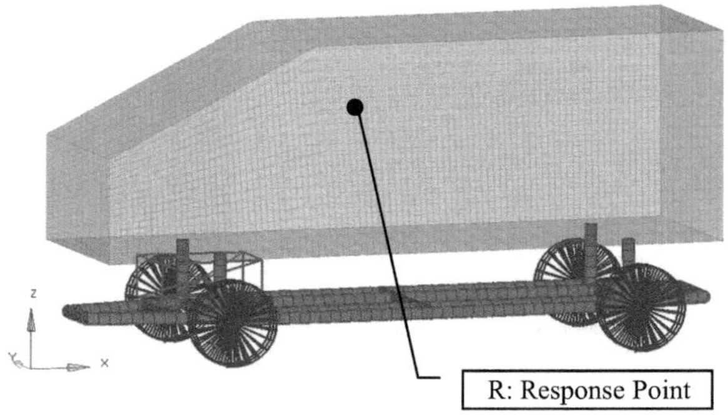

Fig. 4 Simple vehicle model for standard method

Fig. 5 Simple vehicle model for substructure method

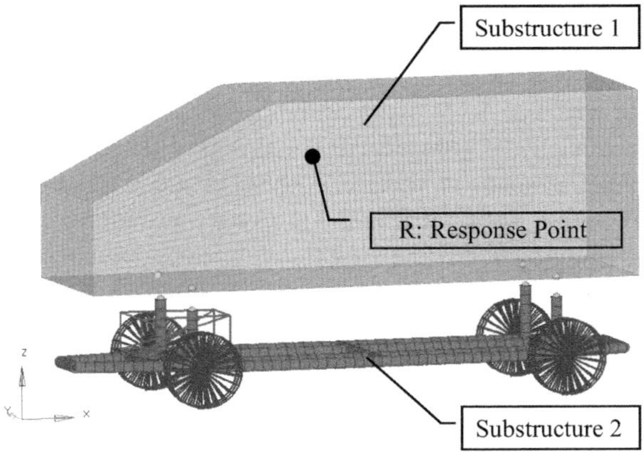

through the multi-dimensional springs with stiffness and damping parameters. The entire FRF matrix of the coupled structure is obtained in accordance with Eqs. (23) to (31). Then, the noise response of this model is analyzed by the substructure method.

With the acoustic characteristic at point *R* as an example, a comparison between the analysis results of two methods is shown in Fig. 6. The acoustic response curve based on the substructure method absolutely retains the features of the curve based on the standard method. Their frequency amplitude almost no deviation and the consistency of the results further verifies the accuracy of substructure method applied to the complex structure model and noise analysis. As a result, the substructure method can be used in automotive engineering perfectly.

Fig. 6 Acoustic characteristic at point *R*

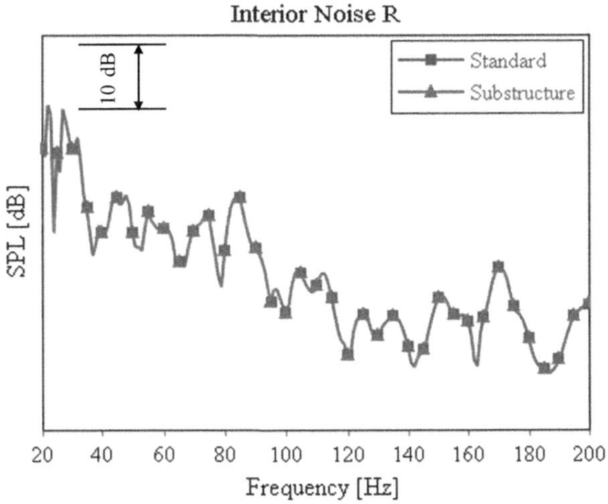

4 Full Vehicle Model

In terms of NVH analysis, body structure is a significant transfer path. To a great extent, the response of body structure determines the road noise level [10]. The finite element model for body in white is set up, as shown in Fig. 7a. The free modal is calculated, and the results are compared with test. Through repeatedly examining and debugging the finite element model, the deviation of simulation and test results should be controlled within 5%, and then the final model of body in white is confirmed usable for full vehicle model [11]. Figure 7b is modal test of the body in white. A comparison between the simulation and test results is shown in Table 1.

In Table 1, it shows that the free modal results of finite element and test are very consistent. The deviations of the first torsion mode and second bending mode are very small, are −3.70 and −3.68%, which in the acceptable range of 5%, and the global torsion and bending modal is significant to the interior noise in the NVH analysis [12]. Therefore, the finite element model of body in white is finalized, and it can exactly reflect the vibration characteristic of the actual vehicle. The free modal shape of body in white is shown in Fig. 8.

Based on above accurate model, the full vehicle finite element model is established for road noise analysis [13] and optimization, as shown in Fig. 9. The point *R* is the response point of the driver's right ear.

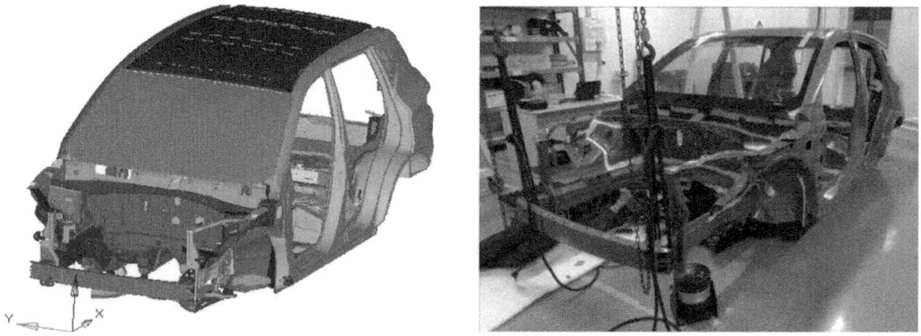

(a) Finite Element Model (b) Modal Test

Fig. 7 Body in white

Table 1 Modal results of body in white

Modal	Test results (Hz)	Finite element results (Hz)	Deviation (%)	Mode
1	41.91	40.36	−3.70	Torsion
2	56.23	54.16	−3.68	Bending

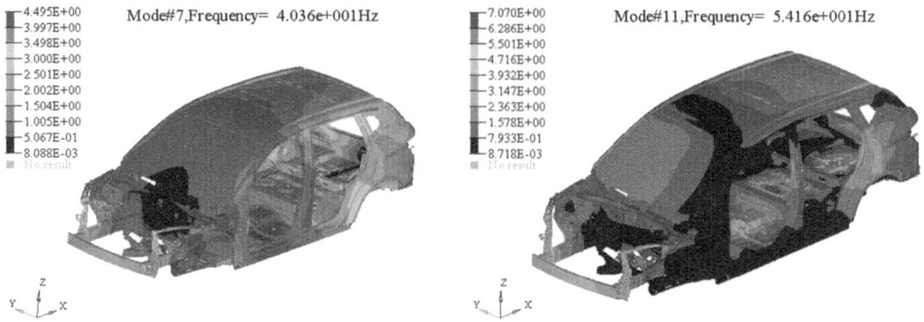

(a) Torsion 40.36Hz (b) Bending 54.16Hz

Fig. 8 Modal shape

Fig. 9 Finite element model of full vehicle

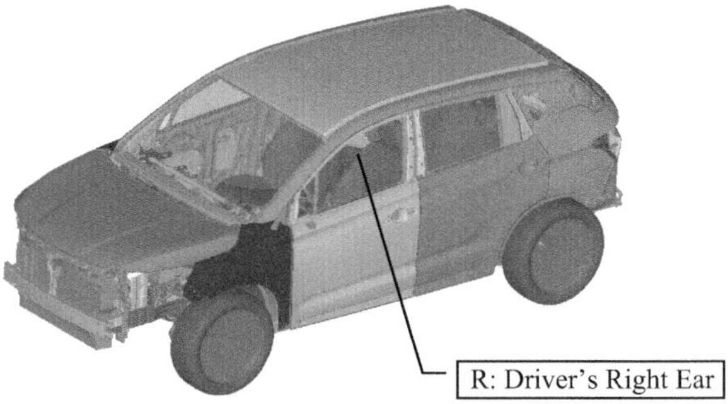

5 Simulation and Test

5.1 Standard Frequency Response Analysis Method

Based on the finite element model in Fig. 9, the experimental road excitation at the center of tires is applied. By standard frequency response analysis method, the road noise from 40 to 200 Hz is analyzed.

5.2 Substructure Frequency Response Analysis Method

Substructure 1 consists of trimmed body and acoustic cavity, and substructure 2 consists of chassis assembly. The two substructures are connected through the 22 vibration isolators, as shown in Fig. 10. By substructure frequency response analysis method, the loads and constraints, which are identical to those in standard method, the road noise is obtained afterward.

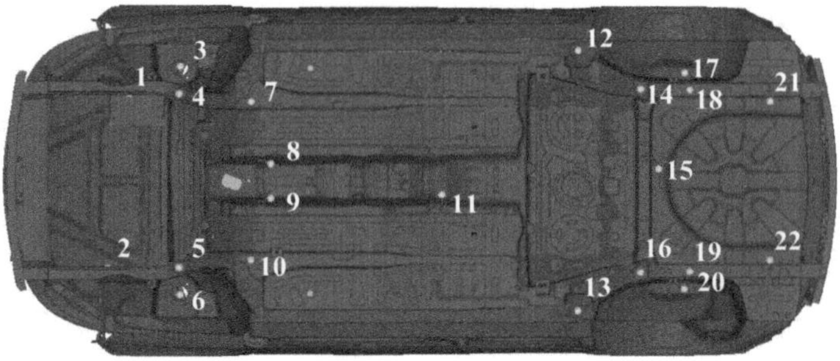

Fig. 10 Substructure connection point

Based on the NVH finite element model, the substructure method is employed to obtain the road acoustic response, which compares with the result obtained through standard method, as shown in Fig. 11. Two sets of curves are of high consistency. In 40–200 Hz whole frequency range, the acoustic characteristic curve retains peak frequency features of standard frequency response result to a great extent. There is little difference between noise peak values of two results, which proves that the result of substructure method is sufficiently accurate. 78, 100, 111, and 170 Hz are potential frequencies corresponding to noise peak. The road noise response curves peak at around 78, 100, 111, and 170 Hz, and it becomes most distinct at 110 Hz. At the location of driver's right ear, the noise exceeds the target value of 6 dB.

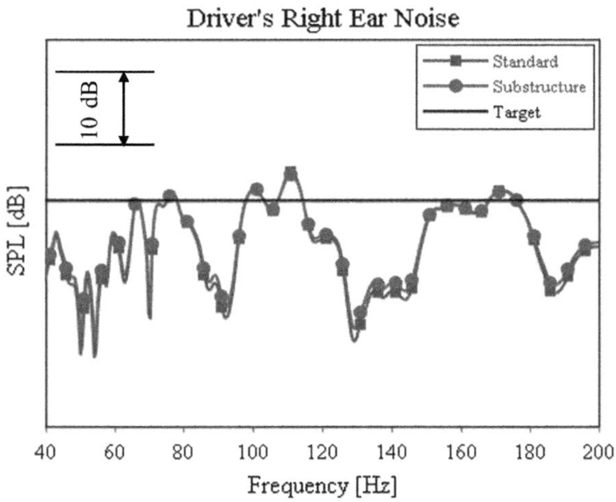

Fig. 11 Characteristics of road noise response

5.3 Test

A vehicle to 40 km/h uniform speed in the bad road is shown in Fig. 12a. The test results of road noise are shown in Fig. 12b. The "Base Model" curve represents substructure result. At 100, 110, and 120 Hz, the peak is higher, respectively, beyond the target value of 2, 3, and 4 dB. Comparison with the substructure method results, the overall trend of the curve is basically the same. At 100–120 Hz, the simulation results are in good agreement with the peak characteristic of the test, the peaks exceeding the target value of 3 dB. Therefore, the road noise level in the 100–120 Hz needs to be optimized. Thus, it is further verified that the substructure simulation model and method is trustworthy.

6 Optimization

The multi-objective optimization model is established. The bushing stiffness of each direction of each connection point is the design variable. The FRF matrix files of substructures are used as the constraint functions. Comparison with the above results of road noise test, the noise peaks of common characteristics are determined, and the 100–120 Hz peaks are used as optimization objectives.

The optimal Latin hypercube method is used to optimize the above-mentioned optimization model. And then, the optimal parameters are introduced into the finite element analysis model to verify the results, until all objectives satisfy the target; otherwise, the variable range is expanded, repeat a new round of optimization analysis.

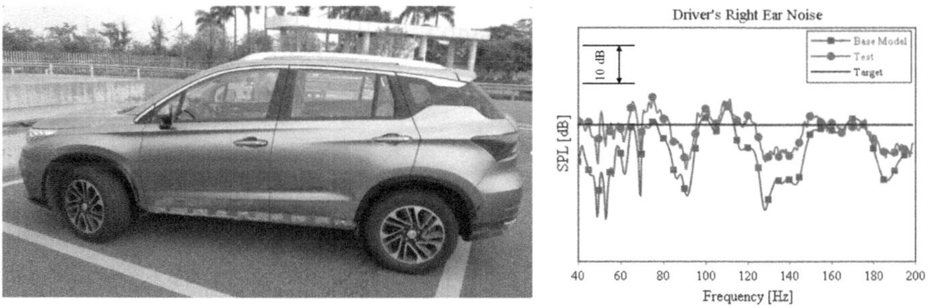

(a) Scene (b) Results Comparison

Fig. 12 Road noise test

6.1 Single-Objective Optimization

From the simulation results, it shows that at 111 Hz, the noise peak is the largest, exceeding the target value of 5 dB, which is an urgent problem to be optimized. Taking the stiffness of each direction of the bushing as the optimization variable, the noise peak at 111 Hz is optimized for the objective. The optimized bushing parameters are introduced into the substructure model to obtain the 40–200 Hz road noise curve, as shown in Fig. 13: 111 Hz peak decreases significantly, lower than the target value of 2.5 dB; 78, 100, and 170 Hz peak also decreases, and they are lower than the target value, but this phenomenon is unknown, with chance; 59 Hz peak deterioration, exceeding the target value of 1.5 dB, road noise is still not up to the target.

Obviously, it is not advisable to simply use the 111 Hz noise peak as the optimization objective, since the 59 Hz frequency peak does not make any constraint, and the unpredictability of its peak deterioration leads to the inefficiency of the optimization scheme. Therefore, single-objective optimization causes a single-frequency peak to drop, but may cause other frequency peaks to deteriorate. It cannot take into account the performance of the whole frequency range. So, it is necessary to introduce multi-objective optimization.

6.2 Multi-objective Optimization

In the last single-objective optimization scheme, the decrease of 111 Hz peak causes the peak of 78, 100, and 170 Hz to decrease, which shows that there are some common relations among the peaks of these four frequencies. Therefore, in the multi-objective optimization, the four peaks are regarded as an objective

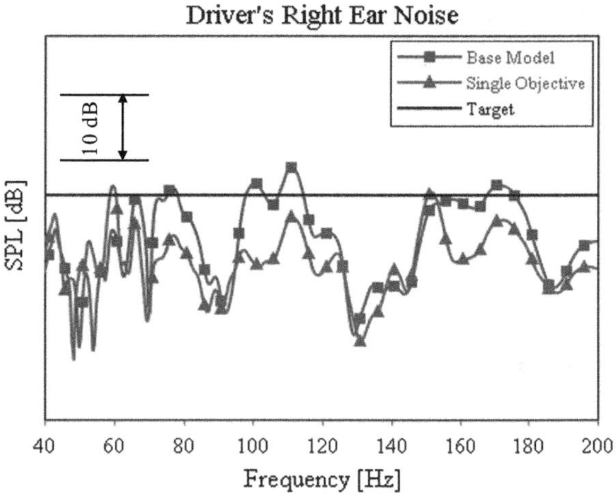

Fig. 13 Single-objective optimization results

function, and the 59 Hz peak is regarded as another objective function. The multi-objective of the optimization model is also carried out with the multi-island genetic algorithm.

The iterative process of K23 and K26 is shown in Fig. 14. Sample parameters are evenly distributed in the value range. Table 2 shows the stiffness of the bushings of the rear subframe point 14 in Fig. 10. KX_{21}, KX_{22}, KX_{23}, KX_{24}, KX_{25}, and KX_{26} are K21, K22, K23, K24, K25, and K26 initial base values. Obviously, the stiffness changes by +0.73, +9.27, −34.00, +15.83, +18.34, and −44.47% in each direction. It shows that in the vehicle NVH performance optimization, all the six directions bushing changes should be focused, and their trends are both positive and negative, rather than just consider a certain direction.

The iterative process of objectives R59 and R111 is shown in Fig. 15. In the 58th sample, they achieve the minimum value at the same time. The K23 and the K26 are the translational stiffness and the rotational stiffness of the *Z* direction of the front attachment of the rear subframe, respectively. R59 and R111 are the amplitudes of the road noise at 59 and 111 Hz. The optimized bushing parameters

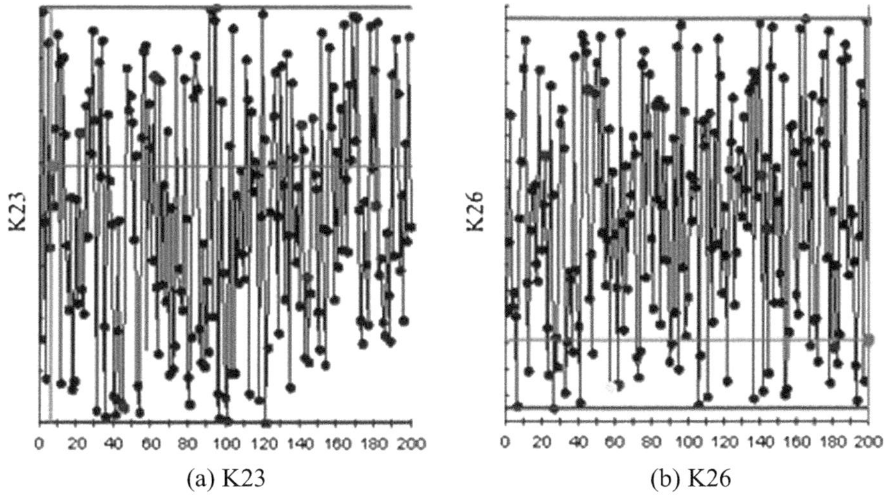

Fig. 14 Iterative process of variable K23 and K26

Table 2 Optimization of stiffness

	Base model (N/mm)	Optimization (N/mm)
K21	KX_{21}	$(1 + 0.73\%) * KX_{21}$
K22	KX_{22}	$(1 + 9.27\%) * KX_{22}$
K23	KX_{23}	$(1 - 34.00\%) * KX_{23}$
K24	KX_{24}	$(1 + 15.83\%) * KX_{24}$
K25	KX_{25}	$(1 + 18.34\%) * KX_{25}$
K26	KX_{26}	$(1 - 44.47\%) * KX_{26}$

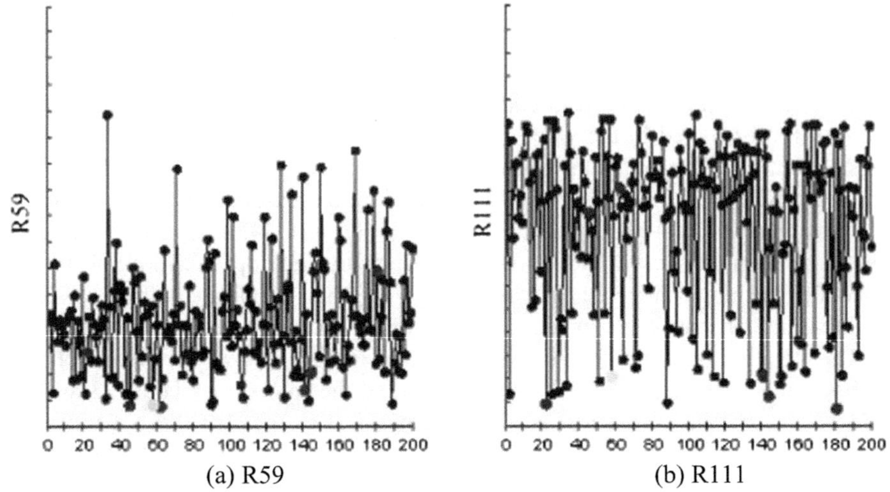

Fig. 15 Iterative process of objectives R59 and R111

are introduced into the substructure model to obtain the 40–200 Hz road noise curve, as shown in the following Fig. 16: 111 Hz peak decreased significantly, meet the target value; 59, 78, 100, and 170 Hz peak also decreased, both meet the target value, and road noise level meets the target.

Obviously, the optimization is effective. For the optimization analysis of the road noise, the multi-objective optimization can be used to constrain some noise peaks in the whole frequency range, which can avoid some unpredictable peak deterioration, and finally make the noise level in the 40–200 Hz band to meet the target.

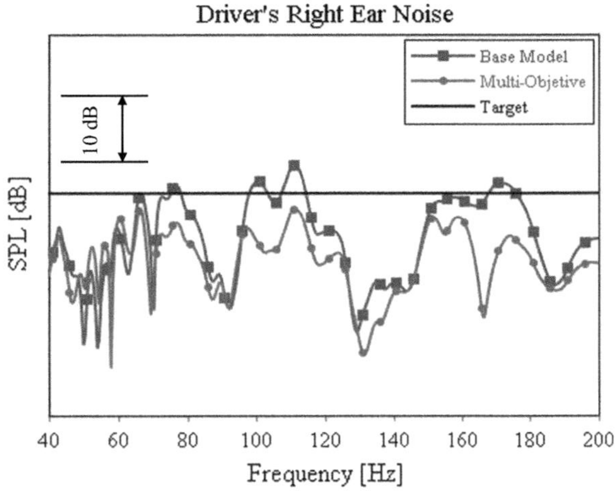

Fig. 16 Multi-objective optimization results

7 Efficiency

Importantly, the above 200 iterations of optimization are very fast, because of the high efficiency of the substructure method. There is a total of about 13 million DOFs to be solved. The model is submitted to workstation, which is equipped with double 2.4 GHz Intel Xeon E5-2620 CPUs and 32 GB memory. The computational time is about 5.5 h for the road noise analysis by standard finite element method.

On the other hand, the chassis substructure model contains about 3.5 million DOFs need to be solved, solving its FRF takes about 1 h. Meanwhile, the body substructure model contains about 9.5 million DOFs need to be solved; solving its FRF takes about 3.5 h. And then, it only takes nearly 4 min to invoke the FRF matrixes and carry out the full vehicle substructure frequency response analysis, which is equivalent to 0.07 h. The time-consuming of the standard method and the substructure method is displayed in Table 3.

As for substructure method, the FRFs of body substructure and chassis are analyzed, respectively; the total time needed to solve the full vehicle FRF is about 4.57 h. Comparison with the standard method, 56 min are saved, indicating that the efficiency increases by 17%. Apparently, the application of the method in the NVH model greatly improves the efficiency, especially for the great number of DOFs model, whose changed structure schemes require repeated calculations. Applying the method can substantially shorten the calculation time.

In this paper, the bushing stiffness of each direction of each connection point is the design variable, only the external program needs to be rewritten and recomputed, which takes just about 0.07 h. In comparison with the 5.5 h required by the standard method, the bushing stiffness optimization scheme takes only 4 min, and efficiency increases by 98%, as shown in Fig. 17a. It should be emphasized that the optimization analysis, only 4 min to complete a full vehicle-level iterative calculation, and it can greatly shorten the time for optimizing iterative calculations.

In addition, as for structure optimizations, such as body and chassis, substructure method also can improve efficiency. For the body thin plate parts, by the substructure method, the analysis model retains the FRF matrix file of the chassis, only the FRF of the optimized body substructure needs to be recomputed. Subsequently, the FRF matrixes of each substructure are combined and frequency response analysis is carried out, which takes about 3.57 h in total. Comparison with the

Table 3 Time-consuming

Method	Analysis project	DOFs (million)	Time-consuming (h)	Total time-consuming (h)
Standard	Frequency response of full vehicle	13	5.5	5.5
Substructure	Chassis FRF analysis	3.5	1	4.57
	Body FRF analysis	9.5	3.5	
	Substructure frequency response	–	0.07	

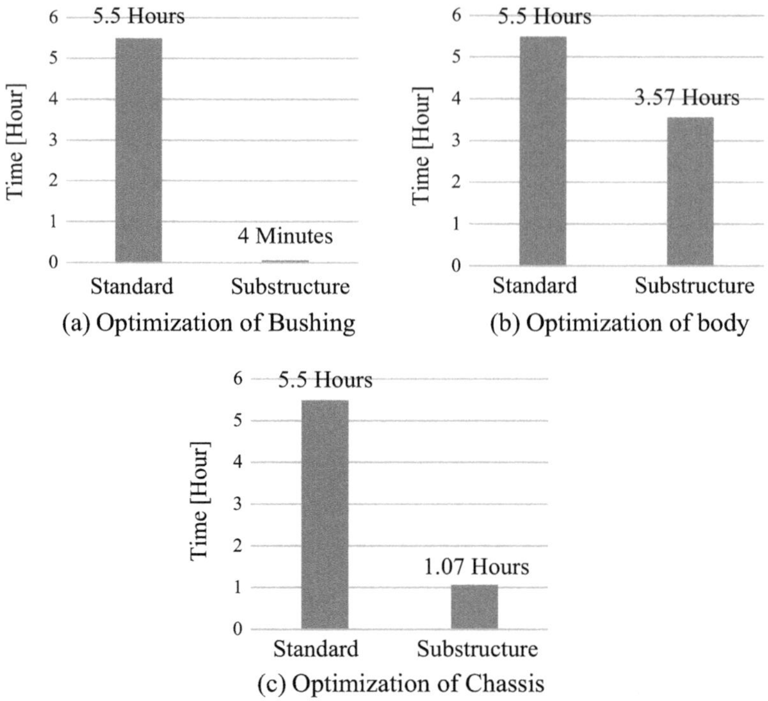

Fig. 17 Time-consuming

standard method, the efficiency increases by 35%, as shown in Fig. 17b. For the subframe or transmission structure, by the substructure method, only the FRF of optimized chassis substructure needs to be recomputed, which takes about 1.07 h in total. The efficiency increases by 80%, as shown in Fig. 17c.

To sum up, as for bushing optimization scheme, the efficiency increases by 98%. In addition, as for chassis structure optimization scheme, the efficiency increases by 80%; as for body optimization scheme, the efficiency increases by 35%. Details are displayed in Table 4. In the optimization analysis with many schemes, the superiority of substructure method is palpably reflected. For 200 iterations of bushing optimization, it takes 1100 h to calculate models by standard method. Nevertheless, by substructure method, it just takes 13 h to complete the calculation of 200 schemes, and efficiency increases by more than 80 times.

Table 4 Time-consuming of optimization

	Standard method (h)	Substructure method (h)	Efficiency increase (%)
Bushing optimization	5.5	0.07	98
Body optimization	5.5	3.57	35
Chassis optimization	5.5	1.07	80

8 Conclusions

It is verified that the substructure FRF method is feasible according to theoretical formula, and the FRF matrix of the whole structure S is deduced according to the FRF matrix of the substructure M and the substructure N.

It is verified that the substructure FRF method is accurate from the simulation perspective by comparison with the results of standard frequency response analysis and substructure method. And then, by the substructure method, the full vehicle model is confirmed to be realistic by comparison with the test results of road noise. So, substructure FRF method can replace the standard method for engineering research.

Multi-objective is more suitable for solving complex engineering problems. Single-objective optimization only guarantees a single peak frequency, but cannot take into account the dynamic performance of the whole frequency band, which may cause other frequency peaks to be deteriorated. Multi-objective optimization can make all the peaks in the whole frequency reach a balanced optimization state; that is to say, all peaks are meeting the target values.

It is efficiency of the substructure method that is superior. Substructure method can increase the computational efficiency by 98% when it comes to the optimization of the bushing stiffness in this paper. In addition, as for others structure optimization scheme, such as the thin plate parts of body, the efficiency increases by 35%; such as subframe or transmission structure of chassis, the efficiency increases by 80%.

References

1. Hampl N (2010) Advanced simulation techniques in vehicle noise and vibration refinement. Veh Noise Vibr Refinement, pp 174–188
2. Kim KC, Kim CM (2005) A study on the body attachment stiffness for the road noise. J Mech Sci Technol 19(6):1304–1312
3. Park SY, Park DC, Yoon KS et al (2009) A study of front subframe system optimization for improving vehicle nvh performance. SAE Technical Paper
4. Douville H, Masson P, Berry A (2006) On-resonance transmissibility methodology for quantifying the structure-borne road noise of an automotive suspension assembly. Appl Acoust 67(4):358–382
5. Kido I, Ueyama S (2005) Coupled vibration analysis of tire and wheel for road noise improvement. SAE Technical Paper
6. Gholizadeh S, Salajegheh E, Torkzadeh P (2008) Structural optimization with frequency constraints by genetic algorithm using wavelet radial basis function neural network. J Sound Vib 312(1):316–331
7. Liu CQ, Liu X (2004) A new method for analysis of complex structures based on FRF's of substructures. Shock Vibr 11(1):1–7
8. Liu CQ (2008) Combination of an improved FRF-based substructure synthesis and power flow method with application to vehicle axle noise analysis. Shock Vibr 15(1):51–60

9. Mahmoudi H, Zimmermann H (2015) On optimal latin hypercube design for yield analysis of analog. Microelectronics (Austrochip), 2015 Austrian Workshop on - IEEE, pp 46–49
10. Sakhaei B, Durali M (2014) Vibration transfer path analysis and path ranking for NVH optimization of a vehicle interior. Shock Vibr
11. Ahlersmeyer T (2010) Advanced experimental techniques in vehicle noise and vibration refinement. Veh Noise Vibr Refinement, pp 189–216
12. Liu CG (2011) Research on car interior noise control and structure optimization. Hunan University
13. Cameron CJ, Wennhage P, Göransson P (2010) Prediction of NVH behaviour of trimmed body components in the frequency range 100–500 Hz. Appl Acoust 71(8):708–721

Driving Cycle Construction of Xi'an Bus Line Based on Cluster Analysis

Yaohua Li, Zhongyu Li, Qizhi Gou, Tianyuan Ren, Pandeng Shao and Peng Liu

1 Introduction

Driving cycle of vehicle is represented by speed versus time and is supposed to represent typical driving patterns [1, 2]. Driving cycle is also called driving schedule, velocity profile, or speed profile. Driving cycle plays an important role in transportation analysis, vehicle economic analysis, vehicle's power system parameter matching, emissions testing, and optimization of the control strategy [1, 3–5]. City bus is the main carrier of passenger transport. The transportation characteristics of city bus are affected by many factors, such as road, climatic conditions, population density. Thus, it is a complex and difficult process to develop the driving cycle based on the actual data collection.

There are many methods to construct the driving cycle. Cluster analysis and the Markov are the main methods. The document [6] and [7] constructed the driving cycle based on the cluster analysis. The advantage of this method is that it divides different classes of working conditions, so that the synthetic conditions are closer to the actual traffic characteristics. The document [8] and [9] use the Markov method to build the cycle. The advantage is that the theory is stronger. In these documents above, the driving cycles are developed by data collection. But they are not judged whether the data is enough or not. It is a lack of theoretical support of the number of data collections.

The selection of sampling road is a complicated task, because the city transportation is very complicated. Another problem is how to judge the sample size. If the sampling data is too much, lots of resources are wasted. If the sampling data is too less, the developed driving cycle is blind and random, and the result is incorrect. So how to select the sampling road and how to determine the sampling size will be

Y. Li · Z. Li (✉) · Q. Gou · T. Ren · P. Shao · P. Liu
School of Automobile, Chang'an University, Xi'an, People's Republic of China
e-mail: lzyjsxz0623@163.com

Society of Automotive Engineers of China (SAE-China) (ed.), *Proceedings of the 19th Asia Pacific Automotive Engineering Conference & SAE-China Congress 2017: Selected Papers*, Lecture Notes in Electrical Engineering 486, https://doi.org/10.1007/978-981-10-8506-2_48

discussed in this paper. In the end, the driving cycle of a bus line in Xi'an was constructed based on cluster analysis. Two methods are used to evaluate whether the developed cycle is reasonable or not.

2 Driving Cycle Data Collection

2.1 The Sampling Route Analysis

Area coverage should be considered in data collection, basically including Xi'an business district and the suburbs. In addition, the proposed route should have a large enough running density. The concept of "site-line density" is proposed to evaluate the bus running density. Occurrence frequency of the bus stop needs to be calculated in "site-line density" method. In theory, the bus line being used is directly proportional to the number of its occurrence frequency; the more of the route being used, the more it is important and typical. Specific steps are as follows. Firstly, each site occurrence frequency must be calculated and replace the domain name with the number of frequency. Secondly, the frequency of each route site is respectively added (here called "site frequency"). Finally, the total frequency of the route divides the number of its sites, the site strength of the line being obtained. The density of the line (here called "line density") is a theoretical basis for selecting lines. The equation of "site frequency" ψ and "site density" λ can be described as follows.

$$\psi_i = \sum_{j=1}^{m} \varphi_{ij} \tag{1}$$

$$\lambda_j = \frac{1}{n} \sum_{i=1}^{n} \psi_i \tag{2}$$

Where φ_{ij} is the occurrences times of the bus station i in the road j, m is the total number of bus lines in Xi'an, and n is the total bus station in road j. When the station occurs, $\varphi_{ij} = 1$, or $\varphi_{ij} = 0$.

There are more than 200 bus lines in Xi'an. Considering the "site-line density" and the coverage area, one line is chosen as an example. This city bus is: Tang chengqiangjiuqu transit hub–Hong qixilu transit hub, which is shown in Fig. 1. It has 36 bus stations. This bus line can well reflect transportation state in Xi'an traffic arteries. It represents the main direction of north and south, and covers the most important areas.

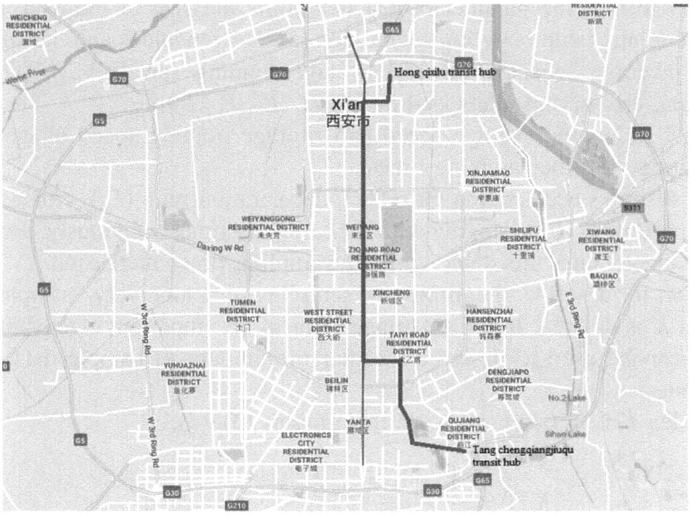

Fig. 1 This representative line running route map

2.2 *The Amount of the Sampling Data Analysis*

In this paper, the experimental data is collected by consecutive weeks gathering from 6: 00 to 23: 00 which is including traffic information on rush hours, medium peak hours, and low peak hours on weekends and weekdays. Taking the different weather conditions (cloudy, sunny, rainy, etc.) into account, we have done continuous sampling until the main characteristic parameter values get stable. The GPS acquisition equipment is installed in the urban and suburban buses. The sampling frequency is set at 1 Hz [10].

Theoretically, the more data is collected, the more accurate of the results. However, when the amount of collected data reaches a certain extent, even the amount of data increases, the accuracy will not continue to be greatly improved as shown in Fig. 2.

Fig. 2 Relation between accuracy and data size

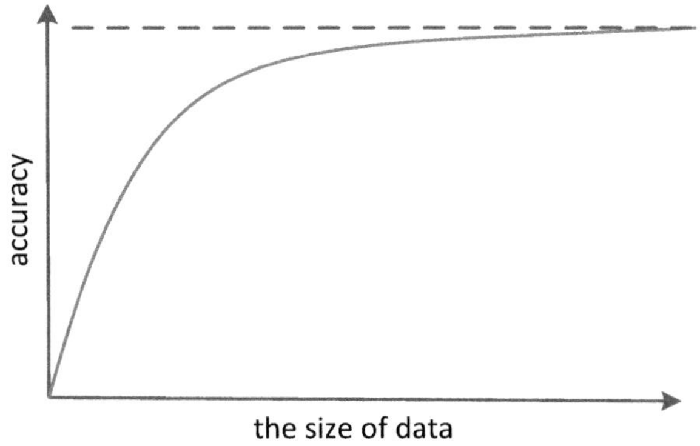

In order to find the parameters that determine the size of the data, this paper studies the relationship between several important eigenvalues and the size of data, including average speed(V_{mean}) and average running speed (V_{mr}), acceleration, deceleration, cruise, and idle speed ratios (P_a, P_d, P_c, P_i).

Figure 3 shows the average speed and average running speed of the sampling data changing with the sampling times.

Figure 4 shows acceleration ratio (P_a), deceleration ratio (P_d), cruise ratio (P_c), and idle speed ratio (P_i) changing with the sampling times. Obviously, these several important eigenvalues are stable with the sampling times increasing.

In order to describe the times of sampling quantitatively, sample stability coefficient K is defined. It is the average error of six important eigenvalues above. K values can be used to describe whether the amount of sample data meets the requirements.

The index K is defined as shown in formulas (3)–(4).

$$K_i(j) = \frac{\overline{A_i}(j) - \overline{A_i}(j-1)}{\overline{A}_i(j)} \quad (i = 1 \sim 6) \tag{3}$$

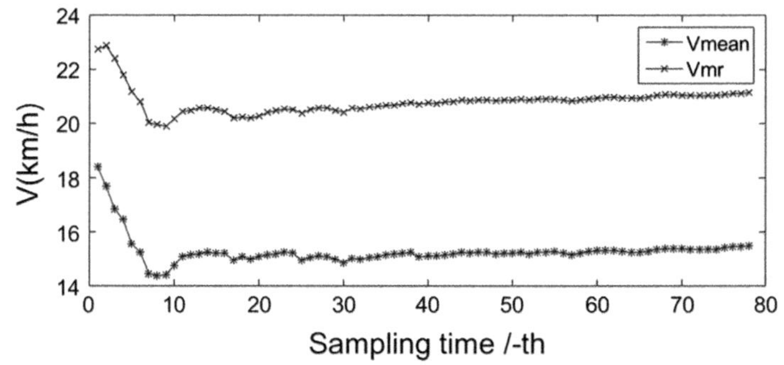

Fig. 3 Average speed and average running speed changing with sampling times

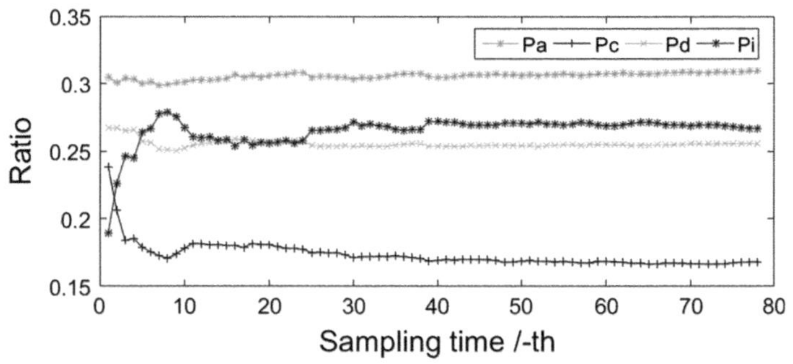

Fig. 4 P_a, P_d, P_c, and P_i changing with sampling times

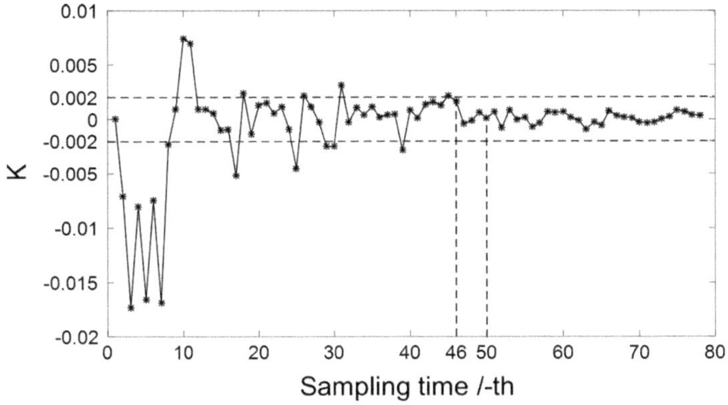

Fig. 5 Comprehensive index *K* changing with sampling times

$$K(j) = \frac{1}{n}\left[\sum_{i=1}^{n} K_1(j)\right] \quad (n = 6) \tag{4}$$

Where $\overline{A_1}(j), \overline{A_2}(j), \overline{A_3}(j), \overline{A_4}(j), \overline{A_5}(j), \overline{A_6}(j)$ represent the *j*th cumulative acceleration ratio (P_a), deceleration ratio (P_d), cruise ratio (P_c), idle ratio (P_i), the average speed (V_{mean}), and the average running speed (V_{mr}) of the sampling data, respectively. Thus, K_1–K_6 represent the acceleration ratio stability, deceleration ratio stability, cruise ratio stability, the proportion of idling stability, the average speed stability, and the average running speed stability with the sampling times.

The relation between index *K* and sampling times is shown in Fig. 5.

Figure 5 shows that the sample stability coefficient *K* converges to 0. The range of the *K* can change according to different actual requirement. In this paper, if the *K* meets that $|K| < 0.002$ and keep it for five times, the sample data can be considered stable. Figure 5 shows that after 45 times, the sampling data has been stable. So, 50 times will meet our demand and keep the data sufficient.

3 Driving Cycle Construction

For the city bus, taking its frequent start and brake into account, a micro-trip can be defined as a process from one idle point to the next adjacent idle point [11, 12]. If the length of the fragment is less than 20 s, this fragment will not be divided and add it to the fragment before as shown in Fig. 6.

By dividing the 50 sets of data, 3141 micro-trips are obtained. The flowchart of developing Xi'an city bus driving cycle based on cluster method is shown in Fig. 7.

The eigenvalues are characteristic values of the running vehicle that can fully describe the driving condition and behavior. There are 24 eigenvalues that are

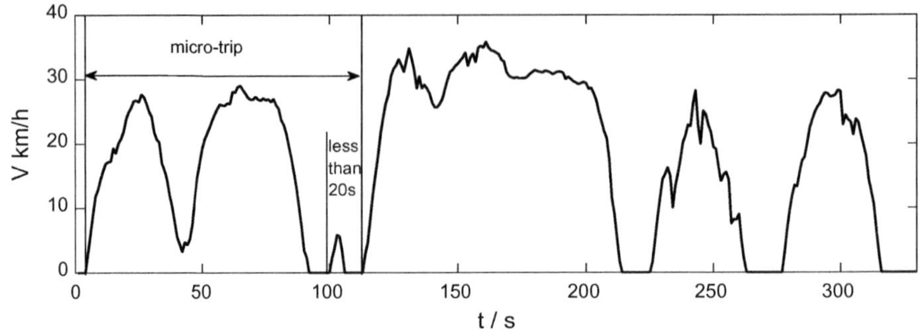

Fig. 6 Micro-trip partitioning method

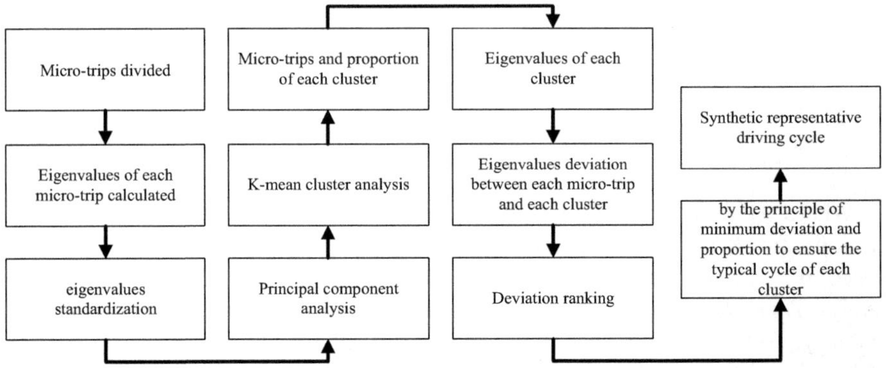

Fig. 7 Flowchart of developing Xi'an city bus driving cycle

defined to describe the characteristic values of the micro-trips and sampling data. These eigenvalues can be divided into two categories, 15 of them are descriptive eigenvalues and the others are statistic eigenvalues. The descriptive eigenvalues are defined including travel distance (L), total time (T), acceleration time (T_a), idle time (T_i), deceleration time (T_d), cruise time (T_s), maximum speed ($V_{\max}$), average speed (V_{mean}), the average running speed (V_{mr}), speed standard deviation (V_{sd}), maximum acceleration ($A_{\max}$), average acceleration (A_{mean}), deceleration average (D_{mean}), maximum deceleration ($D_{\max}$), and the acceleration deviation (A_{sd}), which are mainly used to do principal component analysis and cluster analysis. The calculation of each eigenvalue is shown in formulas (5)–(19).

$$L = \sum_{i=1}^{n} V_i \Delta t = \sum_{i=1}^{n} V_i, \quad i = 1, 2, \ldots, n \tag{5}$$

$$T = \mathrm{t} \tag{6}$$

$$Ta = \{t|a \geq 0.15\ \text{m/s}^2\} \tag{7}$$

$$Td = \{t|a \leq -0.15\ \text{m/s}^2\} \tag{8}$$

$$Ti = \{t|V = 0\} \tag{9}$$

$$T_c = T - T_a - T_d - T_i \tag{10}$$

$$V_{\text{max}} = \max\{V_i, i = 1, 2, \ldots, n\} \tag{11}$$

$$V_{\text{mean}} = \frac{L}{T} \tag{12}$$

$$V_{\text{mr}} = \frac{L}{T - T_i} \tag{13}$$

$$V_{\text{sd}} = \sqrt{\frac{1}{n-1}\sum_{i=1}^{n}(V_i - V_m)^2}, \quad i = 1, 2, \ldots, n \tag{14}$$

$$A_{\text{max}} = \max\{a_i, i = 1, 2, \ldots, n\} \tag{15}$$

$$A_{\text{mean}} = \frac{\text{sum}\{a_i|a_i \geq 0.15, \quad i = 1, 2, \ldots, n\}}{T_a} \tag{16}$$

$$D_{\text{max}} = \min\{a_i, i = 1, 2, \ldots, n\} \tag{17}$$

$$D_{\text{mean}} = \frac{\text{sum}\{a_i|a_i \leq -0.15, i = 1, 2, \ldots, n\}}{T_d} \tag{18}$$

$$A_{\text{sd}} = \sqrt{\frac{1}{n-1}\sum_{i=1}^{n} a_i^2}, \quad i = 1, 2, \ldots, n \tag{19}$$

There are nine statistical eigenvalues that are defined as follows: acceleration ratio (P_a), deceleration ratio (P_d), cruise ratio (P_c), idle ratio(P_i), and each speed segment velocity distribution ratio ($P_{0\text{–}10}$, $P_{10\text{–}20}$, $P_{20\text{–}30}$, $P_{30\text{–}40}$, $P_{40\text{–}50}$). The calculation of eigenvalues is shown in formulas (20)–(28).

$$P_{0-10} = \frac{\{t|0 < V \leq 10\ \text{km/h}\}}{T} \tag{20}$$

$$P_{10-20} = \frac{\{t|10 < V \leq 20\ \text{km/h}\}}{T} \tag{21}$$

$$P_{20-30} = \frac{\{t|20 < V \leq 30\ \text{km/h}\}}{T} \tag{22}$$

$$P_{30-40} = \frac{\{t|30 < V \leq 40\ \text{km/h}\}}{T} \tag{23}$$

$$P_{40-50} = \frac{\{t|40 < V \leq 50\ \text{km/h}\}}{T} \tag{24}$$

$$P_a = \frac{\{t|a \geq 0.15\ \text{m/s}^2\}}{T} \tag{25}$$

$$P_d = \frac{\{t|a \leq -0.15\ \text{m/s}^2\}}{T} \tag{26}$$

$$P_i = \frac{\{t|V = 0\}}{T} \tag{27}$$

$$P_c = 1 - P_a - P_d - P_i \tag{28}$$

The eigenvalues of each micro-trip are shown in Table 1.

There are 20 eigenvalues that are defined to describe the characteristics of the general experimental data. They are maximum speed (V_{max}), average speed (V_{mean}), the average running speed (V_{mr}), speed standard deviation (V_{sd}), maximum acceleration (A_{max}), average acceleration (A_{mean}), deceleration average (D_{mean}), maximum deceleration (D_{max}), acceleration deviation (A_{sd}), acceleration, deceleration, cruise, idle ratios (P_a, P_d, P_c, P_i), each speed segment velocity distribution ratio (P_{0-10}, P_{10-20}, P_{20-30}, P_{30-40}), etc. These 20 eigenvalues are utilized to define the objective functions in the next section and evaluate the error of the cycle we have developed. The eigenvalues of the total sampling data are shown in Table 2.

Table 1 Eigenvalues of each micro-trip

No.	T/(s)	T_a/(s)	T_d/(s)	…	P_c	P_i
1	43	15	10	…	0.04	0.62
2	60	19	21	…	0.39	0.09
…	…	…	…	…	…	…
3140	86	35	28	…	0.116	0.162
3141	62	25	19	…	0.083	0.028

Table 2 Eigenvalues of general experimental data

V_{mean}	V_{mr}	V_{sd}	V_{max}	A_{mean}
15.22	20.86	13.08	49.14	0.519
A_{max}	D_{mean}	D_{max}	A_{sd}	P_{0-10}
2.98	−0.629	−3.93	0.511	0.152
P_{10-20}	P_{20-30}	P_{30-40}	P_{40-50}	P_a
0.178	0.229	0.157	0.013	0.306
P_d	P_c	P_i (P_0)		
0.254	0.169	0.271		

Because of the coupling between eigenvalues, the principal component analysis (PCA) is used to reduce the dimension [13]. Because the distance (L), total running time (T), acceleration time, deceleration time, idle time, cruise time (T_a, T_d, T_i, T_c), maximum speed (V_{max}), maximum acceleration (A_{max}), and maximum deceleration (D_{max}) are the cumulative eigenvalues, does not have statistical significance. In this paper, the eigenvalues above are ignored, and the other 15 eigenvalues are calculated in principal component analysis. Through the principal component analysis, the cumulative contribution rates of the first four principal components have reached 84.903%. According to the statistical theory, the cumulative contribution rate of 80% of the main component can be selected. Therefore, the first four principal components are selected to do cluster analysis. After K-means cluster analysis, the number of micro-trips of each cluster is shown in Table 3, and the average characteristic values in each cluster are shown in Table 4.

As can be seen from Table 4, the micro-trips in the first cluster have highest average speed (V_{mean}), at an average speed of 18.22 km/h. The first cluster represents the smooth driving condition. The average speed of the second cluster is 12.77 km/h, and the idle ratio is 0.28, but the average acceleration rate and the average deceleration rate are between others. The second cluster represents the normal conditions. The average speed of the third cluster is 14.46 km/h, and the idle ratio is 0.33. The third cluster represents the congestion road driving condition. As Table 3 shows, the proportion of the number of three cluster fragments is about 4:4:2. Some micro-trips are selected with the smallest degree of deviation from each cluster correspondingly. Thus, this representative driving cycle is combined (in this paper, we call it XA-HBUS). The following formula can be used to describe the deviation between each micro-trip and the general experimental data.

$$d = \sum_{i=1}^{n} \left| 1 - \frac{\overline{y}_i}{y_j} \right| \tag{29}$$

Table 3 Results of *K*-means cluster analysis

	Custer 1	Cluster 2	Cluster 3
No.	1199	1281	601

Table 4 Average eigenvalues of each cluster

Eigenvalues	V_{mean}	V_{mr}	A_{mean}	D_{mean}	$P_{0\text{–}10}$	$P_{10\text{–}20}$	$P_{20\text{–}30}$
Smooth	18.22	23.53	0.58	−0.77	0.12	0.15	0.24
Normal	12.77	17.7	0.46	−0.55	0.21	0.20	0.21
Congestion	14.46	21.43	0.51	−0.58	0.12	0.17	0.14

Eigenvalues	$P_{30\text{–}40}$	$P_{40\text{–}50}$	P_a	P_d	P_c	P_i
Smooth	0.24	0.02	0.35	0.27	0.17	0.23
Normal	0.10	0.01	0.29	0.25	0.20	0.28
Congestion	0.14	0.01	0.27	0.24	0.18	0.33

Where y_j is the average eigenvalue of cluster j, and $\overline{y}_i$ is the eigenvalue of each micro-trip. The result is shown in Table 5 by sorting the deviation degree.

According to the proportion and the results above, the driving cycle is constructed by selecting the fragments of 2798, 1258, 2201, 2990, 2045, 780, 2973, 805, 1036, and 3053, which is shown in Figs. 8 and 9. Refer to the documents at home and abroad, and the length of most construction cycles is between 15 and 30 min [11]. The length of the constructed driving cycle is about 18 min. The characteristic values of XA-HBUS are shown in Table 6.

Table 5 Deviation of each micro-trip in three clusters

Smooth	Deviation	2.8374	2.9203	3.1361	3.2838	…	6.1021
	No.	2798	1258	2201	2990	…	2885
Normal	Deviation	3.7900	4.0465	4.0722	4.1989	…	7.9664
	No.	2045	780	2973	805	…	2686
Congestion	Deviation	3.2856	3.4742	3.5133	3.8108		9.8832
	No.	1036	3053	2292	1432	…	1873

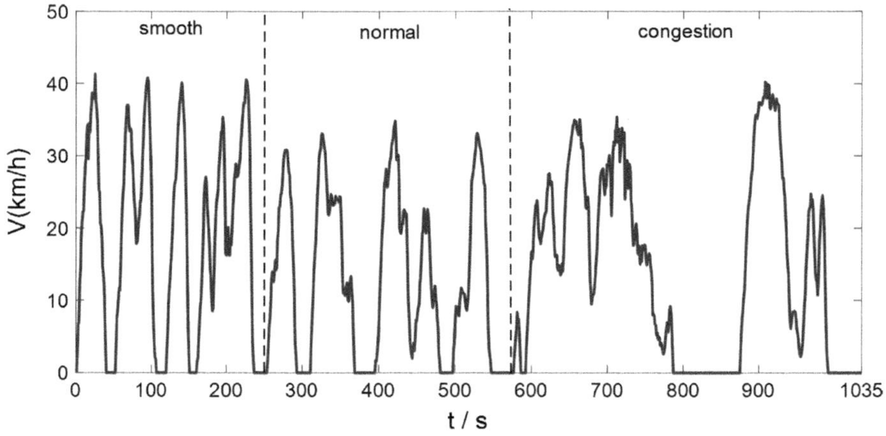

Fig. 8 Velocity of XA-HBUS

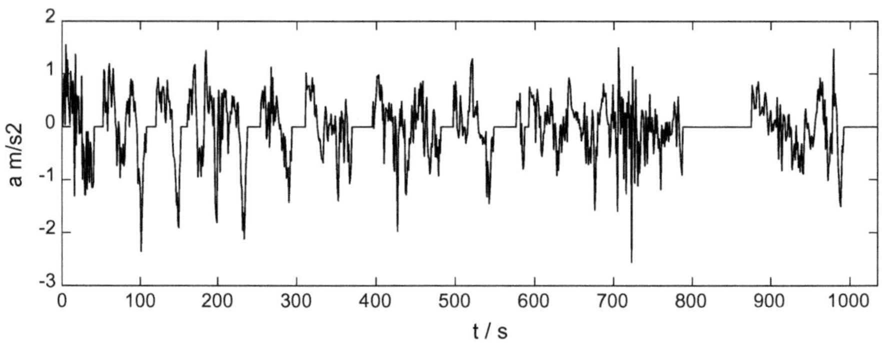

Fig. 9 Acceleration of XA-HBUS

Table 6 Eigenvalues of XA-HBUS

T/(s)	T_a/(s)	T_d/(s)	T_c/(s)	T_i/(s)	L/(km)	V_{max}	V_{mean}	V_{mr}
1035	333	263	149	290	4.26	41.23	14.83	20.60
V_{sd}	A_{max}	A_{mean}	D_{max}	D_{mean}	A_{sd}	$P_{0–10}$	$P_{10–20}$	$P_{20–30}$
12.88	1.55	0.52	-2.56	-0.66	0.53	0.15	0.19	0.22
$P_{30–40}$	$P_{40–50}$	$P_{50–60}$	$P_{60–70}$	P_a	P_d	P_c	P_i (P_0)	
0.15	0.01	0	0	0.32	0.25	0.14	0.28	

The speed interval distribution of sampling data and driving cycle is shown in Figs. 10 and 11.

As can be seen from the figures and tables above, the distribution of the speed intervals and the eigenvalues of the developed driving cycle are basically the same as the sampling data. Different methods are used to analyze the error of the construction condition further.

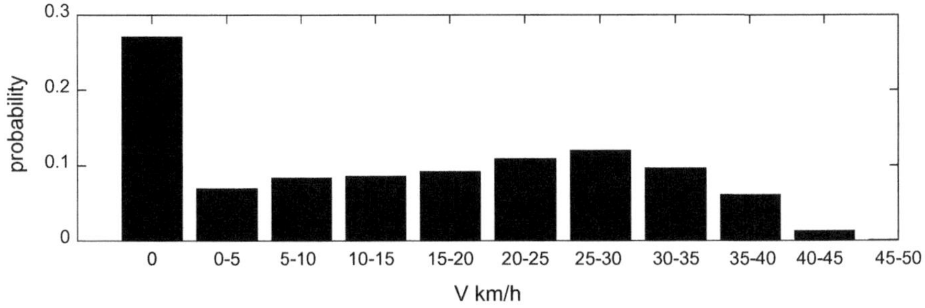

Fig. 10 Speed interval distribution of sampling data

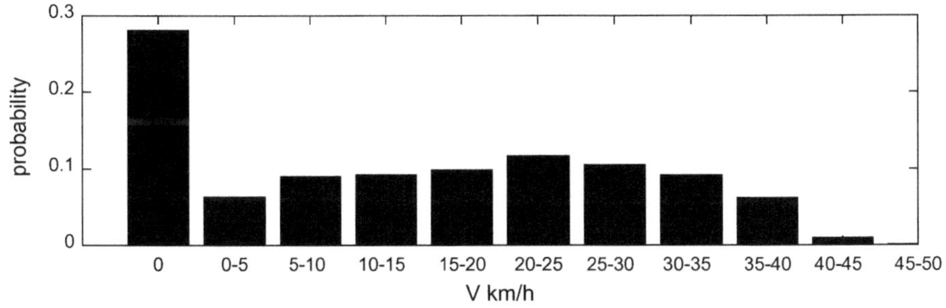

Fig. 11 Speed interval distribution of driving cycle

4 Driving Cycle Error Analysis

Average speed (V_{mean}), the average operating speed (V_{mr}), speed standard (V_{sd}), average acceleration (A_{mean}), deceleration average (D_{mean}), the acceleration (A_{sd}), acceleration, deceleration, cruise, idle ratios (P_a, P_d, P_c, P_i), and a total of ten important eigenvalues are selected as indicators. Equations (30–32) are defined as error evaluation to evaluate the error of XA-HBUS.

$$\delta_i = \frac{|A_i - A|}{A} \times 100\% \tag{30}$$

$$\delta = \sum_{i=1}^{n} \delta_i \tag{31}$$

$$\overline{\delta} = \frac{1}{n}\sum_{i=1}^{n} \delta_i \tag{32}$$

The errors between XA-HBUS and the overall sample data are shown in Table 7.

Table 7 suggests that the average error between developed XA-HBUS driving cycle and sampling data is 2.93%. Each error is relatively small and closest to the sampling data.

The velocity and acceleration probability distribution matrix is adopted to verify the accuracy of the developed driving cycle further from different aspects. The matrix of the sampling data and the developed driving cycle data is shown in Tables 8 and 9. The speed range is [0, 50] km/h, and the acceleration is [−3, 2] m/s^2.

Euclidean distance is used to verify the accuracy of cycle based on the *V*-*A* matrix analysis. There are two *n*-dimensional vectors, a[X_1, X_2, X_3,…, X_n] and b [Y_1, Y_2, Y_3,…, Y_n]. The distance is shown in formulas (38) and (39) that can be used to measure the similarity of matrix.

$$d = \sqrt{(x_1 - y_1)^2 + (x_2 - y_2)^2 + (x_3 - y_3)^2 + \cdots + (x_n - y_n)^2} \tag{38}$$

$$e = d/n \tag{39}$$

Where *d* is the Euclidean distance, *e* is the deviation rate, and *n* is dimensional vectors.

Table 7 Error between XA-HBUS and the sampling data %

Cycles	δ_{Vmean}	δ_{Vmr}	δ_{Vsd}	δ_{Amean}	δ_{Dmean}	δ_{Asd}	δ_{Pa}	δ_{Pd}	δ_{Pc}	δ_{Pi}	δ	$\overline{\delta}$
XATD-HBUS	4.64	2.26	3.17	0.14	0.89	1.56	2.47	1.27	7.84	5.02	29.3	2.93

Table 8 Velocity and acceleration matrix of the sampling data %

V\a	[−3, −2.5)	[−2.5, −2)	[−2, −1.5)	[−1.5, −1)	[−1, −0.5)	[−0.5, 0)	[0, 0.5)	[0.5, 1)	[1, 1.5)	[1.5, 2]
[0, 5)	0.01	0.05	0.19	0.55	1.72	28.01	1.98	1.35	0.11	0.00
[5, 10)	0.02	0.07	0.24	0.52	1.05	1.83	2.47	1.91	0.19	0.01
[10, 15)	0.02	0.08	0.21	0.50	1.05	2.03	2.51	1.80	0.34	0.03
[15, 20)	0.01	0.05	0.18	0.52	1.24	2.28	2.71	1.73	0.42	0.04
[20, 25)	0.01	0.02	0.13	0.42	1.40	2.96	3.68	1.85	0.36	0.06
[25, 30)	0	0.01	0.05	0.27	1.20	3.65	4.95	1.58	0.22	0.04
[30, 35)	0	0	0.02	0.10	0.67	3.32	4.52	0.90	0.11	0.02
[35, 40)	0	0	0	0.02	0.23	2.33	3.07	0.37	0.04	0.01
[40, 45)	0	0	0	0	0.03	0.47	0.71	0.06	0.01	0
[45, 50]	0	0	0	0	0	0.02	0.04	0	0	0

Table 9 Velocity and acceleration matrix of the driving cycle data %

V\a	[−3, −2.5)	[−2.5, −2)	[−2, −1.5)	[−1.5, −1)	[−1, −0.5)	[−0.5, 0)	[0, 0.5)	[0.5, 1)	[1, 1.5)	[1.5, 2]
[0, 5)	0	0	0	0.48	1.45	29.44	1.93	0.68	0.19	0
[5, 10)	0	0.10	0.19	0.58	0.87	2.51	2.22	2.32	0.10	0
[10, 15)	0	0	0.19	0.58	1.06	2.61	2.80	1.64	0.10	0.10
[15, 20)	0	0.10	0.29	0.48	1.93	2.03	2.41	1.83	0.68	0
[20, 25)	0.10	0	0.39	0.39	1.06	2.70	4.25	2.41	0.29	0
[25, 30)	0	0	0	0.77	1.16	2.03	4.44	1.64	0.39	0
[30, 35)	0	0	0	0.19	0.87	2.70	3.86	1.25	0.19	0
[35, 40)	0	0	0	0.10	0.48	1.93	2.90	0.68	0	0
[40, 45)	0	0	0	0	0	0.29	0.39	0.19	0	0
[45, 50]	0	0	0	0	0	0	0	0	0	0

The Euclidean distance between the sampling data matrix and the driving cycle matrix is $d = 0.00075$. The deviation rate is 7.5×10^{-6}. According to the error analysis, the constructed cycle conforms to the actual driving condition.

The accuracy and similarity of the driving cycle are verified from two aspects: The one is statistical characteristic average error, and the other is *V-A* matrix of distribution. The first aspect analyzes the errors based on the important statistical eigenvalues. The second aspect analyzes the errors from different distributions of velocity and acceleration between sampling data and developed cycle. According to the error analysis which is calculated by two methods, this paper verifies the effectiveness of the constructed driving cycle.

5 Conclusion

In this paper, sampling data collection and analysis processes of Xi'an city bus driving cycle are discussed. Firstly, the concept of "site-line density" is proposed as a basis for route selection. The selection of typical important lines for data collection can greatly improve efficiency and reduce the costs in the case of construction. Secondly, based on the analysis of relation between some typical eigenvalues and the sampling times, the integrated index K is defined. The index K is used to describe whether the amount of the sample data meets the requirements or not. The index K can avoid collecting excessive data, so that the data collection is more targeted and make full use of resources. Thirdly, this paper constructs the driving cycle based on the cluster analysis. Finally, this paper uses two methods verifying the accuracy and rationality of the diving cycle from different angles. According to the error analysis which is calculated by two methods, this paper verifies the effectiveness of the constructed driving cycle. However, there are some deficiencies in this article. This paper constructs driving cycle based on one line. The problem of interline coupling is not mentioned, but it can be solved by the rule of "site-line density." In the future, based on the construction of line cycle, we will further study line coupling and construction of urban conditions accurately.

References

1. Nyberg P, Frisk E, Nielsen L (2017) Driving cycle equivalence and transformation. IEEE Trans Veh Technol 66(3):1963–1974
2. Wang Q, Huo H, He K et al (2008) Characterization of vehicle driving patterns and development of driving cycles in Chinese cities. Transp Res Part D 13(5):289–297
3. Pavlovic J, Marotta A, Ciuffo B (2016) CO_2, emissions and energy demands of vehicles tested under the NEDC and the new WLTP type approval test procedures. Appl Energy 177:661–670
4. Jeong NT, Yang SM, Kim KS et al (2016) Urban driving cycle for performance evaluation of electric vehicles. Int JAutomot Technol 17(1):145–151

5. Liu J, Wang X, Khattak AJ (2016) Customizing driving cycles to support vehicle purchase and use decisions: fuel economy estimation for alternative fuel vehicle users. Transp Res Part C 67:280–298
6. Gao X, Zhang B, Xiong X et al (2015) Construction and analysis of the Dalian driving cycle. Int J Control Autom, 8
7. Zaccardi JM, Berr FL (2013) Analysis and choice of representative drive cycles for light duty vehicles—case study for electric vehicles. Proc Inst Mech Eng Part D J Autom Eng 227 (4):605–616
8. Shi S, Lin N, Zhang Y et al (2016) Research on Markov property analysis of driving cycles and its application. Transp Res Part D 47:171–181
9. Li L, Huang C, Lu B, et al (2013) Study on the design method of time-variant driving cycles for EV based on Markov Process. Veh Power Propul Conf IEEE, pp 1277–1281
10. Bishop JDK, Axon CJ, Mcculloch MD (2012) A robust, data-driven methodology for real-world driving cycle development. Transp Res Part D Transp Environ 17(5):389–397
11. Lai J, Yu L, Song G et al (2013) Development of city-specific driving cycles for transit buses based on VSP distributions: case of beijing. J Transp Eng 139(7):749–757
12. Günther R, Wenzel T, Wegner M et al (2017) Big data driven dynamic driving cycle development for busses in urban public transportation. Transp Res Part D Transp Environ 51:276–289
13. Shi S, Wei S, Kui H et al (2009) Improvements of the design method of transient driving cycle for passenger car. IEEE Veh Power and Propul Conf IEEE pp 1581–1586

Driver Fatigue Detection Based on SVM and Steering Wheel Angle Characteristics

Li Zhang, Dongmei Yang, Hongfei Ni and Tao Yu

1 Introduction

Statistics show that [1] 57% of traffic accidents are related to fatigue driving. Therefore, the study of driving fatigue monitoring methods is of great significance to reduce traffic accidents and improve the road traffic safety environment. In recent years, domestic and foreign scholars have proposed different driving fatigue monitoring methods from different research perspectives. These methods have unique opinions on the description and evaluation of driver's fatigue state, but there are some shortcomings, such as the use of physiological signals as a driving fatigue evaluation index [2, 3], requires the measuring instrument to be in direct contact with the driver, which could interfere with the driving behavior, and affects the judgement of driving fatigue state [4, 5]; choosing the physiological response characteristics as the evaluation index could avoid interference in direct contact with the instrument, but the applied instruments would have a negative impact on driver's psychological mood, and is not conducive to the accurate evaluation of the driver's fatigue state.

This paper proposed driving fatigue detection method based on the characteristics of steering wheel angle and chose the steering wheel angle characteristics as fatigue evaluation basis, which had no influence on driver's driving behavior and psychological emotion. The detection model was established by support vector machines and introduced the cross-validation method to optimize the parameters, which could effectively avoid the shortcomings of the above methods.

L. Zhang (✉) · D. Yang · H. Ni · T. Yu
Dongfeng Commercial Vehicle Technical Center, Dongfeng Commercial Vehicle Co., Ltd., Wuhan 430056, China
e-mail: yf-zhangl@dfcv.com.cn

Society of Automotive Engineers of China (SAE-China) (ed.), *Proceedings of the 19th Asia Pacific Automotive Engineering Conference & SAE-China Congress 2017: Selected Papers*, Lecture Notes in Electrical Engineering 486, https://doi.org/10.1007/978-981-10-8506-2_49

2 Basic Theory Introduction

2.1 Support Vector Machines—Introduction

Support vector machine (SVM) was firstly proposed by Vapnik in the 1990s and was based on the statistical learning theory [6]. The algorithm can better achieve the idea of structural risk minimization; that is, the input vector of the sample is mapped to the feature space of high dimension by kernel function, so as to construct the optimal classification surface in the feature space and realize the function of linear classification sample [6, 7]. SVM theory has become the current research hot spot and is widely used in engineering field [8, 9]. The principles and steps of solving the linear separable problem of SVM are as follows:

(1) Assume known training set: $T = \{(x_1, y_1), \cdots, (x_l, y_l)\} \in (X \times Y)^l$, where $x_i \in X = R^n$, $y_i \in Y = \{1, -1\}(i = 1, 2, \ldots, l)$, x_i as the eigenvector; then select a kernel function $K(x, x')$ and the appropriate parameters C, construct and solve the optimization problem, and finally, obtain the optimal solution: $\alpha^* = \left(\alpha_1^*, \ldots, \alpha_l^*\right)^T$.

The specific formula for the optimization problem is $\min\limits_{\alpha} \frac{1}{2} \sum\limits_{i=1}^{j} \sum\limits_{j=1}^{l} y_i y_j \alpha_i \alpha_j K\left(x_i, x_j\right) - \sum\limits_{j=1}^{l} \alpha_j$

$$s.t. \quad \sum_{i=1}^{l} y_i \alpha_i = 0, \quad 0 \le \alpha_i \le C(i = 1, \ldots, l)$$

(2) Select a positive component of α^*, and calculate the threshold: $b^* = y_j - \sum\limits_{i=1}^{l} y_i \alpha_i^* K\left(x_i, x_j\right)$

(3) Construct the decision function, and the specific formula is $f(x) = \text{sgn}\left(\sum\limits_{i=1}^{l} \alpha_i^* y_i K(x, x_i) + b^*\right)$

For the nonlinear separable problems, the Gaussian radial basis function used in this paper is: $K(x, x_i) = \exp\left(-\frac{\|x - x_i\|^2}{2\sigma^2}\right)$.

2.2 Cross-validation Method Introduction

Cross-validation (CV) is a statistical analysis method used to verify the performance of classifiers [10, 11]. There are three commonly used cross-validation methods, namely hold-out method, K-fold cross-validation (K-CV), and leave-one-out cross-validation (LOO-CV). In this paper, the K-CV method is used to optimize the relevant parameters of the support vector machine to obtain the best parameters, so as to improve the recognition rate of the recognition model.

The specific content of the K-CV method is that the original data is divided into K groups (usually equalized), and each subset of the data were made as a validation set, while the rest of the $K - 1$ groups of sub-data as a training set; then, the K models are obtained, and the average of the classification accuracy of the final verification set of K models is used as the performance index of the classifier under K-CV. K is generally greater than or equal to 2. In the actual operation, K starts from 3 and is only equal to 2 when the original amount of data is small. K-CV method can effectively avoid overlearning and the occurrence of poor learning state and finally get the results more reasonable.

3 Driving Fatigue Detection Model Establishment Based on SVM

In this paper, the driving fatigue detection model uses SVM as the classifier, the steering wheel angle characteristic as the characteristic vector (input vector), and the driver's fatigue state as the output vector.

3.1 Classification of Driving Fatigue Status

The driver's subjective fatigue experience can provide a reference for the driving fatigue classification test. This chapter used the Stanford sleepiness scale [12] to divide the driving fatigue into three states: normal driving state, quasi-fatigue driving state, and fatigue driving state. Table 1 shows the contents of the Stanford sleepiness scale. In the table, 1–2 levels represent the normal driving state, 3–4 stand for quasi-fatigue state, and 5–7 represent the fatigue driving state. The relevant testers have agreed to test and publish test data. The relevant testers include two drivers and two testing engineers.

Table 1 Stanford sleepiness scale

Scale	Meaning
1	Extremely alert
2	Alert, attention can be focused
3	Attention cannot be focused
4	Some signs of sleepiness
5	Sleepy, no effort to stay awake
6	Sleepy, some effort to stay awake
7	Very sleepy, great effort to keep awake

3.2 Steering Wheel Angle Analysis Based on Timing

Steering wheel angle data from the actual road test changes over time as shown in Fig. 1. Figure 1a represents the normal driving state, Fig. 1b represents the quasi-fatigue driving state, and Fig. 1c stands for the fatigue driving state. From Fig. 1, it is difficult to observe the internal variation between steering wheel angle and driving fatigue from its external changing. Therefore, it is necessary to use information processing technology to characterize and extract, and find its internal changes.

In this paper, AR model was used to do timing analysis for the three states of the steering wheel angle. The fourteen AR model parameters were used as the eigenvectors of the steering wheel angle in the normal driving state, the quasi-fatigue driving state, and the fatigue driving state and then acted as the mode vector of the driving fatigue detection based on the steering wheel angle characteristic, which can be expressed as follows: $Xi = \{\varphi i1,, \varphi i2, \varphi i3, \varphi i4, \varphi i5, \varphi i6, \varphi i7, \varphi i8, \varphi i9, \varphi i10, \varphi i11, \varphi i12, \varphi i13, \phi i14\}$; Xi is eigenvector of the i th signal; φij represents j th $(j = 1, 2, \Lambda, 14)$ parameter of the eigenvector corresponding to the ith state; the AR model 14 order is derived from the FPE criterion [13].

3.3 The Driving Fatigue Detection Model Establishment

3.3.1 Design of Input and Output Vectors

The fourteen-dimensional eigenvector of the steering wheel angle extracted by the AR model was used as the input vector of the detection model. The classification status of the driving fatigue was identified, which is used as the output vector. The normal driving status is 1, the quasi-fatigue driving status is 2, and the fatigue driving status is 3.

As the fourteen-dimensional eigenvector had higher dimension and larger the data volume, this paper chose PCA method [14] to reduce the sample data dimension and eliminate the above bad effects, and improve the accuracy of the

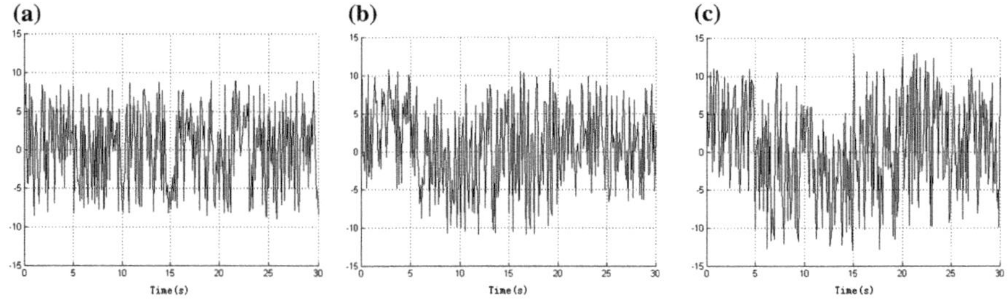

Fig. 1 Steering wheel angle timing diagram of three driving fatigue states

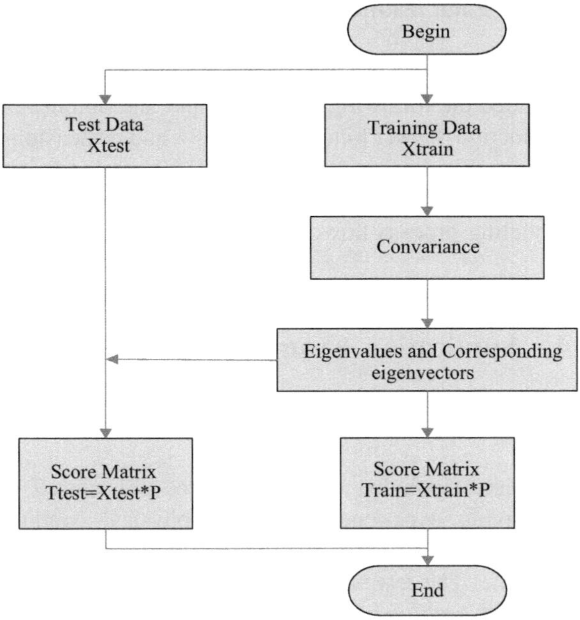

Fig. 2 PCA dimensionality reduction flowchart

model. The percentages of the principal component of dimensionality reduction in this paper were determined by the actual simulation experiments. The process flowchart was shown in Fig. 2.

3.3.2 Selection and Optimization of Classifiers

As mentioned above, this paper chose SVM as the classifier of the detection model. SVM used different kernel functions to obtain different SVM algorithms [6]. At present, the most common kernel functions in practical problems include linear kernel function, polynomial kernel function, Gaussian radial basis function, and sigmoid kernel function. This paper's kernel function was determined by the results of actual simulation experiments.

As mentioned above, this chapter also used the cross-validation method to optimize the SVM. The main steps of the optimal parameter selection based on the cross-validation method were as follows: (1) First, the penalty parameters c and the kernel function parameters g were taken values within a certain range; (2) then, for a certain group of c and g, the sample set was used as the original data set to obtain the classification accuracy of the classifier by applying the above K-CV method. That is the classification accuracy of the classifier in the CV sense; (3) finally, the final set of c and g which had the highest classification accuracy of the classifier acted as the best parameters.

3.3.3 Establishment of SVM-Based Model

In summary, the modeling process based on SVM driving fatigue model mainly included the following steps: ① input and output vectors; ② sample collection and production, kernel function selection, and dimension reduction processing; ③ SVM parameter optimization based on *K*-CV method; ④ training SVM, identifying the test set, solving the recognition accuracy, and choosing the best parameters. Modeling process flowchart was shown in Fig. 3.

4 Application Examples

4.1 Experimental Design

The testers included a driver and three assistants. The driver has more than 10 years of driving experience and good physical condition; the experiment time was selected from 9:00 am to 12:00 pm, 13:00 to 17:00 pm, with the middle interval of 1 h time. The test was divided into A and B groups. At 12 o'clock to 13 o'clock, driver of A group could go to sleep and have a rest, but driver of B group would not be allowed to sleep.

Fig. 3 SVM-based model flowchart

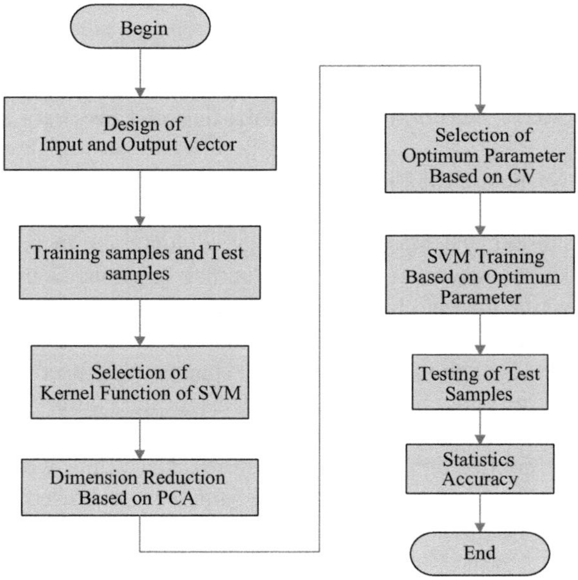

4.2 Collection and Preprocessing of Sample Data

Using the above timing analysis method, the 60 sets of steering wheel angle signal in the above three states were extracted, and the feature vector and the corresponding state number were taken as samples of the model. Finally, 180 groups were obtained. According to a general principle of testing samples (two-thirds of the data as training samples and one-third of the data as test samples), 120 groups of training samples and 60 sets of test samples were randomly selected. Part of the sample data was shown in Table 2.

The preprocessing of the sample as described above refers to the reduced dimension processing of the PCA method. The percentage of the principal component was 85–99%. When the percentage of the main component was 95%, the accuracy of the model was up to 70.00%. The cumulative contribution of 95% of the percentage of the principal component was shown in Fig. 4.

4.3 Simulation Results and Analysis

Based on the above method and data, SVM classifier was implemented by the support vector machine toolbox and MATLAB software, and the Gaussian radial basis function was selected by concrete sample simulation. The method of partitioning the coordinate network was used to select the penalty parameters c and kernel function parameters g of the support vector machine, and the cross-validation was taken as 5 which meant the validation set of classifier was divided into five

Table 2 Example of sample data

Sample number	1	2	3	4	5	6
Input	0.0332	0.0435	−0.0929	0.0410	−0.0544	0.0343
	−0.0295	0.0446	−0.0945	−0.0478	−0.0056	−0.0044
	0.0223	−0.0542	−0.0055	0.0201	−0.0553	0.0147
	−0.0456	0.0285	−0.0527	−0.0159	−0.0150	−0.1017
	−0.0415	−0.1220	−0.0187	−0.0389	−0.0148	−0.1097
	−0.0839	−0.0622	−0.0256	−0.1514	−0.0562	−0.0153
	0.0354	−0.0151	−0.0212	−0.0041	−0.0523	−0.0748
	−0.0706	−0.0189	0.1327	−0.0158	0.00511	−0.0518
	−0.0980	−0.0679	−0.0326	−0.0868	−0.1022	−0.0557
	−0.0194	−0.0226	−0.0461	0.0369	0.09712	−0.0632
	−0.0994	−0.0461	−0.0692	−0.1265	0.03874	0.0296
	0.0754	−0.0311	0.0042	−0.1721	−0.0241	−0.0507
	0.0686	0.0411	−0.0929	0.0410	−0.0544	0.0343
	−0.0561	−0.0030	−0.0945	−0.0478	−0.0056	−0.0044
Output	1	1	2	2	3	3

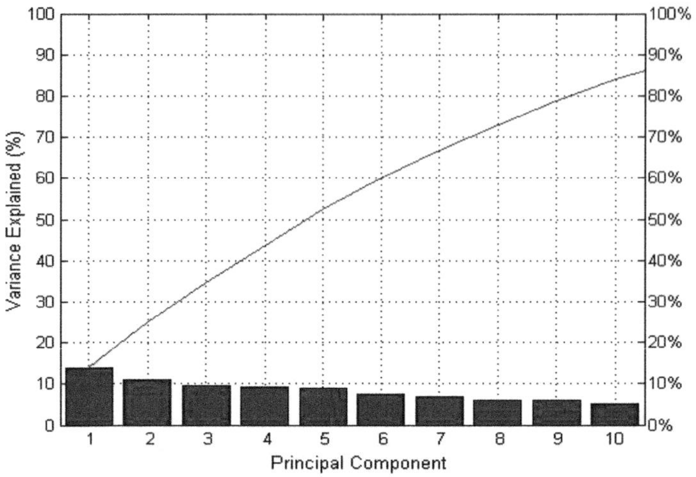

Fig. 4 Cumulative contribution of 95% of the principal component percentage

Fig. 5 3D view of parameter selection results

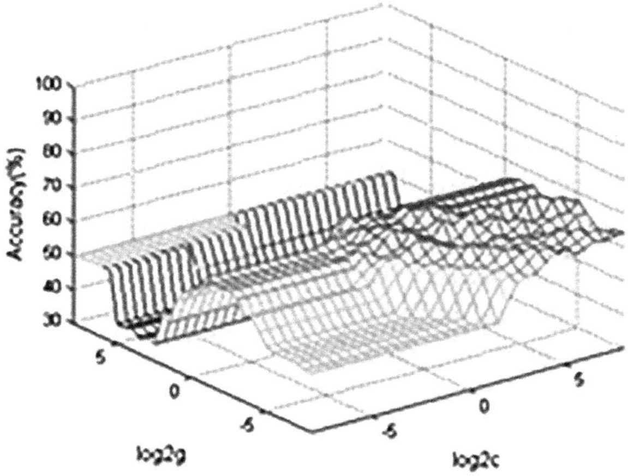

parts. The best penalty parameter was 64.00, the best kernel function was 0.0625, and the final classification accuracy of classifier was 69.17% in CV. The result of the parameter selection was shown in Fig. 5.

The meaning of the graph: x-axis represented the value after c taking the logarithm of 2; y-axis represented the value after g taking the logarithm of 2; the contour of the three-dimensional graph showed the accuracy rate of K-CV method by using the corresponding c and g; In figure, the values range of c and g were $2^{(-8)}$–$2^{(8)}$, and the step of the network division was 0.5, which meant the values of c and g were obtained according to the form $2^{(-8)}, 2^{(-7.5)}, \ldots, 2^{(8)}$.

Table 3 Test results of driving fatigue

	Normal driving state	Quasi-fatigue driving state	Fatigue driving state	Total detection rate for driving fatigue
Number of samples	20	20	20	60
Number of detected	17	20	17	54
Number not detected	3	0	3	6
Detection rate	85.00%	100.00%	80.00%	90.00%

The driving fatigue test model was simulated by using the combination of the best parameters, and the test results of the driver's driving fatigue state were shown in Table 3. The detection rate was 85.00% for the normal driving state, 100.00% for the quasi-fatigue driving state, and 80.00% for the fatigue driving state, and the total detection rate for driving fatigue was 90.00%.

5 Summary

Aiming at the problem of driving fatigue detection, this paper applied the support vector machine to detect the driving fatigue state. Firstly, the steering wheel angle eigenvector was proposed as the input vector by using the timing method. According to the Stanford sleepiness scale, the driving fatigue state was divided into three different state scales which were the output vector. Secondly, according to the support vector machine toolbox and using MATLAB software to compile the program, the best parameters of support vector machine were chosen by using cross-validation method after the reasonable selection of kernel function and dimension reduction degree, and then, the mode using the best combination of parameters was simulated. The results showed that the detection model had a high detection rate, which proved that the driving fatigue detection model could effectively detect the fatigue of driver, and had a good application and promotion.

References

1. Sun W, Zhang W, Zhang X, Chen G (2009) Research progress of fatigue driving early warning system. Auto Electr Parts 1:4–8
2. Yan S, Wei J-Q, Wu Y-H (2005) Study of EEG features extraction for Doze car drivers.Chin. J Biomed Eng 24(1):110–113
3. Fang R-X (2010) Experimental research of effectiveness of pre-warning countermeasure of driving fatigue based on physiological. Beijing University of Technology

4. Kithil PW, Jones RD, John M (1998) Development of driver alertness detection system using overhead capacitive sensor array. SAE Technical Paper Series 982292
5. Guan Y, Song L (2010) Driver eye state detection based on support vector machine-decision tree. J Harbin Univ Sci Technol 15(6):5–8
6. Vapnik V (1998) Statistical learning theory. Wiley, New York, NY
7. Hsu CW, Lin CJ (2002) A comparison of methods for multi-class support vector machines. IEEE Trans Neural Network 13(2):415–425
8. Tan X (2009) The application of support vector machine in wavelet packet De-noising. Shanghai Jiao Tong University
9. Huang J (2011) The research of support vector machine algorithm and the application in meteorological data mining. China University of Mining and Technology
10. Tang B-P, Liu W-Y, Jiang Y (2010) Parameter optimized morlet wavelet de-noising method on cross validation method. J Chongqing Univ 33(1):1–5
11. Liu XY, Li P, Gao C (2011) Fast leave one out cross validation algorithm for extreme learning machine. J Shanghai Jiao Tong Univ 45(8):1140–1144
12. Chalder T, Berelowitz G, Pawlikowska T et al (1993) Development of a fatigue scale. J Psychosom Res 37(2):147–153
13. Yin A-D (2002) Research on the applications of fuzzy classification in the gears fault. HeFei University of Technology
14. Yang WK, Ji SB, Ren MW (2009) Face extraction based on enhance 2D PCA and its application to face recognition. J Image Graph 14(2):227–232

Study on Multiaxial Loading Fatigue Test of Steering Knuckle Based on Road Load Spectrum

Chunpeng Tang, Xianzhen Luo and Yadong Wu

The steering knuckle is the key load-bearing component of automobile suspension. Almost all the ground loads are transmitted to the vehicle body through the steering knuckle, and its fatigue performance is directly related to the safety of automobile and occupant.

There are three main methods for fatigue test in the world, road test, proving ground test and the bench test. Among them, the road test and proving ground test need a long period of testing, cannot be accelerated, with high cost, and are influenced by environmental factors [1, 2]. At the same time, the test boundary conditions of each round are not completely consistent. The bench test has the advantages of low cost, short test time, consistent boundary conditions. But the bench test load is often the uniaxial constant amplitude load [1, 3], and it is different from the automobile's real load.

In this paper, a fatigue test method based on multiaxis test system (MTS) and remote parameter control (RPC) technology [4] is developed. The MacPherson suspension steering knuckle is taken as the experimental object. The research contents include test bench design, the road load spectrum acquisition, processing and compression,load spectrum getting, iteration tests.

C. Tang (✉) · X. Luo · Y. Wu
China Automotive Engineering Research Institute CO., Ltd.,
No. 9 Jin Yu Road, North New District, Chongqing 401122, China
e-mail: tangchunpeng@caeri.com.cn

Society of Automotive Engineers of China (SAE-China) (ed.), *Proceedings of the 19th Asia Pacific Automotive Engineering Conference & SAE-China Congress 2017: Selected Papers*, Lecture Notes in Electrical Engineering 486, https://doi.org/10.1007/978-981-10-8506-2_50

1 Multiaxial Loading Fatigue Test Principle and Bench Design

As shown in Fig. 1, the loads generated by the ground are transferred from the tire, wheel and hub to the steering knuckle when the vehicle is moving. Then, the loads are transferred from the steering knuckle to the body, through the upper column, the steering rod and the controlling arm.

According to the force and the counterforce of interchangeable principle, design of the test bench has two ideas. First is to restrain the steering knuckle body side and apply six-dimensional wheel force at the wheel hub, as shown in Fig. 2; the other is to restrain the steering knuckle tire side and apply force at the body side, as shown in Fig. 3.

In the test by restraint body side, the six-dimensional wheel force transducer is used as the test load control point. The advantage is that the six-dimensional wheel force load spectrum can be used as the test load spectrum directly, which is more intuitive. In the test by restraint tire side, the joint force sensors of steering knuckle and body are used as the test load control point. The test load spectrum should be decomposed from the six-dimensional wheel force load spectrum. But the test by restraint tire side does not require the six-dimensional wheel force transducer involved in the test, can reduce the test cost and avoid damage to six-dimensional wheel force transducer. At the same time, all loading actuator can be arranged horizontally, reducing the difficulty of test bench establishment.

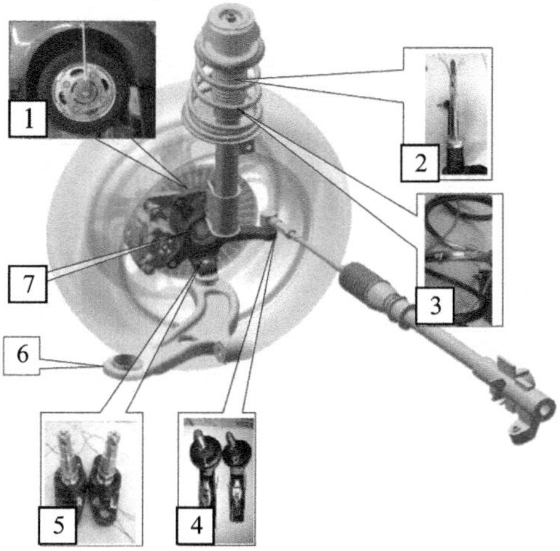

1- wheel hub and six-dimensional wheel force acquisition point;2- shock absorber and Z strain acquisition point;3- spring and shear strain data acquisition point;4- steering rod and Y strain acquisition point;5- ball hinge and XY strain acquisition point; 6- controlling arm;7- steering knuckle.

Fig. 1 MacPherson suspension and road load spectrum data acquisition point

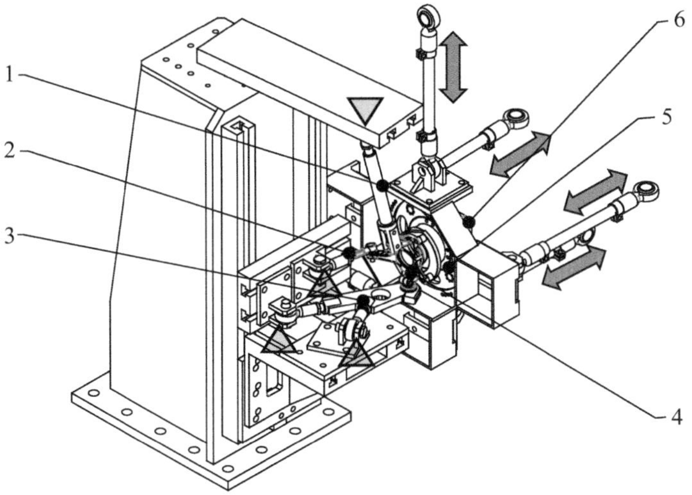

1- Simulated upper column;2- Simulated steering rod;3- Simulated controlling arm;4- Steering knuckle; 5- Adapter for six-dimensional wheel force transducer; 6-Six-dimensional wheel force transducer.

Fig. 2 Test by restraint body side

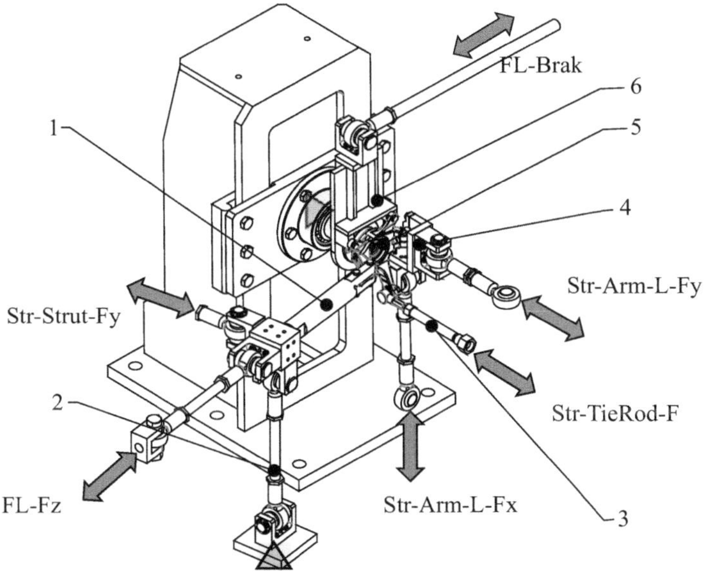

1- Simulated upper column;2- Two-force rod; 3- Simulated steering rod;4- Simulated controlling arm; 5- Steering knuckle; 6- Simulated brake disc.

Fig. 3 Test by restraint tire side

Comparing the two test methods, the test by restrained tire side is adopted in this paper. The final design is shown in Fig. 4. The steering knuckle is mounted on the mounting seat through a core shaft and a pair of tapered roller bearings, as shown in Fig. 5, and $\bar{X}, \bar{Y}, \bar{Z}$ translational degree of freedom (DOF) and $\widehat{X}, \widehat{Z}$ rotational DOF in the vehicle coordinate system are constrained. As shown in Fig. 3, the $\widehat{Y}$ rotational DOF is constrained by the two-force rod. Six electro-hydraulic servo actuators apply vertical and lateral loads on the simulated upper column, apply steering load on the simulated steering rod, apply longitudinal and lateral loads on the simulated controlling arm and apply braking torque on the simulated brake disc. By controlling the magnitude and phase relation of the loads in each channel, the load state of the steering knuckle in the test process will be consistent with that in the road test.

2 Road Load Spectrum Acquisition

The road load spectrum acquisition site is the strengthening test section of the Chongqing Xibu Proving Ground, as shown in Fig. 6, and the full length is 4.4 km. The sections used for data acquisition mainly includes eight-shaped reversing road, body twisting road, washboard road, shoal road, railroad, long wave road, vibration road, seam road, hollow road, gravel road, left ramp road, Belgian road, boulder road, regular long wave road, city residential road, gravel road. The sampling frequency is 512 Hz. The road spectrum's rain flow matrix of combined typical road surfaces should be consistent with the road spectrum rain flow matrix of the user's target road.

The acquired signals include the six-dimensional wheel force signals, the *Z* strain signal of the shock absorber piston rod, the shear strain signal of the spring, the strain signal of the steering rod and the *XY* strain signal of the controlling arm ball joint. The data acquisition points are shown in Fig. 1.

3 Test Load Spectrum

3.1 The Preprocessing of the Six-Dimensional Wheel Force Load Spectrum

The original signals acquired may be mixed with some non-real signals due to interference and so on, resulting in zero drift, trend, high-frequency noise and burrs. In order to ensure the reliability and reality of road load spectrum, it is necessary to preprocess the original signal. The pretreatment process includes not only deburring, trend elimination, low-pass filtering but also stationarity test, distribution characteristic check for the signals acquired in different times.

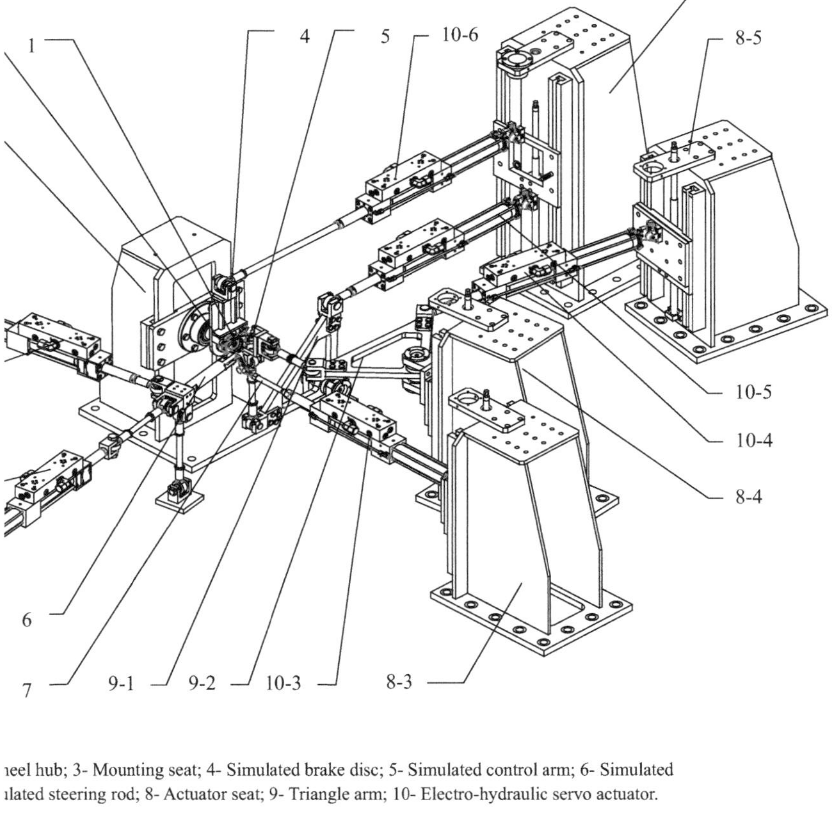

Fig. 4 Steering knuckle six-axle loading fatigue test bench

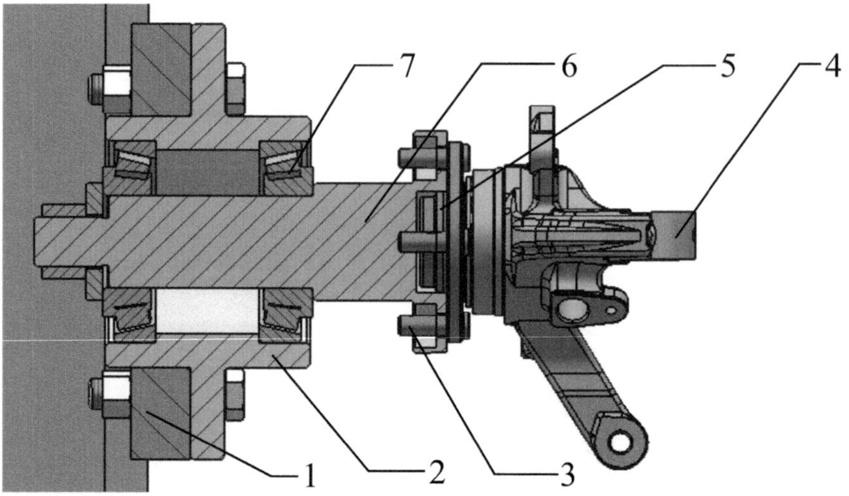

1- Mounting board; 2- Mounting flange; 3- Wheel bolt; 4- Steering knuckle; 5- Wheel hub; 6- Core shaft; 7- Tapered roller bearing.

Fig. 5 Steering knuckle mounting core shaft

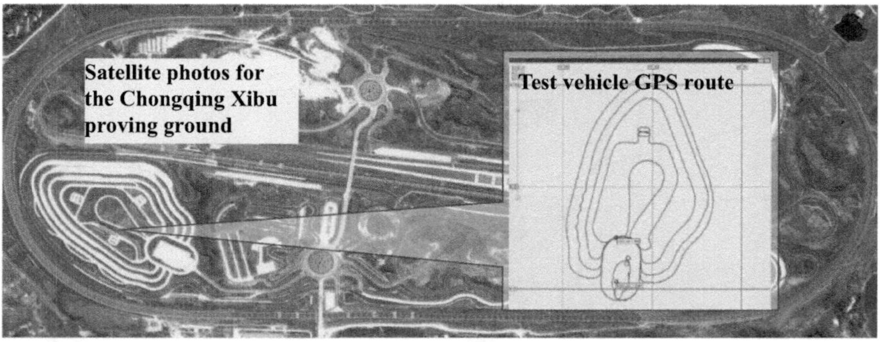

Fig. 6 Road load spectrum acquisition site

3.2 Compression Method of the Six-Dimensional Wheel Force Load Spectrum Based on Time-Domain Damage

3.2.1 Fatigue Damage Calculation Method of the Steering Knuckle

The design life of the steering knuckle is relatively long. If the target load spectrum is not compressed, the bench test will not be accelerated, which will lead to high test cost [6].

The correct compression method used for load spectrum should ensure the damage before and after compression as the same and should meet the following two requirements: (1) the total damage before and after compression is basically the same; (2) the damage distribution and histogram before and after compression are consistent. For different fatigue types, different fatigue analysis methods should be adopted. The fatigue of the steering knuckle belongs to the high cycle fatigue (HCF) and is applicable to the stress–life analysis method (SN method).

Multiaxis damage cumulative rule generally adopts the Dang Van criterion, but the application of this criterion is complicated. The purpose of the damage calculation is to compare the equivalence of the load spectrum before and after compression. Therefore, a simple and practical Miner criterion is used to calculate the total damage.

According to the document [5], the steady-state cyclic stress–strain curve can be expressed as:

$$\varepsilon_a = \varepsilon_{ea} + \varepsilon_{pa} = \frac{\sigma_a}{E} + \left(\frac{\sigma_A}{K'}\right)^{\frac{1}{n'}} \tag{1}$$

The strain–life curve is expressed as:

$$\varepsilon_a = \varepsilon_{ea} + \varepsilon_{pa} = \frac{\sigma_f'}{E} \cdot (2N)^b + \varepsilon_f' \cdot (2N)^c \tag{2}$$

The stress–life curve is expressed as:

$$\lg N = A + B \cdot \lg S \tag{3}$$

Miner linear fatigue cumulative damage expression can be expressed as:

$$N = 1/D = 1 \Big/ \left(\sum_{i=1}^{l} \frac{n_i}{N_i}\right) \tag{4}$$

$$\varepsilon_a = \frac{\sigma_f'^2 - \sigma_m^2}{E \cdot \sigma_f'} \cdot (2N)^b + \varepsilon_f' \cdot (2N)^c \tag{5}$$

Where ε_a strain amplitude; ε_{ea} elastic strain amplitude; ε_{pa} plastic strain amplitude; σ_a stress amplitude; E modulus of elasticity; K' cyclic strength coefficient; n' cyclic strain hardening exponent; σ_f' fatigue strength coefficient; ε_f' fatigue ductility coefficient; b fatigue strength index; c fatigue strength exponent; A and B coefficient; N fatigue life (times); D total fatigue damage; n_i number of cycles at each load level; N_i fatigue life of each load; σ_m stress average.

According to the above formula and the characteristic parameters of materials, the total damage value of any point caused by uniaxial random load can be calculated. For multiaxial loads, ε_a principal strain amplitude; ε_{ea} principal amplitude of elastic strain; ε_{pa} principal plastic strain amplitude; σ_a principal strain amplitude.

3.2.2 Six-Dimensional Wheel Force Load Spectrum Compression Process

Under the given load spectrum, the damage values of each part of the steering knuckle are different, and the fatigue life of the steering knuckle depends on the fatigue life of the most dangerous points. The most dangerous point is taken as the load spectrum compression process's target point, and the process is shown in Fig. 7.

Finite element method is used to calculate the node stress under the action of separated six-dimensional force. After the results are imported into nCode DesignLife software, the principal stress and its direction of each node under loads are calculated by the nCode software, in combination with the six-dimensional wheel force load spectrum and material data. According to the stress of the most dangerous point, the time slice with small damage value in the load spectrum is excised to form a compressive load spectrum. The total damage values of the most dangerous points before and after compression are calculated, respectively. When the total damage value after compression is more than 95% and the compression ratio of load spectrum is greater than 30%, the compressed load spectrum is preserved. Otherwise, the compression parameters are adjusted for the next round of calculation.

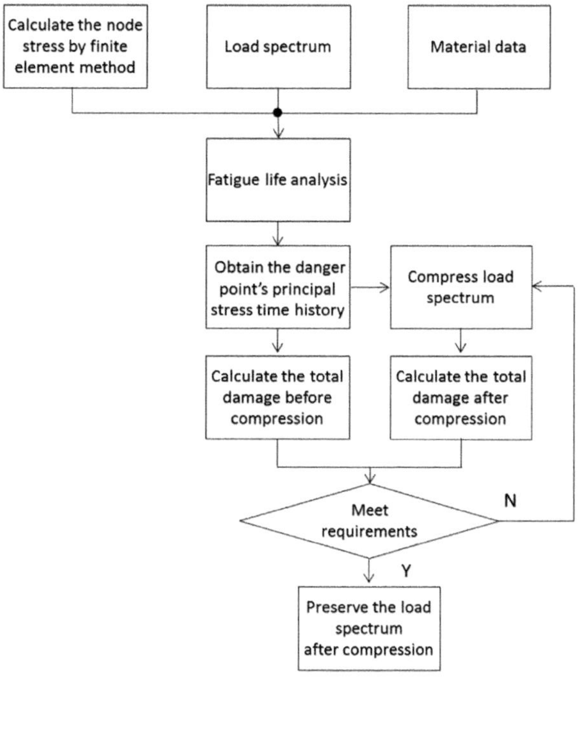

Fig. 7 Flow block diagram of load spectrum compression

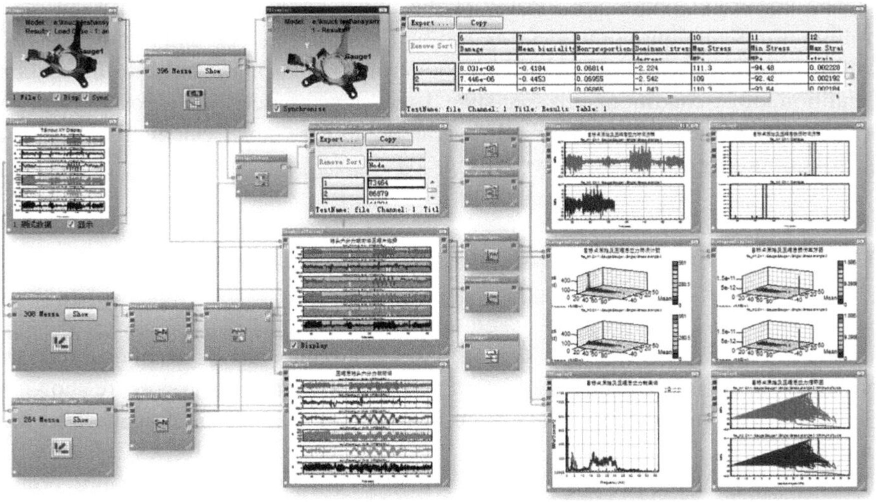

Fig. 8 Load spectrum compression process flow based on nCode DesignLife software

The load spectrum compression process designed by nCode DesignLife software is shown in Fig. 8.

3.2.3 Finite Element Analysis of Steering Knuckle

The steering knuckle is made of cast steel with a yield strength of 230 MPa and a tensile strength of 450–600 Mpa. The 3D model of the steering knuckle is introduced into the ANSYS software to mesh, and the adaptive mesh is used to refine the transition region and the connecting part. The meshing of the steering knuckle results in 111565 units and 187071 nodes, as shown in Fig. 9.

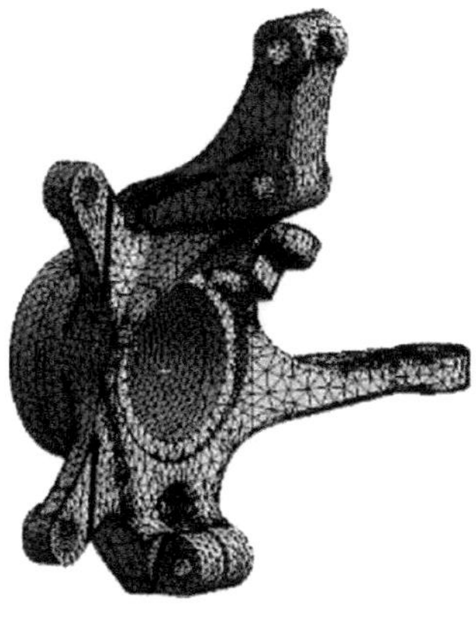

Fig. 9 Mesh of finite element

The constraint points are located at the upper column installation point, controlling arm mounting point, steering rod ball joint installation and hydraulic brake calliper mounting points. The calliper mounting point applies only to the braking condition, other conditions in a free state. Load steps are set to six steps, in turn, *Fx*, *Fy*, *Fz*, *Mx*, *My* and *Mz* are applied separately, and the stress calculation results of each load are recorded, as shown in Fig. 10.

3.2.4 Comparison and Analysis of Load Spectrum Compression Results

Taking the load spectrum of six typical pavement combinations as an example, the principal stress and damage time history before and after compression are shown in Fig. 11. The compression ratio of time history is 53%, and the damage retention ratio is 98%.

The histogram of the main stress rain flow before and after compression is shown in Fig. 12, and the damage histogram is shown in Fig. 13. As can be seen from the diagram, the low-amplitude loads which have no contribution to the damage but have a large number of cycles are removed, and the loads which contribute much to the damage are well preserved. The damage histogram is basically consistent before and after compression.

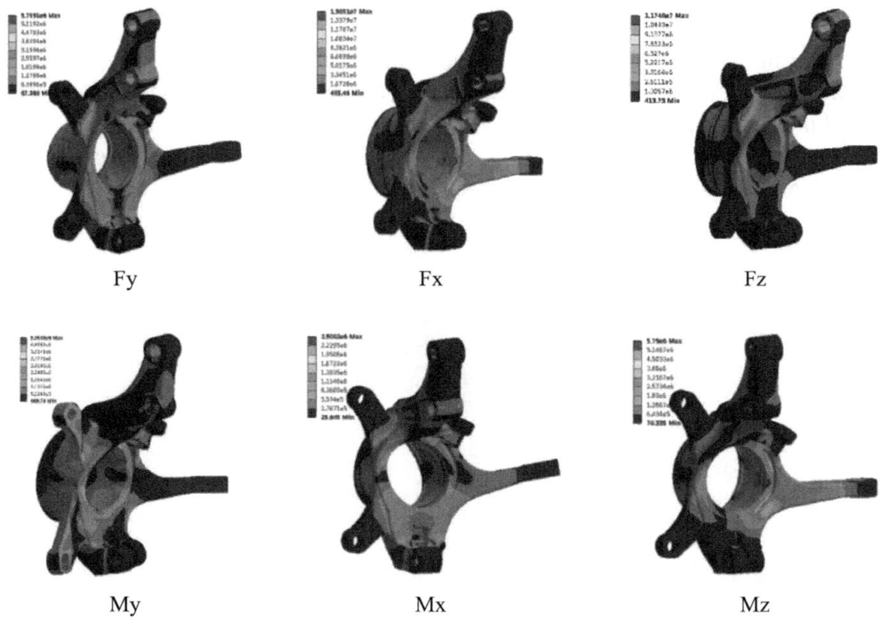

Fig. 10 Stress nephogram of different loads

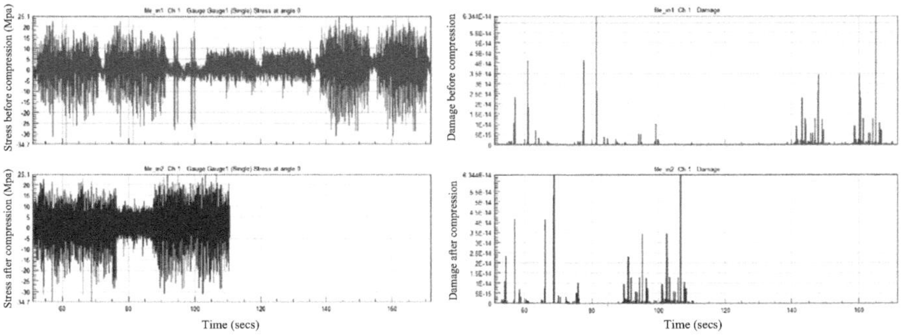

Fig. 11 Principal stress and damage time history before and after compression

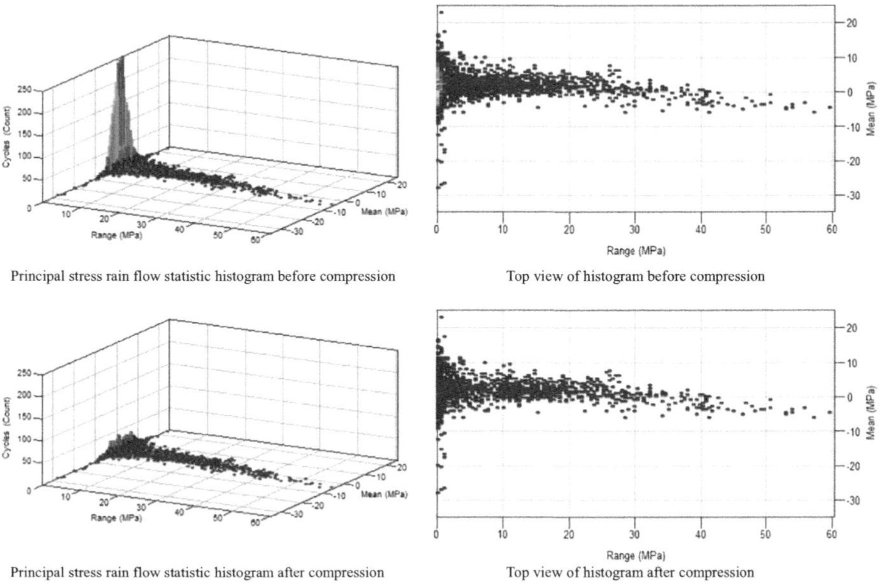

Fig. 12 Principal stress rain flow statistic histogram before and after compression

From the analysis of frequency domain, the power spectrum of the main stress before and after compression is shown in Fig. 14. Except for the 3–6 Hz frequency range, the amplitude, frequency and frequency of the peak are basically the same, and the signal's frequency characteristics have not changed greatly. At the same time, the area under the curve is basically the same, so most of the energy is preserved.

Based on the compression process of the dangerous point stress, the corresponding six-dimensional wheel force load spectrum synchronous compression process is shown in Fig. 15, in which the white is the reserved part and the green is

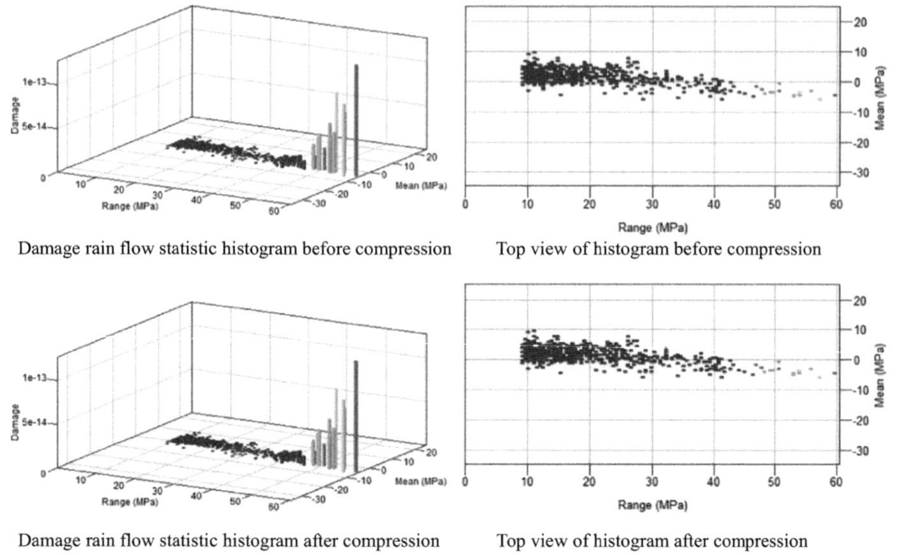

Damage rain flow statistic histogram before compression

Top view of histogram before compression

Damage rain flow statistic histogram after compression

Top view of histogram after compression

Fig. 13 Damage rain flow statistic histogram before and after compression

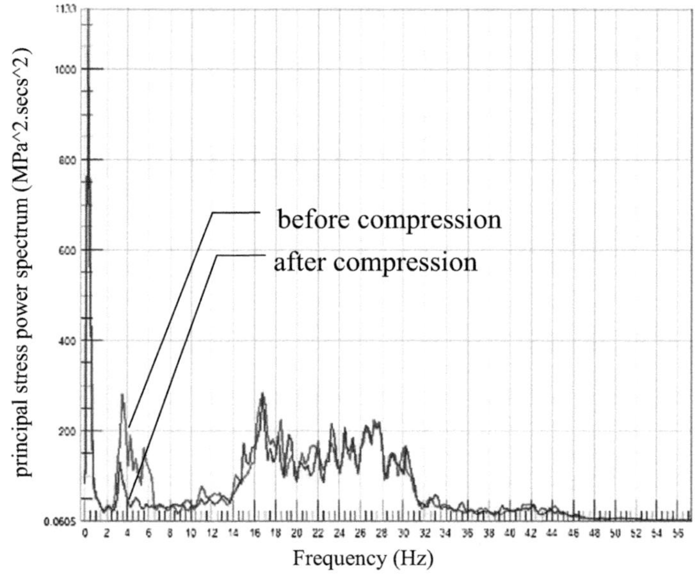

Fig. 14 Principal stress power spectrum before and after compression

the compression and deletion part. The compression ratio of the six-dimensional wheel force load spectrum is the same as the dangerous point, and the compression ratio is 53%.

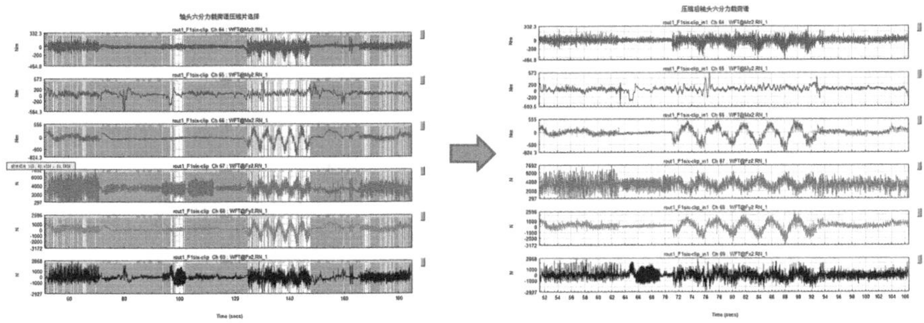

Compressed time slice selection   Load spectrum after compression

Fig. 15 Six-dimensional wheel force load spectrum compression

3.3 *Obtain the Test Load Spectrum*

According to the test principle, the six-dimensional wheel force load spectrum needs to be decomposed to the joints of the steering knuckle and the body, and complete the test load spectrum.

The load decomposition model of steering knuckle is set up for ADAMS, as shown in Fig. 16. In order to ensure the correct geometric relationship, the steering knuckle, the simulated upper column, the simulated controlling arm ball hinge, the simulated brake disc and the simulated steering rod are all imported into the ADAMS by 3D model. The electro-hydraulic servo actuator is simplified as a two-force rod or spring. The six-dimensional wheel force load spectrum is applied at the hub of the steering knuckle to extract the load spectrum at the joint of each body, which is the test load spectrum.

Fig. 16 Load decomposition model

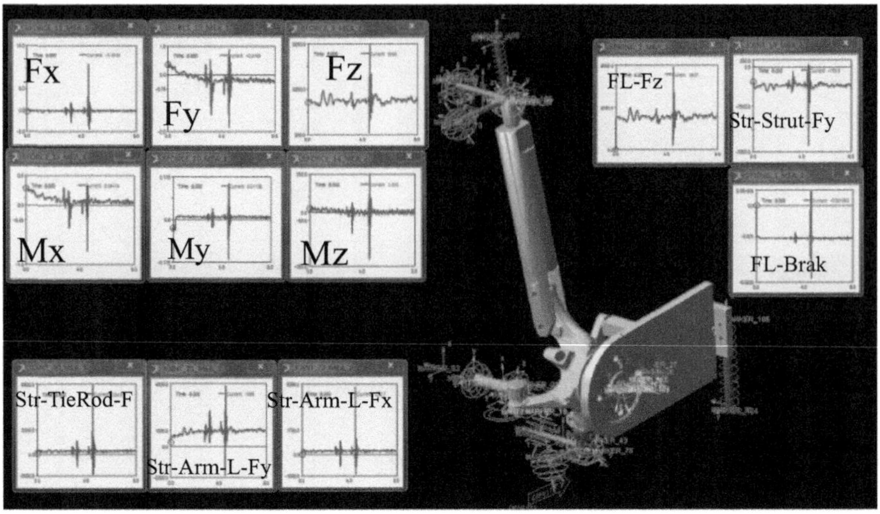

Fig. 17 Decomposition result of hollow road load spectrum

Take a section of hollow road load spectrum for example, the six-dimensional wheel force load spectrum and test load spectrum are shown in Fig. 17 (definition of loads is shown in Fig. 3).

4 Multiaxial Loading Bench Fatigue Test

The multiaxle loading bench fatigue test is shown in Fig. 18. Six electro-hydraulic servo actuators use the test load spectrum as the target response signal to carry out fatigue test. All load channels work at force control mode.

Fig. 18 Multiaxial loading fatigue test bench

The test system is a typical multi-input and multioutput system, and the whole system is nonlinear, including sample, fixture and measurement system. This will result in errors between the response and the desired target obtained from the linear system assumption. In order to eliminate the nonlinear influence, it is necessary to modify the driving signals of six electro-hydraulic servo actuators by the iterative calculation method, so that the response signal of the system approaches the target signal. The main iterative steps by MTS and RPC software are as follows.

(1) Use the broadband white noise signal as the six electro-hydraulic servo actuator driving signal and signals of six force sensors as the response signals, calculate driving signal's auto-power spectrum $S_{AA}(f)$, and calculate cross-power spectrum $S_{AB}(f)$ between the response signal and the driving signal. Finally, the frequency response transfer function $H(f)$ of the test system is:

$$H(f) = \frac{S_{AB}(f)}{S_{AA}(f)} \tag{6}$$

(2) Calculate the first round driving signal of the system according to the target response signal.

$$X_0(f) = [H(f)]^{-1} \cdot \alpha \cdot Y_T(f) \tag{7}$$

$$x_0(t) = IFFT[X_0(f)] \tag{8}$$

$Y_T(f)$: Target response signal in frequency domain; $X_0(f)$: first round frequency-domain driving signal; α: gain; $x_0(t)$: first round time-domain driving signal.

(3) The first round time-domain driving signal is used to excite the system, and the response signal is acquired to calculate the error between the response signal and the target signal. A new driving signal is generated by correcting the original driving signal according to the error, and the process is iteratively repeated until the response signal error RMS meets the requirements, and the final drive signal of the system is obtained.

$$e_i(t) = y_t(t) - y_i(t) \tag{9}$$

$$\Delta Y_i(f) = FFT[e_i(t)] \tag{10}$$

$$\Delta X_i(f) = [H(f)]^{-1} \cdot \beta \cdot \Delta Y_i(f) \tag{11}$$

$$\Delta x_i(t) = IFFT[\Delta X_i(f)] \tag{12}$$

$$x_{i+1}(t) = x_i(t) + \Delta x_i(t) \tag{13}$$

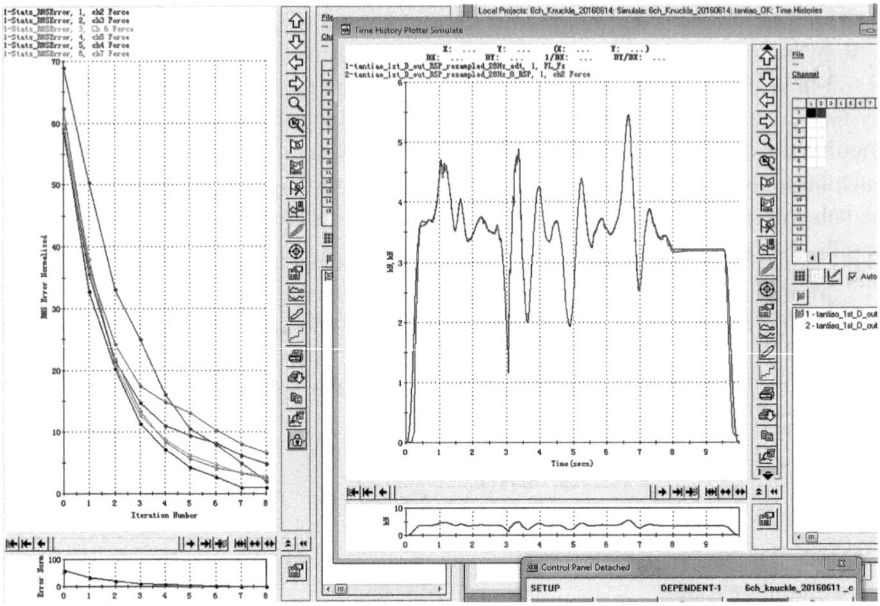

Fig. 19 Error convergence of multiaxial loading fatigue test

$y_t(t)$: Target response signal in time domain; $y_i(t)$: time-domain response signal; $e_i(t)$: time-domain response error signal; $\Delta Y_i(f)$: frequency-domain response error signal; β: gain; $\Delta X_i(f)$: frequency-domain driving error signal; $\Delta x_i(t)$: time-domain driving error signal; $x_{i+1}(t)$: modified time-domain driving signal.

After the N iteration, the response signal error RMS converges percentage is calculated by the following formula:

$$e_n = \frac{\sqrt{\sum_t [y_t(t) - y_n(t)]^2}}{\sqrt{\sum_t [y_i(t)]^2}} \cdot 100\% \quad (14)$$

e_n: The time-domain response error value converges percentage after the N iteration.

The iteration effect is shown in Fig. 19. After nine iterations, two-channel RMS converges to 6 and 8%, and the rest of the channels converge to less than 5%.

5 Conclusion

In this paper, the road test of steering knuckle has been successfully reproduced on the indoor bench. Explore a complete set of test methods, including bench design, road load spectrum acquisition, road load spectrum processing and compression, the road load spectrum is decomposed into the test load spectrum, the bench test iteration stages, provides a high precision, low cost and high efficiency of the automotive components fatigue verification method. Through the study, we can draw the following conclusions:

(1) Based on the software and hardware of MTS company, combined with the design of the platform, the actual road test can be successfully reproduced on the indoor bench, and the convergence accuracy can be stably controlled within 10%.
(2) The accelerated test method based on finite element and time-domain damage analysis can commendably keep the original load spectrum damage to the specimen and greatly shorten the test time and reduce the test cost.
(3) The test bench designed by equivalent principle can greatly reduce the complexity and avoid the risk of damage to the six-dimensional wheel force transducer under the premise of ensuring the quality of the test.

References

1. Shi X, Wang R, Zou X, Qiao K (2013) Design of three-channel road simulator for motorcycle frame. Mach Des Manuf 3:194–196
2. Fan Y, Fanxing M, Shaoming Z (2010) Research on the theoretical basis of vehicle reliability road simulation technology. Sci Times 2:124–125
3. Xueling W, Chuanxing B, Qing L (2010) Method for establishing drive spectrum based on long time history time domain control technology. J Heifei Univ Technol 33(11):1601–1604
4. Xu PJ, Wong D (2005) Pierre leblanc, gerry peticca. Road test simulation technology in light vehicle development and durability evaluation. SAEp5aper2005-01-0584
5. Lee YL, Pan J, Richard B (2011) Hathaway fatague testing and analysis theory and practice. National Defense Industry Press, Beijing
6. Cheng YG, Zheng SL, Feng JZ (2009) A research on the optimization of load spectrum for durability test of a car suspension. Automot Eng 31(5):467–470

Visual Equity Matrix: A New Perspective of Enterprise Operation in the Automotive Industry

Xiai Chen and Jixiang Chen

Nomenclature

BCG Boston Consulting Group
OBC On Board Charger
PEPS Passive Entry Passive Start
VEM Visual Equity Matrix

1 Introduction

Although the domestic automotive industry in China has been a rapidly growing development, there is still considerable disruption going on in automotive now than at any time since cars replaced horses as the dominant mode of long-distance transportation [17]. First, accompanied by the China's Internet Booming [6], many enterprises have announced plans to build "the Internet Auto"; it is reported that Le SEE, NEXTEV, XPENG, SINGULATO, Weltmeister, CHEHEJIA, YOUXIA, and many other Internet car brand will be in production in the year of 2017. The most important technical expand in car with big data, social media, and Internet of things, which become the foundation of business models, is taking place in accompany with the rapid development of China's Internet.

Second, More than 200 companies including Alibaba Group Holding plan to build EVs in China to avoid the exhaust pollution in China's urban areas, derived by the government's allowance plans to the electric vehicles users and companies

X. Chen (✉)
R&D Department, Shanghai Kostal-Huayang Automotive Electric Co., Ltd.,
YuanGao Road no. 77, Anting, Jiading, Shanghai 201814, China
e-mail: jasminchenxiai@hotmail.com

J. Chen
Antai College of Economics & Management, Shanghai JiaoTong
University, Shanghai 200240, China
e-mail: jxchen@sjtu.edu.cn

Society of Automotive Engineers of China (SAE-China) (ed.), *Proceedings of the 19th Asia Pacific Automotive Engineering Conference & SAE-China Congress 2017: Selected Papers*, Lecture Notes in Electrical Engineering 486, https://doi.org/10.1007/978-981-10-8506-2_51

willing to develop them. But only ten new ones win the permit of government rules, such as SAIC Motor and BYD, Beijing Electric Vehicle, Hangzhou Chang Jiang Passenger Vehicle, Wanxiang Group, Zhejiang Geely Holding Group. There are so many entrants in the market of electric cars just like a Bubble [3].

Finally, the Chinese traditional automobile industry was hit hard by that two kinds of companies. Although the industrial relations of the automobile industry are tense and unstable [4], a harmonious and stable labor relationship is also needed. The new-emerging automotive companies exploit the talented person out of the traditional automotive companies by offering higher wages even with double level which induce personnel instability. There has been a common phenomenon of frequent job hopping in the automotive industry recently.

On one hand, as the second most populous country in the world, India's car market has the potential to grow up to 6 million cars by 2020 [18]. It makes the managers in traditional automotive companies scramble to adjust their recruitment budgets so they can confront the losses of staff in China. They are inclined to collaborate with India as a supplier to cut down the cost of operation and talent management in China for the example of KOSTAL China.

On the other hand, Foster said "After decades of trying, China has found a market-based strategy: building scale and experience from the bottom up. In the long run, this is likely to be far more effective than going out to buy foreign companies" [16]. Building internally is effective than external resource. The supplier collaboration management is also inseparable from an effective internal operation management. The paper is aimed to construct a visual equity matrix (VEM) in automotive industry as a new perspective of enterprise operation in the consideration of talent management to help to build scale and experience from the bottom up. The theories of value chain, talent management, BCG matrix, and 4M1E (Man, Machine, Method, Material, Environment) contribute most. VG Ortiz-Gallardo improves the knowledge regarding the influence conditions by identifying and describing the relationship between two variables: key activities and six influential factors, drawing the results from case analysis of the industries from chemicals, oil and gas and biofuels [23]. The method is widely developed by VEM in automotive industry in the paper.

Technology and market-oriented open innovation (OI) modes can positively contribute to performance enhancement for the firm performance in small- and medium-sized enterprises (SMEs) [17]. It becomes a harmonious pursuit and a management, technological innovation when designers and managers learn cybernetics about the operation of electronic machines and human brain through learning and understand the role of modeling in the field of computer collaborative work, business process management, and the impact on people [1]. The five factors of 4M1E contribute most to the firm performance in automotive industry.

Resource limitations between processes are the cause of process interdependent [5]. There is no way to figure out how to decouple resource relationships. The sustainable management of resources is the essential for long-term production in manufacture firms considering resource constrained, and roadmaps is facilitated to self-manage common resources, which the roadmaps are a visual tool, synthesizing the workshop activities into a single page format [7]. The cross-sectional perspective of roadmaps by cutting of a time point is missing. Roles collaboration framework helps to enhance the collaboration in multiple software design projects by coupling tasks and resource [2]. The contribution size of its role is not quantified. These insights widely researched into technology management literature are key to the implementation of VEM.

2 Theory and Research Question

The contents of the section are divided into two parts: research foundations of theory and research questions.

2.1 Theory

From the analysis of the company value network, we can see a large number of cooperation opportunities in the supply chain. Facing the complicated market environment and changeable customer demand, only cooperation can satisfy the demand of the products [8], and the time delay and unnecessary cost expenses can be avoided. So we are concerned about BCG matrix, Porter's Value Chain, 4M1E, and talent management.

2.1.1 Boston Consulting Group (BCG) Matrix

The market difference between a product and another product lies in the difference between the dynamic and static states of market. The transformation of product and product market share, the change of growth rate lies in the interaction of different dynamic and static states.

Boston Consulting Group proposes the BCG matrix or the growth–share matrix from the 1970s. It is widely supported to be a useful company resource analytical tool from different management fields of brand marketing, product, strategic, and portfolio analysis.

The dynamic and static states between question marks (also known as problem children) and cash cows are very different. There is a great difference between the dynamic and static states between dogs and stars. This is the state of nature, and this is Tao of nature. Motion of dynamic and static states is Tao. In KOSTAL China, PEPS is a cash cow changed from original question marks. OBC is a star in all products coming from the original of dogs.

Market share, the market growth rate is one thing between every minute. Dynamic is that the state of the market share; the market growth rate is detected, as the market is analyzed as the Boston matrix which belongs to dynamic. Static is to keep the original state of the market, not to be explored.

It will never be known that the market state is both fast/expansion and slow/contraction at a time. Just as it does not let the manager known exactly of the result of the Boston matrix and known if the strategy is incompatibility to market according to Boston matrix.

It is a shame to move the market, and static market with its original state is cultivation. Moving market is to detect the state of the market; it is still not to do this thing and feel disgrace. Static market that is not to detect the state of the market in accordance with the original state of the market has been maintained.

It is a marketing and strategic planning that made a product market to be a problem. Perhaps the product itself before just in good market prospect of bad market prospects of the ambiguous state, when the company at the wrong time to increase market share, when marketing stimulates it, its market state becomes an established state at a moment, such as the market reduction, become a state of intense competition with other products.

Just as there are four kinds of photon states in quantum mechanics, the state of each moment is uncertain, and once a detection also changes the original state of the photon. Dynamic and static state of Tao means that, in the static state, the market state is not measurable, and the market state can be detected when in the dynamic state, which it also changed market its original state (Fig. 1).

2.1.2 Porter's Value Chain

A value chain helps a company identify a series of activities that operate in a particular industry and valuable products or services. A series of activities to create value need to collaborate among the members of the enterprise. The cooperation is poor, the cost of the value chain is high, and the efficiency is low.

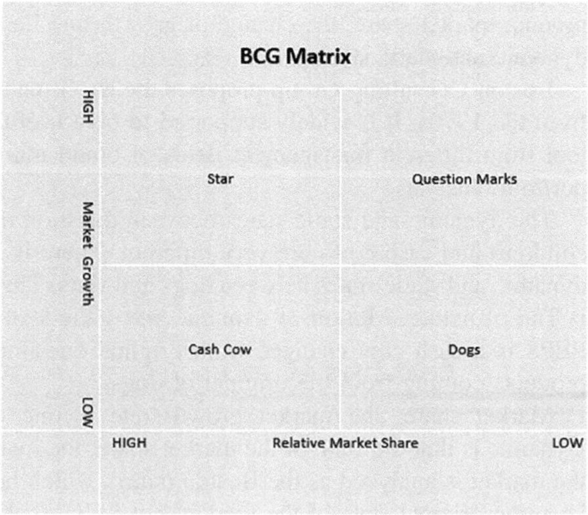

Fig. 1 BCG matrix

In the identification process of company value chain, there are four steps to follow. First, differentiate primary and support activities. Second, assign the company's costs to each individual activity. Third, assign the company's resources to each individual activity. Fourth, identify the contribution of each activity to the value creation process. This activity-based costing is to assess company's cost competitiveness. It is useful method of benchmarking to improve company's cost competitiveness.

The competition of modern enterprises is mainly reflected in the cost competition of internal execution activities of company value chain, and the cost of supplier value chain should not be ignored in the value chain and interests of the whole value chain. It is necessary to identify the different activities in the industry and identify the relations existing between these activities. It is impartment to identify the most profitable activities (Fig. 2).

2.1.3 The Five Factors of Production (4M1E)

The five factors in the management of industrial manufacturing enterprises are man, machine, material, method, and environment.

The loop of man machine material method is the five main factor affecting the quality of product in the theory of total quality management. Man refers to the product manufacturing personnel. Machine is the manufacture equipment used in product. Material refers to the use of raw materials. Method refers to the method used in manufacturing products. Environment refers to surrounding environment during the process of manufacturing. Man is the biggest difficulty in production management and is also the focus of all the management theory.

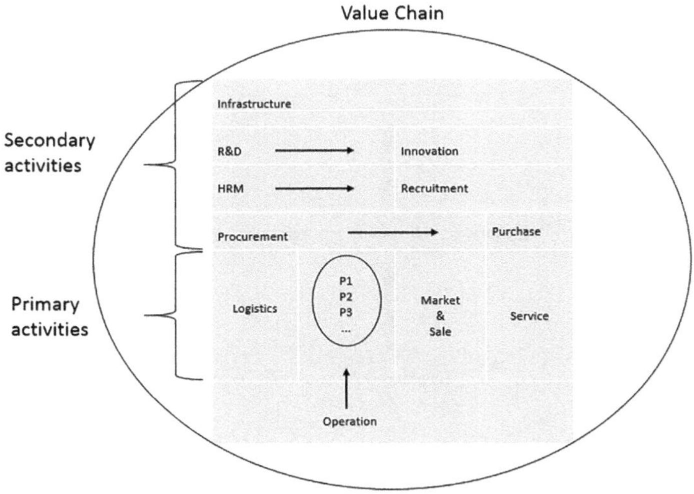

Fig. 2 Company value chain

From the insight of man, machine, material, method, and environment, it is needed to distinguish person and thing.

People not only by the senior animal desire at the mercy of the universe, life and people have broad and inner connection, and all human life have a common and sacred source (mental body).

People are not simply manipulated; people need the initiative, the pursuit of goals.

According to the distinction between platform and application, although in theory is consistent with the development of logic, the logic is conducive to the smooth, but the premise is the body in which people are willing to agree with self and as such identity and role to ongoing work.

At first, for the sake of salary and stability of the company, everyone is willing to accommodate the arrangement of the organization. For a long time, many people find that this is a passive acceptance at work, rather than an active one. This one is very bright.

The reason is: most Chinese has always been concerned only flesh is good to live in this world, but rarely to care for life, the origin of the universe. Life cut the noumenal world, transcendence and spiritual connection himself from the "people" downgraded.

From the point of view of the company's cooperation network, the factors of production provide a variety of economic resources involved in the development of enterprises and the role of mutual cooperation. The differences among industries result from the integration of factors and the difference of competition structure. The integration of advanced or unique factors of production is of great importance in the competition. It is the integration of many factors that cannot be imitated by rivals and form the base of competitive advantage. The integration of production factors plays a very important role (Fig. 3).

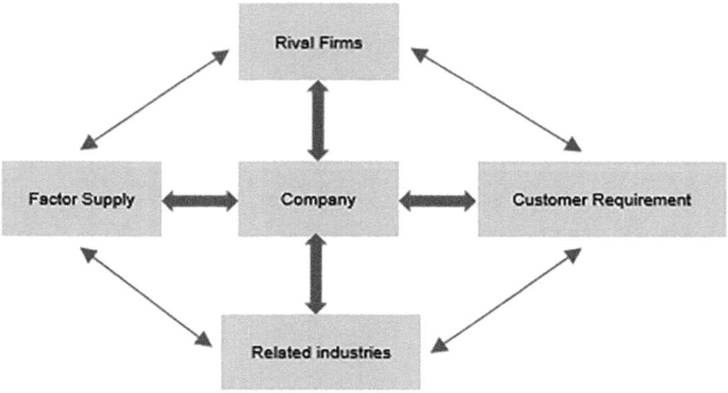

Fig. 3 Cooperation network [14]

2.1.4 Talent Management (TM) and a Company's Performance

Many companies know some employees are strong with ability and capability, who have high-qualified talent and are valuable to an organization, but these employees tend to be covered and freeze up; if employees do not ask, employers do not pay those more.

Most companies assign a lot of work which belongs to old employees who leave to other employees, but do not pay a salary added. The confrontation is that if someone takes a job or not, if someone will be able to carry it. If yes, these employees will have to take more and more work. Its too honest to take responsibility when it comes to problems. Then, they have no choice but to submit a letter of demission.

2.1.5 Factors for Successful Talent Management Implementation

The nature and future of business are to improve people's sense of happiness. But it is impossible for everyone to be rich, because the cash flow of the whole company cannot grow infinitely, and the circulation of money cannot be unlimited. The total amount of flow to everyone is limited, and only a small number of people can attract a large amount of cash flow. Thus, company cannot manage employees; each employee is influenced and managed by their leader.

Leader management is a method of three supervise steps. One is guess, two is check, and three is question. The first task of leader is asking subordinates every day: what do you do today? What's up? If a leader does not know the employee working status, the leader can only guess. In a meeting, leader can check it in meeting minutes. Asking questions is a process of seeking proof.

After a mistake, as the second task, leader uses the different related employees or resources to correct. In this way, leader can know where are wrong.

In all, working by employee themselves first, and don't go to design and record in data at the beginning of working. After working it, someone made a mistake, which was impressed, and learned something. This is the reality model of talent management.

2.1.6 Factors of Operation Management Success and Its Business Impact

Core competences are "An area of specialized expertise that is the result of harmonizing complex stream of technology and work activity". The marked characteristics of core competences are relevance, difficulty of imitation, and breadth of application.

Each team of the company constitutes an intuitive organizational chart; as you know, these organizations are due to the company's business objectives, which projects in the strategic deployment. The objectives or the strategy of the company changes; each project involved in the organizational structure of the members is changing accordingly.

The information from the project to the employee is communicated with the organization information. As each organization's competitiveness is made up of the competitiveness of the employees involved in each project. The type of organization or project determines the type of business information that each employee receives, but each employee's ability determines the type and size of the organization or project. The goal of the company is to determine the nature of each project, and the nature of each project shows the characteristics and business scale of the company (Fig. 4).

Core competences are the ability of a certain kind of employees, which is the concept of a kind of employees because of the difference between the content and the way they work. The reason why enterprises have different core competences is different from the way and quantity they work.

Inside each group of the organization, members will change, each job will change, but as long as the core competences of the company always exists, regardless of how to change the nature of the project, as long as the content of the work will be assigned to each employee and let the employee work effectively to ensure the company can continue to grow the business.

The lifeblood of the company lies in the smooth flow of business information. As well as the number of business units to ensure that business requirements each employee received is accurate. The design of the business unit is the system architecture. The purpose of the architecture is to ensure that the company's project information reaches every employee, so that employees have sufficient information to work.

To make the core competences of the enterprise sustainable development, it is necessary to know how many employees needed to support each core competence, how long training an employee, how long to adjust organization structure, in order to ensure the competitive advantage of enterprises (Fig. 5).

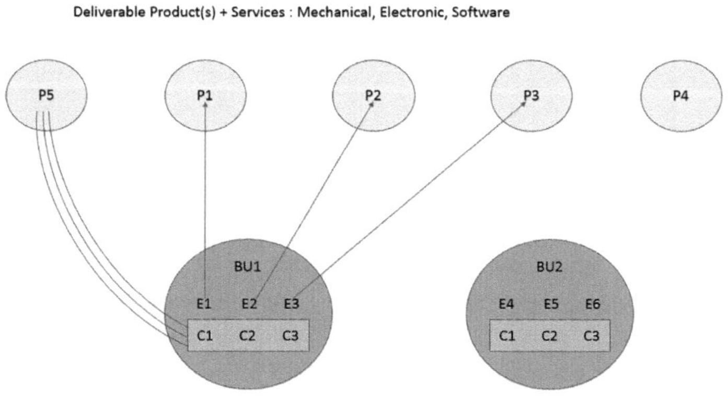

Fig. 4 Deliverable products and services

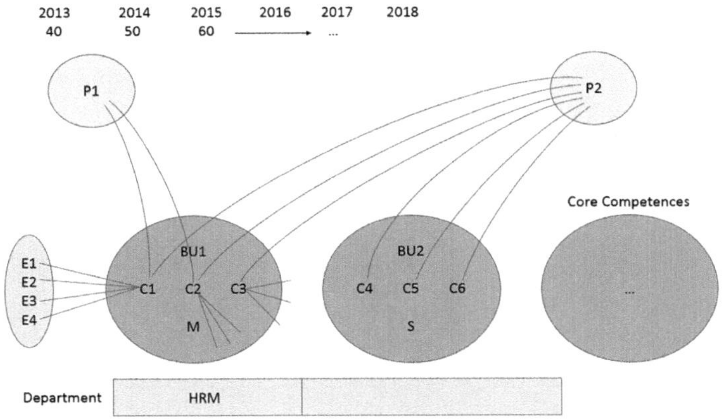

Fig. 5 Core competences

Market is exclusive survival soil for all companies; business information is the market demand. All projects need project management, and the demand is the information that the project management needs to transmit. The size and category of the project determine the direction of enterprise development, which include how much profit, how much market risk resisted, how long growth or development will go on (Fig. 6).

The development and transformation of enterprises are all based on the inherent regularity. Each employee's ability is not the same, but all need to be fully imported business information. Project management is the transfer of information. Information needs to pass effectively, in order to achieve the strongest information utility. Effective refers to the fastest spread of information to employees; the strongest is the strongest combat, which means any business can be overcome with full competitiveness.

The market determines the ability of the staff who need to face. The speed of these employees to carry out a business activity is directly determined by the content and the number of information, and the priority and the delivery path of information. Once each employee receives business information and once resources required is adequate, they begin to work effectively within their own capacity.

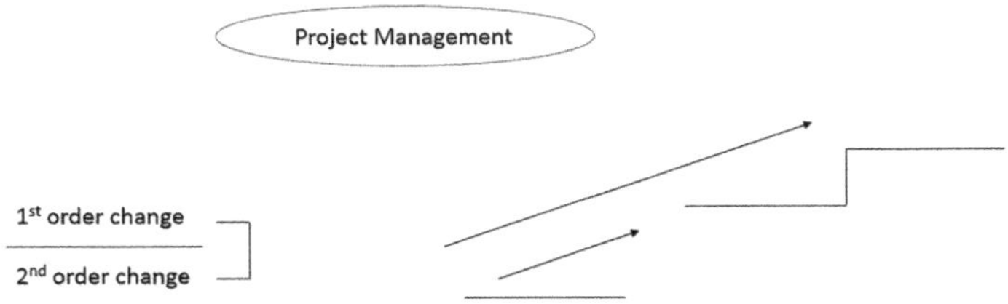

Fig. 6 Project management outline

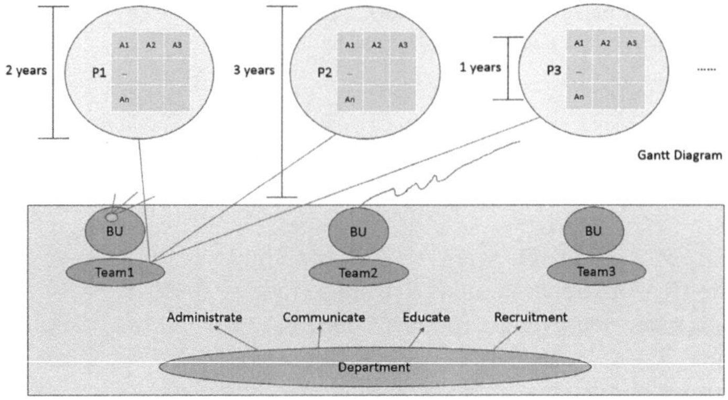

Fig. 7 Activity in Gantt diagram

What kind of people to work for the enterprise determines what kind of products will be produced. The standard of maturity of the company's organizational structure is the ability to accurately and quickly convey the correct amount of business information to the correct employees. The core competence of the organization is to define the path of each business information transmission, the kinds of capabilities which employees participate, in order to complete the extension and development of business. At present, the task of each employee is always known, but there is a missing of visual path information to reflect the business requirements of these tasks (Fig. 7).

2.2 Research Question

The results of the discussion of knowledge gaps between the literature and practice support three relevant research questions.

2.2.1 Research Question 1

How to transform skills or competences into dynamic forces for efficient production? (Fig. 8).

2.2.2 Research Question 2

How to make the process of transforming accurately match to the capture of business opportunities? (Fig. 9).

Fig. 8 Research question 1

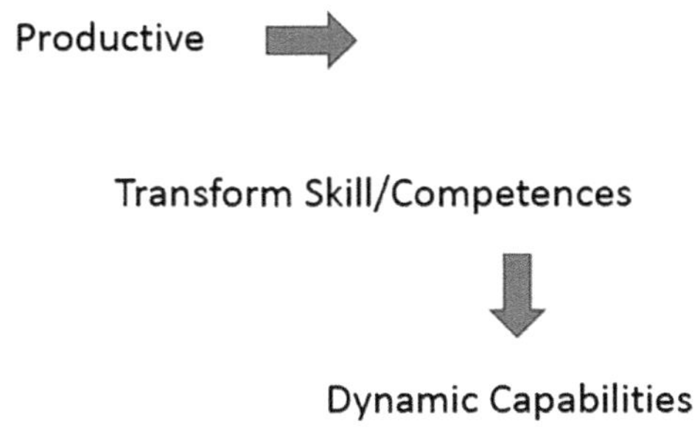

Fig. 9 Research question 2

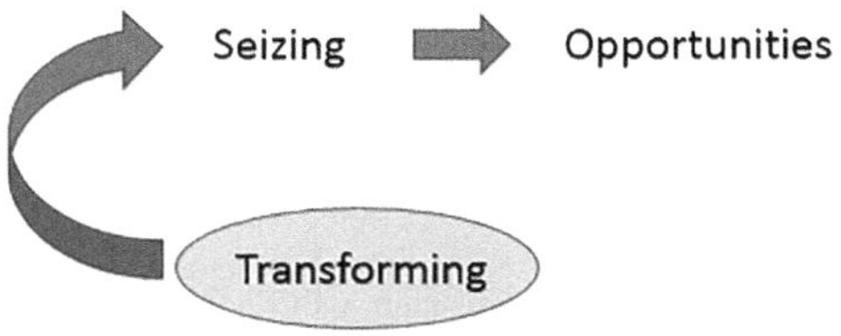

2.2.3 Research Question 3

How to ensure the variety of basic resources used by the staff for the objective are fully satisfied?

3 Methodology

This section describes the methodological approach, hypotheses, and algorithm undertaken in this research, to present a matrix built upon the experience foundation.

3.1 Matrix and Hypotheses

To avoid direct conflict against people, there is a hypothesis that it is not going to complain about other departments or people, because a company's people issue can be attributed to the problem of the process, because there are contradictions in the process of cooperation.

Departments and organizations are the classification and stratification of different activities of all company's employees. The beginning and end of the activity time are the timing problem of project management. Process is the interaction or collaborative issues between activities and activities.

The activity schedule pulled out from VEM is the project schedule.

⊖ 2 years

可视化衡平矩阵 VISUAL EQUITY MATRIX (VEM)					波士顿矩阵BCG Matrix																							
					明星产品Stars						现金牛产品Cash Cow						问题产品Question Marks						瘦狗产品Dogs					
					P1						P2						P3						P4					
					A1	A2	A3	A4	A5	A6	A1	A2	A3	A4	A5	A6	A1	A2	A3	A4	A5	A6	A1	A2	A3	A4	A5	A6
				Mean	1.5	1	0.5	1.5	1.5	1	1.5	1	0.5	1.5	1.5	1	1.5	1	0.5	1.5	1.5	1	1.5	1	0.5	1.5	1.5	1
五大生产要素The Five Factors Of Production	人 Man	奖励点数Award Points	E1	1.5	X								X								X						X	
			E2	1																								
			E3	0.5		X																						
			E4	1.2				X				X																
			E5	1		X																						
			E6	1.3				X								X												
			E7	1																					X			
			E8	1.5	X																							
			E9	1															X									
			E10	0.6		X							X															
			E11	1.5				X							X													
			E12	1		X							X															
	机 Machine	问题概率Defect Probability	Tool1	0	X							X						X						X				
			Tool2	0.2		X				X			X						X									
			Tool3	0				X			X				X		X				X		X					
	法 Mothed		Process1	0.07	X	X						X	X					X						X				
			Process2	0				X			X						X				X		X				X	
			Process3	0					X							X						X						X
	料 Material		Material1	0	X							X						X						X				
			Material2	0.2		X							X															
			Material3	0																								
	环 Environment		Envir1	0.07	X	X						X	X					X										
			Envir2	0																	X						X	
			Envir3	0					X							X						X						X

Fig. 10 Visual equity matrix (VEM)

In addition to people, the other factor has a service evaluation of 11 degree of customers satisfaction, and there is an individual evaluation probability under the activity were recorded, so there is a glance of a problem related to which process, which material, or which environment (Fig. 10).

3.2 Matrix Algorithm

$$Y(\text{reward points}) = aX1 + bX2 + cX3 + dX4 + eX5 \tag{1}$$

$X1$ time elapse
Xx other factors of five factors except man ($x > 1$)

Therefore, the current reward points Y will increase or decrease with the time and the factors of complaints.

First, define an activity start time and release the task entry; then, people enter into it in the beginning period of the activity. Nobody can register an activity when the entry is expired to prevent people from mixing projects or not in the project. The project manager or supervisor can also kick somebody out of the project or withdraw the person's activity registration directly. Staff can also enter the activity again through the complaint.

Define an activity end time, and then compare with the current time, the dynamic display the available reward points is the current activities real points. When the activity is closed, the project staff related to the activity will be unified brush with the same reward point. Closure of the project is carried by the project manager; the reward point has been reduced until the activity closed.

With the elapse of time, at beginning each person's participation rate will rise, rise will get the corresponding points, all points are equal to the highest reward of 150% to points 1.5 when complete half of activity schedule time, afterward, a turning point of subtraction is exist, time over 10% reward points dropped 5%, but the points of activity is inversely proportional to the duration, and the number will fall with time pass. Rise and fall, the activity time of a turning point t is 100%, and the point number is 1. Points of activity have the highest score before the completion of the activity; with the elapse of time, the score given will be reduced; that is, the reward will decline. Bonus from 150% down to 100%, and then to 50%, which the number of points is 0.5.

Each person's wages are not the same, points take all of their own points and divided by the number of points, that is, the average rewards point.

Final salary = position salary × the average rewards point.

From another point of view, the matrix answers the question of what tools will be used by people to do what they do. With time as the axis, it shows what tasks are done at a special time

3.3 Data Analysis

X (Tick) represents the association or the number of 1.

Section analysis of a moment:

More dense in the axis, lower utilization of resources. It is needed to increase the marketing of such projects in order to disperse resources to different projects. Or these resources need to be integrated into a platform.

More dense in the horizontal axis, higher ability.

More intensive under a project, more stable the project is. Or else, it is necessary to recruit or allocate resources to the project to ensure the progress of the project.

One day, no one is going to register or to maintain the project, which means the project lacks life; it is the end of the project or project crisis

The registration of a person at a point has been constant over time, indicating that the person is too professional, there is no change in the arrangement of the work, and the utilization rate is reduced.

The duration of each point to the current time indicates the time at which the work is committed. The standard time of the task can be set as workload. Input time divided by rated operating time is equal to work efficiency. Higher in the *Z* axis (time) of this point, higher of level. Level of 1 is defined as a normal level, less than 1 indicates a low pressure area, with light color of high efficiency, and greater than 1 indicates a high-pressure area, with deep color of low efficiency. So that catch sight of a view that the staff is suitable for what to do. If the vertical axis of the color is dark, indicating that this project is a trouble, indicating that the workload is required to reduce.

As long as each person selects an entry, the system automatically records the start time, and at the same time, the time of the last entry is ceased to be recorded. The same person and the same task of the same project will accumulate calculation.

Each person's entrance is open casually to their own. If there is a lot of red in the horizontal axis, this person's work is too easy.

If the activity is not completed in the standard time, it is necessary to allow other resources to enter; if the vertical axis is red, then the resource is not enough and also need to recruit people. Organizational structure can be integrated by intensity.

The technical work has the characteristic of that less than or equal to three projects which are carried on in the horizontal axis at the same time; the management has the characteristic of that more than three projects at the same time.

Green as the forest is very valuable, which helps to conserve talent resource and prevents the loss of project opportunities; the project is completed in advance only 80% of the normal range

Yellow as the residential area is very safe, indicating that once the house of the residence was built, it is well as progress continued. The project needs to be completed between 120 and 80% of the normal range, marking the project safety.

Red as the desert is very difficult, if not improved, improper handling, it will lead to desertification, that is, project crisis. Completion of the project requires more than 120% of the normal range.

As time goes by, these points will be different. From the forest to the residential area to the desert, the beginning is always very valuable forest, then to the normal residential areas, and finally may become desert.

For a project, the more residential areas have, the safer to be. Too many forests need to exploit more projects, more deserts need to hire more people.

The number of projects is around a circle. The more the enclosure, the faster growth. The project of less area is about to exit. So VEM is also called the enclosure matrix (circle ground matrix).

Coverage per project = number of man in a project/number of all staff. This value is suitable for 10–50%. Too little is explained as the project resource is not enough; too many is explained that the project will lead to the operation of the organization structure is not stable. The rights of the project manager will exceed that of the executive manager.

4 Findings and Discussion

This section presents the results findings and discussion, respectively.

4.1 *Understanding of Operation Management Phenomenon with VEM*

VEM has three related properties. Elite resources are exclusive, which are nearly 150% level of the reward average point of more than 1. Common resources are very safe, which is around 100% level of the reward average point of about 1. Problem resource needs to be improved, which is close to 50% level of the reward average points below 1.

Star project involves more investment tools than other kind of project, which are costs.

VEM is a fair matrix with a glance of ups and downs for the enterprise performance. The advantage is that you can visually see the balance sheet of activity, staff, project, and other resources. Optimal performance of workforce is alignment to business strategy. Each activity is business information. Hook link is like a contract, the longer link the more professional capabilities. Employees are visually different from professional contribution to the organization.

VEM effectively eliminates three work efficiency problems:

(1) The current leadership style is led by people; that is, if a leader is on a business trip or vacation for a few days, their subordinates are likely to be idling for several days, and nothing is going on.
(2) In progress of all projects, the leader is always to be inquired when a problem arises. But the leader would be not aware of this project and ask suspiciously, "have we done this project?"
(3) The current project group structure obscures the actual link between the project and the project worker, which is actually a window link between the project and the problem. When there is a problem, or there is a sample delivery, the problem is first arrived to the project team as a window, and then a lot of people are used to solve the problem, which is like "Press-gang Able-bodied Men." After the problem of the project is settled, it says that you are an "Able-bodied Men," which you have to hard work and you have to pay the project results; then, you are asked to stay away. The final decision is that you are not officially organized by the project team, and that the project success has no credit to you, but conversely doubt on you with the say of "did you also devote to do the project?"

In brief, in terms of a company, the real leader is the company's goals and objectives of a project. When making the objectives of company and tasks to be open and fair and showed to all employees, employees will consciously work for the company's goals and objectives of the project. That is to say, the independence and consciousness of staff help to improve the work efficiency most.

4.2 Factors for Successful Operation Management Implementation

If the decline causes of the reward point may be the tool issue or may be an unreasonable process to set, or the project schedule question, these factors are used to adjust the number of rate of decline that these are used as adjustment factors. The elapse of time is a direct factor in the decline of reward points.

The factor of man needs to have an element of reward points. The other four factors of machine, material, method, environment, such as tools or processes, need to set a probability index, as a complaint. For example, this tool is not enough or the process is irrational, and then, the defect probability will rise, according to the evaluation of 11 grades, from normal to very satisfied or very dissatisfied. Very dissatisfied is considered as an additional factor of reward points, and satisfied is a subtraction factor. The condition of the evaluation is that the person is in the activity and the tool is used for this activity. This probability will be used as a, b, c, and other values; the activity reward points are directly promoted by the percentage of probability rise. The highest point is 1.5. Points = (1 + defect probability) × points. When the project is closed with the number of points 1.5, the activities of the four factors such as tools and processes related to the defect probability must be cleared by 0, because it is sure that the five factors must be no problem.

5 Conclusions

This article is a preliminary attempt to understand the complexity of operation processes from a team-based approach to VEM. An important task of managers and strategists is to lead the operation processes by crucially identifying and allocating resources in order to develop sustainable competitive advantages.

The visual equity matrix is a map of the strategy of the company's product. Each process needs business information passed to it. The scheduling of the project is considered together with the reward of staff.

Visual equity matrix is a glance of enterprise operation, which helps to quickly diagnose the operation management of enterprises and fix the corresponding solution. The VEM matrix also helps to find the different nature of resources and resource distribution because there is a lot of emotional problems in the evaluation of resources.

VEM is applicable to supply statistics of time recording at the macrolevel of company operation, applicable to supply statistics of tasks arrangement at the microlevel of departments and groups, applicable to supply open and fair on the level identification of employee performance, applicable to supply planning and tracking for project management, and applicable to promote the process improvement of enterprise operation management.

Quantitative characterization is the most effective and fair. It is useful to visually see the number of people around the project and the performance of these people. The quantitative definition of related resources can be carried out. According to the process of identifying the value chain, it is necessary to link each activity with the value chain of the company, so it is necessary for us separating the activities in future. It is short of identification that the type of activity is valuable, and the standard duration of the activity to be completed in a reasonable number, and then the actual accomplishment data to complete or such quantitative information which needs to be collected further.

Acknowledgements This paper results from the support and guidance of numerous individuals. Firstly and foremost, I would like to my sincere thanks to Ji-xiang Chen, Frédéric Le Roy, Saïd Yami for their constant guidance and tolerance in my work. I would like to thank to my KOSTAL colleagues for many helpful comments and suggestions. I am also immensely grateful to my family for their encouragement. Finally, I would like to thank all the people for their openly share information and insights into their experience. Their devotion contributed significantly to complete this paper.

References

Journals

1. Maier AM, Wynn DC, Andreasen MM, Clarkson PJ (2012) A cybernetic perspective on methods and process models in collaborative designing. Int Des Conf 5:233–240
2. Xu B, Yang X, Ma A (2008) Role based cross-project collaboration in multiple distributed software design projects. Int Conf Comput Support Coop Work Des, 177–182
3. Bloomberg LP (2016) China's latest bubble: electric cars. Bloomberg Businessweek 4489:19–20
4. Lüthje B (2014) Labour relations, production regimes and labour conflicts in the Chinese automotive industry. Int Labour Rev 153(4):535–560
5. Wynn DC, Eckert CM, Clarkson PJ (2011) Simulating intertwined design processes that have similar structures: a case study of a small company that creates made-to-order fashion products. Int J Prod Develop 14(1–4):118–143
6. Jung E (2016) China's internet boom. MIT Technol Rev 119(4):10–12
7. More E, Gungorb ZE, Phaal R, Probert D (2015) Addressing resource over-exploitation via cooperative institutions: examining how technology roadmapping could contribute. Int J Prod Develop 26:173–178
8. Pellegrin-Boucher E, Le Roy F, Gurău C (2013) Coopetitive strategies in the ICT sector: typology and stability. Technol Anal Strateg Manag 25(1):71–89
9. Le Roy F, Yami S (2007) The development of an innovation strategy in the SME context. Int J Entrepreneurship and Small Business 4(4):431–448
10. Le Roy F (2016) Wojciech Czakon managing coopetition: the missing link between strategy and performance. Ind Mark Manage 53:3–6
11. Iles P (2008) Book review. Human Resour Develop Int 11(2):215–218
12. Stedt J, Westport CT (2006) Talent balancing: staffing your company for long-term success. Praeger, USA

13. Chen J, Wang M (2009) A review of the latest research on destructive innovation theory. Sci Technol Prog Policy 26(11):155–160
14. Chen J, Huo P, Wang Z (2000) Enterprise stragegic synergy under the hypercompetition. J Sjtu (Socia L Sci Ed) 8(4):86–89
15. Ahn JM, Minshall T, Mortara L (2015) Open innovation: a new classification and its impact on firm performance in innovative SMEs. J Innov Manag 2:33–54
16. Lammers D (2016) What is your China strategy? Solid State Technol 59(7):20–22
17. Putre L (2016) How the auto industry is positioning itself for a decade of change. Ind Week 265(5):10–14
18. Rao C, Huang Y, Mayumi W (2016) India's auto industry drives polymer demand. Popular Plast Packag 61(8):29–31
19. Schiemann WA (2006) People equity: a new paradigm for measuring and managing human capital. Human Resour Plann 29(1):34–44
20. Schlegelmilch BB, Diamantopoulos A, Kreuz P (2003) Strategic innovation: the construct, its drivers and its strategic outcomes. J Strateg Mark 11(2):117–132
21. Sun X, Wang Y, Zheng H (2014) The industrial heterogeneity of R&D impacts on total factor productivity based on the empirical test of Chinese manufacturing industry. J Ind Eng Eng Manag 28(3):33–41
22. Ferrier WJ, Smith K, Grimm CM (1999) The role of competitive action in market share erosion and industry dethronement: a study of industry leaders and challengers. Acad Manag J 42(4):372–388

Monograph

23. Ortiz-Gallardo VG (2013) Technology acquisition: sourcing technology from industry partners. University of Cambridge, New York

Effect of Cooling Rate on Microstructure Evolution of Hot Forming High Strength Steel Based on Non-Isothermal Constitutive Model

Wenfeng Zhu, Guoliang Wang, Chun Xu and Xiaojuan Li

1 Introduction

Automotive products are faced the great demand for lightweighting, safety and crashworthiness improvement. It causes the great need to manufacture automobile structural components by ultra high strength steels (UHSS) [1, 2]. However, the deformation resistance of UHSS increases significantly, making it unsuitable to be processed by conventional cold stamping, which could cause big problem of big spring-back, fracture and so on. When heated to high temperature, high strength steels can reduce their deforming resistance and spring-back. Swedish company Plannja firstly developed and patented an innovative hot stamping process in 1977 for manufacturing sheet steel parts with extraordinary high strength [3].

Saab Automobile AB Hardtech firstly used a boron-alloyed steel to manufacture vehicle component in the Saab model 9000 in 1984 [4, 5]. Later, hot stamping attracted much attention in automobile industry and more and more components were fabricated such as A/B-pillar reinforcements, bumpers, door beams, roof rails and other structural members [6, 7]. The number of automotive parts produced by the hot stamping was increased 41 times in 23 years (from 3 million parts/year in 1987 to 124 million parts/year in 2010) and most likely go up to approximate 350 million parts/year in 2015 [8].

Hot stamping numerical simulation is getting more and more important. From the mechanical and metallurgical standpoints, the modelling of hot flow stress and

W. Zhu (✉) · G. Wang
School of Mechanical Engineering, Tongji University, Shanghai 201804, China
e-mail: zhuwenfeng@tongji.edu.cn

C. Xu · X. Li
School of Material Science and Engineering, Shanghai Institute of Technology, Shanghai 201418, China

Society of Automotive Engineers of China (SAE-China) (ed.), *Proceedings of the 19th Asia Pacific Automotive Engineering Conference & SAE-China Congress 2017: Selected Papers*, Lecture Notes in Electrical Engineering 486, https://doi.org/10.1007/978-981-10-8506-2_52

the prediction of flow curves, together with constitutive equations, are important in metal-forming mechanism research. As a result, considerable researches have been focused on this subject in recent years.

Conventionally, isothermal deformation was assumed to obtain these constitutive relationships. Merklein and Svec conducted experiments on the effect of isothermal deformation during quenching of Usibor 1500P [9]. However, during actual industrial process, the forming part is heated up first in furnace, transferred into the mould and then formed and quenched subsequently in the closed tool. Because of cooling water in mould, there exists some instant heat exchange, which would induce energy loss and temperature decreasing for forming panels. In other side, it also undergoes complicated thermal–mechanical phase coupling effect and phase transformation during forming process, introducing non-negligible latent heat, which wound increasing temperature.

Consequently, Naderi et al. [10] carried out non-isothermal numerical simulations of hot stamping of boron-alloyed heat-treated steels. A fully coupled thermomechanical FE model was proposed by Cui et al. for a non-isothermal hot forming of boron alloy steel [11]. P. F. Gao studied the microstructure evolution and processing microstructure relationship in the non-isothermal local loading forming of TA15 titanium alloy [12]. Isothermal and non-isothermal multi-stage hot compression tests were carried out by Fulin Jiang et al. [13] on 7150 aluminium alloy. This shows that hot stamping is accompanied simultaneously by temperature changing and phase transformation. It is in essence a non-isothermal deformation process.

Regarding the four steps of hot stamping, Merklein et al. [14] stated that if a diffusionless martensitic transformation wanted to be induced, cooling rate must exceed a minimum cooling rate of approximately 27 K/s at a temperature of around 400 °C, due to the contact between the hot blank and the cold tool. Bardelcik et al. [15] investigated the effects of cooling rate on mechanical properties of hardened boron steel under high strain rate conditions. LIU Wei [16] found that cooling system by ejecting water into channel in final step can obtain high cooling rate of hot stamped component, which facilitates the martensite phase transformation and improve the UTS of hot stamped component.

This paper firstly presented the boron steel BR1500HS non-isothermal tensile tests on servo-hydraulic Gleeble1500 thermal simulator. True stress–strain curves and microstructures at the relative conditions are gained under different actual cooling speed and with different cooling rates. Different from conventional isothermal scheme, high-temperature work-hardening stress versus strain curves were obtained, and new non-isothermal constitutive equations were regressed and attained by the regression analysis of experimental data.

Then, numerical simulation of hot forming on typical automotive box-shaped parts of metal sheet was carried out, in which the obtained non-isothermal constitutive equations were adopted. The die tool's cooling water flow rates varied at 0.1, 0.2 and 0.3 m/s, by which the practical non-isothermal deformation environment were simulated. The evolution curves of martensite, bainite and pearlite and also their contour during hot forming process were quantitatively obtained.

Thirdly, in order to verify the numerical analysis results under non-isothermal constitutive relationships, hot stamping experiments on same box-shaped plate were conducted. Influence of cooling rate on workpieces' temperature and microstructure evolution was thereby physically monitored and researched. The experimental data had good consistency with numerical results. It shows that the non-isothermal deformation can provide better control in hot stamping design of box-shaped automotive components. It can provide guidance for microstructure optimization in the non-isothermal hot forming of high strength steel.

2 Non-isothermal Tensile Experiments

2.1 Experimental Materials

To obtain new constitutive equation, the material used for this study is BR1500HS, which is pre-coated with a protective layer of aluminium–silicon coating in order to avoid surface oxidation and decarburization. The chemical composition and mechanical properties of the investigated steel are listed in Table 1. The quenched blanks were CNC machined into the miniature dog-bone style samples, on which the uniaxial tension tests were conducted. The sample geometrical dimensions are shown in Fig. 1. The blanks used in this study have a nominal thickness of 1.8 mm, and rolling direction was longitudinal. The non-isothermal tensile scheme of high strength steel BR1500HS is shown in Fig. 2.

Table 1 Composition and mechanical properties of BR1500HS

Composition/mass (%)				
C	Mn	Si	P	S
2.0–2.5	11.0–14.0	1.50–4.00	<0.25	<0.05
Al	Cr	B	Ti	N
0.20–0.60	1.00–3.50	0.02–0.05	0.20–0.50	<0.09
Mechanical properties				
Hardness (HV)	R_m (MPa)	$R_{p0.2}$ (MPa)		$A_{11.3}$ (%)
172	515	354		34

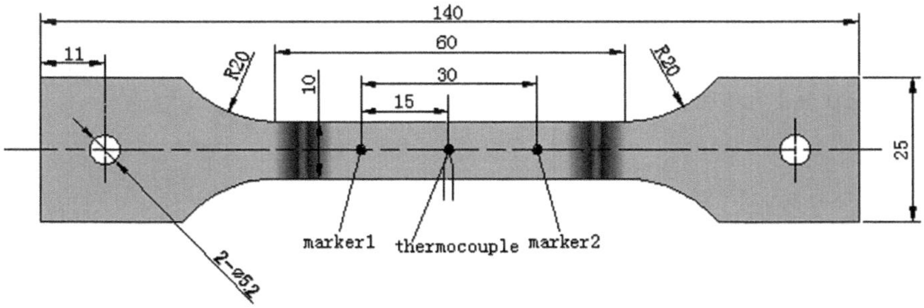

Fig. 1 Specimen geometrical dimensions

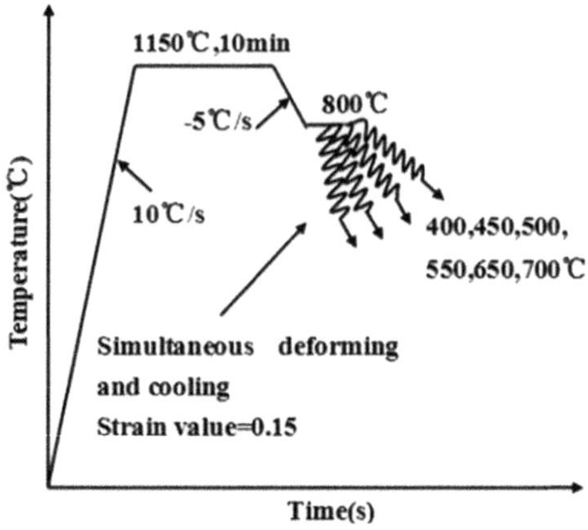

Fig. 2 Non-isothermal tensile scheme

2.2 *Experimental Procedure*

In practical industrial process, hot stamping blank is firstly heated to a temperature between 850 and 950 °C in furnace and held for 5 min and then transferred to cooling die within 3 s. Due to the time required to transfer the samples, as well as the contact to mould with built-in cooling waterway, the deforming temperature dropped rapidly to approximately lower than 850 °C prior to quenching. So the hot stamping constitutive equations need to be obtained by non-isothermal scheme in order to simulate the actual production with higher precision on microstructure prediction, as stated in Reference that hot stamping or press hardening is a non-isothermal sheet metal-forming process, in which the forming and quenching procedure were combined in a single step.

Non-isothermal uniaxial tension tests were carried out on the Gleeble1500 thermal simulation machine. Experimental data of forces, displacements, flow stress has been recorded in order to make regression for new non-isothermal constitutive relationships. During tensile experiments, samples were heated up to the temperature of 1150 °C at rate of 10 °C/s and held for 10 min to attain a homogeneous austenitic microstructure. Thereafter, they were cooled down to 800 °C at rate of 5 °C/s. After dwelling of 2 min, samples were uniaxially extended in non-isothermal way with specific strain and strain rate until the largest deforming degree reaches 15%. To study the influence of difference deformation temperature, extended samples were ultimately cooled to different temperatures of 400, 450, 500, 550, 650 and 700 °C, respectively.

2.3 Non-isothermal Constitutive Model

2.3.1 Non-isothermal Stress–Strain Curves

Figure 3 presents the stress–strain curves of both the non-isothermal and isothermal uniaxial tension tests under different temperatures. It can be found that there exists obvious difference. For non-isothermal one, the flow stress of all six non-isothermal deformation samples increases with the increasing of deformation degree from the beginning to the end. These results show that the non-isothermal tension of high strength steel has strong work-hardening characteristics at different final deformation temperatures.

Comparatively, for isothermal deformation samples with final deformation temperature of 800, 900, 1000 and 110 °C at strain rate of 10/s, flow stress firstly increases with the increasing of strain and then reaches a steady-state stage, and finally, they gradually decrease with the increasing of strain. Isothermal deformation samples with different cooling rates have dynamic recovery and dynamic recrystallization characteristics.

Fig. 3 Stress–strain curves of non-isothermal (**a**) and isothermal (**b**)

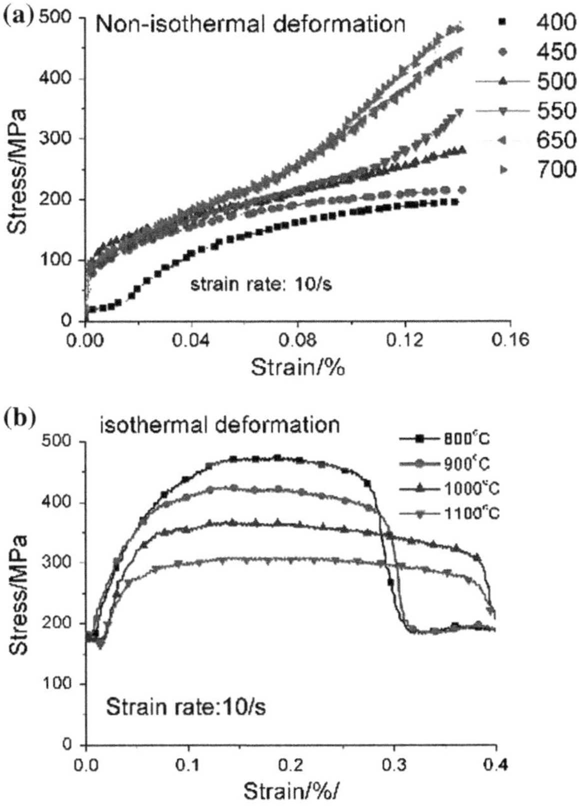

2.3.2 The Modified Constitutive Model

In hot forming die tool, deformation is accompanied simultaneously by cooling water system, causing the flow stress in non-isothermal tension test significantly different from those in isothermal deformation. Experimental results indicated that the temperature is the most prominent factor affecting the flow stress. Hardening index K and the strain rate sensitivity coefficient m also vary with temperature during non-isothermal hot forming.

To better describe the hot stamping non-isothermal forming process, this paper takes equations $K = a + bT$, and $m = c + dT$ to amend the conventional equation and establishes the mathematical material model under non-isothermal condition. The new high-temperature constitutive equation is shown in Formula 1.

$$\sigma = C\varepsilon^{a+bT}\dot{\varepsilon}^{c+dT}\exp\left(\frac{\beta}{T}\right) \tag{1}$$

By experimental results, these coefficients are further obtained using multiple linear regression fitting: $C = 412$, $k = 0.20$, $m = 0.06$, $\beta = 985.2$. The final experiment data consists of 3611 points in the aggregate to guarantee the high precision, and the correlation coefficient of the fitting curve R2 is 0.98386.

$$\sigma = 412\varepsilon^{0.20}\dot{\varepsilon}^{0.06}\exp\left(\frac{985.2}{T}\right) \tag{2}$$

3 Application and Verification

Deep drawing of box-shaped part is the typical forming process in automotive sheet metal manufacturing. Therefore, hot forming of box-shaped part has been numerically modelled by commercial software code DEFORM in this paper. Applying the aforementioned non-isothermal high-temperature constitutive equation, the simulation studied the effects of cooling rate on the box-shaped parts' microstructural evolution. Volume fraction of martensite, bainite and ferrite during microstructural evolutions of non-isothermal hot forming at each cooling water flow rate has been measured and compared with later optical and field scanning electron microscope. Therefore, hot forming of box-shaped part has been numerically modelled by commercial software code DEFORM in this paper. Applying the aforementioned non-isothermal high-temperature constitutive equation, the simulation studied the effects of cooling rate on the box-shaped parts' microstructural evolution. Volume fraction of martensite, bainite and ferrite during microstructural evolutions of non-isothermal hot forming at each cooling water flow rate has been measured and compared with later optical and field scanning electron microscope.

3.1 Geometrical Modelling and Meshing

Considering that the box-shaped part is symmetrical with both x- and y-axes, only 1/4 model is needed in order to reduce computation resources and time. The generated tetrahedral meshes are shown in Fig. 4, the total mesh number is 24,874 and mesh type is Solid45.

3.2 Thermal–Mechanical Boundary Conditions

In practical production line, hot forming starting temperature is around 800 °C when the whole blank is fully austenitized. Therefore, forming starting temperature of 800 °C and austenite microstructure are applied as initial conditions in numerical simulation. Since the machined die mould has unilateral gap of 1.8 mm, the selected blank holder's thickness is 1.8 mm. The ambient heat coefficients of both mould and the surrounding environment are both 0.7 W/(m K), while the convective heat transfer coefficient changes with flow velocity in the range of 2600–5000 W/(m^2 K). Moulding and forming procedures last for 3 s, and the dwelling time occupies the next 4 s. The cooling water flow rate v was 0.1, 0.3, 0.5 m/s, respectively. More detailed parameters are shown in Table 2.

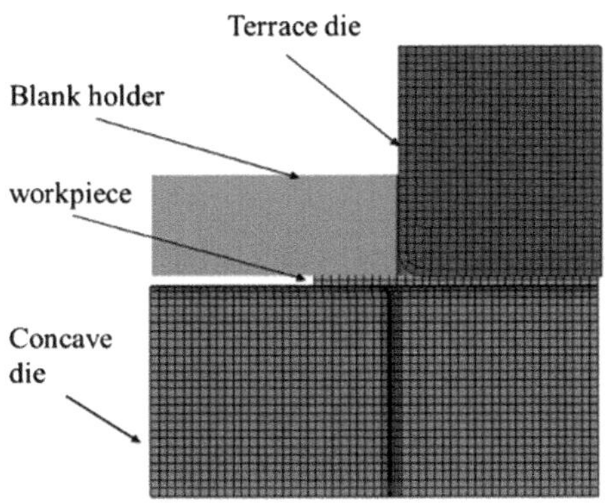

Fig. 4 Geometrical model and meshing of cube-box forming

Table 2 Hot forming parameters

Forming time	Dwelling time	Clamping force	Stamping velocity	Cooling water flow velocity	Friction coefficient
3 s	4 s	100,000 N	25 mm/s	v	0.4

3.3 *Different Cooling Water Flow Rate*

3.3.1 Flow Rate of 0.1 m/s

Different part of box-shaped workpiece has varied contact status in the cooling mould tool, causing different non-isothermal situation during hot forming microstructural evolution. In order to trace this, three typical positions from different places on workpiece, which are flange, sidewall and bottom wall, marked by P1, P2 and P3, are measured. Their microstructural evolution information during numerical analysis is retrieved and analysed.

Figures 5, 6 and 7 show the microstructure evolution of martensite, bainite and ferrite in terms of volume fraction under cooling water flow rate of 0.1 m/s.

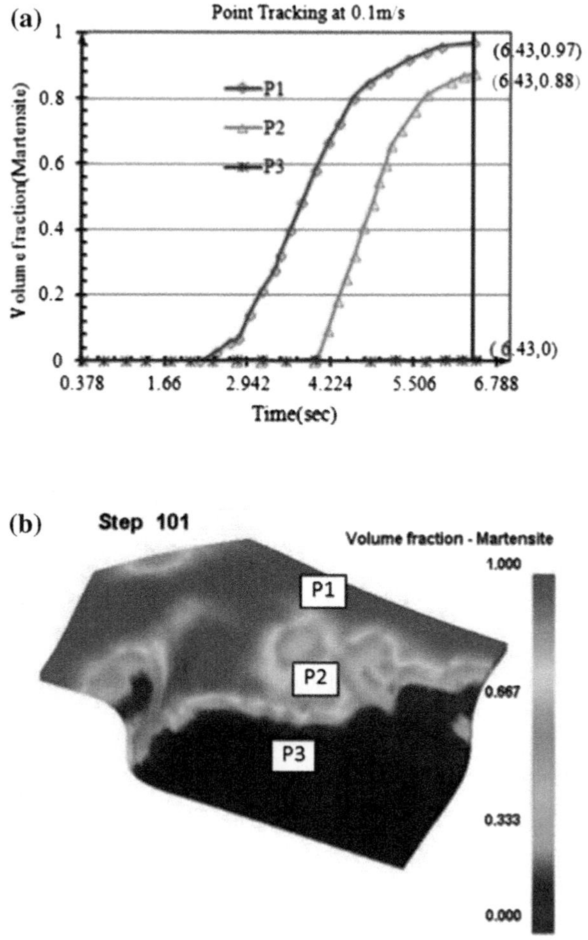

Fig. 5 **a** Transformation curve of martensite, **b** transformation contour of martensite

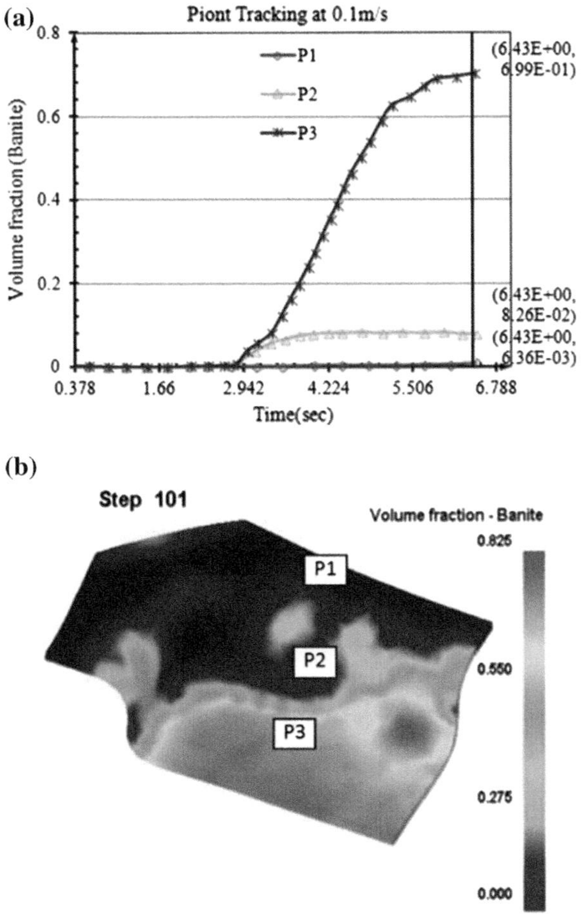

Fig. 6 **a** Transformation curve of martensite, **b** transformation contour of bainite

At this cool rate, the flange area P1 is basically composed of the martensite with its volume fraction as high as 96.9%. The sidewall area P2 consists of martensite (87.7%), bainite (8.26%) and ferrite (2.61%). The bottom area P3 is mainly composed of bainite (69.9%) and ferrite (22.2%).

Martensite plays dominant contribution for hot forming produce strength. It shows that the flange area P1 has the highest volume fraction of martensite. This is because when contacting with mould tool at relatively low cooling water flow rate, this area is the first portion in close contact with the blank holder and die. It causes flange area firstly satisfying the required minimum cooling rate and starting martensite transformation earlier, compared with other areas like sidewall and bottom.

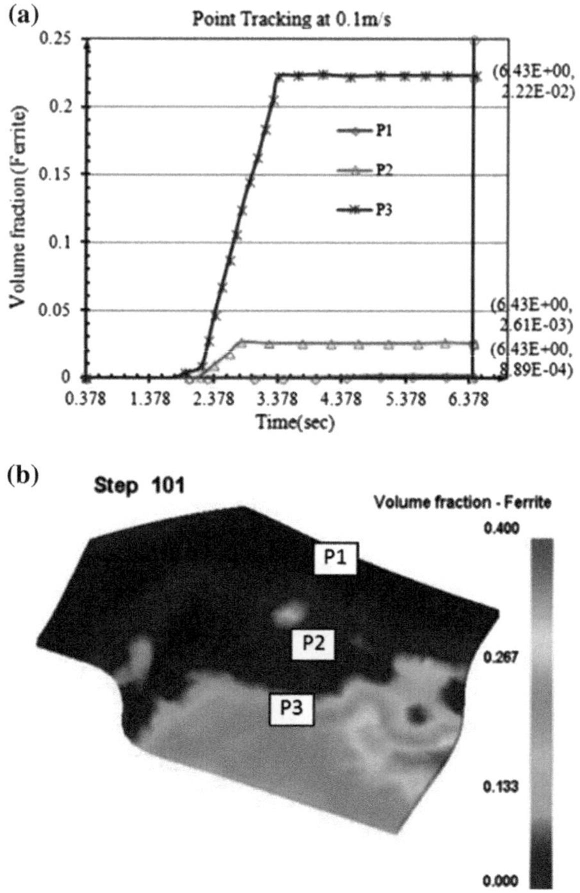

Fig. 7 **a** Transformation curve of ferrite, **b** transformation contour of ferrite

3.3.2 Flow Rate of 0.3 m/s

When the cooling water flow rate is increased to 0.3 m/s, the microstructure transformation of martensite, bainite, ferrite and austenite is shown in Figs. 8, 9, 10 and 11, respectively.

It can be seen that both flange area P1 and sidewall area P2 this time are primarily composed of martensite, with P1 94.9% and P2 91.6%. Since the P1 has earlier contact with cold mould, it has more time for martensite evolution history and reached higher volume fraction of martensite than that of P2, as shown in Fig. 8a. However, the P2 has significantly greater transformation speed than that of P1. This is because on flange area P1, the material flow is not in the same direction with punching force and gravity. At sidewall area, its material flow parallel with the punching force and gravity, causing faster material flow rate. This could make more mass of material to realize martensite transformation in same time. It finally causes the martensite transformation to be realized at relatively less time with high speed.

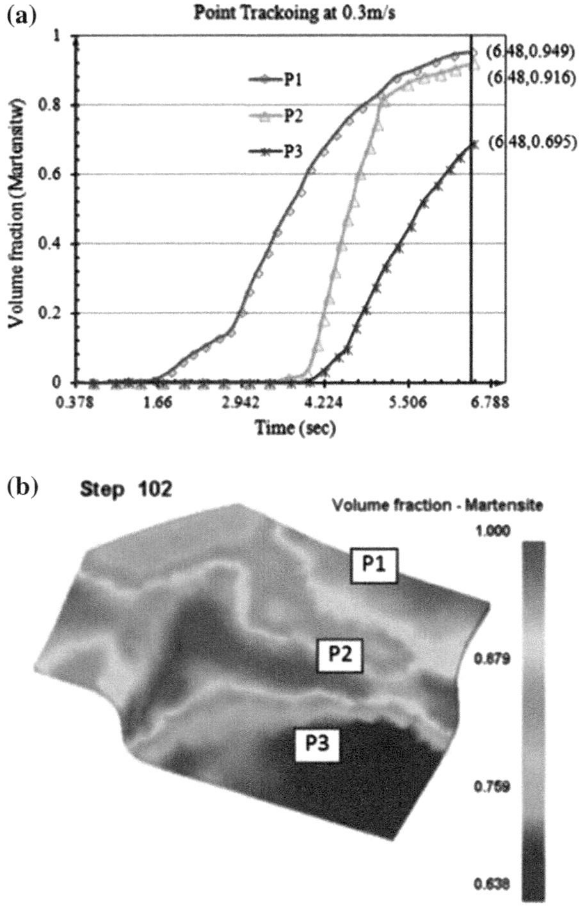

Fig. 8 **a** Transformation curve of martensite, **b** transformation contour of martensite

From Fig. 11b, there exists some residual austenite in flange and the bottom wall area. It may due to small portion of austenite failing to transform into martensite without sufficient time at this cooling rate and contact status.

3.3.3 Flow Rate of 0.5 m/s

Figures 12, 13 and 14 are the microstructure evolution of martensite, bainite and ferrite at water flow velocity of 0.5 m/s, respectively. Figure 12 indicates that flange area P1 is basically composed of fully martensitic microstructure, with a highest volume fraction of almost 98% due to faster cooling rate than the previous 0.1 and 0.3 m/s cases. Since flange area is always in close and tight contact with the blank holder and die, it takes away a large amount of heat from the workpiece, causing the largest amount of martensitic microstructural evolution.

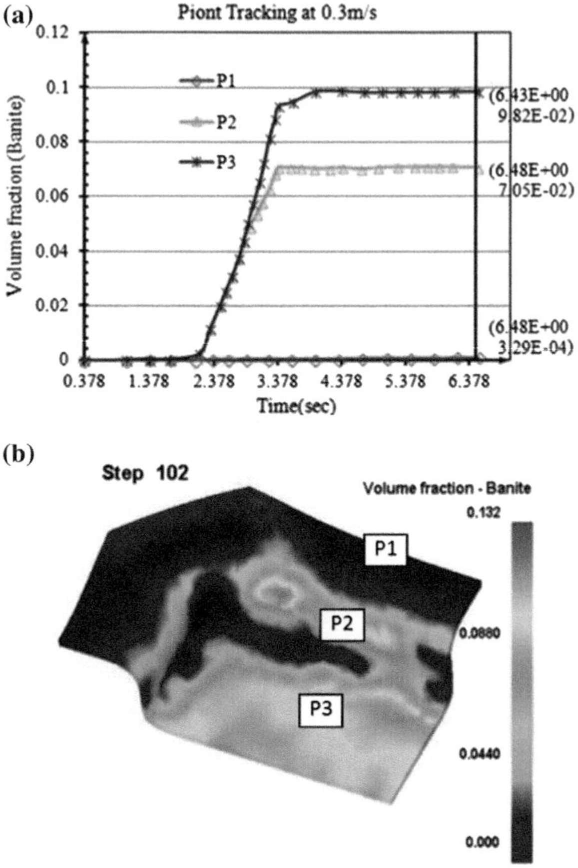

Fig. 9 **a** Transformation curve of bainite, **b** transformation contour of bainite

The sidewall area P2 consists of martensite (96.1%, P2 in Fig. 12a) and bainite (2.72%, P2 in Fig. 13a) and some tiny ferrite (0.211%, P2 in Fig. 14a). The bottom wall area P3 is composed nearly no martensite (0%, P3 in Fig. 12a), but of bainite (62.2%, P3 in Fig. 13a) and ferrite(27.3%, P3 in Fig. 14a).

Overall, the cooling water flow rate influences the composition of sidewall area P2 significantly. At 0.1 m/s, its volume fraction of martensite is only 87.7%. For 0.3 and 0.5 m/s, it increases much higher with 91.6 and 96.1%, respectively. This is because it contacts directly with the dies where the cooling water circulates. The effect of cooling rate on the microstructure of the box-shaped stamping parts can be obviously seen from the three different martensite proportions.

Fig. 10 **a** Transformation curve of ferrite, **b** transformation contour of ferrite

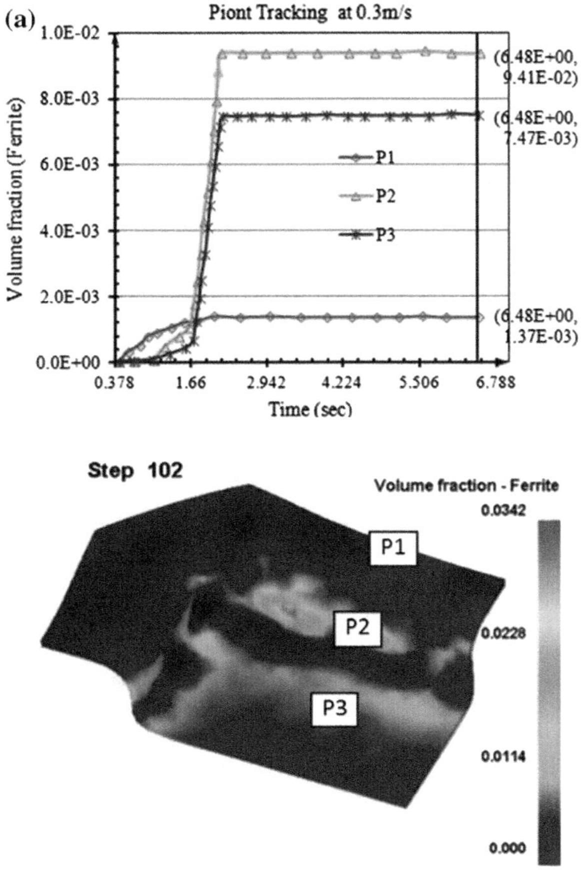

4 Experiment Verification and Analysis

To verify the correctness of the numerical analysis of microstructural evolution based on non-isothermal constitutive equation, hot forming of experimental measurement is performed subsequently and a microstructure prediction model of box part can be established to enhance the hot forming quality of sheet metal.

4.1 Experiment Device

To prevent high-temperature oxidation, the uniaxial tension experiment is conducted on workpiece of boron steel BR1500HS (high strength steel with Al–Si coating). In order to avoid surface oxidation and decarburization, most sheet metal blanks are pre-coated with a protective layer. The widespread protection is an Al–Si

Fig. 11 a Transformation curve of austenite, **b** transformation contour of austenite

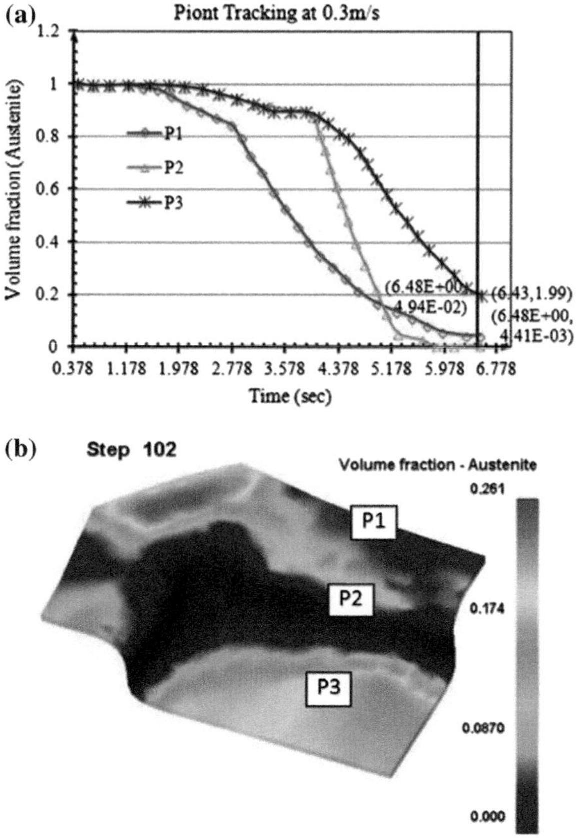

layer preventing scale formation on the steel during the direct hot stamping operation. The geometrical dimension of stamping workpiece is marked in Fig. 15. The steel's original microstructure composition is shown in Fig. 15, which is mainly composed of coarse grains of ferrite and pearlite. And pearlite distributes in ferrite grain boundary in the form of flake.

Initially, the material exhibits a ferritic–pearlitic microstructure with a tensile strength of about 600 MPa. After the hot stamping process, the component finally has a martensitic microstructure with a total strength of about 1500 MPa.

In industrial practical process, the workpiece is firstly heated to 950 °C and held for 5 min so that workpiece can be fully austenitized. Then it is transferred to the designed stamping mould with a water cooling system, as shown in Fig. 16. Real-time temperature changes during hot stamping process is measured with K-type thermocouple welded to the bottom, and data is recorded with the acquisition equipment.

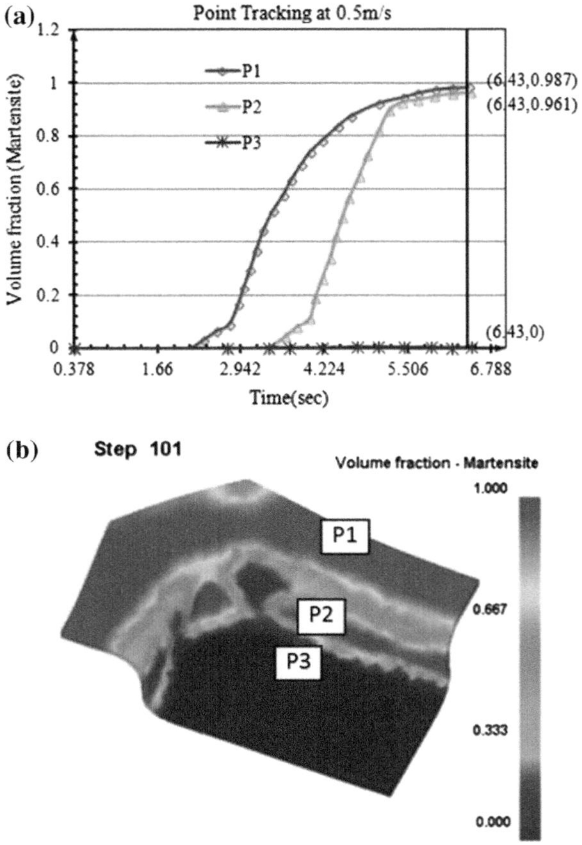

Fig. 12 a Transformation curve of martensite, **b** transformation contour of martensite

4.2 Experimental Overall Results

The circulating water takes away large quantity of heat, causing forming temperature decreasing quite quickly, which eventually results in poor plasticity and cracks. Figure 17 displays the overall appearance of three real deep drawings at different cool rates of 0.1, 0.3 and 0.5 m/s.

It can be seen that there is no crack when the cooling water flow velocity is just 0.1 m/s. However, tiny cracks appeared at 0.3 m/s rate and much longer one occurs at 0.5 m/s rate. Zoom-in show of cracking area, shown in Fig. 17a, b reveals that with the increasing of flow velocity, the crack line becomes longer and wider.

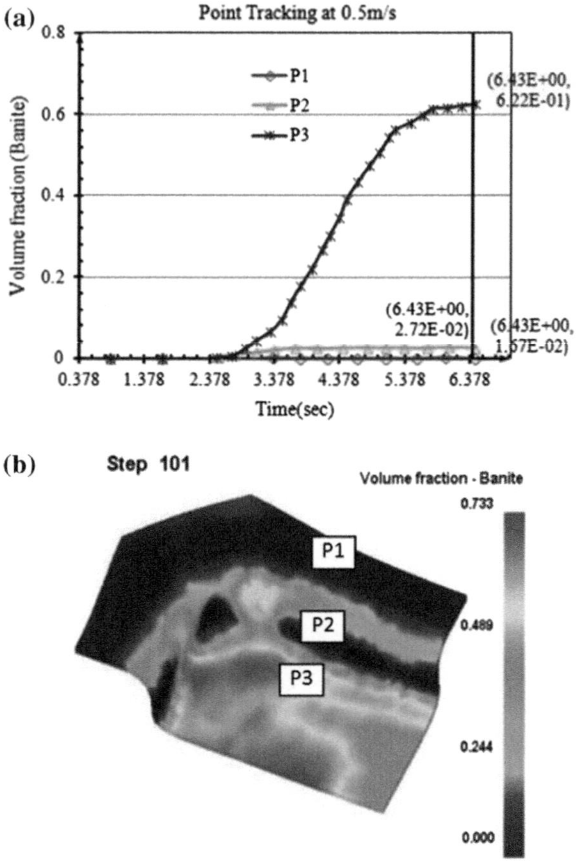

Fig. 13 **a** Transformation curve of bainite, **b** transformation contour of bainite

4.3 *Optical Metallurgy of Microstructures*

Different parts of the stamped box-shaped workpieces, including flange, bottom wall and sidewall, are cut by wire-electrode cutting machine and sampled to make microstructure constitution comparison after hot forming process. Their microstructure at different cooling rates is shown in Fig. 18.

4.3.1 Flange Area P1

Figure 18g, h, i shows the detailed microstructure of flange area under three cooling conditions. When water flows at 0.1 m/s rate, its microstructure is composed of martensite (nearly 95%) and small amount of ferrite. At 0.3 m/s, its constitutions are martensite (96%) and retained austenite (3%). Only martensite appears at 0.5 m/s.

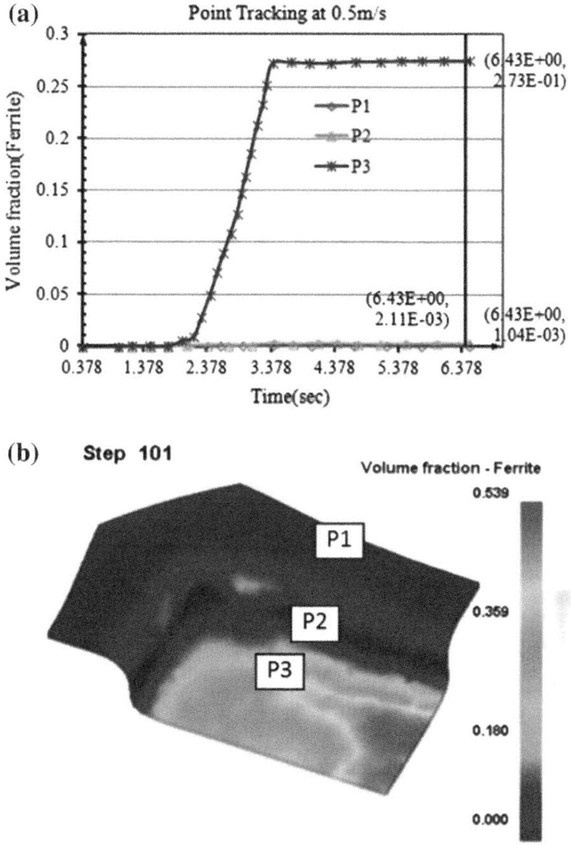

Fig. 14 **a** Transformation curve of ferrite, **b** transformation contour of ferrite

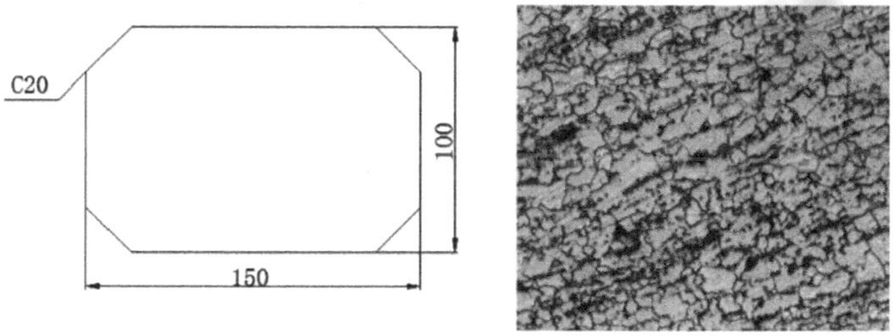

Fig. 15 Workpiece geometrical dimensions and original microstructure composition before hot stamping

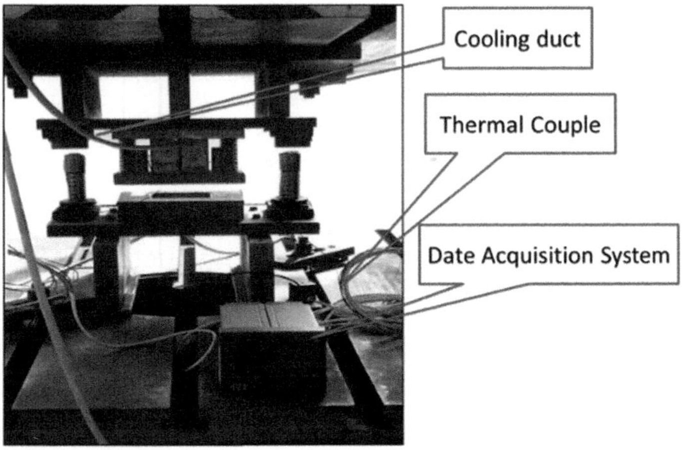

Fig. 16 Actual stamping mould tools and its diagram

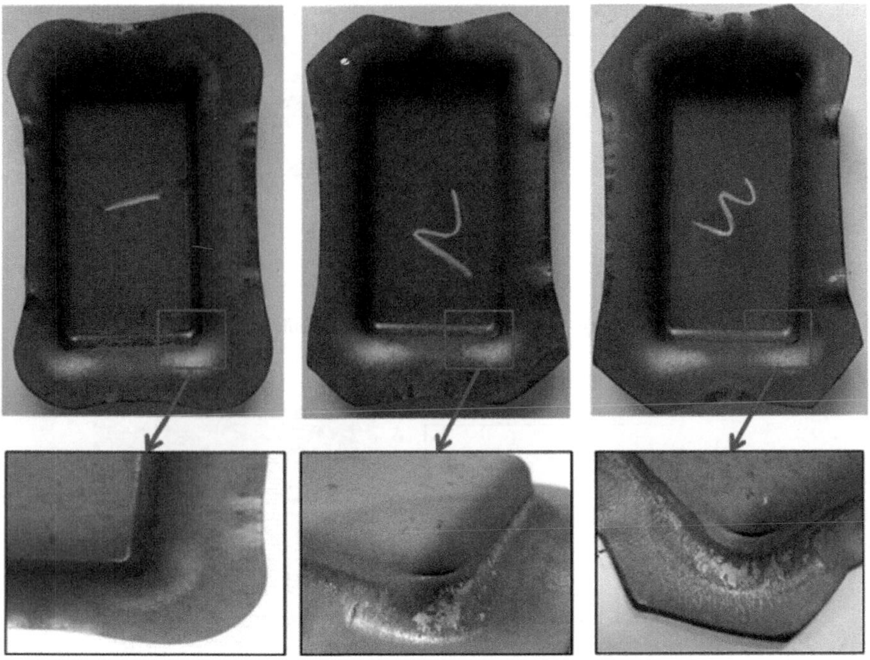

Fig. 17 Experiment results at different cooling water flow rate

This is because the much better close contact with both blank holder and die cooled the workpiece with higher speed. Greater width and density full of martensite banded structure are transformed accordingly. Since different flow

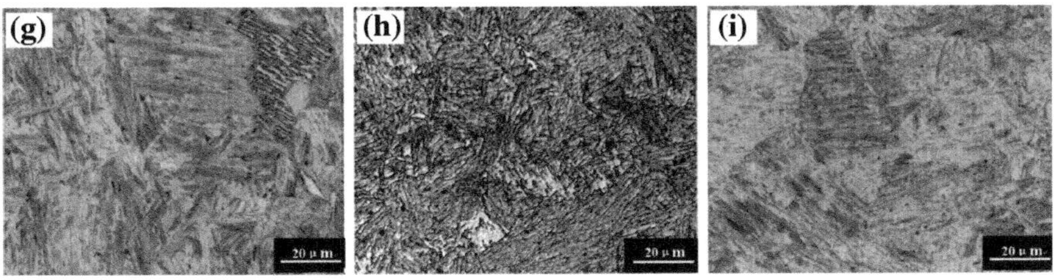

Fig. 18 Microstructures of flange area P1 at flow rates of 0.1, 0.3 and 0.5 m/s

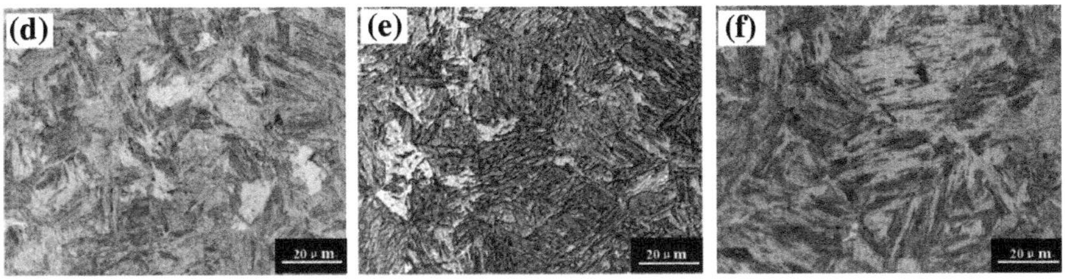

Fig. 19 Microstructures of sidewall area P2 at flow rates of 0.1, 0.3 and 0.5 m/s

velocity means different cooling rates and final forming temperatures, the grain size of the final microstructure differs. At the relative lower cooling rate of 0.1 and 0.3 m/s, the phase transformations are companied by stamping deformation. Hence, the final microstructure is composed of relative tiny grains on the same site. When the cooling rate is higher in 0.5 m/s case, austenite is transformed into martensite directly, retaining its original coarse appearance, shown much clearly in Fig. 18i.

4.3.2 Sidewall Area P2

Figure 19d, e, f shows that sidewall compose of quantities of martensite, some bainite and minor amount of retained austenite. Data shows that cooling mould can effectively guarantee the cooling rate during hot forming process and make workpiece fully be martensized with high strength. Also, quantity of heat has been taken away through the sidewall by the cooling water circulating in the moulds, which accelerated the cooling rate and formed a large numbers of martensite eventually. The martensite banded structures are widened and densified obviously, and Fig. 19d and f can clearly show that.

4.3.3 Bottom Area P3

Figure 20a, b, c shows that the bottom area composes of large quantities of bainite, some ferrite and minor amount of martensite.

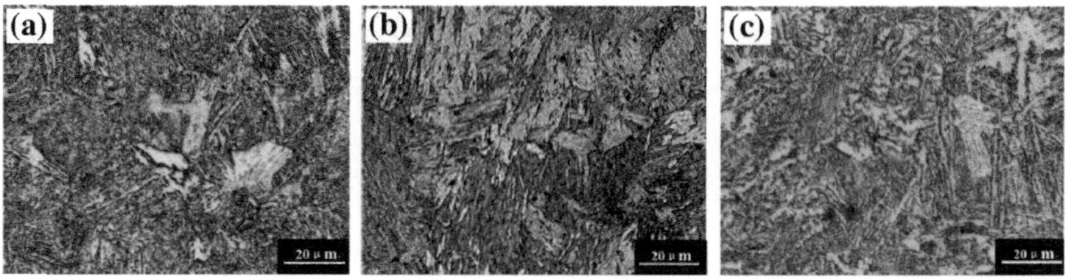

Fig. 20 Microstructures of bottom area P3 at flow rates of 0.1, 0.3 and 0.5 m/s

Only minor amount of martensite was transformed. This is because a large number of heat was taken away quickly when the bottom wall area firstly contacts with the punching die while the transient cooling rate reaches the highest. With the stamping processing, the bottom sheet metal bends gradually and loses contact with terrace die finally, causing big clearance between die and workpiece, leading to an insufficient cooling rate and heat conduct. It finally causes a number of bainite eventually when the cooling rate decreases in short time.

When the cooling water circulates at 0.3 m/s, the volume fraction of martensite is greater than the other two values. This is because that it has better cooling effect when the sampling region is closer to sidewall area.

4.4 *The Overall Comparison*

The comparison of experiment with simulation result is listed detailed in Table 3. It shows a difference of less than 10%. Considering the simulation accuracy and optical microscope measurement error, it can be conclude that good consistency exists between experiment and simulation. The new non-isothermal constitutive equation can provide better instruction in mould clearance design of hot stamping for high strength steel at industrial application.

Table 3 Composition of microstructure between numerical analysis and experiment

Flow rate(m/s)		0.1		0.3		0.5	
Flange area P1	Simulation	M	97%	M + A	95 + 4%	M	99%
	Experiment	M + F	95 + 4%	M + A + B	96 + 3 + 1%	M	98%
Flange area P2	Simulation	M + B	88 + 8%	M + B	92 + 7%	M + B	96 + 3%
	Experiment	M + B	85 + 10%	M + A + B	87 + 3 + 8%	M + B	95 + 3%
Flange area P3	Simulation	B + F	70 + 22%	M + A + B	70 + 19 + 9%	B + F	62 + 27%
	Experiment	B + F	78 + 20%	M + A + B	65 + 12 + 5%	B + F	70 + 25%

M Martensite; *A* Austenite; *B* Bainite; *F* Ferrite

5 Conclusions

(1) Because of close contact with moulds in large area, industrial hot stamping for high strength steel is accompanied with rapid temperature decreasing and making it actually a non-isothermal process.
(2) To study the effects of cooling rate on final phase constitution, non-isothermal deformations at three cooling water flow rates of 0.1, 0.3 and 0.5 m/s are simulated and experimentally measured. Better comprehensive performance appears at 0.1 m/s than in the other two.
(3) The good agreement of numerical simulation results with the experimental data manifests the effectiveness of non-isothermal high-temperature constitutive equation for hot stamping.

Acknowledgements This research is supported by the National Natural Science Foundation of China (NSFC-China) Project (No. 51275359).

References

1. Åkerström P (2006) Modeling and simulation of hot stamping. Doctoral Theses, Department of Applied Physics and Mechanical Engineering, Lule University of Technology, Sweden
2. Bardelcik A, Salisbury CP, Winkler S et al (2010) Effect of cooling rate on the high strain rate properties of boron steel. Int J Impact Eng 37(37):694–702
3. Patent GB1490535. Manufacturing a hardened steel article. NorrbottentensJaernverk AB Sweden; 1977
4. Berglund G (2008) The history of hardening of boron steel in northern Sweden. In: Proceedings of the 1st international conference on hot sheet metal forming of high-performance steel, pp 175–177
5. Jonsson M (2008) Press hardening, from innovation to global technology. In: Proceedings of the 1st international conference on hot sheet metal forming of high-performance steel, pp 253–265
6. Macek B (2007) Developing a deep drawn hot stamped fuel tank guard. In: Bentlerautomotive, great designs in steel seminar
7. Karbasian H, Tekkaya AE (2010) A review on hot stamping. J Mater Proc Technol 210: 2103–2118
8. Neugebauer R, Schieck F, Polster S, Mosel A, Rautenstrauch A, Schönherr J et al (2012) Press hardening-an innovative and challenging technology. Arch Civil MechEng 12:113–118
9. Merklein M, Svec T (2010) Transformation kinetics of the hot stamping steel 22MnB5 in dependency of the applied deformation on the austenitic microstructure. In: IDDRG 2010 International conference, Graz, Austria, pp 71–80
10. Naderi M, Ketabchi M, Abbasi M et al (2011) Analysis of microstructure and mechanical properties of different high strength carbon steels after hot stamping. J Mat Proc Technol, vol 211, pp 1117–1125
11. Cui J, Lei C, Xing Z, Li C (2012) Microstructure distribution and mechanical properties prediction of boron alloy during hot forming using FE simulation. Mater Sci Eng A 535: 241–251

12. Gao PF, Yang H, Fan XG et al (2012) Microstructure evolution in the local loading forming of TA15 titanium alloy under non-isothermal condition. J Mater Process Technol 212: 2520–2528
13. Jiang F, Zhang H, Su J et al (2015) Constitutive characteristics and microstructure evolution of 7150 aluminum alloy during isothermal and non-isothermal multistage hot compression. Mat Sci Eng A636:459–469
14. Merklein M, Lechler J (2008) Determination of material and process characteristics for hot stamping processes of quenchable ultra-high strength steels with respect to a FE-based process design. SAE World congress: innovations in steel and applications of advanced high strength steels for automobile structures, Paper No. 0853
15. Bardelcik A, Salisbury C, Winkler S, Wells MA, Worswick MJ (2010) Effect of cooling rate on the high strain rate properties of boron steel. Int J Impact Eng 37:684–702
16. Wei L, Liu H, Xing Z et al (2012) Effect of tool temperature and punch speed on hot stamping of ultra-high strength steel. Trans. Nonferrous Met. Soc. China 22 s534–s541

A Simulation Method for Predicting Connecting Rod Buckling Stress and Fatigue Safety Factor of a Four-Cylinder Small Displacement Engine

Yongming Zheng, Hongyu Wang and Zhaoshuo Wu

1 Introduction

As we all know, taking simulation measures to improve success rate of product design and shorten development period and cost during concept phase or definitely phase of an engine development program is an essential approach. Connecting rod is one of the key dynamic parts of an engine whose function is to turn the reciprocating motion of piston into rotating motion of crankshaft. During engine operation, connecting rod mainly sustains combustion gas pressure, inertial pull force, and lateral force. Because of the complex load condition and connecting rod boundary condition, it is difficult to capture accurate deformation or stress. A little simplification should be made to reduce calculation time, but inappropriate simplification may cause unrealized results.

This paper takes a four-cylinder small displacement engine connecting rod as an example to illustrate a new simulation method to predict connecting rod buckling and fatigue safety factor accurately. Operation loading of an engine comes from crank mechanism MBD code ENGDYN.

Because this is a turbocharged engine, we need to predict low-speed maximum torque case, maximum cylinder pressure case, rated power case, and no-load overspeed case. In each case, 20 key operating crank angle need to be considered including top dead center, bottom dead center, maximum lateral force, maximum big end lateral force, and so on.

Y. Zheng (✉) · H. Wang · Z. Wu
Jiangxi Chang He Automobile Co. Ltd, Jiu Jiang 332005, China
e-mail: zhengyongming@changhe-suzuki.com

Society of Automotive Engineers of China (SAE-China) (ed.), *Proceedings of the 19th Asia Pacific Automotive Engineering Conference & SAE-China Congress 2017: Selected Papers*, Lecture Notes in Electrical Engineering 486, https://doi.org/10.1007/978-981-10-8506-2_53

2 Analysis Model

2.1 Meshed Model

3D geometry models include connecting rod, big end bearing cap, big end bearing shells, big end bearing bolts, simplified crankpin, piston pin, and piston pin bush. Finite element model in key areas such as big end transition, small end transition, and cap bolts fillets should get refinement (see Fig. 1). Connecting rod and big end cap are meshed as stress quality tetrahedral. Remaining parts are meshed first-order combination of hexahedral and pentahedra. Table 1 shows the details of the mesh.

2.2 Joining Strategy [1]

Components in connecting rod assembly interact with each other. Thus, joints need to be built to simulate the interactions. This part is done in FEARCE software.

Sets for all the contact part of the components are required to be selected in FEARCE (or other preprocess software), which will be used to build the joint in the

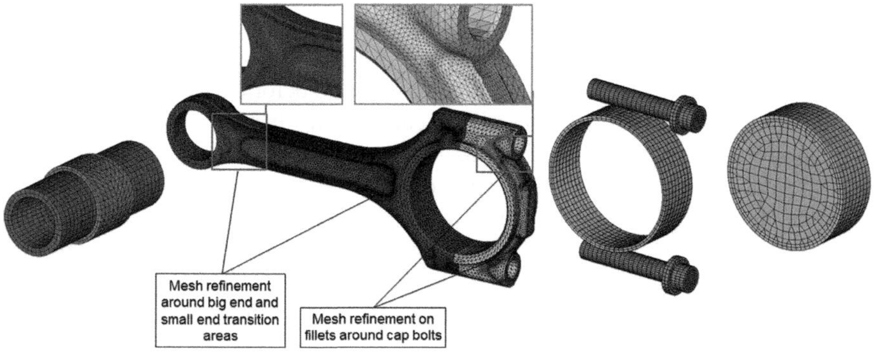

Fig. 1 Finite element models

Table 1 Mesh details

No.	Component	Type of mesh	Size (mm)
1	Big end bearing cap	High-order stress quality tetrahedral	0.3–3
2	Connecting rod	High-order stress quality tetrahedral	0.3–3
3	Big end bearing shells	First-order hexahedral	2
4	Big end bearing bolts	First-order hexahedral and pentahedral	2
5	Crank pin	First-order hexahedral and pentahedral	3.5
6	Piston pin	First-order hexahedral and pentahedral	2
7	Piston pin bush	First-order pentahedral	2

further step. Figure 2 shows some sets of the conrod which contact to other components in the assembly. The green area is conrod to cap, the yellow and purple area is conrod to bolts, and the blue area is conrod to upper shell.

All the joints in the connecting rod assembly are described below; Fig. 3 shows all components assembly relationship:

Bolt threads (bolt to conrod): Tied in radial directions with thread angle definition.
Shell to crankpin: Abaqus contact, set friction coefficient and radial clearance according to benchmark product or drawings.
Shell to cap: Abaqus contact, set friction coefficient.
Bolts to cap: Tied in all directions.
Upper shell to lower shell: Abaqus contact, set friction coefficient.
Cap to conrod: Abaqus contact, infinite friction used to model fracture-split connecting rod.
Bush to piston pin: Abaqus contact, set friction coefficient and radial clearance according to benchmark product or drawings.
Bush to conrod: Abaqus contact, set friction coefficient.

Fig. 2 Sets of conrod

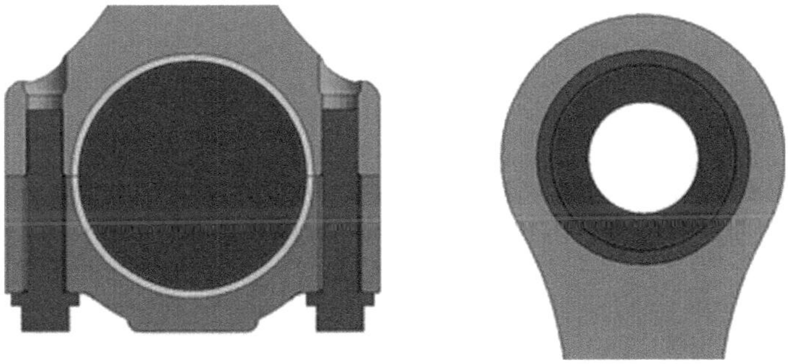

Fig. 3 Interaction components

2.3 Boundary Conditions

2.3.1 Global Restraints

To conduct a FEA analysis, restraints must be defined and should show motion as the connecting rod works in the engine. First, the whole connecting rod body should move as planar motion perpendicular to the crankshaft axis, some nodes in the middle plane of the rod body must be restrained in *X*-direction to prevent *X*-direction motion. Because the small end moves in the liner of the block, *Z*-direction should be restrained to prevent other direction motion. Crankpin is restrained in all directions. See Fig. 4.

2.3.2 Loading Regime

The connecting rod sustains both the assembly load and engine operation load. Assembly load includes bolts preload and interference load of bearing shell and bush. Engine operation load is the combustion load (cylinder pressure) which is varied with the crank angle.

(1) Assembly loads

(1) Bolts preload

The preload of bolts could be calculated as follow empirical formulas:

$$F_{\text{pre}} = k \cdot \sigma_b \cdot A_t$$

$$A_t = \frac{\pi}{4} \cdot (D_{\text{bolt}} - 0.9382 \cdot p)^2$$

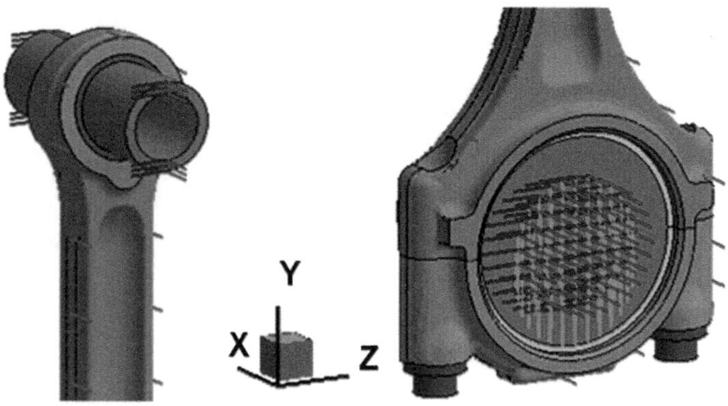

Fig. 4 Global restraints

where

D_{bolt} Nominal diameter of bolt;
p Bolt thread pitch;
A_t Tensile stress area of bolt;
k Fraction of bolt stress, depending on the tightening methods;
σ_b Bolt stress—Proof stress, Rp02 or proofing load depends on tightening methods.

(2) Interference load

Interference exists between big end of connecting rod and shells, small end, and bush. To simulate the crush load results from interference, a directional temperature expansion method is used. That is using temperature orthotropic material with expansion in tangential direction only. This method could avoid unrealistic expansion of the shell and the bush in axial and radial direction. In each direction, expansivity is defined as shown in Fig. 5.

(2) Operation loads

(1) Piston pin loads [2]

Operating loads of connecting rod are gained from MBD code ENGDYN; piston pin is the key part to sustain combustion gas load, so "pin-loaded" method has been used.

Rigid rod model is used in ENGDYN for load calculation. Crank pin and big end journal section are added to the rod FE assembly (Fig. 6).

- Model restrained at the big end journal.
- Restrained to move only in the bore axis at the pin.
- Structural gaps equivalent to nominal clearance applied to interface between bearings and journal faces.

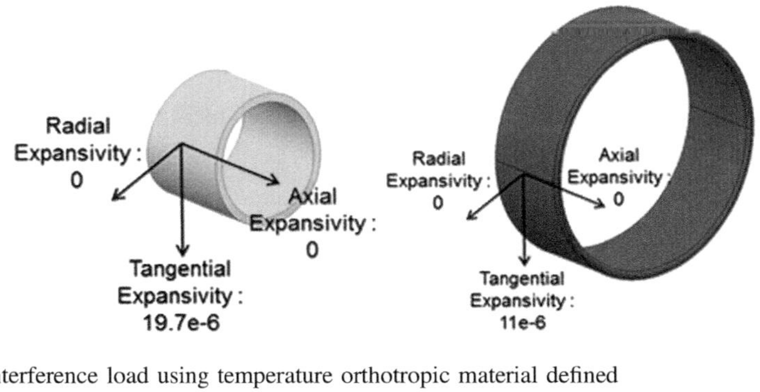

Fig. 5 Interference load using temperature orthotropic material defined

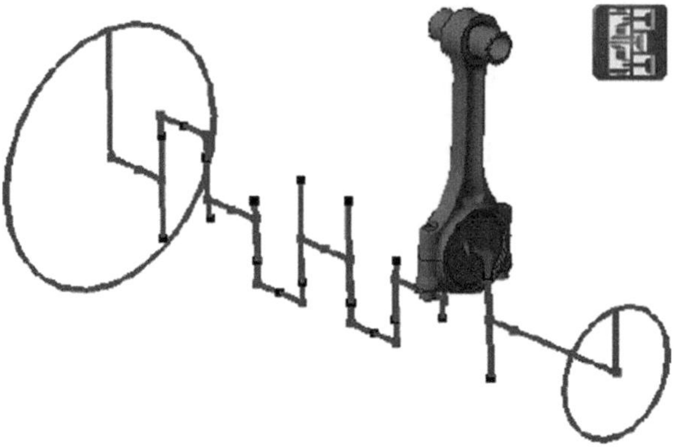

Fig. 6 Calculated load in ENGDYN

Connecting rod FEA was loaded using the ENGDYN model; journal bearing force is conducted using parabolic distribution in axis direction while cosine distribution circumferential in 180° (see Fig. 7).

(2) Operating load cases

To assess the strength of connecting rod, several critical engine cases should be evaluated: low-speed max torque, max cylinder pressure, rated power, and no-load overspeed, which based on engine performance curve showed in Fig. 8.

Following speeds are analyzed for high cycle fatigue:

- Low-speed max torque: 1500 rpm.
- Max cylinder pressure: 5000 rpm (110 bar).
- Rated power: 6000 rpm.
- No-load over speed: 6900 rpm.

130 bar is assumed to be the maximum cylinder pressure allowing for manufacturing and combustion variation.

According to ENGDYN calculated result (see Fig. 9), crank angles shown in Table 2 are selected by assessing axial force as well as lateral force diagram.

3 Stress Result

After running the FEA model under the critical cases mentioned above, the stress result can be available.

3.1 Stress and Deformation Under Critical Cases

The results show quite similar stress distribution and deformation tendency under low-speed max torque (1500 rpm) case, max cylinder pressure (5000 rpm) case,

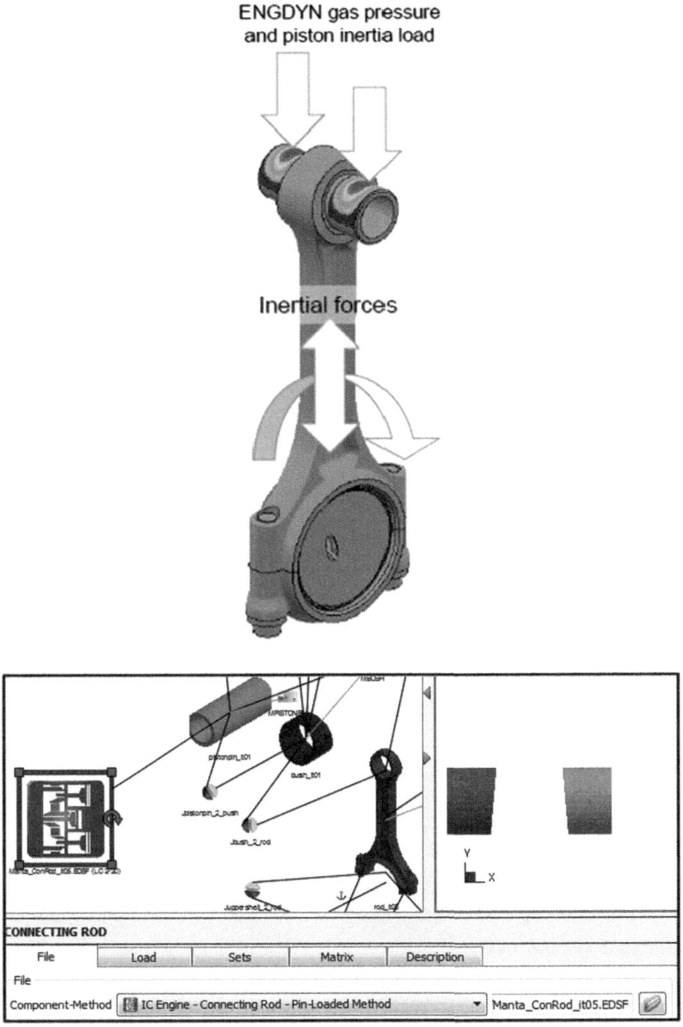

Fig. 7 Piston pin loaded

and rated power (6000 rpm) case, as the connecting rod sustains max compression load at TDC firing and max pull load at TDC no firing. But the result is more critical under max cylinder pressure (5000 rpm) condition. Figure 10 shows the highly loaded crank angles for the connecting rod; the crankshaft is under compression at TDC + Firing, whereas under tension at TDC non-firing where maximum pull force is applied.

At no-load over speed condition, TDC non-firing and lateral loading crank angles (when crank movement is perpendicular to shank axis) become more critical, conrod stress; see Fig. 11.

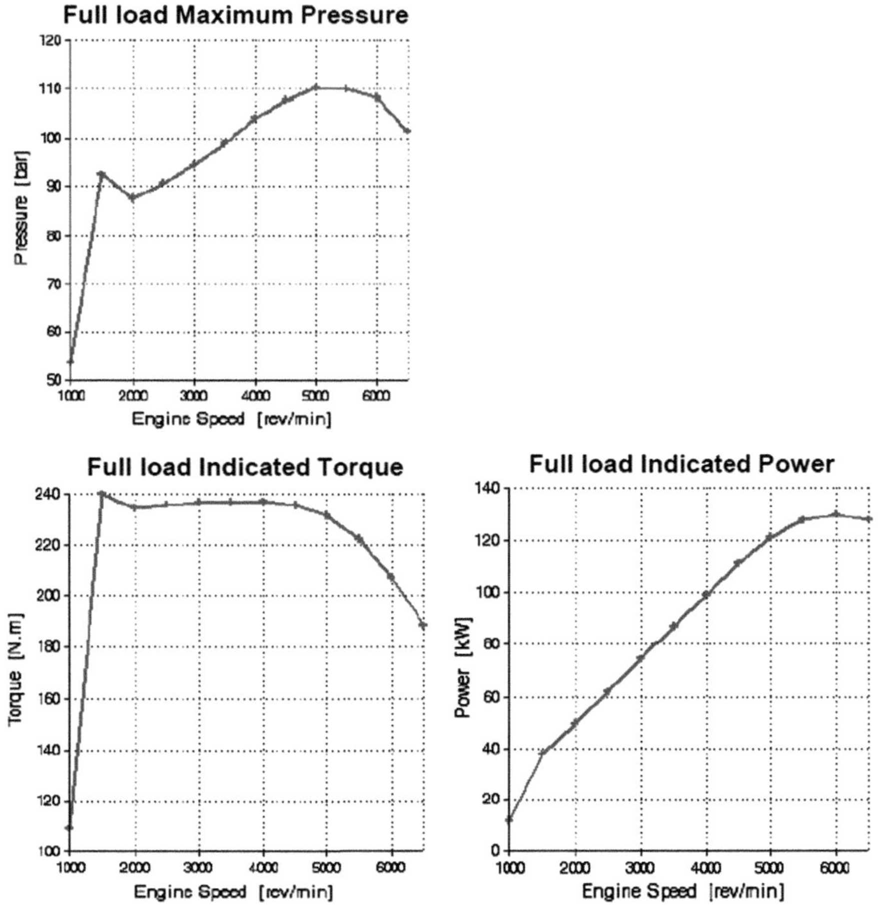

Fig. 8 Engine performance curve

3.2 *Yield Cover Factor*

Signed von Mises stress distribution at peak compressive speed (5000 rpm) TDC firing and peak tensile overspeed (6900 rpm) TDC non-firing become more critical. Extract yield cover factor from connecting rod stress distribution cloud diagram (see Fig. 12), and shown in Table 3.

The minimal yield cover factor on the connecting rod and bearing cap is above recommended value of 1.2 except P_{max} 130 bar at 5000 rpm. Lowest yield cover factor is predicted on "Small End Transition 3 = 1.13" location for P_{max} 130 bar, which experiences high compressive load. Low cover factor at small end transition is affected by the pessimistic value of compressive yield strength—against guideline of 580 MPa, the yield cover factor would be above 1.2.

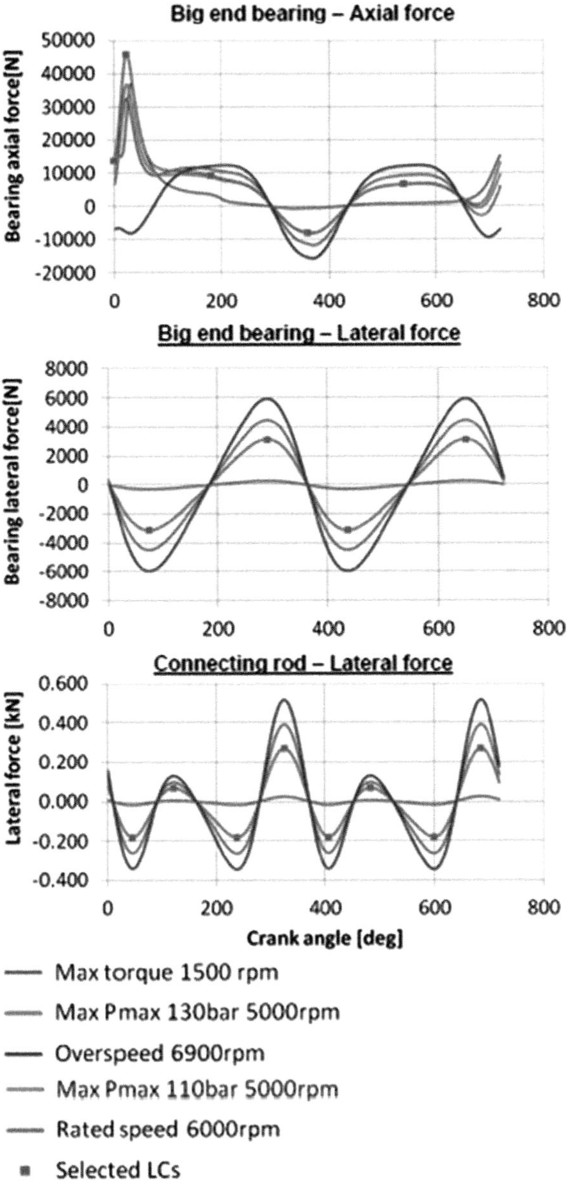

Fig. 9 ENGDYN calculated results

In order to give greater confidence in the design, plastic collapse analysis is conducted to assess yielding in compression.

Table 2 Critical crank angle

Load case	Crank angle	Description
1	NA	Assembly (Bolt pretension + Bearing Shell crush)
2	0	TDC
3	23	TDC firing
4	46	Max lateral force
5	76	Max big end lateral force
6	123	Max lateral force
7	180	BDC
8	239	Max lateral force
9	291	Max big end lateral force
10	325	Max lateral force
11	360	TDC non-firing
12	371	Axial force
13	406	Max lateral force
14	436	Max big end lateral force
15	483	Max lateral force
16	540	BDC
17	599	Max lateral force
18	651	Max big end lateral force
19	685	Max lateral force
20	720	TDC

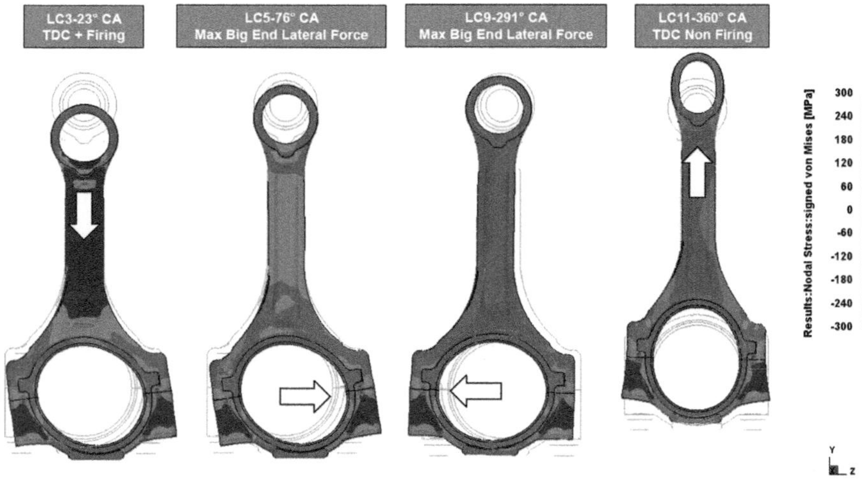

Fig. 10 Stress and deformation under max cylinder pressure condition

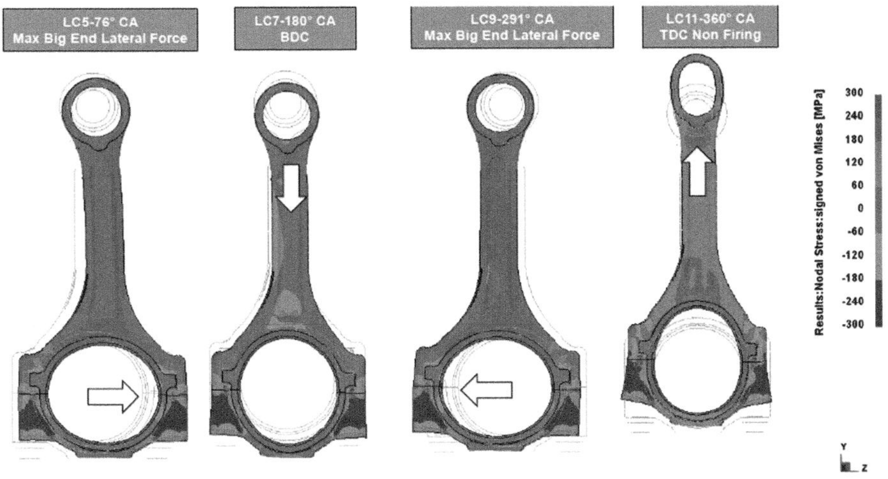

Fig. 11 Stress and deformation at no-load overspeed condition

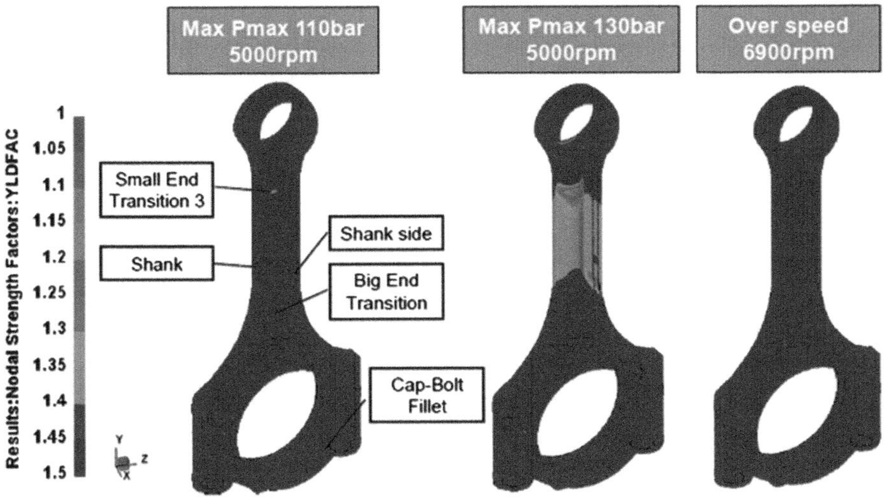

Fig. 12 Yeild cover factor

4 Buckling Mode Predicting

A buckling analysis must be undertaken to predict the accidental unstable plastic collapse. When buckling analysis proceeds, elastic–plastic material data of connecting rod should be assumed. In this analysis, model is a carryover, and a load is

Table 3 Yeild cover factor

Location	Max P_{max} 110 bar, 5000 rpm, LC3: TDC + Firing		Max P_{max} 130 bar, 5000 rpm, LC3: TDC + Firing		Overspeed, 6900 rpm, LC11: TDC non-firing	
	von Mises stress [MPa]	Yield cover factor	von Mises stress [MPa]	Yield cover factor	von Mises stress [MPa]	Yield cover factor
Small end transition 3	−398	1.38	−486	1.14	85	6.21
Shank	−334	1.65	−410	1.34	73	7.26
Shank side	−356	1.54	−436	1.26	100	5.29
Big end transition	−302	1.82	−369	1.49	54	9.73
Cap bolt fillet	261	2.04	259	2.05	405	1.31

applied to the rod in increments to a level which is high enough to buckle the component. When the load increases enough, mode of collapse could be seen. This is the point of buckling. Loads are applied to the piston pin using a distributed force that is representative of a bearing load. The small end is restrained to allow movement in the bore axis while restricting large movement orthogonally.

4.1 Elastic–Plastic Material Data

The material of connecting rod and cap is C70S6 which could be derived based on Ramberg–Osgood equation. Connecting rod material stress–strain is shown in Fig. 13. The Ramberg–Osgood equation was created to describe the nonlinear relationship between stress and strain—that is, the stress–strain curve—in materials near their yield points [3]. It is especially useful for metals that harden with plastic deformation, showing a smooth elastic–plastic transition.

$$\varepsilon = \frac{\sigma}{E} + 0.002 \cdot \left(\frac{\sigma}{\sigma_0}\right)^n$$

where

E The elastic Young's modulus;

σ_0 Yield strength;

n Ramberg–Osgood coefficient, describing the hardening behavior of the material, fitting of tensile (or compressive) experimental data, through test $n = 7.66$.

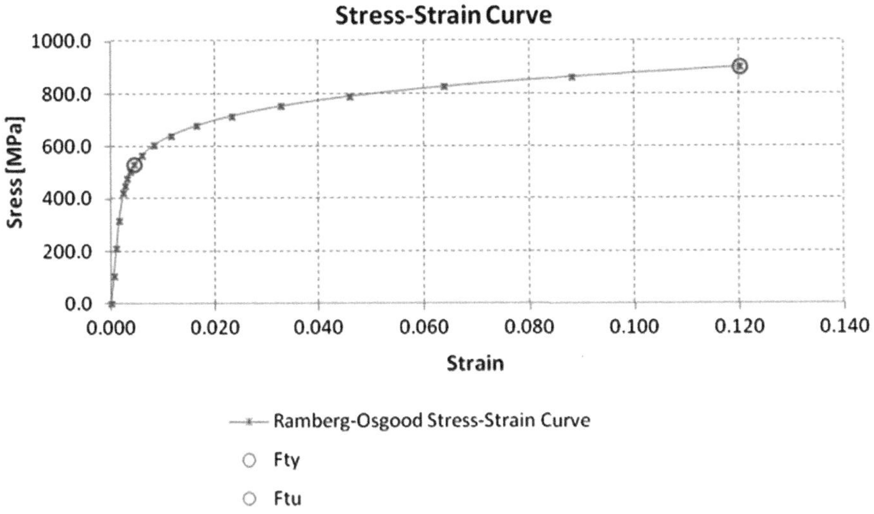

Fig. 13 Connecting rod material (C70S6) stress–strain curve

4.2 Buckling Analysis Loading Method

Riks method is used to undertake buckling analysis. It is generally used to predict unstable, geometrically nonlinear collapse of a structure which includes nonlinear materials and boundary conditions. This method follows an eigenvalue buckling analysis to provide complete information about a structure's collapse. It also can be used to speed convergence of ill-conditioned or snap-through problems that do not exhibit instability.

If the Riks step is a continuation of a previous history, any loads that exist at the beginning of the step and are not redefined are treated as "dead" loads with constant magnitude. A load whose magnitude is defined in the Riks step is referred to as a "reference" load. All prescribed loads are ramped from the initial (dead load) value to the reference values specified [4].

$$P_{total} = P_0 + \lambda(P_{ref} - P_0)$$

where P_0 is the "dead load," P_{ref} is the reference load vector, and λ is the "load proportionality factor."

5 Simulation Results

5.1 Yield Cover Factor

The minimal yield cover factor on the connecting rod and bearing cap is above recommended value of 1.2 except P_{max} 130 bar at 5000 rpm (Fig. 14). Lowest yield cover factor is predicted on “Small End Transition = 1.16” location for P_{max} 130 bar, which experiences high compressive load.

5.2 Buckling Analysis

In order to give greater confidence in the design, plastic collapse analysis is conducted to assess yielding in compression. Maximum axial force is 41 kN at 5000 rpm 110 bar, while maximum axial force is 50 kN at 5000 rpm 130 bar (Fig. 15). Buckling analysis shows that maximum force onset of plastic buckling is 62 kN (Fig. 16), which derived that the buckling cover factor is 1.51 at 110 bar and 1.24 at 130 bar, and both are passed guideline 1.2 for normal operating.

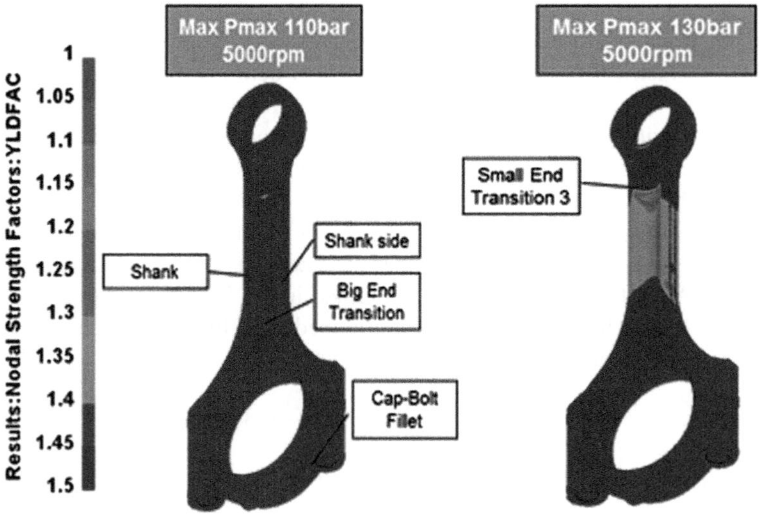

Fig. 14 Connecting rod yield cover factor

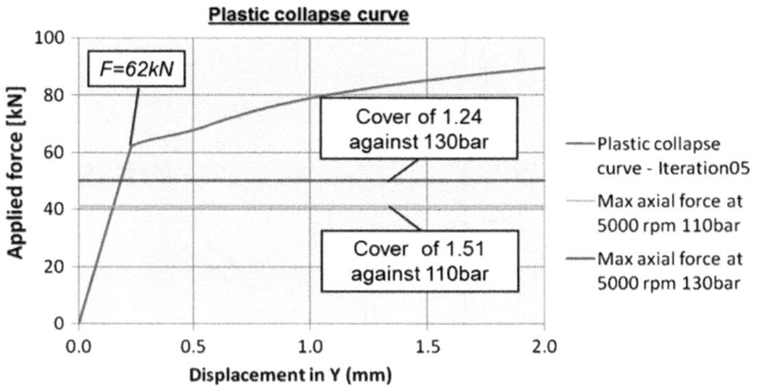

Fig. 15 Connecting rod plastic buckling curve

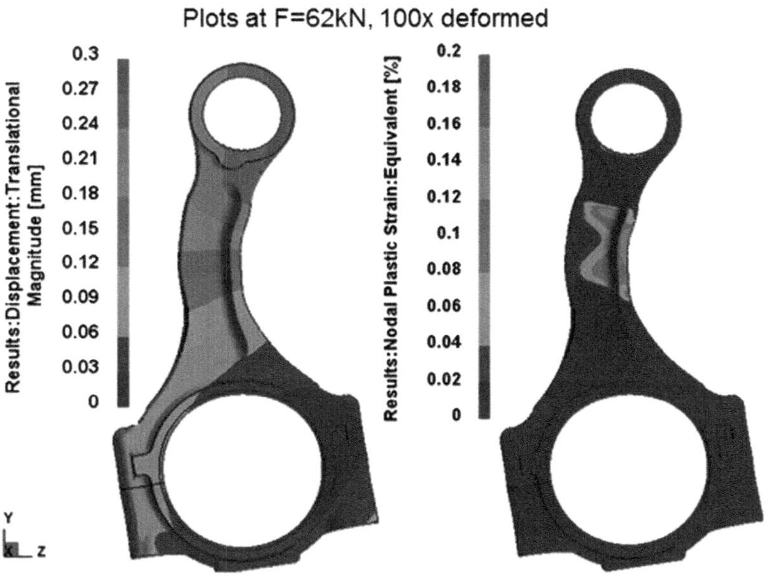

Fig. 16 Connecting rod buckling collapse mode

5.3 *High Cycle Fatigue Prediction [1]*

Haigh diagram has been used for high cycle fatigue prediction. The Haigh diagram is constructed using the basic material strengths as shown (Fig. 17):

σ_{yt} Tensile yield strength;
σ_{yc} Compressive yield strength;
σ_{uts} Ultimate tensile strength;
σ_{yt} Fatigue strength.

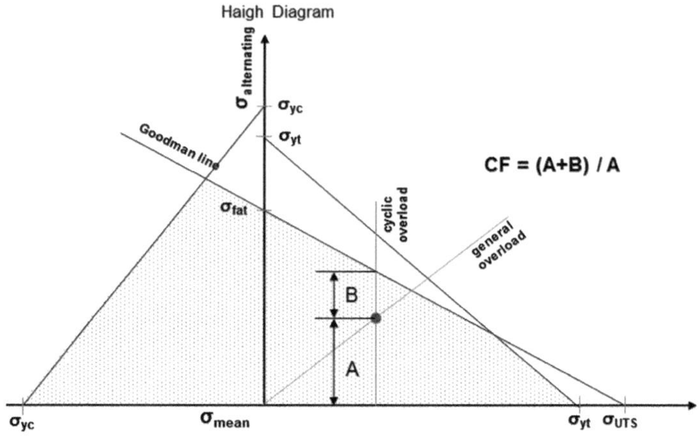

Fig. 17 Haigh diagram

The operating point for each node (red spot) is plotted on the Haigh diagram. Cover factor is calculated in the following way:

$$\text{CF} = (A + B)/A$$

In this, cover factor = 1 when $B = 0$, i.e., when the operating point lies on the limit line.

Uniaxial fatigue tests are normally either conducted in a tensile or bending mode. However, the two types of tests will produce different results. Some of this difference can be attributed to the different stress gradients induced by the two loading regimes (Fig. 18). It is then difficult to decide, for a particular stress state, which the fatigue limit (tension or bending) is most appropriate to use.

Panoskaltsis stress gradient correction allows the calculation to automatically vary the fatigue strength in relation to the intensity of the stress gradient. Low gradients use the Axial Fatigue Strength and higher gradients use the Bending Fatigue Strength.

For connecting rod durability, 6900 rpm no-load overspeed case is the most critical case; the minimum fatigue safety factor is 1.51 (Fig. 19).

6 Experimental Validation

Rig testing is an effective method for establishing component durability limits within a compressed timeframe, and without requiring engine builds and test activity. In the majority of cases, the rig testing is implemented to investigate and validate particular load cases identified as significant during simulation, or to

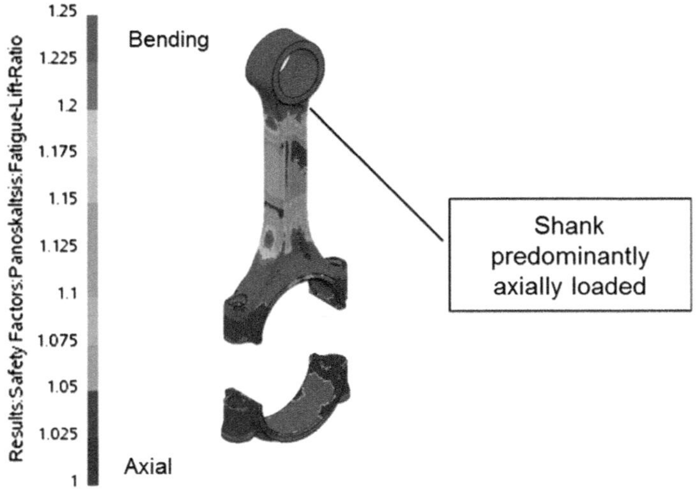

Fig. 18 Connecting rod loading

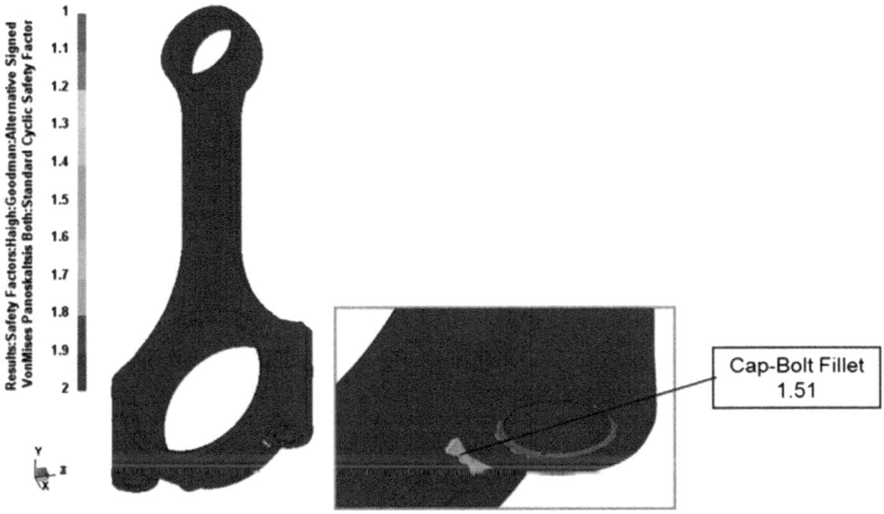

Fig. 19 No-load overspeed case fatigue safety factor

validate a simulation at one design level prior to further simulation at other design levels. This test procedure describes a staircase method for determining the buckling and the fatigue limit of a connecting rod.

6.1 Buckling Load Test

In order to detect the buckling load limit, Test Rig Instrumentation should be set up to monitor the following parameters: axial load, axial displacement between small end and big end. The rig load should be increased from zero until yield is observed. Increase axial load on connecting rod, measuring displacement, until yielding or buckling of the connecting rod is observed; this is the point at which the force/displacement gradient ceases to be linear (Fig. 20).

Tests have been undertaken for seven samples (a minimum of seven test samples is recommended to ensure statistical validity, but a minimum of ten test samples is recommended if possible), the buckling loads vary from 82 kN to 89 kN, mean load could be derived: 86 kN, and test results are shown in Table 4.

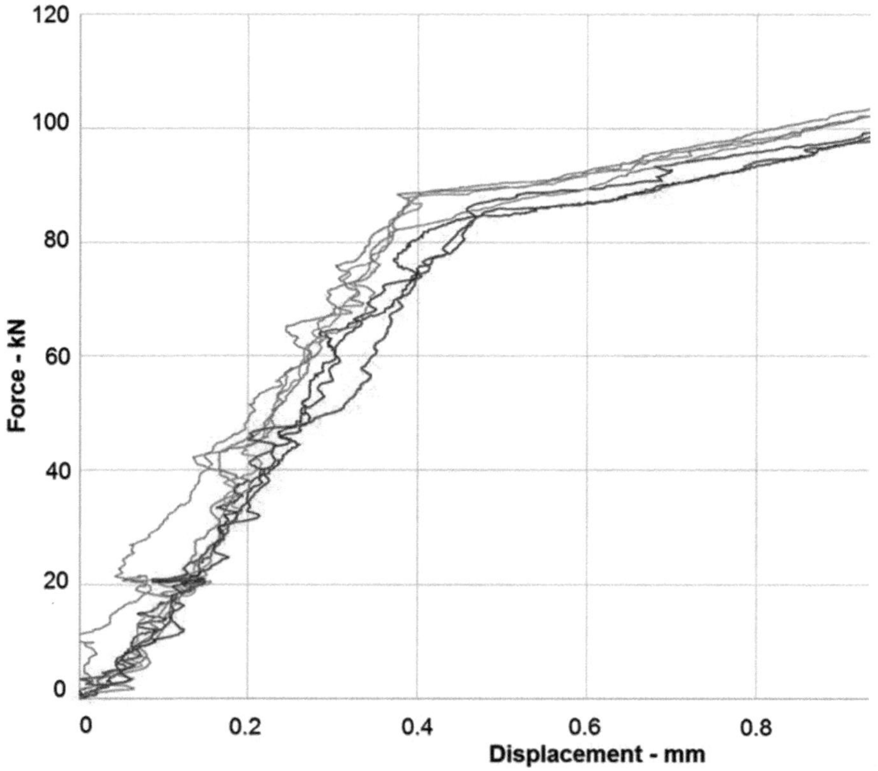

Fig. 20 Plastic collapse curve during connecting rod test

Table 4 Buckling test results

Yield load scatter	
Specimen number	Yield load (kN)
S1	84
S2	88
S3	89
S4	88
S5	87
S6	86
S7	82

6.2 Fatigue Assessment

The fatigue limit of a connecting rod determined by a Power Swing Rig. This Instrumentation works by testing components at a fully or partially reversed load and times the changes to excite the test piece's natural frequency. The rig detects when the natural frequency changes above a given threshold, indicating a failure has occurred, and stops the test automatically. For each test, a run out criteria of 10,000,000 cycles without failure is applied.

The rig should alternately apply worst-case compressive and tensile loading, which are calculated based on engine operating conditions.

The results are processed using a Dixon–Mood "staircase" method for determining safe operating loads; the flow chart for the staircase testing methodology is shown in Fig. 21.

If test sample fails before 10^7 cycles are completed, replace with a new sample and test at a decreased load level. If the test sample completes 10^7 cycles without failure, replace with a new test sample and test at an increased load level.

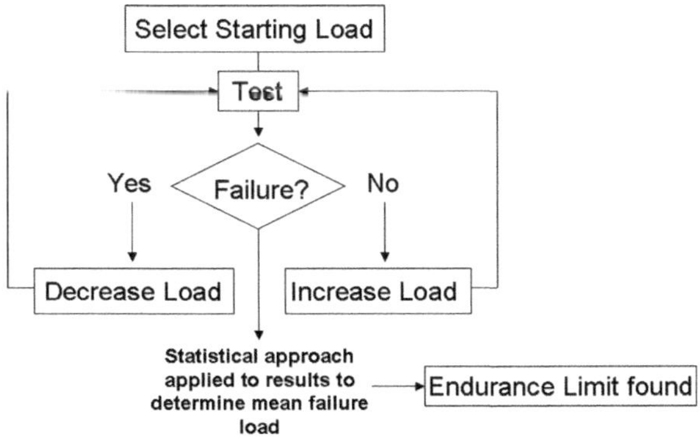

Fig. 21 Staircase testing process flow

Table 5 Fatigue loading test regime

Tensile load (kN)	Compressive load (kN)	Amplitude (kN)	Mean load (kN)	Safety factor
10.0	−53.0	31.5	−21.5	1.00
10.5	−55.7	33.1	−22.6	1.05
11.0	−58.3	34.7	−23.7	1.10
11.5	−61.0	36.2	−24.7	1.15
12.0	−63.6	37.8	−25.8	1.20
12.5	−66.3	39.4	−26.9	1.25
13.0	−68.9	41.0	−28.0	1.30
13.5	−71.6	42.5	−29.0	1.35

In order to determine appropriate loadings for the fatigue testing, a slider-crank derivation was implemented in conjunction with the simulation analysis, using mass and geometry data; worst-case compressive and tensile loads were alternately applied, determined as follows; and the test result is shown in Table 5:

(1) Compressive loads—maximum gas loading, with inertial relief corresponding to the lowest speed at which peak has occurred.
(2) Tensile loads—inertial loading corresponding to max engine speed.

6.3 *Results from Fatigue Durability Testing*

Eight connecting rod samples were tested, and result is shown in Table 6. First connecting rod went through multiple tests at different loads to find a suitable starting level. Each test was only run for 10 million cycles. The first connecting rod failed at 130% of initial starting load; initial load is the load at safety factor 1.00. Four of the samples run out at 10 million cycles which at 120% load level, while three samples failed run out for 10 million cycles at 125% load level. The operating frequency is a function of connecting rod stiffness; dropping frequency signifies

Table 6 Fatigue loading test results

Sample No.	Load level %	Outcome
1	Variable	Failed at 130%
2	120	Run out at 10×10^6
3	120	Run out at 10×10^6
4	125	Failed at $\sim 7 \times 10^6$
5	120	Run out at 10×10^6
6	125	Failed at $\sim 5.5 \times 10^6$
7	120	Run out at 10×10^6
8	125	Failed at $\sim 1.4 \times 10^6$

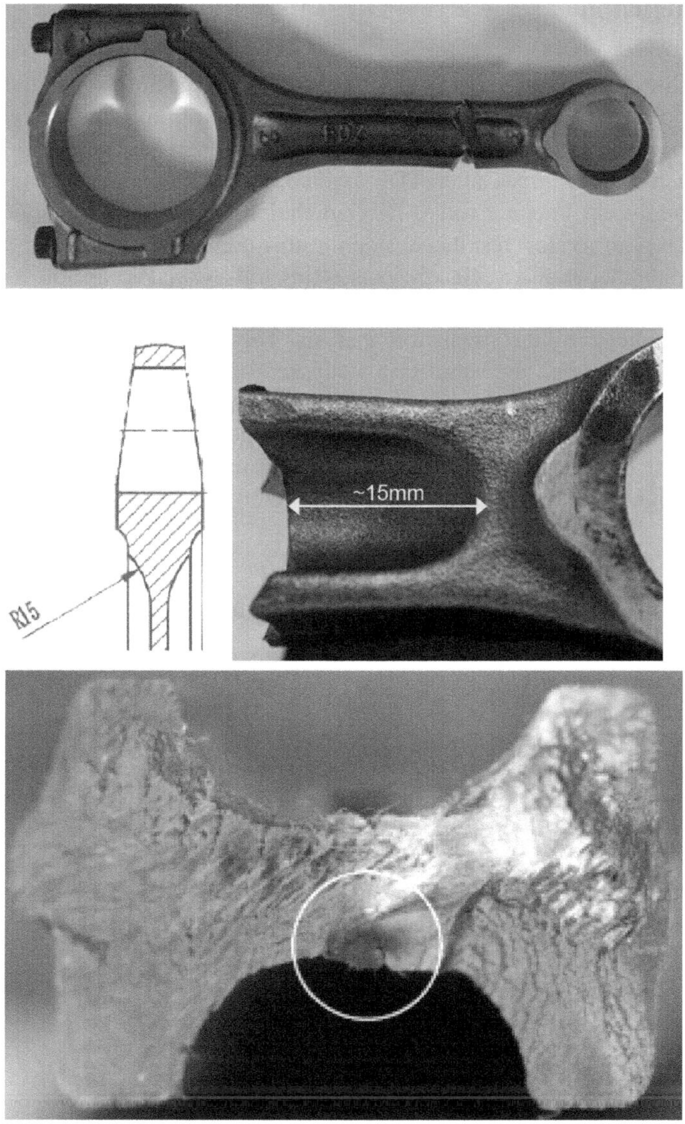

Fig. 22 Fracture point of these connecting rod

crack propagation. These failed samples' fraction places were at the small end transition corresponding to the simulation analysis results. The small end transition has a radius of 15 mm which relates approximately to the fracture point on the shank of the connecting rod (Fig. 22).

Based on the Dixon–Mood analysis, the fatigue life for the connecting rod is estimated to be a safety factor of 1.18, corresponding to a compressive load of 62.27 kN and a tensile load of 11.75 kN.

7 Conclusion

The buckling results showed an improved parity with the analysis predicting using this three-dimensional simulation, 22 kN improvement from simulation, but the rig test is performed with the rod essentially at TDC, whereas peak buckling load in the engine occurs 15 degrees after TDC, which reduces the buckling strength and would give the rig a greater value. This simulation method basically should show a well match with the test results considering this difference.

Regarding the fatigue results, the connecting rod showed acceptable durability in agreement with this analysis method. Test returns a safety factor of 1.18, similar to the 1.16 predicted by the simulation analysis. The failure initiation points showed decent agreement with the simulation analysis.

A method for three-dimensional connecting rod strength simulation has been developed that makes it possible to calculate buckling cover factor and fatigue safety factor of connecting rod. This simulation method can predict the buckling load and high cycle safety factor accurately which was verified by connecting rod experiment.

References

1. Ricardo FEARCE Software Manual
2. Ricardo ENGDYN Software Manual
3. Ramberg W, Osgood WR (1943) Description of stress–strain curves by three parameters. Technical Note No. 902. National Advisory Committee For Aeronautics, Washington DC
4. SIMULIA ABAQUS Software Manual

A Study of Misfire Detection

Tonghao Song, Tingwei Liu and Song Yan

1 Introduction

A misfire refers to incomplete combustion of an air/fuel mixture in a cylinder of the engine. For gasoline engine, misfire can result from a number of causes, such as defective ignition plugs, improperly timed spark, inaccurate measurements or improperly timed delivery of fuel, insufficient of intake air, or malfunction of valve system. The misfire can cause exceeded emissions, catalyst damage, reduced fuel economy, and engine performance. In recent years, along with the introduction of stringent government emission regulations, the requirements of misfire detection have arisen for equally. Misfire detection has been an essential ingredient of the OBD system.

Various techniques have been applied for misfire detection. One way to detect misfire is to directly monitor in-cylinder physical parameters, such as in-cylinder pressure or ionization current [1]. However, installing a sensor inside each cylinder is prohibitively expensive. An alternative is to measure exhaust pressure by pressure sensors and use it for misfire detection by monitoring abnormal exhaust pressure changes [2]. Again, this is not feasible from a production viewpoint due to high sensor system cost. Other work [3] describes the use of oxygen sensor methods to detect misfire. This class of methods monitors the abnormal exhaust air/fuel ratio changes to detect misfire. These are also sensitivite to the performance of oxygen sensor and catalyst in the case of oxygen sensor failure or aging catalyst. The most frequently used methods for misfire detection are based on engine crankshaft angular speed measurements, which utilize the engine speed fluctuation reflected by the signal of crankshaft angular sensor to detect misfire. This is feasible from a production viewpoint due to equipped sensor reuse, low system cost, and high accuracy.

T. Song · T. Liu (✉) · S. Yan
China FAW Corporation Limited R&D Center, Changchun, China
e-mail: liutingwei@faw.com.cn

Society of Automotive Engineers of China (SAE-China) (ed.), *Proceedings of the 19th Asia Pacific Automotive Engineering Conference & SAE-China Congress 2017: Selected Papers*, Lecture Notes in Electrical Engineering 486, https://doi.org/10.1007/978-981-10-8506-2_54

In general, there are three ways to detect misfire based on crankshaft speed. One way is based on an evaluation of the statistical distribution of the angular speed. Another possibility is to compare the angular speed data from one cylinder to another [4, 5]. Misfire would be detected, while the differences exceed a threshold value. These methods are easy to implement and are necessary to predetermine a certain speed- and load-dependent threshold. Another way is to analyze the “engine roughness” [6], which is defined as change of the crankshaft angular acceleration of the engine. Misfire would be detected if the engine roughness exceeds a threshold value. Another approach is to estimate combustion torque from the crankshaft speed of the engine by the crankshaft dynamic model to detect misfire. This method needs a Kalman filter to estimate the combustion torque; thus, the algorithm of which is complicated. Otherwise, at higher engine speeds, the performance of the Kalman filter decreases, and the estimation limit of this approach is between 3000–4000 rpm.

In practice, the first two classes of methods are commonly used for misfire detection. This article will focus on the engineering application on a vehicle equipped with a three-cylinder 1.0T engine of the two classes of methods. In this article, the misfire detection performance of the two classes of methods is compared, and an improved method is presented by the analysis of the experimental results. Then, the improved method is also compared with the conventional methods, and discussion about the result is demonstrated.

2 Basic Analysis of Misfire Detection

In this section, the analysis of the misfire detection based upon crankshaft speed is proposed. Furthermore, the aspects which affect misfire detection results are also discussed.

The signal of crankshaft speed is measured by EMS with a certain sample interval T. There are gear teeth on the flywheel, and C corresponding to a nominal crankshaft gear tooth interspacing of A crank angle degrees is given by:

$$C = \frac{A}{6\omega T} \tag{1}$$

Here, ω is the engine speed (RPM). And the differential of C is:

$$\begin{aligned} \mathrm{d}C &= \frac{\partial C}{\partial A} \cdot \mathrm{d}A + \frac{\partial C}{\partial \omega} \cdot \mathrm{d}\omega \\ &= \frac{1}{6\omega T} \cdot \mathrm{d}A - \frac{A}{6\omega^2 T} \cdot \mathrm{d}\omega \\ &= \frac{1}{6\omega T}\left(\mathrm{d}A - \frac{A}{\omega} \cdot \mathrm{d}\omega\right) \end{aligned} \tag{2}$$

Equation (2) can be used to analyze the change of C due to the change of engine speed and gear tooth interspacing. For misfire detection, the former may be the result of the misfire event, and the latter may result due to tooth machining errors.

It is also shown in Eq. (2) that in order to detect misfire correctly, the machining errors must be negligible. That is,

$$|\mathrm{d}A| \ll \left|\frac{A}{\omega} \cdot d\omega\right| \tag{3}$$

A first-order crankshaft dynamic model is used to simulate the drop in engine crankshaft speed with the occurrence of a misfire, and the aspects which affect the misfire detection results are analyzed. The crankshaft dynamic model is:

$$J \cdot \frac{\mathrm{d}\omega}{\mathrm{d}t} = -\beta \cdot \omega + M_{\text{load}} \tag{4}$$

where J denotes the rotational inertia, β is a coefficient representing engine friction, and M_{load} represents net engine torque. Consider the simple that the net engine torque during the time interval of one engine firing event remains approximately constant. The time interval is:

$$T_{\text{interval}} \approx \frac{720/N_c}{6\omega_0} = \frac{120}{N_c\omega_0}$$

where N_c is the number of cylinders, and ω_0 is the engine crankshaft speed just prior to the misfire event. For the occurrence of a misfire event, the crankshaft speed can be simplified to be stable at ω_0. Thus, the solution of Eq. (4) is:

$$\omega(t) = \omega_0 \cdot \mathrm{e}^{-\frac{\beta}{J}t} + \frac{M}{\beta}\left(1 - \mathrm{e}^{-\frac{\beta}{J}t}\right) \tag{5}$$

The crankshaft speed drop over the course of one cylinder event is given by:

$$\begin{aligned} \Delta\omega &= \omega(t = 0_-) - \omega(t = T_{\text{interval}}) \\ &= \omega_0 - \left(\omega_0 \cdot \mathrm{e}^{-120\beta/(\omega_0 N_c J)} + \frac{M}{\beta}\left(1 - \mathrm{e}^{-120\beta/(\omega_0 N_c J)}\right)\right) \\ &= \omega_0\left(1 - \frac{M}{\beta\omega_0}\right)\left(1 - \mathrm{e}^{-120\beta/(\omega_0 N_c J)}\right) \end{aligned} \tag{6}$$

By Eqs. (3) and (6), it is clear that to be able to detect a misfire easily, we need:

$$\left|A \cdot \left(1 - \frac{M}{\beta\omega_0}\right)\left(1 - \mathrm{e}^{-120\beta/(\omega_0 N_c J)}\right)\right| \gg |\mathrm{d}A| \tag{7}$$

The fluctuation of engine crankshaft speed is used for misfire detection:

$$F = \left|\frac{\Delta\omega}{\omega_0}\right| = \left|\left(1 - \frac{M}{\beta\omega_0}\right)\left(1 - e^{-120\beta/(\omega_0 N_c J)}\right)\right| \tag{8}$$

It is common that under the occurrence of a misfire event, the net engine torque is negative. Since the second term in Eq. (8) decreases exponentially with increasing engine speed, F decreases rapidly with increasing engine speed. By an argument similar to that for high engine speed, misfire detection is harder for engines with higher inertia J, and a larger number of cylinders make it more difficult to detect misfire as well.

It is also clear that it is harder to detect under low load conditions, because the net engine torque would be less negative and would result in a smaller fractional change in engine speed.

3 Misfire Detection Method

The basic workflow of the crankshaft speed-based misfire detection is shown in Fig. 1.

As shown in Fig. 1, the first step is to measure the segment time by using the crankshaft angular sensor (240 °C A interval time for three-cylinder engine), and then learn and correct the tooth machining errors to compensate for disturbances (shown in Fig. 2). The next step is to calculate misfire characteristic according to the diagnosis algorithm. Finally, the characteristic result is compared to a threshold (derived from a speed- and load-dependent threshold map) in order to make normal/misfire decisions.

The key point of the crankshaft speed-based misfire detection method is to calculate the misfire characteristic. The calculation formula of the misfire characteristic by comparing the segment time from cylinder to cylinder is [4]:

$$\delta = T_i - \hat{T}_i \tag{9}$$

Here, δ denotes the misfire characteristic, T is the segment time, $\widehat{T}$ is the segment time estimation, and i is the sample data. For the engine with N_c cylinders, the $\widehat{T}$ can be denoted as:

$$\hat{T}_i = f(T_{i-N}, T_{i-N+1}, \ldots, T_{i+N}), \quad N \in [0, N_c] \tag{10}$$

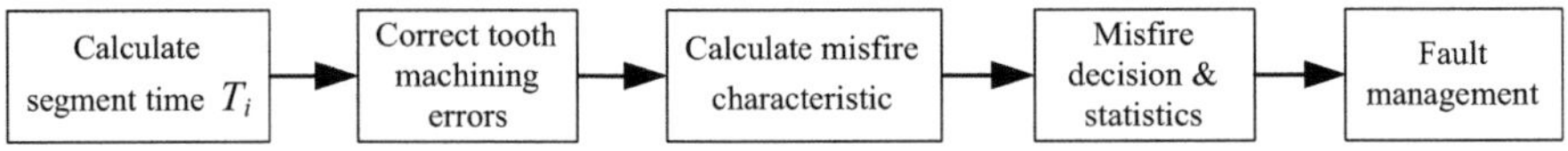

Fig. 1 Basic workflow of the crankshaft speed-based misfire detection

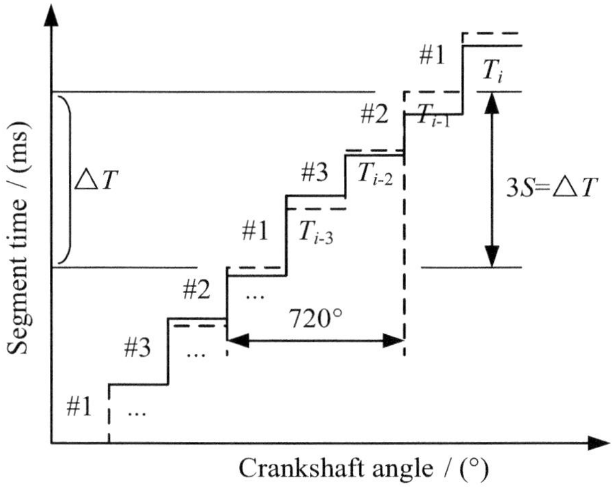

Fig. 2 Segment time calculation and compensation

In order to detect the misfire event, the function f should distinguish the statistical distribution of the misfire characteristics under normal event from the one under misfire event; i.e., there is no overlap between the two distributions. However, the engine operations are various and complicated, and the optimal function f cannot be derived by theoretical method easily. In engineering practice, the misfire detection algorithm uses multiple functions to make detection, and the combination of the multiple functions is determined by the experiments.

After the calculation of the misfire characteristic is determined, the ECU will calculate the misfire characteristic of each cylinder in real time. When the misfire characteristic is larger than the threshold, it is considered that the misfire event is detected.

For a three-cylinder engine, the $\widehat{T}$ can be evaluated with the following methods:

Method 1: Assume that there is no misfire event and the engine speed is stable. Thus, the T_i can be estimated by T_{i-1}, and then $\widehat{T} = T_{i-1}$ and

$$\delta = T_i - \hat{T}_i = T_i - T_{i-1} \tag{11}$$

Method 2: Assume that there is no misfire event and the engine speed is stable, and the nonlinear relationship of the engine speed and segment time is negligible. Thus, the T_i can be estimated by T_{i-1} and T_{i-2}, and then $\widehat{T} = 2T_{i-1} - T_{i-2}$ and

$$\delta = T_i - \hat{T}_i = T_i - 2T_{i-1} + T_{i-2} \tag{12}$$

Method 3: Assume that there is no misfire event and the engine speed is stable, and the nonlinear relationship of the engine speed and segment time is negligible.

Thus, the T_i can be estimated by T_{i-1} and T_{i+2}, and then $\hat{T} = 2/3 \cdot T_{i-1} + 1/3 \cdot T_{i-2}$ and

$$\delta = T_i - \hat{T}_i = T_i - \frac{2}{3}T_{i-1} - \frac{1}{3}T_{i-2} \tag{13}$$

Method 4: Assume that there is no misfire event and the acceleration the engine speed change rate is stable, and the nonlinear relationship of the engine speed and segment time is negligible. Thus, the T_i can be estimated by T_{i-2}, T_{i-1}, and T_{i+1}, and then $\hat{T} = -1/3 \cdot T_{i-2} + T_{i-1} + 1/3 \cdot T_{i-2}$ and

$$\delta = T_i - \hat{T}_i = \frac{1}{3}T_{i-2} - T_{i-1} + T_i - \frac{1}{3}T_{i-2} \tag{14}$$

In a certain engine load and speed operation, the misfire characteristic δ with normal event can be treated as statistical sample S_1, and under the same operation, the misfire characteristic with misfire event can be treated as statistical sample S_2. Consider the assumption that S_1 and S_2 obey the normal distribution; thus, misfire detection algorithm with a good performance should meet the following equation:

$$D = \frac{|\mu_1 - \mu_2|}{|\sigma_1 + \sigma_2|} > 3 \tag{15}$$

where D is a measurement of the detection performance, μ_1 is the average of the misfire characteristic under misfire events, σ_1 is the standard deviation, μ_2 is the average of the misfire characteristic normal events, and σ_2 is the standard deviation.

Figure 3 illustrates the distinction between the misfire characteristics under normal and misfire events. A higher D translates to lower mis-detect (actual misfire events that go un-detected) as well as lower false alarm (normal misfire events that get flagged as misfire events) probabilities. There is no overlap between the distributions of S_1 and S_2 when $D > 3$, and the misfire threshold can be set as any value between the $\mu_1 + 3\sigma_1$ and $\mu_2 - 3\sigma_2$. When $D < 3$, the probability distributions of S_1 and S_2 are overlapped, and no matter how the misfire threshold is chosen, the misfire event cannot be completely separated from the normal event.

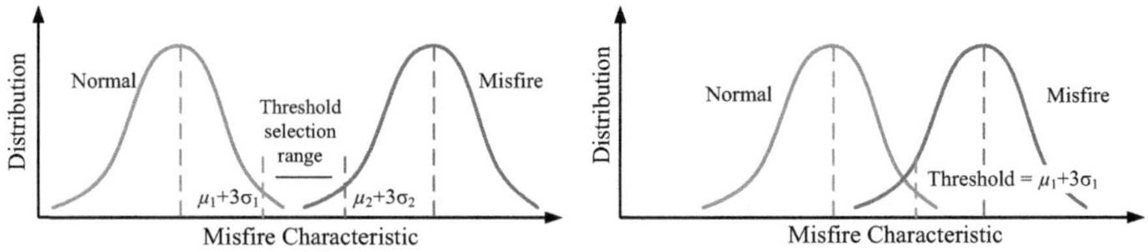

Fig. 3 Misfire detection performance measure

4 Misfire Detection Experiment

In order to compare the performance of current detection methods and optimize algorithm, a misfire detection experimental system is set, and misfire detection experiments have been performed. A misfire generator is designed and embedded in the ECU software to simulate the misfire events. In the experiment, the misfire generator can generate continual and random misfire signal to induce engine misfire events, and the misfire signal and engine speed and load can be measured at the same time. As is illustrated in Fig. 4, the misfire detection experimental system is divided into functional blocks.

The experiments are performed on an A-class vehicle equipped with a three-cylinder engine and a DCT transmission. The ignition order of the engine is #1 $\rightarrow$ #2 $\rightarrow$ #3, the rated power is 88 kW, and the maximum torque is 170 N m. The misfire generator is a program embedded in the ECU and simulates the misfire event by prohibiting fuel injection or ignition. The misfire generator can set misfire occurrence cylinder and generate single cylinder continual misfire, random misfire, periodic misfire, and multi-cylinder continual misfire signal. And the CANape tool is used to measure the segment time, engine speed, engine load, and misfire signals.

During the experiment, the first step is to set the transmission gear and make the engine operation stable at a certain speed and load. The next step is to simulate continual and random misfire event at each cylinder, respectively. After the completion of misfire information measurement at a certain operation point, pause the misfire generator and wait for the engine stable at a subsequent operation point, and then continue the misfire detection experiment.

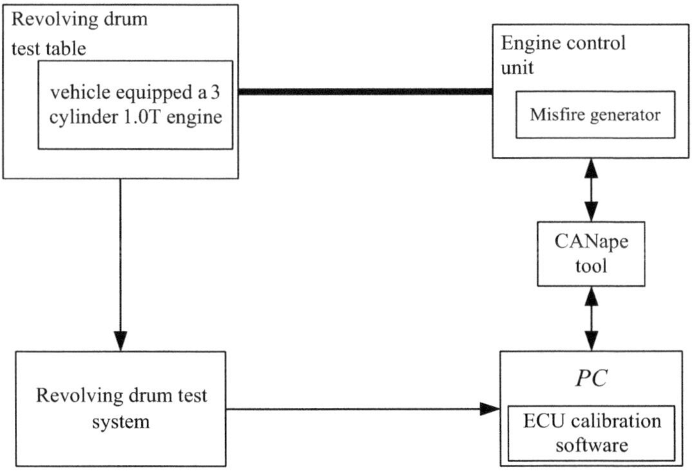

Fig. 4 Misfire detection experimental system

5 Results and Discussion

Figure 5a–d shows the misfire experimental data at 3800 rpm engine speed and 20% maximum engine load.

The performance of the four misfire detection methods is shown in Fig. 5. By evaluating the δ calculated by each method, it is shown that the sample data δ of all the methods meet the normal distribution law well. Thus, it means that the normal distribution assumption is satisfied, and the misfire detection performance measure D is reasonable.

Figure 5 shows that, no matter with which method, the averages of the δ distribution under normal events of the third cylinder are always larger than the second

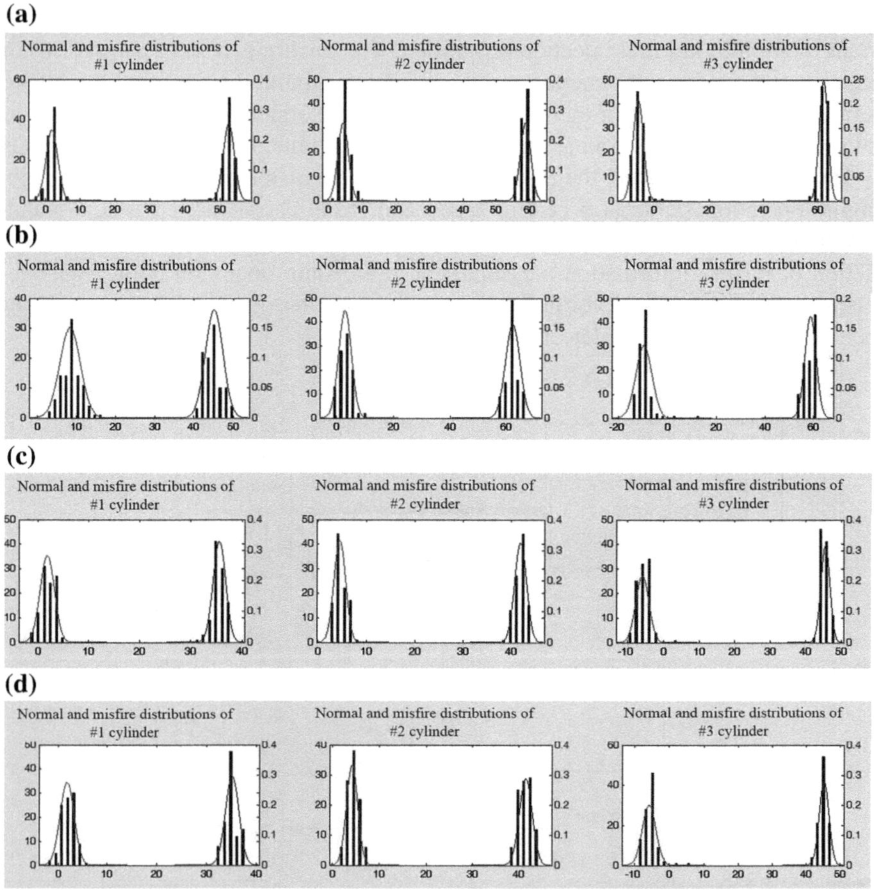

Fig. 5 **a** The detection performance of Method 1 at 3800 rpm and 20% load, **b** the detection performance of Method 2 at 3800 rpm and 20% load, **c** the detection performance of Method 3 at 3800 rpm and 20% load, and **d** the detection performance of Method 4 at 3800 rpm and 20% load

cylinder, as well as which of the second cylinder is always larger than the first cylinder. The averages of the δ distribution under normal events of the third cylinder of all four methods are all less than 0, and the averages of the δ distribution under normal events of the first and second cylinders of all four methods are larger than 0. This indicates that there is a stability error in the engine speed measurement or the inhomogeneity of each cylinder.

As shown in the tables of misfire detection performance *D*, It can be found that the *D* values of the four methods are greater than 3 in the positive torque range specified by the OBD regulations, which indicates that the four methods can meet the requirements of the OBD regulations (Table 1).

The *D* values map of the four methods is plotted in the same three-dimensional graphs (stereoscopic and top view, respectively). The following results can be found in Fig. 6:

Table 1 **a** The misfire detection performance *D* of Method 1, **b** the misfire detection performance *D* of Method 2, **c** the misfire detection performance *D* of Method 3, and **d** the misfire detection performance *D* of Method 4

(a)

	850	1200	1400	1600	1800	2000	2200	2400	2600	3000	3400	3800
10			3.2	3.3	10.8	10.2	10.1	11.7	6.1	10.6	4.9	3.7
15	9.8	6.1	4.8	7.7	10.7	11.7	4.5	13.9	14.4	12.6	12.5	8.6
20		12.6	10.5	9.3	17.5	14.4	13.7	17.1	19.2	15.5	15.4	17.4
25		14.2	11.5	11.9	21.8	16.5	16.7	16.8	22.7	16.0	17.4	17.3
30		15.2	12.8	14.2	20.5	17.0	17.8	20.6	23.1	17.7	19.4	15.9
35		14.8	14.2	28.8	22.1	19.0	19.2	21.0	22.1	19.7	18.0	19.7
40		11.4	16.1	29.5	26.4	23.2	19.4	22.3	27.0			
45				21.2								
50				15.3								

(b)

	850	1200	1400	1600	1800	2000	2200	2400	2600	3000	3400	3800
10			4.2	4. 6	10.6	10.4	10.6	8.3	8.5	7.4	1.5	2.2
15	6.3	7.9	7.1	10.0	11.3	14.0	4.9	11.9	11.6	10.5	10.0	4.5
20		11.3	15.2	12.4	17.0	15.6	15.6	14.9	14.5	13.2	11.5	10.7
25		12.9	13.7	14.7	19.9	18.5	17.3	15.9	16.5	14.1	14.0	12.8
30		13.8	12.4	16.0	19.2	18.5	17.0	17.9	18.2	15.8	14.9	11.6
35		12.4	10.9	16.1	19.0	20.1	17.7	16.3	18.1	17.0	13.8	14.6
40		9.1	12.3	18.1	20.3	22.9	19.5	17.7	22.0			
45				12.7								
50				8.0								

(continued)

Table 1 (continued)

(c)

	850	1200	1400	1600	1800	2000	2200	2400	2600	3000	3400	3800
10			3.3	3.6	11.4	12.0	10.9	6.1	8.6	9.3	3.9	3.0
15	6.5	7.5	6.5	9.8	11.5	14.6	5.5	15.4	14.1	12.3	12.8	7.5
20		17.4	12.1	13.5	19.2	17.1	17.8	20.4	18.6	15.2	16.3	15.0
25		18.8	14.4	16.4	25.8	19.4	21.6	20.9	23.4	15.8	16.9	15.5
30		18.7	14.6	18.5	26.1	20.4	23.1	23.4	22.0	17.9	19.8	14.4
35		17.5	15.1	28.3	26.4	23.1	25.0	22.8	21.0	18.7	18.1	17.8
40		15.0	15.7	25.6	29.9	26.4	24.5	24.9	23.3			
45				17. 1								
50				12.4								

(d)

	850	1200	1400	1600	1800	2000	2200	2400	2600	3000	3400	3800
10			5.1	5.9	11.9	12.3	13.7	10.9	10.9	8.2	2.1	2.9
15	7.5	14.1	7.8	12.1	15.6	22.3	4.5	16.7	16.1	13.3	11.9	5.1
20		17.7	15.4	16.5	22.7	19.6	25.0	22.4	20.1	17.5	14.6	13.8
25		19.5	16.7	20.4	31.9	22.2	26.6	20.1	26.0	17.0	16.0	14.6
30		20.2	16.0	22.3	27.2	21.6	22.9	22.4	22.7	18.3	17.9	12.7
35		17.5	15.3	22.5	25.8	23.3	23.1	20.1	20.7	19.1	16.0	16.1
40		14.6	17.0	22.0	26.4	28.4	24.4	22.0	24.1			
45				16.7								
50				11.0								

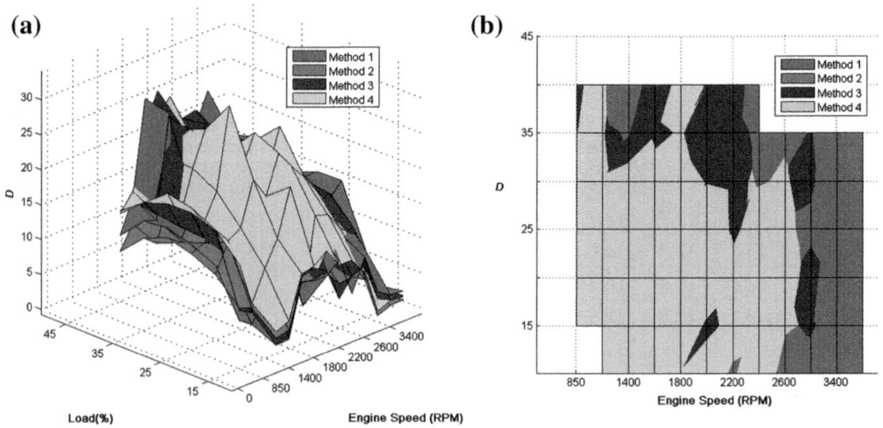

Fig. 6 **a** The *D* values map of the four methods (stereoscopic view) and **b** the *D* values map of the four methods (top view)

1. The D values of Method 1 are overall low, which indicates that the misfire detection capacity of Method 1 is lower than that of the others.
2. Method 2 has maximum D values in some local areas, but the local areas are discontinuous. This indicates the stability of detection performance of Method 2 is worse than that of the others.
3. The D values of Method 3 are overall large in the low-speed area, indicating that this method has a better detection performance in the low engine speed area.
4. The D values of Method 4 are overall large in the high area, indicating that this method has a better detection performance in the high engine speed area.

6 Conclusion

This paper introduces a process of development of the misfire detection strategy. Four misfire characteristic calculation methods and detection performance measure are introduced with this strategy. And then, the experiments are used to compare the effect of the four algorithms in the whole engine operating range. The experimental results show that even if there is cylinder non-uniformity or segment measurement error, the four methods are still able to independently meet the requirements of OBD regulations. The Method 3 has a better detection performance in the low engine speed area, and the Method 4 has a better detection performance in the high engine speed area. Thus, the combination of the Method 3 and Method 4 can have a better detection performance. The combination strategy is robust and can be used in engineering.

References

1. Asano M et al (2001) Further development of an ion current combustion control system (ICCS). No. 2001-01-0266. SAE Technical Paper, 2001
2. Chiatti G, Chiavola O (2002) Misfire diagnosis in spark ignition engines. No. 2002-01-1670. SAE Technical Paper
3. Chung Y, Bae C, Choi S et al (1999) Application of a wide range oxygen sensor for the misfire detection. SAE Technical Paper
4. Rizzoni G (1989) Diagnosis of individual cylinder misfires by signature analysis of crankshaft speed fluctuations. SAE Technical Paper
5. Ribbens W B, Rizzoni G (1989) Applications of precise crankshaft position measurements for engine testing, control and diagnosis. SAE Technical Paper
6. Plapp G, Klenk M, Moser W (1990) Methods of on-board misfire detection. SAE Technical Paper

Preparation and Application of Copper-Based Small-Pore Molecular Catalysts for Diesel Vehicle

Lina Yu, Bin Zhang, Long Cui, Kejin Zhang, Jian Han, Xinyan Mi and Zhonghui Su

1 Introduction

Selective catalyst reduction (SCR) is a commercially proven technology for removing NO_x [1, 2] from diesel engine exhaust to meet more strict international emission standard. Although urea-SCR has recently been commercialized, the window of this technology is severely limited by the decomposition temperature of urea into NH_3 and the SCR catalyst work temperature. Therefore, the catalyst should have good low-temperature performance for cold-starting, but also have good hydrothermal stability, since the diesel exhaust temperature can rise periodically well above 600 °C when the DPF is frequently regenerated to remove PM.

Recent years, it seems like the SCR of NO_x in the presence of ammonia using small-pore molecular sieves with a chabazite (CHA) structure (e.g., Cu/SSZ-13 [3] and Cu/SAPO-34) [4] has drawn much attention as a highly promising commercial catalyst for the urea-SCR system due to their excellent low-temperature activity and robust hydrothermal stability in the automotive industry. However, there are still many problems to be improved in order to realize the industrial application of copper-based zeolite catalysts, such as unable enlarge production of the ion exchange and the aging deactivation of the impregnation. In this work, Cu–Ce/SAPO-34 and Cu–Ce/SSZ-13 were prepared by once ion exchange–impregnation methods, which were investigated of NO_x activity, hydrothermal aging properties, and sulfur-resistant properties. The Cu–Ce/SSZ-13 was a promising technology for De–NO_x with NH_3, which showed good low-temperature activity of NH_3–SCR performance.

L. Yu (✉) · B. Zhang · L. Cui · K. Zhang · J. Han · X. Mi · Z. Su
China FAW Corporation Limited JieFang Business Division, Changchun, China
e-mail: yulina@rdc.faw.com.cn

Society of Automotive Engineers of China (SAE-China) (ed.), *Proceedings of the 19th Asia Pacific Automotive Engineering Conference & SAE-China Congress 2017: Selected Papers*, Lecture Notes in Electrical Engineering 486, https://doi.org/10.1007/978-981-10-8506-2_55

2 Experimental

2.1 The Preparation of Cu–Ce/Sapo-34 and Cu–Ce/Ssz-13

SAPO-34 zeolites were ion-exchanged by using an aqueous solution of copper nitrate containing the required amount of Cu. The solid-to-liquid ratio for the ion-exchanged method was fixed at 10 g/L, and the resultant mixture was maintained under stirring for 4 h at 70 °C. The ion-exchanged zeolites were then filtered, washed once with de-ionized water, dried at 105 °C and were continuously stirred using cerium acetate solution for 3 h, dried at 105 °C and calcined for 5 h at 550 °C, and obtained Cu–Ce/Sapo-34. Cu–Ce/SSZ-13 was also prepared in the same way according to the methods of Cu–Ce/Sapo-34.

2.2 Catalyst Characterization

The prepared catalysts were characterized as follows: crystalline structure of zeolites by X-ray diffraction (XRD), surface morphology scanning electron microscopy (FE-SEM), surface area and porosity by N_2 adsorption–desorption at −196 °C, acidic properties by NH_3 temperature programmed desorption (NH_3–TPD), and reducibility of Cu species by Hydrogen temperature programmed reduction (H_2–TPR).

2.3 SCR Performance

The SCR performance was performed on a fixed-bed quartz tube reactor (inner diameter 20 mm) under atmospheric pressure. The reaction temperature was monitored by a programmable temperature controller (150–450 °C). Reactions were carried out under the conditions: [NO] = [NO_2] = 500 ppm, [NH_3] = 1000 ppm, [O_2] = 4.5 vol%, balance N_2, GHSV = 40,000 h^{-1}, and total flow rate 200 L/h;

The sulfur-resistant treatment was carried out under the condition: [SO_2] = 100 ppm, balance N_2, 250 °C for 32 h;

The hydrothermal aging was carried out under the condition: [H_2O] = 10%, balance N_2, 650 °C for 48 h.

The sulfur-resistant and hydrothermal aging test conditions were done according to our own standards, and the experiment test system was shown in Fig. 1.

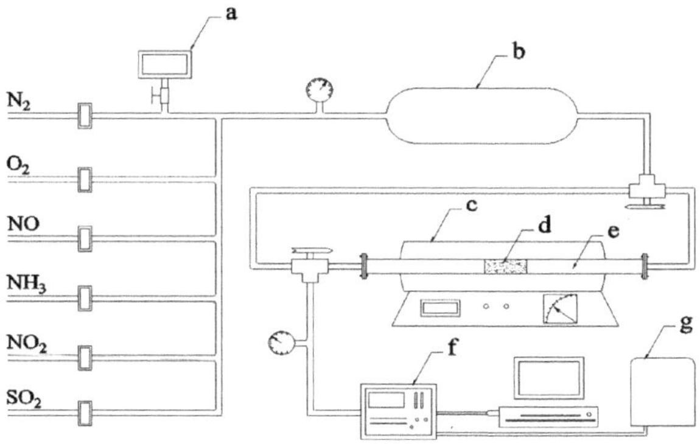

Fig. 1 Experiment test system (**a** water vapor generator; **b** mixer; **c** tube furnace; **d** catalyst; **e** quartz tube; **f** analyzer; **g** waste gas treatment)

3 Results and Discussions

3.1 Structure Characterization

The crystal structure of the as-prepared samples was studied by using powder X-ray diffraction (XRD), and the result was shown in Fig. 2. The zeolite frameworks were well maintained after the ion exchange–impregnation process, as Fig. 2a, b clearly showed that the typical structure diffraction peaks of initial products were observed in both of these catalysts. It should be noted that the contents of copper and cerium were too low and could not be detected in these XRD profiles, which suggested that the dispersion effect of copper and cerium was homogeneous and the crystallinity of initial products would not be destroyed under the treatment we described for both Cu–Ce/SAPO-34 and Cu–Ce/SSZ-13. In general, the excellent copper species dispersion was beneficial to the redox ability, which contributed to enhance the superior activity.

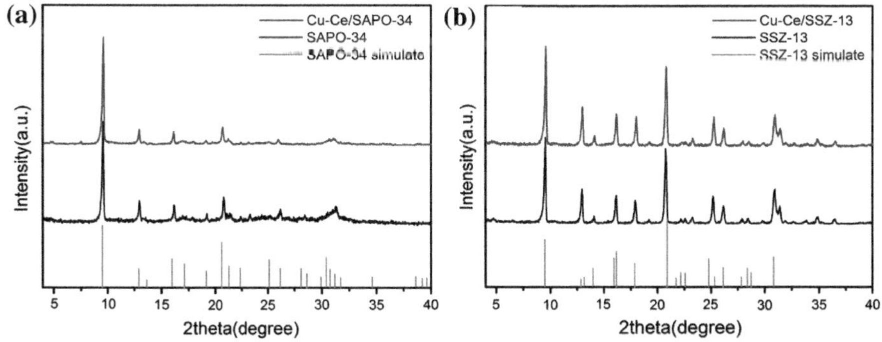

Fig. 2 XRD patterns of the initial and the as-prepared samples: **a** SAPO-34 initial zeolites, Cu–Ce/SAPO-34, SAPO-34 simulated; **b** SSZ-13 initial zeolites, Cu–Ce/SSZ-13, SSZ-13 simulated

3.2 Scanning Electron Microscope

The resultant materials were also characterized by using SEM to study their crystal morphology. SSZ-13 showed cubic crystals with similar average size (0.2–1 μm, Fig. 3a), and SAPO-34 also showed cubic crystals with similar average size (3–6 μm, Fig. 3c). The SEM images also confirmed that Cu–Ce/SAPO-34 (Fig. 3b) and Cu–Ce/SSZ-13 (Fig. 3d) samples had high crystallinity, for that the structure did not change obviously compared with SAPO-34 and SSZ-13.

3.3 Bet Surface Areas

Figure 4 displays the N_2 adsorption isotherms of the SAPO-34, Cu–Ce/SAPO-34, SSZ-13, and Cu–Ce/SSZ-13 samples. All of the four samples exhibited the typical Langmuir-type adsorption, indicating the presence of uniform micropores. It should be noted that the sample Cu–Ce/SAPO-34 had a BET surface area of 550 m^2/g, which was very close to that (596 m^2/g) of a SAPO-34 zeolite, the BET surface area of Cu–Ce/SSZ-13 (580 m^2/g) was also close to that (623 m^2/g) of a SSZ-13 (see Table 1), and both Cu–Ce/SAPO-34 and Cu–Ce/SSZ-13 demonstrated a little lower micropore volume than do a regular SAPO-34 and SSZ-13. It was clearly

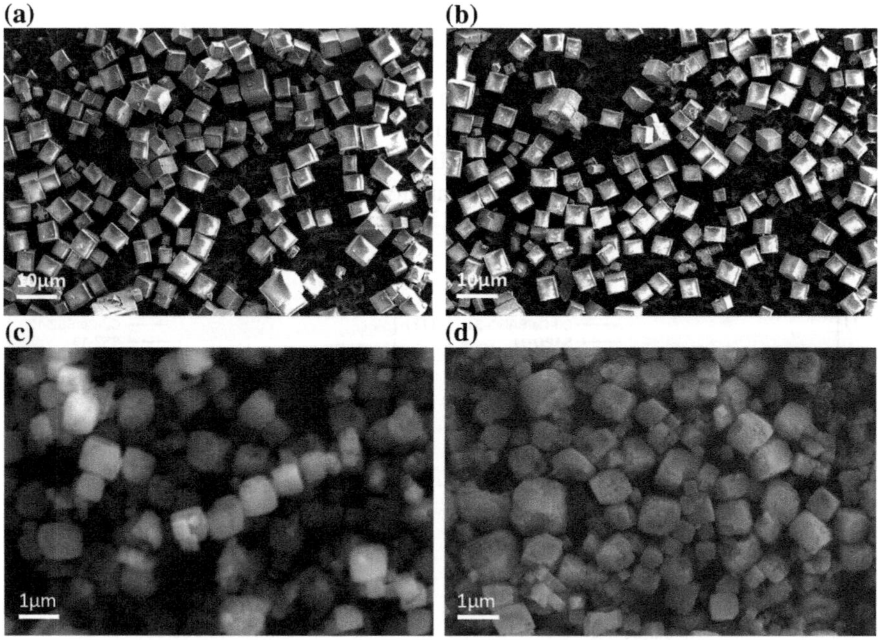

Fig. 3 SEM images of **a** SAPO-34, **b** Cu–Ce/SAPO-34, **c** SSZ-13, and **d** Cu–Ce/SSZ13

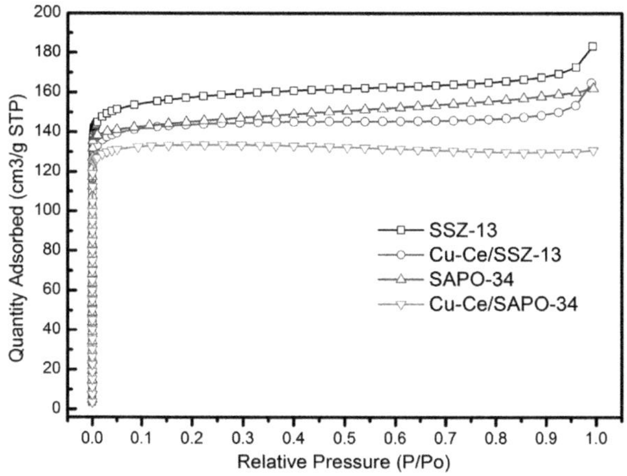

Fig. 4 N_2 adsorption isotherms of SAPO-34, Cu–Ce/SAPO-34, SSZ-13, and Cu–Ce/SSZ-13

Table 1 BET surface areas and pore volumes of the initial and as-prepared samples

Sample	BET surface area (m^2/g)	Median pore width (nm)	Micropore volume (cm^3/g)
SSZ-13	623.13	0.5860	0.2161
Cu–Ce/SSZ-13	580.22	0.5827	0.2070
SAPO-34	596.57	0.5893	0.2019
Cu–Ce/SAPO-34	552.65	0.5875	0.2021

displayed the microporosity structure of initial products retained in the microscale, and the results were fairly consistent with the conclusion from XRD.

3.4 *Analysis of H_2–TPR*

To further investigate the chemistry of the Cu–Ce/SSZ-13 and Cu–Ce/SAPO-34, H_2–TPR tests were conducted and the results were shown in Fig. 5. The H_2 consumption signal of Cu–Ce/SAPO-34 from 100 to 350 °C was divided into three peaks centered at 160, 160–200, and 250 °C. The H_2–TPR curves of Cu–Ce/SSZ-13 appeared three peaks centered at 110–150, 345, and 600 °C, respectively. The Cu–Ce/SSZ-13 showed lower initial reduction temperature than Cu–Ce/SAPO-34 by comparing two kinds of catalyst reduction temperature. Corroborating the findings from NH_3–SCR results, the Cu–Ce/SSZ-13 catalyst had a lower initial reduction temperature, which could be one of the reasons for its high catalytic activity at lower temperature. As a general trend, the lower the initial reduction temperature was, the easier it was to oxidize NO to NO_2 at low temperature. It was

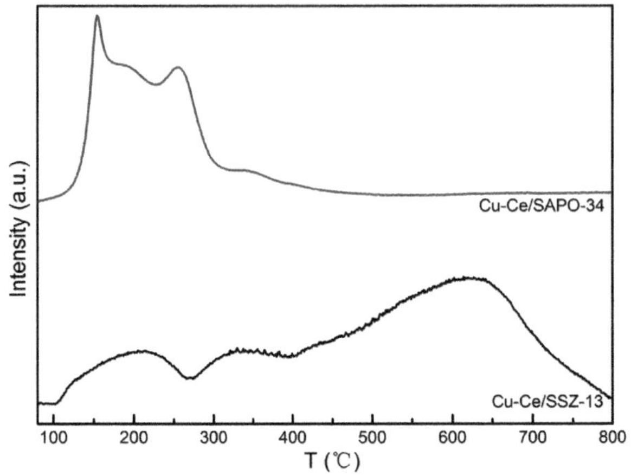

Fig. 5 H_2–TPR spectra of Cu–Ce/SAPO-34 and Cu–Ce/SSZ-13

obvious that we could see all the peaks of Cu–Ce/SAPO-34 had desorption at 350 °C, while the desorption peaks of Cu–Ce/SSZ-13 were completed at 800 °C, which could explain the NO_x activity of Cu–Ce/SSZ-13 showed better at the lower temperature and was slightly worse at the higher temperature. In addition, the coordination environment and the reducing ability of the metal species on the SSZ-13 and SAPO-34 were also different under the same preparation method.

3.5 Analysis of Surface Acid Sites

For the NH_3–SCR reaction, the acid sites of catalysts were beneficial to the adsorption and activation of NH_3, which was one of the most important steps for the reduction of NO by NH_3. Usually, the desorption peaks at 100–250 °C were attributed to the weak acid sites of adsorption NH_3, and 250–700 °C were attached to the strong acid sites adsorption of NH_3. As shown in Fig. 6, the samples of Cu–Ce/SAPO-34 and Cu–Ce/SSZ-13 appeared three peaks near 160, 350, and 515 °C, which were the weak desorption peak of NH_3 (physical adsorption and Lewis acid adsorption at 160 °C), strong acid sites at 350 °C (copper ion induced Lewis acid strongly acidic), and strong Brönsted acid near 515 °C.

It was clear that the Brönsted and Lewis acid of Cu–Ce/SSZ-13 was far more than Cu–Ce/SAPO-34 through the desorption amount of ammonia. There were some differences between the effects of different acid sites on the conversion of NO_x. More sites and isolated Cu^{2+} could improve NH_3–SCR activity at low temperature, while the strong oxidation of Brönsted acid amount could accelerate the oxidation of NH_3 at higher temperature. Combined with NH_3–SCR results, it was reasonable to explain that Cu–Ce/SSZ-13 showed better low-temperature activity than Cu–Ce/SAPO-34, while Cu–Ce/SAPO-34 showed higher activity than Cu–Ce/SSZ-13 at high temperature (above 400 °C).

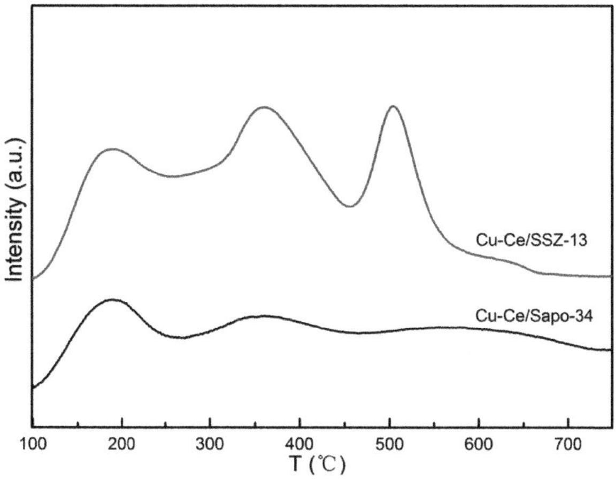

Fig. 6 NH_3–TPD profiles of Cu–Ce/SAPO-34 and Cu–Ce/SSZ-13

3.6 NH_3–SCR Performance

Figure 7 displays the NO_x conversions over Cu–Ce/SAPO-34 and Cu–Ce/SSZ-13 catalysts before and after hydrothermal and sulfur-resistant treatment. It was shown that the T50 (the temperature at which NO_x conversion reached 50%) of fresh Cu–Ce/SAPO-34 was 175 °C, its effective reaction window (T90, temperature range at which NOx conversion was over 90%) was in a temperature range of 300–450 °C, and the maximum NO_x conversion was about 97%. Compared with Cu–Ce/SAPO-34 catalyst, Cu–Ce/SSZ-13 exhibited better NH_3–SCR performance in the whole temperature range of 150–450 °C. The T50 of fresh Cu–Ce/SSZ-13 was lower than 150 °C, and its effective reaction window was widened from 250 to 425 °C; moreover, its maximum NO_x conversion was promoted to 98%.

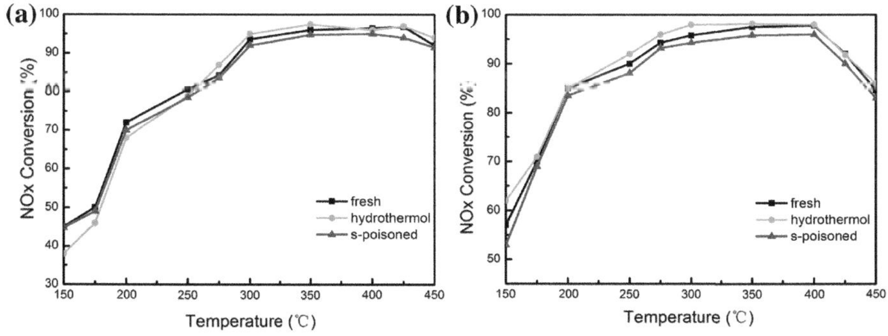

Fig. 7 NO_x conversions of **a** Cu-SAPO-34 and **b** Cu-SSZ-13 catalysts for fresh, hydrothermal aging and s-poisoned at different temperatures

Comparing the hydrothermal stability of Cu–Ce/SSZ-13 and Cu–Ce/SAPO-34 catalysts (Fig. 7), the two catalysts maintained the best performance after aging at 650 °C for 48 h in simulated exhaust gases containing water. However, a significant difference between the two samples appeared under 250 °C. Cu–Ce/SSZ-13 catalyst still had better performance with even a slight increase in NO_x conversion in low-temperature range (150–200 °C), while a remarkable NO_x activity loss was observed with the Cu–Ce/SAPO-34 sample.

As shown in Fig. 7, both Cu–Ce/SAPO-34 and Cu–Ce/SSZ-13 catalysts decreased slightly after s-poisoned conditions at 250 °C kept for 32 h in the absence of 100 ppm SO_2. It implied that the two catalysts possessed excellent resistance to the SO_2. Combined with SEM, XRD, BET, H_2-TPR, NH_3-TPD analysis, Cu–Ce/SSZ-13 had better thermal stability and sulfur resistance, which could be highly stability dispersion, starting with lower reduction temperature and the maximum of acid sites.

4 Conclusions

The Cu–Ce/SAPO-34 and Cu–Ce/SSZ-13 catalysts had been prepared and investigated for NH_3–SCR of NO_x. The results showed that Cu–Ce/SSZ-13 catalyst exhibited the better catalytic activity, excellent resistance to H_2O and SO_2, good dispersion, lower reducing temperature, and the stronger surface acidity. The SCR activity of the Cu–Ce/SSZ-13 catalyst was closely related to physicochemical properties. T50 < 150 °C; especially, over 70% of NO_x conversion was achieved at the temperature of 175–200 °C and the maximum NO_x conversion was about 90% in the range of 250–425 °C, showing robust low-temperature and high-SCR activity. Due to the low cost, high thermal stability, sulfur resistance and simplicity, Cu–Ce/SSZ-13 catalysts was a promising technology for the preparation of the ion exchange–impregnation route.

References

1. Granger P, Parvulescu VI (2011) Catalytic NO_x abatement systems for mobile sources: from three-way to lean burn after-treatment technologies. Chem Rev 111:3155–3207
2. Roy S, Hegde MS, Madras G (2009) Catalysis for NO_x abatement. Appl Energy 86(11): 2283–2297
3. Yang X, Wu Z, Moses-Debusk M (2012) Heterometal incorporation in metal-exchanged zeolites enables low-temperature catalytic activity of NO_x reduction. Phys Chem C 116(44): 23322–23331
4. Raquel M-F, Manuel M, Cristina F, Arkady K, Avelino C (2012) Rational direct synthesis methodology of very active and hydro-thermally stable Cu-SAPO-34 molecular sieves for the SCR of NO_x. Appl Catal B 127:273–280

End-to-End Adaptive Cruise Control Based on Timing Network

Zhong Cao, Diange Yang, Kun Jiang, Tinghan Wang, Xinyu Jiao and Zhongyang Xiao

1 Introduction

Intelligent vehicle technology has developed rapidly in the last 20 years, and some advanced driver assistant system (ADAS) products have already set up in the new version cars. Since the Google driverless car released in 2012, autonomous vehicle technology has developed much faster. It has been proved that the potential of this technology on the safety [1], economy [2], and efficiency will create great value in society.

Obviously, creating the autonomous vehicle needs a huge and complex system typically including environment perception, motion planning, and vehicle dynamic control with both hardware and algorithm. Nowadays, each of these three portions has its own problems. The environment perception needs sensors to collect information, such as lidar and cameras [3]. The outputs of this part should suggest not only the spatial information such as the moving objects' distance, speed, direction, and acceleration, but also the objects' properties, moving intention, motivation, and some rules. As for computer vision system, some researches focus on detection [4] and have breakthroughs these years. When getting the spatial information, the more accuracy required, the more computing resource it needs. Because of the sensor error and weather factors [5], it is difficult to improve the accuracy of environment perception. On the other hand, the motion planning framework is still discussed and mainly has four portions, which are route planning, path planning, maneuver choice, and trajectory planning [6]. There are many researchers working in these

Z. Cao · D. Yang (✉) · K. Jiang (✉) · T. Wang · X. Jiao · Z. Xiao
State Key Laboratory of Automotive Safety and Energy, Collaborative Innovation Center of Intelligent New Energy Vehicle, Department of Automotive Engineering, Tsinghua University, Beijing 100084, China
e-mail: ydg@tsinghua.edu.cn

K. Jiang
e-mail: jiangkun@tsinghua.edu.cn

Society of Automotive Engineers of China (SAE-China) (ed.), *Proceedings of the 19th Asia Pacific Automotive Engineering Conference & SAE-China Congress 2017: Selected Papers*, Lecture Notes in Electrical Engineering 486, https://doi.org/10.1007/978-981-10-8506-2_56

areas and proposing kinds of strategies. Many results of them are generated based on the ideal environmental assumptions. Various models will constitute multi-target framework, and the evaluation system has not formed a consensus yet.

Since 2016, NVIDIA proposed the "end-to-end" controller system [7], proving a new technology road map to achieve driverless vehicle. In this system, the monocular image is the only input information, while the outputs contain steering angle, throttle, and brake signal. The whole system is a trained neural network by deep learning framework, and the intermediate variable cannot export. Even though the results cannot illustrate that it can achieve the autonomous vehicle, "end-to-end" controller system is still a significant breakthrough in this area. Similar to this thought, Xu et al. [8] established the "end-to-end" driving learning models with the large-scale video datasets. This work focuses on how to build the learning model with different scale dataset. Koutník et al. [9] built the reinforcement learning neural network and achieved "end-to-end" driving in simulation platform. This framework can only run in the racing game.

Adaptive cruise is a typical situation while driving a car, and there are already some researches in this area. Similar to other control procedures, adaptive cruise control system also has three portions: environment perception, motion planning, and vehicle control (throttle and brake). In environment perception system, the relative distance, speed, acceleration between the preceding vehicle and the following vehicle are required. Computer vision and radar [10] are main approaches to measure the parameters. Although radar is a simple way to get the distance information, it limits the range of the aim objects and cannot comprehend the environment. As a result, most of the intelligent vehicle choose cameras as the sensors for these situations [11]. They show that in order to get the spatial parameters, computer vision needs to be calibrated precisely [12], and both intrinsic and extrinsic parameters have direct impact on the results. In fact, when the driverless vehicle driving in the street, the intrinsic and extrinsic parameters cannot be observed accurately in real time. In addition, many factors will cause the error of the distance information.

Some control strategies are applied on the single-lane cruise especially car following. The main differences between strategies are the control models and optimization targets. Control models mainly use PID, or some specific models [13]. The parameters are adjusted manually, and it would cause some bias while the application scenarios change. On the other hand, optimization targets have more adjusted parameters, which makes it more various. The definition of safe distance, vehicle model, and the driving target such as high-efficiency and low fuel consumption would generate plenty of systems that can be used in the specific scene. Furthermore, when both vehicle following and lane keeping exist, which kinds of strategies should be applied is also a problem.

From the review above, the ideal implementation approach for the adaptive cruise control can achieve that the input data can get from a camera only which has not calibrated yet. And the output data is control signal that has the features the drivers need in the last driving process. Here we are trying to apply the "end-to-end" controller in adaptive cruise control situation. Machine learning is used in the system and big driving data, which are generated based on some special strategies, are collected. Neither the parameters of the camera nor the strategies are given to the whole model. A timing neural network should be used in this model.

The main contribution of this method is to achieve the end-to-end adaptive cruise control based on the NARX net. The remainder of this paper is organized as follows. Section 2 overviews the "end-to-end" system including application condition, input and output data. The network based on a timing neural network is described in Sect. 3. Section 4 illustrates the collected dataset. Some training details and the simulation results are introduced, respectively, in Sects. 5 and 6. At last, the summary is given in Sect. 7.

2 Overview System

Figure 1 shows the "end-to-end" adaptive cruise control system. The image from the camera installed in the vehicle is the only input information, and the system will give the control of the throttle and brake. Because the situation in this paper is adaptive cruise control (on the straight), the steering angle should be set to approximately zero. The decision module is based on the machine learning and should be trained with big driving data. It should be noticed that the camera could not be calibrated in the vehicle so that the distance between of the vehicles cannot be obtained directly. In addition, the control model and strategies of the training generated data cannot be obtained.

In this paper, in order to make the whole structure clear, Fig. 2 shows the structure of this paper. With the development of computer vision, the detection of the image can get the lane and the following vehicle position in the image. Some deep learning approaches such as faster RCNN [14] make it easy to get. On the other hand, some researches on the vehicle dynamic control can achieve the control from the desired acceleration, and because of the whole system is made up in mechanics, the dimension in this problems leads to the unnecessary of using learning to finish this process. So in this paper, the main contribution is the focus on the neural network from the image which has detected the vehicle and lane, to the desired acceleration. Furthermore, a vehicle dynamic model is applied to generate the control signal.

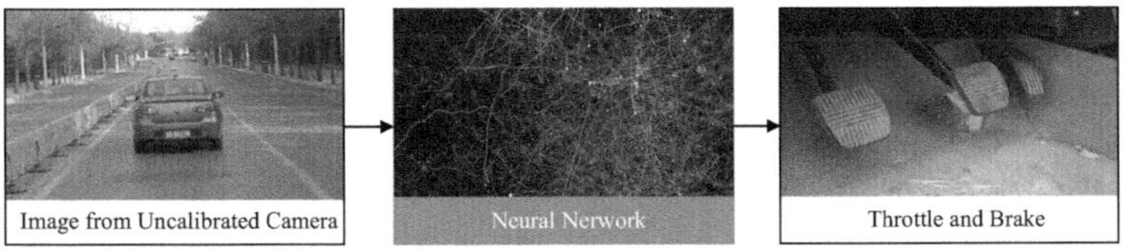

Fig. 1 End-to-end adaptive cruise control system framework

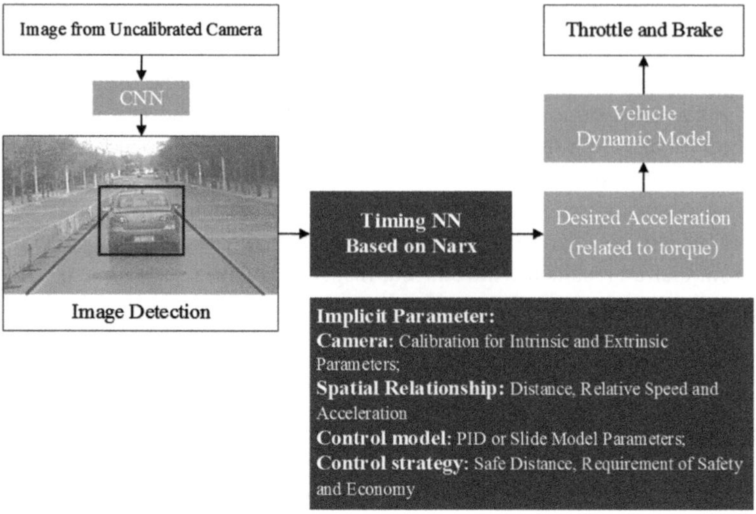

Fig. 2 System decomposition structure

Because the neural network has the characteristics of high-dimensional fitting, it will find the relationship between some parameters which cannot solve in the situation lack of subjective information. In this model, the implicit parameters have four portions showed in Fig. 2. We cannot get the intrinsic and extrinsic parameters of the uncalibrated camera. As a result, the distance could not calculate either, similar to the relative speed and acceleration. The data can be generated in a specific control model with some parameters, which is not known, and some strategies like the safe distance and driving target are also invisible. All of these parameters have relationship with the input image and the output control signal. Therefore, the neural network needs to finish the function fitting in this situation.

3 Network Architecture

3.1 Input Data Format

As described in Sect. 2, the input image is detected by some mature algorithms that can get the position of the preceding vehicle and lane on the image. To make the network focus on the decision, the input should show the detection results directly. As for the vehicle, a circumscribed rectangle represents the position of the preceding vehicle on the image. The upper and lower edges of the rectangle are parallel to the edge of the image, so that it needs two points to represent. In this paper, the upper left corner and the lower right corner are chosen and the four parameters are as the input of the network. In the single-lane cruise system, the only vehicle which influences the result is preceding vehicle in the lane. So, four parameters for the preceding vehicle can meet the requirement.

The input data format is shown in Table 1. The controllers in this paper are the desired acceleration. Therefore, it applies on the straight lane. In order to get the lane position on the image, two points represent one lane. The first point is the intersection of the lane line and low edge of image. In addition, another one is the infinity point of the lane line position on the image, which should be the intersection between the optical axis of the camera and the imaging plane. Both left lane and right lane have the same point as the axis and imaging plane intersection. So, four parameters can present the lane position on the image. Once the image size is settled, these eight parameters are the complete input data to present the image.

3.2 Output Data Format

The output data of the system is the control signal including throttle and brake, while the acceleration of the following desired has the mapping relation with the control signal in the specific vehicle model. So, the only parameter is the desired acceleration which is equivalent to the control signal. As the consequence, the output neural network has one parameter which is the desired acceleration measured in m/s^2.

3.3 Neural Network Structure

In the control model, the control signals need to be generated according to process while it also accords with people's cognition. On the other hand, one image can show the incomplete state of a moment. Obviously, the "end-to-end" neural network needs memory function and the timing data. NARX net is a classical network applied for dynamic neural network and event sequence prediction. The neural network in this paper is designed based on the NARX net [15]. The structure is shown in Fig. 5.

Table 1 Input data format

Preceding vehicle (pixel)	Circumscribed rectangle upper left corner u_{vl}
	Circumscribed rectangle upper left corner v_{vl}
	Circumscribed rectangle lower right corner u_{vr}
	Circumscribed rectangle lower right corner v_{vr}
Lane line (pixel)	Axis and imaging plane intersection u_0
	Axis and imaging plane intersection v_0
	Left lane and image low edge the intersection v_{ll}
	Right lane and image low edge the intersection v_{lr}

The neural network has four layers, which are input layer, timing delay layer, hidden layer, and output layer. The input layer has eight points got from the detected image, which has been described in Table 1. In this situation, the results need to consider the last decision it made, and in order to get this information, the last output should be given as the feedback of another input data. They are the orange points in Fig. 3. In general, the input layer has nine points, consisting of a feedback point and eight detected points observed from the images.

Timing delay is the feature of the NARX net structure, because both the speed and the acceleration are the differentiation and second derivatives of the distance to time. The network needs to memorize the timing data, and the timing delay will help to finish this function. The parameters I settled are 1:5, which means every calculation has five times data to generate the results. After the timing delay, the intermediate node will save all of these data. At the same time in this layer, the feedback data will also be delayed and generate other points. The orange node in timing delay layer represents this process. A constant node also exists in this layer. Until now, all of the data that needs to be considered are saved in this layer and we can use the classical BP neural network to train the model and get the machine learning system.

Hidden layer is the important layer in BP neural network, and the weight of each point has significant influence on the output. As we tested via our data, the layer with 10 nodes is a suitable setting for the network. The training details will also have effect on this layer which will be explained in Sect. 5.

The output layer has only one node, which is the desired acceleration we need to calculate for the control signal in the vehicle model. The only thing needed to explain is that the output value should be the feedback to the input layer. Here the whole network architecture has been described, and it would apply in the training dataset.

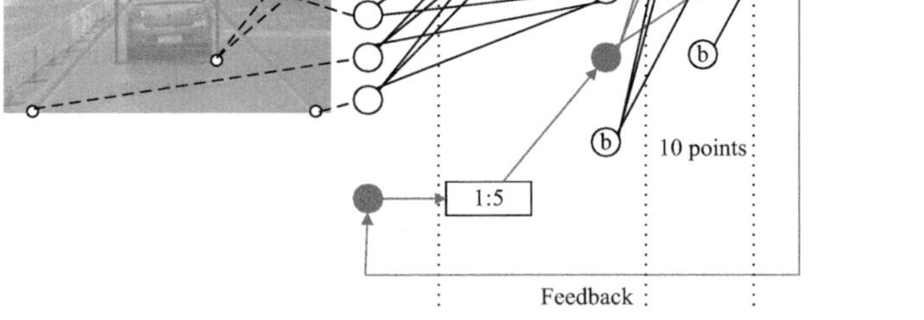

Fig. 3 Neural network structure in this paper

4 Data Collection

4.1 Data Situation Description

The training and testing dataset is generated by "expert", who will always follow an internal driving policy accurately. The input parameters of the dataset includes car-following distance, lane width, relative speed and accelerations between the preceding and following vehicle. These parameters imitate the real scene containing various conditions. They are shown in Fig. 4. It is supposed that both vehicles have the same driving direction. With these parameters, firstly we set the camera model to reconstruct image. In addition, expert-like control model and control strategies are applied in these conditions to generate the desired acceleration. With these data, the training dataset and test dataset are generated.

4.2 Image Data

4.2.1 Traffic Parameters

In order to rebuild the image and finish the whole simulation, some traffic parameters need to be set according to experience. These parameters are not the input data to the neural network in this paper. So, these parameters can only influence the generated image, which are the preceding vehicle external dimensions including the width and height, and the lane width. Table 2 shows the traffic parameters' values.

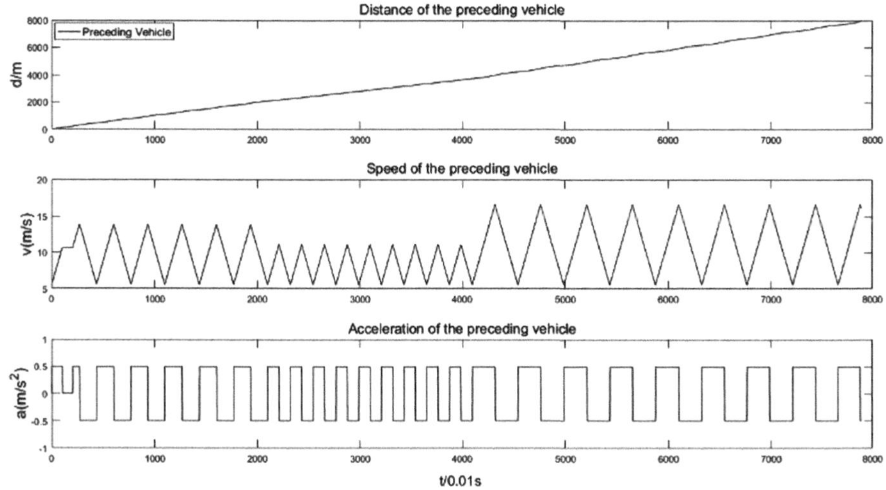

Fig. 4 Preceding vehicle motion situation

Table 2 Traffic parameters

Preceding vehicle width pv_w	3 m
Preceding vehicle height pv_h	2 m
Lane width l_w	3.5 m

4.2.2 Camera Parameters

The camera parameters are also the significant factors while rebuilding the image, including the intrinsic and extrinsic parameters. In this paper, we assume that the camera model is ideal without the distortion. The parameters are shown in Table 3.

4.2.3 Image Rebuilding

The neural network input data needs rebuilding image, and the basic model used in this paper refers to the model in [16]. The main idea is that with the camera parameters and the size of the traffic factors, some feature points in the real world will be projected to the image plane. With the car-following distance and camera's paramters, the front vehicle's circumscribed rectangle postion can be calculated in Eq. 1. After that, the vehicle input data would be expressed as in Eq. 1.

$$\begin{bmatrix} v_{vr} \\ v_{vl} \\ u_{vl} \\ u_{vr} \end{bmatrix} = \begin{bmatrix} v_0 + \frac{f \cdot c_h}{d} \\ v_0 + \frac{f \cdot c_h}{d} - \frac{f \cdot pv_h}{d} \\ u_0 - \frac{f \cdot pv_w}{2d} \\ u_0 + \frac{f \cdot pv_w}{2d} \end{bmatrix} \tag{1}$$

As for the lane line, the intersection of optical axis and image plane is represented by (u_0, v_0). The intersection of left and right lane with image low edge needs to be calculated with the lane width and the parameters of the camera. Firstly, the distance of the lane at the low edge of the image with the camera needs to be calculated. After that, with the lane width the width of the lane on the image low edge can be calculated. The other two parameters are calculated in Eq. 2.

Table 3 Camera settled parameters

Image plane size	320 × 160 pixel
Focal length f	500 pixel
Intersection of optical axis and image plane (u_0, v_0)	(160, 80)
The height of the camera c_h	1.5 m
Optical axis angle	Parallel to the ground
Optical axis direction	Same to the velocity direction

$$\begin{bmatrix} v_{ll} \\ v_{lr} \end{bmatrix} = \begin{bmatrix} v_0 - \frac{l_w \cdot v_0}{2c_h} \\ v_0 + \frac{l_w \cdot v_0}{2c_h} \end{bmatrix} \quad (2)$$

4.3 Control Signal

The control signals are generated by "expert," which has the specific control model and strategies. With the current vehicle status and the environment information, the controller generates the desired acceleration at the next time. The vehicle model will receive the desired acceleration, which represents the torque kindly. After that, both of the vehicle spatial relationship is shown in Fig. 5. This result is considered as the driving process of an experienced driver. The desired acceleration is shown in Fig. 6.

5 Training Details

Some training details are given in this section. Most of them need to adjust by experience, and they are shown in Table 4.

With these parameters, the network can be trained, and the results are shown in Sect. 6.

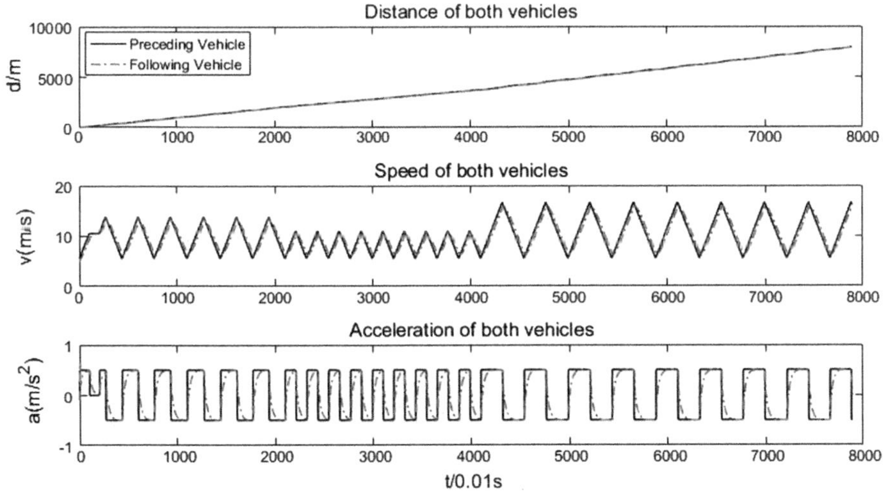

Fig. 5 Control results of both preceding vehicle and following vehicle

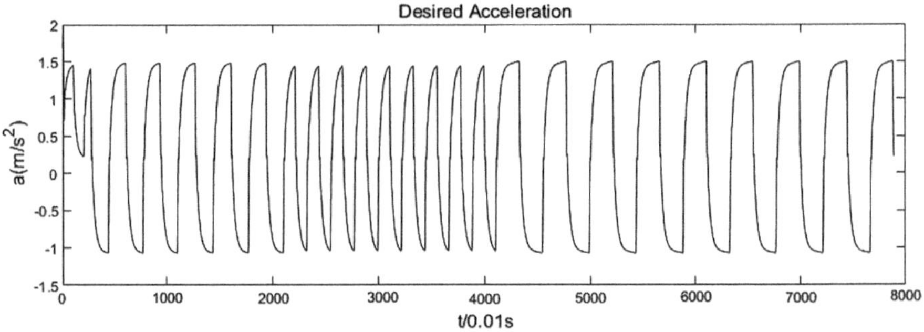

Fig. 6 Control signals represented by desired acceleration

Table 4 Training details

Input and feedback delays	1:5
Hidden layer size	10
Data division	Random
Data pretreatment	Remove constant rows
Training algorithm	Levenberg–Marquardt(LM)
Performance	Mean squared error
Calculations	MEX
Epoch	1000
Max MU	10
MIN gradient	10^{-7}
Train data divide	40%
Validate data divide	30%
Test data divide	30%

6 Simulation Results

The simulation platform in this paper was chosen as the software MATLAB 2016b, which has some basic functions for neural network. With the training parameters settled in Sect. 5, it can get the result. The results are shown and analyzed in this section.

The settled epochs are 1000, which is enough to finish the training process. In most time, the training may stop at approximately 500 epochs, while the validation results show that it has finished the training process. The results shown in this section are some of them, and the stop epoch is 487. The results are shown in Table 5, which is measured by mean squared error (MSE).

In order to visualize the results directly, the output data, the expert data, and the preceding vehicle acceleration are shown in Fig. 7. The figure is divided into three parts, which are the training data, validation data, and the test data. In fact, all of

Table 5 Training results measured by the mean squared error with both output data and target data

Train performance MSE	2.7885e−05
Validation performance MSE	8.9530e−06
Test performance MSE	1.2595e−05

this data are divided randomly, so this is just the expression to show the proportion of these kinds of data.

In general, the neural network completes the task of the control signal prediction, which can be verified by the test data performance and the figure. Because both the frequency and amplitude of the training data change, and the preceding vehicle acceleration is not included in the input data, the neural network cannot just memorize the results, which means it is not overfitting. To make this point more clear, the A and B points on Fig. 7 are enlarged and shown in Fig. 8.

Although the output data is close to the target data, it is clear in Fig. 8 that some details are still different. Points A and B have both different frequencies and amplitudes of the target desired acceleration tendency, and the output data has discrepant results. While points B and C have almost the same input, the output data still have difference. As a result, it can verify that the neural network finishes the prediction task without overfitting.

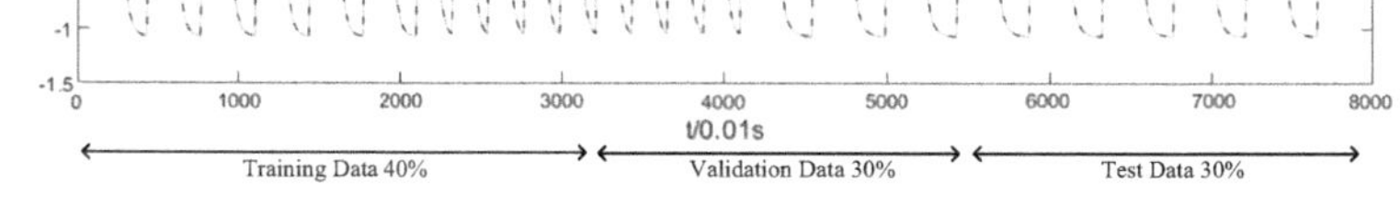

Fig. 7 Comparison between target data and the output data

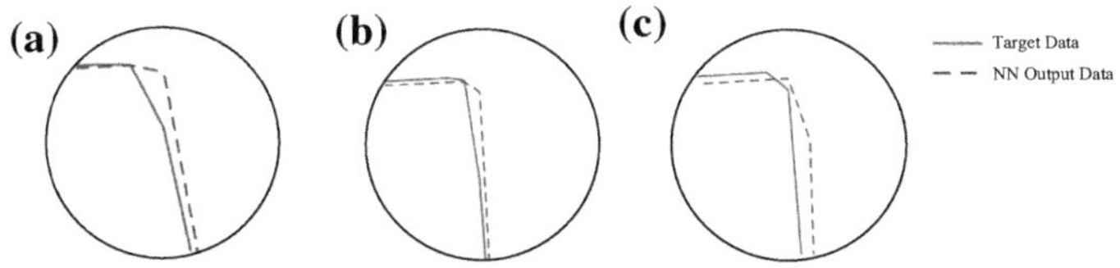

Fig. 8 Some details that show the difference of the target and output data in the results

The training process is shown in Figs. 9 and 10, which show the MSE in every epoch and the change of gradient and validation check. This neural converges to a small error in a short time and with the test performance tendency. It can also verify that the system is not overfitting.

From the simulation results, the neural network can realize the delay and feedback function. Compared with the target data, which is generated by the specific control model and strategies, the neural network gives an accurate output, which is close to the target. According to the output data tendency in details and the performance of the training data and validation data, the network model is not overfitting and can achieve the desired acceleration prediction. The simulation

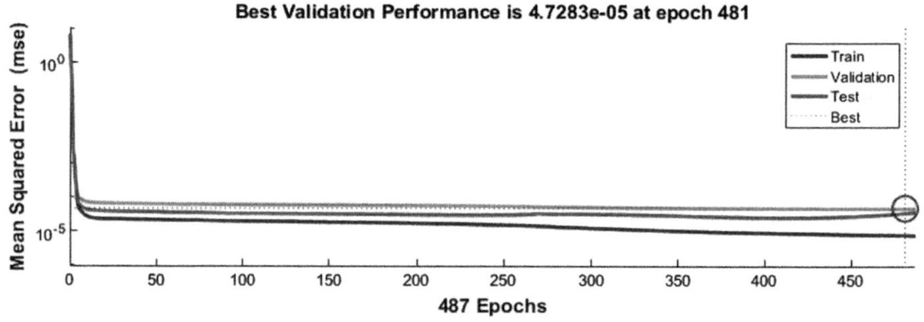

Fig. 9 MSE change tendency measured with the epochs

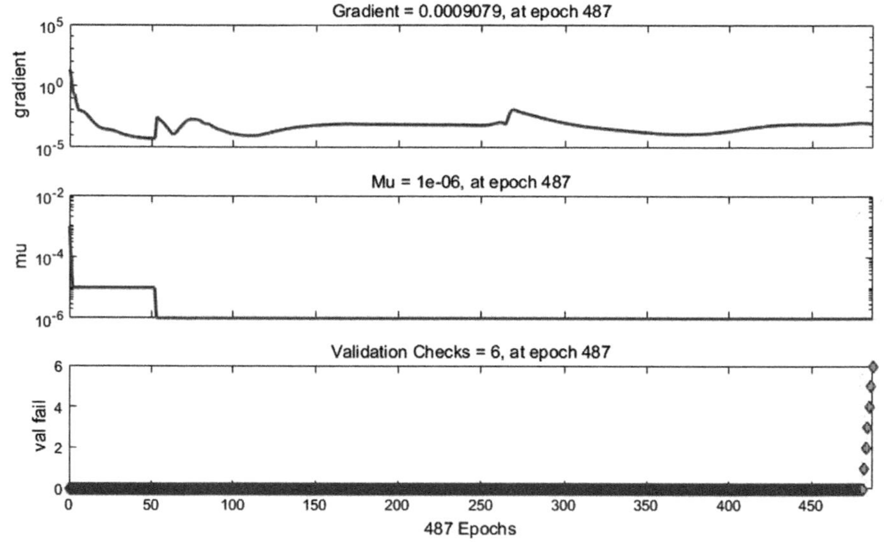

Fig. 10 Some train parameters change tendency to show the training process

results show that the neural network and the training process proposed in this paper are effective on the condition of the car following in a single straight lane.

7 Summary

In this paper, we propose an approach with machine learning to predict the control signal from the uncalibrated monocular camera directly. The camera parameters, the spatial relationship between both vehicles, and the control model are as neither the input data nor a priori information. The main contribution of this paper is that the neural network structure proposed in this paper based on the timing NARX net has the time delay and the feedback function. This structure can memorize the history image feature and the output data in the past time. Because of this structure, the history information can be used to predict the next control signal and make the results more accurate. The training data used in this paper is generated by the "expert" control strategies, and we described the camera projection model to rebuild the image from the training data. The training data format and some training details are also described in this paper. The simulation results show that the trained net can predict the control signal from the abstract image. The mean squared error is within the acceptable limits. The outputs and training process can also verify the network is not overfitting. From all above, we can see that the approach proposed has the satisfactory performance for the prediction of the control signal from the abstract image taken by the uncalibrated monocular camera. In the future, we will apply the "end-to-end" model to the actual experimental scenery to verify the validity of this approach further. We will also compare the results with other methods, such as stochastic process.

Acknowledgements This work is supported by International Science & Technology Cooperation Program of China under contract No. 2016YFE0102200.

References

1. Jarašūniene A, Jakubauskas G (2007) Improvement of road safety using passive and active intelligent vehicle safety systems. Transport 22(4):284–289
2. Luo Y, Chen T, Zhang S et al (2015) Intelligent hybrid electric vehicle ACC with coordinated control of tracking ability, fuel economy, and ride comfort. IEEE Trans Intell Transp Syst 16(4):2303–2308
3. Zhou L, Deng Z (2014) LIDAR and vision-based real-time traffic sign detection and recognition algorithm for intelligent vehicle. IEEE, Int Conf Intell TranspSys IEEE, 578–583
4. Nieto M, Laborda JA, Salgado L (2011) Road environment modeling using robust perspective analysis and recursive Bayesian segmentation. Mach Vis Appl 22(6):927–945
5. Ditze M, Golatowski F, Laum N et al (2010) A survey on intelligent vehicle safety systems for adverse weather conditions. Fisita World Automot Congr

6. Katrakazas C, Quddus M, Chen WH et al (2015) Real-time motion planning methods for autonomous on-road driving: State-of-the-art and future research directions. Transp Res Part C Emerg Technol 60:416–442
7. Bojarski M, Testa DD, Dworakowski D et al (2016) End to end learning for self-driving cars
8. Xu H, Gao Y, Yu F et al (2016) End-to-end learning of driving models from large-scale video datasets
9. Koutník J, Cuccu G, Schmidhuber J et al (2013) Evolving large-scale neural networks for vision-based reinforcement learning. Mach Learn 541–548
10. Li SE, Jia Z, Li K et al (2015) Fast online computation of a model predictive controller and its application to fuel economy-oriented adaptive cruise control. IEEE Trans Intell Transp Syst 16(3):1199–1209
11. Alonso IP, Llorca DF, Gavilan M et al (2012) Accurate global localization using visual odometry and digital maps on urban environments. IEEE Trans Intell Transp Syst 13(4): 1535–1545
12. Stein GP, Mano O, Shashua A (2003) Vision-based ACC with a single camera: bounds on range and range rate accuracy. In: Intelligent Vehicles Symposium, 2003. Proceedings. IEEE. IEEE Xplore, pp 120–125
13. Yang HH, Peng H (2010) Development of an errorable car-following driver model. Veh Sys Dyn 48(6):751–773
14. Ren S, He K, Girshick R et al (2015) Faster R-CNN: towards real-time object detection with region proposal networks. In: International conference on neural information processing systems. MIT Press, Cambridge, pp 91–99
15. Sani MG, Abdul Wahab N, Sam YM et al (2014) Comparison of NARX neural network and classical modelling approaches. Appl Mech Mater 554(554):360–365
16. Stein GP, Mano O, Shashua A (2003) Vision-based ACC with a single camera: bounds on range and range rate accuracy. In: Intelligent vehicles symposium, 2003. Proceedings. IEEE, pp 120–125

Design Optimization of Multi-material Battery Enclosure for Electric Vehicle

Xiaokai Chen, Mengqiang Li, Siqi Li, Jiawei Jin and Cheng Zhang

1 Introduction

The battery enclosure protects and supports the power battery; it plays a key role in ensuring the safety of electric vehicles. The improper design of the battery enclosure will result in many serious problems, such as cracking, causing noise, or battery damage [1, 2]. At the same time, the weight of the battery enclosure is large; in order to improve the driving range of the electric vehicle and reduce the influence of the battery on the vehicle dynamic performance and acceleration performance, it is necessary to carry out the lightweight design of the battery enclosure. Zhao et al. [3] used topology optimization method to design the battery enclosure structure; after optimization, the weight loss effect of the battery enclosure is obvious. Hartmann et al. [4] used the finite element method to optimize the structure of the battery enclosure, reducing the thickness of the battery enclosure to achieve the goal of 20% weight loss, while improving the low-order natural frequency of the enclosure. Schubert et al. [5] adopted a method of improving the original structure to realize the lightweight design. Although these studies had optimized the structure of the battery enclosure, reducing the weight of the battery enclosure, most of the vehicle battery enclosure was welded by steel plate, which causes heavy weight. Therefore, some studies used carbon fiber composite materials for battery enclosure

X. Chen (✉) · M. Li · S. Li · J. Jin · C. Zhang
School of Mechanical Engineering, Beijing Institute of Technology, Beijing 100081, China
e-mail: chenxiaokai@263.net

M. Li
e-mail: limengqiangbit@163.com

X. Chen · M. Li · S. Li · J. Jin · C. Zhang
Collaborative Innovation Center of Electric Vehicles in Beijing, Beijing 100081, China

Society of Automotive Engineers of China (SAE-China) (ed.), *Proceedings of the 19th Asia Pacific Automotive Engineering Conference & SAE-China Congress 2017: Selected Papers*, Lecture Notes in Electrical Engineering 486, https://doi.org/10.1007/978-981-10-8506-2_57

design. Zhang et al. [6] used carbon fiber in the battery enclosure and analyzed the strength of it, which proved the feasibility of carbon fiber applications; Wang et al. [7] applied carbon fiber composites to the box of the battery enclosure; compared with the steel battery enclosure, the carrying capacity of the battery enclosure was not reduced on the premise of reducing the weight of the battery box by 64%. At present, the multi-material design of battery enclosure is relatively rare, but many scholars put forward the multi-material design of the body; Sakundarini et al. [8] proposed the application of a variety of new materials to achieve structural lightweight, to improve product performance, and to reduce costs; Yan Hui [9] pointed out that compared to using steel solely in body design, applicating a variety of materials can make use of various materials' characteristics, so that the materials can be used efficiently.

To sum up, multi-material structure design is one of the development trends of structure design; at present, there is no perfect design technology in multi-material battery enclosure area. With the promotion of energy efficiency, using the new materials to carry out lightweight design of the battery enclosure has become a trend. In this paper, the multi-material optimization design of the battery enclosure is presented. Using an electric vehicle battery enclosure as an example, under the bumpy condition and turning condition, the finite element analysis method is used to analyze the strength and sensitivity of the battery enclosure. Through reasonable allocation of materials, the carbon fiber composite materials and aluminum alloy materials are used to reduce battery weight and to ensure the mechanical properties of the battery enclosure.

2 Battery Enclosure Finite Element Modeling

The battery enclosure studied in this paper is composed of covers, floor, front and rear air inlet plate, side circumference, bracket and battery separator. The length, width, and height of the battery enclosure are 1.45 m × 1.20 m × 0.21 m, and the total mass of it is 41.13 kg, and the overall weight of the battery pack is 321.1 kg. The geometric model of the battery enclosure created by CATIA is shown in Fig. 1.

During the battery enclosure meshing, on the premise of not affecting the precision of calculation, the unimportant geometric characteristics such as the chamfer, the small hole, and the non-bearing parts are simplified. Because the structure of the battery enclosure is made of plates and shells, the component's scale is much smaller in the thickness direction than the other two directions. And it has the characteristics of thin wall structure. So that the PSHELL element was used to simulate the enclosure structure. The battery was replaced by the PSHELL element with the appropriate thickness and density. Because the elastic modulus of the battery is much lower than that of the steel, its mechanical response is ignored in this paper during the finite element analysis process. There are 480 beam units and

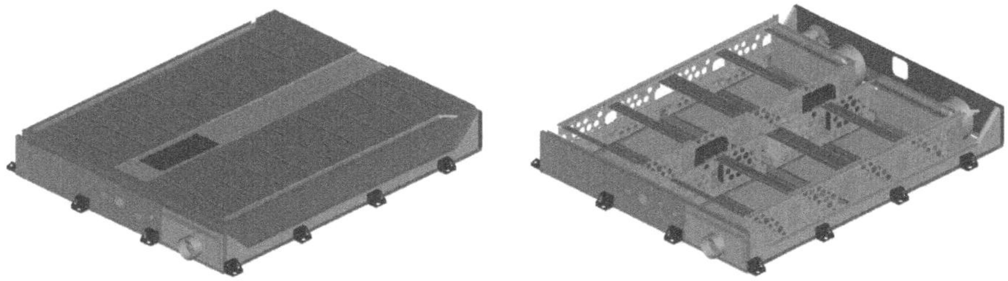

(a) Global geometric model of the battery enclosure

(b) Inner structure model of the battery enclosure

Fig. 1 External geometric model and inner structure model of the battery enclosure

113,678 tetrahedral elements in the model, and the number of the triangular elements is 2415. The quality of the FEA model achieve a reasonable level, in which the quality index of elements(comp. QI) is 0.02. The finite element model of the battery enclosure and the constraint condition of the enclosure are shown in Fig. 2.

In this paper, the materials mainly used in the battery enclosure modeling and optimization process are steel, aluminum alloy, and CFRP. And the material properties of them are shown in Tables 1 and 2.

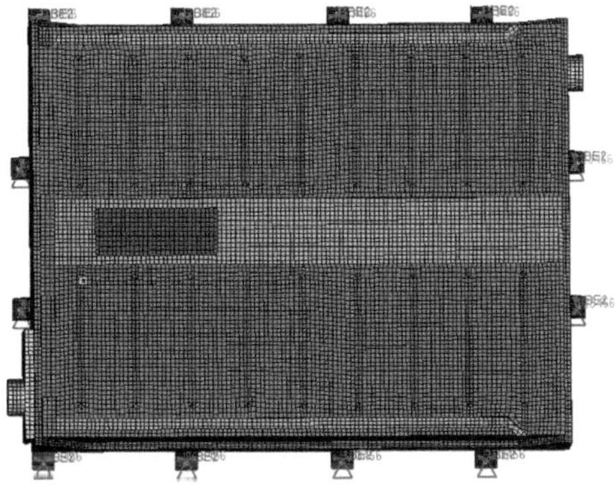

Fig. 2 FEA model and constraint of battery enclosure

Table 1 Metallic material properties

Material	Density (t/mm^2)	Young modulus (MPa)	Poisson ratio
Steel	7.9E−9	2.10E+5	0.3
Aluminum alloy	2.7E−9	7.2E+4	0.3

Table 2 T300/5208 composite parameters

E_1 (Gpa)	E_2 (Gpa)	NU_{12}	G_{12} (Gpa)	G_{1z} (Gpa)	G_{2z} (Gpa)
181	10.3	0.28	7.17	7.17	3.78
X_t (Mpa)	X_c (Mpa)	Y_t (Mpa)	Y_c (Mpa)	S (Mpa)	
1500	1500	40	246	68	

3 Battery Enclosure Prototype Model Analysis

3.1 Battery Enclosure Prototype Modal Analysis

The modal analysis of the battery enclosure is an effective method to study the structural dynamics of the battery enclosure. The analysis method of the modal characteristic is to analyze the dynamic performance of the structure by applying the basic theory of the system vibration. Each mode has its own natural frequency, natural modal shape, modal mass, modal stiffness, modal damping, and so on. The high-order vibration of the battery enclosure has little damage to the dynamic performance of the enclosure, and the high-order frequency intensity is serious; the coupling between each frequency is very powerful, and it is difficult to correctly identify the modal parameters. At the same time, in order to ensure the consistency of the analysis results and the actual situation, this paper studies the battery enclosure constraint mode under the low-order natural frequency. If the low-order natural frequency is high enough, resonance with vehicle vibration source could be avoided. The numerical value of the first four-order constraint mode of the battery enclosure and the corresponding vibration shape obtained are shown in Table 3, and the modal shapes of the first-order modes are shown in Fig. 3.

The vibration frequency range of the vehicle vibration source is in the range of 18–30 Hz. The results show that the first-order vibration frequency of the battery enclosure is slightly lower, and the vibration spread over the whole enclosure body. In order to avoid resonance with the vehicle vibration source, the battery enclosure should be optimized.

Table 3 First four-order restrained frequencies and corresponding shapes of vibration

Order	Frequency (Hz)	Shape
1	30.23	Whole first-order bending
2	34.21	Whole first-order torsion
3	37.33	Whole second-order torsion
4	39.13	Whole second-order bending

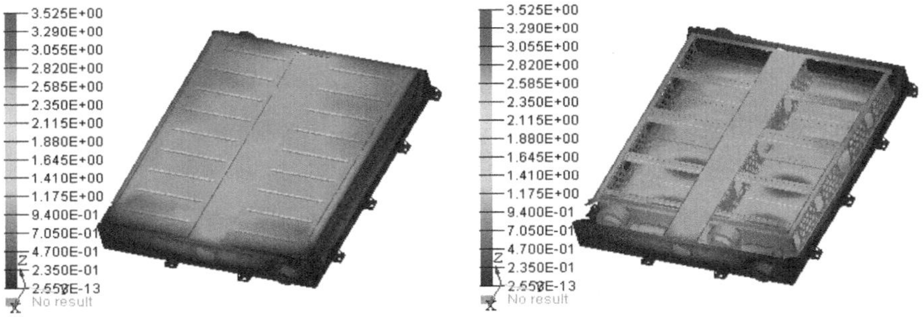

(a) Main view of the vibration mode (b) Inner view of the vibration mode

Fig. 3 Vibration mode of the first-order frequency

3.2 The Strength and Stiffness Analysis of the Basic Battery Enclosure

Under the vertical bump condition and left sharp turn condition, analyze the strength and stiffness of the battery enclosure. Under the two conditions, the impact of the battery on the battery enclosure can be equivalent to the static load evenly distributed on the inner wall of the battery enclosure. When the vehicle is through the potholes, the vertical acceleration is 2.5 g, and when the vehicle turns sharply to the left, the lateral acceleration is 0.75 g. The stress nephogram and strain nephogram of the battery enclosure in the case of the vertical bump and the left sharp turn are shown in Fig. 4.

It can be seen from the results, when vehicle bumping through potholes, the maximum stress value of the power battery enclosure is 172.4 MPa, located on both sides of the supporting legs and on the bottom near the junction of the battery enclosure; the maximum deformation value is 1.56 mm, located on the middle of the bottom surface of the lower enclosure. Under the bumpy condition, the performance of the battery enclosure meets the mechanical requirements. The safety factor is 1.36, which is relatively low, indicating that the structure of the battery enclosure should be strengthened. When the vehicle turns to the left, the maximum stress value of the battery enclosure is 51.87 MPa, which is located on the left side of the lower enclosure. The maximum deformation value of the battery enclosure is 0.435 mm, which is located on the bottom of the enclosure. The safety factor of the battery box is 8.13, which indicates that there is redundancy in the design of the battery enclosure in the left turn condition. Although the mechanical properties of the battery enclosure are meeting the requirements, but in order to reduce the weight of the battery enclosure and improve the mechanical performance of the battery enclosure, we need to use new lightweight materials for battery enclosure design. Carbon fiber composites and aluminum alloy materials are the lightweight materials with high recognition in the structure design. Although the cost of these materials is high now, the multi-material integration structure design is a future development trend.

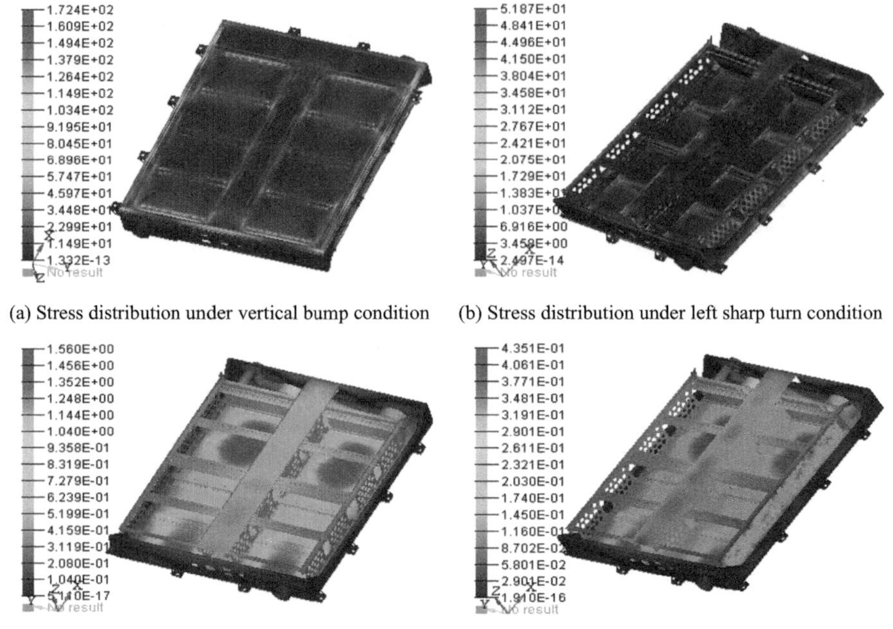

(a) Stress distribution under vertical bump condition (b) Stress distribution under left sharp turn condition

(c) Strain distribution under vertical bump condition (d) Strain distribution under left sharp turn condition

Fig. 4 Stress and strain distribution under vertical bump condition and the left sharp turn condition

3.3 Sensitivity Analysis of Battery Enclosure's Thickness

The contribution of the battery enclosure component to the lightweight and the first-order natural frequency is different. In order to make CFRP and aluminum alloy to be used in the appropriate parts, in this paper, the sensitivity analysis method was used to select the components of main contribution to the targets [10]. The sensitivity analysis results are shown in Table 4.

Table 4 Design variables of sensitivity analysis model

Variable	Name	Mass sensitivity	First-order natural frequency sensitivity
x_1	Cover	1.43E−02	3.52E+00
x_2	Floor	1.32E−02	4.28E+00
x_3	Apron	3.68E−03	1.44E−01
x_4	Air intake plate	1.61E−03	2.43E−01
x_5	Air outlet plate	2.94E−03	4.75E−01
x_6	Side wall	2.61E−03	1.05E+00
x_7	Beam set	5.87E−03	5.61E+00
x_8	Battery separator	1.97E−03	9.30E−01

Through the sensitivity analysis, in order to ensure the integrity of the structure design, the design variables x_1, x_2, x_4, x_5, x_6 represent the relatively variables of CFRP structures. And the design variables x_3, x_7, x_8 represent the relatively variables of aluminum alloy material structures.

4 Optimization and Analysis of Multi-material Battery Enclosure

4.1 *Multi-material Battery Enclosure Optimization Design*

In this paper, we take the minimum of the weight and the maximum of the first-order natural frequency as the targets. In order to simplify the optimization problem, the battery enclosure was optimized by the main objective method [11]. The idea is transforming the multi-objective optimization problem into a constrained optimization problem. According to the sensitivity analysis results, the weight of battery enclosure was chosen as the objective to minimize, the other objectives of the original problem were transformed to the design constraints. While the first-order natural frequency and the box stress serve as constraints. The expression of the optimization problem is shown as follows:

$$\text{Minimize } m, \qquad \text{Subject to}: \begin{cases} 0.5\text{ mm} \le xi \le 5\text{ mm}\ (i = 1, 2, \ldots, 8) \\ S_{\text{bump}} \le 422\text{ Mpa} \\ S_{\text{left}} \le 422\text{ Mpa} \\ f(1) \ge 32\text{ Hz} \end{cases} \tag{1}$$

where m is the total mass of the battery enclosure; $f(1)$ is the first-order natural frequency constraint; S_{bump} and S_{left} are, respectively, the global maximum von Mises stress of the battery enclosure in the case of vertical bump and turn sharply to left.

The initial value and range of the design variables of the battery enclosure are shown in Table 5.

Considering the manufacturing feasibility, the optimization results are approximated, and the optimization results are shown in Table 6.

4.2 *Static Analysis of Multi-material Battery Enclosure*

In order to ensure the reliability of the optimization results, static analysis of the optimized structure was carried out. The design variables of it were modified to the value of the optimized results. The first mode shape and the stress and strain distribution of the multi-material enclosure are shown in Figs. 5 and 6.

Table 5 Initial value and range of the design variables

Variable	Name	Upper limit (mm)	Initial value (mm)	Lower limit (mm)
x_1	Cover	5.0	2.4	0.5
x_2	Floor	5.0	2.4	0.5
x_3	Apron	5.0	1.0	0.5
x_4	Air intake plate	5.0	2.4	0.5
x_5	Air outlet plate	5.0	2.4	0.5
x_6	Side wall	5.0	2.4	0.5
x_7	Beam set	5.0	1.2	0.5
x_8	Battery separator	5.0	3.0	0.5

Table 6 Optimization results of the design variables

Variable	Name	Initial value (mm)	Final value (mm)	Selected value (mm)
x_1	Cover	2.4	1.172	1.19
x_2	Floor	2.4	1.047	1.05
x_3	Apron	1.0	0.50	0.50
x_4	Air intake plate	2.4	1.394	1.40
x_5	Air outlet plate	2.4	1.279	1.40
x_6	Side wall	2.4	2.219	2.10
x_7	Beam set	1.2	3.15	2.60
x_8	Battery separator	3.0	2.651	2.65

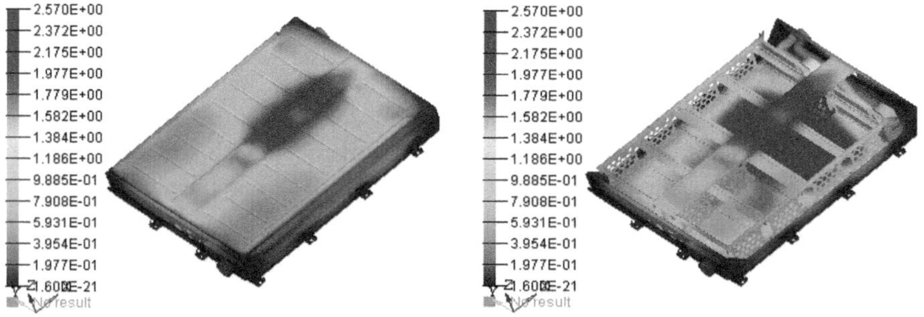

(a) Main view of the vibration mode (b) Inner view of the vibration mode

Fig. 5 Vibration mode of the first-order frequency

As can be seen from the analysis results of Figs. 5 and 3, the vibration position distribution of multi-material battery enclosure is stable, it is symmetric along the X-axis, the maximum amplitude is reduced, and the first-order natural frequency is significantly improved, from the initial 30.23–32.03 Hz, which can avoid the resonance between the battery enclosure and the vehicle vibration source.

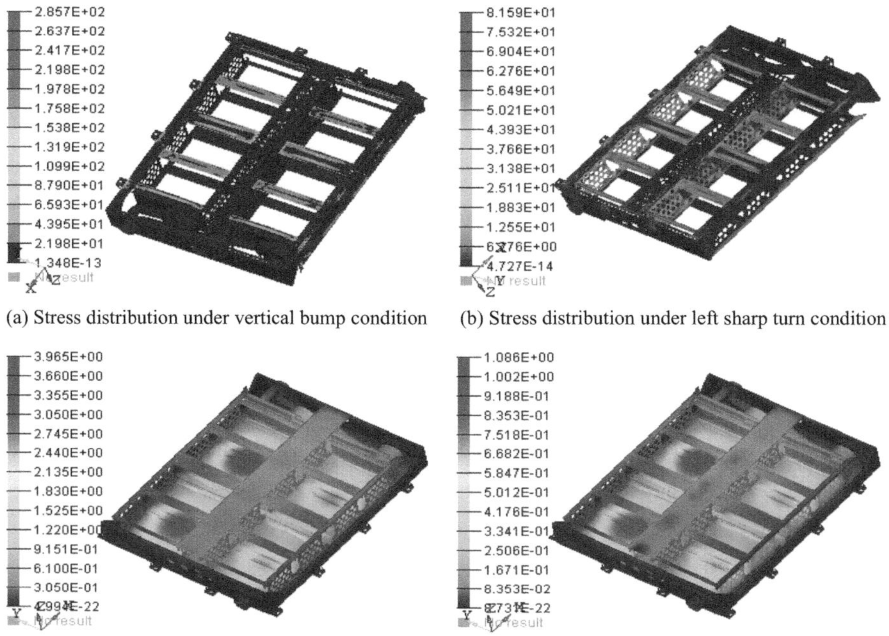

(a) Stress distribution under vertical bump condition (b) Stress distribution under left sharp turn condition

(c) Strain distribution under vertical bump condition (d) Strain distribution under left sharp turn condition

Fig. 6 Stress and strain distribution of the multi-material battery enclosure under the condition of the bump and the sharp turn

As shown in Figs. 6 and 4, the stress of the multi-material battery enclosure under the conditions of vertical bump and left sharp turn is increased relative to the stress of the battery enclosure prototype. But the distribution is more uniform, and the stress concentration is reduced. At the same time, the deformation of the multi-material battery enclosure is larger than the deformation of the battery enclosure prototype, but it still meets the rigidity requirement of the battery enclosure. The safety factor of the multi-material battery enclosure in the bumpy condition is 1.47, the mechanical performance of the prototype enclosure is strengthened, the safety factor of the left sharp turn condition is 5.14, the mechanical performance of the prototype enclosure is reduced, but it still meets the security requirement.

4.3 Battery Enclosure Strength Check

The mechanical mechanism of carbon fiber composite material is complex, and this paper uses Tsai–Wu standard [12] to check the strength; its expression is as follows:

$$f(\sigma) = F_i\sigma_i + F_{ij}\sigma_i\sigma_j < 1 \;(i,j = 1,2,\ldots,6) \tag{2}$$

In the formula above, F_i and F_{ij} are the strength parameters, which can be expressed by the basic strength of CFRP.

The formula (2) can be derived as follows:

$$f(\sigma) = F_{11}\sigma_1^2 + 2F_{12}\sigma_1\sigma_2 + F_{22}\sigma_2^2 + F_{66}\tau_{12}^2 + F_1\sigma_1 + F_2\sigma_2 < 1 \tag{3}$$

CFRP parameters are shown in Table 3. Put them into the formulas:

$$\begin{cases} F_1 = \frac{1}{X_t} - \frac{1}{X_c} = 0 \\ F_{11} = \frac{1}{X_t X_c} = 0.444\ \text{GPa}^{-2} \\ F_2 = \frac{1}{Y_t} - \frac{1}{Y_c} = 20.93\ \text{GPa}^{-2} \\ F_{22} = \frac{1}{Y_t Y_c} = 101.6\ \text{GPa}^{-2} \\ F_{66} = \frac{1}{S^2} = 212.2\ \text{GPa}^{-2} \\ F_{12} = -\frac{1}{2}\sqrt{\frac{1}{X_t X_c Y_t Y_c}} = -3.36\ \text{GPa}^{-2} \end{cases} \tag{4}$$

From Fig. 7, we can get the maximum principal stress and maximum shear stress of the composite structure respectively: $\sigma_1 = 82.99$ Mpa, $\sigma_2 = 28.94$ Mpa, $\tau_{12} = 15.63$ Mpa, these values are used to calculate the strength of composite structures. The result of formulas (4) and σ_1, σ_2, τ_{12} is brought into formula (3), which can be calculated:

$$f(\sigma) = 0.73 < 1$$

The strength of the CFRP was checked by Tsai–Wu criterion, and the maximum stress of the structure of the CFRP was determined to meet the strength requirement. From Fig. 6, we can get the maximum stress of the battery enclosure, which was 285.7 Mpa, less than the yield strength of aluminum alloy material. To sum up, the strength of the multi-material battery enclosure can meet the requirements in the bump condition and the left sharp turn condition. The comparison of the

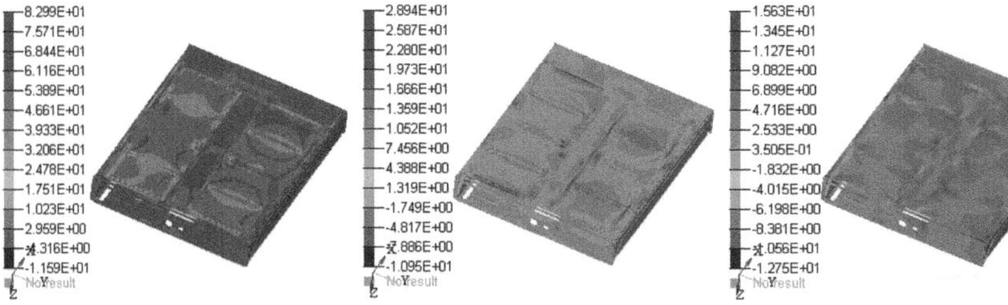

(a) σ_1 stress nephogram (b) σ_2 stress nephogram (c) σ_3 stress nephogram

Fig. 7 Stress nephogram of CFRP structure

Table 7 Performance comparison of multi-material battery enclosure and battery enclosure prototype

Condition name	Battery enclosure prototype	Multi-material battery enclosure
Stress in bump (Mpa)	172.4	285.7
Strain in bump (mm)	1.56	3.96
The left turn stress (Mpa)	51.87	81.59
The left turn strain (mm)	0.435	1.06
First-order natural frequency (Hz)	30.23	32.03
Mass (kg)	41.13	22.13

performance index of the multi-material battery enclosure and the battery enclosure prototype is shown in Table 7.

The stress and strain of the multi-material battery enclosure have been analyzed previously. It can be concluded from Table 7 that the weight of the multi-material battery box is reduced by 48.86% compared with that of the battery enclosure prototype. The first-order natural frequency of the multi-material battery enclosure can avoid resonance with the vibration source of the vehicle. And the stress and strain of the multi-material battery enclosure in the bump and the left sharp turn condition are higher than those of the battery enclosure prototype, but the requirements are met.

5 Conclusion

In this paper, the multi-material battery enclosure was designed and optimized under multiple loading cases. The strength and stiffness were obtained through the static analysis performed on the battery enclosure prototype under the condition of bumping and turning, which provided the basis for the optimization of the multi-material battery enclosure. The first-order mode of the battery enclosure was analyzed, and the first-order natural frequency of the enclosure needed to be improved.

Sensitivity analysis on the battery enclosure prototype was executed. This method is an effective method to determine the influence of the composition of the battery enclosure on the total mass and the first-order natural frequency of the enclosure. CFRP and aluminum alloy were used to design parts. CFRP and aluminum alloy materials are the lightweight materials with high recognition in the structure design. Although the cost of these materials is high now, the multi-material integration structure design is a future development trend.

Taking the weight minimum and the first-order natural frequency maximum as the goals, the optimization of the components design with the main objective method was implemented. Then, the static analysis was applied to validate the optimization results. Compared to the original design, the mass of the new multi-material battery enclosure can be decreased 48.86%, and the first-order natural frequency can be increased up to 32.03Hz at the same time.

Acknowledgments This research is supported by National Natural Science Foundation of China (Grant No. 51675044).

References

1. Mohammadian SK, Yuwen Z (2015) Thermal management optimization of an air-cooled Li-ion battery module using pin-fin heat sinks for hybrid electric vehicles. J Power Sources 273(273):431–439
2. Lin S, Zhiming L, Cheng P (2011) Research on vibration characteristics of the electric vehicle battery enclosure. Electron Test 12:1–4
3. Hongwei Z, Xiaokai C, Yi L (2009) Topology optimization of power battery cabin in electric vehicle. J Jilin Univ 39(4):846–850
4. Hartmann M, Roschitz M, Khalil Z (2013) Enhanced battery enclosure for electric vehicle: noise reduction and increased stiffness. Mater Sci Forum 765:818–822
5. Schubert E, Klassen M, Zerner I et al (2001) Lightweight structures produced by laser beam joining for future applications in automobile and aerospace industry. J Mater Process Technol 115:2–8
6. Xiaohong Z, Feng Z, Qi F et al (2014) Design and research on carbon fiber box of vehicle power battery. Shanghai Automot 09:60–62
7. Jianong W, Xiaoyu Z (2016) Study on the lightweight of carbon fiber epoxy resin composite battery box. FRP composites (12): 99–102 + 33
8. Sakundarini N, Taha Z, Abdul-Rashid SH et al (2013) Optimal multi-material selection for lightweight design of automotive body assembly incorporating recyclability. Mater Des 50 (17):846–857
9. Hui Y (2013) Automotive body lightweight material application research. Mech Res Appl 01:140–142
10. Zhiyuan H, Gengqiang P, Yunkai G (2003) Sensitivity analysis and optimization for Light-bus body stiffness. J Mech Strength 25(1):70
11. Honggang L, Pengfei W et al (2008) A method for reducing and prolonging life of liquid propellant rocket engine based on main objectives. J Aerosp Power 08:1530–1535
12. Guanlin S, Gengkai H (2004) Mechanics of composite materials. Tsinghua University Press, Beijing, p 9

Improving the Dynamics of a Vehicular Platooning by Taking Account of Individual Vehicle Dynamics

Xiujian Yang and Yayong Chen

1 Introduction

Vehicular platooning which is typically defined as a number of interconnected vehicles moving in a group autonomously has been an attractive topic for a few decades. Much more effort has been done since 90 s when the California PATH program was initiated to develop an automated highway system [1, 2]. The core issue of vehicular platooning is related to autonomous formation control for interconnected or cascaded systems. The aim consists of several aspects such as increasing traffic throughput, improving safety or reducing fuel consumption [3–5]. A great deal of research effort has been put in the related issues such as string stability and control policy which are mainly investigated in automatic control community [6].

String stability of a platoon can be roughly interpreted as the property of spacing error attenuation toward the upstream direction of the platoon. Up to now, string stability is usually viewed as a performance criterion rather than a stability property to evaluate the control performance of a platoon [6]. The control of vehicular platooning can roughly be classified into decentralized and centralized formation controls [7]. In previous studies, each vehicle in a platoon is commonly controlled based on the relative spacing or/and speed information from its neighboring vehicles obtained by a radar or other range sensors, e.g., adaptive cruise control (ACC) system. This kind of scheme corresponds to the decentralized formation control. The centralized control case generally needs communicating information with other vehicles besides the neighboring members. Though the centralized control method presents some superior properties in theoretical studies, it still has severe challenges such as the difficulty and cost of implementation, the limited

X. Yang (✉) · Y. Chen
Faculty of Transportation Engineering, Kunming University of Science and Technology, Kunming 650500, China
e-mail: yangxiujian2013@163.com

Society of Automotive Engineers of China (SAE-China) (ed.), *Proceedings of the 19th Asia Pacific Automotive Engineering Conference & SAE-China Congress 2017: Selected Papers*, Lecture Notes in Electrical Engineering 486, https://doi.org/10.1007/978-981-10-8506-2_58

communicating range, the communicating delay, and the drop of packet [8]. In contrast, the decentralized control method is more practical and much easier to implement since communication is not needed. Spacing policy as a typical decentralized control scheme is often found in the literature, and in this policy the space between adjacent vehicles maybe constant, speed-dependent, or other forms [1, 7, 9–12].

Though the vehicular platooning or cascaded systems are extensively discussed in the literature, in most of the studies, linear vehicle models, e.g., single integrators, double integrators, double integrator with first-order actuator dynamics, are commonly assumed and used to discuss the platooning stability or control issues theoretically [12–14]. Since these simple linear models do not encompass the factors of actual vehicle dynamics such as wheel rotational dynamics and the property of nonlinear tire force, the theoretical policies reported in the literature would face other challenges in real application. Especially, the wheel rotational dynamics in critical situations, e.g., emergent braking or intense driving on slippery road, would greatly affect the dynamics and stability of the entire string. In this paper, we take the wheel rotational dynamics as the input disturbance and discuss the platooning control method which can compensate the effect of this disturbance before large wheel slip appears to improve the entire platooning dynamics. Therefore, we propose a novel time headway platooning control policy (speed-dependent spacing policy) which involves feedforward of wheel rotational velocity in the platooning control law.

2 Vehicle Model

In order to evaluate the platooning control performance when taking into account the effect of wheel dynamics and the nonlinear tire force property, a 4DOF single-track nonlinear vehicle model is developed in this section based on Lagrange's approach. The four degrees of motion consist of the longitudinal motion, body pitch motion, and rotational dynamics of the front and rear wheels. Figure 1 illustrates the coordinate system and the schematic of body motion of this model.

Fig. 1 Vehicle coordinate system and the schematic of body motion

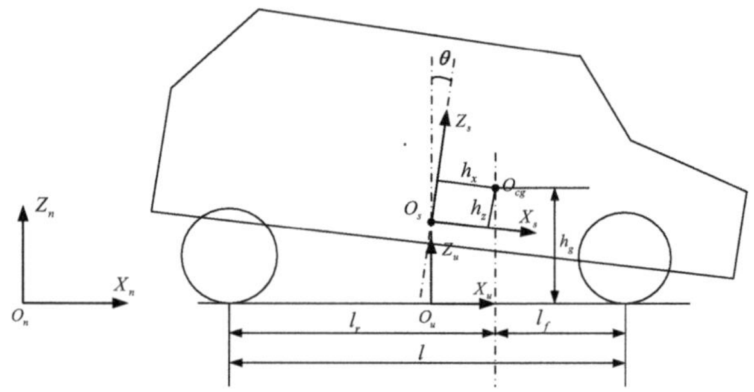

Three coordinates, such as the global coordinate $X_nO_nZ_n$ fixed on the ground, chassis coordinate $X_uO_uZ_u$ fixed on the chassis, and sprung mass coordinate $X_sO_sZ_s$ fixed on the center of the pitch motion of the sprung mass, are defined here to facilitate the model derivation. The equations of motion are derived here based on Lagrange's approach. The nonlinear dynamical equation depicting the longitudinal dynamics and the pitch motion is respectively given in Eqs. (1) and (2) as follows:

$$m\dot{v}_x + m(h_x \sin\theta + h_z \cos\theta)\ddot{\theta} + m(-h_z \sin\theta + h_x \cos\theta)\dot{\theta}^2 = F_{xf} + F_{xr}, \quad (1)$$

$$\begin{aligned} m(h_z \cos\theta + h_x \sin\theta)\dot{v}_n + (mh_x^2 + mh_z^2 + I_y)\ddot{\theta} + mgh_x = {} & F_{xf}(h_x \sin\theta - l_f \sin\theta) \\ & + F_{zf}(h_x \cos\theta - l_f \cos\theta) \\ & + F_{xr}(l_r \sin\theta + h_x \sin\theta) \\ & + F_{zr}(l_r \cos\theta + h_x \cos\theta) \end{aligned} \quad (2)$$

where θ denotes the pitch angle, h_x and h_z are used to identify the relative position between the pitch center O_s and the center of gravity O_{cg}, and I_y denotes the moment of inertia of the pitch motion. The tire normal force for the front and rear wheels is respectively estimated as

$$F_{zf} = \frac{mgl_r}{l_f + l_r} - \frac{ma_xh_g}{l_f + l_r}, \quad F_{zr} = \frac{mgl_f}{l_f + l_r} + \frac{ma_xh_g}{l_f + l_r}. \quad (3)$$

The longitudinal tire force F_{xf}, F_{xr} produced when braking or accelerating, respectively, for the front and rear wheels is calculated from pure longitudinal slip 'Magic Formula' tire model [15], which is formulated as

$$F_x = \mu D_x \sin(C_x \arctan(B_x \Phi_x)), \quad (4)$$

With $\Phi_x = (1 - E_x)\lambda + (E_x/B_x)\arctan(B_x\lambda)$, $E_x = a_{x6}F_z^2 + a_{x7}F_z + a_{x8}$, $B_x = \left(a_{x3}F_z^2 + a_{x4}F\right)_z/(C_xD_xe^{a_{x5}F_z})$, $C_x = 1.65$ $D_x = a_{x1}F_z^2 + a_{x2}F_z$. In tire model (4), μ denotes road adhesion coefficient, λ denotes longitudinal wheel slip, and $a_{x1} \sim a_{x8}$ are empirical constants. The wheel slip is modeled as

$$\lambda = \frac{\omega_w r_w - v_x}{\omega_w r_w} \text{ when accelerating,} \quad \lambda = \frac{\omega_w r_w - v_x}{v_x} \text{ when braking.} \quad (5)$$

The wheel rotational angular velocity can be obtained from the wheel rotational dynamics equation

$$I_w \dot{\omega}_w = T_w - F_z f - F_x r_w, \quad (6)$$

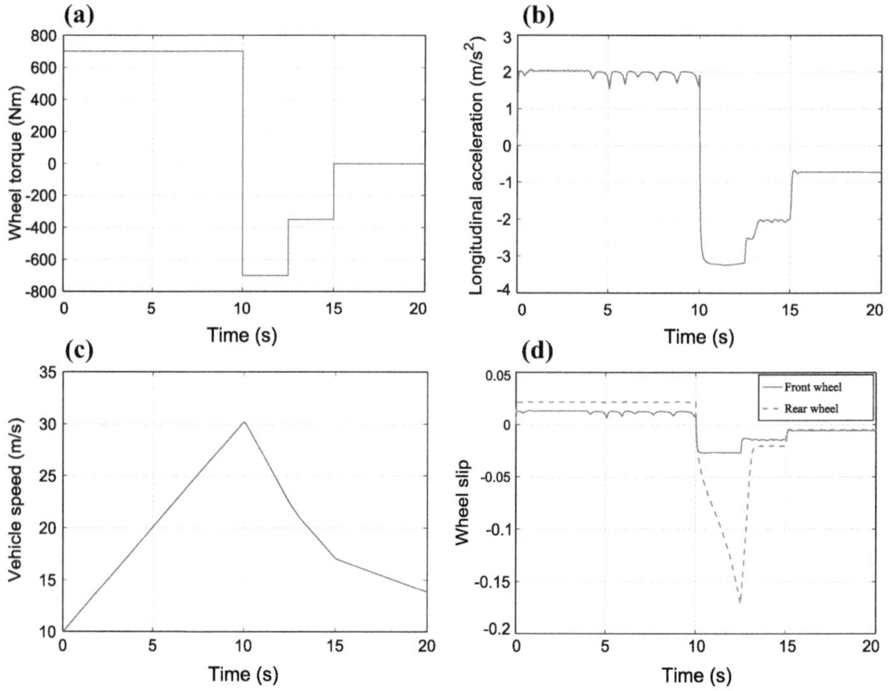

Fig. 2 Illustration of the response of the 4DOF single-track nonlinear vehicle model. **a** Input wheel torque; **b** response of longitudinal acceleration; **c** response of vehicle speed; **d** response of wheel slip

where T_w is the braking/accelerating torque of the wheel and f is the rolling resistance coefficient. The effect of the power train system dynamics is represented by a first-order transfer function which will be presented in the following sections when designing the platooning control strategy.

As an illustration, the response of the formulated 4DOF single-track nonlinear vehicle model to the wheel torque input given by Fig. 2a is presented in Fig. 2b–d. It is noted that the response conforms to the basic characteristic of a typical response of vehicle longitudinal dynamics and the nonlinear tire slip property can also be embodied. Hence, we will use this model in simulation of the platooning dynamics to evaluate the platooning control strategies in the following sections.

3 Spacing Error Dynamics

Figure 3 presents the schematic of a longitudinal vehicle platoon which is comprised of n identical vehicles. The lead vehicle is identified by number 1, and the number for the followers increases toward the tail of the string, therefore the last vehicle is

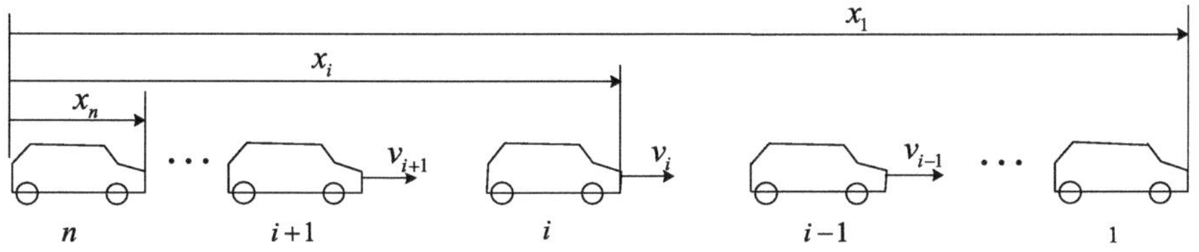

Fig. 3 Schematic of a typical homogeneous vehicle platoon

labeled as n. In the figure, x_i and v_i, respectively, denotes the position and velocity of vehicle i, and similar definitions for other vehicles are defined accordingly.

The actuator dynamics is represented by a first-order transfer function and is mainly used to characterize the power train dynamics property. The vehicle model is given by

$$P(\mathrm{s}) = \frac{1}{s^2(\tau\,\mathrm{s}+1)}, \tag{7}$$

where τ is a time constant and s is a complex variable. Let $X_i(\mathrm{s})$, $U_i(\mathrm{s})$, and $D_i(\mathrm{s})$ be the Laplace transform of the position $x_i(t)$, control input acceleration $u_i(t)$, and disturbance acceleration $d_i(t)$ of vehicle i in time domain. We assume that the lead vehicle runs freely disregarding the motions of its followers. It is also assumed that all the vehicles start from rest that is the initial velocity for all the vehicle is zero. For one of the following vehicles i, $2 \le i \le n$, its position can be derived as

$$X_i(\mathrm{s}) = P(\mathrm{s})(U_i(\mathrm{s}) + D_i(\mathrm{s})) + \frac{x_i(0)}{s}, \tag{8}$$

where $x_i(0)$ is the initial position of vehicle i. Assuming the initial position of the lead vehicle is $x_1(0) = 0$ and the initial spacing between two adjacent vehicles is δ, hence the initial position of vehicle i, $2 \le i \le n$, is $x_i(0) = -(i-1)\delta$. Then according to Eq. (8), the position of vehicle $i-1$ and vehicle i, (here $3 \le i \le n$) can, respectively, be rewritten as

$$\begin{aligned} X_{i-1}(\mathrm{s}) &= P(\mathrm{s})(U_{i-1}(\mathrm{s}) + D_{i-1}(\mathrm{s})) - \frac{i-2}{s}\delta, \\ X_i(\mathrm{s}) &= P(\mathrm{s})(U_i(\mathrm{s}) + D_i(\mathrm{s})) - \frac{i-1}{s}\delta. \end{aligned} \tag{9}$$

The time headway policy with constant headway time is adopted here [6, 11], and then the control law for vehicle $i-1$ and i is respectively

$$\begin{aligned} U_{i-1}(\mathrm{s}) &= K(\mathrm{s})E_{i-1}(\mathrm{s}) = \left(k_p + k_v s\right)E_{i-1}(\mathrm{s}), \\ U_i(\mathrm{s}) &= K(\mathrm{s})E_i(\mathrm{s}) = \left(k_p + k_v s\right)E_i(\mathrm{s}). \end{aligned} \tag{10}$$

The spacing error for vehicle i following its immediate predecessor is defined as

$$e_i(t) = x_{i-1}(t) - x_i(t) - \dot{x}_i(t)h_w - \delta_d, \tag{11}$$

where h_w is the time headway constant and δ_d is also a constant representing the desired standstill distance. The spacing error in Laplace domain is given as

$$E_i(\mathrm{s}) = X_{i-1}(\mathrm{s}) - (1 + h_w s)X_i(\mathrm{s}) + h_w x_i(0) - \frac{\delta_d}{s}. \tag{12}$$

Substituting (8)–(10) into (12) and collecting appropriately yields

$$E_i(\mathrm{s}) = G_e(\mathrm{s})E_{i-1}(\mathrm{s}) + G_{d1}(\mathrm{s})D_{i-1}(\mathrm{s}) - G_{d2}(\mathrm{s})D_i(\mathrm{s}) + G_s(\delta - \delta_d), \tag{13}$$

with $G_e(\mathrm{s}) = \frac{P(\mathrm{S})K(s)}{\mathrm{Den}(s)}$, $G_{d1}(\mathrm{s}) = \frac{P(\mathrm{S})}{\mathrm{Den}(s)}$, $G_{d2}(\mathrm{s}) = \frac{P(\mathrm{S})(1+h_w s)}{\mathrm{Den}(s)}$, $G_s(\mathrm{s}) = \frac{1}{\mathrm{Den}(s)}\frac{1}{s}$,

$\mathrm{Den}(s){=}1 + (1 + h_w s)P(s)K(s)$. The transfer function $G_e(\mathrm{s})$, G_{d1}, G_{d2}, G_s respectively depicts the effect on spacing error $\mathrm{E_i(s)}$ from the tracking error of the immediate predecessor, the acceleration disturbance of the immediate predecessor, the input acceleration disturbance of the vehicle itself and the initial spacing with respect to the desired standstill distance. Obviously, all these factors can affect the spacing error. Note that if the initial spacing equals the desired standstill distance that is $\delta - \delta_d = 0$, then the initial spacing has no effect on the spacing error. We assume that the second vehicle (immediate behind the lead vehicle) causes a spacing error $e_2(t)$, and then the spacing error dynamics is expressed as

$$\tilde{E}(\mathrm{s}) = H_E(\mathrm{s})E_2(\mathrm{s}) + H_D(\mathrm{s})\tilde{D}(\mathrm{s}) + H_S(\mathrm{s})(\delta - \delta_d), \tag{14}$$

with the definitions as follows:

$$\tilde{E}(\mathrm{s}) = [E_3(\mathrm{s}) \quad \ldots \quad E_n(\mathrm{s})]^{\mathrm{T}}, \tilde{D}(\mathrm{s}) = [D_2(\mathrm{s}) \quad \ldots \quad D_n(\mathrm{s})]^{\mathrm{T}},$$

$$H_E(\mathrm{s}) = \begin{bmatrix} G_e(\mathrm{s}) & G_e(\mathrm{s})^2 & G_e(\mathrm{s})^3 & \ldots & G_e(\mathrm{s})^n \end{bmatrix}^{\mathrm{T}},$$

$$H_D(\mathrm{s}) = \begin{bmatrix} G_{d1}(\mathrm{s}) & -G_{d2}(\mathrm{s}) & & & \\ G_e(\mathrm{s})G_{d1}(\mathrm{s}) & G_{d1}(\mathrm{s}) - G_e(\mathrm{s})G_{d2}(\mathrm{s}) & -G_{d2}(\mathrm{s}) & & \\ \vdots & \vdots & \ddots & \ddots & \\ G_e(\mathrm{s})^{n-3}G_{d1}(\mathrm{s}) & G_e(\mathrm{s})^{n-4}(G_{d1}(\mathrm{s}) - G_e(\mathrm{s})G_{d2}(\mathrm{s})) & \cdots & G_{d1}(\mathrm{s}) - G_e(\mathrm{s})G_{d2}(\mathrm{s}) & -G_{d2}(\mathrm{s}) \end{bmatrix},$$

$$H_S(\mathrm{s}) = \left[1 \; G_e(\mathrm{s}) + 1 \ldots \; \sum_{k=3}^{n} G_e(\mathrm{s})^{n-k}\right]^{\mathrm{T}} G_s(\mathrm{s}).$$

Note that $G_e(s)$ is the transfer function governing the propagation of the spacing error. We can rewrite the spacing error for vehicle i as

$$E_i(s) = G_e(s)^{i-2}E_2(s) + G_e(s)^{i-3}\bar{D}_2(s) + \sum_{k=3}^{i-1}\left(G_e(s)^{i-1-k}\bar{D}_{1k}(s) - G_e(s)^{i-k}\bar{D}_{2k}(s)\right) - \bar{D}_i(s) + \sum_{k=3}^{i} G_e(s)^{i-k}\bar{\delta}, \quad (15)$$

where $\bar{D}_2(s) = G_{d1}(s)D_2(s)$, $\bar{D}_{1k}(s) = G_{d1}(s)D_k(s)$, $\bar{D}_{2k}(s) = G_{d2}(s)D_k(s)$, $\bar{D}_i(s) = G_{d2}(s)D_i(s)$, $\bar{\delta} = G_s(s)(\delta - \delta_d)$. The spacing error transfer function $G_e(s)$ in Eq. (13) can be further formulated as

$$G_e(s) = \frac{k_v s + k_p}{\tau s^3 + (1 + k_v h_w)s^2 + (k_v + k_p h_w)s + k_p}. \quad (16)$$

In order to obtain a stable string, it is required the maximum magnitude of $G_e(s)$ to be less than 1 for all frequencies, that is

$$\|G_e(s)\|_\infty := \sup_\omega |G_e(j\omega)| < 1. \quad (17)$$

Following this criterion, the design parameters need to satisfy

$$\begin{cases} \frac{(1+k_v h_w)^2}{2(k_p h_w + k_v)} \geq \tau \\ k_p h_w^2 > 2 \end{cases} \text{or} \begin{cases} \frac{(1+k_v h_w)^2}{2(k_p h_w + k_v)} < \tau \\ 4\tau^2(k_v^2 + 2k_p k_v h_w + 2k_p) - 4\tau(1+k_v h_w)^2(k_v + k_p h_w) + (1+k_v h_w)^4 < 0 \end{cases}. \quad (18)$$

Obviously, the string stability property closely relates to all the design parameters. Recall the actual vehicle dynamics that the time constant τ is typically below 1 and the constant h_w in time headway policy is generally around 1 as an alternative to expect a relatively satisfying platooning dynamics [1]. In fact, the first condition of (18) is very easy to satisfy by appropriately designing the control gains while the second condition is not.

According to Eq. (8), if the vehicle is subjected to input disturbance acceleration $D_i(s)$, the actual control effort to the vehicle is not the commanded $U_i(s)$ but $\tilde{U}_i(s) = U_i(s) + D_i(s)$. Seen from the spacing error dynamics (15), on the other hand, the disturbance acceleration $D_i(s)$ can also be made use to help mitigate the spacing error if the platoon is string stable. In practical case, any factor directly affecting vehicle acceleration can be viewed as a disturbance such as wind resistance, road grade, wheel slip. It is known that wheel rotational dynamics which is closely related to wheel slip is rather important for vehicle longitudinal dynamics since a small variation of wheel slip would generally cause a large variation of vehicle acceleration especially in low adhesion condition. So, if wheel rotational dynamics is involved in the platooning controller and designed appropriately, the spacing error is expected to be well attenuated.

4 Improving Platooning Dynamics by Regulating Wheel Rotational Dynamics

4.1 The Role of Regulating Wheel Dynamics

Take vehicle i as an illustration, the schematic is depicted in Fig. 4. The upper platooning controller of vehicle i generates its desired acceleration command $\bar{a}_i$ based on the velocity and position information from its immediate predecessor and itself. The brake/drive torque controller is designed based on PID control method to force the vehicle to track the commanded acceleration $\bar{a}_i$. The wheel slip regulator which is also a PID controller is designed to prevent large wheel slip by tracking the desired wheel slip $\bar{\lambda}_{i1} \ldots \bar{\lambda}_{i4}$. The nominal value of the parameters in the platooning control law is $\tau = 0.3$, $h_w = 1$, $k_p = 10$, $k_v = 10$, and the initial speed of the platoon is 25 m/s. The string stability can be guaranteed with these nominal parameters according to the condition (18). In the simulation, the vehicle parameter values are the same as those given in Sect. 2.

First, the spacing error response without wheel slip regulating respectively on a dry road ($\mu = 0.8$) and on a slippery road ($\mu = 0.4$) is presented in Fig. 5. As shown in Fig. 5a, the spacing errors for all the following vehicles can converge rapidly from an initial value of 10 m to zero on a dry road. However, the spacing errors for vehicle 6 and its followers would diverge and cannot converge to zero any more on the slippery road; see Fig. 5b. It is no doubt that a collision would not be avoided in this situation. This is confirmed from the position response given by Fig. 6. Though the string is theoretically stable according to the condition (18), if the road adhesion is relatively low, the converging of spacing errors for the vehicles

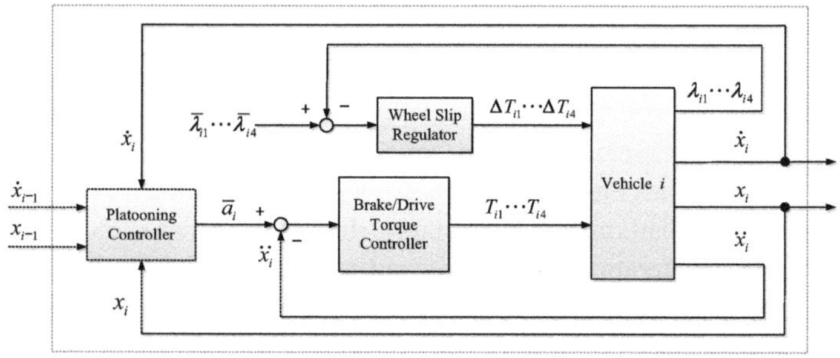

Fig. 4 Platooning control by regulating wheel dynamics. (take vehicle i as an illustration here)

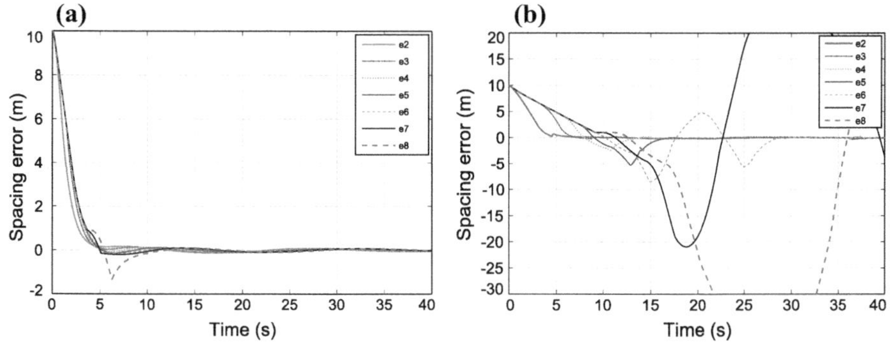

Fig. 5 Spacing error responses without wheel slip regulating. Road adhesion: **a** $\mu = 0.8$, **b** $\mu = 0.4$

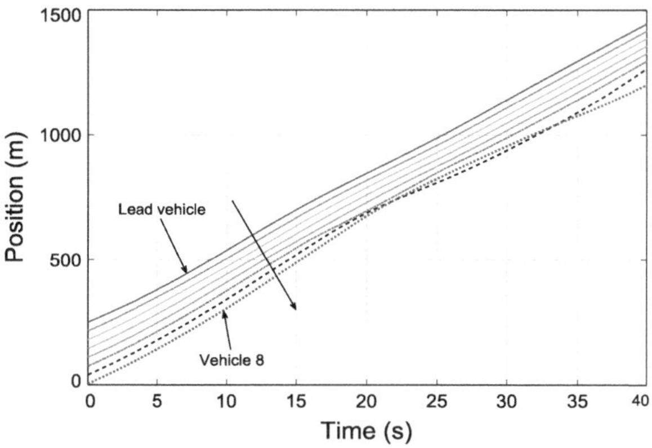

Fig. 6 Vehicle position responses without wheel slip regulating on slippery road ($\mu = 0.4$)

far from the leader becomes more and more difficult and collisions would occur. From the corresponding variations of wheel slip given in Fig. 7, it is noted that in this case large wheel slip would propagate toward the upstream direction of the platoon, which may have much relation to the adopted speed-dependent predecessor–follower spacing policy since a vehicle only follows its immediate predecessor. Thus, this predecessor–follower spacing policy faces much difficulty to control the platoon in low adhesion conditions, and this problem would become even severe as the platoon size grows.

Compared with Fig. 5a, if the wheel slip is regulated, the spacing error for all the following vehicles can converge to zero (see Fig. 8a) though the transient overshoot is still propagated, and no collision happens which can be found from the position responses given by Fig. 8b. However, remarkable wheel slip propagation still appears which can be seen from Fig. 9. It is not difficult to imagine that there must

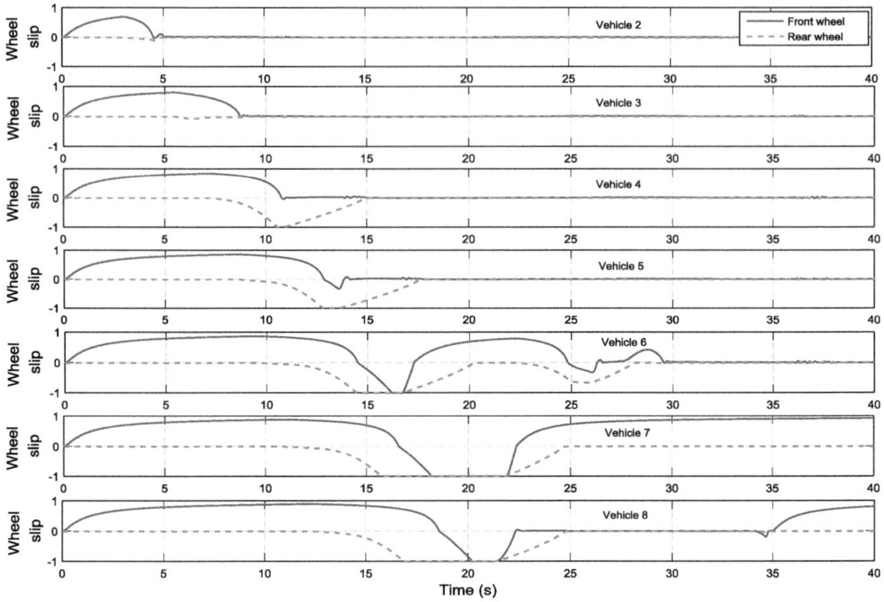

Fig. 7 Variations of wheel slip if it is not regulated on slippery road ($\mu = 0.4$)

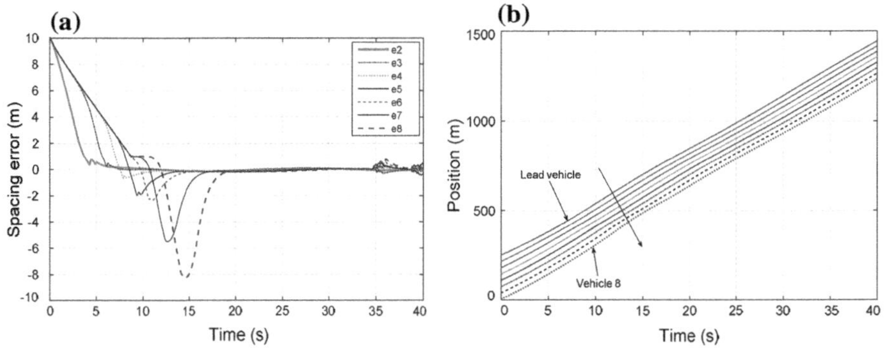

Fig. 8 Response of spacing error and vehicle position with wheel slip regulating on slippery road ($\mu = 0.4$). **a** Spacing error; **b** vehicle position

be a vehicle the spacing error corresponding to which and the followers would diverge as the size of the platoon grows. Hence, if a vehicle only follows its immediate predecessor, this kind of control policy may be not suitable for long platoon case. As a result, in order to compensate the deficiency in robust property when using this kind of platooning control policy, we propose to involve the wheel speed (disturbance) in the upper platooning control law based on the feedforward compensation control method to improve the platooning dynamics performance.

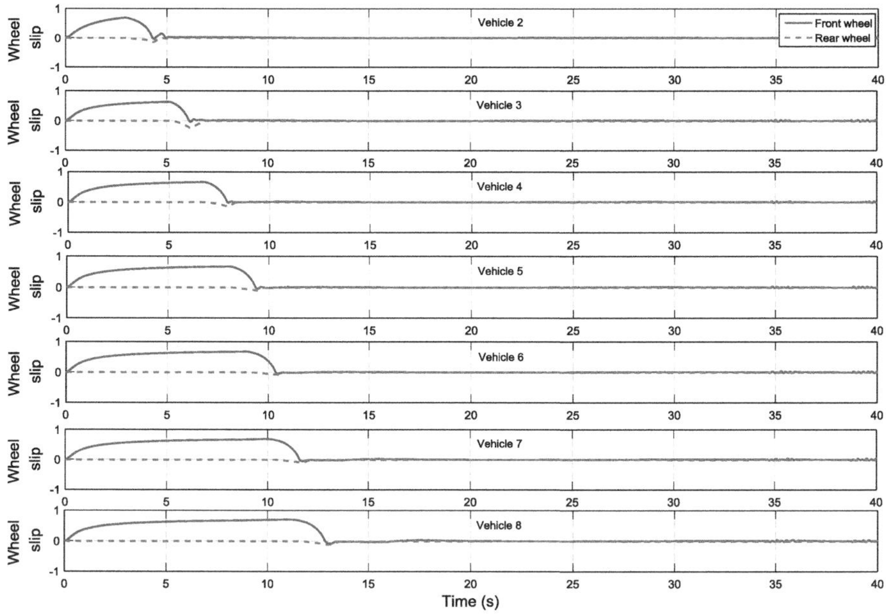

Fig. 9 Variations of wheel slip with wheel slip regulating on slippery road ($\mu = 0.4$)

4.2 *Feedforward of Wheel Speed in the Platooning Controller*

The schematic of the proposed platooning control law with feedforward of wheel rotational speed is illustrated in Fig. 10.

On the basis of the conventional time headway policy (10), this novel control law is defined in time domain as follows:

$$u_i(t) = k_p e_i(t) + k_v \dot{e}_i(t) + k_w(\dot{x}_i(t) - v_{wi}(t)), \tag{19}$$

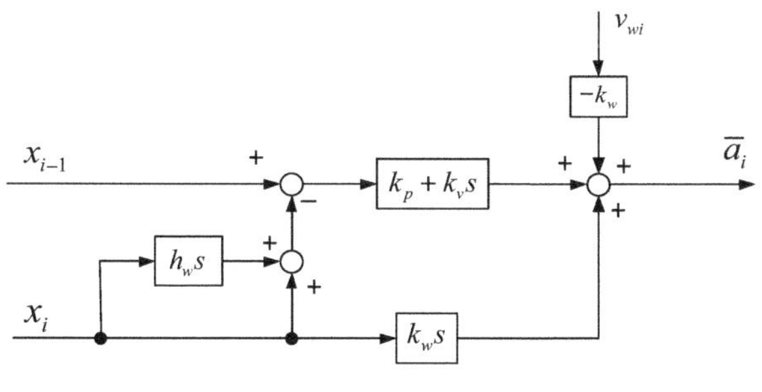

Fig. 10 Schematic of the time headway platooning control law with feedforward of wheel rotational speed

where $v_{wi}(t)$ is the wheel rotational speed of vehicle i and it is viewed as the disturbance of feedforward. Since there are several wheels for a vehicle and the wheel speed may be different, we assume $v_{wi}(t)$ to be the maximum wheel speed among all the wheels, that is $v_{wi}(t) = \omega_{w\max}(t)r_w$, where $\omega_{w\max}(t)$ denotes the maximum wheel angular velocity. k_w denotes the feedforward gain of wheel rotational speed and $k_w > 0$. The spacing error $e_i(t)$ is the same as the definition given by Eq. (11). We assume that all the initial spacing errors are zero and the initial position for vehicle i, $2 \leq i \leq n$, is $x_i(0) = -(i-1)\delta$, then the control law (19) can be represented in Laplace domain as

$$U_i(\mathrm{s}) = \left(k_p + k_p s\right)E_i(\mathrm{s}) + k_w s X_i(\mathrm{s}) - k_w V_{wi}(\mathrm{s}) + k_w(i-1)\delta. \tag{20}$$

Similarly, for vehicle $i-1$, it is

$$U_{i-1}(\mathrm{s}) = \left(k_v s + k_p\right)E_{i-1}(\mathrm{s}) + k_w s X_{i-1}(\mathrm{s}) - k_w V_{wi-1}(\mathrm{s}) + k_w(i-2)\delta. \tag{21}$$

The position of vehicle $i-1$ and vehicle i is respectively represented in Laplace domain as

$$X_{i-1}(\mathrm{s}) = P(\mathrm{s})U_{i-1}(\mathrm{s}) - \frac{i-2}{s}\delta,\ X_i(\mathrm{s}) = P(\mathrm{s})U_i(\mathrm{s}) - \frac{i-1}{s}\delta, \tag{22}$$

According to the definition of spacing error in Laplace domain by Eq. (12) and combining Eqs. (20)–(22), the spacing error for vehicle i is formulated as follows:

$$E_i(\mathrm{s}) = G_{ew}(\mathrm{s})E_{i-1}(\mathrm{s}) - G_{dw1}(\mathrm{s})V_{wi-1}(\mathrm{s}) + G_{dw2}(\mathrm{s})V_{wi}(\mathrm{s}) + G_{sw}(\delta - \delta_d), \tag{23}$$

With $G_{ew}(\mathrm{s}) = \frac{k_v s + k_p}{\mathrm{Den}(s)}$, $G_{dw1}(\mathrm{s}) = \frac{k_w}{\mathrm{Den}(s)}$, $G_{dw2}(\mathrm{s}) = \frac{(1 + h_w s)k_w}{\mathrm{Den}(s)}$, $G_{sw}(\mathrm{s}) = \frac{\tau s^2 + s - k_w}{\mathrm{Den}(s)}$,
and $\mathrm{Den}(s) = \tau s^3 + (1 + k_v h_w)s^2 + \left(k_p h_w + k_v - k_w\right)s + k_p$, where the transfer function $G_{ew}(\mathrm{s})$, $G_{dw1}(\mathrm{s})$, $G_{dw2}(\mathrm{s})$, and $G_{sw}(\mathrm{s})$ respectively depicts the effect on spacing error from the spacing error of the immediate predecessor, the wheel speed disturbance of the immediate predecessor, the wheel speed disturbance of the vehicle itself, and the initial spacing between two adjacent vehicles. Since the final value of the step response of $G_{sw}(\mathrm{s})$ is not zero but tends to $-k_w/k_p$, the effect of the initial spacing on the spacing error can be mitigated if the gain k_w is designed much smaller than k_p. The final values of the step response of $G_{dw1}(\mathrm{s})$ and $G_{dw2}(\mathrm{s})$ are equal and both tend to k_w/k_p, so spacing error can be reduced by coordinating the wheel speed of the two adjacent vehicles. Hence, we are provided an alternative to reduce the spacing error when involving wheel speed in the upper platooning control law.

Since the string stability is governed by the spacing error propagating transfer function $G_{ew}(\mathrm{s})$, it is primarily necessary to guarantee the maximum magnitude of $G_{ew}(\mathrm{s})$ to be less than 1 for all frequencies, that is

$$\|G_{ew}(\mathrm{s})\|_{\infty} := \sup_{\omega} |G_{ew}(j\omega)| < 1. \tag{24}$$

Following this criterion, the design parameters need to satisfy

$$\begin{cases} \frac{(1+k_v h_w)^2}{2(k_p h_w + k_v - k_w)} \geq \tau \\ \frac{1}{2k_s}\left(k_p^2 h_w^2 - 2k_p k_w h_w + k_w^2 - 2k_p\right) > k_v \end{cases}, \quad \begin{cases} \frac{(1+k_v h_w)^2}{2(k_p h_w + k_v - k_w)} < \tau \\ 4\tau^2\left(k_v^2 + 2k_p k_v h_w + 2k_p\right) \\ -4\tau(1+k_v h_w)^2\left(k_v + k_p h_w - k_w\right) + (1+k_v h_w)^4 < 0 \end{cases} \tag{25}$$

For the left string stability condition of (25), the combination of $\tau = (1+k_v h_w)^2/2(k_p h_w + k_v - k_w)$ and $k_v = \left(k_p^2 h_w^2 - 2k_p k_w h_w + k_w^2 - 2k_p\right)/2k_w$ corresponds to the critical string stability situation where the platoon is near string instability. Based on these two equalities, take $h_w = 1$ and then the left condition of (25) can be interpreted as the illustration of Fig. 11, where the left and right plots respectively correspond to the second and the left inequality of (25). Hence, the platooning control gains can be designed straightforwardly with the help of this figure. On the contrary, there are very limited parameters which can satisfy the right condition of (25), so left condition of (25) is the primary consideration here.

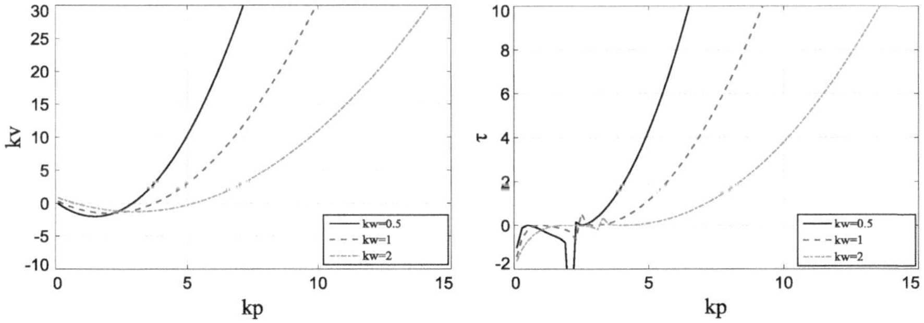

Fig. 11 Design of the platooning control parameters for string stability according to the left condition of (25). For a certain curve in the left plot, it is suggested to choose the value of k_v below the curve, and then the calculated value of τ on the corresponding curve of the right plot with the chosen value of k_v should be not less than the actual time constant of the vehicle for string stability

The proposed platooning control law is evaluated on the same eight-vehicle platoon as used in previous sections on a slippery road with the adhesion coefficient $\mu = 0.4$. We still take the nominal values for the parameters, that is $\tau = 0.3$, $h_w = 1$, $k_p = 10$, $k_v = 10$, and the initial speed is also 25 m/s. The parameter values are set to be identical to those used in Sect. 4.1 to facilitate the comparison. Note that the string stability can be theoretically guaranteed with these nominal parameters.

The response of spacing error and vehicle position with the proposed novel platooning control law is presented in Fig. 12. Compared with the spacing error response without feedforward of wheel speed illustrated in (Fig. 8a), the transient overshoot is greatly attenuated and all the spacing error can converge much more rapidly to zero (see Fig. 12a) as expected with our proposed platooning control law. See from the vehicle position response Fig. 12b, with this control law the spacings between two adjacent vehicles are more uniform comparing with the case without feedforward compensation given in Fig. 8b. Especially, see Fig. 13, there is no obvious large wheel slip propagation with the proposed feedforward compensation control law comparing with Fig. 9. This is very important since the propagation of large wheel slip would lead the spacing error to diverging or even cause collisions. Hence, the platooning control law is more robust to the disturbance and relatively large platooning size is allowed.

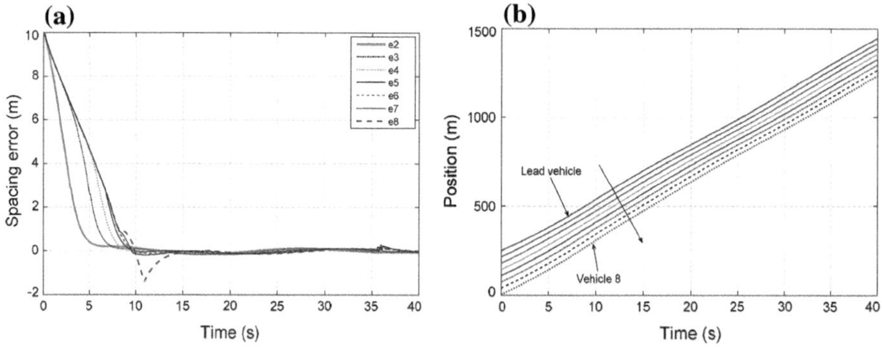

Fig. 12 Response of spacing error and vehicle position with feedforward and regulating of wheel speed on slippery road ($\mu = 0.4$). **a** Spacing error; **b** vehicle position

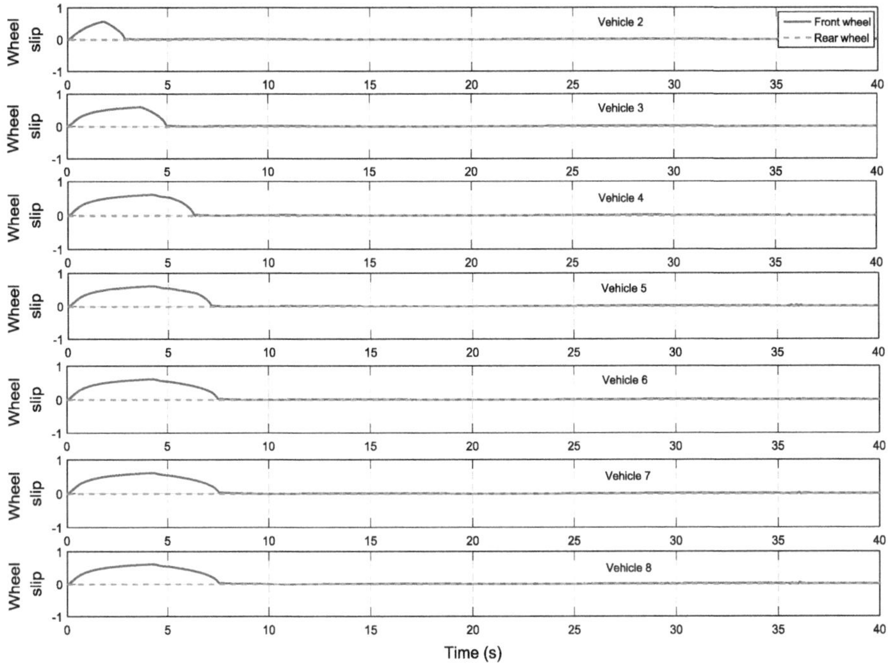

Fig. 13 Variations of wheel slip with feedforward and regulating of wheel speed on slippery road ($\mu = 0.4$)

5 Conclusions

Vehicular platooning control is a typical instance of the control of a cascaded system. In order to mitigate the effect of vehicle dynamics, in this work the wheel rotational speed is viewed as an input disturbance to compensate the effect of individual vehicle dynamics. Simulations based on nonlinear vehicle models indicated the effectiveness of the proposed control scheme. It is worth noting that the platooning control policy is rather important for the vehicular platooning. If the commanded acceleration of a vehicle only depends on the information from its immediate predecessor, once the dynamics of a vehicle drastically decays the tracking performance of the following vehicles is doomed to get worse and worse toward the upstream direction of the platoon. Fortunately, the platooning control law proposed in this work which is realized by feedforward of wheel speed can remarkably compensate the deficiency of this kind of platooning control policy. Nevertheless, a platooning control policy in which the desired motion of a vehicle does not only depend on its immediate predecessor is expected to improve the robust property of the platooning. In the future work, coordinated control of individual vehicle dynamics and the platooning dynamics would be further investigated.

References

1. Rajamani R (2011) Vehicle dynamics and control. Springer Science & Business Media, New York
2. Swaroop D, Hedrick J, Chien C, Ioannou P (1994) A comparison of spacing and headway control laws for automatically controlled vehicles. Veh Syst Dyn 23:597–625
3. Ellis M, Gargoloff J, Sengupta R (2015) Aerodynamic drag and engine cooling effects on class 8 trucks in platooning configurations. SAE Int J Commercial Veh (2015-01-2896) 8:732–739
4. Alam A, Mårtensson J, Johansson K (2015) Experimental evaluation of decentralized cooperative cruise control for heavy-duty vehicle platooning. Control Eng Pract 38:11–25
5. Alam A, Besselink B, Turri V, Mårtensson J, Johansson KH (2015) Heavy-duty vehicle platooning for sustainable freight transportation: a cooperative method to enhance safety and efficiency. IEEE Control Syst 35:34–56
6. Ploeg J, Van De Wouw N, Nijmeijer H (2014) Lp string stability of cascaded systems: application to vehicle platooning. IEEE Trans Control Syst Technol 22:786–793
7. Shaw E, Hedrick JK (2007) String stability analysis for heterogeneous vehicle strings. In: Proceedings of 2007 American control conference, pp 3118–3125
8. Orosz G (2016) Connected cruise control: modelling, delay effects, and nonlinear behavior. Veh Syst Dyn 54:1–30
9. Lestas I, Vinnicombe G (2007) Scalability in heterogeneous vehicle platoons. In: Proceedings of 2007 American control conference, pp 4678–4683
10. Yanakiev D, Kanellakopoulos I (1996) A simplified framework for string stability analysis in AHS. In: Proceedings of the 13th IFAC world congress, 1996, pp 177–182
11. Liang CY, Peng H (1999) Optimal adaptive cruise control with guaranteed string stability. Veh Syst Dyn 32:313–330
12. Seiler P, Pant A, Hedrick K (2004) Disturbance propagation in vehicle strings. IEEE Trans Autom Control 49:1835–1842
13. Jovanovic MR, Bamieh B (2005) On the ill-posedness of certain vehicular platoon control problems. IEEE Trans Autom Control 50:1307–1321
14. Talukdar S (2015) Control of vehicular platoons using nearest neighbor interactions. In: SAE 2015 world congress & exhibition. Paper No. 2015-01-0299
15. Bakker E, Pacejka H B, Lidner L (1989) A new tire model with an application in vehicle dynamics studies. In: Autotechnologies conference and exposition. Paper No. 890087

Analysis on Energy Flow of Electric Vehicle

Yanxin Nie, Tian Yang, Dongyue Huang, Tao Li and Peiliang Yu

1 Introduction

Only improving driving range has not been satisfied with technology trend. Promotion of energy utilization is becoming the focus. Besides, ambient temperature and drive loads have a significant impact on electric vehicle, which should be considered in the process of electric vehicle evaluation. For example, reduction rate of the range in cold winter has to be controlled in an acceptable scope.

The energy flow graph can show energy utilization and dissipation clearly, so building an accurate energy flow is the key to understand energy utilization of a vehicle and propose optimizations. In order to obtain vehicle's overall energy consumption characteristics, a Smart ED was tested under different temperatures and driving cycles. After the test, main systems and components were monitored and quantitatively analyzed. Based on energy flow graph, evaluation indexes were proposed and calculated. Then, energy flow graphs in different test conditions were analyzed, and the optimization suggestions were put forward.

2 Build Energy Flow Model

The vehicle driving energy flow model was built based on vehicle force analysis and electrical schematic. The chassis dynamometer is used to emulate the loads that vehicle experiences on the road. The propulsion power at the wheel overcomes change in momentum and road load force caused by wind resistance, rolling

Y. Nie · T. Yang (✉) · D. Huang · T. Li · P. Yu
Automotive Engineering Research Institute, China Automotive Technology & Research Center, Tianjin 300300, China
e-mail: yangtian@catarc.ac.cn

Society of Automotive Engineers of China (SAE-China) (ed.), *Proceedings of the 19th Asia Pacific Automotive Engineering Conference & SAE-China Congress 2017: Selected Papers*, Lecture Notes in Electrical Engineering 486, https://doi.org/10.1007/978-981-10-8506-2_59

resistance, friction losses, and other vehicle level losses [1]. So demanded power at the wheel when vehicle is driving can be mathematically described as follows:

$$P_{\text{propulsion}} = \left(m \times \frac{\partial(V)}{\partial t} + F_{\text{roadload}}\right) \times V$$

where m is the vehicle test mass; V is the vehicle speed; F_{roadload} is the road load force. The road load force can be described as follows:

$$F_{\text{roadload}} = A + BV + CV^2$$

where A, B, C are road load coefficients.

The Smart ED has a high integration, and PDU, DCDC, MCU are integrated into a power electronic control unit. In driving part, electric energy converts into mechanical energy which drives the vehicle. Mechanical energy overcomes rolling resistance and air resistance, eventually converts into kinetic energy [2]. In braking process, through mechanical braking and electrical braking, kinetic energy converts into heat energy and electric energy. Most of electric energy flows into battery, and the others flow into air conditioner and accessories. Figure 1 shows the energy flow of Smart ED. The energy flow values were obtained from time integration for power under the driving cycles.

3 Test Method

Hioki's power analyzer was used to acquiring data of energy flow process. According to energy flow model, the acquisition scheme was developed [3]. Under driving conditions, power analyzer took in charge of monitoring current and voltage

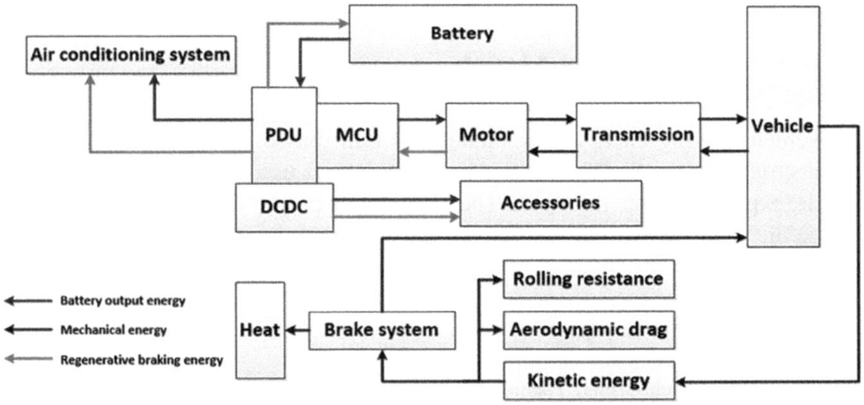

Fig. 1 Test vehicle energy flow model

of motor, high-voltage battery, PTC, AC, DCDC, 12 V battery. Under charging condition, power analyzer took in charge of monitoring current and voltage of high-voltage battery, OBC, DCDC, 12 V battery [4]. The current clamps and voltage taps of AC and PTC remained unchanged for convenience. Figures 2 and 3 present the data acquisition scheme.

The vehicle was tested on a 4WD chassis dynamometer test cell with a thermal chamber of CATRAC. Choose NEDC as test driving cycle. NEDC includes city condition and suburb condition, which can reflect the real operation of vehicle. Furthermore, NEDC has a relatively high quantity of braking processes, which is suitable as a driving and braking energy flow test cycle.

Vehicle condition, ambient temperature, and other requirements referred to the Electric vehicle-Energy consumption and range-Test procedures (GB/

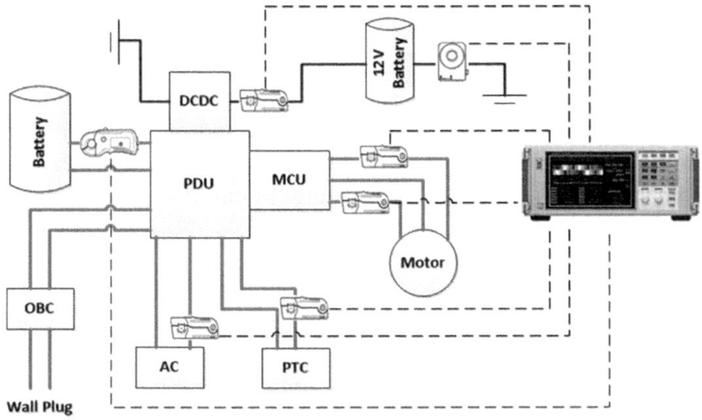

Fig. 2 Data acquisition scheme under driving condition

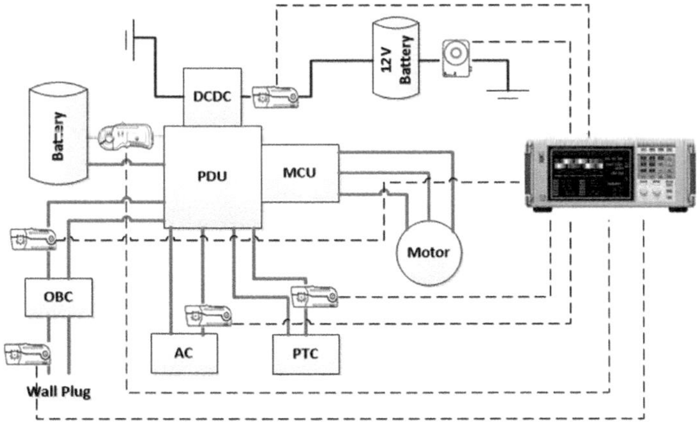

Fig. 3 Data acquisition scheme under charging condition

T18386-2005). The tolerance of velocity is ±2 km/h, and test procedures are as follows:

- In order to set the dynamometer parameters, coast-down tests were carried out. Firstly, testing coast downs at the track, and deriving road load coefficients. Secondly, matching the coast downs on the dynamometer. Figure 4 illustrates the driving resistance curve. From the curve, we derived the parameters.
- Before the test, the vehicle was soaked at test temperature for 8 h. Then, test equipment was connected to the vehicle and debugged. Ensure that vehicle and equipment maintain appropriate state.
- When in normal temperature tests, the vehicle climate control was turned off, and the driver's window left open, for all tests at an ambient temperature of 25 °C. When in low temperature tests, the climate control was turned on and set to 25 °C in automatic mode at ambient test temperatures of −10 or −20 °C, and vehicle windows were closed. When in high temperature tests, the climate control was set to 25 °C in automatic mode at ambient test temperatures of 40 °C. All the vehicle windows were closed. During all 40 °C tests, the solar emulation lamps were turned on and set to the level of 850 W/m^2.
- When the highest velocity could not reach 70 km/h or meet the trace, end the test. Then, the equipment was changed as shown in Fig. 3. Charging test begun in an hour at the same ambient temperature. Figure 5 shows the details of tests.

4 Test Analysis

4.1 Energy Flow Analysis on Standard Test Drive Cycles

It is important to choose reasonable evaluation indexes. Range, capacity consumption per kilometer, energy consumption per kilometer are commonly used as evaluation indexes for energy consumption and utilization of electric vehicle.

Fig. 4 Driving resistance curve of Smart ED

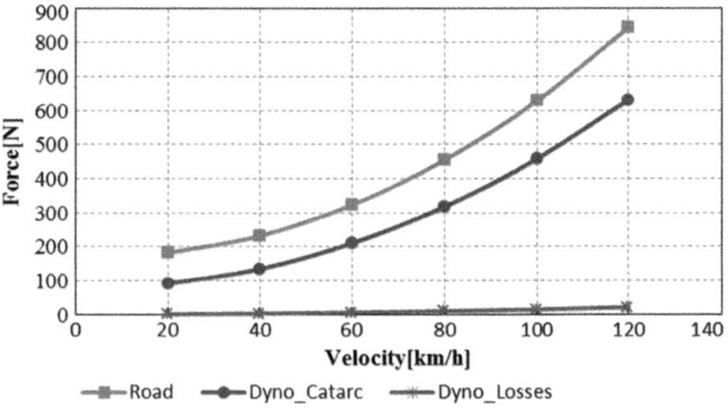

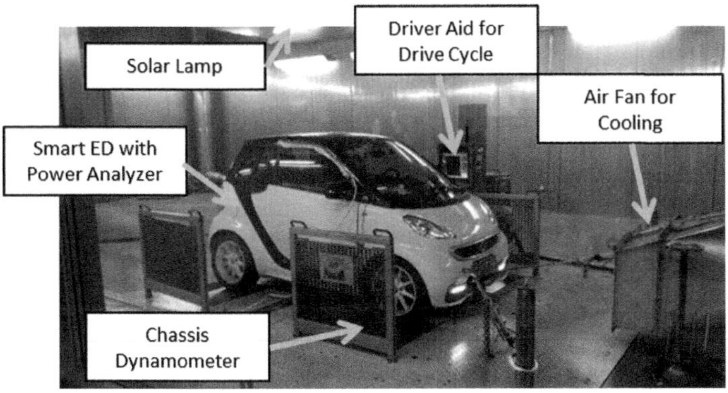

Fig. 5 Test vehicle in 4WD chassis dynamometer test cell with a thermal chamber

But range and capacity consumption per kilometer depend on battery capacity and voltage platform, which cannot apply to assess vehicle economy accurately. Energy consumption per kilometer eliminates the impacts of battery capacity and voltage platform, which can be divided into alternating current energy consumption and direct current energy consumption [5]. The AC energy consumption per kilometer represents energy consumption from power grid. The DC energy consumption per kilometer represents energy consumption from battery. Table 1 shows evaluation indexes and values.

Figure 6 illustrates energy distribution under 25 °C test temperature. Wall plug delivered energy as overall trip energy input. Fully charged battery depleted the

Table 1 Evaluation indexes of energy flow

Evaluation indexes	Value
Energy consumption [DC kWh/100 km]	13.16
Energy consumption [AC kWh/100 km]	15.61
Overall trip energy utilization	64.8%

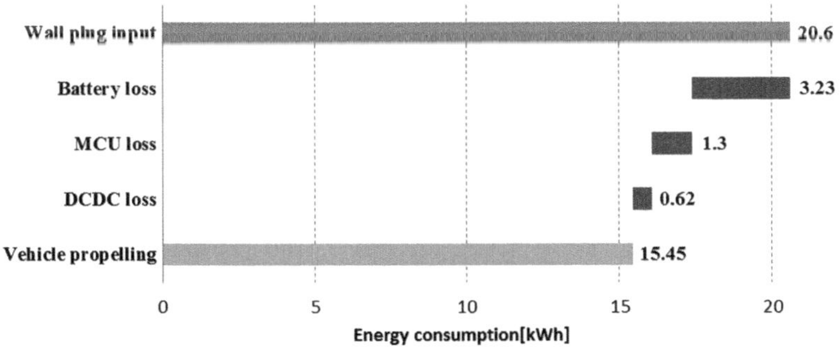

Fig. 6 Energy distribution under test cell temperature of 25 °C

energy as vehicle powertrain energy input. After a series of losses, energy was used to propel the vehicle. In this test method, motor loss was included in vehicle propelling energy. So when calculating powertrain efficiency, motor efficiency should be considered.

Above-mentioned evaluation indexes are on vehicle level, which cannot be used to describe energy consumption characteristics of system or component. In order to get system- or component-level energy flow, an analysis method was proposed.

The vehicle system was divided into power supply module, energy distribution and conversion module, energy conversion and transmission module. Energy supply module is high-voltage battery that provides DC power for powertrain. Energy distribution and conversion module includes PDU, MCU, DCDC, which are integrated together. This module is responsible for distributing energy from high-voltage battery. The majority of the energy is converted into three-phase alternating current through MCU. Then, energy inputs to motor. Other energy is converted into low voltage through DCDC which is used for accessories. Energy conversion and transmission module function is converting electrical energy into mechanical energy that propels vehicle.

Energy flow was divided into two parts for the analysis. When power at the wheel was positive, energy flow accessed driving part. When power at the wheel was negative, energy flow accessed braking part. Figure 7 illustrates the system- and component-level energy flow. Figures 8 and 9 show the important data acquired during the NEDC for energy flow analysis, including power transient variation and energy accumulation of road load, inertia, motor, and battery.

Based on the above method, the sub-evaluation indexes were calculated for one NEDC. Table 2 presents the sub-evaluation indexes of energy flow.

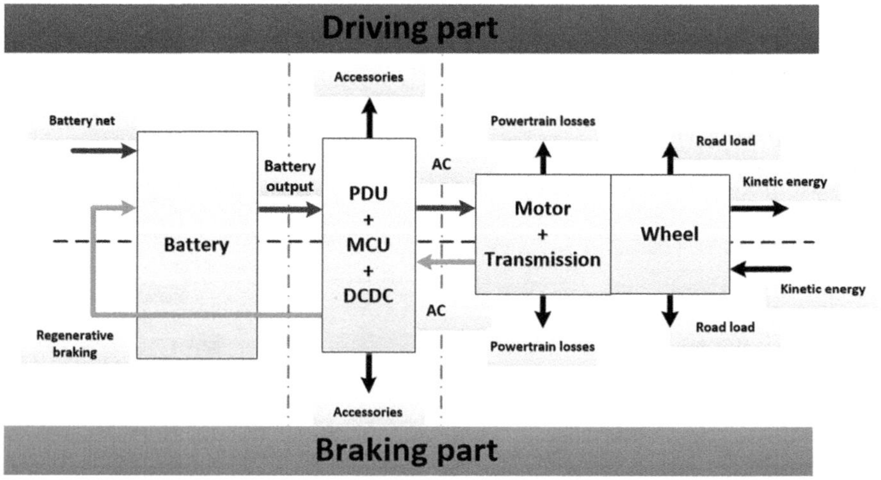

Fig. 7 Energy flow at system and component level

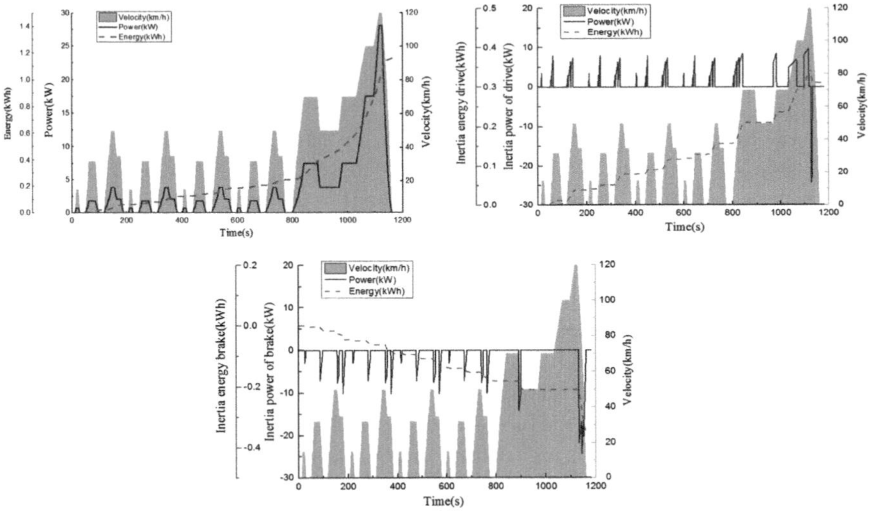

Fig. 8 Power transient variation and energy accumulation of road load and inertia

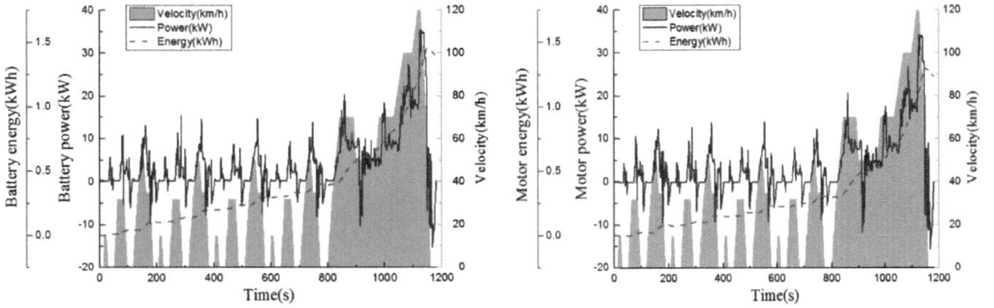

Fig. 9 Power transient variation and energy accumulation of battery and motor

Table 2 Sub-evaluation indexes of energy flow

Sub-evaluation indexes	Value (%)
Energy supply module efficiency	88.6
Energy distribution and conversion module efficiency	92.6
Energy conversion and transmission module efficiency	87.7
Regenerative braking contribution rate	13.6
Battery pack utilization	80.4
Charger efficiency	88.6

4.2 *Energy Flow Analysis Under Nonstandard Test Conditions*

Driving conditions have a significant influence on energy flow. Driving conditions mainly consist of ambient temperature and driving cycle. Quantitative analysis was executed as follows. Figure 10 shows ambient temperature and driving cycle impact on energy consumption. In low ambient temperature, the energy consumption is more than twice that observed at 25 °C in NEDC. High temperature increases the energy consumption by 50%. When in ECE cycle, the trend is more obvious, especially in −10 °C, the energy consumption is three times that observed at 25 °C. So we can confirm that the most severe increment of energy consumption occurs in a traffic jam on a very cold winter.

There are many reasons for this phenomenon. Figure 11 presents AC or heater energy consumption in a series of driving conditions. The additional load of the air-conditioning system or heater has a largest impact on energy consumption. It is observed that the heater power doubles the electric compressor.

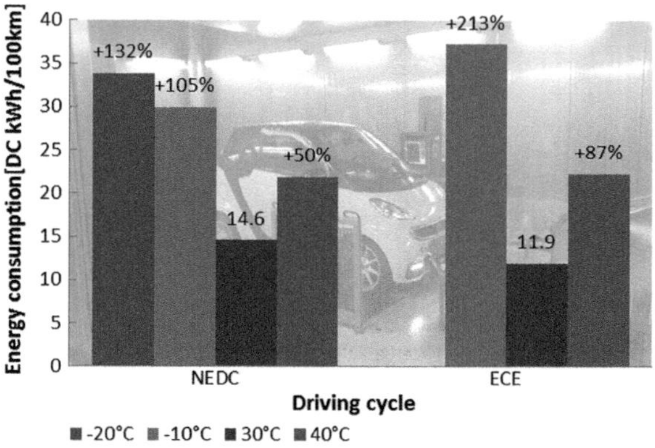

Fig. 10 Ambient temperature impact on energy consumption under NEDC and ECE

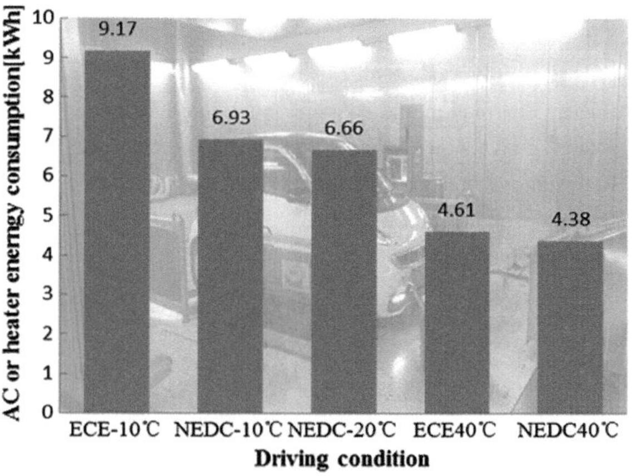

Fig. 11 AC or heater energy consumption in a series of driving conditions

Accessory load is a less significant fraction of the total energy consumption. Figure 12 quantifies the impact of test cycles and ambient temperatures. Longtime test cycles yield higher energy consumption in the same ambient temperature. High ambient temperature conditions yield higher energy consumption, which means the accessory load used to heat dissipation is much more than warming. According to the proportion, it concludes that accessory load has a more pronounce effect during less aggressive, lower-speed driving.

Changing ambient temperature has a slight impact on regenerative contribution rate. As shown in Fig. 13, the lower temperature yields lower regenerative contribution rate in the same test cycle. It appears that temperature limits the amount of regenerative braking power it can absorb at low or high temperature. In the same ambient temperature, the regenerative braking energy absorbed on city-type driving is almost twice that absorbed in comprehensive test cycle.

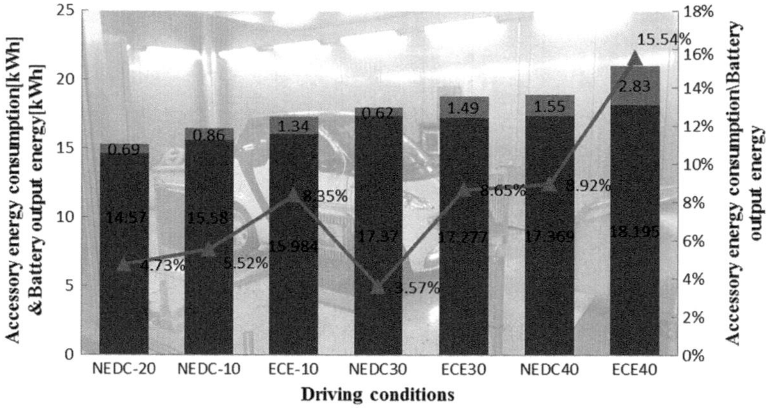

Fig. 12 Accessory energy consumption for different test cycles and ambient temperatures

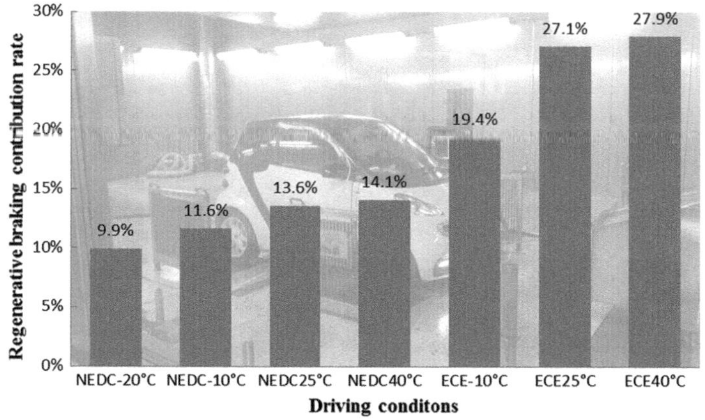

Fig. 13 Regenerative braking contribution rate

Ambient temperature has a significant influence on battery capacity. In high temperature, as warmer environment can improve activity of reactants, actual capacity is higher than rated. But in low temperature, capacity attenuates sharply, which is one of the main reasons electric vehicle has an obvious decline on a cold winter. Figure 14 shows the amount of battery output energy on different ambient temperatures. Additionally, resistances were derived and plotted, as shown in Fig. 15. The slope of the polarization curves is the resistance of battery at different ambient temperatures. Using first cycle data for plotting curves as vehicle cooling or heating system may active, which may cause the characteristics of battery are

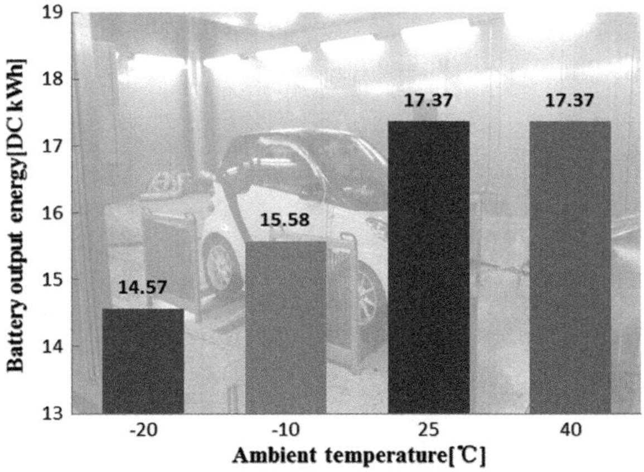

Fig. 14 Ambient temperature impact on battery output energy

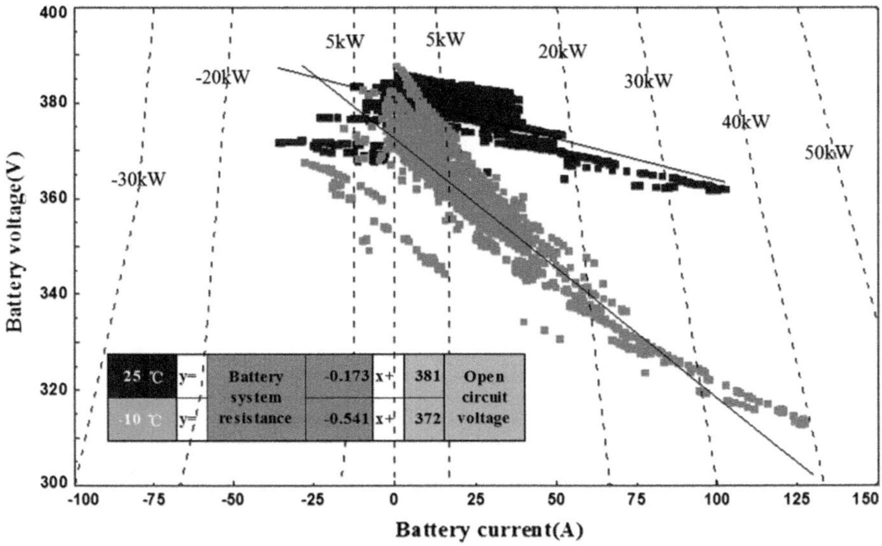

Fig. 15 Resistance curves at different ambient temperatures

changed in specific temperature. It appears that the resistance of battery increases obviously in low ambient temperature.

Charging efficiency was considered in energy flow analysis. The test results were shown in Fig. 16. It appears that properly lower ambient temperature yields higher charging efficiency. But at −20 °C, low temperature influenced components' working state, and charging efficiency decreased. This exercise also shows that high ambient temperature makes heat dissipation worse.

Through above analysis, evaluation indexes were calculated as shown in Table 3, and sub-evaluation indexes were calculated as shown in Table 4.

Through the energy flow analysis method as shown in Fig. 7, energy consumption was decomposed [6]. Tables 5 and 6 show decomposition of energy consumption in propelling part and braking part. The percentage in tables is defined as the amount of energy consumption that divides the amount of net battery energy output for completing the drive cycle. From the Tables 5 and 6, it concludes that the energy used to overcome road forces holds the largest part at room temperature in propelling part. Yet with the test temperature changed, the proportion of overcome road force energy decreases. Particularly at low temperature, accessories' consumption energy holds 44.4%, larger than all parts.

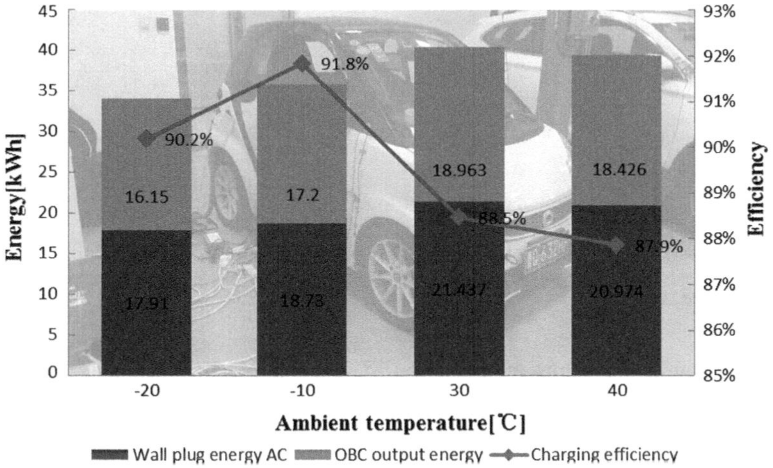

Fig. 16 Ambient temperature impact on charging efficiency

Table 3 Evaluation indexes at different temperatures

Evaluation indexes	−20 °C	−10 °C	40 °C
Energy consumption [DC kWh/100 km]	27.54	24.89	17.87
Energy consumption [AC kWh/100 km]	33.85	29.97	21.86
Overall trip energy utilization	37.2%	40.1%	48.3%

Table 4 Sub-evaluation indexes at different temperatures

Sub-evaluation indexes	−20 °C (%)	−10 °C (%)	40 °C (%)
Energy supply module efficiency	74.3	79.5	88.6
Energy distribution and conversion module efficiency	96.3	96.8	94.9
Energy conversion and transmission module efficiency	84.3	85.5	86.8
Regenerative braking contribution rate	9.9	11.6	14.1
Battery pack utilization	67.5	72.1	80.4
Charger efficiency	90.2	91.8	87.9

Table 5 Decomposition of energy consumption in propelling part

Energy consumption items	−20 °C (%)	25 °C (%)	40 °C (%)
Overcome road load forces	34.0	73.1	53.5
Overcome inertial forces	10.3	22.2	16.2
Accessories	44.4	8.7	31.9
Transmission loss	8.2	6.0	4.4

Table 6 Decomposition of energy consumption in braking part

Energy consumption items	40 °C	25 °C	-20 °C
Overcome road load forces	6.9	9.4	4.4
Overcome inertial forces	−18.0	−24.5	−11.4
Accessories	5.1	2.4	4.2
Transmission loss	−5.8	−12.3	−1.1

5 Suggestions

Energy flow analysis points the direction of vehicle optimization. The main aspects were presented below:

1. The large power of PTC obviously reduces the energy efficiency and produces severe range anxiety. Use heat pump air-conditioning instead of PTC to reduce the depletion rate of range. By calculation, it shows that range increases 3.5 km per 0.5 kW of air-conditioning power declines.
2. In cold winter, range has a sharp decline, and battery pack utilization is half that observed at room temperature. So on the one hand, battery low-temperature properties should be promoted. On the other hand, vehicle cooling circuit should be optimized. Only by ensuring the activity of reactants can enhance battery pack utilization to support the range.

3. This exercise concludes that regenerative braking contribution rate is less than 15%, which has a potential to further improve. When regenerative braking contribution rate is reaching at 15%, the 150 km range is approached.
4. Efficiency of energy distribution and conversion module is high enough in all ambient temperatures, but the DCDC efficiency remains low. Besides, the ambient temperature has a negative impact on DCDC efficiency, especially in charging condition. So optimization of DCDC needs to be put on the agenda.

6 Summary

This paper presents a methodology on how to analyze energy flow of an electric vehicle. The methodology has considered test difficulty caused by trend of automotive integration. Employing this approach, we can ensure high precision and low testing difficulty. Energy flow can analyze vehicle efficiency in different levels, and a series of evaluation indexes were reasonably selected and calculated which can appraise energy utilization of vehicle clearly. These are very important to build advanced vehicle evaluation system. In addition, energy flow was divided into driving part and braking part for the analysis, by which energy consumption of every aspect was shown fully. In view of changeable weather and traffic of vehicle in daily life, energy flow on different ambient temperatures and driving cycles was analyzed. Through above energy flow analysis, energy consumption characteristics of systems and components were confirmed, and optimization direction comes naturally.

References

1. Yu Z (2008) Automotive theory, 4th edn. China Machine Press, Beijing
2. Chu L (2014) Research on brake energy regenerative evaluation and test method of pure electric vehicle. Huazhong Univ Sci Technol 42(1):18–22
3. Alley R (2012) VTool: a method for predicting and understanding the energy flow and losses in advanced vehicle powertrains. MS thesis, Virginia Tech
4. Argonne National Laboratory (2013) Ambient temperature (20°F, 72°F and 95°F) impact on fuel and energy consumption for several conventional vehicles, hybrid and plug-in hybrid electric vehicles and battery electric vehicle. SAE International
5. Ehsani M, Gao Y, Emadi A (2009) Modern electric, hybrid electric, and fuel cell vehicles: fundamentals, theory, and design. CRC press
6. Wang Z (2004) Analysis of energy consumption distribution and factors of influence in electric vehicles. Trans Beijing Inst Technol 24(4):306–310

Gearbox Noise Optimization of Electric Bus Based on Gear Modification

Yong Chen, Miao Yu, Hai Liu, Koucheng Zuo, Zhanjiang Li, Yuan Yang and Zhongzhe Li

1 Introduction

With the improvement of the electric vehicles' motor and battery performance, the pure electric bus becomes more popular because of its environmental protection and energy-saving advantages. Because the dynamic performance of the electric bus is becoming more and more perfect, the performance of NVH becomes people's focus of attention. Although there is no internal-combustion engine noise in the pure electric bus, it also loses the masking effect of the internal-combustion engine [1]. Thus, the new NVH problem seriously affects the riding comfort.

The gearbox is not an absolutely rigid body in real work, and its parts have the processing error, the assembling error, and the deformation. All these factors cause the NVH problem of the power train [2]. According to the cause and the transmission route of the gearbox noise, at present, through the micro-modification of gears, which reduces the gear dislocation amounts and the meshing error, thereby reducing the amplitude and the fluctuation of the transmission error, avoiding the occurrence of gear noise from the source is the basic idea to solve the problem. Many scholars have studied about the theoretical research and the test on the gear modification of the gearbox. At home, Wang [3] who comes from Chongqing University combined the theoretical formula with the Romax software to calculate the gear modification parameters and optimized the parameters of the filtering gear reducer to improve the transmission performance; Chen [4] who comes from

Y. Chen · M. Yu · H. Liu (✉) · Y. Yang · Z. Li
School of Mechanical Engineering, Hebei University of Technology, Tianjin 300130, China
e-mail: 750409101@qq.com

K. Zuo
Romax Technology (Beijing) Ltd., Beijing 100027, China

Z. Li
Nanjing YueBoo Power System Co. Ltd., Nanjing 210019, China

Society of Automotive Engineers of China (SAE-China) (ed.), *Proceedings of the 19th Asia Pacific Automotive Engineering Conference & SAE-China Congress 2017: Selected Papers*, Lecture Notes in Electrical Engineering 486, https://doi.org/10.1007/978-981-10-8506-2_60

Central South University built a single gear pair dynamic model, which considered the meshing stiffness and the transmission error, and then analyzed the influence of different modification parameters for gear pair dynamic motion, thus determining the appropriate modification parameters. In foreign countries, Lin [5] did the computer simulation under different working conditions and determined the best profile modification program to minimize the dynamic load of the gear system; Kahraman [6] set up the nonlinear gear contact finite element model, took the bending stress as the optimization objective, and then designed the gear profile modification. At the same time, the contrast experiment of the profile modification between before and after was carried out. The results show that the gear modification can prolong the service life and reduce the vibration noise of the gear.

The transmission system of the electric bus which has higher load torque is different from the traditional passenger vehicle, so the deformation of the gear is more serious. The quantities and the law of gear modification are different from the gear of traditional passenger vehicles and the transmission gear used in laboratory. In this study, the model is built and the corresponding load spectrum is set. The influence law of the gear modification parameters of the pure electric bus gearbox is summarized by orthogonal experimental method, and then the plans according to the law are made so as to improve the noise problem.

2 Gear Modification and Optimization

2.1 Transmission Error

Due to the manufacturing error, the assembly error, and other reasons, the passive gear will be along the meshing line direction ahead or behind the theoretical position in the actual gear meshing process, and the displacement of the offset is the transmission error [7, 8] (TE), such as the formula (1). The transmission error is the excitation source of vibration and noise of the gear system. The gear pair is affected by the dynamic meshing excitation and the generated whine noise. Especially the tooth numbers of the gearbox constant mesh gears are 30 and 46. Because they have common divisor, it is easy to wear and increase the transmission error.

$$\mathrm{TE} = r_{b2}\theta_2 - r_{b1}\theta_1 \tag{1}$$

θ_1 is the theoretical rotation angle of driving gear; θ_2 is the actual rotation angle of driven gear; r_{b1}, r_{b2} are the base circle radius of driving and driven gears.

2.2 Gear Micro-Modification

The gear micro-modification is to remove the part of the tooth surface material during grinding, and it reduces the gear misalignment due to the system error and the tooth deformation. Thus, it makes the gear contact deformation under load and keeps tooth surface contact stress distribution uniform for avoiding the phenomenon of bias load [9]. Figure 1 is a three-dimensional sketch map of four kinds of the micro-modification.

Setting up the gear modification parameters as the optimization variables, the gear strength as the constraint condition, the transmission error as the main optimization goal, and considering the contact stress and the tooth surface load nephogram synthetically, then the multi-objective optimization is realized. Ultimately, the purpose that reduces the transmission error and the internal dynamic excitation is achieved, and then the problem of gear whine is improved. The technical route is shown in Fig. 2.

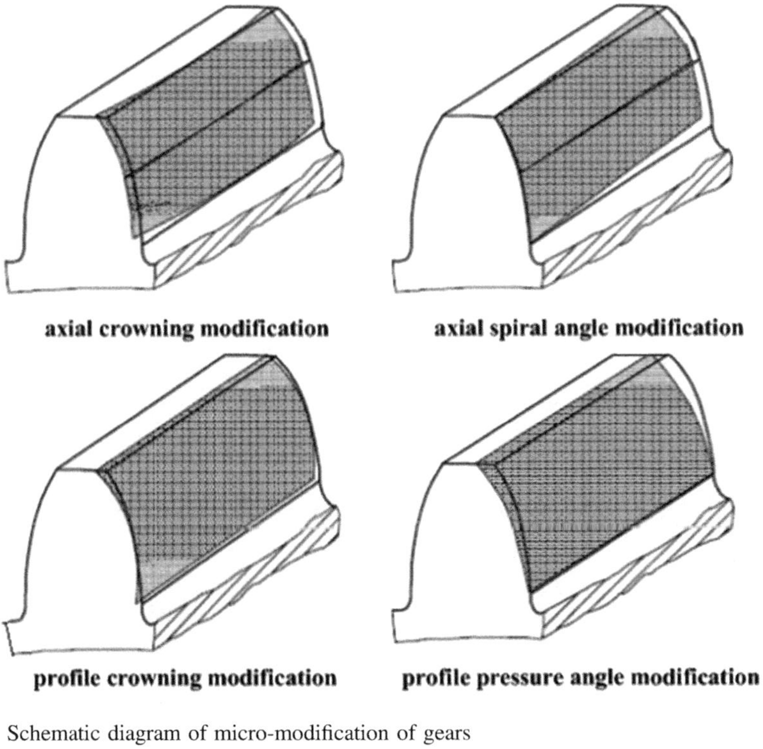

Fig. 1 Schematic diagram of micro-modification of gears

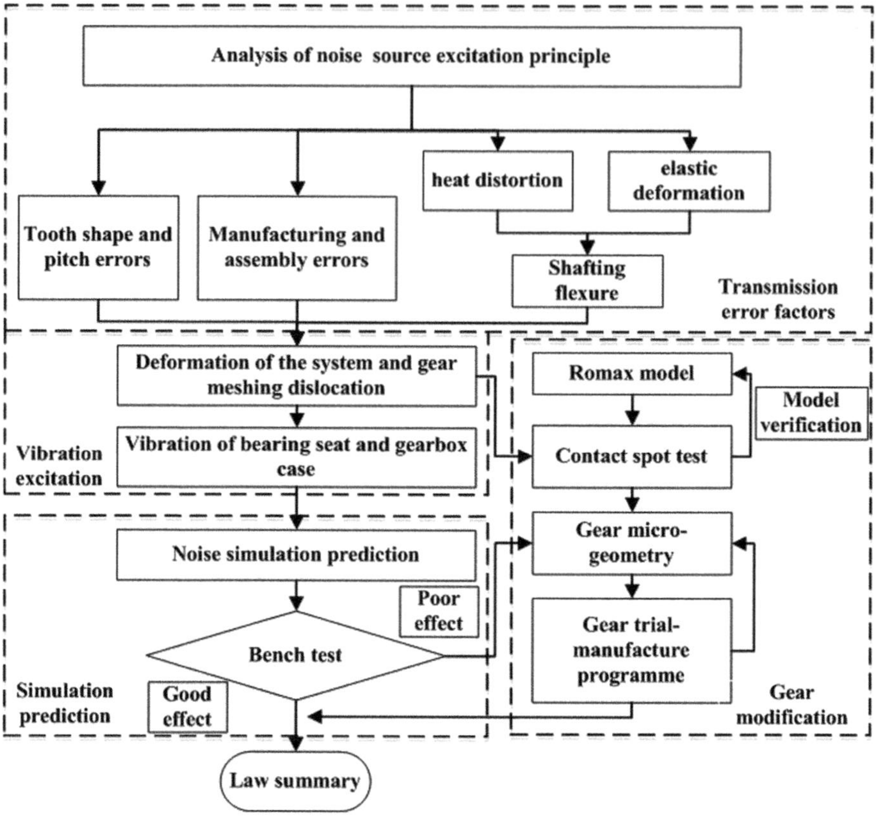

Fig. 2 Technical road map

3 Gearbox Model Establishment and Verification

3.1 Dynamic Model Construction

Romax Designer software is used in parametric modeling of the gearbox, and the model of the gear shaft transmission system is established, as shown in Fig. 3. The gear is driven by the input shaft through the input of the constant gear pair, and the power is transmitted to the output shaft corresponding to the gear through the intermediate shaft gear. At last, it exports by the output shaft.

The constant mesh gear pair of the four gearbox always meshes when it works. The fourth gear is direct gear, and the first, the second, and the third gears all transmit load by the constant mesh gears and the corresponding gear. The gear parameters are as shown in Table 1.

The casing, the bearing, the shaft, and the gear wheel hub of the gearbox all have the microscopic deformation because of the force and the vibration in working condition. In order to simulate the actual working condition of the gearbox, the finite

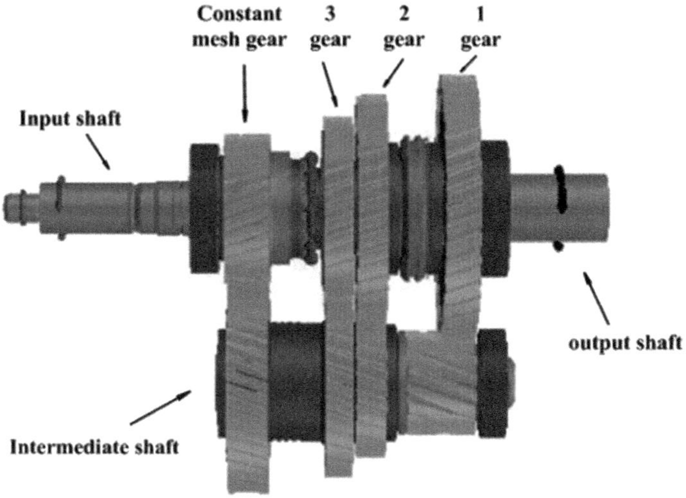

Fig. 3 Model of the gear shaft system of the four gearboxes

Table 1 Parameters of the third gear transmission system

Parameter	Input stage (constant mesh)		Output stage (the third gear)	
	Driving	Passive	Driving	Passive
Teeth number	30	46	31	30
Modulus	3.7		4.5833	
Pressure angle	20°		20°	
Spiral angle	20°		20°	

element mesh (FEM) of the gearbox casing and the gear wheel hub are conducted by using Hypermesh software. And the gear shaft system assembly is connected with the bearing and the bearing seat after meshing by software. The dynamic model of rigid flexible coupling is constructed. In order to simulate the actual working state of the gearbox, the motor rotor is added at the input end by the spline, and then the output shaft is added to the analog load, as shown in Fig. 4.

3.2 *Kinetic Model Validation*

It needs the gear contact spot test to reduce the model's error. The gearbox is installed on the comprehensive performance test bed of gearbox, and both ends are respectively connected with the driving motor and the load motor, and it coated with a layer of the red lead powder (Pb_3O_4) in the corresponding gear surface. The conditions of the bench test are set as the torque 400 N m and the speed 150 r min^{-1}. The model is modified according to the comparison of the experimental spots and the simulation of the shape and the position of the spots (Fig. 5).

Fig. 4 Gearbox rotor-load model of a certain electric bus

Fig. 5 Comparison of meshing speckle test and simulation nephogram

As shown in Fig. 6, the meshing surface of the test gear is basically the same as that of the model tooth surface nephogram after correction, which proves the accuracy of the model meeting the requirements. Then the model can be further modified on the model.

4 Analysis on Influence Law of Gear Modification

4.1 Research on Gear Modification Parameters Based on Orthogonal Experimental Method

The influence of the gear modification parameters on transmission error is different for various gearboxes, and especially the size and the direction of the modification parameters. So it is necessary to clear the influence law and the influence weight of the modification parameters for the transmission error before making the modification scheme. Because of the interaction between the gear shaping factors,

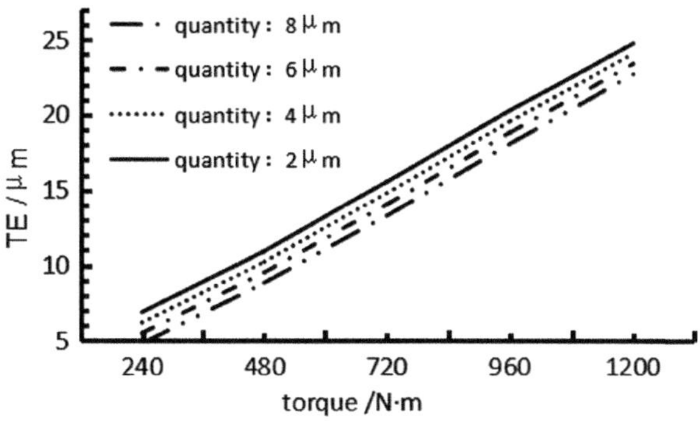

Fig. 6 TE trend chart of the third gear profile crowning modification

a 7-factor and 2-level experiment table is designed first [9], and the A, B, C, D, respectively, represent axial crowning modification (quantity is 2 μm or 4 μm), profile crowning modification (quantity is 6 μm or 8 μm), profile pressure angle modification (quantity is −5 μm or 0 μm), and axial spiral angle modification (quantity is 5 μm or 10 μm). The load spectrum is set in the Romax (i.e., commonly used steady-state conditions: input torque 600 N m and speed 1200 r min^{-1}). The upper and lower limit deviations of tooth surface modification are selected 10 and 90%, respectively, and the results are given in Table 2. (In the table, "1" represents the first modification amount, "2" represents the second modification amount; T_{i1} and T_{i2}, respectively, represent the sum of corresponding column "1" or "2"; K_{i1} and K_{i2}, respectively, represent the average value of corresponding column "1" or "2"; R represents the difference between the corresponding column K_{i1} and K_{i2}.)

Table 2 L_8 (2^7) orthogonal test table

Serial number	1	2	3	4	5	6	7	TE
	A	B	A × B	C	–	A × D	D	
1	1	1	1	1	1	1	1	7.73
2	1	1	1	2	2	2	2	8.10
3	1	2	2	1	1	2	2	6.63
4	1	2	2	2	2	1	1	6.23
5	2	1	2	1	2	1	2	8.37
6	2	1	2	2	1	2	1	8.17
7	2	2	1	1	2	2	1	6.63
8	2	2	1	2	1	1	2	6.90
T_{i1}	28.69	32.37	29.36	29.36	29.43	29.23	28.76	
T_{i2}	30.07	26.39	29.40	29.40	29.33	29.53	30.00	
K_{i1}	7.173	8.093	7.340	7.340	7.358	7.308	7.190	
K_{i2}	7.518	6.598	7.350	7.350	7.333	7.383	7.500	
R	0.345	1.495	0.010	0.010	0.025	0.075	0.310	
Order	B > A > D > A × D > C = A × B							

A conclusion can be got according to the R value: The influence of the transmission error for the gear which from large to small are profile crowning modification, axial crowning modification, axial spiral angle modification and profile pressure angle modification in turn. The influence weight of profile crowning modification is greater than the others. Through the orthogonal experiment method, the influence law of each gear modification factor on the transmission error can be obtained. The optimization scheme is put forward based on this rule.

4.2 Determination of the Modification Amount

The transmission error is calculated respectively after modification of four single factors which include axial crowning modification, profile crowning modification, profile pressure angle modification, and axial spiral angle modification. Taking profile crowning modification of the third gears as an example, the 20, 40, 60, 80, and 100% of the peak torque whose value is 1200 N m are taken as the working conditions. The error is calculated after setting the third gear passive gear profile crowning modification, respectively, as 2/4/6/8 μm, and the data is drawn into weight trends. As shown in Fig. 6, the transmission error decreases with the increasing of the profile crowning modification amounts.

The load torque of the pure electric bus is large, so the four modification factors weight trend chart of the third gear is made in turn, and the electric bus transmission gear modification is got which has the following rules in medium-high torque conditions: (1) The more axial crowning modification, the more of the transmission error; (2) the more profile crowning modification, the less of the transmission error; (3) the more axial spiral angle modification, the transmission error decreases first and then increases; (4) the profile pressure angle modification has little effect on the transmission error.

According to the experience, in order to save the cost of modification, usually it chooses the gear whose radius is smaller in a pair of meshing gears, or unified repairs the passive wheel or the driving wheel. Through the calculation and the comparison, it shows that for the electric bus transmission gear, compared with just a pair of gears, taking two gears the suitable amount of modification at the same time, the maximum contact stress and the transmission error both decrease, and the transmission error decreases significantly, so it is better for vibration and noise reduction, as shown in Table 3.

Table 3 Comparison before and after modification

	Original gear	One gear	Two gear	Unit
Maximum contact stress	1207	1190	1172	MPa
Transmission error	13.92	11.73	9.47	μm

5 Optimization Scheme and Noise Prediction

5.1 Micro-Modification Scheme

The third gears of the pure electric bus gearbox have the 10 μm axial crowning modification originally. The transmission error is calculated on this basis, and the optimization scheme is formulated, as shown in Table 4.

Because of the limited machining accuracy, the gear will be put into the gear three-coordinate measuring machine to measure after gear grinding. The actual gear modification scheme is shown in Table 5 according to the measurement report.

Because it has the machining error, the three gears are processed in a total of 10 groups (10 driving wheels and 10 driven wheels). The combination of all these driving wheels and passive wheels are freed (10 × 10, a total of 100 combinations). Then the combined parameters are calculated in Romax, and the results are shown in Table 6. The best five groups are selected and assembled in the five gearboxes for testing according to the transmission error, the maximum contact stress, and the meshing tooth surface nephogram.

5.2 Optimization Effect Analysis

First of all, the theoretical optimal solution is calculated. According to the comparison of the transmission error, the peak-valley value of the transmission error is reduced from 13.97 to 9.47 μm, which decreases 32.21%. The contact stress maximum unit length load is reduced from 711.068 to 669.409 N mm, which decreases 41.659 N mm. From Fig. 7, longitudinal load distribution is more

Table 4 Modification of the three gears

	Axial modification		Profile modification	
	Crowning	Spiral angle	Crowning	Pressure angle
3 Stop output shaft gear (driven wheel)	7 ± 1	9	15 ± 1	−5
3 Stop counter shaft gear (driving wheel)	7 ± 1	No modification	10 ± 1	No modification

Table 5 Actual modification of the three gears

	Axial modification		Profile modification	
	Crowning	Spiral angle	Crowning	Pressure angle
3 Stop output shaft gear (driven wheel)	8.63	8.55	16.8	−2.9
3 Stop counter shaft gear (driving wheel)	8	4.35	10.6	−1.1

Table 6 Transmission error of the three-gear combination

3 Gear (No modification 13.97 μm/theoretical scheme 9.47 μm, decline 32.21%)										
TE	1	2	3	4	5	6	7	8	9	10
1	3.84	3.97	3.95	**3.83**	3.9	6.23	6.08	6.02	6.04	6.23
2	**6.71**	6.81	6.87	6.72	6.83	9.08	8.95	8.89	8.91	9.08
3	4.29	**4.43**	4.43	4.29	4.39	6.7	6.56	6.5	6.52	6.7
4	8.04	8.17	8.19	8.03	8.15	10.4	10.3	10.2	10.2	10.4
5	7.81	7.91	7.95	7.82	7.92	10.11	10.01	9.93	9.94	10.11
6	7.43	7.53	7.58	7.43	7.55	9.78	9.65	9.59	9.61	9.78
7	7.05	7.15	7.21	7.05	7.17	9.4	9.27	9.21	9.22	9.4
8	7.86	7.97	8.01	7.87	7.97	10.2	10.1	10	**10.0**	10.2
9	8.05	8.17	8.17	8.02	**8.13**	10.4	10.3	10.2	10.2	10.4
10	7.66	7.77	7.81	7.67	7.78	10.0	9.93	9.82	9.81	10.0

Bold indicates rough number is the transmission error of a preliminary selection scheme

uniform than the initial modification, especially the addendum and the dedendum loading problems have been improved significantly. It not only improves the gear carrying capacity and service life, but also reduces the noise and the vibration caused by the tooth deformation. According to the calculation results of the micro-gear modification, the effect of decreasing the transmission error fluctuation is obvious. And decreasing the peak-valley value of the transmission error can improve the vibration characteristics of the gear system, which will effectively solve the problem of squeal noise. Table 6 shows the actual processing, the best optimal combination of the gear transmission error fluctuations value is reduced to 3.84 μm, and the stress distribution equilibrium degree is significantly better than that without modification of the gear. The maximum contact stress is 1204 MPa, and it also shows the contact stress which does not exceed the acceptable range of gears' material ($20CrMnTaH_3$).

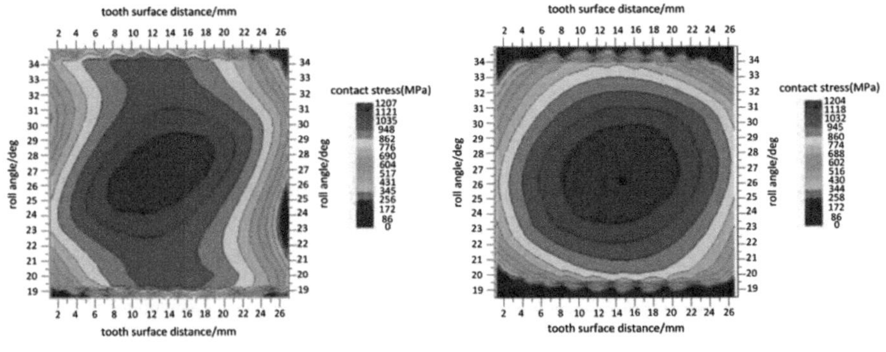

Fig. 7 Comparison of tooth surface nephogram before and after gear modification

5.3 Gearbox Radiation Noise Prediction

The first step is to simulate the radiation noise before the trial-producing gear. A transmission acoustic boundary element model (BEM) is built by LMS. Virtual. Lab is an acoustic simulation software of symmetry, selecting sites, and establishing the sound field at a distance of 300 mm shell, as shown in Fig. 8 [10]. The vibration response of the box body is calculated in Romax, whose simulation condition corresponds to the common working condition of gear modification. The dynamic excitation of the bearing seat is mapped as a boundary condition to the model before and after modification, and the numerical simulation of the noise radiation is carried out according to the vibration response calculation function [11]. Then the optimization results of gear modification by noise simulation are evaluated intuitively.

Taking the top point of the gearbox as an example, the frequency response curve in the prediction of radiation noise is analyzed and compared before and after the optimization, as shown in Fig. 9. The sound pressure level decreased significantly (Fig. 10).

Fig. 8 Radiation noise model and sound field

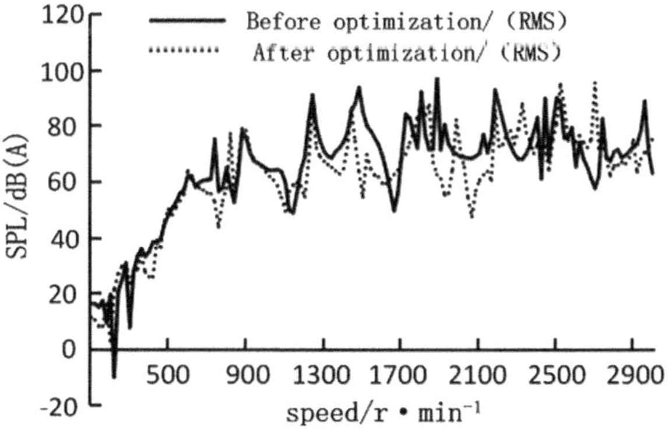

Fig. 9 Prediction of sound pressure level comparison before and after modification

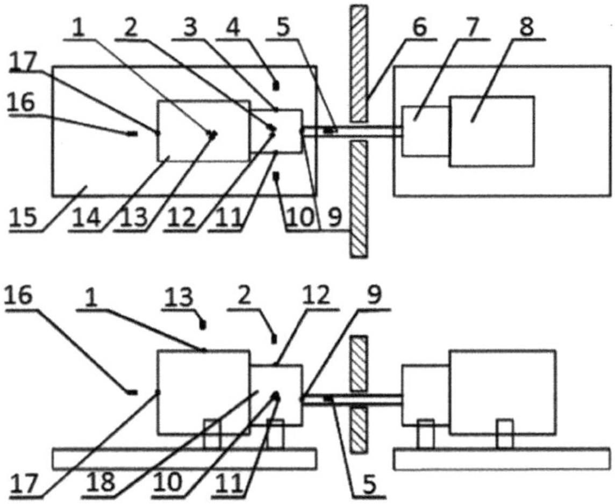

Fig. 10 Arrangement of the test points of the bench. 1,3,9,11,12,17—three-component acceleration sensors; 2,4,5,10,13,16—acoustic sensor; 6—wall of anechoic chamber; 7—gearbox; 8—load motor; 14—driving motor; 15—bench; 18—gearbox to be tested

6 Test and Analysis

6.1 Test Data Collection

In order to avoid the influence of the ambient noise and the wind speed, the bench test is carried out in anechoic chamber. And then the sound absorbing cotton and lead cover are used to avoid motor noise interference. There are six noise sensors which are arranged at the distance of the shell 300 mm, and six three-component acceleration sensors are arranged on the shell corresponding to the acoustic sensor [12]. The measuring points are arranged as shown in Fig. 11.

Using the front-end(DataRec4-DIC 24 data) of Beijing Launder Company to acquire the gearbox near field noise and vibration acceleration of the points on the gearbox shell. In the process of data acquisition, the load spectrum was tested according to the load spectrum in the common working condition and compared with the background noise test of the anechoic chamber. Figure 11 was the test site photograph. In the process of data acquisition, the load spectrum was tested according to the load spectrum in the common working condition and compared with the background noise test of the anechoic chamber. Figure 11 is the test site photo.

6.2 Data Analysis

Using software (LMS .Test. Lab) to collect the data for post-processing, the comparison of the gearbox third gear of the modification schemes of 1 and 2 before and after the modification under the different working conditions was made. The noise contrast before and after modification is shown in Table 7 (unit/dB).

The comprehensive comparison and the analysis of experimental data show that the noise of the first modification scheme decreases 2–3 dB, and the noise of the

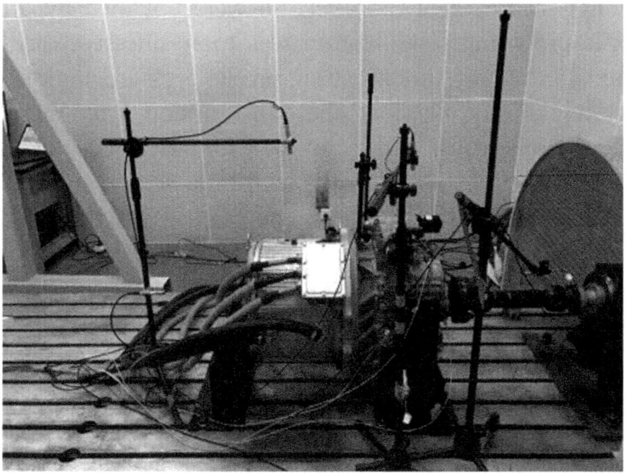

Fig. 11 Gearbox bench test site

Table 7 Noise value of gearbox under the common working condition (3 dB)

500 N m					
Transmission number	NO-1		NO-2		Original
Speed (rpm)	RMS value	Difference	RMS value	Difference	RMS value
1000	86.39	0.70	86.58	0.51	87.09
1200	82.87	6.49	83.51	5.85	89.36
1400	87.18	2.41	87.06	2.52	89.59
1600	91.28	2.29	88.88	4.68	93.57

second modification scheme decreases 3–4 dB. The modification effect makes the radiation noise of gearbox decrease by 3.18 dB averagely, and the improvement effect is ideal.

7 Conclusion

(1) The transmission error increased linearly with the increment of the load torque, and the gear transmission error of the conventional passenger vehicle gearbox can be reduced to less than 2 μm after the gear modification. However, the transmission error of pure electric bus was greater than 2 μm in most conditions as a result of the large load torque. In a pure electric bus gearbox, the modification of a pair of meshing gears was more effective than single gear in the reduction of transmission error.

(2) The following rules of gear modification were obtained by orthogonal experiment: Profile crowning modification was more effective on the transmission error than axial crowning modification, profile pressure angle modification, and axial spiral angle modification; the transmission error decreased with the increase of the tooth profile and increased with the increase of the teeth' amount under the load spectrum of the pure electric bus.
(3) After bench test data analysis in the anechoic chamber, the noise of the gearbox was declined by 2–3 dB; the test results remained good consistent with the simulation results, which proved that the gear micro-modification could significantly improve the power train performance and noise control.
(4) Taking a pure electric bus gearbox as the research object, the application method of "Modeling—Verification—Modification—Forecast—Test" could effectively achieve the expected noise reduction target. The feasibility of solving the problem of gear whine noise by simulation was proved. The cost and prototype optimization of time were saved compared with the past simply relying on the experience of modification.

References

1. Moore BCJ, Glasberg BR, Baer T (1997) A model for the prediction of thresholds, loudness, and partial loudness. J Audio Eng Soc 45(4):224–240
2. Umezawa K, Suzuki T (1986) Vibration of power transmission helical gears: approximate equation of tooth stiffness. Bull JSME 29(251):1605–1611
3. Guan H, Wang J, Shi Z (2013) Tooth profile modification of filtering gear reducer research based on Romax software. J Mech Des Res 29(3):28–31
4. Chen S, Tang J, Wang Z (2014) Effect of modification on dynamic characteristics of gear transmissions system. J Mech Eng 50(13):59–65
5. Lin HH, Oswald FB, Townsend DP (1994) Dynamic loading of spur gears with linear or parabolic tooth profile modifications. Mech Mach Theory 29(8):1115–1129
6. Ding H, Kahraman A (2008) Interactions between nonlinear spur gear dynamics and surface wear. J Sound Vib 130:1–5
7. Smith JD (2003) Gear noise and vibration. CRC Press
8. Munro RG, Morrish L, Palmer D (1999) Gear transmission error outside the normal path of contact due to corner and top contact. Proc Inst Mech Eng Part C J Mech Eng Sci 213(4): 389–400
9. Ge R, Jiang X, Yang W (2009) A study on the application of micro-modifications on gear tooth profile to noise reduction of automotive transmission. Autom Eng J 31(6):557–560
10. Wu X, Zhang B, Liu Y (2013) Research on prediction of radiated noise for sump based on LMS virtual lab. Mod Veh Power 3:38–43
11. Beacham MR, Bell DJ, Powell NN et al (1999) Development of transmission whine prediction tools. SAE Technical Paper
12. Yang B, Chu C, Tang H (2012) Research on effects of gear modification parameters with gear-box transmission performance. J Mech Transm 36(9):8–11

Lyapunov-Based Feedback Control of Auxiliary Power Unit for Series Hybrid Electric Vehicles

Munan Hong, Zhenhui Yao, Ling Su, Bo Liu and Wei Liang

Nomenclature

APU	Auxiliary power unit
ICE	International combustion engine
p_m	Intake pressure
ω	Combustion engine speed
c_1, c_2, c_3	Intake air model coefficients
τ_e	Net torque of the combustion engine
L, τ_d	Rotational dynamic model coefficients
J_e	Equivalent inertial of crankshaft
τ_l	Load torque
a_1, a_2, a_3	Rotational dynamic model coefficients
c_{1n}, c_{2n}, c_{3n}	Nominal model coefficients
a_{1n}, a_{2n}, a_{3n}	Nominal model coefficients
d_p	Intake air model errors
d_ω	Rotational dynamic model errors
P_r	Power reference
ω_r	Speed reference
e_p	Tracking error of intake pressure
e_ω	Tracking error of engine speed
k_1, k_2	Proportional control gains
k_{p1}, k_{p2}	Proportional control gains
k_{i1}, k_{i2}	Integral control gains
α	Throttle position
p_0	Ambient pressure

M. Hong (✉) · Z. Yao · L. Su · B. Liu · W. Liang
New Energy Vehicle R&D, Chongqing Changan Automobile Co., Ltd.,
No.9, Danhu Road, Shuangfengqiao, Yubei District, Chongqing 401120, China
e-mail: yunzhongtian@163.com

Society of Automotive Engineers of China (SAE-China) (ed.), *Proceedings of the 19th Asia Pacific Automotive Engineering Conference & SAE-China Congress 2017: Selected Papers*, Lecture Notes in Electrical Engineering 486, https://doi.org/10.1007/978-981-10-8506-2_61

T_0	Ambient temperature
κ'	Specific heat ratio
α_0, c_t	Throttle model coefficients
D, $\dot{m}_{at0}$	Throttle model coefficients

1 Introduction

In series hybrid vehicle, engine and generator are connected together as an auxiliary power unit (APU) [1, 2].

The APU supplies the battery electricity and offers the driving energy for the traction motor. If the battery can be charged from the grid, the key function of the APU is to increase the vehicle's range when the battery is depleted which can help alleviate range anxiety. In this case, the APU is used as a range extender.

Since both the APU and the battery can output power, there must be a supervisor which determines the distribution of the output power. The engine speed target should be also calculated to optimize vehicle fuel consumption and emission. Therefore, the supervisor is at a high hierarchical control level, in which the power and the target speed of the APU are decided based on a suitable energy management strategy [3–5]. At a lower hierarchical level, the combustion engine and generator must be controlled in order to track the target speed and provide the power demand [6]. This is not very easy due to different response characteristics of the combustion engine and the generator.

In the APU system, the generator control provides a predetermined load torque and the combustion engine control tries to track a given engine speed reference. Most research solves this control problem by introducing an APU system model. Tian takes attention to detailed generator model, while the internal combustion engine model is simplified [1]. Jiao [2] and Fiengo [6] give more complicated approach by modeling the dynamics of both the combustion engine system and the generator system. Actually, the combustion engine response is much slower because there is time delay in the air flowing into the cylinders and the crankshaft rotation. Therefore, the combustion engine dynamics are more significant in the APU control. For the sake of simplicity, the time delay of the generator is omitted in this paper. The torque output of the generator is supposed to be instantaneous. Thus, the generator system is simplified to a static system and the generator torque (load torque of the APU) is seemed as a control variable.

The organization of this paper is as follows. In Sect. 2, the APU system architecture and model are described. Then, a basic controller and a robust controller of the APU system are proposed in Sect. 3. Simulations are conducted, and results are analyzed in Sect. 4 to verify the controllers. The conclusion is summarized in Sect. 5 finally.

2 APU System Architecture and Modeling

The APU system architecture is shown in Fig. 1. The high hierarchical controller determines the APU power demand and the target ICE speed. The low hierarchical control consists of the generator controller and the APU controller where the ICE speed control is included.

As it is mentioned, this paper only focuses on the APU controller design. The control objective is to determine the throttle position of the ICE and the required load torque from the generator such that the ICE power output tracks a given reference at a preset speed. The generator control is not considered due to its fast response, and only the ICE model is necessary for the APU controller design.

The engine system is a nonlinear and time-varying dynamic system. It is difficult to build a very exact engine model. Lots of literatures have investigated engine modeling issue from different viewpoints. It is found that a physic-based model from fundamental thermodynamic, fluid mechanics, and rigid body mechanics is helpful for online controller design. The engine system's working principle is shown briefly in Fig. 2. It can be simply described as follows: The air is induced into the manifold passing the throttle and fills the cylinders; the fuel is injected to the manifold or the cylinders directly; the mixture in the cylinders is ignited by the spark and generates the driving torque, which rotates the crankshaft. Usually, it is very quick for the fuel injection and the combustion process, so they can be modeled by algebraic function, while the air filling dynamic and the rotational dynamic should be modeled using differential equation. A compact engine model is given in [7]. Since it is very simple and suitable for controller or observer design, it is adopted here to design the APU controller.

The intake air filling dynamic can be represented by the following differential equation

$$\dot{p}_m = c_1 u - c_2 \omega p_m + c_3 \omega \tag{1}$$

where p_m is the intake pressure, ω is the ICE speed, c_1, c_2, c_3 are positive constants, which can be available from engine tests, and u denotes the air mass flow passing through the throttle position.

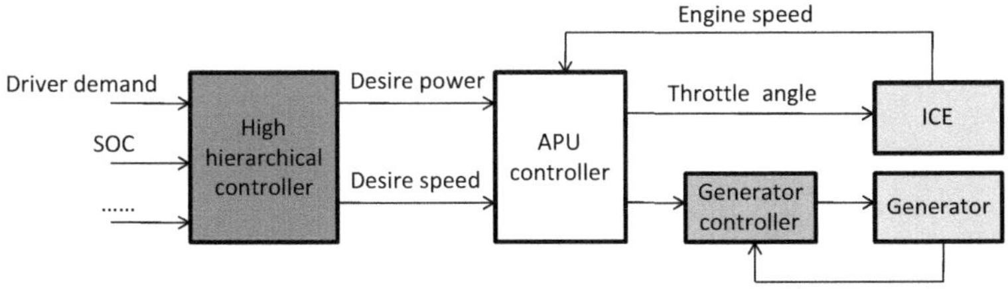

Fig. 1 APU system architecture

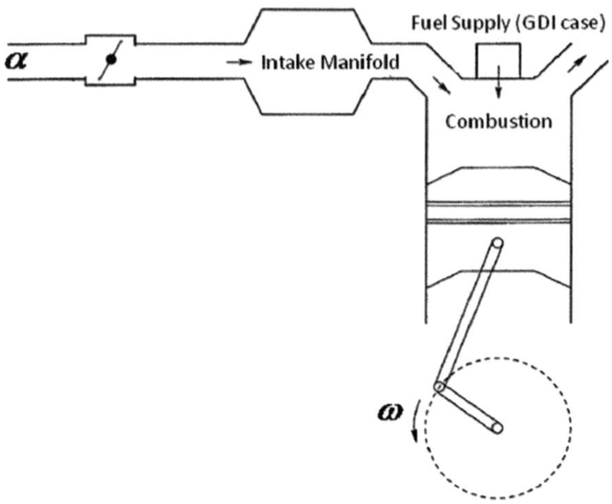

Fig. 2 Combustion engine system

The net torque generated by the combustion engine can be calculated by

$$\tau_e = Lp_m - \tau_d \tag{2}$$

where τ_e is the net torque of the engine, and L and τ_d are coefficients related to the engine speed, the air–fuel equivalence ratio, the spark timing, and the intake temperature. They are approximately constants over a bounded engine operating region.

According to Newton's second law, the rotational dynamic of the crankshaft can be derived directly

$$J_e\dot{\omega} = \tau_e - \tau_l \tag{3}$$

where J_e is the equivalent inertial of crankshaft which includes pistons and engine flywheel and τ_l is the engine load.

Substituting (2) to (3) yields

$$J_e\dot{\omega} = Lp_m - \tau_d - \tau_l \tag{4}$$

By using the following notations, $a_1 = L/J_e, a_2 = \tau_d/J_e, a_3 = 1/J_e$, the rotational dynamic equation can be written as

$$\dot{\omega} = a_1p_m - a_2 - a_3\tau_l \tag{5}$$

Thus, the engine dynamic model can be described by

$$\begin{cases} \dot{p}_m = c_1u - c_2\omega p_m + c_3\omega \\ \dot{\omega} = a_1p_m - a_2 - a_3\tau_l \end{cases} \tag{6}$$

3 APU Controller Design

In this section, an APU controller based on the engine model will be shown and the stability of the control system will be analyzed. Generally, tracking control problem can be solved by making the tracking error dynamics be asymptotically stable at the origin. So, the first step is to change the engine system to a tracking error dynamic system where the state is the error between the measured and the reference signal. Then, to apply the linear control technologies in the controller design, the error system should be linearized by selecting appropriate feedback control law. This is possible because the system (6) is exactly feedback linearizable if the parameters of the system are known and measurable. Moreover, much complicated condition that the model parameters are not exactly known is also considered in the controller design.

3.1 *Basic Controller*

If the desired power and speed of the engine are given as P_r, ω_r, the required engine torque should be P_r/ω_r. From Eq. (2) and the definition of a_1, a_2, a_3, the desired intake pressure is

$$p_{mr} = \frac{P_r}{L\omega_r} + \frac{\tau_d}{L} = \frac{a_3 P_r}{a_1 \omega_r} + \frac{a_2}{a_1} \tag{7}$$

Thus, the error system of (6) is

$$\begin{cases} \dot{e}_p = -c_2\omega e_p + c_1 u - c_2 \omega p_{mr} + c_3\omega - \dot{p}_{mr} \\ \dot{e}_\omega = a_1 e_p + a_1 p_{mr} - a_2 - a_3\tau_l - \dot{\omega}_r \end{cases} \tag{8}$$

where $e_p = p_m - p_{mr}$, $e_\omega = \omega - \omega_r$ are the tracking errors. The error system (8) can be rewritten as

$$\begin{cases} \dot{e}_p = -c_2\omega e_p + v_1 \\ \dot{e}_\omega = a_1 e_p + v_2 \end{cases} \tag{9}$$

where $v_1 = c_1 u - c_2\omega p_{mr} + c_3\omega - \dot{p}_{mr}$,

$v_2 = a_1 p_{mr} - a_2 - a_3\tau_l - \dot{\omega}_r$ are the intermediate variables which are determined by the control input u and τ_l, respectively.

Construct the Lyapunov function candidate as

$$W_1 = \frac{1}{2}e_p^2 + \frac{1}{2}e_\omega^2 \tag{10}$$

The time derivative of (10) along (9) is

$$\begin{aligned}\dot{W}_1 &= \dot{e}_p e_p + \dot{e}_\omega e_\omega \\ &= -c_2 \omega e_p^2 + (v_1 + a_1 e_\omega) e_p + v_2 e_\omega\end{aligned} \tag{11}$$

By Lyapunov theorem, the error system is stable if $\dot{W}_1 \leq 0,\ \forall\, t \geq 0$. Therefore, the control law can be chosen as follows

$$\begin{cases} v_1 = -a_1 e_\omega \\ v_2 = -k_2 e_\omega \end{cases} \tag{12}$$

where k_2 is a positive gain. It decides the response time of the engine speed and can be adjusted properly. While the convergency rate of the intake air pressure depends on the coefficient $c_2\omega$, which is physically determined. The controller gain k_2 does not contribute directly to the convergence of e_p. That means the system transient performance is limited. To overcome this deficiency, another positive gain k_1 is introduced to the control law as follows

$$\begin{cases} v_1 = -a_1 e_\omega - k_1 e_p \\ v_2 = -k_2 e_\omega \end{cases} \tag{13}$$

It is easy to get the derivative of the Lyapunov function candidate

$$\dot{W}_1 = -c_2 \omega e_p^2 - k_1 e_p^2 - k_2 e_\omega^2 \leq 0 \tag{14}$$

Thus, the error system (8) is globally stable according to Lyapunov stability theorem.

Furthermore, from the definition of v_1, v_2 and the control law (13), the control variable is determined by

$$\begin{cases} u = \frac{1}{c_1}\left(\dot{p}_{mr} + c_2 \omega p_{mr} - c_3 \omega - a_1 e_\omega - k_1 e_p\right) \\ \tau_l = \frac{1}{a_3}\left(a_1 p_{mr} - a_2 - \dot{\omega}_r + k_2 e_\omega\right) \end{cases} \tag{15}$$

The air mass flow u can be calculated with the information of throttle position α by a well-known function [8, 9]:

$$u = h_0 \sqrt{p_r^{\frac{2}{\kappa}} - p_r^{\frac{\kappa+1}{\kappa}}}\,[1 - \cos(\alpha - \alpha_0)] + \dot{m}_{at0} \tag{16}$$

where $p_r = \max\left\{\frac{p_m}{p_0}, \left(\frac{2}{\kappa+1}\right)^{\frac{\kappa}{\kappa-1}}\right\}$, $h_0 = c_t \frac{\pi}{4} D^2 \frac{p_0\sqrt{\kappa'}}{RT_0}$, $\kappa' = \frac{2\kappa}{\kappa-1}$, p_0, T_0 are the ambient pressure and temperature, κ' is the specific heat ratio, and α_0, c_t, D and $\dot{m}_{at0}$ are constants. Therefore, the throttle position is decided by

$$\alpha = \alpha_0 + \cos^{-1}\left[1 - \frac{u - \dot{m}_{at0}}{h_0\sqrt{p_r^{2/\kappa} - p_r^{\kappa+1/\kappa}}}\right] \quad (17)$$

3.2 Extension of the Controller

The exact engine model parameters are usually not available in real applications. Also, the engine model is built with the assumption that the air–fuel ratio is maintained by a separate controller at the stoichiometric value and the spark advance is appropriately chosen. If the air–fuel ratio and the spark advance deviate from the ideal values, there will be errors between the model and the real engine system. These errors produced in the modeling and the parameter identification can be dealt with as the uncertainty of the model parameters. Use unknown functions d_p, d_ω to represent the model errors

$$\begin{cases} d_p = (c_1 u - c_2 \omega p_m + c_3 \omega) - (c_{1n} u - c_{2n} \omega p_m + c_{3n} \omega) \\ d_\omega = (a_1 p_m - a_2 - a_3 \tau_l) - (a_{1n} p_m - a_{2n} - a_{3n} \tau_l) \end{cases} \quad (18)$$

where $c_{1n}, c_{2n}, c_{3n}, a_{1n}, a_{2n}, a_{3n}$ are known nominal model parameters.

Then, the engine dynamics (6) can be changed to

$$\begin{cases} \dot{p}_m = c_{1n} u - c_{2n} \omega p_m + c_{3n} \omega + d_p \\ \dot{\omega} = a_{1n} p_m - a_{2n} - a_{3n} \tau_l + d_\omega \end{cases} \quad (19)$$

And the tracking error dynamics are

$$\begin{cases} \dot{e}_p = -c_{2n} \omega e_p + d_p + v_1 \\ \dot{e}_\omega = a_{1n} e_p + d_\omega + v_2 \end{cases} \quad (20)$$

where $e_p = p_m - p_{mr}$, and the calculation of p_{mr} is similar to Eq. (7), but nominal parameters are used because the real ones are not exactly known.

$$p_{mr} = \frac{a_{3n} P_r}{a_{1n} \omega_r} + \frac{a_{2n}}{a_{1n}} \quad (21)$$

Compared to the error system (9), the system (20) brings in two unknown terms which represent the model errors. Corresponding to this, the control law is improved as follows

$$\begin{cases} v_1 = -a_{1n} e_\omega - k_{p1} e_p - k_{i1} \int e_p dt \\ v_2 = -k_{p2} e_\omega - k_{i2} \int e_\omega dt \end{cases} \quad (22)$$

where $k_{p1}, k_{i1}, k_{p2}, k_{i2}$ are positive gains.

The error system (20) is then changed to

$$\begin{cases} \dot{e}_p = -c_{2n}\omega e_p + d_p - k_{i1}\int e_p dt - k_{p1}e_p - a_{1n}e_\omega \\ \dot{e}_\omega = a_{1n}e_p + d_\omega - k_{i2}\int e_\omega dt - k_{p2}e_\omega \end{cases} \quad (23)$$

Define $e_1 = d_p - k_{i1}\int e_p dt$, $e_2 = d_\omega - k_{i2}\int e_\omega dt$, and supposing that d_p, d_ω are slowly time varying yields

$$\begin{cases} \dot{e}_p = -c_{2n}\omega e_p + e_1 - k_{p1}e_p - a_{1n}e_\omega \\ \dot{e}_1 = -k_{i1}e_p \\ \dot{e}_\omega = e_2 - k_{p2}e_\omega + a_{1n}e_p \\ \dot{e}_2 = -k_{i2}e_\omega \end{cases} \quad (24)$$

Construct the Lyapunov function candidate as

$$W_2 = \frac{1}{2}e_p^2 + \frac{1}{2}e_\omega^2 + \frac{1}{2k_{i1}}e_1^2 + \frac{1}{2k_{i2}}e_2^2 \quad (25)$$

The time derivative of (25) along (24) is

$$\begin{aligned} \dot{W}_2 &= \dot{e}_p e_p + \dot{e}_\omega e_\omega + \frac{1}{k_{i1}}\dot{e}_1 e_1 + \frac{1}{k_{i2}}\dot{e}_2 e_2 \\ &= -c_2\omega e_p^2 - k_{p1}e_p^2 - k_{p2}e_\omega^2 \end{aligned} \quad (26)$$

By Lyapunov stability theorem, the error system (24) is globally stable. Finally, the control variable is determined by

$$\begin{cases} u = \frac{1}{c_{1n}}\left(\dot{p}_{mr} + c_{2n}\omega p_{mr} - c_{3n}\omega - a_{1n}e_\omega - k_{p1}e_p - k_{i1}\int e_p dt\right) \\ \tau_l = \frac{1}{a_{3n}}\left(a_{1n}p_{mr} - a_{2n} - \dot{\omega}_r + k_{p2}e_\omega + k_{i2}\int e_\omega dt\right) \end{cases} \quad (27)$$

4 Simulations and Results Discussion

To illustrate the effectiveness of the proposed controllers, simulations are conducted and results are discussed. The nominal model parameters are identified from a production engine, and the values are given as follows:

$c_1 = 1.01 \times 10^8$, $c_2 = 0.2236$, $c_3 = 425$, $a_1 = 0.0022$, $a_2 = 40.0107$, $a_3 = 0.7143$.

4.1 Basic Controller Verification

The basic controller is verified first by simulation. Figures 3, 4, 5, and 6 show the results in which the desired power is a step curve and the engine speed is expected to be kept at a constant point. Figure 3 gives the APU system's response at 2000 r/min with $k_1 = 0,\ k_2 = 5$. The intake air pressure increases to have more air in the cylinders; therefore, more torque is generated. The engine speed deviates a little at the beginning, but finally goes back to the set point. From the inequality (14), the convergence of engine speed is determined by k_2. Figure 4 compares the speed tracking performance using different values of k_2. It is observed that the engine speed convergences faster as the gain k_2 increases.

Another conclusion from the inequality (14) is that the convergence of the intake pressure relies on the engine speed and the gain of k_1. To validate this conclusion, the convergence rate of the intake pressure is investigated at different engine speeds and control gains of k_1. The effect of the engine speed is illustrated in Fig. 5. It is

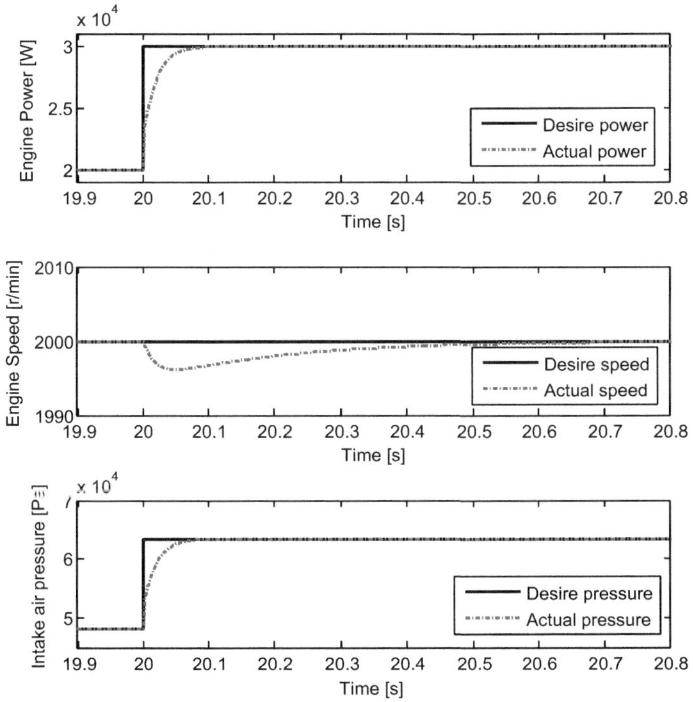

Fig. 3 Response at 2000 r/min

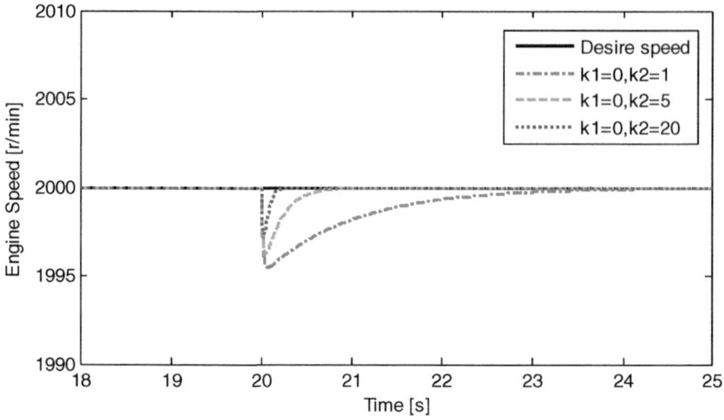

Fig. 4 Response with different k_2

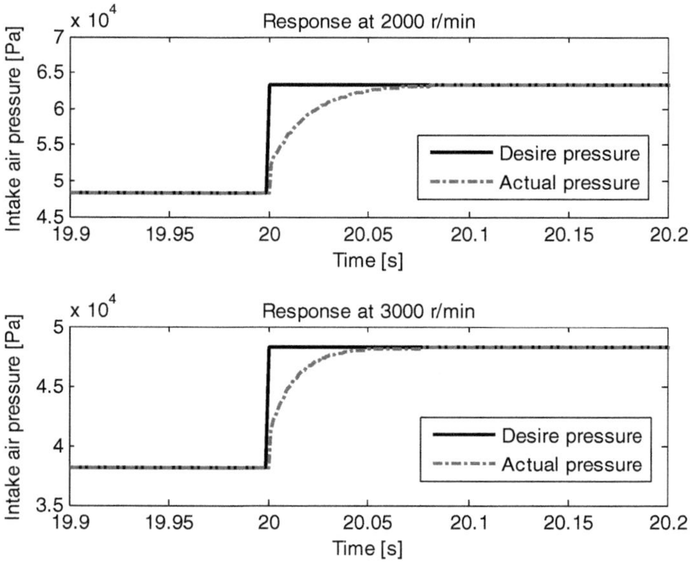

Fig. 5 Response at different engine speeds

shown that the intake pressure follows the target better at higher engine speed. And besides, large k_1 gets better tracking performance as Fig. 6 shows.

Furthermore, to show the effects of the model parameters, three cases are applied in the simulation. In Case A, the correct model parameters are used in the controller, while model errors are implemented in Case B and Case C.

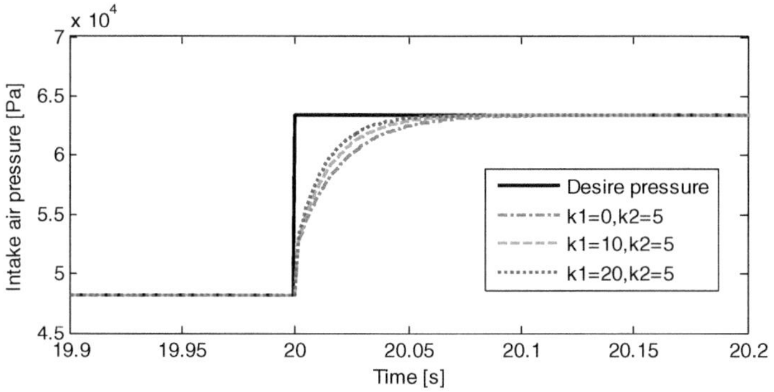

Fig. 6 Response with different k_1

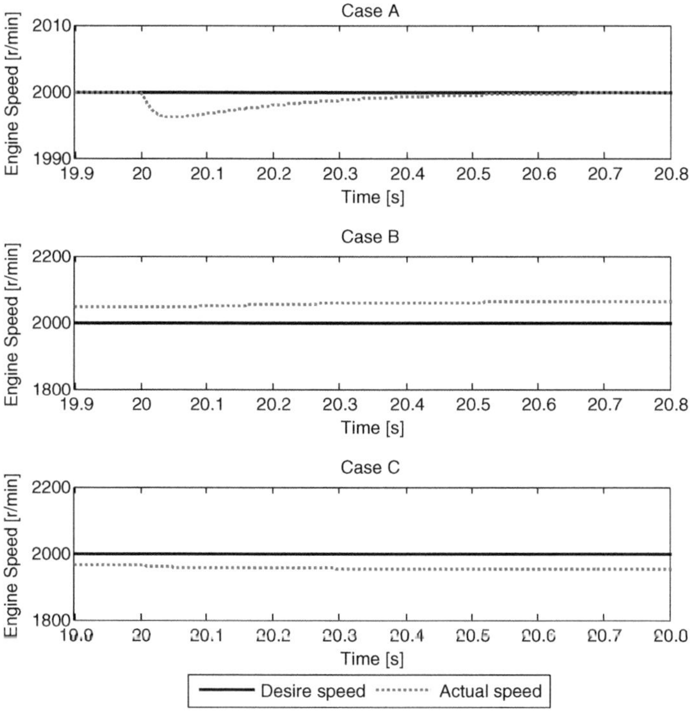

Fig. 7 Engine speed response in different cases

Case A: All model parameters are correct.
Case B: $c_{1n} = 0.9c_1, a_{1n} = 0.9a_1$, and others are correct.
Case C: $c_{1n} = 1.1c_1, a_{1n} = 1.1a_1$, and others are correct.

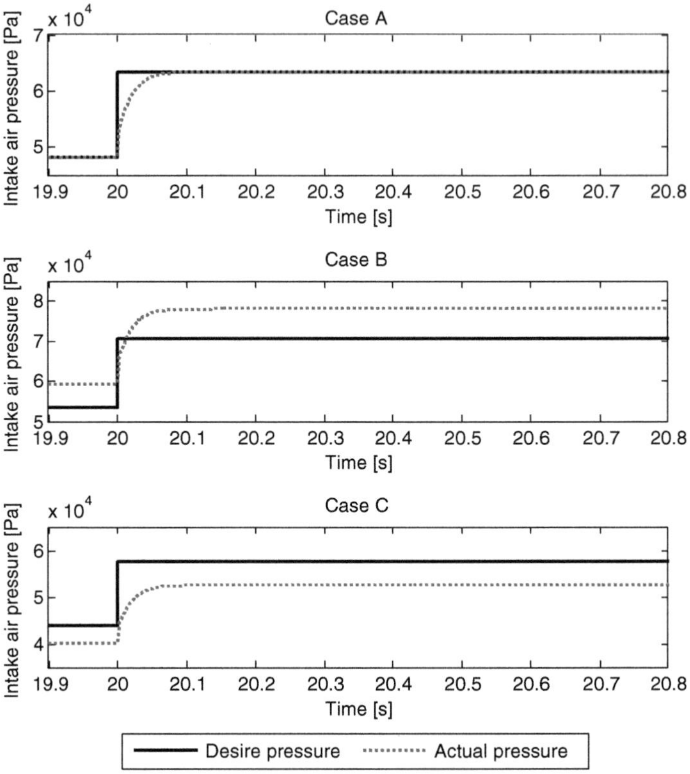

Fig. 8 Intake pressure response in different cases

It is shown from Figs. 7 and 8 that the engine speed and the intake air pressure cannot follow their target values anymore in Case B and Case C. As is discussed in Sect. 3.2, the improved controller can overcome the shortage. Simulation using the improved controller will be detailed introduced in the following.

4.2 Improved Controller Verification

The improved controller is used to confirm its robustness to the model uncertainty. Figures 9 and 10 give the results using the improved controller in different cases of parameters deviation. In the simulation, the control gains are given as $k_{p1} = 30,\ k_{i1} = 50,\ k_{p2} = 5,\ k_{i2} = 5$. It can be found that the engine speed and the intake pressure follow their targets well in all cases.

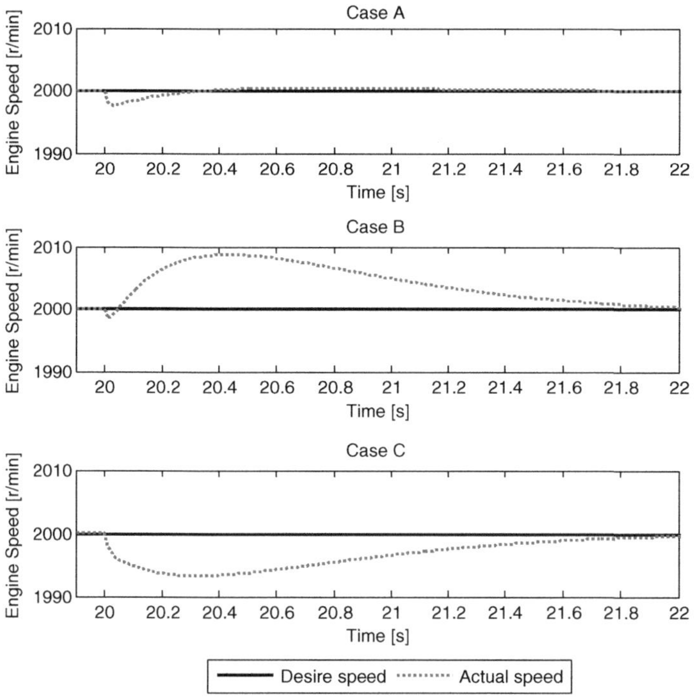

Fig. 9 Engine speed response with improved controller

To verify the controller's effectiveness in wider situation, the varying targets both in speed and power are given. The engine speed target is set as a periodical signal, and the desired power is chosen as a step curve from 0 to 20 kW. The response of the APU system with the control gains of $k_{p1} = 30$, $k_{i1} = 5$, $k_{p2} = 5$, $k_{i2} = 1$ is shown in Fig. 11, and the control input is illustrated in Fig. 12. It can be seen that the intake air pressure is well controlled by the air mass flow. When the target speed changes, the generator torque is adjusted quickly to achieve good speed tracking performance. In the simulation, the generator torque is limited to ± 200 N m. Positive torque means the generator is in generation mode, while negative torque denotes it is in the motor mode. It can be seen that the generator works in motor mode as a driving force of the crankshaft when the target speed goes up. When the target speed falls down, the generator works in generation mode as a load torque.

It should be noted that the PI gains $k_{p1}, k_{i1}, k_{p2}, k_{i2}$ should be properly chosen even though any positive number can ensure the control system's stability. Figure 13 shows the tracking performance and the control inputs using different PI gains. The normal gains are chosen as $k_{p1} = 30$, $k_{i1} = 5$, $k_{p2} = 5$, $k_{i2} = 1$, which

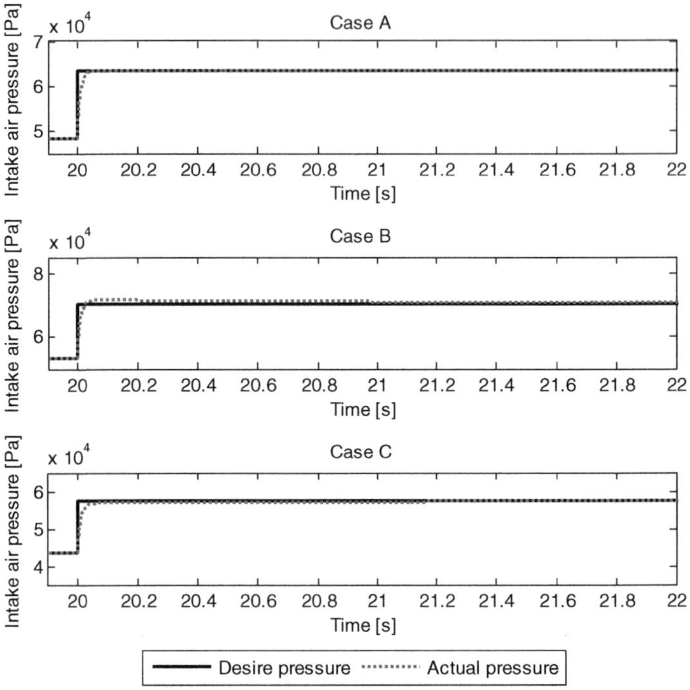

Fig. 10 Intake air pressure response with improved controller

is denoted as middle gains in the figure. "Large gains" and "small gains" are 200 and 50% of the middle gains, respectively. It is observed that larger gains get better speed tracking performance. It is also found that the generator torque may go to the limit if too big gain is used. In this case, increasing the gains further is not so effective to boost the speed response.

5 Conclusions

A basic controller and a robust controller of auxiliary power unit (APU) are proposed based on a mean value engine model. Lyapunov theory is applied to construct the controllers and ensure their stability. Simulations demonstrate that both controllers can track the engine speed and intake air pressure well if the model parameters are exactly known. Only the improved controller is able to keep good performance if the model parameters are not very exact.

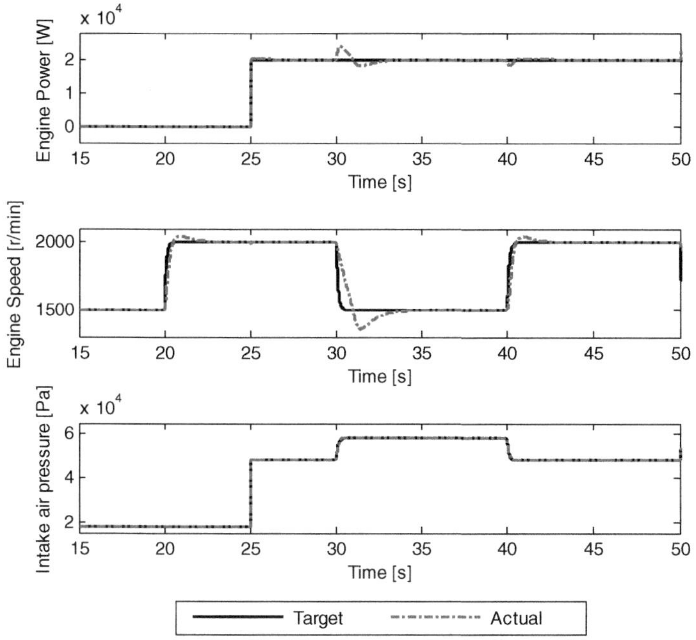

Fig. 11 Speed and power tracking performance

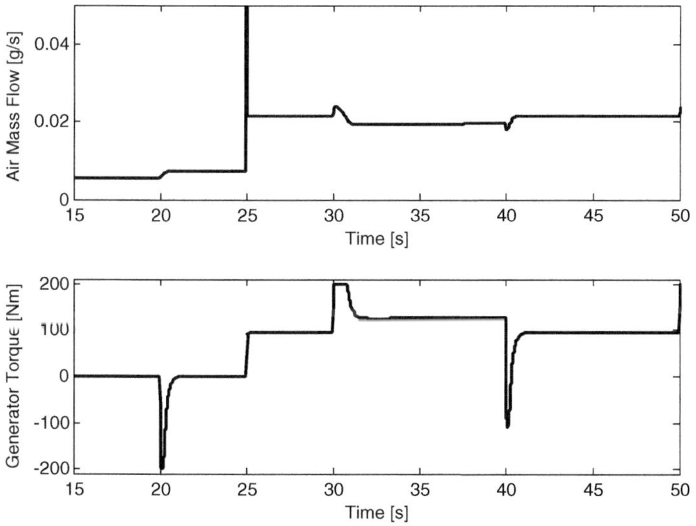

Fig. 12 Air mass flow and generator torque

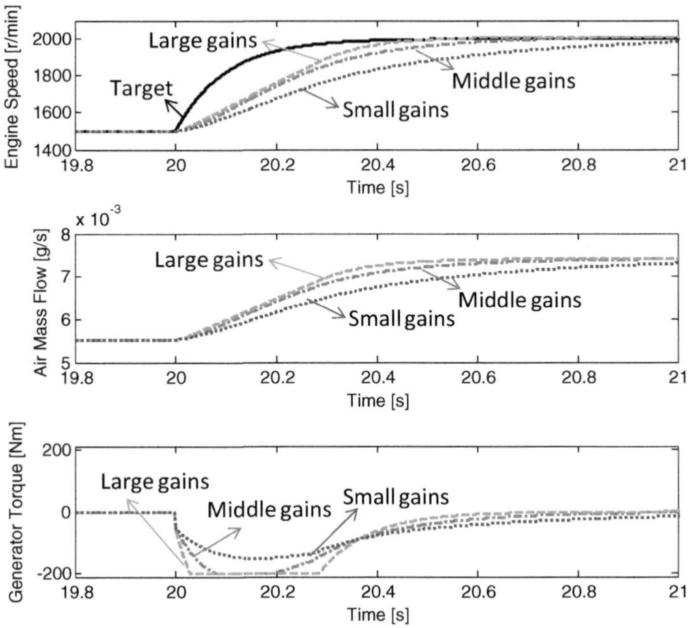

Fig. 13 Results comparison with different PI gains

References

1. Tian S, Cao G, Han Q, Li J, Yang M (2006) Modeling and decoupling control of ICE APU with uncontrolled rectifier in series hybrid vehicle. IEEE Veh Power Propul Conf 1–6
2. Jiao Xh, Xg Zhao (2011) Adaptive control for auxiliary power unit of series hybrid electrical vehicle. Electr Mach Control 15(10):75–81
3. Liu W, He H, Wang Z (2016) A comparison study of energy management for a plug-in serial hybrid electric vehicle. Energy Procedia 88:854–859
4. Chen Z, Xia B, You C, MI CC (2015) A novel energy management method for series plug-in hybrid electric vehicles. Appl Energy 145:172–179
5. Hu X, Murgovski N, Johannersson L, Egardt B (2013) Energy efficiency analysis of a series plug-in hybrid electric bus with different energy management strategies and battery sizes. Appl Energy 111:1001–1009
6. Fiengo G, Glielmo L, Vasca F (2007) Control of auxiliary power unit for hybrid electric vehicles. IEEE Trans Control Syst Technol 15(6):1122–1130
7. Hong M, Shen T, Ouyang M, Kako J (2011) Torque observers design for spark ignition engines with different intake air measurement sensors. IEEE Trans Control Syst Technol 19 (1):229–237
8. Hendricks E, Sorenson SC (1990) Mean value modeling of spark ignition engines. SAE World Congress, Paper 900616
9. Benvenuti L, Di Benedetto MD, Di Gennaro S, Sangiovanni-Vincentelli A (2003) Individual cylinder characteristic estimation for a spark injection engine. Automatica 39:1157–1169

Hardware in Loop Simulation Platform for Power Train System in Series Hybrid Electric Tracked Vehicles

Chao Liu, Yuan Zou, Jinlong Wu and Bing Wu

1 Introduction

The customer demands, environment requirements, and market competition drive new HEV technologies forward at a faster pace, which enforce the company to react more promptly and to reduce research cost [1]. Nowadays, HEV is in higher demand as the fuel price soars and the environmental regulation gets tighter. However, the research cost on HEV is much higher than that of the conventional vehicle due to its complex power train systems, which integrates the conventional internal combustion engine (ICE), electric motors (EM), and battery management system (BMS). Any updates to power train architecture will lead to adjustments and redesign of its control system. In order to remain competitive on HEV techniques, reduce the research cost, and make the development process reusable, hardware in loop (HIL) technology is widely utilized in power train development cycles, where control engineers can simultaneously work on design, test, and validation of the control system to hardware prototyping stage.

In the implementation and testing of control systems HIL of the HEV power train system design, the target controller is real hardware-based, while all other relevant components are model-based, or the target controller and some of the key components in the control loop are real hardware-based, while the rest are model-based. Usually, a plant is simulated because it is not available to simultaneously build up physical plant [2, 3]. Even though it is feasible to build the physical plant, it is very

C. Liu (✉) · Y. Zou · J. Wu
National Engineering Laboratory for Electric Vehicle,
Beijing Institute of Technology, Beijing 100081, China
e-mail: liuchao1028@bit.edu.cn

B. Wu
SAIC Motor Corporation Limited, Shanghai, China

Society of Automotive Engineers of China (SAE-China) (ed.), *Proceedings of the 19th Asia Pacific Automotive Engineering Conference & SAE-China Congress 2017: Selected Papers*, Lecture Notes in Electrical Engineering 486, https://doi.org/10.1007/978-981-10-8506-2_62

time-consuming and costly. It is especially meaningful in the case of the controller being a real hardware and all other plants being mathematical models, because this combination can achieve the design objectives with less design iterations and better performance of the HEV power train system at lowest cost and shortest development period [4].

This chapter builds an 'Engine-VCU' hardware in loop platform based on the series hybrid tracked vehicle which includes the real engine, real-time simulator *RT-LAB*, *AVL* dyno, and VCU. In this platform, the electronic throttle and engine speed in loop control system based on cascade PI theory is delivered, the power-followed multi-point speed switching control strategy of the engine is tested, and the feasibility and effectivity of the whole vehicle control algorithm in controlling engine speed and the vehicle's power chain coordination is verified.

2 'Engine-VCU' Hardware in Loop Platform

2.1 The Architecture of Hybrid Electric Tracked Vehicle

The architecture of the hybrid power train system is shown in Fig. 1. An engine-generator-rectifier set and a battery pack supply the electricity to dual motors, propelling the sprocket independently. The engine gives 50 kW maximum power and 93 Nm maximum torque within the speed range from 1200 to 6200 r/min. The generator offers 107 Nm maximum output torque and 40 kW maximum power within the speed range from 0 to 6400 r/min. The 37.6 Ah lithium-ion battery pack gives 307 V rated voltage. Due to the limitation of battery capacity, the main power source of vehicle is engine-generator set; battery is used to provide better power, attain better economy and achieve ultra-quiet driving when necessary.

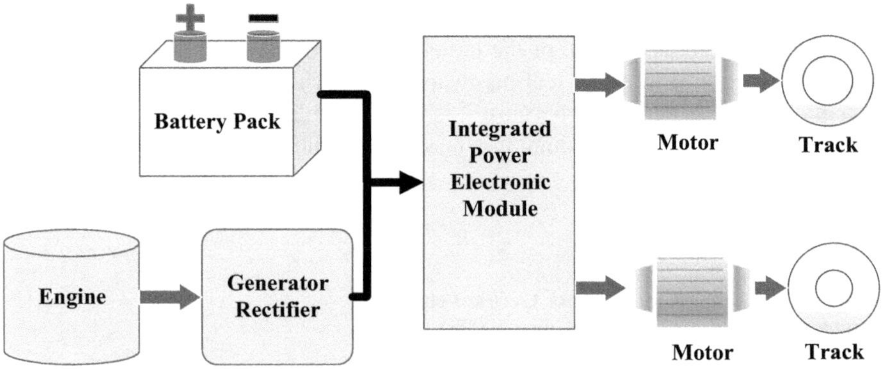

Fig. 1 Architecture of the hybrid power train system

2.2 Platform Design

The 'Engine-VCU' hardware in loop platform is shown in Fig. 2 which consists of mechanical systems, electrical systems, signal systems, hardware in loop real-time control systems, and peripheral auxiliary equipments. The platform includes a real vehicle controller, a real engine, virtual vehicle components, and dynamic models. Among them, the vehicle controller is based on automatic code generation *MotoTron* in order to achieve engine-generator system control, model-based vehicle dynamics control, state monitoring, and diagnosis of the platform. *Suzuki* three-cylinder *K10B* engine is used and its output shaft is mechanically connected with AVL power dynamometer to accomplish real-time synchronous loading and monitoring. Real-time simulation system *RT-LAB* can communicate with vehicle controller and *PUMA* power dynamometer through CAN bus. To be specific, it can send motor speed signals and operating instructions calculated by motor, generator and kinetic models while receiving the torque command of motor, the actual speed of engine, the target and actual throttle opening from VCU.

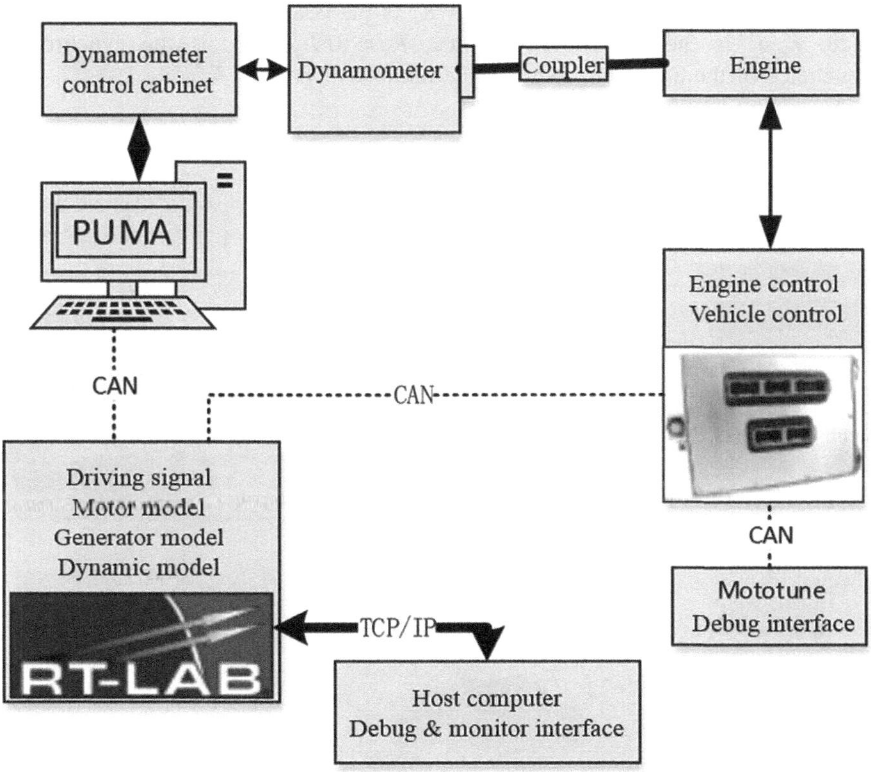

Fig. 2 'Engine-VCU' hardware in loop platform

2.3 Modeling

2.3.1 Engine-Generator Set (EGS) Model

In the series hybrid tracked vehicle, engine drives the generator directly to generate three-phase alternating current, and then it is rectificated into direct current which is provided for the battery or both sides of the motor through rectifier bridge. The rated power of generator is 40 kW. Figure 3 illustrates the equivalent electric circuit of the engine-generator set (EGS) [5].

According to electrical relations, the current and output voltages of the generator, I_g and U_g, respectively, are computed by [6]

$$U_g = K_e\omega_g - K_x\omega_g I_g \tag{1}$$

$$T_g = K_e I_g - K_x I_g^2 \tag{2}$$

$$K_x = \frac{3PL^g}{\pi} \tag{3}$$

where T_g is the electromagnetic torque, K_e is the coefficient of the electromotive force, $K_e\omega_g$ is the electromotive force, $K_x = 3PL_g/\pi$, L_g is the synchronous inductance of the armature, and P is the number of poles.

According to mechanical relations,

$$\left(\frac{J_e}{i_{eg}^2} + J_g\right)\frac{\mathrm{d}n_g}{\mathrm{d}t} = \frac{\left(\frac{T_{eng}}{i_{eg}} - T_g\right)}{\frac{2\pi}{60}} \tag{4}$$

$$n_g = i_{eg} n_{eng} \tag{5}$$

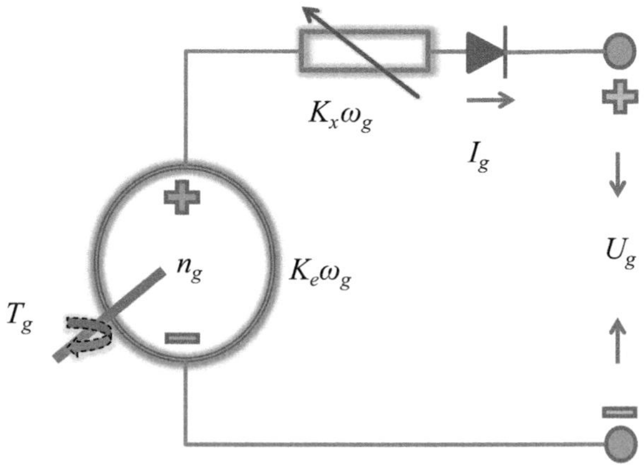

Fig. 3 Equivalent electric circuit of the EGS

where n_g is the rotational speed of the generator, n_{eng} is the engine speed, i_{eg} is the fixed gear ratio connecting the engine and generator, and J_e and J_g are the moments of the inertia. T_{eng} is the torque of engine and the sole control variable is engine throttle, which determines the power split between engine-generator set (EGS) and battery.

2.3.2 Battery Model

An OCV-internal resistance model is used which is shown in Fig. 4 [7].

The state of charge (SOC) of the battery can be calculated by the following formula:

$$\dot{SOC} = -I_{bat}(t)/C_{bat} \tag{6}$$

where C_{bat} is the rated capacity of the battery, and I_{bat} is the current.

According to the Kirchhoff's Law and power balance,

$$U_{bat} = V_{oc} - I_{bat}R_{int} \tag{7}$$

$$P_{bat} = V_{OC}I_{bat} - I_{bat}^2R_{int} \tag{8}$$

According to the above formula, the value of I_{bat} can be calculated by

$$I_{bat} = \left(V_{OC} - \sqrt{V_{OC}^2 - 4R_{int}P_{bat}}\right)/2R_{int} \tag{9}$$

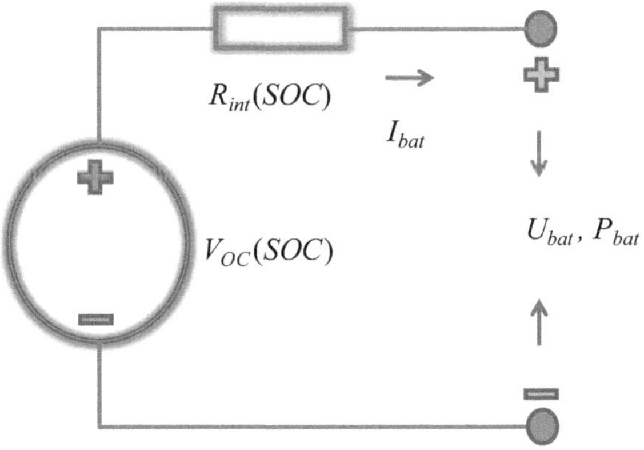

Fig. 4 An OCV-internal resistance model

2.3.3 Electric Coupling Between Engine-Generator Set (EGS) and Battery

This section mainly analyzes the power–balance relationship of series hybrid tracked vehicles which means the balance between front and rear power chains. Front power chain reflects the power parts of engine-generator set and battery, and on the other hand, rear power chain refers to the power parts of motor controllers, motors, and road loads. The simplified model for HEV power train is shown in Fig. 5.

To satisfy the driver power demand, Eq. (10) must be satisfied all the time

$$P_{\text{dem}} = U_{\text{dc}}(I_g + I_{\text{bat}}) \tag{10}$$

According to the charging characteristics of battery, the following constrains are existed:

$$\begin{aligned} & I_{\text{bat,min}} \le I_{\text{bat}} \le I_{\text{bat,max}} \\ & n \cdot u_{\text{b,min}} \le U_{\text{dc}} \le n \cdot u_{\text{b,max}} \\ & \text{SOC}_{\text{min}} < \text{SOC} < \text{SOC}_{\text{max}} \end{aligned} \tag{11}$$

where $I_{\text{bat,min}}$ and $I_{\text{bat,max}}$ represent the maximum charge current and maximum discharge current, with the value of -180 and 180 A, respectively. n is the number of series cells in the battery pack, and SOC_{min} and SOC_{max} represent the minimum and maximum state of charge, with the value of 0.6 and 0.9, respectively.

Depending on the match results of battery and engine-generator, the following constrains are existed:

$$\begin{aligned} & n_{g,\text{min}} < n_g < n_{g,\text{max}} \\ & 0 < I_g < I_{g,\text{max}} \end{aligned} \tag{12}$$

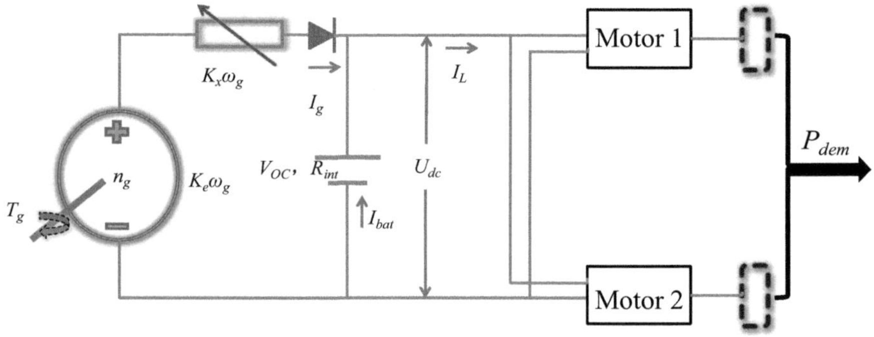

Fig. 5 Simplified model for HEV power train

where $n_{g,\min}$ and $n_{g,\max}$ represent the minimum and maximum of the generator speed, with the value of 1200 and 4700 r/min, respectively. $I_{g,\max}$= 300 A is the maximum of DC bus current.

The electric bus voltage U_{dc} is determined by

$$\begin{cases} \begin{cases} U_{dc} = U_{bat} > U_g \\ I_g = 0 \end{cases}, & V_{OC} \geq K_e\omega_g \\ \begin{cases} U_{dc} = U_g = U_{bat} \\ I_g > 0 \end{cases}, & V_{OC} < K_e\omega_g \end{cases} \tag{13}$$

I_g= 0 occurs when $K_e\omega_g$ is lower than V_{OC}, and thus, the generator is not producing any electric power. In this case, all the power is provided by the battery. When I_g> 0, generator supplies electric power alone or together with the battery.

Based on the power demand balance and some relevant derivations, U_{dc}, I_g, and I_{bat} are computed by

$$\begin{cases} \begin{cases} U_{dc} = U_{bat} = V_{OC} - I_{bat}R_{int} > U_g \\ I_{bat} = I_L = \frac{P_{dem}}{U_{dc}} = \frac{V_{OC} - \sqrt{V_{OC}^2 - 4R_{int}P_{dem}}}{2R_{int}} \\ I_g = 0 \end{cases}, & V_{OC} \geq K_e\omega_g \\ \begin{cases} U_{dc} = U_g = U_{bat} = \frac{-b + \sqrt{b^2 - 4ac}}{2a} \\ I_{bat} = \frac{V_{OC} - U_{dc}}{R_{int}} \\ I_g = I_L - I_{bat} = \frac{P_{dem}}{U_{dc}} - I_{bat} \end{cases}, & V_{OC} < K_e\omega_g \end{cases} \tag{14}$$

$$a = \frac{1}{R_{int}} + \frac{1}{K_x n_g}, b = -\left(\frac{V_{OC}}{R_{int}} + \frac{K_e}{K_x}\right), c = P_{dem}$$

2.3.4 Vehicle Dynamic Model

The vehicle is taken as a rigid body with a concentrated mass and planar rotational movement. The demand power consists of the heading and steering power is calculated by

$$P_{dem} = \left(ma + mg\ \cos\theta f + \frac{C_D A v_{ave}^2}{21.15} + mg\ \sin\theta\right) v_{ave} + M\omega \tag{15}$$

$$v_{ave} = (v_1 + v_2)/2 \tag{16}$$

$$\omega = (v_1 - v_2)/B \tag{17}$$

where P_{dem} is the power demand. The former part of the first equation is the heading power, including the power from acceleration resistance, aerodynamic resistance, rolling friction, and road slope. v_{ave} is the speed of vehicle, and m is the curb weight; a is the vehicle acceleration; g means the gravity acceleration; θ is the road slope angle, and f denotes the rolling resistance coefficient; C_D represents the aerodynamic coefficient; A is the front area of the vehicle. The latter part computes steering power, and M is the resisting yaw moment; ω means the rotational speed of vehicle. v_1 and v_2 are the speeds of outside and inside tracks. B denotes the tread of the vehicle. The value of M is calculated as

$$M = \frac{1}{4} u_t mgL \tag{18}$$

where u_t is the steering resistance coefficient; L is the track contacting length.

3 Control Strategies of the Platform

3.1 Engine-Generator Set Speed Closed-Loop Control

3.1.1 Electronic Throttle Control

The electronic throttle is modified from the pull-type throttle of *Suzuki* engine, including DC drive motor, reduction gear, cable, pull back spring, and throttle valve as shown in Fig. 6. Incremental discrete multi-parameter PI closed-loop control [8] is used in the control system. The input parameter is target opening of throttle, the control output is PWM duty cycle, and the feedback is the actual opening of the throttle acquired from TP sensor.

The dynamic characteristics of the throttle system after multi-PI parameter tuning are shown in Fig. 7, and the output duty cycle is shown in Fig. 8. It is clearly that at 6.6 s, the object throttle opening percentage rises from 20 to 30%, resulting in the PWM duty cycle rise from 26 to 80% quickly. Then, the real throttle percentage reduces to the lowest point 15% abnormally in about 0.02 s which leads to the PWM duty cycle rise to the highest point 95%. After that, the throttle percentage

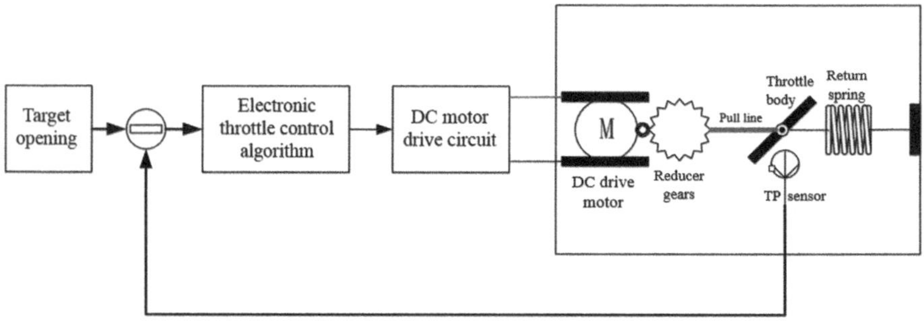

Fig. 6 Electronic throttle control structure

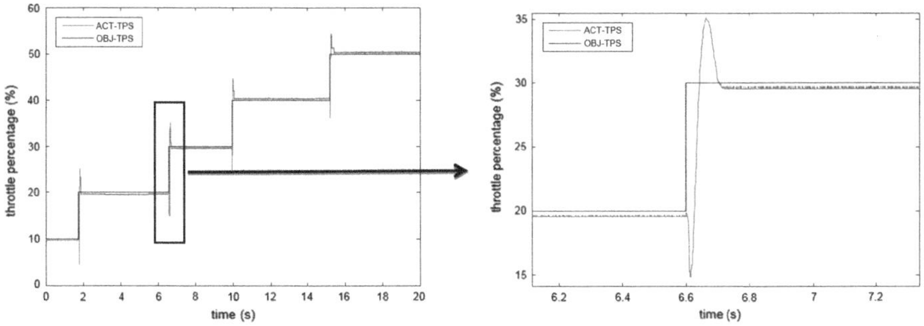

Fig. 7 Electronic throttle opening percentage results

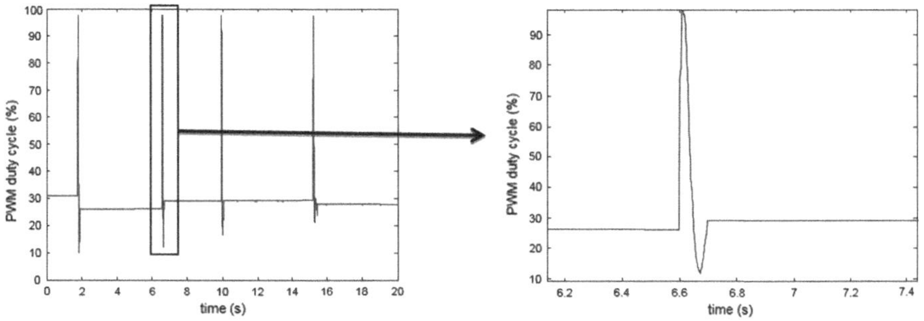

Fig. 8 Electronic throttle control output signals

goes up and after a 5% overshoot, it finally becomes stable to 30%. The whole process lasts about 0.1 s and the abnormal decline of throttle percentage may result from the nonlinear characteristics of the pull line. The correctness of control algorithm in electronic throttle percentage control system is verified and the better dynamic follow performance of electronic throttle percentage is proved from the experimental results.

3.1.2 Engine Speed Control

Cascade PI algorithm is used in the whole engine speed control closed-loop system which is shown in Fig. 9 while incremental discrete multi-PI control is adopted in this system. As for engine speed control, target throttle percentage, engine speed is served as output and feedback, respectively. Target throttle percentage, which is the control output of this stage, is also the input of throttle opening control at the next stage. Engine target speed which is used for the input value depends on the requirements of rear power chain.

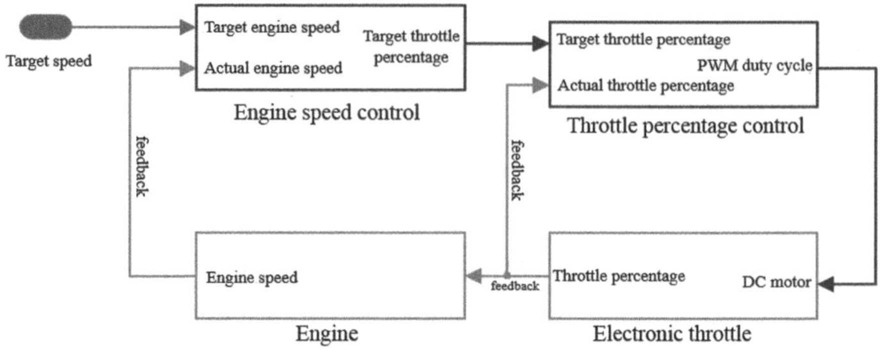

Fig. 9 Whole engine speed control closed-loop system

In order to increase the efficiency and economy of engine, the multi-point speed switching control mode is delivered. In this mode, the engine speed is divided into several different values according to power demands. The demand power under different working conditions of the tracked vehicle used in this chapter is shown in Table 1.

Thus, engine speed is divided into three different values according to different ranges of power demands as shown in Table 2. In order to avoid frequent speed switching, each area has a power hysteresis band of 5 kW. For example, if the power rises from 18 to 24 kW, the speed will maintain 2400 r/min.

3.2 *Vehicle Dynamics Closed-Loop Control*

Driving control strategy of the tracked vehicle is shown in Fig. 10. In a certain control cycle, driving signals such as accelerator pedal, brake pedal, steering wheel, and stall mode are parsed to a motor target speed. Based on the actual speed of the

Table 1 Demand power under different working conditions

Working conditions	Power demand (kW)
32 km/h rapid acceleration	37
32 km/h cruising	13.8
Climbing	39
Pivot steering	5.2
32 km/h steering	34

Table 2 Multi-point speed switching area division scheme

Area	Power range (kW)	Speed (r/min)
I	0–20	2400
II	20–30	3200
III	30–40	4500

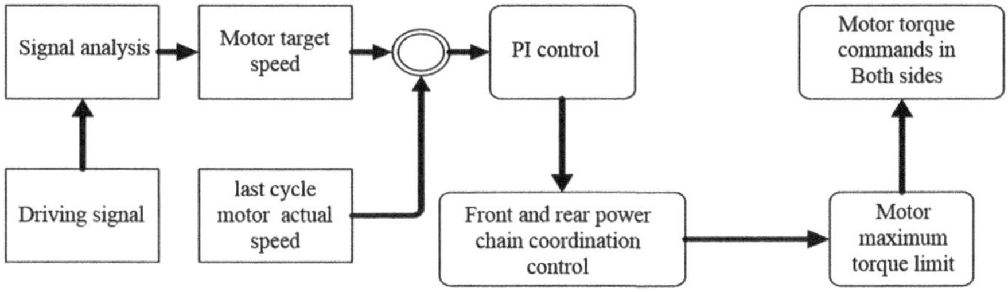

Fig. 10 Driving control strategy of the track

motor on the last cycle which is regarded as negative feedback of the system, the target motor speed in this cycle is calculated through PID control algorithm. Due to the coordination of front and rear power chains, motor torque commands acquired from PI algorithm are not directly sent to the actual motor. When the power demand of rear power chain can be satisfied from the front, the torque command of drive motor is the final result; on the contrary, if the power demand of rear power chain cannot be satisfied from the front, in order to protect the 'engine-generator' system, the torque will be formulated on the basis of the maximum power provided by the front power chain.

4 'Engine-VCU' Hardware in Loop Test

4.1 Driver Operating Instructions

The 'Engine-VCU' hardware in loop test lasts about 200 s. At 10 s, the driving mode signal switches from 0 to 1 which means the tracked vehicle begins to move. At 30 s, the accelerated pedal signal switches from 0 to 1 which means the vehicle begins to accelerate rapidly. At 40 s, the vehicle begins to swerve until 130 s. At 180 s, the speed drops to 0 in 2 s and at 190 s the driving mode signal switches from 1 to 0.

4.2 Experimental Results Without Battery

The aim of this experiment was to prove whether engine speed control is accurate according to rear power demand under different driving instructions and verify if the dynamic and steady-state performance of the engine can meet the requirements of entire electric power train when 'engine-generator' is the only power source. Multi-point speed switching control of engine is used in this experiment.

The results are shown in Figs. 11 and 12. At first, the vehicle is in a parking state and the engine runs at its own idle speed. Then, at 10 s, the vehicle starts to move so that the engine speed rises to 2400 r/min while the bus voltage increases. At 30 s, the vehicle starts to accelerate rapidly and both sides of the target motor speed reach to the maximum value. Engine speed rises to 4500 r/min due to the increasing of power demand. According to the requirements of vehicle dynamics, the motor torque command on both sides sent by VCU increases rapidly from 0 and the bus current increases swiftly which leads to the quick increasing of electromagnetic torque at the generator side. At the same time, both sides of the real target motor speed will rise to the maximum value. The electronic throttle percentage rises from 5 to 20%. At 40 s, the vehicle starts to swerve and the target motor speed reaches to 4000 and 1500 r/min each side, respectively. Due to the high-resistance moment when swerving, both sides of the motor demand power are very large. To avoid the insufficient power supply of engine, the motor's output torque is limited based on the engine's maximum output power. That is why both sides of the motor speed decline at 40 s and then attain to the object speed. At 130 s, the steering instruction ends. The aligning strategy is that both sides of the motor speed tends to the average value at first and then reaches to the target, which results in the decrease of demand power. Then, the bus current and engine electromagnetic torque decrease. At 180 s, the driver's instruction ends, and both sides of the motor speed are reduced to zero and the engine speed drops.

The effectiveness of engine multi-point control strategy is verified through the whole experiments. The engine speed can response quickly with the change of rear power chain based on the default control logic. When the engine maintains a certain speed and the generator electromagnetic torque changes due to the alter of rear

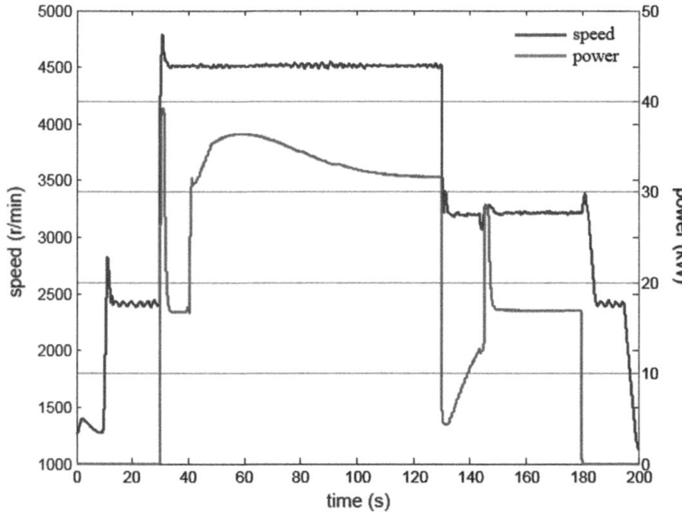

Fig. 11 Multi-point switching of engine speed

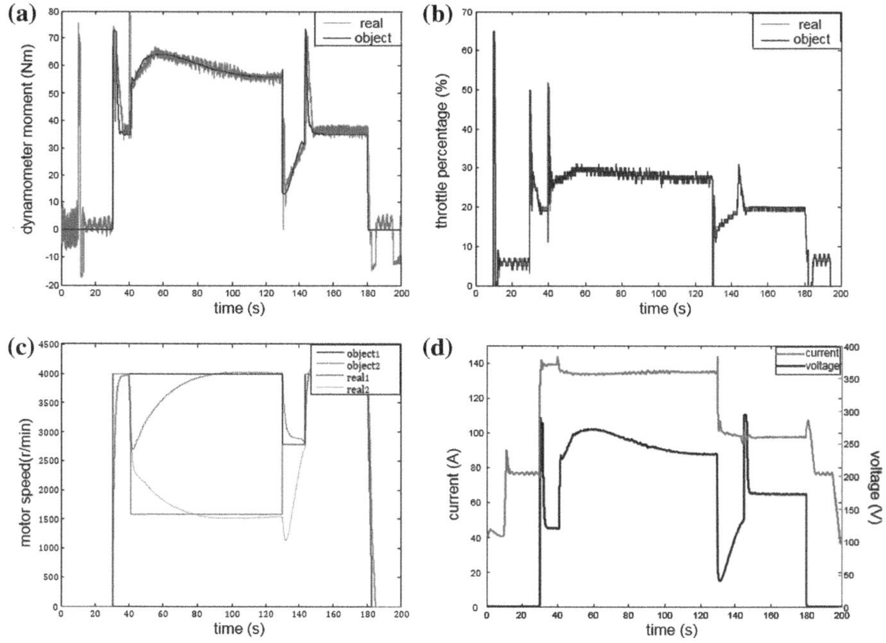

Fig. 12 Hardware in loop test results

power chain, the engine speed control is stable and has a high anti-interference ability. The driver instructions are analyzed accurately while the driving motor is controlled well according to the front power chain by the vehicle controller which can meet the requirements of the dynamic performance.

4.3 Experimental Results with Battery

When battery is added into power source, the fluctuations of bus voltage should be considered in order to avoid overcharge and over discharge. In the meantime, engine-generator and battery can discharge together when the rear power chain is under heavy demand. Also, energy can be recycled by battery under braking condition.

The electrical characteristics of the system are shown in Fig. 13a, and the bus voltage is stabilized at about 307 V and fluctuates slightly only when the demand power of rear power chain has a significant change. At 30 s, the vehicle starts to accelerate rapidly and the rear power chain requires more power; both generator and battery have a high current discharge and the maximum discharge current of battery is about 110 A. At 40 s, the vehicle starts to swerve, and power demand of the rear power chain is negative which means battery is at a charging state with the maximum charge current 70 A.

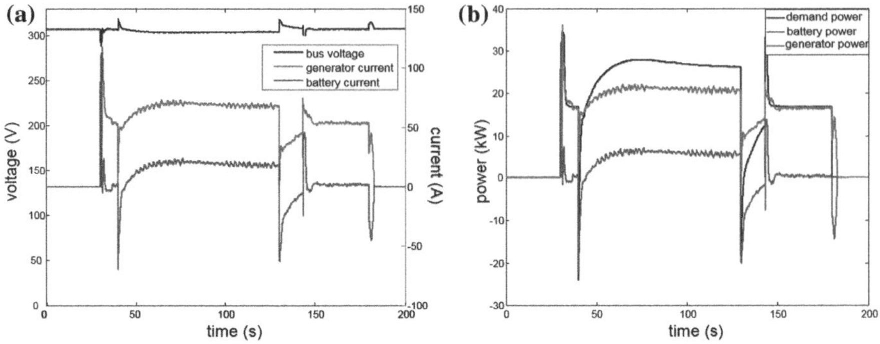

Fig. 13 Voltage (**a**), current (**a**), and power (**b**) characteristics of the system

Engine-generator set, battery pack, and the demand power of system are shown in Fig. 13b. In the case of rapid acceleration, engine-generator and battery produce the power together to meet the requirements; in the case of brake or steering conditions, engine-generator goes into no-load mode with no power output while battery enters into the energy recovery mode with negative power.

The results indicate that the main power source of the system is engine-generator set; battery plays an important role in the energy recovery and auxiliary output of high power demand. The validity and effectiveness of the control strategies are verified.

5 Conclusion

This chapter builds an 'Engine-VCU' hardware in loop platform and carries out the researches of control strategies based on the series hybrid electric tracked vehicle. The results are listed as follow:

1. Build an 'Engine-VCU' hardware in loop platform based on the series hybrid tracked vehicle including the real engine, real-time simulator RT-LAB, AVL dyno, and VCU. Create a detailed mathematical model of the hybrid tracked vehicle consisting the engine-generator set, battery pack, and whole vehicle dynamics model.
2. Complete the electronic throttle percentage control based on the incremental discrete multi-parameter PI closed-loop control. Finish the engine speed control based on cascade PI algorithm. Put forward the driving control strategy of the tracked vehicle. Deliver an engine multi-point speed switching control mode.
3. When engine-generator is the only energy source of the power train system, effectiveness of engine multi-point control strategy is verified. When battery is added into the power source, it plays an important role in the energy recovery and auxiliary output of high power demand.

Acknowledgements Supported by National Natural Science Foundation of China (51375044) and National Defence Basic Scientific Research program of China (B20132010).

References

1. Yan QZ, Williams JM, Li J (2002) Chassis control system development using simulation: software in the loop, rapid prototyping, and hardware in the loop. In: SAE 2002 Automotive Dynamics & Stability Conference and Exhibition. Singapore
2. Hardware-in-the-loop simulation for the design and testing of engine-control systems. Control Eng Prac 5(7):643–653 (1999)
3. Hanselmann H (1993) Hardware-in-the-loop simulation as a standard approach for development, customization, and production test. Int Congr Exposition. Chicago, USA
4. Wu H, Zhang H, Vahid M et al (2013) Hybrid electric vehicle powertrain controller development using hardware in the loop simulation. SAE Technical Paper. Detroit, Michigan, USA
5. Zou Z, Sun F, Hu X et al (2012) Combined optimal sizing and control for a hybrid tracked vehicle. Energies 5(12):4697–4710
6. Zou Y, Kong Z, Liu T, Liu D (2017) A real-time markov chain driver model for tracked vehicles and its validation: its adaptability via stochastic dynamic programming. IEEE Trans Veh Technol 1(1):99
7. Plett GL (2004) Extended Kalman filtering for battery management systems of LiPB-based HEV battery packs: part 2. Model Iden 2(134):262–276
8. Wan L, Chen Y (2015) Automatic control principle. Wuhan University Press, Wuhan

Study on Control Strategy of Dual-Motor Electric Power Steering System for Commercial Vehicle

Yaohua Li, Qianlong Feng, Yangsen Zhang, Pengfei Ou and Peng Xu

1 Introduction

The performance of the steering system directly affects the stability and safety of the vehicle. Compared with the hydraulic power steering system, the electric power steering system is assisted by the motor to provide assist torque, reducing the complexity of the power system, and has high reliability [1]. Commercial vehicles have large changes in steering torque at different vehicle speeds and different loads. Especially when the loading is full and the speed is low, the vehicle needs a large steering torque [2]. Ordinary EPS-assisted motor is no longer able to meet the requirements of the commercial vehicle for steering torque, so most commercial vehicles still use hydraulic power steering system. In order to use the EPS on commercial vehicle, the designed EPS must be provided with sufficient torque and the torque can be changed over a wide range. Therefore, a dual-motor EPS for commercial vehicles is proposed in this paper. Based on ADAMS/Car, the multi-body dynamic model of bus is established, and the basic assist characteristic curve of commercial vehicle is given. At the same time, two kinds of electric current distribution and fault-tolerant control strategies of dual-motor EPS system are proposed. Finally, the control model of EPS is established based on MATLAB/Simulink; the feasibility and fault tolerance of current distribution strategies proposed are verified by co-simulation.

Y. Li · Q. Feng (✉) · Y. Zhang · P. Ou · P. Xu
School of Automobile, Chang'an University, Xi'an, China
e-mail: fengql2009@qq.com

Society of Automotive Engineers of China (SAE-China) (ed.), *Proceedings of the 19th Asia Pacific Automotive Engineering Conference & SAE-China Congress 2017: Selected Papers*, Lecture Notes in Electrical Engineering 486, https://doi.org/10.1007/978-981-10-8506-2_63

2 Dual-Motor Coupling Devices

In this paper, the dual-motor-coupled electric power steering device uses the planetary gear mechanism to achieve the torque coupling of the two motors. The structure is shown in Fig. 1, the carrier 6 is fixed, and the ring gear 4 is the output device. The power provided by the motor coupling is transmitted to the worm gear reducer, and then output to the gear rack steering gear, to provide assistance for the steering of the vehicle. The vehicle steering assist torque can be provided separately by the two motors, or by the coupling of the two motors' torque. The dual-motor power steering modes are shown in Table 1, where "1" indicates that the clutch is in the closed state and "0" indicates that the clutch is in the open state; "L" indicates that the synchronizer slides to the left and engages with the torque coupling gear; "R" means that the synchronizer slides to the right and engages with the auxiliary motor gear; "M" indicates that the synchronizer is in the middle position.

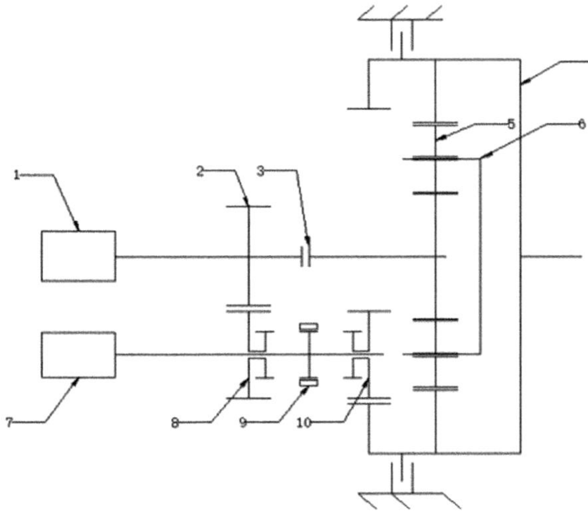

Fig. 1 Double-motor structure. 1. main motor 2. main motor gear 3. electromagnetic clutch 4. ring gear 5. planetary gear 6. planet carrier 7. auxiliary motor 8. torque coupling gear 9. Synchronizer coupling sleeve 10. auxiliary motor gear

Table 1 Dual-motor power steering modes

Modes	Main motor	Auxiliary motor	Electromagnetic clutch	Synchronizer position
The main motor provides assist alone	ON	OFF	1	M
Auxiliary motor provides assist alone	OFF	ON	0	R
Main and auxiliary motor torque coupling assist	ON	ON	1	L

The dual-motor EPS designed in this paper has three kinds of power-assisted mode. When the vehicle is running at high speed, the required steering assist torque is smaller, so the main motor works alone to provide power. At this time, the electromagnetic clutch is closed and the synchronizer combination sleeve is in the middle position, and the main motor is used to transmit the assisted torque through the ring gear. When the vehicle is running at low speed, then the steering assist torque is larger. The main motor and the auxiliary motor work simultaneously to provide assist power for steering. At this time, the electromagnetic clutch is closed and the synchronizer is engaged with the torque coupling gear, and the torque is output through the sun gear and the ring gear. When the ECU detects the main motor failure, then the main motor should stop working, the assist power provided by the auxiliary motor. The electromagnetic clutch is disconnected, and the synchronizer combination sleeve slides to the right and engages with the auxiliary motor gear ring. Then, the power of the auxiliary motor output is transmitted to the ring gear to provide the assist torque.

3 Multi-body Dynamic Model of Commercial Vehicle

In order to analyse the steering dynamics of the vehicle, Adams/Car software is used to build the multi-body dynamic model including EPS steering system, suspension system, transmission system and tyres [3]. The model parameters are shown in Table 2, and the multi-body dynamic model of the bus is shown in Fig. 2.

4 Design of EPS Assist Characteristic Curve

The assist characteristic curve of electric power steering system is the law that the assist torque varies with the change of the vehicle's movement state. A good assist characteristic can solve the contradiction between the steering handiness and the

Table 2 Some parameters of a commercial vehicle

Total quality (kg)	15,000	Suspension	Double wishbone suspension
Axle load (kg)	7000/8000	Steering gear reduction ratio	22.2–26.2
Length × width × height (mm)	7710 × 2500 × 3385	Positive efficiency of recirculating ball steering gear	80%
Tread (mm)	2070	Steering gear output shaft angle range	±47°
Wheelbase (mm)	4300	Maximum output torque (N*m)	5312–6269
Drive form	4 × 2	Tyre pressure (Mpa)	0.72 Mpa

Fig. 2 Vehicle model

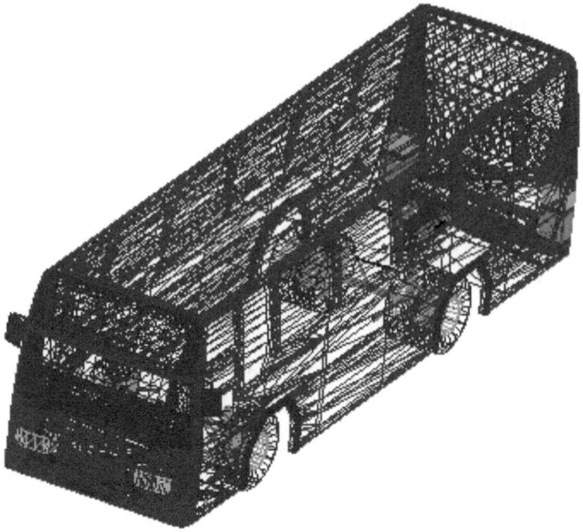

driver's road sense [4]. The assist characteristic curves used in EPS include the linear assistance characteristics curve, the polygonal curve assistance characteristic curve and the curved assistance characteristic curve [5]. The linear assistance characteristic curve is adopted in this paper.

For ordinary passenger cars, when the steering power system is operated, the torque applied to the steering wheel is generally 1.5 N m to 7 N m. Considering that the diameter of the steering wheel of a commercial vehicle is larger than that of a passenger car, when the EPS system works, the minimum torque T_{d0} applied to the steering wheel is 3 N m, and the maximum torque $T_{d\max}$ is 25 N m. According to the corresponding formula, the maximum assisting current is 65.8435 A, so $I_{\max} = 66$ A is selected in this paper. The determination of the speed coefficient should ensure the vehicle has a good steering handiness at low speed and has a good road sense at high speed, so the ideal assist torque should be reduced with the increase of speed. The speed coefficient R_v formula is as follows:

$$R_v = K_a / \left(i_j \times K_i\right) \quad (1)$$

$$K_a(v) = \frac{T_{\exp}}{T_{d\max} - T_{d0}} \quad (2)$$

where i_j is the assist current, K_i is the current coefficient. $T_{\exp}$ is the ideal assist torque of the power system. As described in the literature [6], the State Key Laboratory of Automotive Simulation and Control (Jilin University) studied the preferred steering torque of the domestic passenger car driver and obtained the steering torque of the driver at different lateral acceleration. The experimental results are shown in Fig. 3.

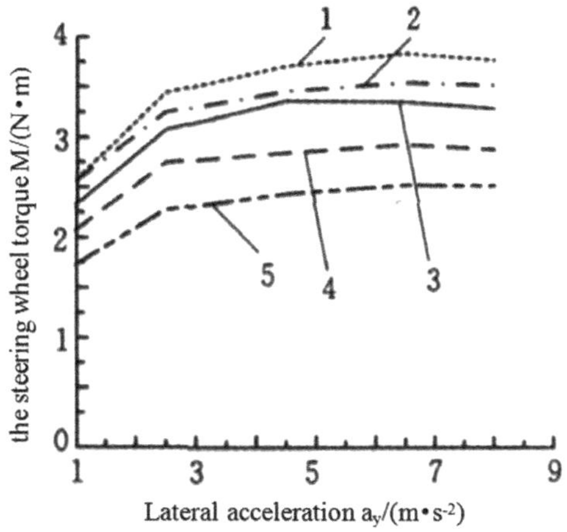

1. v=120km/h 2. v=100km/h 3. v=80km/h
4. v=60km/h 5. v=40km/h

Fig. 3 Characteristic curve of the Chinese drivers' favourite steering torque varying with the lateral acceleration

According to the reference [7], select the steering wheel torque of the lateral acceleration 0.3 g as the ideal hand torque of the driver. Because the data above is based on the test of passenger cars, so the torque is multiplied by the coefficient 1.25 as the ideal torque of the bus. In Adams, a step simulation test is used and set the speed to 10, 20, 30, 40, 50, 60, 70 km/h, then turn the steering wheel at a uniform speed and ensure that the lateral acceleration is 0.3 g, and the maximum steering wheel torque at each vehicle speed is obtained, as shown in Fig. 4. According to formula (1) and formula (2), the speed coefficient of characteristic speed can be obtained, as shown in Table 3.

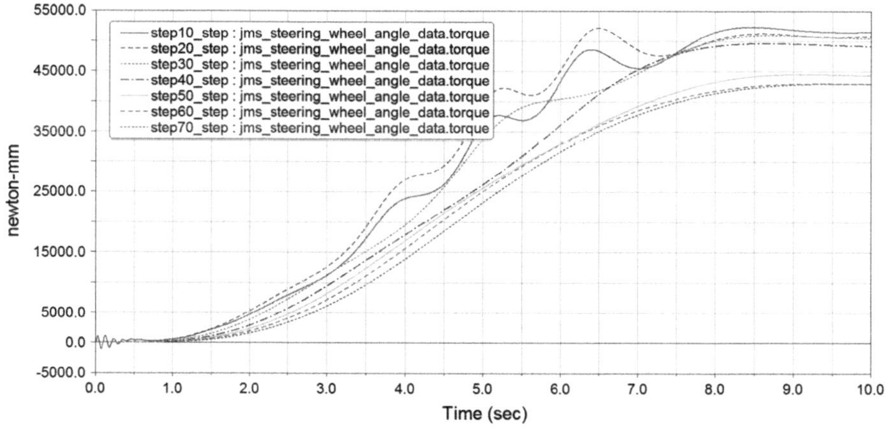

Fig. 4 Maximum steering wheel torque at each vehicle speed

Table 3 Speed coefficient at characteristic speed

v (km/h)	0	10	20	30	40	50	60	70
R	1	0.7663	0.5529	0.4104	0.3433	0.2648	0.1850	0.1576

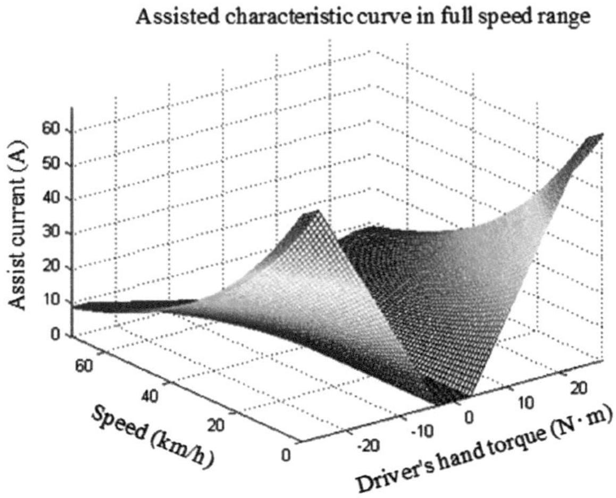

Fig. 5 Assist characteristic curve in the whole vehicle speed range

In order to ensure a continuous change in the vehicle speed, the change in the assist torque is also continuous, so the assist characteristic curve is fitted over the entire vehicle speed range, thereby improving the assist effect. In this paper, the speed coefficient is fitted according to the fitting formula (3), and the assist characteristic curve in the whole vehicle speed range is obtained, as shown in Fig. 5.

$$y = e^{-0.000034162x^2 - 0.027092x} \tag{3}$$

5 Control Strategy of Dual-Motor EPS System

5.1 Current Distribution Strategy

In this paper, two kinds of current distribution strategies are proposed. Scheme 1: when rotating the steering wheel, if the required torque is smaller, the target current I_{goal} is not higher than the rated current I_{rated} of the main motor, then the target current is entirely borne by the main motor, and the auxiliary motor current is zero. When the required torque is large, the target current I_{goal} is higher than the rated current of the main motor I_{rated}, the main motor works at rated current, and the more part is taken by the auxiliary motor, as shown in Fig. 6a. Scheme 2: the main motor and the auxiliary motor are each assigned half of the target current, when the power system work, the two motors provide the same size of the torque, as shown in Fig. 6b.

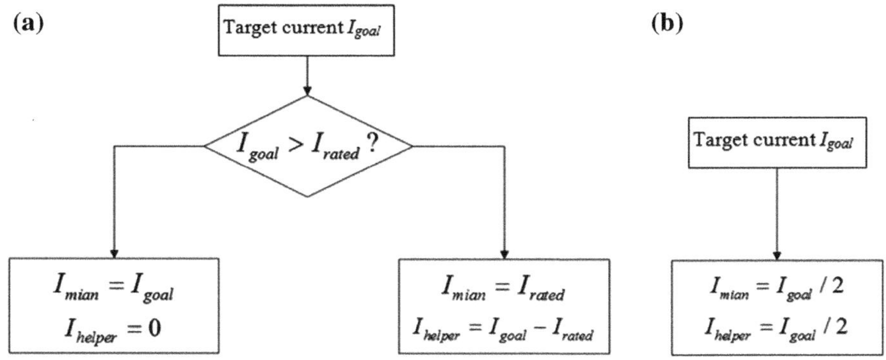

Fig. 6 Current distribution control strategy

5.2 *Fault-Tolerant Control Strategy*

The dual-motor assist steering system not only provides a large assist torque, but also when one of the motors fails, the other motor can still provide all or part of the steering torque to improve the reliability of the dual-motor electric power steering system. In this paper, two fault-tolerant control strategies are designed for two current distribution schemes.

The fault-tolerant control strategy for scheme 1 is shown in Fig. 7. When the bus is running at low speed, the steering system needs large torque, and the two assisted motors are all working. The main power-assisted motor works at rated current, and the target current of the excess part is taken by the auxiliary motor. What's more, the ECU detects whether the main motor and auxiliary motor are working properly; if there is any motor working abnormal, then disconnecting this motor's electromagnetic clutch, the other motor operates at rated current, and the power system

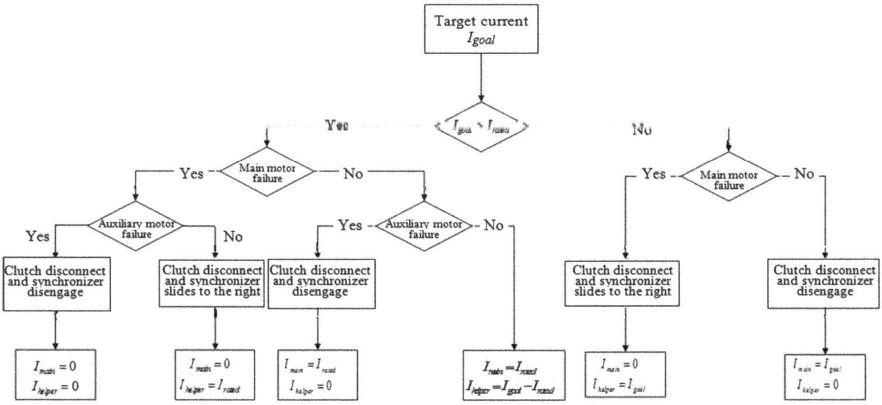

Fig. 7 Fault-tolerant control strategy of scheme 1

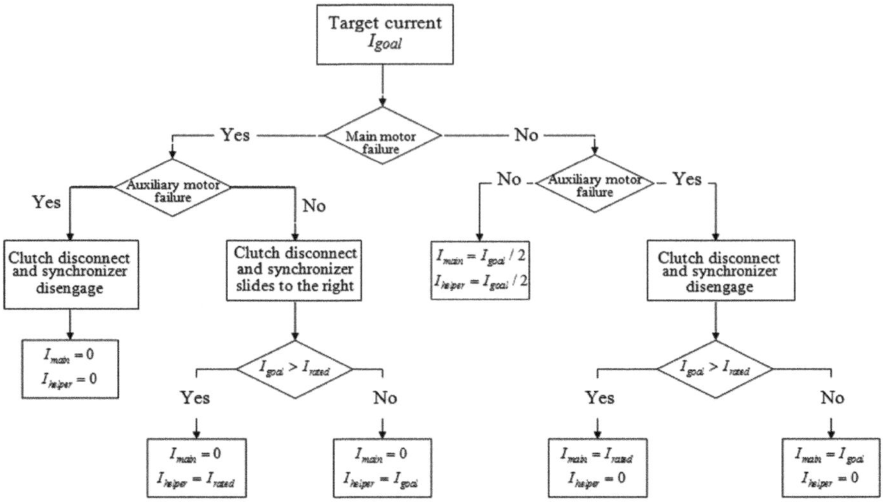

Fig. 8 Fault-tolerant control strategy of scheme 2

cannot provide all the assist torque. When the bus is running at high speed, the steering system needs small torque, and the main motor provides the assist torque alone. If the main motor is working abnormally, disconnect the main motor electromagnetic clutch and the main motor current is set to zero, and the auxiliary motor will bear all target current. If two motors are all faulty, disconnect the total electromagnetic clutch, and the required steering torque is fully supplied by the driver.

The fault-tolerant control strategy for scheme 2 is shown in Fig. 8. When the bus is steering, the main motor and the auxiliary motor bear half of the assist current. When ECU detects one motor failure, the motor electromagnetic clutch should be disconnected, and the assist current is entirely borne by another motor. If the target current is higher than the rated current of the motor, in order to ensure the normal operation of the motor, the motor should work at rated current. If two motors are all faulty, disconnect the total electromagnetic clutch and the target current is set to zero, and the required steering torque is fully supplied by the driver.

6 Co-simulation of Dual-Motor EPS System

6.1 Establishment of Co-simulation Model

According to the control strategy of EPS, the control model of EPS system is established in MATLAB/Simulink, and the bus dynamic model and the dual-motor EPS control model are connected by ADAMS/Control. In the bus dynamic model, the input signal is the power torque acting on the steering rack. The output signals are the bus speed, the steering wheel torque and the rotation angle. The input signals in MATLAB/Simulink are bus speed, steering wheel torque and rotation angle, and the output signal is the power torque. This software realizes the co-simulation by linking the input signal and the output signal. The co-simulation model is shown in Fig. 9.

6.2 Verification of Current Distribution Control Strategy

In this paper, the current distribution control strategy is verified by step test. Figure 10 shows the dual-motor current distribution in high-speed driving condition of scheme 1, and the required steering torque is small. Set the simulation time as 3 s, turn the steering wheel at 1 s, and keep the steering wheel stable at 1.5 s. As the hand torque is larger than T_{d0}, the power system starts to provide assist torque for the car steering through the dual-motor EPS. But at this time, the target current does not exceed the rated current, so only the main motor works.

Figure 11 shows the dual-motor current distribution in low-speed driving condition of scheme 1. Set the simulation time to 3 s, turn the steering wheel at 0.5 s, and keep the steering wheel stable at 2.5 s. Due to the need for larger steering assist torque, when the simulation reaches 1.5 seconds, the main motor current reaches the rated value, and the assist torque is not enough to reach the required assist torque for steering, and then the auxiliary motor starts to work. The assist torque is coupled by the two motors to provide power for the steering system.

Figure 12 shows the dual-motor current distribution of scheme 2. Set the simulation time to 3 s, turn the steering wheel at 1 s, and keep the steering wheel stable at 1.5 s. The driver's hand torque is larger than T_{d0}, and then the main motor and the auxiliary motor bear half of the assist current to provide assist torque for the steering system.

The simulation results show that the two current distribution strategies designed in this paper can play the performance of the main motor and the auxiliary motor, which can provide the required assist torque for the steering system. And the actual current has better tracking to the target current, improving the response speed of the power-assisted system.

6.3 Verification of Fault-Tolerant Control Strategies

This paper verified the fault-tolerant control strategies by single-lane change test. Keep the speed at 13.286 m/s, and when the simulation is 5 s, the main motor fails and stops working. Figures 13, 14 and 15 show the simulation results of fault-tolerant control strategy with scheme 1. At the beginning of the simulation, the main motor bears all of the power-assisted tasks. When the simulation time reaches 5 s, the main motor fails and stops working. According to the fault-tolerant

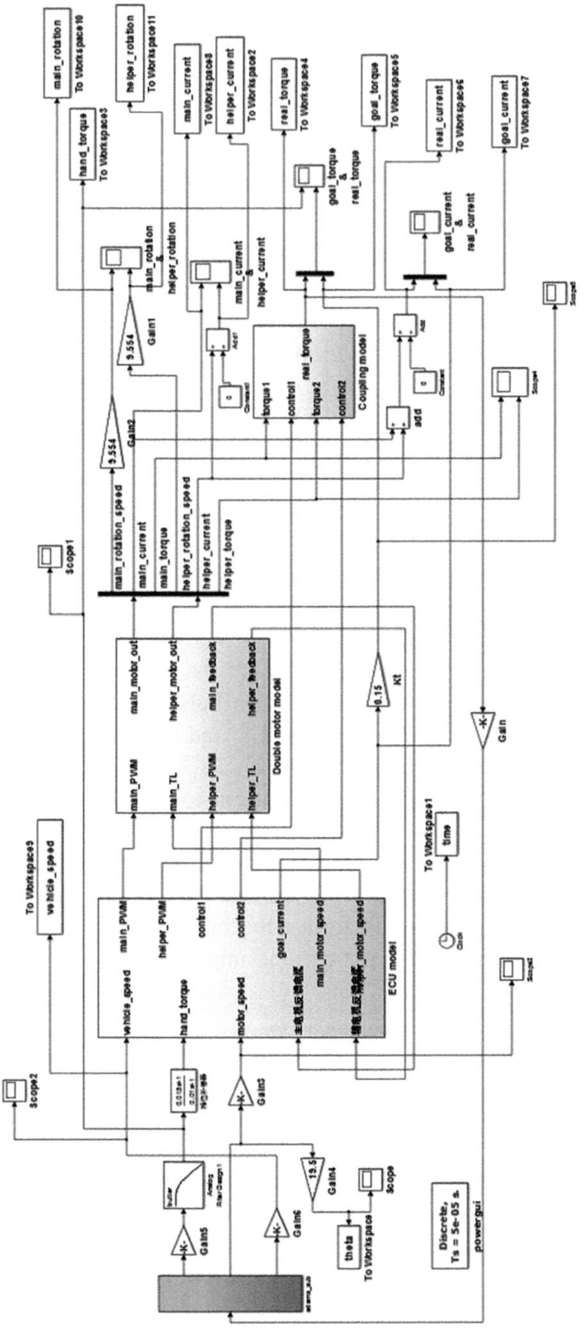

Fig. 9 Co-simulation model of dual-motor EPS

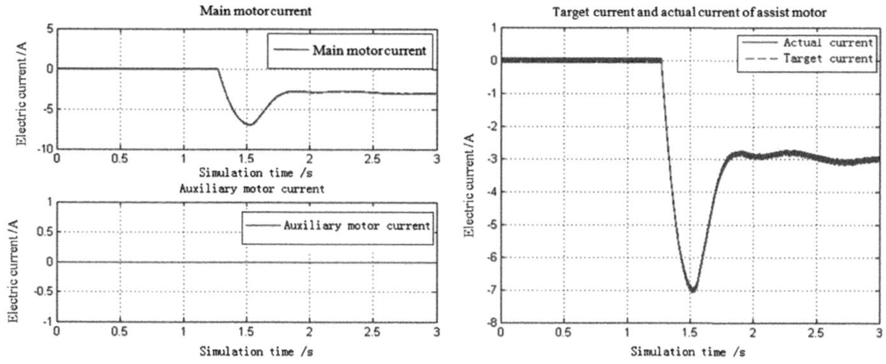

Fig. 10 Current distribution of a bus at high speed

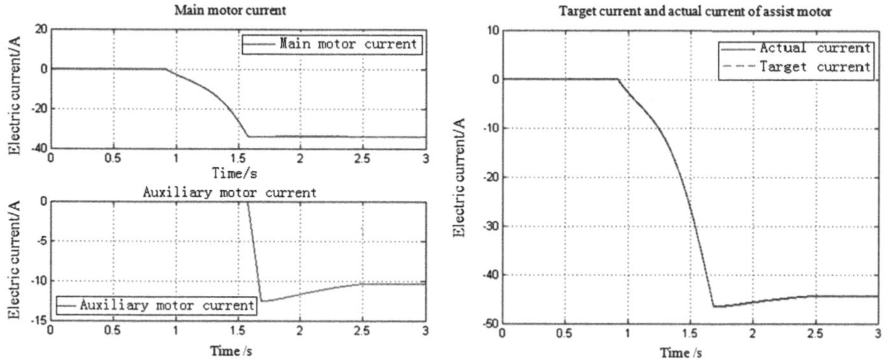

Fig. 11 Current distribution of a bus at low speed

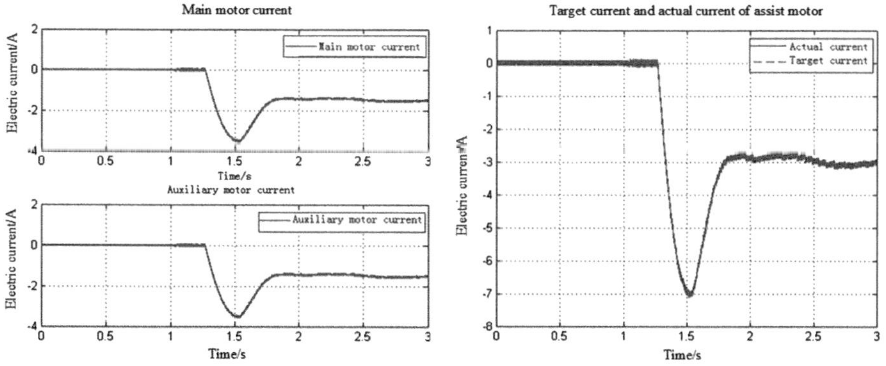

Fig. 12 Current distribution of scheme 2

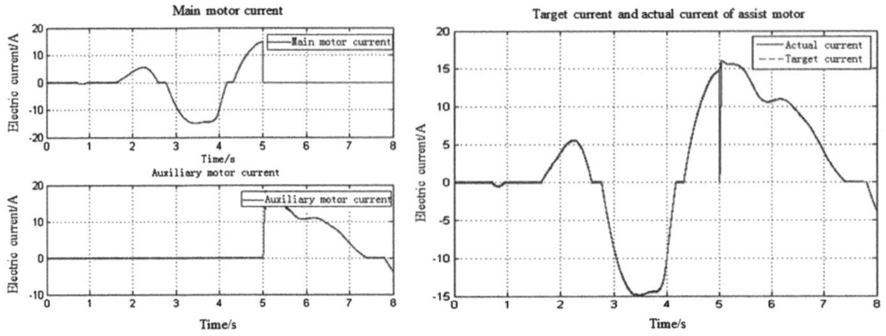

Fig. 13 Current simulation of fault-tolerant control strategy results in scheme 1

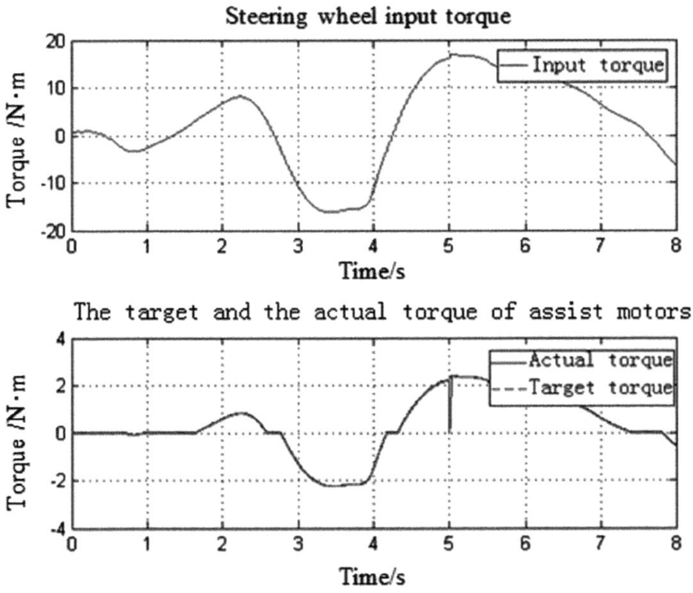

Fig. 14 Assist torque change of scheme 1

control strategy, the main motor no longer receives the reference voltage, and the auxiliary motor takes the power steering task and outputs the assist torque.

Figures 16, 17 and 18 show the simulation results of fault-tolerant control strategy with scheme 2. At the beginning of the simulation, the main motor and the auxiliary motor bear half of the assist current. When the simulation time reaches 5 s, the main motor fails and stops working. Then, the auxiliary motor takes on all the power-assisted tasks, and the output torque increases.

The simulation results show that the two fault-tolerant control strategies have better fault-tolerant control effect. When one motor fails, the other motor can

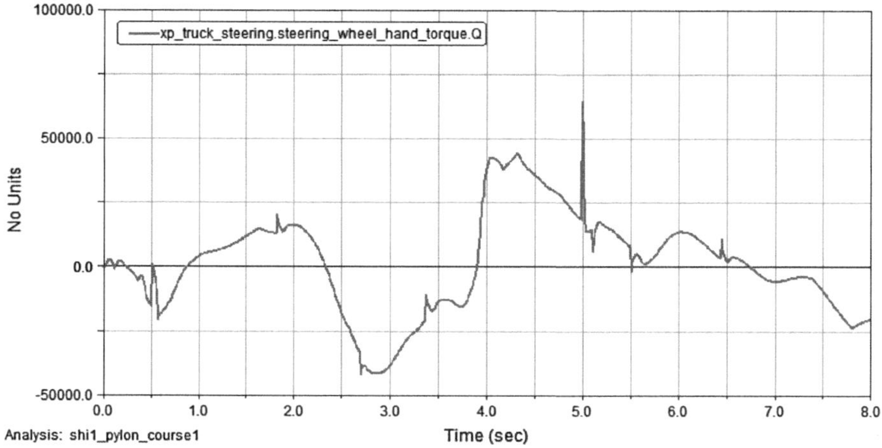

Fig. 15 Steering wheel torque changes of scheme 1

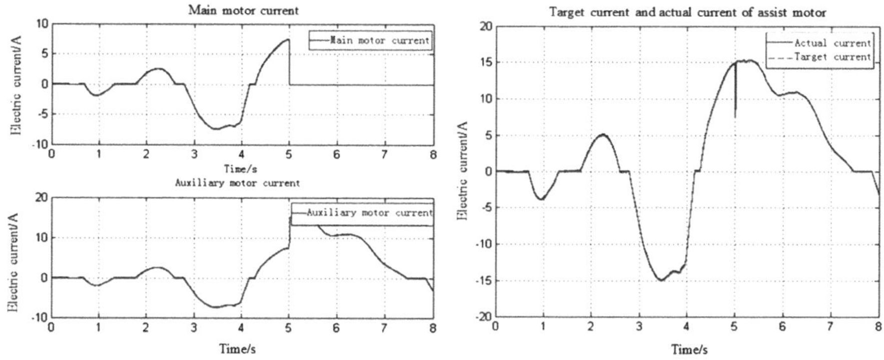

Fig. 16 Current simulation of fault-tolerant control strategy results in scheme 2

provide the required torque according to the control strategy. At the time of 5 s in the simulation, the main motor fails and stops working, and the auxiliary motor began to take on all power steering tasks. Since the motor takes a period of time from start-up to normal operation, there is a large fluctuation in torque and current. The simulation results of the two schemes show that the torque fluctuation of scheme 2 is smaller than that of scheme 1, and the two motors work together at scheme 2 which can make the motors fully utilized. Through the comprehensive comparison, the control strategy of scheme 2 is better than that of scheme 1.

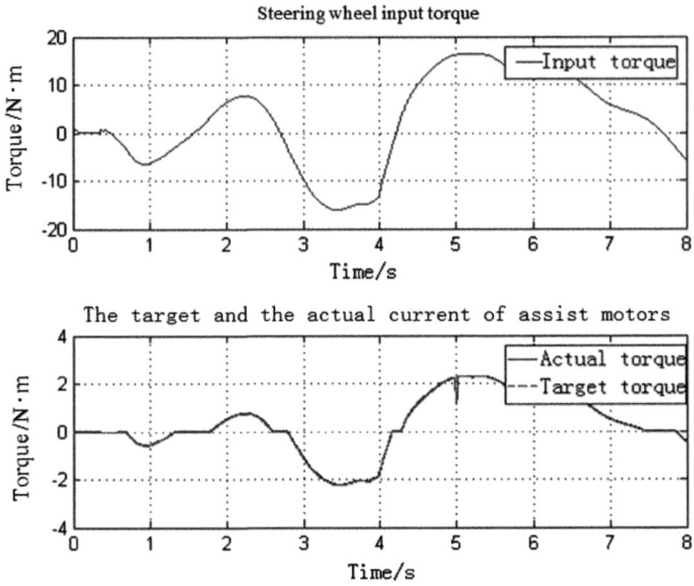

Fig. 17 Assist torque change of scheme 2

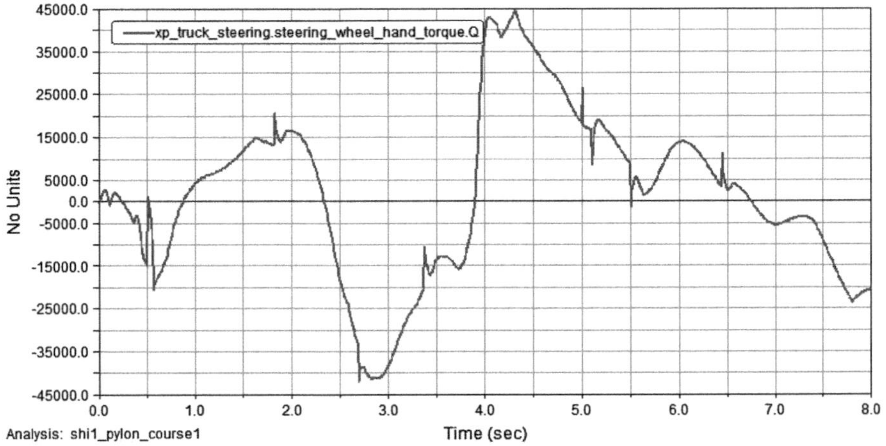

Fig. 18 Steering wheel torque changes of scheme 2

7 Conclusion

This paper presents a dual-motor electric power steering system for commercial vehicles. The multi-body dynamic model of the vehicle is established in ADAMS/Car. The assist characteristic curve of the commercial vehicle is given, and two kinds of current distribution and fault-tolerant control strategies of dual-motor EPS system are proposed. Finally, the control model of the dual-motor EPS is built based on MATLAB/Simulink, and the co-simulation is performed. The simulation results show that the two current distribution strategies can make the double-motor EPS have better assist characteristics and fault tolerance, and the current distribution of scheme 2 has better stability.

References

1. Zhang SY, Wang N, Wang Y et al (2015) Study and design of ECU for automobile electric power steering system. Electron Des Eng
2. Ning C (2012) Study on control strategy of electric power steering system of large-size electric bus. Wuhan University of Technology
3. Li S (2016) Based on The Joint simulation of Adams and MATLAB of racing car handling stability study. Hunan University
4. Zhang Y, Wang Y, Chen D et al (2016) Study on design and evaluation of assisted characteristics curve for EPS. J Chongqing Univ Technol (Natural Science)
5. He Q, He Z, Fang H et al (2014) Study on the design and evaluation of speed correlation coefficient for a Minivan EPS. Mech Sci Technol Aerosp Eng 33(3):404–408
6. Zong C, Na HH, Hu D et al (2007) Study on steering effort preference of drivers based on driving simulator. China Mech Eng
7. Bertollini GP, Hogan RM (1999) Applying driving simulation to quantify steering effort preference as a function of vehicle speed. Sae Technical Papers

A Study of Calibration Method for Fastener Dynamic Torque in Vehicle Durability Tests

Haochong Tang, Yu Wang, Zhuo Cheng, Jiaxing He and Chuanqin Chen

1 Introduction

The most essential function of a chassis suspension system is to alleviate the impact of an uneven road and support the vehicle body. Therefore, a bolt connection failure in the suspension system is a high-level problem. So the thread connection technology is one of the most important parts of technology for the guaranteed reliability in the chassis suspension system [1].

The various complicated forces loaded onto the suspension system, the friction between the fastener and the connected members, the different driving cycle and vehicle load and the unavoidable simulation errors are all factors that make the target torque, set in the conceptual design stage, do not satisfy the vehicle durability test requirements [2–4]. Yang judges the bolt looseness based on the attenuation characteristics of the stress wave which passed through the bolt [5]. Xiang proposed a method of controlled tap detection to detect the bolt loosing state [6]. Junker proposed that the fastening system would be easier to self-loosen by using a static transverse load [7]. The machine and experiment named after him have been widely used to test the anti-loosening property of screw thread connections. Simulation analysis is often used to solve the failure caused by bolt fractures in chassis [8, 9].

Designers previously would solve bolt connection problems as they appeared, so a quantitative analysis and calibration method to test the fastener's required torque are needed to be investigated.

In order to find a calibration method to measure the dynamic required torque in a vehicle durability test, this chapter has built in a calibration and amendment method of required tightening torque based on strain–stress. The essential parameter for bolt connections was defined as the "fastening factor." In order to figure out the drive

H. Tang (✉) · Y. Wang · Z. Cheng · J. He · C. Chen
Guangzhou Automobile Group Co., Ltd, Automotive Engineering Institute, Guangzhou 511434, China
e-mail: tanghaochong@gaei.cn

Society of Automotive Engineers of China (SAE-China) (ed.), *Proceedings of the 19th Asia Pacific Automotive Engineering Conference & SAE-China Congress 2017: Selected Papers*, Lecture Notes in Electrical Engineering 486, https://doi.org/10.1007/978-981-10-8506-2_64

cycle and the dominant frequency that a maximum load can apply onto a bolt, analysis was used from the tension and compression stress of suspension link rods from real-time data collected from the vehicle durability tests. The maximum load can be deconstructed into six component forces and applied in CAE analysis to estimate the rationale behind the bolt selection. The fastening factor of a specific drive cycle (with maximum load applied) can be determined by a simple bench test. The final aim is to determine the upper limit of tightening torque through a torque test [10]. This chapter will state the method of torque amendment of the toe link bolt connection in a certain vehicle.

2 Essential Theory

2.1 Measuring Principle and Application Criteria

The torque amendment and calibration method are mainly applied on two-point link. In order to prevent the bolt from rotating loose or slackening in the durability test or suspension abuse test, the tightening torque may be calculated as follows [11]:

$$M_{\mathrm{t}} = K_{\mathrm{t}} \cdot \left[\frac{F_{ty}}{q_F \cdot \mu_T} + \frac{M_{ty}}{q_M \cdot r_a \cdot \mu_T} + (1 - \Phi) \cdot F_{tz} + \Delta F_{\mathrm{Vth}} \right] \cdot d \quad (1)$$

where M_t—the minimum tightening torque required in a certain time; M_{ty}—torsional moment in the direction along the bolt axis between the members connected by bolt in a certain time; F_{ty}—transverse load between the members connected by bolt in a certain time; F_{tz}—axial load between the members connected by bolt in a certain time; ΔF_{Vth}—change in the preload caused by temperature variation; K_t—tightening factor of connected system in a certain time; d—outside diameter of thread (nominal diameter); r_a—friction radius at the interfaces of the acting part; q_F—number of friction interfaces which transmit shearing force; q_M—number of friction interfaces which transmit torsional moment; μ_T—friction coefficient at the interface; Φ—load factor and relative resilience factor.

The loads on each connecting point in durability test or suspension abuse test can be measured by strain gauges. The influence of bending load and torsional load can be eliminated by proper alignment of strain gauge [12]. Figure 1 shows the circuit principle of load calibration method.

The inner workings of load calibration method are as follows:

1. Two orthogonal strain gauges are pasted on both opposite surfaces;
2. The alignment of strain gauge will make R_1 and R_3, R_2 and R_4 as two set of module.

The relationship between output voltage and resistance variation can be stated as in Eq. 2:

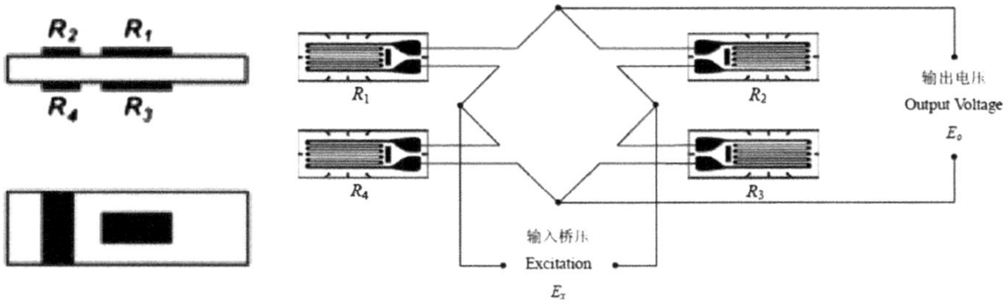

Fig. 1 Circuit principle of load calibration method

$$E_0 = \frac{1}{4}\left(\frac{\Delta R_1}{R_1} - \frac{\Delta R_2}{R_2} + \frac{\Delta R_3}{R_3} - \frac{\Delta R_4}{R_4}\right) \cdot E_x \tag{2}$$

where E_0—the output voltage of the circuit; E_x—the input voltage of the circuit; R_1, R_2, R_3, R_4—resistance of the strain gage; ΔR_1, ΔR_2, ΔR_3, ΔR_4—resistance variation of the strain gage.

As the equation shown above, a reasonable alignment of strain gauges can eliminate the effects of bending load and torsional load and thus measure the pure tensile load.

Equations 3, 4 show relationships of the strain, stress, and force in the measured rod. The strain is linearly related to the force in the case of tension-only, so the "strain–force" curve can be plotted through the calibration test. Then measured object and strain gauge can be assembled as an independent force sensor to measure the axial load along the direction of the rod.

$$\varepsilon = \frac{\sigma}{E} \tag{3}$$

$$\sigma = \frac{F}{A} \tag{4}$$

where ε—the strain of strain gauges; σ—the stress of the strain gauges; F—the force acting on the rod; A—the cross-sectional area of the rod; E—the elasticity modulus of the rod.

Figure 2 shows that comparison of a toe link on a car between calibration and theoretical calculation.

If the toe link made of 45#, then the elasticity modulus is 205 MPa. The resultant curve and the measuring error of the load calibration method is 0.81%. Chart 1 shows the fitting formula and error in a calibration test.

Reasonable alignment of strain sensor can measure the total tension and compression strain variation of the toe link, and the influence of bending and torsional load can be eliminated. So the sensor measured and calibrated is the resultant force of toe link which can be expressed as in Eq. 5:

Fig. 2 Comparison of calibration and theoretical calculation

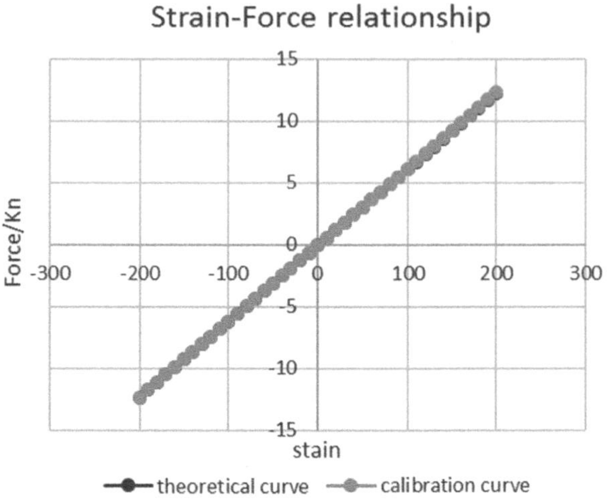

Chart 1 Fitting formula and error

Item	Fitting formula	Error (%)
Calibration value	y = 0.0615*x*−0.0552	0.81
Theoretical calculation value	y = 0.063*x*	

$$F_{tz} = a\sqrt{F_{tz}^2 + F_{ty}^2} \tag{5}$$

where *a*—the factor of axial force.

If Eq. 6 can be achieved, then the error range of tightening torque measurement method can be reduced to $\pm 100 \times n\%$.

$$-n \leq \frac{1 - \sqrt{1 - a^2} - (1 - \Phi) \cdot q_F \cdot \mu \cdot a}{\sqrt{1 - a^2} + (1 - \Phi) \cdot q_F \cdot \mu \cdot a} < n \tag{6}$$

where *n*—the maximum error of tightening torque measurement method.

In order to meet the requirement of Eq. 6, it should meet Eqs. 7 and 8 in the meantime or meet Eq. 9 independently. If the requirement cannot be meet, the method will probably not achieve the expected results that the errors are within $100 \times n\%$.

$$\begin{aligned} a < {} & (q_F \cdot \mu_T + ((\Phi \cdot n \cdot q_F \cdot \mu_T)^2 + 2 \cdot n \cdot (\Phi \cdot q_F \cdot \mu_T)^2 + (q_F \cdot \Phi \cdot \mu_T)^2 \\ & - 2 \cdot \Phi \cdot (n \cdot \mu_T \cdot q_F)^2 - 4 \cdot \Phi \cdot n \cdot (\mu_T \cdot q_F)^2 - 2 \cdot \Phi \cdot (q_F \cdot \mu_T)^2 \\ & + (n \cdot q_F \cdot \mu_T)^2 + n^2 + 2 \cdot n \cdot (\mu_T \cdot q_F)^2 + 2 \cdot n + (\mu_T \cdot q_F)^2)^{0.5} \\ & - \Phi \cdot \mu_T \cdot q_F)/(n + (\mu_T \cdot q_F)^2 - 2 \cdot \Phi \cdot (\mu_T \cdot q_F)^2 + n \cdot (\mu_T \cdot q_F)^2 \\ & + (\Phi \cdot \mu_T \cdot q_F)^2 + n \cdot (\Phi \cdot \mu_T \cdot q_F)^2 - 2 \cdot \Phi \cdot n \cdot (\mu_T \cdot q_F)^2 + 1) \end{aligned} \tag{7}$$

$$
\begin{aligned}
a > {} & (\mu_T \cdot q_F + ((\Phi \cdot n \cdot \mu_T \cdot q_F)^2 - 2 \cdot n \cdot (\Phi \cdot \mu_T \cdot q_F)^2 + (\Phi \cdot \mu_T \cdot q_F)^2 \\
& - 2 \cdot \Phi \cdot (n \cdot \mu_T \cdot q_F)^2 + 4 \cdot \Phi \cdot n \cdot (\mu_T \cdot q_F)^2 - 2 \cdot \Phi \cdot (\mu_T \cdot q_F)^2 \\
& + (n \cdot \mu_T \cdot q_F)^2 + n^2 - 2 \cdot n \cdot (\mu_T \cdot q_F)^2 - 2 \cdot n + (\mu_T \cdot q_F)^2)^{0.5} \\
& - \Phi \cdot \mu_T \cdot q_F)/(-n + (\mu_T \cdot q_F)^2 - 2 \cdot \Phi \cdot (\mu_T \cdot q_F)^2 - n \cdot (\mu_T \cdot q_F)^2 \\
& + (\Phi \cdot \mu_T \cdot q_F)^2 - n \cdot (\Phi \cdot \mu_T \cdot q_F)^2 + 2 \cdot \Phi \cdot n \cdot (\mu_T \cdot q_F)^2 + 1)
\end{aligned}
\tag{8}
$$

$$
\begin{aligned}
a < {} & -((\Phi \cdot n \cdot \mu_T \cdot q_F)^2 + 2 \cdot n \cdot (\Phi \cdot \mu_T \cdot q_F)^2 + (\Phi \cdot \mu_T \cdot q_F)^2 \\
& - 2 \cdot \Phi \cdot (n \cdot \mu_T \cdot q_F)^2 - 4 \cdot \Phi \cdot n \cdot (\mu_T \cdot q_F)^2 - 2 \cdot \Phi \cdot (\mu_T \cdot q_F)^2 \\
& + (n \cdot \mu_T \cdot q_F)^2 + n^2 + 2 \cdot n \cdot (\mu_T \cdot q_F)^2 + 2 \cdot n + (\mu_T \cdot q_F)^2)^{0.5} \\
& - \mu_T \cdot q_F + \Phi \cdot \mu_T \cdot q_F)/(n + (\mu_T \cdot q_F)^2 - 2 \cdot \Phi \cdot (\mu_T \cdot q_F)^2 \\
& + n \cdot (\mu_T \cdot q_F)^2 + (\Phi \cdot \mu_T \cdot q_F)^2 + n \cdot (\Phi \cdot \mu_T \cdot q_F)^2 \\
& - 2 \cdot \Phi \cdot n \cdot (\mu_T \cdot q_F)^2 + 1)
\end{aligned}
\tag{9}
$$

If the requirements of Eq. 6 can be met, then the required tightening torque of the bolt can be calculated by Eq. 10 (considering the temperature variation of suspension while driving, so $\Delta F_{\mathrm{Vth}} = 0$):

$$
M_t = \frac{K_t}{q_F \cdot \mu_T} \cdot \left(F_t + \frac{M_{ty}}{r_a} \right) \cdot d \tag{10}
$$

where F_t—the axial force of toe link.

Each two-point link in the suspension system has two ball joints or rubber bushing joints on both ends [13].

While the two-point link mounts two ball joints, then the M_{ty} do not need to be taken into account. While the two-point link features two rubber bushing joints, the M_{ty} has a very small weight factor in the calculation results as the torsional stiffness of rubber bush is small in general. The maximum M_{ty} in the driving cycle could be calculated by ADAMS [14].

2.2 Calibration Principle

Some research suggests that the classical friction law is not completely correct as the friction coefficient is related to load, relative velocity, and actual contact area [3]. Almost all the fasteners in suspension system work in secular cyclical loading, impulsion and vibration force condition, the limitation of classical friction law may lead to a wrong result in the tightening torque calculation, so it is necessary to measure the characteristic parameters of joint surface.

It takes a lot of time and effort to measure the K, q_F, and μ_T independently [15, 16], but consider that the K, q_F, and μ_T are only affected by bolt selection, material and surface treatment of connected member, loading rate, and pre-tightening force, so the K, q_F, and μ_T is a definite value while the drive cycle and type of connection are determined. The efficiency can be lifted by calibrating these factors together, and we will define a symbolic "η" to express the effects of K, q_F, and μ_T.

The fastening factor η can be calculated as Eq. 11:

$$\eta = \frac{K_t}{q_F \cdot \mu_T} \tag{11}$$

where η—the fastening factor.

So Eq. 10 can be simplified as in Eq. 12:

$$M_t = \eta \cdot \left(F_t + \frac{M_{ty}}{r_a} \right) \cdot d \tag{12}$$

As the fastening factor η is a collection of parameters K, q_F, and μ_T, it is also affected by bolt selection, material and surface treatment of connected member, loading rate, and pre-tightening force.

The maximum required torque can be calculated as given in Eq. 13:

$$M'_A = \eta_{\max} \cdot \left(F_{t\max} + \frac{M_{ty\max}}{r_{a\min}} \right) \cdot d \tag{13}$$

where M_A'—the minimum tightening torque required in a certain time; $\eta_{\max}$—the maximum fastening factor in a certain time; $F_{t\max}$—the maximum axial force of toe link; $r_{a\max}$—the friction radius that considers the maximum tolerance; $M_{ty\max}$—the maximum torsional moment in the direction along the bolt axis between the members connected by bolt in a certain time.

Equation 13 shows that the required tightening torque is proportional to shear force and the nominal diameter of thread, so the fastening factor η can be calibrated easily by the shearing test which aims at measuring the shear sliding force and tightening torque. In the shearing test, the initial torque, bolt, material and surface treatment of connected member, loading rate, and pre-tightening force should be the same as the original connecting point, the maximum loading should be larger than the maximum load in the drive cycle, and the loading frequency should be the same with the dominant frequency in the maximum load case.

The required tightening torque can only be calibrated by measuring the deformation ε of the control arm in a durability test and the fastening factor η. The test not only simplifies 33% of work, while taking the materials, pre-tightening force, and loading rate into consideration, but also reduces the error caused by reliance on simulation.

3 Torque Calibration of Toe Link on a Car

3.1 Applicability Assessment

The calculation application of a bolted joint between knuckle and toe link in multi-link rear vehicle suspension system can be set as an example. The toe link in this case is a two-point link with two rubber bushing joints at each end in the multi-link rear suspension system that applied on a car. The tightening torque for the bolt (M12, in the case of TTJs [11], strength grade 10.9 and pitch of the thread 1.25 mm) is determined as 100 ± 10 N m. Figure 3 shows the detailed structure.

In concept design stage of suspension system development, the loads of each fastening point can be calculated with the application of CAE analysis, so the connection type, tightening specification, and tightening torque can be determined. Figure 4 shows the maximum load of the toe link in different impacting conditions.

The toe link on the car is responsible for transferring lateral load to the vehicle's body, and the component force along the bolt axial direction is small [12]. According to the calculation result, Eq. 6 can be met, so the torque calibration method can be used in such a condition. The bushings at each end of the toe link have small magnitude of torsional stiffness and the $M_{ty} < 13883$ N m that is obtained by dynamics simulation analysis, which is equal to 698 N of shear force that is much less than resultant force F_t. Figure 5 shows the relationship of proportion of axial force-friction coefficient error.

3.2 Load Measured and Calibration

The key point of torque calibration is obtaining accurate dynamic load of toe link in road durability test. While calibrating the toe link tightening torque, Fig. 6 shows the load–strain curve.

Do zero-drift and filtering treatment to data collected by strain gauge, and the axis load in different drive cycles can be found with the help of bench test, shown as in Fig. 7.

The M_{ty} in the corresponding location of the maximum load can be calculated by ADAMS.

Fig. 3 Detailed structure

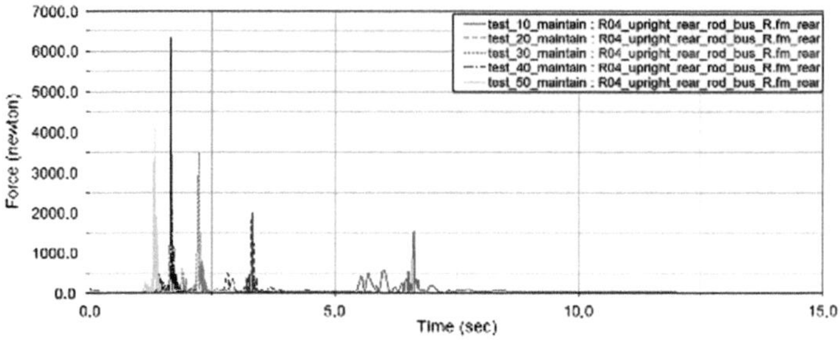

Fig. 4 Maximum load of the toe link in different impacting cases

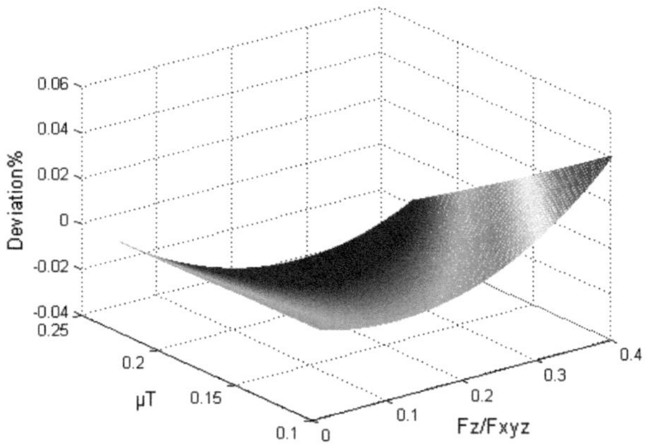

Fig. 5 Relationship of proportion of axial force-friction coefficient error

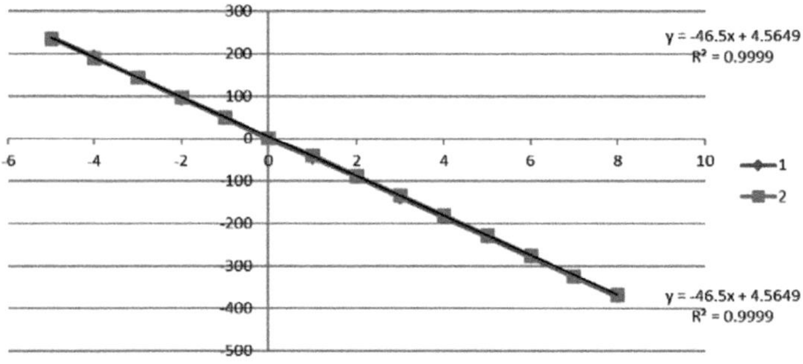

Fig. 6 Load–strain curve of toe link

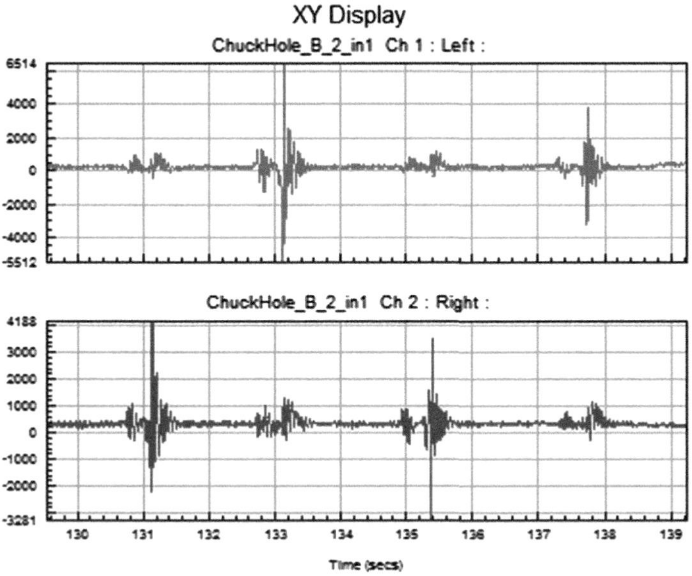

Fig. 7 Axis load of the toe link in durability test

3.3 *Required Tightening Torque Calibration*

The dominant frequency of the maximum axis load can be calculated by using a frequency analysis (Fourier transform), and Fig. 8 shows the frequency spectrum of the axis load.

The fastening factor (η) needs to be calibrated under the dominant frequency of 15 Hz because the fastening factor changes with loading rate. So the fastening factor of the fastening system (knuckle and toe link) is measured by shearing in the condition of 90 N m initial tightening torque and 15 Hz sine loading. Figure 9 shows the measurement of fastening factor in a test cycle.

The statistical analysis of collected data shows the fastening factor distributed in a nearly normal fashion. The value range of fastening factor is from 0.94 to 1.0987. Figure 10 shows the statistical analysis of the fastening factor.

Input the $\eta_{\max}$ into Eq. 13:

$$M'_A = \eta_{\max} \cdot \left(F_{\max} + \frac{M_{Ty}}{r_a} \right) \cdot d = 94.32 \text{ N m} \tag{14}$$

The maximum torque is:

$$M_{\max} = \frac{1 + m\%}{1 - m\%} \cdot M'_A = 115 \text{ N m} \tag{15}$$

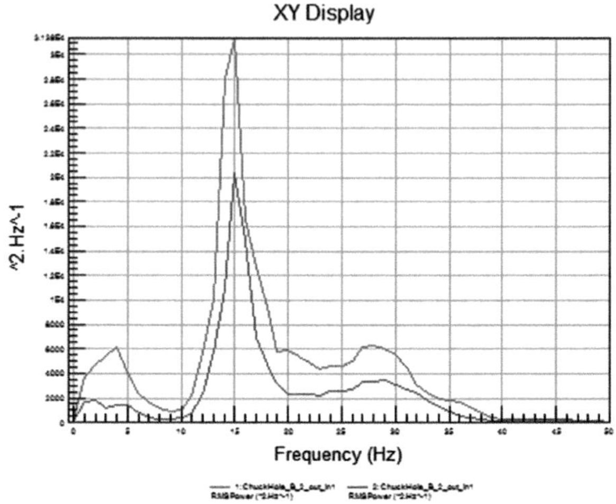

Fig. 8 Spectrum analysis of the maximum load in durability test

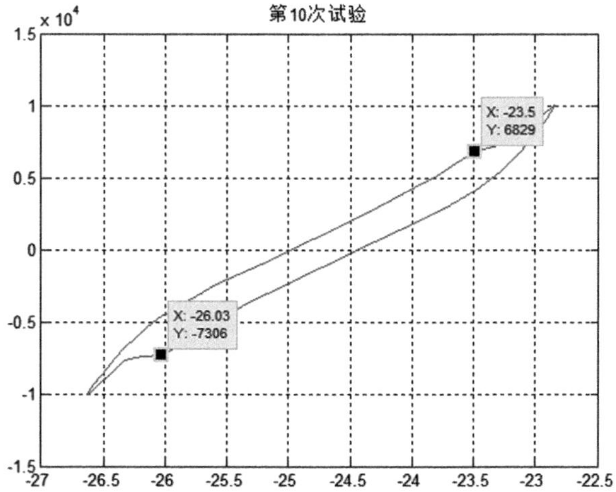

Fig. 9 Fastening factor measurement

where m—the maximum tightening error based on tightening method.

$M_{\max}$ is maximum torque, and m is instrument error. The tightening torque is determined as 110 ± 10 N m according to component tightening test. Figure 11 shows the component tightening test result of this part.

3.4 *Result Analysis*

For fastening connection in suspension system, the bolt slippage will generate self-rotation torque which leads to bolt looseness and torque attenuation [1]. The

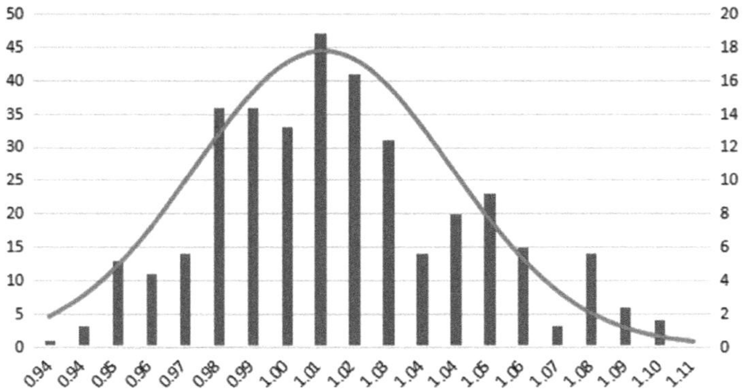

Fig. 10 Statistical analysis of the fastening factor

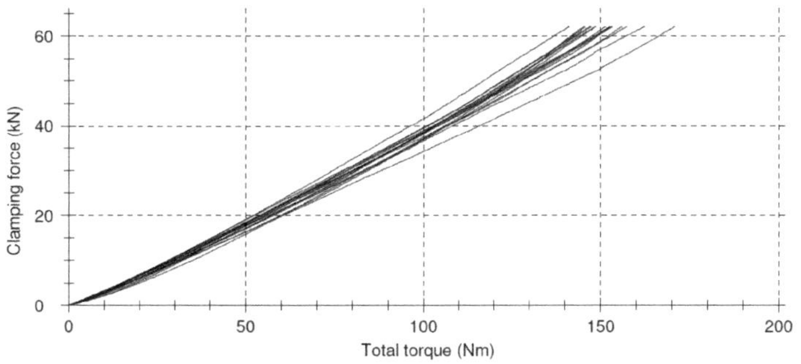

Fig. 11 Component tightening test result

bench durability test and road durability test have shown the reliability of this torque calibration method that there is no bolt looseness and torque attenuation while the modified torque applied. Figure 12 shows the comparison between bench test result of initial torque and modified torque.

4 Conclusion

1. The torque calibration method presented in this chapter is adapted from domestic and international standards. The required tightening torque can only be calibrated by measuring the deformation ε of the control arm in a durability test and the fastening factor η. The test not only simplifies 33% of the work, while taking the materials, pre-tightening force, and loading rate into consideration, but also reduces the error caused by reliance on simulation.

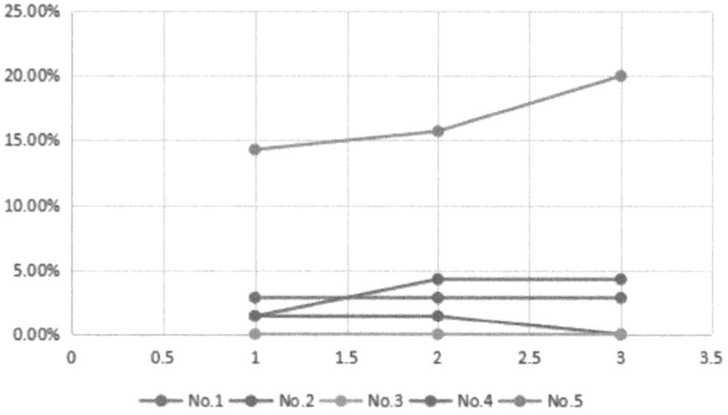

(a) Result of initial torque

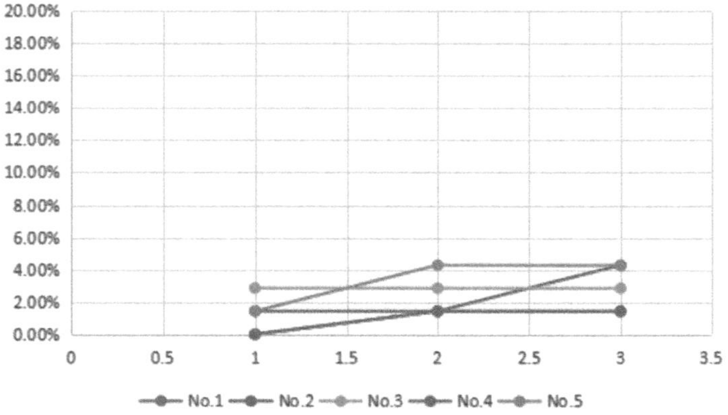

(b) Result of modified torque

Fig. 12 Bench test result

2. The torque calibration method presented in this chapter has a specific range of applications to ensure accuracy, especially for the control arm in the suspension system.
3. The torque calibration method presented in this chapter also can be used to modify the parameter settings in simulation—which is a further direction for investigation.

References

1. Chai Z (2016) Threaded fastener connecting engineering. China Machine Press, Beijing, p 1
2. Liu Y (2009) Principles of tribology. Higher Education Press, Beijing, pp 60–61

3. Bhushan B (2001) Modern tribology handbook. CRC Press
4. Sun S (2015) Discussion on the effects of diameter and type of bolt-hole on anti-sliding coefficient based on principles of tribology. Steel Constr 7:19–23
5. Yang J, Chang FK (2006) Detection of bolt loosening in C-C composite thermal protection Panels: I. Diagnostic principle. Smart Mater Struct 15(2):581–590
6. Xiang Z, Huang J (2012) A controlled tap detection method for bolt tightness. J Exp Mech 27 (5):546–551
7. Junker GH (1969) New criteria for self-loosening of fastener under vibration. SAE Trans 78:314–335
8. Zhou G, Men Y, Rong B et al (2013) Analysis and improvement for the failure of powertrain mount based on simulation and experiment. In: Proceedings of 2013 SAE-CHINA CONGRESS, pp 749–753
9. Li L (2016) Research on moment attenuation of mount system bolts of one passenger car. Agric Equip Veh Eng
10. ISO16047 (2005) Fasteners-torque/clamp force testing
11. Verein Deutscher Ingenieure (2015) Systematic calculation of high duty bolted joints, joints with one cylindrical bolt. VDI 2230 Part 1
12. Rong B, Xiao P, Zhou J et al (2016) Dynamic load cascading and fatigue analysis of rear axle based on road load spectrum. Chin J Automot Eng 6(3):203–211
13. Heißing B, Ersoy M (2011) Chassis handbook. Mercedes Druck, Berlin, pp 289–292
14. Li M, Li M et al (2013) Load decomposition method based on wheel spindle six-component force. Comput Aided Eng (22):61–62
15. Mı´nguez JM, Vogwell J (2006) Effect of torque tightening on the fatigue strength of bolted joints. Eng Fail Anal 13:1410–1421
16. Eccles W, Sherrington I, Arnell RD (2010) Investigation into the self-loosening behavior of bolted joint subjected to rotational loading. In: Eccles W et al. Tribology Int 43:700–707

Study of Wheel Vibration Energy Harvesting for Intelligent Tires

Bing Zhu, Jiayi Han and Jian Zhao

The most direct contact between an automobile and the environment is the interaction between the tires and the road surface. The control of the vehicle which is from the driver or the controller is achieved by the interaction forces between the tires and the road surface ultimately. Therefore, the tire is one of the most important parts of the vehicle. The tractive and brake performance, handling stability, vehicle safety and economy are directly determined by the four tire contact patches which is palm-sized each [1–3]. In recent years, the sensors have developed quickly, which makes the sensors higher integrated, smaller and more inexpensive. Under this circumstances, it is possible to gain the tire and road information directly by embedding sensors in the tire. These kinds of methods which make the tires more intelligent based on tire sensors have been extensively studied. This type of tire is known as intelligent tire [4–8].

However, the problem of energy supply has been widely concerned after the intelligent tire [9]. By reason of the tire rotating independently of the vehicle body, how to power the sensors in the tire has been a bottleneck of intelligent tire development. At present, there is a traditional energy supply method that uses battery to power the sensors in the tire, such as tire pressure monitor system (TPMS). But this method cannot meet energy requirements of the advanced intelligent tire system; besides, replacing the batteries frequently may undermine the practicability and result in environmental pollution. In this case, it will be a new approach to solve this problem by studying energy harvesting of the intelligent tire [10, 11]. The wheel is always vibrating excited by the road roughness when the vehicle is moving [12]. Therefore, the self-power of the intelligent tire will be achieved, if install a cantilever beam which is made of piezoelectric materials on the rim to transform vibration energy into electric energy and then power the sensors in the tire. The precondition of this study is to learn about the characteristics of the

B. Zhu · J. Han · J. Zhao (✉)
State Key Laboratory of Automotive Simulation and Control, Jilin University, Changchun 130022, China
e-mail: zhaojian@jlu.edu.cn

Society of Automotive Engineers of China (SAE-China) (ed.), *Proceedings of the 19th Asia Pacific Automotive Engineering Conference & SAE-China Congress 2017: Selected Papers*, Lecture Notes in Electrical Engineering 486, https://doi.org/10.1007/978-981-10-8506-2_65

wheel vibration and then design the energy harvester according to the characteristics. In this way, a higher energy transform efficiency can be approached.

In this paper, firstly a wheel vibration measurement system is set up for measuring the wheel vibration under different speeds. Then, the wheel vibration characteristics are analysed. Finally, a prototype of piezoelectric cantilever for intelligent tire is designed and tested.

1 Wheel Vibration Measuring System

The wheel vibration measuring system designed in this paper is shown in Fig. 1. A tri-axis accelerometer 3133A2 from DYTRAN is glued to the outside of the hub. The position where the accelerometer is glued corresponds to the position of the energy harvester to guarantee that the measurement results will reflect the excitation condition of the energy harvester. The output signals of the accelerometer are transmitted to the dSPACE DS1103 for recording by slip ring and other devices.

The dSPACE DS1103 uses IBM PowerPC750 GX processor running at 1 GHz. There are 16 16-bit A/D channels, 8 16-bit D/A channels, 5 PWM outputs and 32 digital IO. Besides, it supports CAN and RS232/422 two communication modes. The DS1103 communicates with the upper computer by ISA bus. The three vibration acceleration signals from tri-axis accelerometer are converted and recorded by the DS1103.

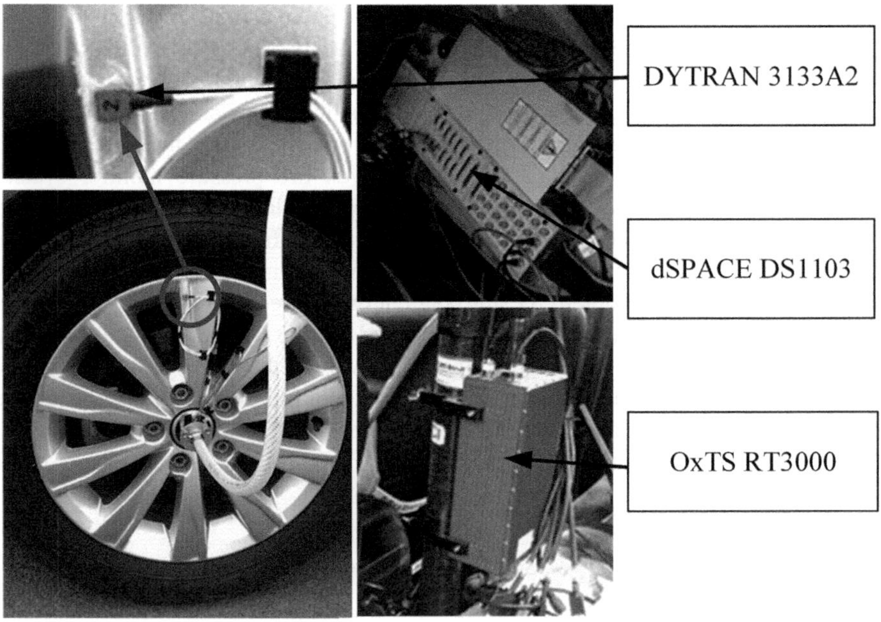

Fig. 1 Wheel vibration measuring system

OxTS RT3000 is an advanced and precision INS/GPS inertial navigation system for measuring movement, position and direction. It can measure the velocity, accelerated velocity, angular velocity, angular acceleration and other kinematic parameters of the vehicle. The RT3000 communicates with the DS1103 by CAN bus. These kinematic parameters of the vehicle can help analysing the relationship between the wheel vibration and vehicle kinestate.

2 Analysis of Wheel Vibration

A road test for measuring the wheel vibration is carried out using the wheel vibration measuring system as introduced earlier. The measuring results are shown in Fig. 2. The test vehicle drives in a straight line at 10, 20, 30, 40 and 50 km/h. The tire pressure is 300 kPa. The vibration acceleration signal lasting for 5 s for

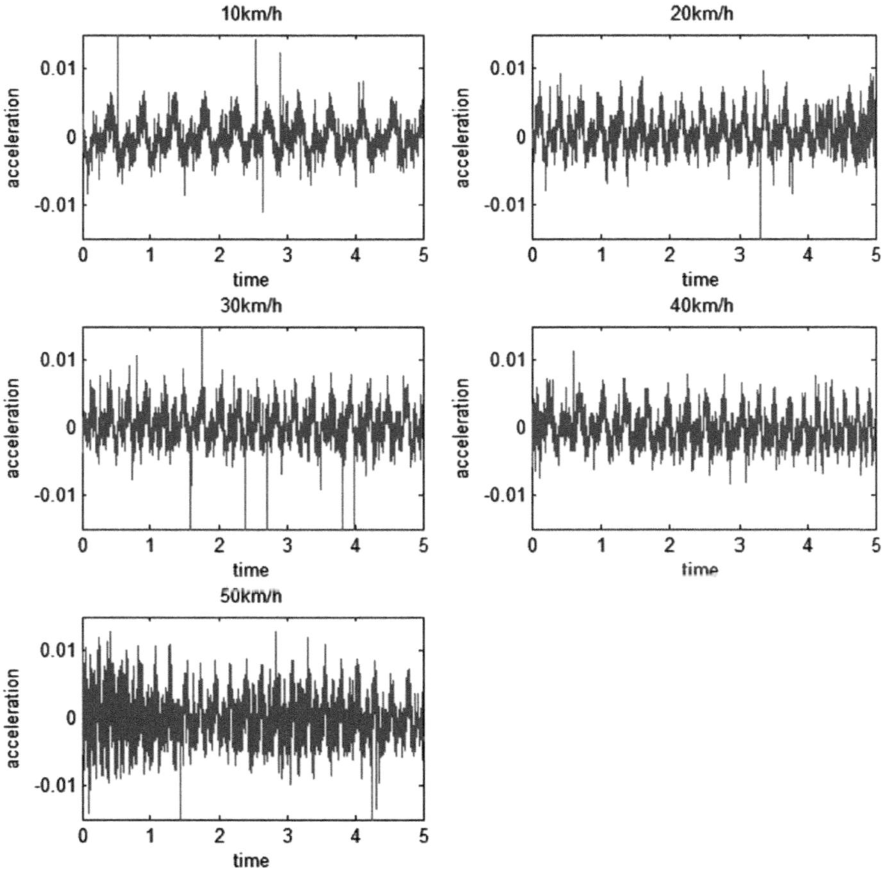

Fig. 2 Wheel vibration at different speeds

each speed is acquired. The sampling frequency is 1 kHz. The vibration acceleration in circumference direction is recorded and analysed emphatically.

A spectrum estimation method based on Yule–Walker AR model is used to calculate the power spectral density (PSD) of the wheel vibration at each speed. The results show that there are two peak values in the power spectral density. One of the two peak values is below 30 Hz, and the other is between 30 and 60 Hz. Emphasis is given to analyse the peak value which is below 30 Hz firstly. The PSD data at each speed below 30 Hz is extracted by a low-pass filter as shown in Fig. 3. It can

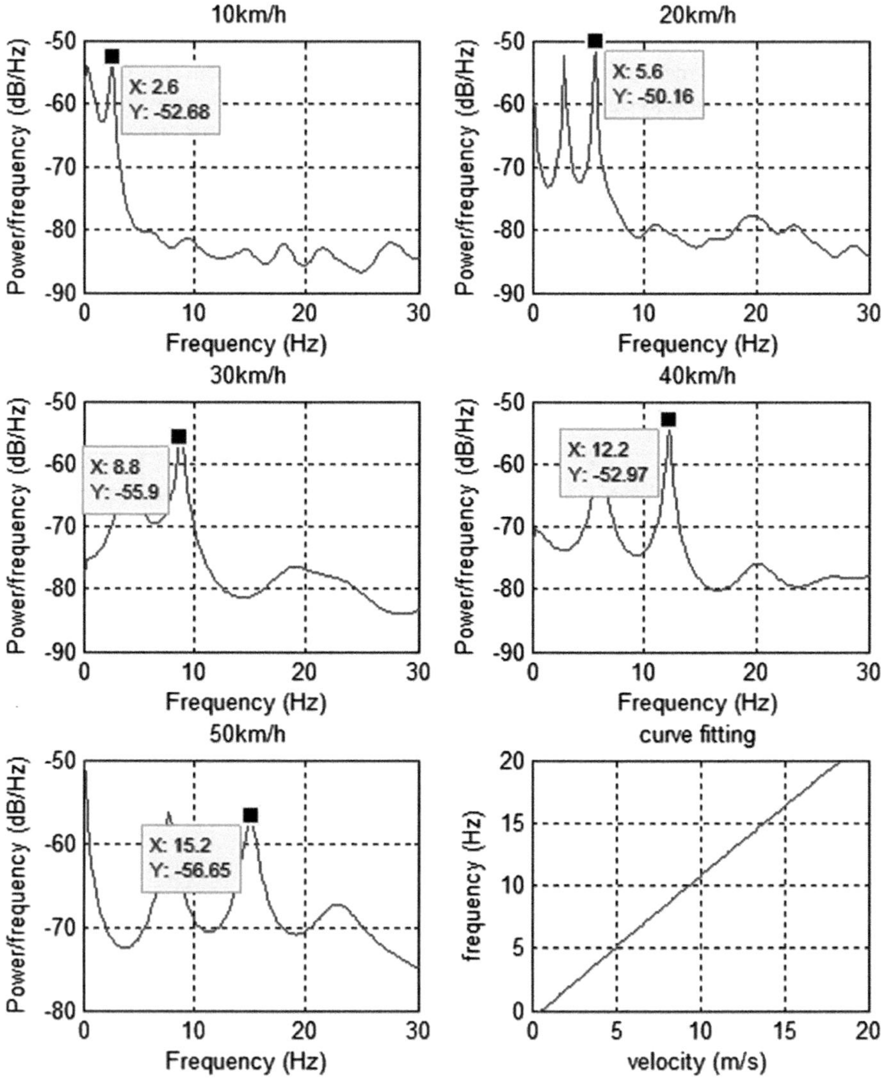

Fig. 3 PSD between 0 and 30 Hz at different speeds

be inferred that the frequency of the peak value at different speeds varies with the vehicle speed. The frequency rises with the increase of the vehicle speed. The relationship curve between the frequency of the peak value and the vehicle speed is fitted according to the principle of least square as shown in the bottom right corner of Fig. 3, and the function is shown below.

$$y = -0.0020x^2 + 1.1633x - 0.6891 \tag{1}$$

Here, y is the frequency of the peak value (Hz) and x is vehicle speed (m/s).

Next, the peak value between 30 and 60 Hz is analysed. A band-pass filter is used to extract the data between 30 and 60 Hz. The spectrum estimation method based on Yule–Walker AR model is also used here. The results show that the frequency of the peak value does not change much with the speed. The frequency is always around 45 Hz as shown in Fig. 4.

3 Test for Energy Harvester

An energy harvester which is based on piezoelectric cantilever beam is designed and made according to the characteristics of the wheel vibration as shown in Fig. 5. Its inherent frequency is 45 Hz. A test for electricity generating capacity of the energy harvester is carried out through the testing system designed in this paper as shown in Fig. 6. The load resistance is 10 kΩ. The vibration exciter excites from 5 to 80 Hz with an interval of 1 Hz. The output voltage of the energy harvester under different excitation frequencies is recorded. The generated power is calculated by the formula below.

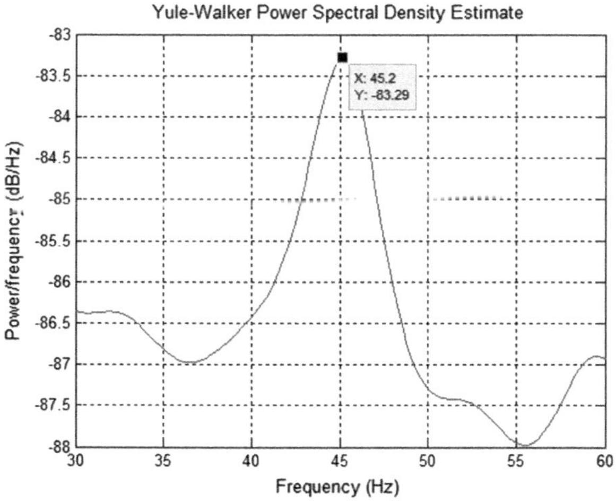

Fig. 4 PSD between 30 and 40 Hz at different speeds

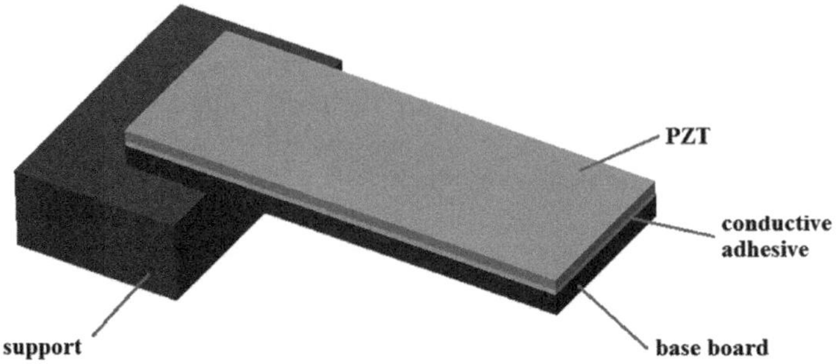

Fig. 5 Piezoelectric cantilever beam

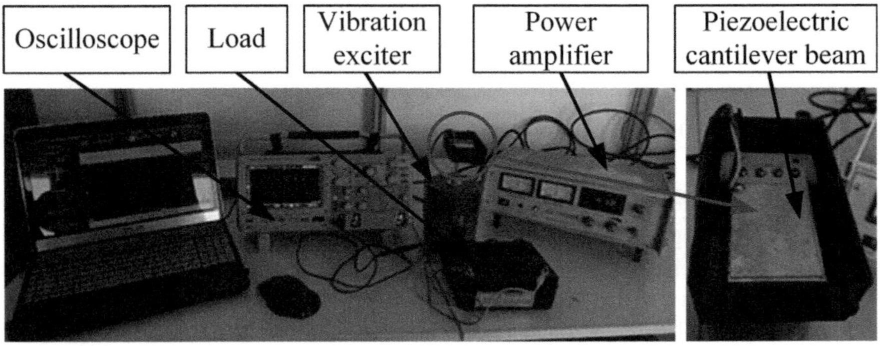

Fig. 6 Energy harvester testing system

$$P = \frac{\int \frac{U^2}{R} \mathrm{d}t}{T} \tag{2}$$

Here, U is the output voltage of the energy harvester, R is the load resistance, and T is recording time.

The result is shown in Fig. 7. It can be observed that the generated power is the highest which reaches 12 mV when the excitation frequency corresponds to the inherent frequency of the energy harvester. With the excitation frequency gradually away from the inherent frequency, the output power decreases. The inherent frequency is designed as 45 Hz. The test indicates that the energy harvester has the highest output when the excitation frequency is between 42 and 47 Hz. That is to say, the energy harvester designed in this paper complies with the design requirement.

Furthermore, the output of the piezoelectric cantilever beam is rectified to direct current with low impedance, low voltage and high current by a nanopower energy-harvesting power supply LTC3588. That is because the output of the piezoelectric cantilever beam is alternating current with high impedance, high

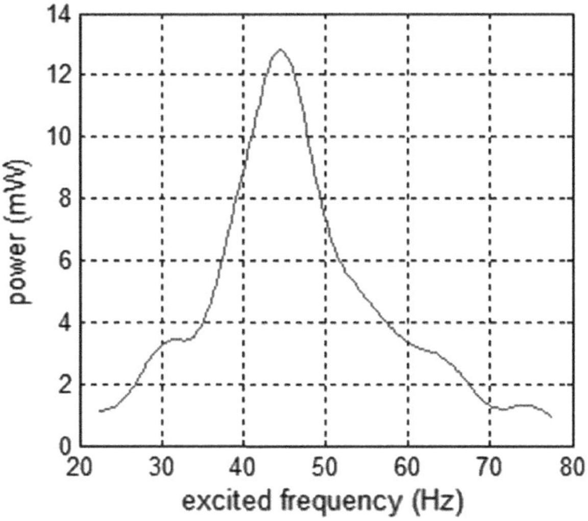

Fig. 7 Output power of energy harvester

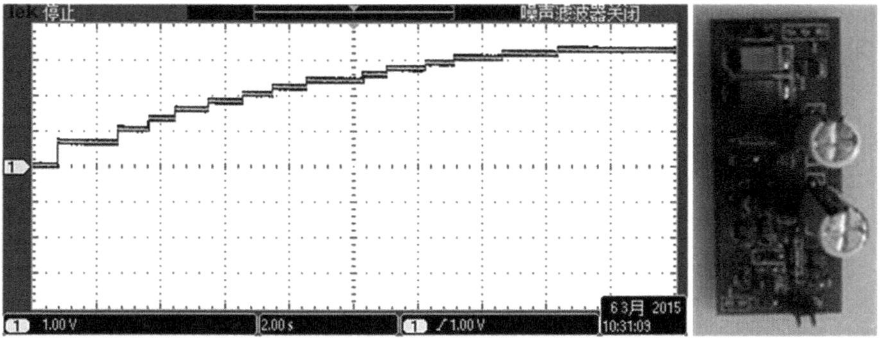

Fig. 8 Energy converting result and energy-harvesting circuit

voltage and low current. The rectification is required to allow the output to fit the electric power loads in the tire. LTC3588 is an energy-collecting circuit from Linear Technology Corporation, which integrates with a low-loss full-wave bridge rectifier and a high efficiency-integrated hysteretic buck. It can collect the output power of the piezoelectric structures and convert it into excellent output to power micro-controllers, sensors or wireless transmitting equipment. The test result is shown in Fig. 8. The figure shows that the output power of the piezoelectric cantilever beam can be collected and converted to a steady direct current at 3.3 V.

4 Conclusion

(1) There are two peak values in the power spectral density of the wheel vibration. One of the two peak values is below 30 Hz. The frequency of this peak value at different speeds varies with the vehicle speed. The frequency rises with the increase of the vehicle speed.
(2) The other is between 30 and 60 Hz. The frequency of this peak value does not change much with the speed. The frequency is always around 45 Hz.
(3) The piezoelectric cantilever beam, which is designed according to the characteristics of the wheel vibration, can harvest the vibration energy effectively. The test shows that the energy harvester can generate more than 12 mW of power when the cantilever beam resonates at about 45 Hz.
(4) The energy-harvesting circuit based on LTC3588 can convert and rectify the output of the cantilever beam to steady direct current at 3.3 V to power the electric equipment.

Acknowledgement This work is partially supported by National Natural Science Foundation of China (51475206, 51575225, U1564211) and Jilin Province Science and Technology Development Plan Projects (20170101138JC).

References

1. Hong S, Erdogan G, Hedrick K et al (2013) Tyre-road friction coefficient estimation based on tyre sensors and lateral tyre deflection: modelling, simulations and experiments. Veh Syst Dyn 51(5):627–647
2. Rajamani R, Phanomchoeng G, Piyabongkarn D et al (2012) Algorithms for real-time estimation of individual wheel tire-road friction coefficients. IEEE-Asme T Mech 17(6):1183–1195
3. Yu Z, Zuo J, Zhang L (2006) A summary on the development status quo of tire-road friction coefficient estimation techniques. Automot Eng 28(6)
4. Niskanen AJ, Tuononen AJ (2014) Three 3-axis accelerometers fixed inside the tyre for studying contact patch deformations in wet conditions. Veh Syst Dyn 52:287–298
5. Singh KB, Arat MA, Taheri S (2013) An intelligent tire based tire-road friction estimation technique and adaptive wheel slip controller for antilock brake system. J Dyn Syst-T Asme 135(3)
6. Reinaa G, Gentileb A, Messina A (2015) Tyre pressure monitoring using a dynamical model-based estimator. Veh Syst Dyn 53(2):1–19
7. Eom J, Lee H, Choi B (2014) A study on the tire deformation sensor for intelligent tires. Int J Precis Eng Man 15(1):155–160
8. Arat MA, Singh K, Taheri S (2013) Optimal tire force allocation by means of smart tire technology. SAE Int J Passeng Cars—Mech Syst 6(1):163–176
9. Singh KB, Bedekar V, Taheri S et al (2012) Piezoelectric vibration energy harvesting system with an adaptive frequency tuning mechanism for intelligent tires. Mechatronics 22(7):970–988
10. Erturk A, Inman DJ (2008) A distributed parameter electromechanical model for cantilevered piezoelectric energy harvesters. J Vib Acoust 130(4)
11. Wischke M, Masur M, Kroner M et al (2011) Vibration harvesting in traffic tunnels to power wireless sensor nodes. Smart Mater Struct 20(8)
12. Lu J, Wu S (2008) Real-time road surface identification based on wheel vibration. J Vib Shock 27(4):19–22

Inter-functional Design and Simulation of Vehicular Electronic Control Units

Pan Song, Xingliang Liu, Jingang Dai and Bolin Gao

1 Introduction

1.1 Background

In recent years, with the electronics, computer science and information technology in widespread use, the automotive intelligence level is getting higher and higher and the automotive electronic control technology has made rapid development and breakthrough. With more and more functions becoming more and more complex, the number of the automotive electronic control systems shows exponential growth. An electronic control system can automatically analyse, compare, calculate and process the signals fed back by various sensors according to the predefined procedures, and then generate new command signals transmitting to the actuators to execute the control action, such that the dynamic performance, active safety, ride comfort and fuel economy of the vehicle can be greatly enhanced [1]. Some of the complex functions may involve a combination of different controllers and actuators outside the master control unit. Figure 1 shows the interactive relationships between the Advanced Driver Assistance Systems (ADAS) and the chassis electronic control units, where different component suppliers are responsible for the realization of the basic functions of all the control units and the OEM is responsible for the design of their interactive logic. However, if any of the sub-systems exert dysfunctional behaviours or their interactive logic is improperly designed, the whole vehicle cannot work in the best way, which has a negative effect on the driving performance and even greatly endangers the safety of the vehicle in some severe cases.

P. Song (✉) · X. Liu · J. Dai · B. Gao
China Automotive Technology and Research Center Co., Ltd., 300300 Tianjin, China
e-mail: songpan@catarc.ac.cn

Society of Automotive Engineers of China (SAE-China) (ed.), *Proceedings of the 19th Asia Pacific Automotive Engineering Conference & SAE-China Congress 2017: Selected Papers*, Lecture Notes in Electrical Engineering 486, https://doi.org/10.1007/978-981-10-8506-2_66

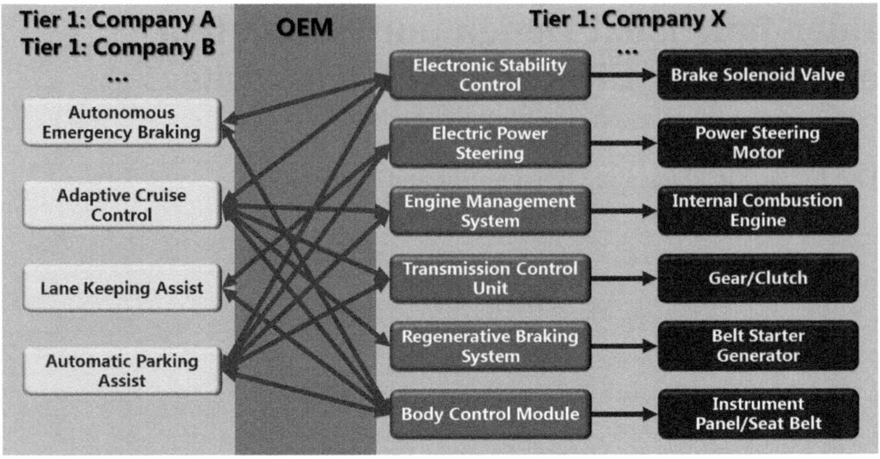

Fig. 1 Interactive relationship between ADAS and chassis ECUs

The traditional step-by-step approach for the development of automotive electronic control units (ECUs) does not work in facing the increasingly complex and huge electronic control system and the short development cycle, mainly reflected in:

A. The relationships among various sub-systems are complex, so the design requirements become hard to reflect and it is difficult to quickly determine whether the final design can meet the requirements.
B. It is very possible that the development cycle of the products may get prolonged, the project risk may get increased, and the products will be unable to meet the rapidly changing market demands.
C. The amount of the software codes are larger and larger, which is time-consuming and laborious to be done by human labours.

In order to resolve the contradiction between the demand of the functional diversity and the rapid development of the products, it is necessary to adopt the computer simulation approach in company with the scientific test case design, such that the detailed interaction definition among systems and the design quality verification can be finished in the early stage of development [2]. The contradiction can be well resolved by utilizing the model-based design method, the modular design concept of which can facilitate labour division and cooperation and improve the scalability and reusability of the simulation models for the secondary development. As a result, the developers of the component suppliers can recognize potential defects in the running process of the products and correct them in a timely manner, such that a complex, stable, safe and reliable product can be developed in a short time. Meanwhile, the electronic and electrical engineers of the OEM are capable of solve the problems that are difficult to find, to analyse, to test and to solve in the development stage [3]. At present, the model-based design method is only employed in the functional safety evaluation [4], electronic and electrical architecture design [5], functional safety verification [6] and so forth. No research has been made to consider and test the inter-functional logic from the whole vehicle-level.

On account of the above-mentioned research status, the simulation platform for the inter-functional testing of vehicular ECUs is built based on the model-based design method. The simulation experiments are conducted for a certain vehicle, where the multi-system coordination manoeuvres under complex functionalities and the correct responding manoeuvres to abnormal interrupts are set up to validate the rationality of its inter-functional logic.

1.2 Purpose

1.2.1 Sorting Out Logical Relationships Among ECUs

The inter-functional relationships of automotive electronic control systems are multithreaded and require the simulation design at the vehicle-level, such that the all the systems can work in a collaborative manner. The specific logical relationships include:

A. Highest-Priority-First: If the Autonomous Emergency Brake (AEB) system is active during the Adaptive Cruise Control (ACC) operation, the ACC should quit immediately.
B. Serial Execution: In the ACC mode, the Automatic Parking Assist (APA) system must not be activated when the Lane Keeping Assist (LKA) system is activated, and vice versa.
C. Parallel Execution: Both the LKA and the Lane Departure Warning (LDW) systems can be activated at the same time. Likewise, both the Regenerative Braking System (RBS) and the ACC system can be activated at the same time.

Through the use of the simulation platform for the inter-functional testing of vehicular ECUs, simulation tests can be conducted in the early stage of development to verify the logical relationships of ECUs, to sort out the logical sequence under the multi-complex functions and to optimize the overall architecture design.

1.2.2 Finding Out Missing Inter-functional Definitions

There exist functional overlaps between each ECU, which need to be coordinated and managed on the simulation testing platform so as to find out the missing inter-functional definitions and avoid the management vacuum in the vehicle research and development. For instance, the engineers responsible for the ACC system do not care about the RBS while the engineers responsible for the RBS do not pay attention to the ACC system. As a consequence, the vehicle may autonomously brake in the ACC mode without the RBS participation, which wastes a good opportunity of energy regeneration and fuel consumption reduction.

1.2.3 Improving Functional Safety Through Interactive Logic Design

Because complex functions require the participation of multiple electronic control systems, when a participant system fails, it is necessary to design and specify on the simulation testing platform how other normal-running systems are coping to meet the functional safety requirements. For instance, in the APA process, if either the APA or the Electric Power Steering (EPS) system fails, the Electric Stability Control (ESC) system needs to self-brake the vehicle and control the vehicle into a safe state.

1.2.4 Achieving Functional Collaboration Between Cross-Domain ECUs

The Internet of Vehicles (IoV), Autonomous Driving Vehicles (ADV) and New Energy Vehicles (NEV) introduce a variety of new features, which cross the traditional functions and complement each other in a way. It is necessary to make full use of this opportunity to enhance the commodity value of the whole vehicle and speed up the deployment process of new products via the use of the simulation testing platform. For instance, by introducing the GPS signals into the ESC controller, the performance of the integrated Tyre Pressure Monitor System (iTPMS) can be greatly improved.

2 Implementation Plan

2.1 Functional Principle

Figure 2 shows the software architecture of the simulation platform for the inter-functional testing, which includes the general vehicle simulation model and its human–machine interfaces, the functional algorithm models of the real ECUs, the real vehicle CAN network and the gateway models, and the test case management module.

Each of the functional algorithm models of the ECUs is the system model developed using the related software tool according to the system requirements, including the definition of the control logic, the model development and the simulation verification. When the whole system is relatively large, it can be divided into a number of small sub-systems, which are then modelled respectively. This executable system model is a modification of the functional algorithm model and

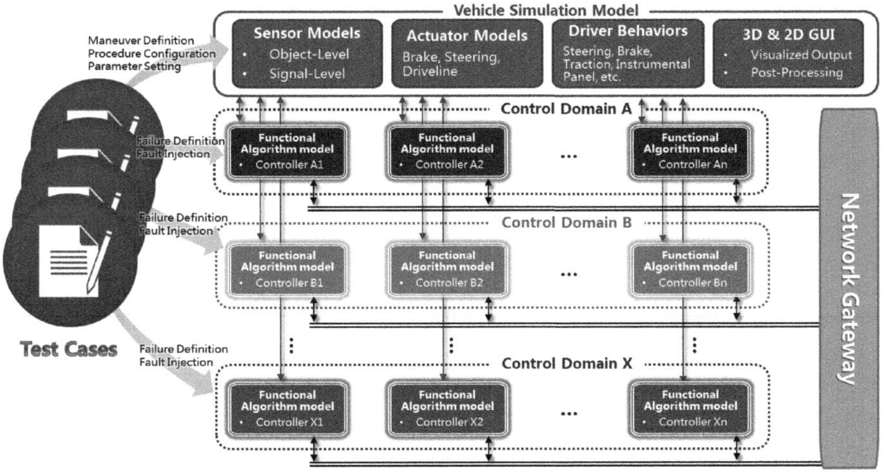

Fig. 2 Architecture of inter-functional simulation test platform of vehicular electronic control units

ensures that the system model can meet the operating standards of the hardware simulation platform. Here, the vehicle simulation model should reserve all the signals imported and exported by the system models for integral simulation of the vehicle behaviours. The functional algorithm models of the real ECUs should be provided by the individual component suppliers, requiring that some critical variables (such as the internal states) indicating the internal functions are open to the vehicle communications.

This kind of the model-based design enables closed-loop feedback in all aspects of the whole system, such that any errors can be corrected in a timely manner. For the errors found during the test phase of the functional algorithm models, the developers could easily correct the related system requirements in time, which prevents the accumulation of the errors afterwards and increases the development efficiency. The developers should enact the test method and procedures in advance and test the whole system accordingly. The test method and procedures should be constantly updated based on the experiences of every several tests. Hence, the interaction and coordination logic can be well improved to make sure the vehicle run in the optimum operation mode and to greatly enhance the handling stability, functional safety, dynamic performance and fuel economy of the vehicle.

After the development of the functional algorithm models is finished, they can be tested immediately. Note that the tests at this time can be either the tests for a single functional module or the integration tests of the whole system. In addition to conducting the inter-functional tests in accordance with the test outlines, the engineers should also perform the coverage tests to detect whether every logical branch of the models is correct and whether there is an unreachable branch. The system requirements should be corrected according to any errors found during the tests, the so-called dual trace of the requirements and the models.

2.2 Software Selection

In this paper, the functional algorithm models as well as the CAN network communication and the gateway models are developed in the MATLAB/Simulink, which provides a complete tool chain to support the complete flow of the development process from the conceptual design to the final implementation.

By adding the interfaces to the vehicle simulation program, the simulation testing platform is capable of not only testing the functionalities of the ECUs but also evaluating their performance. There exist a variety of vehicle simulation programs for co-simulation with MATLAB/Simulink on the market. Figure 3 compares different functions of the mainstream vehicle simulation programs [7]. It can be seen that different vehicle simulation programs focus on different functional aspects. However, no program has been built in functional algorithm models with enough sophistication for supporting the design, development, testing and evaluation of the ECUs. Moreover, in order to build a general platform compatible with the mainstream vehicle simulation programs, the interfaces of the vehicle simulation program should be specifically designed, including: the system-level sensor models, the high-precision actuator models, and the complete driver behaviour model. Thus all these sub-systems should be modelled in detail. In practice, the selection of a vehicle simulation program for the run-time environment of the vehicle itself and the scenarios is determined based on synthetic consideration of its computational efficiency, performance precision, program scalability and use cost.

MATLAB SIMULINK	TASS PreScan	MSC CarSim	dSPACE ASM	IPG CarMaker
Vehicle Dynamics Precision	× Low	✓ High	✓ High	✓ High
Model Expansibility	✓ Strong	× Weak	✓ Strong	× Weak
ADAS Sensor Precision	✓ High	× Low	○ Medium	○ Medium
Traffic Scene Modeling Capability	✓ Good	× Bad	○ Medium	✓ Good
ECU Sophistication	× Bad	× Bad	○ Medium	× Bad
Test Case Management	○ Medium	× Bad	✓ Good	✓ Good
Real-Time Platform Compatibility	× Bad	✓ Good	× Bad	○ Medium
Purchasing Prise	○ Medium	✓ Low	× High	○ Medium

Fig. 3 Functional comparison of mainstream vehicle simulation programs

2.3 Program Structure

Figure 4 shows the MATLAB/Simulink block diagram of the simulation platform for the inter-functional testing. The communication architecture of the real vehicle CAN network helps simulate the interactive behaviours among the ECUs and is the core part of the simulation testing platform, which includes five sub-networks: (1) powertrain; (2) chassis; (3) comfort; (4) information; (5) hybrid. Additionally, the CAN gateway model is developed by using the MATLAB—Vehicle Network Toolbox and connects five sub-networks through bus communications, where the CAN-db protocol, transmit counter, communication delay, bus load and failure simulation are also modelled.

Figure 5 lists the constitution of the ECU modules on the five sub-networks. These modules should be the quasi product-level codes integrally demonstrating the inter-functional logic while their hardwires not directly related to the interactive logic can be moderately simplified. Their coding, compiling, encapsulating and testing all follow the general software development process and supports are needed from each component suppliers.

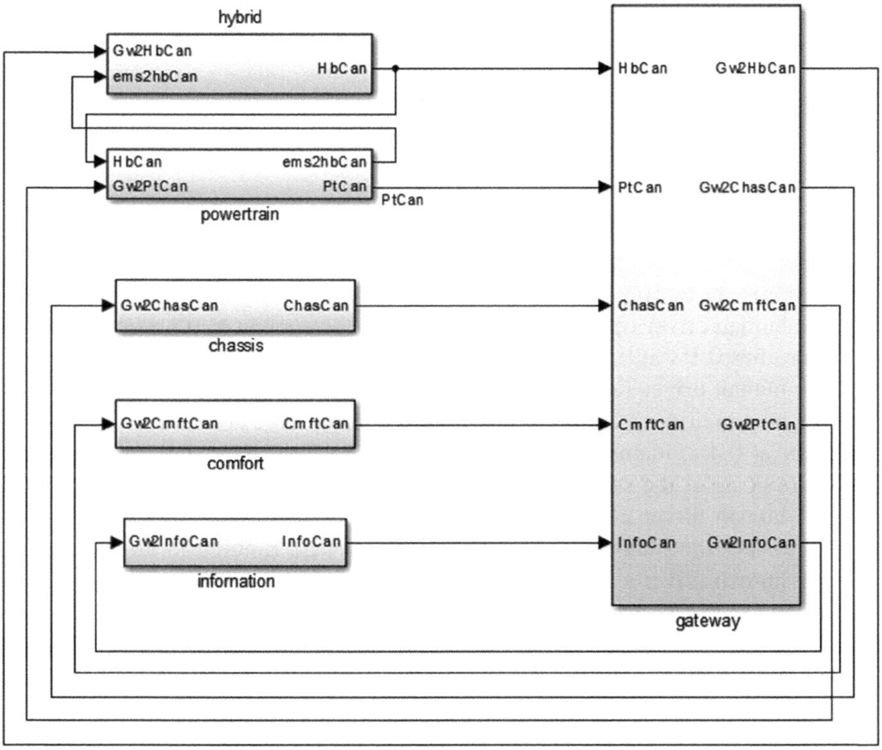

Fig. 4 Block diagram of inter-functional simulation test platform of vehicular electronic control units

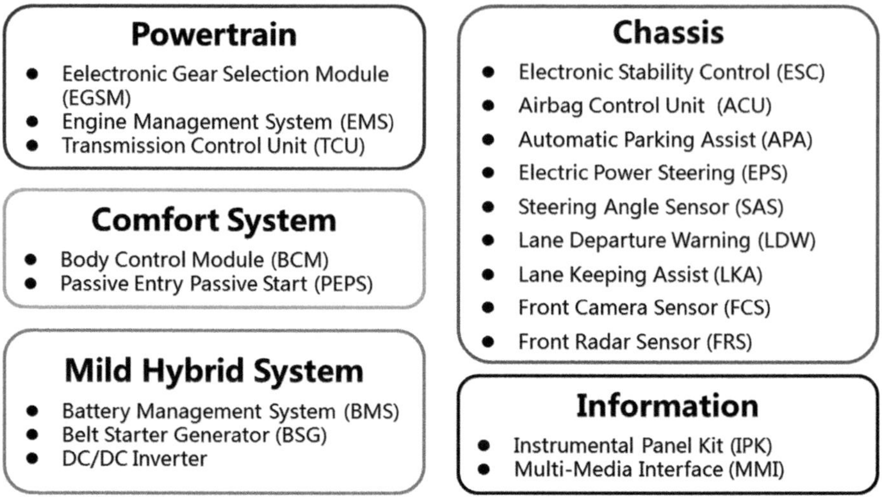

Fig. 5 Module constitution of inter-functional simulation test platform of vehicular electronic control units

3 Simulation Examples

3.1 Multi-system Coordination Manoeuvres in ACC Process

3.1.1 Operation Timing

A. Before the start of this simulation, the initial value of the battery state of charge (SOC) is set to 50%.
B. The human driver opens the door and enters the cockpit at 0.1 s and closes the door after 0.1 s at 0.2 s.
C. The human driver fastens the seatbelt at 0.3 s.
D. The human driver starts to apply the brake pedal force at the rate of 100 N/s to 50 N at 0.4 s, maintaining this value until 4 s, and then starts to release the brake pedal at the rate of −500 N/s at 4 s.
E. The human driver presses the ignition button at 0.5 s and holds it for 1.5 s at 2 s to start the Internal Combustion Engine (ICE), running in idle operation.
F. The human driver starts to release the Electric Parking Brake (EPB) button at 2 s and then lets it return to its original position after 1 s at 3 s.
G. The human driver presses the release button of the Electronic Gear Selection Module (EGSM) and switches the EGSM position from Parking (P) to Drive (D) at 4 s. The ICE is still operating in idle mode.

H. The human driver steps on the gas pedal at 5 s to an openness of 20% and then maintains this value. The vehicle begins to accelerate until the human driver releases the gas pedal at 7 s.
I. The human driver presses the ACC button at 5 s and then presses the confirm button prompted by the Multi-Media Interface (MMI) at 8 s. The vehicle enters the ACC mode.
J. The human driver switches the EGSM position from D to P at 37 s.
K. The human driver pulls the EPB button at 37 s and releases it after 1 s at 38 s.
L. The human driver presses the ignition button at 38 s, holds it for 1.5 s and releases it at 38.5 s.
M. The human driver unfastens the seatbelt at 38.7 s.
N. The human driver opens the door, leaves the cockpit at 38.9 s and closes the door after 0.1 s at 39 s.

3.1.2 Test Results

See Figs. 6, 7, 8, 9, 10 and 11.

This test is the inter-functional simulation between the full-speed range ACC and the RBS. The initial battery SOC is set to 50% before the test begins. After the vehicle accelerates to get close to the target vehicle in front, the host vehicle enters the ACC following mode and decelerates in response to the target vehicle. At this time, the BSG motor starts charging the battery. The test results show that the RBS is operating correctly while the ACC system functioning.

Fig. 6 Animation screenshot of ACC simulation

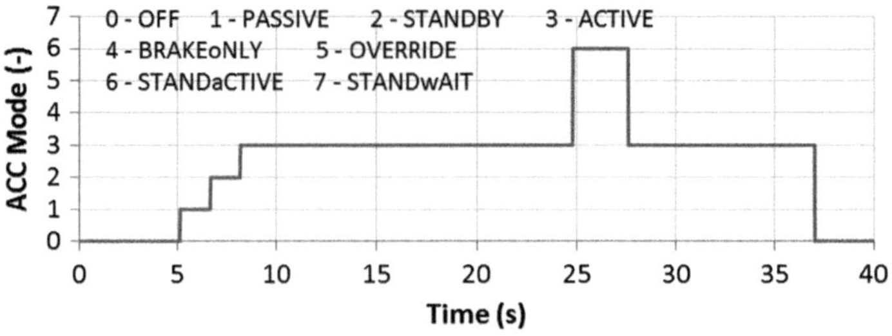

Fig. 7 ACC running state

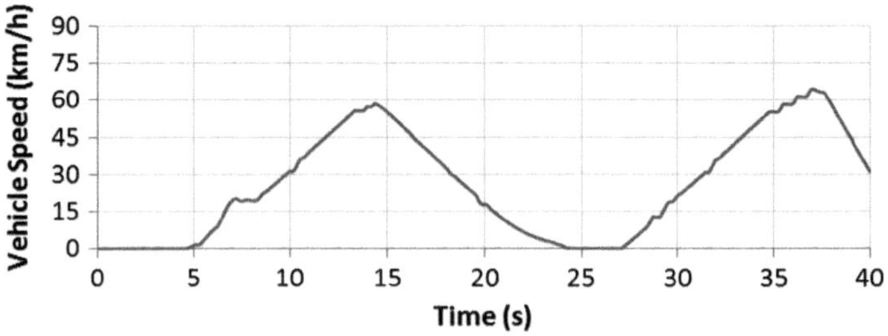

Fig. 8 Vehicle speed

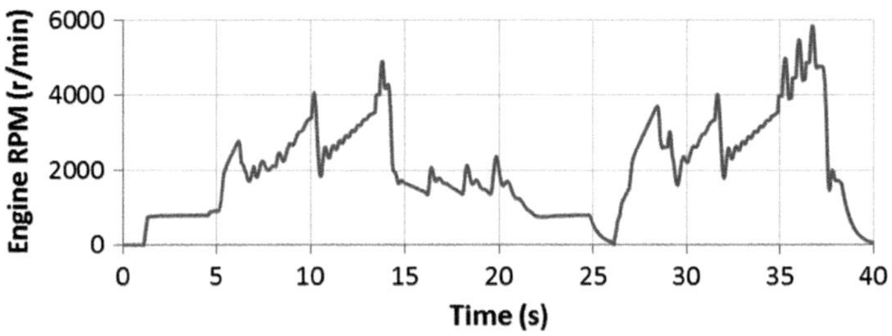

Fig. 9 Engine speed

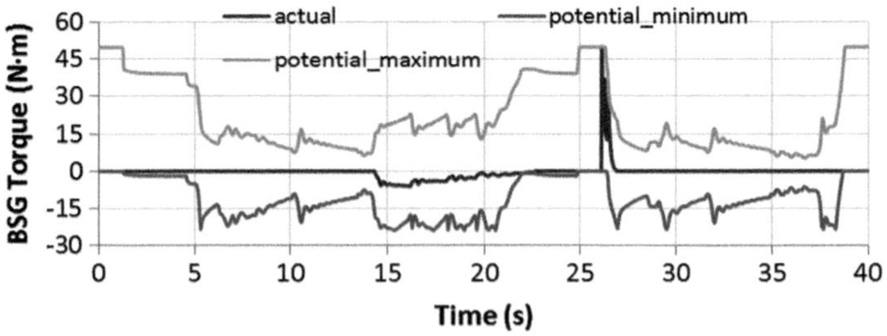

Fig. 10 Actual torque and torque potential of BSG motor

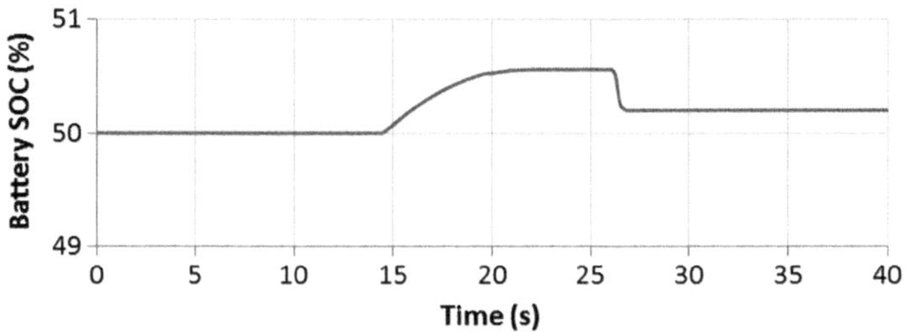

Fig. 11 Battery state of charge

3.2 *Response to Abnormal Interrupt in APA Process*

3.2.1 Operation Timing

A. The human driver opens the door, enters the cockpit at 0.1 s and closes the door after 0.1 s at 0.2 s.
B. The human driver fastens the seatbelt at 0.3 s.
C. The human driver starts to apply the brake pedal force at the rate of 100 N/s to 50 N at 0.4 s, maintaining this value until 4 s, and then starts to release the brake pedal at the rate of −500 N/s at 4 s.
D. The human driver presses the ignition button at 0.5 s and holds it for 1.5 s at 2 s to start the ICE, running in idle operation.
E. The human driver starts to release the EPB button at 2 s and then lets it return to its original position after 1 s at 3 s.
F. The human driver presses the release button of the EGSM and switches the EGSM position from P to D at 4 s. The ICE is still operating in idle mode.
G. The human driver presses the APA button at 10 s and holds it for 1 s. The APA function is activated, and then the human driver operates the vehicle by following the instructions prompted by the MMI until the APA manoeuvres are finished.
H. The human driver steps on the gas pedal at 23 s and then releases it at 30 s.
I. The human driver switches the EGSM position from D to P at 37 s.
J. The human driver pulls the EPB button at 37 s and releases it after 1 s at 38 s.
K. The human driver presses the ignition button at 38 s, holds it for 1.5 s and releases it at 38.5 s.
L. The human driver unfastens the seatbelt at 38.7 s.
M. The human driver opens the door, leaves the cockpit at 38.9 s and closes the door after 0.1 s at 39 s.

3.2.2 Test Results

See Figs. 12, 13, 14, 15 and 16.

Fig. 12 Animation screenshot of APA simulation

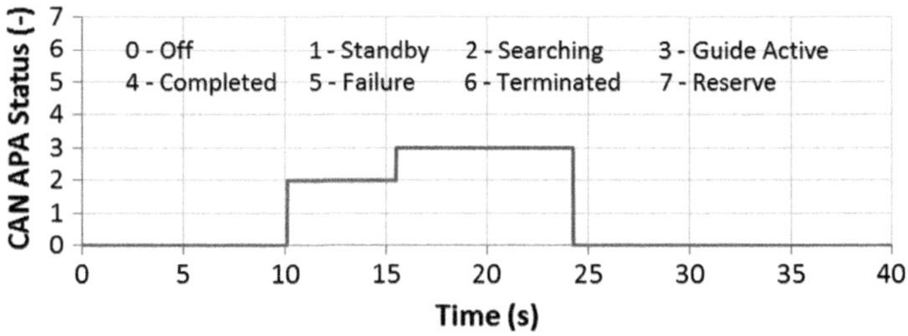

Fig. 13 APA running state

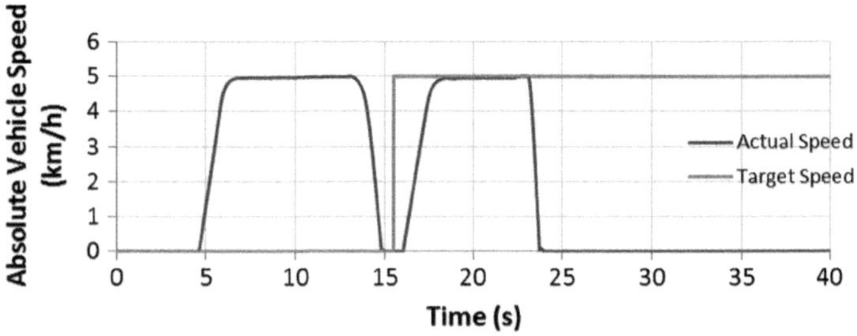

Fig. 14 Vehicle speed and APA target speed (absolute value)

This test simulates the functional response and performance of the APA system when the human driver suddenly steps on the gas pedal during the APA process. After the test begins, the vehicle is in normal operation and parks itself with the APA function. The human driver steps on the gas pedal at 23 s causing the

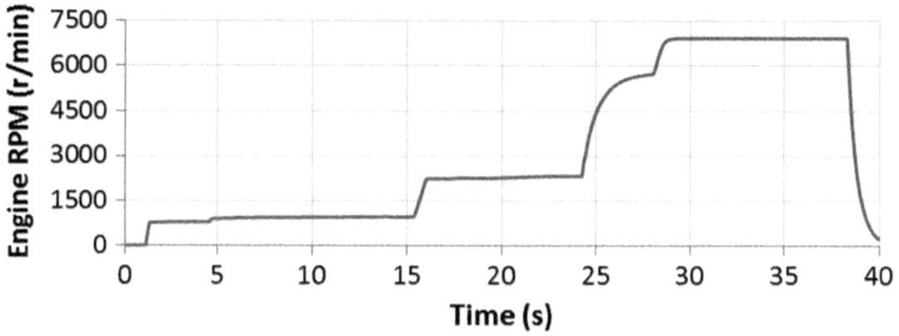

Fig. 15 Engine speed

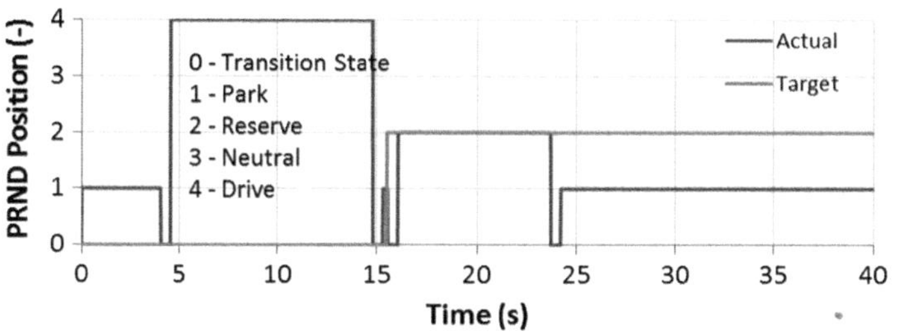

Fig. 16 Actual gear and APA target gear

cessation of the APA function and then releases the gas pedal to allow the APA process to be continued at 30 s. After analysis, the APA and the related systems interact with each other as expected.

4 Conclusions and Prospect

4.1 Research Conclusions

A. Based on the model-based design method, the simulation platform for the inter-functional testing of vehicular ECUs is built, where the interfaces of the vehicle simulation program are specifically designed so as to be compatible with the mainstream vehicle simulation programs, the vehicle sub-systems are modelled in detail so as to support the design, development, testing and evaluation of the vehicular ECUs, and the vehicle network communication is simulated in an actual way so as to achieve the development and integration of the ECU functional modules on each sub-network.

B. The simulation experiments are conducted for a certain vehicle, where the multi-system coordination manoeuvres under the complex ACC function and the correct responding manoeuvres to abnormal interrupts in the APA process are set up to validate the rationality of its inter-functional logic. Furthermore, the early testing platform as well as the test cases is ready for preparing the new complex functions for the subsequent upgrade of the vehicle product.

4.2 Research Prospect

With the increasing application of electronic control systems in the vehicle, the interactions between vehicular ECUs are becoming more and more complex and difficult to manage. As the whole control system is large enough to a certain extent that the engineers are unable to finish coordination of all the sub-systems at the early design stage, the OEM will eventually face irreparable flaws in their close-to-production product. At this point, the OEM needs to accomplish the coordination control of the whole electronic control system through embedded programming in a specific controller. The coordination program should be calibrated and verified on a real test vehicle [8]. By doing so, the component suppliers only need to design and deliver the controllers according to the interfaces and specifications defined by the OEM. Meanwhile, the OEM only needs to design the logic of the interaction and coordination in the Integrated Vehicle Control System (IVCS) in order to achieve the vehicle-level calibration. As shown in Fig. 17, the inter-system logic would be much clearer in the IVCS architecture.

In fact, the IVCS is a code snippet embedded in the ESC controller, similar to the EPBi implementation [9]. The IVCS manages and supervises other control systems through the vehicle network communication. Here, the ESC supplier needs to provide the OEM with the Integrated Development Environment (IDE) and clarify the IVCS operating conditions. The developers of the OEM compile and download the IVCS module to the ESC controller with the assistance of the supplier. The simulation testing platform would also help the OEM design, develop, test and verify the IVCS functions, which would be done in future work.

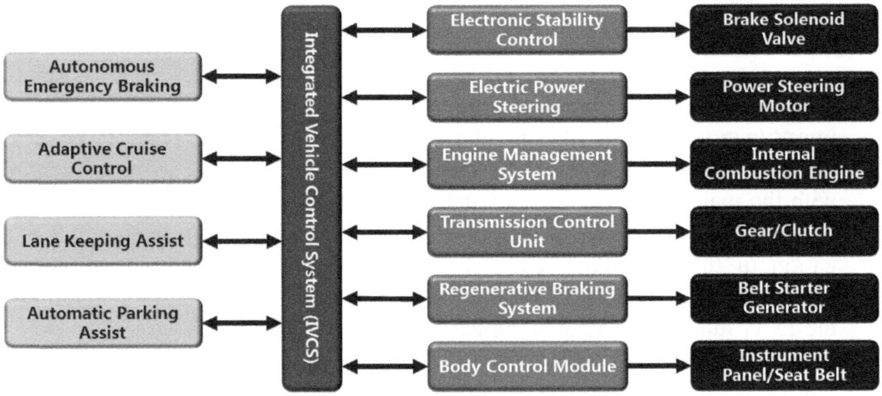

Fig. 17 Integrated Vehicle Control System (IVCS)

References

1. Wang S, Li J, Xia Q (2011) Automotive electronics, 2nd edn. Tsinghua University Press (in Chinese)
2. Yan Jiong, Wang Ji, Chen Huowang (2004) Survey of model-based software testing. Comput Sci 31(2):184–187 (in Chinese)
3. Yang D (2002) System real-time simulation development environment and application. Tsinghua University Press (in Chinese)
4. Nair S, De La Vara JL, Sabetzadeh M et al (2014) An extended systematic literature review on provision of evidence for safety certification. Inf Softw Technol 56(7):689–717
5. Habli I, Ibarra I, Rivett RS, et al (2010) Model-based assurance for justifying automotive functional safety. SAE Technical Paper
6. Liu J, Guo H, Li J (2011) Functional safety standards for automotive electrical and electronic systems ISO 26262. Shanghai Automot 10:57–61 (in Chinese)
7. Helmer T (2015) State of scientific and technical knowledge on pre-crash evaluation. Development of a Methodology for the Evaluation of Active Safety using the Example of Preventive Pedestrian Protection. Springer International Publishing, pp 17–48
8. Mattsson M, Mehler R (2015) Optimal vehicle speed control using a model predictive controller for an overactuated vehicle
9. Zhang Z, Liu W, Xu Y et al (2014) Integrated EPB driving device for parking braking, CN 203404328 U[P] (in Chinese)

Study on Engine Control Software Testing Based on Hardware-in-the-Loop Simulation Platform

Wenwen Zeng, Ying Huang, Xuelong Zheng and Wenqiang Zhao

1 Introduction

Traditional ECU development methodologies have been gradually eliminated, and V-cycle development mode, a model-based development mode mainly including control function design, rapid control prototyping, target code generation, HIL, and calibration as shown in Fig. 1, is recently most favored by vehicle developers [1–3]. Typically, off-line simulation and HIL are used as development methodologies according to different test requirements in different test stages. Normally, the control logic and control algorithm as well as the switching and coordination between various engine operation conditions can be verified through off-line simulation [4]. Different from logical connection in off-line simulation, corresponding interface programs are used in HIL to connect the real ECU and the digital engine model in the processor [5]. The virtual engine, which provides actuator signals for ECU, precisely simulates various sensor signals and to the greatest extent makes the ECU run in a real working environment of engine.

Notice that the dominant merit of HIL is that it can be used for software testing, and it is an efficient methodology for testing software functions and achieving parameter pre-calibration. Besides, software testing is extremely necessary, while the engine control system becomes increasingly complex [6–8].

Referring to the mature software testing methodology classification, engine control software testing technique is classified into many types as shown in Fig. 2 [9]. Typically, it is divided into static test technique and dynamic test technique

W. Zeng · Y. Huang (✉) · X. Zheng · W. Zhao
College of Mechanical and Vehicle Engineering, Beijing Institute of Technology, 100081 Beijing, China
e-mail: hy111@bit.edu.cn

W. Zeng
e-mail: 936754111@qq.com

Society of Automotive Engineers of China (SAE-China) (ed.), *Proceedings of the 19th Asia Pacific Automotive Engineering Conference & SAE-China Congress 2017: Selected Papers*, Lecture Notes in Electrical Engineering 486, https://doi.org/10.1007/978-981-10-8506-2_67

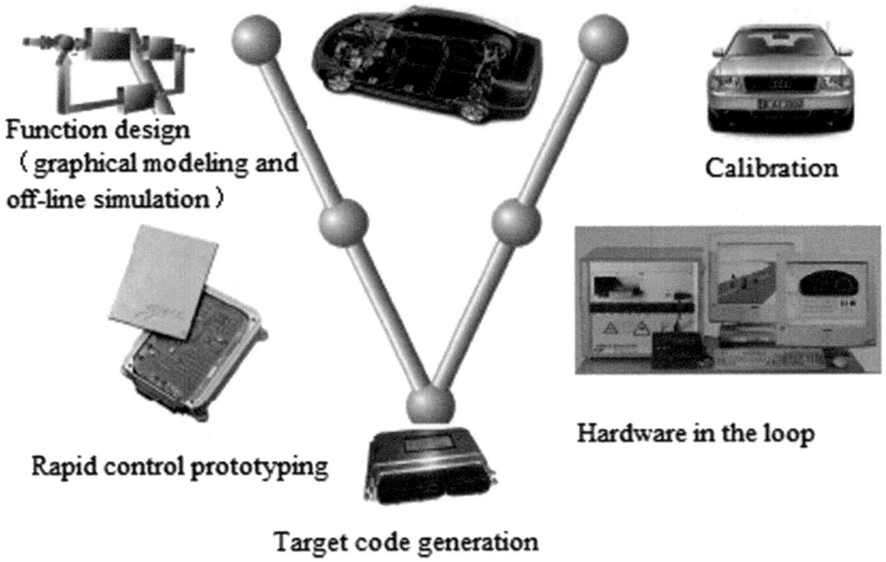

Fig. 1 V-cycle mode in vehicle development

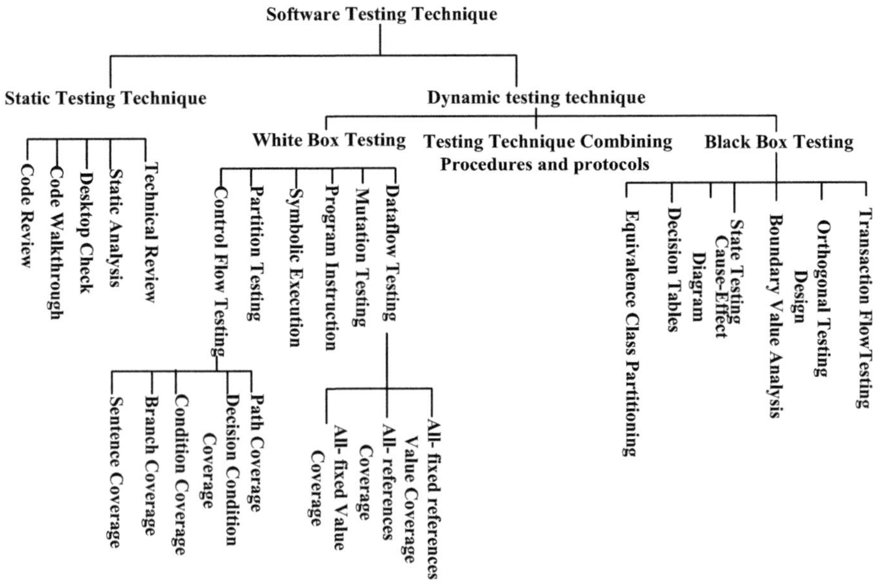

Fig. 2 Classification of software testing technique

depending on whether the software runs dynamically, and dynamic test technique is the most commonly used test methodology [10, 11]. As one of the branches of dynamic testing technique, black-box testing is divided into methodology of equivalence class partitioning and methodology of boundary value, etc. [12–14]. When the black-box testing is performed, the whole tested program is regarded as a sealed black box. To be more specific, the only information available for testers is the software function specification, and the testing can be accomplished by comparing software function specification and the testing results [15, 16].

In this paper, the HIL-based test system was developed by Beijing Institute of Technology. The engine control software function requirements and test requirements were analyzed, and the test items were extracted through considering the coverage ratio and difficulties such as the limiting and dangerous conditions which are hard to be achieved in real engine test bench. In the test case distribution strategy, the reasonable test case distribution of the engine control software involves two procedures: At the first stage, the first-time test case distribution was conducted based on the matrix distribution methodology, and then the test results were analyzed to guide the redistribution of the test cases. For the design of test cases, appropriate black-box test methods such as methodology of equivalence class partitioning, methodology of boundary value, and methodology based on scenario were used. In addition, for the automatic testing of software, the test scripts were designed according to different test cases. Finally, several test results were analyzed in detail.

2 HIL-Based Test System

In order to implement HIL-based software testing, AutoBox, which was developed by dSPACE Company, was selected as the real-time platform to run the engine model. When the software testing is performed, the real-time engine model and electronic control unit (ECU) exchange data through I/O interfaces, which replace the logical interfaces used in off-line simulation. The HIL system supports the implementation of various scheduled ECU software testing in a virtual environment, and the testing results are close to or even equivalent to what is achieved in engine's real working environment.

Figure 3 shows the schematic diagram of HIL-based software testing system. The hardware environment includes PC1, PC2, emulator AutoBox, sensor signal processing board, actuator signal processing board, and the engine controller. The software environment includes engine mean-value model, engine control software, ControlDesk, AutomationDesk, and CANape.

CANape, a software which is used to download the engine control software into the controller, is installed on PC1. Both software ControlDesk and automatic testing software AutomationDesk are installed on PC2. ControlDesk is used to download the engine mean-value model into the real-time emulator, and AutomationDesk is

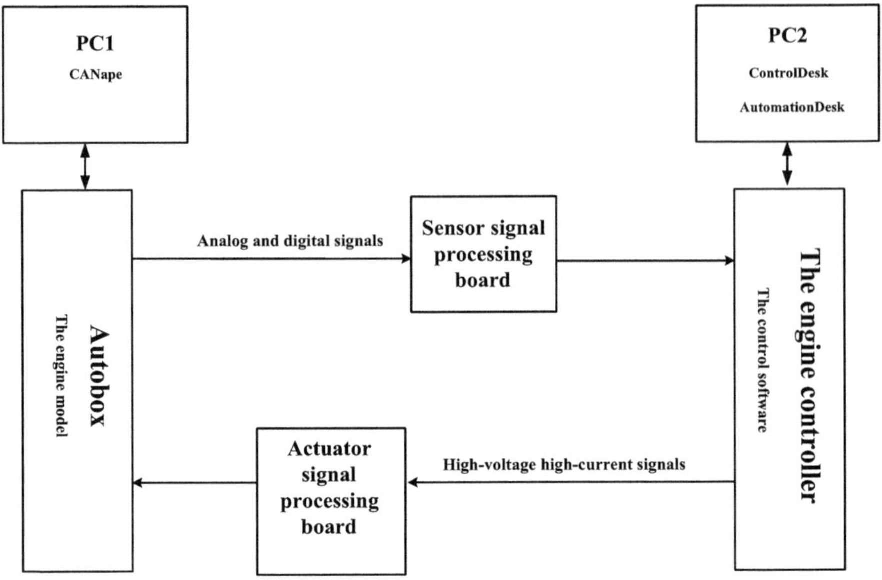

Fig. 3 Schematic diagram of HIL system

used to design and implement test script files. Emulator AutoBox and the engine controller are used to run engine model and the control software in real time, respectively. Analog and digital signals generated by AutoBox are transformed into signals that can be identified by the controller through sensor signal processing board; meanwhile, the high-voltage and high-current driving signals generated by the engine controller are transformed into signals that can be identified by AutoBox via actuator signal processing board. The engine mean-value model, which has a reasonable fidelity and excellent real-time performance, was built by MATLAB/Simulink. So did the engine control software. They interact via I/O interfaces and CAN bus to constitute a real-time closed-loop test system.

3 Test Item Extraction

Test items play an important part in software testing; they correspond to the software function points and can be extracted through the test requirement analysis. Besides, test items ensure the integrity of the software function testing as well as the test coverage ratio. Typically, the test coverage ratio is divided into test code coverage ratio and test requirement coverage ratio, and the test coverage ratio is equivalent to test requirement coverage ratio in terms of software function testing.

3.1 Extraction Strategy

In general, test requirement coverage ratio is determined by congruent relationship between software function requirements and test requirements, if all the software function requirements have been established congruent test requirements, the test requirement coverage ratio is regarded as 100%. Because the test items are extracted through test requirement analysis, then the test item coverage ratio will be 100% too.

3.2 Extraction

Figure 4 illustrates the control logic, depending on which the control software can realize the control functions. Basic control function points include stop condition control function, start condition control function, idling condition control function,

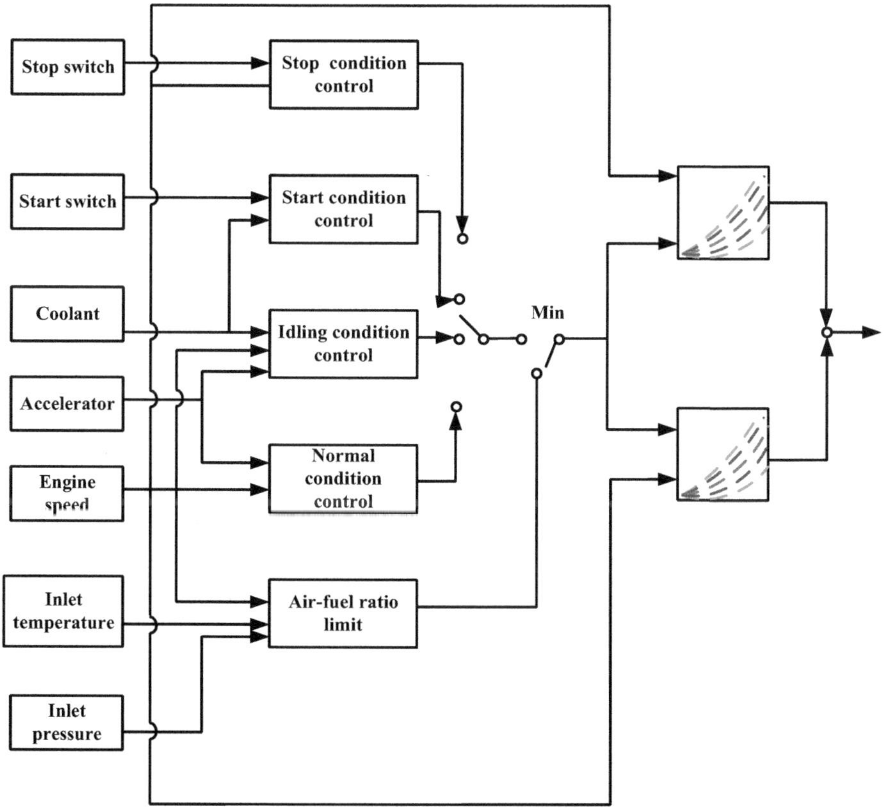

Fig. 4 Diagram of control software function

normal condition control function, and air–fuel ratio limit function. Besides, analog signal acquisition function, digital signal acquisition function, fuel injection pulse function, and data transmission function are also taken into consideration in this study.

Through full analysis for the control software function requirements, the corresponding relationship between software function requirements and test requirements is established to ensure the full coverage of the all function points. Nevertheless, test cases extracted directly on test items could hardly satisfy the detailed test requirements of large-scale or complex function points; most possibly, there are testing omissions that can affect the test adequacy.

Here, only the signal acquisition function and data transmission function are relatively simple and no complex control logic is included, so one test point is enough. However, some large-scale software function points such as start control function and idling control function need to be subdivided into more test points; refer to Table 1.

4 Test Case Distribution and Design

After the extraction of test items, specific test cases for each test point need to be designed, according to which software testing is to be implemented afterward. Notice that the number of test cases should be reasonable; too many test cases could result in a waste of labor and financial resources of programmers, but not enough test cases could lead to testing omissions; thereby, reasonable distribution and design of test cases is the key to improve testing efficiency. Usually, the optimal number is determined by the scale of the software as well as the demands of testers and test time.

Table 1 Test item extraction

Test Items	Test points
Analog signal acquisition	Data acquisition
Digital signal acquisition	Data acquisition
Start condition	Data acquisition, look-up table function, logical judgement
Idling condition	Data acquisition, look-up table function, logical judgement, complex calculation
Normal condition	Data acquisition, look-up table function, logical judgement, complex calculation
Stop condition	Data acquisition, logical testing function
Air–fuel ratio limit	Data acquisition, look-up table function, logical judgement, complex calculation
Fuel injection pulse	Data acquisition, look-up table function
Data transmission	Data acquisition

4.1 Test Case Distribution Strategy

In this study, assuming that the number of test cases grows linearly with the software scale, the complexity and other factors of the engine control software are neglected, only the scale of the software is taken into account. The distribution of the test case number is based on matrix distribution methodology. Specifically, when test cases are distributed preliminarily, test items such as data transmission, fuel injection pulse, air–fuel ratio limit, idling condition control, and signal acquisition are taken as rows of the matrix, and the basic test points such as look-up table function, logic judgement, complex calculation, and data acquisition are taken as columns of the matrix. The total number of test cases required is calculated using Eq. (1):

$$Q_A = \sum S_i * R_i \tag{1}$$

where Q_A is the total number of test cases, i (i = 1,…,n) is the number of function modules after software partition, S_i is the scale of the ith function module, and R_i is the test case density of the ith function module. In general, test case density varies with number of testing participants, permitted time as well as testing scales. The number of test case predicted by the cells in row i and column j of the matrix is calculated using Eq. (2):

$$QC_{ij} = S_i * R_i * W_{ij} \tag{2}$$

where QC_{ij} is the number of test cases in row i and column j of the matrix, W_{ij} is the weight of basic test point in column j for row i of the matrix. Notice that the sum of the weight of all basic test points for each row is 100%, and each weight of the basic test points in each row can be different. In addition, values of S_i and R_i are equivalent to the values defined in the model. Through matrix distribution methodology, the number of test cases can be distributed reasonably.

4.2 Test Case Distribution

The first-time distribution of test cases was shown in Table 2, the total number is calculated by matrix distribution methodology according to software scale and test case density of each test item.

In this study, after the implementation of test cases designed according to test points, the failed test cases were filled into the failed cases matrix, which has the same structure with the first-time distribution matrix. Figures 5 and 6 show the proportion represented by failed test cases based on the rows and columns of the matrix, respectively.

Table 2 First-time distribution of test cases

Test items	Scale	Density	Total number	Data acquisition	Logical judgement	Look-up table function	Complex calculation
Accelerator pedal	10	0.1	1	1			
Coolant temperature	10	0.1	1	1			
Inlet temperature	10	0.1	1	1			
Inlet pressure	10	0.1	1	1			
Crankshaft signal	150	0.01	2	2			
Camshaft signal	100	0.01	1	1			
Start condition	50	0.1	5	2	2	1	
Idling condition	100	0.09	9	3	4	1	1
Normal condition	100	0.07	7	2	2	1	2
Stop condition	30	0.2	6	3	3		
Air–fuel ratio limit	60	0.1	6	2	2	1	1
Fuel injection pulse	80	0.05	4	2		2	
Data transmission	120	0.01	2	2			

The first-time distribution of test cases is based on software scale and test case density; however, this simple distribution is not enough for a full-scale software testing. The redistribution of test cases is based on the error distribution characteristics which determine whether to continue testing after the first-time distribution. If the test case increase is necessary, the redistribution strategy based on the analysis results can be utilized to determine the distribution of increased test cases.

Taking the engine speed acquisition function testing as an example, the engine speed acquisition is based on a wide engine speed range, from low speed to high speed. In essence, in the process of engine speed acquisition, the speed is calculated by only one CPU frequency division in the bottom program of the engine controller, resulting in obvious accuracy variant in different speed segments. Thus, for the detailed testing of speed accuracy calculation function in each speed segment, extra test cases are introduced to speed acquisition function to make the test cases cover all speed segments.

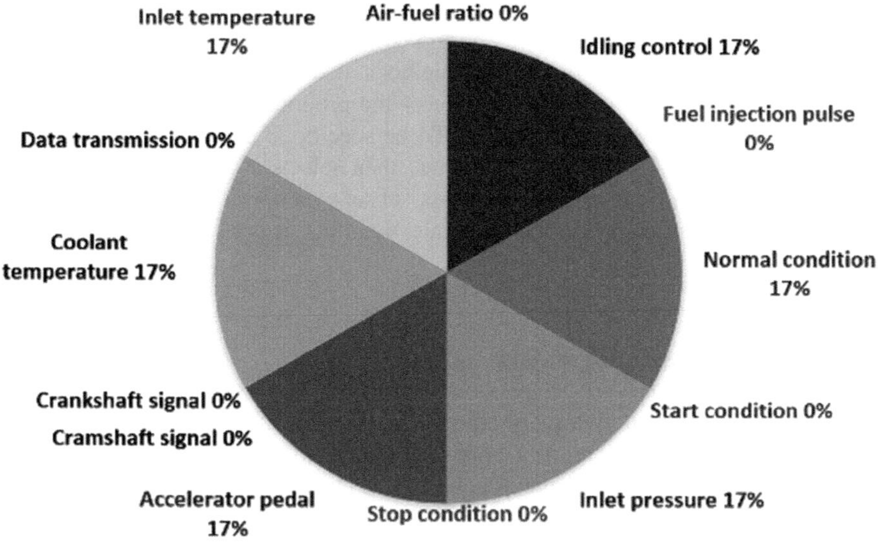

Fig. 5 Failed test case distribution based on test items

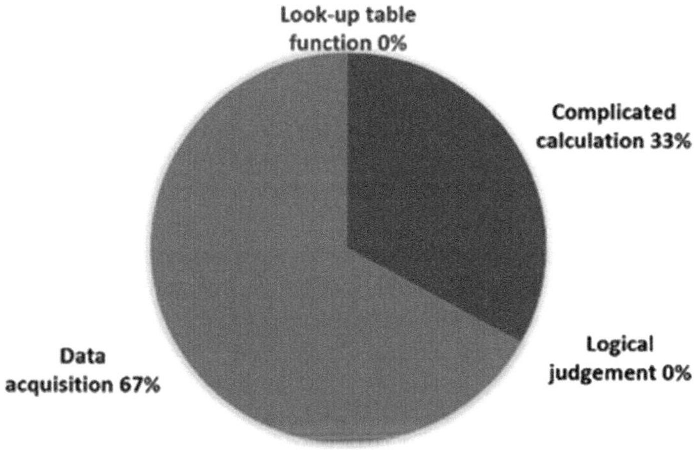

Fig. 6 Failed test case distribution based on test points

4.3 Test Case Design Strategy

Since the engine control software testing is a kind of function testing, the detailed realization process of software function could be completely ignored, so the black-box testing technique is adopted in this paper. Three typical testing methods of black-box testing technique used in this paper are illustrated as follows.

4.3.1 Methodology of Equivalence Class Partitioning

As one of the typical black-box testing methods, methodology of equivalence class partitioning ignores the internal structure of the program and partitions the inputs according to the software specification. To be specific, it divides all possible input data into a number of disjoint subsets, and then selects a small number of representative data from each equivalence class subset as test cases. In this manner, the test points can be remained in a small scale, avoiding heavy workload for software testers.

4.3.2 Methodology of Boundary Value

Different from the methodology of equivalence class partitioning, the methodology of boundary value, which aims at boundary values instead of any internal element within the input and output range of equivalence classes, selects the boundary value of the equivalent class to design the test cases. Besides, it is a supplement of equivalence class partitioning since errors often occur near boundary values. Figure 7 shows the schematic of methodology of equivalence class partitioning and methodology of boundary value.

4.3.3 Methodology Based on Scenario

The scenario is actually the flow of affairs, including elementary streams and alternative streams as shown in Fig. 8. The transaction-driven approach is used to drive the software, and the scenario is developed once the scene is triggered by an event.

From the view of software users, testers simulate ideas of designers by analyzing the design, imagine the operation order to use the software in practice, and list such possibilities as test points one by one; in this way, defects of the software are easier to be found out.

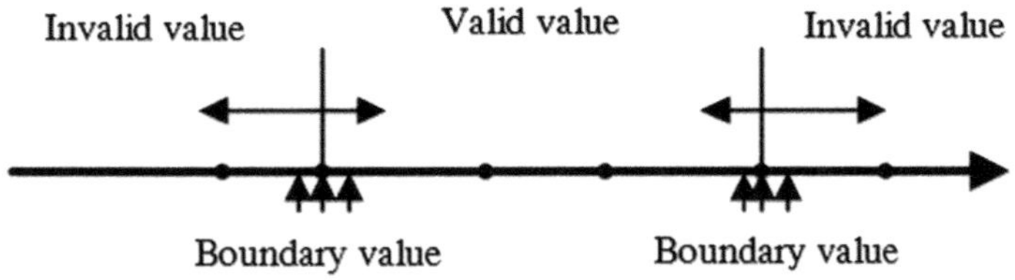

Fig. 7 Schematic of equivalence class partitioning and boundary value

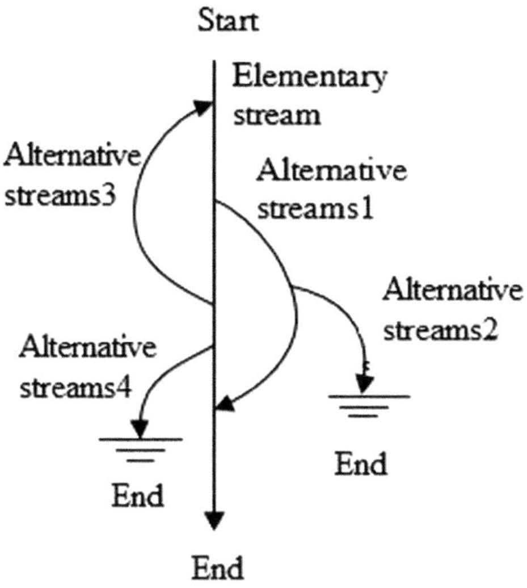

Fig. 8 Schematic of methodology based on scenario

4.4 Test Case Design of Idling Control Function

A complete test case contains the test item name, test methodology, test content, test input, expected output, and judgement criteria. Combined with the test requirement analysis, the test case design of idling control is taken as an example.

The control strategy for idling condition control function is relatively complicated as shown in Fig. 9. The controller judges whether the engine is running at the idling condition according to the accelerator pedal position signal collected from accelerator pedal position sensor and the engine speed signal collected from engine speed sensor. The basic values of the target idling speed, which decrease as the coolant temperatures increase, is obtained by looking up MAP using coolant temperature. Furthermore, the application layer obtains the difference between engine speed and target idling speed via speed contrast, and then inputs it into the

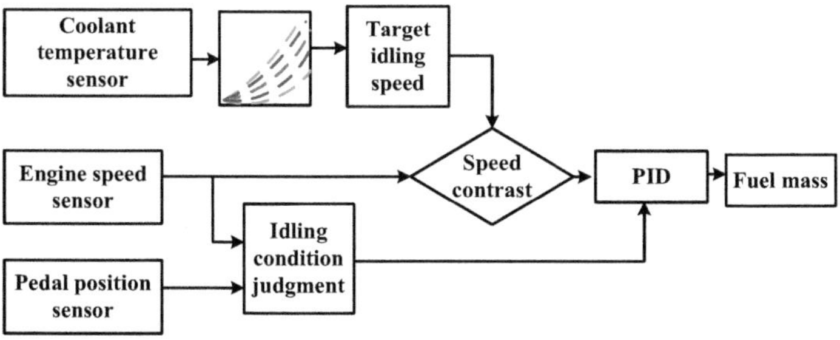

Fig. 9 Control strategy for idling conditions

PID module, through which can calculate out the fuel mass and adjust the engine speed stabilized near the target idling speed.

In this case, the methodology based on scenario is suitable, and test case scenario of idling condition control function is determined by the combination of elementary streams and alternative streams. Fig. 10 shows the flowchart of the test case. First, start the engine, check if the engine status is 2 (i.e., the engine is running at idling condition) when the speed is higher than 500 r/min; if so, check if the target idling speed value is consistent with the expected idling value; and if so, check whether the fuel mass calculated by the PID algorithm is correct; and if so, check whether the error between engine speed and the target idling speed is reasonable when the engine speed is stabilized. Notice that the allowable error between engine speed and the target idling speed is less than 10 rpm, and the test failed if any of the steps above not passed.

As for the detailed testing of look-up table function, both methodology of equivalence class partitioning and methodology of boundary value are indispensable. Notice that the range of input coolant temperature is 0–100 °C in the MAP, and nine values are selected, namely −50, −1, 0, 1, 50, 99, 100, 101 and 150 °C, which include the invalid values, valid values, and near boundary values, covering all the possibilities. In the process of testing, the inputs are set as the values defined above, respectively, and then the output target idling speed values are measured whether they are consistent with expected idling values.

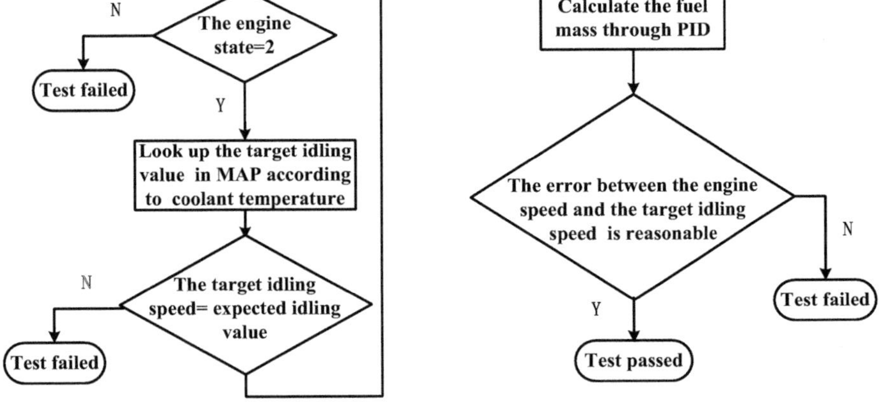

Fig. 10 Flowchart of idling control based on scenario methodology

5 Automatic Software Testing

HIL testing is a critical step in the development and validation of controller,and the aim of function testing is to find out defects in software system. Usually, manual and automatic software testing methods are both used; however, because the constantly improving complexity and frequently updating versions of the software bring heavy and repetitive workload to manual testing, the automatic software testing is used more.

5.1 Development of Automatic Test Scripts

AutomationDesk, a software tool based on dSPACE, is used to design the test scripts. This software tool can not only ensure the consistency of test cases in the regression test but also avoid the repetitive workload in manual testing.

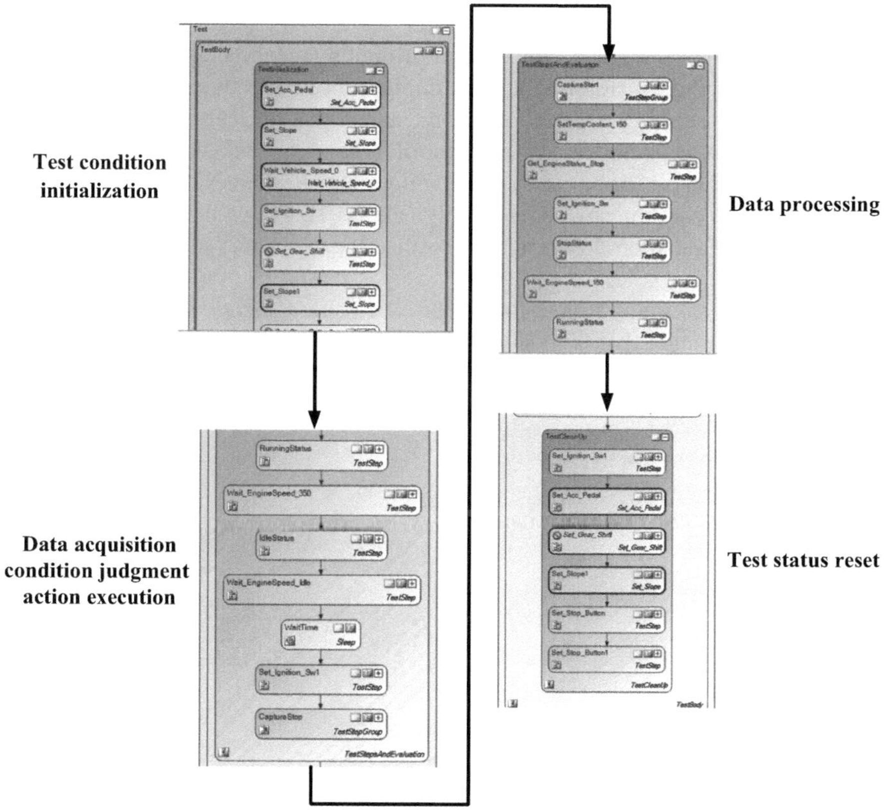

Fig. 11 Test script of automatic testing

Figure 11 shows the test script of idling condition control function, and each test script basically includes test condition initialization, data acquisition, condition judgement, action execution, data processing, and test status reset.

5.2 *Automatic Testing and Result Analysis*

On the basis of the automatic test scripts developed, the automatic testing is implemented and testing reports are generated. The detailed testing and analysis of digital signal acquisition function, analog sensor signal acquisition function, idling condition control function, and normal condition control function are introduced as follows. And the testing analysis contains all the four categories of test points mentioned in Chap. 3.

5.2.1 Testing and Result Analysis of Digital Signal Acquisition Function

The engine speed, which is a typical kind of digital signal, is taken as an example for digital signal acquisition function. Engine speed is obviously a significant parameter for describing engine conditions, and it can be seen from Chap. 4 that most of the control parameters in the control strategy are related to engine speed. Besides, the engine speed accuracy directly determines whether the relevant engine control function can be achieved correctly. Figure 12 shows the testing results of engine speed acquisition function; notice that the allowable acquisition error of speed in this testing is less than $\pm 3\%$.

It is shown in Fig. 12 that the speed acquisition errors are all within the allowable error range; that is to say, the test passed.

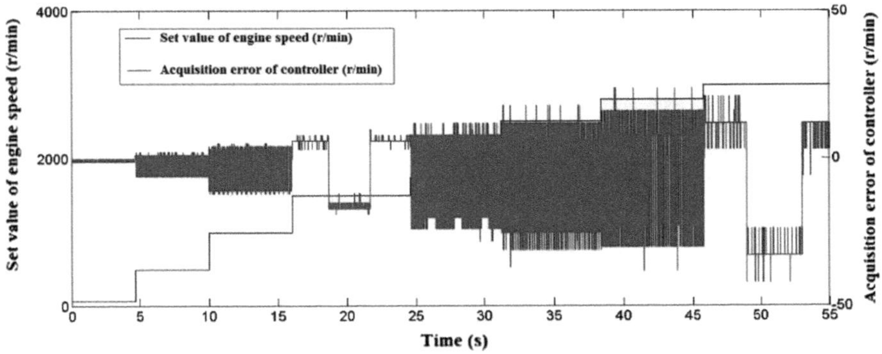

Fig. 12 Function testing of engine speed acquisition

5.2.2 Testing and Result Analysis of Analog Signal Acquisition Function

Likewise, the analog signal acquisition is important in the realization of engine control function. In this paper, the accelerator pedal position signal acquisition is taken as an example; also, the allowable acquisition error is less than ±3%. The judgement criterion is filled in the automatic testing script; the test passes if the collected signal meets the criterion. However, the test failed when the test script was executed for the first time. The results show that the error is a constant. Given this, the reason for the failure emerges: There is a zero drift in the acquisition circuit of controller, which was out of consideration in the development process of software.

In order to solve the problem, the zero drift of the hardware is corrected at the software layer. The correction algorithm is shown in Fig. 13. The voltage signal (0–5 V) collected by voltage signal input module is converted to digital signal (0–5000). The value of Constant1 is set to 50 to eliminate linearity deviation of zero drift. The function of the remaining part is to turn digital signal (0–5000) into the corresponding throttle opening (0–100%).

Figure 14 shows the contradistinction between set value and collected value of the corrected accelerator pedal position. Apparently, the two lines coincide perfectly, particularly when the throttle opening is small; meanwhile, the maximal error is less than 3%, and the test passed.

5.2.3 Testing and Result Analysis of Idling Condition Control Function

(1) Target idling speed control function

The target idling speed control function is taken as an example for look-up table function. As was analyzed in Chap. 4, in order to test the target idling speed control function, the input coolant temperature was divided to nine values, namely −50, −1, 0, 1, 50, 99, 100, 101 and 150 °C; thereafter, corresponding test cases and test scripts were designed.

When the testing was implemented, the input coolant temperature was set as the eight values in turns, and the output target idling speed values were all equal to the expected values; the test passed.

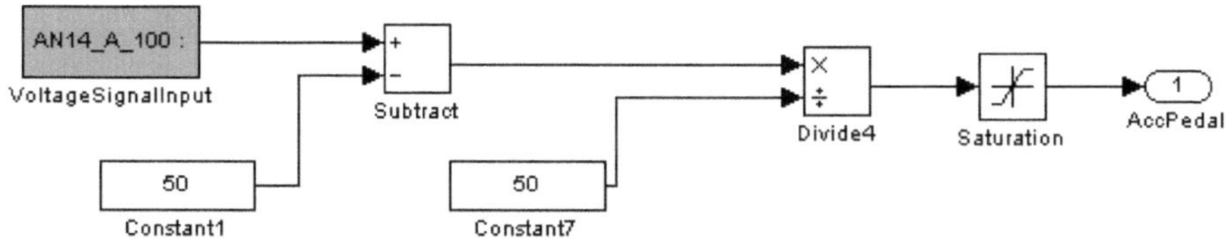

Fig. 13 Correction algorithm of accelerator pedal position sensor

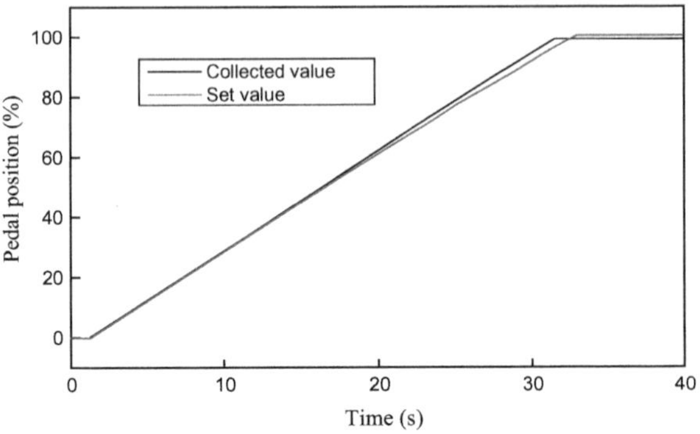

Fig. 14 Contrast diagram after correction of zero drift

(2) Idling speed control function

Likewise, idling condition is important for engine control, and it is rather complex. In Chap. 4, the test item of idling condition is divided into four test points. The PID control is used in the idling speed control function and is taken as an example for complex calculation. Several representative target idling speed values, namely 600, 800, 1000, 1500, and 2000 r/min, were selected for PID control function testing. The test passes only when the error between speed calculated by PID and the set speed is less than 10 r/min. Despite the fact that the PID control function passed the testing in the off-line simulation, it did not work correctly in the HIL. Figure 15 shows the testing result of speed at 800 r/min; the test failed due to the high fluctuation of approximately 150 r/min when the engine speed was stabilized 40 s later.

The reason is that the real-time performances of the model vary in different platforms, and the parameters calibrated before may not be suitable for the HIL platform. As a consequence, CANape, the online calibration software produced by Vector, is used to re-calibrate parameters in PID module. Figure 16 shows the online calibration interface.

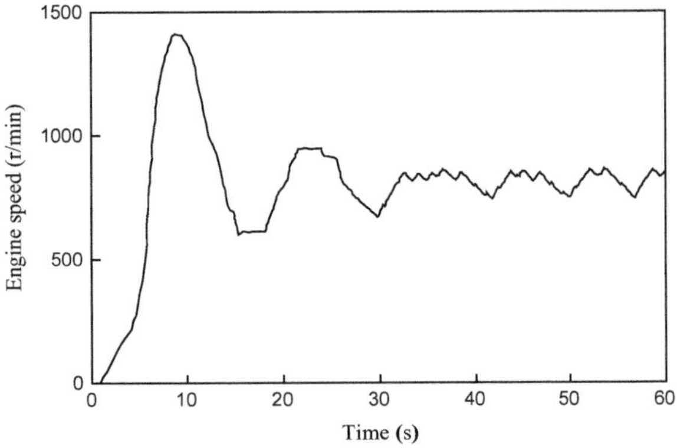

Fig. 15 Testing result of PID control function

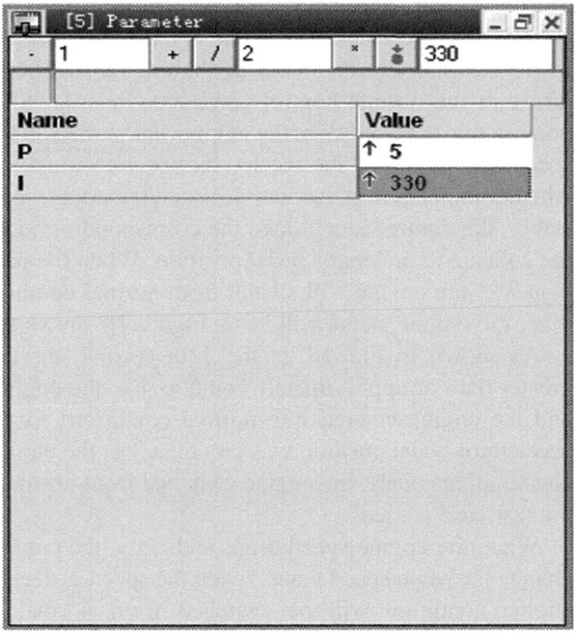

Fig. 16 Online calibration interface

Moreover, regression testing is executed after re-calibration; Fig. 17 shows the testing result; apparently, after re-calibration, the idling speed stabilizes near the target speed (800 r/min) in about 2 s with a speed error of less than 2 r/min, and the test case passed.

5.2.4 Testing and Result Analysis of Normal Condition Control Function

The correct judgement of engine status determines whether the controller could calculate out the right control signals. The judgement of engine status in normal condition function is taken as an example for logical judgement function.

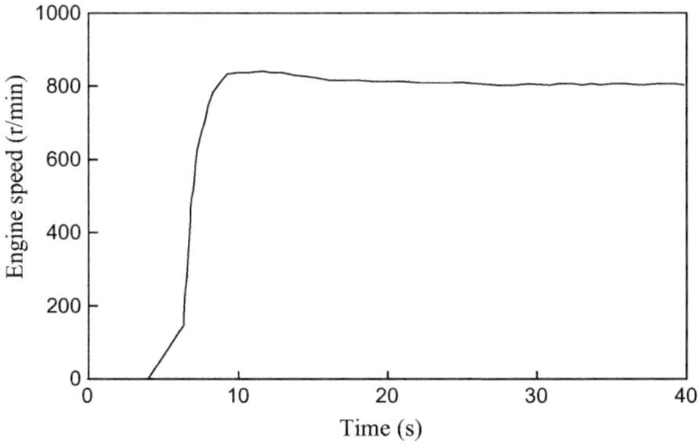

Fig. 17 Testing result of PID control function after re-calibration

The controller determines whether the engine is running at normal conditions according to the accelerator pedal signal. When the position of the accelerator pedal is greater than 4%, the status of the engine will be 3; meanwhile, the engine will turn into the normal control condition. In order to test this logic judgement test point in normal condition, the engine was started to enter idling condition, when the speed was stable and the accelerator pedal opening was increased to over 4%, check whether the engine status was 3. When the engine was running at normal condition stably, the controller will take the corresponding judgements and control based on the collected accelerator pedal position. When the accelerator pedal position is less than 3%, the engine will switch from normal condition to idling condition; at that time, the engine status will be 2. Figure 18 shows the test result.

As shown in Fig. 18, at the 13th second, the accelerator pedal position was greater than or approximately equal to 4%, the engine status switched from 2 to 3, and the engine entered into normal condition. At probably the 35th second, the accelerator pedal position was less than 3%, the engine status switched from 3 to 2, and simultaneously, the engine switched from normal condition to idling condition; the test case passed.

When the engine speed drops suddenly, the controller will make corresponding change based on speed value, when the speed is decreased to less than 100 rpm, the engine condition will be switched from normal condition to stop condition, meanwhile the engine status will switch from 3 to 0, and the fuel supply will be cut off. The test result was shown in Fig. 19.

When the engine was running in normal condition, the load torque suddenly increased at the 43rd second, leading to a sudden drop of engine speed. At the 43rd second, the engine speed decreased to 100 rpm, the engine status switched from 3 to 0, and the engine turned into stop condition. As shown in Fig. 19, this control function was realized correctly, and the test case passed.

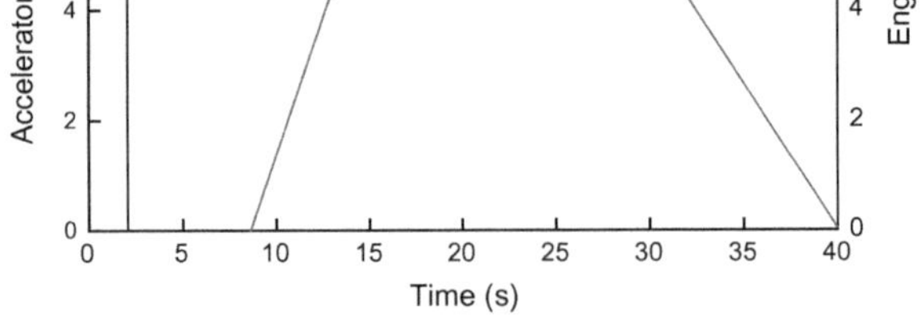

Fig. 18 Logical judgement testing result

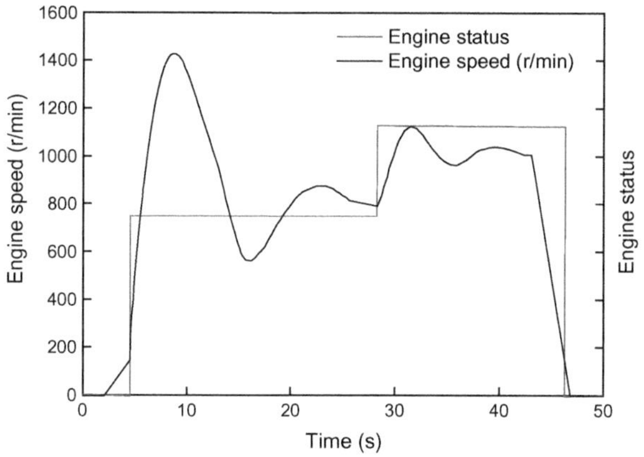

Fig. 19 Engine status testing result

6 Conclusions

In this study, based on the HIL simulation technology, the integrated test platform of hardware and software was developed for the engine ECU; meanwhile, the automatic testing of various functions of the engine control software was accomplished in the semi-physical simulation environment. Moreover, the test case distribution was optimized based on the matrix distribution methodology, and the test case redistribution strategy was applied based on the error distribution characteristics which contributed to a tremendous enhancement of testing efficiency. Besides, the engine control software test cases were designed and the automatic testing scripts were developed using the methodology of equivalence class partitioning, methodology of boundary value, and the methodology based on scenario. The developed automatic testing script library lays the foundation for reuse in subsequent software testing and improvement of testing efficiency. Application on engine control software testing shows that the HIL-based platform can not only test the software function and real-time performance, but also find the further software problems related to the hardware. It was noticed that problems found in the engine control software testing mainly exist in the software relating to hardware circuit such as the zero drift in the controller acquisition circuit. This will be valuable experience for future software testing using HIL-based platform.

References

1. Roche M, Mammetti M (2015) An innovative vehicle behaviour modeling methodology for model-based development. SAE Technical Papers
2. Xu H, Niimi Y, Ono T (2013) Virtual development of engine ECU by modeling technology. Lect Notes Electr Eng 67:66–71

3. Haukap C, Röpke K, Barzantny B (2006) Hardware-in-the-Ioop simulation (HIL) for production engine development. Elsevier Inc.
4. Piscaglia F, Ferrari G (2007) Development of an offline simulation tool to test the on-board diagnostic software for diesel after-treatment systems. SAE Technical Papers
5. Li JW, Dong HG, Wang Y et al (2014) Research on the dynamic model and ECU HIL simulation system of electronic-controlled engine. Adv Mater Res 889–890:962–969
6. Allen J, Dhaliwal A, Warra J (2011) A novel approach to implementing HIL systems for ECU validation and verification for commercial vehicle applications. Chem Nat Compd 7(6): 840–841
7. Mazhari SA, Nampoothiri S (2013) PSO tuned vehicle climate system model for HIL based ECU testing. SAE Technical Papers, 2
8. Ni J, Li X, Shi X et al (2014) Design of host program for engine ECU HIL system based on NI PXI platform. Automobile Technology
9. Vegas S, Basili V (2005) A characterisation schema for software testing techniques. Empirical Softw Eng 10(4):437–466
10. Gupta A, Jalote P (2008) An approach for experimentally evaluating effectiveness and efficiency of coverage criteria for software testing. Int J Softw Tools Technol Transfer 10(2): 145–160
11. Chen TY, Kuo FC, Liu H et al (2013) Code coverage of adaptive random testing. IEEE Trans Reliab 62(1):226–237
12. Katyal RK, Srinath S (2011) Virtualization for ECU platform software testing in automotive embedded. SAE Technical Papers
13. Conrad M, Sadeghipour S, Wiesbrock HW (2005) Automatic evaluation of ECU software tests. SAE Technical Papers, No. 4, pp 595–599
14. Barrett S, Bouchez M (2015) Addressing engine ECU testing challenges with FPGA-based engine simulation. SAE Technical Papers
15. Bryce R, Kuhn R (2014) Software testing. Computer 47(2):21–22
16. Godefroid P, De Halleux P, Nori AV et al (2008) Automating software testing using program analysis. IEEE Softw 25(5):30–37

Dynamic Modeling and Coordinate Control for an Engine-Generator Set

Wei Zhou, Changfu Zou, Junqiu Li and Guangyao Li

1 Introduction

Extended-range electric vehicle (EREV) has been attracting significant attention from both academia and industry since recent years. EREV is generally equipped with a big battery pack and a small range extender, configured in a series-hybrid topology. It works in pure electric mode as long as the battery contains enough energy and in charge-sustaining mode when the battery is depleted and with the range extender supplies the energy needed for propulsion. Therefore, EREV is able to eliminate the so-called range anxiety, which is commonly suffered by pure electric vehicles, and is considered to be an intermediate yet promising solution to the ultimate goal of vehicle electrification in the automotive industry [1].

Range extender is a critical component for EREVs, and engine-generator set is one of the most commonly used range extenders. Considerable research on the engine-generator set control has been carried out for the past several years. For example, Cheng et al. designed dynamic controllers respectively for the engine-generator set and tested their performance of the controllers via experimental approaches [2–4]. Hu et al. synthesized a distributed control system for an EREV, and then implemented this on an experimental test system [5]. Wang et al. proposed a gain adaptive controller for the engine-generator set in construction machinery

W. Zhou (✉) · G. Li
College of Mechanical and Vehicle Engineering, Hunan University,
Changsha, China
e-mail: zhouwei_bit@sina.com

C. Zou
Department of Electrical Engineering, Chalmers University of Technology,
Gothenburg, Sweden

J. Li
School of Mechanical Engineering, Beijing Institute of Technology,
Beijing, China

Society of Automotive Engineers of China (SAE-China) (ed.), *Proceedings of the 19th Asia Pacific Automotive Engineering Conference & SAE-China Congress 2017: Selected Papers*, Lecture Notes in Electrical Engineering 486, https://doi.org/10.1007/978-981-10-8506-2_68

applications [6]. To realize the referred control methodologies, appropriate engine-generator models are desired, which, in general, should capture the key system characteristics but have as simple as possible mathematical structures.

A number of publications have been emerged to address the problem of dynamic modeling for the engine-generator system. Early research practices have led to the production of two well-known modeling software in the field of hybrid electric vehicle simulations, i.e., PSAT and ADVISOR. It is noted that both the two software packages rely on quasi-static modeling approaches, and in other words, the generator and converter characteristics are represented by static maps [7, 8]. This is perhaps sufficient for validating vehicle-level energy management control strategies, but is likely to fail in supporting the component-level controller design and calibration. This is due to the fact that the quasi-static models intrinsically lack the ability to capture system dynamics which can be crucial in real-time control such as in the electrical domain of an engine-generator set. As one step further, Bin et al. proposed an average model of a diesel engine-generator system with excited magnetic synchronous generator and uncontrollable rectifier [9]. Therein, this model was compared with its high-order counterparts which explicitly consider the transient dynamic behaviors of power electronic devices. It showed that the proposed model is capable of effectively predicting the average dynamic behaviors of the generator and rectifier. The average model has a good trade-off between the accuracy and complexity, and thus may be a candidate for dynamic coordinate control of an engine-generator set.

In this paper, we aim to compare the performance of two typical dynamic coordinate control strategies for a diesel engine-generator set with permanent magnetic synchronous generator (PMSG) and controllable AC/DC converter onboard, based on a semi-physical model that captures necessary system dynamics. The remainder of the paper is organized as follows: Section 2 introduces briefly the overall structure of the considered diesel engine-generator system; Sect. 3 presents detailed information on how to semi-physically model each component and the entire system; Sect. 4 validates the model by experiment; Sect. 5 compares two typical dynamic coordinate control strategies by the developed system model. Finally, key conclusions and insights are made in Sect. 6.

2 Overall Architecture of the Investigated Engine-Generator Set

In this study, a distributed scheme is adopted to realize the control purpose of a diesel engine-generator set for an extended-range electric school bus. Figure 1 illustrates the overall topology of this engine-generator set and its distributed controllers. As can be seen, in EREV applications, the engine-generator set is primarily used to provide a certain amount of power dictated by the vehicle-level energy management controller. The energy management controller determines to

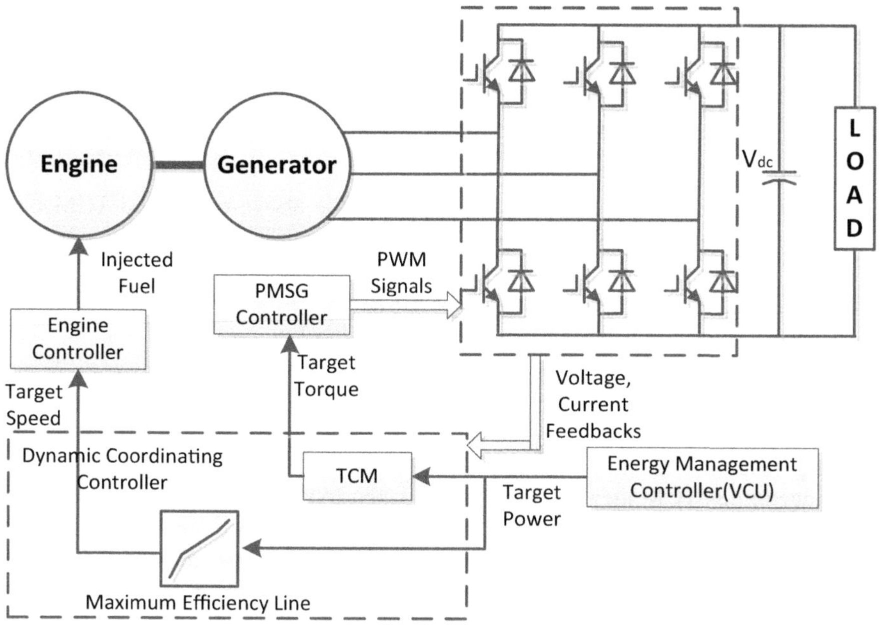

Fig. 1 Overall structure of the investigated engine-generator system and its controllers

split the demanded propulsion power between the engine-generator set and battery pack using heuristic or optimization algorithms [10, 11]. To ensure that the power dictated by the energy management controller is achieved in an energy-efficient, dynamically fast, and stable manner, the engine and generator need to be regulated cooperatively by a dynamic coordinating controller (DCC). As shown in the figure, the task of the DCC is to generate speed and torque command, respectively, for the lower-level engine and generator controller. Since accurate closed-loop torque control is not as easy to achieve by the engine as that by the generator, we choose closed-loop speed control for the engine and closed-loop torque control for the generator in this study. To guarantee optimal energy efficiency, the range extender is desired to operate along its maximum efficiency line under any propulsion loads. Therefore, the target engine speed is determined by using the dictated power and looking up a speed-power table representing the maximum efficiency line of the entire engine-generator system. The maximum efficiency line is found by searching the overall efficiency map of the engine-generator system. One the other hand, for the closed-loop torque control, the commanded generator torque is obtained by a torque coordinating module (TCM), which is responsible for ensuring that the dictated engine-generator power is well tracked, and at the same time, a good dynamic and NVH performance is fulfilled.

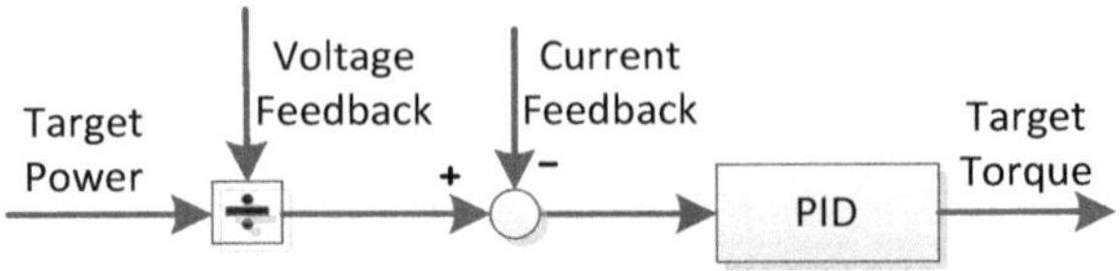

Fig. 2 Closed-loop current control scheme for the TCM

There are potentially three methods to realize the above-mentioned TCM control. The first one is based on the following direct open-loop torque computation:

$$T_g^* = 9549 \frac{P_{\mathrm{RE}}^*}{n_e^* \eta_{\mathrm{RE}}\left(n_e^*, P_{\mathrm{RE}}^*\right)} \tag{1}$$

where T_g^* is the target generator torque; P_{RE}^* is the target power dictated by the energy management controller; n_e^* is the target engine speed; η_{RE} represents the engine-generator efficiency, which is a function of its speed and power. The advantage of this TCM idea is that, due to its open-loop nature, it has fast and stable dynamic performance. In addition, this method needs no current and voltage feedbacks, meaning hardware cost is saved. However, as the efficiency cannot be predicted accurately especially during transient period, this method lacks good power tracking capability, which could lower down the system efficiency. The second way to approach the TCM control can be achieved by closed-loop current control, as shown in Fig. 2. The target generator torque is established by a PID controller aiming to follow the target power using real-time current and voltage feedbacks in a closed-loop manner.

3 Control-Oriented Dynamic Modeling of Diesel Engine and PMSG

The above three DCC strategies all have several control parameters to tune, e.g., PID gains, torque control delay constants, commanded power rate limitations. As the engine-generator set is a complex nonlinear system with coupled multi-domain and multi-time scale dynamics, a large amount of effort will be needed to calibrate these parameters, especially for closed-loop DCC design. In order to accelerate the calibration and tuning process, we utilize a model-based approach, in which a model capturing critical dynamics of all the components and their coupling relationships with high fidelity and appropriate computation burden is desired. As stated in Sect. 2, the main function of the engine and generator in a range extender application is to deliver, respectively, desired engine speed and generator torque. Therefore, this section will elucidate specifically how to dynamically model the diesel engine and PMSG system from the speed and torque control point of view.

3.1 Diesel Engine Model

Since the engine speed control is our main concern in the APU application, an average model that has been successfully used in [12] for engine speed controller design is adopted here in this paper. In this model, the average engine torque produced from fueling can be expressed as follows:

$$T_e(t) = K_1 m_f(t - \tau) \tag{2}$$

where K_1 can be assumed to be constant for a specific engine, and m_f is the injected fuel mass at each instant; τ represents the fueling transport delay and is a nonlinear function of the engine speed:

$$\tau(\omega_e) = \tau_1 + \frac{\tau_2}{\omega_e} \tag{3}$$

where τ_1 and τ_2 are constants.

Through the Laplace transformation, Eq. (2) yields:

$$T_e(s) = K_1 m_f(s) \mathrm{e}^{-\tau s} \tag{4}$$

Using the first-order Pade approximation, the fueling delay can be approximated as follows:

$$\mathrm{e}^{-\tau s} \approx \frac{1}{\tau s + 1} \tag{5}$$

Consequently, the dynamics of the produced torque with transport delay can be described by the following ordinary differential equation:

$$\tau \frac{\mathrm{d}T_e(t)}{\mathrm{d}t} + T_e(t) = K_1 m_f(t) \tag{6}$$

The engine speed dynamics is governed by:

$$T_e(t) - T_d(t) - B\omega_e^2 = (J_e + J_g)\frac{\mathrm{d}\omega_e(t)}{\mathrm{d}t} \tag{7}$$

where T_d is the disturbance load torque, which is equal to the generator electromagnetic torque; B is the damping coefficient; J_e denotes the inertia of the engine, and J_e denotes the inertia of the generator.

The overall diagram of the model in the Simulink environment is shown in Fig. 3. The engine plant model uses injected fuel and disturbance load torque as the inputs, and engine angular speed as output. The injected fuel is established by a PI controller to regulate the engine speed to a reference value. A low-pass filter with a time constant of 100 ms is added after the reference speed command to avoid rapid chattering of the engine speed.

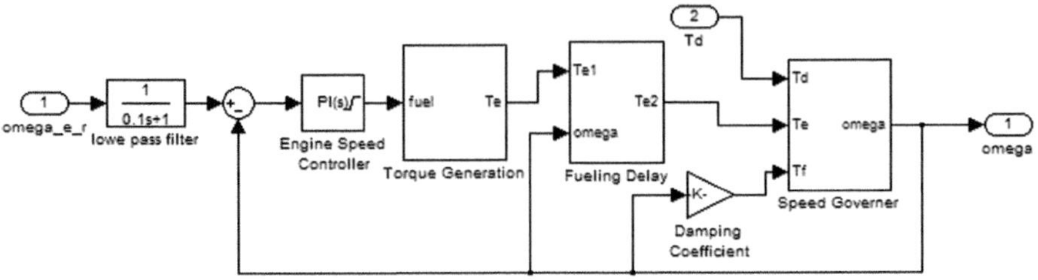

Fig. 3 Model diagram of the diesel engine in Simulink environment

3.2 PMSG Model and Its Controller

(1) PMSG Plant Model

For electric machine control, the dynamic behavior of the electric machine is usually described in the rotor *d*/*q* synchronous frame, where the *d*-/*q*-axis current of the generator is governed by two coupled nonlinear ordinary differential equations with the sign of variables being the generator convention [13–15]:

$$L_d \frac{\mathrm{d}i_d}{\mathrm{d}t} = u_d - Ri_d + \omega_s L_q i_q \tag{8}$$

$$L_q \frac{\mathrm{d}i_q}{\mathrm{d}t} = u_q - Ri_q - \omega_s L_d i_d - \omega_s \lambda_f \tag{9}$$

where L_d, L_q are *d*-axis and *q*-axis components of the stator inductance, respectively. u_d, u_q are *d*-axis and *q*-axis components of the stator control voltage, respectively, R is the stator resistance, λ_f is the flux linkage due to the permanent magnet, and ω_s is the electrical angular speed. For a surface-mounted PSMG, the following relationship holds:

$$L_d = L_q = L \tag{10}$$

p is the number of pole pairs. The generator mechanical speed relates to the electrical angular speed through the following equation:

$$\omega_s = p\omega_g \tag{11}$$

The electromagnetic torque produced by the generator can be calculated from:

$$T_g = \frac{3}{2} p \left[\lambda_f i_q + \left(L_d - L_q\right) i_d i_q\right] \tag{12}$$

As there is no gearbox between the engine and generator in our case, the generator mechanical speed ω_g is the same as the engine speed ω_e governed by Eq. (7). If a fixed-gear-ratio gearbox exists, the generator and engine speed are related by multiplying a gear ratio i_g:

$$\omega_g = i_g \omega_e \tag{13}$$

The engine speed governing equation is modified to be:

$$T_e - i_g(T_g + T_f) = \left(J_e + i_g^2 J_g\right) \frac{d\omega_e}{dt} \tag{14}$$

(2) Decoupled *d*-/*q*-Axis Current Controller

It has been noted in Sect. 2 that the function of the generator is to deliver appropriate torque such that the desired APU output power can be tracked. Therefore, torque control for the generator is our major concern. Maximum torque control is normally achieved by *d*/*q* synchronous frame current control approach. In order to get around the nonlinear coupled terms to simplify controller design, we use the method of feedback linearization. To do so, Eqs. (8)–(9) are rewritten as:

$$L_d \frac{di_d}{dt} + Ri_d = u_d + \omega_s L_q i_q = u'_d \tag{15}$$

$$L_q \frac{di_q}{dt} + Ri_q = u_q - \omega_s L_d i_d - \omega_s \lambda_f = u'_q \tag{16}$$

In this way, the original coupled nonlinear generator model is separated to two independent first-order linear ones. For the obtained linear system, tracking problems can be readily performed under the hood of PID controllers. Specifically, two controllers can be designed to individually regulate the *d*-/*q*-axis components of the stator current:

$$\frac{I_d(s)}{V'_d(s)} = \frac{1}{L_d s + 1} \tag{17}$$

$$\frac{I_q(s)}{V'_q(s)} = \frac{1}{L_q s + 1} \tag{18}$$

u'_d and u'_q are the equivalent stator control voltages for the decoupled linear model. To restore the stator control voltages u_d and u_q, compensation terms on the feed-forward paths need be added. This technique is claimed to have faster transient performance for the purpose of *d*-/*q*-axis current control than traditional synchronous frame current control approaches [16]. A general scheme illustrating the *d*-/*q*-axis current control approach is shown in Fig. 4. In this approach, the reference *d*-axis current is set to zero, such that a unit power factor can be achieved. The reference *q*-axis current value is obtained from the desired generator torque using Eq. (12).

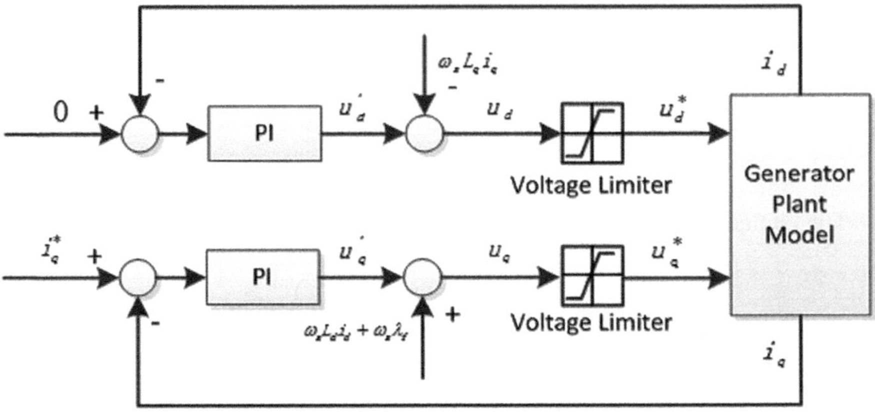

Fig. 4 *d-/q*-axis current control for the PMSG system

In order to ensure that the power electronic devices work in the linear modulation region and also due to the physical restriction of the DC bus voltage (i.e., the maximum voltage that the AC/DC converter can build is physically limited by the DC bus voltage), proper voltage saturation is necessary for the *d*-axis and *q*-axis voltages. There are multiple ways to implement the voltage saturation, among which a commonly used one is the constant-phase angle method [17]. The principle of the constant-phase angle method is illustrated in Fig. 5. It can be seen that if the amplitude of the control voltage exceeds the maximum limited value V_{lim}, the phase angle θ_v is kept unchanged, and the amplitude is reduced to V_{lim}, namely:

$$u_d^* = V_{\mathrm{lim}}\cos\theta_v = V_{\mathrm{lim}} \times \frac{u_d}{\sqrt{u_d^2 + u_q^2}} \tag{19}$$

$$u_q^* = V_{\mathrm{lim}}\sin\theta_v = V_{\mathrm{lim}} \times \frac{u_q}{\sqrt{u_d^2 + u_q^2}} \tag{20}$$

Assuming that the power electronic devices (i.e., IGBTs) work in the linear modulation region, V_{lim} can be calculated by:

$$V_{\mathrm{lim}} = \sqrt{\frac{3}{2}}\frac{V_{\mathrm{dc}}}{2} \tag{21}$$

where V_{dc} is the DC bus voltage which, for the EREV application, is established by the battery system and is equal to the battery terminal voltage.

(3) DC Bus Link

In this study, detailed switching mechanisms can be ignored without introducing large modeling errors due to two reasons: (1) The major concern of this study is to

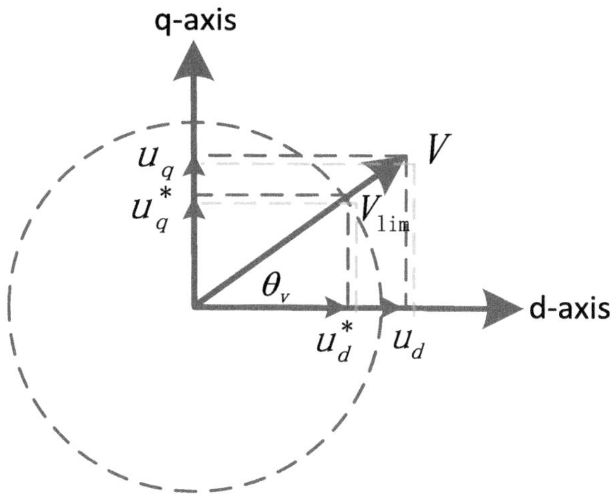

Fig. 5 Constant-phase voltage limiting

model the average behavior of the AC/DC converter, and (2) the switching frequency of the power electronics is much faster than powertrain dynamics. From an average perspective, the *d*-/*q*-axis components of the stator control voltage and the DC bus voltage satisfy the following relationships:

$$u_d^* = D_d V_{dc} \tag{22}$$

$$u_q^* = D_q V_{dc} \tag{23}$$

where D_d and D_q are, respectively, the continuous *d*-axis and *q*-axis switching duty cycles time-averaged from the discrete switching functions for each switching period of the converter.

According to energy balance relationship, the *d*-/*q*-axis currents are related to the DC-side current through:

$$V_{dc} i_{dc} = \frac{3}{2}\left(u_d^* i_d + u_q^* i_q\right)\eta_{con} \tag{24}$$

By substituting u_d^*, u_q^* defined in Eqs. (22)–(23) from Eq. (24), we can obtain:

$$i_{dc} = \frac{3}{2}\left(D_d i_d + D_q i_q\right)\eta_{con} \tag{25}$$

where η_{con} is the average efficiency of the AC/DC converter and is assumed to be constant here.

The overall Simulink model of the engine-generator system is schematically shown in Fig. 6. As can be seen, the PMSG model highlighted uses the target *d*-axis current, target *q*-axis current, DC voltage, actual *d*-axis current, actual *q*-axis current, and the actual engine speed as inputs, and outputs the electromechanical

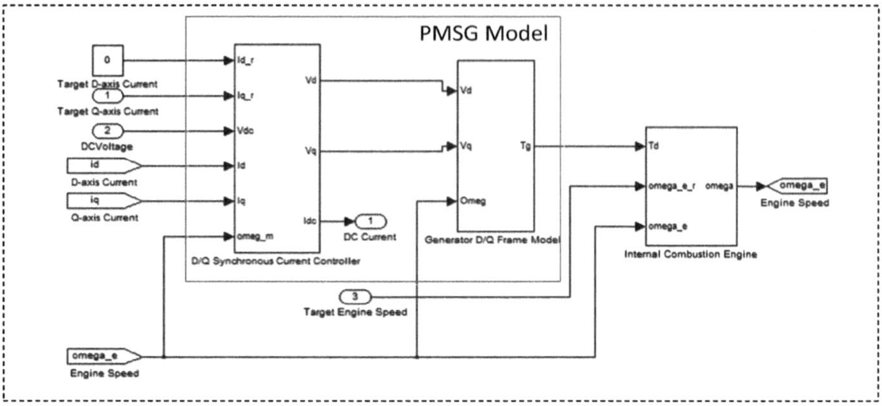

Fig. 6 Model diagram of the PMSG in Simulink environment

Table 1 Values of related parameters

Parameter	Symbol	Value	Unit	Parameter	Symbol	Value	Unit
d-axis inductance	L_d	0.0045	mH	Generator inertia	J_g	0.489	kg m^2
q-axis inductance	L_q	0.0045	mH	Pole pairs	p	6	–
d-axis resistance	R_d	0.017	Ω	Rotor flux linkage	φ_f	0.2125	Wb
q-axis resistance	R_q	0.017	Ω	Converter efficiency	η_{con}	0.94	–
Engine inertia	J_e	0.37	kg m^2	Damping coefficient	B	0.05	Nms

torque and DC current. The electromechanical torque is fed to the engine model as a disturbance load torque. Values of parameters used in the model are listed in Table 1.

4 Experimental Validation

4.1 *Experimental Test Bench Installation*

To validate the proposed model, a full-sized test bench has been built. The overall configuration of the real test bench is presented in Fig. 7. The two battery packs are serially connected to provide an appropriate voltage to the DC bus and also to dynamically compensate the load on the DC bus through charging and discharging. The state of charge (SOC) of the battery is initialized at 30%. The DC-side load is established by a water resistor, which is outside of the laboratory and thus invisible

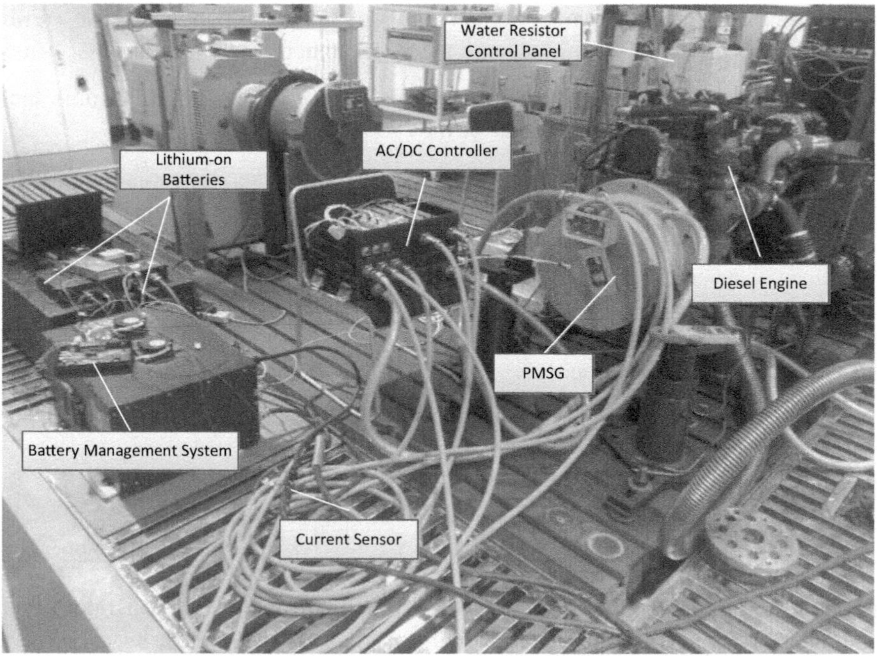

Fig. 7 Configuration of the test bench

Table 2 Parameters of components used in the experiment

Component	Description
Lithium-ion battery	Specification: manganese lithium-ion Nominal capacity: 35 Ah Nominal voltage: 624 V (156 cells in series)
Diesel engine	Displacement: 2.9 L Maximum power: 96 kW@3600 rpm Maximum speed: 4200 rpm
Permanent magnetic synchronous generator	Rated speed: 2800 rmp Rated power: 75 kW
AC/DC converter	Output DC voltage range: 500 700 V Rated capacity: 80 kW

in the figure. By adjusting the area of the plates immersed into the water, i.e., the resistance of the water resistor, the DC voltage can be controlled within a shallow range. During the experiment, the engine speed and generator torque commands are set up on a host computer using LabVIEW-based monitoring software. Afterward, the dynamic coordinate controller interprets these command signals and sends them to the lower-level engine and AC/DC converters. All system signals are recorded by the monitoring software. Critical parameters of the components in Fig. 7 are listed in Table 2.

In order to validate the model under different operating conditions, we carried out the following two dynamic scenarios and implemented them on the test bench:

- Constant Speed (@2000 rpm) Variable Torque Test (CSVT@2000 rpm) and
- Variable Speed Variable Torque Test (VSVT).

As the control parameters in the engine speed and generator torque controllers are unknown, we used the data obtained from the first test to tune the control parameters (i.e., d-axis current PI controller parameters, q-axis current PI controller parameters, and engine speed PI controller parameters) in the Simulink model. Then, we performed simulation tests under the same conditions as those in the second and third tests on the tuned model. Simulation and experimental results are presented in the next section.

4.2 Experimental Validation and Discussions

(1) CSVT@2000 rpm

In this scenario, the engine speed is regulated to 2000 rpm and the generator torque command is changed according to a profile in Fig. 8. The simulation and experimental results are compared. Note that as the model requires DC voltage as the input, we applied the same DC voltage as that in the experimental test to the model. As a result, the same DC voltage condition is guaranteed for both the simulation and experimental tests. It is shown in Fig. 8 that the target generator torque has

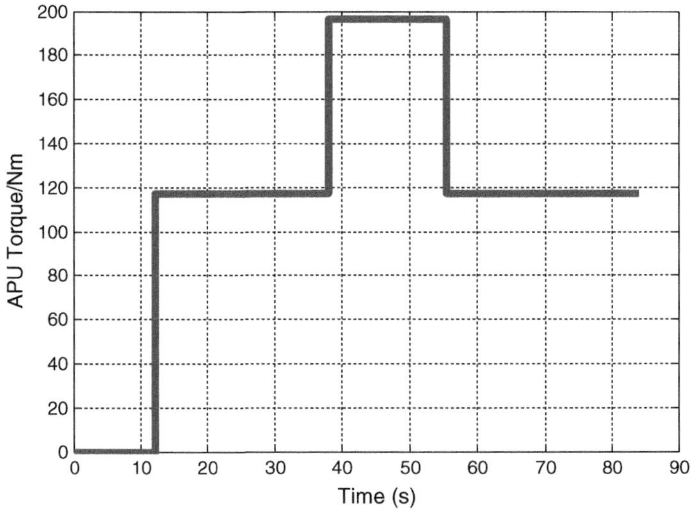

Fig. 8 Target generator torque profile for Test 1

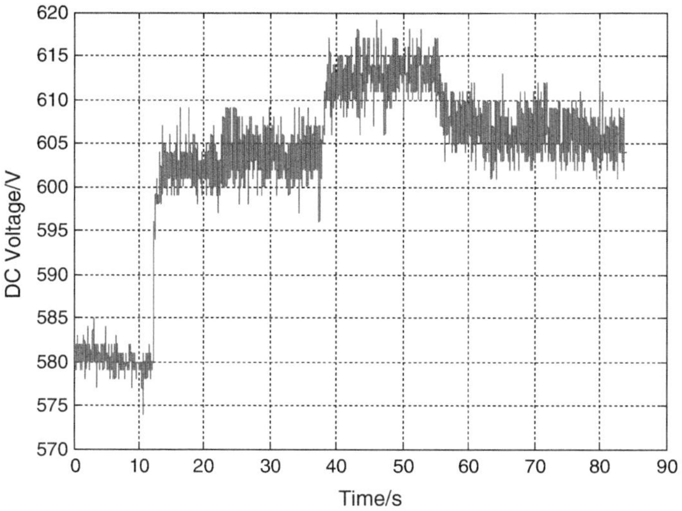

Fig. 9 DC bus voltage for Test 1

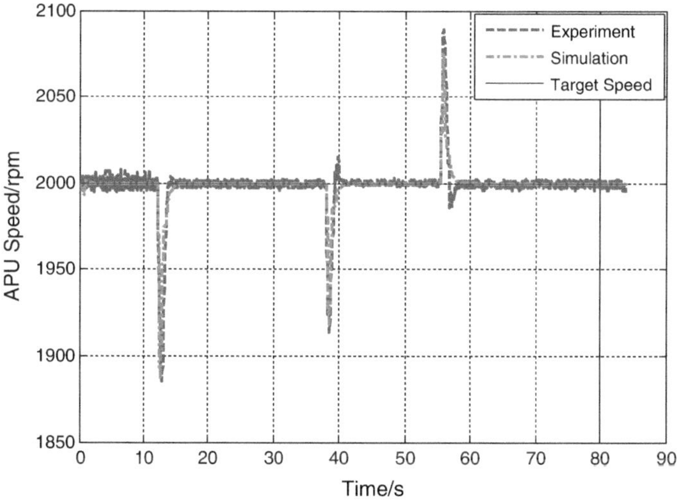

Fig. 10 Engine speed comparison for Test 1

three step changes at 12.15, 38, and 55.4 s, respectively. The DC voltage is presented in Fig. 9. As the DC-side load is a pure resistive one, the DC voltage increases when the DC current increases, and vice versa. However, since the water resistance is adjustable, we are able to control the DC voltage to operate within a narrow range. Figures 10 and 11 show that the engine speed and DC current results obtained from simulation and experiment can well match with each other under

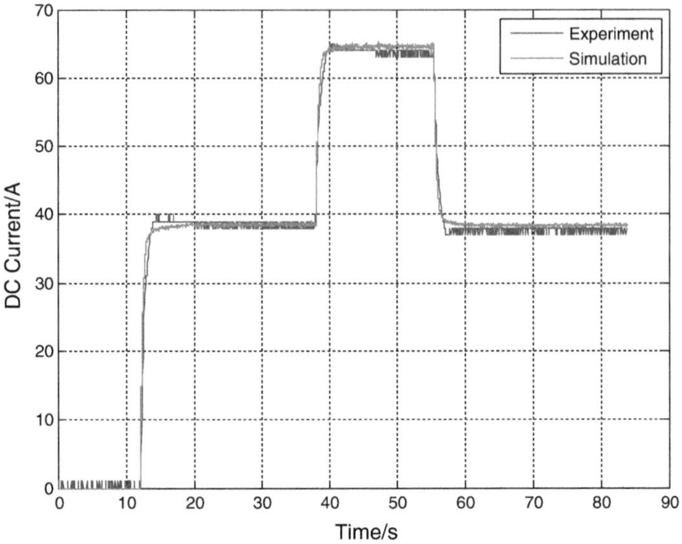

Fig. 11 Generator output DC current comparison for Test 1

appropriate tuning actions. Generally, the simulation results reproduce the average behavior of the experimental results with a high fidelity. At 12.15 and 38 s, the APU speed has two obvious drops due to the sudden increases of the target APU torque. At 55.4 s, the APU speed considerably increases considerably due to a sudden decrease of the APU torque. The DC current varies accordingly with the target APU torque. As can be seen, even during the dynamic process, a good agreement between the simulation and experimental results is achieved.

(2) VSVT

In this test, we changed the APU torque and speed simultaneously. The reference profiles of the target torque, DC bus voltage and speed are shown in Figs. 12, 13 and 14 respectively. It can be observed from Fig. 15 that the simulation and experimental results of the DC current exhibit a very good agreement. For the APU speed, Fig. 14 shows that the simulation result at steady state can be validated by the experiment as well, but noticeable differences exist during transient state. The reason might be that the engine model only considers the average torque production part and neglects the transients during the combustion process. Therefore, engine speed variation dynamics among different values cannot be reflected with good accuracy in the model.

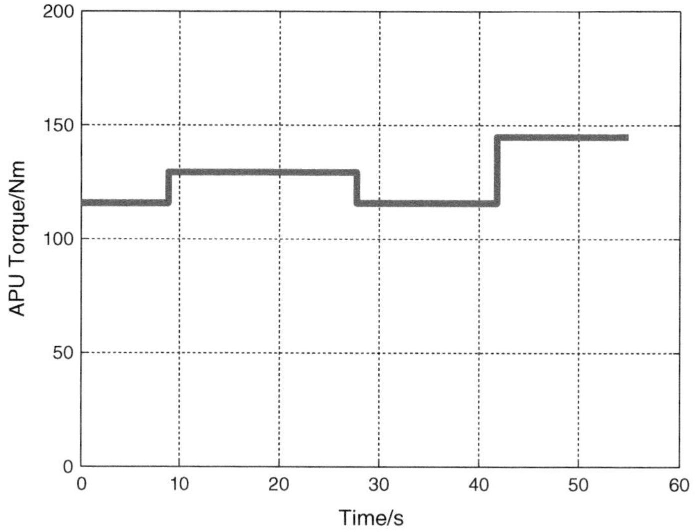

Fig. 12 Target generator torque profile for Test 2

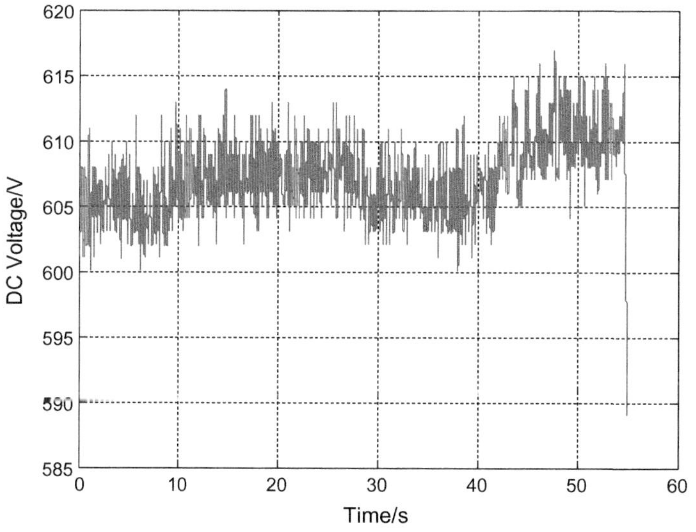

Fig. 13 DC bus voltage for Test 1

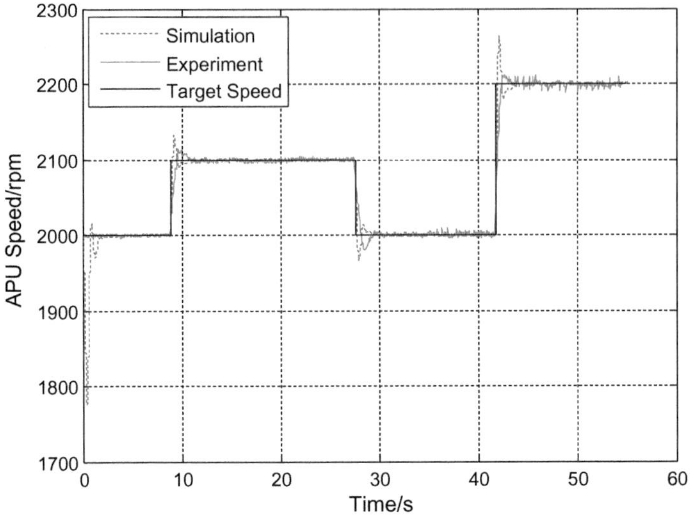

Fig. 14 Engine speed comparison for Test 2

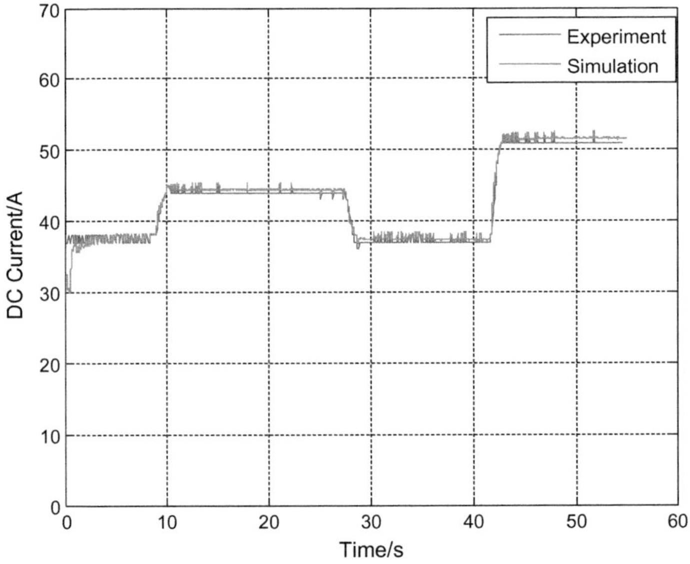

Fig. 15 Generator output DC current comparison for Test 2

5 Comparison of Dynamic Coordinate Control Performance

Dynamic coordinate control of the engine-generator system regulates the engine and generator simultaneously and dynamically to achieve stable and efficient operation. In the coordination control, the engine is usually controlled by speed, and the generator can be controlled either by torque or current. In this section, we compare the performance of these two control strategies, by using the above established system model. Two scenarios, i.e., big load changing rate (the changing rate of the target power is ±10 kw/s) and small load changing rate (the changing rate of the target power is ±5 kw/s), are investigated. Figures 16 and 17 present the comparison results. As can be seen, in both scenarios, the current control gets smaller difference between desired power and actual power, which means generator current control could achieve higher system efficiency. However, as can be seen from the speed trajectory, speed fluctuation in the current control is larger than that in torque control, which means that torque control has better stability. This can be seen more clearly when the load changing rate is big. Therefore, a simple note to guide the dynamic controller design would be: If fast and stable operation is mostly

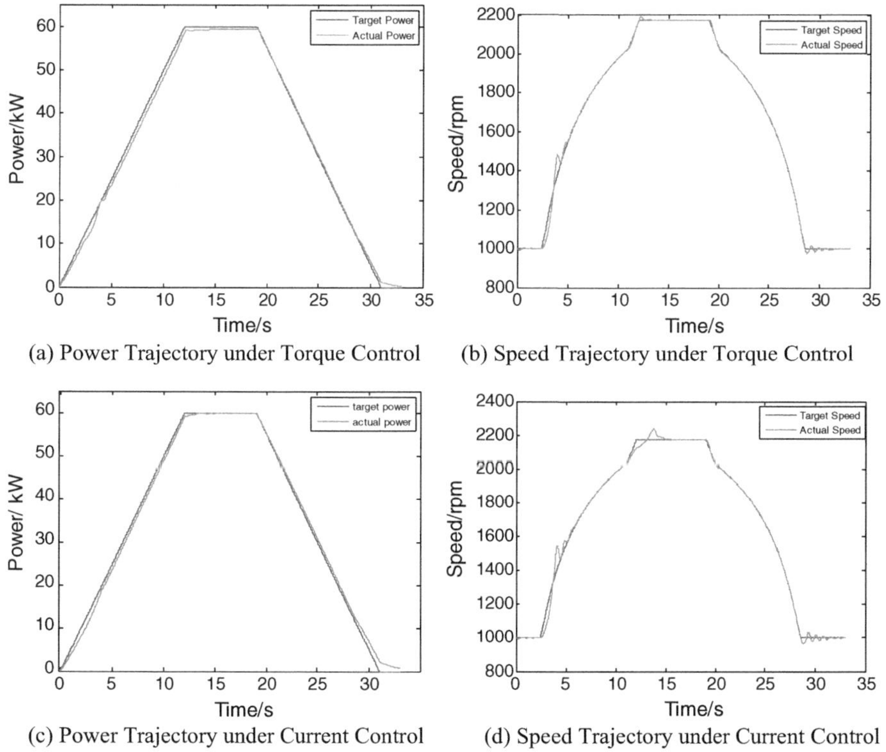

Fig. 16 Simulation results under load changing rate of ±5 kw/s

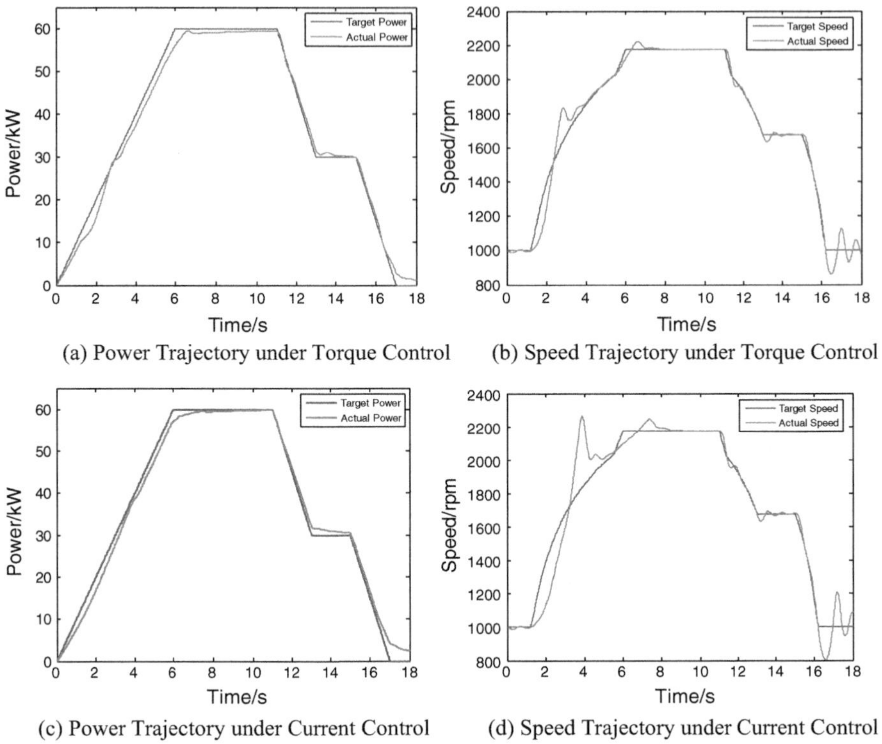

(a) Power Trajectory under Torque Control
(b) Speed Trajectory under Torque Control
(c) Power Trajectory under Current Control
(d) Speed Trajectory under Current Control

Fig. 17 Simulation results under load changing rate of ±10 kw/s

desired, the torque control is preferred; if system efficiency is the most concerned control objective, the current control should be chosen and big load changing rate should be avoided.

6 Conclusion

This paper presents a semi-physical average model for a diesel engine-generator system with PMSG and controllable AC/DC converter onboard in EREV applications. The diesel engine model focuses on speed control by considering the average engine torque production mechanism. The PMSG and AC/DC converter model are developed on the *d*/*q* synchronous frame according to first principles. Detailed switching mechanisms are ignored, as the model is mainly for component-level and vehicle-level control purposes. Experiments are carried out to validate the proposed model. The comparison between simulation and experiment shows that the model generally has good accuracy under different operating conditions. Based on the established and validated model, two dynamic coordinate control strategies are compared in terms of efficiency and stability performance under two load changing rate scenarios. Heuristic guide on how to appropriately design the dynamic coordinate controller according to different control objectives and load conditions is given.

References

1. Tate ED, Harpster MO, Savagian PJ (2008) The electrification of the automobile: from conventional hybrid, to plug-in hybrids, to extended-range electric vehicles. SAE Technical Paper, No: 2008-01-0458
2. Seng C, Lijun H, Hongwen H, Yebao S, Fengchun S (2003) Auxiliary power systems for series hybrid electric vehicles. Chin J Automot Eng 25(3):243–245 (in Chinese)
3. Qu X, Qingnian W, Yuanbin Y (2013) Study on APU control strategy for extended range electric vehicles. Chin J Automot Eng 35(9):763–768 (in Chinese)
4. Bin H, Ximing C, Guijun CAO, Jiaqun X, Minggao O (2006) Dynamic control for auxiliary power unit of series hybrid electric vehicles. Chin J Automot Eng 28(1):12–16 (in Chinese)
5. Hu M, Yang F, Ouyang M, Xu L, Yang X Fang C (2012) Study on distributive control systems for extended hybrid electric vehicles. Automot Eng 34(3):197–202 (in Chinese)
6. Wang T, Wang Q (2014) Control system of permanent generator for boom potential energy recovery. Chin J Agric Mach 10(9):1–6 (in Chinese)
7. Markel T, Brooker A, Hendricks T, Johnson V, Kelly K, Kramer B, O'Keefe M, Sprik S, Wipke K (2002) ADVISOR: a systems analysis tool for advanced vehicle modelin. J Power Sources 110(2):255–266
8. Rousseau A, Sharer P, Pasquier M (2001) Validation process of a HEV system analysis model: PSAT. SAE Technical Paper 2001-01-0953
9. He B, Ouyang M (2005) Average modeling of diesel auxiliary unit for series hybrid electric vehicle. Veh Technol Conf 4:2372–2376
10. Zhou W, Zhang C, Li J (2014) Analysis of optimal power management for series plug-in hybrid electric vehicles via dynamic programming. In: IEEE transportation electrification conference Asia-Pacific, Beijing, China
11. Moura SJ, Fathy HK, Callaway DS, Stein JL (2011) A stochastic optimal control approach for power management in plug-in hybrid electric vehicles. IEEE Trans Control Syst Technol 19(3):545–555
12. Song Q, Grigoriadis KM (2003) Diesel engine speed regulation using linear parameter varying control. In: Proceedings of the American control conference, 2003, pp 779–784
13. Li S, Haskew TA, Xu L (2010) Conventional and novel designs for direct driven PMSG wind turbines. Electr Power Syst Res 80:328–338
14. Pillay P, Krishnan R Modeling, simulation and analysis of permanent-magnet motor drives. Part I: the permanent-magnet synchronous motor drive. IEEE Trans Ind Appl 25(3):265–273
15. Yin M, Li G, Zhou M, Zhao C (2007) Modeling of the wind turbine with a permanent magnet synchronous generator for integration. In: Power engineering society general meeting, IEEE, pp 1–6
16. Rowan TM, Kerkman RJ A new synchronous current regulator and an analysis of current-regulated PWM inverters. IEEE Trans Ind Appl IA-22(4):678–690
17. Lerdudomsak S, Doki S, Okuma S Novel voltage limiter for fast torque response of IPMSM in voltage saturation region. Electr Eng Jpn 175(12):1346–1356

Study on High-Cycle Shear Fatigue Behavior of Two Kinds of Spot-Welded 980 Mpa Ultra-High-Strength Steels

Zhang Ding, Yang Bing, X. G. Qin, J. C. Duan, W. C. Huang, Zhang Mao and Qin Liang

1 Introduction

Resistance spot welding, with the advantage in low cost and high efficiency, is still the primary joint method used for the construction of the automotive body. However, as the development of automotive lightening and increasing demands for safety, the strength of steels used in white body is also higher than before. Ultra-high-strength steels (UHSS) with yield strength no less than 550 Mpa have been widely applied all over the world [1]. Dual-phase steel DP980 [2] and quenching and partitioning steel QP980 [3] are two typical UHSS.

The majority of spot-weld failures in automobiles, however, can be attributed to fatigue loading. During its life cycle, the automotive structure is subjected to random fatigue loading in the form of vibrations, which initiated by the road and resulting in a relative planar motion tendency among steel blanks. In this way, the welding spot would bear alternate shear fatigue load. This is one of the typical working conditions for the welded spot of white body. Compared to the regular high-strength steels, the UHSS would be added more C, Mn, Si or Cr, Mo, Nb, etc., alloy elements to strengthen the microstructure of steels [4], which increase the carbon equivalent of the material itself. And the RSW for the steel plate would be influenced, such as narrowing the welding lobe and more likely spatter occurrence, etc. [5]. Therefore, it is necessary to investigate the shear fatigue performance of the welding spot formulated by the resistance spot welding. In this paper, the welding spot fatigue performance of high frequency for DP980 and QP980 steels would be experimentally investigated and analyzed.

Z. Ding · Y. Bing (✉) · X. G. Qin · J. C. Duan · W. C. Huang · Z. Mao · Q. Liang
Dongfeng Motor Corporation Technical Center, Wuhan, China
e-mail: yangbing@dfmc.com.cn

Society of Automotive Engineers of China (SAE-China) (ed.), *Proceedings of the 19th Asia Pacific Automotive Engineering Conference & SAE-China Congress 2017: Selected Papers*, Lecture Notes in Electrical Engineering 486, https://doi.org/10.1007/978-981-10-8506-2_69

2 Materials and Experimental Procedure

2.1 Materials

The two UHSS used in this study were DP980 and QP980 with 1.8 mm thickness. The chemical composition for the two steels of DP980 and QP980 is shown in Table 1. It could be observed that the carbon equivalent of QP980 is higher than that of DP980. Table 2 shows the mechanical properties of DP980 and QP980.

A multiple-impulse welding with high welding current of 9250 A would be applied (Fig. 1), and the resistance spot-welding test samples were fabricated using a MFDC with chromium–zirconium–copper electrodes (tip diameter 6 mm). The spot-weld joints were welded with an electrode force of 4.5 kN, satisfying the minimum nugget diameter criteria ($4.25\ \sqrt{t}$, t = sheet thickness). Then, the welded samples were set aside for microstructure analysis.

The spot-weld joints of DP980 and QP980 steels with the same welding parameter were sectioned through the centerline, mechanically ground, polished,

Table 1 Chemical composition by percent weight

Steels	Ceq	C	Si	Mn	P	S	Cr	Mo	Al	Ti	Nb
DP980	0.243	0.091	0.279	2.393	0.0094	0.0011	0.557	0.216	0.029	0.0211	0.006
QP980	0.362	0.184	1.48	1.94	0.0128	0.0015	0.03	<0.001	0.0452	0.0032	0.007

Remark Ceq = C + Si/30 + Mn/20 + 2P + 4S

Table 2 Mechanical properties of DP980 and QP980

Steels	Thickness (mm)	Rp0.2 (MPa)	Rm (MPa)	Agt (%)	$A_{50}t$ (%)	n90	r90
DP980	1.8	682.46	1006.72	7.21	14.28	0.08	0.83
QP980	1.8	729.77	991.81	11.78	18.68	0.10	0.99

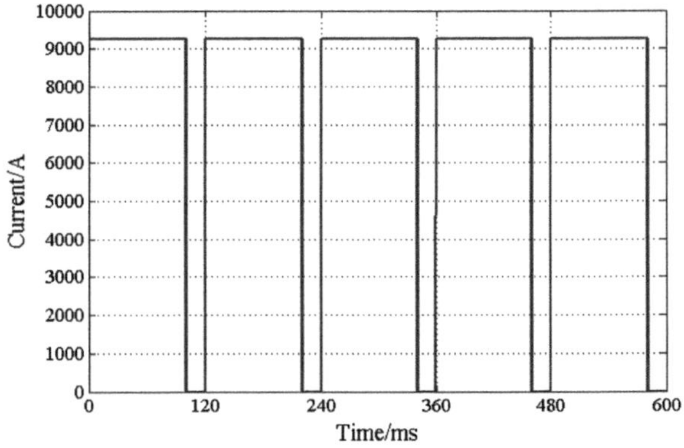

Fig. 1 Schematic drawing of welding current pattern

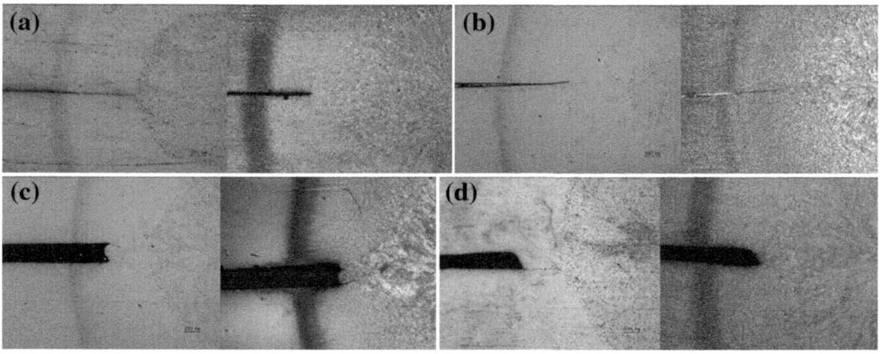

Fig. 2 Typical appearances of pre-cracks near nugget (left: 2% nital acid; right: picric acid)

and etched using 2% natal acid and picric acid alcohol solution. The microstructures show that all spot welds exhibit a notch along the circumference of the weld nugget where the two overlapping sheets are unbonded. The typical cracks initiate at the notch corner and propagate through the HAZ, as shown in Fig. 2. It could be observed that there exists single micro-crack stretch into nugget zone (Fig. 2a), and there are also more than one cracks at the interface between plates (Fig. 2b–d). A part of crack ends are located at the critical position between the fine-grained zone and coarse-grained zone, and some of them even stretch into the nugget zone. The length and shape of those initial micro-cracks exhibit a random characteristic at a certain extent.

2.2 *Experiment Procedure*

Test samples for shear fatigue test of spot-weld joint were prepared by welding two overlapping coupons (150 * 50 mm) with an overlap of 50 * 50 mm as per Standard JIS Z 3138. Two spacers of dimension 30 * 50 mm were placed at both ends of the test samples to induce pure shear and to avoid initial realignment during testing (Fig. 3). All specimens were tested in the as-welded condition. The fatigue testing was carried on 5000 N electromagnetic fatigue testing machine with a frequency of 40 Hz. The samples were subjected to sinusoidal load with $R = 0.1$. The fatigue limit was set to be 10^7 cycles. When a significant load attenuation occurs before the fatigue limit arrives, the experiment would be ended and the circulation times would be recorded.

After the fatigue experiments of the spot-weld joints, the welding spot would be cut in the middle along the tensile direction, and then be etched by 2% nitric acid alcohol solution. Samples would be evaluated the origins and propagations of the fatigue cracks by SEM.

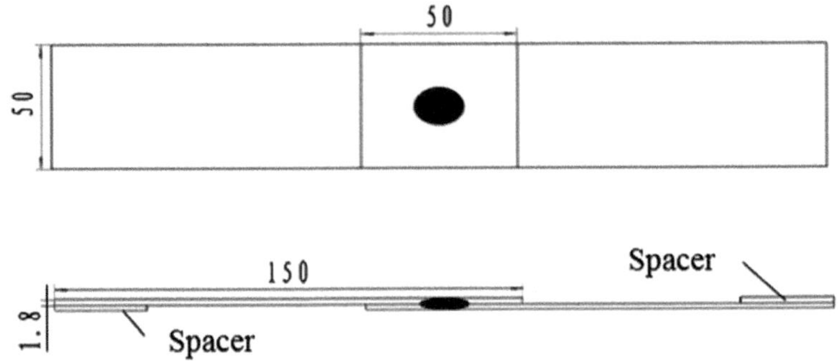

Fig. 3 Schematic of shear fatigue test samples

3 Results and Discussion

3.1 High-Cycle F-N Curves

The load range is plotted as the ordinate and endurance as the abscissa, both of which use double logarithmic scales as shown in Fig. 4. The F-N curves for the welding spots of QP980 and DP980 steels are achieved by fitting linearly those spatter plots as per Standard JIS Z 3138. It could be observed that the fatigue life of welding spot of QP980 is higher than that of DP980. At maximum loading of 2.75 kN, all the welding spot samples of QP980 steel completed 10^7 cycles, while three welding spots' samples of the DP980 steel failed to complete 10^7 cycles at the maximum loading of 2 kN. When the up-and-down threshold measurement method was adopted at 95% confidence degree and 10% possibility of failure, the fatigue limit load of DP980 welding spots is 1.29 kN, and the fatigue limit load of QP980 arrives to 2.54 kN for the fixed cycles 1.00×10^7. Therefore, the welding spot of QP980 has better high-frequency shear fatigue performance compared to that of DP980.

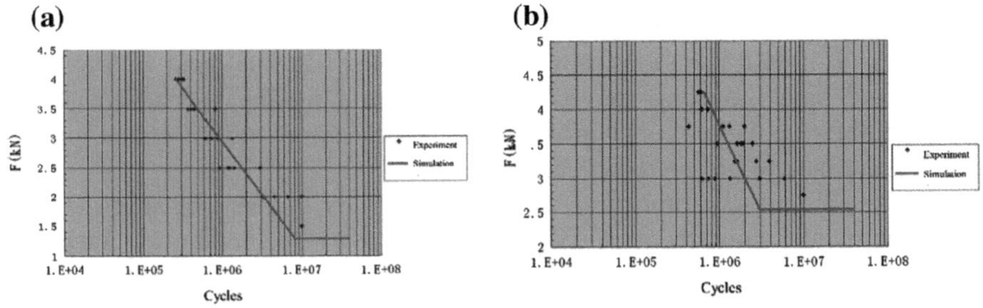

Fig. 4 F-N curves of RSWs after high-cycle shear-tensile fatigue test of DP980 (**a**) and QP980 (**b**)

3.2 Fracture Mode and Crack Initiation

It could have observed that all the cracks of failure samples after fatigue testing were lateral expanded along the plate width direction (Figs. 5 and 6), and the interfacial fracture was never happened. The fatigue cracks for some samples only appeared at one side of the plate, while others at both sides. The fatigue crack of samples experimented high amplitude loading exhibited higher degree of expansion and propagated throughout the whole plate width. The failure mode for the welding spot under high-frequency shear fatigue loading condition was quite different from that under static tensile or low-cycle fatigue conditions [6]. Then the relative locations of the fatigue cracks at the welding spot region were observed for the typical failure samples of those two materials. The information of those samples is presented in Table 3, and the fatigue crack initiation position and propagation direction are shown in Fig. 7. It can be observed that different from the pull-out failure in the quasi-static shear-tensile test, the fatigue cracks of high-cycle fatigue

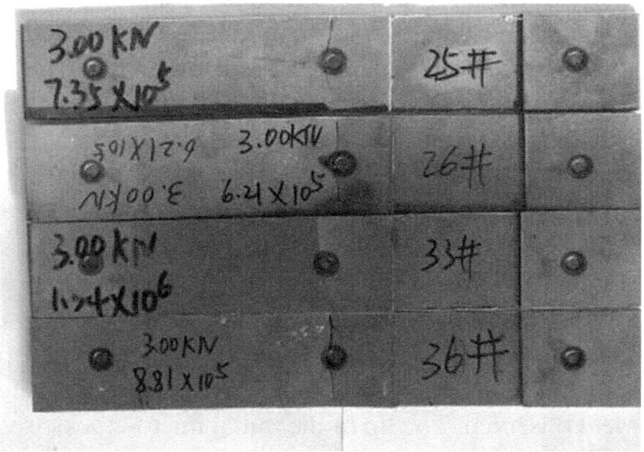

Fig. 5 Typical appearance of spot-welded DP980 after high-cycle shear fatigue test

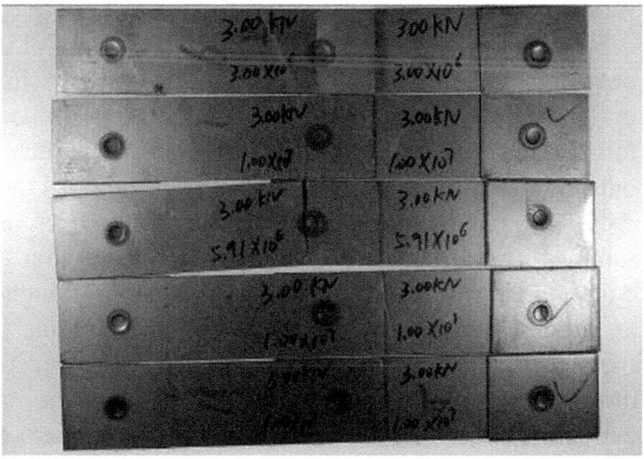

Fig. 6 Typical appearance of spot-welded QP980 after high-cycle shear fatigue test

Table 3 Information of samples for investigate position of crack initiation

Sample No.	Steel	Max load (kN)	Fatigue life	Spatter	Macro-crack location	Figure No.
43#	DP980	2.5	9.19×10^5	Negative	Single	Figure 7d
14#	DP980	4	2.61×10^5	Negative	Both	Figure 7a
8#	QP980	3.5	9.54×10^5	Negative	Single	Figure 7c
13#	QP980	3.75	1.97×10^6	Negative	Both	Figure 7b
3#	QP980	4	7.31×10^5	Positive	Both	Figure 7e

testing were not initiated from the subcritical heat-affected zones, but mainly from the tips of initial micro-cracks between base metals, and the position of fatigue crack initiation is related to the form and location of initial micro-cracks.

When the macro-cracks of the failure specimens occurred in both sides, the macro-cracks originated from the tip of initial micro-cracks even though they were in different forms. At one side of the welding spot of No. 14 sample (Fig. 7a), "tongue" like protrusion has been formed because the softened metal around the nugget zone would flow and squeezed out under the electrode pressure, which caused two initial micro-cracks at this zone. The fatigue cracks would originate from the initial micro-cracks location between those protuberances and base metals and then expand. While at the other side of No. 14 sample where there are no "tongue"-like shape extrusions exist, the initial micro-crack is a crack stretching into the spot-weld nugget zone. The fatigue cracks are originated from the tip of initial micro-cracks and extended along the blank thickness direction, even cross through the spot-weld nugget zone. There was no evidence from the SEM observation results that the initial micro-cracks would propagate firstly along the interface of two base metals into the nugget zone and then along the blank thickness direction. At both sides of the No. 13 samples (Fig. 7b), the initial micro-cracks were observed. The tip of the initial micro-cracks is very close but not stretched into the weld nugget zone. During the propagation process of the cracks along the plate thickness direction at both sides, the fatigue crack at one side partly crosses over the weld nugget zone.

There are two statuses when the macro-cracks of the failure specimens located at only one side. For the first situation, such as the No. 8 sample (Fig. 7c), the crack initiation and propagation characteristics are similar to No. 14 sample, while the initial micro-crack of No. 8 sample does not stretch into the welding nugget zone, but originated from the coarse-grained region. As the fatigue crack does not totally cross through on thickness direction, the cracks are only found at one side of samples in macroscale. It could be speculated that under the fatigue loadings, the crack initiations are more likely generated at "tongue" regions under cyclic strain, as the directions of local stress are different. While for the other situations such as the No. 43 specimen (Fig. 7d), fatigue cracks that propagated along the thickness direction are only found at one side, where the "tongue" zones formed by the extrusion of the melted metal, and the initial micro-cracks propagate into the weld

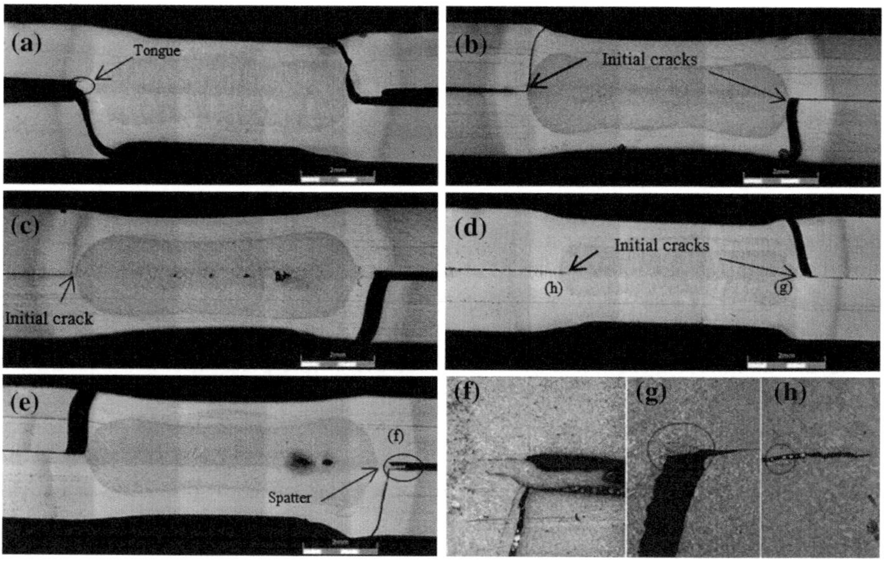

Fig. 7 Fatigue crack initiation position and propagation direction of samples in Table 3 plus typical local features

nugget zone at the other side, which can be found in Fig. 7h that there have been existed a "tongue" zone before fatigue testing at this side. Since the cracks at only one side are not found under the other high-load conditions, it could be calculated that the initial micro-cracks tend to propagate from the tip to the welding nugget zone slowly under sustained low shear fatigue loading.

No. 3 sample of QP980 (Fig. 7e) has a spatter at one side. The fatigue cracks of the spatter side were not initiated from the tip of micro-crack between the spatters and base metals, but originated from two locations of HAZ and propagated until intersected in HAZ (Fig. 7f). The reason may be that one side of the base metal was not strictly parallel to the loading direction and then the stress direction of the heat-affected zone would be influenced when the spatter occurs.

3.3 SEM Analysis of Fractures

In this part, the fracture surfaces of the shear fatigue specimens for those two materials would be analyzed by SEM. The specimen information is presented in Table 4, No. 23 sample has a "tongue" like zone similar to Fig. 7d, and No. 5 specimen with an initial micro-crack as Fig. 7b was selected for QP980. All the fatigue cracks are originated from the inner surface of based material and show a ring initiation (Fig. 8a). The characteristics of fatigue initiation could be observed at the 1/6 circle around the spot-welding joint orientated to the clamp end by

Table 4 Information of samples for fracture analysis

Sample No.	Steel	Max load (kN)	Fatigue life	Spatter	Figure No.
23#	DP980	3.5	4.11×10^5	Negative	Figure 8
5#	QP980	3.5	1.62×10^6	Negative	Figure 9

comparing the macro-cracks (Fig. 8b). With the help of SEM, the morphology of the "tongue" region (Fig. 8c), a benched multi-segment structure, could be clearly observed. Since the oxidation occurs under high temperature, the layer oxide could be observed. Cracks are originated from the tip of initial micro-cracks (the interface between the "tongue" and plate surface). The fatigue cracks would firstly propagate along the thickness direction until cross over the whole plate thickness. And then cracks turn to propagate laterally until throughout the whole plate width. The micro-crack initiation and propagation characteristics for different directions are not observed around the tip of micro-cracks (Fig. 8d), which means that the cracks do not propagate firstly into the welding nugget region and then along the thickness direction. The features of intergranular fracture could be observed around crack initiation positions (Fig. 8d). While as the fatigue cracks propagate, they disappear at the following regions.

The QP980 fatigue fracture surface of the spot-welding joint is presented in Fig. 9. It could be observed from Fig. 9a that there is no "tongue" area similar to Fig. 8c between two based blanks. The cracks are originated from the tips of micro-cracks, and their propagation processes are similar to that of DP980. It could be observed from Fig. 9b that the surface morphology around the initiation zone is different from that of the base blanks, which means that this zone is the surface of initial micro-cracks. No initiation and propagation characteristics of cracks are found at this zone. Intergranular fracture characteristics could be found around the initiation zone of fatigue cracks.

Figure 10 shows the microstructure of the fatigue ridges near crack initiation area for those two materials. Although the ridge morphologies are different at the

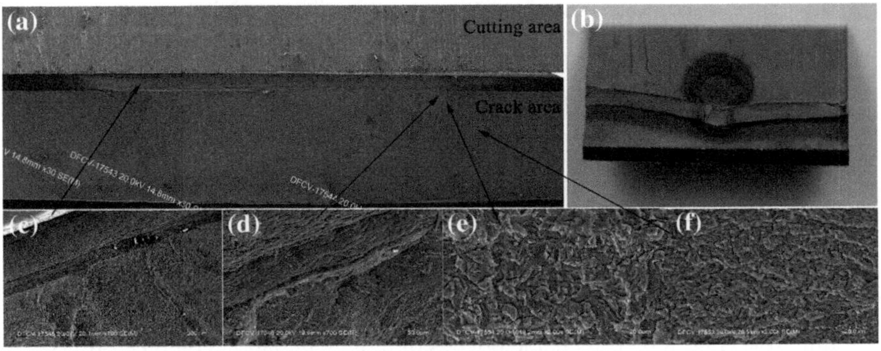

Fig. 8 Fracture appearance of spot-welded DP980 with a 'tongue" area after fatigue test

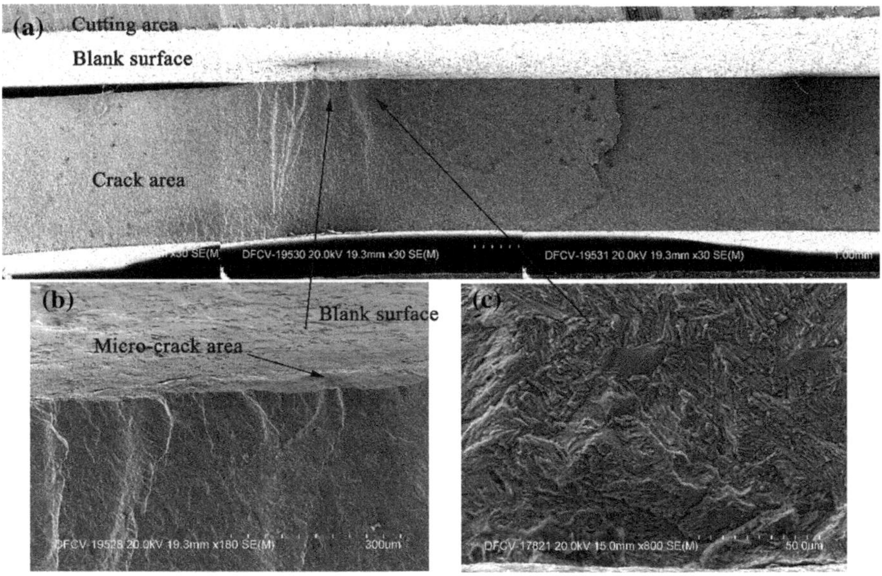

Fig. 9 Fracture appearance of spot-welded QP980 without a 'tongue" area after fatigue test

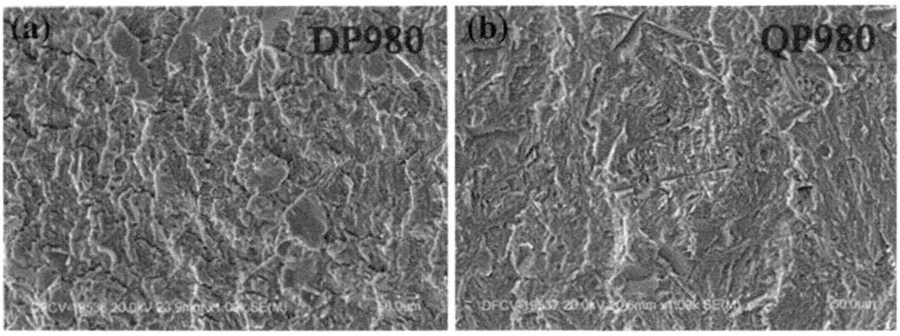

Fig. 10 Appearance of fatigue ridges of spot-welded DP980 (**a**) and QP980 (**b**) near crack initiation area

crack surface of DP980 and QP980, their intervals are almost the same, which means that their crack propagation velocities are at the same level. The crack initiation areas for those two welding spots are all located around the coarse-grained zones. At these two zones, the structures of those two materials are coarse martensites, while the hardness of the martensitic structure in the heat-affected zone of QP980 is higher. It could be observed from the ridge intervals of those two materials that the crack propagation processes are only related to microstructures and independent on the hardness, which is accordance to previous investigation [7]. At the maximum loading of 3.5 kN, the fatigue life of No. 5 sample of QP980 is 4 times of that of No. 23 sample of DP980, because QP980 is more difficultly initiated during the fatigue crack initiation stage. This might be that the hardness of the martensitic in the heat-affected zone (even in the welding nugget zone) of QP980 is higher, and it is more difficult to reach the critical value of plastic strain for the crack initiation which is perpendicular to the board direction under the fatigue loading.

4 Conclusions

(1) The spot-welding joint of QP980 has better anti-high-cycle shear crack performance than DP980.
(2) The initiation and propagation behavior of high-cycle fatigue cracks of the spot-welding joints are similar, for QP980 and DP980 steels. Different from the tensile test, the fatigue cracks are not initiated from the subcritical heat-affected zone, but mainly from the tip of initial micro-cracks. And then they would propagate along the plate thickness direction from the based metal. When the cracks cross over the whole plate thickness, they would propagate laterally until the load significantly decrease and failure.
(3) At low loading situation, there also exist the initial micro-cracks propagating into the welding nugget zone, which results in the cracks at only one side. When spatter occurs, the fatigue cracks may not originate from the tips of initial micro-cracks but from the heat-affected zones.
(4) The crack propagation velocities of these two materials, according to the gap between the fatigue ridges of the crack propagations, are almost the same. It means that for the spot-welding joints of QP980, as the hardness of the tip microstructure of initial micro-cracks is higher, the birth of fatigue cracks would be more difficult compared to the spot-welding joint of DP980.

References

1. Ultralight steel auto body final report. American Iron and Steel Institute, 1998
2. Kang YL, Kwang S, Yin X, Yu H, Liu RD (2006) Research on progress and development of dual phase sheet of automobiles. Automobile Technol Mater 5:1–5
3. Speer J, Matlock DK, Cooman BCD, Schroth JG (2003) Carbon partitioning into austenite after martensite transformation. Acta Mater 51(9):2611–2622
4. Zhang ZG (2016) Research progress and present technique application situation of advanced high-strength automobile steel sheet. HEBEI Metall 241(1):1–7
5. Wang M (2005) The latest development of resistance welding technology in foreign countries. In: The proceedings of the 11th national welding conference (Book 1). China Welding Association
6. Rathbun RW, Matlock DK, Speer JG (2003) Fatigue behaviour of spot welded high-strength sheet steels. Weld J 8:207–218
7. Barsom JM, Rolfe ST (1999) Fracture and fatigue control in structures. ASTM, West Conshohocken, PA, pp 200–204

Effect of Unbalanced Vertical Force of IWM on Comfort and Stability in Four-Wheel Drive Electric Vehicle

Zhe Li, Ling Zheng, Yinong Li, Yusheng Li and Zhoubin Xiong

1 Introductions

IWM-EV has changed the traditional power output channel and canceled the mechanical transmission system. By means of integrated IWM as power source, reducing the transmission system quality, chassis layout is more flexible. Passive safety systems, e.g., anti-lock braking system (ABS), acceleration slip regulation (ASR), and electronic stability program (ESP), can be easily integrated and controlled due to precisely and rapidly longitudinal force response. It is considered as the final form of electric vehicle and has become research hot spots in recent years [1].

Switched reluctance drive (SRD), as one of the most optimal schemes for the motor drive system in IWM-EVs, has a series of advantages such as simple structure, high reliability, and low maintenance. It can also be easy to provide torque and power and meet vehicle dynamic demand. In addition, SR motor phase is independent; it can run under lack of phase, which improves the fault tolerance and robustness in the driving system that is critical to improve the stability in IWM-EVs.

Key project of Chongqing Natural Science foundation (cstc 2015 jcyjBX 0097).
Key technology innovation projects of key industries in Chongqing (csts 2015 zdcy-ztzx 30001).
Key research program of the Ministry of science and technology (2016 YFB 0100900).

Z. Li · L. Zheng (✉) · Y. Li
College of Automotive Engineering, State Key Laboratory of Mechanical Transmission, Chongqing 400044, China
e-mail: zling@cqu.edu.cn

Y. Li · Z. Xiong
Changan Automotive Engineering Institute, Chongqing 401120, China

L. Zheng
College of Automotive Engineering, Chongqing University, 400044 Chongqing, People's Republic of China

Society of Automotive Engineers of China (SAE-China) (ed.), *Proceedings of the 19th Asia Pacific Automotive Engineering Conference & SAE-China Congress 2017: Selected Papers*, Lecture Notes in Electrical Engineering 486, https://doi.org/10.1007/978-981-10-8506-2_70

However, for a SRD system, it is natural to produce a series of torque ripples in operation because of harmonic components in current, salient characteristic of the stator and rotor as well as measurement error. Vertical vibration and torque ripple may deteriorate dynamic performance in vehicles. Once the torque ripple increases, the vehicle suspension resonance may also occur. On the other hand, the unsprung weight is increased due to the special arrangement and structure and the negative effect is caused. Therefore, it is an important task for IWM-EV to alleviate these two negative effects.

In order to suppress the vibration in SRM, optimal methods are studied. Inagaki et al. [2] proposed a H∞ control to suppress the vibrations induced by radial force and torque ripple of the motor. Max Santos et al. [3, 4] used a physical field analysis to predict the radiation characteristics of sound field in EV by introducing parallel paths in windings. Li et al. [5] have reduced the radial force imbalance. It is found that the current can be balanced and unbalanced force can be reduced by parallel paths in windings.

Some new methods are proposed to promote the ride comfort and handling stability of IWM-EVs. Rojas et al. [6] have investigated the relationships between IWM configurations and suspension systems in which the control strategy ensures the safety and comfort in vehicle. Shini et al. [7] have used finite element analysis and optimization method to improve torque ripple. Furthermore, rubber bushings are used by Luo et al. [8] as an efficacious method to isolate the vibration in body caused by the unbalanced force. Lu et al. [9] developed an IWM-EVs dynamic controller to improve the stability in vehicle. The proposed controller is able to improve stability effectively in vehicle. Nam et al. [10], Lu et al. [9], and Lu et al. [11] proposed some methods to enhance the maneuver stability and ride comfort performance in IWM-EVs. However, the influence of vibration on EVs is ignored. Sun et al. [12] proposed genetic algorithm (GA) to optimize parameters in active suspension system, and a series of Pareto optimal solution are obtained.

Most of the researches focus on maneuver stability and ride comfort in IWM-EVs, without taking account of vibration and torque ripple in wheel motor. Also, the coupling between the vibration of IWM and dynamic performance of IWM-EV is rarely considered overall. Therefore, in this paper, a conjoint analysis for vertical stabilities in IWM-EV is carried out and the resonance phenomenon caused by vertical force is investigated.

2 In-Wheel Motor and Vehicle Modeling

Figure 1 shows IWM structure, and it consists of the following parts: shaft, motor bearing, tire, and hub bearing. Two sub-models were established: One is SR-in-wheel motor model which presents vertical force in wheel, and the other is a quarter vehicle model in IWM-EVs which emphasizes the coupling effect of IWM on EVs.

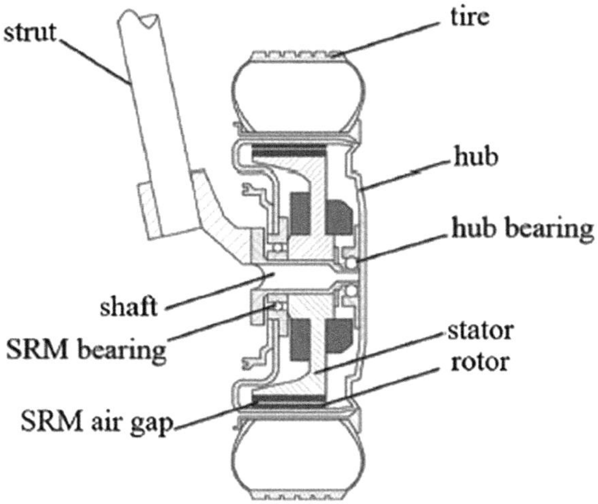

Fig. 1 In-wheel motor structure

2.1 SRM Model

Typical 8/6 SR motor is shown in Fig. 2. The mutual inductance between each phase is ignored in SRM. For a particular phase, e.g., A, phase voltage of pole winding is obtained through the corresponding phase current as:

$$v_a = r_a i_a + \frac{\mathrm{d}\lambda_a(i_a,\theta)}{\mathrm{d}t} = r_a i_a + N\frac{\mathrm{d}\phi_a(i_a,\theta)}{\mathrm{d}t} \text{ or } v_a = r_a i_a + l_a\frac{\mathrm{d}i_a}{\mathrm{d}t} + e_a \quad (1)$$

where $l_a = \partial\lambda_a(i_a,\theta)/\partial i_a$ is winding inductance, $e_a = w\partial\lambda_a(i_a,\theta)/\partial\theta$ is self-inductance and A-phase back electromotive force. Electromagnetic co-energy is expressed as:

$$W'_a(i_a,\theta) \triangleq \int_0^{i_a} \lambda_a(\varsigma,\theta)d\varsigma \quad (2)$$

Individual phase participates in torque independently. The torque developed by each phase is a function of the corresponding electromagnetic co-energy. For a particular phase, the output torque is written as:

$$T_a(i_a,\theta) = \frac{\partial W'_a(i_a,\theta)}{\partial\theta_m} = P_r\frac{\partial W'_a(i_a,\theta)}{\partial\theta} \quad (3)$$

For multi-phase switched reluctance motor, the total developed torque is the sum of the torque produced by all individual phases. The output torque is written as:

Fig. 2 Typical SR motor

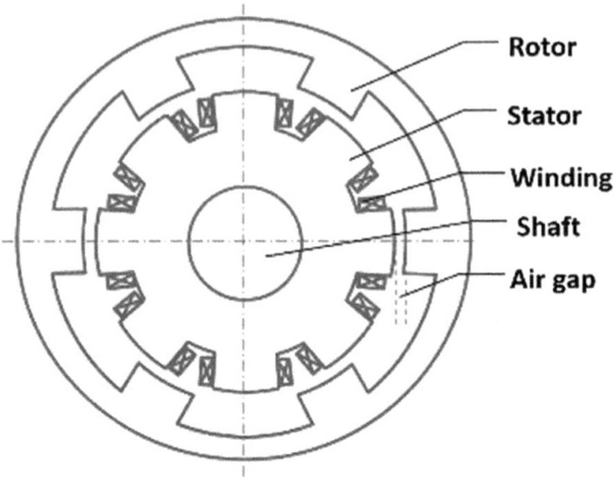

$$T = T_a(i_a, \theta) + T_i(i_i, \theta) + , \cdots , + T_n(i_n, \theta) \quad (4)$$

The mechanical equation of IWM is described as:

$$T = J\frac{\mathrm{d}\omega}{\mathrm{d}t} + B\omega + T_L \quad (5)$$

where ω is the rotor speed, B is the torque friction coefficient, J is the moment of inertia, and T_L is the load torque determined by operating condition of IWM-EV.

Electromagnetic force generated by IWM could be divided into two parts: One is in radial direction, and the other is in tangential direction. Radial force which could not be avoided in rotating electrical machine does not contribute to output power. However, it is the source of noise and vibration. In general, the radial force may be zero under ideal geometrical condition. In fact, it is difficult to ensure machine accuracy and symmetric structure. Therefore, it is unavoidable to produce unbalanced radial force in rotate process. This unbalanced radial force may further deteriorate dynamic performance in EVs.

The rate of relative eccentricity in SRM is written as:

$$\varepsilon = \frac{\Delta g}{g_m} \quad (6)$$

where g_m is the air gap in SRM and Δg is the relative displacement in rotor. When Δg is zero, unbalanced radial force does not exist. The rate of relative eccentricity in 30–60% does not reduce the function of motor [13, 14].

As shown in Fig. 3, radial force on the opposite direction of stator poles can be written as:

Fig. 3 Vertical force of SRM

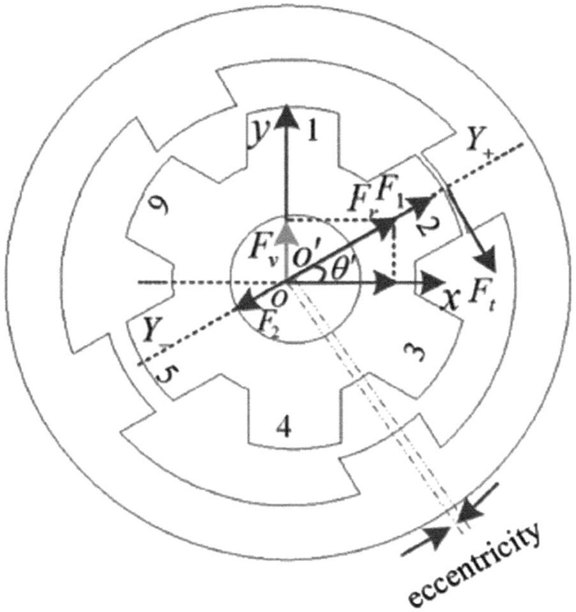

$$F_1 = -\frac{r\,\sin(\theta_o)}{g_m - \Delta g} F_t \tag{7}$$

$$F_2 = -\frac{r\,\sin(\theta_o)}{g_m + \Delta g} F_t\,\Delta g = 0 \tag{8}$$

An imbalanced radial force is the difference of radial force between two forces, and it is written as:

$$F_r = F_1 - F_2 \tag{9}$$

The vertical force exerts on the wheel is yielded:

$$F_v = F_r\,\sin(\theta') \tag{10}$$

where r is the outer radius of rotor, θ_0 is the overlap position between the stator and the rotor pole, and θ' is the angle between vehicle drive direction and Y-coordinate axis.

The SRM and vertical force model are shown in Fig. 3. Radial force F_1 in the positive Y direction decreased at the same time F_2 on the opposite direction increased owing to air gap eccentricity, F_v is the component of the unbalanced radial force in the vehicle vertical direction. In this paper, we take $\theta_0 = 90°$ to investigate the influence of this vertical force on vehicle dynamics (Fig. 4).

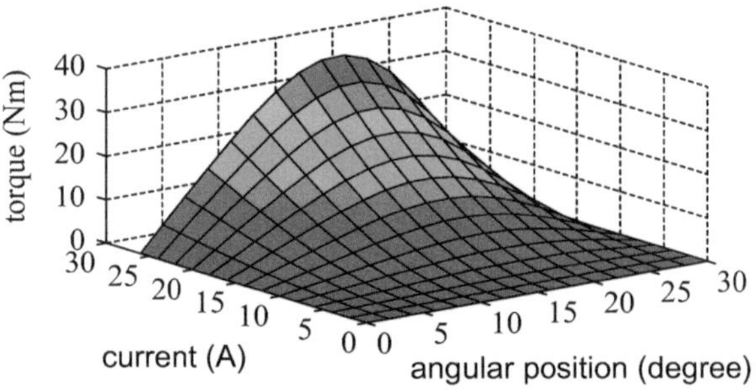

Fig. 4 Output torque related to rotor position and winding current

The radial force and the unbalance radial force of individual phase obtained by Eq. (7) and (8) are shown in Figs. 5 and 6. It can be noted that even a small current excitation could produce extremely large radial force. According to Fig. 6, when winding current is about 20 A, radial force reached 8 kN to the peak. While current excitation increases or decreases, radial force is decreased with a considerable amplitude. In addition, the imbalanced force is the difference between the radial forces generated by anti-poles in the SR-motor and is inversely proportional to the air-gap. Meanwhile, due to the eccentricity, an increase in the radial force from one direction may lead to a decrease in radial force from the opposite direction, which further increases the magnitude of the unbalanced radial force. Furthermore, relative eccentricity is a key factor to unbalanced radial force which increases rapidly with the relative eccentricity rises, and smaller air gap leads to larger unbalanced radial force.

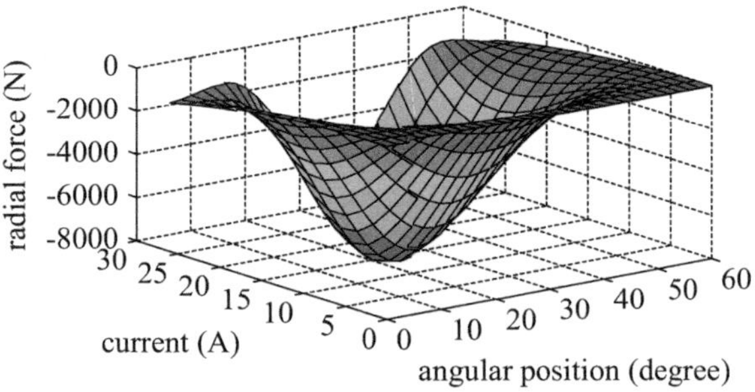

Fig. 5 Phase radial force related to rotor position and winding current

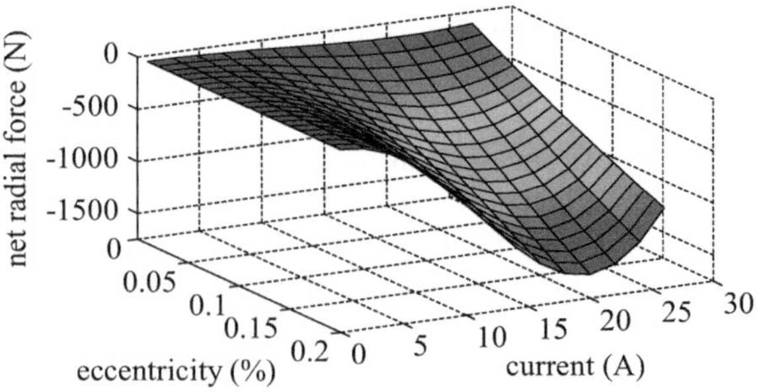

Fig. 6 Peak value of unbalanced radial force related to winding current and eccentricity

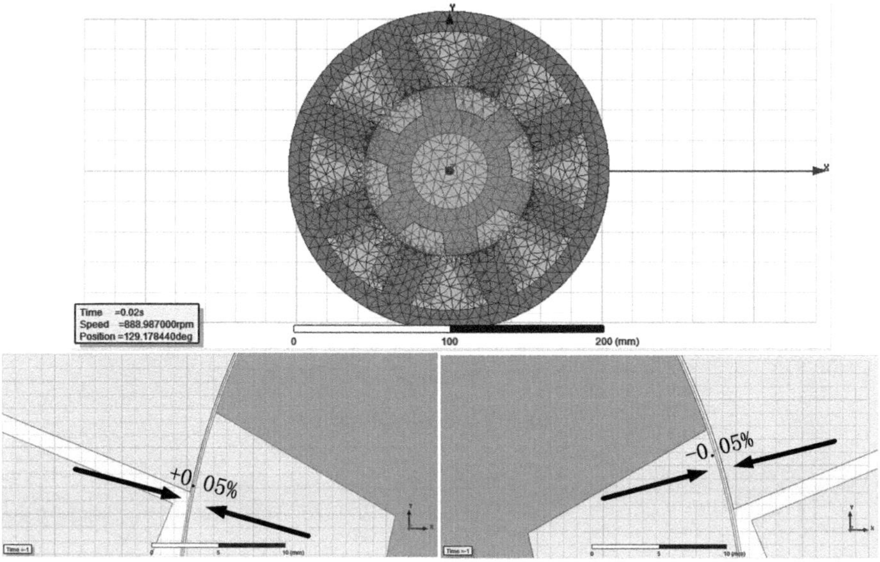

Fig. 7 Finite element model of vertical force

The accuracy of above analysis and simulation is confirmed by a finite element model of eccentric SRM. Rotor eccentricity in this finite element model is set to 0.05%. FE model and partial enlarged detail are shown below (Fig. 7).

Analytical and finite element results are shown in Fig. 8. It is found that two waveforms have the same shape and amplitude. Therefore, analytical method can be verified.

2.2 *Quarter IWM-EV Model*

A quarter vehicle vibration model is chosen to investigate the influence of unbalanced radial force on the vehicle ride comfort and other dynamic response. In addition, it is assumed that eccentricity of the SRM exists only at the stator salient pole in order to simplify the problem. Thus, an additional excitation is present in vertical direction of the vehicle, as shown in Fig. 9.

When rotor is driving wheel rotation, eccentricity of rotor could alter air gap, and there would be a relative displacement between rotor and stator. This will produce imbalanced disturbance force exerted on rotor and stator [15–17]. It is noteworthy that in actual operating conditions, this disturbance force and some other external excitation in motor will aggravate the effect of eccentricity. Rotor eccentricity is assumed to be a fixed value in order to facilitate presentation, as shown in Fig. 9, and the mass of IWM is separated into two sections: One is motor stator and shell

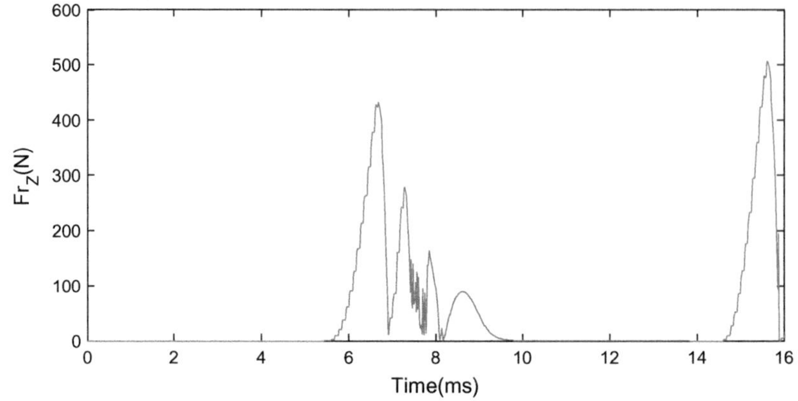

(a) Analytical solution

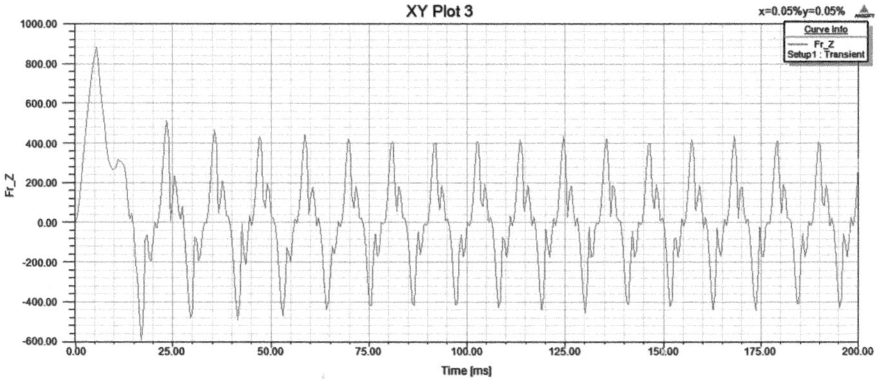

(b) Finite element solution

Fig. 8 Simulink and finite element simulation results

m_s, and the other is tire mass including motor rotor and hub m_u; these two masses are fitted together by hub bearing and motor bearing.

Tire damping is neglected because it is comparatively small in derivation. F_{r_z} acting on m_u and m_s would affect dynamic performance of vehicle. The influence is similar to that caused by dynamic imbalance of an engine or excitation from a road surface.

Wheel rotation balance equation is generated:

$$\frac{I_t \dot{w}}{i_g} = \eta i_g T_e - F_{tx} R_e - M_t \tag{11}$$

Where, I_t is the equivalent inertia of wheel and body mass, $\dot{w}$ is the wheel angular velocity, η is the transmission efficiency, i_g is the transmission ratio, M_t is

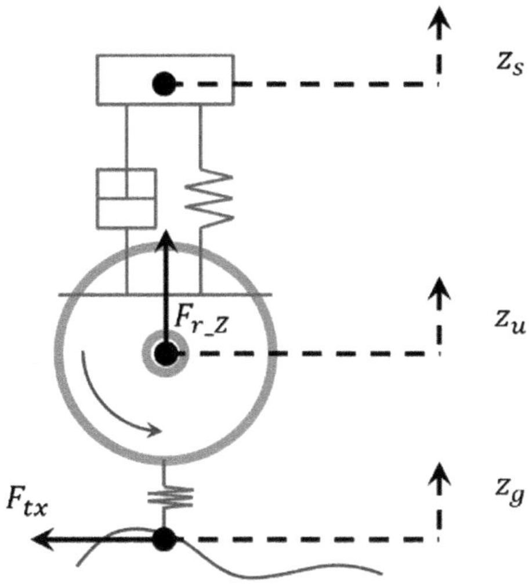

Fig. 9 Quarter vehicle model with SRM driving

the rolling resistance moment generated by the tire, F_{tx} the reaction force between the tire and the road which can be obtained by magic formula [18], T_e is the motor driving torque.

Spring mass equation and non-spring mass vibration equation of quarter vehicle model are as follows:

$$\begin{cases} m_s\ddot{z}_s = k_s(z_u - z_s) + c_s(\dot{z}_u - \dot{z}_s) \\ m_u\ddot{z}_u = k_t(z_g - z_u) - k_s(z_u - z_s) + c_t(\dot{z}_g - \dot{z}_u) - c_s(\dot{z}_u - \dot{z}_s) + F_{r\_Z} \end{cases} \quad (12)$$

Where, m_u and m_s are the non-sprung mass and sprung mass, k_s and k_t are the suspension stiffness and tire vertical stiffness, c_s and c_t are the suspension damping and tire damping, respectively. F_{r_Z} is vertical component of radial force of IWM, z_g is road surface excitation obtained by filtered white noise, z_u and z_s are the vertical displacement of vehicle unsprung and sprung masses.

3 Joint Modeling and Preliminary Analysis

At the present time, there are mainly three kinds of control technologies used for SRM drive, including current chopping control (CCC), angle position control (APC) and voltage chopping control (VCC) or pulse width modulation control (PWMC). Although the implementation of this three driving technologies are different, they are designed to ensure motor windings to generate the desired current waveform. Figure 10 is a typical phase asymmetrical bridge power converter for the SRM, which converts the DC voltage to the four-phase AC drive signal required by the SRM. In this paper, the third method is used to achieve motor speed control,

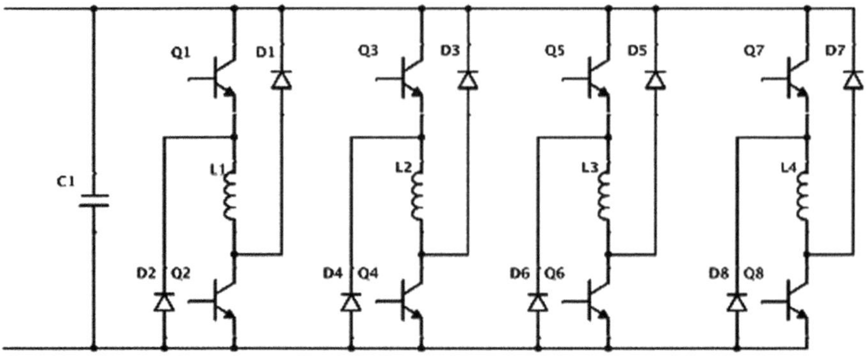

Fig. 10 Asymmetric bridge converter for SRM drives

conduction angle is fixed to 5.5°, and turnoff angle is about 25°. According to required driving torque, PID controller is used to calculate the appropriate PWM duty cycle to control the winding voltage. Then, vertical excitation of the motor obtained by in-wheel motor is applied to quarter vehicle model with SRM driving, and the response of the vehicle to the unbalanced radial force is calculated. Control system diagram of in-wheel model is shown in Fig. 11.

The vibration transmission characteristics and vibration mode energy distribution of electric vehicle with in-wheel switched reluctance motor can be obtained by Eq. (11), as shown in Table 1 and Figs. 12 and 13. According to Table 1 and Fig. 12, it can be seen that vibration mode of quarter vehicle model is mainly the second-order vibration of the sprung mass and the unsprung mass. As shown in Fig. 13, the effect of unbalanced radial force acting on in-wheel motor on vehicle dynamic system is analogous to the effect of road excitation or of wheel unbalance moment of inertia. The unbalanced radial force causes resonance of vehicle on two degree of freedom. Results show that the effect of random road excitation is interrelated to vehicle tire damping and stiffness, while unbalanced radial force is only a matter of eccentricity and the vehicle speed. Consequently, higher eccentricity will lead to greater unbalanced radial force amplitude. In addition, the higher the vehicle speed is, the higher the frequency of the unbalanced radial force. On the other hand, road excitation is generally random excitation, but unbalanced radial force excitation is generally related to wheel speed. So vibration effect of motor on vehicle and excitation effect of road surface are slightly different. The main vibration mode of vehicle system on road excitation is concentrated in the first-order resonance, namely low frequency resonance around 1 Hz, but the resonance of spring mass on exciting force of motor is mainly concentrated in middle and high frequency of second-order resonance, which is about 10 Hz, basically same to vibration response of suspension dynamic jounce in second-order resonance frequency. It can be seen that exciting force of in-wheel motor is related to the speed of vehicle, because motor speed is relatively high, and its influence on vibration response of vehicle is mainly reflected in medium and high frequency.

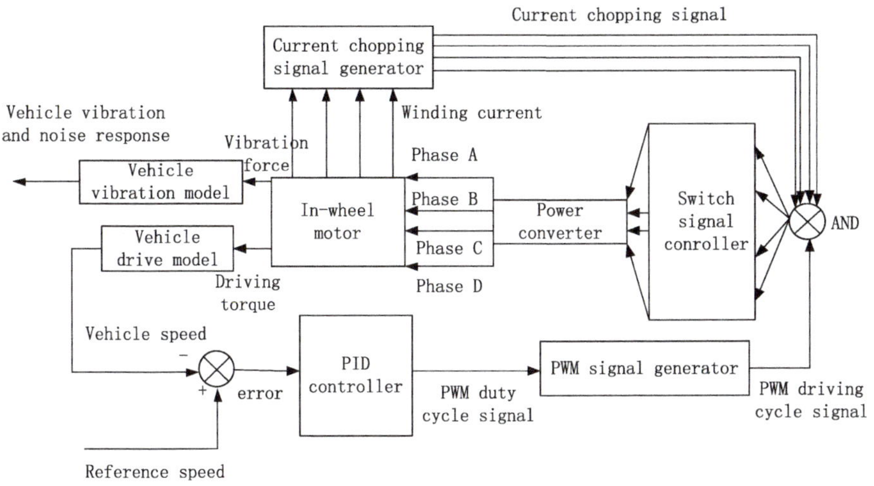

Fig. 11 Control system diagram of in-wheel model

Table 1 Distribution of vibration energy

	Natural frequency (Hz)	Vibrational energy of sprung mass (%)	Vibrational energy of no-sprung mass (%)
First order	1.2439	99.8650	0.1350
Second order	10.3118	0.1350	99.8650

Furthermore, frequency of exciting force can be obtained by

$$f_e = i_g \frac{n_v}{60} N_r = i_g \frac{\omega}{2\pi} N_r \cong i_g \frac{v}{2\pi R_e} N_r = i_g \frac{v_{(\mathrm{km/h})}}{2\pi R_e} \frac{N_r}{3.6} \tag{13}$$

The relationship between exciting frequency and transmission ratio in vehicle is shown in Fig. 14. It can be seen that when the vehicle speed is low (30 km/h), the motor excitation frequency can be as low as the resonance frequency range of the vehicle. Therefore, the unbalanced radial force has more significant influence on the ride comfort of the vehicle at low speed or starting conditions. While in the speed range vehicle commonly used (which is 30–80 km/h), the excitation frequency of motor varies widely depending on the transmission ratio, it can produce 30–80 Hz low-frequency vibrational excitation to vehicle system, which is in the resonance range of tire [19] and will eventually cause great damage to the vehicle dynamic system.

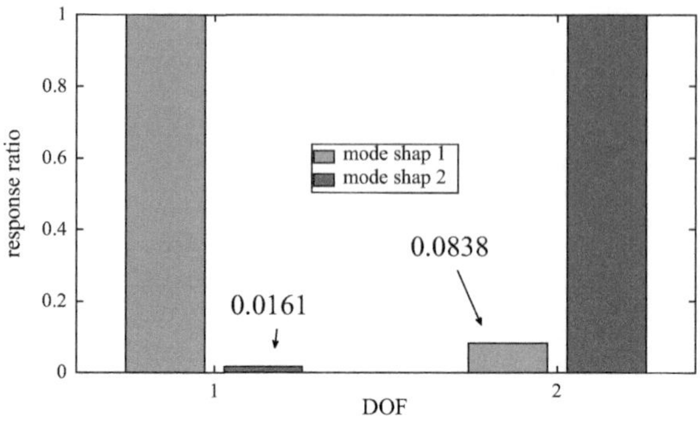

Fig. 12 Quarter vehicle vibration modes

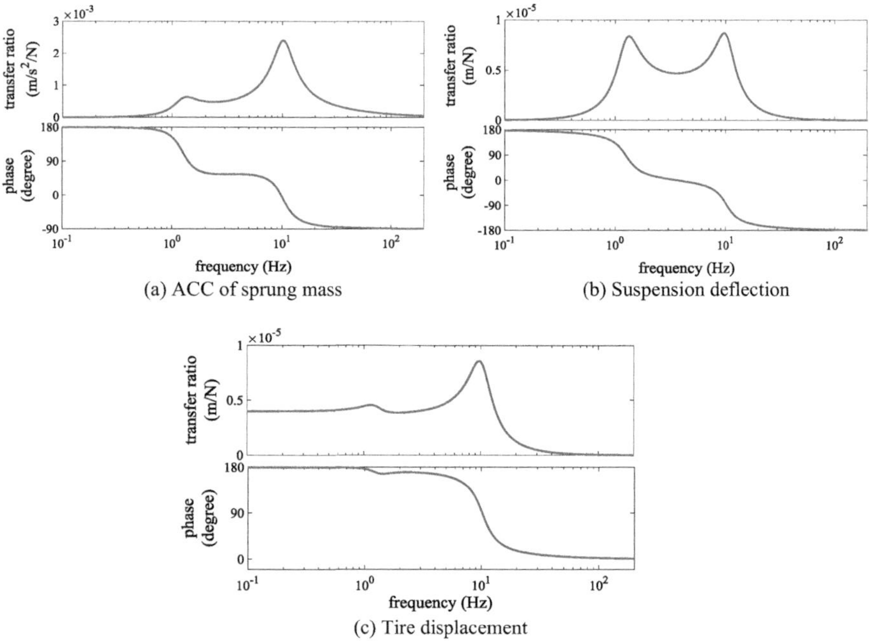

(a) ACC of sprung mass

(b) Suspension deflection

(c) Tire displacement

Fig. 13 Vibration transfer characteristics

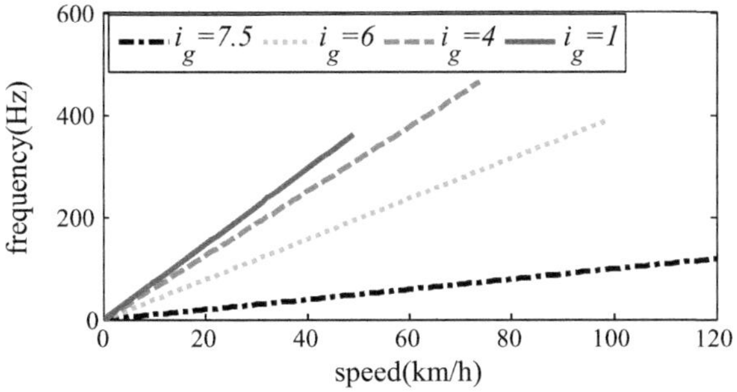

Fig. 14 Vehicle speed and IWM excitation frequency

4 Effect of Vertical Force of In-Wheel Motor on Stability Analysis

Simulation results of vehicle dynamic response in the speed of 5, 10, 20, 40, and 80 km/h are listed in Table 2. It can be noted that when vehicle speed is less than 20 km/h, ride comfort and handling stability are severely damaged by unbalanced electromagnetic excitation. When vehicle speed is greater than 40 km/h, the influence of vertical vibration decreases rapidly, and it weakens with the increase of vehicle speed. When the vehicle is travelling at the speed of 80 km/h, the adverse effects caused by the torque ripple of the motor on the vertical vibration of the vehicle have been very small.

Therefore, at low speed, vertical unbalance force of in-wheel motor has a significant negative impact, resulting in the destruction of vehicle ride and handling stability. With the increase of vehicle speed and over a certain range, the influence of vertical force on the vehicle can be reduced to a relatively low level.

When excitation frequency of in-wheel motor is 1.3 Hz, it is near to natural frequency of vehicle bodywork, which can easily result in body resonance, corresponding speed of 1.83 km/h [20]. Simulation results with C grade road surface excitation are shown in Fig. 15a–d, respectively. As shown in the figure, in the case of taking vertical excitation of the motor into consideration, spring mass acceleration and the tire displacement are greatly increased. In the frequency domain, in case of body resonance, spring mass acceleration in the whole frequency band has increased and spring mass acceleration in low frequency (<8 Hz) segment increased, while in high frequency band (>30 Hz), change is not obvious. To sum up, body resonance caused by vertical force of in-wheel motor increases vibration of car body, which seriously affects comfort of electric vehicle.

When excitation frequency of in-wheel motor is 6.5 Hz, it is close to natural frequency of wheel [21], which can lead to the resonance of wheel. At this time, the corresponding speed of vehicle is 10 km/h, C grade pavement is selected as road input. Under the condition of wheel resonance, vibration response of the vehicle is shown in Fig. 16a–d.

Table 2 RMS of performance indices for different rotate speeds of SRM

		5 km/h	10 km/h	20 km/h	40 km/h	80 km/h
ACC of sprung mass	With Fr_Z	0.2357	0.3334	0.4715	0.6668	1.0429
(m/^2)	Without Fr_Z	0.3670	0.5429	0.6478	0.7417	1.1057
	Amplification (%)	55.70	62.84	37.39	10.01	6.02
Suspension deflection	With Fr_Z	0.0023	0.0033	0.0046	0.0065	0.0092
(m)	Without Fr_Z	0.0028	0.0039	0.0048	0.0066	0.0094
	Amplification (%)	21.74	18.18	4.35	1.52	2.12
Tire dynamic load	With Fr_Z	302.740	389.473	550.7993	778.9478	1121.6
(N)	Without Fr_Z	468.064	570.957	651.2914	884.3341	1197.4
	Amplification (%)	54.61	46.60	18.24	8.39	6076

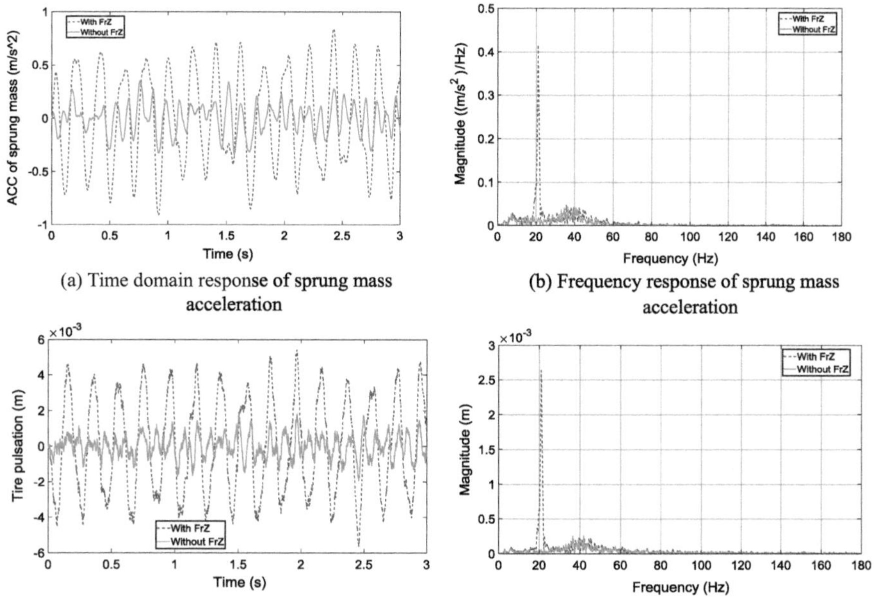

(a) Time domain response of sprung mass acceleration

(b) Frequency response of sprung mass acceleration

(c) Time domain response of tire pulsation

(d) Frequency response of tire pulsation

Fig. 15 Vehicle body resonance

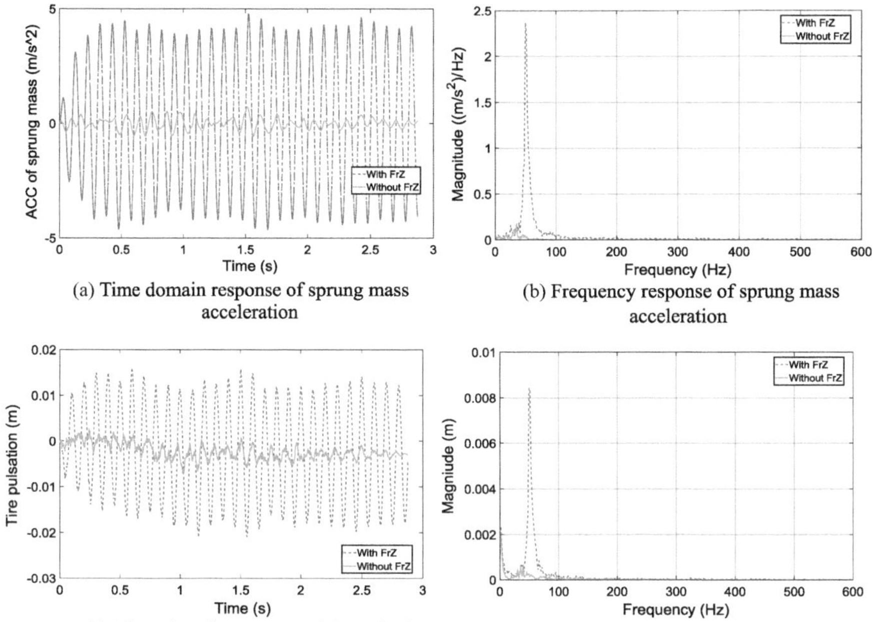

(a) Time domain response of sprung mass acceleration

(b) Frequency response of sprung mass acceleration

(c) Time domain response of tire pulsation

(d) Frequency response of tire pulsation

Fig. 16 Wheel resonance

It can be seen that in case of wheel resonance, vibration acceleration of vehicle body and dynamic load of tire are greatly increased at the resonance frequency. To sum up, vertical excitation of SRM will lead to the resonance of non-spring load, which has a significant effect on vehicle comfort.

5 Conclusion

In this paper, the influences of unbalance vertical force of in-wheel electric on vehicle ride comfort and road holding are discussed. The results are given as follows,

(1) The vertical force of switched reluctance motor (SRM) is an important factor affecting comfort and stability of vehicle. Its excitation frequency covers the resonant frequencies of sprung mass and unsprung mass, which makes it easy to cause the resonance of body and wheel, especially in starting and low-speed running conditions. Therefore, the influence of the vertical force on the vehicle dynamic performance cannot be ignored.

(2) The effect of unbalanced radial force acting on in-wheel motor on vehicle dynamic system is analogous to effect of road excitation or of wheel unbalance moment of inertia. The unbalanced radial force causes resonance of vehicle on 2° of freedom.
(3) The magnitude of imbalance radial force of in-wheel motor is related to the speed of vehicle, and its influence on vibration response of vehicle is mainly reflected in medium and high frequency.
(4) In resonance condition, vertical force caused by switched reluctance motor has a significant influence on comfort of vehicle. Under the condition of body and wheel resonance, acceleration of sprung mass and vibration of the tire are significantly increased.

References

1. Youchuan S, Guodong J (2004) Types and characteristics of electric wheel. Urban Public Transp 4:16–18
2. Inagaki H, Kato H, Kuzuya H (1961) Drive train vibration and acoustic noise reduction control of switched reluctance motor for electric vehicle. SAE paper 6183176
3. Marques dos Santos FL, Anthonis J, Van der Auweraer H (2012) Multiphysics thermal and NVH modeling: integrated simulation of a switched reluctance motor drive train for an electric vehicle. In: 2012 IEEE international electric vehicle conference, Greenville, South Carolina, USA, 4–8 March 2012, IEEE, New York, pp 1–7
4. Marques dos Santos F, Anthonis J, Naclerio F et al (2014) Multiphysics NVH modeling: simulation of a switched reluctance motor drivetrain for an electric vehicle. IEEE Trans Ind Electron 61(1):469–476
5. Li J, Cho Y (2010) Dynamic reduction of unbalanced magnetic force and vibration in switched reluctance motor by the parallel paths in windings. Math Comput Simulation 81 (2):407–419
6. Rojas AER, Niederkofler H, Willberger J (2010) Comfort and safety enhancement of passenger vehicles with in-wheel Motors, SAE Technical Paper, 2010, ⟨http://papers.sae.org/2010-01-1146/⟩
7. Shini GN, Kamaraj V, Giridharan E, Kannan S (2013) Electromagnetic and vibration analysis of in-wheel switched reluctance motor with notches and projections for electric vehicle applications,. In: Wei D, Qi L(eds), Applied mechanics and materials, pp 34–39
8. Luo Y, Tan D (2012) Study on the dynamics of the in-wheel motor system. IEEE Trans Veh Technol 61:3510–351.https://doi.org/10.1109/tvt.2012.2207414
9. Lu X, Yu Z, Wang Y et al (2013) Vehicle dynamics control of four in-wheel motor drive electric vehicle using gain scheduling based on tyre cornering stiffness estimation. Veh System Dynamics 50(6):831–846
10. Nam K, Fujimoto H, Hori Y (2012) Lateral stability control of in-wheel-motor-driven electric vehicles based on sideslip angle estimation using lateral tire force sensors. IEEE Trans Veh Technol 61(5):1972–1985
11. Lu D, Ouyang M, Gu J, Li J (2014) Instantaneous optimal regenerative braking control for a permanent-magnet synchronous motor in a four-wheel-drive electric vehicle. Proc IMechE Part D J Automobile Eng 228(8):894–908
12. Sun W, LI Y, Huang J, Zhang N (2017) Efficiency improvement of vehicle active suspension based on multi-objective integrated optimization. J Vib Control 23(4):539–554

13. Ayari S, Besbes M, Lecrivain M et al. (1999) Effects of the airgap eccentricity on the SRM vibrations. In: IEEE international electric machines and drives conference, IEMDC 1999—Proceedings. IEEE, pp 138–140
14. Garrigan NR, Soong WL, Stephens CM et al (1999) Radial force characteristics of a switched reluctance machine. In: Conference record of the 1999 ieee industry applications conference. thirty-forth IAS annual meeting (Cat. No.99CH36370). IEEE, 4:2250–2258
15. Chen YS, Cheng YD, Liao JJ, Chiu CC (2009) Electromagnetic–rotor dynamic integrated vibration analysis for electrical machines. Proc IMechE Part D J Automobile Eng 223(11): 2549 2259
16. Qiao XL, Zhu CS (2014) The active unbalanced vibration compensation of the flexible switched reluctance motorized spindle. J Vibr Control 20(13):1934–1945
17. Ha K-H, Hong J-P, Kim G-T, Chang K-C (2000) Orbital analysis of rotor due to electromagnetic force for switched reluctance motor. IEEE Trans Magn 36(4):1407–1411
18. Pacejka H (2005) Tire and vehicle dynamics. Elsevier
19. Yang S, Mu L (1989) A study on the mechanism of functional dysphonia. J Voice 3(4):337–341
20. Wang Y-Y, Li Y-N, Sun W, Zheng L (2015) Effect of the unbalanced vertical force of a switched reluctance motor on the stability and the comfort of an in-wheel motor electric vehicle. Proc IMechE Part D J Automobile Eng 229(12):1569–1584
21. Lu S-B, Li Y-N, Choi S-B (2012) Contribution of chassis key subsystems to rollover stability control. In: Proceedings of the institution of mechanical engineers. Part D J Automobile Eng 226(4):479–493

The Load and Centroid Location Estimation of Tractor–Semitrailer

Bing Zhu, Yao Feng and Jian Zhao

Along with the economic development, commercial semi-trailer has been widely used in the road freight transportation, the commercial vehicle's traffic safety has also gained more and more attention, ABS and EBS systems are also designed to improve the safety of commercial vehicles. However, due to the commercial semi-trailer load and centroid position changes, it becomes difficult for the commercial vehicle control system to play its best control effect to ensure the stability of the vehicle. Therefore, domestic and foreign scholars have done a lot of research on the centroid estimation of commercial vehicles: Verner Kober et al. use vehicle air suspension to obtain vehicle axle load and identify the vehicle's centroid through the dynamic model of the vehicle under its roll conditions [1], Rajamani et al. designed a method for estimating the height of the vehicle's center of gravity based on accurate observation of the vehicle roll angle. The estimation method uses recursive least squares (RLS) with variable forgetting factor and can realize the center-of-gravity position estimation both in steady-state and transient steering conditions [2], Cheng et al. use the double Kalman filter to estimate the vehicle parameters and the vehicle state at the same time. Based on the given longitudinal center of gravity, the yaw moment of inertia was obtained, and obtain the vehicle's centroid height under the stable steering condition where the axle load of the vehicle was known [3], Li based on the simplified model of linear four-degree-of-freedom tractor-semitrailer, estimating the longitudinal centroid

National Natural Science Foundation of China (51475206, 51575225, U1564211), and Jilin Province Science and Technology Development Plan Projects (20170101138JC).

B. Zhu · Y. Feng (✉) · J. Zhao (✉)
State Key Laboratory of Automotive Simulation and Control, Jilin University, Changchun 130022, China
e-mail: fengyaojlu@163.com

J. Zhao
e-mail: zhaojian@jlu.edu.cn

Society of Automotive Engineers of China (SAE-China) (ed.), *Proceedings of the 19th Asia Pacific Automotive Engineering Conference & SAE-China Congress 2017: Selected Papers*, Lecture Notes in Electrical Engineering 486, https://doi.org/10.1007/978-981-10-8506-2_71

position of semi-trailer using the recursive least squares method, based on the linear six-degree-of-freedom tractor-trailer, the centroid height of the semi-trailer was estimated under stable steering conditions [4], Solmaz and Akar proposed a multi-model switching identification strategy, the principle of which is through the search for the most appropriate cost function parameter set, making the vehicle center-of-gravity position estimation to achieve the smallest error [5], Fang of Jilin University estimated the vehicle load based on the vehicle's longitudinal dynamic model, and the known axle load is used to estimate the longitudinal position of the load when the vehicle is driven at a constant speed; finally, under the stable steering condition, the height of the vehicle's load center is estimated [6], Momiyama developed the center-of-gravity estimation scheme using the least squares method based on the transfer function from the steering angle to the roll angle [7], Huang and Wang of Ohio State University studied the mathematical relationship between the vertical load variation and the real-time effective radius of the wheel, and based on this, it is proposed to use the axle load transfer to estimate both the height of the vehicle's center of gravity and the wheel radius change at the same time by the Kalman filter algorithm [8]. The above estimation method is based on the air suspension, so the scope of application is limited in the vehicles equipped with air suspension, and the moment of inertia of the load, which is based on the known load center-of-gravity longitudinal position and assuming uniform load condition, is used in the estimation of the center of gravity, making it not applicable to the actual operating conditions of the vehicle. In this paper, the wheel model and the wheel inverse model are used to calculate the wheel force, and the vehicle load and the centroid position of the semi-trailer are obtained by the longitudinal dynamic analysis of the vehicle under the uniform running speed condition and the braking condition. The estimation method does not need external sensors, and the algorithm is simple and accurate.

1 Tire Force Estimation

When the vehicle is under the condition of straight running, where the transverse force of the wheel can be neglected, the longitudinal braking force of the brake wheel under the vehicle braking condition can be estimated by the wheel model, and the longitudinal force of the driving wheel is estimated by the transmission system model. By combining the estimated wheel longitudinal force and the measured tire slip rate, the tire vertical force can be estimated by the tire inverse model.

1.1 Longitudinal Tire Force Estimation

Estimation of longitudinal force of the driving wheel:

When the vehicle is running on a horizontal road with a good road surface, the engine output distortion conveys to the drive wheel through the transmission system, meanwhile the rolling resistance of the wheel is also applied to drive wheel, the formula is as follows:

$$F_t = \frac{T_{tq} i_g i_0 \eta_T}{r} - f \cdot F_z \tag{1}$$

where f is the wheel rolling resistance coefficient, which can be found in the manual:

$$f = 0.0076 + 0.000056 u_v \tag{2}$$

F_z is the drive shaft vertical load; T_{tq} is the engine output torque; i_g is the transmission ratio; i_0 is the main reducer transmission ratio; η_T is the efficiency of the transmission system; r is the wheel effective radius.

Estimation of longitudinal force of the braking wheel:

Figure 1 shows the wheel model, through which the relationship between the brake pressure and the wheel braking force can be derivated. When the brake pressure is measured by the pressure sensor mounted on the wheel cylinder, the wheel braking force can be then calculated. In this figure, T_b stands for the brake torque of the brake, F_x stands for the ground brake force, R stands for the wheel radius, and ω stands for the wheel angular velocity. According to the working principle of the brake:

$$T_b = K \cdot P \tag{3}$$

where K is the brake effectiveness factor.

To analyze the wheel force:

$$T_b - F_x \cdot r = J \cdot \frac{\mathrm{d}\omega}{\mathrm{d}t} \tag{4}$$

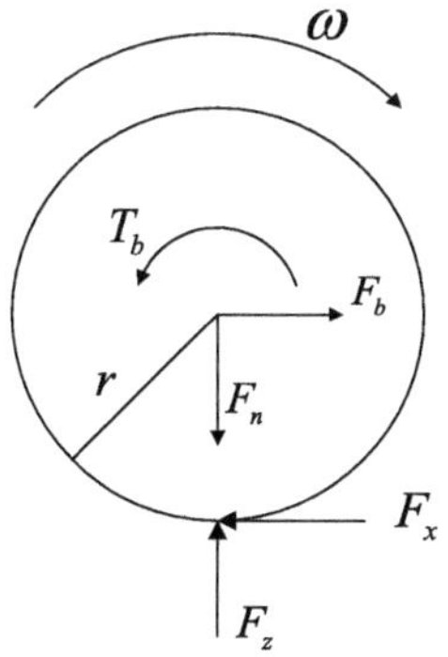

Fig. 1 Force applied to the brake wheel

where J is the rotational inertia of the wheel, and combining Eqs. (3) and (4), F_x can be calculated:

$$F_x = \frac{1}{r}\left(K \cdot P - J \cdot \frac{d\omega}{dt}\right) \tag{5}$$

1.2 Vertical Tire Force Estimation

When estimating the center of the mass of the vehicle, firstly, it is necessary to determine the vertical force of each wheel, which can be derived from the wheel slip rate and the corresponding longitudinal tire force. Different from the conventional tire model, the wheel model used here is called the inverse tire model.

Trucksim provides a virtual tire test platform that can be used to obtain tire strength data under different vertical loads easily. Figure 2 shows the longitudinal wheel force curves on dry asphalt pavement when the wheel is under different vertical loads. Based on these data, the vertical wheel force can be calculated when the slip rate and the longitudinal wheel force are given.

When the vehicle is decelerating, by measuring the slip rate of the wheel when the vehicle is braking, along with the estimated value of the longitudinal braking force of the wheel, the vertical load of the wheel can be obtained through the lookup table. When the vehicle is driven under a constant speed, by measuring the slip rate of the drive wheel, along with the estimated value of the longitudinal force of the wheel, the vertical load of the drive wheel can be obtained through the lookup table.

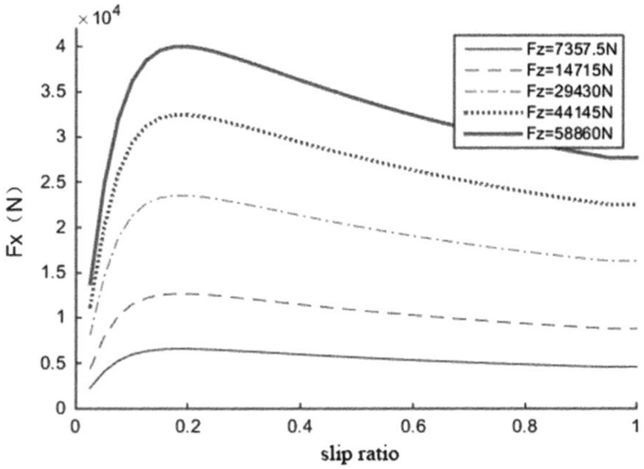

Fig. 2 Vertical tire force table

2 Longitudinal Dynamic Model of the Vehicle

2.1 Dynamic Analysis of the Vehicle Running at Constant Speed

In order to analyze the vehicle force, the vehicle model is simplified as follows: The vehicle is running in a straight line on a well-balanced road surface; the vehicle center of gravity is on the lateral symmetry plane of the vehicle; ignoring the rolling resistance; assuming the point of the air resistance is in the middle of the vehicle's windward position; ignore the body and suspension deformation, the tractor and semi-trailer are rigidly connected through the hinge.

In the vehicle force analysis, the tractor is decoupled from the semi-trailer, both of which are connected by forces acting on the hinges. This paper firstly analyzes the force of the tractor.

When the road surface is horizontal and the semi-trailer rolling resistance is not taken into account, the force of the tractor in the horizontal direction is only the longitudinal force F_t on the drive wheel and the air resistance force F_w acting on the front, and the force of the tractor in the vertical direction is the vertical force F_{z1}, F_{zt} on the wheel, the gravity of the tractor itself W_1 and the vertical force F_{zh} acting on the hinge of the semi-trailer, among them:

$$F_w = \frac{1}{2} C_D A \rho u_r^2 \tag{6}$$

where C_D is the air resistance coefficient; ρ is the air density, and $\rho = 1.2258\,\mathrm{N} \times \mathrm{s}^2 \times \mathrm{m}^{-4}$; A is the wind area; u_r is the relative speed between the car and the air, when ignoring the wind speed, $u_r = u_v$, u_v is the speed of the vehicle. Vehicle force is shown in Fig. 3:

First of all, analysis the force is applied to the tractor:

$$F_{zt} \cdot l_1 = F_w \cdot h_w + W_1 \cdot c_1 + F_{zh} \cdot (l_1 - b) \tag{7}$$

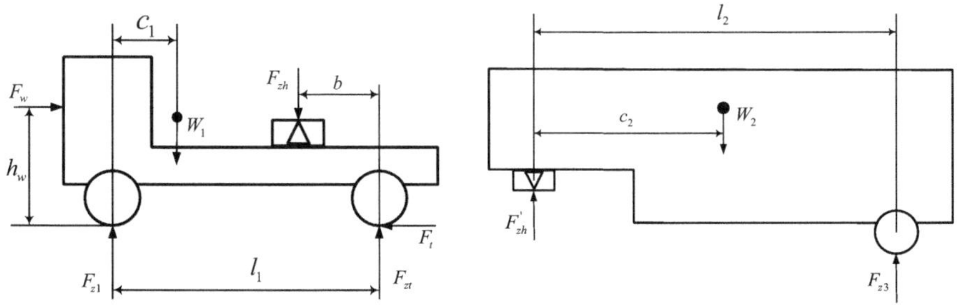

Fig. 3 Force applied on the tractor

where l_1 is the axle spacing of the tractor; W_1 is the tractor weight; h_w is the vertical distance between the air resistance point and the ground; c_1 is the longitudinal distance between the tractor centroid and the front axle, b is the horizontal distance between the hinge point and the rear axle of the tractor. When the vehicle is driven under a constant speed on a horizontal road, the vertical force in the connection hinge can be obtained:

$$F_{zh} = \frac{1}{l_1 - b}(F_{zt} \cdot l_1 - F_w \cdot h_w - W_1 \cdot c_1) \tag{8}$$

Then analyze the force applied to the semi-trailer:

$$W_2 \cdot (l_2 - c_2) = F'_{zh} \cdot l_2 \tag{9}$$

where F'_{zh} is the vertical force of the hinge acting on the semi-trailer, $F'_{zh} = F_{zh}$; W_2 is the total weight of the cargo and the semi-trailer, l_2 is the horizontal distance between the axle of the semi-trailer and the hinge; c_2 is the horizontal distance between the center of gravity of the semi-trailer and the hinge.

2.2 Dynamic Analysis of the Vehicle Under Constant Deceleration

When the vehicle is in a stable braking conditions, the force applying on the tractor and semi-trailer is shown in Fig. 4:

To analyze longitudinal force:

$$F_{x1} + F_{x2} + F_w = W_1 \cdot Z + F_{x4} \tag{10}$$

$$F_{x4} + F_{x3} = W_2 \cdot Z \tag{11}$$

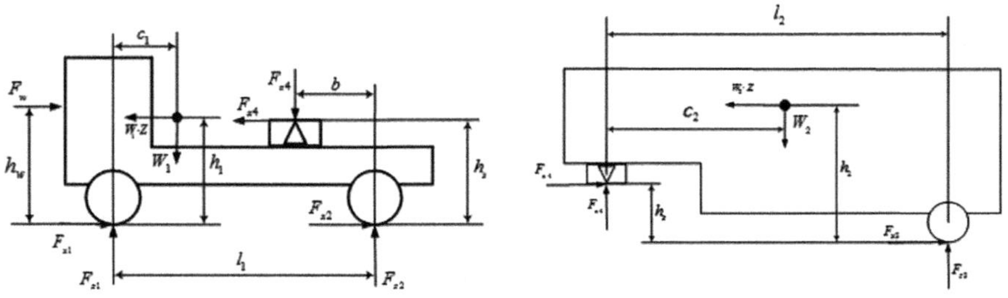

Fig. 4 Force applied on vehicle under braking condition

To analyze vertical force:

$$F_{z1} + F_{z2} = W_1 + F_{z4} \tag{12}$$

$$F_{z3} + F_{z4} = W_2 \tag{13}$$

Moment analysis of the tractor:

$$F_{z2} \cdot l_1 + F_{x4} \cdot h_s + W_1 \cdot Z \cdot h_1 = F_w \cdot h_w + W_1 \cdot c_1 + F_{z4} \cdot (l_1 - b) \tag{14}$$

Moment analysis of the semi-trailer:

$$F_{z3} \cdot l_2 + F_{x3} \cdot h_s + W_2 \cdot Z \cdot (h_2 - h_s) = W_2 \cdot c_2 \tag{15}$$

where Z is the deceleration of the vehicle; h_s is the vertical distance between hinge point and the ground; F_{x1}, F_{x2}, and F_{x3} are the longitudinal braking force of the front axle, rear axle of the tractor, and the axle of the semi-trailer, respectively. F_{z1}, F_{z2}, and F_{z3} are the vertical force applied to the axles; F_{x4} and F_{z4} are the longitudinal force and vertical force acted at the hinge.

2.3 The Load and Centroid Estimation of the Semi-trailer

According to the analysis of the force applied on the vehicle which is under a constant speed or a stable deceleration, the mass of the semi-trailer can be calculated as well as its centroid. The calculation process is as follows:

Combining Eqs. (10) and (11), the mass of the semi-trailer can be obtained:

$$W_2 = \frac{1}{Z}(F_{x1} + F_{x2} + F_{x3} + F_w) - W_1 \tag{16}$$

Combining Eqs. (9) and (14), the height of the centroid of the semi-trailer can be obtained:

$$h_2 = \frac{1}{W_2 \cdot Z}\left[(W_2 - F'_{zh} - F_{z3})l_2 + (W_2 \cdot Z - F_{x3})h_s\right] \tag{17}$$

Combining Eqs. (8), (9), and (16), the longitudinal position of the centroid of the semi-trailer can be obtained:

$$c_2 = l_2\left(1 - \frac{F'_{zh}}{W_2}\right) \tag{18}$$

3 Simulation Verification

The simulation conditions are set as follows: The vehicle load is set to 5000 kg; the coordinates of the load center are $C2$ = 5221 mm, $H2$ = 1935 mm.

3.1 Estimation of Tire Force in Brake Process

The road is horizontal, and its adhesion coefficient is 0.85, the vehicle initial speed is 90 km/h, and when the vehicle is running, set the brake pressure to 0.4 Mpa, so that the vehicle will slow down till parking; the result of the tire force estimated is shown in Figs. 5 and 6, and they represent the tire force of the front axle and rear axle of the tractor, respectively. Table 1 shows the estimated tire force and the real value of it.

In the first two seconds of the simulation, the braking force of the wheel is zero, so that the vehicle's vertical load estimated which is based on the braking force is zero too. When the vehicle is braking, the longitudinal braking force and the estimated value of the vertical load of the wheel are very close to the actual value.

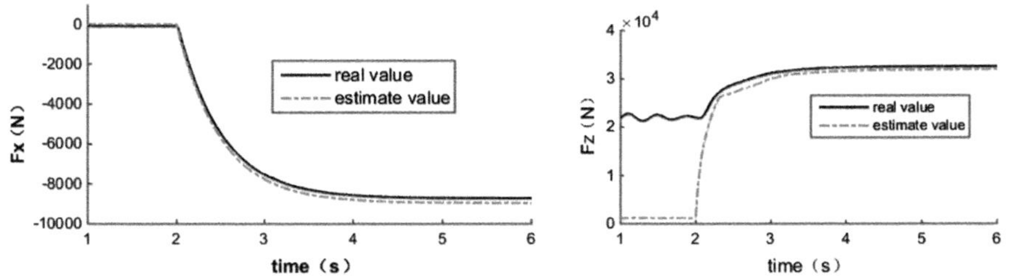

Fig. 5 Tire force estimation results

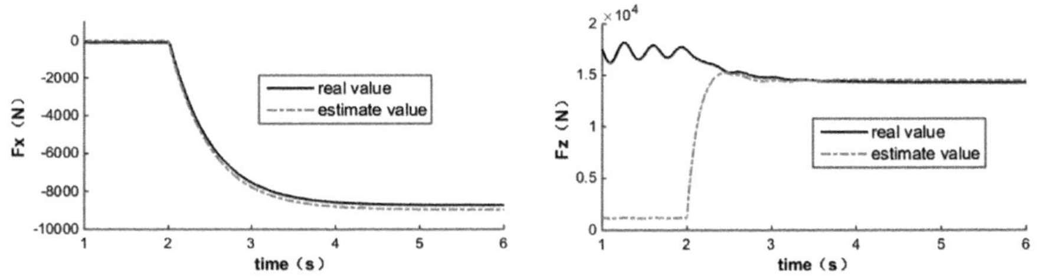

Fig. 6 Tire force estimation results

Table 1 Tire force estimation value and real value

Parameter	F_{x1}	F_{x2}	F_{x3}	F_{z1}	F_{z2}	F_{z3}
Estimation value	17,440	8714	8717	31,990	11,760	13,560
Real value	17,890	8644	8665	32,540	11,470	13,380
Error (%)	2.50	0.81	0.60	1.70	2.53	1.35

3.2 *Estimation of Tire Force When the Vehicle Is Evenly Running*

Simulation conditions are set as follows: Gearbox stalls are fixed, accelerator pedal is fixed, and the vehicle is running at 90 km/h on a horizontal road. Figure 7 shows the estimation results of the longitudinal and vertical loads of the drive wheel. The jitter of the tire force at the initial stage of the simulation is due to the slight jitter of the vehicle speed at that period. Table 2 shows the estimated value and the actual value of the drive wheel tire force, the vertical force at the hinge and the wind resistance.

3.3 *Mass and Centroid Estimation of the Semi-trailer*

According to the tire forces estimated in 3.1 and 3.2, and the Eqs. (16), (17), and (18), the vehicle load and center-of-mass estimation results are given in Table 3:

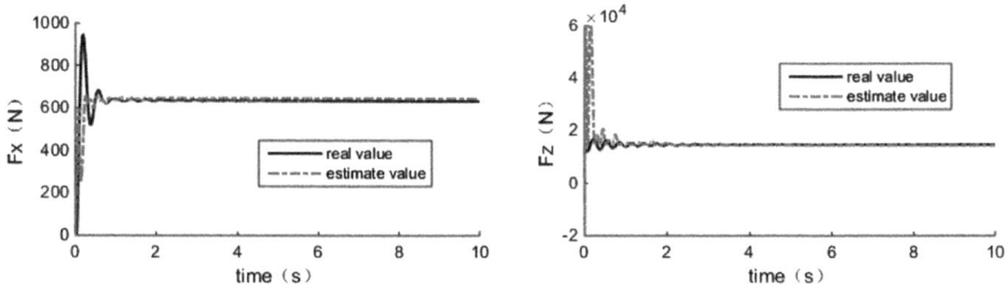

Fig. 7 Tire force estimation results of the drive wheel

Table 2 Tire force estimation value and real value

Parameter	F_x	F_z	F_w (风 阻)	F_{z_h} (铰链处纵向力)
Estimated value	642.6	14,480	1505	50,670
Real value	629.9	14,750	1481	51,560
Error (%)	0.16	1.83	1.62	1.73

Table 3 Mass and centroid estimation value and real value of semi-trailer

Parameter	W_2	C_2	H_2
Real value	107,800	5221	1935
Estimation value	108,400	5330	2042
Error (%)	0.56	2.09	5.53

4 Conclusion

The vehicle load and center-of-gravity position estimation method based on tire force is easy to be calculated and implemented. There is no need to add additional load sensor, and it is suitable even when the load distribution is uneven. The simulation results show that the load and center-of-gravity estimation method is with high accuracy and is reliable. The load and center-of-gravity estimation results can be further applied to the vehicle braking force distribution strategy, thereby improving vehicle braking performance.

References

1. Kober W, Hirschberg W (2006) On-board payload identification for commercial vehicles. In: Mechatronics, 2006 IEEE international conference on IEEE, pp 144–149
2. Rajamani R, Piyabongkarn D, Tsourapas V et al (2011) Parameter and state estimation in vehicle roll dynamics. IEEE Trans Intell Transp Syst 12(4):1558–1567
3. Cheng C, Cebon D (2011) Parameter and state estimation for articulated heavy vehicles. Veh Syst Dyn 49(1–2):399–418
4. Li H (2015) Study on structural parameters and state estimation of semi-trailer. Chinese J Traffic Eng
5. Solmaz S, Akar M, Shorten R et al (2008) Real-time multiple-model estimation of centre of gravity position in automotive vehicles. Veh Syst Dyn 46(9):763–788
6. Fang Y (2010) Research on stability control algorithm of three-axle semi-trailer. Jilin University
7. Momiyama F, Kitazawa K, Miyazaki K et al (1999) Gravity center height estimation for the rollover compensation system of commercial vehicles. JSAE Rev 20(4):493–497
8. Huang X, Wang J (2013) Center of gravity height real-time estimation for lightweight vehicles using tire instant effective radius. Control Eng Practice 21(4):370–380

Study on Co-simulation of a Dual-Motor EPS for Bus Based on ADAMS and MATLAB

Yaohua Li, Yangsen Zhang, Qianlong Feng, Pengfei Ou and Peng Xu

1 Introduction

Compared with the hydraulic power steering system, the EPS has the advantages of adjustable traction characteristics, good following ability, simple structure, high reliability, and so on [1]. Due to the large weight of the commercial vehicle, greater steering assist torque is required, and because of the limitation of automobile electric voltage and installation size, the ordinary single-motor EPS only can provide smaller torque, and it cannot meet the demand of commercial vehicle for normal steering torque, so most commercial vehicles still use hydraulic power steering system; however, the hydraulic power steering system has many disadvantages, such as large power consumption, easy to leak, and difficult to control the steering assist torque [2]. In order to use the EPS on buses, the power-coupling scheme, the power-assisted current allocation scheme, and the fault-tolerant control algorithm of the dual-motor electric power steering system are designed. The whole vehicle dynamics model is built by using ADAMS, and the EPS simulation model of dual motor is established by using Matlab/Simulink. The performance of dual-motor electric power steering system of commercial vehicle is verified by joint simulation [3].

Y. Li (✉) · Y. Zhang (✉) · Q. Feng · P. Ou · P. Xu
Chang'an University, Automotive Institute, Xi'an 710064, China
e-mail: nuaaliyaohua@126.com

Y. Zhang
e-mail: 971415335@qq.com

Society of Automotive Engineers of China (SAE-China) (ed.), *Proceedings of the 19th Asia Pacific Automotive Engineering Conference & SAE-China Congress 2017: Selected Papers*, Lecture Notes in Electrical Engineering 486, https://doi.org/10.1007/978-981-10-8506-2_72

2 Commercial Vehicle Dual-Motor EPS System

The key of dual-motor EPS design lies in the arrangement of the booster and the torque coupling scheme of the two motor. The overall structure diagram of the dual-motor EPS system is shown in Fig. 1. ECU collects speed signals, steering wheel torque and corner signals and inputs them into the control unit, and the target current of the power-assisted motor is obtained by comparing with the data of the driving characteristic curve [4]; then, it assigns the proper current to the main motor and the auxiliary motor and realizes the dual-motor power steering by the dual-motor torque coupler. The main motor or auxiliary motor may malfunction during operation. Therefore, a fault-tolerant control scheme is designed; when a power-assisted motor is broken, the electromagnetic clutch of this motor is disconnected, and the auxiliary power is completely borne by another motor. When the two motor fault, ECU will disconnect the total electromagnetic clutch, and the steering torque will be provided by the driver.

The structure of the dual-motor coupling electric power steering device is shown in Fig. 2. The dual-motor system determines the engagement or disconnection of the unidirectional clutch in accordance with the control signals issued by the ECU. If the engagement of the unidirectional clutch is needed, the dual-motor system will enter the dual-motor-assisted steering mode. At the same time, ECU can also control the current and voltage of the dual-motor system to achieve the purpose of regulating output torque. If not required, the dual-motor system will enter the main power-assisted mode.

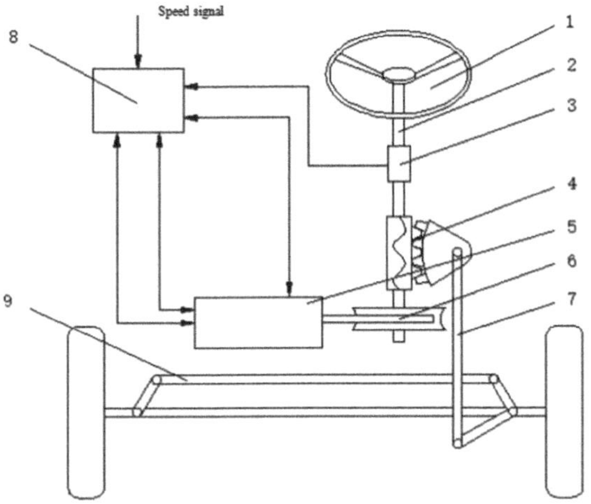

1.Steering wheel 2.Steering shaft 3.Torque and angle sensor 4. Recirculating ball type steering gear 5. Double motor structure 6. Worm gear reducer 7. Steering tie rod 8. ECU 9.Steering trapezium

Fig. 1 Overall structure diagram of dual-motor electric power steering system

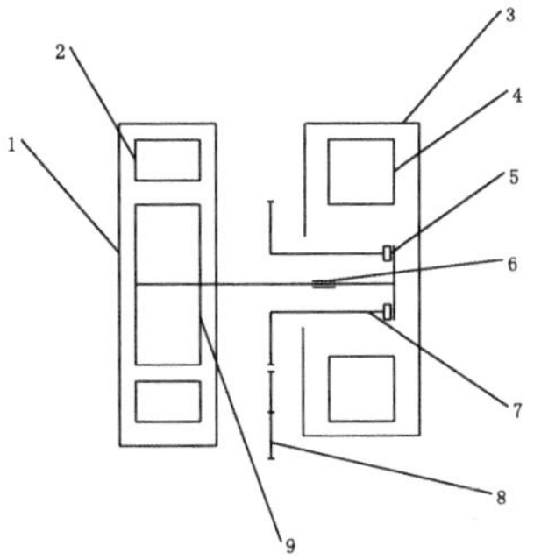

1. Auxiliary power assisted motor 2. Auxiliary booster motor stator 3. Main power assisted motor 4. Main auxiliary motor stator 5. One-way clutch 6. Support device 7. Main power assisted single rotor 8. Power output terminal 9. Auxiliary power assisted motor rotor

Fig. 2 Dual-motor construction

3 Dual-Motor EPS Control Strategy

3.1 EPS Control Strategy

The control strategy of the electric power steering system is shown in Fig. 3. ECU collects the speed, steering wheel torque, and corner signal, selects the appropriate control pattern, compares the acquired signal with the characteristic data in the assistance characteristic curve, obtains the target current, and then the ECU assigns the target current to the two power-assisted motors, so that the two motor outputs the appropriate power moment, and eventually, the coupling power is transmitted to the steering shaft by a torque coupler.

3.2 Current-Sharing Strategy

The dual-motor EPS controls the output torque by controlling the total input current, and then the input currents of the two motors are assigned to ensure that the output torque provides the proper torque for the steering of the car [5]. The control strategy of the dual-motor electric power steering system mainly solves the reasonable distribution of the target current of the two motors. In this paper, the current-sharing strategy of main and auxiliary motors is proposed; the main motor and auxiliary motor assume half of the target current, as shown in Fig. 4.

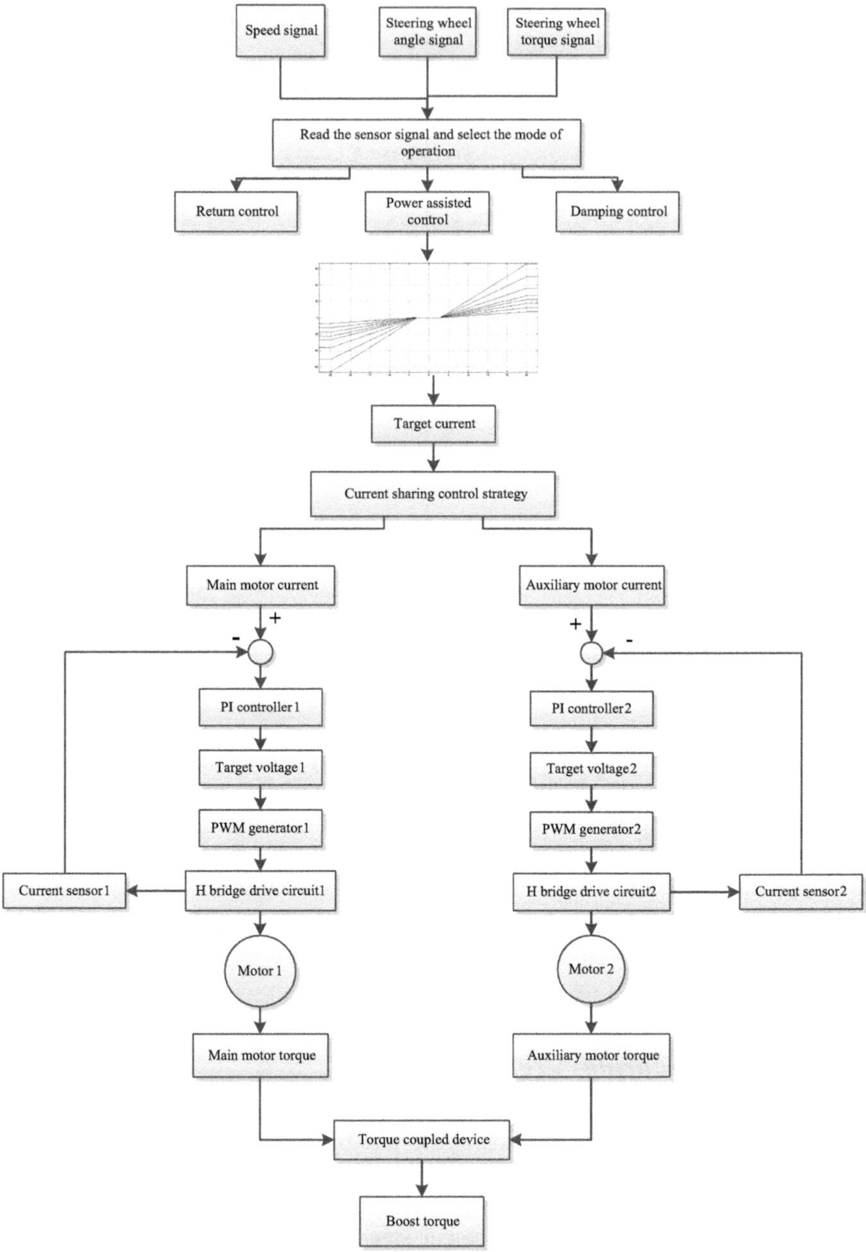

Fig. 3 EPS control strategy

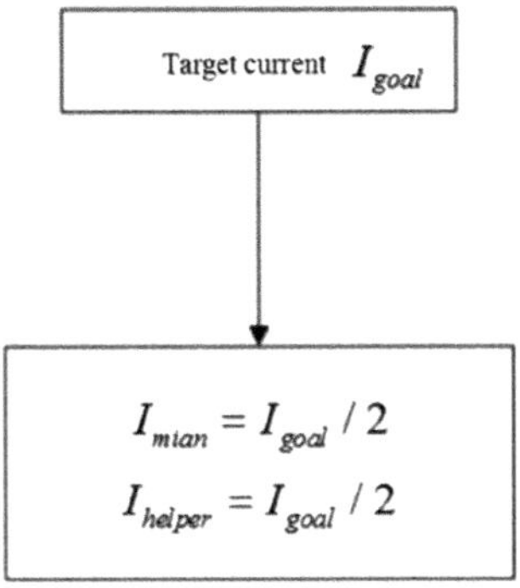

Fig. 4 Current-sharing control strategy

3.3 *Fault-Tolerant Control Strategy*

Due to the failure of the main motor or auxiliary motor during operation [6], this paper proposes a fault-tolerant control strategy, as shown in Fig. 5. Under normal circumstances, the main motor and auxiliary motor always bear half of the auxiliary current, when a motor failure, disconnect the motor electromagnetic clutch; the auxiliary current is entirely borne by another motor; if the target current is higher than the rated current of the motor, the motor works on the rated current.

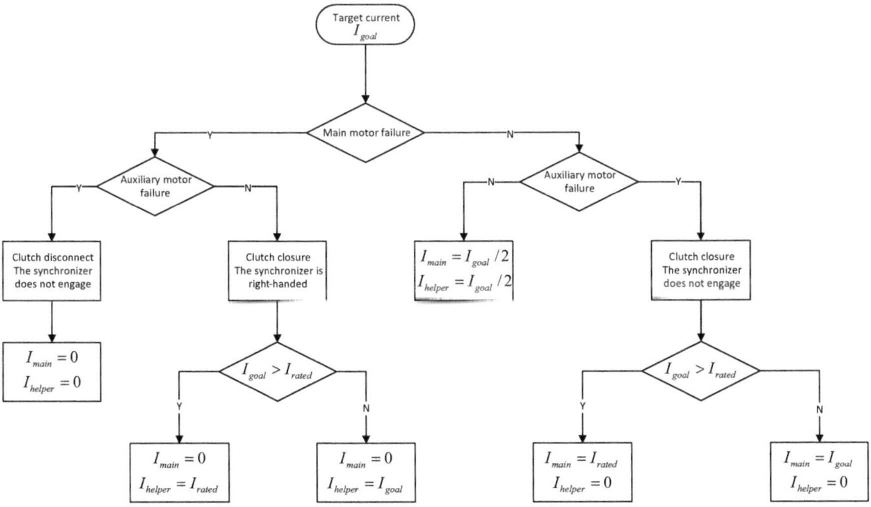

Fig. 5 Fault-tolerant control strategy

4 Simulation Verification

4.1 Joint Simulation Model of Dual-Motor EPS System

In this paper, a dual-motor EPS model is established in MATLAB/Simulink, and the whole vehicle dynamics model is established by ADAMS/Control, and then the joint simulation is carried out. The joint simulation model is shown in Fig. 6.

4.2 Current Distribution Control Strategy Validation

In this paper, a step test is used to validate the current allocation strategy. Set the simulation time of the system to 3S, turn the steering wheel to 180° at 1 s, last 0.5 s, and then keep the steering wheel stationary. The simulation results are shown in Figs. 7 and 8. As we can see from the picture, when the driver's hand torque is greater than T_{d0}, the system provides power, and the main and auxiliary motors share the target current and provide the output of the power moment.

4.3 Fault-Tolerant Control Strategy Test

Fault-tolerant control is used to simulate the fault of the main motor. When the main motor fails, the auxiliary motor still can bear all or part of the torque, so that the power system will not fail. Test driver files use the single-lane change test [7], and the speed is maintained at 13.286 m/s, and when the simulation time reaches 5S, the main motor is broken and stops working. The simulation results are shown in Figs. 9, 10, and 11. The simulation results show that the target auxiliary current is

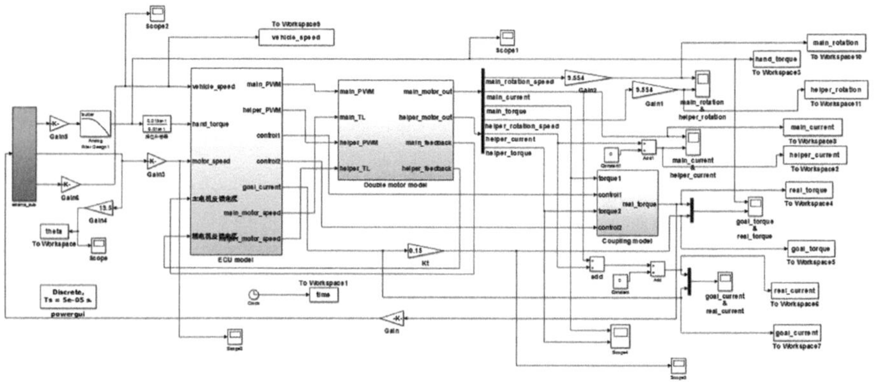

Fig. 6 Joint simulation model of dual-motor EPS system

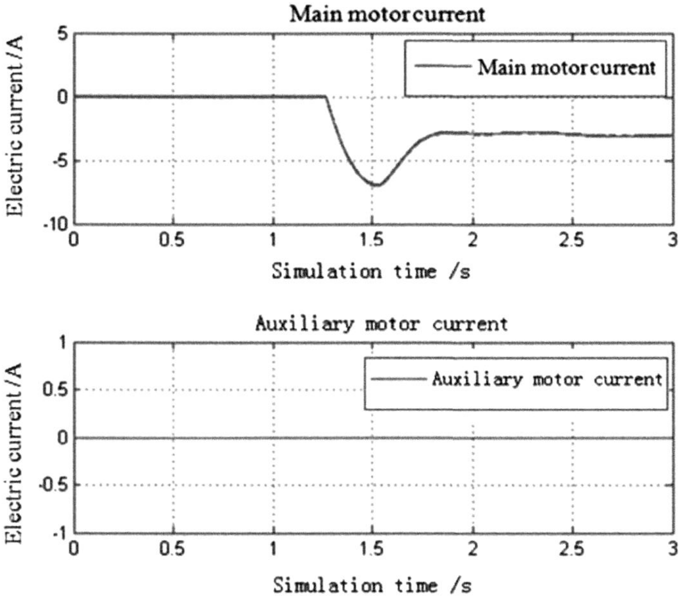

Fig. 7 Main and auxiliary motor current

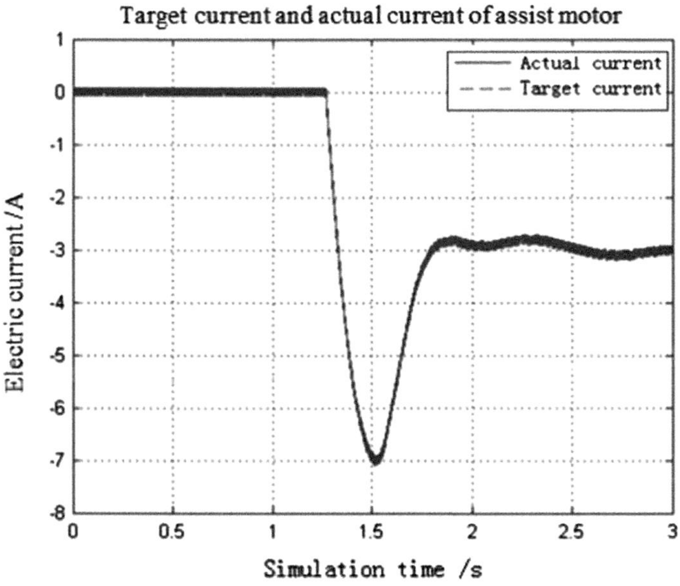

Fig. 8 Current simulation

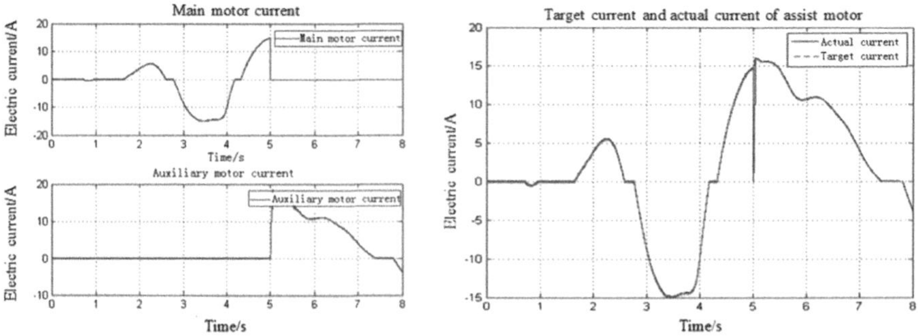

(a) Current distribution of main and auxiliary motors

(b) Following situation of total current

Fig. 9 Simulation results of fault-tolerant control current

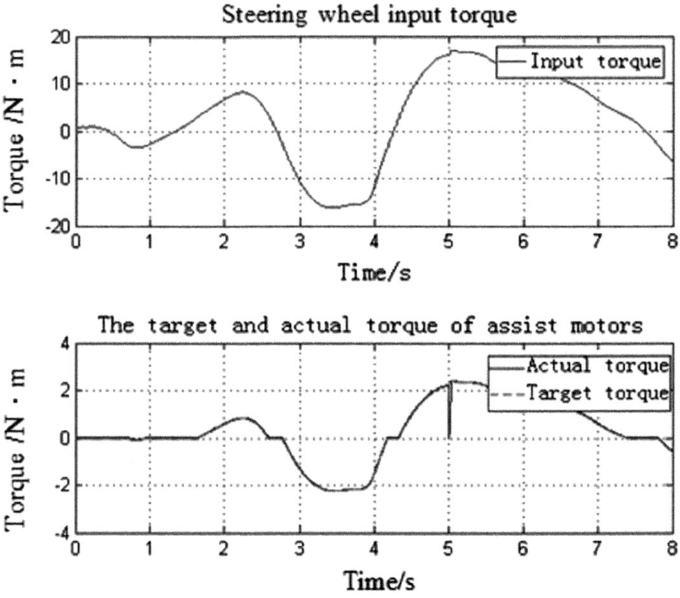

Fig. 10 Torque simulation

divided equally to the main and auxiliary motors under normal conditions. When the main motor is broken and stopped working, the ECU allocates the target auxiliary current to the auxiliary motor according to the control strategy formulated, and all power-assisted tasks are undertaken by auxiliary motors.

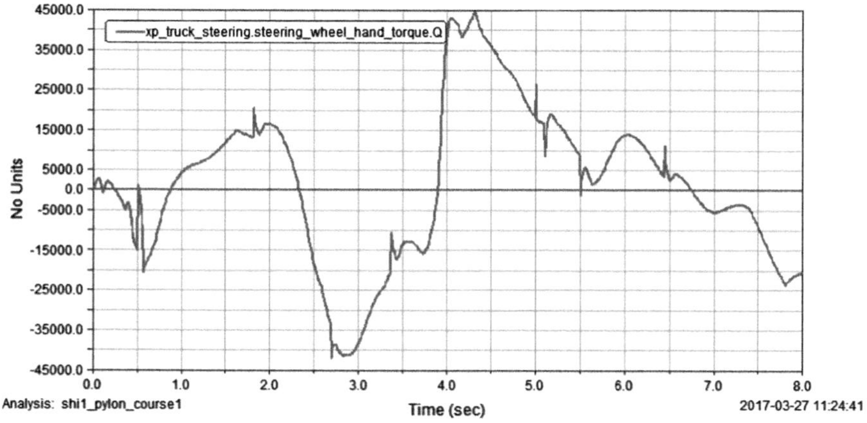

Fig. 11 Steering wheel torque

4.4 *Steering Portability Test*

The main purpose of EPS is to improve the steering agility of the car at low speed and maintain a good road feel at high speed. This paper verifies the steering handiness of the designed dual-motor EPS by double-curve test. The double track of the passenger train is constructed in the ADAMS, and then the bus runs along the trajectory at a constant speed of 10 km/h. The simulation results are shown in Figs. 12, 13, and 14. Figure 12 is the trajectory of a double curve. Figure 13 is the change simulation of the hand torque and the power torque of the steering wheel, in which the dashed line is the hand torque, and the solid line is the power moment.

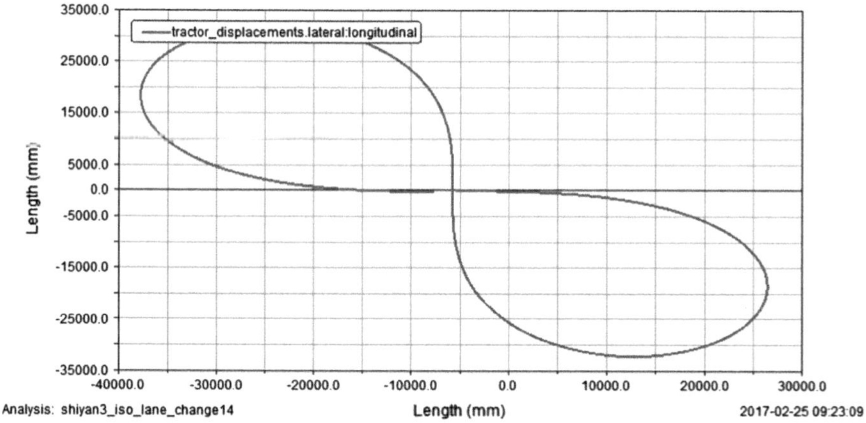

Fig. 12 Kinematic track of double-curve test

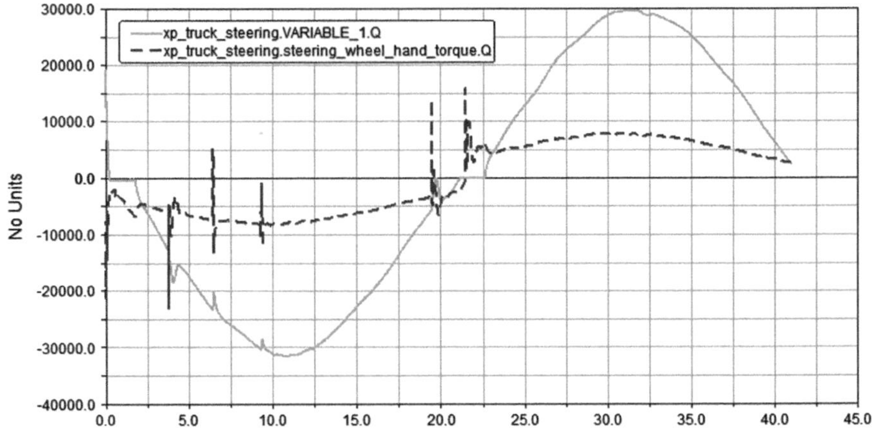

Fig. 13 Steering wheel hand torque and assist torque of the double-curve test

Figure 14 shows the relationship between the steering wheel torque and the steering wheel corner of the double-wire test, where the relationship between the steering wheel torque and the corner when the dotted line is helpless, and the relationship between the steering wheel torque and the corner when the solid line is assisted; the relationship without assist torque between the steering wheel torque and the speed is shown by the dotted line; the relationship with assist torque between the steering wheel torque and the angle is shown by the solid line. From the data shown in the figure, we can see that the maximum torque of the steering wheel is 8.7 N m, and the maximum power moment is 31.42 N m. The steering wheel diameter of the car is 0.5 m, and the maximum tangential force which is applied to the steering wheel is 43.5 N. The maximum steering force rating standard as indicated in the

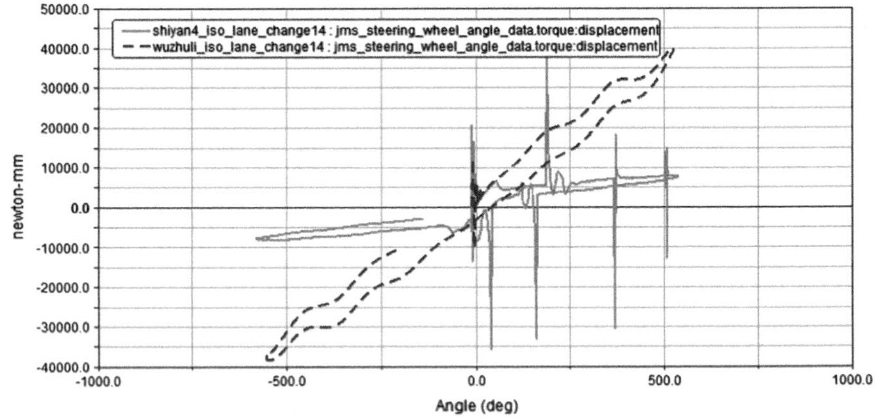

Fig. 14 Steering wheel torque and corner relation of the double-curve test

formula (1), N_{Fm} = 107.3. If the score is over 100, the score is 100. Therefore, the dual-motor electric power-assisted system designed in this paper has better steering handiness.

$$N_{Fm} = 60 + \frac{40}{F_{m60} - F_{m100}} \times (F_{m60} - F_m) \tag{1}$$

5 Conclusion

A dual-motor electric power steering system for commercial vehicle is proposed in this paper. The dynamic coupling scheme, power-assisted current distribution scheme, and fault-tolerant control algorithm of dual-motor electric power steering (EPS) system are designed, The whole vehicle dynamics model is built in ADAMS, and the EPS simulation model of dual motor is established by using Matlab/ Simulink. Finally, the performance of dual-motor electric power steering system of commercial vehicle is verified by joint simulation.

References

1. Zhang S,Yu Q, Yan G, Zhang Q, Zhang P (2013) A research on the power-assistance control strategy for the EPS in a commercial vehicle. Automot Eng 35(9)
2. Liu J (2016) Design of capacitive torque sensor and research of electric power steering system. University of Electronic Science and technology of China
3. Chen W, Shi P, Gao L et al (2007) Co-simulation of ADAMS and Matlab EPS and the vehicle system. J Agricu Mach 38(2):22–25
4. Deng L, Zhang J, Wang S (2010) EPS system assist characteristic and control strategies of J. Agricu Equip Veh Eng (10):19–22
5. Saifia D, Chadli M, Karimi HR et al (2015) Fuzzy control for Electric Power Steering System with assist motor current input constraints[J]. J Franklin Inst 352(2):562–576
6. Sheng De, Li H (2015) Joint simulation and parameter analysis for ADAMS and MATLAB for. In: 2015 proceedings of the annual meeting of China Automotive Engineering Society, vol 2
7. He Q, He Z, Fang H et al (2014) Design and evaluation of speed induction coefficient of a minibus EPS system. Mechanical Sci Technol 33(3):404–408

Technology of Squeeze Casting for Aluminium Alloy Automobile Triangle Arms

Juncheng Ren, Zezhong Wang, Liqiong Heng, Xinghong Zhang, Chaoliang Shi, Qiulai Wang, Bing Yang, Wanghong Xia and Chuandong Ren

1 Background

In recent years, due to the requirements of energy conservation and emission reduction, aluminium alloy lightweight materials in the application of automotive chassis develop very quickly. Squeeze casting combines the characteristic of casting and forging, and it works to minimize both solidification shrinkage and gas entrapment [1], resulting in fine structure and excellent mechanical properties. It is a precise casting process with good application prospect for aluminium alloy.

Automobile triangular arm is one of the important safety parts on the chassis. The prototype of the triangular arm mentioned in the article was a stamping-welding product; the weight of which was 2.7 kg, and it is expected to be reduced by more than 20% using the squeeze casting technology combined with structure optimization.

2 The Structure Design and Material Selection

The material of stamping-welding triangular arm (see Fig. 1) is QSTE460TM steel plates and 20 steel pipes. Considering the reliability and replaceability, as well as the characteristics of aluminium alloy, the interface between the steering knuckle

J. Ren (✉) · Z. Wang (✉) · L. Heng · X. Zhang · C. Shi · Q. Wang · B. Yang
Dongfeng Motor Coporation Technical Center, Wuhan 430058, China
e-mail: renjc@dfmc.com.cn

Z. Wang
e-mail: wangzz@dfmc.com.cn

W. Xia · C. Ren
Guangzhou KINBON Liquid Mold Forging Technology Co., Ltd, Guangzhou, China

Society of Automotive Engineers of China (SAE-China) (ed.), *Proceedings of the 19th Asia Pacific Automotive Engineering Conference & SAE-China Congress 2017: Selected Papers*, Lecture Notes in Electrical Engineering 486, https://doi.org/10.1007/978-981-10-8506-2_73

Fig. 1 Stamping-welding triangular arm

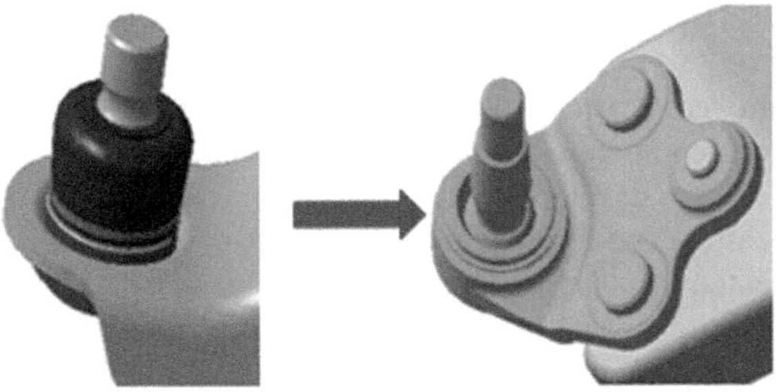

Fig. 2 Change of the interface between the steering knuckle and triangular arm

and triangular arm changes from column pin to taper pin (see Fig. 2), the other interfaces of the triangular arm remain unchanged.

A356 alloy is a widely used aluminium alloy material with good comprehensive mechanical properties, the price of which is relatively cheaper than other high-strength aluminium alloy. The mechanical properties of A356 are stipulated in Table 1.

According to the mechanical properties of A356, the environment of triangular arm, the characteristics of squeeze casting and CAE analysis results, we redesign and optimization the structure of the aluminium alloy triangular arm. Figure 3 shows the preliminary model of the aluminium alloy triangular arm. In order to minimize the volume and the weight of the product, we use the topology optimization technology as well as multi-body dynamics simulation analysis method to redesign triangular arm. Figure 4 shows the optimization model, and there are several hollow areas exist in the optimization triangular arm; as a result, the weight of the aluminium alloy triangular arm is only 2.05 kg. Compared to the steel triangular arm (2.7 kg), the optimization model reduces 24% weight.

Table 1 Mechanical properties of A356

Rm/MPa	Rp 0.2/MPa	A/%	Hardness/HB
≥ 303	≥ 230	≥ 10	>90

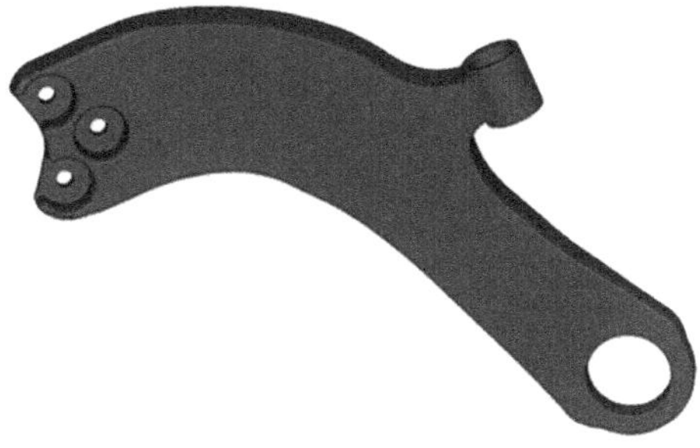

Fig. 3 Preliminary model of the aluminium alloy triangular arm

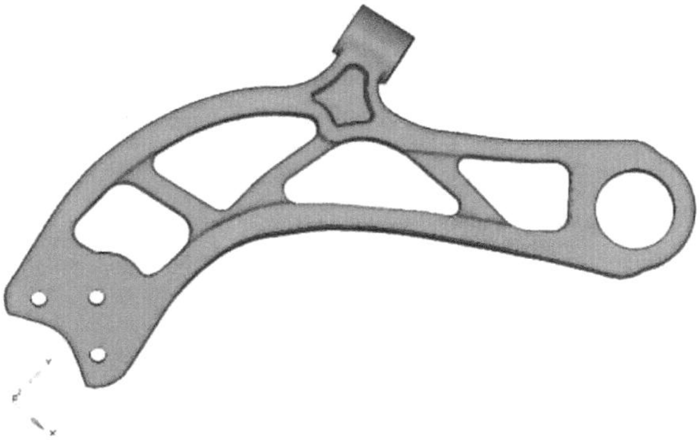

Fig. 4 Optimization model of the aluminium alloy triangular arm

3 CAE Strength Analysis for the Aluminium Alloy Triangular Arm

The results (see Table 2) of strength analysis for the optimization model of the aluminium alloy triangular arm show that the maximum stress of seven working conditions (such as turning, braking and speeding up) is less than the material yield strength (230 MPa), the stress distribution of each working conditions; see Fig. 5.

Using the measured wheel load spectrum as the input load, we calculated the fatigue cumulative damage value of the aluminium alloy triangular arm as well as the steel one. The damage distributions of the two kinds of triangular arm are listed in Table 3. The cumulative damage value of the aluminium alloy is 1.54e-6, which is much smaller than the value of the steel one, and to a certain extent, it proves that the fatigue properties of the aluminium alloy triangular arm are better than the steel one.

Table 2 Analysis results of different working conditions(MPa)

Working condition	Case 1	Case 2	Case 3	Case 4	Case 5	Case 6	Case 7	Target requirement
Max stress	82	144	104	209	99	44	22	≤ 230

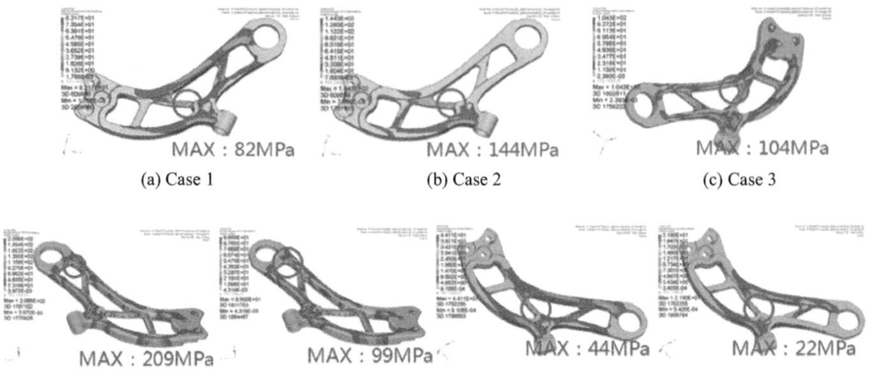

(a) Case 1 (b) Case 2 (c) Case 3

(d) Case 4 (e) Case 5 (f) Case 6 (g) Case 7

Fig. 5 Stress distribution of the triangular arm

Table 3 Damage distributions of the two kinds of triangular arm

Items	The steel triangular arm	The aluminium triangular arm
Material	QSTE460TM	A356
Weight	2.740 kg	2.047 kg
Damage distribution	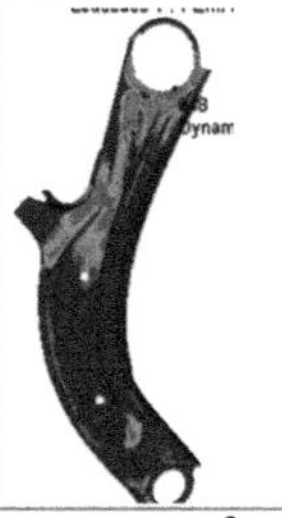	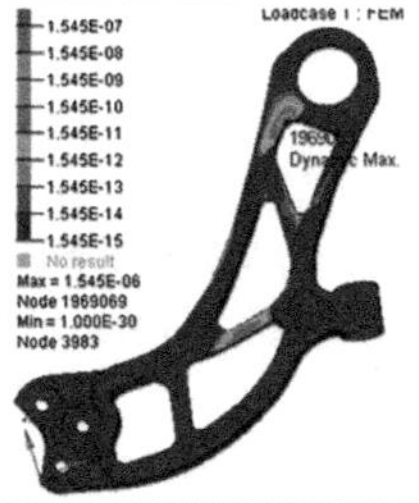
Max damage value	9.86×10^{-2}	1.54×10^{-6}

4 Simulation Analysis of Squeeze Casting Process

The basic parameters, injection parameters and local Squeeze parameters for simulation analysis of aluminium alloy casting process see Tables 4, 5 and 6.

The results of simulation analysis of squeeze casting process see Fig. 6, the filling effect is relatively good, some hot spots exist, but the value is low,point cooling can be adopted to minimize the hot spots in the casting.

Table 4 Basic parameters

Mould material	Mould temperature	Casting temperature
SKD61	180 °C	690 °C

Table 5 Injection parameters

Stroke/mm	0	20	230	250	280	300	420
Injection speed/(m/s)	0	0.15	0.15	0.08	0.08	0.1	0.1

Table 6 Pressurization parameters and local squeeze parameters

Pressurization parameters				Local Squeeze parameters		
Initial pressure	Pressure building time	Pressurization pressure	Pressure keeping time	Squeezing time	Pressure keeping time	Squeezing pressure
90 MPa	0.05 s	100 MPa	15 s	0.4 s	40 s	100 MPa

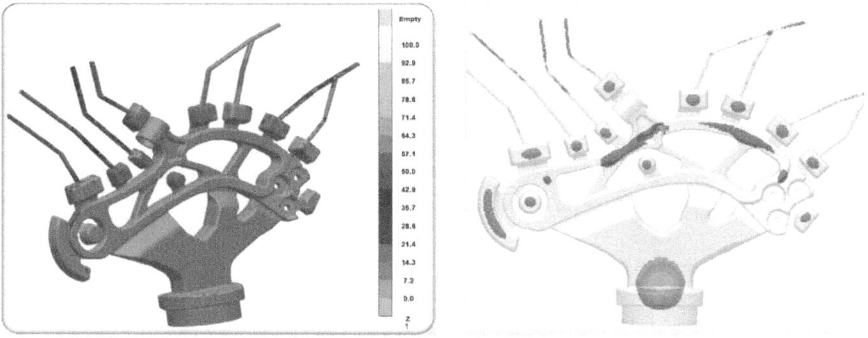

(a) Gas content simulation results (b) hot spots simulation results

Fig. 6 Results of simulation analysis of squeeze casting process

5 Trial Problems and Solutions

In the trial process of the aluminium alloy triangular arm, it is easy to produce shrinkages and bubbles (after heat treating). In order to solve these problems and ensure product stability, there are several main steps, such as core pulling, local squeeze, point cooling (see Fig. 7), changing the gate size and the speed.

After the heat treatment, it is easy to generate bubbles and micro-cracks; the reason is that the die casting speed was too high, which led to the reflow of the molten aluminium. The solution is to reduce the filling speed (Fig. 8).

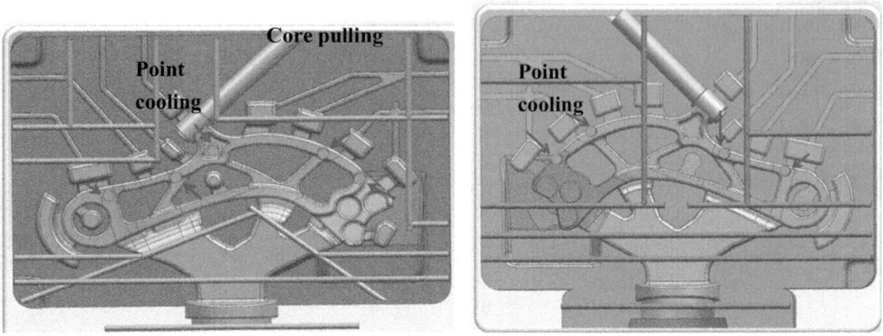

(a) Ejector die (b) Cover die

Fig. 7 Several steps to ensure product stability

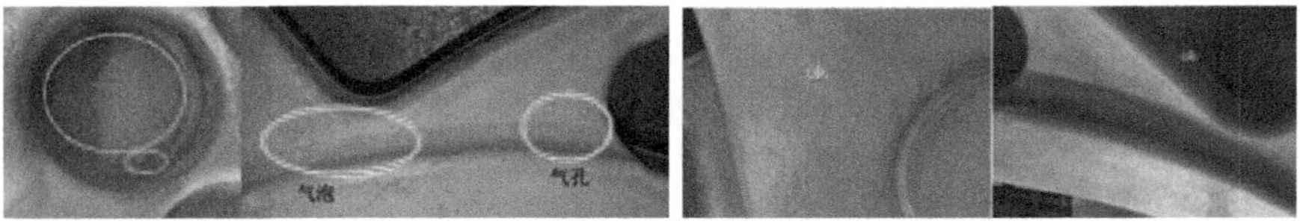

(a) before improvement (b) after improvement

Fig. 8 Effect of reducing the filling speed

6 Material Tests, Bench Tests and Vehicle Tests

The material tensile tests (see Table 7) show that the properties can meet the expected requirements; the microstructure is α phase + granular eutectic silicon + a small amount of acicular Fe phase, and no shrinkage exists; see Fig. 9.

We do bench tests to the aluminium alloy triangle arms, and all the requirements of the tests keep the same with the steel triangle arms, such as the value, frequency and the direction of the load; one of the bench tests is durability test(see Fig. 10), and the aluminium alloy triangle arms endure 1.5 million cycles without deformation or fracture; the performance is better than the steel triangle arms.

The aluminium alloy triangle arms were equipped into the vehicle, and there was no deformation or fracture after the vehicle ran 80,000 km.

Table 7 Tensile test results

Mechanical properties	Rm	Rp0.2	A
Sample 1	323	253	13.4
Sample 2	310	240	10.7
Sample 3	316	253	10.6
Sample 4	310	251	10.5
Average	314.75	249.25	11.3
Requirement	⩾303	⩾230	⩾10

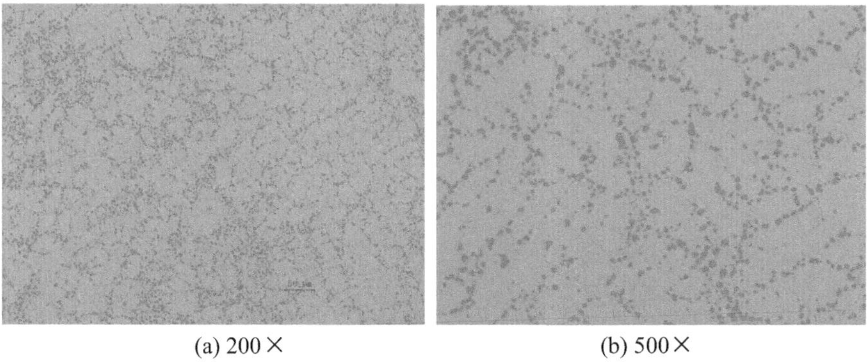

Fig. 9 Microstructure photograph

Fig. 10 Durability test

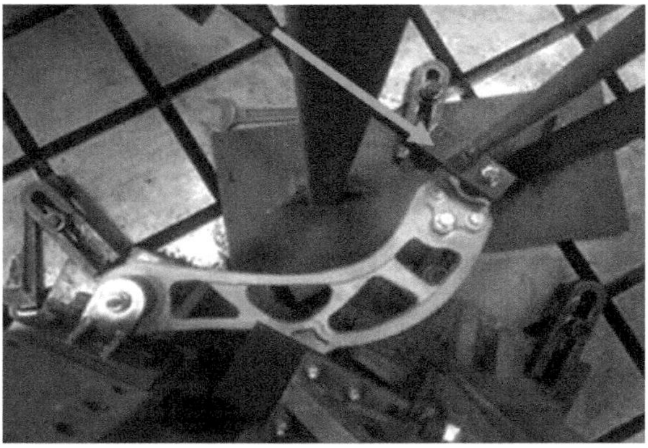

7 Conclusions

The technology of squeeze casting to make aluminium alloy automobile triangle arms to replace steel triangle arms can reduce 24% weight combined with topology optimization technology and multi-body dynamics simulation analysis.

Bench tests and vehicle tests show that the aluminium alloy automobile triangle arms can fulfil the requirements of this product. The paper provides valuable experiences for lightweight of automobile chassis.

Reference

1. Vinarcik EJ (2002) High integrity die casting processes. Wiley, USA, pp 51–52

Application of NX Innovative Design Model in the Fuzzy Front End Period of NEV Development

Ning Xiao and Menghan Tao

1 Introduction

At present, the NEV industry begins to flourish worldwide. The development of NEV industry provides an opportunity for China to get rid of technological dependence to foreign countries and to improve its industrial competitiveness. It also helps shorten the technological gap between China and foreign countries in vehicles industry and push the industry in China into the fast lane for "corner overtaking." NEV is a technology and knowledge-intensive industry, and much innovative investment is needed in its development.

In 2016, about 517,000 NEVs (including passenger vehicles and commercial vehicles) were produced in China, among which 507,000 were sold; a 51.7 and 53% increase from the previous year. The sales volume of NEV has surpassed USA and ranked first worldwide [1]. Despite the fact that the NEV industry is emerging in China, manufacturers fail to produce ground-breaking and innovative Tesla-ish products. As a result, Chinese-made NEVs are in a backward situation among global vehicle industries. Furthermore, NEV-related patents are basically owned by foreign enterprises in China. In a global perspective, Japan and USA are taking lead in patents application of NEV-related technologies (shown in Fig. 1).

Therefore, the development of NEV in China should focus and invest more on the breakthrough and innovation of products. How to develop and create a breakthrough NEV product? The key point is the development during FFE period, as this period is the first step of product development and the process of innovative opportunity growing out of nothing. Research has shown that the success during the

N. Xiao · M. Tao (✉)
Gac-Toyota Motor Co., Ltd, Guangzhou, China
e-mail: menghan_tao01@gtmc.com.cn

N. Xiao
e-mail: ning_xiao@gtmc.com.cn

Society of Automotive Engineers of China (SAE-China) (ed.), *Proceedings of the 19th Asia Pacific Automotive Engineering Conference & SAE-China Congress 2017: Selected Papers*, Lecture Notes in Electrical Engineering 486, https://doi.org/10.1007/978-981-10-8506-2_74

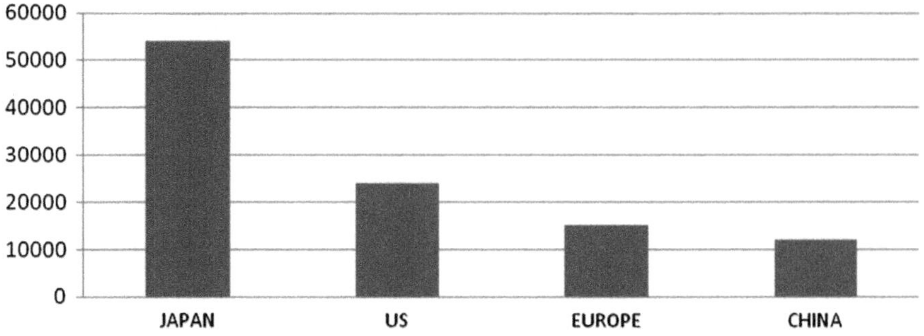

Fig. 1 Distribution of patent applications worldwide [2]

FFE period can save 30% of new product development time, and the rate of product ideas passing into specific development (forming product prototype) is only 0.47%, while only 7.14% product for mass production can achieve business success [3]. The reason can be concluded as follows: the uncertainty and unreliability of innovative opportunities, technology, and available resources. This paper tried to establish an innovative model to apply to the FFE period of NEV development in order to ensure the innovativeness and breakthrough of product based on the premise of integrated resources and technology.

2 Establishing NX Innovative Design Model for FFE Period of Vehicle Product Development

According to comprehensive research report on innovative design strategy issued by Chinese Academy of Engineering in 2016 [4], innovative design (design in a broad sense) refers to "human"-oriented innovative and creative activities that serve the industry. It is a combination of engineering design, industrial design (design in a narrow sense) and service design, etc. Ezio Manzini, an Italian scholar, put forward the concept of "design in the broad sense" in design, when everybody designs [5]: Design in the broad sense is to construct meaning and to solve problems. It breaks away from the skills and fixed principles that design in the narrow sense depends on. Be it professional designers or related experts, be it officials from government or decision makers from enterprises, they all are the participants and implementer of designs. In conclusion, the specific functions of creative design lie in its ability to evolve into corporate strategies (including business model, service design, and experience design) as well as guidance and methods for FFE period of a new product development.

The question is, how could enterprise make innovative designs? Based on the understanding and analysis of the guidelines above, this article puts forward an NX innovative design model that can serve product development and evolve into corporate strategy (or strategic design) of vehicle industry. Then comes the next question. What is the core of this model? How is it formed and how can it play a part? First, we need to understand two core concepts of NX innovative design model by delving into current theories: "integrated New Product Development" and "Design-Driven Development."

2.1 *Principle and Significance of Integrated New Product Development*

Integrated new product development is an innovative product development method in which developers truly understand target users and turns this understanding into quantitative analysis method that can be upgraded into product standards. It puts an emphasis on the fact that a development team should possess indispensable forces technology and design, and should focus on aspects that are the users' concern.

The significances of this method are as follows: The integrated development team must discard their own standpoint and try to form a new product concept prototype centered on users. Whenever the products are accepted, regarded as useful, handy, and something to yearn for, it means the success of products. This kind of development method enjoys a high successful rate and can easily evolve into ground-breaking product concepts.

Traditional FFE period of new product development works as follows: the marketing personnel decide what kind of new vehicle is needed according to their past experience. Then engineers and designers also design products based on their own experiences and lifestyles. The iNPD method, however, emphasizes that explorers and decision makers who participate in FFE period should build an in-depth connection with users because this essentially helps improve certain aspects of vehicle development for creating more breakthrough product.

For example, in the NX innovative design model, personnel from engineering, marketing, and design areas must discard their own standpoints before the concepts of products are formed. Marketing personnel should not think about sale volume; designers should not only think about product modeling, and engineers should not think about feasibility and reliability. As shown in Fig. 2, personnel from these three parts form an integrated team and try to study users with a user-oriented perspective. They put themselves into the roles of users when conducting analysis, which perfectly matches the purpose of innovative design that must cover various fields.

2.2 *Principle and Significance of Design-Driven Innovation*

The design-driven innovation (DDI) [7] strategy discovered and put forward by the Italian scholar Roberto Verganti states that to focus on the real meaning of products (the reasons why users choose to buy a product) to innovate, can deliver brand new concepts or visions. These concepts and visions are always expected by users but

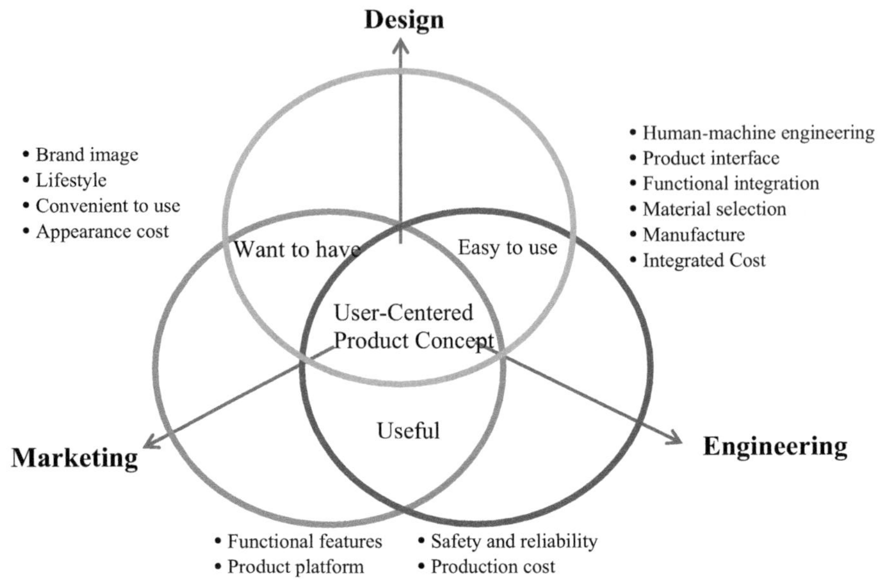

Fig. 2 User-centered integrated new product development (iNPD) [6]

have not come into being, and will be accepted and liked by the users. In short, it is an innovation in which the message of the product and design language has a greater novelty than its functionality and technology.

As shown in Fig. 3, there are two traditional innovation strategies: 1. market-driven innovation. As an incremental innovation, it is a forecast for the next two or three years based on the sales situation in the past; 2. technological innovation. It shows the leap of product performance with the help of technology breakthrough. However, innovation is not just about optimizing the performance of products or meeting the needs of consumers. Design-driven innovation emphasizes on adopting the connotative and innovative methods and endows the products with a core value that users really care about.

2.3 Establishing NX Innovative Design Model

The reason why researchers select iNPD and DDI as the theoretical support of NX innovative design model is that both concepts are user-oriented, matching the core thinking of innovative design. The key difference and the deficiencies are as follows: DDI is a guiding strategy for product innovation, without reference to specific implementation methods as well as new business models; iNPD, however, provides specific implementation methods for FFE period of new product development and

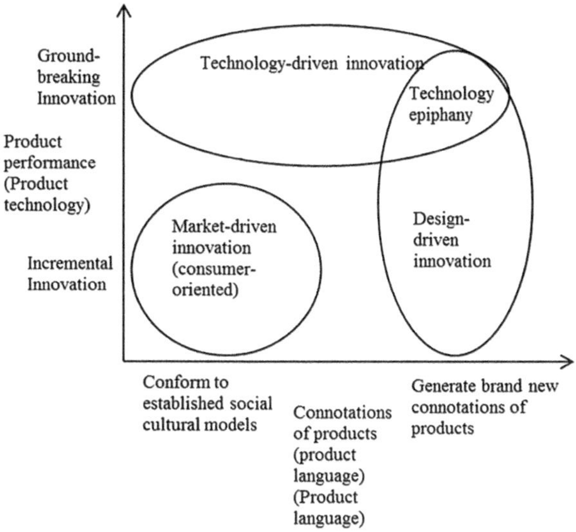

Fig. 3 Product positioning under the guidance of different innovation strategy

is not helpful for formulating enterprise strategy. Therefore, NX innovative design model combines both and makes them complement each other and can evolve into a three-dimensional model based on different mechanisms of action.

Firstly, based on the iNPD, actually applying cross-domain team to implement the product/service prototype innovation during the FFE period of new product development. The original point is definitely the user (human). Thereby, a prototype of NX innovative design model is formed (shown in Fig. 4).

The forming of the second step is with the guidance of DDI strategy, which is also user-oriented. It can link to the iNPD through the "user" central axis. While in the actual business activity, DDI strategy is not only limited to "prototype" level, but strategically provide guidance to the iNPD's vertical process, downing to form the product concept "prototype" at the same time, upping can also form a product strategy, which can also proceed to transform to enterprise strategy. This process is seen as strategic design. It is the key step for NX innovative design model. It

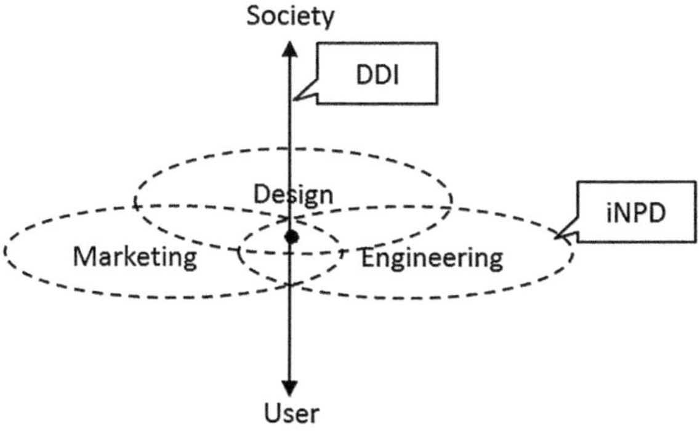

Fig. 4 First/second step of forming an NX innovative design model

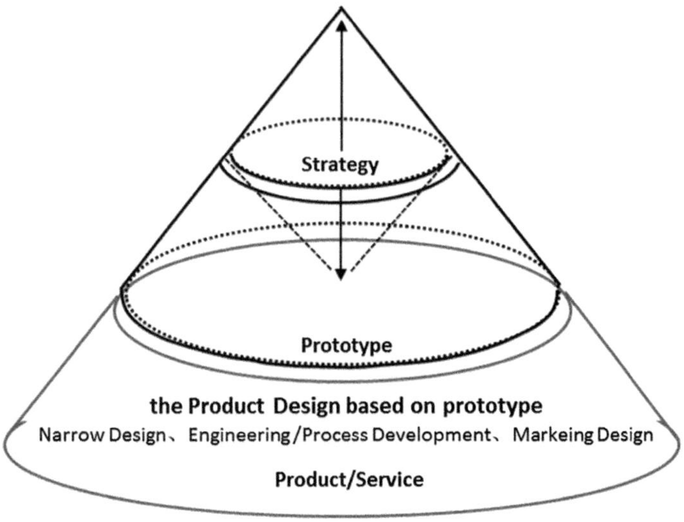

Fig. 5 Third step of forming an NX innovative design model

provides a link between enterprise strategy which is based on product strategy and the specific development process of prototypes; in the integrated innovative process, it serves as an important coordinative agent and guidelines.

The third step is the implementation of product concept, or the process to turn concepts into products (problem solving). This step is composed of design in a narrow sense (industrial design), engineering research and development, technique development, implementation of commercialized marketing design (shown in Fig. 5). The above steps are the basic level that Chinese enterprises should be familiar with.

The fourth and last step, the strategy attributes of constructing meaning and solving problems are added for building the model according to the concept of design in the broad sense.

As shown in Fig. 6, the trend of the two axes also represents the responsibilities that innovative design need to assume, or, in other words, to construct meaning and value from users to company with regards to social responsibilities.

The central axis of the cone is "human." Because iNPD and DDI both put emphasis on "human," they can fit each other on the central axis. Therefore, developers must pay attention to the central axis of "human-oriented" when conducting innovative design. Developing teams must abandon their original standpoint on marketing, engineering, and design in the narrow sense so as to function strategically and coordinately in the model.

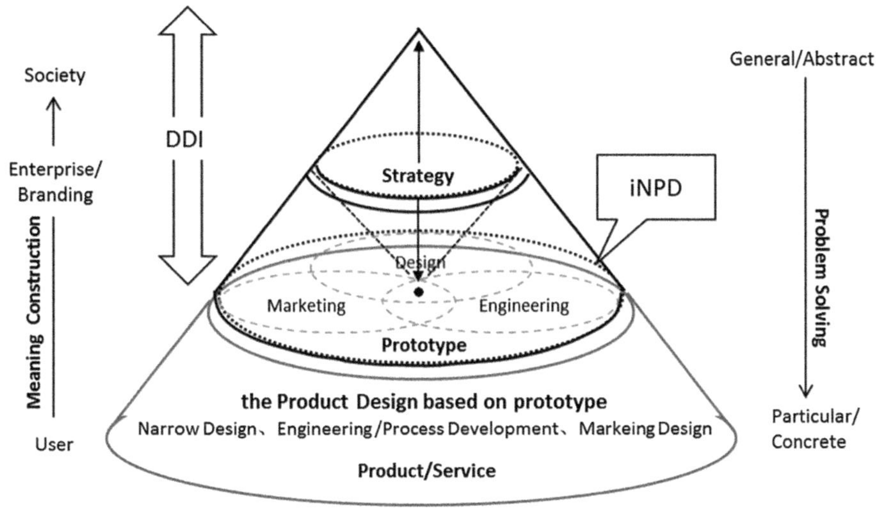

Fig. 6 NX innovative design model

3 Application of NX Innovative Design Model in FFE Period of Nev Development

3.1 Demand Identification in the FFE Period of New Product Development

3.1.1 Opportunity Identification

After the company has decided to research on and develop a new product, the first step is to build a team as shown in Fig. 7 of professionals in engineering, marketing, and design areas.

The first step is to make use of socioeconomic technological factors to analyze strategic background and to look for opportunity gap. In terms of social factors,

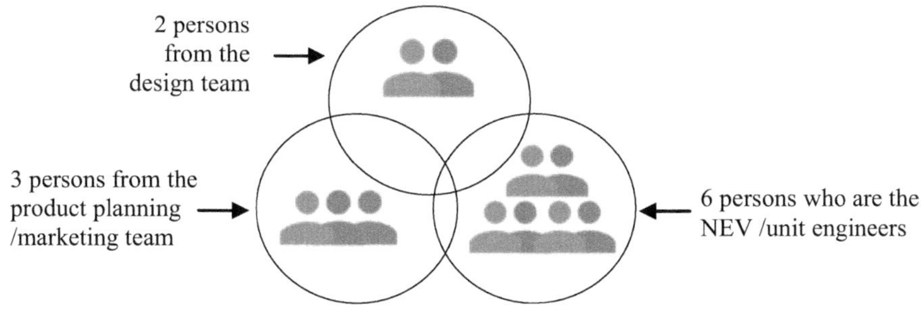

Fig. 7 NX innovative design group

because of the environmental friendliness of NEV and government support, NEV has become a focus of vehicle industry. In terms of economic factors, on one hand, the car allowance for NEV is decreasing year by year; on the other, the production cost of NEV is still high, so it becomes a problem to persuade users to purchase expensive NEV. In terms of technological factors, the physical features of NEV mean totally different usage features from ordinary vehicles, and users may encounter many inconveniences in their usage. Here, an opportunity gap can be discovered: The NEV product design and development should pay attention to solving such inconveniences in the process of popularizing NEV usage.

The second step is to conduct research focusing on NEV users, with the aim to fully understand them and their pain points. Currently, the users of NEVs are mainly early adopters. Users with in-depth user experience are selected as research objects in line with the aim of the research—solving user inconveniences and improving product force. Quantitative research method is used to study the main quality of users (Fig. 8).

Information and materials collected in user research are analyzed and sorted into 50 typical scenes of using NEVs and two persona prototypes of users (Fig. 9).

Then, through the workshop method, based on two user personas, to analyze pain points in different scenes. With the criteria of attractiveness/feasibility/innovation, the most typical scenes (pain points of most users) are drawn: NEV

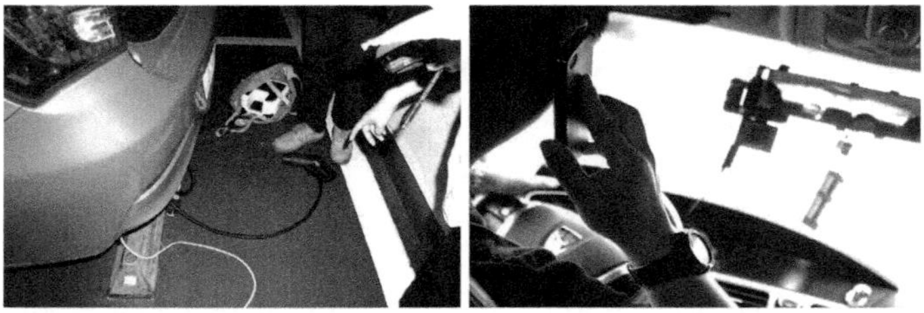

Fig. 8 Researchers are observing how user charges the car and drive

user persona		Life Style	Personality
Experienced Semi-Car-Expert		Career:rising and stable stage Family:married,extended family	Take car as brother or friend,take efforts to do maintanance and refitting; Impulsive and rational coexistence, dare to win the freshest things; Believe in that NEVs are the future development trend
Pioneering Eco-Living Practitioner		Career:rising stage Family:single,small family	Take car as a means of transport, used to enhance the quality of life Brave,trendy, rational consumption; Perfer Win-win products, like to show off to friends

Fig. 9 Two user persona prototypes

Where/ when/ how	Slow down				
	Pedestrian–vehicle mixed traffic				
Who/ whom	Charging	Start-up	Quiet	Extremely quiet	Noisy
Pedestrian	• Light (Warning: do not get close) •Sound	•Anticipation	To remind car is approaching		
			• Sound (normal) #Not disturbing/ easy to notice/easy to distinguish/ warning	• Volume(small) #Same features as on the left (more joyful, more of a warning) • Light (nighttime)	• Sound (louder, more of a warning)
Analysis	○, **Charging indicator lamp** ×, Warning tone is disturbing	○, **Sound of low-speed driving**	**Responsive sound of low-speed driving: unable to identify current situation due to limiting technology**		
Driver	• Note: Successful connection • Note: Starting charging •Note: Charging/ full battery • Fault reminder	• Feedback on start-up (The driver can perceive whether it is ready to drive whether it is ready)	Same as on the right	• Convenient (Controllable and audible like car horns) • The driver needs to know whether the VSP system is working	The same as on the left
Analysis	※ Pilot lamp on dashboard	○, **Ready reminder/Start-up reminder**	○, **When driving at low speed, the driver can click twice the switch. An alternative strategy may be needed when the car is at a full stop**		

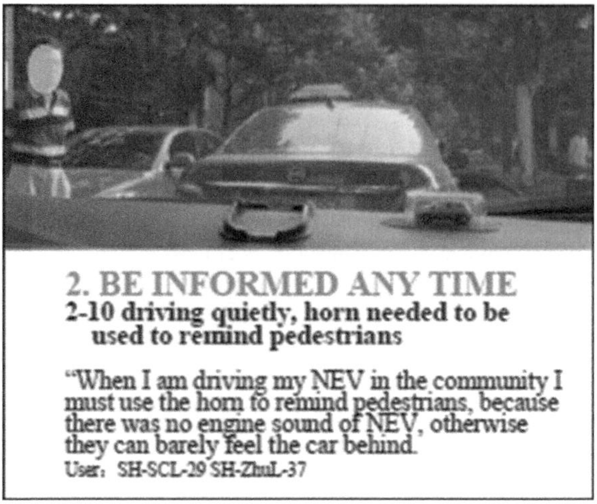

Fig. 10 One of the typical scenes of NEV

(EV and PHEV in EV mode) are too quiet to drive, and it is not easy for passengers to note that an NEV is passing by. Users are particularly bothered by this (Fig. 10).

3.1.2 Analysis of Requirements

- Scenario segmentation analysis:
- Supplementary scene mining:

In addition, according to the discovery that pain points can be solved through using acousto-optic signals under different subdivision scenes, team members start brain-storming again to mine possible usage scenes on the premise of achieving such functions, and try to give more meanings to production. In line with the attractiveness/feasibility/creativeness of products, the following applicable scenes and pain points can be determined: When the driver/online car-hailing passenger is searching cars in parking lot and looking for appointed car-hailing vehicles, they are not sure about the specific location of the car and cannot recognize it. After the appointed cars has come to the designated destination, the driver still has to contact the passenger through telephone, which results in low recognition efficiency and hidden danger of driving.

- Supplementary scene requirement analysis:

If an acousto-optic signaling device is installed in the car, the scene will be different. Online car-hailing users can define the acousto-optic signals of the cars on their APP; when the car is near the passenger, it can give out the defined acousto-optic signal for the passengers' identification. If the passenger has not defined acousto-optic signals, the device can send a series of driver-defined signals to the APP on the passengers' cell phone, so as to remind them the appointed car is there (Fig. 11).

Fig. 11 Supplementary of scenario requirements

3.2 Formulating Product Plans

User demands that need to be solved and analysis on these requirements can evolve into respective product strategic prototypes based on different application scenes.

- Development goal: 1. Vehicle sound for pedestrians; 2. VSP + applied in NEV car-hailing vehicles
- Product design specification (Fig. 12).
- Architectural design (Fig. 13).

3.3 Preparation Phase for Production

Next, the team starts the preparation work from product plan to production. They finish the establishment of product prototype and testing with the focus on VSP and VSP + as well as demonstration running on NEVs.

3.4 Evaluation of Application Results

Through feasibility analysis, the application result is evaluated so as to judge whether the concept prototype can be turned into corporate strategies and guidance for production. In the post-phase of development, our team discovered that GB/T 32694-2016 issued on January 1, 2017, recommend VSP device on EV/PHEV,

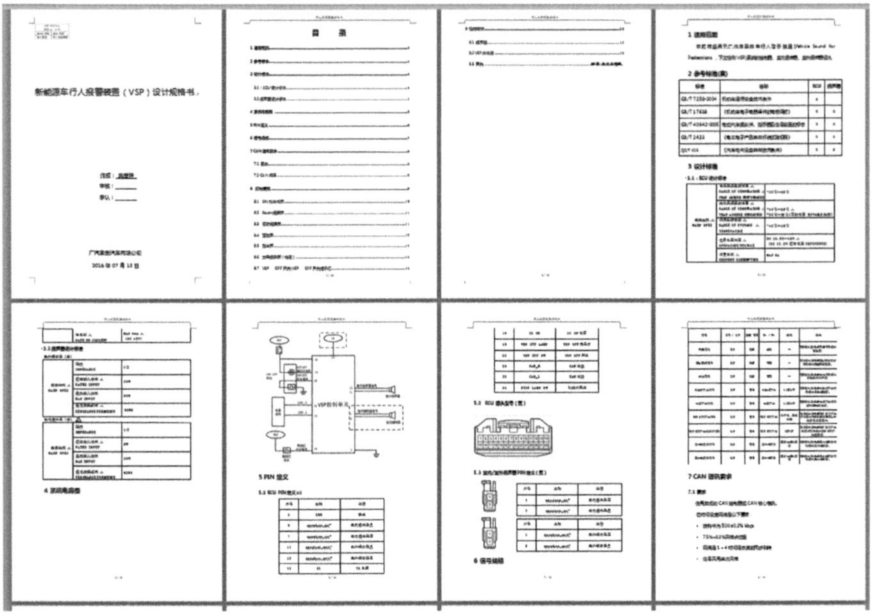

Fig. 12 Product design specification

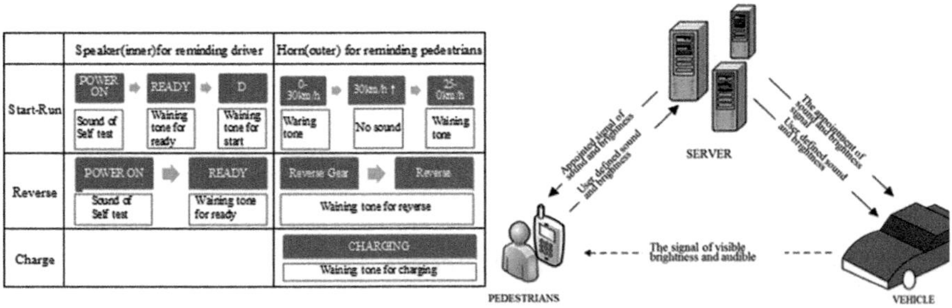

Fig. 13 Concept map of VSP and VSP+ architectural design

which means VSP product will be installed on all NEVs before long. On the occasion, VSP will be meaningless to users and become a cost-adder rather than a selling point.

As shown in Fig. 3.11, the upgrade from VSP to VSP+ is a new meaning for products and a change of product positioning without changing technological level (Figs. 14, and 15).

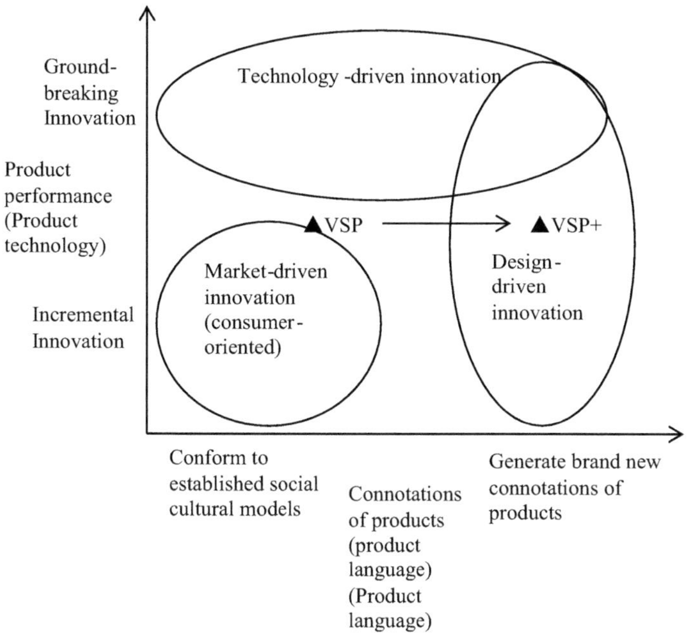

Fig. 14 Application achievements of NX innovative design model

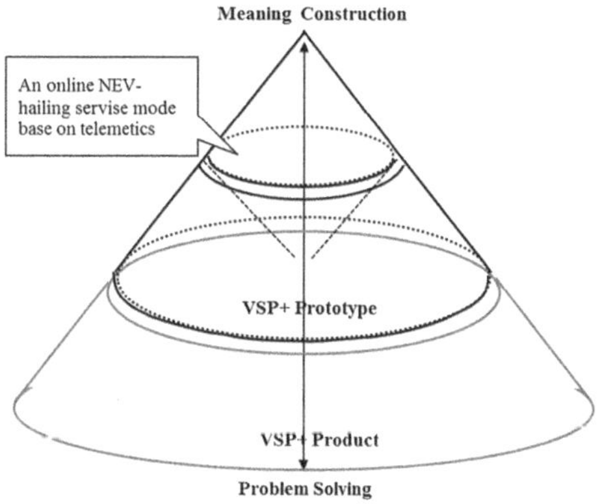

Meaning	1.Brand image of the company is injected with internet gene
	2.Increasing cooperation opportunity between the enterprise and car hailing companies/internet companies
	3.Building a new connective technological barrier
	4.Increasing the business opportunity in internet service sector
	5.Reminding the enterprises of the user-oriented and meaning-driven concepts
Problems	1.Pain points: NEV is too quiet when driving
	2.Driver and passenger cannot find each other in car-hailing service
	3. The parking lot is near, but I can find it.

Fig. 15 Application achievements of NX innovative design model

4 Conclusions and Application Prospect

Thronging above exploration and practice, upgrading simple VSP just for legal purpose to multi-functional, user-oriented VSP+ can help increasing product force and creating new value with just a slightly higher cost. Because of such benefits,

VSP+ is expected to be installed on specialized online hailed cars volume production with volume production.

The exploration proves that NX innovative design model is a set of new guidelines applied in the FFE period of NEV development. It successfully improved the NEV product technology's inner meaning, meanwhile produced breakthrough innovation and made the product attractive to the users. In addition, attractive NEV products are created and new strategic model which specifics to the product/service innovation of online hailed car. Through the above application, the practical value of the model has been validated.

The next step will focus on the transformation from technological "pointed innovation" to the "crowd-outsourcing innovation" of the enterprise, and to expand into the fields including interactive innovation, experience innovation, service innovation and collect the domestic successful breakthrough cases of different enterprises to compare with NX innovative design model, to see if there is any similarity, in order to further verify the universality and boundary conditions of NX innovative design model.

In terms of the follow-up application of the innovative design model, the issue also exists; for instance, the traditional vehicle development process of Chinese enterprises and the closed bureaucratic culture of the traditional automobile industry are obviously not conducive to innovation. If the enterprise wants to create a team with the NX innovative design model as the core, the only solution is to start all over again or make independent authorization. The common outsourcing approach cannot solve the core issue of strategy plan. On the contrary, if the internal team mastered the correct method of product innovation, the more practical solutions can be made out for the enterprise; however, the outsourcing resource can be reasonably applied under the enterprise internal team's strategic guidance, to form a large number of prototype innovation and supplement product concepts. In other word, enterprises can flexibly use the external resources. At the same time, the creativity of external team also becomes a reference for the internal innovation team.

The vehicle industry is about to revolution. User-oriented iNPD method and DDI strategy that create new meanings for products are spreading rapidly throughout the vehicle industry. The core value of the both is "human" oriented. The combinative appliance of the both will create new values that users are willing to "pay for" and evolve into corporate strategies which constantly improve the image of enterprises. This is the fundamental purpose of NX innovative design model to integrate iNPD and DDI.

References

1. CAAM Economic operation of automobile industry in 2016 [EB/OL]. http://www.miit.gov.cn/newweb/n1146312/n1146904/n1648362/n1648363/c5466622/content.html. 2017-01-17
2. SIPO State intellectual briefing of patent [R/OL]. http://www.sipo.gov.cn/tjxx/zltjjb/201603/t20160323_1254508.html. 2016-03-02

3. Spencer E (2016) Ante, Ethnographic Hits, Business Week online [EB/OL]. http://images.businessweek.com/ss/06/05/ethnography/index_01.htm?chan=search
4. Project Group on Innovative Design Development Strategy (2016) Comprehensive report on innovative design development strategy. China Science and Technology Press
5. Ezio M, Zhong F (2016) Ma J. Trans. design, when everybody designs: an introduction to design for social innovation. Publishing House of Electronics Industry
6. Jonathan C, Craig MV, Xin XY, Pan L (2003) Trans. creating breakthrough products. Beijing: China Machine Press
7. Roberto V, Ma YJ Edit DS (2014) Trans. design driven innovation: changing the rules of competition by radically innovating what things mean. China Renmin University Press

Vehicle Fuel Consumption Prediction Based on Least Squares Support Vector Machine Optimized by Improved Particle Swarm Optimization Algorithm

Yu Liu, Mengliang Li, Yueyun Xu and Guo Jinwei

1 Introduction

As one of the main body of energy consumption, the automobile industry faces serious challenge with the increasing pressure of energy and environment. China has established the public system of the fuel consumption for light-duty vehicles in 2010, and clearly stipulates that the vehicle comprehensive fuel consumption must reach to the level of 5 L/100 km in 2020.

The vehicle comprehensive fuel consumption is obtained through type-approval test, and it is about 34% lower than the actual fuel consumption [1]. There are two main reasons for this difference: The first one is the test cycle; in China, the new European driving cycle (NEDC) is used for the type-approval test; however, the average speed, idle ratio, acceleration, and deceleration characteristics have significant differences with the actual operating characteristics of Chinese vehicles. The other reason is that the test procedures are inconsistent with the actual situation. To solve these problems, the Ministry of Industry and Information Technology (MIIT) has entrust China Automotive Technology Research Center to develop China automobile test driving cycles (CATC), which can fully reflect the actual operating conditions of Chinese vehicles, and it is planned that CATC and the new test procedures will be applied to the new fuel consumption test standards together. However, it will take a long time to develop CATC and put it into practice. Therefore, a method for predicting the actual fuel consumption of vehicles is very necessary before the implementation of CATC.

Tolouei [2], Ericsson [3], and Lapointe [4] have analyzed the main affected factors of vehicle fuel consumption from different points of view, but they only use a qualitative description. Wu [5] and Liu [6] have quantitatively predicted the

Y. Liu (✉) · M. Li · Y. Xu · G. Jinwei
China Automotive Technology & Research Center, Tianjin 300300, China
e-mail: liuyu2016@catarc.ac.cn

Society of Automotive Engineers of China (SAE-China) (ed.), *Proceedings of the 19th Asia Pacific Automotive Engineering Conference & SAE-China Congress 2017: Selected Papers*, Lecture Notes in Electrical Engineering 486, https://doi.org/10.1007/978-981-10-8506-2_75

vehicle fuel consumption with the neural network, but the neural network method has many inherent problems, such as over learning and lack of learning.

This paper summarized the vehicle operation characteristics and design parameters that have some influence on vehicle fuel consumption and determined the fuel consumption sensitive characteristics with the mutual information method. Then, with sensitive features and fuel consumption of type approve test as the input parameters, and the actual vehicle fuel consumption as output parameters, a vehicle fuel consumption prediction model based on least squares support vector machine (LSSVM) and improved particle swarm optimization algorithm (IPSO) [9] were established. Finally, the IPSO-LSSVM model was applied in the fuel consumption prediction of vehicles.

2 Correlation Analysis for the Influence Factors of Vehicle Fuel Consumption

2.1 Influence Factors of Vehicle Fuel Consumption

The vehicle operation characteristics and the design parameters have an important impact on fuel consumption. Operation characteristics include average speed and idle ratio, design parameters include auto-body parameters, engine parameters, and transmission system parameters. [7]. The auto-body parameters include the vehicle body structure and the vehicle weight; the engine parameters include the number of cylinders, the cylinder displacement, the cylinder arrangement, the air inlet mode, the maximum power, and the power output per litre; transmission system characteristic is the type of transmission. Table 1 summarizes the actual vehicle operation characteristics and the design parameters, and their quantitative results are also given.

2.2 Determine the Sensitive Influence Factors of Vehicle Fuel Consumption

In this section, the mutual information theory was used for feature sensitivity evaluation. For two discrete random variables X and Y, the mutual information is defined as follows:

$$MI(X;\ Y) = H(X) - H(X|Y) \tag{1}$$

where $H(\cdot)$ is the entropy, and $H(\cdot|\cdot)$ is the conditional entropy. The mutual information $MI(X;\ Y)$ represents the additional information about X provided by Y, so it can be used to reflect the correlation between measured variables.

Table 1 Quantitative results of vehicle actual operation characteristics and the design parameters

	Characteristics	Specific description
Operation characteristics	Average velocity f_1	Unit: km/h
	Idle ratio f_2	Unit: %
Auto-body parameters	Vehicle body structure f_3	2 stand for hatchback, 3 stand for sedan
	Vehicle weight f_4	Unit: kg
Engine parameters	Number of cylinders f_5	
	Engine displacement f_6	Unit: L
	Cylinder arrangement f_7	0 stand for V type arrangement, 1 stand for in-line arrangement
	Air inlet mode f_8	0 stand for naturally aspirated engines, 1 stand for supercharged engine
	Maximum power f_9	Unit: kw
	Power output per litre f_{10}	Unit: kw/L
Transmission system characteristic	Gearbox type f_{11}	1 stand for MT, 2 stand for AT 3 stand for DCT, 4 stand for CVT

Some papers state that the mutual information is biased in favor of features with more values. In addition, the values should be normalized to ensure that they are comparable [8]. Targeting these problems, symmetrical uncertainty (*SU*) is defined as follows:

$$SU(X,Y) = \frac{2 \times MI(X;Y)}{H(X)+H(Y)} \tag{2}$$

In this paper, the data of 300 light-duty vehicles (refer to Sect. 4 for specific) are selected as the research samples, and *SU* is used to measure the correlation between the parameters in Table 1 and actual vehicle fuel consumption. The result is shown in Fig. 1.

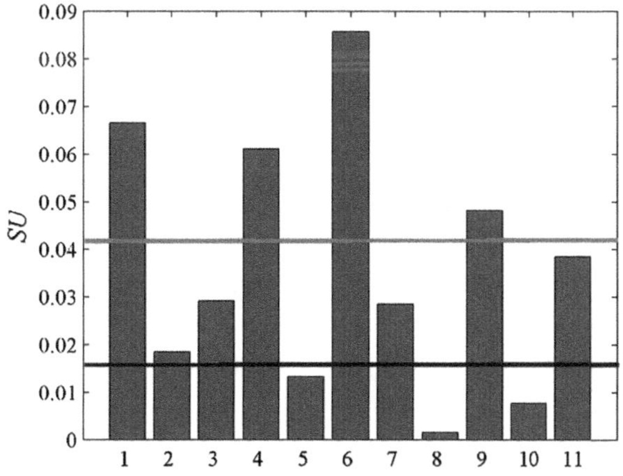

Fig. 1 Fuel consumption sensitive feature parameters selection by *SU*

Figure 1 shows that the average speed, engine displacement, vehicle weight, maximum power, and transmission types are the critical parameter that affects the vehicle fuel consumption; the vehicle body structure, idle ratio, and cylinder arrangement also have some influence on the fuel consumption; the other design parameters have less influence on the fuel consumption. Therefore, we can choose the sensitive operation characteristics and the design parameters to construct the vehicle fuel consumption prediction model.

3 Lssvm

Given a training sample set $\{\boldsymbol{x}_i, y_i\}_{i=1}^{N}$, where $\boldsymbol{x}_i \in R^n$ is the ith training sample and $y_i \in (0, 1)$ is the class label of $\boldsymbol{x}_i$. To class the training sample set, the LSSVM [9] approach aims at constructing the optimal separating hyperplane

$$y(\boldsymbol{x}) = \operatorname{sign}\left(\boldsymbol{w}^T \varphi(\boldsymbol{x}) + b\right) \tag{10}$$

where $\varphi(\cdot)$ is the nonlinear function, $\boldsymbol{w}$ is the weight vector, and b is the basic term.

The parameters $\boldsymbol{w}$ and b can be got by solving the optimization problem as follows:

$$\begin{aligned} &\min_{w,e} J(\boldsymbol{w}, \xi) = \frac{1}{2}\boldsymbol{w}^T\boldsymbol{w} + \frac{\gamma}{2}\sum_{i=1}^{N}\xi_i^2 \\ &s.t \quad y_i\left[\boldsymbol{w}^T\varphi(\boldsymbol{x}_i) + b\right] = 1 - \xi_i, \quad i = 1, \ldots, N \end{aligned} \tag{3}$$

where γ is the regularization parameter, and ξ_i is the nonnegative error for misclassification. The first term of Eq. (3) stands for the minimization of the Vapnik–Chervonenkis (VC) dimension, while the second term minimizes the training errors. Define the Lagrangian function as follows:

$$L(\boldsymbol{w}, b, \xi, \alpha) = J(\boldsymbol{w}, \xi) - \sum_{i=1}^{N} \alpha_i\left\{y_i\left[\boldsymbol{w}^T\varphi(\boldsymbol{x}_i) + b\right] - 1 + \xi_i\right\} \tag{4}$$

where α_i is the Lagrange multiplier, and only the α_i of support vectors equals to zero. To find the optimal solution, Eq. (4) should satisfy $\frac{\partial L}{\partial w} = 0$, $\frac{\partial L}{\partial b} = 0$, $\frac{\partial L}{\partial \xi_i} = 0$, and $\frac{\partial L}{\partial \alpha_i} = 0$. After elimination of $\boldsymbol{w}$ and $\boldsymbol{\xi}$, the following linear system is obtained:

$$\begin{bmatrix} 0 & \boldsymbol{Y}^T \\ \boldsymbol{Y} & \boldsymbol{Z}\boldsymbol{Z}^T + \gamma^{-1}\boldsymbol{I} \end{bmatrix} \begin{bmatrix} b \\ \boldsymbol{\alpha} \end{bmatrix} = \begin{bmatrix} 0 \\ \vec{\boldsymbol{I}} \end{bmatrix} \tag{5}$$

where $\boldsymbol{Z} = \left[\varphi(x_1)^T y_1; \ldots; \varphi(x_N)^T y_N\right]$, $\boldsymbol{\alpha} = [\alpha_1; \ldots; \alpha_N]$, $\boldsymbol{Y} = [y_1; \ldots; y_N]$ and $\vec{\boldsymbol{I}} = [1; \ldots; 1]$.

The kernel function is then used

$$K(\boldsymbol{x}_i, \boldsymbol{x}_j) = \varphi(\boldsymbol{x}_i)^T \varphi(\boldsymbol{x}_j) \tag{6}$$

Finally, the classification function of LSSVM is obtained, and the RBF kernel function is adopted as the kernel function of LSSVM.

$$y(\boldsymbol{x}) = \operatorname{sgn}\left[\sum_{i=1}^{N} \alpha_i y_i K(\boldsymbol{x}, \boldsymbol{x}_k) + b\right] \tag{7}$$

4 Parameter Optimization of LSSVM Based on IPSO

Two parameters have important influence on the performance of LSSVM. One is the regularization parameter γ, by which the trade-off between the training error minimization and smoothness of the classification function can be determined. The other is the kernel parameter σ, which has an important influence on the distribution of the data in high-dimensional space.

In recent decades, research on determining the best parameters of the LSSVM has been a hot topic. Commonly used parameter optimization algorithms include GA [10], particle swarm algorithm (PSO) [11]. GA was the most widely used parameter optimization algorithm in the past decades. However, its computation cost is relatively high for adopting the bit-string encoding strategy. In addition, GA often falls into the local optimal solution. The PSO algorithm simulates the bird flocks foraging, and it is easy to compute and is simple to implement. Nevertheless, it also suffers from the local optimal solution problem [12]. To get a satisfactory prediction model of LSSVM, in this section, a novel IPSO algorithm is proposed for the parameters optimization of LSSVM.

For an optimization problem in a D-dimensional space, the PSO algorithm begins with a group of random particles, and each particle is a possible solution. Then each particle adjusts its searching direction and velocity according to experiences of the particle population and its own experience, and this process will be repeated until a terminal condition is met. The velocity and position updating formulas are as follows [13]:

$$\boldsymbol{v}_i^{m+1} = w \cdot \boldsymbol{v}_i^m + c_1 \cdot r_1 \cdot (\boldsymbol{p}_i^m - \boldsymbol{x}_i^m) + c_2 \cdot r_2 \cdot (\boldsymbol{p}_g^m - \boldsymbol{x}_i^m) \tag{8}$$

$$\boldsymbol{x}_i^{m+1} = \boldsymbol{x}_i^m + \boldsymbol{v}_i^{m+1} \tag{9}$$

where i is the label particle; m is the number of iteration; $\boldsymbol{x}_i^m = (x_{i1}^m, x_{i2}^m, \ldots, x_{iD}^m)$ is the particle position; $\boldsymbol{v}_i^m = (v_{i1}^m, v_{i2}^m, \ldots, v_{iD}^m)$ is the particle velocity; $\boldsymbol{p}_i^m = (p_{i1}^m, p_{i2}^m, \ldots, p_{iD}^m)$ is the optimal position that the ith particle has reached; $\boldsymbol{p}_g^m = (p_{g1}^m, p_{g2}^m, \ldots, p_{gD}^m)$ is the optimal position that the population has reached; w is inertia weight factor; c_1 is the cognition learning factor; c_2 is the social learning factor; and r_1 and r_2 are two random numbers in the range [0, 1]. Previous studies have shown that the inertia weight factor w, which is used to control the impact of the previous velocity on current iteration, is a very important parameter for the PSO algorithm [11]. A large w can improve the global searching ability but with worse local searching ability and low convergence speed, while a relatively small w can improve the local searching ability and convergence speed but with more likelihood of falling into a local optimum. Here, a variable inertia weight is used to ensure strong global searching ability in the prophase of the whole searching process and strong local searching ability and high convergence speed in the anaphase of the whole searching process:

$$w_m = w_{\max} - \left(w_{\max} - w_{\min}\right) \cdot \left(1 - \mathrm{e}^{-(5m/t)}\right) \quad (10)$$

where m is the number of the current generation, t is the maximum number of generations, $w_{\max}$ and $w_{\min}$ are the maximum and minimum inertia weight factor, respectively.

5 Vehicle Operation Data Collection

In the experiment, the vehicle operation data of 300 light vehicles were collected by the free driving method (the drivers without any particular instruction to drive their cars), the acquisition time is from May 1, 2016, to January 31, 2017. The total mileage of this experiment is 1.5 million km.

The test system consists of two parts: vehicle data acquisition terminal (sampling frequency 1 Hz) and data management platform. The vehicle data acquisition terminal can collect the GPS signal and the onboard diagnostic system (OBD) signal according to the unified data protocol and transmit it to the data management platform through GPRS network; the data acquisition schematic diagram is shown in Fig. 2. The parameters that are used for fuel consumption prediction include GPS

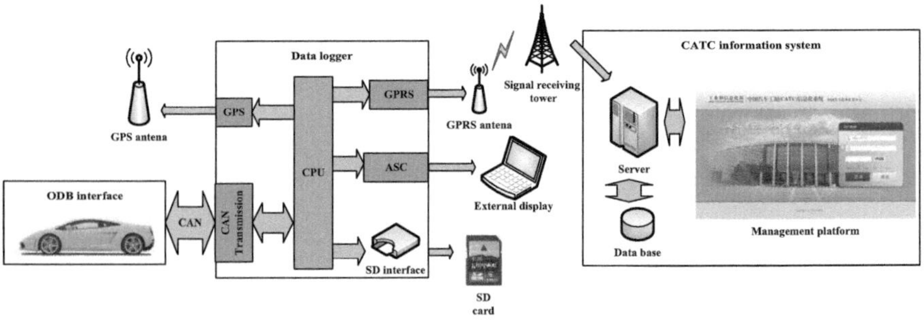

Fig. 2 Data acquisition schematic diagram

speed, engine speed, fuel consumption. The fuel consumption is calculated by the inlet mass flow rate and air–fuel ratio as follows:

$$\dot{m}(t) = q_m(t)/\lambda(t) \tag{11}$$

where $\dot{m}(t)$ is the instantaneous fuel consumption, $q_m(t)$ is the air intake instantaneous mass flow, $\lambda(t)$ is the air–fuel ratio. Part of the acquisition data and vehicle characteristic parameters is shown in Table 2.

6 Vehicle Fuel Consumption Prediction Model Based on IPSO-LSSVM

In order to establish the mapping relationship between the fuel consumption sensitive characteristics and the actual fuel consumption, the nonlinear prediction method LSSVM is introduced in this section. Figure 3 shows the flowchart of the proposed fuel consumption prediction method.

Table 2 Part of the acquisition data and vehicle characteristic parameters

Samples	1	2	3	4	5	6	7	8
Average velocity (km/h)	42.04	34.18	33.48	29.43	33.37	30.61	37.7	31.94
Idle ratio (%)	17	18	31	31	27	24	21	20
Displacement (L)	2	1.8	1.5	1.4	2	2.5	3	1.6
Inlet type	0	1	0	0	0	0	0	0
Number of cylinders	4	4	4	4	3	4	6	4
Cylinder arrangement	1	1	1	1	1	1	0	1
Maximum power (kw)	114	118	96	86	123	120	199	89
Power/litre (kw/L)	57	65.56	64	53.75	61.5	48	66.33	55.63
Vehicle weight (kg)	1499	1520	1058	1180	1475	1766	1999	1380
Vehicle body structure	3	3	2	2	3	3	3	3
Gearbox type	4	1	1	1	2	2	2	1
Type-approval fuel consumption (L/100 km)	7	7.9	5.7	6.5	7	9.6	10.7	6.9
Actual fuel consumption (L/100 km)	7.97	9.12	7.00	13.46	11.24	12.7	13.47	9.42

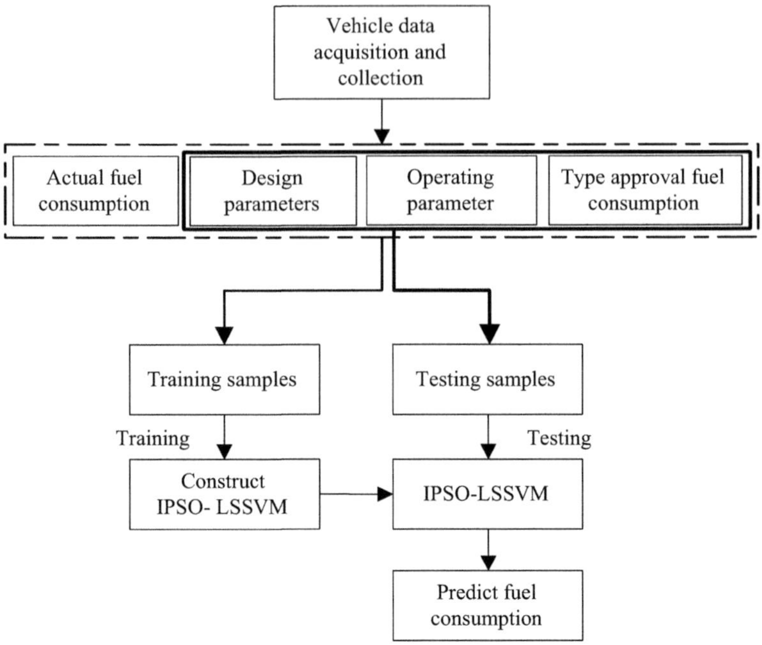

Fig. 3 Flowchart of the proposed fuel consumption prediction method

6.1 Construct the Prediction Model

The data of 200 vehicles were randomly selected from the experiment data to train the IPSO-LSSVM, and the remaining data of 100 vehicles were used to test the trained model. The input parameters of the model include: average speed, engine displacement, vehicle weight, maximum power, transmission types, the vehicle body structure, idle ratio, cylinder arrangement, and the fuel consumption of type approve test.

The average prediction error rate was selected as the fitness function of IPSO algorithm. The searching range of γ and σ was set as (0, 100]. After implementing a number of different trials on the same processor, the number of individuals was set as 10, the maximum number of generations was set as 100, and the learning factors c_1 and c_2 were set as 1.5 and 1.7, separately. The range of inertia weight factor was set as [0.4, 0.9]. The fitness curve of IPSO algorithm is given in Fig. 4.

6.2 Prediction Results Analysis

After the IPSO-LSSVM model trained, the rest testing samples were tested with the trained model and the results are given in Fig. 5. Besides, in order to further verify

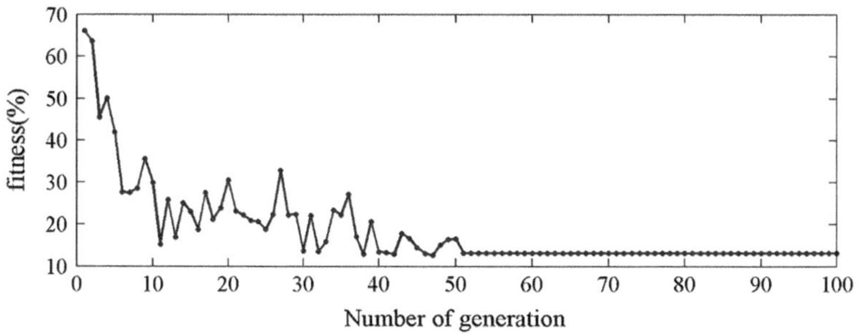

Fig. 4 Average fitness curve of IPSO-LSSVM

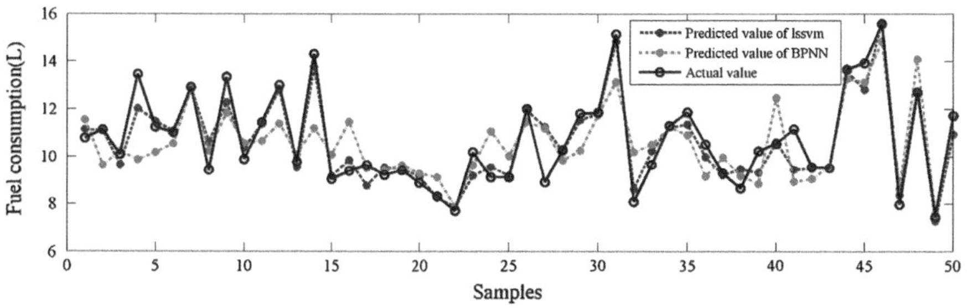

Fig. 5 Part of the prediction results

the effectiveness of the fuel prediction model, the BPNN method was selected to replace the IPSO-LSSVM model in fuel prediction. There are 9 input nodes, 10 hidden nodes, and 1 output node in BPNN; the learning was 0.1, the maximum iteration was 100, and the target error was 0.0004. The prediction results of BPNN and the actual fuel consumption value are also given in Fig. 5.

From Fig. 5, we can find that the prediction accuracy of IPSO-LSSVM is obviously higher than BPNN. The average relative error of IPSO-LSSVM is 4.2%, and the average relative error of BPNN is 9.2%. Therefore, it can propose fuel prediction model in this paper is better than the traditional BPNN method, and it can be used to describe the mapping relationship between vehicle running characteristics, design parameters, and actual fuel consumption characteristics. The proposed fuel prediction model is very suitable for the vehicle fuel consumption prediction.

7 Conclusions

(1) This paper summarized the vehicle operation and design parameters, which are directly related to the fuel consumption, and *SU* is used to select the sensitive parameters of fuel consumption. The results show that the average speed,

engine displacement, vehicle weight, maximum power, and transmission types are the critical parameter that affects the vehicle fuel consumption; the vehicle body structure, idle ratio, and cylinder arrangement also have some influence on the fuel consumption;

(2) The advanced nonlinear prediction methods LSSVM is introduced to predict the vehicle fuel consumption, and an IPSO algorithm is used to optimized the kernel function parameter of LSSVM;

(3) The proposed fuel consumption prediction model can accurately predict the actual fuel consumption and can effectively make up for the gap between the actual fuel consumption and fuel consumption of type approve test.

Acknowledgements We gratefully acknowledge the research and development of China new energy vehicle driving cycle development project and also thanks for the driver's permission to using their driving data.

References

1. http://www.cnautonews.com/xw/201702/t20170221_523057.htm
2. Tolouei R, Titheridge H (2009) Vehicle mass as a determinant of fuel consumption and secondary safety performance. Transp Res Part D Transp Environ 14(6):385–399
3. Ericsson E (2001) Independent driving pattern factors and their influence on fuel-use and exhaust emission factors. Transp Res Part D Transp Environ 6(5):325–345
4. LaPointe C (1973) Factors affecting vehicle fuel economy. SAE Technical Paper
5. Wu JD, Liu JC (2012) A forecasting system for car fuel consumption using a radial basis function neural network. Expert Syst Appl 39(2):1883–1888
6. Wu JD, Liu JC (2011) Development of a predictive system for car fuel consumption using an artificial neural network. Expert Syst Appl 38(5):4967–4971
7. Cheng XJ, Han QL, Quan CG (2016) Vehicle fuel consumption forecast models of optimized BP neural network based on grey relation and genetic algorithm. 46(8):43–51
8. Yu L, Liu H (2004) Efficient feature selection via analysis of relevance and redundancy. J Mach Learn Res 5:1205–1224
9. Suykens JAK, Vandewalle J (1999) Multiclass least squares support vector machines. In: Neural networks, 1999. IJCNN'99. International joint conference on IEEE, 1999, 2, pp 900–903
10. Xie Z, Shepard S Jr, Woodbury KA (2009) Design optimization for vibration reduction of viscoelastic damped structures using genetic algorithms. Shock Vibration 16(5):455–466
11. Xu H, Chen G (2013) An intelligent fault identification method of rolling bearings based on LSSVM optimized by improved PSO. Mech Syst Signal Process 35(1):167–175
12. Ardia D, Boudt K, Carl P et al (2011) Differential Evolution with DEoptim: An application to non-convex portfolio optimization. R J 3(1):27–34
13. Eberhart RC, Kennedy J (1995) A new optimizer using particle swarm theory. Proc Sixth Int Symp Micro Mach Hum Sci 1:39–43

Production and Engine Performance and Emission Evaluation of Karanja and Jatropha-Based Biodiesel

Patel Akashkumar, Chougule Nagesh, J. M. Babu, Jain Aatmesh and K. C. Vora

1 Introduction

The depletion of fossil fuels due to over-usage and many serious environmental threats associated with burning fossil fuels forced many researchers to find ways of using alternative sources of energy instead of conventional fuels such as gasoline and diesel. There are many alternative fuels available, and among them, biodiesel seems very promising due to various technical and non-technical advantages: It is highly biodegradable and very low toxic in nature. Blends of biodiesel can be used with existing diesel engines without major modifications and only minimal decrease in performances is reported, almost zero emissions of sulfates and aromatic compounds, a small increment in carbon dioxide (CO_2) when the whole life cycle is considered (including cultivation, production of oil and conversion to biodiesel). It also supports rural economic in country like India [1–4] Jatropha and Karanja are non-edible oils and are abundantly available in India. These oils can easily be converted into biodiesel using one-step or two-step transesterification process, depending on their FFA content. Biodiesel is defined as mono-alkyl esters of long chain fatty acids derived from vegetable oils or animal fats and alcohol with or without a catalyst. Compared to diesel fuel, biodiesel produces no sulfur, no net

P. Akashkumar · C. Nagesh · J. M. Babu · J. Aatmesh (✉) · K. C. Vora
ARAI Academy, The Automotive Research Association of India (ARAI), Pune 411038, India
e-mail: aatmeshjain@gmail.com; jain.pga@araiindia.com

C. Nagesh
College of Engineering Pune, Pune, India

J. M. Babu
Veltech Dr. RR & Dr. SR University, Chennai, India

J. Aatmesh · K. C. Vora
The Automotive Research Association of India, Pune, India

Society of Automotive Engineers of China (SAE-China) (ed.), *Proceedings of the 19th Asia Pacific Automotive Engineering Conference & SAE-China Congress 2017: Selected Papers*, Lecture Notes in Electrical Engineering 486, https://doi.org/10.1007/978-981-10-8506-2_76

carbon dioxide, less carbon monoxide, particulate matters, smoke, and hydrocarbons emission. Moreover, biodiesel is itself oxygenated fuel which further helps in the complete combustion of fuel and reduced emissions [5, 6].

2 Biodiesel Production Process and Experimental Procedure

2.1 Oil Expelling

Motorized oil expeller setup was used to expelling oil from jatropha and karanja seeds as shown in Fig. 1. It consists of a screw which rotated with the help of a motor. The screw crushes and chews the seeds which result in expelling of the oil from the seeds. The shell part of jatropha seeds is comparatively hard, and thus, it causes excess friction between seed-cake paste and screw. Due to this excess friction, the expeller may choke. Dehulling of jatropha seeds was done before the expelling in order to reduce the effort and chocking of expeller. Dehulled seeds are known as kernals. The optimum dehulling ratio is found to be 70/30% (30% seeds with 70% kernals) for motorized oil expelling setup [7, 8]. As karanja seeds are comparatively soft in nature, they can be used to extract the oil, without dehulling. The preheat treatment of the seeds before expelling shows significant effect on its oil color. Also, preheating reduces the effort of expelling.

Effect of preheat treatment of Jatropha and Karanja seeds at 50 and 60 °C on quantity of oil extracted is compared with sundried condition of seeds in Table 1.

Fig. 1 Motorized operated oil expeller used for expelling oil from vegetable seeds

Table 1 Oil expelling process parameters of jatropha seeds

Parameter	Jatropha seeds			Karanja seeds		
	Sundried	50 °C	60 °C	Sundried	50 °C	60 °C
Weight of seeds (kg)	1	1	1	1	1	1
Oil extracted (ml)	280	255	250	270	260	240

2.2 *Titration*

Titration process is used to find out the free fatty acid (FFA) content present in the oil. In stoichiometric transesterification process, oil does not contain any free fatty acids (FFAs). But in actual scenario, oil contains some FFA in it. Oils having FFA value less than 2% can be trans-esterified directly. However, if FFA content of oil is higher than 2%, oil must be neutralized first before undergoing transesterification process [9]. In that case, acid pretreatment is suggested before transesterification process. Direct transesterification of oils, having FFA content higher than 2%, will result in reduction of biodiesel yield. So, it is necessary to first check the FFA content present in oil by following titration process.

2.2.1 Titration Procedure

Titration is a process of neutralizing unknown amount of acids with known amount of base in the presence of neutral alcohol solvent. Steps followed to obtain FFA% of oil, in this study, are as follows:

1. Prepare a titration solution by adding 1gm of KOH in 1 L of distilled water.
2. Take 10 ml of isopropyl alcohol in a beaker and add 1 ml of oil in it.
3. Add 4–5 drops of phenolphthalein solution in that beaker.
4. Add drop by drop titration solution in it and measure the quantity of titration in ml solution required to change the color of mixture to pink.

Divide the quantity of titration in ml required to titrate the solution by 1.8. This will give approximate FFA% of oil. This process also helps to find out amount of KOH required during transesterification process. For a sample of 1 L of oil, whatever amount of titration solution is required to neutralize the acid present in 1 ml of oil add equal amount of extra KOH in grams with stoichiometric amount of KOH [9].

Titration process is used to find out FFA values of extracted jatropha and Karanja oil. It is observed that FFA value of jatropha oil is less than the FFA value of karanja oil. Table 2 shows the FFA values of extracted jatropha and karanja oil.

Table 2 FFA % value of extracted Jatropha and Karanja oil

	Jatropha	Karanja
FFA % obtained from Titration	1.51	3.77
Further process required	Transesterification	Pre-esterification

2.3 *Pre-esterification Process (FFA >2%)*

When FFA content of oil is greater than 2%, it is always suggested to do pre-esterification process which helps to reduce it below 2%. By neutralizing oil with acid as a catalyst, FFA content can be reduced. In this process, oil is reacted with methanol in the presence of acid catalyst H_2SO_4, D5081, or A46. H_2SO_4 is used as acid catalyst in the present study. Methanol to oil ratio has been taken as 6:1 and heated to 55 °C for 1 h in the presence of H_2SO_4 and is stirred continuously. Reaction of the same is as follows (Fig. 2) [10, 11].

Pre-esterification process has been used for Karanja oil as its FFA % is higher than 2%. After pre-esterification process, FFA % of Karanja oil has been reduced to 1.7%. Table 3 shows the FFA value of Karanja oil after pre-esterification process.

2.4 *Transesterification Process*

Transesterification process can be used to convert oil into biodiesel with the help of alcohol and catalyst. It is the process of exchanging the organic group R″ of an ester with the organic group R′ of an alcohol. It consists of many consecutive reversible reactions. In these reactions, the triglyceride is converted stepwise into diglycerides, then monoglyceride and finally into glycerol, being heavier, sinks to the bottom and biodiesel, being lighter, floats on the top of glycerol. Catalyst is usually used to improve the reaction rate and biodiesel yield. Since the transesterification reactions are reversible, excess amount of alcohol is used to shift the equilibrium to the products side. Methanol and ethanol are most commonly used alcohols for trans-esterification process, especially methanol because of its low cost and its physical and chemical advantages. Currently, homogeneous alkaline catalysts are usually favored over acid catalysts because of the higher reactivity and milder process conditions such as the lower temperature required. After transesterification of triglycerides, the products are a mixture of esters, glycerol, alcohol, catalyst and tri-,

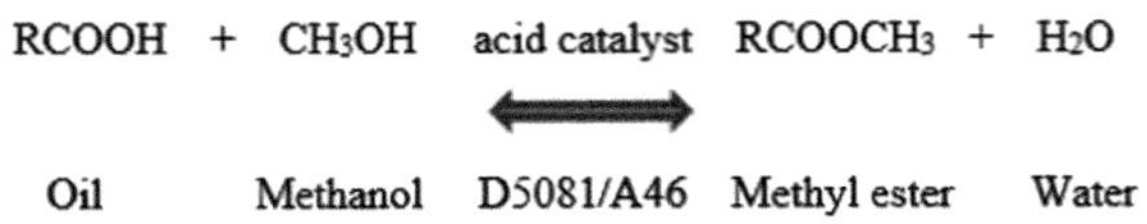

Fig. 2 Pre-esterification process reaction

Table 3 FFA % of Karanja oil after pre-esterification process

	Karanja
FFA % obtained after pre-esterification process	1.7
Further process required	Transesterification

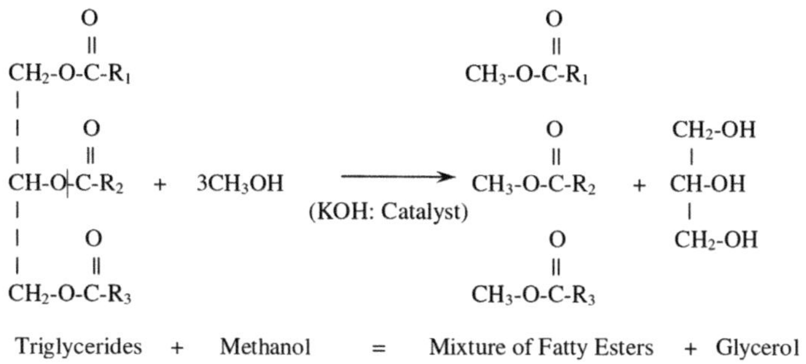

Fig. 3 General chemical reaction of transesterification process

di- and monoglycerides. Transesterification is one of the best methods among other approaches due to its low cost and simplicity. Transesterification reaction is shown in Fig. 3 [12, 13].

In the present study, various process parameters like reaction temperature, reaction duration, stirring speed, amount of catalyst, and volume of methanol were optimized for Karanja and jatropha oil. Oil is reacted with methanol in the presence of base catalyst KOH. Methanol to oil ratio has been taken as 6:1 and heated to 60 °C for 2 h in the presence of KOH and is stirred continuously. The final products obtained are biodiesel and glycerol. Biodiesel being heavier appears on the top and glycerol stays on the bottom. Obtained biodiesel is separated from glycerol manually. Further, glycerol is taken to separate it from excess methanol and glycerine is obtained.

Optimized process parameters for jatropha and karanja biodiesel are shown in Table 4. Figure 4 shows karanja and jatropha biodiesel samples with glycerol at

Table 4 Optimized process parameters for Jatropha and Karanja oil

	Jatropha oil	Karanja oil
FFA %	1.51	3.77
Process	Single-step transesterification	Two-step transesterification
Materials required/Optimized parameters	• 100-ml oil • 1 gm KOH • 25-ml methanol • 60 °C reaction temperature • 90-min reaction time	*First Step* • 100-ml oil • 1-ml H_2SO_4 • 33-ml methanol • 55 °C reaction temperature • 90-min reaction time *Second Step* • 100-ml oil • 1-gm KOH • 30-ml Methanol • 60 °C Reaction temperature • 60-min Reaction time
Biodiesel yield efficiency (%)	95	86

Fig. 4 **a** Jatropha biodiesel with glycerol **b** karanja biodiesel with glycerol

bottom. The optimized parameters were used for large quantity production of biodiesel in 5 litres capacity per batch biodiesel pilot plant.

2.5 *Washing and Heating*

Washing of biodiesel is necessary to separate extra methanol, catalyst, and soap suspended in the fuel. For washing, 2–3 times of water is required to wash the biodiesel and stirred. After washing and settling, the water along with the impurities has been drained from the bottom of the reactor. 3–4 wash cycles are recommended to make biodiesel clean. Heating is necessary after washing to remove remaining water from the biodiesel. Biodiesel is heated above 100 °C for the same.

After washing and heating, physiochemical properties of the jatropha and karanja biodiesel were evaluated as per the ASTM standards, as shown in Table 6.

3 Experimental Validation

To evaluate the performance and emission characteristics of obtained biodiesel, experiments were conducted on a three-cylinder, naturally aspirated, water-cooled, tractor DI diesel engine as per 8-mode emission test cycle. The engine specifications and experimental setup is shown in Table 5 and Fig. 5, respectively.

The experimental test setup consists of a three-cylinder diesel engine, of which specifications are listed in Table 5, eddy current dynamometer and electronic data acquisition and controls with various automatic measuring systems. Equipment setup is capable to measure the performance parameters like brake specific fuel consumption (BSFC) and brake thermal efficiency (BTE) at full throttle as well as part throttle position at various engine speeds.

Tests were conducted with diesel fuel to establish a performance and emission baseline values. Tests were then conducted using obtained Jatropha and Karanja

Table 5 Test Engine Specification

Naturally aspirated water-cooled DI diesel engine	
Max Power output	50 HP @ 2150 rpm
Max Torque	180 Nm @ 1200 rpm
Bore and stroke	108 mm and 120 mm
Compression ratio	18.5:1
Connecting rod length	208 mm
Engine cubic capacity	3.3 l
Inlet valve diameter	48 mm
Maximum valve lift	10.92 mm

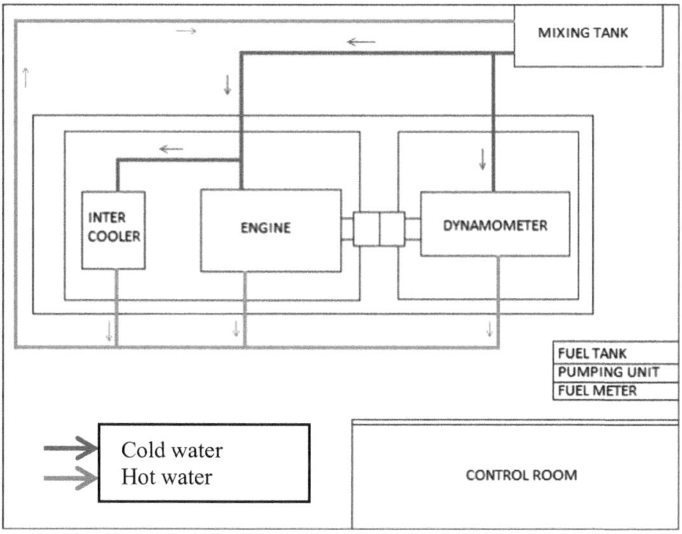

Fig. 5 Schematic diagram of test setup

biodiesel blended with diesel fuel in the blend ratio of B10 and B20 [14]. Eight mode test cycle from ISO 8178 Type C1 standard is adopted for the engine emission test [15] (Refer Fig. 6).

4 Results and Analysis

4.1 Fuel Properties

Various important properties of biodiesel are evaluated as per ASTM test procedures and compared with IS-15607 biodiesel standards. All the properties of biodiesel tested gives a very satisfactory result as all are conforming to the IS 15607 2005 Standard.

Table 6 Properties of Jatropha and Karanja biodiesel

Properties	Test method	IS 15607 2005	JB	KB	Diesel
Density at 15 °C (kg/m^3)	IS 1448 (P:32)	860-900	876.7	888.4	833
Kinematic viscosity @ 40 °C (cSt)	ASTM D445	2.5–6	3.849	5.521	2.246
Sulfur content	ASTM D129	50 max	16 (ppm)	0.11 (% by wt.)	–
Water content (% by Vol)	ASTM D95	0.05 max	<0.05	<0.05	–
Flash point (°C)	ASTM D93	120 min	134	166	70
Copper corrosion	ASTM D130	No 1	No. 1	No. 1	No. 1
Acid number (mg KOH/g)	ASTM D664	0.50 max	<0.50	<0.50	<1
Carbon residue (% mass)	ASTM D524	0.05 max	0.022	0.020	0.015
Sulphated ash (% mass)	ASTM D874	0.02 max	0.01	0.01	0.01
Gross calorific value (Cal/g)	ASTM D4868	–	9660	9780	10,432
Cetane index	ASTM D4737	–	61	57	52

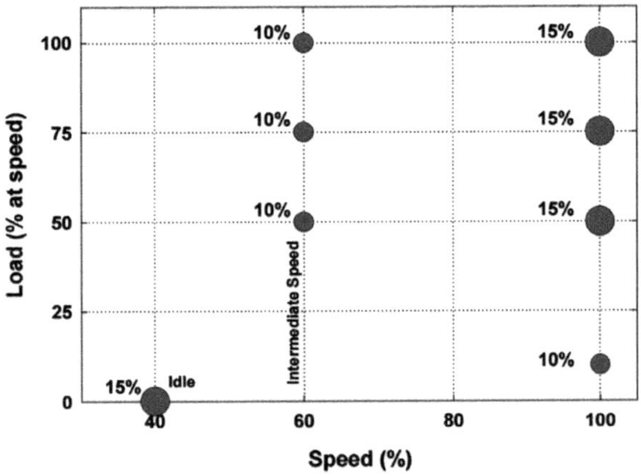

Fig. 6 ISO 8178 test cycle with weighting factors [9]

The calorific value of jatropha and karanja biodiesel is lower than that of diesel. Oxygen content present in biodiesel is higher than diesel value. The self-oxygenated ability of biodiesel helps in better and more complete combustion of fuel in the engine. The flash point of biodiesel is less than biooil, but it is still higher than that of diesel. Thus, it is safer to store biodiesel–diesel blends as compared to diesel alone.

4.2 *Engine Performance and Emissions*

Figure 7 shows percentage change in CO, HC, NO_x, BSFC and power with various blends of jatropha and karanja biodiesel as compare to diesel. Negative sign indicates reduction in value and positive sign indicates increment in value. Maximum CO emission reduction has been observed for JB20 blend of biodiesel. Maximum HC emission reduction has been observed for KB20 blend of biodiesel. Maximum NO_x emission reduction has been observed with JB10 blend of biodiesel. Maximum engine break power has been observed with KB20 blend of biodiesel (Table 7).

4.2.1 Engine Performance Parameters

Brake-specific fuel consumption (BSFC), power and break thermal efficiency (BTE) are considered as the engine performance indicator.

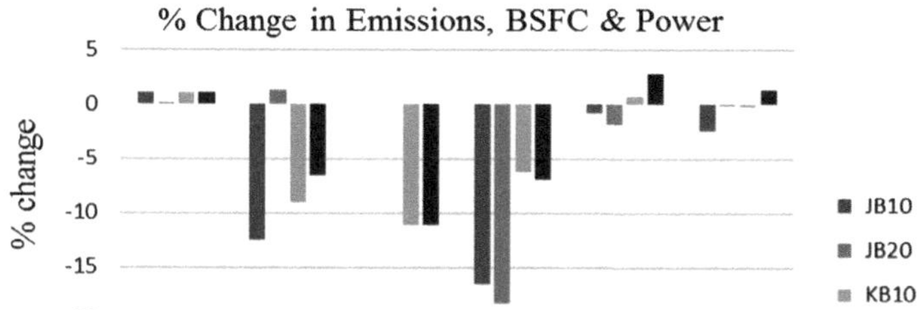

	POWER	BSFC	HC	CO	NOx	CO2
JB10	1.140549	-12.4767	0	-16.5519	-0.81901	-2.3887
JB20	0.118807	1.303538	0	-18.2815	-1.80019	-0.03147
KB10	1.072235	-8.95717	-11.1111	-6.22593	0.771164	-0.20794
KB20	1.143519	-6.57356	-11.1111	-6.93444	2.790302	1.364263

Fig. 7 Change in emissions and performance for various biodiesel blends

Table 7 Overall performance parameters and engine emissions values according to eight-mode engine test cycle for different biodiesel blends

	Diesel	JB10	JB20	KB10	KB20
Power (kW)	16.834	17.026	16.854	17.014	17.026
BSFC (kg/kW-hr)	0.537	0.470	0.544	0.4889	0.5017
HC (g/kW-hr)	0.018	0.018	0.018	0.016	0.016
CO (g/kW-hr)	2.70	2.2531	2.2064	2.5319	2.5127
NO_x (g/kW-hr)	12.332	12.231	12.11	12.427	12.676
CO_2 (g/kW-hr)	746.85	729.01	746.61	745.29	757.03

1. Break Power
   Figure 8 illustrates the engine brake power at different mode numbers. There is overall small amount of increment in break power has been observed for bio-diesel blends with compare to diesel fuel. At lower speed, all the fuels show same trend because diesel contains a larger portion of all the fuels and time available for combustion is relatively higher and the combustion temperature is lower. However, at higher speed, time available for combustion is very short and the combustion temperature is relatively high. At this stage, cetane number of the fuel plays a key role. Moreover, higher viscosity of biodiesel, which enhances fuel spray penetration, and thus improves air-fuel mixing which explains recovery in power for biodiesel relative to diesel. With this, higher power indicates better and complete combustion of fuel.
2. BSFC
   Figure 9 shows the BSFC of the blends of the biodiesels and diesel at different mode number. With increase in load, BSFC of biodiesel decreases. One probable reason for this trend could be the higher percentage increase of break power

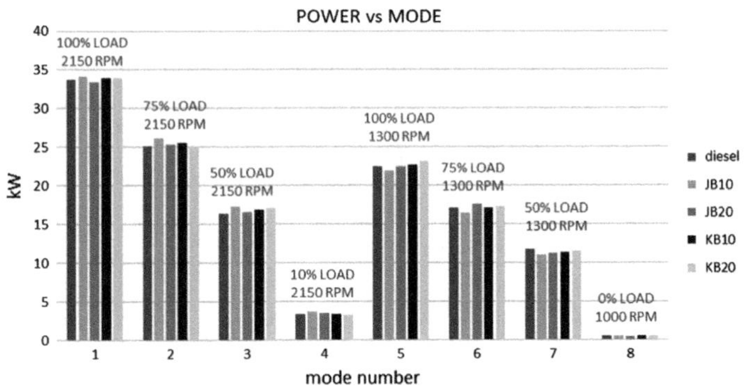

Fig. 8 Variation of power with mode number

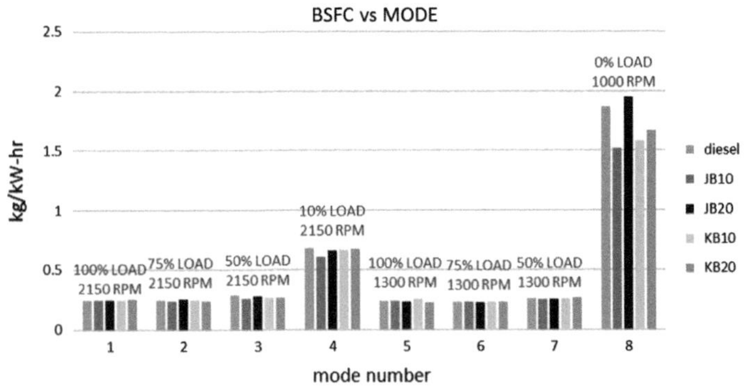

Fig. 9 Variation of BSFC with mode number

for biodiesel blends as compared to percentage increase in fuel consumption with load. Moreover, at higher speed, higher combustion temperature and additional oxygen content of biodiesel facilitate better combustion and reduces BSFC (compare mode 4 and 8).

3. Break Thermal Efficiency
   The brake thermal efficiency characteristics at different mode numbers are as shown in Fig. 10. As break thermal efficiency is the ratio of energy output (break power) to the energy input (product of mass of fuel consumption and CV), higher break power, lower fuel consumption, and lower calorific value of jatropha and karanja biodiesel explains higher break thermal efficiency for all the biodiesel blends.

4.2.2 Engine Emission Parameters

1. HC
   The results for HC emission are shown in Fig. 11. HC emission is negligible in case of diesel engine. At higher load condition (100 and 75%), jatropha biodiesel shows higher hydrocarbon emission then the diesel fuel. Lower exhaust gas temperature at higher load can explain this increment in HC emissions. Lower HC emissions have been observed at lower load because of complete combustion (higher AFR). Overall, HC emission for jatropha biodiesel is almost equal to diesel.
   It has been observed that the hydrocarbon emission is lower for all blends of karanja biodiesel. Higher exhaust gas temperature and higher oxygen contained in karanja biodiesel explains this reduction in HC emissions (complete combustion).

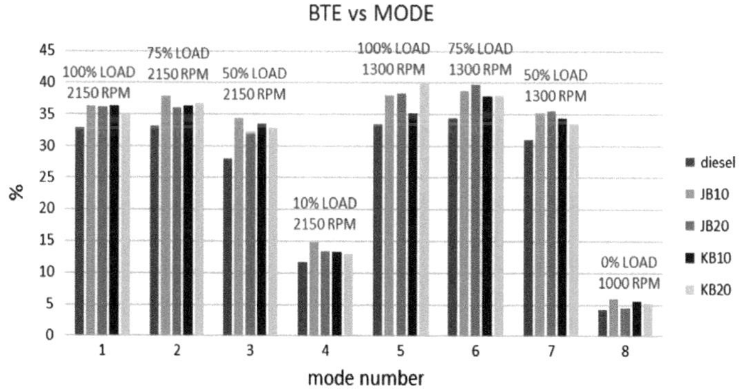

Fig. 10 Variation of BTE with mode number

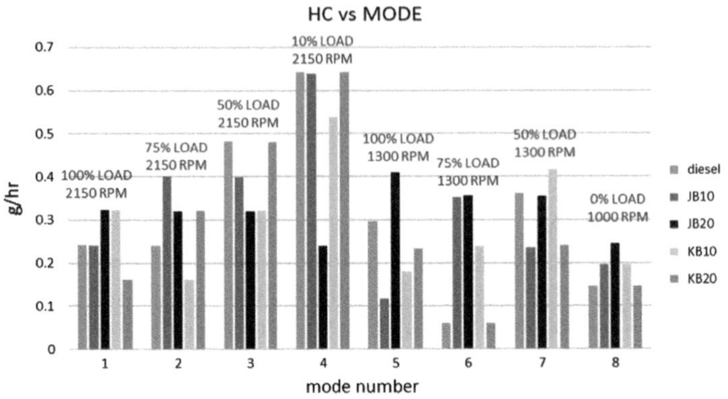

Fig. 11 Variation of HC emissions with mode number

2. CO
   The CO exhaust emission for different biodiesel blends at different mode numbers is shown in Fig. 12. It has been observed that overall CO emissions are lower for all the biodiesel blends with compare to diesel fuel. At mode number 1 and 2, because of higher speed and higher load (incomplete combustion), there is not significant reduction in CO emissions has been observed for all the blends of jatropha and karanja biodiesel. But for all the other modes, lower CO emissions have been observed (complete combustion) for all the blends of jatropha and karanja biodiesel. Emission of CO is greatly dependent on the AFR relative to the stoichiometric proportion. This explains the trend of CO emissions of all the biodiesel blends.
3. NO_x
   The NO_x emission characteristics are shown in Fig. 13. It has been observed that overall NO_x emissions for jatropha biodiesel blends are lower than the pure diesel case. Incomplete combustion at higher load (lower exhaust gas temperature) explains this trend of jatropha biodiesel blends.

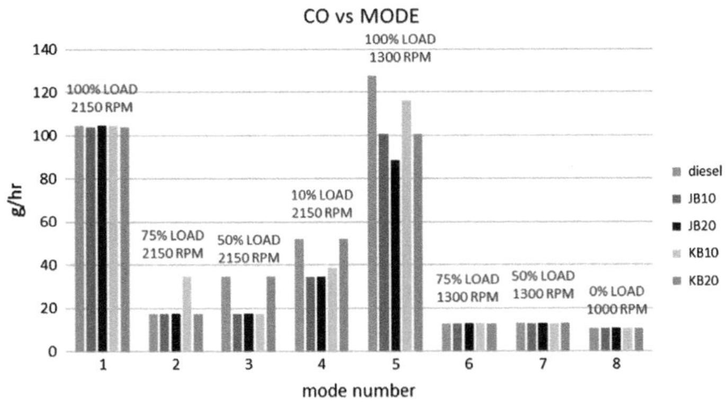

Fig. 12 Variation of CO emissions with mode number

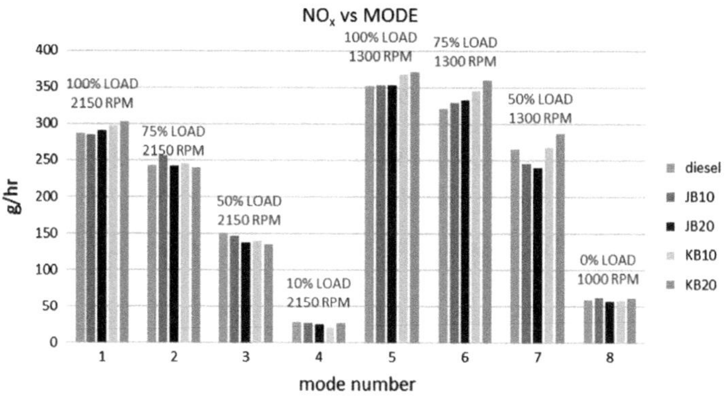

Fig. 13 Variation of NO_X emissions with mode number

Overall, NO_x emissions with karanja biodiesel blends have been observed higher than the diesel fuel. At higher load, because of more oxygen contained in biodiesel (around 10% more) and higher temperature, NO_x emissions are higher for karanja biodiesel blends.

Higher NO_x emissions have been observed at lower speed because of larger time available for NO_x formations. NO_x emissions have been strongly depending on the peak cylinder temperature.

4. CO_2

The results for CO_2 emission are shown in Fig. 14. For all the biodiesel blends except KB20, the overall CO_2 emission was lower than diesel. Higher oxygen contained in biodiesel can explain this reduction in CO_2 emissions. With KB20 blend, there is slight increment in CO_2 emissions which has been observed.

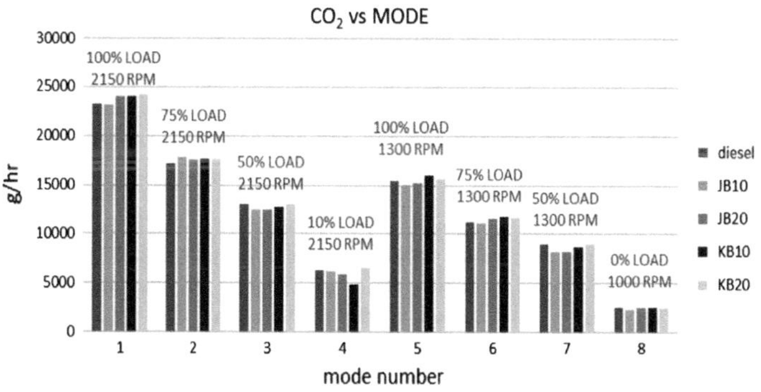

Fig. 14 Variation of CO_2 emissions with mode number

5 Conclusion

Process for biodiesel production from jatropha and karanja seeds has been optimized to produce optimized quality biodiesel. Results show that biodiesel from Karanja and Jatropha feedstocks can be used in CI engine up to B20 blend, as an alternative to conventional diesel fuel.

- 270 and 280 ml of oil are extracted from 1 kg of sundried karanja and sundried jatropha seeds respectively.
- Titration of oil is the best way to find out FFA contain of oil.
- For Jatropha oil, single-step transesterification process is recommended as FFA content of oil was less than 2%. For 100 ml of oil, 25 ml of methanol and 1gm KOH is suggested for quality biodiesel production.
- For Karanja oil, two-step transesterification process is used. Acid esterification followed by transesterification process is followed as FFA content of oil was 3.5%. For 100 ml of oil, 33 ml methanol and 1 ml H_2SO_4 was used for acid esterification (first step). After that 29-ml methanol and 1gm KOH was used for transesterification process.
- Among JB10 and JB20 biodiesel blends, JB10 is better as it reduces 16.55% CO emission, 0.8% NOx emission and 2.38% CO_2 emission with 12.47% reduction in BSFC. Also, 1.14% increment in power has been observed with JB10 biodiesel blend.
- Among KB10 and KB20 biodiesel blends, KB10 is better as it reduces HC emission by 11%, CO emission by 6.22%, with only 0.77% increment in NO_x emission. Also, reduction in BSFC is 8.95% with 1.02% increment in power output. Higher power is obtained with KB20 blend at the cost of higher emission.
- Among JB10 and KB10 blends, JB10 is proved to better fuel as CO and NOx emission are lower than KB10. Power and BSFC are also higher with JB10 blend. However, HC emission is lower in KB10 blend.
- In general, the emission characteristics like CO, HC, and CO_2 were observed lower at all engine load conditions compared to diesel fuel. NO_x emission was observed slightly higher than diesel fuel. Performance was found much better at higher loads.

Acknowledgements The authors are grateful to the ARAI, Pune (The Automotive Research Association of India) for providing oil expelling and biodiesel production facility and for sponsoring the whole project. Authors are also grateful to VELTECH University, Chennai, for providing engine testing facility for research work. Authors are also grateful to the team of Automotive Materials Laboratory, ARAI to help us in testing for fuel properties as per standards.

References

1. Dufey A Biofuels production, trade and sustainable development: emerging issues. Environmental economics program sustainable markets group. ISBN: 978-1–84369-643-8. September
2. Nag AS, Bhattacharya MS, DE KB (1995) Utilization of vegetable oils. J Am Oil Chemist Soc 72:1591–1593
3. Keith O (2000) A review of Jatropha Curcas: an oil plant of unfulfilled promise. Biomass Bioenerg 19(1):1–15
4. Alam R Studies on the properties of karanja oil for probable industrial application. Master's thesis, National institute of technology, Rourkela
5. Panigrahi N, Mohanty MK, Pradhan AK (2012) Non-edible Karanja biodiesel-a sustainable fuel for C.I. engine. Int J Eng Res Appl 2(6):853–860
6. Chauhan B, Kumar N, Haeng M, Chang L (2013) A study on the performance and emission of a diesel engine fueled with Karanja biodiesel and its blends. Energy 56:1–7
7. Aatmesh J, Vinod SS, Vora KC (2016) Optimisation of parameters for the production of biodiesel from Jatropha Oil. SAE Technical Paper 2016-28-0144, 2016. https://doi.org/10.4271/2016-28-0144
8. Darshan J, Sarvanan K, Vora KC (2015) Parametric optimization for biodiesel production from Jatropha Curcus. SAE Technical Paper 2015-26-0047, 2015. https://doi.org/10.4271/2015-26-0047
9. Mettler T Basics of titration
10. Amit S Biodiesel production and properties. International standard book number: 978-1-84973-470-7
11. Gajendra babu MK, Subramanian KA Alternative transportation fuels. International Standard Book Number-13: 978-1-4398-7281-9. Chapter 5, pp 232–244
12. Kurl M Biodiesel production. International standard book number: 978-1-63240-078-9. Chapter 1, pp 12–17
13. Kumar R, Sharma M, Ray SS, Sarpal AS, Gupta AA, Tuli DK, Sarin R (2004) Biodiesel from Jatropha Curcas and Pongamia Pinnata. SAE Technical Paper 2004-28-0087, 2004
14. Biodiesel Blending Guide by BC Transits Canada. BBI Biofuels International–Canada. 2008
15. Indian standard reciprocating internal combustion engines, exhaust emission measurement part 4 steady-state test cycles for different engine applications. Bureau of Indian standards. ISO 8178-4: 2007

Experimental and Simulation Study on Vehicle Ride Comfort and Parameter Optimization

Junnan Chen, Jiqing Chen, Fengchong Lan, Jun Fang and Dexing Liu

1 Introduction

With the improvement of living standards, automobile manufacturers, researchers, and consumers are increasingly demanding on the vehicle ride comfort. The automobile is a complex multi-mass vibration system, which causes complex vibration of the system due to the impact of the road injustice, the inertial force during acceleration or deceleration, and the exciting force generated by the vibration of the engine and the drive shaft, resulting in car ride comfort down [1]. The average vehicle speed and transport efficiency of the vehicle with poor ride comfort will be reduced. The dynamic load generated by vibration accelerates the wear of the parts and reduces the service life of the automobile. Vibration will also cause additional energy consumption so that the poor economic performance of the car [2].

How to improve car ride comfort, how to optimize those factors that affect the car ride comfort, these are the actual need to focus on solving the problem. Nearly a decade of development of the virtual incentive method, it will be a smooth response to the calculation of simple response to the calculation of simple response. It simplifies the actual calculation of stochastic vibration theory by converting the calculation of nonstationary stochastic response into ordinary stepwise integral calculation [3–5]. Shen Xiaoan and others studied the vehicle ride comfort model. They established two-degree-of-freedom single-wheel model, four-degree-of-freedom monorail model and seven-degree-of-freedom vehicle model according

J. Chen · J. Chen · F. Lan · J. Fang · D. Liu
School of Mechanical & Automotive Engineering, South China University of Technology, Guangzhou 510641, China
e-mail: tong_fang@126.com

J. Chen · J. Chen (✉) · F. Lan · J. Fang · D. Liu
Key Laboratory of Automotive Engineering of Guangdong Province, South China University of Technology, Guangzhou 510641, China
e-mail: chjq@scut.edu.cn

Society of Automotive Engineers of China (SAE-China) (ed.), *Proceedings of the 19th Asia Pacific Automotive Engineering Conference & SAE-China Congress 2017: Selected Papers*, Lecture Notes in Electrical Engineering 486, https://doi.org/10.1007/978-981-10-8506-2_77

to stochastic vibration theory. By calculating the model, it was found that the total weighted acceleration root mean square of the three models is very close to each other, which verified that the method of ride comfort analysis with a single or monorail model was feasible [6]. Peng Xian and others established a three-mass vehicle-car system based on the vertical dynamic model of the vehicle. They used the road simulation as the input model, the use of virtual excitation method for the variable speed of the vehicle vibration analysis. By calculating the instantaneous power spectrum of the nonstationary response of the vehicle, the vehicle ride comfort of the variable speed was investigated [7]. Li Jie and others using the 1/4-vehicle 2-DOF vibration model, using virtual excitation method, constructing the virtual pavement excitation, deduced the frequency response characteristic of the system, and then established the power spectral density formula of the system vibration response [8]. Xia Jianman was based on multi-body system dynamics, established a vehicle model in the ADAMS environment and carried out dynamic simulation. In addition, they also discussed the impact of vehicle suspension system parameters on vehicle ride comfort. Through the analysis of the test and simulation results, the optimal design parameters of the vehicle suspension system were determined and the simulation analysis was carried out to achieve the purpose of improving the vehicle ride comfort [9]. Wu Yingxi and others analyzed the vehicle ride comfort based on the ten-degree-of-freedom vehicle model. They simulated the vibration characteristics of the vehicle by MATLAB/Simulink and discussed the influence of the tire stiffness and the mounting rigidity of the power assembly on the ride comfort [10]. Zhang Liping and others built the eight-degree-of-freedom vehicle model. Under the condition that the road excitation was unknown and the wheel vertical vibration response was known, the vertical acceleration of the body, the dynamic deflection of the suspension and the dynamic load of the wheels were calculated by the inverse virtual excitation method. The ride comfort of the road surface and the unknown road surface were compared by using the MATLAB calculation. The results showed that it was feasible to study the ride comfort of the vehicle by using the inverse virtual excitation method on the unknown road surface. The analytical method did not need to construct the pavement incentive model, which was convenient for popularization and application [11].

However, the above research is more focused on theoretical research, and it is not convenient for engineering application. Therefore, based on the three-degree-of-freedom system, the virtual incentive method is used to establish the road surface incentive model, and the ride comfort model is tested, simulated and optimized to improve the vehicle ride comfort.

2 Theoretical Analysis

The essence of studying vehicle ride comfort is to study its vibration problem. Road roughness and speed are the inputs of the automotive vibration system, while automobile tires, suspension, seats, spring mass and sprung mass constitute the vehicle vibration system. After the input of the vibration system is processed, the spring mass or the vibration acceleration of the body, the dynamic deflection of the suspension and the dynamic load of the wheels are output. For the use of the environment, road roughness and vehicle speed have been determined, so the main purpose of comfort research is dynamic characteristics of output characteristics and system design of the control system, to ensure that in the common road and speed, the driver and passengers comfort can be at a reasonable level or above.

2.1 Virtual Incentive Method

Virtual excitation method is to use simple harmonic excitation instead of smooth random excitation, through the dynamic analysis, to solve the frequency response function, so as to establish the link between the incentive and response. In addition, the virtual incentive method is widely used in the field of random vibration research, which is simple and efficient compared with the traditional algorithm [12, 13].

For a linear vibration system with only one stationary random excitation source, when the power spectrum density $G_{xx}(\omega)$ of the stationary random excitation is known, the virtual excitation x can be constructed:

$$x = \sqrt{G_{xx}(w)}\mathrm{e}^{iwx} \tag{1}$$

Let the response be z, the power spectral density is $G_{xx}(\omega)$, the transfer function is $H(w)$, and the response z can be expressed as:

$$z = \sqrt{Gzz(w)}|H(w)|\mathrm{e}^{iwx} \tag{2}$$

According to the virtual excitation method, the relationship between the response power spectral density and the excitation power spectral density is:

$$G_{zz}(w) = |H(w)|^2 G_{xx}(w) \tag{3}$$

2.2 Pavement Incentive Model

According to the virtual incentive method, the road surface excitation model is established. GB/T 7031-2005 mechanical vibration road pavement spectrum measurement data report recommended road displacement power spectral density $G_q(n)$ with the following formula (see formula 4) fitting:

$$G_q(n) = G_q(n_0)(n/n_0)^{(-W)} \tag{4}$$

$G_q(n)$ Road surface power spectral density, also known as road roughness coefficient;
n Spatial frequency, unit: m^{-1}, indicates the number of wavelengths per meter length;
n_0 Reference spatial frequency $n_0 = 0.1\,m^{-1}$
W Frequency index General value $W = 2$.

When the car with a certain speed v (unit: m/s) through the spatial frequency of n road roughness, the input time frequency f and the spatial frequency n relationship is:

$$f = v \cdot n \tag{5}$$

According to the frequency range and the calculated vehicle speed, the range of the spatial frequency can be obtained, and the power spectral density of the corresponding road surface excitation can be obtained. The relationship between the displacement power spectral density and the pavement roughness power spectral density is:

$$G_q(f) = G_q(n)/v \tag{6}$$

Pavement incentives are a function of time; therefore, the virtual road excitation can be constructed as follows according to the displacement power spectral density:

$$q(t) = \sqrt{G_q(f)} \cdot e^{j2\pi ft} \tag{7}$$

$$\dot{q}(t) = j2\pi f \cdot \sqrt{G_q(f)} \cdot e^{j2\pi ft} = j2\pi f \cdot q(t) \tag{8}$$

$$\begin{aligned} \ddot{q}(t) &= j2\pi f \cdot j2\pi f \cdot \sqrt{G_q(f)} \cdot e^{j2\pi ft} \\ &= -(2\pi f)^2 \cdot q(t) \end{aligned} \tag{9}$$

2.3 *Calculation Model of Ride Comfort and Calculation of Vibration Response*

In order to facilitate the analysis of ride comfort theory, automotive vibration systems generally need to be simplified. At present, the car around the suspension of the quality of close, you can approximate that the car around the suspension of vibration is not linked to each other. If you further ignore the left and right rutting differences, you can simplify the vehicle model to single-wheel model. Ride comfort is mainly concerned with the low-frequency vibration, modeling generally ignore the high-frequency engine excitation and aerodynamic incentives [14].

The seat stiffness is nonlinear and the seat stiffness approximation is calculated as a constant. When driving at constant speed, the vibration in the *X* and *Y* directions is much smaller than that in the *Z* direction. Therefore, only the *Z* direction should be taken into account in modeling [15].

In this paper, the seat system is added on the basis of the commonly used two-degree-of-freedom single-wheel ride comfort model, and the vibration system of the target model is simplified as the constant-parameter linear system of the human body-sprung mass-reed mass three degrees of freedom (Fig. 1) to carry out the study of ride comfort, and follow-up vibration response calculation will be based on this model for theoretical analysis.

m_1 *Suspension unsprung mass*
m_2 *Suspension sprung mass*
m_3 *The seat body load*
k_1 *Tire radial stiffness*
k_2 *Suspension spring stiffness*
k_3 *Seat stiffness*
c_2 *Shock absorber damping*
q *The road height*

According to the mathematical model established in Fig. 1, Newton's second law is applied to establish the differential equation of motion:

$$\begin{cases} m_3\ddot{z}_3 + k_3(z_3 - z_2) = 0 \\ m_2\ddot{z}_2 + c_2(\dot{z}_2 - \dot{z}_1) + k_2(z_2 - z_1) = k_3(z_3 - z_2) \\ m_1\ddot{z}_1 + k_1(z_1 - q) = c_2(\dot{z}_2 - \dot{z}_1) + k_2(z_2 - z_1) \end{cases} \quad (10)$$

According to the stochastic vibration theory, for a linear system subjected to a smooth random excitation, the frequency response is a function of the dynamic characteristics of the system in the frequency domain. Setting the frequency response of the system is *H*_ (*z*1 − *q*) (*f*), *H*_ (*z*2 − *q*) (*f*), *H*_ (*z*3 − *q*) (*f*); for zero mean stationary random road input, we can calculate the frequency response function (see Eqs. 11, 12, 13, 14, 15, 16, 17):

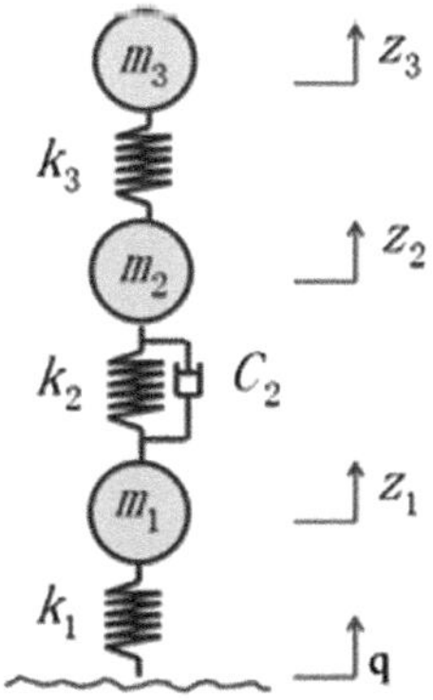

Fig. 1 Target vehicle ride comfort model

$$H_{z1-q}(f) = \frac{k_1 k_3^2 - k_1 A_2 A_3}{k_3^2 A_1 - A_1 A_2 A_3 + A_3 A_4^2} \tag{11}$$

$$H_{z2-q}(f) = \frac{k_1 A_3 A_4}{k_3^2 A_1 - A_1 A_2 A_3 + A_3 A_4^2} \tag{12}$$

$$H_{z3-q}(f) = \frac{k_1 k_3 A_4}{k_3^2 A_1 - A_1 A_2 A_3 + A_3 A_4^2} \tag{13}$$

Among them:

$$A_1 = -(2\pi f)^2 \cdot m_1 + j2\pi f \cdot c_2 + k_1 + k_2 \tag{14}$$

$$A_2 = -(2\pi f)^2 \cdot m_2 + j2\pi f \cdot c_2 + k_2 + k_3 \tag{15}$$

$$A_3 = -(2\pi f)^2 \cdot m_3 + k_3 \tag{16}$$

$$A_4 = -j2\pi f \cdot c_2 - k_2 \tag{17}$$

2.4 Method for Evaluating Ride Comfort of Target Vehicle

The evaluation of ride comfort can refer to the GB/T 4970-2009 vehicle ride test method, which measures a total of 9 lines of vibration in the three directions of the hip, back, and foot parts *X*, *Y*, and *Z*. Generally, recommended body sitting posture vibration model is shown in Fig. 2.

The weighted acceleration root mean square is to calculate the acceleration signal in vibration direction and the sensitivity of human body to different vibration frequencies. To study the effect of vibration on the comfort of the human body, the total weighted acceleration root mean square values of the seat cushion, seat back, and foot floor are generally evaluated:

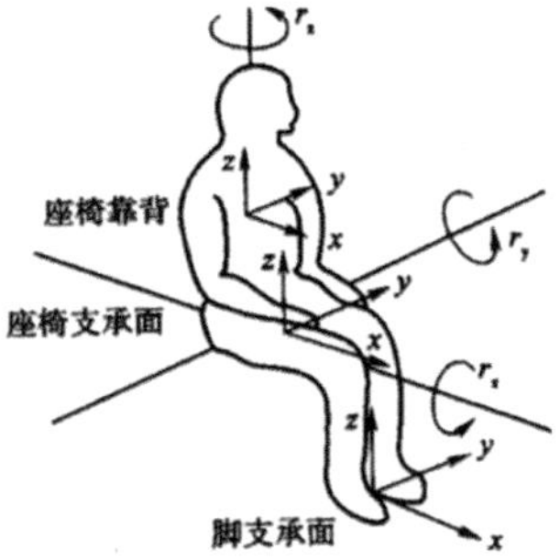

Fig. 2 Vibration model of human sitting posture

$$\bar{a}_v = \left(\bar{a}_{v1}^2 + \bar{a}_{v2}^2 + \bar{a}_{v3}^2\right)^{\frac{1}{2}} \tag{18}$$

Since the root mean square range of the total weighted acceleration range is large, in order to facilitate the description, the general use of the total weighted vibration level as a car at a speed of the ride comfort evaluation indicators:

$$L_{aw} = 20\,\lg\frac{\overline{a_v}}{a_0} \tag{19}$$

L_{aw} Total weighted vibration level, unit: decibel (dB)
a_0 Reference acceleration, $a_0 = 10^{-6}$ m/s^2

Car ride comfort is expressed by the relationship between the total weighted vibration level and the vehicle speed curve. In the MATLAB environment, the data processing program is programmed to calculate the total weighted acceleration root mean square value and convert to the total weighted vibration level.

Taking the speed as the *X*-axis and the total weighted vibration stage as the *Y*-axis, the curve is drawn, which is the evaluation result of ride comfort.

3 Test Scheme and Numerical Model

The ride comfort test of the target vehicle model can confirm the ride comfort level of the target vehicle and the accuracy of the numerical model. Determined the test plan, selected the appropriate instrument, you can carry on the ride test of the target vehicle type. The corresponding data are processed by MATLAB, and the total weighted acceleration root mean square values of each speed are plotted into curves with speed, and the final experimental results of ride comfort are obtained.

3.1 Test Plan

Car test is generally carried out in the automotive test site; according to the ride test requirements, select the performance test road test. GB/T 7031-2005 Appendix C: General description of road spectrum using statistical parameters, given the road classification, is generally A, B, C, D, the corresponding degree of roughness were 16×10^{-6} m^3, 64×10^{-6} m^3, 256×10^{-6} m^3, 1024×10^{-6} m^3. The domestic expressway is mainly in the B, C grade, the test site performance road in the A level.

Before the test, we should make sure that the test vehicle is in good working condition, and then install the sensor, the test equipment, hardware, and software settings. After confirming the equipment normal, we can adjust the parameters according to the provisions of loading. Finally, we should confirm whether the weather conditions meet the test requirements, the relevant data to be recorded archive. The test shall be carried out in an automotive test site.

Measurement of the acceleration of the sensor, according to the working principle can be divided into piezoelectric, piezoresistive, capacitive, inductive, and photoelectric. In general, we often use piezoelectric acceleration sensor. It has a wide range of measurement frequency, large range, small size, lightweight, ease of installation, and other advantages, widely used in the field of vibration testing.

The sensitivity of the acceleration sensor is critical to obtaining accurate and reliable test data. Therefore, the sensitivity of the acceleration sensor should be calibrated using a dedicated acceleration calibrator before testing. Select the calibrator to match the sensor. It is recommended to use the model recommended by the sensor supplier.

According to the sampling theorem, the sampling frequency of the data collector should be determined according to the highest frequency of the signal to be analyzed by the vibration test when selecting the data collector. For the ride test, the sampling frequency of the data collector should be greater than 180 Hz.

The models used in the test are shown in Table 1.

According to the human body by the vibration model, and the ride test requirements, we need to measure the human body buttocks and back of the acceleration. In order to facilitate the installation, while not affecting the normal driving operation, we need to produce the acceleration sensor fixture, as shown in Figs. 3 and 4.

According to the national standard for the sensor layout requirements, the test should be arranged in 6 axis acceleration sensor, taking up 18 channels of data acquisition. In addition, the front windshield of the car should install speed display. Finally, we connect the data collector and car power, connect the computer, start the software, set the hardware and software, the initial confirmation of system work. The connection of the test system is shown in Fig. 5.

After the calibration of the sensor, the experimental data can be collected. Respectively, at different speed conditions 40, 50, 60, 70, 80, 90, 100, 110,

Table 1 Test vehicle parameters

Models	Class a car
Engine capacity	1497 mL
Maximum power	77 kW
Maximum power speed	5800 rpm
Maximum torque	136 N·m
Maximum torque speed	4800 rpm
Front suspension type	*Mcpherson-independent suspension*
Rear suspension type	*H torsion beam type semi-independent rear suspension*
Curb weight	1022 kg
Full weight	1463 kg
Maximum speed	178 km/h
Front track	1455 mm
Rear track	1455 mm

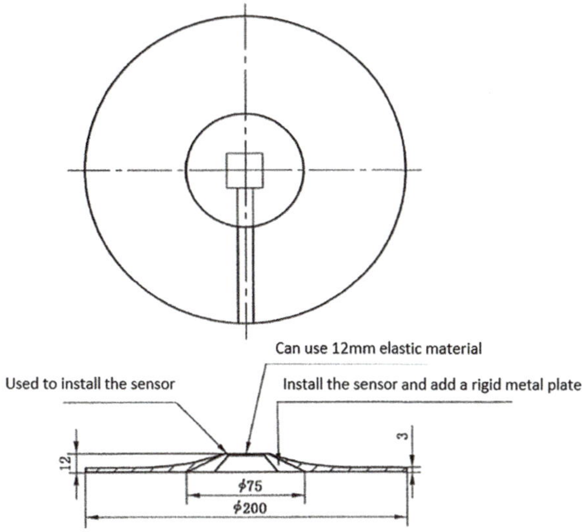

Fig. 3 Cushion sensor fixture diagram

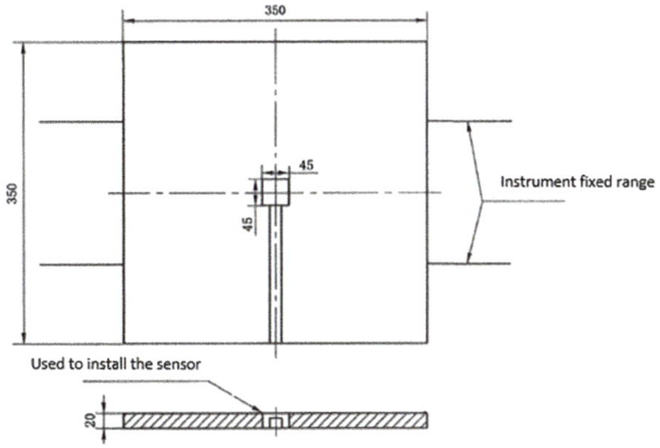

Fig. 4 Backrest sensor fixture diagram

120 km/h, in the driver and the crew, we measure *X*, *Y*, *Z* three-way foot, cushion, backrest acceleration value. Then, the self-power spectrum calculation, the frequency weighting calculation, and the shaft weight calculation are carried out to obtain the final total weighted acceleration root mean square value.

(1) MATLAB using Welch method to calculate the power spectral density of the syntax is as follows:

[Pxx, f] = pwelch(x, window, noverlap, nfft, fs)
Pxx: Results of power spectral density vector f: Frequency vector corresponding to power spectral density
x: Random signal sequence window: Window function options
noverlap: The number of overlapping data of random signal sequence

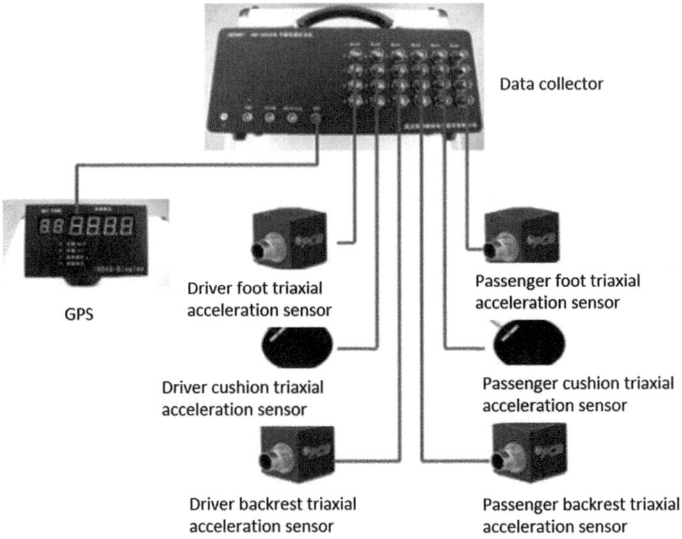

Fig. 5 Experimental instrument connection diagram

nfft: The number of data calculated by Fourier transform
fs: The frequency of the random signal sequence

(2) Calculate parameters

Fs = 512(Accorded to the test data sampling frequency to determine)
Window = *hann*(4096) (*According to the minimum segment* 8 s *to determine*)
Nfft = 4096 (*According to the minimum segment* 8 s *to determine*)
Noverlap = 2048 (*By default overlap rate* 50%)

(3) Calculation results

Calculate the frequency range returned by the result: 0 ~256 Hz; frequency resolution: 0.125 Hz;
Frequency point: 2049.
Power spectral density: the power spectral density of the frequency point is calculated in the frequency range.

3.2 Numerical Model

The calculation model is shown in Fig. 1.According to the formula (11), the formula (12), the formula (13) analysis results, The input/output relationship of the power spectrum density of the target model is shown in formula (20), formula (21) and formula (22):

$$G_{(\ddot{Z}_1)}(f) = \left|H_{(\ddot{Z}_1-q)}(f)\right|^2 G_q(f) = (2\pi f)^4 \cdot \left|H_{(Z_1-q)}(f)\right| G_q(f) \tag{20}$$

$$G_{(\ddot{Z}_2)}(f) = \left|H_{(\ddot{Z}_2-q)}(f)\right|^2 G_q(f) = (2\pi f)^4 \cdot \left|H_{(Z_2-q)}(f)\right| G_q(f) \tag{21}$$

$$G_{(\ddot{Z}_3)}(f) = \left|H_{(\ddot{Z}_3-q)}(f)\right|^2 G_q(f) = (2\pi f)^4 \cdot \left|H_{(Z_3-q)}(f)\right| G_q(f) \tag{22}$$

$G_{(\ddot{Z}_1)}(f), G_{(\ddot{Z}_2)}(f), G_{(\ddot{Z}_3)}(f)$, respectively, represent the acceleration power spectral density of the foot Z direction, the cushion Z direction, and the backrest Z direction. According to the above formula, we can calculate the $G_{(\ddot{Z}_1)}(f), G_{(\ddot{Z}_2)}(f), G_{(\ddot{Z}_3)}(f)$, and then calculate the ride comfort level.

The calculation program comply with MATLAB, calculating the calculation speed by using MATLAB, changing the calculated vehicle speed, calculating the power response spectrum of human body under the vehicle speed of 40, 50, 60, 70, 80, 90, 100, 110, 120 km/h, respectively, then perform frequency weighting and axis weighting, and finally calculate the total weighted acceleration mean square root response to vibration in the driver. The calculation parameters are used in Table 2.

4 Analysis and Optimization of Results

4.1 *Result Analysis*

First, it is necessary to check the calculation model established by the numerical analysis method by comparing the test value with the simulation value. Figure 6 shows the comparison result between the test value and the simulation value.

Table 2 Calculation parameters of the numerical model

Parameter	Name	Design value
m_3	The seat body load	62.0 kg
m_2	Suspension sprung mass	272.5 kg
m_1	Suspension unsprung mass	80.0 kg
k_3	Seat stiffness	8.0 kN/m
k_2	Suspension spring stiffness	20 kN/m
k_1	Tire radial stiffness	1000 kN/m
c_2	Shock absorber damping	3750 N·s/m

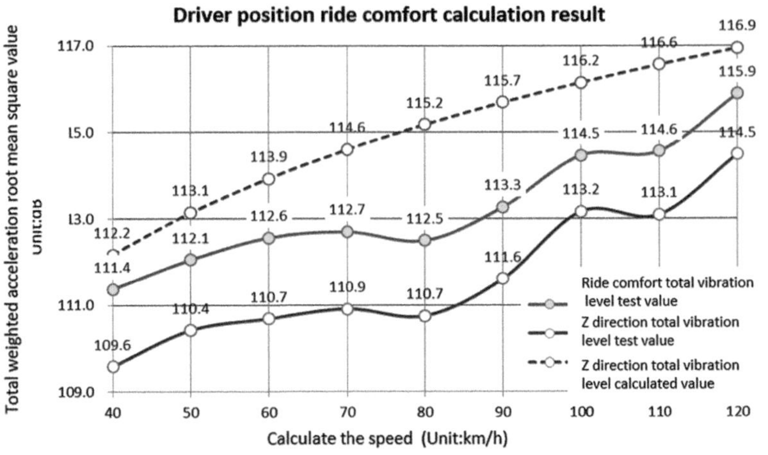

Fig. 6 Driver position ride comfort calculation result

From Fig. 6, we can see: Compared with the test value, the calculated value is high, and there are differences in the trend of the speed of the vehicle; but from the order of magnitude, the calculated results and test results in the same range. Therefore, it can be considered that the calculated model has sufficient computational accuracy, and the model parameters can be optimized based on the simulation results.

4.2 *Ride Comfort Model Optimization*

The results show that the target vehicle ride comfort and vehicle speed is approximately linear. The higher the speed, the worse the ride. In the model analysis, the ride comfort at 120 km/h speed can be selected as the research object, and its research results are instructive to optimize the ride comfort at other speeds. The parameters that affect the ride comfort of the vehicle mainly include the quality parameters, the shock absorber parameters, the tire and seat parameters and the ride comfort model. The method of changing the variables by MATLAB can optimize the ride comfort model.

Figure 7 shows the effect of changing the quality parameters on vehicle ride comfort. From the lightweight performance, quality parameters can be decreased by 30%; from the emission control requirements, quality parameters can go up 10%. According to the above range, the variability index with the same quality parameter is used as the reference value, and the variation of the ride comfort index is calculated by MATLAB. The calculation result is shown in Fig. 7.

It can be seen from Fig. 7 that the ride comfort of this model is beneficial to reduce unsprung mass for each reduction of 5%, the average ride comfort index down 0.1 dB; ride comfort is beneficial to increase sprung mass for each additional

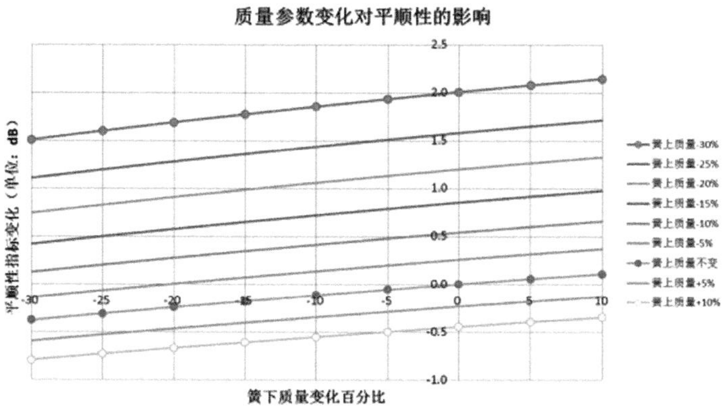

Fig. 7 Influence of quality parameters on ride comfort

of 5%, the average ride comfort index down 0.3 dB; this model uses the adjustment of quality parameters to optimize the ride comfort, and ride comfort indicators can be reduced by up to 0.8 dB or so.

Figure 8 shows the effect of the damper parameter change on the car ride comfort. In order to observe the effect of the damper parameter on the ride comfort, the variation range of the stiffness and damping of the shock absorber is set to −50 ~ +50%, and the ride comfort parameter of the damper parameter is used as the reference value. The change degree of ride comfort index is calculated by MATLAB and the results are shown in Fig. 8.

From Fig. 8, we can see that this model shock absorber damping reduced, favorable for ride comfort, each reduced by 10%, the average ride comfort to reduce the average 0.5 dB; each spring stiffness reduced by 10%, the average ride reduces the average 0.1 dB; this model to adjust the shock absorber parameters to optimize the ride comfort, ride comfort indicators can be reduced by up to 3.2 dB or so.

Figure 9 shows the impact of tire and seat parameter changes on vehicle ride comfort. In order to observe the influence of the radial dynamic stiffness parameter

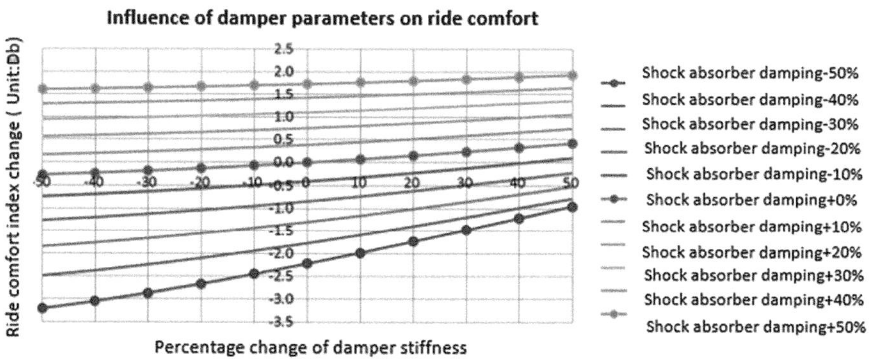

Fig. 8 Influence of damper parameters on ride comfort

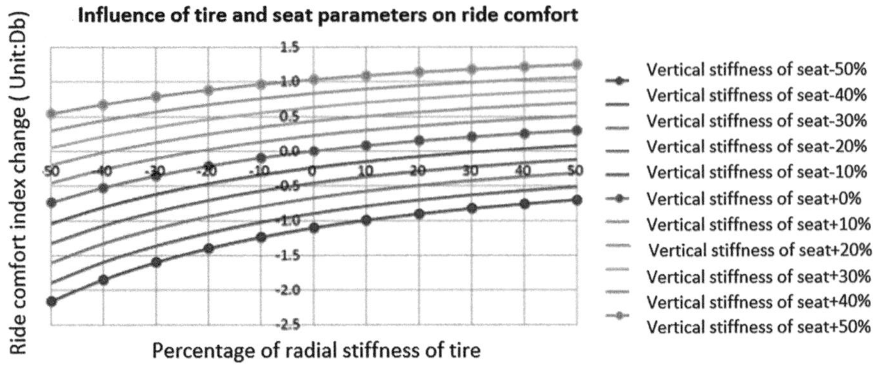

Fig. 9 Influence of tire and seat parameters on ride comfort

and the variation of the vertical stiffness parameters on the ride comfort, the variation range of the tire radial dynamic stiffness parameter and the seat vertical stiffness parameter is set from −50 to +50%. Based on the treadness of the tire radial dynamic stiffness parameter and the vertical stiffness parameter of the seat as the reference value, the variation of the ride comfort index is calculated by MATLAB. The calculation result is shown in Fig. 9.

Figure 9 shows that this model tire radial stiffness reduction, favorable for ride comfort, each reduced by 10%, the average ride comfort index reduced by 0.2 dB; this model seat vertical stiffness reduction, favorable for the ride, each reduction small 10%, the ride comfort down the average 0.2 dB; in this model, the ride comfort is optimized by adjusting the radial stiffness of the tire and the vertical stiffness parameters of the seat, and the ride comfort index can be reduced by up to 2.2 dB.

According to the result of the above analysis, because the quality of the changes in power, emissions, fuel consumption, noise, control and other major performance and development time and cost are larger, the effect of optimization is not suitable as an optimization program. Tires and seats as important general-purpose components; the design is very mature, generally do not do custom development, while this is not conducive to vehicle manufacturers to reduce costs, and therefore not suitable as an optimization program. Therefore, it is the best solution to adjust the suspension parameters for ride comfort optimization. Considering the best effect of adjusting the damper damping to ride comfort, the damper damping parameters are adjusted as the final scheme of ride comfort optimization. To adjust the damper damping parameters, we must consider the suspension dynamic $f_d(\mathrm{t})$, defined as follows:

$$f_d(t) = Z_2(t) - Z_1(t) \tag{23}$$

The power spectral density of the dynamic deflection is defined as follows:

$$G_{f_d}(f) = |f_d(t)|^2 \tag{24}$$

According to the theoretical results, we can conclude that

$$\begin{aligned} f_d(t) = {} & H_{z2-q}(f) \cdot q(t) \\ & - H_{z1-q}(f) \cdot q(t) \end{aligned} \tag{25}$$

$$\begin{aligned} G_{f_d}(f) = {} & \left| H_{z2-q}(f) - H_{z1-q}(f) \right|^2 \\ & \cdot G_q(f) \end{aligned} \tag{26}$$

According to the results of the theoretical calculation, you can locate the damping range $-50 \sim +50\%$, using MATLAB program to calculate the results shown in Fig. 10:

Figure 10 shows that the maximum value of the dynamic deflection spectral density is about $0.9 \times 10^{-5}\ \mathrm{m}^2 \cdot \mathrm{s}$, and the maximum dynamic deflection is about 3 mm at the corresponding excitation frequency. In the range of 0–90 Hz road excitation frequency, the maximum dynamic deflection of the initial phase of 0 and the peak value of MATLAB is calculated in Fig. 11.

The design of the suspension deflection value is 11 cm; according to the damping parameters on ride comfort and dynamic deflection of the calculation results, only considered from the ride comfort, the shock absorber damping dynamic deflection can be reduced by 10%, and deflection increased by about 3.8%, ride comfort index decreased 0.5 dB. After optimization, the results are as follows (Fig. 12).

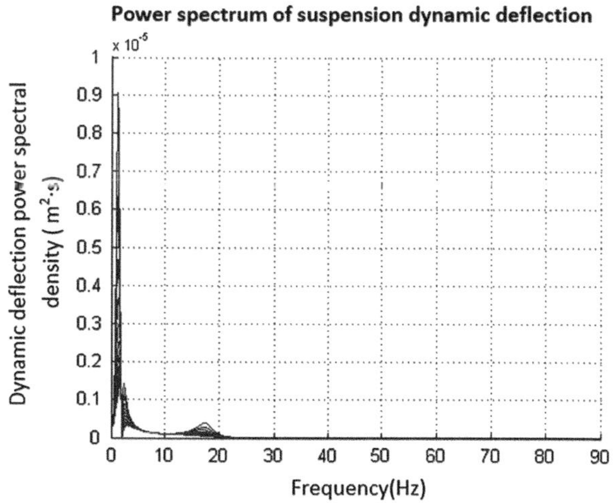

Fig. 10 Power spectrum of suspension dynamic deflection

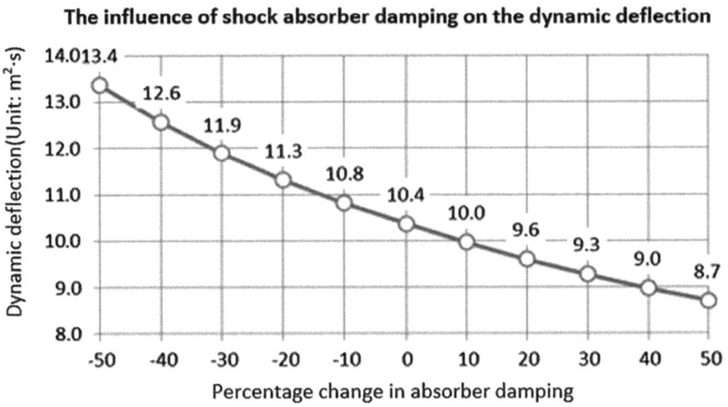

Fig. 11 Influence of shock absorber damping on the dynamic deflection

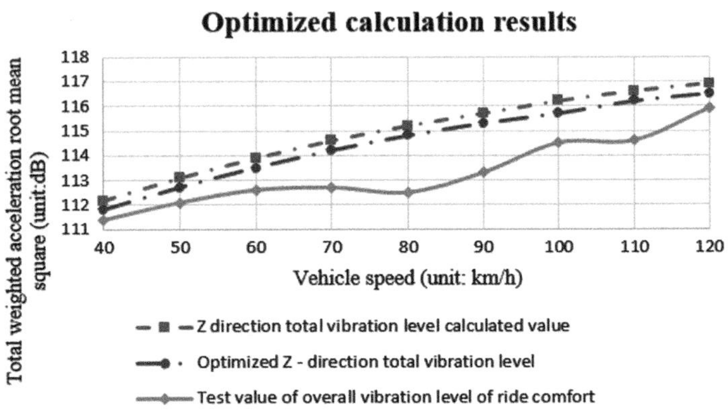

Fig. 12 Optimized calculation results

5 Conclusions

Based on the two degrees of freedom, the seat system is used to establish the three-degree-of-freedom single-wheel model. Using virtual excitation method, the experimental analysis and modeling of random pavement excitation are carried out to verify the accuracy of numerical calculation model. The optimization measures and suggestions are given from the aspects of quality parameters, parameters of damper, tire and seat parameters and model optimization. Finally, the suspension parameters are selected as the best scheme for ride comfort optimization. This paper quantitatively analyzes the influence degree of the key technical parameters on the ride comfort from the practical point of view and provides the reference for the designers to avoid the repeated design and experiment, thus reducing the development cost and shortening the development cycle. However, the optimization design of the damper can improve ride comfort and will also affect the handling stability. If we want to do further research, we can establish a unified model to study ride comfort and handling stability.

Acknowledgments This work was supported by Science and Technology Planning Project of Guangdong Province (2015B010137002/2016A050503021).

References

1. Zhisheng Yu (2009) Automobile theory. Mechanical Industry Press, Beijing, pp 203–206
2. GB/ T 4971-2009 Terms and definitions of vehicle ride comfort
3. Lin J, Zhang Y, Zhao Y Fuzzy vibration virtual excitation method and engineering application. Supported by the National Natural Science Foundation of China (No. 10072015)
4. Jiahao L, Yahui Z (2004) Virtual excitation method for random vibration. Science Press, Beijing
5. Lin JH, ZhangYH (2005) Vibration and shock handbook, Chapter 30. Scismic random vibration of long-span structures. CRC Press, Boca Raton, U.S.A
6. Xiaoan S (2006) Modeling and Simulation of vehicle ride comfort. Zhejiang University of Technology, Zhejiang
7. Xian P, Xiaohui L, Guilin W (2007) Vibration analysis of variable speed vehicle based on virtual excitation method. J Hunan Uni Nat Sci Edition 34(11):37–41
8. Jie Li, Yuying Qin, Qi Zhao (2009) A new method for the analysis of random vibration of vehicle. Mechan Design 26(4):14–17
9. Jianman X, Chenghua L, Wei Z (2012) Vehicle ride comfort test and optimization design of suspension system. J Shenyang Agri Uni 43(5):571–575
10. Yingxi W, Hong Z, Erbing W, Xianyin L (2011) Analysis of vehicle ride comfort based on 10-DOF vehicle model. J Jiamusi Uni 29(5):684–688
11. Liping Z, Lixin G (2012) Ride comfort of vehicle based on reverse virtual excitation. J Northeastern Uni (Nat Sci Edition) 33(5):719–721
12. Hao L, jian Z (1998) A mark on the virtual excitation method and the random response of structures. J Comput Mech 15(2):217–223
13. Yan Z, Yahui Z, Hao L (2013) Virtual excitation method for response spectrum analysis of vehicle random vibration. Appl Mathe Mech 34(2):107–117
14. Lili L, Yongde G (1993) Establishment of linear model in the study of vehicle ride comfort. J Taiyuan Inst Mech Eng 14(4):336–343
15. Lanzhu Z, Jun W (2008) Effect of vehicle suspension on ride comfort. J East China Uni Sci Technol (Nat Sci Edition) 34(6):908–912

Intelligent Laser Solutions for Automotive Manufacturing

Jianghua Ji and Houxue Cui

1 Introduction

Laser is a flexible and universal tool which allows for typical applications in the automotive industry [1–3], e.g. laser brazing and welding of car body, laser cutting of hot-forming components and carbon fibre-reinforced plastic (CFRP) for car body, laser welding of power train components, laser hardening of engine parts, laser metal deposition for mould, laser marking for traceability, [4, 5]. See Fig. 1 for an overview of typical laser applications in automotive manufacturing. Due to these more and more laser applications in the automotive industry, intelligent laser solutions fit into the age of Industry 4.0. Intelligent laser solutions include intelligent laser equipment, intelligent laser process and eventually the intelligent factory.

The intelligent laser equipment, sensors for collecting data, interface for transferring data, and software for processing data are necessary. To process the data correctly and precisely, we must have all the data synchronized, which needs Precision Time Synchronization Protocol. And all these features are equipped in TRUMPF new TruDisk Laser, which is a newly developed intelligent version based on the proved Disk Laser technology.

To make a complete manufacturing process, sensors for process monitoring are also very important and necessary components. Thus, TRUMPF also developed its own sensor technology for important laser processes, including laser welding, cutting, seam tracking, which make the laser processes more transparent and reliable.

Finally, these data could be transferred to an intelligent platform which is provided by Axoom, an independent company initiated by TRUMPF. Axoom is an open plat-

J. Ji (✉)
TRUMPF (China) Co., Ltd, Taicang, Jiangsu 215400, China
e-mail: Joe.ji@cn.trumpf.com

H. Cui
Dongfeng Motor Cooperation, Wuhan, China

Society of Automotive Engineers of China (SAE-China) (ed.), *Proceedings of the 19th Asia Pacific Automotive Engineering Conference & SAE-China Congress 2017: Selected Papers*, Lecture Notes in Electrical Engineering 486, https://doi.org/10.1007/978-981-10-8506-2_78

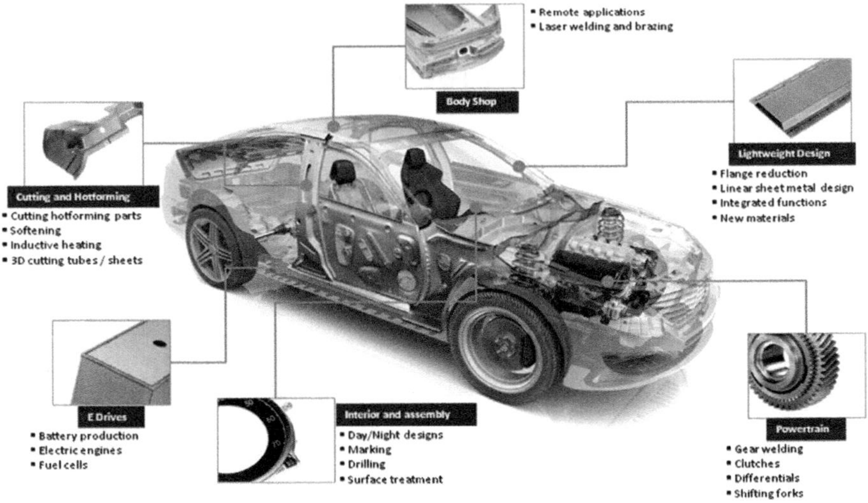

Fig. 1 Laser applications in automotive manufacturing

form for the intelligent factory, which could collect and process all manufacturing data, including status of equipment, production and process of each station. By processing these data, the complete laser system makes the production more flexible and robust, thus more intelligent. Equipment availability and productivity are being improved, which finally reduces cost and helps to make you even more competitive in the market!

2 Intelligent Laser Equipment

A laser system for manufacturing (welding, cutting …) normally consists of a laser source, laser light cable, processing optics and a robot (or another motion unit). All these components in an intelligent system must be also intelligent.

The laser source, providing the energy for manufacturing, is the core component of the system. In TRUMPF TruDisk laser, a few dozen sensors are installed all over the laser device. Figure 2 gives an overview of some exemplary sensors that are used within TRUMPF latest disk lasers. Sensors in power supply unit for measuring voltage and current keep the input power supply within a well-defined range for stable production. Sensors in the cooling unit measure water quality, pressure and flow rate to ensure proper cooling of the optical components in the laser device; sensors in optical unit for measure temperature, humidity, scattered light and laser power make sure optical components are working in proper environment, and then the laser power is generated according to the settings and transmitted in the right way without extra loss. With all these sensors, we know that the laser is working in right conditions, output right laser power and finally transmitted to the workpiece.

The values measured by the sensors are not independent from each other, such as scattered light is normally related to the output laser power. So we should analyse

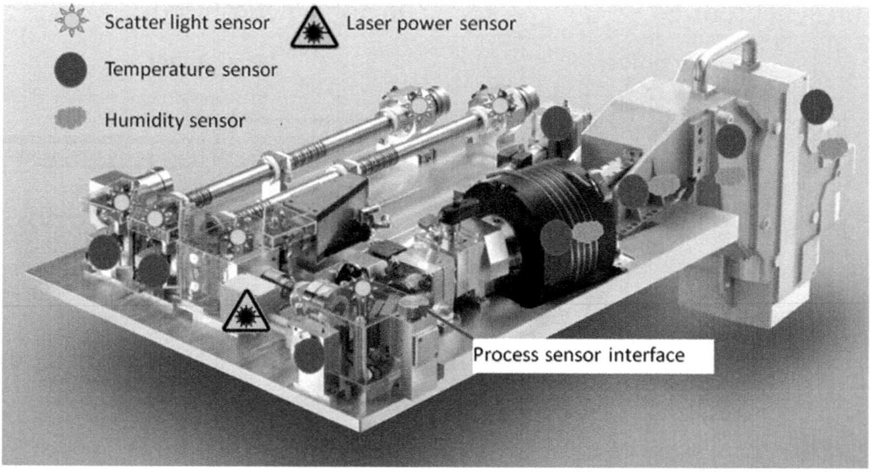

Fig. 2 Sensors in optical unit of TRUMPF TruDisk laser. (*Source* TRUMPF)

all these data simultaneously, which requires synchronization of these data records. Since values can change very fast, e.g. during pulsed mode operation, signals need to by time synchronized. Thus, TRUMPF's new TruDisk laser uses the standard Precision Time Protocol for data record and synchronization. Figure 3 gives an example of synchronized versus not synchronized sensor values.

2.1 Intelligent Laser Process

To make the laser process intelligent, intelligent processing optics and sensors are necessary.

I-PFO is the intelligent generation of the TRUMPF scanner welding head Programmable Focusing Optics (PFO). In the I-PFO, there is a processor and interface for connecting the laser with a robot, and coordinate them with each other automatically. Thus, they can be operated and programmed as one unit to realize welding on the fly. Welding on the fly further increases the productivity of laser welding significantly by reducing the positioning time of the robot. An exemplary set-up can be seen in Figs. 4 and 5.

Offline programming software TruTops I-PFO is also provided together with the I-PFO. In this programming tool, the robot, I-PFO, could be considered as one system, and then makes a welding program for all base on the 3D model of the workpiece. During the welding process, the I-PFO receives the real-time position of the robot, calculates its own position and then automatically shoots the laser beam to the right position of the workpiece. You don't have to worry about the welding position as long as your workpiece is put in the right position.

But normally the position of the workpiece, especially the fillet joint seam position is not accurate enough for laser welding. Thus, a sensor for detecting the joint seam

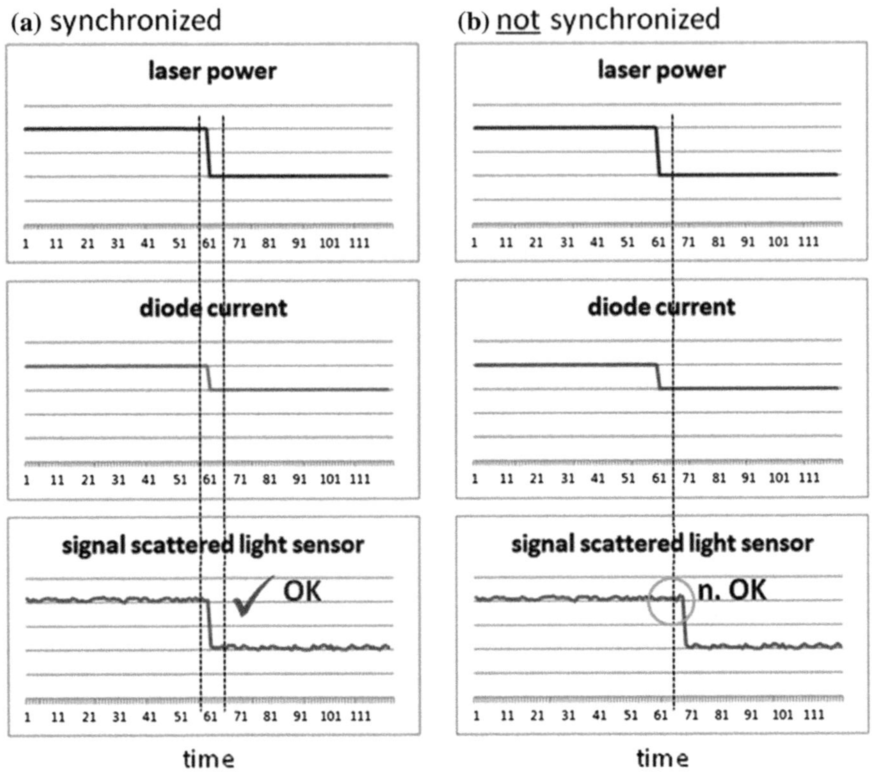

Fig. 3 Importance of time synchronization. (*Source* TRUMPF)

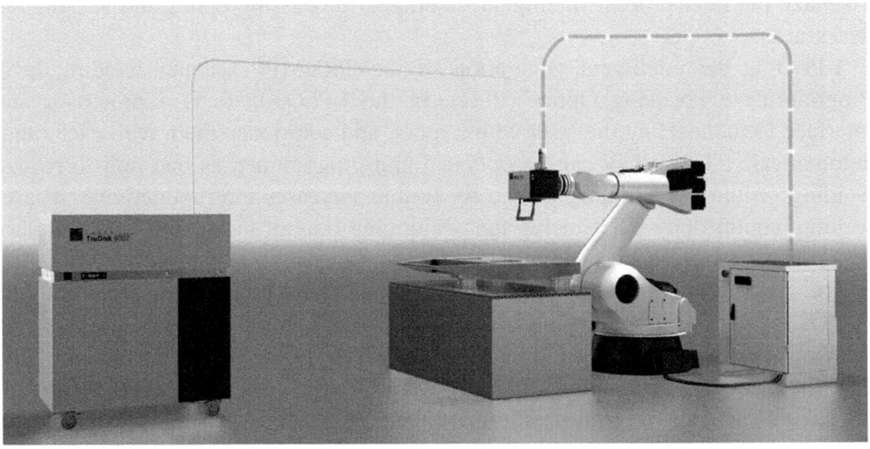

Fig. 4 Welding on the fly system: laser, I-PFO and robot. (*Source* TRUMPF)

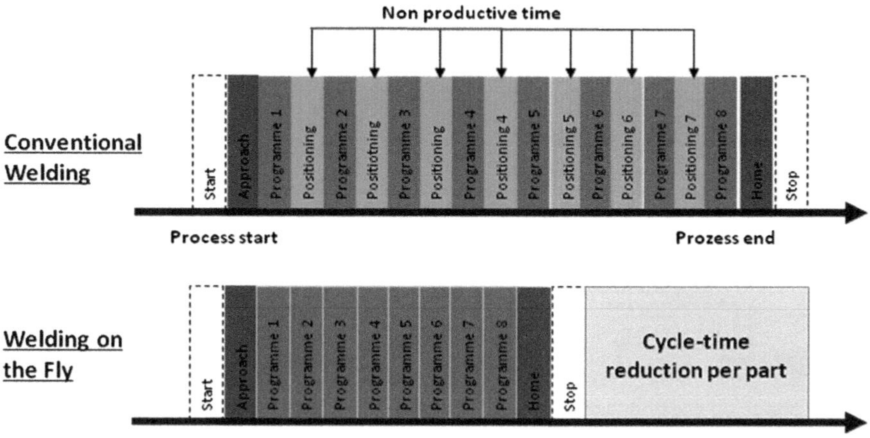

Fig. 5 Cycle time reduction with "welding on the fly". (*Source* TRUMPF)

has been developed, which is called SeamLine Remote; see Fig. 6. The SeamLine Remote system projects three laser lines onto the joint seam of the workpiece. A camera is used to take the pictures. By analysing the pictures, the angle of the projected laser lines on the joint seam is determined, and the exact seam position is calculated, as well as the gap between the two sheets. With these data, the I-PFO could adjust the programmed position accordingly and shoot the laser beam exactly onto the seam, with proper focus diameter and laser power adapt to the gap. This system allows for highest robustness and perfect welding seams. With the same system, the welding quality is double-checked online. The camera will also take a picture after the welding simultaneously and analyse the seam profile to assess the resulting welding quality. The data of each welding seam is stored in a Quality Assessment Software, which is also an option of the SeamLine Remote system. The Quality Assessment Software will analyse the quality of each seam and give out the assessment of the workpiece welding quality completely, according to the preset conditions by the user.

Other intelligent processing optics and sensors are also developed for each different laser process in automotive, such as dynamic laser cutting head with distance control sensor, plastic laser welding head with pyrometer sensor.

2.2 *Smart Factory*

With intelligent laser equipment and process, we have a vertical integrated intelligent workstation, which can improve productivity, reliability and availability of this process itself. The smart factory needs to connect and coordinate all workstations and processes, thus we have horizontally connected and integrated all stations into an intelligent system.

Axoom is an open platform initiated by TRUMPF as a solution for the smart factory. As an open platform, the production data of all stations, not only TRUMPF

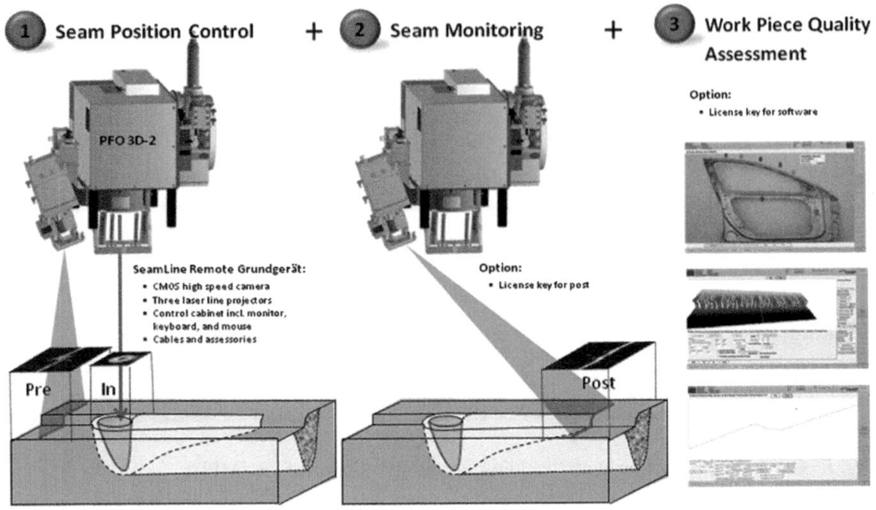

Fig. 6 SeamLine Remote for seam tracking and welding quality check. (*Source* TRUMPF)

devices can be connected, transferred and stored with the Axoom platform. All stations are synchronized, thus minimizing waste of material and time.

TRUMPF will also provide "Condition Monitoring Services", e.g. based on the Axoom platform. The customer can transfer the data of TRUMPF devices to TRUMPF Service Centre, and then all these devices and their processes are analysed and shown on dashboards, monitored and analysed by TRUMPF experts. Once there is any risk in the production, the system will give warning or even alarm messages.

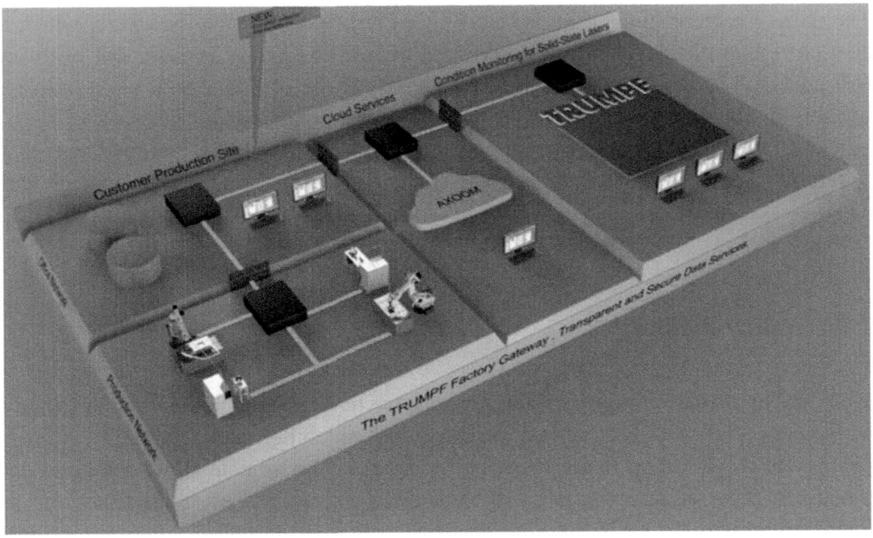

Fig. 7 TRUMPF Axoom platform. (*Source* TRUMPF)

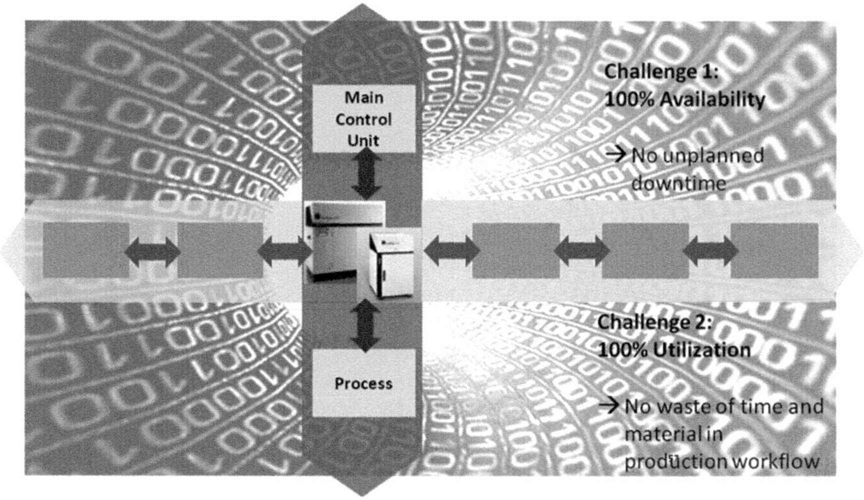

Fig. 8 Smart factory with 100% availability and utilization. (*Source* TRUMPF)

TRUMPF experts will check the data and provide solutions to customers. By this procedure, unplanned production interruption will be minimized (Figs. 7 and 8).

3 Conclusion

With intelligent features like equipment monitoring, process monitoring and process control, laser processes will become even more important for future automotive manufacturing. The laser as a highly versatile tool will play the key role in acquiring, gathering and pre-processing data for the intelligent factory. Thus, lasers and laser processes will pave the way to future automotive designs and concepts, eventually being the cornerstone of reliable and robust automotive production lines.

References

1. Bernhardt R, Schubert H Laser welding applications in the new Mercedes-Benz C-class. In: 8th European conference and exhibition, european automotive laser application
2. Brockmann R, Mann K (2007) Fertigungstechnologie bei Audi: Laserscannerschweißen an den Tueren des neuen Audi A4. Laser Technik J 2
3. Brockmann R, Mann K, Schlueter H, Havrilla D (2008) High-power disk laser. In: Photonics West
4. Giesen A, Speiser J (2007) Fifteen years of work on thin-disk lasers: Results and scaling laws. IEEE J Sel Top Q Electron 13:598
5. Gottwald T, Kuhn V, Schad S-S, Stolzenburg C Recent developments in high power thin disk lasers at TRUMPF Laser. In: Proceedings of SPIE 8898, 88980P

Noise Detection Technology of Automotive Powertrain and Application

Wei Xia, Xiangjun Cao and Xinyu Duan

1 Preface

With the rapid development of automotive industry, people pay more and more attention to the automotive comfort and environmental protection and energy saving, also require more strictly on automotive noise control. The major sources of noises that affect automotive comfort are: engine and gearbox (GMP), air intake, exhaust, tires (moving noise), aerodynamic noise, equipment noise, interference noise. While engine from idle to stop, GMP noise constitutes 100% user-perceived noise; at low speed (30 km/h), GMP noise constitutes 40–60% user-perceived noise; at a high speed (>100 km/h), GMP noise accounts for 20–30% of user-perceived noise. Obviously, vibration and noise of powertrain is one of main sources of automotive overall noise, which is one of key performance indicators to measure the quality of the powertrain, and its overall quality will directly affect the driver's feelings, and therefore became the target of research. The purpose of noise detection is, from the complex sound source system, with the use of reasonable means of testing, to find abnormal sound source location, frequency characteristics, etc., and to provide guidance for troubleshooting.

Usually in manufacture of engine and transmission, on the assembly line, or on the automobile, in the professional laboratory, detecting vibration and noise, analysis and verification processing compliance and assembly accuracy of engine or transmission parts and correcting anomalies, this has a crucial role to ensure the quality of assembly. This paper focuses on noise detection method commonly used in the mass production of engine and transmission.

W. Xia (✉) · X. Cao · X. Duan
Dongfeng Peugeot Citroen Automobile Ltd. Company, Wuhan, China
e-mail: xiawei@dpca.com.cn

Society of Automotive Engineers of China (SAE-China) (ed.), *Proceedings of the 19th Asia Pacific Automotive Engineering Conference & SAE-China Congress 2017: Selected Papers*, Lecture Notes in Electrical Engineering 486, https://doi.org/10.1007/978-981-10-8506-2_79

2 Basic Concept of Acoustics

Sound is a mechanical vibration of matter; it is a longitudinal wave phenomenon that propagates in the air or in solids and liquids, characterized by frequency, amplitude, and phase. A pure tone is the simplest form of sound; it is a sine wave. The vast majority of sound perceived daily is composite sound of complex wave form; it may be regarded as compounded by pure-tone audiometry. Sound frequency ($f = 1/T$) is the number of sound vibration cycles finished within one second; its unit is Hertz (Hz). Low frequency corresponds to the bass, and high frequency corresponds to the treble. Human average hearing can only perceive sounds within a certain frequency range, approximately (depending on age, race, etc.) between 16 and 20 kHz. Below the audible sound is called infrasound, and higher than audible sound is called ultrasound. The distance in space that sound waves propagate within a time period is called a wavelength λ, and the relationship of wavelength with sound speed and frequency is $\lambda = c/f$. Propagation speed of sound in air at 20 °C is 344 m/s.

2.1 *Sound Pressure*

Sound spreads in the form of sound pressure change around the average pressure. Sound pressure is the atmospheric pressure changes after the disturbance, the fluctuation of atmospheres pressure, its unit Pascal (Pa). $P = P$ average $+\Delta P$ sound (is equivalent to the atmospheric pressure plus the perturbation pressure increment). On the average human hearing, the average hearing limit and the average hearing pain limit have been estimated. The average hearing limit: $P_0 = 2 \times 10^{-5}$ Pa (audibility limit 2000 Hz); the average hearing pain limit: $P_{MAX} = 60$ Pa (1000 Hz). In view of the large difference between the two limit values, we use a logarithmic scale, from the perspective of the sensation of sound that better corresponds to the human ear perception.

Sound pressure level: Pressure level P (dB) or Lp is used to express sound pressure.

$$P_{dB} = 20 \log\left(\frac{P}{P_0}\right)$$

Average human audibility limits expressed in decibels (dB):

$$P_{0(dB)} = 20 \log\left(\frac{P_0}{P_0}\right) = 20 \log(1) = 0\,\mathrm{dB}$$

Fig. 1 Relation of DB and pressure

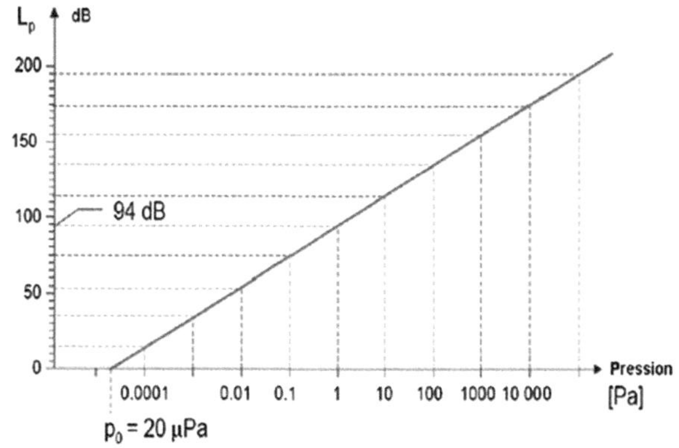

Average human hearing pain limits expressed in decibels:

$$P_{\mathrm{MAX}}(\mathrm{dB}) = 20\log\left(\frac{P_{\mathrm{MAX}}}{P_0}\right) = 20\log(3\times 10^6) = 130\,\mathrm{dB}$$

DB and pressure conversion can be done with help of Fig. 1.

2.2 *Sound Intensity and Sound Power*

Noise can also be used to describe the sound intensity level and sound power level.

Sound intensity: the energy per unit time that passes through a unit area that is perpendicular to the propagation direction of sound waves is acoustic energy density per unit, w/m^2.

Sound intensity level: $I_{\mathrm{dB}} = 10\log\left(\frac{I}{I_0}\right)$. In the air $I_0 = 10^{-12}$ w/m^2

The relationship between sound pressure and sound power depends on the form of sound waves. If the size of the sound source is much smaller than the wavelength, it can be considered as point simple harmonic spherical sound waves, this sound intensity:

$$I = \frac{P^2}{2\rho c}$$

Sound intensity I_{dB}, sound pressure P (Pa), medium density ρ(kg/m^2), sound speed c (m/s).

For a spherical sound wave, if the measuring point distance from the sound source to r, this sound power:

$$W = 2\pi\frac{P^2}{\rho c}r^2$$

Propagation of a sound source in air, at temperatures of 20 °C, medium density ρ = 1.21 kg/m^2, sound speed C = 344 m/s. A 100 kW of engine under full load, sound pressure is 100 dB (2 Pa), the sound propagate 1 m, sound power is

$$W = 2\pi \frac{P^2}{\rho c} = 2\pi \frac{2^2}{1.21 \times 344} = 60\,\mathrm{mW}$$

In summary, $P_{\mathrm{dB}} = 20\ \log(P/P_0) = 10\ \log(P^2/P_0^2) = 10\ \log(W/W_0)$.

Average human audibility limits: Reference sound pressure is $P_0 = 2 \times 10^{-5}$ Pa, and reference sound power is $W_0 = 1 \times 10^{-12}$ w.

2.3 Loudness Level

Loudness level is a method of assessing environmental noise based on the human auditory characteristics. Usually, the human ear can hear the sound frequency range of 20–2000 Hz; higher than 20 kHz is called ultrasound, less than 20 Hz is called infrasound. Ear's perception of sound not only depends on pressure but also frequency. Compare two kinds of sounds that have the same sound pressure level but different frequencies, the higher frequency sound sounds more loudly than the lower. If the measured sound compares with the reference sound that is pure tones of frequency 1000 Hz, sound like the same, then the reference sound pressure levels (dB) is called the loudness level of the measured sound; its unit is (PHON). This is compared with the reference pure tone method, and you can get the entire range of pure tones' loudness level.

Pure tone equal loudness curves obtained by a large number of tests are shown in Fig. 2. Each curve in the graph corresponds to these sounds that have different frequencies and sound pressure levels but sound the same, which have the same loudness level (PHON).

Fig. 2 Equal loudness curve

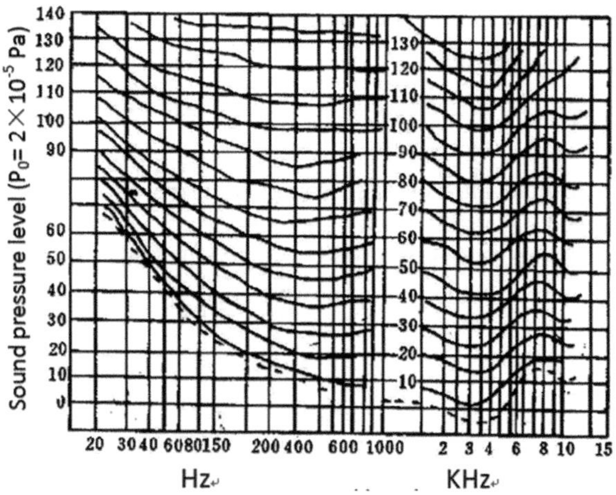

Showed by the equal loudness curve, the human ear is sensitive to high frequency sound, not sensitive to low frequency sound. Sounds of loudness level 70 PHON, if the frequency is 100Hz ,the sound pressure level is 76dB; if the frequency is 1000Hz ,the sound pressure level is 70dB; if the frequency is 4000Hz, the sound pressure level is 62dB, but they are all in the 70 PHON curve; they sound the same.

2.4 Common Detection Methods for Vibration and Noise

Noise is composite sound that is made up of many frequencies and is non-periodic vibration. It has all the characteristics of acoustic wave. The lowest frequency is called the fundamental acoustic wave, and acoustic wave in the higher frequencies is called overtones or harmonics. If overtones frequency is an integer multiple of the fundamental frequency is called harmonics.

The overall function of the acoustic wave $f(x)$ can be expressed by the sum constants and the sine function (sine and cosine):

$$f(x) = a_0 + \sum_{n=1}^{\infty} [a_n \cos(nx) + b_n \sin(nx)]$$

Use sound pressure level, noise intensity level, and sound power level to measure the strength of the sound, use frequency, and spectrum to measure sound characteristic and analyze and judge location of the noise source. While noise propagates, it can be transmitted directly to the receiver or indirectly to the receiver, according to whether the noise propagation process is encountered obstacles, can be divided into primary airborne and secondary airborne sound.

Sound in the transmission process, if received directly by the receiver without any hindrance, this sound is called primary airborne sound. If when noise spread, it encounters other obstacles, which occur a certain vibration, and becomes a source of encouragement; then the receiver receives the noise which is not just the initial sound source issued, but also mixed with the noise generated by the vibration of the obstruction, and this noise is known as secondary airborne sound. In indirect way to spread, there may be multiple conversions of solid, liquid, air, and resonance between individual components.

Common methods for noise and vibration detection and analysis:

(1) Sound pressure measurement method—weighted sound level

Weighted sound level has *a*, *b*, *c*. They are noise values measured by acoustic measurement that is set weight network "*a*", "*b*", or "*c*", do remember dB (*a*), dB (*b*), and dB (*c*). The A—Weighted network is designed to simulate the human ear in accordance with 40 PHON equal loudness curve. When it receives sound passing through, low frequency sound that is not sensitive to the human ear has a greater attenuation, followed by medium-frequency attenuation, high frequency does not decline. The B—weighted network is designed according to 70 PHON equal loudness curve, and there are only some attenuation in low frequency band. The C—weighted network is designed based on 100 PHON equal loudness curve; almost no attenuation over the entire audible range of frequencies. In all three weight sound levels, A—weighted sound level that best reflects the human auditory characteristics and be currently used widely.

Sound pressure measured with a sound level meter is susceptible to the effects of background noise and acoustic reflex and will produce a sound superimposed (also known as dB synthesis). Precise sound pressure measurements are made in anechoic room and semi-anechoic room.

The same frequency noise superimposes $P(t) = P1\ (t) + P2\ (t)$

- Note weather superimposed phases are the same: This rule is adapted to the fundamental frequency (and 2, 4, 6… times the overtone.)
  Except in special cases (e.g., in the opposite phase of the sine wave), with linear Watts are needed to calculate the increase in sound power.
  For example, we have two 60 dB loudspeakers.
  A 60 dB sound source: 60 dB $\Leftrightarrow$ 10^{-6} W/m^2 ,
  Two kinds of 60 dB sound source: 2 × 10^{-6} W/m^2 $\Leftrightarrow$ 63 dB.
  A 60 dB sound source: 60 dB $\Leftrightarrow$ 10^{-6} W/m^2,
  A 40 dB sound source: 40 dB $\Leftrightarrow$ 10^{-8} W/m^2,
  A total of 1.01 × 10^{-6} w, which is, 60.04 dB.
  Different frequency noise superimposes $P^2 = P_1^2 + P_2{}^2$
- The rules to be used to calculate the treble peak:
  For example, different frequency noise superimposes
  $N1–N2 = 0$ (See Fig. 3, abscissa value 0)
  Find the ordinate 3 dB,
  After superimposed $N = N1 + 3$ dB = 77 Db (Tables 1 and 2).

(2) Sound intensity measurement method

Sound intensity measurement technology rapidly develops and matures with advances electronic technology and signal processing technology in the year of the seventies or eighties in twentieth century. It has unique advantages in sound power determination and main noise source identification. Sound intensity is vector, and using this method to measure sound power, it can eliminate the effect of other sound sources outside the enclosed surface and environmental impact on the measurement

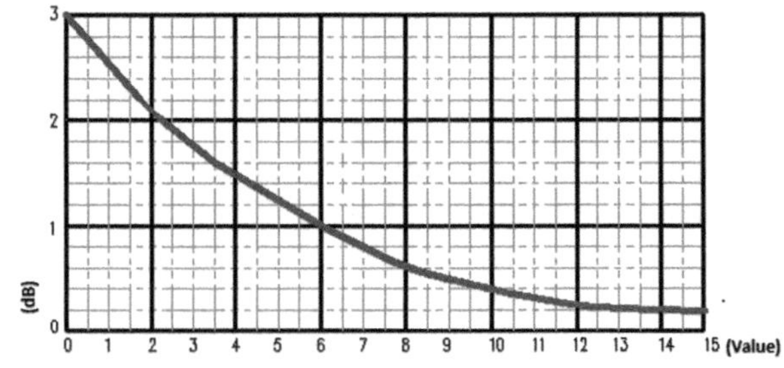

Fig. 3 Noise superimpose (dB)

Table 1 Same frequency noise

W_{14} (dB)	W_2 (dB)	Total (dB)
60	60	63.0
60	59	62.5
60	57	61.7
60	50	60.4
60	40	60.04

Table 2 Examples of noise superimposed

		Same frequency noise		Different frequency noise	
	N1	74 dB		74 dB	
	N2	74 dB	68 dB	74 dB	68 dB
N1N2 same phase	N	80 dB	77.5 dB	77 dB	75 dB
N1N2 opposite phase		0 dB	68 dB		

results. Using sound intensity measurement can detect location of the surface radiation noise source and display visually. Mainly, there are two types of sound intensity measurement methods: Direct method is the combination of microphone and direct measurement of particle velocity sensor, referred to as *p-u*; indirect method is dual-pressure response of the microphone, referred to as *p-p*. Use two microphones that have the same phase response and the straight amplitude-frequency characteristic curve to realize indirect measurement. Measuring distance *r* is much smaller than when the acoustic wavelength, and the sound intensity value of intermediate point can approximate the two probes of the arithmetic mean.

$$\text{Instantaneous sound intensity}\, \mathrm{I(t)} = \frac{\mathrm{P1(t) + P2(t)}}{2\rho 0\, \Delta \mathrm{r}} \int [\mathrm{P1(t) - P2(t)}]\mathrm{dt};$$

$$\text{time} - \text{averaged sound intensity}\, \mathrm{Ir} = \frac{\mathrm{P1(t)P2(t)}}{2\rho 0} \int \frac{[\mathrm{P1(t) - P2(t)}]}{\Delta \mathrm{r}}.$$

T—time for test, Δr—distance entre two microphone, ρ_0—air density, *P*1(t)—Microphone 1 sound pressure components, *P*2(t)—Microphone 2 sound pressure components,

(3) Vibration velocity measurement method

Structure-borne sound can be characterized by surface vibration velocity level (or acceleration level) and its radiation efficiency. For a fixed medium, a fixed point in the sound field whose mean square value of sound pressure and sound power are proportional to mean square value of vibration velocity of the surface. As long as you know the proportional coefficient of δ, the sound pressure level and sound power of noise can be calculated by measuring the particle velocity level. This conclusion has a very important meaning to application vibration velocity measurement method. This method is more used in the detection of gearbox vibration and noise.

$$P^2 = \frac{1}{2\pi r^2}\delta\rho^2 c^2 SV^{\,2};\ W = \delta\rho c SV^{\,2}$$

P^2 Mean-square sound pressure level, W sound power, δ coefficient of noise radiation, ρC (air density × sound speed), V^2 mean square value of vibration speed, *r* the distances between measuring point and the sound source.

The vibration velocity measurement method is measurement of particle velocity levels of noise radiation on surface. The vibration velocity level Lv (dB). $Lv = 10\log(V^2/V_0^{2)}$.

V effective vibration speed in effective frequency range; structure-borne sound baseline speed is $V_0 = 5 \times 10^{-8}$ m/s.

Sound pressure level of noise radiation: (dB) $Lp = 10\log\left(P^2/P_0^2\right) = Lv + 10\log\frac{\delta\rho^2 c^2 SV_0^{\,2}}{2\pi r^2 P_0^2}$

Sound power levels of noise radiation: (dB) $Lw = 10\log\left(W_2/W_0^2\right) = Lv + 10\log\frac{\delta\rho c SV_0^{\,2}}{2\pi r^2 W_0^2}$

Average hearing limit: Reference sound pressure $P_0 = 2 \times 10^{-5}$ Pa, reference sound power $W_0 = 1 \times 10^{-12}$ w.

(4) Spectrum analysis.

Noise is compounded by different frequencies of sound, in order to identify the composition and distribution of noise range, it is necessary to determine the sound pressure level according to different frequencies. Spectrum is a graph showing the relationship between noise pressure and vibration amplitude and frequency, the abscissa is marked according to the logarithmic scale of frequencies, and the ordinate is sound pressure levels or sound power levels in decibels or vibration amplitude.

The spectrum analysis of noise signal is to transform the signal in time domain into frequency domain for analysis, the theoretical basis is the Fourier transform.

The periodic noise signal can be decomposed into a Fourier series to obtain its spectrum–discrete spectral lines, for example, the fundamental frequency of the engine intake and exhaust noise, fan noise and gear noise. The aperiodic signals obtained its spectrum by Fourier transform and the continuous spectrum. The spectrum of the usual vibration noise is a discrete type, continuous type, or both mixed type.

Usually, the audible frequency range 20–20,000 Hz is divided into several frequency bands, that is often said the frequency bands. Each band has a central

Table 3 Frequency range of octave

Central frequency	31.5	63	125	250	500	1000	2000	4000	8000	16,000
Frequency range	22–45	45–90	90–180	180–355	355–710	710–1400	1400–2800	2800–5600	5600–11,200	11,200

frequency, maximum and lower frequencies. Difference between the maximum and lower frequencies is called frequency bandwidth. The most commonly used is the octave (see Table 3 for center frequency and the frequency range) and 1/3 times octave. Octave are the frequency bands in which the ratio of maximum and lower frequencies is 2:1; the above-mentioned one frequency band is divided into 3 points to get a finer frequency division called 1/3 times octave.

Usually, the logarithmic scale of central frequencies of the octave or 1/3 times octave is the abscissa, sound pressure levels (or sound intensity level and sound power level) in decibels or vibration amplitude is the ordinate, make a spectrum, in order to understand the distribution of situation of sound pressure level of frequency bands in different frequency ranges, and accordingly determine composition and nature of noise, this is spectrum analysis.

3 Application of Engine Vibration and Noise Detection in Production

3.1 Engine Noise Classification

According to the mode of noise generation, the noise of engine can be divided into combustion noise, mechanical noise, and gas dynamic noise. The combustion noise is the gas pressure in the cylinder due to the combustion of the cyclical changes; mechanical noise is a noise caused by the impact between moving parts and between moving and fixed parts or caused by parts surfaces vibration that caused by the parts elastic deformation produced while internal forces changing periodically. The combustion noise and mechanical noise propagate outward through the engine surface, it is also known as the engine surface noise; noise and intake and exhaust fan, exhaust gas turbocharger noise is radiated directly to the atmosphere.

It is hard to distinguish the combustion noise and mechanical noise. Some combustion machinery noise is indirect excitation in the cylinder of the engine, such as the formation of combustion pressure inside the cylinder through the vibration of cylinder head, piston-connecting rod-crank—noise radiated by burning the body is excited. The piston of cylinder liner, piston ring friction, a timing gear and a belt, chain transmission, gas distribution mechanism, oil supply system, noise excitation and vibration between the other such as generator, air compressor, cooler, hydraulic pump and other mechanical moving parts produced by the impact that the mechanical noise.

In vibration theory, the noise on the surface of an engine arises from the vibrations that occur during its operation. Often use multiple harmonic vibration to describe the vibration of the engine, same of the engine speed and vibration frequency named vibration (*H*1), the frequency is two times of the engine speed is two-order vibration (*H*2), and so on, there are three-order (*H*3) and four-order (*H*4) vibrations. The higher the vibration frequency is, the smaller the amplitude is. In general engineering, three or more vibrations can be neglected.

The first noise source of engine, the vibration force produced by combustion pressure and inertial force: vertical vibration force, lateral oscillation force, side roll vibration force, left and right oscillation forces.

For a straight line 4-cylinder engine, the vertical and vertical vibration forces, the transverse oscillation vibration, and the vibration forces are derived from the two-order inertial forces.

The second noise sources of engine. The phenomenon of a second vibration noise source produced by the impact of a first vibration noise generator; a system driven by an engine; a synchronous phenomenon associated with engine rotation; vibration and noise produced by the system itself. Second vibration and noise sources are timing mechanism, timing drive, electric injection system, accessory drive system (generator, compressor, power steering pump, water pump, and oil pump), the impact of the inherent gap, etc…

3.2 General Method of Engine Noise Detection in Batch Production

Because there are many factors affecting noise in the working cycle of the engine, noise detection is very important. While the frequency structure of each sound source is very complex, the delivery channels vary the mutual interference between the sound source and sound source, resulting in the process of fault diagnosis; it is difficult to effectively distinguish between different sound sources, here in the engine batch production process, a method of noise detection.

3.2.1 Heuristic Method

According to experience, use ears hearing to distinguish the sources of noises and the primary and secondary order. The method can be understood the basic situation of noise sources and provide a preliminary scheme for noise control. It is requires the operator to have some experience in noise recognition. DPCA has a new engine in the production process, before the engine off assembly line, they are going on bench test and noise test 100%, bench test member has a wealth of practical experience, and the abnormal engine cannot escape their ears. However, due to the diversity of operator perception, especially when new candidates are required to provide guidance in document assistance, it is possible to prepare inspection instructions as shown in Table 4.

Table 4 Operation instructions for noise perception inspection

No	Fault location	Cause	Fault phenomenon description
1	Main bearing part	The main bearing part is loose	A loud noise; a muffled noise
		All the main bearings are loose	Loud and short strokes. The knock is regular and varies with the speed
		The axial clearance of crankshaft is too large	Intermittent sharp strikes; irregular noises
		The crankshaft is too deep to balance the hole	Crankshaft squeal
2	Connecting rod bearing parts	Excessive clearance of connecting rod	Knock in middle of cylinder block
		The connecting rod is not screwed	Apparent patter
3	Piston position	Piston looseness	Double click percussion
		Piston pin loose	On the wall of metal patter
		Piston reverse	Very obvious pops, acceleration is more obvious
4	Timing drive system	Timing belt interference	The sharp crack, squeak, hissing
		The covers are not packed right now	Interference with accessories pulley gives plastic friction
5	Valve drive part	Hydraulic tappet holds or leaks	The valve in the cylinder head has a muffled sound or a metal beating
		Camshaft profile defect	Camshaft squeal
		The spark plug fell off	Loud, loud fires, and acceleration
6	Cooling system	Abnormal water pump bearing	A sharp hiss
7	Intake manifold	Intake manifold abnormality	Slight intake sound, no obvious abnormal sound

3.2.2 Step-by-Step Operation

Some parts can be respectively running, if the noise radiation were measured in motion or disconnect these parts of the case, it can cause noise apart in order to identify the main noise source. But sometimes the operations of each part are interrelated, and a part of the work stops the operation of the other parts. Therefore, it is often difficult to allow only one part of the work without affecting the operation of other parts, and thus the use of this method is limited. In the present production, the method of replacing parts is used when analyzing faults, and we also think that the method of step-by-step operation is adopted. For example, a suspect in the engine balance box with pad gap is abnormal, and the suspected problem balance box is installed in another cylinder, test hearing.

3.2.3 Spectrum Analysis Method

According to the frequency spectrum characteristic of noise source, the method of determining the main noise source is called spectrum analysis. Through the noise spectrum, on the one hand, we can understand the frequency distribution of the noise source, such as whether the noise source is mainly low frequency, middle frequency, or high frequency. On the other hand, it can determine the source of some peak noise. The spectrum analysis method can be used to estimate the sound energy source of noise evaluation points, but it cannot estimate its influence degree accurately. DPCA in batch producing, engine assembly except in the assembly line 100% bench test, when the monthly output is more than 1000 engines, take one to laboratory for performance sampling test every month. When sampling, we should also evaluate the noise.

Use spectrum analysis method, the noise of a new type of engine parts are measured.

Test equipment consists of analyzer 1, 3 direction acceleration sensor, microphone, host 1, used to connect acceleration sensors, microphones, power supply. The installation of the sensors and microphones on the engine's detected locations is shown in Table 5.

The test shall be done step by step as follows:

Table 5 Installation method of test device

No	Detected site	Microphone arrangement	Acceleration sensor placement	Icon
1	Balancing box noise	In the 90 degree direction of the balance box, the distance between them is 15 cm (cold test)	1 Y acceleration sensors attached to the oil pan (near the 1# cylinder side of the balance box, the flywheel side of the engine exhaust side)	
2	Pump noise	Distance timing side (water pump) 25 cm	The Y is attached to the water pump to the acceleration sensor	
3	Timing noise (idler pulley, belt tensioner wheel, exhaust gear)	Timing side 25 cm	The Y + Z is attached to the oil pan (timing side) to the acceleration sensor	

Engine noise detection is carried out under cold, hot, and full load conditions, respectively:

(1) Cold cycle measurement (balancing box)

- The idle speed is 30 s, and then accelerates from 960 to 3600 RPM with 2 min, and the idle speed is 2 min.
- Cold, partial load, that is, torque 40 N m, speed accelerated from 960 to 3600 RPM, time 2 min.

(2) Thermal cycle measurement

After the cold test is finished, the microphone of the balancing box is moved to the position of the detecting water pump, and the hot water temperature is controlled at 90 °C and the oil temperature is controlled at 110 °C.

- Idle for 30 s, and then accelerate from 970 to 4000 RPM in 2 min, idle for 2 min.
- Thermal partial load, that is, torque 40 N m, speeds from 970 to 4000 RPM, time 2 min.

(3) Measurement under full load

- Hot, full load, speed from 1200 to 5000 RPM, time 2 min

When cold only detect balance box noise, at this time only in the balance box microphone at work, check out the detection of balance box microphone moved to the pump test position, hot state detection timing mechanism, water pump, nozzle microphone, and each operation ends, according to difference curve, compliance immediately checking the monitoring points.

3.2.4 Method for Determining Tolerance Range Line of Noise Detection

Tolerance range is the standard to evaluate the vibration frequency of the engine at different speeds. A noise evaluation tolerance curve, according to the above detection in a longer period of time within the laboratory bench measurement engine, using subjective judgment, step-by-step operation, the method of frequency spectrum analysis, combined with the sensory evaluation, the initial formation of the range curve, according to the revised and improved the tolerance curve of market user evaluation of noise information. Set up in line with the company's products meet the requirements, scope of customer demand curve for sound quality evaluation of engine manufacturing process.

Noise criteria

The long-term monitoring and evaluation of the noise can be carried out in the 10 points. The acceptable value is 7 points, and the following 7 points are not acceptable. It is similar to a bench deduction card that used in the factory. The specific rules are shown in Table 6.

Table 6 Noise ratings scoring rules

User evaluation	Unbearable		Unacceptable		Dissatisfied		Acceptable	Satisfied		Excellent
	1	2	3	4	5	6	7	8	9	10
	Return goods		Asking for replacement		Strong Dissatisfaction	Dissatisfied	Reluctantly satisfied	More satisfactory	Very satisfied	Very satisfied
	1	2	3	4	5	6	7	8	9	10
Notes	Rejection interval				Critical interval		Acceptance interval			
determine	Non-deliverable				Improvements must be made in the short term	Improve over a long period of time	Deliverable			

4 Online Detection Technology for Vibration and Noise of Automotive Gearbox

Online detection of vibration and noise of automotive gearbox is getting more and more attention. In producing process scientifically and reasonably set gearbox vibration and noise test, to discover a gearbox problem and correct in time. DPCA's five-speed manual gearbox test stands as an example to illustrate.

More than 90% of vibration and noise of gearbox mechanical transmission occurs in the gears, shafts, and bearing operations. Gearbox is a closed structure, part of gear mesh impact noise and the bearing vibration through the box spread into the air to form primary airborne sound, and the other part of the vibration of the box wall propagates into air to form secondary airborne sound. Usually, the greater the area of noise reception is, the relatively greater the contribution to noise radiation. Select the upper surface of gearbox to detect vibration and noise of gearbox on this test bench.

Gearbox test rig consists of a driving motor load motor and two closed systems. Quantitative oiling of the gearbox before the test removed the lubricants from the gearbox after the test, and the whole detection process is done automatically by the device (Fig. 4).

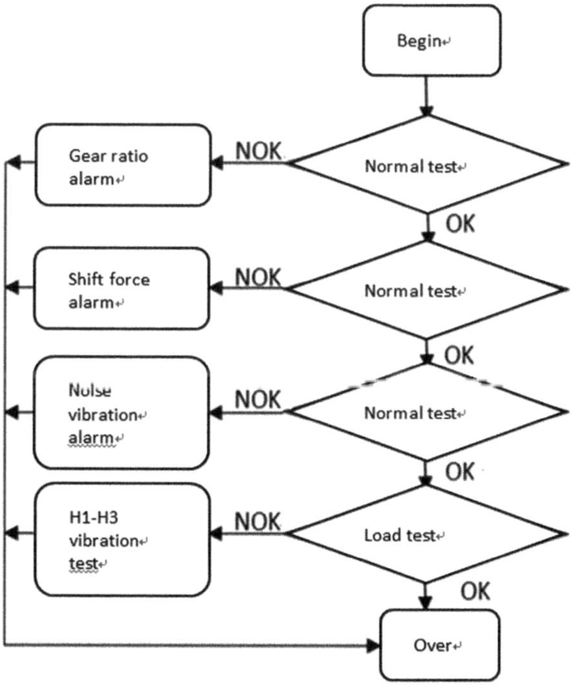

Fig. 4 Gearbox test flow

When the gearbox in place, the motor rotates through the splined drive shaft, the spline shifts on both sides of the differential are transferred to the gearbox for a certain load, simulating conditions on the vehicle, the contact type vibration probe mounted on the upper surface of gearbox detect and collect data throughout the test cycle, and data collected is analyzed by software to identify the quality of the gearbox. The whole process is divided into two parts of normal test and load test. The normal test is to carry out the conventional shift test on the gearbox at different speeds; load test aims to detect the gearbox quality under loaded conditions. The difference between normal and loading tests is that the evaluation criteria are different. The loading test is a supplement and a strict addition to normal test, increase detections of first-order, second-order and third-order vibrations of gearbox.

Normal test is to detect the vibration curve of acceleration/deceleration at every gearbox shift ($5 \rightarrow 4 \rightarrow 3 \rightarrow 2 \rightarrow 1 \rightarrow$ Neutral $\rightarrow$ R $\rightarrow$ Neutral $\rightarrow 1 \rightarrow 2 \rightarrow 3 \rightarrow 4 \rightarrow 5$) and compare with standard vibration curve, and the main evaluation index is the points of vibration and vibration density. Quality of gearbox is affected by the production batch of transmission parts, precision and stability of machine, quality of operation and other factors, from the subtle perspective of each gearbox assembly quality are different, and this difference is inevitable and acceptable. By contrast, we found that even if the same gearbox was subjected to repeated bench tests, each of the measured values was different. In order to identify whether the difference in gearbox test affects the quality, in addition to the “standard curve”, “alarm curve” and “stop curve” are introduced in the test cycle. Figure 5 curves monitoring in the testing cycle.

The loading test is the transmission under large torque, completed $2 \rightarrow 3 \rightarrow 4 \rightarrow 5 \rightarrow$ Neutral the shifting process. The main aim of loading test is to monitor the gearbox vibration in acceleration and deceleration cycles. Monitoring projects are mainly as follows: first/second/third-order frequency of gears and main reduction gear. After test of each speed ratio, you can get $H1/H2/H3$ frequency corresponding to the three acceleration curves and three deceleration curves.

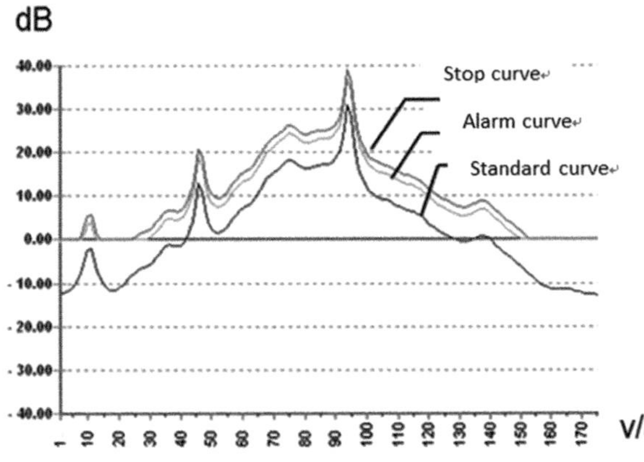

Fig. 5 Curve monitoring in the test cycle

4.1 Failure Analysis of Gearbox Vibration and Noise

Various parts or components in the machines have certain functions, such as passing motion, force, or energy, achieving the provisions of the actions, maintaining a certain geometric shapes. When the mechanical components lose its original defined function under load (including mechanical load, thermal load, corrosion, and integrated load), it is known as failure.

Determine the principles of mechanical failure or not based on three points: (a) not work, (b) the specified function is not completed according to the specified specification; (c) cannot be reliably and safely used. Online test bench for gearbox distinguishes the third type failure, which cannot be reliably and safely used (Fig. 6).

The gearbox structure is complex, and parts' failure may cause the overall gearbox failure. According to production statistics, the proportions of the different parts in the overall failure of the gearbox are shown in Table 7.

There are three main parts that cause the gearbox failure: gear, bearing, and shaft. Proportion of these three parts of the failure accounted for nearly 90%, and the failure of the gear is also accounted for the total failure ratio of almost 60%.

The vibration and noise generated by a pair of gears during the meshing process are mainly due to the generation of pitch and meshing forces in the gear operation. A pair of gear in meshing process, the gear manufacturing error, and periodic and non-periodic load changes and other factors will cause the gear pitch impact and meshing force.

(1) The vibration and noise caused by the impact of the pitch on the gear

Figure 7 is a pair of gear meshing simulation diagram. A pair of gears in the meshing of process and the meshing points are on the meshing line, between the teeth of the two gears, not only the existence of rolling friction; there is a certain of relative sliding, so that at the same time at joint point there will be rolling friction and sliding friction, and it will be a similar pulse power, which is the impact of the pitch.

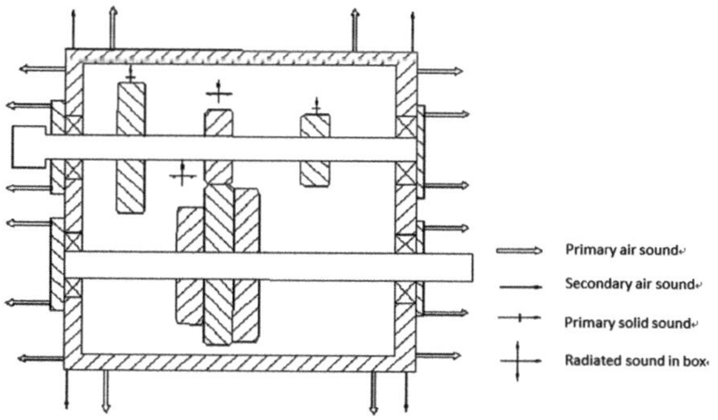

Fig. 6 Gearbox noise transmission diagram

Table 7 Specific past failure ratio

Failure parts	Gear	Bearing	Axis	Box	Fastener	Oil seal
Failure ratio (%)	60	19				1

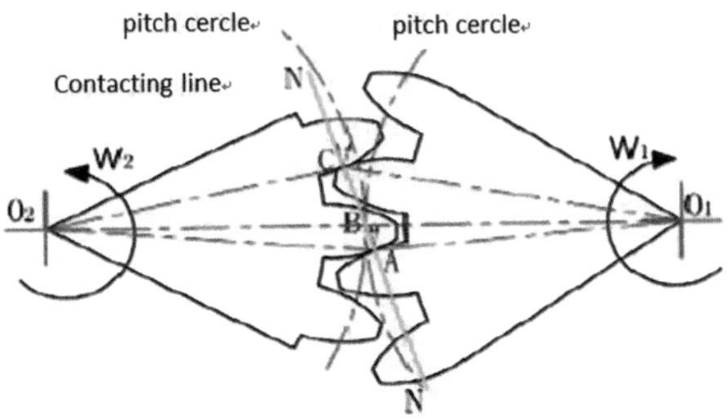

Fig. 7 A pair of gear meshing simulation diagram

(2) The vibration and noise caused by gear meshing impact

No error and unimpeded gear is ideal gear, in an ideal gear, meshing force does not exist, but there is no absolute ideal gear. Gear meshing process will inevitably produce force and deform. Due to the existence of gear manufacturing errors and installation errors, the gear meshing produces the impact force of the collision.

Gear noise sources can be divided into four parts: design, manufacturing, installation, use, and maintenance.

The design factors have (a) improper precision grade selection; (b) improper gear parameter selection (such as for gearbox that fit the passenger car, usually used smaller series of modules and gear width larger, to make gear meshing noise smaller also to ensure gear strength); (c) structure selection (such as: to use the web stiffener structures as far as possible, to reduce inherent frequency of gear itself, to reduce gear meshing noise).

Manufacturing factors are: (a) gear accuracy error (such as tooth profile and tooth pitch error and helix error); (b) the tooth surface roughness.

Installation factors are mainly reflected in assembly gear centricity and balance, especially for high-precision gears; its balance will have a direct impact on the gear transmission quality.

Maintenance factors are: (a) gear cleanliness and (b) proper lubrication of gears.

(3) Bearing noise

The failure of the bearing to the overall failure rate of the gearbox is 19%, which is one of the main sources of noise for the gearbox. According to statistics, the rotating machinery with bearings has about 30% fault caused by bearing, and quality of bearing has a great impact on the working conditions of the machine.

The main sources of bearing noise are of three parts: raceway, cage, and fault noises.

Table 8 Summary of abnormal noise problems in gearbox test

Abnormal phenomena	Abnormal expression	Possible cause
Noise abnormal	Noise: When gears rotating at high speed, there are a certain frequency of percussion sounds and clicks	Gear bump, gear tooth surface roughness, tooth indentation or without top trimming, gear is partially unshaven, gear run out…
	Whistler-type noise: Sound like a screaming voice of whistle or wind-blown, when gears rotating at high speed	Bad gear accuracy (tooth profile, pitch error, and helix); backlash is too tight or too large…
	Other type noise: Other abnormal sounds, such as chirping, whining	Driven gear synchronous cone grinding defects, sliding sleeve fork groove is rough or bumped, gear faces bumped, fork interference…

(a) Raceway noise

When the bearing is working, the bearing rolling element and the raceway are relatively rolling. The contact between the bearing rolling element and the raceway actually forms a flexible nonlinear vibration system. Raceway will inevitably exist some manufacturing errors, and so there will be some rough corrugations on the surface, which caused raceway noise when bearing rotates.

(b) Cage noise

During the bearing work, there must be some friction between the rolling elements and cage, especially when the lubrication situation is bad, cage will make some noise.

(c) Fault noise

When the bearing is of high quality, good lubrication conditions, bearing noise energy is very small. However, when the bearing sound failure, both failures between rolling elements and the cage or between rolling elements and raceway will be directly produce some noise (Table 8).

5 Development Trend of Transmission Online Diagnosis Technology

Along with the rapid development of modern information technology and the development of computer technology and artificial intelligence technology, online fault diagnosis technology of vehicle transmission has been in rapid development.

Traditional online detection technology of noise and vibration has occurred changes:

(a) Intelligent test: Transmission fault diagnosis intelligent system will be the main development trends;
(b) Multi-functional integrated: Transmissions online test is only a single check in the past, and now online detection technology not only detects and diagnoses the gearbox, but also develops toward the direction of integration of management and scheduling;
(c) Functional distribution: Function centralized steering function distributed, with the developments of single-chip microcomputer and innovation of hardware and software of computer systems, and the diagnosis system of the future will be more reliable.

It is noteworthy that differences of the product structure, equipment structure, parts status, process parameters, and a production environment will give a larger challenge to the accuracy of online detection system of gearbox vibration and noise. This requires us to take the necessary technical means and management measures to control the above factors in every period, such as process planning, equipment investment and installation, mass production. Apply reliable online diagnostic technology to ensure transmission assembly quality.

6 Conclusion

Based on the above mentioned, according to products features and design requirements, in production of automobile engine and gearbox, may use method of subjective judgment, step-by-step operation, the method of frequency spectrum analysis and the tolerance range curve for evaluation establish noise ratings scoring rules based on user perception. In the automotive powertrain assembly line, use more mature intelligent and multi-functional integrated noise diagnosis system, to effectively monitor, judge, and exclude various noise defects that are produced in powertrain mass production, also provide direction to continuous products improvement, to meet consumer requirements for driving pleasure more and more.

The Application of Film Broken Alarm Detection Technology in Polishing Machine

Xuesong Liu, Rui Chen and Zhiyong Han

1 Introduction

Polishing machines are used for the final machining of crankshaft and camshaft parts. In the drill or reamer processing, there are some mature tool broken alarm monitoring software, such as ARTIS, Digitalway, and so on, to monitor the power or torque of the spindle to judge whether the tool is damaged or not. In the polishing machine, film broken is possible; polishing tile will directly touch the parts if it is not found timely and accurately and thus will cause scratch or no-polishing of the parts' surface; then, the product is not qualified. At the same time, because the parts have been grinded before polishing, it is difficult for operators to identify the parts which were not polished, and there is a great risk of quality, which had happened ever in the crankshaft line of our factory. Therefore, it is necessary to study the principle of the film break alarm and improve the alarm effect.

2 The Application of the Principle of Film Broken Alarm Detection Technology

There are three different kinds of polishing machines used in our factory. It is said that the effect of film broken detection devices is not good and has been frequent false alarms. That is to say, the film is good but the machine said it is broken, so we have to shield the alarm device to continue the production.

X. Liu (✉) · R. Chen · Z. Han
Dongfeng Peugeot-Citroën Automobile Co., Ltd, Wuhan 441004, China
e-mail: liuxuesong@dpca.com.cn

Society of Automotive Engineers of China (SAE-China) (ed.), *Proceedings of the 19th Asia Pacific Automotive Engineering Conference & SAE-China Congress 2017: Selected Papers*, Lecture Notes in Electrical Engineering 486, https://doi.org/10.1007/978-981-10-8506-2_80

2.1 *The Principle of Film Broken Alarm for IMPCO Polishing Machine*

As shown in Fig. 1, the American IMPCO polishing machine film broken alarm principle diagram, this polishing machine is mainly used for polishing the CAM and oil seal of camshaft in our plant, for four-cylinder engine's camshaft, and it needs eight films for polishing eight CAMS and one for the oil seal. In figure, the polishing film 1 walks through the lower edge of the dam-board 3, and the direction is shown as the arrow. If the polishing film is not broken, the film takes in a state of tension, because of the film walking toward left on the edge of dam-board, and makes the dam-board in position 1, and thus signal from the sensor 2 will not be blocked with dam-board; it can receive the signal on the other side, so the polishing machine thinks film is not broken. If the film was broken, it is in loose condition, and then the dam-board will be in position 2 due to its own weight; the signal from the sensor will be blocked by the dam-board, and it could not get the signal at the receiving end; thus, the machine will alarm as film broken.

This method has simple structure, ease to implement machine design, and lower manufacturing cost. No matter how many polishing films are tied together, only one sensor is needed. But the drawbacks are: Firstly, the machine cannot know which film is broken when alarming broken, and the operators need to identify it by themselves. Secondly, when the dam-board is placed in position 2, it needs to overcome the friction of the films around, and the weight design of the board should also be explored through practical experience. When the film walks more complicated with more friction, it is difficult to overcome those friction only by weight. From the actual use of production, as shown in the picture, the film walks through a low concave, and in actual production, it often happens that the film is broken without alarming.

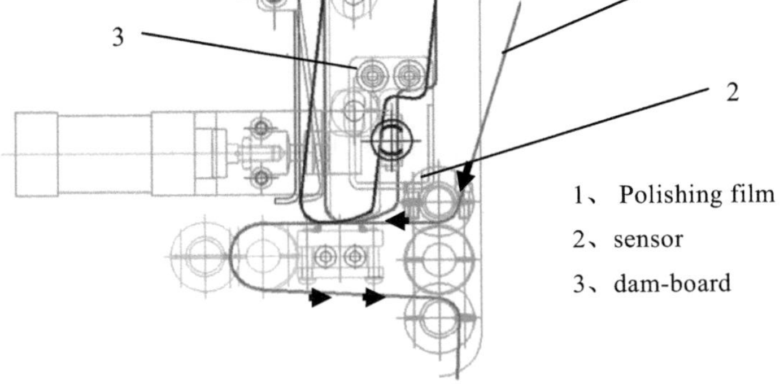

Fig. 1 Alarm principle of IMPCO polishing machine

2.2 The Principle of Film Broken Alarm for SUPFINA Polishing Machine

Fig. 2 shows the SUPFINA polishing machine film broken alarm principle diagram. Number 1 is the polishing film; the walking direction is shown as the arrow. Number 2 is the spring. When the film is not broken, it is same in a tight state. This tension force acts on block 3, causing the spring 2 to be compressed. At this point, the signal from the sensor 4 will not be blocked by block 3, and the signal can be received at the other end; thus, the polishing machine thinks that the film is not broken. When the polishing film breaks, film gets into loose condition, and the spring not subjected to external forces will spread outward from the original state of compression, in a free state; then, the block 3 will move toward left, blocking the signal from the sensor, and the signal could not be obtained at the receiving end, so the machine will alarm polishing film broken.

This structure and principle of film broken alarm are similar to that of IMPCO polishing machine; both use the block which will be in two different positions when the film breaks or not. Monitoring is realized by judging whether the signal sent by the sensor can be received. Similarly, the machine cannot automatically know which film is broken. The difference is that one uses self-weight and the other one uses pressure spring. The selection of pressure spring's force also needs to be explored, but it is easier to explore than the self-heavy choice design. At the same time, with the pressure spring, the spring is permanently compressed, which cannot affect its ability to recover to the free state after long normal use. You do not have to change the spring regularly. This equipment has been used in our factory for nine years, and it has worked well in actual production. There is no case that the polishing band is broken but there is no alarm.

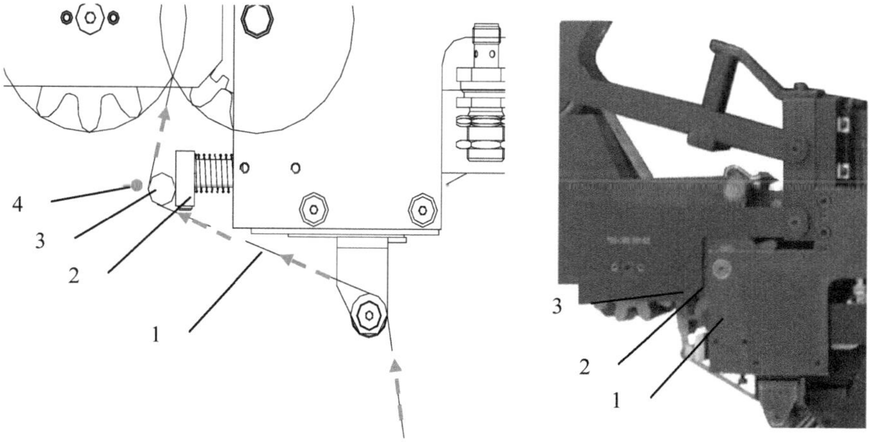

1、Polishing film 2、spring 3、block 4、sensor

Fig. 2 Principle of film broken alarm for SUPFINA polishing machine

2.3 *The Principle of Film Broken Alarm for Grind Master Polishing Machine*

Fig. 3 shows the film broken alarm principle diagram for Grind Master polishing machine. 1 stands for the polishing film, and the walk direction is indicated by the arrow and walking by signal round 2; 3 is the sensor. The film broken alarm principle mainly uses the friction between wheel and film, which can drive the wheel to rotate, through the sensor sensing black and white signals on the wheel, raising the pulse signal. At the same time, a proper pulse number is set in the system. When the pulse number of the sensor sensing is more than that setting in the system, it is considered that the polishing film walks with a right length. On the contrary, when the pulse number of the sensor sensing is less than that setting in the system, it is considered that the length of the film walk is not enough, it may be the film is broken, and then the machine will alarm polishing film broken.

This structure design has set a signal wheel and a sensor for every strip of polishing film, and the machine can automatically find out which film is broken when alarming and the operator should not judge by himself. This polishing machine was only introduced in recent years and added the function to recognize which polishing film is broken when compared to the former two kinds of polishing machines. However, in the actual production, there is also a false alarm situation; that is, the polishing film is not broken, but the machine alarms broken. Combined with the principle of film broken alarm to analyze the reason, it is not hard to find the main reason is only rely on the friction between the film and signal wheel to drive the wheel rotation, there is no guarantee that it is absolutely rolling friction, existing the phenomenon of sliding friction, thus the pulse number of the sensor sensing will be less than that of the machine set, therefore, the alarm break band appears, while the polishing band is not actually broken.

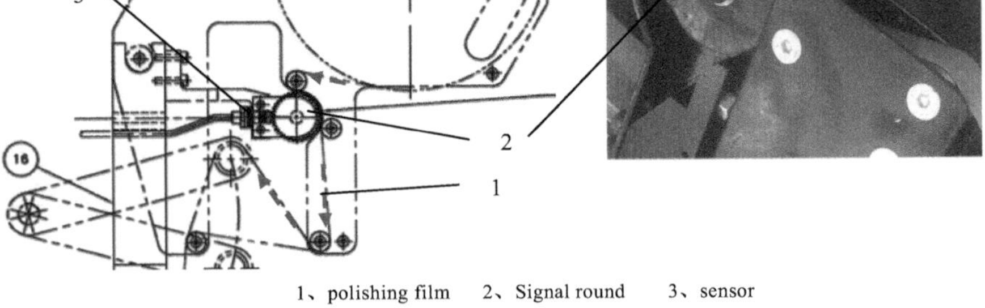

Fig. 3 Principle of film broken alarm for Grind Master polishing machine

3 Optimization Design of Polishing Film Broken Alarm Device

In order to solve the problem of false alarm of Grind Master polisher, we optimized the device. Because the main reason is the friction between the signal wheel and film is not enough, we set out from here, using a spring, one end of which connects with polishing arm body, signal wheel, and sensor as a whole hinged connect with the polishing arm, and the other end of the spring connects the whole of signal wheel and sensor. This makes use of the spring to push the signal wheel against the polishing film, increasing the friction between the polishing film and the signal wheel.

To make the tension of the spring absolutely passed to the force of the signal wheel for polishing film, it needs the tension of the spring and the force of the signal wheel for film in the same direction, and the choice here is horizontal. As shown in Fig. 4, the spring is installed in the horizontal direction; the center of the signal wheel and the rotary center of the film are on the same horizontal line. The horizontal tension of the spring on the signal wheel is then transmitted to the horizontal left tension of the signal wheel on the belt. When the polishing belt moves, it produces vertical downward friction on the signal wheel. Other alarm principles remains the same.

You will find that it is line contact between the film and the signal wheel, and there is a doubt that the line contact will have less friction than the previous surface contact. No, when the spring tension is constant, the friction between the signal wheel and the film will not change; no matter whether it is line contact or surface contact. (It can be demonstrated by the synthesis and resolution of forces.)

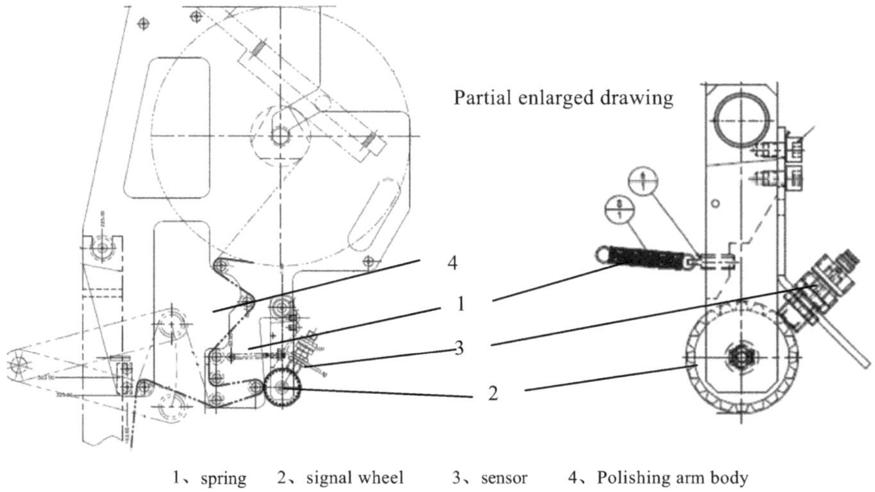

Fig. 4 Optimization design scheme for Grind Master polishing machine film broken alarm

Therefore, the design of the line contact here will not affect its effect. The signal wheel can rotate with the film if you pull the film by hand.

As the SUPFINA polishing machine used the pressure spring, improved Grind Master polishing machine used the tension spring which will be in the tensile state for a long time; tension of the spring will weaken after used for a long time, there is a risk of flase alarm.

There was a good effect after improving within one year; however, the false alarm appeared then again. There may be two main reasons, of which, the first was the black dirt on the surface of signal wheel resulting in that sensors could not receive optical signal and generate impulse, and the other may be the failed tension spring causing the signal wheel unable to revolve with the polishing belt. Therefore, after cleaning the signal wheel and replacing with a new spring, the false alarm never appeared again when used in practical production. On a separate note, cleaning the signal wheel and changing the tension spring in good time was required.

4 Development Direction

With the rapid development of manufacturing industry and the need for high-efficiency production, users are increasingly demanding for machine. The polishing machine on the TU/EW/EC project could only give a film broken alarm but could not tell which film was broken accurately. However, on the EB/EP project, it is able to identify exactly which film is broken. Relatively speaking, the film broken alarm detection technology was improved. In actual production, the problem of the false alarm in polishing machine was widespread, and the detection technology of broken band alarm is not mature enough and needs to be improved.

The tension of the normal polishing film was utilized when the methods for film broken alarm were summarized. At the same time, the distance between each adjacent film is too close owing to the very close axial distance of shaft parts during polishing, resulting in that most of the manufacturers chose to install alarm device on the back of the film surface, instead of the lateral face of the film. Then, can we try to install the alarm device on the lateral of the film? Can we install the alarm device in the position of the polishing film installed?

Alternatively, a pneumatic detection device is designed to be installed on the back of the film surface where the film will walk by. In normal conditions, the polishing film is tight and close to the stomata, and the blow pressure is high. When the polishing film is broken, the tensioning force becomes smaller, the air vent blowing can blow up the polishing film, and the pressure of the gas will become smaller. This method references the alarm device of whether the part loading position is correct or not. It still needs to be verified whether it can be successfully used in polishing machine.

5 Conclusion

By comparing different principles of polishing film broken alarm detection, the advantages and disadvantages of the detection principle are found, which is helpful for polishing machine manufacturers to improve the alarm detection technology and improve alarm accuracy. At the same time, understanding the alarm principles is also helpful for the technician to analyze the problems on site and optimize the existing problems. In the new projects' equipment selection, you can identify the alarm reliability of the equipment early, choose suitable machine, and improve investment performance.

Bibliography

1. IMPCO polishing machine manual
2. SUPFINA polishing machine manual
3. Grind Master machine manual

Xuesong Liu female, was born in November 1988 and obtained her bachelor's degree in July 2012. Mechanical parts processing technology in EB processing office, technical center of power assembly chassis, DPCA. LTD. Communication address: EB processing office, xiangyang factory of DPCA, XiangYang city, Hubei province. Zip code: 441004, Tel: (0710) 3389609, Email: liuxuesong@dpca.com.cn.

Rui Chen male, was born in March 1981 and obtained his bachelor's degree in 2001. He is the team leader of the crankshaft processing and the crankshaft processing technology in DPCA Ltd, working in the process for 15 years. Communication address: EB processing office, xiangyang factory of DPCA, XiangYang city, Hubei province. Zip code: 441004, Tel: (0710) 3389600, Email: chenrui@dpca.com.cn.

Zhiyong Han male, was born in November 1973 and obtained his bachelor's degree in July 1994. Mechanical parts processing technology in EB processing office, technical center of power assembly chassis, DPCA LTD. Communication address: EB processing office, xiangyang factory of DPCA, XiangYang city, Hubei province. Zip code: 441004, Tel: (0710) 3389600, Email: hanzhiyong@dpca.com.cn.

GPSR Compliance

The European Union's (EU) General Product Safety Regulation (GPSR) is a set of rules that requires consumer products to be safe and our obligations to ensure this.

If you have any concerns about our products, you can contact us on ProductSafety@springernature.com

In case Publisher is established outside the EU, the EU authorized representative is:

Springer Nature Customer Service Center GmbH
Europaplatz 3
69115 Heidelberg, Germany

Batch number: 10755075

Printed by Printforce, the Netherlands